Hot and Dense Nuclear Matter

NATO ASI Series

Advanced Science Institutes Series

A series presenting the results of activities sponsored by the NATO Science Committee, which aims at the dissemination of advanced scientific and technological knowledge, with a view to strengthening links between scientific communities.

The series is published by an international board of publishers in conjunction with the NATO Scientific Affairs Division

A	Life Sciences	Plenum Publishing Corporation
B	Physics	New York and London
C	Mathematical and Physical Sciences	Kluwer Academic Publishers
D	Behavioral and Social Sciences	Dordrecht, Boston, and London
E	Applied Sciences	
F	Computer and Systems Sciences	Springer-Verlag
G	Ecological Sciences	Berlin, Heidelberg, New York, London,
H	Cell Biology	Paris, Tokyo, Hong Kong, and Barcelona
I	Global Environmental Change	

Recent Volumes in this Series

Series B: Physics

Hot and Dense Nuclear Matter

Edited by

Walter Greiner and
Horst Stöcker

Johann Wolfgang Goethe University
Frankfurt-am-Main, Germany

and

André Gallmann

Louis Pasteur University of Strasbourg
Strasbourg, France

Springer Science+Business Media, LLC

Proceedings of a NATO Advanced Study Institute on
Hot and Dense Nuclear Matter,
held September 26–October 9, 1993, in Bodrum, Turkey

NATO-PCO-DATA BASE

The electronic index to the NATO ASI Series provides full bibliographical references (with keywords and/or abstracts) to more than 30,000 contributions from international scientists published in all sections of the NATO ASI Series. Access to the NATO-PCO-DATA BASE is possible in two ways:

—via online FILE 128 (NATO-PCO-DATA BASE) hosted by ESRIN, Via Galileo Galilei, I-00044 Frascati, Italy

—via CD-ROM "NATO Science and Technology Disk" with user-friendly retrieval software in English, French, and German (©WTV GmbH and DATAWARE Technologies, Inc. 1989). The CD-ROM also contains the AGARD Aerospace Database.

The CD-ROM can be ordered through any member of the Board of Publishers or through NATO-PCO, Overijse, Belgium.

Library of Congress Cataloging-in-Publication Data

Hot and dense nuclear matter / edited by Walter Greiner and Horst
  Stöcker and André Gallmann.
      p.   cm. -- (NATO ASI series. Series B, Physics ; v. 335)
    "Published in cooperation with NATO Scientific Affairs Division."
    "Proceedings of a NATO Advanced Study Institute on Hot and Dense
  Nuclear Matter, held September 26-October 9, 1993, in Bodrum,
  Turkey"--T.p. verso.
    Includes bibliographical references and index.
    ISBN 978-1-4613-6071-1   ISBN 978-1-4615-2516-5 (eBook)
    DOI 10.1007/978-1-4615-2516-5
    1. Nuclear matter--Congresses. 2. Heavy ion collisions-
  -Congresses. I. Greiner, Walter, 1935-   . II. Stöcker, Horst.
  III. Gallmann, André. IV. North Atlantic Treaty Organization.
  Scientific Affairs Division. V. NATO Advanced Study Institute on
  Hot and Dense Nuclear Matter (1993 : Bodrum, Turkey) VI. Series.
  QC793.3.N8H68   1994
  539.7'4--dc20   94-39369
                                                             CIP

Additional material to this book can be downloaded from http://extra.springer.com.

ISBN 978-1-4613-6071-1

©1994 Springer Science+Business Media New York
Originally published by Plenum Press in 1994

PREFACE

Ladies and Gentlemen,
dear colleagues,

Welcome in Bodrum to the NASI on
Hot and Dense Nuclear Matter!

Welcome also to Mrs. Governor Dr. Lale AYTAMAN. We are very honored, that you, Governor of the Mugla-State, came here to greet us. We are particularly grateful to you that you offered help and assured us to do everything that we can enjoy two <u>safe</u> weeks in Bodrum, in this wonderful area of your country.

I have chosen Bodrum as the place for our NASI because I like this historic region where many cultures meet (e.g., Oriental and European (Greek, Roman) culture) and where you find numerous places which played a role in ancient science and in early Christianity- I mention Milet (Thales) and Ephesus (Apostle Paulus), both of which are close by.

Our NASI will exhibit the most recent developments in high energy heavy ion physics. The meeting is both a school and a conference: A school, because there are very many advanced students, who frequently are themselves already top researchers, attending the lectures of distinguished scientists and leading researchers. It is also a conference because new material, new results of this exciting and wonderful field - our field - high energy heavy ion physics will be presented. It is the topic of hot and dense nuclear matter, which we are focusing on. Here we investigate how matter may have been formed and what its properties were just fractions of seconds after the big bang, after the inflationary phase of the creation of our universe is over. In other words we are interested in the ur-question of how our world was created, how baryo-synthesis and nucleo-synthesis might have happened, how stars are built (in particular such exotic objects as neutron-stars, quark stars), how supernova explosions occur, etc. Hot and dense nuclear matter may have completely new phases as compared to ordinary matter: It is believed that at high densities and temperatures the QGP with dissolved quarks and gluons appears. Recent theoretical investigations, based on QCD (Rischke, Gorenstein) indicate that it is most likely - if not certain - that the QGP is a cluster plasma. Baryons and mesons are not completely - in fact only 20% - dissolved. There are

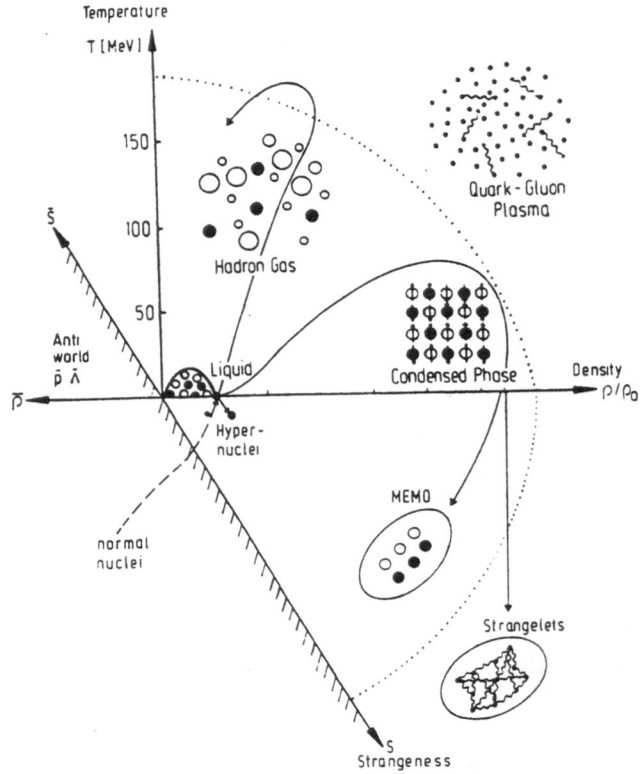

Matter, temperature, antimatter, strangeness, antistrangeness. Hot and dense nuclear matter obtained through the shock wave mechanism leads to the creation of new forms of matter (nuclei, mesons, ...).

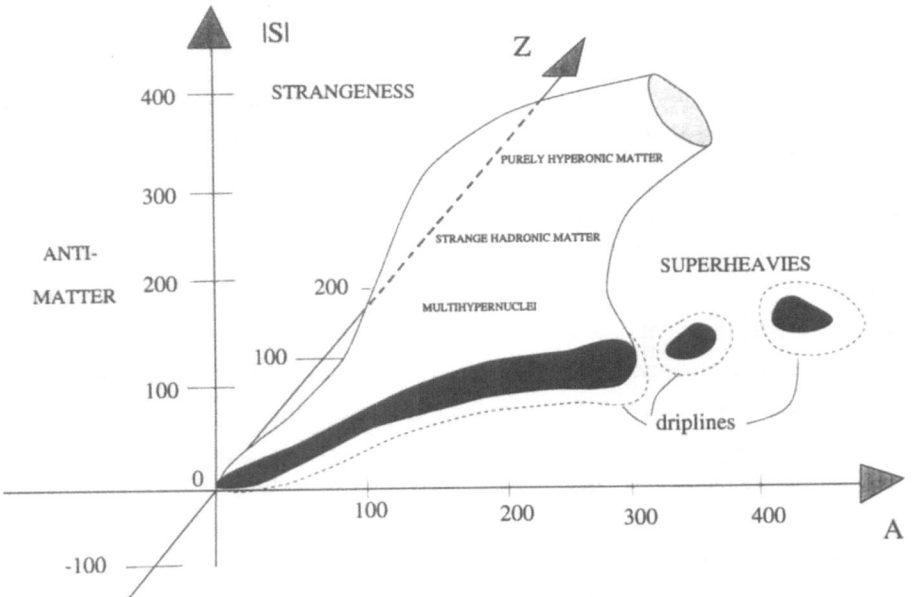

The periodic system and its extension into the strangeness sector. The darkness indicates the degree of stability (lifetime). One can easily imagine its extension into the antimatter domain, which is not shown in this figure.

also the exciting prospects of forming very new forms of matter, strange matter (strangeletts, multi - Λ - hypernuclei, multiply strange exotic mesonic objects (Memos), antinuclei, multipionic atoms, etc.) The preceding two figures illustrate these predictions and expectations. This opens the door to an extension of our periodic system into completely new directions, i.e. strange, and antimatter directions, really exciting prospects!

What we need is hot and dense nuclear matter, and this is achieved through the shock wave mechanism in its various forms. When nuclei collide at high energies they compress and heat up tremendously.

We will have plenty of discussions, I hope, both here in the lecture hall as well as outside in this pleasant garden. I have choosen the hotel KARIA PRINCESS for our meeting, because it is a first class, wonderful place. According to my taste it's the best hotel in town. It seems to me particularly suited because it is small enough so that we can stay compactly together. The whole hotel is ours for the next two weeks. As it turned out, we had far too many applications - over 350 and we had to turn down most of them. Nevertheless we still have an unusually high number of physicists here for such a NASI. In addition, and I must say not surprisingly, there are many wifes and companions who joined their husbands and friends.

Well, before I go on, let me make a couple of technical remarks concerning the organization of our meeting here. I would like to thank my colleagues and friends Prof. A. GALLMANN and Prof. H. STÖCKER for helping tremendously in the organisation. NATO's scientific program has been very generous with its support for this school. It deserves our warmest thanks. The German Ministry for Science, in particular Dr. D. HARTWIG, has also been very helpful with its support. The same is true for the Institut National de Physique Nucléaires et de Physique Particules (IN2P3), Paris, France, and for GSI (Darmstadt).
All of them deserve our "thank you" for their financial support, without which this meeting could not take place in this form.
Finally I appreciate the help of Mrs. Nicole GROSS, Mrs. Margarita OLLIG and Mrs. Daniela RADULESCU in secretarial and technical matters which was quite important.

A particular "thank you" goes to Dipl.-Physiker Jens KONOPKA, who helped extraordinarily with contacting the speakers, reminding them to deliver their contribution to the proceedings in time and organizing most of the editorial work.

The next two weeks will be full of hard work, but I hope that there will also be joy. Then our NASI will be remembered!

Now, lets begin with the work! I wish us a very pleasant, interesting and successful meeting!

Walter Greiner

CONTENTS[1]

---

[1]The proceedings volume is organized in the very same way as the conference itself was. In particular, the ordering of the contributions corresponds to the sequence of the talks delivered during the conference.

PAST AND PRESENT OF NUCLEAR MATTER

Hans Georg Ritter

Nuclear Science Division, Lawrence Berkeley Laboratory
Berkeley, CA 94720

INTRODUCTION

The subject of nuclear matter is interesting for many fields of physics ranging from condensed matter to lattice QCD. Knowing its properties is important for our understanding of neutron stars, supernovae and cosmology.

Experimentally, we have the most precise information on ground state nuclear matter from the mass formula and from the systematics of monopole vibrations.[1] This gives us the ground state density, binding energy and the compression modulus κ ($\kappa \approx 210 MeV$) at ground state density. However, those methods can not be extended towards the regime we are most interested in, the regime of high density and high temperature. Additional information can be obtained from the observation of neutron stars and of supernova explosions.[2] In both cases information is limited by the rare events that nature provides for us.

High energy heavy ion collisions, on the other hand, allow us to perform controlled experiments in the laboratory. For a very short period in time we can create a system that lets us study nuclear matter properties. Density and temperature of the system depend on the mass of the colliding nuclei, on their energy and on the impact parameter. But this control has a price. The system created in nuclear collisions has at best about 200 constituents, not even close to infinite nuclear matter, and it lasts only for collision times of $\approx 10^{-22} sec$, not an ideal condition for establishing any kind of equilibrium. Extended size and thermal and chemical equilibrium, however, are a priori conditions of nuclear matter. As a consequence we need realistic models that describe the collision dynamics and non-equilibrium effects in order to relate experimental observables to properties of nuclear matter.

With those constraints, heavy ion collisions give us access to the study of nuclear matter over a wide range of temperature and density. New phenomena occur and boundaries are crossed opening up new degrees of freedom. At low temperature

Hot and Dense Nuclear Matter, Edited by
W. Greiner *et al.*, Plenum Press, New York, 1994

and density we observe multifragmentation where the nucleus possibly undergoes a first or second order transition to the gas phase. At temperatures of the order of $100\,MeV$ and densities of a few times ground state density, excited nucleons start to play an important role and we enter the regime of resonance matter, while at yet higher energies partonic degrees of freedom become important. At the highest energy available now or in the near future, we expect a new form of matter, the Quark-Gluon plasma, to be formed. Thus as we increase the energy available with accelerators we push our limits of observation closer to the origin of the universe, the big bang. All these exciting subjects are the topic of this NATO Advanced Study Institute.

The study of high energy nuclear collisions started at the Bevalac. The lessons learned and the concepts developed are important and relevant for the work at lower and higher energies. I will try to summarize the results from the Bevalac studies, the highlights of the continuing program, and extension to higher energies without claiming to be complete.

THE PAST

Progress in the understanding of high energy heavy ion collisions at the Bevalac came as an interplay between experiments, theory development, and accelerator improvements. In the early stage, spectra were compared with simple thermal models, like the Fireball model,[3] with cascade models[4, 5] that treat the nuclear collision as a superposition of nucleon-nucleon collisions, and with hydrodynamical models.[6, 7] All the models reproduced the data within a factor of two to four, but there was no conclusive evidence, favoring or excluding any particular model.

Substantial improvements were made when the experiments became more sophisticated. The first 4π detectors, the Streamer Chamber and the Plastic Ball, were able to detect and identify most of the emitted charged particles and to perform global event analysis. This novel type of analysis made the identification of collective behavior possible. In a bulk matter picture part of the energy available in the center of mass is converted into thermal energy and part is converted into compressional (potential) energy. The relationship between thermal (E_{th}) and compressional (E_c) energy as a function of density is shown in Fig. 1 for temperature $T = 0$. This relationship is sometimes called the equation of state. A more detailed discussion can be found in reviews.[8, 9]

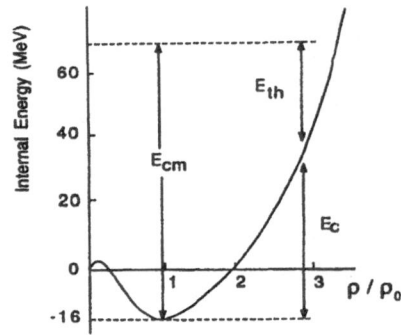

Figure 1. Energy per baryon as a function of nuclear matter density.

In the expansion phase the compressional energy leads to collective flow. For events with finite impact parameter the flow should have a directed (non-spherical) component. This directed collective flow had been predicted by hydrodynamic model calculations[10] and was observed first by the Plastic Ball experiment[11] in $Nb + Nb$ collisions at $400 MeV$ per nucleon.

The Streamer Chamber group followed a different approach. Their reasoning was that if there is potential energy in the form of compressional energy then it is not available for particle production, not at the point of maximum compression, nor during the expansion phase where it leads to collective flow. The group performed a systematic study of pion production,[12] and the basic idea to measure the available energy by studying particle production and sub-threshold particle production is still the basis for many experiments at GSI, at the AGS and at CERN. However, it is important to do careful comparisons with realistic models before any information about the reaction dynamics or about the property of nuclear matter can be extracted from those data.

In addition to collective flow the experiments revealed that in the most central collisions of the heaviest systems there were many indications that some kind of thermal equilibrium is achieved. The rapidity distributions[13] peak near mid-rapidity, as seen in Fig. 2 for the most central collisions, leaving no indication for the presence of

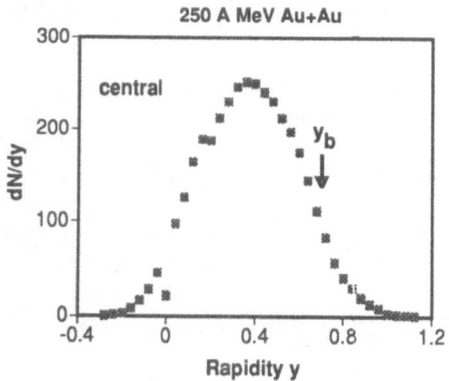

Figure 2. Rapidity distribution for baryons from Au+Au at 250 MeV per nucleon.

projectile spectators. Another indication for near equilibrium conditions comes from the measurement of the stopping ratio R, the ratio of the mean longitudinal to the mean transverse momenta of all the measured particles. A value of $R = 1$ indicates that the originally longitudinal momentum is distributed equally over longitudinal and transverse degrees of freedom. For the example of $Nb + Nb$ at $400 MeV$ per nucleon[14] Fig. 3 shows that this condition is reached for the most central events.

Despite the great progress that was made on the experimental side it was not possible to unambiguously determine the equation of state. The concepts were too simple minded. Flow does not only depend on the static properties of nuclear matter (equation of state) but as well on the transport properties (viscosity). In addition, hydrodynamic or simple thermal models did not take into account non-equilibrium or finite-size effects. New microscopic models (VUU,[15] BUU,[16] and QMD[17]) including mean field effects, the Pauli principle, and momentum dependent interactions,

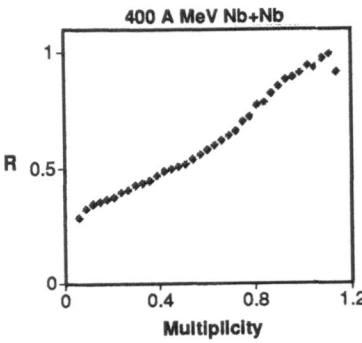

Figure 3. Stoping ratio R as a function of normalized charged particle multiplicity for Nb+Nb at 400 MeV per nucleon.

became available. This made a dramatic difference, since the shortcomings of the other models were overcome and information about the form of the equation of state is included via density dependent potentials. Now it was possible to predict a broad range of phenomena over the complete energy range of the Bevalac. As an example, Fig. 4 shows VUU predictions[18] for the energy dependence of the mean transverse momentum in the reaction plane[19] as a function of rapidity.

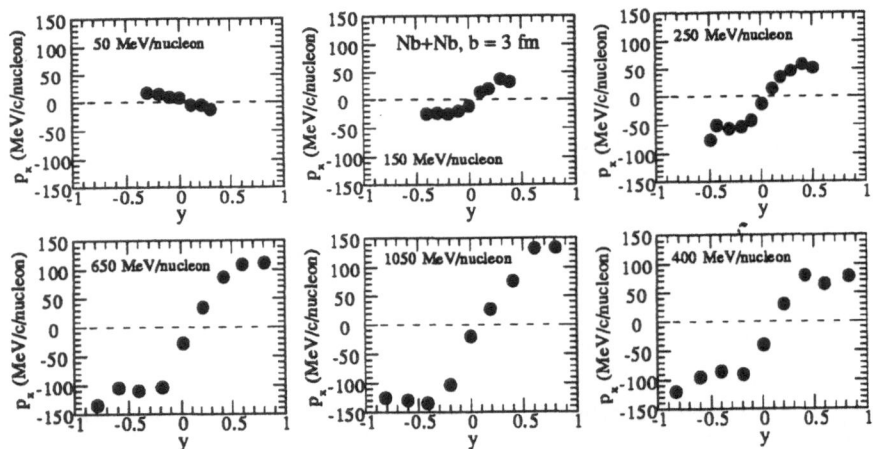

Figure 4. VUU predictions for the energy dependence of the mean transverse momentum in the reaction plane versus rapidity for Nb+Nb.

The very encouraging result from the rapid development in theoretical models is that the data of high energy heavy ion collisions are sensitive to the equation of state. However, observables are only sensitive to less than a factor of two at best, and the uncertainties in other parameters, like the modification of the in-medium cross sections or even the technical implementation of well known effects into the models, make a clear statement difficult.[20] It was very obvious that for the next step progress had to come from the experimental side with more precise and more complete data covering a large variety of observables over a large range of energy and mass.

THE PRESENT

In the past years a very active program has developed at many places. SIS at GSI has replaced the Bevalac as the main accelerator in what is now called intermediate energy ($\approx 100 MeV$ up to $2 GeV$ per nucleon). This new facility has made a big impact with three new second generation experiments, FOPI, ALADIN and KAOS. At lower energies MSU and GANIL complement the program.

Multifragmentation

At low energies we observe in the exit channel of nuclear collisions sequential binary decay. Above $\approx 35 MeV$ (depending on the mass of the system) this mechanism changes[21] and many intermediate mass fragments are emitted,[22] a process that is called multifragmentation. Especially the ALADIN experiment has established the systematics of multifragmentation as a function of mass and energy.[23] The data have been used to test microscopic models that have been combined with statistical multifragmentation codes.[24] Many ideas have been proposed in an effort to explain the multifragmentation mechanism, among them the development of surface instabilities[25] and the formation of bubbles and rings.[26] It was suggested to see multifragmentation as a critical phenomenon quite a while ago[27]. The new data from the EOS experiment have been used in an attempt to extract critical exponents[28] by analyzing the data in terms of moments proposed by Campi.[29] All those new possibilities and ideas make multifragmentation one of the most interesting subjects with the potential to determine the nuclear equation of state at densities below ground state density and to study phase transitions in small systems. Such studies could lead to tools and methods that can be applied in the search for the Quark-Gluon plasma and in other areas of physics.

Flow

Flow is caused by the short range nuclear repulsion and its strength decreases with decreasing energy. At very low energy the attractive part of the nuclear potential becomes more important and the direction of the flow is reversed. This was first seen in the VUU calculations[18] shown in Fig. 4. This behavior is called disappearance of flow and has been observed experimentally.[30] This subject is covered in several contributions.[21, 31] The point where the flow is zero is called balance energy. It can be determined very precisely and is quite independent of detector biases. Detailed and systematic measurements of the balance energy as a function of mass have been performed. Fig. 5 shows a summary of these studies[21] and a comparison with BUU calculations. Fig. 5 indicates that the balance energy does not depend strongly on the equation of state. This is not astonishing, given the low density at those energies, but the effect is expected to depend strongly on the in-medium cross sections.[21, 32]

The detailed study of directed flow has been considered the best way to pin down some of the parameters of the nuclear matter equation of state. As a consequence systematic measurement have been performed in many experiments at many accelerators. As an example, the excitation function of the flow is shown in Fig. 6 where new results from EOS[33] are compared with the Plastic Ball data[34]. The EOS data show the increase of flow with energy much more clearly pronounced, differences especially at the high energies being most probably due to missing acceptance corrections in the Plastic Ball data. This trend is further highlighted by the analysis of the data

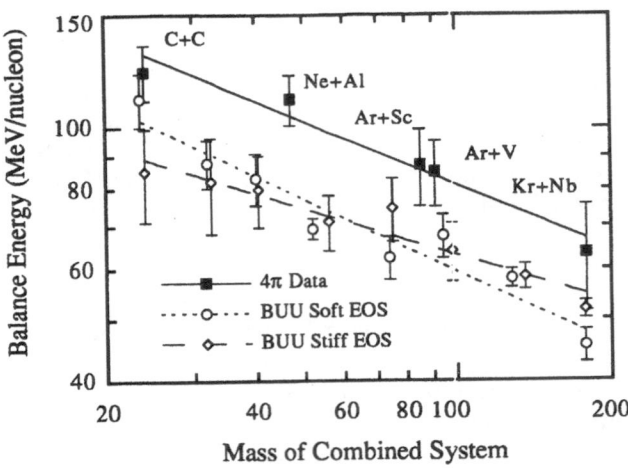

Figure 5. The balance energy as a function of the mass of the combined system is compared with BUU calculations with a soft and stiff equation of state.

in terms of hydrodynamic scaling.[35] In addition to the systematic study of the magnitude, new data and new methods allow us to ask if the flow signal is an effect of focussing the particles in the reaction plane, or of those particles having higher momenta, or a combination of both.[33] Another encouraging sign of progress is that systematic comparisons with different models now are being done routinely.[36]

Composite particles show larger flow effects.[22] In a thermal picture, the mean kinetic energy of all particles is the same, which means that the mean energy per nucleon decreases with increasing mass. The collective boost from flow, however, is independent of fragment mass and therefore becomes more important with increasing mass. The FOPI collaboration has emphasized the importance of the fragments in their systematic analysis of directed fragment flow[37] and in the exciting new evidence for "radial" flow.[38]

The energy contained in directed flow is only a few percent of the total available kinetic energy.[11, 36] From entropy considerations[39] and from general energy estimates,[40] we would expect to see a much larger fraction of the total energy contained in collective flow. With a large dynamic range in fragment mass coverage, the FOPI collaboration was able to plot the mean kinetic energy per nucleon as a function of fragment charge as shown in Fig. 7. The data (open squares) do not follow the pattern expected from a purely thermal source with Coulomb corrections (triangles). Instead, they follow closely the predictions from a thermal model with isotropic radial flow (circles), and for large masses the mean kinetic energy per nucleon seems to saturate. The large difference between the data and the thermal expectations can be taken as an indication for the presence of a large amount of collective flow.

The data presented in Fig. 7 are from the most violent events with very stringent cuts on centrality and are taken within a limited range of emission angle in the center of mass system. Therefore, it is not possible to distinguish experimentally between radial (isotropic) flow and azimuthally symmetric flow perpendicular to the beam axis, the limiting case of directed flow for very central events. Since the energy contained in this new type of flow is much larger than the energy contained in directed flow,

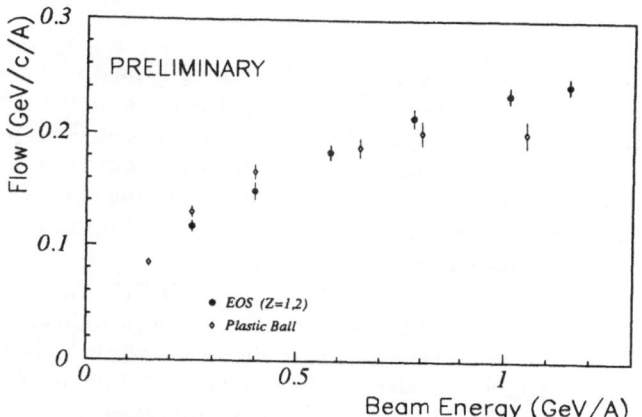

Figure 6. Flow results from Plastic Ball and EOS data as a function of beam energy for Au+Au.

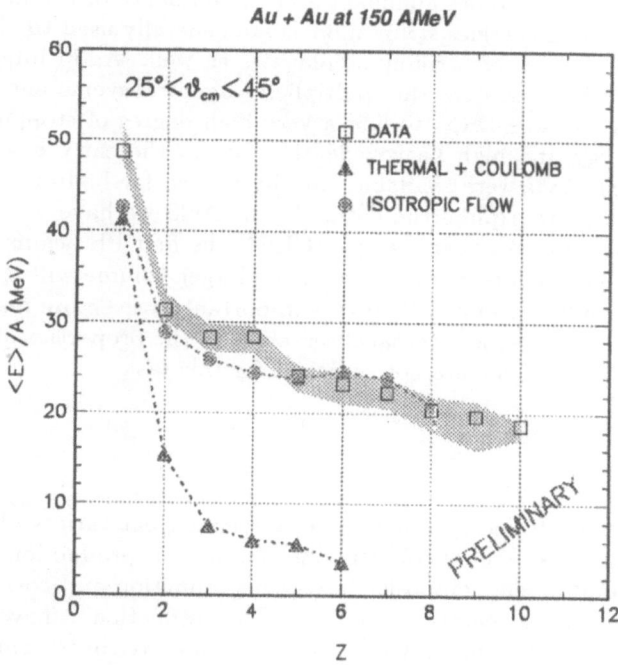

Figure 7. Mean energy per nucleon as a function of fragment charge for Au+Au at 150 MeV per nucleon. The open squares show the FOPI data, the triangles the results from a thermal model, and the circles the expectation for isotropic flow.

it should be expected that this is radial flow, predicted a long time ago.[41] It should be seen at all impact parameters and it should agree well with model calculations[42] that reproduce the directed flow since it is generated by the same mechanism.

Higher Energies

The energy dependence of the flow and the scaling behavior[33] suggest that there will be a large amount of flow at energies higher than one or two GeV per nucleon. With measurements at AGS energies between two and ten GeV per nucleon with the EOS experiment[43] it will be possible to extend the range towards higher density. This is very important since there are new studies suggesting that the density reached at top SIS/Bevalac energies might be lower than expected up to now.[44] Flow at the highest AGS energy has been observed recently.[45]

The energy frontier has moved to the AGS and to CERN. An excellent overview of the physics and the experimental program can be found in topical reviews.[46] At the higher energies the most important goal is to discover and to investigate the Quark-Gluon plasma, a new phase of nuclear matter predicted by lattice-QCD calculations. Of course if one wants to study a new phase it is important to know the properties of the phase from which such a transition is to occur. Thus the much broader goal of AGS and CERN physics is to understand the properties of hot and dense hadronic matter. At higher energies the nature of the nucleon-nucleon interactions changes, resonance production becomes abundant and the influence of the mean field less important. At very high energies string models are generally used to describe the basic interactions. But there are striking similarities as well. As in intermediate energy collisions geometry determines the multiplicity and transverse energy distributions and at the AGS and at CERN we find a very high degree of stopping,[45, 47] indicative of high energy and high baryon density. As in the early days of the Bevalac research only light ions were available initially and so far hadronic effects generally have been sufficient to explain the data. At the AGS we have seen the first results from experiments with Au beams and at CERN the first Pb beams will be available end of 1994. It will be interesting to see if the larger volume will lead to qualitative changes in the results. Also it will be very important to see more realistic models developed that relate the experimental observables to the properties of nuclear matter. Superposition models alone are not sufficient for this task.

SUMMARY

In the intermediate energy range we have seen a great variety of new experimental data. Systematic studies of collective effects, particle production and correlations, sub-threshold production, composite fragment production and correlations, rapidity distributions, multifragmentation, and di-lepton production[48] have been performed during the last years. Individually most of those data have been compared to dynamical models and their sensitivity to static and dynamic properties of nuclear matter have been tested and established. Often the uncertainties in the model predictions are larger than the sensitivity of the experimental observables to the nuclear matter properties. However, the large body of diverse experimental results gives us the chance to test the ability of the models to describe all the data for a given system at a given energy. Only as a result of such a comprehensive comparison will it be possible to determine the properties of nuclear matter at high density and temperature. Mapping the equation of state in this way over a large range of energies[49] does not depend on the concept of a quadratic form and one parameter, the compression modulus κ, but it will leave open the possibility to find new effects and processes[8] that manifest themselves in the shape of the equation of state.

REFERENCES

1. D.H. Youngblood, C.M. Rozsa, J.M. Moss, D.R. Brown, and J.D. Bronson, Phys. Rev. Lett. 39, 1188 (1977); G.E. Brown, Nucl. Phys. A 488, 689c (1988) and references therein.

2. J.R. Wilson and H.A. Bethe, Astrophys. J. 295, 14 (1985).

3. G.D. Westfall, J. Gosset, P.J. Johansen, A.M. Poskanzer, W.G. Meyer, H.H. Gutbrod, A. Sandoval, and R. Stock, Phys. Rev. Lett. 37, 1202 (1976).

4. Y. Yariv and Z. Fraenkel, Phys. Rev. C 20, 2227 (1979).

5. J. Cugnon, Phys. Rev. C 22, 1885 (1980).

6. W. Scheid, H. Müller, and W. Greiner, Phys. Rev. Lett. 32, 741 (1974).

7. A.A. Amsden, G.F. Bertsch, F.H. Harlow, and J.R. Nix, Phys. Rev. Lett. 35, 905 (1975).

8. H. Stöcker and W. Greiner, Phys. Rep. 137, 277 (1986).

9. K.-H. Kampert, J. Phys. G 15, 691 (1989).

10. G. Buchwald, G. Graebner, J. Theis, J. Maruhn, W. Greiner, and H. Stöcker, Phys. Rev. Lett. 52, 1594 (1984).

11. H.A. Gustafsson, H.H. Gutbrod, B. Kolb, H. Löhner, B. Ludewigt, A.M. Poskanzer, T. Renner, H. Riedesel, H.G. Ritter, A. Warwick, F. Weik, and H. Wieman, Phys. Rev. Lett. 52, 1590 (1984).

12. R. Stock, R. Bock, R. Brockmann, J.W. Harris, A. Sandoval, H. Ströbele, K.L. Wolf, H.G. Pugh, L.S. Schroeder, M. Maier, R.E. Renfordt, A.D. Dacal, and M.E. Ortiz, Phys. Rev. Lett. 49, 1236 (1982).

13. H.H. Gutbrod, K.H. Kampert, B.W. Kolb, A.M. Poskanzer, H.G. Ritter, and H.R. Schmidt, Z. Physik A 337, 57 (1990).

14. H.A. Gustafsson, H.H. Gutbrod, B. Kolb, H. Löhner, B. Ludewigt, A.M. Poskanzer, T. Renner, H. Riedesel, H.G. Ritter, A. Warwick, and H. Wieman, Phys. Lett. B 142, 141 (1984).

15. H. Kruse, B.V. Jacak, and H. Stöcker, Phys. Rev. Lett. 54, 289 (1985).

16. G.F. Bertsch, H. Kruse, and S.D. Gupta, Phys. Rev. C 29, 673 (1984).

17. J. Aichelin and H. Stöcker, Phys. Lett. B 176, 14 (1986).

18. J.J. Molitoris and H. Stöcker, Phys. Lett. B 162, 47 (1985).

19. P. Danielewicz and G. Odyniec, Phys. Lett. B 157, 146 (1985).

20. G. Peilert, H. Stöcker, W. Greiner, A. Rosenhauer, A. Bohnet, and J. Aichelin, Phys. Rev. C 39, 1402 (1989).

21. G. Westfall, these proceedings.

22. K.G.R. Doss, H.A. Gustafsson, H. Gutbrod, J.W. Harris, B.V. Jacak, K.H. Kampert, B. Kolb, A.M. Poskanzer, H.G. Ritter, H.R. Schmidt, L. Teitelbaum, M. Tincknell, S. Weiss, and H. Wieman, Phys. Rev. Lett. 59, 2720 (1987).

23. C.A. Ogilvie et al., Phys. Rev. Lett. 67, 1214 (1991); J. Pochodzalla, these proceedings.

24. M. Begemann-Blaich et al., Phys. Rev. C 48, 610 (1993).

25. L.G. Moretto, K. Tso, N. Colonna, and G.J. Wozniak, Phys. Rev. Lett. 69, 1884 (1992).

26. W. Bauer, G.F. Bertsch, and H. Schulz, Phys. Rev. Lett. 69, 1888 (1992).

27. R.W. Minich, S. Agarwal, A. Bujak, J. Chuang, J.E. Finn, L.J. Gutay, A.S. Hirsch, N.T. Porile, R.P. Scharenberg, B.C. Stringfellow, and F. Turkot, Phys. Lett. B 118, 458 (1982).

28. J.Elliott and P. Warren, these proceedings.

29. X. Campi, J. Phys. A 19, L917 (1986).

30. D. Krofcheck, W. Bauer, G.M. Crawley, C. Djalali, S. Howden, C.A. Ogilvie, A. Vander Molen, G.D. Westfall, W.K. Wilson, R.S. Tickle, and C. Gale, Phys. Rev. Lett. 63, 2028 (1989).

31. J. Péter, these proceedings.

32. B.A. Li, Phys. Rev. C 48, 2415 (1993); D. Klakov, G. Welke, and W. Bauer, Phys. Rev. C 48, 1982 (1993).

33. D. Keane, these proceedings.

34. H.H. Gutbrod, A.M. Poskanzer, and H.G. Ritter, Rep. Prog. Phys. 52, 1267 (1989).

35. A. Bonasera and L.P. Csernai, Phys. Rev. Lett. 59, 630 (1987).

36. T. Wienold, these proceedings.

37. J.P. Coffin, these proceedings.

38. K.D. Hildenbrand, these proceedings.

39. K.G.R. Doss, H.A. Gustafsson, H.H. Gutbrod, B. Kolb, H. Löhner, B. Ludewigt, A.M. Poskanzer, T. Renner, H. Riedesel, H.G. Ritter, A. Warwick, and H. Wieman, Phys. Rev. C 32, 116 (1985).

40. R. Stock, Phys. Rep. 135, 259 (1986).

41. P.J. Siemens and J.O. Rasmussen, Phys. Rev. Lett. 42, 880 (1979).

42. P. Danielewicz and Q. Pan, Phys. Rev. C 46, 2002 (1992).

43. G. Rai et al., AGS Proposal E895 (1993).

44. J. Jänicke and J. Aichelin, Nucl. Phys. A547, 542 (1992).

45. P. Braun-Munzinger, these proceedings.

46. J. Stachel and G.R. Young, Annu. Rev. Nucl. Part. Sci. 42, 537 (1992); H.R. Schmidt and J. Schukraft, J. Phys. G 19, 1705 (1993).

47. J. Bächler et al., Phys. Rev. Lett. in press.

48. G. Roche, these proceedings.

49. G. Peilert, J. Randrup, H. Stöcker, and W. Greiner, Phys. Lett. B 260, 271 (1991).

THE HYDRODYNAMIC MODEL FOR HIGH–ENERGY HEAVY ION REACTIONS

Y. Pürsün, U. Katscher, A. von Keitz, D.H. Rischke,
B. Waldhauser, J.A. Maruhn* and W. Greiner

Institut für Theoretische Physik, J.W. Goethe–Universität
60054 Frankfurt am Main, Germany

Introduction

Hydrodynamic models[1-3] have the great advantage of allowing for a simple relativistic formulation and explicit use of the equation of state of excited hadronic matter. On the other hand, when applying one–fluid hydrodynamics to heavy ion collisions, one assumes local thermodynamic equilibrium, which is probably not a good approximation at high bombarding energies. Nevertheless, at relatively low bombarding energies the hydrodynamic model has been successfully used to describe heavy ion collisions[4-6] and even at CERN energies it seems not to contradict experimental data[7]. Note that also the Landau model[8] and related models[9-12] are successfully applied to describe certain observables of multiparticle production in hadron–hadron, hadron–nucleus and nucleus–nucleus collisions.

In the following we will first examine the lower energy hydrodynamics and in more detail the equation of state and the approach of viscosity in nonrelativistic situations. Then we discuss the relativistic formulation of the hydrodynamic model and introduce a model equation of state with a deconfinement phase transition. Finally we give a short exposition of the fundamental problem of viscous relativistic hydrodynamics.

Lower Energy Hydrodynamics

Equations of Motion

The equations of motion of non-viscous hydrodynamics simply correspond to conservation conditions for the macroscopic fields ρ (density), $\rho\vec{v}$ (momentum density),

---

*Invited speaker

and ρE (energy density)

$$
\begin{aligned}
\partial_t \rho + \nabla \cdot (\rho \vec{v}) &= 0 \quad, & (1) \\
\partial_t (\rho \vec{v}) + \nabla \cdot (\rho \vec{v} \vec{v}) &= -\nabla p \quad, & (2) \\
\partial_t (\rho E) + \nabla \cdot (\rho E \vec{v}) &= -\nabla \cdot (\vec{v} p) \quad. & (3)
\end{aligned}
$$

Here $\vec{v}$ is the flow velocity, E the energy per particle (containing kinetic and internal energy), while p stands for the pressure, which should be determined from ρ and E using the equation of state. The equation of state being essentially unknown and heuristic functions were usually employed.

Typical Heuristic Equations of State

A widely used but in principle unjustified assumption is that of separability of the equation of state into a compressional and a thermal part. Formally the internal energy $W(\rho, T)$ enters into the total energy per nucleon via

$$
E = \tfrac{1}{2} m v^2 + W(\rho, T) \quad, \tag{4}
$$

and the latter is assumed to split according to

$$
W(\rho, T) = W_0(\rho) + W_{\text{th}}(\rho, T) \tag{5}
$$

with $W_0(\rho) = W(\rho, T = 0)$. This function is often simply referred to as the equation of state, because it contains the unknown compressional behaviour of nuclear matter, while W_{th} is usually assumed to be given by the Fermi gas expressions.

The known properties of nuclear matter impose some constraints on the function $W_0(\rho)$. The equilibrium value must be $W_0(\rho_0) = -B_0$ and the incompressibility should be $K = (9 \partial^2 W_0 / \partial \rho^2)_{\rho_0}$. In principle vacuum properties should also be reproduced, i. e. $W_0(\rho = 0) = 0$ with a quadratic rise for low densities, where nuclear matter should be close to a gas of noninteracting nucleons. In reality, however, nuclear matter is unstable in this region with respect to a breakup into nucleons and light nuclei, so that the equation of state is not required to describe this region correctly. In practical calculations the behaviour of nuclear matter in the breakup region is described by different models, so that the functional form of $W_0(\rho)$ does not matter there. The breakup density is expected to be between one third and one half of ρ_0.

Another condition is provided by causality. Relativistically, the speed of sound is given by the formula

$$
c_s^2 = \left. \frac{\partial W(\rho, s)}{\partial e} \right|_s \quad, \tag{6}
$$

which in the case of zero entropy reduces to $c_s^2 = d W_0(\rho) / de$, where e is the total internal energy density including the rest mass, $e = \rho(mc^2 + W)$. Now c_s should not exceed the speed of light. This leads to the condition that $W_0(\rho)$ should not rise more rapidly than linearly with ρ for large ρ. However, since c_s^2 is related to the curvature of $W_0(\rho)$, there may also be problems if this function contains regions of larger curvature.

Widely used parametrizations are (to abbreviate the expressions we use the *compression ratio* $x = \rho / \rho_0$):

- The quadratic equation of state

$$W_0(x) = B_0 + \frac{K}{18}(x-1)^2 \quad , \quad p_0(x) = \frac{K}{9}\rho_0 x^2(x-1) \quad , \tag{7}$$

$$c_s^2 = \frac{2x(3x-2)}{18(mc^2+B_0)/K + 3x^2 - 4x + 1} \quad , \tag{8}$$

with the standard choices $K = 210$ MeV, $B_0 = -16$ MeV and $mc^2 = 938$ MeV, causality becomes violated at all densities above $x \approx 5.2$.

- The linearized equation of state

$$W_0(x) = B_0 + \frac{K}{18x}(x-1)^2 \quad , \quad p_0(x) = \frac{K}{18}\rho_0(x^2-1) \quad , \tag{9}$$

$$c_s^2 = \frac{x}{9(mc^2+B_0)/K + x - 1} \quad , \tag{10}$$

which for reasonable values of the parameters never becomes acausal.

Thermal Part

The thermal part is, as has been mentioned, usually assumed to be given by the expressions for an ideal gas. For many qualitative studies it is sufficient to use the relation $p_{th} = \frac{2}{3}\rho W_{th}$ which is the only property of the equation of state needed for inviscid hydrodynamic simulations and is exactly valid for any mixture of nonrelativistic noninteracting particles. However, sometimes this is not sufficient. If the temperatures reach values above about 50 MeV, the creation of pions becomes quite important, and these cannot be treated as nonrelativistic particles because of their small mass. In this case, and if the temperature of the system is needed for the equations of motion, one has to explicitly evaluate the Fermi integrals for the nucleon and its excited states (Δ–resonance etc.) and the Bose integrals for the pion and heavier mesons. The resulting calculation is so involved that the results are usually inserted into the hydrodynamic simulations in tabular form.

Viscosity

Investigations concerning viscosity were carried out by Schmidt et al.[13]. In the equations of motion, one has to add a Newtonian viscous stress tensor, i. e. replace the scalar pressure by the tensor σ_{ij}, which can be written as

$$\sigma_{ij} = -p \cdot \delta_{ij} + \eta \cdot \left(\frac{\partial v_i}{\partial x_j} + \frac{\partial v_j}{\partial x_i} - \frac{2}{3}\delta_{ij}\frac{\partial v_k}{\partial x_k} \right) + \zeta \cdot \delta_{ij}\frac{\partial v_k}{\partial x_k} \tag{11}$$

with η and ζ being the coefficients of shear and bulk viscosity. These are, in general, functions of density and temperature[14]. In our model, however, they are set constant.

To allow a comparison of our results with the Quantum Molecular Dynamics (QMD) and to show the influence of the equation of state two different values for the incompressibility K were used: the *soft* equation of state, where $K = 160$ MeV, and the *hard* equation of state, where $K = 400$ MeV.

Let us now examine just a few selected results. Figure 1 shows the influence of viscosity and the equation of state on the compression and entropy achieved. Apparently viscosity affects both quantities considerably, and this shows that not only the associated broadening of the shock front is responsible, because in that case the conditions in the compressed zone would not be affected at all. Even more interesting, however, is the effect on the observables compared to that of the equation of state.

As one can see from figure 1, the influence of the equation of state on the entropy is low. Here we compare calculations of Au+Au at 400 A MeV with $\eta = 0$, $\zeta = 0$ and $\eta = 60$, $\zeta = 0$ MeV/fm^{2}c for both the hard and the soft equation of state. Particularly in the case of a viscous calculation, the equation of state does not show *any* influence. A comparison with experimental data shows that the larger value of the viscosity may be more appropriate.

Also it turns out that the flow angle is almost independent of either the equation of state and viscosity[13]. It does not even vary significantly for different impact energies; however, it depends strongly on the impact parameter. This implies that the flow angle is a purely geometrical quantity. In summary, it is therefore not yet clear at present to what extent the equation of state can be untangled from the effects of dissipative terms, which in principle are no better known than the equation of state itself.

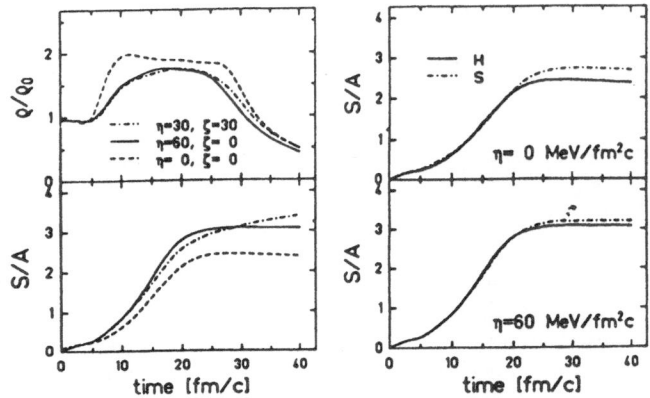

Figure 1. Compression (top,left) and entropy per baryon in a Au+Au reaction at 400 A MeV, impact parameter $b = 3$ fm. Left: influence of the viscosity on compression and entropy (hard equation of state); right: influence of the equation of state (H: hard, S: soft) for two different viscosities. The viscosity value is given in units of MeV/fm^{2}c.

In microscopic models, of course, both effects are also present. Roughly speaking, one may liken the equation of state to the mean field, and the dissipative terms to the influence of collisions. The difference is that the collisional cross sections are better known than the viscosity in hydrodynamics; however, there is the problem of whether they are the same in nuclear matter as in free nucleon–nucleon collisions, where they are measured.

Relativistic Formulation

Equations of Motion

We formulate the equations in an arbitrary inertial frame, for which in the calculations usually the laboratory–fixed frame or the equal velocity frame are taken. For the notation we follow[15], *i.e.* four–vectors $x^\mu = \{t, \vec{x}\} = \{t, x^k\}$ represents the space–time coordinates, with Greek indices running from 0 to 3 and Latin indices from 1 to 3 and choose the natural units $\hbar = c = 1$. The inner product is defined via the metric tensor $g_{\mu\nu}$ which is diagonal with diagonal elements $g_{00} = 1 = -g_{ii}$.

The equations of motion of relativistic hydrodynamics can be obtained from conservation conditions for the four–current and energy–momentum tensor

$$\partial_\mu j^\mu = 0 \quad , \quad \partial_\mu T^{\mu\nu} = 0 \quad . \tag{12}$$

The four–current density j^μ is defined in terms of the fluid density in the rest frame, n, as $j^\mu = nu^\mu$, where $u^\mu = \gamma(1, \vec{v})$ denotes the four–velocity with the familiar relativistic factor $\gamma = (1 - \vec{v}^2)^{-1/2}$ and is normalized to $u_\mu u^\mu = 1$. For a perfect fluid the energy-momentum tensor is given by[16]

$$T^{\mu\nu} = (e + p)u^\mu u^\nu + p\,g^{\mu\nu} \quad , \tag{13}$$

where e is the energy density measured in the local rest frame (including the rest mass of the particles), and p the hydrodynamic pressure. To rewrite these equations in a more familiar form, the individual parts of $T^{\mu\nu}$ in the laboratory frame that correspond to classical quantities are defined as the energy and momentum density

$$E = T^{00} = \gamma^2(e + p) - p \quad , \tag{14}$$
$$M_k = T^{0k} = \gamma^2(e + p)v_k \quad . \tag{15}$$

In the same way, the four–current can be decomposed into the density in the laboratory frame,

$$\rho = j^0 = \gamma n \tag{16}$$

and the spatial part given by $j^k = \rho u^k = \gamma \rho v^k$. Utilizing these definitions, the equations of motion can be rewritten in three–dimensional notation as

$$\partial_t \rho + \nabla \cdot (\rho \vec{v}) = 0 \quad , \tag{17}$$
$$\partial_t \vec{M} + \nabla \cdot (\vec{v}\vec{M}) = -\nabla p \quad , \tag{18}$$
$$\partial_t E + \nabla \cdot (\vec{v}E) = -\nabla \cdot (\vec{v}p) \quad . \tag{19}$$

These equations are identical in form to the nonrelativistic ones, so that many of the numerical methods of standard fluid dynamics can be taken over. The big difference, however, is in the application of the equation of state, which is given *in the local rest frame* as $p = p(n, e)$. The fields ρ, $\vec{M}$, and E, which are all defined in a fixed system of reference such as the laboratory frame, have been propagated to a new time, the rest frame quantities n and e as well as the velocity $\vec{v}$ have to be found *at each point* such as to satisfy the equations (14)–(16).

The full set of equations, together with the equation of state $p = p(n, e)$ is solved with the SHASTA algorithm[17]; for details see Ref.[18].

Equation of State

For high energies it is also necessary to take into account relativistic effects. In this section we present a model equation of state for nuclear matter with a deconfinement phase transition. To describe a strongly interacting matter covariantly we use the mean–field approximation[19, 20]. Following Ref.[21] we obtain a thermodynamical self consistent formulation of the hadronic phase. The thermodynamical quantities denote as functions of the temperature T and baryochemical potential μ:

$$n = \frac{\gamma}{(2\pi)^3} \int d^3k(f^+ - f^-) \quad , \tag{20}$$

$$p = \frac{\gamma}{3(2\pi)^3} \int d^3k \frac{k^2(f^+ + f^-)}{(k^2 + m_N^{*2})^{1/2}} + P(m_N^*) + P_1(n) \quad , \tag{21}$$

$$e = \frac{\gamma}{(2\pi)^3} \int d^3k(k^2 + m_N^{*2})^{1/2}(f^+ + f^-) + U(n)n - P(m_N^*) - P_1(n) \quad , \tag{22}$$

where $\gamma = 4$ is the internal degree of freedom of nucleons and $f^\pm$ are the Fermi distribution functions for (anti–)nucleons with effective mass m_N^* in the scalar field $U(n)$. The functions $P(m_N^*)$ and $P_1(n)$ describe additional mesonic field contributions to the pressure and energy density.

The relation

$$P_1(n) = U(n)n - \int_0^n dn'U(n') \tag{23}$$

determines P_1 and is a thermodynamic self–consistent condition for the equations (20)–(22). The equation of state is then parameterized by choices of the functions U and P. In our calculations we choose

$$U(n) = C_v^2 n - C_d^2 n^{1/3} \quad , \quad P(m_N^*) = -\frac{1}{2C_s^2}(m - m_N^*)^2 \quad , \tag{24}$$

proposed in Ref.[21]. For $C_d^2 = 0$ in (24) one obtains the Walecka model[19] where C_s^2 and C_v^2 are parameters related to the coupling constants of the baryon fields with the scalar and vector fields. An additional contribution to the hadronic phase comes from an ideal relativistic Bose gas of thermally excited massive pions, further baryonic and mesonic resonances are neglected. We choose the model parameters[21] $C_s^2 = 300.01\,\text{GeV}^{-2}$, $C_v^2 = 242.30\,\text{GeV}^{-2}$, $C_d^2 = 0.184$ to ensure the nuclear matter ground state values $n_0 = 0.16\,\text{fm}^{-3}$ for baryonic density, $W_0 = -16\,\text{MeV}$ for binding energy per particle and $K_0 = 300\,\text{MeV}$ for incompressibility constant.

The quark–gluon plasma phase is described by the MIT bag model equation of state[25]. For two flavours of massless quarks (u and d) and gluons the thermodynamic functions in terms of the temperature T and baryochemical potential μ are

$$n_Q(T,\mu) = \frac{2}{9}\mu(T^2 + \frac{1}{2\pi^2}\mu^2) \quad , \tag{25}$$

$$p_Q(T,\mu) = \frac{37}{90}\pi^2 T^4 + \frac{1}{9}\mu^2 T^2 + \frac{1}{162\pi^2}\mu^4 - B \quad , \tag{26}$$

$$e_Q(T,\mu) = 3p_Q(T,\mu) + 4B \quad , \tag{27}$$

for the nonperturbative vacuum pressure B we take the value $B = (235\,\text{MeV})^4$. Then the whole phase diagram in the (T,μ)–plane is constructed according to the Gibbs criterion for systems with a first order phase transition

$$p_H(T,\mu) = p_Q(T,\mu) \quad . \tag{28}$$

The critical temperature with the mentioned parameters is $T_c = 169\,\text{MeV}$ and we get a large value for the latent heat density $l = 4B = 1.59\,\text{GeV/fm}^3$.

Shock Model

The compressional shock model for nucleus–nucleus collisions relies on the conservation of energy–momentum flow and baryon current. It seems that the nuclear stopping power at weak relativistic laboratory energies is large enough to justify this approach at least for central collisions of heavy nuclei.

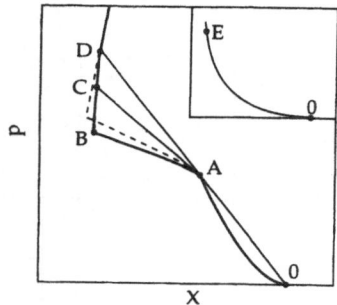

Figure 2. The generalized RHTA for an equation of state with thermodynamically anomalous regions (A–D) and the RHTA with stable single shock solutions (smaller part).

It is therefore of interest to calculate the compressional shock adiabat, the so-called Rankine–Hugoniot–Taub adiabat (RHTA), which links the states in the unperturbed incoming nuclear matter with the compressed zone

$$n^2 X^2 - n_0 X_0^2 - (p - p_0)(X + X_0) = 0 \quad , \tag{29}$$

where $X = (e + p)/n^2$ is the generalized specific volume and the parameters $p_0 = 0$, $X_0 = 5.8\,\text{GeVfm}^3$ correspond to the nuclear matter ground state. The points reached on the RHTA is defined by the bombarding energy according to

$$E_{\text{lab}} = 2m_N \left[\left(\frac{e/n}{e_0/n_0} \right)^2 - 1 \right] \quad . \tag{30}$$

The form of the RHTA is further determined by the equation of state. Generally for equations of state without a phase transition the solutions of Eq. (29) have the form shown by the inlay of Fig. 2 which corresponds just to single shock wave solutions. For equations of state with deconfinement phase transition thermodynamically anomalous regions may be reached, especially if the compressed zone is in the quark–hadron mixed phase[22]. In this case the single shock can be unstable and stable configurations of several shock waves occur. Figure 2 shows the result for the introduced equation of state. The points of the section A–D correspond to unstable single shock transitions (dashed line). The stable physical solution (full line) is the so called generalized RHTA[23]. Instead of single shocks the solutions in the section A–D are series of shock waves which are shown in Figure 3. One interesting part A–B of the generalized RHTA yields the configuration shock plus compressional simple wave (Fig. 3.b). The shock wave is fixed for all states in this section, namely a single shock from the centre of the RHTA 0 to A. The simple wave solution according to poisson adiabat from A to a point between A–B has a constant density s/n. This leads to a constant value of the total entropy of the system in some interval of bombarding energy. In our calculation the deconfinement transition occurs in the energy region of $E_{lab} = 1.4 - 4.2$ A GeV and we reach the poisson adiabat for bombarding energies of $E_{lab} = 2.4 - 3.8$ A GeV.

The existence of a mixed phase and therefore a thermodynamically anomalous region of the equation of state causes a plateau like structure in the excitation function of the total entropy and temperature. The constant value of the total entropy leads also to a plateau like structure of the pion multiplicity[24]. In summary the results with the generalised RHTA shows signature of the deconfinement transition.

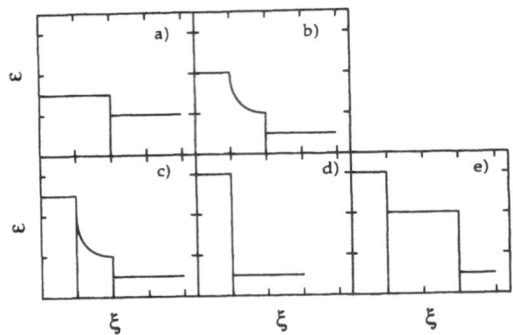

Figure 3. Energy density profiles of compression shock wave configurations corresponding to the generalized RHTA in Fig. 2: between the points 0–A and above D stable shock waves appear (a,d), (b) shows a superposition of a compressional simple wave and a shock wave for the region A–B, (c) is a simple wave between two shock waves corresponding to the points between B–C and (e) is a double shock wave configuration for the region C–D.

Relativistic Heavy Ion Collisions

In the (3+1)–dimensional calculation we chose the collision O+Au at 200 A GeV, corresponding to a Lorentz–factor of $\gamma = 10$, and a impact parameter $b = 0$. In the hydrodynamical simulation of a heavy ion collision, the SHASTA works most effectively on a computational grid which is fixed with respect to the equal velocity frame of the colliding nuclei. To increase the longitudinal resolution of the colliding nuclei (which appear as "pancakes" at the respective energies), it seems attractive to contract the calculational grid along the collision axis (in our case, the x–axis) by the γ–factor of the colliding nuclei in the equal velocity frame. Although the conservation laws are not violated, the feature of a Lorentz–contracted grid causes the acausal propagation of compressional shock waves in the transverse (y–z–) direction and thus a too fast deceleration of the colliding nuclei.

To avoid such errors a cubic grid was used in our calculations. In Fig. 4 we show the calculation with a cubic grid. One clearly observes that the oxygen nucleus drills a hole into the gold nucleus. The transverse shock wave travels with subluminous velocity into the gold nucleus, so that a large amount of spectator nucleons remains at the initial rapidity until the late stage of the reaction[7]. We note that the flow geometry suggests the very intuitive conical shock wave model for asymmetric head-on collisions, which was proposed in[26] and extensively discussed in[27]. The left part of the same figure shows the rapidity distribution, which are in good agreement with data[28].

We mention that up to now we have only considered *fluid* rapidity distributions, i.e., the "freeze-out" of the fluid has been neglected. Our distribution will be "smoothed" and smeared out by this effect.

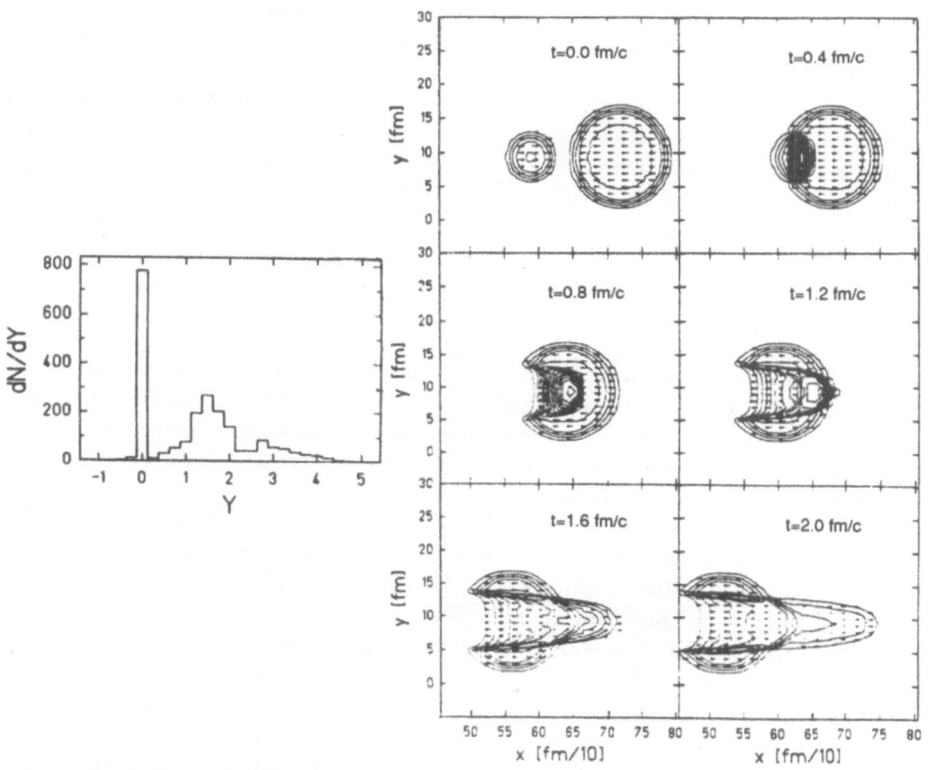

Figure 4. Right part: contour plots of the baryonic density in the equal-velocity frame for a central collision of O+Au at 200 A GeV. The beam direction corresponds to the horizontal axis and is stretched by $\gamma = 10$. Left part: fluid rapidity distribution at $t = 5.2$ fm.

These results show that the SHASTA exactly fulfills the conservation laws of nuclear fluid dynamics at ultrarelativistic energies and reproduces the correct propagation of shock waves[7]. The reasonable agreement with the experimental rapidity distribution for O+Au at 200 A GeV was also found by the authors of Ref.[29], with another numerical scheme to solve Euler's equations. Furthermore a comparison between the SHASTA and a principally different numerical scheme, the Basis Spline Collocation Method[30], shows well agreement[31]. These results are important for establishing the validity of the numerical methods used.

More–Fluid Models

In order to take into account nonequilibrium effects in high–energy heavy ion collisions Amsden et al. proposed the first more–fluid formulation, a two–fluid model[32]. In this model target and projectile nucleons are considered as two separate fluids coupled via interaction terms. This leads to a partial mutual stopping of the fluids during the collision. The two–fluid model was further developed in Refs.[2, 6, 33]. In Ref.[34] a nonrelativistic three–fluid model was developed, introducing aa additional fluid of stopped baryons. Depending on the strength of the interaction between target and projectile fluids this model contains the one–fluid model (strong coupling) as well as the two–fluid model (weak coupling) as limiting cases. The relativistic form of this three–fluid model[35] was restricted to heavy ion collisions with energies below 1 A GeV, since only elastic nucleon–nucleon collisions were taken into account in the coupling terms.

An ultrarelativistic formulation of the three–fluid model in (3+1)–dimensions can be found in this volume. Here separately two baryonic fluids, formed of projectile and target nucleons, and a hadronic fluid, which is produced as a result of inelastic nucleon–nucleon collisions, will be considered. The results obtained agree well with observed particle spectra. Also in a simplified form the model seems to work reasonable[36].

Fundamental Problems

It may appear strange that in the nonrelativistic case it was shown that viscosity is quite important for the hydrodynamic model, yet no attempt has been made up to now to incorporate viscosity into the relativistic formulation. The reason for this is the lack of a consistent formulation of of relativistic viscous hydrodynamics. Let us briefly explain the state of this field.

The non–viscous theory as presented and used is not known to have any problems. It is stable, preserves causality, and has been used extensively in astrophysical and cosmological investigations. Unfortunately there is no more direct experimental proof of its validity, such as is possible with classical fluids, and one may have doubts, e. g., whether the assumption of a separation of time scale between collective and single–particle motion can at all be fulfilled in an ultrarelativistic situation. Nevertheless, the equations of motion are a good starting point for relativistic many-body physics.

For the viscous case the natural generalization is the extension of Newtonian viscosity to a covariant form. Such equations have been applied occasionally, but were later shown to lead to acausal propagation of effects[37] and to also contain serious instabilities[38], which could conceivably make a water drop explode within 10^{-34}s! The latter problem is still being discussed, some[39] argue that the instability may only occur in regions where the hydrodynamic approximation is not valid anyway.

Israel and Stewart[40] proposed an extended theory, which is similar to Grad's 13-moment equations in the nonrelativistic case. This theory was believed to be causal and stable until Hiscock and Olson[41] discovered similar problems, leading to the statement that "the theory is significantly less pathological when formulated in the Landau-Lifshitz frame" – a formulation we do not want to explain in detail here, but which makes the uncertain state of affairs quite obvious. The principal result is that the Israel-Stewart theory is always acausal and sometimes unstable. In addition, there are problems with strong shocks, in the sense that the regular shock structure breaks down for an upstream Mach number larger than 1.65 even in the nonrelativistic case[42, 43].

At present one may thus conclude that using one of the formulations of viscous relativistic hydrodynamics entails the danger of testing the shortcomings of such a theory and not the equation of state when applied to high–energy heavy ion collisions. There is, of course, a similar problem in the presently used classical microscopic theories, since there is also no universally accepted and experimentally proven formulation of the classical relativistic many-body system with interacting particles. In our opinion this situation adds to the challenge of high–energy heavy ion physics by giving an opportunity for the study of relativistic many–body physics on a very fundamental level, going beyond the properties of nuclear matter, that may not be possible in the laboratory anywhere else!

Acknowledgements

This work was supported by the Gesellschaft für Schwerionenforschung (GSI) and the Bundesministerium for Forschung und Technologie (BMFT).

REFERENCES

1. J.A. Maruhn and W. Greiner in *Treatise on Heavy–Ion Science*, vol. 4, ed. D.A. Bromley, Plenum Press, NY and London:565 (1985).

2. R.B. Clare, D. Strottman, Phys. Rep. 141:177 (1986).

3. I.N. Mishustin, V.N. Russkikh, L.M. Satarov in *Relativistic Heavy–Ion Physics*, vol. 6, eds. L.P. Csernai, D. Strottman, World Scientific, Singapore:179 (1991).

4. G. Buchwald, G. Graebner, D. Barthel, T. Rentzsch, J.A. Maruhn, W. Greiner, H. Stöcker, Nucl. Phys. A428:537c (1984).

5. W. Schmidt, B. Waldhauser, H. Stöcker, University Frankfurt preprint UFTP 295 (1992).

6. I.N. Mishustin, V.N. Russkikh, L.M. Satarov, Sov. J. Nucl. Phys. 48:454 (1988); Sov. J. Nucl. Phys. 54:260 (1991).

7. B. Waldhauser, D.H. Rischke, U. Katscher, J.A. Maruhn, H. Stöcker, W. Greiner, Z. Phys. C54:459 (1992).

8. L.D. Landau, S.Z. Belen'kii, Izv. Akad. Nauk SSSR, Ser. fiz. 17:51 (1953).

9. I.L. Rozental, Y.A. Tarasov, Sov. Phys. JETP 58:892 (1983).

10. K.S. Lee, E. Schnedermann, J. Sollfrank, U. Heinz, Nucl. Phys. A525:523 (1991).

11. U. Ornik, R.M. Weiner, Phys. Lett. B263:503 (1991).

12. J. Stachel, P. Braun-Munzinger, Phys. Lett. B216:1 (1989).

13. W. Schmidt, PhD–thesis, University Frankfurt (1989).

14. P. Danielewicz, Phys. Lett.B146:141 (1984).

15. C.W. Misner, K.S. Thorne, J.A. Wheeler, "Gravitation" (W.H. Freeman and Co., San Francisco 1973).

16. L.D. Landau and E.M. Lifshitz, "Fluid Mechanics" (Pergamon Press, London 1959).

17. J.P. Boris and D.L. Book, J. Comp. Phys. 11:38 (1973).

18. V. Schneider, U. Katscher, D.H. Rischke, B. Waldhauser, J.A. Maruhn, C.D. Munz, J. Comp. Phys. 105:92 (1993).

19. J.D. Walecka, Ann. Phys. 83:491 (1974); B.D. Serot and J.D. Walecka, Phys. Lett. 27B:172 (1979).

20. J. Theis, G. Graebner, G. Buchwald, J.A. Maruhn, W. Greiner, H. Stöcker, J. Polonyi, Phys. Rev. D28:2286 (1983).

21. M.I. Gorenstein, D.H. Rischke, K.A. Bugaev, H.Stöcker, W. Greiner, University Frankfurt preprint UFTP 239 (1989).

22. K.A. Bugaev, M.I. Gorenstein, V.I. Zhdanov, Z. Phys. C43:365 (1988).

23. K.A. Bugaev, M.I. Gorenstein, B. Kämpfer, V.I. Zhdanov, Phys. Rev. D40:2903 (1989).

24. K.A. Bugaev, M.I. Gorenstein and D.H. Rischke, Phys. Lett. B255:18 (1991).

25. J. Clemens, R.V. Gavai, E. Suhonen, Phys. Rep. 130:217 (1986).

26. H.G. Baumgardt, J.U. Schott, Y. Sakamoto, E. Schopper, H. Stöcker, J. Hofmann, W. Scheid, W. Greiner, Z. Phys. A273:359 (1975); G.F. Chapline, A. Granik, Nucl. Phys. A459:681 (1986); Nucl. Phys. A511:747 (1990).

27. D.H. Rischke, H. Stöcker, W. Greiner, Phys. Rev. D42:2283 (1990).

28. H.R. Schmidt and the WA80 collaboration, Z. Phys. C38:109 (1988).

29. T.L. McAbee, J.R. Wilson, J.A. Zingman, C.T. Alonso, Mod. Phys. Lett. A4:983 (1989).

30. C. Bottcher and M.R. Strayer, Ann. of Physics 175:175 (1987).

31. D.J. Dean, C. Bottcher, M.R. Strayer, J.C. Wells, A.v. Keitz, Y. Pürsün, D.H. Rischke, J.A. Maruhn, (submitted to Phys. Rev. E, 1993).

32. A.A. Amsden, A.S. Goldhaber, F.H. Harlow, J.R. Nix, Phys. Rev. C17:2080 (1978).

33. V.N. Russkihk and Y.B. Ivanov, Nucl. Phys. A543:751 (1992).

34. L.P. Csernai, I. Lovas, J.A. Maruhn, A. Rosenhauer, J. Zimanyi, W. Greiner, Phys. Rev. C26:149 (1982).

35. A. Rosenhauer, J.A. Maruhn, W. Greiner, L.P. Csernai, Z. Phys. A326:213 (1987).

36. U. Katscher, D.H. Rischke, J.A. Maruhn, W. Greiner, I.N. Mishustin, L.M. Satarov, Z. Phys. A346:209 (1993).

37. I. Müller, Z. Physik 198:329 (1967).

38. W.A. Hiscock and L. Lindblom, Phys. Rev. D31:725 (1985).

39. Ch.G. van Weert, in: *Relativistic Fluid Dynamics*, ed. A. Anile and Y. Choquet-Bruhat (Lecture Notes in Mathematics 1385, Springer-Verlag, Berlin and Heidelberg): 290 (1989).

40. W. Israel and J.M. Stewart, Ann. Phys. 118:341 (1979).

41. W.A. Hiscock and T.S. Olsen, Phys. Lett. a141:125 (1989).

42. H. Grad, Commun. Pure Apll. Math. 5:257 (1952).

43. A.M. Anile and A. Majorana, Meccanica 16:149 (1981).

ELECTROMAGNETIC PROBES OF
HOT AND DENSE NUCLEAR MATTER

Karl-Heinz Kampert

University of Münster
Institut für Kernphysik
D-48149 Münster
and
WA80-Collaboration, CERN

INTRODUCTION

The primary motivation for studying high-energy heavy-ion collisions is to investigate nuclear matter under conditions of extremely high densities and temperatures. The response of the nuclear medium to changes of its temperature and density is described by the nuclear matter equation of state which is subject of quantitative discussions in the energy regime of 1 GeV per nucleon. Many new results from experiments adressing these questions at the Berkeley Bevalac and SIS in Darmstadt have been discussed during this meeting. Among the ultimate goals of ultrarelativistic nuclear collisions ($E_{proj} \gtrsim 10\,\mathrm{AGeV}$) is the formation and observation of a quark-gluon plasma (QGP) which has been predicted by QCD lattice calculations (for a recent review on this subject the reader is referred to Ref.[1]). In such a novel state of matter quarks and gluons are deconfined over an extended volume and chiral symmetry may (partially) be restored. A vigorous experimental programme is now under way at the CERN-SPS and Brookhaven-AGS accelerators with additional preparations being carried out for Brookhaven's Relativistic Heavy-Ion Collider (RHIC) and CERN's Large Hadron Collider (LHC). The creation and observation of a QGP represents a considerable challenge, both in its experimental realization and also in the theoretical interpretation of the experimental results. The lifetimes involved are of the order of $\approx 10\,\mathrm{fm/c}$ and the detailed dynamics of the collision process may furthermore play an important rôle, because the subsequent hadronization tends to mask the signal from the QGP phase, thus complicating the extraction of a clear signal. Nevertheless, much progress has been made both in theory and experiment, and even in the absence of a genuine QGP the study of hot and dense hadronic systems is a fascinating subject from which a great deal can be learned[2].

Electromagnetic probes are very well suited for studies of strongly interacting particles under conditions from free space to nuclear matter at high densities and temperatures. They carry away virtually unaltered information about the reaction since they do not suffer strong interaction rescattering. This results in mean free paths much bigger than typical sizes of the nuclear systems. Since production rates are rapidly increasing functions of temperature and density, electromagnetic signals provide valuable information on the early hot and dense phases of the reaction and should constitute precious aids in the process of analyzing the behavior of a hot QGP. Early theoretical work on this subject can be found in Refs.[3, 4, 5, 6, 7, 8]. For hadronic particles the situation is quite different, since their abundances and phase space distributions are changed by the collisions as the final matter expands. However, as with any possible experimental signature of the QGP, a great deal of care must go into the calculation of a corresponding "purely hadronic" signal that is a contribution to the same experimental observables from sources other than a deconfined chiral-symmetric phase. Thus, one must learn if contributions from sources other than the QGP either are dominant in some part of phase space or have so distinct properties so that they can be separated.

To establish an overall picture of the electromagnetic emission, we shall first consider the basic production processes of real and virtual photons in a QGP as compared to a hadron gas and perform rate calculations for static systems at different temperatures. More interestingly for heavy-ion collisions, however, are absolute cross sections. They allow to judge, whether such probes are observable above background in a real experiment. We shall assess the various yields in a rather simple framework of ideal gases by assuming entropy conservation and a one-dimensional Bjørken expansion. Preliminary experimental data from WA80 on direct photon production will be presented and possible interpretations be discussed. Finally, an outlook to future experiments at collider energies will be given.

SOURCES OF REAL AND VIRTUAL PHOTONS

A system formed during relativistic heavy ion collisions consists of many charged objects moving in close proximity and thus emitting radiation in form of real and virtual photons with the latter being observed as lepton-pairs (e^+e^- or $\mu^+\mu^-$). In a QGP those charged objects are represented by (anti-)quarks, whereas in a hadron gas they are represented by mesons and baryons. Photons and lepton-pairs are thus emitted during the hole collision process. This is schematically depicted in Fig. 1 where the invariant mass spectrum of lepton pairs is shown for high energy nuclear collisions.

When the nuclei start to interpenetrate at very high collision energy, their partons will interact non-thermally. In case of real photons the primary production processes (order $\alpha \alpha_s$) are quark-gluon Compton scattering, $qg \rightarrow q\gamma$, and quark-antiquark annihilation $q\bar{q} \rightarrow g\gamma$ (Fig. 2a, b, respectively). The particular primary process (order α^2) of lepton-pair production is the Drell-Yan process of quark-

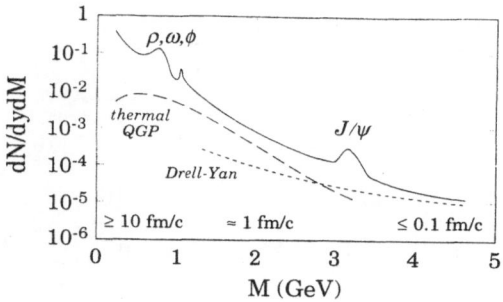

Figure 1. A schematic presentation of the di-lepton mass distribution in high energy nuclear collisions. Typical time scales of their emission are indicated.

antiquark annihilation; $q\bar{q} \rightarrow \gamma^* \rightarrow \ell^+\ell^-$ (Fig. 2c). All of these hard processes are rather well understood and can be treated as interactions of point-like particles in the framework of perturbative QCD. Its comparison to experimental data has provided a great deal of information about the quark- and gluon structure functions[9, 10, 11] of nucleons and nuclei and furthermore serves as a reference for the study of the suppression of the J/ψ and higher charmonium and bottonium states in a QGP. As sketched in Fig. 1, lepton-pair spectroscopy allows to identify the decay of various vector mesons, like $J/\psi, \phi, \rho$, and ω and to study the production properties of these particles. We will return to this important aspect below.

As mentioned above, the differential cross sections of the diagrams in Fig. 2 are well understood and can after carrying out color-sums and spin-averages for the Compton and annihilation graph be expressed as follows[12]:

$$
\frac{d\sigma}{dt}(qg \rightarrow \gamma q) = \pi\alpha\alpha_s e_q^2 \frac{1}{6}\frac{-2}{\hat{s}^2}\left(\frac{\hat{u}}{\hat{s}} + \frac{\hat{s}}{\hat{u}}\right) \tag{1}
$$

and

$$
\frac{d\sigma}{dt}(q\bar{q} \rightarrow \gamma g) = \pi\alpha\alpha_s 2e_q^2 \frac{4}{9}\frac{2}{\hat{s}^2}\left(\frac{\hat{u}}{\hat{t}} + \frac{\hat{t}}{\hat{u}}\right) \tag{2}
$$

with the quark charge e_q and the usual Mandelstam invariants $\hat{s}, \hat{t}$ and $\hat{u}$ for the constituent subprocess ($\hat{s} = x_a x_b s$).

Thermal photon radiation from a QGP

The thermal component in Fig. 1 arises from collisions among quanta which overlap after the primary collision and have a momentum spread characterized by the temperature. Thus, to calculate rates of thermal photon and lepton-pairs in

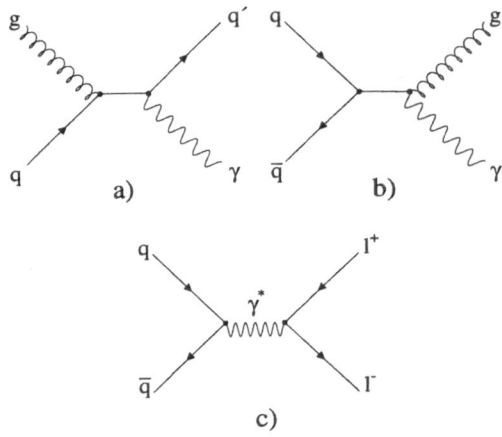

Figure 2. QCD Feynman diagrams of leading order real and virtual photon production in a QGP; (a) quark-gluon Compton $qg \rightarrow q\gamma$, (b) quark-antiquark annihilation $q\bar{q} \rightarrow g\gamma$, (c) lepton pair production by quark-antiquark annihilation $q\bar{q} \rightarrow \gamma^* \rightarrow \ell^+\ell^-$.

a QGP one assumes Fermi-Dirac and Bose-Einstein distributions of a certain temperature for the quarks and gluons, respectively. Sometimes, both distributions are approximated for simplicity by a Maxwell-Boltzmann distribution[13, 14]. Perturbative QCD at finite temperature suffers, however, from serious problems leading to infrared singularities and gauge dependence for many observables calculated in this way. The reason for this is the fact that naive perturbation theory of finite temperature is incomplete, i.e. an expansion in the number of loops is not equivalent to an expansion in the coupling constant g. This problem can be circumvented by using effective propagators and vertices based on a resummation of the so-called hard thermal loop diagrams, as was shown by Braaten and Pisarki[15]. In this way, medium effects, e.g. Debye screening, are included and improve the infrared behavior of the results drastically. At the same time, effective perturbation theory leads to consistent results for observables, i.e. gauge independent results which are complete to leading order in g.

The resummation technique has been applied by Baier *et al.*[16] and Ruuskanen[14]. In both cases the thermal invariant emission rate of photons is related to the imaginary part of the photon self-energy[17, 7]. For real photons the consistent lowest order (Fig. 2a, b) calculation of emission rate at fixed temperature T can then be written as [16]

$$E\frac{\mathrm{d}R_\gamma}{\mathrm{d}^3p\mathrm{d}^4x} \cong \left(\sum Q_f^2\right) \frac{\alpha\,\alpha_s}{2\pi^2}\, T^2\, e^{-\frac{E}{T}} \ln\left(\frac{c}{\alpha_s}\frac{E}{T}\right) \tag{3}$$

with the constant $c \cong 0.23$ and where the bare quark mass has already been replaced by the thermal mass $m_\beta = \frac{2\pi}{3}\alpha_s T^2$. A slightly different result is obtained by Ruuskanen[14], mainly because of the approximation of the thermal distributions by a Boltzmann law. A quantitative comparison[18] shows that the rate at a temperature of $T = 200\,\mathrm{MeV/c}$ is thereby enhanced compared to Baier *et al.*[16] by

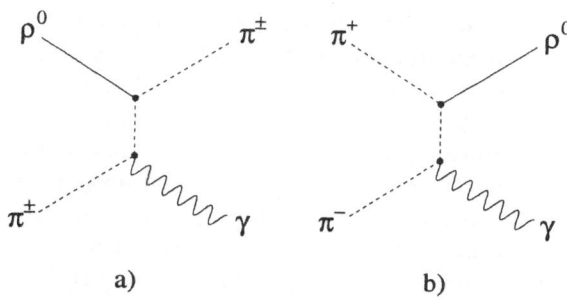

Figure 3. Feynman diagrams of photon production in a hadron gas; (a) $\pi\rho$ Compton with an intermediate virtual π, ρ, or $A_1(1260)$ meson, (b) $\pi^+\pi^-$ annihilation.

approximately a factor of two. Almost identical emission rates to Baier's calculations, on the other hand, were found by Hwa and Kajantie[8], who have evaluated the Feynman graphs in an effective manner by an approximation to high energy e^+e^- data.

Thermal photon radiation from a Hadron-Gas

Only recently it has been pointed out by Kapusta et al.[19] that a hot hadronic gas radiates thermal photons with similar emission rates as a QGP of the same temperature. The dominant contribution comes from the reactions $\pi\pi \to \rho\gamma$ and $\pi\rho \to \rho\gamma$ (Fig. 3). Xiong et al.[20] later pointed out that the latter of the two processes proceeds in a hot hadron gas not only via an intermediate virtual π or ρ, but also via the broad $A_1(1260)$ resonance.

If form-factors are neglected, i.e. if particles are treated again as point-like objects, the evaluation of the Feynman graphs proceeds in an almost identical fashion as sketched above for the diagrams of Fig. 2. A quantitative comparison of the different contributions shows that for a thermal hadron gas of $T = 150 \, \text{MeV/c}$ the pion annihilation dominates for $p_\perp \lesssim 0.7 \, \text{GeV/c}$, while the $\pi\rho$ channel dominates for large $p_\perp$. Here, the photon production via the intermediate A_1 amounts to approximately twice the yield compared to the intermediate π and ρ state.

Future calculations should take into account also form factors. A rough estimate shows that this will reduce the photon rate at $p_\perp \approx 2.5 \, \text{GeV/c}$ by a factor of approximately 2.8[19].

Lepton-Pair Spectroscopy of a Hadron-Gas

Lepton-pair emission is, similar to that of photons, not only expected for a thermalized QGP but also for a thermal hadron gas. As discussed above, the electromagnetic radiation couples to hadronic (mesonic) matter via vector mesons. These vector mesons carry the same quantum numbers as the photon and are dominated by the ρ, ω, and ϕ. They decay with a branching ratio of $\sim 10^{-4}$ into e^+e^- or $\mu^+\mu^-$ pairs. This picture is well known as the vector dominance model (VDM) and provides a means to calculate by the aid of experimental form factors the expected rates.

The lifetime of the ρ-meson of approximately $1\,\text{fm/c}$ combined with the decay into lepton pairs opens an exciting diagnostics tool for hot and dense nuclear matter. Because the lifetime is very short compared to typical time-scales of the collision, ρ mesons decay almost instantaneously after their formation in the interior of the hot and dense matter. Since lepton pairs will escape almost unaltered by the surrounding matter their spectroscopy allows to test the properties of those particles inside the medium. Changes of the width and mass may be expected according to the density and temperature of the system[21, 22]. Of course, such a diagnostics tool is not provided by the dominant $\pi\pi$-decay channel, since those pions will strongly be affected by rescattering before reaching the detector.

An experiment, however, has to cope with several competing sources of lepton pairs. At high mass this is the perturbatively treated Drell-Yan process. However, the mass region $M \lesssim 1\,\text{GeV}$ constitutes the major part of the measured yield (cf. Fig. 1), but the present knowledge and understanding of the intermediate and low mass data in hadronic collisions is rather limited. Recent studies[23, 24, 22] show that one has to consider bremsstrahlung contributions from $NN, NN\pi, NN\pi\pi$, etc., radiative decays of the Δ-resonance, direct $\pi^+\pi^-$ annihilation, and Dalitz decays of the η and π^0 in the e^+e^- channel and Dalitz decays of the ω and η in the $\mu^+\mu^-$ channel. The relative importance of each of these channels depends furthermore on the global conditions of the system, i.e. on the projectile energy, the mass of the fireball, its density, temperature, etc., so that precise predictions seem hard to be achieved at present. However, such a knowledge is mandatory for any firm conclusions on thermal lepton-pair production.

EXPECTED YIELDS IN NUCLEAR COLLISIONS

Up to now, only the photon and di-lepton radiation of *static* systems with fixed temperature was discussed. Nuclear collisions are, however, highly *dynamic* systems where even the question about a thermal equilibrium may be raised. Based on results from kinetic theories and experimental data, which provide at least some evidence for a local equilibrium, we will in the following proceed with the assumption of a thermalized system and calculate in a rather idealized way the real photon production in nuclear collisions. We shall assume that the system starts with a QGP of temperature T_i being formed after some initial thermalization time τ_i. Then it cools isentropically due to longitudinal Bjørken expansion. Furthermore, we assume a first order phase transition to a mesonic gas at temperature $T_c \cong 160\,\text{MeV/c}$. The calculation is stopped after the hadron gas has cooled to a freeze-out temperature of $T_f \approx 100\,\text{MeV/c}$. Figure 4 visualizes the time evolution and its relevant parameters.

The photon yield from the QGP is largely dominated by the initial temperature T_i which in turn is linked to the initial time τ_i via the entropy density of the system. The entropy density ζ can be estimated from the experimental particle multiplicities dN/dy and the projectile radius $\propto A_p^{2/3}$. Using the uncertainty principle $\tau_i \gtrsim \hbar/\langle E \rangle$, we may write (compare Ref.[25, 26])

$$\tau_i \approx 0.93 \sqrt{\frac{A_p^{2/3}}{dN/dy}} \ (\text{fm/c}) \tag{4}$$

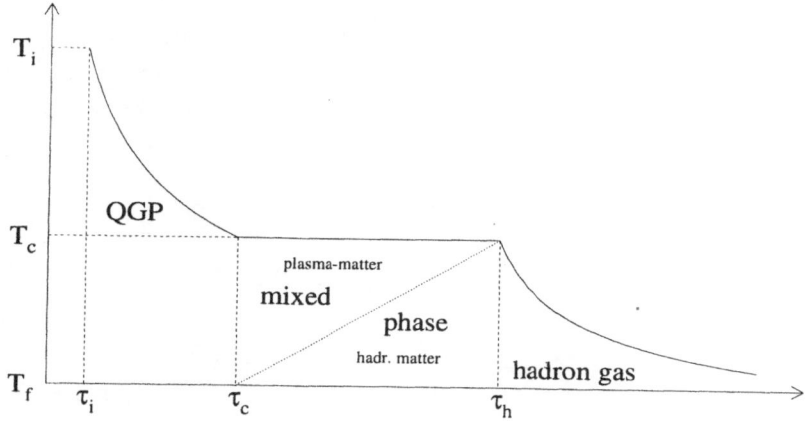

Figure 4. Time evolution of the temperature in a nuclear reaction with a first order phase transition.

which represents a lower limit for the initial time. Now we can use this relation and estimate the initial temperature from the entropy density of an ideal gas[8], $\zeta = d\varepsilon/dT = 4aT^3$ (a = degrees of freedom) and find

$$T_i \approx 81 \sqrt{\frac{dN/dy}{A_p^{2/3}}} \; (\text{MeV/c}). \tag{5}$$

Entropy conservation, $\zeta_i \tau_i = \text{const} = \zeta\tau$, and the ideal gas equation of state $P = \varepsilon/3$ then determines the cooling rate $T(\tau) = T_i \cdot (\tau_i/\tau)^{1/3}$ until the critical temperature is reached at time $\tau_c = \tau_i \cdot (T_i/T_c)^3$. Now the mixed phase starts and plasma matter is converted into hadronic matter. Entropy conservation still holds and the end of the mixed phase is then given by $\tau_h = r\tau_c$ with r being the ratio of the degrees of freedom in the QGP over the hadron gas. In analogy to above, the system then cools again according to $T(\tau) = T_c \cdot (\tau_h/\tau)^{1/3}$ until the end of the reaction is marked by the freeze-out temperature T_f.

The total yield is obtained by the time integration over the three phases. The initial temperature for central S + Au collisions has been assumed $T_i = 336\,\text{MeV/c}$ and $228\,\text{MeV/c}$, both of which are lower than expected from eqn. 5. The number of degrees of freedom was assumed to be $r = 42.25/6.6$ in order to take into account contributions from quarks heavier than the u and d and mesons heavier than the π.

Experimentally, the relevant quantity is the γ/π^0-ratio rather than the cross section of thermal photons alone. This is because an experiment has to cope with a large background of photons from the π^0 decay. The γ/π^0-ratio thus represents a measure of the surplus of single gammas compared to gammas from meson decays and tells, whether the amount of single photons is sufficient to be detected by an experiment. Using the experimental π^0-spectra, the $\gamma_{\text{thermal}}/\pi^0$ can now be calculated and its result is presented in Fig. 5 for central S + Au reactions[18]. According to these calculations, photons from a hadron gas dominate the region $p_\perp \lesssim 2.5\,\text{GeV/c}$. The spectral slope, $d\sigma/dp_\perp(\gamma_{\text{HG}})$, reflects in this scenario basically the temperature

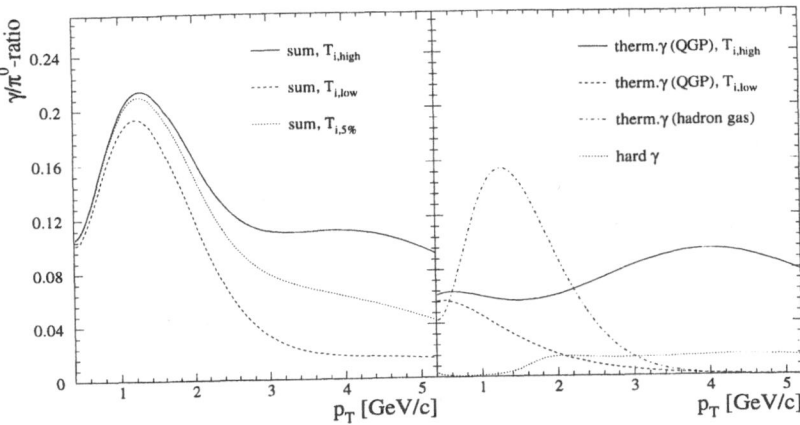

Figure 5. Calculated γ/π^0-ratio for central S + Au collisions at 200 AGeV. The left panel shows the total yield and the right panel its differential contributions. The three different initial temperatures are $T_i = 338$, 310, and 228 MeV/c.[18]

of the mixed phase. Photons from the QGP exhibit the expected strong variation with the initial temperature. For the lowest of the three temperatures, they would remain unobserved within the present experimental uncertainty, but would be clearly visible for the upper temperature. The figure also shows that hard direct photons from the initial state of the collision do not play a significant rôle at SPS fixed target energies.

Finally, it should be stressed again that there are certain simplifications in these calculations; (*i*) thermal equilibria are assumed, something which should be checked and verified by parton kinetic simulations[27], (*ii*) the rate calculations were performed for zero chemical potentials $\mu_q \cong \mu_B/3 \cong 0$ which are unlikely to be reached in present experiments. The inclusion of baryons is expected to reduce the photon yield at fixed energy density[28]. (*iii*) Scale invariance according to Bjørken expansion is assumed, but not supported by present experiments. (*iv*) Transverse expansion effects are neglected and would tend to reduce the photon yield in the late phase of the reaction, i.e. in the hadron gas. Obviously, there are large uncertainties in such calculations and the results should only be regarded a rough estimate. Much more work is still needed and many theoretical efforts are underway particularly at collider energies. Independent of the quantitative results of such calculations experiments may investigate scaling properties of single photons to learn about its origin. The most important characteristics is the expected scaling with $(dN/dy)^2$, since quark-antiquark and quark-gluon collisions produce the photons.

EXPERIMENTAL MEASUREMENTS

Summarizing the results from the previous two sections, thermal photons and lepton-pairs are both considered unique probes of a quark gluon plasma and hot hadronic gas. Lepton-pair experiments require hadron-blind detection techniques in order to suppress signals from the $\sim 10^4$ more abundantly produced hadrons. In $\mu^+\mu^-$ experiments this is usually achieved by massive absorber materials behind the target so that only muons can penetrate and be detected in wire chambers. Their momenta are usually measured by their curvature in a magnetic field. Measurements of e^+e^--pairs require detectors inherently blind to hadrons. An example of such kind of experiment is NA45 (CERES) at CERN[29]. It employs ring imaging Cherenkov detectors (RICH) with a radiator such that only electrons and very high energetic pions exceed the Cherenkov threshold of of $\gamma_{th} \cong 32$. Such experiments have succeeded to identity vector mesons, but the questions of thermal lepton-pair radiation was not yet addressed.

Single photon radiation is studied experimentally by two different approaches, called the 'direct' and 'conversion' method. The former one uses a thin converter behind the target (with a thickness of some percent of a radiation length, which may be the target itself) to convert a small amount of photons into e^+e^--pairs. These may then be investigated by a lepton pair experiment as discussed above. An example of such an experiment is again NA45. The direct method, on the other hand, employs finely granulated electromagnetic calorimeters to measure the photons themselves. The calorimeter is chosen in such a way that π^0- and η-mesons can be identified by their two-photon decay. The important advantage the direct method over all of the aforementioned techniques is that the *background of the thermal radiation is measured* within the same experiment so that there is no need to rely on (imprecise) calculations. Such kind of technique is employed by the WA80 and WA93 experiments at CERN. The following discussion will concentrate on data from this experiment. Results from the photon conversion experiment NA45 have been presented by Specht[29] on this conference.

Results from WA80

In order to allow for the subtraction of photons from neutral meson decays measured within the same experiment, the detector is primarily designed for high precision π^0 and η reconstruction[30, 31]. This is achieved by a 3,800 modules lead-glass calorimeter covering the pseudorapidity range $2.1 \leq \eta \leq 2.9$ at a distance of 9 m to the target. To allow tagging of charged particles, a 40,000 pad streamer-tube detector is placed in front of the lead-glass. With that detector the accessible $p_\perp$-range in 200 AGeV S + Au reactions is $0.2 \leq p_\perp \leq 4.5$ GeV/c for π^0-mesons and $0.5 \leq p_\perp \leq 2.5$ GeV/c for η-mesons. The experimental trigger allows selection of different ranges of centrality to allow for comparisons of spectra from central and peripheral events.

Among the most important experimental difficulties in measuring single photons in high energy nuclear collisions is the precise determination of the photon and π^0 reconstruction efficiencies, ε_γ and ε_{π^0}, respectively. Their calculation is based on the actual experimental data and is performed by superimposing single hadronic and electromagnetic showers on a raw-data level to the measured heavy-ion events. The additional showers are generated either by the GEANT simulation package and assuming phase space distributions of the various particles according to experimental data or are, for reasons of consistency checks, taken from very peripheral S + Au reactions. The artificial events are processed with the same chain of shower re-

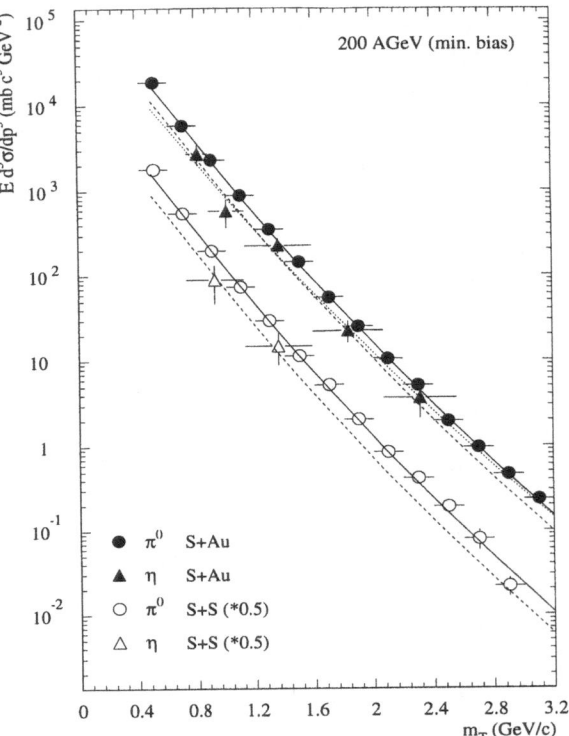

Figure 6. Invariant cross sections of π^0- and η-mesons as a function of transverse mass for 200 AGeV S + Au and S + S minimum bias data[32]. The S+S data are scaled by a factor 0.5 for a better presentation. The different lines represent fits to a function $C \cdot [p_0/(p_0 + m_\perp)]^n$.

construction routines as is used for real data events. The photon reconstruction efficiency is then defined by the ratio of the reconstructed photon spectrum divided by the known input photon spectrum and is found to depend both on the local particle density and on the transverse momentum of the photon. As a typical result for central S + Au events, the correction according to $\varepsilon_\gamma(\varrho_{\text{local}}, p_\perp,)$ leads to a change of the inverse slope parameter in the inclusive photon $p_\perp$-distributions from $T_{\text{raw}} = 210\,\text{MeV}$ to $T_{\text{corr}} = 207\,\text{MeV}$. The systematic and statistical uncertainty in the photon (and π^0) reconstruction efficiency is found to be approximately 3 %.

Differential cross sections of π^0 and η mesons are obtained by measuring their yields above the combinatorial background in invariant mass distributions of $\gamma\gamma$ pairs selected according to the variable of interest. In order to determine the precise shape (and yield) of the background under the $\pi^0(\eta)$ peak, a mixed event method has been developed. Here, the invariant mass distribution is constructed by combining photons from one event with those from another one of the same global characteristics.

Acceptance and efficiency corrected invariant cross sections of π^0 and η are shown in Fig. 6 as a function of transverse mass $m_\perp = \sqrt{p_\perp^2 + m_0}$ for minimum bias,

S + S and S + Au reactions. In addition to the statistical errors, systematic errors of the acceptance and efficiency correction are added in quadrature. In such a representation the π^0-mesons of both systems exhibit a concave shaped spectrum with a similar shape observed within the experimental uncertainties for the η-mesons. The data are thus compatible with the so called phenomenological m_T-scaling behavior. The η/π^0-ratio in bins of equal width in $m_\perp$ or for very large transverse momentum is found to be 0.66 ± 0.09 and 0.57 ± 0.18 for S + Au and S + S data, respectively[32]. A corresponding analysis of p + p, π +p, and p + C reactions found in the literature and taken at $\sqrt{s} = 24.3$ - $62\,\text{GeV}$ yields 0.55 ± 0.02.

The single photon yield is generally expressed in terms of the γ/π^0-ratio and can now be extracted from the measured γ, π^0, and η yields according to the expression 9, 33.

$$\frac{\gamma}{\pi^0} = \frac{N_\gamma}{N_{\pi^0}} \cdot \frac{\varepsilon_{\pi^0}}{\varepsilon_\gamma} \cdot A_{\text{geo}} - (R_{\pi^0} + R_\eta + R_X) \tag{6}$$

where ε_{π^0} and ε_γ denote the photon and π^0 reconstruction efficiencies, respectively. A_{geo} is the geometrical acceptance of the detector for π^0's, and R_{π^0}, R_η, and R_X are the Monte-Carlo calculated ratios of observed background photons from measured π^0 and η mesons, and higher (non-measured) resonances, respectively. Such a presentation of data has the advantage that certain experimental errors, e.g. the absolute cross section normalization, cancel out thus allowing a total experimental sensitive in the γ/π^0 ratio of up to 5 %.

In case of high heavy-ion reactions where the π^0 and η production cross sections are in principle unknown it is very important to measure those data in the same experiment and in the appropriate $p_\perp$ region. Heavier resonances are then to a large part automatically taken already into account, since the major fraction of these resonances decays via the π^0 or η branch (e.g. $\eta' \rightarrow \pi^0 \pi^0 \eta$) and is therefore already contained in the measured spectra. The different hadronic photon contributions as calculated from the experimental π^0 and η yields and complemented by assumed relative production ratios of heavier particles are displayed in Fig. 7 as a function of $p_\perp$. Besides the π^0 decay the most important hadronic photon background is the $\eta \rightarrow \gamma\gamma$ decay. These photons amount to approximately 10 % compared to those from the π^0 while the sum of all heavier resonances contributes on the 1–2 % level. Possible inherent systematic uncertainties when assuming their production cross-sections from different experiments and reactions are thus suppressed by approximately the same factor.

The single photon yield, obtained after subtracting the calculated decay contributions from Fig. 7, is shown in Fig. 8 for central S + Au reactions. Different than in the peripheral data the central sample shows a positive signal and a slight increase towards low $p_\perp$. The error bars represent the sum of all collected statistical and systematic uncertainties. Assuming a gaussian distribution, we arrive at an uncertainty of 7-8 % and 8-15 % for $p_\perp$ bins below and above $2\,\text{GeV/c}$, respectively. The excess of photons in central data over the expected yield from the hadronic background for $p_\perp \lesssim 2\,\text{GeV/c}$ thus reaches a level of approximately 2 standard deviations in each point. Data from peripheral collisions are on the other hand well described by known hadronic sources. To allow for comparisons with other experiments and to demonstrate the variation of the

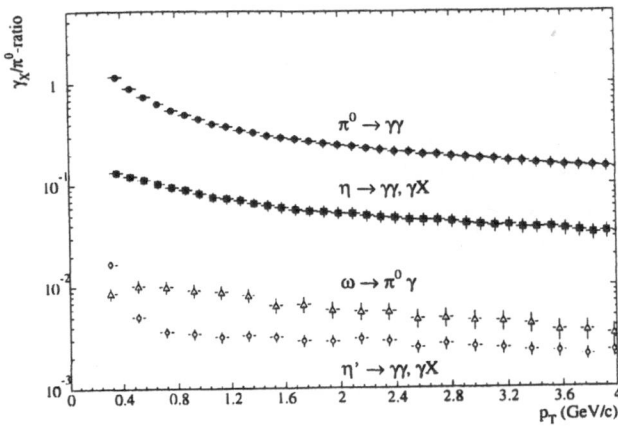

Figure 7. Different hadronic photon contributions normalized to the π^0 yield as a function of $p_\perp$ for central S + Au reactions.

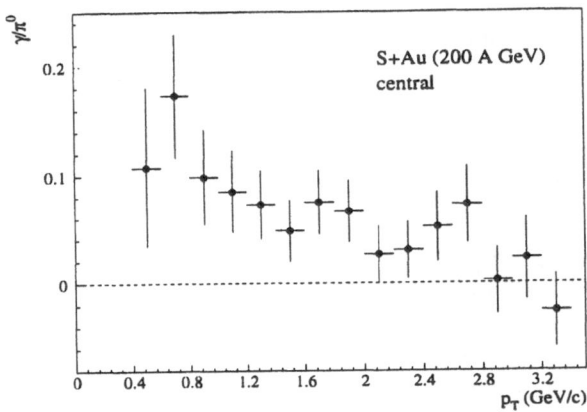

Figure 8. Preliminary background subtracted γ/π^0 ratio for central 200 AGeV S + Au reactions as a function of $p_\perp$.

signal between central and peripheral data one may integrate the γ/π^0 ratio in $0.4 \leq p_\perp \leq 2.0$ GeV/c to find $\gamma/\pi^0|_{\mathrm{int}}^{\mathrm{cen}} = 0.091 \pm 0.016(\mathrm{stat}) \pm 0.058(\mathrm{syst})$ in central events and $\gamma/\pi^0|_{\mathrm{int}}^{\mathrm{per}} = 0.013 \pm 0.019(\mathrm{stat}) \pm 0.058(\mathrm{syst})$. NA45[29] finds for central collisions of the same reaction $\gamma/\pi^0 = 0.075 \pm 0.005(\mathrm{stat}) \pm 0.110(\mathrm{syst})$ and NA34[34] reports no evidence for an excess within an experimental uncertainty of $\sigma(\gamma/\pi^0) = (0.04 - 0.11)(\mathrm{stat}) \pm 0.09(\mathrm{sys})$. All results are thus compatible with each other, but the uncertainty of the latter two experiments is still too large to allow for a sensitive comparison.

The spectral shape observed in Fig. 8 resembles very much the shape expected from the hadronic interactions shown in Fig. 5. At this stage of the analysis, however, no quantitative comparison to calculated photon spectra will be made. This is because of both large uncertainties in the calculations, as stated above, and the preliminary character of the experimental data. A more obvious link to the temperature of the system may be given instead by the invariant cross section of single

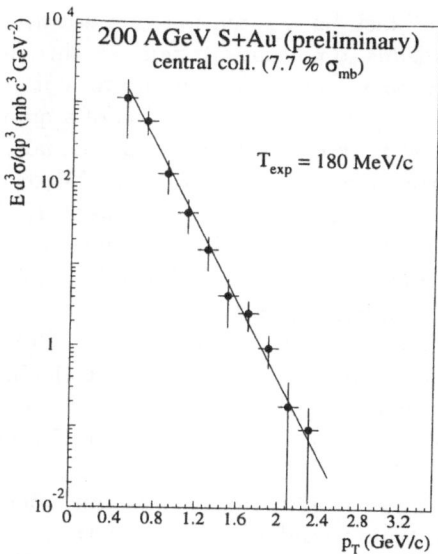

Figure 9. Preliminary invariant cross section of single photons from central S + Au data (WA80). The line shows an exponential distribution of $T_{\exp} = 180$ MeV/c.

photons which is obtained from Fig. 8 by multiplying with the experimental π^0 distribution. Such a preliminary distribution is shown in Fig. 9. Fitting an exponential distribution to the data points yields a slope parameter of $T_{\exp} \cong 180$ MeV/c which would – after taking into account the decay kinematics of the Feynman diagrams of Fig. 3 – translate into a hadronic gas temperature of $T_{\mathrm{HG}} \approx 165$ MeV/c.

Before any firm conclusions can be drawn from the data, more work is still needed to verify the significance of the photon excess. The most time consuming and critical work is the Monte-Carlo procedure to determine the γ, π^0, and η reconstruction efficiencies. Therefore, an independent re-analysis is planned in order to gain extra confidence in the estimated systematic errors.

SUMMARY AND CONCLUSIONS

The measurement of photons and lepton-pairs provides a powerful tool to learn about the properties of hot and dense matter over a wide range of bombarding energy. Informations extracted from these signals are to a large part complementary with some common features. An important and unique aspect of lepton-pair spectroscopy is the possibility to infer properties of the excited medium from modifications observed in the leptonic decay-width and mass of vector mesons. After pioneering work on this subject performed by the DLS-Collaboration[35] is being terminated due to the recent shutdown of the Bevalac, a challenging experimental program is now under discussion by the HADES-Collaboration at GSI. At CERN-

SPS energies the thermal 'black-body radiation' of a hot fireball is of primary interest, because it provides means to measure its temperature in the early phase of the collision. High temperatures, i.e. large photon and di-lepton yields, combined with high entropy densities are expected for a formation of a quark-gluon plasma. From a theoretical point of view, both of these probes are equally well suited. However, experimentally real photons measured by the 'direct' detection technique with π^0 and η identification are advantageous, because the hadronic background to the signal can be subtracted based on the experimental data alone without any relevant theoretical assumption. Such an experiment allows for a precision in the γ/π^0-ratio of 5 - 6 % as has been demonstrated by WA80[36, 37].

Preliminary results show for the first time an excess of photons in central nuclear collisions at transverse momenta $p_\perp \lesssim 2.5$ GeV/c. Calculations of single photons from various sources have been reviewed and suggest an interpretation of the experimental excess and its observed shape in terms of a hadronic gas at a temperature of approximately 165 MeV/c.

Assuming a first order phase transition, a two component structure observed in the single photon $p_\perp$-spectrum would allow for an extraction of the critical temperature T_c and signal the formation of a QGP and its initial temperature by a hard component at larger values of $p_\perp$.

Prospects for single photon measurements at RHIC and LHC collider experiments seem rather encouraging. Recently, it has been pointed out by Shuryak and co-workers[38] and confirmed by parton-kinetics[27] that the thermal equilibration of gluons happens very fast leading to a 'hot-glue' scenario with very high gluon temperatures. Quark production on the other hand is much slower, so that a chemical non-equilibrium situation is expected. It is however found that the smaller quark number is more than compensated by the fact that they are embedded into the hotter glue. The predicted yields are thus larger than considered before.

In summary, the experimental detection of electromagnetic emission from the plasma is not going to be easy but the chances for the observation of the signals are realistic. The fact that they are the only direct signals from the quark-gluon plasma stage justifies a major experimental effort in this area.

REFERENCES

1. B. Petersson, Nucl. Phys. (Proc. Suppl.) **B30** (1993) 66–80.

2. Proceedings to Quark-Matter Conference, Borlänge 1993, Nucl. Phys. **A**, 1993.

3. E. L. Feinberg, Nuovo Cimento **34** (1976) 391–412.

4. E. V. Shuryak, Phys. Lett. **78B** (1978) 150–153.

5. K. Kajantie and H. I. Miettinen, Z. Phys. **C9** (1981) 341–345.

6. F. Halzen and H. C. Liu, Phys. Rev. **D25** (1982) 1842–1846.

7. L. D. McLerran and T. Toimela, Phys. Rev. **D31** (1985) 545–563.

8. R. C. Hwa and K. Kajantie, Phys. Rev. **D32** (1985) 1109–1118.

9. T. Ferbel and W. R. Molzon, Rev. Mod. Phys. **56** (1984) 181–221.

10. J. F. Owens, Rev. Mod. Phys. **59** (1987) 465–503.

11. R. M. Turnbull, J. Phys. **G14** (1988) 135–161.

12. L. Cormell and J. F. Owens, Phys. Rev. **D22** (1980) 1609–1616.

13. M. Neubert, Z. Phys. **C42** (1989) 231–242.

14. P. V. Ruuskanen, Nucl. Phys. **A544** (1992) 169c–182c.

15. E. Braaten and R. D. Pisarski, Nucl. Phys. **B337** (1990) 569–634 and Nucl. Phys. **B339** (1990) 310–324.

16. R. Baier, et al., Z. Phys. **C53** (1992) 433–438.

17. H. A. Weldon, Phys. Rev. **D26** (1983) 2789–2796.

18. D. Bucher, Diploma thesis, University of Münster. 1993.

19. J. Kapusta, P. Lichard, and D. Seibert, Phys. Rev. **D44** (1991) 2774–2788.

20. L. Xiong, E. Shuryak, and G. E. Brown, Phys. Rev. **D46** (1992) 3798–3801.

21. M. Herrmann, B. L. Frieman, and W. Nörenberg, Z. Phys. **A343** (1992) 119–120.

22. G. Wolf, W. Cassing, and U. Mosel, Nucl. Phys. **A552** (1993) 549–570.

23. P. Koch, Z. Phys. **C57** (1993) 283–303.

24. K. Haglin and C. Gale, Preprint McGill/93-9, 1993.

25. J. Kapusta, L. McLerran, and D. K. Srivastava, Phys. Lett. **B283** (1992) 145–150.

26. S. Chakrabarty, et al., Phys. Rev. **D46** (1992) 3802–3806.

27. K. Geiger and B. Müller, Nucl. Phys. **B369** (1992) 600–654.

28. A. Dumitru, et al., Mod. Phys. Lett. **A8** (1993) 1291.

29. H. Specht et al., CERES-Collaboration, Proceedings to this conference, 1993.

30. K.-H. Kampert et al., WA80-Collaboration, Proceedings to *2. International Conference on Physics and Astrophysics of the Quark-Gluon Plasma*, Calcutta, January 19.-23., 1993, World Scientific, Singapore, 1993.

31. R. Santo et al., WA80-Collaboration, Proceedings to Quark-Matter Conference 1993, Borlänge, Nucl. Phys. **A**, 1993.

32. T.C. Awes et al., WA80-Collaboration, to be submitted to Phys. Lett. **B**, 1994.

33. R. Albrecht et al., WA80-Collaboration, Z. Phys. **C51** (1991) 1–10.

34. T. Åkesson et al., NA34-Collaboration, Z. Phys. **C46** (1990) 369–375.

35. G. Roche et al., DLS-Collaboration, Proceedings to this conference, 1993.

36. G. Clewing, Doctoral Thesis, University of Münster, 1993.

37. G. Hölker, Doctoral Thesis, University of Münster, 1993.

38. E. Shuryak, Phys. Rev. Lett. **22** (1992) 3270–3272 and E. Shuryak and L. Xiong, Phys. Rev. Lett. **70** (1993) 2241–2244.

MULTIFRAGMENTATION AND THE QUEST
FOR THE LIQUID-GAS PHASE TRANSITION

Josef Pochodzalla for the ALADiN - Collaboration

Gesellschaft für Schwerionenforschung Darmstadt, Germany

INTRODUCTION

The determination of fundamental properties of nuclear matter far from the ground state is one of the principal motivations for studying energetic collisions between two complex nuclei. At densities slightly above the normal nuclear density transverse flow phenomena are regarded as fingerprints of the equation of state [1,2]. At higher densities of typically 2-3 times normal density particle production - in particular particle production via multistep processes - is believed to be a sensitive probe of nuclear matter [3]. At subsaturation densities on the other hand, the fragmentation pattern of the nuclear system is expected to be influenced by the bulk properties of nuclear matter.

The special interest in this low density regime , however, arouse from the conjecture that the nuclear many-body system might undergo a phase-transition during the disassembly. This idea [4,5] was inspired by the Van der Waals - like behaviour of the nucleon - nucleon interaction. Of course, such a phenomenon is not only fascinating by itself but it is also of more general interest: despite many quantitative differences - also in ultra relativistic heavy ion studies we face the problem to identify a phase transition in a finite, transient system.

THE RISE AND FALL OF MULTIFRAGMENTATION

For a nucleus at low excitation energies, $E_x/A \leq 1$ MeV, the dominant decay channels are the evaporation of light particles and fission. If it is possible to excite a nucleus far beyond its total binding energy, a complete vaporization into individual nucleons is anticipated. Between these two limits, we expect a regime where the production of several intermediate mass fragments (IMFs) is the dominant exit channel.

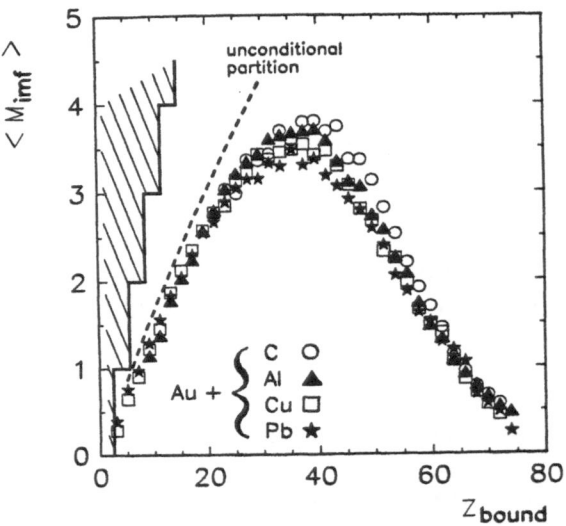

Figure 1. $\langle M_{IMF} \rangle - Z_{bound}$ correlation measured for Au + C, Al Cu and Pb reactions at E/A = 600 MeV [6].

It is obvious that events with multiple fragment production may provide the expected glimpse of a nuclear liquid-gas phase transition since they are the favorite candidates for a true multifragmentation.

Departure from the Evaporation Regime

In the first experiment at the ALADIN we, therefore, investigated which collisions lead to the emission of several IMFs. We have measured the reactions of Au on targets of C, Al, Cu and Pb at 600 MeV/nucleon and have studied the impact parameter dependence of multi-fragment emission from the excited projectile spectator. In order to characterize the events a quantity called Z_{bound}, measuring the number of charges contained in the projectile fragments of $Z \geq 2$, has been introduced. Z_{bound} gives the atomic number of the excited projectile spectator reduced by the number of hydrogen nuclei emitted from it. It thus depends on the primary abrasion *and* the secondary deexcitation stages and reflects the violence of the collision in a two-fold way. The most striking result of this first analysis was that by ordering the collisions according to the summed charge Z_{bound} a remarkable target independence of the charge distributions was found [7].

From a nearly complete measurement of the resulting projectile fragments we have observed the different modes of nuclear disassembly, ranging from spallation and evaporation to multi-fragment emission and finally to complete vaporization of the projectile spectator (Figure 1). It was found for the C target that the largest number of IMFs is observed for central collisions. For heavier targets the maximum of the IMF multiplicity occurs for more peripheral reactions. On all three targets this maximum of 3 to 4 IMFs is observed at approximately the same size of the spectator of 150 nucleons.

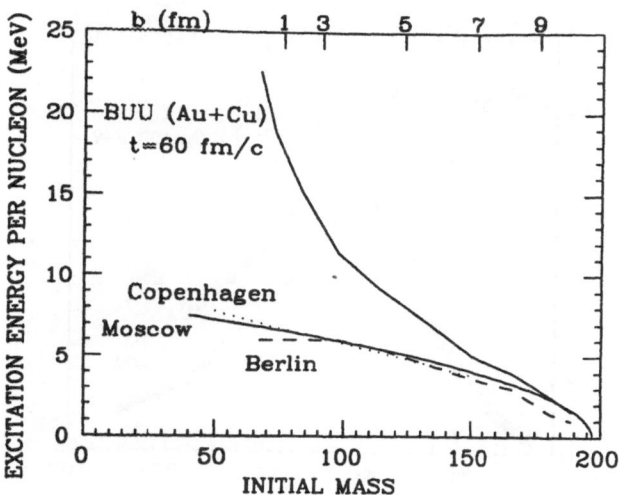

Figure 2. Excitation energy per nucleon of the decaying projectile spectator as a function of its mass. The upper solid line represents the deposited energies predicted by BUU simulations of Au+Cu collisions at E/A = 600 MeV[8]. The solid, dashed and dotted lines are the theoretical inputs required by the MOSCOW [18], BERLIN [9] and COPENHAGEN [10] multifragmentation codes in order to reproduce the observed charge distributions.

The Fall towards Vaporization

For a more quantitative interpretation of the data displayed in Figure 1, one has to keep in mind that the mass of the pre-fragment decreases whereas the excitation energy per nucleon increases when going from peripheral to the most violent reactions. For illustration, the solid line in Figure 2 shows the relation between the size and the specific excitation energy of the projectile spectator as predicted by microscopic BUU simulations.

To disentangle size and excitation energy effects and explore the evolution from multifragmentation towards a complete vaporization, central collisions of symmetric target-projectile combinations are ideally suited. In such collisions the (not necessarily thermal) excitation energy can be controlled by the incident beam energy thereby keeping the size of the excited system rather constant. In a joined experiment of the ALADIN/LAND and the MINIBALL/WALL collaborations we, therefore, investigated the system Au+Au at E/A = 100, 250 and 400 MeV. Fragments with charge Z $\geq$ 2 and rapidity close to the projectile rapidity were detected in the ALADIN forward spectrometer. Fragments originating from the target spectator and the participant region were detected in the MSU MINIBALL/WALL array which was complemented by the Si-CsI hodoscope of the ALADIN facility.

Figure 3 shows the observed mean multiplicity of intermediate mass fragments with $3 \leq Z \leq 30$ as a function of the total charged particle multiplicity N_c detected in the MINIBALL/-WALL for the three beam energies [11]. Provided that N_c is a monotonous function of the impact parameter, the data show that at E/A = 100 MeV the most central collisions are associated with the highest average multiplicity of

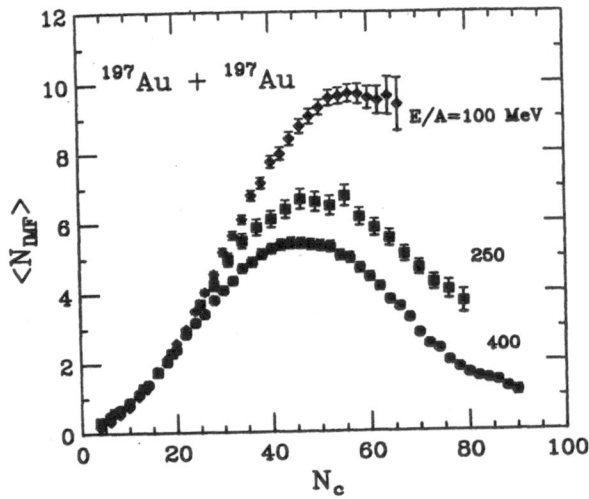

Figure 3. Mean multiplicity of intermediate mass fragments $\langle N_{IMF} \rangle$ as a function of the charged particle multiplicity N_c observed in the MINIBALL/-WALL for Au+Au collisions at E/A = 100, 250 and 400 MeV [11].

IMFs. With increasing energy a larger number of free nucleons and/or light particles and a corresponding smaller fragment multiplicity is observed in central collisions, thus signaling the onset of vaporization of a system that is too highly excited to produce significant numbers of fragments. Confirming our qualitative interpretation of the data presented in figure 1, the maximal IMF production is shifted towards larger impact parameters.

Despite the fact that the charge distributions and the maximum observed multiplicity of IMF's normalized to the total size of the decaying system are rather similar in central ($\approx 10/400$) and peripheral ($\approx 3.5/150$) fragmentation processes, it is important to realize a potential difference between these two types of reactions. The large compression and excitation energy reached in central collisions between two heavy nuclei and the resulting large radial collective flow [12, 13] may lead to fast explosion of the system. Because of the short time scale only short range fluctuations will therefore determine the fragmentation pattern. On the other hand, the projectile spectator produced in peripheral collisions are presumably not significantly compressed (see also below) and the expansion of the system is predominantly driven by the thermal pressure. As a consequence a relatively slow expansion might take place and the system might have enough time [14] to develop instabilities in a low density phase or even to undergo a liquid-gas phase transition.

CHARGE AND MASS PARTITION IN MULTIFRAGMENTATION

As mentioned already above, a given value of Z_{bound} corresponds to different impact parameters [15] if different target nuclei are involved. Therefore, the observed universal behavior seen in Figure 1 provided a necessary – although not sufficient – condition

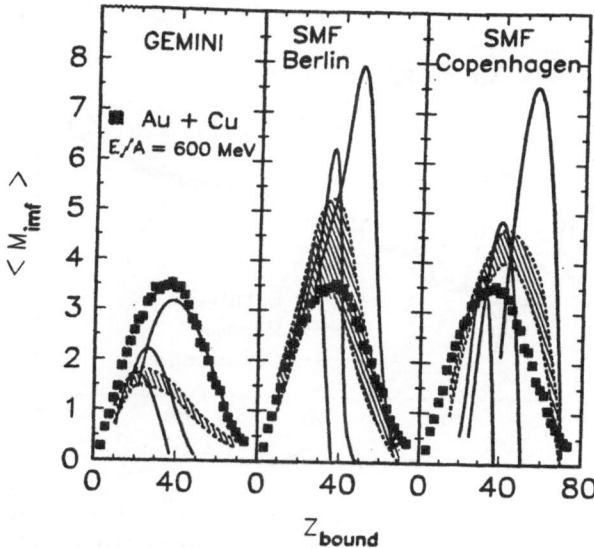

Figure 4. Comparison of the $\langle M_{IMF} \rangle$ – Z_{bound} correlation observed for Au + Cu reactions (squares) with predictions of the sequential decay code GEMINI (left part) and of the SMF models of Gross and co-workers (center part) and of Bondorf and co-workers (right part). The lines represent excitation functions for a given initial systems $(A_1, Z_1) = (100,40), (131,54)$ and $(190,75)$. The shaded bands are predictions based on initial conditions provided by BUU simulations.

for a chemical equilibrium being established during the fragmentation process. It seemed, therefore, worthwhile to confront the observed fragment distributions with predictions of statistical models. In particular, we performed calculations with the sequential evaporation code GEMINI [16], which is based on a transition state theory [17], and the COPENHAGEN [10] and BERLIN [9] statistical multifragmentation (SMF) models which both describe the simultaneous decay of an expanded nuclear system.

The shaded areas in Figure 4 mark the results of calculations using the primary projectile spectators predicted by the BUU simulations (see solid line in Figure 2). Qualitatively, all calculations exhibit a rise and fall of the IMF multiplicity. Whereas the BERLIN [9] and COPENHAGEN [10] multifragmentation codes predict rather similar relations between Z_{bound} and $\langle M_{IMF} \rangle$ which are in reasonable agreement with the data, the IMF multiplicities deduced from the sequential decay program GEMINI are lower by a factor of 2 – 3. Clearly, the observed charge distributions favor statistical decay calculations which assume an expanded decaying system [6].

In the mean time, these results have triggered a series of theoretical activities, where the complete ALADIN data (i.e. not only $\langle M_{IMF} \rangle$ but also higher moments of the charge distributions) where used to fix the input parameters of SMF models [18, 9, 10]. Despite several minor differences in the calculations, all three groups agree that a significantly lower excitation energy of the decaying projectile spectator than expected from the microscopic BUU simulations is required in order to reproduce the measured charge distributions (see Figure 2). Qualitatively, such excitation energies are consistent with predictions of the QMD-model [19] for the energy in the pre-fragment at a later time of about 200 fm/c and may, therefore, signal the energy loss via particle emission during the expansion of the system prior to the breakup [20].

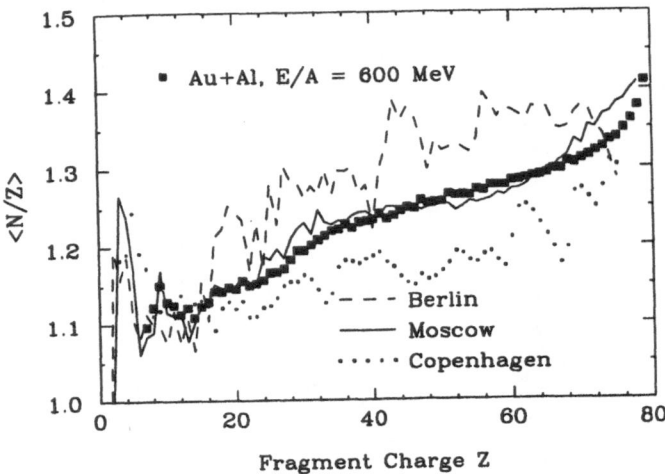

Figure 5. Mean N/Z ratio as a function of the fragment charge [22]. The solid, dashed, dotted lines represent the predictions of the MOSCOW, BERLIN and COPENHAGEN SMF codes, respectively. The large fluctuations in the latter two calculations are due to the discrete steps for the input mass and excitation energy corresponding impact parameter steps of 0.5 fm.

Isotope Distribution

The isotopic composition of complex fragments reflects the neutron-to-proton ratio of the decaying system [21]. In addition, the isotope distributions depend strongly on the sequential decay processes after freeze-out. Therefore, isotope distributions represent a stringent quantitative test ground for statistical decay models.

The masses and momenta of the fragments were determined by measuring their trajectories with the TP-MUSIC detector located behind the ALADIN magnet. Combining the TOF and the TP-MUSIC information, the mass of fragments with $Z \geq 8$ was evaluated with a relative resolution of $\Delta A/A \approx 3\%$ (FWHM). Figure 5 shows the measured mean neutron-to-proton ratio of the fragments as a function of the fragment charge measured for the reaction Au+Al at E/A = 600 MeV. Generally the fragments are on the more neutron poor side of the valley of stability. (Note that for fragments with charges greater than about 65, this ratio may be distorted by our experimental trigger condition.)

Predictions of the MOSCOW, BERLIN and COPENHAGEN statistical multifragmentation models are given by the solid, dashed and dotted lines, respectively. Despite the fact that these three models give similar charge distributions [18, 9, 10] they show clear differences in the isotopic distributions. Whereas the MOSCOW code agrees rather well with the experimental observations, the COPENHAGEN and the BERLIN codes underestimate resp. overestimate the average N/Z ratios significantly. It was checked, that these deviations cannot be compensated by any reasonable variation of the N/Z ratio of the decaying system nor by an increase or fluctuation of the excitation energy. Most likely, the different theoretical N/Z ratios reflect the different treatment of the final sequential decay stage after freeze out [23]. In turn, the observed discrepancies with the data call for a more detailed treatment of this secondary stage.

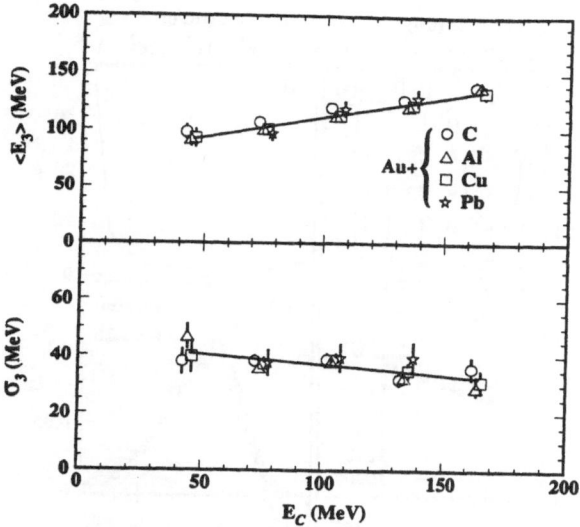

Figure 6. Mean center-of-mass total kinetic energy $\langle E_3 \rangle$ (top) and the standard deviation σ_3 (bottom) as a function of the nominal Coulomb energy for the three largest projectile fragments emitted in Au + C, Al, Cu and Pb collisions at E/A = 600 MeV.

DYNAMICAL OBSERVABLES

Breakup Configurations from Proximity Correlations

In the previous section indirect evidence for a simultaneous breakup of excited nuclear systems has been accumulated by comparing measured fragment yields with predictions based on statistical decay models [24, 25, 6]. More characteristic fingerprints of the disintegration dynamics may show up in kinematic correlations which are governed by the long-range Coulomb force[26-35]. In the following we will consider only those events where exactly three fragments were tracked in the TP-MUSIC. For this event class Z_{bound} varies between 45 and 70 and peaks at $Z_{bound} \approx 55$. In the present analysis the momenta $\vec{p}_{c.m.,i}$ of the three fragments in their center-of-momentum (c.m.) frame are used to determine the summed c.m. - kinetic energy

$$E_3 = \sum_{i=1}^{3} p_{c.m.,i}^2/(2 \cdot m_0 \cdot A_i) \tag{1}$$

where m_0 is the atomic mass unit. In Figure 6 the average value $\langle E_3 \rangle$ (top part) and the standard deviation σ_3 of E_3 (bottom part) are displayed as a function of the sum of a nominal Coulomb repulsion

$$E_c = e^2 \cdot \sum_{i<j} Z_i Z_j/(1.4 \cdot (A_i^{1/3} + A_j^{1/3})). \tag{2}$$

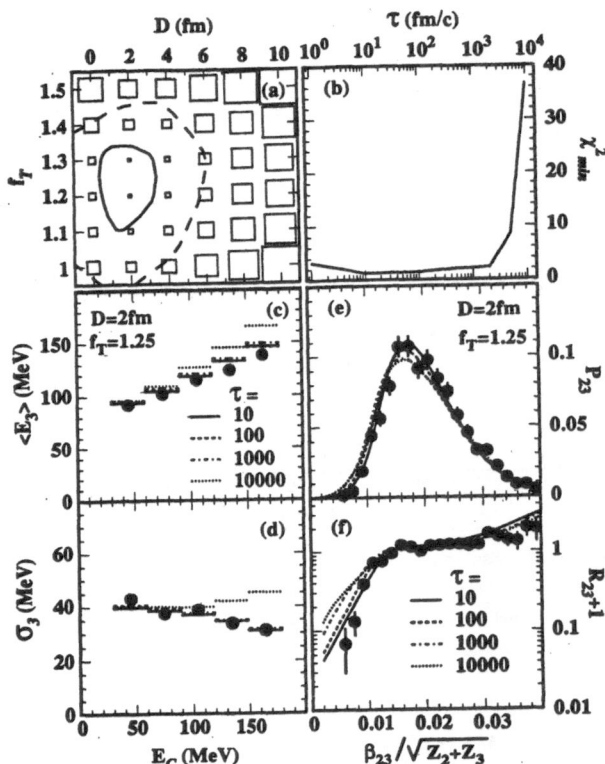

Figure 7. Summary of the sequential decay calculations. Part a: Two-dimensional display of χ^2 as a function of the initial fragment-fragment separation D and the scaling factor f_T of the temperature for a life time $\tau = 100$ fm/c. The area of the largest box corresponds to $\chi^2 = 50$. Part b: Minimum χ^2 as a function of the lifetime τ. Parts c and d: Optimal predictions for $\langle E_3 \rangle$ and σ_3 for different lifetimes τ. Parts e and f: Coincidence yield and two-fragment correlation function of fragment 2 and 3 as a function of their scaled relative velocity for the optimum parameter set with D=2 fm and $f_T = 1.25$ and the indicated values of τ.

Within the statistical uncertainties no target dependence is apparent and the data depend linearly on E_c.

In order to test the sensitivity of our observations to the breakup geometry we performed classical three-body trajectory calculations. We examined two schematic disintegration configurations which mimic the main categories of breakup scenarios discussed in the literature [24, 25, 6]:

(i) The first class simulates two sequential, binary splittings of the initial nucleus. In both steps, the surfaces of the two produced fragments were initially separated by a fixed distance 2·D. The time delay t between the two decays was chosen from an exponential distribution $P(t) \propto exp(-ln2 \cdot t/\tau)$. The initial relative energy of the two fragments was selected from a distribution

$$P(E) \propto E^\alpha \cdot exp(-E/T) \tag{3}$$

where for the exponent α values of 0.5 or 1 were used. The direction of the corresponding initial relative velocity was isotropically distributed.

(ii) With the second type of simulations a simultaneous emission out of a given volume is modelled. The centers of the three non-overlapping fragments are distributed randomly within a sphere of radius R. To each fragment an isotropically distributed initial velocity was assigned. Constrained by total momentum conservation, these velocities were selected according to the energy distribution described by Eq. 3. In addition to this random motion, an initial radial flow velocity $\vec{v}_{f,i} = (2\epsilon_f/m_0)^{1/2} \cdot \vec{d_i}/R$ was taken into account. Here, $\vec{d_i}$ is the position of the center of the fragment i with respect to the center-of-momentum, and the parameter ϵ_f denotes the flow energy per nucleon for fragments located at $d_i = R$.

The charges and masses of the fragments were obtained by a Monte Carlo sampling of the measured events, thus reducing the uncertainties associated with the fragment distribution. For each event the temperature parameter T was chosen according to the experimental value of Z_{bound} from the relation $T = f_T \cdot (2 \cdot (79 - Z_{bound}))^{1/2}$, where f_T is a free parameter. For $f_T = 1$ this relation describes within the relevant range of Z_{bound} the impact parameter dependence of the temperature of the initial projectile spectators as predicted by microscopic transport calculations [6] reasonably well. The paths of the fragments were calculated under the influence of their mutual Coulomb field and two-fragment proximity forces according to Ref. [36]. Since for the further analysis those trajectories were rejected for which the fragments did overlap during the propagation, the influence of the proximity force turns out to be rather small (see also below). In order to account for the recoil from light particles emitted sequentially from the initial fragments, the measured charges and masses were increased prior to the trajectory calculations in accordance with the initial temperature. After the fragment-fragment interaction has ceased, the sequential emission of light particles leading to the observed masses and charges was assumed to take place.

In order to quantify the agreement between the simulations and the experimental observations we define a reduced χ^2

$$\chi^2 = \frac{1}{5} \sum_{i=1}^{5} (x_i - y_i)^2/\delta_i^2 . \tag{4}$$

Here, x_i are the four coefficients characterizing the linear fits to the data in Figure 6 and, in addition, the mean scaled [31] relative velocity between the two lighter fragments 2 and 3. δ_i and y_i denote the experimental uncertainties of these quantities and the corresponding model predictions, respectively.

The results of the sequential type of calculations are summarized in Figure 7. The box-plot in part (a) displays for a given lifetime of $\tau = 100$ fm/c the variation of χ^2 as a function of the two other model parameters f_T and D. For orientation, contour lines for $\chi^2 = 4$ and 10 are shown. A distinct minimum at D = 2 ± 0.5 fm and f_T = 1.25 ± 0.05 can be discerned. Within the quoted uncertainties, similar minima were found at the *same* values of D and f_T for all considered values of $\tau \leq 10^4 fm/c$. However, the height of this minimum, χ^2_{min}, increases strongly for large values of τ (Figure 7b) and clearly excludes emission times $\tau > 2000$ fm/c. The good agreement of the simulations with the data for smaller values of τ is illustrated in parts c and d of Figure 7. Whereas the E_c dependences of $\langle E_3 \rangle$ and σ_3 provide no further constraint on τ, the probability distribution P_{23} of β_{23} is best reproduced by calculations using small $\tau \sim 10$ fm/c (solid line in Figure 7e). This preference for very small relative

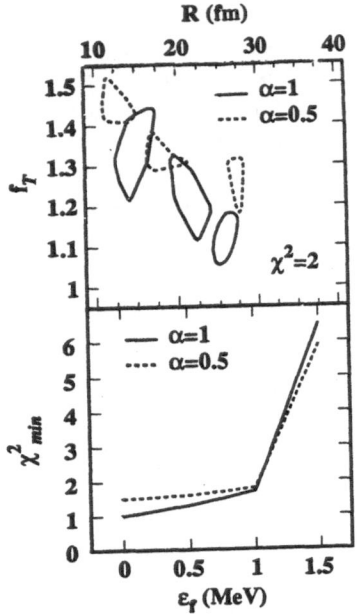

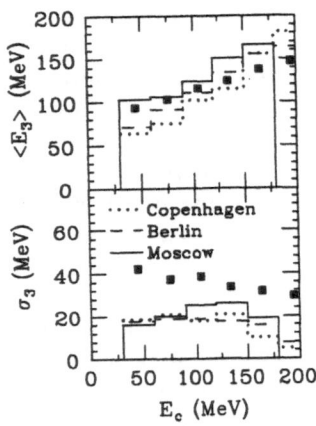

Figure 8. Summary of calculations assuming a volume emission. Top part: contour lines for $\chi^2 = 2$ in a plane defined by the volume radius R and the temperature parameter f_T for flow parameters $\epsilon_f = 0$, 0.5 and 1 MeV (increasing from left to right). Bottom part: minimum χ^2 as a function of ϵ_f.

Figure 9. Mean center-of-mass total kinetic energy $\langle E_3 \rangle$ (top) and the standard deviation σ_3 (bottom) as a function of the nominal Coulomb energy averaged over the 4 reactions Au + C, Al, Cu and Pb at E/A = 600 MeV. The dotted, dashed and solid histograms present predictions of the COPENHAGEN [10], BERLIN [9] and MOSCOW [18] multifragmentation models, respectively.

emission times is even more evident in Figure 7f where we compare the measured and predicted correlation functions [28] of fragment 2 and 3, $R_{23}(\beta_{23})+1$.

Results of calculations modelling the simultaneous volume emission of the three fragments are compiled in Figure 8. Similar to the case of the sequential simulations a clear minimum of χ^2 can be determined for each given flow parameter ϵ_f by varying independently the other two model parameters R and f_T. The upper part of Figure 8 shows in a R - f_T plane the contour lines with $\chi^2 = 2$ for $\epsilon_f = 0$ (at R $\approx$ 15 fm), 0.5 (R $\approx$ 22 fm) and 1 MeV (R $\approx$ 26 fm) and for two different values for the exponent α. The corresponding minimum values of χ^2 are displayed in the lower part as a function of ϵ_f. Values of ϵ_f larger than 1 MeV are ruled out whereas smaller values of ϵ_f show no significantly different χ^2_{min}. There exists, for each given value of α, a rather unique relation between ϵ_f, R and f_T. For large emission radii R > 15 fm the data can only be described if a collective radial velocity is taken into account. This radial motion can be provided by the Coulomb repulsion if the breakup occurs out of a volume with R = 15 fm. In the latter case, the average initial distance between fragment 2 and 3 is about $\langle d_{23} \rangle \approx 18$ fm and is comparable to a value of $\langle d_{23} \rangle \approx 16$ fm (21 fm) obtained for the sequential scenario using $\tau = 10$ fm/c (100 fm/c). Again, the relative velocity distributions and the correlation function $R_{23} + 1$ are reasonably well described with the optimum parameter sets and $\epsilon_f \leq 1$ MeV.

Kinematic Observables from Multifragmentation Models

Taken at their face values, the trajectory calculations suggest a fast disintegration process of an highly excited ($f_T = 1.1 - 1.35$) and - at the same time - rather extended ($R \geq 15$ fm) nuclear system. If comparative source parameters are utilized in statistical multifragmentation codes charge distributions are predicted which are clearly inconsistent with the experimental observations. Using in turn input parameters for the statistical models which fit the experimental charge distributions [18, 9, 10], $\langle E_3 \rangle$ is indeed reasonably well reproduced (see top part in figure 9). Similar results are obtained by the schematic trajectory calculations if corresponding values $f_T \approx 0.6$-0.8 and $R \approx 7$ - 9 fm are adopted, thus illustrating the compensation of a reduced thermal contribution to E_3 ($\sim f_T$) by an increased Coulomb repulsion ($\sim 1/R$). However, the width σ_3 is significantly underpredicted in this case (lower part in Figure 9). In addition to thermal fluctuations, σ_3 reflects fluctuations due to the position sampling within the breakup volume. Therefore, the lower temperatures *and* the smaller radius as compared to the fit results of the trajectory calculations lead to a significant reduction of σ_3. It has been verified, that this underprediction of σ_3 can not be compensated by allowing reasonable fluctuations of the initial excitation energy or the system size. (Note, that the MOSCOW [18] multifragmentation code takes also fluctuations of the initial excitation energy into account.) Thus, it seems that within the framework of these models it is not possible to describe simultaneously the fragmentation phase-space and the fluctuative dynamics in the final disassembly stage.

For any further attempt to reconcile the kinetic observables and the fragment yields within the statistical decay models it might be important to realize that all dynamical calculations presented hitherto ignore the nuclear interaction during the breakup dynamics: First, in the statistical models the interaction between the fragments is limited to the Coulomb repulsion. Secondly, although the schematic three-body trajectory calculations include a nuclear proximity potential, its influence is strongly suppressed by the requirement that the fragments do not overlap. One might speculate that in case of a stronger overlap of the fragments in an earlier stage of the breakup, the nuclear attractive force between the fragments may partially compensate the Coulomb repulsion. Thus, smaller radii would not necessarily lead to an overestimation of the kinetic energies. At the same time large fluctuations - perhaps akin to dissipative phenomena or shape fluctuations known to be important in binary fission [37]- may arise. Indeed, in the reactions studied here binary events also show fluctuations of the relative kinetic energy which can only be described by a rather high - and probably unrealistic - thermal contribution if in addition only the Coulomb interaction is considered. First steps to add the nuclear interaction between the fragments to statistical decay models in a consistent manner have indeed already been undertaken [38]. Although such calculations emphasize the importance of the nuclear interaction for the multifragmentation process, a quantitative understanding of fluctuations and their development during the disassembly phase clearly requires dynamical transport models which include a realistic treatment of fluctuations on a microscopic level.

Acknowledgements

I wish to thank all collaborators of the ALADIN group who contributed to this work, especially S. Aiello, M. Begemann-Blaich, A. Cosmo, A. Ferrero, T. Hofmann, J. Hubele, G. Imme, I. Iori, P. Kreutz, G.J. Kunde, W.D. Kunze, V. Lindenstruth, U. Lynen, A. Moroni, W.F.J. Müller, B. Ocker, C.A. Ogilvie, V. Pappalardo, G. Raciti, H. Sann, R. Scardoni, A. Schüttauf, W. Seidel, V. Serfling, L. Stuttge, W. Trautmann, A. Tucholski, A. Wörner, and B. Zwieglinski. Enlightening discussions are gratefully acknowledged with D.H.E. Gross and A.S. Botvina. This work was supported by the Deutsche Forschungsgemeinschaft under the grant no. Po 256/2-1.

REFERENCES

1. For a recent reviews see, e.g., *The Nuclear Equation of State*, NATO ASI Series B: Physics Vol. 216A, ed. W. Greiner and H. Stöcker (Plenum Press, New York 1989); K.-H. Kampert, J. Phys. **G15**, 691 (1989); H.H. Gutbrod, A. Poskanzer, and H.G. Ritter, Rep. Prog. Phys. **52**, 1267 (1989).

2. W. Schmidt *et al.*, Phys. Rev. C **47**, 2782 (1993).

3. For a recent review see Gy. Wolf, GSI-preprint, GSI-93-55 (1993).

4. P.J. Siemens, Nature (London) **305**, 410 (1983).

5. G.F. Bertsch and P.J. Siemens, Phys. Lett. B **126B**, 9 (1983).

6. J. Hubele *et al.*, Phys. Rev. C **46**, R1577 (1992).

7. P. Kreutz *et al.*, Nucl. Phys. **A556**, 672 (1993).

8. C.A. Ogilvie *et al.*, Phys. Rev. Lett. **67**, 1214 (1991).

9. Bao-An Li, A.R. DeAngelis, D.H.E. Gross, Phys. Lett. B **303**, 225 (1993); for a recent review see D.H.E. Gross, Rep. Prog. Phys. **53**, 605 (1990).

10. J. Bondorf *et al.*, Nucl. Phys. **A443**, 321 (1985); H.W. Barz *et al.*, Nucl. Phys. **A561**, 466 (1993).

11. M.B. Tsang *et al.*, Phys. Rev. Lett. **71**, 1502 (1993).

12. W. Bauer *et al.*, Phys. Rev. C **47**, R1838 (1993).

13. N. Hermann, Nucl. Phys. **A553**, 739c (1993).

14. D. Idier *et al.*, Phys. Rev. C **48**, R498 (1993).

15. J. Hubele *et al.*, Z. Phys. A **340**, 263 (1991).

16. R.J. Charity *et al.*, Nucl. Phys. **A483**, 371 (1988).

17. L.G. Moretto, Nucl. Phys. **A247**, 211 (1975).

18. A.S. Botvina, A.S. Il'inov, and I.N. Mishustin, Yad. Fiz. **42**, 1127 (1985) [Sov. J. Nucl. Phys. **42**, 712 (1985)]; A.S. Botvina and I.N. Mishustin, Phys. Lett. B **294**, 23 (1992) and private communication.

19. J. Konopka *et al.*, Prog. Part. Nucl. Phys. **30**, 301 (1993).

20. Z. He, J. Wu, and W. Nörenberg, Nucl. Phys. **A489**, 421 (1988).

21. For an overview see O.V. Lozhkin and W. Trautmann, Phys. Rev. C **46**, 1996 (1992) and references therein.

22. V. Lindenstruth *et al.*, to be published.

23. D.H.E. Gross and K. Sneppen, HMI-Preprint, HIM 1993/P1-Gros 4 (1993).

24. D.R. Bowman *et al.*, Phys. Rev. Lett. **67**, 1527 (1991).

25. K. Hagel *et al.*, Phys. Rev. Lett. **68**, 2141 (1992).

26. P. Glässel *et al.*, Z. Phys. A - Atoms and Nuclei **310**, 189 (1983).

27. D. Pelte *et al.*, Phys. Rev. C **34**, 1673 (1986).

28. R. Trockel *et al.*, Phys. Rev. Lett. **59**, 2844 (1987).

29. D.H.E. Gross *et al.*, Phys. Lett. B **224**, 29 (1989).

30. R. Bougault *et al.*, Phys. Lett. B **232**, 291 (1989).

31. Y.D. Kim *et al.*, Phys. Rev. Lett. **67**, 14 (1991).

32. Y.D. Kim *et al.*, Phys. Rev. C **45**, 338 (1992)

33. G. Bizard *et al.*, Phys. Lett. B **276**, 413 (1992).

34. D.R. Bowman *et al.*, Phys. Rev. Lett. **70**, 3534 (1993).

35. E. Bauge *et al.*, Phys. Rev. Lett. **70**, 3705 (1993).

36. J. Lopez and J. Randrup, Nucl. Phys. **A503**, 183 (1989).

37. For a recent review see F. Gönnenwein, *The Nuclear Fission Process*, Edt. C. Wagemans, CRC Press, Boca Raton (1991), pg. 287 ff.

38. L. Satpathy *et al.*, Phys. Lett. B **237**, 181 (1990); A. Das *et al.*, J. Phys. G **19**, 319 (1993).

STRANGE MATTER: A NEW DOMAIN OF NUCLEAR PHYSICS

C. Greiner

Department of Physics, Duke University,
Durham NC 27708-0305, U.S.A.

A. Diener, J. Schaffner and H. Stöcker

Institut für Theoretische Physik, J.W. Goethe-Universität,
D-60054 Frankfurt, Germany

1. INTRODUCTION

Perhaps the only unambiguous way to detect the transient existence of a temporarily created quark gluon plasma (QGP) might be the experimental observation of exotic remnants, like the formation of strange quark matter (SQM) droplets [1]. First studies in the context of the MIT-bag model predicted that sufficiently heavy strangelets might be absolutely stable [2] or smaller ones at least metastable [1]. The reason for the possible stability of SQM lies in introducing a third flavour degree of freedom, the strangeness, where the mass of the strange quarks is considerably smaller than the Fermi energy of the quarks, thus lowering the total mass per unit baryon number of the system. According to this picture, SQM should appear as a nearly neutral and massive state because the number of strange quarks is nearly equal to the number of massless up or down quarks and so the strange quarks neutralize that hypothetical form of nuclear matter.

Still, on the other side, strangeness remains also an experimentally as theoretically largely unexplored degree of freedom in strongly interacting baryonic matter [3]. This lack of investigation reflects the experimental task in producing nuclei containing (weakly decaying) strange baryons, which is conventionally limited by replacing one neutron (or at maximum two) by a strange Λ-particle in scattering experiments with pions or kaons.

Hot and Dense Nuclear Matter, Edited by
W. Greiner *et al.*, Plenum Press, New York, 1994

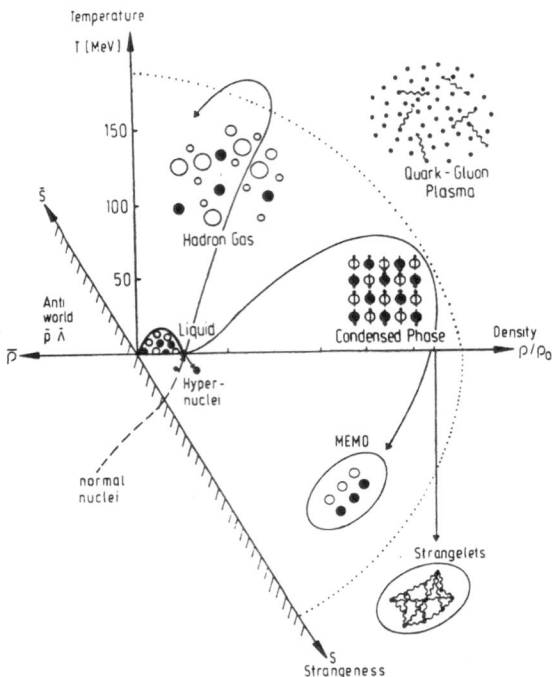

Figure 1. Phase diagram of (hot) nuclear matter including the strangeness degree of freedom – MEMOs and possibly also strangelets establish as stable multistrange configurations.

However, central relativistic heavy ion collisions provide also a source for the formation of multi-hypernuclear objects, consisting of nucleons, Λ's, Σ's and Ξ's. To be more specific, the RQMD model [4] predicts on the average the occurence of 20 Λ's, 10 Σ's and 1 Ξ's per event for Au(11.7 AGeV)Au and of 60 Λ's, 40 Σ's and 5 Ξ's per event for Pb(160 AGeV)Pb. (The number of Ξ's are solely obtained by string fragmentation – rescattering effects are not taken into account.) By employing a relativistic meson-baryon field theory (RMF), which gives a rather excellent description of normal nuclear and single Λ-hypernuclear properties [5], it was found that such configurations may exist as small metastable objects [6]. From a more general point of view, based on these theoretical observations one is now tempted to ask about their principle existence also as much *larger* objects.

In Figure 1 we depict qualitatively what we want to emphasize in the following: Customarily the equation of state of hot and dense hadronic matter is characterized by means of a phase diagramm ($\rho_B \leftrightarrow T$), where at some critical temperature and/or nonstrange baryon density eventually a phase transition to a deconfined QGP-state does occur. However, the EOS to be passed through during a heavy ion collision incorporates also a new degree of freedom, the *net* strangeness (counting a surplus of strange over antistrange quarks). Like the occurence of bound nonstrange nuclear matter, multihypernuclear matter or small droplets (MEMOs) of this new state, may be revealed. In addition, also the phase transition to the deconfined state is affected by the possible conglomeration of the strangeness degree of freedom. In particular, if the strangelet does exist in principle, it has to be regarded as a cold, stable and bound manifestation of that phase being a remnant or 'ash' of the originally hot QGP-state.

In chapter 2 we summarize the reasons for the existence of both this novel and exotic states, particularly we describe our findings for stable multihypernuclear configurations in more detail. In chapter 3 the mechanism of strangeness separation out of a QGP and the possible distillation of a strangelet are reviewed. Also we give some arguments that a somewhat similar scenario might hold solely in the hadronic sector, assembling and localizing more hyperons in phase space. In chapter 4 we more critically discuss the detection possibilities of these states by their properties and lifetimes, also in respect to the present experimental undertaking at Brookhaven and at CERN.

2. STRANGE MATTER

2.1 Strange quark matter

Let us briefly summarize how a stable or metastable strangelet might look like: Think of bulk objects, containing a large number of quarks ($u...u$, $d...d$, $s...s$), so-called multiquark droplets. Multiquark states consisting only of u- and d-quarks must have a mass larger than ordinary nuclei, otherwise normal nuclei would be unstable. However, the situation is different for droplets of SQM, which would contain approximately the same amount of u-, d- and s-quarks. If the mass of a strangelet were smaller than the mass of the corresponding ordinary nucleus with the same baryon number, the strangelet would be absolutely stable and thus be the true groundstate of nuclear matter [2]. Presently such a scenario cannot be ruled out.

On the other hand, it is also conceivable that the mass per baryon of a strange droplet is lower than the mass of the strange Λ- baryon, but larger than the nucleon mass. The droplet is then in a *metastable* state, it cannot decay into Λ's [1, 7]. For phenomenological bag parameters $B^{1/4}$ lower than 190 MeV strange quark droplets can only decay via weak interactions. For larger B-values strangelets are instable. Due to the strong finite size effects [1] and the wider range of the employed model parameters, smaller strangelets are much more likely to be metastable than being absolutely stable. (Witten's idea [2] of absolutely stable SQM droplets does work in this approach only for bag parameters around 145 MeV.)

2.2 Strange hadronic matter – MEMOs

The existence of metastable exotic multihypernuclear objects (MEMOs) has been predicted just recently [6]. A classification scheme for metastable combinations of nucleons and hyperons exhibits that combinations of nucleons, Λ's and Ξ's are favoured compared to combinations with Σ's due to their Q-values in vacuum. The substantial energy release $Q \approx 75 - 80$ MeV in reactions like $\Sigma N \to \Lambda N$ or $\Sigma \Xi \to \Lambda \Xi$ can not be overcome by binding effects. The situation is less clear for the reaction $\Xi N \to \Lambda \Lambda$ with the smaller energy release of $Q = 23$ MeV for Ξ^0 and $Q = 28$ MeV for Ξ^-. Here the reaction can be prohibited by Pauli-blocking effects in bound strange hadron matter (SHM). An example is shown in Figure 2, where the single particle levels of a strange nucleus consisting of two of each proton, neutron, Λ, Ξ^0, Ξ^- are plotted.

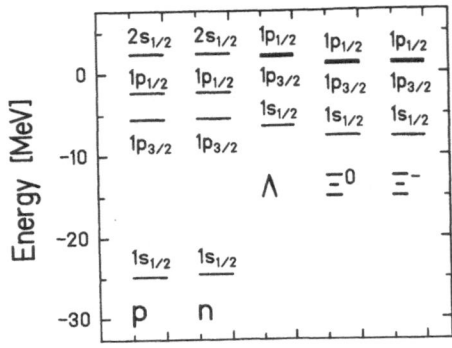

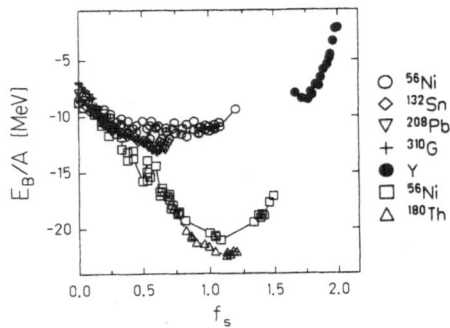

Figure 2. Single particle energy of a MEMO consisting of two of each baryon of the baryon octet except the Σ's. The binding energy difference cancels the mass difference of the strong reaction channels so that the whole 'nucleus' is metastable.

Figure 3. The binding energy versus the strangeness fraction f_s for several sequences of superstrange hypernuclei based on various nuclear cores as indicated. The calculations for the upper three cores employed the model without strange meson exchange, whereas the calculation for the lower one includes the strong YY interaction. Y denotes purely hyperonic matter.

Note that each baryon sits in the 1s-state. The reaction $\Xi N \to \Lambda\Lambda$ can not induce a strong decay because the two Λ's sitting in the 1s-level cause the produced Λ's to escape in vacuum. But this is energetically unfavoured resulting in an overall metastable compound system. The calculation was carried out in a relativistic mean field model taking care of the nucleon-nucleon and nucleon-hyperon interaction. Extrapolation to heavier systems shows a minimum binding energy of $E_B/A = -13$ MeV with a strangeness fraction of $f_s = |S|/A \approx 0.6$ and a charge fraction around $Z/A \approx 0.1$ nearly independent of the total mass. (Here and in the following f_s counts the amount of net strangeness per baryon in the system.) The most important effect for this raising binding energy and stability comes from the vanishing Coulomb repulsion due to the addition of the negatively charged Ξ^-. In contrast to normal nuclei one can easily think of baryon numbers of the order of $A > 500$ because the fission process will be removed since the ordinary Coulomb repulsion generated by the protons can be compensated by a comparable number of Ξ^-'s. Thus SHM is predicted to be metastable even in the bulk limit. Interestingly, multi-Λ hypernuclei are generally unstable against the reaction $\Lambda\Lambda \to \Xi N$ for some critical number of Λ's, i.e. pure multi-Λ hypernuclei get more bound by adding Ξ's. E.g. ^{56}Ni collapses for more than 14 Λ's to a multi-$\Lambda\Xi$ hypernucleus.

An extension of this model also implements the scarce information about the hyperon-hyperon interaction. This is done by introducing two new meson fields into the theory, σ^* and ϕ, which couple to strange baryons only [8]. Note that the weak $\Lambda\Lambda$ interaction in the usual relativistic mean field model is inconsistent with the experimental known values. After the implementation of these two new meson exchanges one can easily get strong enough hyperon-hyperon potentials. The binding energy for MEMOs will be enhanced due to this additional interactions. Indeed, binding energies of -21 MeV and more have been found with a strangeness fraction of $f_s \approx 1$ as is visualized in Figure 3. Even negatively charged strange nuclear systems are possible without losing stability. The filled circles in Figure 3 represent stable aggregates of purely hyperonic Λ, Ξ^0, Ξ^- matter with E_B/A as large as -9 MeV for $f_s \approx 1.7$ and $f_q \approx -0.3$. The lightest stable object of this type is likely to be $(2\Lambda + 2\Xi^0 + 2\Xi^-)$. Purely hyperonic matter, in contrast to SHM in general, is not stable in the bulk limit, because of the Coulomb repulsion generated by the Ξ^-'s.

Therefore finite MEMOs share the two features highlighted as a possible "smoking gun" for the existence of SQM, namely $f_s \approx 1$ and $q/A < 0.1$. This analogy goes even further if one considers the baryon densities of these multihypernuclear objects, which range from 2 to 3 times that of normal nuclear matter similar to SQM calculations. Nevertheless, SQM might be absolutely stable, while SHM will be unstable against weak interactions, particular for $\Delta S = 1$ nonmesonic decays.

3. STRANGENESS SEPARATION

Multiple collisions per hadron ensure that a system starts to equilibrate which might be suited to search for the most interesting collective effects. In particular all exotic multistrange objects need for their formation *large strange particle numbers, high degree of equilibration* and *large densities*. These requirements should enlarge their production probabilities at least due to simple coalescence arguments. In order to get this kind of states one should use high energies to produce enough strangeness and energy density, and heavy nuclei to gain as much equilibration as possible. If the degree of stopping and thermalization at high densities is large enough, as predicted by the RQMD model, the argumentation with more simple models like e.g. thermodynamic rate equations seems to be reasonable. In the following we want to sketch two related arguments why the production of SQM clusters, if they do exist in principle, is very likely, if a baryon rich and hot QGP is created in such collisions.

The net strangeness of the QGP is zero from the onset, although an equal, however large, number of strange and antistrange quarks has been produced by gluon fusion [9]. Yet, there is a physical mechanism which separates the strange quarks from their antiparticles [1]. Herefore consider the phase transition of the QGP to the hadron gas at some critical temperature. There is no reason why these different quarks should hadronize in the same manner and time, especially if one thinks of a baryon rich system. It is 'simple' for the antistrange quarks to materialize in kaons $K(q\bar{s})$ because of the lots of light quarks as compared to the s-quarks which could only move into the suppressed antikaons (K^-) or the heavy hyperons. Hence, during an equilibrium phase transition a large antistrangeness builds up in the hadron matter while the QGP retains a large strangeness excess.

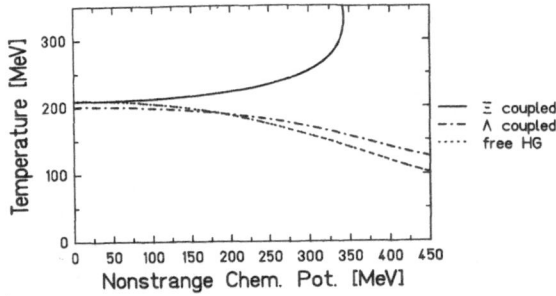

Figure 4. The strangeness separation out of a QGP occurs if the phase transition takes place below the curves presented here. In the case of interacting baryons in a RMF approach, the border is shifted to higher temperatures, in particular if the Ξ-particle is coupled to the mean field. Including also the additional interactions among the hyperons (model 2 of section 2.2), the border vanishes and the separation will always occur.

Figure 4 shows the physical region where the above outlined separation occurs [10]. For a free gas of hadronic particles and resonances the critical curve below which the mechanism starts working lies already at rather high temperatures or baryochemical potentials. If one includes interaction among the baryons like in the RMF approach outlined in section 2.2, this 'border' moves even to higher temperatures. (For the RMF-model 2 there indeed exist no such border, the mechanism should work inside the whole $(T \leftrightarrow \mu)$-plane.) In any case, it is expected that the phase transition from a QGP to an (interacting) hadronic gas lies below these presented curves so that essentially the strangeness separation should always take place.

Furthermore, rapid kaon emission leads similarly to a finite *net* strangeness of the expanding system [1]. This, in turn, results in an even stronger enhancement of the s-quark abundance in the quark phase. This prompt kaon (and, of course, also pion) emission may cool the quark phase, which then condenses into metastable or stable droplets of SQM.

3.1 Strangelet distillation

In order to model the evolution of an initially hot fireball a two phase equilibrium description between the hadron gas and the QGP was combined with the nonequilibrium radiation by incorporating the rapid freeze-out of hadrons from the hadron phase surrounding the QGP droplet during the phase transition [1].

Two scenarios may describe the evolution to the final state: The quark droplet may remain unstable until the strange quarks have clustered into Λ-particles and other strange hadrons to carry away the strangeness and the plasma has completely vanished into standard particles. This scenario is customarily accepted. However, if SQM exists at low temperatures in configurations having a mass per baryon lower than the mass of the Λ-particle, the hot SQM droplet would remain at the phase transition boundary much longer. As shown in [1], producing SHM like Λ particles is energetically more expensive and therefore less likely than producing SQM like

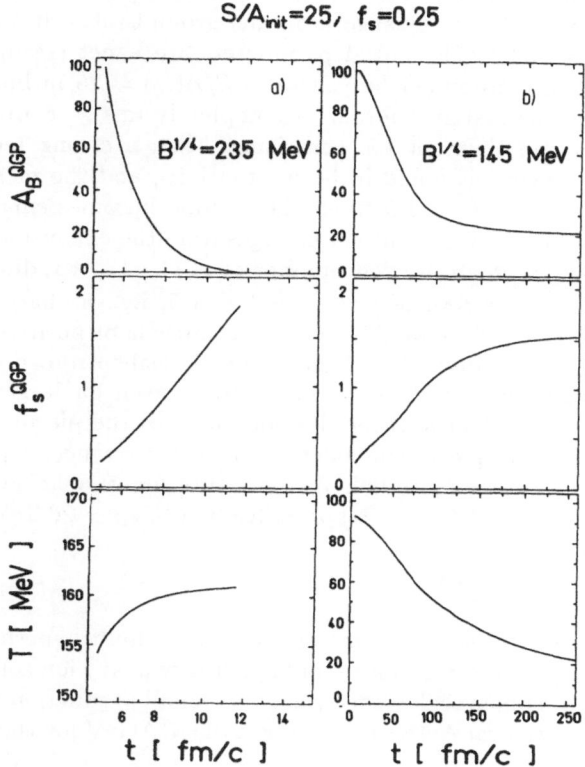

S/A$_{init}$=25, f$_s$=0.25

Figure 5. a. Baryon number, strangeness content and temperature of the quark glob during complete hadronization as a function of time for a very large bag constant $B^{1/4} = 235$ MeV. The initial values are an initial baryon content of $A_B(t_0) = 100$, an entropy per baryon ratio of $S/A(t_0) = 25$ and an initial net strangeness fraction of $f_s(t_0) = 0.25$. Note the strong increase of the strangeness content with time.

b. The same situation as in a., however, for a small bag constant $B^{1/4} = 145$ MeV, when a strangelet is *distilled*. One observes a strong decrease in the evolving temperature.

strangelets. Towards the end of the evolution only baryons are allowed to escape from the droplet, since at this point all of the antiquarks are gone. The baryons will be mostly nucleons, since the hyperons are heavier and require more energy for formation. These nucleons remove energy but they do not carry away any strange quarks, so the ratio of strange to nonstrange quarks increases further, refining the distillation of strangeness. With a reasonable but small probability, the hot strange matter cools down to cold lumps of size $A \sim 5 - 50$, depending on the original baryon content of the plasma.

Figure 5 gives an impression how the hadronization proceeds for a large bag constant ($B^{1/4} = 235$ MeV – no strangelet in the groundstate) and a small bag constant ($B^{1/4} = 145$ MeV). The initial parameters are a net strangeness content of $f_s(t_0) = 0.25$ and an entropy per baryon ratio $S/A(t_0) = 25$ in both cases: For the large bag constant the system hadronizes completely in $t \sim 8$ fm/c, which is customarily expected and thus not too surprising. Yet, a strong increase of the net strangeness of the system is found in both situations, and the plasma drop reaches a strangeness fraction of $f_s \sim 1.5$ when the volume becomes small. For the small bag constant, however, a *cold* strangelet emerges from the expansion and evaporation process with an approximate baryon number of $A_B \sim 17$, a radius of $R \sim 2.5$ fm, and a net strangeness fraction of $f_s(t \to \infty) \stackrel{>}{\sim} 1.5$, i.e. a charge to baryon ratio $Z/A = (1 - f_s)/2 \sim -0.2$! It would comprise a nucleus of positive baryon number, but negative charge. Also it was found that a high initial entropy does not necessarily prohibit strangelet formation. The distillery works even for larger initial entropies $S/A = 50$ or 100. Abundant kaon production enriches the plasma rapidly with net strangeness at high entropies. This offers to look for strangelet production at the highest bombarding energies available in the future for very heavy systems, e.g. at the CERN SPS ($E_{LAB} \sim 200$ GeV/N) or at RHIC ($E_{LAB} \sim 20$ TeV/N).

3.2 Purely hadronic fireball

It may be possible to produce some of the lightest multihypernuclear objects envisaged here in the laboratory by means of high energy heavy ion collisions. Estimates based on grounds of simple coalescence yield very small production rates, for instance 3×10^{-9} events per central Au+Au collisions at 11.7 AGeV for the lightest bound Ξ system $^{7}_{\Xi^0 \Lambda\Lambda}$He.

One is tempted to ask whether a separation could also manifest if initially there is no QGP at all, i.e. the fireball is only made out of hadronic constituents. Obviously, any separation could only happen by nonequilibrium effects. For a more or less global chemical equilibrium the strange and antistrange particles are distributed rather homogenously across the fireball, equalizing locally on the average. It is the interior QGP phase which due to its different affinity assembles the strange quarks if both phases stay in chemical equilibrium.

Potential non equilibrium effects turn out to be interesting and important for the following two reasons [11]: The mesons like the pions and the kaons are much lighter than the baryons, especially compared to the moderately heavy hyperons. Accordingly their average velocity should be much higher, so that a tremendous amount of the mesons should decouple reasonably earlier from the surface of the fireball before an overall freeze-out of the hadrons in the deep inside of the system takes place. Secondly in a baryonrich fireball the kaons are much more abundant than the antikaons, however they possess a much smaller annihilation cross section compared to the antikaons, e.g. $\sigma_{KN} \ll \sigma_{\bar{K}N}$. In principle this means that the kaons decouple earlier from the fireball compared to the antikaons.

To demonstrate the consequences of this inspired picture we simulate the evolution of an initially very hot system by allowing only the pions and kaons (case 1) and additionally also the antikaons (case 2) to evaporate from the outer layer. The initial parameters are specified by the entropy per baryon content $(S/A(t_0) = 40; 25; 15)$ and a fixed baryon density (ρ_B equals two times normal nuclear matter density – the

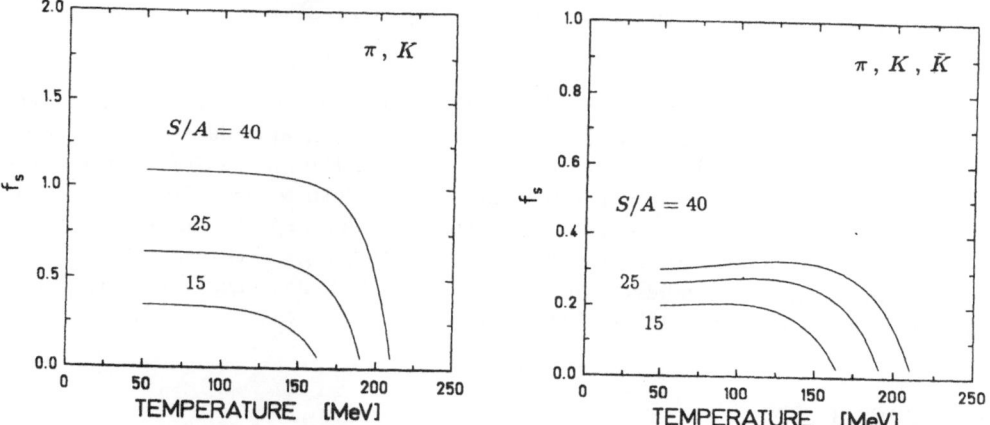

Figure 6. a. Strangeness enrichment in a purely hadronic fireball. Pions and kaons escape from the outer layer (case 1). The initial configuration is specified by the entropy and a baryon density of two times normal nuclear matter.

b. In addition, to show the other 'extreme' also antikaons are allowed to escape from the system (case 2).

results are only minorly sensitive to this choice). The results for the net strangeness enrichment of the remaining and cooling fireball are shown in Figure 6 as a function of the decreasing temperature. Indeed, in case 1 we observe a reasonable to strong increase of the net strangeness fraction f_s in the range $0.3 - 1.1$, depending on the initial entropy content. This increase occurs in the early evolution at still high temperatures $\sim 150 - 170$ MeV. In the second case this enrichment also takes place, but being strongly suppressed, reaching only values $f_s \sim 0.2 - 0.3$. This last scenario, however, should be seen with some caution, because it does not pay respect to the

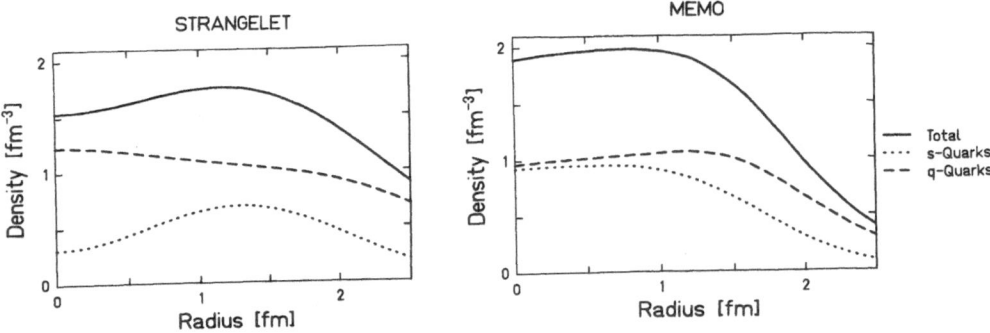

Figure 7. a. The spatial quark distribution of the light quarks and the strange quarks in a hypothetical metastable strangelet with a baryon number of $A_B = 30$ and $f_s = 1$ (the bag parameter $B^{1/4} \approx 170$ MeV was fixed in such a way that the binding energy is the same as for the multihypernuclear object in b.).

b. The same for a multihypernuclear metastable object $6\left\{\, pn\Lambda\Xi^0\Xi^-\,\right\}$.

fact that the kaons decouple much earlier and that the antikaons may be annihilated. Also, the number of produced antikaons are here nearly as abundant as the number of kaons, which is not observed by experiment where the number of antikaons are suppressed. So to speak, the 'truth' may lie between both scenarios presented. The strange quarks or the hyperons are more localized in phase space. Thinking in terms of a coalescence picture for producing exotic multistrange clusters, such an early increase in the net strangeness content, whether a QGP has formed or not, should affect the production probability of these clusters.

4. DETECTION

It is important to note that these objects are a *new form of matter*, not a specific new particle. The strange droplets produced in these reactions do not come in the form of a single type of particle. Many different sizes of droplets may be produced, spanning a range in mass, charge, and strangeness content. The experimental task of finding the new form of matter is therefore challenging. Here any detected particle having an unusual charge to mass ratio is a potential *strangelet* candidate.

To identify a particle or cluster, its charge and mass need to be measured. To determine that the particle is a new form of strange matter, its strangeness content must also be revealed. The experimental approach is first to find 'objects' having a peculiar or new charge/mass ratio. (The strangeness might be seen by interaction with a secondary nucleus : multiple production of Λs, Σs, Ξs and $\bar{K}$s in such a secondary reaction would signal its existence.) The key idea here is that the charge/mass ratio will be unlike that of any normal nuclear isotope (the ^{8}He with a $Z/A = 0.25$ would be the isotope candidate with the smallest ratio). Strangelets or MEMOs would have a charge ~ 0, being slightly positively or negatively charged. In particular in the range $-0.25 < Z/A < +0.25$ there exists no quasistable form of nuclei or antinuclei. Such a range will be covered by the E864 experiment taken next year at Brookhaven. E878, using a focussing spectrometer at zero degree, is seizing a much smaller selected range, which in respect to cover still the full range of interest can be steadily adjusted. A similar technique like in E878 will also be employed next year by the Newmass collaboration (NA52 experiment) at CERN at much higher energies, and later possibly also with the heavy Pb-beam.

However, the global properties of a strangelet or a MEMO are likely to be identical (see Figure 7): a similar small charge $|Z|$ and nearly the same average baryon density $\sim 3\rho_0$. In principle, to distinguish experimentally between both one has in addition to resolve the mass E/A very accurately. A MEMO is only bound in the order of $E_B/A \sim 10$ MeV whereas the strangelet may be bound from $10 - 200$ MeV (which is, of course, speculation). In Figure 7 the spatial quark distribution of a MEMO and a strangelet are depicted. Due to the strong overlap of the corresponding distributions, the MEMO would decay into a strangelet, if the latter is energetically more favourable and stable.

Employing TOF-techniques to reveal the velocity and thus the charge to mass ratio, the experimental setup sets a natural time scale $\sim 10^{-7}$ sec. So, an important question we finally have to adress are the lifetimes of these objects. The lifetime of a MEMO should be similar to the Λ's lifetime, i.e. $\sim 10^{-10}$ sec. Thus an open geometry detectional device will be needed to discover their existence. If a produced strangelet is absolutely stable the only energetically possible decay mode is the weak leptonic decay ($s \rightarrow d$, $Q \rightarrow Q' + e + \bar{\nu}$), which will turn the strangelet to its minimum value in energy at $f_s \sim 0.8$. The time-scale for this weak process has been estimated [7] to be $\sim 10^{-4}$ sec. Hence, the strangelet would remain in its initial condition.

The situation turns out to be more complicated, if the strangelet is metastable. The weak conversion rate $u + s \leftrightarrow u + d$ for almost cold SQM was calculated [12] to be $\sim 10^{-6} - 10^{-7}$ sec. This process changes an s-quark into a d-quark or vice versa, whichever is energetically possible, and thus drives a SQM droplet to $f_s \sim 0.8$, where the Fermi energies are equal ($\mu_s = \mu_d$). This estimated rate would be sufficiently small to permit detection. However, there might be another similar important weak decay channel, the weak nucleon decay: $s \rightarrow d$, $Q \rightarrow Q' + n$. The conversion of a strange quark is accompanied immediately by nucleon emission. This decay is energetically possible in the range of $0.5 < f_s < 1.7$. For $f_s > 1$, accompanying nucleon emission drifts the strangelet to a higher net strangeness content (and hence to a larger chemical potential μ_s for the strange quarks). Subsequent weak decay processes, as described here, 'heat' up the droplet (on a scale of a few MeV). The droplet may be cooled by γ-radiation or by nucleon emission. Both processes may compete in magnitude. Still, the lifetime is a matter of debate. At least for energetical reasons it should be lower than the one of a hypernuclear object. An open geometry device is needed too if the strangelets decay in times of the same order as the MEMOs.

REFERENCES

1. C. Greiner, P. Koch and H. Stöcker, Phys. Rev. Lett. **58**, 1825 (1987);
   C. Greiner, D.H. Rischke, H. Stöcker and P. Koch, Phys. Rev. D **38**, 2797 (1988);
   C. Greiner and H. Stöcker, Phys. Rev. D **44**, 3517 (1992)

2. E. Witten, Phys. Rev. D **30**, 272 (1984);
   E. Farhi and R. L. Jaffe, Phys. Rev. D **30**, 2379 (1984)

3. C. B. Dover, D.J. Millener and A. Gal, Phys. Rep. **184**, 1 (1989)

4. R. Mattiello and H. Sorge, private communication

5. M. Rufa, J. Schaffner, J. A. Maruhn, H. Stöcker, W. Greiner and P.-G. Reinhard, Phys. Rev. C **42**, 2469 (1990)

6. J. Schaffner, C. Greiner and H. Stöcker, Phys. Rev. C **46**, 322 (1992)

7. S. A. Chin and A. K. Kerman, Phys. Rev. Lett. **43**, 1292 (1979)

8. J. Schaffner, C. B. Dover, A. Gal, C. Greiner and H. Stöcker: submitted to Phys. Rev. Lett.

9. P. Koch, B. Müller, and J. Rafelski, Phys. Rep. **142**, 167 (1986)

10. A. Diener, C. Greiner, J. Schaffner and H. Stöcker, publication in preparation

11. C. Greiner, publication in preparation

12. P. Koch, Nucl. Phys. B (Proc. Suppl.) **24B**, 255 (1991);
    H. Heiselberg, J. Madsen and K. Riisager, Phys. Scri. **34**, 556 (1986)

THE ANALYSIS OF PHASE-SPACE
TEXTURES IN ULTRA-RELATIVISTIC
MULTIHADRON PRODUCTION

Peter A. Carruthers

Department of Physics
University of Arizona
Tucson, Arizona, 87521

and

Institut für Theoretische Physik
Johann Wolfgang Goethe-Universität
Robert-Mayer Str. 8-10
60054 Frankfurt am Main
Germany

ABSTRACT

In this paper we discuss the progress made recently in the analysis of higher order correlations. Although we mostly consider multihadron distributions, the methods are of wide applicability. In addition, concepts related to fractals and wavelet analysis are discussed.

INTRODUCTION

It has become clear in recent years that an understanding of high order correlations among particles produced in nuclear collisions is crucial for the purpose of answering several fundamental questions in this interesting field of research [1].

There are at least two crucial aspects to the required analyses. First of all, the methods employed cannot depend on uncontrolled approximations. Rigorous statistical methods are essential[2,3]. The second aspect to keep in mind is the continued absence of theoretical methods that can give a good description of complex multiparticle dynamics, let alone the still more complicated hadronization process at energies for which production of hadronic secondaries dominates the statistical analysis of final states.

Because of this the attitude my own research group has increasingly focussed on defining a rigorous statistical analysis that is independent of modelistic bias. Our method is to work from the outside to the inside rather than the usual inside to outside approach in theoretical physics. In a sense, we develop the analogue of a phase shift analysis, to precisely characterize data which can then be studied for their dynamical content. It turns out that very interesting systematics are emerging from this approach. The next step is to understand their meaning.

The key problem[4] now to be faced is simple. Can we really learn anything of fundamental significance from work in this field? A typical collision in the relativistic regime produces a variety of complex products, from "uninvolved" nucleons in the projectile and target, which jump into phase space, evading equilibration, as well as generating additional complications due to impact parameter effects. The difficulty is how to separate the components of the reaction products in a clean way that allows the proper separation of the physical components.

In lower energy collisions, proton/neutron liberation dominates, along with collective flows and quasiparticle excitation. At higher energies with particle production and secondary-dominated ultrarelativistic collisions, it is also essential to have unambiguous tools of analysis. In the latter case the dynamical situation is especially ambiguous. In the former case it has been difficult to incorporate correlations and fluctuations into the dynamical models.

The procedures described here are based mathematically on point distributions of a single species of particle. New experimental results which discriminate between like-sign and unlike-sign or all charge mixtures are extremely important and must be dealt with. Some results are already available. and have become more interesting in the light of new results on Hanbury-Brown/Twiss (Bose Einstein) correlations in various reactions.

Historically, the earliest efforts to describe large multiplicity events in cosmic rays led to statistical methods and the absorption of relativistic fluid mechanics. We call this approach the SHM (statistical-hydrodynamics model), summarizing a large body of interesting work by Fermi, Heisenberg and especially the work of the Russian (Landau) and Japanese "schools". This methodology exploits the fact that the lowest order description merely expresses energy- momentum conservation and an equation of state which allows LTE (local thermodynamic equilibrium) to close the equations of motion. If the reader is interested in the details of the history of the subject, I recommend my lectures at previous NASI meetings held at Peñiscola in 1990[5] and Il Ciocco[6] in 1992.

An important feature absent from these early works is a proper description of fluctuations[2]. These can be geometrical, stereotypically depicted by the concept of irregular, perhaps fractal, event-by-event hadronization geometries. Or they can (as commonly measured) be count probabilities or moments thereof of number distributions in various patches of phase space. These days we are subjected to many ideas about "the" equation of state, without a careful discussion about those parts of phase space (including of course time and spatial coordinates) for which the equilibrium results should be relevant.

Although pessimism is premature, it is essential to think clearly about the many complications that must be addressed if we are to find decisive evidence for "new" physics. In order to evade such complications various "signals" have been conceived, typically emphasizing unusual properties of hot nuclear matter which should indicate the onset of, especially, the quark-gluon plasma or other new phases of hadronic matter. Here we can mention dilepton pairs, photons, strangeness and other flavor enhancements J/psi "supression", etc. (As far as I can see, no "new physics" is yet in sight.)

Despite the suggestive nature of these alleged phenomena, there remains no unambiguous indication that something really new is going on in high energy nuclear collisions. What is missing, we believe, is a serious and systematic analysis of correlations and fluctuations in nuclear collisions. We also believe that many of these deficiencies have been remedied in recent work.

Closely related is the issue of quasiparticles in excited hadronic matter. Hydrodynamic schemes, or one-body style kinetic equations need extension in this direction to capture the physics.

WHAT IS THE BEST WAY TO DESCRIBE EVENT BY EVENT MULTIPARTICLE FLUCTUATIONS?

Honesty requires some candor about the experimental debris that can easily conceal important systematics of bulk matter that everybody hopes to see. Among these observables, many are explained by non-deep physics, laws of large numbers, statistical averaging, etc. The crucial point here is to segregate those phenomena which are trivial (after the fact!) from those suggestive of something more interesting. In typical collisions we expect the production of resonances, fragments, diffractive components (strongly dependent on collision geometry) not to mention the presence of many hadrons. The inter-relation of all these mechanisms requires sophisticated techniques of analysis.

The need for black-box Monte Carlo codes remains important for the "predictions" of and analysis of experiments we need to avoid the illusion that such methodologies are predictive theories. In addition "corrected" data sets inherit the intellectual errors of the codes. For this reason we emphasize here recent developments which are based on mathematical identities. Despite the existence of a substantial mathematical literature, the particular needs of current high energy and nuclear data analysis have required special ideas as the precision of the experiments has increased.

In order to formulate the problem clearly we consider collisions of hadrons, described as

$$h + h, \qquad h + A, \qquad A + A' \tag{1}$$

where h is an "elementary" hadron such as p, pi, K, and A a nucleus.

We note the incomplete nature of the experimental analysis, without considering acceptance and other tricky detector issues. Assuming (wrongly) a unique center of mass collision energy, we know that for high energies there will be a substantial dispersion in the number of produced particles. In addition, most contemporary detectors are not sensitive to neutrals, a fact which has to be considered. In fact, even the plus and minus aspect is often ignored, being replaced by the question of whether the particle is charged or not. The influence of the cuts made in the data samples, made with respect to the above analysis, is scarcely trivial.

Immediately we can identify some important issues:

1) What are the best observables? Besides mere counts, the further labels such as momentum, flavor, etc. will be crucial. In particular, the behavior of collective flow phenomena depends on the correct choice of dynamical variables. This should be a productive area of research.

2) What information best describes the geometry of the typical collision? For example, it is natural of identify high multiplicity with small impact parameter, but this is only a gross measure. Further, it is reasonable to expect fluctuations in many variables, in order to consider the statistical variation of the phase space geometry of the emitting source.

3) What is the interplay involved in the construction of ensembles and their analysis? In particular, one needs to consider the strong effect of rare and concentrated "spike" events on the statistical accuracy of the moment analysis. These non-trivial problems have led us to modify bin-oriented data analyses to improve statistical accuracy for high resolution.

Now, we have sets of points. Take one example, with density

$$\hat{\rho}(x, s) = \sum_{i=1}^{N} \delta(x - s_i) \tag{2}$$

where x locates the measurement and s_i range over locations labeled 1 to N.

Of course, one example is not enough, in general. The particular example[2] is only that, an example, and must be complemented by many others. We have repeatedly explained the methods necessary to formulate these problems.

1) Fundamental problems in probability theory. In contemporary experimental reality, there is no true statistical ensemble, even for $10^4 - 10^5$ events. The reason for this is that rare but large fluctuations corrupt the data analysis at high resolution, at least for the currently popular variables.

2) As mentioned before, we must find the most insightful physical variables. History has led to a natural emphasis on number counts and their correlations[7]. The further breakdown into flavors and other indicators given by distinct particles is well known to be important. Energy and transverse momentum spectra and correlations among these variables may become a dominant theme in the near future.

COUNTS AND THEIR ANALYSIS

First we recall the connection between moments and correlations: in a patch of phase space it is not really difficult to measure counts and their moments. This is in contrast to the difficulty of measuring correlation functions. As we have emphasized, averages of correlation functions identically yield count moments. The inverse process is more difficult.

In order to understand the nature of fluctuations it is crucial to subtract background terms. This leads from density to "cumulant" correlations. The prototype is given by the 2-particle example

$$\rho_2(x, y) \rightarrow \rho_2(x, y) - \rho_1(x)\rho_1(y) \tag{3}$$

Corresponding to this is the number count:

$$< n(n-1) > \rightarrow < n(n-1) > - < n >^2 \qquad (4)$$

(These formulae assume that the count distribution relates to one species of "particle".) Now we can note that this and higher order analogues provide a precise test of statistical independence, in the sense of Poissonian statistics.

The advance which caught the attention of experimentalists was made by Bialas and Peschanski in 1986. They partitioned phase space (then just rapidity) into bins and averaged over them in order to improve statistics. Although discussions among experts have occurred about the best way to assess the data we simply express the basic formula for the "bin-averaged factorial moment" F_p:

$$F_p = \frac{1}{M} \sum_{i=1}^{p} \frac{< n_i(n_i - 1) \cdots (n_i - p + 1) >}{< n_i >^p} \qquad (5)$$

averaged over M bins of equal size, labeled n_i.

According to elementary arithmetic, Poissonian counts lead to $F_p = 1$.

The second idea proposed was that there could be a scaling behavior, originally conceived to be present in the rapidity variable y. The initial writings over-represented these speculations, as we have explained elsewhere.

Clearly if the correlation functions scale (whether for densities, or better, cumulants), the F_p moments, or their cumulant analogues K_p, will also scale.

The history of this subject is clouded by the premature zeal to detect scaling. We can see this by looking at a standard graph of F_2.

As is clear[2], the form of $F_2(\delta y)$ is a simple manifestation of ρ_2. This follows from the identity (here written for a single bin labeled i)

$$F_2^i \equiv \int_{\omega_i} dy_1 dy_2 \frac{\rho_2(y_1 1 y_2)}{\rho_i^2 (\delta y)^2} \qquad (6)$$

where ρ_i is the average single particle density in the chosen bin ω_i of size δy. The dependence of F_2 on the resolution δy is then determined by the experimental knowledge of ρ_2. It turns out that although ρ_2 is not translation invariant, a simple exponential or Gaussian with a single correlation length gives as good a description of F_2 as more sophisticated parametrizations of the data. Fig. 1 indicates the behavior of F_2 as a function of resolution. The decrease for large bin size is a direct consequence of the cutoff due to the correlation length. The straight lines indicate the scaling fits to the high resolution domain. For the analysis of higher correlations and the cumulant decomposition, we refer to ref. 2.

There is no space here to discuss speculations about the structure of higher order correlations, and their connection with the negative binomial count distribution which is pervasive in the count statistics of multihadron production. (The validity of this approach in nucleus-nucleus collisions is not yet confirmed.) It seems true that the linked pair approximation for the hierarchy of cumulant correlations with appropriate negative binomial coefficients works quite well. Unfortunately there is no unique dynamical explanation at present.

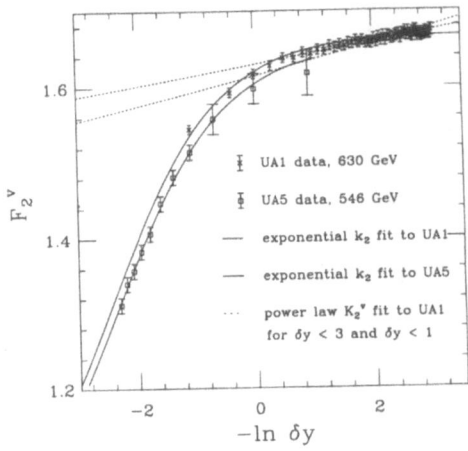

Fig. 1. Best fits for F_2^y moments for exponential correlation $k_2 = \gamma exp(-|y_1 - y_2|/\xi)$ for UA1 [$\gamma = 0.669$, $\xi = 3.24$] and UA5 [$\gamma = 0.656$, $\xi = 2.95$] (solid lines). Dashed lines show best fits using a power law correlation $k_2 = x(y_1 - y_2)^{-\nu}$

PROBLEMS WITH THE BIN ANALYSIS
AND IMPROVEMENTS THERETO

The big advance permitted by the BAFM (bin averaged factorial moment) approach was that high order correlations could be sampled, though not directly measured, because of precise new data on number fluctuations. However, even in 2nd order correlations one can detect procedural problems. The moment analysis is especially interesting in several respects. The dependence on resolution is perhaps the most important feature. This gives an indication of whether there is a scaling regime. Second, the trivial fact that Poissonian noise components are removed in the BAFM procedure needs little further discussion. However the usual approach forgets that it is necessary to subtract lower order background correlations, which obscures the meaning of the results. For this reason we reminded people of the necessity to use cumulant correlations[2]. As a consequence, it became clear that contemporary data involving nuclei have almost no correlations beyond second order at present energies. The real issue is simple. Are there interesting fluctuations to be found in RHIC? Will a new phase appear at RHIC energies? Does multiple scattering destroy most of the correlations that contain information about dynamical fluctuations?

A more serious problem created by binning arises from rare but especially interesting "spike" events. The mathematical problem is obvious once pointed out. The counting procedure displays a major defect with regard to amplifying moment fluctuations for high resolution. Suppose that we improve the bin resolution by a factor of two, for a particular event with 12 points in a bin of size $2\delta y$. Further suppose that we have 5 and 7 points in the split bins of size δy. Consider the third moment F_3. For the $2\delta y$ bin we have $12 \cdot 11 \cdot 10 = 1320$ for the un-normalized moment. Splitting the bins we have $5 \cdot 4 \cdot 3 + 7 \cdot 6 \cdot 5 = 270$. Despite normalization issues, it is clear that the bin moments are sensitive to fluctuations due to procedure rather than nature. These difficulties can be removed[8] by modifications of the moment analysis.

By redefining the moment definitions it is possible to reduce the noise at high resolution. In lowest (second) order this coincides with the well-known analysis of Grassberger-Hentschel-Procaccia[11]. It must be admitted that devilish problems associated with normalizations remain. In higher orders there are details not to be brought up here.

For this meeting I would point to the problem of CPU time required to digest events of high multiplicity and geometric complexity. Thanks to the hard work of my collaborators, Eggers and Lipa[9,10,12,13], still another modification of the sampling of points needed for the correlation analysis, which they call the "star" integral, allows realistic data analysis techniques for relativistic nuclear collisions in which hundreds of produced particles will be typical. The required CPU time becomes manageable in this framework.

The papers in press need to be consulted for detailed computational procedures.

MORE REMARKS

I have not spoken much about fractals, multifractals, and so forth although these concepts are always on my mind. Many of these techniques are based on models without background noise. Therefore the applications of standard formulas can be dominated by noise in phenomena like nuclear collisions. After many meanders, I came to appreciate a different approach, the wavelet method which seems to have been properly formulated especially by Belgian and French scientists. The wavelet transform acts as a microscope, localizing and magnifying the texture of the histograms. A good introduction to wavelet theory can be found in Doubechies' ten lectures[16].

I consider this to be the tool of choice for texture analysis, for data compression and the simplification of many body dynamics. In addition I predict that in a year everybody will recognize the possibility to improve and economize on detector design using such ideas. In a recent paper Hakioglu and I discussed[14] power spectrum methods of data analysis. However, despite my pleasure in constructing that work, the inability of Fourier techniques to simultaneously localize in space and time is a symptom of its weakness.

The lecture by Martin Greiner[15] at this meeting gives a nice introduction to the ways in which wavelet transforms could be useful in nuclear and particle physics.

ACKNOWLEDGEMENTS

In addition to my collaborators in my research group, I have had the great pleasure to work with European colleagues in many countries. The support of the Alexander von Humboldt Stiftung, the GSI Darmstadt, the Institute for Theoretical Physics in the J-W Goethe University of Frankfurt and the Deutsche Forschungsgemeinshaft, have been especially helpful. For this I thank Prof. Dr. Walter Greiner and his many talented associates. In the USA my activities have been supported by the Department of Energy Divisions of Nuclear and High Energy Physics.

REFERENCES

1. XXIV International Symposium on Multiparticle Dynamics, Aspen, Colorado, September 12-16, 1993. Many key references can be traced from the proceedings, eds. M. Block and A. White, (World Scientific, Singapore, to be published)

2. P. Carruthers, H. C. Eggers and I. Sarcevic, Physics Letters $\underline{B254}$ 258 (1991). From this one can find references to earlier work by the Tucson group.

3. P. Carruthers, monograph on "Correlations and Fluctuations in Complex Systems", to be published.

4. L. Neise, "Quantum Effects in the Phase Space Description of Particle Production and Multiple Scattering", these proceedings.

5. P. Carruthers, "Scale Transformations, the Energy-Momentum Tensor and the Equation of State" p. 513 in the Proceedings of the NATO Advanced Study Institute "The Nuclear Equation of State", ed. W. Greiner and H. Stocker (Plenum Press, NY, 1990)

6. P. Carruthers, "The Quark-Gluon Plasma: Search for the Holy Grail", p. 669 in the NATO Advanced Study Institute "Particle Production in Highly Excited Matter", Il Ciocco, Italy, eds. H. H. Gutbrod and J. Rafelski, Plenum Press, New York (1993)

7. P. Carruthers and C. C. Shih, Intl. Journal Modern Physics A, 1447 (1987)

8. P. Lipa, P. Carruthers, H. C. Eggers and B. Buschbeck, Physics Letters $\underline{285B}$ 300 (1992)

9. H. C. Eggers, P. Lipa, P. Carruthers and B. Buschbeck, Phys. Rev. D (in press)

10. H. C. Eggers, P. Lipa, P. Carruthers and B. Buschbeck, Physics Letters $\underline{301B}$ 298 (1993)

11. P. Grassberger and I. Procaccia, Phys. Rev. Letters $\underline{50}$, 346 (1983); H. G. E. Hentschel and I. Procaccia, Physica $\underline{D8}$ 435 (1983)

12. P. Lipa and H. C. Eggers, "Unbiased estimators for correlation measurements", preprint, September, 1993

13. H. C. Eggers and P. Lipa, "Star integrals and unbiased estimators", preprint, September 1993 (Cracow Workshop on Multiparticle Production, April 1993)

14. P. Carruthers and T. Hakioglu, Phys. Rev. $\underline{D45}$ 4046 (1992)

15. M. Greiner, these proceedings.

16. I. Daubechies, "Ten Lectures on Wavelets", Soc. for Industrial and Applied Mathematics (1992)

RECENT RESULTS FROM LATTICE QCD

Jochen Fingberg

Supercomputer Computations Research Institute
The Florida State University
Tallahassee, FL 32306-4052, USA

INTRODUCTION

Quantum Chromodynamics (QCD), the theory of strong interactions, exhibits a phase transition from hadronic matter to a plasma of quarks and gluons at high temperature, $T \simeq 150\text{MeV}$, or density, $n \simeq (5 - 10)\ n_0$, where $n_0 = 0.15\text{fm}^{-3}$ is the density of ordinary nuclear matter. Not only does this further our understanding of the early universe, but also it gives us hope that a quark gluon plasma (QGP) may be created in ultra-relativistic heavy-ion collisions planned at Brookhaven (RHIC) and CERN (LHC)[1].

Lattice gauge theory allows us to calculate the consequences of QCD for any energies. Questions that can be studied in numerical simulations include the existence of a phase transition, the order of the transition, and, of course, the temperature of the transition. The plan of these lectures is to give our current view of the structure of QCD at high temperature from the viewpoint of lattice QCD calculations. Lattice calculations must be done with a finite cutoff and also in a specific fermion scheme. Therefore we have to control the continuum limit in order to extract physical results. Significant progress in this direction has been achieved recently in the methodology of lattice gauge theory calculations in the pure gauge sector.

In the following sections the previous questions will be discussed based on recent results from lattice studies. A summary which illustrates the advances in lattice size and quark mass is contained in Table 1.

The outline of the lectures is the following: after a brief discussion of the finite temperature formalism we will present some recent results in pure gauge theories. We will then come to simulations of full QCD: here the focus will be on the flavour dependence of the transition and on the nature of the high temperature phase.

For a more general overview on finite temperature QCD we recommend the excel-

Hot and Dense Nuclear Matter, Edited by
W. Greiner *et al.*, Plenum Press, New York, 1994

Table 1. Lattice studies of QCD thermodynamics with staggered quarks.

N_f	lattice size	m_q/T	group, reference
4	$12^3 \times 4$	0.1	Fukugita et al, (1990)[2]
	$16^3 \times 6$	0.06	Columbia, (1990)[3]
	$16^3 \times 8$	0.08	MTc, (1990)[4]
2+1	$16^3 \times 4$	.1, .1 to 2	Columbia, (1990)[3]
2	$12^3 \times 4$	0.05	Fukugita et al, (1990)[2]
	$12^3 \times 6$	0.075	MILC, (1992)[5]
	$12^3 \times 8$	0.1	Gottlieb et al. (1990)[6]
	$16^3 \times 8$	0.1	HTMCGC, (1993)[7]
	$16^3 \times 8$	0.05	HTMCGC, (1993)[7]
	$16^3 \times 8$	0.032	Columbia, (1993)[8]
	$32^3 \times 8$	0.032	Columbia, (1993)[9]

lent review talks given at previous Lattice and Quark Matter conferences[10] as well as the comprehensive summary talks by F. Karsch and S. Gottlieb[11].

QCD THERMODYNAMICS

The basic quantity for numerical studies of Quantum Field Theories at finite temperature is the Euclidean path integral for the partition function[12]

$$Z(V,T,\mu) = \int \mathcal{D}A \, \mathcal{D}\bar{\psi} \, \mathcal{D}\psi \, e^{-\int_V d^3x \int_0^{1/T} dx_0 \mathcal{L}_{QCD}(\mu)} \tag{1}$$

$$V \; - \; \text{volume}$$
$$T \; - \; \text{temperature}$$
$$\mu \; - \; \text{chemical potential}$$

with the well known QCD lagrangian

$$\mathcal{L}_{QCD}(\mu) = \mathcal{L}_G + \mathcal{L}_F \tag{2}$$
$$\mathcal{L}_G = \tfrac{1}{4} F^A_{\mu\nu} F^A_{\mu\nu} \tag{3}$$
$$\mathcal{L}_F = \bar{\psi}_{j,a}(-\gamma_\mu D_\mu + m_j - i\mu_j)^{ab} \psi_{j,b} \tag{4}$$
$$A = 1,...,N^2-1 \quad \text{(adjoint colour)}$$
$$a,b = 1,...,N \quad \text{(fundamental colour)}$$
$$j = 1,...,N_f \quad \text{(flavour)}$$

The gluonic field strength tensor and the covariant derivative are defined as

$$F^A_{\mu\nu} = \partial_\mu A^A_\nu - \partial_\nu A^A_\mu + g f^A_{BC} A^B_\mu A^C_\nu \tag{5}$$
$$D^{ab}_\mu = \left(\partial_\mu - i\frac{g}{2} A^A_\mu \lambda_A\right)^{ab} \tag{6}$$

Non-zero temperature requires periodic boundary conditions, $A_\mu(0, \vec{x}) = A_\mu(1/T, \vec{x})$. for bosonic fields and antiperiodic boundary conditions, $\psi(0, \vec{x}) = -\psi(1/T, \vec{x})$ for fermionic fields.

The real world corresponds to $N = 3$ colours and either $N_f = 2$ or 3, if one believes that the dynamics is controlled by 2 light flavours or 2 light and 1 heavier quarks.

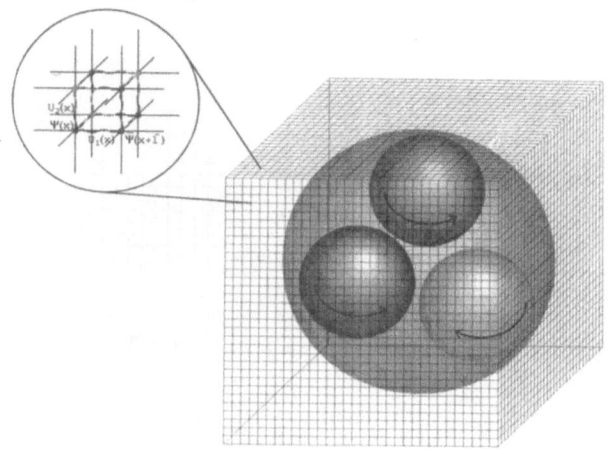

Figure 1. Attempt to picture a proton on the lattice.

The lattice regularization is introduced by replacing the 4 dimensional space time continuum by a finite lattice.

$$\int_V d^3 x \int_0^{1/T} dx_0 \qquad \longrightarrow \qquad a^4 \sum_{n_1, n_2, n_3 = 1}^{N_\sigma} \sum_{n_0 = 1}^{N_\tau}$$

lattice spacing: a temperature: $T = (N_\tau a)^{-1}$
lattice size: $N_\sigma^3 \times N_\tau$ volume: $V = (N_\sigma a)^3$

The plaquette operator is defined as the trace of a product of 4 link variables $U_{n,\mu} \in SU(N)$ around an elementary square on the lattice.

$$P_{n,\mu\nu} = 1 - \frac{1}{N} \, \text{Re Tr} \, U_{n,\mu} U_{n+\hat\mu,\nu} U^\dagger_{n+\hat\mu+\hat\nu,\mu} U^\dagger_{n+\hat\nu,\nu} \tag{7}$$

The Wilson action is used almost exclusively and is given by

$$S_G^{SU(N)} = \frac{2N}{g^2} \sum_n \sum_{0 \le \mu < \nu \le 3} P_{n,\mu\nu} \tag{8}$$

It approximates the continuum action up to corrections which vanish quadratically with the lattice spacing. Integrating out the fermion fields the partition function for zero chemical potential can be written in the form

$$Z(T, V) = \int \prod_{n,\mu} dU_{n,\mu} \prod_{i=1}^{n_f} \det Q^i \; e^{-S_G^{SU(N)}} \tag{9}$$

where $dU_{n\mu}$ is the invariant Haar measure and Q the fermion matrix. The temperature and the volume are adjusted by the parameters N_τ, N_σ and g. The continuum limit, $a \to 0$ at fixed temperature, is controlled by the $SU(N)$ β-function

$$\beta_f(g) = -a\frac{dg}{da} = -b_0 g^3 - b_1 g^5 + \mathcal{O}(g^7) \; . \tag{10}$$

$$g_1, a_1 \qquad \xrightarrow[\;g_1 \to g_2\;]{\;a_1 \to a_2\;} \qquad g_2, a_2$$

The running coupling $g(1/a)$ implicitly defines a mass scale Λ_L

$$\Lambda_L = \frac{1}{a} \, (b_0 g^2)^{b_1/2b_0^2} \, e^{-1/2b_0 g^2} + \mathcal{O}(\ln a) \tag{11}$$

which allows to remove the dependence on the lattice spacing in the continuum limit. The functional dependence $a(g)$ given by equation (11) without corrections this is called asymptotic scaling (AS). The weaker condition of scaling only implies that ratios of physical quantities become constant, as a goes to zero.

UNIVERSALITY AND THE CONTINUUM LIMIT

The concept of an order parameter is an extremely powerful tool to study phase transitions. An order parameter reflects the symmetry of the system. The QCD lagrangian has an exact symmetry only in the limit of infinite or vanishing quark mass. The corresponding order parameters are the Polyakov loop $\langle L \rangle$ and the chiral condensate $\langle \bar\psi\psi \rangle$; they are related to the $Z(N)$ and chiral symmetry of the corresponding action. Universality arguments based on the symmetry of the order parameter and the dimensionality of the system are used to classify a phase transition. Different universality classes can be labeled by critical exponents, which describe the singular behaviour of thermodynamic quantities at the transition point in the infinite volume limit. Powerful finite-size scaling (FSS) methods have been developed to extrapolate from finite lattices to this limit.

Svetitsky and Yaffe[13] made the conjecture that $(3+1)$-dimensional $SU(N)$ gauge theories are in same universality class as 3-dimensional $Z(N)$ spin models with ferromagnetic short ranged interactions. If the transition is second order for $SU(2)$, then the critical exponents of the 3-dimensional Ising model should be found. $SU(3)$ gauge theory is expected to be in the same universality class than the 3-dimensional Potts model which has a first order phase transition. Numerical results[14, 15] calculated for $N_\tau = 4$ show that the $SU(2)$ deconfinement phase transition is second order with the critical exponents of the 3-dimensional Ising model[16]. For $SU(3)$ the first order nature of the transition could be confirmed[17].

Table 2. Comparison of critical exponents.

	SU(2)	3-d Ising	SU(3)	3-d $Z(3)$ Potts
β/ν	0.545(30)	0.518(7)		
γ/ν	1.931(15)	1.970(11)	3.02(14)	3
ν	0.65(4)	0.6289(8)	0.339(13)	1/3

The question is whether this persists for all higher values of N_τ, i.e., in the scaling limit. For $SU(2)$ a FSS method which includes the N_τ dependence[18] has given consistent results up to $N_\tau = 16$. For $SU(3)$ the latent heat has been calculated recently[19], and it has been shown to scale for $N_\tau = 4$ and 6, when it is expressed in terms of the energy density of a free gluon gas, corrected for finite size effects.

IMPROVED COUPLINGS AND ASYMPTOTIC SCALING

It has been noticed that the bare lattice coupling shows large violations of AS. However, dimensionless ratios are found to scale[20, 21]. This implies that a possible universal part of AS violating terms can be absorbed in effective coupling schemes.

Several methods[22, 23] have been developed to improve the situation. The general strategy is to define a new scheme, which allows us to extract continuum results from coarser lattices. For comparison the continuum modified minimal subtraction ($\overline{\text{MS}}$) scheme can be used as a reference. The perturbative relation between the bare lattice coupling g and the $\overline{\text{MS}}$ coupling (the numbers given are for $SU(3)$) shows a large first order correction

$$g^{-2} = g_{\overline{\text{MS}}}^{-2} + 0.3087 + \ldots \quad . \tag{12}$$

The idea of Lepage and Mackenzie[22] was to define an improved expansion parameter g_V^2 that approximates the coupling strength of a gluon with momentum q. A possible definition uses the heavy-quark potential, such that

$$V_{q\bar{q}}(q) = -\frac{C_2(N)\, g_V^2(q)}{q^2} \tag{13}$$

is valid without any higher-order corrections. The next step is to determine the typical gluon momentum q^* so that $g_V(q^*)$ is the appropriate expansion parameter. This can be done by identifying q^* with the average momentum circulating in the Feynman diagrams. This procedure for defining a renormalized coupling with proper scale enables us to convert from the bare lattice coupling g to the improved coupling $g_V(q^*)$. Investigations of the interquark potential have shown that the coupling defined in this way indeed almost behaves perturbatively[21, 24]. However these measurements are highly involved and expensive.

A straightforward way to define a new coupling is to use a measured quantity from a simulation. A very simple choice is the plaquette operator $U_p = \frac{1}{N}\text{Tr}U_1 U_2 U_3^\dagger U_4^\dagger$. If we expand the plaquette operator in terms of g we get

$$U_p = 1 - c_1 g^2 - c_2 g^4 + \mathcal{O}(g^6) \quad . \tag{14}$$

An empirical concept, the plaquette scheme[25], is obtained by truncation of the power series of the plaquette to lowest order. This defines an effective coupling g_e with reduced corrections,

$$g_e^{-2} = \frac{c_1}{1 - \langle U_p \rangle} = g_{\overline{\mathrm{MS}}}^{-2} + 0.2070 + \ldots \quad . \tag{15}$$

It has been successfully applied in a large number of studies. An example[18] shown in Figure 2 is the critical temperature in units of $\Lambda_{\overline{\mathrm{MS}}}$ for $SU(3)$ pure gauge theory. It demonstrates that the effective coupling g_e substantially reduces the dependence on the lattice spacing.

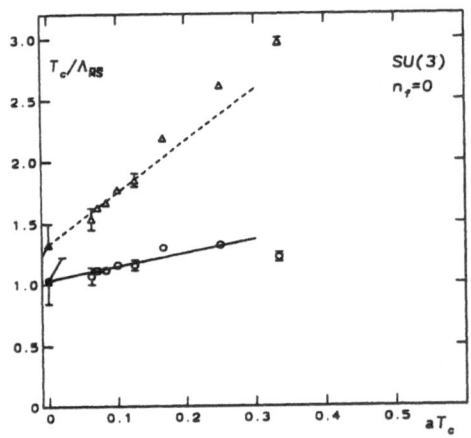

Figure 2. Extrapolation of the critical temperature to the continuum limit in the bare (triangles) and the effective (circles) coupling scheme.

If we expand the logarithm of the plaquette $T_U = -\ln \langle U_p \rangle$ instead and again truncate the series to first order we get an even smaller first order correction.

$$g_V^{-2}(1/a) = \frac{c_1}{T_U} = g_{\overline{\mathrm{MS}}}^{-2} + 0.0403 + \ldots \tag{16}$$

It has been verified extensively that this choice improves the approach to asymptotic scaling considerably[22].

A different approach is based on a direct measurement of a non-perturbative β-function[26]. Most recently the QCD-TARO collaboration[27] has performed a Monte Carlo Renormalization Group (MCRG) study of SU(3) lattice gauge theory in the interval $5.85 < \beta < 7.60$ on 32^4 lattices. The MCRG method determines the coupling shift $\Delta\beta$ induced by blocking transformations with a scale factor s from a matching procedure of a set of Wilson loops. An integral equation relates $\Delta\beta$ to the β-function

$$\ln s = \int_{g(\beta)}^{g(\beta - \Delta\beta)} \frac{dg}{-\beta_{f_u}(g)}$$

$$\beta_{f_u}(g) = -b_0 g_u^3 - b_1 g_u^5 - 0.0119(8) \ b_0 \ g_u^7$$

$$g_u^{-2} = g_{\overline{MS}}^{-2} - 0.442(4)$$

The authors found that violations of AS, shown in Figure 3a, can be removed by a next-order correction to the 2-loop β-function together with a shifted coupling g_u.

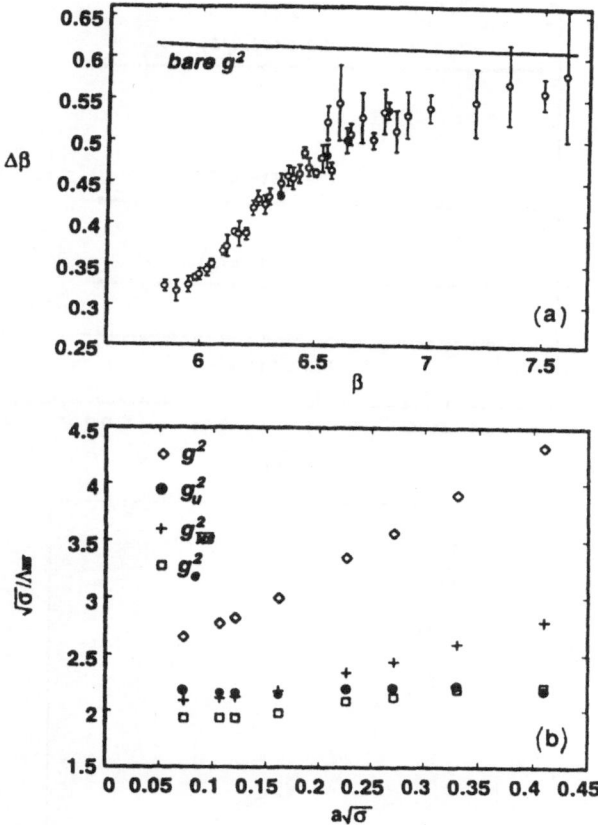

Figure 3. (a): Discrete β-function. Solid curve shows 2-loop asymptotic scaling with respect to the bare coupling constant. (b): Scaling of the string tension in different coupling constant schemes.

A similar approach[28], the step scaling function, determines a renormalized coupling $\bar{g}$ by recursive finite-size techniques. It is defined non-perturbatively as the response of the system to a constant colour electric background field which is parameterized by a constant η.

$$\frac{k}{\bar{g}^2} = - \left. \frac{\partial \ln Z}{\partial \eta} \right|_{\eta=0} \tag{17}$$

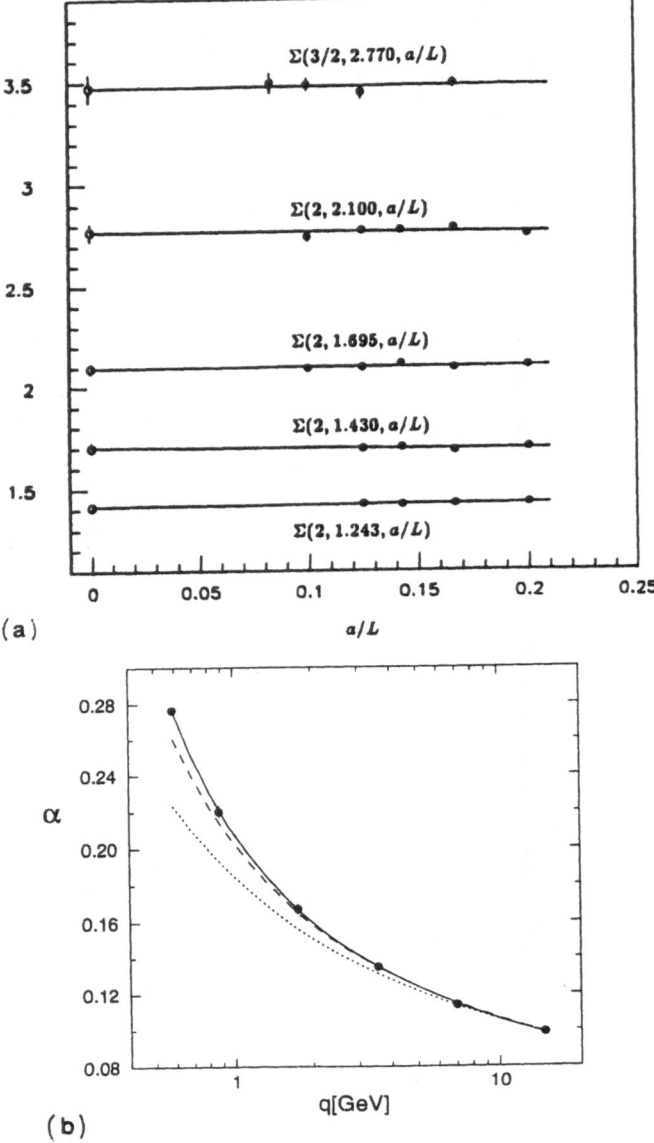

Figure 4. (a): Extrapolation of the lattice step scaling function $\Sigma(a/L)$ to the continuum limit a/L=0. (b): Comparison of the running coupling $\alpha = \bar{g}^2/4\pi$ with perturbation theory. The full line represents an effective 3-loop β-function, while the dashed (dotted) curve used the 2-loop (1-loop) function.

The small coefficient in the perturbative relation relation shows that this scheme is close to the $\overline{\text{MS}}$ scheme.

$$\bar{g}^{-2} = g_{\overline{\text{MS}}}^{-2} - 0.0999 - \dots \tag{18}$$

The lattice step scaling function gives the renormalized coupling after a scale transformation $L \rightarrow s L$

$$\Sigma\left(s, \bar{g}^2(L), a/L\right) = \bar{g}^2(sL) \tag{19}$$

One has to extrapolate Σ linearly to $a/L = 0$ to get the continuum step scaling function $\sigma(s, \bar{g}^2)$. Then the coupling has to be evolved back to low energies

$$\text{low energy} \quad \left\{ \begin{array}{ccccc} g_0^2 & \dots & g_i^2 = \sigma(s, g_{i-1}^2) & \dots & g_{max}^2 \\ L_0 & \dots & L_i = s\, L_{i-1} & \dots & L_{max} \end{array} \right\} \quad \text{high energy}$$

to convert the scale $L_{max} = L(g_{max}^2)$ to physical units using an alternative reference scale r_0. The conversion factor can be obtained from the the heavy quark force $F(r)$ calculated through numerical simulation on large symmetric lattices.

$$L_{max}/r_0 = 0.674(50), \quad r_0 \simeq 0.5\text{fm} \tag{20}$$

In Figure 4a we see that the lattice step scaling function is nearly independent of the lattice size and can be extrapolated to the continuum limit with great accuracy. The resulting running coupling shown in Figure 4b behaves almost perturbatively.

From Figure 3b and Figure 4b we see that all effective coupling schemes considered here work remarkably well. The deviations from asymptotic scaling seen in the bare lattice coupling are drastically reduced. However small deviations between different improved schemes remain and this uncertainty certainly needs to be removed. A special problem is the ambiguity in the extrapolation to the continuum limit: this appears to be under control best in the method of Lüscher and collaborators. However, for this method a generalization to full QCD is not obvious while it is straightforward for the other ones.

THE NATURE OF THE HIGH TEMPERATURE PHASE

Our present knowledge of the high temperature phase of QCD tells us that the plasma most certainly is not a simple gas of nearly free quarks and gluons. Directly above T_c, energy density and pressure differ from their ideal gas value[14]. We know that interactions are strong at arbitrary high temperature and different phenomena are expected at various length scales.

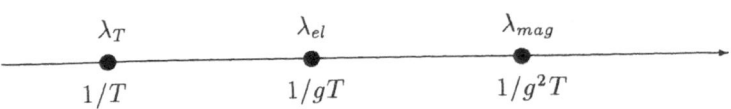

To understand the features of the quark gluon plasma a detailed knowledge on all length scales is necessary. In the electric sector interactions of colour charges are screened. This is the mechanism of Debye screening giving rise to the famous J/ψ-suppression, which has been proposed as an experimental signal for the QGP[29]. Numerical simulations of a pure gluon plasma[30] show good agreement of the electric gluon mass with the leading order perturbative value, $m_{el} = \sqrt{N/3}\, gT$, for $T > 3.5 T_c$. However perturbation theory breaks down at λ_{mag} and little is known at this scale. It has been speculated some time ago by Linde[31], that the generation of a magnetic gluon mass $m_{mag} = cg^2 T$ might cure the infrared divergences of perturbative QCD.

Many detailed studies of hadronic correlation functions performed recently[32] have shown, that there are strong spatial correlations between quarks even when the temperature is increased beyond T_c. A recent computation[33] relates these correlations to the observation[34, 35] that spatial Wilson loops exhibit an area law, $W(R, S) \sim \exp{(-\sigma_s RS)}$, at all temperatures. The spatial string tension, σ_s, is related to the magnetic interaction between two colour currents in the high-temperature limit and is expected to give information about the magnetic gluon mass.

In a pioneering study of $SU(3)$ gauge theory on a $10^3 \times 6$ lattice Manousakis and Polonyi[34] found a value of the spatial string tension that is essentially constant across the transition, $\sigma_s(2T_c) \simeq \sigma_s(T = 0)$. A recent study however revealed a temperature dependence, $\sigma_s = \sigma_s(T)$[36]. A detailed Monte Carlo investigation of $SU(2)$ gauge theory on $32^3 \times N_\tau$ lattices at two different values of the coupling $\beta = 2.5115, 2.74$[37] resolved the temperature scaling of σ_s. The spatial string tension was determined from a fit to temperature dependent pseudo-potentials from spatial Wilson loops

$$V_T(R) = \lim_{S \to \infty} \ln \frac{W(R, S)}{W(R, S + 1)} \stackrel{\text{fit}}{=} V_0 - e/R + \sigma_s R \qquad (21)$$

Figure 5b shows that the results at the two couplings $\beta = 2.5115$ and 2.74 are consistent. This means that the spatial string tension is scaling and thus non-vanishing in the continuum limit. It was found constant below T_c, $\sigma_s = \sigma(T = 0)$ and increasing with temperature above T_c. For $T \geq 2T_c$ a fit of the form

$$\sqrt{\sigma_s(T)} = c\, g^2(T/\Lambda_T)\, T \qquad (22)$$

gave the values $c = 0.369(14)$ and $\Lambda_T = 0.076(13)\, T_c$. The corresponding result for the 3-dimensional $SU(2)$ gauge theory[38] is $\sqrt{\sigma} = 0.3340(25)\, g_3^2$. The good agreement of the 3 and 4-dimensional coefficients indicates that the pure gauge part seems to give the dominant contribution.

We see that the spatial string tension shows indeed a temperature dependence above T_c and in this way does not connect the low and high temperature phase in a smooth way. Its scaling behaviour, given by equation (22), supports the existence of a magnetic scale $\lambda_{mag} = 1/g^2 T$.

Further evidence for dynamically realized electric and magnetic mass scales comes from numerical results for the finite temperature excitation spectrum in the gluon sec-

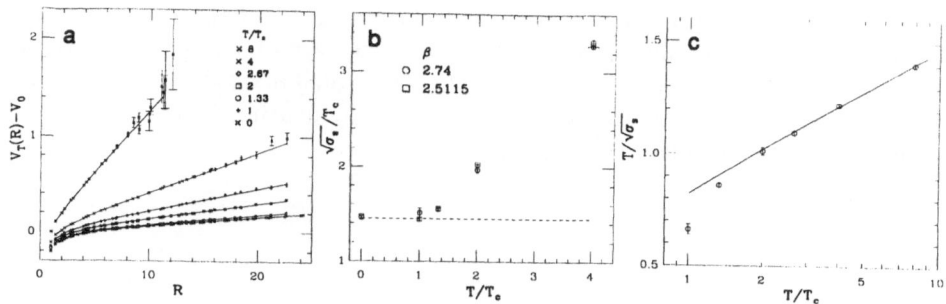

Figure 5. (a): spatial potentials at β=2.74. (b): scaling of the spatial string tension. (c): temperature dependence of σ_s.

tor of $SU(3)$ gauge theory[39]. Below the phase transition the authors found that the zero temperature symmetries are dynamically realized whereas above the transition the spectrum distinguishes between operators coupling to electrically and magnetically polarised gluon fields. In the limit of high temperature the measured screening masses can be identified with twice the electric and magnetic mass of the gluon.

$$m_{A_1^{++}} = 2.8(4)T \approx 2m_{el} \qquad m_{A_1^{++\prime}} = 5.8(4)T \approx 2m_{mag}$$

These observations are consistent with deconfinement. No bound states but rather gluons with dynamically generated mass m_{el} and m_{mag} are found.

DYNAMICAL FERMIONS

The biggest challenge to the simulation of QCD on a lattice is the inclusion of dynamical fermions. There are two popular implementations, staggered and Wilson fermions

$$S_F^S = m\sum_x \bar{\chi}_x\chi_x + \frac{1}{2}\sum_{x,\mu}[\bar{\chi}_x\eta_\mu U_{x,\mu}\chi_{x+\mu} - \bar{\chi}_{x+\mu}\eta_\mu U_{x,\mu}^\dagger\chi_x] \tag{23}$$

$$S_F^W = \sum_x \bar{\psi}_x\psi_x - \kappa\sum_{x,\mu}[\bar{\psi}_x(1-\gamma_\mu)U_{x,\mu}\psi_{x+\mu} + \bar{\psi}_{x+\mu}(1+\gamma_\mu)U_{x,\mu}^\dagger\psi_x] \tag{24}$$

Staggered fermions have a remnant of chiral symmetry even at finite lattice spacing. For them the chiral limit is easy to find: $m_q a \to 0$. An exact algorithm, Hybrid Monte Carlo[40], however exists only for $N_f = 4$. Wilson fermions on the other hand have an exact algorithm for any even number of flavours. Here one has to search the chiral limit, $\kappa = \kappa_c$, where the pion becomes light. So far lattice simulations with Wilson fermions at non-zero temperature show a strange behaviour for $N_\tau \leq 6$ where the transition occurs when the pion mass is still quite high. Therefore the results quoted in the rest of this lecture all correspond to staggered fermions. Finally of course we have to convince ourselves that both formulations give the same results in the continuum limit.

Monte Carlo simulations have shown that the dynamics of the phase transition strongly depends on the number of flavours. For the case of $N_f = 4$ there is clear evidence for a first order chiral transition. The MT_c-collaboration[4] found long-lived metastable states and an abrupt change in the entropy density. This was confirmed by a detailed finite-size analysis[2].

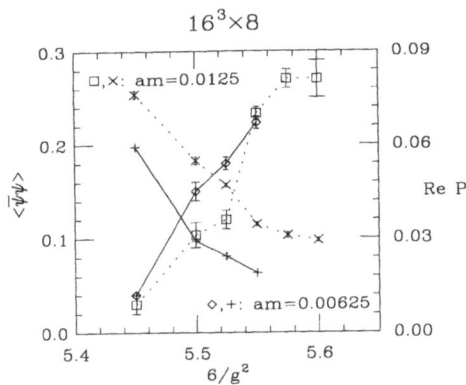

Figure 6. Deconfinement and chiral order parameter for 2 flavours of dynamical fermions.

The case $N_f = 2$ is still under discussion. No evidence for a first order phase transition has been found so far, rather a rapid crossover which is consistent with a nearby second order phase transition at zero quark mass[7, 41].

For $N_f = 3$ there is a first order transition for small enough quark mass[3]. In nature however the s-quark may be too heavy for a first order transition.

In the pure gauge sector, $N_f = 0$, the first order deconfinement phase transition has been analyzed in great detail up to large values of N_τ (see Figure 7) and the results can be used to set the scale. The strong flavour dependence manifests itself in a decrease of the critical temperature with increasing number of flavours[42].

Table 3. Flavour dependence of the critical temperature.

N_f	T_c [MeV]	
	$\sqrt{\sigma} = 440\mathrm{MeV}$	$m_\rho = 770\mathrm{MeV}$
0	251(13)	231(40)
2	-	147(7)
4	166(22)	130(7)

Figure 7 shows the ratio T_c/m_ρ and we see that the behaviour of the data points for $N_f = 0$ is consistent with scaling. To draw a conclusion for dynamical fermions we clearly need results for larger values of N_τ.

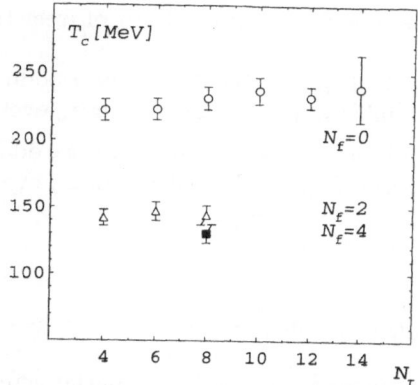

Figure 7. The critical temperature calculated from the rho mass.

HADRONIC EXCITATIONS IN HOT QCD

In early studies non-trivial correlations of hadronic modes[43] have been interpreted as possible indication for bound states in the plasma.

If we want to identify the relevant excitations in the plasma phase we can investigate the spectrum directly through the study of hadronic correlation functions

$$G^H(x_0, \vec{x}) = \langle H^\dagger(x_0, \vec{x}) H(0, \vec{0}) \rangle_T \tag{25}$$

The operator H carries the desired hadronic quantum numbers. Two types of functions have been used, temporal

$$
\begin{aligned}
G_t(x_0, \vec{p}) &= \frac{1}{L^3} \sum_{\vec{x}} e^{i\vec{p}\vec{x}} G^H(x_0, \vec{x}) \\
\vec{p} &= (p_1, p_2, p_3), \quad \text{hadron momentum}
\end{aligned}
\tag{26}
$$

and spatial correlators

$$
\begin{aligned}
G_s(x_3, \tilde{p}) &= \frac{T}{L^2} \sum_{\tilde{x}} e^{i\tilde{p}\tilde{x}} G^H(x_0, \vec{x}) \\
\tilde{x} &= (x_0, x_1, x_2) \\
\tilde{p} &= (p_0, p_1, p_2), \quad \text{momentum in rotated space}
\end{aligned}
\tag{27}
$$

$$
p_0 = \begin{cases} 2\pi T n & \text{for bosons} \\ 2\pi T(n + 1/2) & \text{for fermions} \end{cases}
\tag{28}
$$

The correlation functions become simple in the limit of free staggered quarks

$$G_t(x_0, \vec{0}) = \begin{cases} A_t \cosh[E_t(x_0 - N_\tau/2)] & \text{for } x_0 \text{ odd} \\ B_t \sinh[E_t(x_0 - N_\tau/2)] & \text{for } x_0 \text{ even} \end{cases} \tag{29}$$

$$G_s(x_3, \tilde{p}_{min}) = \begin{cases} A_s \sinh[E_s(x_3 - N_\sigma/2)] & \text{for } x_3 \text{ odd} \\ B_s \cosh[E_s(x_3 - N_\sigma/2)] & \text{for } x_3 \text{ even} \end{cases} \tag{30}$$

$$E_{t,s} = \ln\left(\omega_{t,s} + \sqrt{\omega_{t,s}^2 + 1}\right)$$

$$\omega_t^2 = m_{eff}^2$$

$$\omega_s^2 = \sin^2(p_{0,min}) + m_{eff}^2 = \sin^2(\pi/N_\tau) + m_{eff}^2$$

We see that already free fermions lead to a large spatial screening mass. Since the lowest Matsubara frequency is known in principle information on the particle mass m_{eff} can be extracted from spatial as well as temporal correlators. At large distance we expect an exponential fall-off

$$G_{s,t}(x_{3,0}) \sim e^{-\omega_{s,t}\, x_{3,0}} \quad . \tag{31}$$

If the temporal direction is short it is difficult to measure the particle mass which is determined by the long distance behaviour from temporal correlation functions. The spatial correlators on the other hand can be completely dominated by a large fermionic Matsubara frequency $p_{0,min} = \pi T$. For a bosonic bound state the spatial correlator will be insensitive to changes in temporal boundary conditions. For fermionic modes however a large drop in screening masses is expected when changing from antiperiodic to periodic boundary conditions.

The free fermion correlator has been compared with results from a Monte Carlo study of the quark propagator[44] with $N_f = 4$ staggered fermions of mass $ma = 0.01$ ($m/T \approx 0.08$) on a lattice of size $16^3 \times 8$. The critical coupling for this lattice size is $\beta_c(N_\tau = 8) = 5.15(5)$. Across the phase transition Figure 8 shows that ω_s is indeed

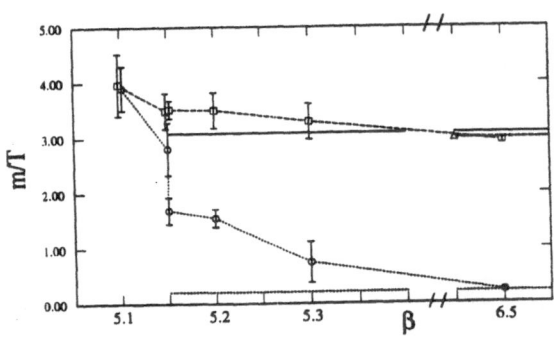

Figure 8. The temporal (circles) and spatial (squares) effective quark masses in units of the temperature versus the coupling.

close to πT and m_{eff} drops rapidly above T_c showing no strong deviation from free quark behaviour for $T > T_c$.

Mesons are composite objects that carry an internal momentum sum $\vec{P} = \vec{p}_q + \vec{p}_{\bar{q}}$.

The meson correlator is obtained according to equation (25) from an operator H which carries mesonic quantum numbers. In the free fermion limit and for $\vec{P} = 0$ we get the following expression

$$
\begin{aligned}
G_t^H(x_0, \vec{0}) &= \sum_{\vec{p}} \begin{cases} C_H \dfrac{\cosh^2\left[E(x_0 - N_\tau/2)\right]}{\cosh^2(EN_\tau/2)\cosh^2(E)} & \text{for } x_0 \text{ odd} \\[3mm] f_H(m) \dfrac{\sinh^2\left[E(x_0 - N_\tau/2)\right]}{\cosh^2(EN_\tau/2)\sinh^2(2E)} & \text{for } x_0 \text{ even} \end{cases} \\[3mm]
E &= \ln\left(\omega + \sqrt{\omega^2 + 1}\right) \\[2mm]
\omega^2 &= \sum_{k=1}^{3} \sin^2(p_k) + m^2
\end{aligned}
\tag{32}
$$

A similar formula holds for the spatial correlator. For large distances we expect again an exponential fall-off but now with a different screening mass for mesons, $\mu_M(T) \simeq 2\pi T$, and for a baryons, $\mu_B(T) \simeq 3\pi T$. Numerical results for four flavours

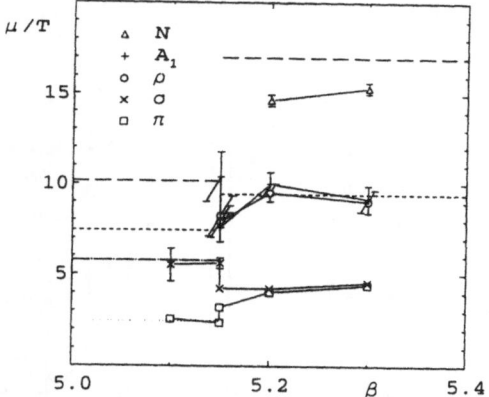

Figure 9. Spatial screening masses for 4 flavours. Lines on the right correspond to free quark propagation in the meson and baryon channel.

of dynamical quarks[45] gave the main result that the screening masses for the ρ and nucleon channnels are close to their free gas values already at $T \geq 1.2T_c$. This is shown in Figure 9, where a screening mass for propagation of two or three weakly interacting quarks is seen in the vector, pseudo-vector and baryon quantum number channels. Only the scalar channel shows large deviations. Here internal quark momenta obscure the contribution of the low momentum excitations. Wall source operators, which project on zero internal quark momentum, can be used to improve the situation. From Figure 10 we see that the lowest mesonic mode is compatible with twice the effective quark mass. The conclusion is that the pion is not a bound state in the QGP, but interactions between the two unbound quarks are still strong. In Figure 11 we see that the correlation functions in in the hadronic phase are unaffected by a change in boundary conditions while the corresponding results in the plasma phase differ substantially. This is in accordance with the expectation that the fundamental excitations in the low temperature phase are bosonic bound states in the low temperature phase and unbound quarks in the QGP.

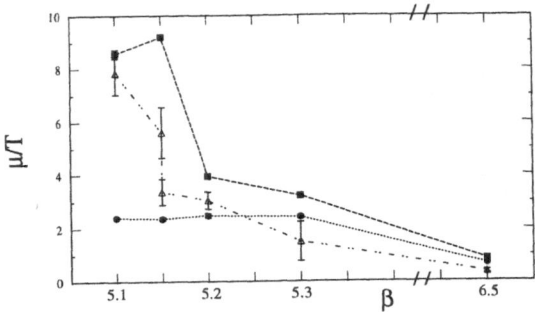

Figure 10. Masses from temporal wall source operators for the ρ-meson (squares) and the pion (circles). Also shown is twice the effective quark mass (triangles).

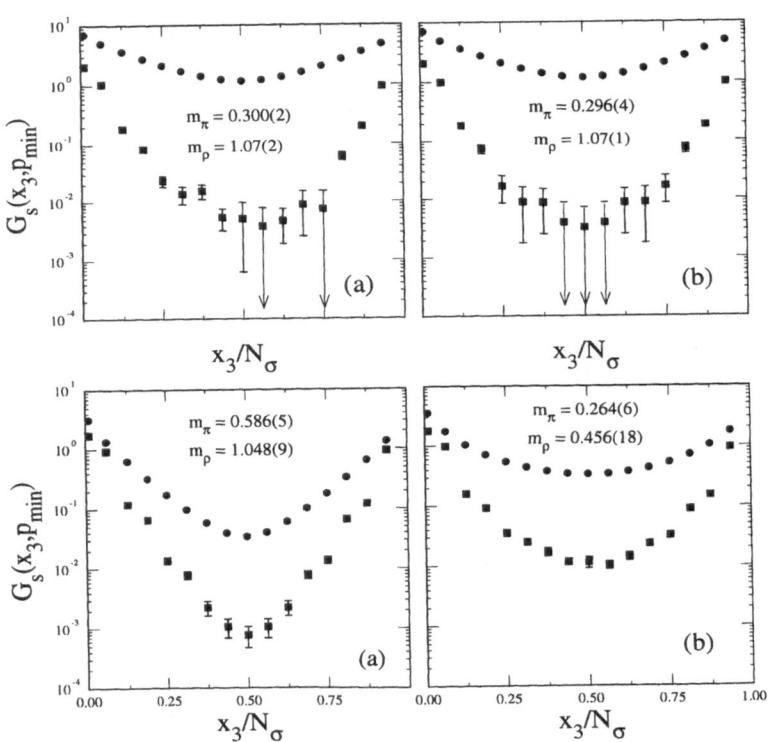

Figure 11. The spatial meson correlator at β=5.1 (a) and 5.3 (b) with periodic (top) and antiperiodic (bottom) boundary conditions in the temporal direction for the ρ-meson (squares) and the pion (circles).

THE CHIRAL TRANSITION

For massless quarks it is generally believed that the phase transition is driven by the restoration of chiral symmetry which is spontaneously broken at low temperature.

The order of the chiral transition has first been discussed by Pisarski and Wilczek[46, 47] on the basis of a renormalization group analysis. They concluded that the transition is of first order for 3 or more flavours. For $N_f = 2$ the transition can be first or second order depending on the fate of the anomaly in the axial baryon number current. Recently Wilczek[47] has argued in favour of a second order transition. In this case universality relates the chiral transition in QCD to 3-dimensional chiral spin models. If the hypothesis is true one should find the critical exponents of the $O(4)$ model[48]. For the specific heat the prediction is

$$C_v(T) \rightarrow A_\pm \, |T - T_c|^{0.21(2)} + \text{less singular terms} \qquad (33)$$

with a dip at the transition point. This is in contrast with the existing lattice data which suggest a peak in this region. The observation of a dip on top of this peak would strongly support the hypothesis. The expected scenario will depend on the relative contribution to the total energy density from chiral condensation and deconfinement. It depends on the microscopic details of the theory and also on the value of T_c.

The agreement of this scenario with previously calculated lattice data has been tested recently[49] in a first FSS analysis of the order parameter which is the chiral condensate $\langle \bar{\chi} \chi \rangle$.

The point of largest slope of $\langle \bar{\chi} \chi \rangle$ as a functions of quark mass defines a pseudo-critical coupling $g_c^2(m_q a, N_\tau)$. For this coupling a scaling law can be derived[49, 50]

$$6/g_c^2(m_q a, N_\tau) - 6/g_c^2(0, N_\tau) = c \, (m_q a N_\tau)^{1/\beta\delta} \quad . \qquad (34)$$

In Figure 12 we see, that all data show a scaling behaviour in terms of $h = m_q a N_\tau$ which is consistent with $O(4)$ exponents. This gives additional support for the as-

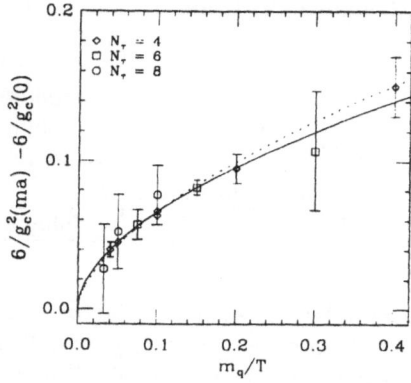

Figure 12. Scaling of the pseudo-critical couplings. The solid (broken) line corresponds to a fit with $O(4)$ ($O(2)$) critical exponents.

sumption of a second order phase transition. However the precision is still to low to exclude other possibilities like for instance $O(2)$ critical exponents.

SUMMARY AND CONCLUSIONS

The purely gluonic calculations are in good shape. Improved coupling schemes and finite-size scaling methods allow accurate quantitative studies.

In the preceding sections we have seen that lattice results are converging towards a consistent picture of the high temperature phase. A non-perturbative quantity in the gauge sector, the spatial string tension has been identified as a candidate for the strong spatial correlations between quarks above T_c. For $SU(2)$ gauge theory the spatial string tension persists the continuum limit. It is constant below T_c and temperature dependent in the form $\sqrt{\sigma_s} = c\, g^2(T)\, T$ for $T > 2T_c$. This gives evidence for the dynamical generation of a magnetic gluon mass $m_{mag} \sim g^2 T$. The origin of the spatial string tension is in the gauge sector of the underlying dimensionally reduced theory. The picture of a hierarchy of non-perturbative length scales gets further support from glueball-like screening masses which show indications for the existence of 2 different length scales $\lambda_{el} \sim 1/(gT)$ and $\lambda_{mag} \sim 1/(g^2T)$ in the plasma.

Simulations including sea quark effects show that the dynamics of the phase transition strongly depends on the number of flavours as is seen from $T_c(N_f)$.

The order of the transition in the physical case $N_f = 2$ is still not clear. So far the numerical results show a rapid but smooth crossover which is consistent with a nearby second order phase transition for $m_q = 0$. However quark masses and lattice spacings are still relatively large and a first order transition at smaller values can not be excluded.

Hadronic correlators above T_c clearly reflect properties of the fermionic substructure. This suggests the absence of light quark bound states.

Present results from Wilson fermions for non-zero temperature are still unsatisfactory. There is a need to go to smaller pion mass and still larger lattices to reach the quality of staggered simulations. In order to make predictions for full QCD as

Table 4. Status of QCD Thermodynamics on the Lattice

Theory	Lattice Size	CPU time
$SU(2)$	$48^3 \times 16$	$\sim 1000\text{h} \times 1\,\text{GFlops}$
$SU(3)$	$24^3 \times 16$	$\sim 5000\text{h} \times 1\,\text{GFlops}$
QCD $m_q/T = 0.01$	$16^3 \times 8$	$\sim 5000\text{h} \times 1\,\text{GFlops}$
What do we want?		
QCD $m_q/T = 0.01$	$48^3 \times 16$	$\sim 50000\text{h} \times 1\,\text{TFlops}$

accurate as the pure gauge simulations we need a factor $\sim 10^4$ increase in computing and algorithmical power. This can be achieved only on a Teraflop computer. Already today the fastest machines achieve a speed of ~ 400 GFlops and in 1995/6 we can be on the level of TFlops. With this optimistic prediction the basic questions, including the critical temperature and the equation of state, can be answered before RHIC or LHC produce heavy ion data.

ACKNOWLEDGEMENT

It is a pleasure to thank F. Karsch and U. M. Heller for discussions and for giving me their results prior to publication. I am indepted to the organizers of this meeting for their generous support. This work was supported in part by the DOE under grants #DE-FG05-85ER250000 and #DE-FG05-92ER40742.

REFERENCES

1. H. Satz, Nucl. Phys. A544 (1992) 371.

2. M. Fukugita, H. Mino, M. Okawa and A. Ukawa, Phys. Rev. Lett. 65 (1990) 816; Phys. Rev. D42 (1990) 2936.

3. F.R. Brown, F.P. Butler, H. Chen, N.H. Christ, Z. Dong, W. Schaffer, L.I. Unger and A. Vaccarino, Phys. Lett. B251 (1990) 181; Phys. Rev. Lett. 65 (1990) 2491.

4. R.V. Gavai, S. Gupta, A. Irrbäck, F. Karsch, S. Meyer, B. Petersson, H. Satz and H.W. Wyld, Phys. Lett. B241 (1990) 567.

5. C. Bernard, M.C. Ogilvie, T.A. DeGrand, C. DeTar, S. Gottlieb, A. Krasnitz, R.L. Sugar and D. Toussaint (the MILC Collaboration), Phys. Rev. D45 (1992) 3854.

6. S. Gottlieb, W. Liu, R.L. Renken. R.L. Sugar and D. Toussaint, Phys. Rev. D41 (1990) 622.

7. S. Gottlieb, A. Krasnitz, U.M. Heller, A.D. Kennedy, J.B. Kogut, R.L. Renken, D.K. Sinclair, R.L. Sugar, D. Toussaint and K.C. Wang (the HTMCGC Collaboration), Phys. Rev. D47 (1993) 3619.

8. R.D. Mawhinney, Nucl. Phys. B (Proc. Suppl.) 30 (1993) 331.

9. D. Zhu, talk given at the conference Lattice '93 in Dallas.

10. B. Petersson, Nucl. Phys. B (Proc. Suppl.) 30 (1993) 66; D. Toussaint, Nucl. Phys. B (Proc. Suppl.) 26 (1992) 3; S. Gottlieb, Nucl. Phys. B (Proc. Suppl.) 20 (1991) 247; N. Christ , Nucl. Phys. A544 (1992) 81c; B. Petersson, Nucl. Phys. A525 (1991) 237c.

11. F. Karsch, *Deconfinement and Chiral Symmetry Restoration on the Lattice*, workshop on "QCD - 20 Years Later", Aachen, Germany, June 9-13, 1992, Vol. 2, p.717, edited by P.M. Zerwas and H.A. Kastrup (World Scientific 1993); S. Gottlieb, *Finite Temperature Lattice QCD Simulations With Staggered And Wilson Fermions*, published in Upton Hot Summer Daze, 1991.

12. C. Bernard, Phys. Rev. D9 (1974) 3312.

13. B. Svetitsky and L.G. Yaffe, Phys. Rev. D26 (1982) 963; Nucl. Phys. B210 (1982) 423.

14. J. Engels, J. Fingberg and M. Weber, Nucl. Phys. B332 (1990) 737.

15. J. Engels, J. Fingberg and V.K. Mitryushkin, Nucl. Phys. B392 (1993) 493.

16. A.M. Ferrenberg and D.P. Landau, Phys. Rev. B44 (1991) 5081.

17. J. Kuti, S.A. Gottlieb, D. Toussaint, A.D. Kennedy, R.L. Sugar, S. Meyer, B.J. Pendleton, J. Stat. Phys. 43 (1986) 1105; F.R. Brown et al., Phys. Lett. 61 (1988) 2058; P. Bacilieri et al., Phys. Rev. Lett. 61 (1988) 1545; Nucl. Phys. B318 (1989) 553; Phys. Lett. B224 (1989) 333; M. Fukugita, M. Okawa and A. Ukawa, Nucl. Phys. B337 (1990) 181.

18. J. Fingberg, U.M. Heller and F. Karsch, Nucl. Phys. B392 (1993) 493.

19. Y. Iwasaki et al. (the QCDPAX Collaboration), Phys. Rev. D46 (1992) 4657.

20. S.P. Booth, et al. (the UKQCD Collaboration), Phys. Lett. B275 (1992) 424; Nucl. Phys. B394 (1993) 509.

21. G.S. Bali and K. Schilling, Phys. Rev. D46 (1992) 2636; Phys. Rev. D47 (1993) 661.

22. G.P. Lepage and P.B. Mackenzie, Phys. Rev. D48 (1993) 2250.

23. A.X. El-Khadra, G. Hockney, A.S. Kronfeld and P.B. Mackenzie, Phys. Rev. Lett. 69 (1992) 729.

24. G.S. Bali, preprint WUB 93-37, Oct. (1993).

25. G. Parisi, Proceedings of the xxth International Conference on High Energy Physics 1980, Madison, Eds. L. Durand, and L.G. Pondrom, American Institute of Physics, New York (1981) 1531.

26. D. Petcher, Nucl. Phys. B275 (1986) 241; J. Hoek, Nucl. Phys. B339 (1990) 732.

27. K. Akemi et al. (the QCD-TARO Collaboration), preprint HUPD-9317, July (1993).

28. M. Lüscher, R. Sommer, P.Weisz and U. Wolff, Nucl. Phys. B389 (1992) 247; preprint DESY-93-114, Sep. (1993).

29. T. Matsui and H. Satz, Phys. Lett. 178B (1986) 416; S. Gavin, H. Satz, R.L. Thews, R. Vogt, preprint CERN-TH-6644-93, Feb. (1993).

30. M. Gao, Phys. Rev. D41 (1990) 626.

31. A.D. Linde, Phys. Lett. 96B (1980) 289.

32. C. Bernard, M.C. Ogilvie, T.A. DeGrand, C. deTar, S. Gottlieb, A. Krasnitz, R.L. Sugar and D. Toussaint, Phys. Rev. Lett. 68 (1992) 2125.

33. V. Koch, E. Shuryak and G. Brown, Phys. Rev. D46, (1992) 3169.

34. E. Manousakis and J. Polonyi, Phys. Rev. Lett. 58 (1987) 847.

35. C. Borgs, Nucl. Phys. B261 (1985) 455.

36. L. Kärkkäinen, P. Lacock, B. Petersson and T. Reisz, Phys. Lett. B312 (1993) 173.

37. G.S. Bali, J. Fingberg, U.M. Heller, F. Karsch and K. Schilling, accepted for publication in Phys. Rev. Lett.

38. M. Teper, Phys. Lett. B311 (1993) 223.

39. B. Großmann, F. Karsch, S. Gupta and U.M. Heller, preprint HLRZ 61/93, Sep. (1993).

40. S. Duane, A.D. Kennedy, B.J. Pendleton and D. Roweth, Phys. Lett. 195B (1987) 216; S. Gottlieb, W. Liu, D. Toussaint, R.L. Renken and R.L. Sugar, Phys. Rev. D35 (1987) 2531.

41. S. Gottlieb, Phys. Rev. D47 (1993) 3619.

42. F. Karsch and E. Laermann, *Numerical Simulations in Particle Physics*, BI-TP-93-10, March (1993), submitted to Rept. Prog. Phys.

43. C. DeTar and J.B. Kogut, Phys. Rev. D36 (1987) 2828;
S. Gottlieb, W. Liu, D. Toussaint, R.L. Renken and R.L. Sugar, Phys. Rev. Lett. 59 (1987) 1881;
A. Gocksch, P. Rossi and U.M. Heller, Phys. Lett. B205 (1988) 334.

44. G. Boyd, S. Gupta and F. Karsch, Nucl. Phys. B385 (1992) 481.

45. K.D. Born, S. Gupta, A. Irrbäck, F. Karsch, E. Laermann, B. Petersson and H. Satz, Phys. Rev. Lett. 67 (1991) 302.

46. R. Pisarski and F. Wilczek, Phys. Rev. D29 (1984) 338,

47. F. Wilczek, J. Mod. Phys. A7, No. 16 (1992) 3911.

48. F. Wilczek, preprint IASSNS-HEP-93/48, July (1993); K. Rajagopal and F. Wilczek, Nucl. Phys. B399 (1993) 395.

49. F. Karsch, private communication.

50. G. Boyd, J. Fingberg and F. Karsch, Nucl. Phys. B376 (1992) 199.

INVESTIGATIONS WITH THE 4π-DETECTOR SYSTEM FOPI AT SIS : FLOW IN HIGHLY CENTRAL AU ON AU COLLISIONS

K. D. Hildenbrand for the FOPI Collaboration[*]

Gesellschaft für Schwerionenforschung
D-64291 Darmstadt
Germany

INTRODUCTION

A series of lectures on this conference is devoted to 4π-studies; by this term we denote measurements with detectors which are capable of detecting, at least ideally, all emerging products of a heavy ion collision over the whole phase space in each single event. One can fairly state, that our present-day knowledge on collisions in the energy range below 2 A•GeV is based to a large extent on such investigations. This has been demonstrated in a convincing way by an earlier contribution,[1] in which the most important results[2] of the pioneering devices,[3] placed at the BEVALAC, have been reviewed, which have been obtained about ten years ago. Since then not much has been changed in the field, except the results obtained with the DIOGENE detector[4] at Saclay.

Only during the past five years this situation has improved. With the EOS system at LBL a whole series of measurements was carried out[5] before the BEVALAC was shut down. At the new heavy ion accelerator SIS/ESR of GSI[6] a wide-spread investigation program has been initiated. Whereas various detectors have been set up for the measurement of special probes in a limited solid angle, the spectrometer ALADIN[7] and especially the modular 4π-detection system FOPI[8] can be considered as second-generation large-solid-angle devices for studies in the energy range of SIS up to 2 A•GeV. At lower energies of 100 A•MeV and below, the MSU 4π-Ball[9] and the devices at GANIL[10] have been successful in studying light and medium-heavy systems. It is for the first time that representatives of all these devices report about results of their experiments on one conference.

The older investigations had established our present idea about semi-central and central heavy ion collisions: In the geometrical overlap zone of the colliding nuclei a hot and dense transient nuclear state is formed, which exhibits in its subsequent decay several collective features as well as thermal properties. To infer the behaviour of nuclear matter under such extreme conditions from the measured observables needs a close interplay between the experiment and theoretical models, since none of the relevant parameters such as density or temperature is derivable in a model-free manner. A decent understanding of all aspects and a final derivation of the nuclear equation of state can be achieved only by a measurement

Hot and Dense Nuclear Matter, Edited by
W. Greiner *et al.*, Plenum Press, New York, 1994

of all relevant probes and of a large number of experimental signatures in multi-dimensional cross-sections, for systems of different colliding masses and over a wide range of incident energies. Only by confronting such a huge body of results to the state-of-the-art theories we may hope to overcome ambiguities in the parameters employed in the latter. 4π-studies should be able to provide these experimental results with a minimum of experimental bias.

In most of the first-generation experiments the identified probes were restricted to light charged particles; measurements of mesons and strange particles (π, K, Λ) were very scarce, as well as those of heavier fragments with Z > 2, called clusters or intermediate-mass fragments. A common study of all of them in one experiment was not feasible since none of the older devices was suited for those measurements. The FOPI system is more universal in this respect: Due to its modular configuration it is able to identify clusters, the emission of which is the dominating aspect at incident energies up to a few hundred A•MeV, as well as π and K at higher energies. The experiments carried out so far have concentrated on the system Au + Au. In this contribution as well as in two further talks[11−12] results of these experiments are presented which concentrate on various aspects of cluster emission, such as entropy and directed sidewards flow, at incident energies between 100 and 800 A•MeV. A last contribution[13] deals with isotope ratios of light elements in these reactions.

The present contribution describes a novel phenomenon, the flow which is observed in very central collisions in these reactions. Before, as an introduction also for the other contributions, a brief introdution on the experimental method is given, followed by a description of the methods used for a selection of the most central events.

The 4π-DETECTION SYSTEM FOPI AT SIS

The experiments described in the following were carried out with the so-called Phase I of FOPI. It consists, see fig. 1, out of a highly-granular (764 single elements) scintillator wall, which determines the time of a flight (with an in-beam start detector as reference) and gives a ΔE-signal of the penetrating products. The wall is covered by a shell of 188 thin energy-loss detectors delivering a ΔE information for heavy/slow fragments which are stopped in the scintillators. ΔE signals and the TOF (or velocity-) measurement allow an element identification of particles up to Z $\approx$ 12 with thresholds increasing from 15 to 40 A•MeV for Z = 1 and 12, respectively.[8]

The wall covers in full azimuthal acceptance the polar angles from 1.2° to 30°. Larger angles ($\theta \leq 90°$) were covered by a box of multiwire chambers delivering the charged-particle multiplicity in that angular range. A set of telescopes[14], movable in front of the wall or behind a multiwire chamber to any angle was used to identify A and Z of the products by means of the ΔE-E method. With this given set up, the acceptance of the apparatus covers, except the angular regions outside 1.2° and 30°, the forward hemisphere in the Au + Au collision; particle thresholdes stay, for 150 A•MeV beam energy up to Z = 4 below midrapidity (hence not influencing the forward hemisphere). At all measured energies samples of 0.5 - 1 • 10^6 central triggers were accumulated, selected by an adapted hardware threshold on the charged particle multiplicity in the angular range 7°-30°.

The present (1993) lay-out of FOPI[8] comprises in addition to the forward wall a superconducting solenoid, with a central drift chamber (θ = 34°-135°) around the target (cf. fig. 2). This device has been used already for a study of Au + Au at 1.06 A•GeV, where p, d, t and of π^+, π^- have been measured (identified via Bρ - ΔE measurement) together with the forward-wall product identification in Z and v. The data of this experiment are being analyzed at present. The final completion of FOPI will be achieved in 1994 when a second drift chamber between 8° and 30° as well as a barrel-like detector out of scintillator and Cherenkov-strips will be operational.[8]

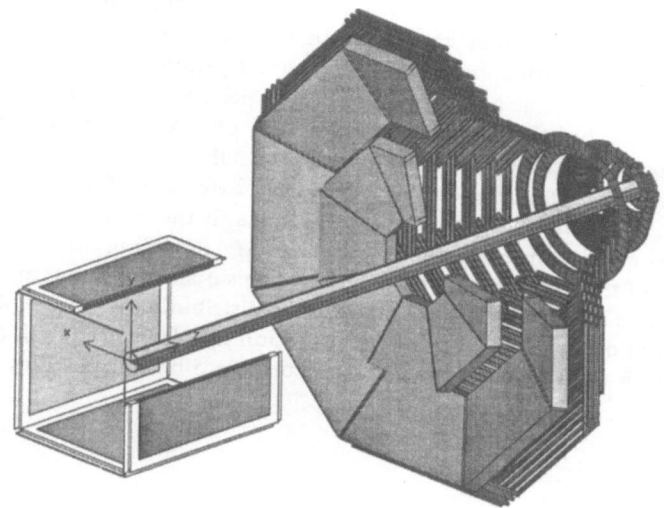

Fig. 1: Phase I of FOPI, the 4π detection system for charged particles at the SIS accelerator of GSI. It comprises a highly granular time-of-flight wall of 764 scintillators distributed over the full azimuth of the polar angles from 1.2° to 30°, which is covered by a shell of cluster detectors (gas filled ionisation chambers and thin scintillator sheets); both are partly cut off in the figure. The target is located at the left side in air, surrounded by a box of multiwire-chambers. The wall diameter is 4 m, the length about 6 m. For details see text and ref. 8. This set-up has been used in the experiments which are described in this article as well as in refs. 11-13.

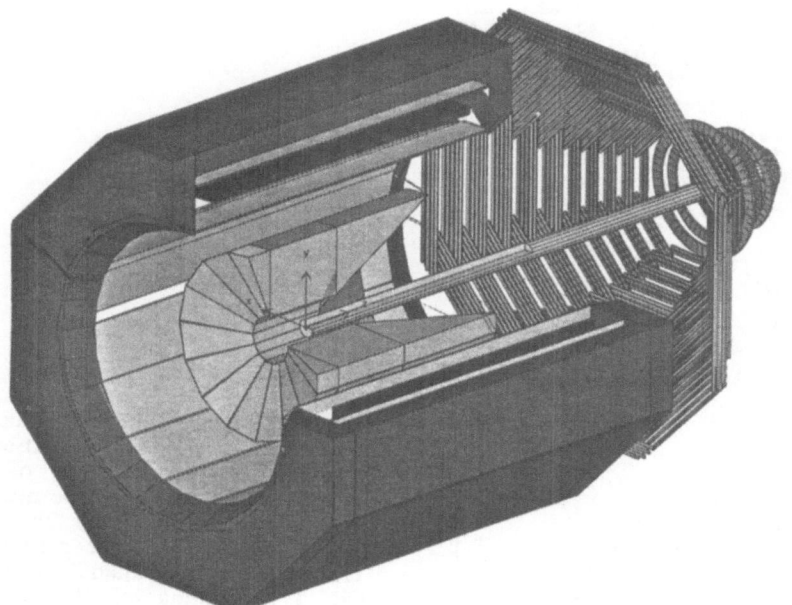

Fig. 2: Existing (fall 93) FOPI set-up. The forward wall has been complemented by a superconducting magnet housing a cylindrical drift chamber which covers the angles backward of 34°. In the final lay-out (FOPI phase II) this chamber will be surrounded by a barrel-like shell of scintillators and Cherenkov-detectors, and another drift chamber will cover the polar angles down to 8° (cf. ref. 8).

SELECTION OF CENTRAL EVENTS

A crucial issue in all experiments is the selection of the impact parameter. In principle any measured quantity which varies with b can be used, but the effectiveness can only be determined in comparison with appropriate model investigations or by a careful inspection of the effect on the data sample analyzed under the respective cut. Care has to be taken to avoid autocorrelations by which one denotes features observed in the sample which are caused in an indirect or direct way by the applied cut itself.

Selection criteria as used in the analyses presented here (as well as in refs. 11-13) are displayed in fig. 3. The most widely used concept (part a) is the impact parameter selection by means of the observed charged-particle multiplicity M_c, assuming that the centrality and hence the violence of the reaction are proportional to this quantity. In our analysis we have adopted the Plastic Ball recipe, where the plateau-like distribution is cut into 4 equally wide M_c bins between 0 and the value, where the distribution has dropped to half of the plateau height. Together with the bin above this value one obtains 5 bins, denoted by PM1 to PM5 in the following.

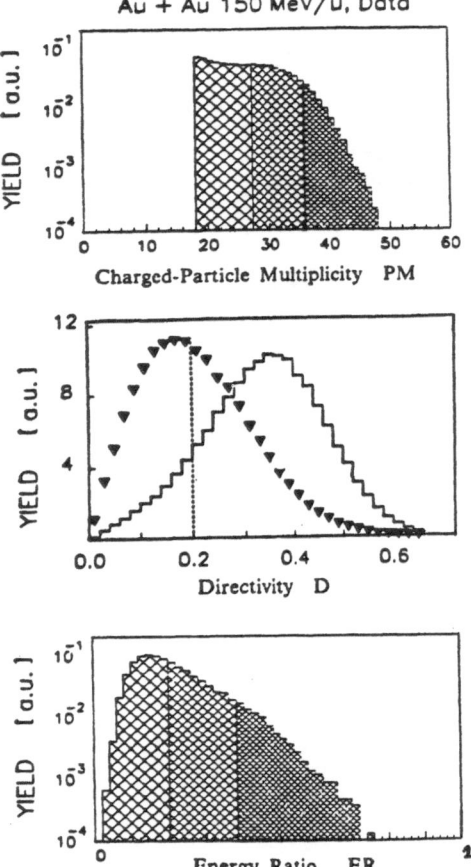

Fig. 3: Experimental observables (Au + Au, 150 A•MeV) which are used in the analysis for the selection of central events: a) Charged-particle multiplicity PM (7° to 30° only), b) Directivity D (solid line, see text for definition), c) Transverse-to-longitudinal energy ratio ER The binning of this quantities, shown by differently hatched areas, is explained in the text.

The effect of these cuts can be depicted from fig. 4, which displays two-dimensional cross sections of $p_\perp/A$ vs. y for $Z=1$ and $Z=3$ fragments under different selection criteria. When going from PM3 to PM5, the distributions change from clear blobs around the beam rapidity (y = 1 in the present notation) to distributions stretching over a wide y-range from mid- to targetrapidity. This phenomenon, especially visible in the $Z=1$ distributions, dem-

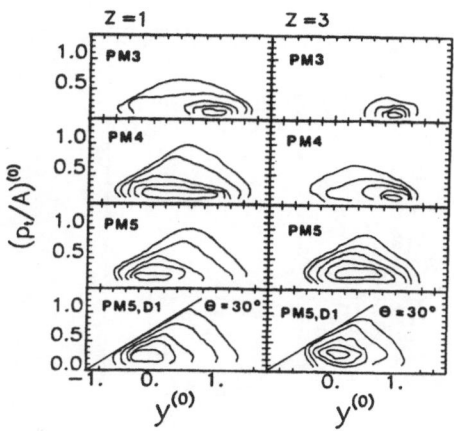

Fig. 4: Experimental $p_\perp$ /A vs. y distributions (Au on Au, 600 A•MeV). Both quantities are normalized to the respective beam quantities in the c.m. system as indicated by the suffix (o). Shown are the distributions for $Z=1$ and $Z=3$ for different cuts on PM and the double cut PM5D1 (cf. fig. 3), demonstrating the presence of a single-source component at midrapidity in the most central collisions. The acceptance cut at 30° is indicated in the lowest two figures only.

onstrates the large fluctuations present: obviously also semi-central collisions lead to high particle multiplicities, or there are, in central collisions, protons and heavier fragments emitted which, after a few first collisions manage to escape hence preserving a seizable fraction of their primary velocity.

A better selection criterion is to require (in addition to a high M_c) a high degree of axial symmetry in the emission pattern of the products. One possibility to quantitatively determine this symmetry is the directivity[15,16]

$$D = \frac{|\Sigma_i \, \vec{p}_{\perp i}|}{\Sigma_i \, |p_{\perp i}|}$$

where $p_{\perp i}$ denotes the transverse momentum component of particle i; the sum runs over the forward c.m. hemisphere only.

Fig. 3b shows the experimental directivity distribution; the triangle distribution shown in addition is obtained after the azimuthal angles of all fragments have been randomized[16], demonstrating that D approaches zero for completely symmetric events (the fact that the randomized distribution is still finite is a finite number effect). An instructive comparison of the momentum distributions of events characterized by different values of D is shown in fig. 5, which displays the $p_\perp$ distributions (p_x vs. p_y plane of two events of the same multiplicity ($M_c = 33$ in our usual 7° - 30° cut) but of rather different directivity. The low-directivity event exhibits an almost isotropic distribution, the other event clearly shows a preferred emission characteristics as a remainder of the reaction plane (only the forward e.m. hemisphere is displayed).

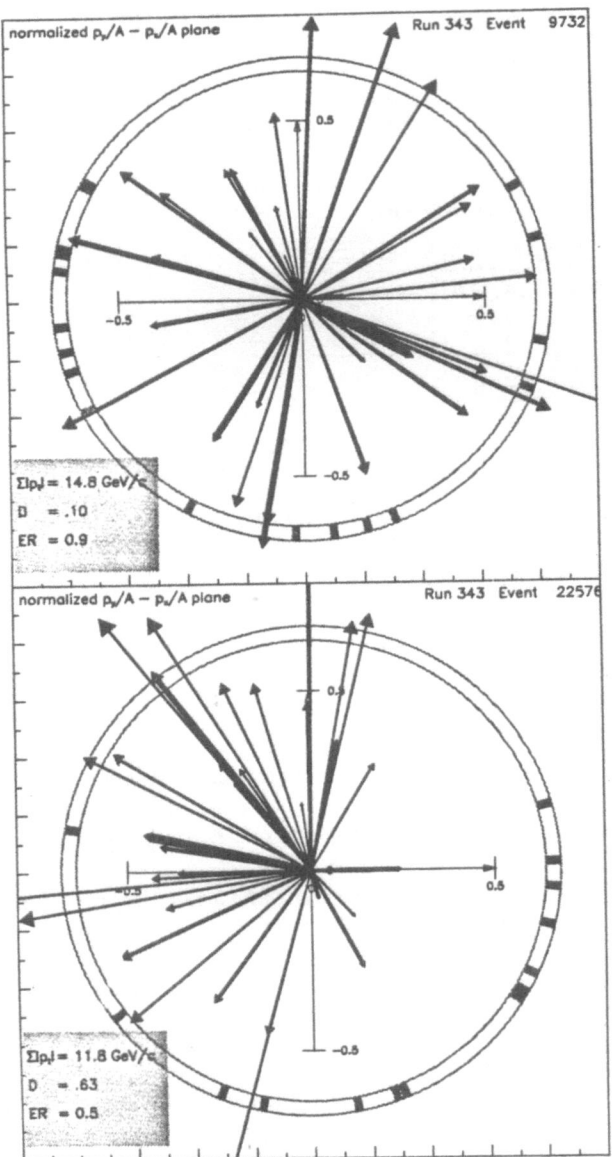

Fig. 5: Display of two events in a normalized p_y vs. y_x plane (only forward hemisphere). Different elements are denoted by the thickness of the momentum arrow, beginning with hydrogen as thinnest line. In both cases 33 charged particles are observed with $Z = 5$ as heaviest fragment. Nonetheless the event patterns are clearly different: The upper event shows an almost perfect azimuthal symmetry, the lower one a preferred sideward emission reminiscent of the reaction plane. This difference is reflected in the directivity which is 0.1 and 0.63 in the two cases, respectively.

The combined effect of a cut on high M_c (PM5) and low directivity D1 (D = 0.2 at 150 A•MeV, as indicated in fig. 3b), called PM5D1 cut in the following, is to be seen in the lower row of fig. 4: A clear component around y = 0 is prepared, which, even more pronounced for Z = 3 and heavier clusters not shown here, demonstrates in a striking way a single midrapidity source for these fragments.

Another way of cutting on centrality is the ratio of transversal to longitudinal energy in an event; the experimental distribution is shown in fig. 3c. The different bins, called ERi are chosen to contain the same cross secton as the bins PMi on the multiplicity. The selective power of an ER5 cut is equivalent to the cut PM5D1, and it prepares a very similar midrapidity component in the $p_\perp$ vs. y plane as shown in the lowest row of fig. 4. Either one of these cuts has been used in the following, if we talk about the most central collisions. Differences found between the two criteria will be mentioned whenever present and important for the analysis. For sake of completeness one should add, that several other se-

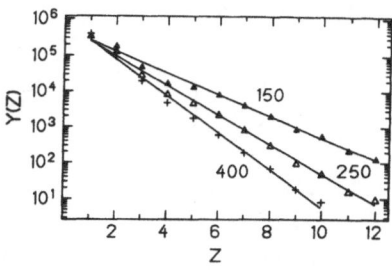

Fig. 6: Element distributions for Au + Au at 150, 250 and 400 A•MeV, selected by the ER5 cut (see text). The two higher energies have been scaled by the factors 0.64 and 0.49 to make the Z = 1 points coincide. The solid lines are exponential fits to the data.

lection criteria have been looked at,[17] a few of which are suited in a similar way for the preparation of the central source.

As summary we can say that they did enable us to clearly isolate a central component exhibiting a midrapidity source of clusters in Au on Au collisions.

CLUSTER DISTRIBUTIONS IN CENTRAL AU ON AU COLLISIONS

Fig. 6 shows element distributions analysed under the ER5 cut for the energies of 150, 250 and 400 A•MeV. They are clearly characterized by exponential fall-offs, ranging down to element numbers as high as 12. Quite similar distributions can be extracted[18-19] by combining cuts on less central collisions such as PM4 or PM5 with (somewhat arbitrary) cuts on central rapidities: This demonstrates that at least the chemical composition of the hot central zone does not or only weekly depend on its size.

When looking on fig. 6 one has to bear in mind that the available c.m. energy in the present cases are way above the average binding energy of cold nuclei (about 37, 60 and 95 A•MeV compared to 8 A•MeV). So the mere existence of such fragments in central collisions is an interesting fact. Indeed, an analysis[19] of the cluster distributions (or the chemical composition) of the central component with the goal to derive the baryonic entropy in these reactions has resulted in apparent temperatures at freeze out as low as 12 Mev at 400 A•MeV. This analysis is described in detail in a separate contribution.[11]

The mean multiplicities of the clusters as function of the measured energies as well as their relative yield are displayed in the left and right part of fig. 7, respectively.[20] Extrapolated into 4π about 13 clusters ($Z \geq 3$) are emitted at 150 A•MeV, the value dropping with increasing energy and finally levelling off at about 5. The total charge bound in these clusters amounts to 40% and 10%, respectively, at the extreme energies.

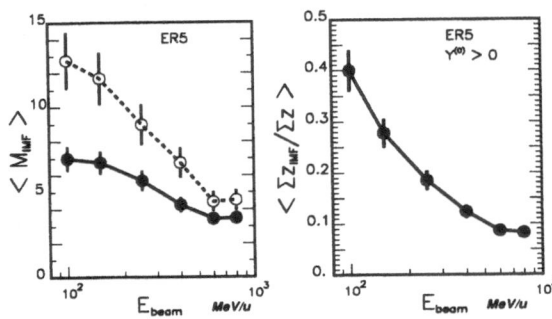

Fig. 7: Mean multiplicity of clusters (left) and their relative share in the observed total charge (right) for incident energies from 150 to 800 A•MeV (central events from ER5 cut). The two curves on the left refer to twice the multiplicity in our experimental acceptance (because only the forward hemisphere is covered in experiment) and to the values extrapolated into 4π.

THE FLOW IN CENTRAL COLLISIONS

The observation of different flow phenomena in collisions at relativistic energies is one of the most interesting findings - if not the most interesting one. It may provide us with information on the interplay between thermal and collective aspects and on nuclear matter properties in general.

In semi-central collisions, where a reaction plane can be determined experimentally, two main signatures have clearly been established: An in-plane flow under a certain polar angle ("side splash"[21]) and an out-of-plane flow ("squeeze-out"[22]) which is maximal perpendicular to this plane.

Only little is known on flow in highly central collisions, where the reaction plane is not well determined. Such collisions are, on the other hand, of particular interest, since stopping and compression are largest. Indeed flow effects have been predicted for these cases both in microscopic and hydrodynamics calculations.[23-24]

As described above we are able to isolate in our data a well-separated central component, which is located at midrapidity; this simple geometry facilitates its investigation.

Fig. 8 shows experimental (Au + Au, 150 A•MeV) rapidity distributions ($p_\perp/A$ vs. y plane, normalized quantities) for $Z = 4-8$, the yields of these fragments, selected via a PM5D1 (upper) and ER5 (lower part) cut, are weighted by their respective charge. The detection boundaries, 1° to 30° in the laboratory, are indicated as well as the lower threshold for Z = 6 (shaded area).

The two distributions are shown to recall that each selection criterion may introduce its special bias. As one can depict from the figures and quantify in a more thorough analysis,[25] the ER cut by its very nature favours somewhat the transverse momenta, whereas MD might have contaminants from spectator-like products at forward angles, resulting in a distribution more elongated along the beam axis. Hence for the analysis shown in the next

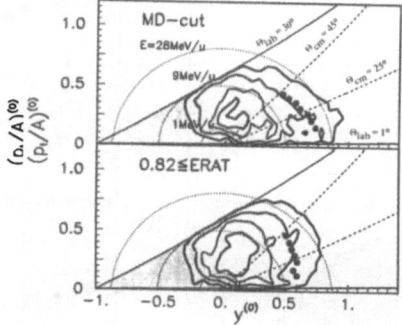

Fig 8: Invariant cross section ($p_\perp/A$ vs. y) of Z = 4-8 fragments produced in central Au + Au collisions at 150 A•MeV, selected via a PM5D1 (upper) and ER5 cut (lower part). Both are divided by the respective beam quantities in the c.m. system. The detector boundaries, 1° and 30°, are delimited by the solid lines, regions affected by detection thresholds are shaded. The dotted circles are contours of constant c.m. energies. The dashed lines, 25° and 45° in the c.m. system, give the region which has been used for the analyses shown in figs. 9,10. The solid points present the mean kinetic energy $<E/A>$ under the respective c.m. angles.

two figures we have used a ER > 0.65 cut, somewhat less stringent than the one in fig. 8b, which leaves a cross section of 200 mb corresponding to b≤2.5 fm in a sharp cut-off approximation. It leads to a pattern intermediate between the two shown in fig. 8 thus averaging to a certain extent the discussed trigger bias. In addition the analysis has been restricted to a c.m. angle bin of 25° to 45°, as indicated by the dashed lines in fig. 8, where one does not suffer from the geometrical acceptance cuts.

Fig. 9 shows the invariant energy spectra of the elements up to Z = 6 in this angular range and fig. 10 the mean kinetic energies/nucleon of these spectra. Both representations show features which are impossible to reconcile with a pure thermal scenario: The heavier elements exhibit rather high velocity tails which deviate drastically from Boltzmann-like distributions, i.e. they are much harder than inferred from the hydrogen spectrum. This goes along with only slightly decreasing, almost constant average velocities of about 20 to 25 A•MeV in fig. 10 for these elements; only for $Z = 1,2$ a thermal behaviour is seen, where $<E/A>$ decreases with $1/A$ as expected from the relation $<E> \propto T$. This fact can be understood by requiring, in addition to thermal motion, a flow velocity common for all particles. It is equivalent to a picture of the midrapidity source exploding under the influence of compression[26] and/or heat; both aspects are nicely discussed in ref. 27.

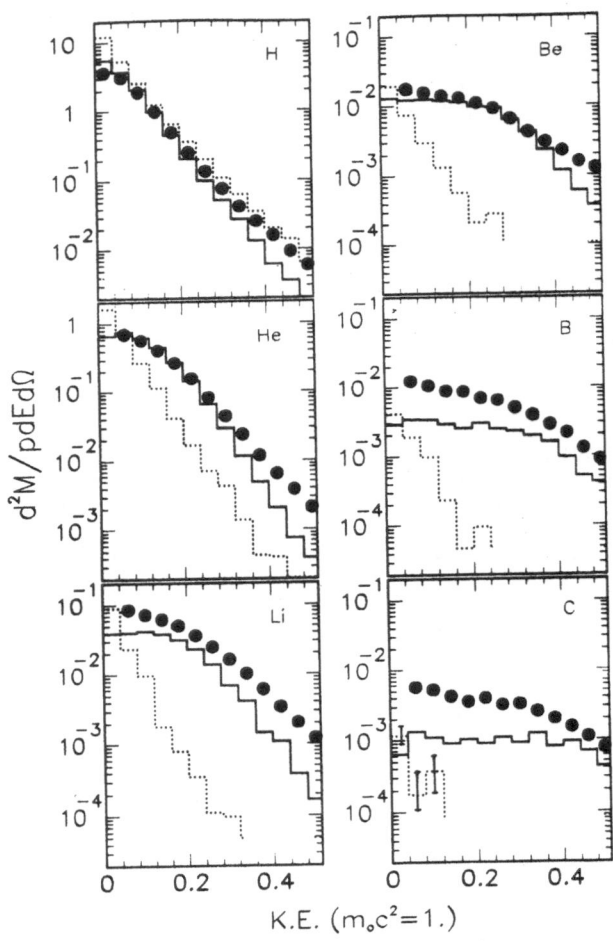

Fig 9: Invariant kinetic energy distributions for the elements $Z = 1$ to 6. The experimental points (solid dots) were obtained by integrating a distribution similar to those of fig. 8 between $25°$ and $45°$ in the c.m. system (the curves are normalized to the number of events). The dotted and dashed lines are results obtained with the code FREESCO[28] for a thermal scenario (including Coulomb affects) and a calculation including an isotropic flow, respectively. For details see text.

In a more quantitative analysis[25] we have used the modified event generator FREESCO[28] to calculate the fragments velocities, both in the pure thermal approach (including Coulomb effects) where the total available energy is converted into heat and by including an isotropic flow pattern, equal in all three directions and common to all elements. The results of the thermal calculations confirm the intuitive conclusions mentioned above (see dotted lines in fig. 9 and triangles in fig. 10): The hydrogen spectrum is described fairly well, but this temperature fails completely in describing the spectra of heavier elements, the discrepancy becoming larger with increasing Z. The average velocities < E/A > of fig. 10 are missed by about 20 A•MeV for the heavier fragments. This amounts to about 55% of the total available c.m. energy of 37 A•MeV. The fraction is

Au + Au at 150 A MeV

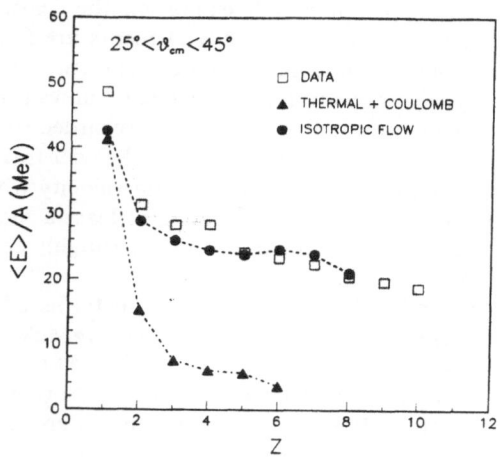

Fig 10: Mean kinetic energy/nucleon < E/A > as function of Z. The experimental points (open squares, the shaded band is to represent the variation observed, when different selection cuts are applied) are obtained from the spectra shown in fig. 9. The full triangles represent the result of the thermal FREESCO calculation, the full dots the calculation with included flow (cf. fig. 9 and the text).

even higher when one considers the lower effective available energy at the moment of freeze-out, where about 9 A•MeV are tied up by nuclear binding energies and by interfragment Coulomb repulsion. Hence only 6 A•MeV remain which correspond, in the model framework, to a freeze-out temperature of about 9 MeV and an entropy of 1.9, which is in very good agreement with the findings of ref. 11.

The share of the collective part in the total available energy has been analysed in a somewhat different mannor[18] also for higher bombarding energies. It was found that the mentioned ≃50% stay rather constant for the beams of 150, 250 and 400 A•MeV.

The results of the calculation including a radial flow with $(E_x, E_y, E_z) = (4,4,4)$ A•MeV are shown as solid lines in fig. 9 and by the full dots in fig. 10. The drastic underestimation of the mean energies by the pure thermal scenario is obviously cured by this calculation which needs only a total flow energy of 12 A•MeV. This is less than the 20 A•MeV read off simply from the difference between the data and the thermal calculation. The calculated value does, however, depend intimately on the details of the model asumptions, for example on the spatial cluster distribution inside the freeze-out volume.[25] Hence we have not tried to vary the flow parameters further in order to improve e.g. the still insufficient description of the spectral shapes in fig. 9. In particular we did not see any reason to use a non-isotropic flow scenario. The importance of the model calculation lies in the demonstration that after a complete failure of the thermal description an explosion scenario, a nuclear blast, is indispensable to explain the gross features seen in the experiment.

SUMMARIZING REMARKS

Highly central events of the reaction Au on Au have been analyzed at incident energies ranging from 150 to 800 A•MeV. Using various selection criteria it has been shown that a well-isolated component can be prepared, exhibiting the features of a single-emitter source located at midrapidity. Interestingly heavy elements are found in this component, the Z-spectra and the multiplicities of which have been shown.

The shape of the energy spectra as well as their mean values (analyzed in a c.m. angle bin of 25° to 46°, where trigger bias and set-up cuts are regarded to be least) clearly exhibit the features of an explosion-like nuclear blast, where all heavier fragments (the lighter ones being less affected) are emitted with an almost common velocity which accounts for about 50% of the total available c.m. energy. This value is at least a factor 10 higher than the energies observed in the directed sidewards flow[12] and is roughly constant for the analyzed energies up to 400 A•MeV.

This magnitude does not allow for an explanation in terms of a simple quasi-eleastic release of the compression energy. In our range of ≤800 A•MeV one may hope to reach densities of 2 times the nuclear ground state density at best. At those densities compression energies of 20 to 40 A•MeV cannot be reconciled with any present idea about the shape of the EOS functional: So there has to occur a reconversion of internal energy into a macroscopic outward flow.[29]

The high energy amount found in collective motion leaves less room for thermal energy; in a model analysis[25] apparent freeze-out temperatures of just 8 MeV were derived at 150 A•MeV. This is in accordance with the results of an analysis of the baryonic entropy in these collisions[11] and explains the survival of the heavy fragments observed in the detectors.

A last but certainly important aspect concerns the isotropy of this nuclear blast. On the theoretical side there are quite a number of predictions (cf. ref. 24 and refs. therein) about the angular emission pattern in central collisions. The most recent calculated angular distributions for the present systems[24] are very sensitive to the selected impact parameter range: A 90°-peaked distribution soon turns into a flat or even forward/backward peaked one when going from 1 to 3 and 5 fm impact parameter. So for a meaningful comparison any experiment needs a very high selectivity on b, but even then the effects might be washed out by b fluctuations.

Because of the acceptance cut of 30° in our present setup the full c.m. angular distribution can be derived only for c.m. energies below a certain threshold (e.g. 13.2 A•MeV at 150 A•MeV beam energy), i.e. by disregarding all faster fragments. We are currently doing a thorough investigation of our tracking and particle identification efficiencies which are non-trivial especially at angles forward of about 30° in the c.m. system (that is why the distributions shown in ref. 11 have to be looked at with caution in that region). This should allow us to see possible non-isotropies in the nuclear blast observed in highly central collisions.

*)Bucharest - Budapest - Clermont-Ferrand - Darmstadt - Dresden - Florence - Heidelberg - Moscow - Strasbourg - Warsaw - Zagreb - Collaboration with (by the time when the data presented here were taken) the following members:

J.P. Alard, Z. Basrak, N. Bastid, I.M. Belayev, M. Bini, T. Blaich, R. Bock, A. Buta, R. Caplar, C. Cerruti, N. Cindro, J.P. Coffin, M. Crouau, P. Dupieux, J. Erö, Z.G. Fan, P. Fintz, Z. Fodor, L. Fraysse, R. Freifelder, S. Frolov, A. Gobbi, Y. Grigorian, G. Guillaume, N. Herrmann, K.D. Hildenbrand, S. Hölbling, O. Houari, S.C. Jeong, M. Jorio, F. Jundt, J. Kecskemeti, P. Koncz, Y. Korchagin, R. Kotte, M.Krämer, C. Kuhn, A. Lebedev, I. Legrand, V. Manko, G. Mgebrishvili, J. Mösner, D. Moisa, G. Montarou, P. Morel, W. Neubert, A. Olmi, G. Pasquali, D. Pelte, M. Petrovici, G. Poggi, F. Rami, W. Reisdorf, Z. Seres, D. Schüll, B. Sikora, V. Simion, S. Smolyankin, U. Sodan, K.M. Teh, R. Tezkratt, M. Trzaska, M.A. Vasiliev, P. Wagner, J.P. Wessels, T. Wienold, Z. Wilhelmi, D. Wohlfahrt, A.V. Zhilin.

REFERENCES

1. H.G. Ritter, contribution to this conference
2. High Energy Nuclear Collisions, GSI-LBL Collaboration at the Bevalac, Papers 1975 - 1987, GSI Report 87-10 (1987)
3. A. Baden et al., NIM 203, 189 (1982) and R. Stock: Heavy Ion Collisions, ed. R.Bock (North Holland, Amsterdam, New York, Oxford, 1979) Vol. I, p. 607
4. D. L'Hote et al., Nucl. Phys. A519, 331c (1990) and references therein
5. H.G. Ritter and D. Keane. contributions to this conference
6. GSI-Reports GSI-03-88 and relevant articles in Annual Reports GSI-01-91, 01-92 and 01-93
7. J. Pochodzalla, contribution to this conference
8. A. Gobbi et al., NIM A324, 156 (1993)
9. G. Westfall, contribution to this conference
10. J. Peter, contribution to this conference
11. J.P. Coffin, contribution to this conference
12. T. Wienold, contribution to this conference
13. G. Poggi, contribution to this conference
14. G. Poggi et al., NIM A324, 177 (1993)
15. P. Beckmann et al., Mod. Phys. Lett. A52, 163 (1987)
16. J.P. Alard et al., Phys. Rev. Lett. 69, 889 (1992)
17. W. Reisdorf, proc. XXth Intern. Workshop on Gross Properties of Nuclei, Hirschegg, Austria (1992) p. 38
18. W. Reisdorf, proc. XXIIIth Masurian Lakes Summer School on Nuclear Physics, Piaski, Poland (1993)
19. C. Kuhn et al., Phys. Rev. C48, 1232 (1993)
20. T. Wienold, Thesis, Heidelberg 1993, GSI-Report GSI-93-28
21. H.A. Gustafson et al., Phys. Rev. Lett. 52, 1590 (1984)
22. D. L'Hote, Nucl. Phys. A488, 457c (1988)
23. P. Danielewicz et al., Phys. Rev. C46, 2002 (1992)
24. W. Schmidt et al., Phys. Rev. C47, 2782 (1993)
25. S.C. Jeong et al., GSI Preprint GSI-93-38 and to be published in Phys. Rev. Lett.
26. W. Scheid et al., Phys. Rev. Lett. 32, 741 (1974)
27. G. Batko et al., Nucl. Phys. A563, 97 (1993)
28. J. Randrup et al., Phys. Lett. B115, 281 (1982) and J. Randrup, LBL Preprint 33865 (1993)
29. J. Cugnon, Phys. Lett. B 135, 374 (1984)

NON–PERTURBATIVE EFFECTS IN THE SU(3)–GLUON PLASMA

D.H. Rischke, M.I. Gorenstein*, A. Schäfer, H. Stöcker,
and W. Greiner

Institut für Theoretische Physik
Johann Wolfgang Goethe Universität
Robert–Mayer–Str. 10
D–60054 Frankfurt am Main
Germany

* perm. address: Institute for Theoretical Physics, Kiev, Ukraine

INTRODUCTION

Quantum chromodynamics (QCD) is supposed to be the fundamental theory of strong interactions. Unfortunately, QCD can be perturbatively solved only in the region of asymptotic freedom, i.e. for high momenta [1]. For small momenta, the expansion parameter of perturbation theory, the strong coupling constant α_S, is of the order of unity, and one is forced to use non–perturbative methods. One of the most successful non–perturbative approaches are lattice gauge calculations [2]. These calculations are especially suitable to study perturbative as well as non–perturbative effects in QCD. The presently available lattice data mostly concern the simulation of $SU(N)$ pure gauge theory [3]. Therefore, we also restrict ourselves to gluons in our present considerations. Moreover, only in the pure gauge case one believes to have lattice artefacts well under control [4]. Thus, only in this case we are able to draw conclusions for the continuum theory from the lattice data.

One of the most interesting features of lattice simulations of $SU(3)$ pure gauge theory is an – apparently first order – phase transition from a phase of confined gluons ("glueballs") to deconfined gluons ("gluon plasma") [3]. This leads to a sharp rise in the energy density as a function of temperature at a phase transition temperature T_c. Interestingly enough, also in the $\sigma - \omega$–model [5] a similar phase transition (from massive to almost massless nucleons) appears at small net baryon densities [6], although this model knows nothing about quark and gluon degrees of freedom.

In a recent work [7], available data on a $(N_\sigma^3 \times N_\tau) = (16^3 \times 4)$–lattice for the energy density and pressure of the $SU(3)$ pure gauge theory were re–analysed using a modified relation for the QCD β–function [8]. In this talk we want to present a phenomenological model equation of state for the gluon plasma which quantitatively reproduces these data. The aim is to understand the lattice data in simple physical terms and to supply dynamical models for the simulation of heavy–ion collisions with a realistic equation of state for deconfined matter. Only then these simulations allow for a reliable study of signals for the deconfinement phase transition in heavy–ion collisions.

The outline of the talk is as follows: in the next section we review lattice data for thermodynamical functions of the SU(3) pure gauge theory and discuss how they can be compared to a continuum model. Then we develop a continuum model equation of state for the gluon plasma and compare it to the lattice data. Finally, we interpret our results, draw conclusions for experimental signatures of the quark–gluon plasma, and give an outlook for further investigations.

LATTICE DATA FOR SU(3) PURE GAUGE THEORY

The energy density and the pressure of $SU(N)$ pure gauge theory can be deduced from the Wilson action [2] on the lattice as functions of $\beta \equiv 2N/g^2$, where g is the coupling constant, provided the QCD β–function $\beta_g \equiv -adg/da$ is known. Assuming that the coupling is weak, β_g can be perturbatively calculated and one obtains

$$a\frac{dg}{da} = b_0 g^3 + b_1 g^5 + \cdots , \tag{1}$$

where $b_0 = 11N/48\pi^2$, $b_1 = 102b_0^2/121$ [1]. However, in numerical studies of the Monte Carlo Renormalization Group one observed violations of this perturbative scaling relation at smaller values of β, near the phase transition [9]. Therefore, in Ref. 8 a QCD β–function was proposed which accounts for these scaling violations, but simultaneously reproduces the perturbative form at large β,

$$a\frac{dg}{da} = b_0 g^3 \frac{(1 - a_1 g^2)^2 + a_2^2 g^4}{(1 - [a_1 + b_1/2b_0] g^2)^2 + a_3 g^4} , \tag{2}$$

where for $SU(3)$ pure gauge theory the values $a_1 = 0.853572$, $a_2 = 0.0000093$, $a_3 = 0.0157993$ were suggested in Ref. 8. In turn, a re–analysis [7] of the lattice data for the energy density and the pressure with this non–perturbative β–function yielded dramatic changes in the Monte Carlo results, especially in the phase transition region.

In order to prepare these improved data for a comparison with a continuum model, we have to perform two steps. First, we have to transform the β–scale into a temperature scale. Integrating the β–function (2) yields a as a function of $\beta = 2N/g^2$, for instance $a \equiv F(\beta)$ [8]. This function is uniquely determined up to a multiplicative constant Λ_L, the lattice scaling parameter. To eliminate this constant, we take $a_c = F(\beta_c)$, the lattice spacing at the well–known phase transition point β_c ($\simeq 5.6925$ for $SU(3)$ on a $(16^3 \times 4)$–lattice), and calculate the ratio a/a_c, where Λ_L cancels out. Since $a/a_c \equiv F(\beta)/F(\beta_c)$ and $T \equiv (aN_\tau)^{-1}$, we obtain

110

$T/T_c = F(\beta_c)/F(\beta)$ [10]. The last equation is a relation between the temperature in units of the deconfinement temperature T_c and β, and thus represents the required scale transformation.

The second step concerns the observation [4] that the calculation of a thermodynamic quantity on a finite lattice yields a different result than in the continuum. For instance, the value of the energy density of an ideal massless Bose–gas (i.e., of a Stefan–Boltzmann gas) on a $(16^3 \times 4)$–lattice is $\epsilon^L_{SB} = 1.4922175\,\epsilon_{SB}$, where $\epsilon_{SB} = \pi^2(aN_\tau)^{-4}/30$ is the corresponding continuum energy density [4]. For the pressure of an ideal massless Bose gas, the relation $p = \epsilon/3$ still holds on the lattice [4], consequently $p^L_{SB}/p_{SB} = \epsilon^L_{SB}/\epsilon_{SB}$. In order to eliminate these finite size effects, it is common practice to normalize lattice data for the energy density ϵ^L and pressure p^L of $SU(3)$–gluons to the corresponding Stefan–Boltzmann values for gluons on a lattice of the same size. We mention, however, that this prescription is accurate only up to the order of 5% [10]. The resulting data for ϵ/ϵ_{SB} and p/p_{SB} [7] are shown in Fig. 1 as functions of T/T_c (full squares and circles).

A MODEL EQUATION OF STATE FOR THE GLUON PLASMA

Let us now construct a phenomenological equation of state for the gluon plasma, which fits the lattice data. One observes two important features of the data above T_c (Fig. 1): (a) For $T/T_c \simeq 1 \div 2.2$ there is a difference between ϵ/ϵ_{SB} and p/p_{SB}. This means that the relationship between energy density and pressure is not that of an ideal gas of massless particles, $\epsilon = 3\,p$. Since $\epsilon_{SB} = 3\,p_{SB}$ one would have $\epsilon/\epsilon_{SB} = p/p_{SB}$ for such a gas. (b) Even above $T/T_c \simeq 2.2$, the energy density and the pressure do not assume the corresponding Stefan–Boltzmann values, we rather have $\epsilon \simeq 0.7\,\epsilon_{SB}$ and $p \simeq 0.6\,p_{SB}$ and the rate of increase with T is small. This behaviour is in striking contrast to the $SU(2)$ case [11].

To explain the first observation (a) we have to assume that, in the vicinity of the phase transition point ($T/T_c \simeq 1 \div 2$), effects from non–perturbative interactions (which render a gas non–ideal) are still large. QCD suggests that gluons with small momenta are subject to confining interactions, which bind them into (colour–less) "glueball" states with a comparatively large mass $M_{gb} \sim$ GeV, while gluons with large momenta are asymptotically free. Following Refs. 11 – 13, we introduce a momentum K which separates these two regions in momentum space. Although an appreciable part of momentum space is occupied by the gluons with small momenta (the glueball constituents, see Fig. 2), their contribution to the thermodynamic functions is far smaller (suppressed due to the large glueball mass M_{gb}) than that of the massless, asymptotically free gluons. Therefore, we may safely omit these states with momentum $k \equiv |\vec{k}| < K$ in the following (a more thorough investigation of the influence of glueballs is given in Ref. 10). Hence K, which we take as a free parameter in our model, is also called "cut–off" momentum in the following. Such a "cut–off model" has been proven to be rather successful in describing lattice data for the thermodynamical functions [11] and the heavy–quark potential in an $SU(2)$–gluon plasma [13].

To implement the second observation (b) in our model equation of state for the gluon plasma, we must reduce the energy density and pressure of an ideal (Stefan–Boltzmann) gluon gas by a factor which is (apart from the T^4–dependence) very slowly varying with temperature. Such a factor is for instance given by including the

first order of the standard perturbative loop expansion of the pressure [14] into our
model equation of state. To see this, we remember that these correction terms to
the pressure are proportional to $\alpha_S \equiv g^2/4\pi$. If we use the expression for α_S as
improved by the Renormalization Group analysis, we obtain a logarithmic dependence
of α_S on the renormalization point [1], and therefore on the temperature of the gluon
gas, if we take this point as the mean thermal momentum in the gluon system [14].
Hence, the first order perturbative corrections to the pressure have the required weak
(i.e. logarithmic) temperature dependence.

However, let us emphasize that a *purely* perturbative treatment of the gluon plasma (with $K \equiv 0$) will not be reasonable. To describe the data in the interval

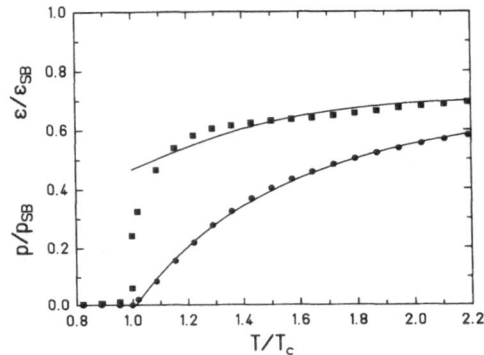

Figure 1. Full lines: fit of p/p_{SB} and ϵ/ϵ_{SB} to the lattice data (full circles and squares),
$\Lambda = 2.1\ T_c$, $K = 4.0\ T_c$, $B^{1/4} = 0.8\ T_c$.

$T/T_c = 1 \div 2.2$, the corrections to the Stefan–Boltzmann values have to be of the
same order as these values themselves. Consequently, the convergence of the series
of perturbative corrections to the Stefan–Boltzmann values is questionable. On the
other hand, for a non–vanishing K we treat *only* the interaction between gluons with
momenta larger than K perturbatively. The perturbative corrections thus apply to
an *already non–perturbative* reference state. As we will see, the corrections are by far
smaller in this case and hence a perturbative treatment more reliable.

Finally, we note that the true QCD vacuum state is non–perturbative, with an
energy density which is about B lower than for the perturbative vacuum. B is
assumed to be constant and can thus play the same role as the bag constant in the
MIT–bag model [15].

Hence, our equation of state reads as follows

$$p(T) = p_0(T) + p_I(T) - B .$$ (3)

p_0 is the pressure of an ideal gluon gas, however, where gluons with momenta smaller than K are excluded,

$$p_0(T) = 2 \, N_g \, T \int \frac{\mathrm{d}^3 \vec{k}}{(2\pi)^3} \, \ln\{1 + N_B(k;T)\} \, \theta(k - K) .$$ (4)

Here N_B is the Bose–Einstein distribution function

$$N_B(k;T) = \frac{1}{\exp(k/T) - 1} .$$ (5)

The term p_I in equation (3) is the first order correction to the ideal gas pressure according to the perturbative loop expansion. In the case of a gluon plasma, only graphs with gluon and ghost lines are to be considered in the calculation of p_I [14]. Since there are neither gluons nor ghosts with momenta less than K, we arrive at the – as compared to Ref. 14 slightly modified – result

$$
\begin{aligned}
p_I(T) \;=\; g^2 N_c N_g \Bigg\{ &-3 \left(\int \frac{\mathrm{d}^3 \vec{k}}{(2\pi)^3} \frac{1}{k} N_B(k;T)\, \theta(k-K) \right)^2 \\
&+ \frac{9}{4} \int \frac{\mathrm{d}^3 \vec{k}_1}{(2\pi)^3} \frac{\mathrm{d}^3 \vec{k}_2}{(2\pi)^3} \frac{1}{k_1 k_2} N_B(k_1;T)\, N_B(k_2;T) \\
&\quad \times \; \theta(k_1 - K)\, \theta(k_2 - K)\, \theta(|\vec{k}_1 + \vec{k}_2| - K) \\
&- \frac{1}{4} \int \frac{\mathrm{d}^3 \vec{k}_1}{(2\pi)^3} \frac{\mathrm{d}^3 \vec{k}_2}{(2\pi)^3} \frac{1}{k_1 k_2} N_B(k_1;T)\, N_B(k_2;T) \\
&\quad \times \; \theta(k_1 - K)\, \theta(k_2 - K)\, \theta(|\vec{k}_1 - \vec{k}_2| - K) \Bigg\} .
\end{aligned}
$$ (6)

The evaluation of (6) is slightly more complicated as compared with the analogous calculation in Ref. 14. Details are presented in Ref. 10.

As mentioned above, we use a Renormalization Group improved expression for the coupling constant α_S ("running coupling constant"). For the sake of simplicity we take the usual expression obtained by solving the Callan–Symanzik equation for β_g at the renormalization point M [1]

$$\alpha_S(M^2) = \frac{g^2}{4\pi} = \frac{4\pi}{11 \, \ln(M^2/\Lambda^2)} .$$ (7)

Since Λ is also taken as a free parameter in our model, we argue that any changes of this relation due to the modified momentum spectrum can be absorbed in this parameter. We now have to relate M to the temperature, i.e., to the independent variable of the equation of state. For the sake of simplicity we decided to choose the relation

$$M^2 = T^2 + K^2 ,$$

(8)

which is motivated by considering that the smallest possible gluon momentum in the cut–off model is of the order of K . To include temperature effects, the most simple way is to make the Ansatz (8). Let us mention that more sophisticated choices [14] lead to quantitatively comparable results [10]. Therefore it is sufficient to use (8).

The energy density can be calculated from the pressure according to the thermo-dynamical relation

$$\epsilon(T) = T \frac{\mathrm{d}p}{\mathrm{d}T} - p(T).$$

(9)

We finally mention that, for a fixed Λ , $\alpha_S(T)$ as given by equations (7), (8) is much smaller than in the common perturbation expansion of p_I . This is due to the fact that M^2 is always larger for finite K than for $K = 0$. Finally, also the coefficients resulting from the evaluation of the terms in (6) are smaller in the case $K > 0$, due to the restricted integration over momentum space. Consequently, a perturbation expansion of p_I in the framework of the cut–off model seems to be more reliable. Higher order perturbative effects, which may be large in the case $K = 0$, are now absorbed in the non–perturbative parameter K .

COMPARISON TO LATTICE DATA

We now adjust the free parameters of our phenomenological model equation of state, K , Λ , and B , to fit the lattice data for the energy density and pressure of $SU(3)$–gluons. The first question is whether all parameters are necessary to produce a reasonable fit to the data or whether some are redundant. To clarify this question we discuss three different cases:

(a) Assuming $K = \Lambda = 0$, we arrive at the standard Bag model equation of state without perturbative corrections. This equation of state cannot reproduce the lattice data, since $\epsilon = \epsilon_{SB} + B > \epsilon_{SB}$, while $\epsilon < \epsilon_{SB}$ for the data. Nevertheless, it is used for most predictive calculations concerning quark–gluon–plasma signatures.

(b) For $B = \Lambda = 0$ we obtain the "cut–off" model in one of its earliest versions [11, 16, 17]. This model reduces the Stefan–Boltzmann energy density and pressure by the energy density and pressure of the low momentum modes. Moreover, thermody-namic consistency requires a term in the pressure which ensures that p is not simply related to ϵ by the usual ideal gas formula. However, a quantitative agreement with the data cannot be obtained, mainly because the correction terms have not the required weak temperature dependence observed in the data at higher T .

(c) For $B = K = 0$ we arrive at the astonishing result that the qualitative behaviour of the lattice data can be quite well reproduced for $\Lambda \simeq 0.5 \, T_c$ [10]. However, the first order perturbative correction to the pressure is too large in this case. Consequently, the loop expansion for p_I is not reliable and we cannot draw physical conclusions from the agreement with the data.

Nevertheless, the discussion of the preceding case has shown that we have to assume $\Lambda \neq 0$ to obtain (at least qualitative) agreement with the data at larger T . Furthermore, K must not be too small, since then the perturbative corrections to the pressure are smaller and the reliability of the loop expansion is improved. In Fig. 1 we show a fit for $\Lambda = 2.1\,T_c$, $K = 4.0\,T_c$, $B^{1/4} = 0.8\,T_c$, which produces reasonable agreement, especially for the pressure. In this case, the perturbative corrections to the pressure are indeed small, especially at lower temperatures, due to the diminished momentum space in the integrals in (6). The reproduction of the energy density is worse in the immediate vicinity of T_c. The reason is that not only the pressure itself, but also its derivative enters the calculation of the energy density, equation (9). Thus, differences between the pressure of the continuum model and the data are enhanced in the energy density. Astonishingly enough, if $T_c \simeq 180$ MeV, the required value for B is of the order of the MIT bag constant as obtained by fitting properties of hadronic particles [15].

Three final remarks are in order:

(i) Up to now we have omitted the contribution of glueballs to our model equation of state. As was shown in Ref. 10, their influence on the thermodynamic functions is – due to the large glueball masses – only of the order of 10% in the respective temperature range. By a suitable adjustment of the parameters of the model, agreement with lattice data can be re-obtained. The respective change in the numerical values of the parameters is also only on the 10% level, which gives some confidence concerning their *general order of magnitude*. However,

(ii) due to the fact that lattice artefacts cannot be completely eliminated from the lattice data (see discussion above), the actually obtained values for K, B, and Λ *must not be taken too literally*.

(iii) Finally, the parameters K and B may not be independent variables. The bag constant is related to the energy density of the vacuum in the absence of colour charges, which is determined by non–perturbative gluonic modes. On the other hand, in our model K defines which modes are perturbative or non–perturbative. Thus, K and B are certainly related, but this relationship is not yet clear to us.

CONCLUSIONS

In conclusion, lattice data for the thermodynamical functions of the $SU(3)$ pure gauge theory suggest that even at temperatures of several T_c the gluon plasma is not at all an ideal gas. In order to understand the underlying physical principles, as well as to enable an implementation of the quark–gluon–plasma equation of state into a dynamical model, we have discussed a phenomenological equation of state for the gluon plasma, which reproduces the lattice data. The basic idea of this equation of state is a separation of the momentum scale into a perturbative and a non–perturbative regime (similar as in the resummation method of Braaten and Pisarski [18], although there the temperature is assumed to be much larger).

From a technical point of view, the introduction of the cut–off momentum K is crucial in the sense that the perturbative loop expansion of the pressure is now performed around a non–perturbative reference state, which renders higher order terms small as compared to those in the conventional expansion around the ideal gas pressure and thus improves the reliability of the expansion.

From the physical point of view, the gluon–plasma phase is an admixture of de-confined gluons with momenta larger than a cut–off momentum K and of massive quasi–particles ("glueballs"), constituted by gluons with momenta smaller than K (due to the large glueball mass their contribution to the pressure is considerably smaller than that of the free gluonic modes). In a thermal excitation spectrum of gluonic momenta there are considerably more gluons with low momenta. Thus, the larger part of all gluons present at a given temperature are bound into glueball states (see Fig. 2). Consequently, the gluon plasma is not an ideal gluon gas but rather a state consisting mainly of "gluon–clusters" and a few deconfined gluons (which nevertheless dominate the thermodynamic properties of the system).

Such a state bears much resemblance to hot and dense hadronic matter. In fact, we could extend our considerations to quark degrees of freedom. Then, the massive states constituted by gluons *and* quarks/antiquarks could be identified with hadrons present even well above the deconfinement transition temperature (similar ideas are quantified

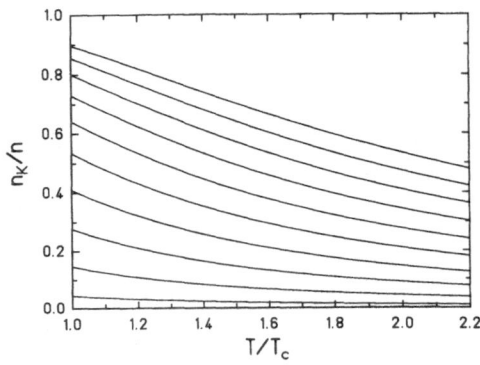

Figure 2. Fraction of gluons with momentum smaller than K as a function of temperature. From below to above: $K = 0.5, 1, ..., 4.5, 5 \, T_c$.

in Refs. 19 and 20 and references therein). Consequently, one might suspect that the physics of the quark–gluon plasma is not very much different from hadronic physics at high temperatures. If this interpretation is correct, an experimental distinction between hot hadronic matter and the quark–gluon plasma will become extremely difficult. In turn, it becomes questionable whether the proposed signals for the quark–gluon plasma (strangeness abundance, dileptons, photons, strangelets,...) will survive.

For instance, it was suggested that, in case a quark–gluon plasma is produced in a heavy–ion collision, the abundance of strange quarks and antiquarks should reach its equilibrium value due to the short equilibration time scale in a hot quark–gluon plasma [21]. In Fig. 3 we show the $s\bar{s}$–creation rate due to gg– and $q\bar{q}$–reactions, if there are no coloured objects with momenta smaller than $K = 4 \, T_c$. As one observes, the

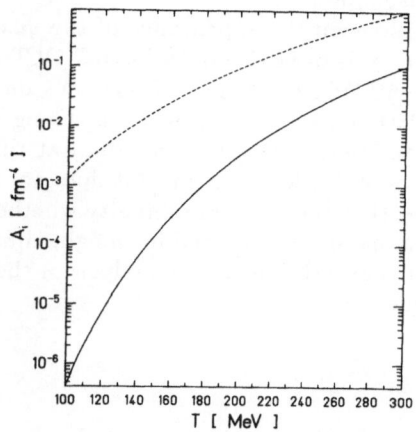

Figure 3. $s\bar{s}$-pair creation rate as function of T, $M_s = 150$ MeV, $\alpha_S = 0.6$. Full line: $K = 4\,T_c$ ($T_c = 180$ MeV), dashed line: $K = 0$.

rate decreases by about two orders of magnitude (at $T = 160$ MeV) as compared to the case $K = 0$. Correspondingly, the equilibration time scale will increase. However, to draw definite conclusions on the role of the strangeness abundance as signature for the quark–gluon plasma is difficult at this stage, since up to now this simple consideration neglects the influence of the hadronic modes on the $s\bar{s}$–creation. Moreover, the lattice data can be also interpreted assuming that the gluons acquire a (thermal) mass instead of having no gluons with momentum below K [22]. In this case, these massive gluons may additionally decay and the equilibration rate even increases.

Nevertheless, supposing the validity of our result that there are no free gluons with low momentum, one would expect a decrease in *any* equilibration rate with gluons in the ingoing channel. Hence, even the *thermal* (not only the chemical [23]) equilibration of gluons in the very early, gluon–dominated stage of an ultrarelativistic (i.e. RHIC or LHC energy) heavy–ion collision [24] is questionable on the available time scale. Further investigations are mandatory to raise our understanding of the dynamics of the initial stage of ultrarelativistic heavy–ion collisions beyond the level reached so far via simple parton cascade models.

Finally, definite conclusions on the appearance of the *quark*-gluon plasma should not be done prior to an analysis of lattice data for *full* QCD, which is – due to poor statistics and difficulties with the treatment of fermions on the lattice [2] – not yet available. However, first lattice results [25] indicate a strong rise in the gluon energy density (even well above the Stefan–Boltzmann value) at the expense of the energy density of the quark degrees of freedom. This result does not yet invalidate our above analysis, it rather indicates that there is an ambiguity whether the interaction energy between quarks and gluons should be included in the gluonic or the quark part of the total energy density [26]. It is exactly the above analysis of the pure gauge theory that may help to decide this question.

Acknowledgements

We thank F. Karsch for kindly providing us with remarks and criticism which enabled us to complete Ref. 10.

REFERENCES

1. H.D. Politzer, Phys. Rep. 14 (1974) 129

2. For an introduction, see e.g.: M. Creutz, "Quarks, gluons and lattices", Cambridge University Press, Cambridge, 1983

3. Y. Deng, Nucl. Phys. B (Proc. Suppl.) 9 (1989) 334
   F.R. Brown, N.H. Christ, Y. Deng, M. Gao and T.J. Woch,
   Phys. Rev. Lett. 61 (1988) 2058

4. J. Engels, F. Karsch, H. Satz, Nucl. Phys. B205 (1982) 239
   H.-Th. Elze, K. Kajantie, J. Kapusta, Nucl. Phys. B304 (1988) 832

5. J.D. Walecka, Ann. Phys. 83 (1974) 491

6. J. Theis, G. Graebner, G. Buchwald, J. Maruhn, W. Greiner, H. Stöcker, J. Polonyi, Phys. Rev. D28 (1983) 2286

7. J. Engels, J. Fingberg, F. Karsch, D. Miller and M. Weber, Phys. Lett. 252B (1990) 625

8. J. Hoek, Nucl. Phys. B339 (1990) 732

9. A. Hasenfratz, P. Hasenfratz, U. Heller, F. Karsch, Phys. Lett. 140B (1984) 76, and Phys. Lett. 143B (1984) 193
R. Gupta, G. Guralnik, A. Patel, T. Warnock, C. Zemach, Phys. Rev. Lett. 53 (1984) 1721, and Phys. Lett. 161B (1985) 352

10. D.H. Rischke. J. Schaffner, M.I. Gorenstein, A. Schäfer, H. Stöcker, W. Greiner, Z. Phys. C56 (1992) 325

11. J. Engels, J. Fingberg, K. Redlich, H.Satz, M. Weber, Z. Phys. C42 (1989) 341

12. T. Hatsuda, invited talk given at the "International Symposium on High Energy Nuclear Collisions and Quark Gluon Plasma", Kyoto, Japan, June 6–8, 1991

13. M.I. Gorenstein, O.A. Mogilevsky and St. Mrówczyński, Phys. Lett. 246B (1990) 200

14. J.I. Kapusta, Nucl. Phys. B148 (1979) 461

15. A. Chodos, R.L. Jaffe, K. Johnson, C.B. Thorn, V.F. Weisskopf, Phys. Rev. D9 (1974) 3471

16. M.I. Gorenstein and O.A. Mogilevsky, Phys. Lett. 228B (1989) 121

17. D.H. Rischke, M.I. Gorenstein, H. Stöcker and W. Greiner, Phys. Lett. 237B (1990) 153

18. E. Braaten, R.D. Pisarski, Nucl. Phys. B337 (1990) 569

19. C. Bernard, M.C. Ogilvie, T.A. DeGrand, C. DeTar, S. Gottlieb, A. Krasnitz, R.L. Sugar, D. Toussaint, Proc. of the 9th Int. Conf. on Ultra–relativistic Nucleus–Nucleus Collisions, Quark Matter '91, to be published in Nucl. Phys. A.

20. I. Mardor, B. Svetitsky, Phys. Rev. D44 (1991) 878

21. J. Rafelski, B. Müller, Phys. Rev. Lett. 48 (1982) 1066

22. T.S. Biró, P. Lévai and B. Müller, Phys. Rev. D42 (1990) 3078

23. T.S. Biro, E. van Doorn, B. Müller, M.H. Thoma, X.-N. Wang, Duke University-preprint DUKE–TH–93–46

24. K. Geiger, Phys. Rev. D47 (1993) 133

25. J.B. Kogut, D.K. Sinclair, Nucl. Phys. B344 (1990) 238

26. V. Koch, G.E. Brown, SUNY–preprint SUNY–NTG–93–5

HADRONIC MATTER AND DILEPTONS

or

IN-MEDIUM PROPERTIES OF HADRONS

Guy Roche

Laboratoire de Physique Corpusculaire
Université Blaise Pascal/IN2P3
63177 Aubière Cedex, France

INTRODUCTION

The topic concerns Relativistic Heavy Ion Physics, i.e., the laboratory energy domain from about 1 to 10 GeV per nucleon, or $\sqrt{s} \sim 2-5$ GeV in the nucleon-nucleon center-of-mass frame. The presentation will be illustrated with experimental data from the DLS (DiLepton Spectrometer) program at the Lawrence Berkeley Laboratory's Bevalac. Older experiments around 10 GeV incident energy[1] had quite low statistics and were concerned with the low-mass low-pt anomaly observed and studied in the late 70's/early 80's. This issue seems to be settled now and will be discussed by H.J. Specht[2]. Some indication on the HADES (High Acceptance DiElectron Spectrometer) project at SIS, Gesellschaft für Schwerionenforschung/Darmstadt, will be given herein. For a more complete description of this project, see the presentation by W. König[3].

The ultimate objective of both programs (DLS, HADES) is the use of dileptons (here e^+e^- pairs) as a probe of the hadronic matter formed in the collision of relativistic heavy ions, and in particular the study of in-medium properties of hadrons. It is obvious that a good knowledge of the probe is needed first. The DLS collaboration has spent a great deal of time to investigate the basic processes of dilepton production in the nucleon-nucleon interaction between 1 and 5 GeV (no data were available in this energy range). Interpretation of the results actually revealed a specific interest of the dilepton probe, namely its sensitivity to electromagnetic form factors in the time-like region.

A brief description of the experimental technique will be given in the next section.

Hot and Dense Nuclear Matter, Edited by
W. Greiner *et al.*, Plenum Press, New York, 1994

The following two sections will be devoted to the basic processes of dilepton production, and to the approach of nuclear matter and medium effects with the dilepton probe, respectively. Conclusions and outlook will end the presentation.

EXPERIMENTAL TECHNIQUE

The DLS collaboration[4] started its dilepton program in 1984 (date of the first proposal) and collected data in p-p and p-d, p-nucleus and nucleus-nucleus (up to Nb-Nb) collisions from 1986 to 1993. The measured dileptons are electron pairs mostly for experimental reasons. Dimuons would present the physics advantage of a lower bremsstrahlung production rate (as far it is considered as a background), but would result in a higher invariant mass threshold (200 MeV) that may be a limitation for studies such as pion dispersion in the nuclear medium where a possible expected effect is the reduction of the threshold below its vacuum valus of $2m_\pi$. In any case, the high energy technique of dimuon measurement that consists in dumping all hadrons in a suitable absorber is not applicable to our energy domain, while identification of electrons in the relevant momentum range of $50 - 100$ MeV/c up to $1 - 2$ GeV/c is very simple (in principle). Using the Cherenkov signal in appropriate gases (with a γ_{th} of $20 - 30$), all electrons are above the Cherenkov threshold, and actually at saturation, all other particles being below threshold.

The main experimental difficulty is the huge hadron flux. There is about one dilepton produced per 10,000 nucleon-nucleon collisions. In the DLS scheme[5], suitable hadron rejection power is achieved with two threshold Cherenkov counters adequatly segmented in each DLS arm. The rest of the system includes two dipole magnets and drift chambers for tracking, scintillator hodoscopes for redundant information and trigger purpose, a multiplicity detector around the target[6], and various detector calibration and beam control components.

A second difficulty, though less serious than the above one, is due to the combinatorial background, commonly referred to as false pairs. These result from two π^0 decays, directly through the Dalitz channel ($\pi^0 \longrightarrow \gamma e^+ e^-$, BR = 1.2 %), or indirectly through the external conversion of the real photons from the main channel ($\pi^0 \longrightarrow \gamma\gamma$, BR = 98.8 %). There are two electron pairs but, in some/many cases, one member in each pair is not seen by the detectors. The false $e^+ e^-$ pairs cannot be identified on an event-by-event basis, but their contribution in the opposite sign sample can be accurately estimated from the measurement of the like sign pairs, $e^+ e^+$ and $e^- e^-$, and then subtracted[7]. Let us point out that real photon measurements are more difficult to correct for combinatorial background[8].

BASIC PROCESSES OF DILEPTON PRODUCTION

Basic processes and electromagnetic form factors

The main basic processes of dielectron production at intermediate energy are:

- two-body decays of vector mesons ($\rho,\omega,\phi \longrightarrow e^+ e^-$),

- three-body Dalitz decays of mesons ($\pi^0, \eta \longrightarrow \gamma e^+ e^-$ and $\omega \longrightarrow \pi^0 e^+ e^-$) and delta and nucleon resonances ($\Delta, N^\star \longrightarrow N e^+ e^-$),

- hadronic bremsstrahlung (pp,pn $\longrightarrow e^+e^- + X$),

- and pion annihilation ($\pi^+\pi^- \longrightarrow e^+e^-$).

Measurement of dileptons from two-body decays yields full information on the vector mesons themselves. Table 1 reproduces masses, widths and branching ratios for the three mesons. The ρ and ω have about the same mass, but their width or life time are quite different. It follows that most of the time the ρ will decay inside the interacting nuclear matter and the dilepton will carry out the information on its in-medium properties, namely any mass shift and/or width change. On the other hand, the ω will be much less affected (except perhaps for production yield) and can be used as a reference. This study is one of the main goals of the HADES project. A consequence is the need for high resolution[1] (in order to clearly see the ω) and high acceptance (cf. branching ratios). For a reason of production rate connected to a higher mass, the ϕ will be more difficult to study.

Table 1. Vector mesons.

	M (MeV)	Γ (MeV - fm/c)	BR
ρ	770	149 - 1.3	4.4×10^{-5}
ω	783	8.4 - 23.5	7.1×10^{-5}
ϕ	1020	4.4 - 46	3.1×10^{-4}

Dileptons from three-body decays do not provide full information on the parent. There are plans to use TAPS behind HADES to measure real photons in coincidence with dileptons and reconstruct the η (M = 549 MeV, Γ = 1.2 keV, BR = 5×10^{-3}). The π^0 Dalitz decay is not of great interest in our case because of the very low dilepton invariant mass (the average e^+e^- mass is 15 MeV) and thus contributes mostly to the combinatorial background discussed in the previous section. Other three-body decays (including the η) can be approached via model calculations, depending upon their contribution yields in the dilepton spectrum.

The pp/pn bremsstrahlung process turns out to be quite difficult to estimate. It will be discussed in some detail in the next subsections. Pion annihilation received much attention a few years ago and may still be an interesting issue related to medium effect as it will be shown later in the talk. This last process is actually connected to the ρ behavior in nuclear matter via the pion electromagnetic form factor that we will discuss just below.

Dilepton production is described as the decay of a virtual/massive photon (labeled as γ^*) and all annihilation or radiative processes involve the coupling hadron-γ^*. The vector dominance model (VDM) says that the coupling is done via a neutral vector meson, ω for the isoscalar part of the electromagnetic current and ρ^0 for the isovector part[9]. Fig. 1 shows the pion annihilation and nucleon radiation graphs drawn under VDM assumption. Of course, the hadron-γ^* coupling is also found in other processes like the η and Δ Dalitz decays.

Available electromagnetic form factor measurements are in good agreement with VDM for pions and in reasonnable agreement for etas[9]. This is not the case for baryons, which form factors have in fact been measured for nucleons only. The recent proton experimental data from LEAR[10] exhibits a significant deviation from the

---

[1]In the previous section, I briefly discussed the interest of dielectrons vs. dimuons at intermediate energy. Let us add here that, for $\rho/\omega/\phi$ studies, even at high energy, dielectrons are preferable because the dimuon detection technique reduces resolution capabilities

VDM calculation in the time-like region for $q^2 > 4m_\rho^2$. Furthermore, no data is available in the region relevant to our q^2 domain, around 1 GeV^2 (which is forbidden in $p\bar{p}$ experiments). Thus, dilepton measurements should bring some information on baryon electromagnetic form factors in the time-like region through pp/pn bremsstrahlung and Δ Dalitz decay as will be seen in the next subsections.

Figure 1. Pion annihilation and nucleon bremsstrahlung under VDM assumption.

The DLS experimental results

The first DLS data[11] in p-Be between 1 and 5 GeV were supposed to produce a suitable information on the basic nucleon-nucleon dilepton processes. They were mostly interpreted in terms of pn bremsstrahlung, Δ decay and pion annihilation[12]. The bremsstrahlung from pp scattering was considered to be suppressed. The calculations indicated that pn bremsstrahlung is a dominant source for dielectrons, and increasingly with beam energy, in sharp contrast with older interpretations of the low-mass low-pt anomaly[1]. A ratio as large as 10 for the dielectron yields in p-n and p-p interactions could be expected at 5 GeV. Furthermore, the pn bremsstrahlung calculations were considering only the elastic channel where the final state would consist of the two nucleons and the dilepton, in contradiction with the fact that, at increasing energy, inelastic channels with produced pions contribute to most of the nucleon-nucleon cross section. These difficulties led the DLS collaboration to undertake studies of p-p and p-d collisions, as a direct measurement of p-n collisions would not be feasible.

Fig. 2 shows the differential dielectron yield ratios pd/pp as a function of invariant mass for the measurements at 1.0, 1.2, 1.6, 1.8, 2.1 and 4.9 Gev incident energies. Obviously, the ratios should go to infinity (except for resolution smoothing) at the p-p kinematic limit because of Fermi momentum, but qualitatively one would expect a rather sharp effect. What is observed instead at the three lowest energies is a continuous increase as a function of mass which might be indicative of different processes at work. The ratio distribution at 2.1 GeV is unexpectedly flat up to the kinematic limit.

Fig. 3 presents the two excitation functions of the integrated yield ratio pd/pp for masses above 0.15 and below 0.10 GeV, respectively. The low mass ratio has a rather flat behavior, going down smoothly from about 3 to 2. This mass range is essentially populated by π^0 Dalitz decay (true) dielectrons, and the DLS ratios about 1 GeV are actually in agreement with an inclusive measurement of the π^0 cross sections in p-p and p-n by A.P. Batson[13] which yields a pd/pp ratio of 4.1 ± 0.8 (the open circle in the figure). The excitation function for m > 0.15 GeV (no π^0 contribution) exhibits a drastic change in between 1 and 2 GeV, going down from about 10 to 2 with increasing energy. The η and ρ-ω nucleon-nucleon thresholds are indicated by arrows on fig. 3. It would be interesting to have another point below 1 GeV to see if the excitation function (m > 0.15 GeV) is really flattening off at low energy.

Finally, associated hadron-dilepton events in the DLS have been analyzed for some of the above dilepton measurements[14]. It is found that there are about 3 ± 1 % pion-dilepton event above 2 GeV beam energy, and a preliminary estimate of the corresponding pion production yield in the whole phase space per detected dilepton amounts to 40 ± 20 %.

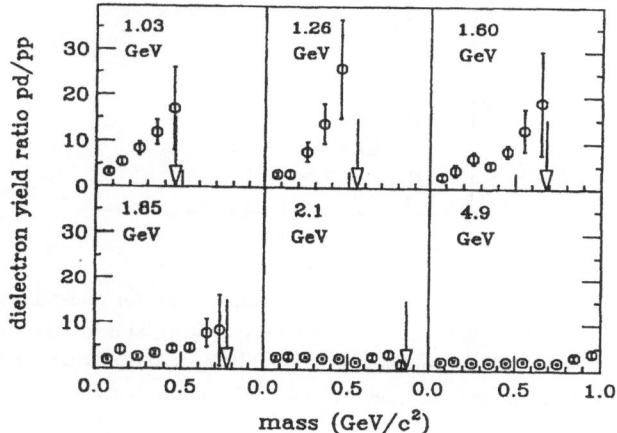

Figure 2. The pd/pp ratio as a function of the dielectron invariant mass. The maximum dielectron mass kinematically allowed in p-p collisions is indicated by an arrow for each beam energy except 4.9 GeV, where the limit, m = 1.7 GeV, is off-scale. The figure is from W.K. Wilson et al.[15].

These experimental results in p-p and p-d clearly establish that the pn bremsstrahlung process as estimated in the first model calculations is not a dominant source, at least above 2 GeV beam energy, and that pp is not suppressed.

Present status of model calculations

In this subsection, we are going to focus on pn/pp bremsstrahlung calculations. The pn process has been first computed in the soft photon approximation[12] that retains only radiation from external charged lines, treats the strong interaction blob as being on shell, and makes use of the elastic nucleon-nucleon cross section. The pp process is neglected considering the non-relativistic dipole limit. Energy conservation is taken care of with the aid of a phase space correction factor. Later on, covariant computations of the pn process were performed in the one-boson-exchange description of the nucleon-nucleon interaction[16]. The results were found to be in reasonable agreement with those from the soft photon approximation and no further attempt of covariant calculations were undertaken. Recently, L.A. Winckelmann et al.[17] pointed out that the forward peaking of the differential elastic pn cross section gives a reduction of about a factor of 3 of the dilepton yield compared to the previously used symmetric parametrization at 4.9 GeV beam energy. B. Kämpfer et al.[18] studied off-shell and VDM form factor corrections and obtained pd to pp ratios consistent with the DLS data at 4.9 GeV. The more recent work by K. Haglin and C. Gale[19] introduces the concept of inelastic nucleon-nucleon bremsstrahlung and we are going to examine some of their results.

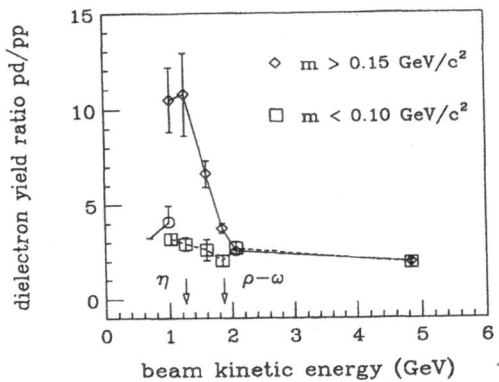

Figure 3. The pd/pp ratio as a function of the beam energy for dielectrons in the mass region dominated by π^0 Dalitz decay, $0.05 < m < 0.10$ GeV (squares), and the rest of the mass spectrum, $m > 0.15$ GeV (diamonds). The beam energy thresholds for η and ρ-ω production are indicated by the two arrows. The lines are only to guide the eyes. The figure is from W.K. Wilson et al.[15].

These authors uses the soft photon approximation that includes radiation from all external charged lines of multi-hadron final states. They first investigate the electromagnetic weightings for the pp and pn elastic interactions and find they are of comparable magnitude at 4.9 GeV, while at 100 MeV pn dominates pp by about a factor of 10, in agreement with the dipole limit. Then accurate parametrizations of the differential elastic cross sections lead to the interesting result that the dilepton yield at 4.9 GeV is larger for pp than for pn, both being much lower than the experimental measurements from the DLS. Contributions from the inelastic channels with one, two and three pions in the final state are found to drastically increase the dilepton yield, with a resulting mass distribution getting steeper than for the elastic channel. Their overall result for p-p at 4.9 GeV, including inelastic bremsstrahlung, η and Δ Dalitz decays, and pion annihilation is reproduced in fig. 4. Agreement with the DLS data is reasonnable. However, the intermediate mass region around 400 MeV is underestimated by about a factor of 3. More work is needed, perhaps the inclusion of form factors in the η Dalitz decay and the bremsstrahlung. Also, the double Dalitz decay of the η may help a little bit[20]. If hadronic processes are not enough, then one might think of parton mechanisms[21].

In conclusion of this section, I'd like to say that the large amount of DLS data and the on-going theoretical effort should result in adequate understanding of the dilepton probe to be used in the study of nuclear matter.

NUCLEAR MATTER AND MEDIUM EFFECTS

Nuclear matter and dileptons - first aspects

In the temperature-baryon density phase diagram of nuclear matter that has already been shown several times at the school, the domain under investigation here covers temperatures about 100 MeV and baryon densities in the range of 3 to 8 times normal density (for beam energies from about 1 to 10 GeV/A). This is the domain of hadronic matter, sometimes referred to as "hadron gas", in which the hadronic degrees of freedom are suitable to describe the medium properties. Nevertheless, there is indication from first estimates that QCD underlying effects (partial restoration of chiral symmetry) could be observable at rather low temperature and smoothly increasing baryon density, while at low baryon density, temperature has to reach its critical value (about 200 MeV) before a sharp discontinuity should occur, a scenario attainable with ultrarelativistic heavy ion beams. We will get to this point in the next subsection.

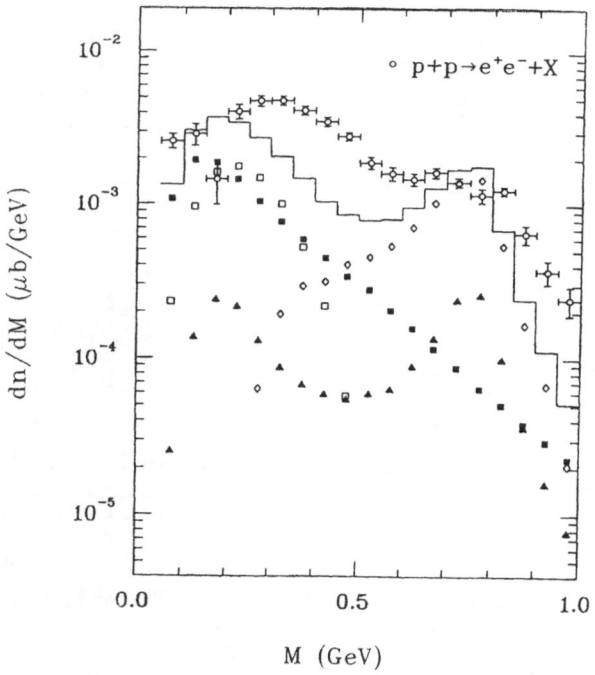

M (GeV)

Figure 4. Invariant mass distributions from various mechanisms of dielectron production in p-p collision at 4.9 GeV: bremsstrahlung is presented as solid squares, radiative decay of the Δ as solid triangles, Dalitz decay of the η as open squares, two-pion annihilation as open diamonds, and finally, the sum of all contributions is shown as the solid histogram. The data points are from the DLS. Notice that the data is not corrected for acceptance and the calculation is weighted by the DLS acceptance. The figure is from K. Haglin and C. Gale[19].

The main reason for using dileptons in the study of hadronic matter is that it is an electromagnetic penetrating probe. The current well established description of relativistic heavy ion collisions, best seen in central collisions of symmetrical systems, introduces an initial stage at high temperature and high baryon density (due to stopping of matter) followed by an expansion and cooling stage that ends with the emission of hadrons (freeze out). Thus, direct information on the initial stage cannot be obtained with hadronic probes, one can speak of a loss of memory, and there is need for a reconstruction of the whole history of the collision with the aid of simulation codes (BUU, QMD,...). In contrast, dileptons can leave the high density interaction region without sizeable disturbance and provide a more unique signal. Analogy can be made with the use of neutrinos instead of visible light or X rays to study the center of the sun.

The first DLS data in p-Be and Ca-Ca[22] are sometimes interpreted without medium effect. Fig. 5 shows a typical model calculation by Gy. Wolf et al.[23] with the following general features: (1) the low mass region below 400 MeV is mostly populated by η Dalitz decay, (2) the bremsstrahlung and Δ decay dominate in the intermediate mass range (400 to 600 MeV) in p-Be at 2.1 GeV which makes this mass region sensitive to electromagnetic form factors and suitable to VDM tests, and (3) the highest mass domain above 600 MeV in Ca-Ca collision is the cleanest with only one dilepton production mechanism at work, pion annihilation, making this mass domain particularly attractive to medium effect studies, as will be discussed in the next subsection. Two remarks have to be done however: (1) this pn bremsstrahlung estimate is probably not very reliable as pointed out in the previous section, and (2) the above conclusions do not apply to p-Be at 1.0 GeV where the bremsstrahlung and Δ contributions dominate over the whole mass range (above 100 MeV).

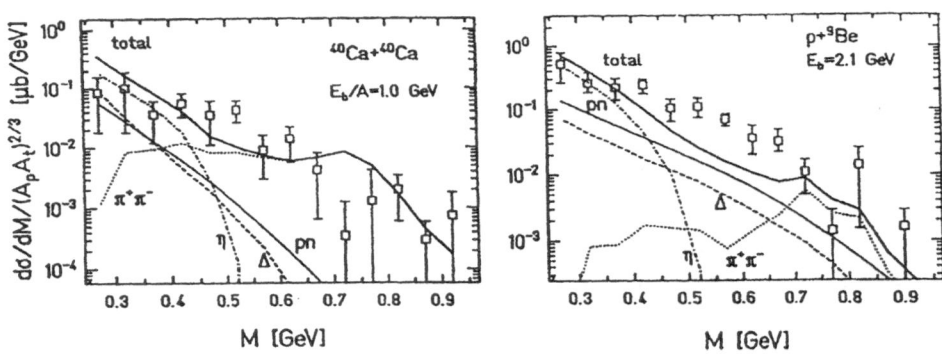

Figure 5. Dilepton invariant mass spectra for Ca-Ca at 1 GeV/A and p-Be at 2.1 GeV bombarding energies. The thick solid line is the sum of all contributions indicated on the graphs. Data points are from the DLS. The figure is from Gy. Wolf et al.[23].

Medium effects

The interest of dileptons to study medium effects was first pointed out by C. Gale and J. Kapusta[24] and made more quantitative by L.H. Xia et al.[24]. At the time, the emphasis was put on the dispersion of pions in nuclear matter, with actually the use of the free space pion electromagnetic form factor. The propagation of pions was described by the following relationship:

$$\omega^2 = \mathbf{k}^2 + m_\pi^2 + \prod(\omega, \mathbf{k}, \rho) \tag{1}$$

where ω is the pion total relativistic energy, $\mathbf{k}$ its vector momentum and ρ the nuclear matter baryon density. These first estimates indicated a drastic enhancement of the dilepton yield close to the mass threshold $(2m_\pi)$ because of the softness of the dispersion relation in the nuclear medium. The mass threshold could actually shift below its value in vacuum (279 MeV) at density high enough when the relation presents a minimum. Later on, C.L. Korpa and S. Pratt[25] realized that inclusion of medium effect on the pion annihilation vertex would strongly reduce the enhancement previously estimated. A more precise calculation by C.L. Korpa et al.[25] taking into account finite dilepton momenta (in the collision rest frame) and delta width still reports a quite substantial enhancement and threshold shift. Finally, the recent paper by W. Ehehalt et al.[25] indicates little sensitivity of the dilepton spectra to the in-medium pion dispersion relation.

Nowadays, the theoretical effort has moved towards the pion electromagnetic form factor and the behavior of the rho meson in the nuclear medium. One class of calculation is based on a hadronic description of the propagation of the rho in the medium. Results[26] are not in precise agreement because of differences in the approximations and the treatment of various corrections. However, common features are the weakening of the resonance and a slight shift of the peak position towards high mass with increasing density. There may be a structure in the spectral function of the meson about a mass value of 400-500 MeV. Gy. Wolf et al.[23] put the medium corrected pion form factor from M. Herrmann et al.[26] in their BUU calculation. Fig. 6 shows their results for Ca-Ca at 1 GeV/A and p-Be at 2.1 GeV. For the Ca-Ca collision, the dashed line labeled "background" is the sum of all contributions other than pion annihilation, the dotted and solid lines are the annihilation contributions with free space and density dependent form factors, respectively. The dilepton yield at the ρ mass of 770 MeV is reduced by a factor of 4 when medium correction is introduced. For p-Be, the authors try to estimate the medium modification of the pn bremsstrahlung component[2] by simply using the same form factor as for pions. In the figure, "background" refers now to all processes other than pn bremsstrahlung, the dotted line is the bremsstrahlung estimate without a form factor, and the dashed and dot-dashed lines show the effects of free space and medium corrected form factors, respectively. One observes a huge increase of the dilepton yield when the form factor is introduced, more than one order of magnitude, and then again a quite sizeable reduction in the rho mass region due to medium correction. The DLS experimental points in the figure qualitatively support medium correction and clearly emphasize the need for high statistics data.

QCD related approaches have also been used[27] and seem to indicate that QCD vacuum modification and partial restoration of chiral symmetry could be observed at low temperature and finite baryon density $(\rho \neq 0)$, which could actually be a better place than the finite temperature system with $\rho = 0$. In a consistent treatment of QCD sum rules in the nuclear medium, T. Hatsuda and S.H. Lee[27] obtain a linear reduction of the vector meson masses valid in the density domain below about 2 times

---

[2]Remark (1) at the end of the preceding subsection still applies

the normal nuclear matter density ρ_0, that can be expressed as

$$m(\rho)/m(0) \simeq 1 - C(\rho/\rho_0) \qquad (2)$$

with $C \simeq 0.18$ for the rho and omega mesons, and $C \simeq 0.15y$ for the phi, y being the strangeness content in the nucleon. Even at moderate density, the mass reduction would be quite impressive. For instance, the rho-omega mass would go down by 200 MeV at $\rho/\rho_0 = 1.5$. At the same density, the phi mass would be below the threshold of its main $K\bar{K}$ decay mode (if the kaon mass does not change in medium), which could be checked by comparison to the dilepton decay mode. In addition to the above mass shift, M. Asakawa and C.M. Ko[27] report a reduction in the rho width as the nuclear density increases. These modifications of the rho meson obviously react on the dilepton mass distributions as will be presented by L.A. Winckelmann[28].

These theoretical calculations all indicate that dilepton invariant mass spectra are sensitive to medium effects, in particular in the ρ/ω mass region. They may still look uncertain, and probably more so for the QCD sum rules approach[3], but they clearly call for high statistics and high resolution experimental data.

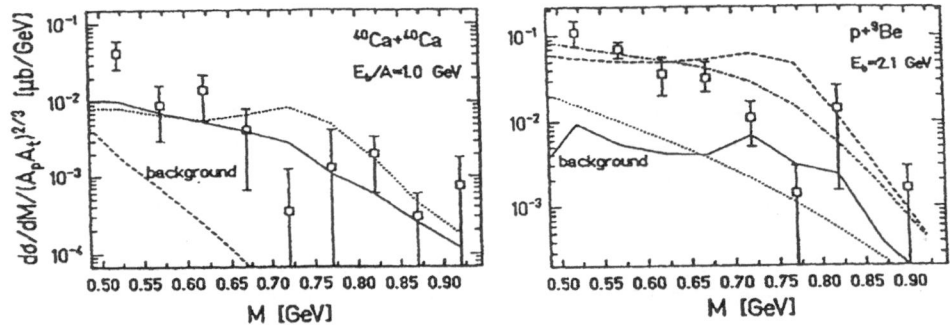

Figure 6. In-medium change of the pion annihilation component for Ca-Ca at 1 GeV/A and effects of the proton form factor on the dilepton spectra for p-Be at 2.1 GeV. See text for an explanation of the different lines. The figure is from Gy. Wolf et al.[23].

CONCLUSIONS AND OUTLOOK

Elementary processes of dilepton production at intermediate energy, decays of particles and resonances or bremsstrahlung and annihilation, should be hadronic. Dalitz decays and bremsstrahlung/annihilation are subject to hadron-virtual photon coupling. Two body decays of vector mesons and pion annihilation seem to be of main interest in relativistic heavy ion physics. On the other hand and already in p-p and p-nucleus collisions, Dalitz decay of the Δ resonance and nucleon-nucleon bremsstrahlung can provide information on the baryon electromagnetic form factors in the time-like region, around $q^2 \sim 1$ GeV^2, where no data exist.

---

[3]During the presentations and discussions at the school, the validity of the four-quark condensate expansion raised some questions.

The large amount of DLS data in p-p and p-d collsions between 1 and 5 GeV and the on-going theoretical effort should result in an adequate understanding of the basic cross sections of dilepton production to be used in the simulation codes of relativistic heavy ion collsions. Sometimes, comparisons of experimental data and models are done for mass spectra only, as these look more promising for medium effect studies. It must be stressed that the measured p_t and y cross sections are also useful to constrain the models. Moreover, any accurate comparison implies that the DLS acceptance filter is included in the cross section calculations.

Dileptons, a penetrating electromagnetic probe, are appropriate to the study of the hot and dense hadronic matter formed in relativistic heavy ion collisions. One may distinguish two regions in their invariant mass spectrum, the one below 500-600 MeV, mostly populated by Dalitz decays and bremsstrahlung, more complex to interpret, and the other above, more simple, where pion annihilation and vector meson two-body decays dominate, more sensitive to medium effects. Calculations of in-medium properties, based on hadronic or QCD related approaches, are rather uncertain and/or present significant discrepancies. There is an obvious need for experimental data.

The first DLS measurements in nucleus-nucleus collisions are in qualitative agreement with hadronic models of the propagation of the ρ vector meson in nuclear matter, and they clearly show the importance of high statistics and high resolution experiments. The recently collected data in d-Ca, α-Ca, C-C and Ca-Ca at 1 GeV/A have much better statistics, actually a factor of 10 better in the Ca-Ca case. They show a significant change in slope of the mass distributions from C-C to Ca-Ca. The first data presented a similar feature, with less accuracy, but from p-Be to Ca-Ca.

The HADES detector is designed to achieve a very high acceptance, 40 % in the collision frame, and good resolution, adequate to unambigously identify the ρ and the ω. Its main goal will be a high statistics study of the vector mesons for projectile-target systems up to Au-Au, at 1 - 1.6 GeV/A.

It would also be of great interest to perform measurements in the energy range around 10 GeV/A, where the vector meson cross sections are higher by more than two orders of magnitude, and the baryon density is estimated to reach values about 7 - 9 times normal nuclear density, compared to about 3 times at 1 GeV/A.

REFERENCES

1. M.R. Adams et al., Phys. Rev. **D27**, 1977 (1983); D. Blockus et al., Nucl. Phys. **B201**, 205 (1982); S. Mikamo et al., Phys. Lett. **B106**, 428 (1981).

2. H.J. Specht, "Direct Photons and Lepton Pairs in pp, pA and AA above 10 GeV/c", these proceedings.

3. W. König, "Subthreshold Antiproton Production and Dilepton Production", these proceedings.

4. The present DLS collaboration - Lawrence Berkeley Laboratory: J. Cailiu, W.G. Gong, L. Heilbronn, H.Z. Huang, G.F. Krebs, A. Letessier-Selvon, H.S. Matis, J. Miller, C. Naudet, R.J. Porter, L.S. Schroeder, P.A. Seidl, W.K. Wilson; University of California at Los Angeles: S. Beedoe, J. Carroll, G. Igo, M. Toy; The Johns Hopkins University: T. Hallman, L. Madansky, R. Welsh; Louisiana State University: P. Kirk, Z.F. Wang; Northwestern University: D. Miller; Université Blaise Pascal/Clermont II: M. Bougteb, F. Manso, M. Prunet, G. Roche; CEBAF: A. Yegneswaran.

5. A. Yegneswaran et al., Nucl. Inst. and Meth. in Phys. Res. **A290**, 61 (1990).

6. S. Beedoe at al., Phys. Rev. **C47**, 2840 (1993).

7. H.Z. Huang et al., Phys. Lett. **B297**, 233 (1992).

8. K.H. Kampert, "Electromagnetic Signals of Hot and Dense Nuclear Matter", these proceedings.

9. See M. Soyeur and W. Weise, 4th Journées des Théoriciens (November 22-23, 1990), Laboratoire National Saturne, 91191 Gif-sur-Yvette, France, and references therein.

10. G. Bardin et al., Phys. Lett. **B257**, 514 (1991).

11. C. Naudet et al., Phys. Rev. Lett. **62**, 2652 (1989); A. Letessier-Selvon et al., Phys. Rev. **C40**, 1513 (1989).

12. L. Xiong et al., Phys. Rev. **C41**, R1355 (1990); C. Gale and J. Kapusta, Phys. Rev. **C40**, 2397 (1989); J. Kapusta and P. Lichard, Phys. Rev. **C40**, R1574 (1989).

13. A.P. Batson, Proc. of the Royal Society **A251**, 233 (1959).

14. F. Manso, Thèse présentée à l' Université Blaise Pascal/Clermont II, numéro DU 538 (May 6, 1993).

15. W.K. Wilson et al., Phys. Lett. **B316**, 245 (1993).

16. K. Haglin et al., Phys. Lett. **B224**, 433 (1989); M. Schäfer et al., Phys. Lett. **B221**, 1 (1989).

17. L.A. Winckelmann et al., Phys. Lett. **B298**, 22 (1993).

18. B. Kämpfer et al., Phys. Lett. **B301**, 123 (1993).

19. K. Haglin and C. Gale, McGill/93-9 Preprint, and Phys. Rev. **C**, in press.

20. P. Bertin, H. Fonvieille and P. Lichard, private discussions.

21. P. Lichard and J.A. Thompson, Phys. Rev. **D44**, 668 (1991).

22. G. Roche et al., Phys. Lett. **B226**, 228 (1989).

23. Gy. Wolf et al., Universität Gießen preprint, and Nucl. Phys. **A552**, 549 (1993).

24. C. Gale and J. Kapusta, Phys. Rev. **C35**, 2107 (1987); L.H. Xia et al., Nucl. Phys. **A485**, 721 (1988).

25. C.L. Korpa and S. Pratt, Phys. Rev. Lett. **64**, 1502 (1990); C.L. Korpa et al., Phys. Lett. **B246**, 333 (1990); W. Ehehalt et al., Phys. Lett. **B298**, 31 (1993).

26. M. Herrmann et al., Nucl. Phys. **A560**, 411 (1993); G. Chanfray and P. Schuck, Nucl. Phys. **A555**, 329 (1993); M. Asakawa et al., Phys. Rev. **C46**, R1159 (1992).

27. M. Asakawa and C.M. Lo, Nucl. Phys. **A560**, 399 (1993); T. Hatsuda and S.H. Lee, Phys. Rev. **C46**, R34 (1992); G.E. Brown and M. Rho, Phys. Rev. Lett. **66**, 2720 (1991).

28. L.A. Winckelmann, "Dielectron Production from 1 to 200 GeV", these proceedings.

MULTIFRAGMENTATION

Claus-Konrad Gelbke

National Superconducting Cyclotron Laboratory
Michigan State University
East Lansing, MI 48824, USA

ABSTRACT

Nuclear collisions at intermediate energies lead to highly excited systems which decay into multi-fragment exit channels. Experimental results addressing impact parameter selection, fragment multiplicity, fluctuations, emission time scales, and expansion dynamics will be reviewed and compared to model calculations.

INTRODUCTION

For bulk nuclear matter at moderate temperatures and low densities ($T \approx 10$ MeV, $\rho < \rho_0$) a liquid-gas phase transition is expected to occur [1,2]. Nuclear collision experiments might yield information about this phase transition. In a possible reaction scenario, a hot system, formed during the initial stages of the reaction, may expand under the action of thermal pressure, possibly aided by a decompression cycle following an initial compression. If the reaction trajectory enters regions of sufficiently low density, instabilities with respect to density fluctuations can cause a prompt multifragment disintegration. Multifragment emission processes may, therefore, signal a liquid-gas phase transition.

Unfortunately, this interpretation of multifragment emission is not unique. Dynamically induced density fluctuations (not directly linked with phase instabilities) could also lead to multifragment final states [3,4], or the system could evolve dynamically into strongly prolate, oblate or even toroidal [5,6] shapes for which Rayleigh instabilities could lead to a multifragment breakup. Moreover, statistical models of compound nucleus decay also predict multiple fragment emission [7-10] that may be strongly enhanced under extreme conditions of temperature and density due to a decrease of the decay barriers. Which of these scenarios are applicable must be answered by careful experimental and theoretical investigations.

In this paper, I will review some of the work recently conducted at MSU with the goal of obtaining a better understanding of the physics underlying intermediate energy nucleus-nucleus collisions and of exploring whether conditions necessary for a liquid-gas transition in hot nuclear systems can be achieved. In particular, I will touch issues with regard to impact parameter selection, relative abundances of intermediate mass fragments ($Z_{IMF} \approx 3\text{-}20$) and light particles ($Z = 1,2$) among the emitted particles, the influence of conservation laws on charged particle multiplicity fluctuations, emission temperatures extracted from the relative populations of states, and information about emission time scales as determined by intensity interferometry.

IMPACT PARAMETER SELECTION

Impact parameter filters are generally constructed from global observables which, on average, exhibit a monotonic dependence on the magnitude of the impact parameter. Common impact parameter filters are based upon the measured multiplicity of charged particles N_C [11-14], the transverse energy E_t [15], or the summed charge of particles emitted at intermediate rapidity Z_y [16]. For collisions with incident energies of a few hundred MeV per nucleon, the summed charge, Z_{bound} (the complement of the hydrogen multiplicity N_1), of particles with atomic number $Z \geq 2$ [17] has also been used.

If one assumes a monotonic relationship between impact parameter b and an observable X, one can define a reduced impact parameter scale via the geometrical relation [14,18]

$$\hat{b}(X) = b(X) / b_{\max} = \sqrt{\frac{1}{\sigma_R} \int_X^\infty \frac{d\sigma(\xi)}{d\xi} d\xi} \, , \qquad (1)$$

where σ_R is the reaction cross section. The reduced impact parameter scale ranges from $\hat{b} = 1$ for glancing collisions to $\hat{b} = 0$ for head-on collisions.

A priori, it is unclear to what extent different observables X select similar or equivalent impact parameters, and whether one technique provides superior resolution as compared to another. For $^{36}Ar + ^{197}Au$ collisions of E/A = 35 - 110 MeV, this question was investigated in refs. [18-20] for impact parameter filters based upon $X = N_C$, E_t, Z_y, or N_1. The overall similarity of these impact parameter filters was established, and quantitative questions concerning the resolution of individual methods were addressed. For the selection of central collisions, filters based upon E_t had slightly better resolution than filters based upon Z_y, N_C, and Z_{bound}.

For central collisions, the particle emission pattern must be axially symmetric. For non-central collisions, on the other hand, the emission pattern is generally not axially symmetric. Non-isotropic azimuthal distributions are readily detected by measuring azimuthal correlation functions,

$$1 + R(\Delta\phi) = \frac{\sigma_2(\phi_1 - \phi_2)}{\sigma_1(\phi_1)\sigma_1(\phi_2)} , \tag{2}$$

where ϕ_i denotes the azimuthal angle of particle i, $\Delta\phi = \phi_1-\phi_2$, and σ_1 and σ_2 are the single and two-particle cross sections for a given class of events. Figure 1 illustrates that central collisions can, indeed be selected, by appropriate cuts on the reduced impact parameter [18,20]: the azimuthal correlation functions become isotropic for $\hat{b} \to 0$.

Impact parameter selection appears to be understood at a sufficiently quantitative level to make comparison of impact-parameters filtered data with model prediction meaningful.

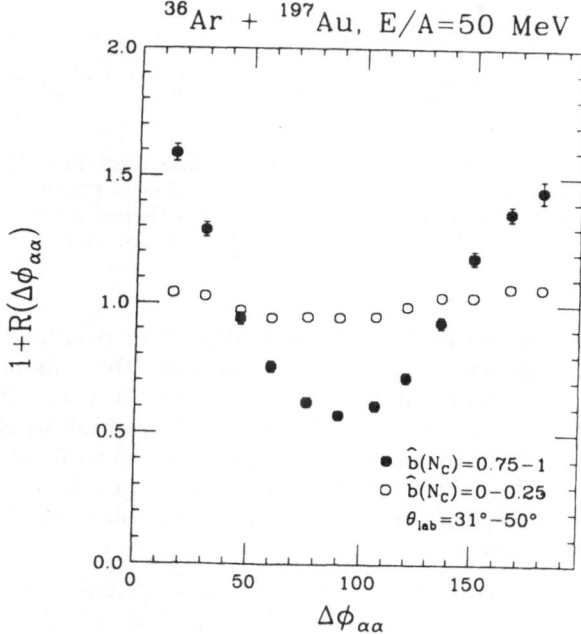

Figure 1. Azimuthal correlation functions for α-particles emitted in impact-parameter selected ^{36}Ar + ^{197}Au collisions; the cuts on reduced impact parameter are indicated in the figure. From refs. [18,20].

FRAGMENT ADMIXTURES

Surprisingly stringent tests of various multifragmentation models can already be obtained from the relative abundance of intermediate mass fragments and light particles among the emitted charged particles [21-24].

The solid points in Fig. 2 show the average number of detected intermediate mass fragments, $<N_{IMF}>$, as a function of the raw charged particle multiplicity, N_C, for the Xe + Au reaction [21] at E/A=50 MeV. Calculations with the microscopic quantum molecular dynamics (QMD) and quasiparticle-dynamics (QPD) transport models of refs. [3,4] predict average fragment multiplicities much smaller than observed experimentally [21]. Both numerical simulations underestimate the growth of density fluctuations leading to fragment emission.

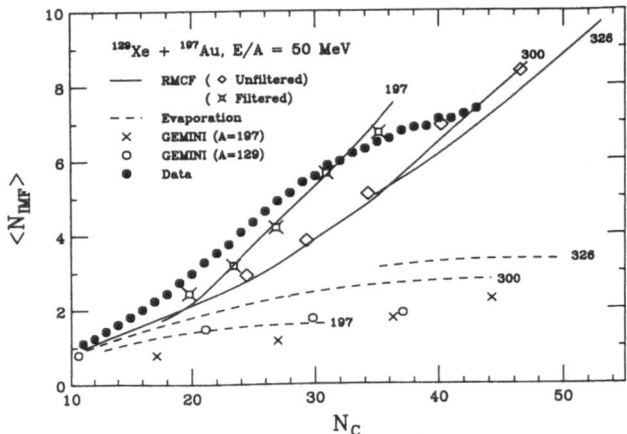

Figure 2. Solid points: relation between $<N_{IMF}>$ and N_C measured for $^{129}Xe + ^{197}Au$ reactions. Calculations are discussed in the text. Diamonds and star-shaped points illustrate instrumental distortions: Diamonds represent unfiltered distributions, star shaped points depict corresponding values filtered by the response of the experimental apparatus. From ref. [21].

Open circles and crosses in Fig. 2 show multiplicities predicted by the statistical compound nuclear decay code GEMINI [4], and the dashed curves depict predictions of the statistical compound nuclear model of ref. [5]. None of these calculations for the decay of excited nuclei of normal nuclear density can reproduce the high IMF multiplicities measured for $N_C > 20$. When hot nuclei are allowed to expand [6], however, significantly larger IMF multiplicities are predicted and reasonable agreement with the data is obtained (solid curves, star- and diamond-shaped points).

Figure 3 shows the mean observed [24] IMF multiplicities as a function of $\hat{b}$ for $^{197}Au + ^{197}Au$ collisions at $E/A = 100$, 250, and 400 MeV. At $E/A=100$ MeV, $<N_{IMF}>$ is largest for small impact parameters, consistent with the trends shown in Fig. 2. Contrary to the incident energy dependence observed at lower incident energies [21-23], the IMF multiplicities in central collisions decrease with incident energy. Such a trend is expected in a statistical picture. At high temperatures, entropy drives the emission of more nucleons and light clusters.

The open squares and circles in Fig. 3 represent IMF multiplicities predicted by the QMD and QPD calculations [24], respectively. Both models predict enhanced fragment multiplicities for central collisions at $E/A = 100$ MeV, but they underpredict the measured peak IMF multiplicities, and they underestimate the shift in the peak fragment multiplicity to larger impact parameters with incident energy. These discrepancies are acerbated when the QMD and QPD calculations are corrected for the detection efficiency, see dashed and dashed-dotted lines [24]. The growth of density fluctuation is clearly not understood at a satisfactory level.

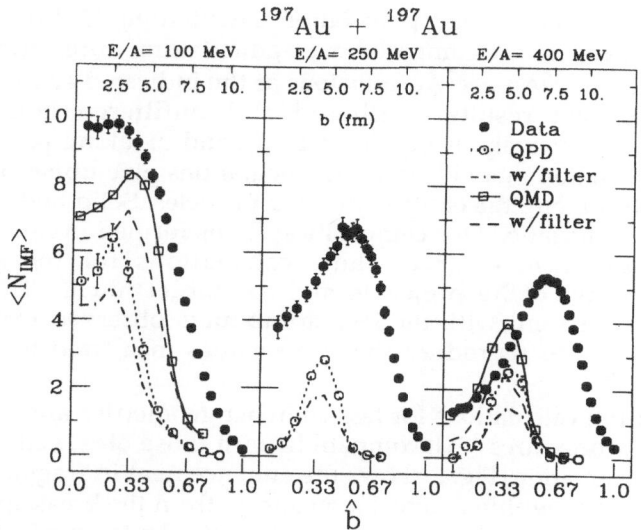

Figure 3. The measured impact parameter dependence of the mean fragment multiplicity observed in ^{197}Au + ^{197}Au collisions is shown by the solid points. The open circles and open squares depict the unfiltered predictions of the QPD and QMD molecular dynamics model, respectively. The dashed and dashed-dotted lines depict the QPD and QMD calculations, filtered through the experimental acceptance. From ref. [24].

A number of theoretical investigations of phase transitions in finite nuclear systems have been based on percolation models [25-28]. Percolation models are attractive, since they exhibit a well-defined phase transition for infinite systems and since they allow straightforward generalizations to finite systems. They have been rather successful [25] in describing the observed [29] power-law behavior of measured fragment mass distributions and in developing techniques to extract critical exponents.

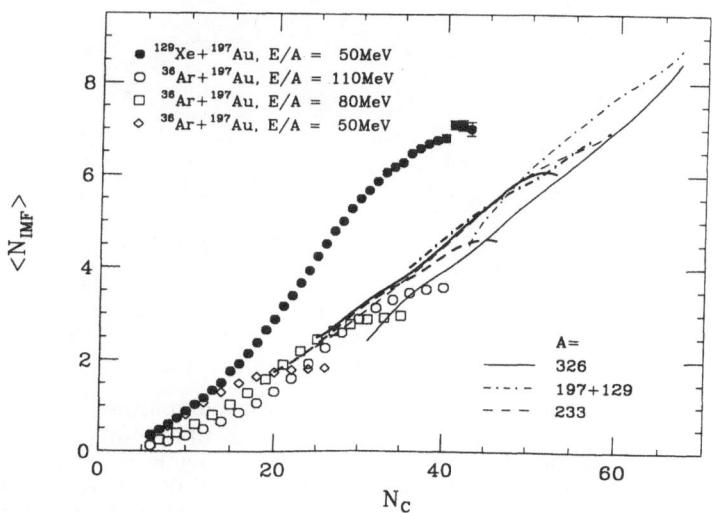

Figure 4. Points show measured relation between $\langle N_{IMF} \rangle$ and N_C for the indicated reactions. Curves depict calculations with the bond-percolation model of ref. [25] as explained in the text. From ref. [30].

In Fig. 4, the results of bond percolation calculations [30] for the decay of spherical nuclear systems are compared to fragment admixtures measured [21,22] for ^{129}Xe + ^{197}Au and ^{36}Ar + ^{197}Au collisions at the indicated energies. Thick and thin curves represent results of filtered and unfiltered bond-percolation calculations [25], respectively, using the critical bond-breaking parameter, p=0.7. For the bond-percolation model [25], these calculations give upper bounds for the admixture of IMFs among the emitted charged particles. Solid and dashed curves show percolation calculations for composite systems formed in Xe + Au and Ar + Au collisions; dot-dashed curves show calculations for the simultaneous multifragment decays of Xe projectile and Au target nuclei. The percolation calculations are consistent with the IMF admixtures observed for the Ar + Au system, but they fail to reproduce the large values measured for the Xe + Au system.

Bond percolation calculations for less compact geometries are able to produce higher fragment admixtures [31], compatible with those observed for the Xe+Au reaction. As an illustration, Fig. 5 shows the enhancement in fragment production which can be obtained within a percolation ansatz from the breakup of bubbles as compared to that of spheres [31]. Percolation model calculations for finite systems indicate a strong dependence of extracted "critical" parameters on the geometrical configuration of the system at breakup. Compilations of power law exponents λ characterizing mass distributions observed for different entrance channels [32], impact parameters [33], or incident energies [32,34,35] may contain samples representing different geometrical configurations rendering a minimum in λ difficult to interpret. For infinite systems, critical exponents govern the scaling laws near critical points. The application of scaling laws to finite systems of potentially complex breakup geometries is likely to be much less straight forward [31] than originally surmised [36].

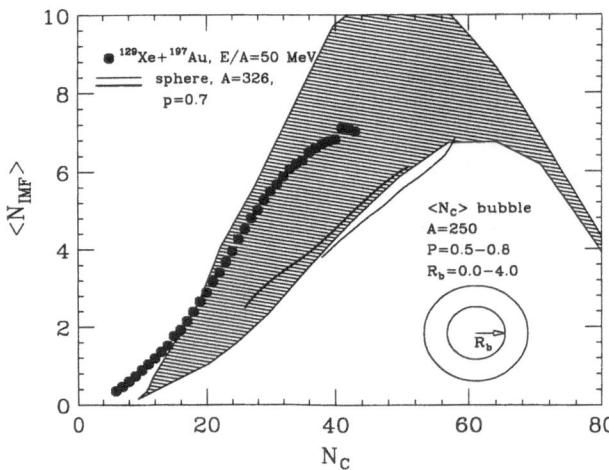

Figure 5. Relation between average IMF and charged particle multiplicities. Solid points represent values measured for ^{129}Xe + ^{197}Au at E/A=50 MeV. Thin (thick) solid line shows the raw (efficiency corrected) percolation calculation for a solid sphere using the near-critical bond-breaking parameter p = 0.7. The hatched area shows the range of average IMF and average charged particle multiplicities predicted by percolation calculations for bubble-shaped breakup configurations.

CHARGED PARTICLE FLUCTUATIONS

The question has been raised whether multifragmenting systems might exhibit large scale fluctuations or evidence for intermittency, which could indicate that fragmentation processes are scale invariant. Intermittency as an indicator of nontrivial physics is generally sought for systems exhibiting larger than Poisson fluctuations. Evidence for such large fluctuations must be sought in events representing similar initial conditions. However, constraints from conservation laws may lead to reduced fluctuations because statistical independence of individual bin occupations is lost.

Figure 6 depicts the relation between $\langle N_c \rangle$ and $\sigma_c^2/\langle N_c \rangle$ extracted [37] for central (b<0.3) Ar + Au collisions at E/A = 35, 50, 80, and 110 MeV. Here, $\langle N_c \rangle$ and σ_c denote the mean value and the variance of the measured charged particle multiplicities. At all energies, the fluctuations of the charged particle multiplicity are considerably smaller than expected for Poisson distributions (for which $\sigma_c^2/\langle N_c \rangle = 1$).

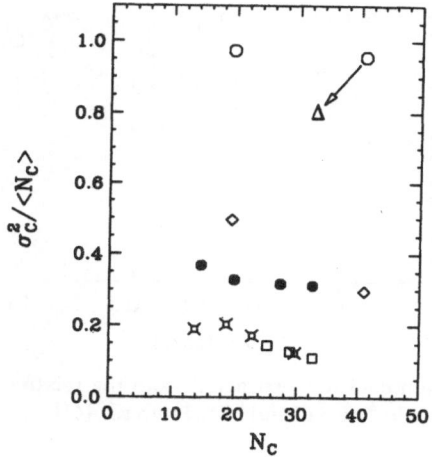

Figure 6. Relation between $\sigma_c^2/\langle N_c \rangle$ and $\langle N_c \rangle$. Solid points show experimental values extracted from near-central ^{36}Ar + ^{197}Au reactions at E/A=35, 50, 80 and 110 MeV. Open points depict results of calculations. Open circles: standard bond-percolation model [37] using p=0.6 and 0.7 (the open triangle illustrates the magnitude of instrumental distortions); open diamonds: percolation model with fixed number of broken bonds; open squares: predictions by GEMINI [7]; star shaped points: predictions by Copenhagen fragmentation model [10]. From ref. [37].

These surprisingly narrow charged particle multiplicity distributions may result from phase space constraints imposed by energy conservation [37]. In Fig. 6, the effect is illustrated by bond percolation calculations in which the number of broken bonds corresponds to excitation energy. In standard percolation calculations, shown by open circular points, the number of broken bonds is allowed to fluctuate from event to event, and the predicted fluctuations in N_C are much larger than observed experimentally. If, on the other hand, one requires a fixed number of broken bonds (i.e. a fixed excitation energy), the widths of the charged particle distributions are strongly reduced (see open diamonds). More realistic statistical model calculations [7,10] which incorporate energy conservation on an event-by-event basis predict even smaller fluctuations. Examples of such calculations are shown by the open square- and star-shaped points which depict predictions of the sequential decay model GEMINI [7] and of the Copenhagen fragmentation model [10], respectively. The experimental N_C-distributions are somewhat broader than predicted by these latter two models, possibly due to imperfect impact parameter selection [37]. Broadening due to poor impact parameter selection and implications for intermittency analyses have been illustrated and discussed in ref. [37].

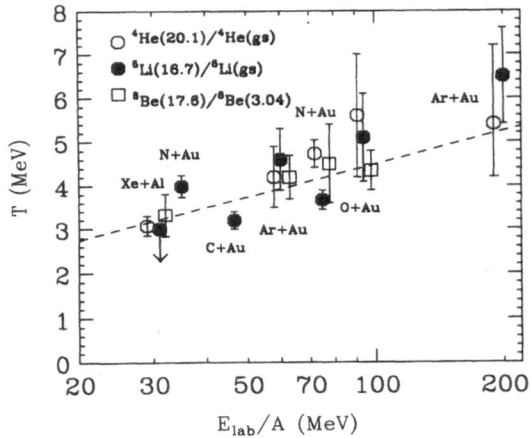

Figure 7. Emission temperatures extracted from the relative populations of widely separated states in ^{4}He, ^{5}Li and ^{8}Be. From ref. [41].

EMISSION TEMPERATURES

Statistical treatments of fragment emission processes require the knowledge of the local excitation energy density (or temperature) at the time of emission. Emission temperatures have been determined from relative populations of states. Particularly suitable for such investigations are widely separated states in ^{4}He and ^{5}Li for which uncertainties due to feeding from higher lying states are of minor importance [38-41]. Figure 7 summarizes emission temperatures determined [38-41] for a number of reactions with incident energies ranging from E/A = 30 - 200 MeV. The extracted temperatures are of the order of 3 - 6 MeV, significantly smaller than "kinetic" temperature parameters which characterize the slopes of the kinetic energy spectra of emitted particles, and they exhibit a small, but significant rise with incident energy [40]. These features can be understood [42] in terms of the expanding compound nucleus model of ref. [9]. Alternative interpretations in terms of nonequilibrium transport models exist [40] and cannot be ruled out. Note however, that these transport models predict too small fragment multiplicities.

EMISSION TIME SCALES

Information about the time scale of multifragment emission processes can be extracted from fragment-fragment correlation measurements [43]. Correlation functions measured [43,44] for the ^{36}Ar + ^{197}Au reaction are shown in Fig. 8. The correlation functions are shown as a function of the reduced relative velocity, v_{red} = $v_{rel}/(Z_1 + Z_2)^{1/2}$, and summed over fragment pairs with $4 \leq Z_{IMF} \leq 9$, see ref. [43] for technical details. The correlation functions exhibit a pronounced minimum at $v_{red} \approx 0$ which is mainly due to the repulsive Coulomb final state interaction between the two coincident fragments. The width of this minimum is sensitive to the source dimension and the time scale of emission.

The curves in Fig. 8 represent calculations with the classical version of the Koonin-Pratt formula. The derivation of this formula is based upon the assumption that the final-state interaction between the two detected fragments dominates, that final-state interactions with all remaining particles can be neglected, and that the correlation functions are determined by the two-body density of states as corrected by the interactions between the two particles. For two-fragment correlations, these assumptions are not strictly valid, and Coulomb distortions in the mean field of the residual system introduce uncertainties for the fragment emission time scales of approximately 50% [43,44].

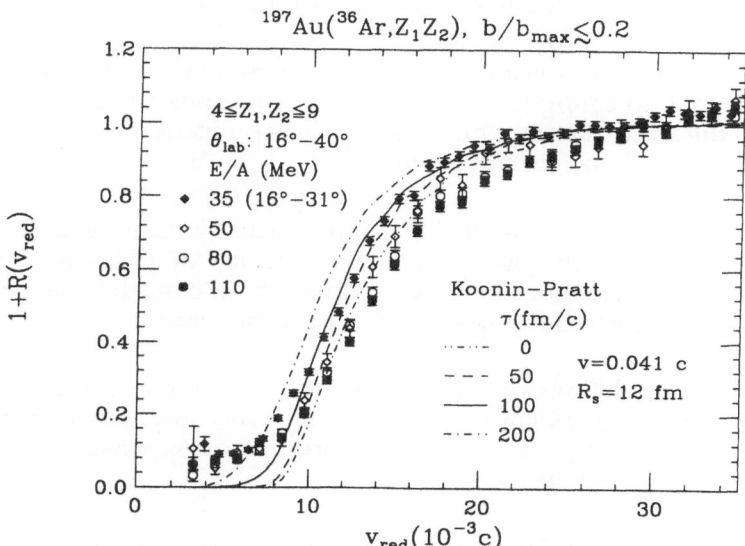

Figure 8. Two-fragment correlation functions measured for the ^{36}Ar + ^{197}Au reaction. From ref. [44].

Time scales extracted from the comparison between experiment and theory are of the order of 100 fm/c or less. These time scales are commensurate with typical time scales of nonequilibrium transport phenomena, and they are well within the realm of current microscopic transport theories. The failure of these microscopic theories to produce sufficient numbers of fragments presents therefore a particularly challenging problem for future investigations.

SUMMARY

Considerable progress has been made in studies of nucleus-nucleus collisions leading to multifragment final states:

• Empirical impact parameter filters can be constructed which allow the selection of small impact parameter collisions, and it is possible to cross-calibrate impact parameter filters based upon different observables and assess their respective resolution.

• Multifragment disintegrations exist over a broad range of incident energies. Microscopic transport theories, explicitly constructed for the treatment of multifragment disintegrations, fail to account for the large number of fragments observed experimentally.

• Statistical models which allow for fragment emission from a low-density phase, on the other hand, are rather successful in reproducing the observed large fragment admixtures. Emission from a low-density regime is a necessary (but not sufficient) condition for a liquid-gas transition.

• Heavy-ion collisions leading to multifragment final states may involve non-compact breakup configurations. Such configurations significantly affect the shape of the mass distributions making the application of scaling laws less straight forward than generally surmised.

• For impact parameter selected reactions, fluctuations in the charged particle multiplicity are much smaller than fluctuations of Poisson distributions, indicating that fluctuations in finite systems exhibit considerable damping due to phase space constraints imposed by conservation laws.

• Relative populations of states, measured over a broad range of incident energies, indicate fragment formation at relatively low temperatures, $T \approx 4\text{-}6$ MeV, again pointing towards the importance of expansion and fragment formation at low average density.

• Studies of two-fragment correlation functions indicate that multifragmentation time scales are short, $\tau \approx 100$ fm/c, and that fragment emission already sets in during the early nonequilibrium stages of the reaction.

Much progress has been made over the last few years, but the fundamental problem of understanding the growth of fluctuations in the nuclear medium is not yet understood at the quantitative level.

All of what I have reported has been learned in collaboration with many friends and colleagues (the authors of refs. [8,9,11,18-24,30,31,37-44]) whose many contributions I wish to acknowledge. This work is based upon work supported by the National Science Foundation under Grant number PHY-92-14992.

REFERENCES

1. G. Bertsch and P.J. Siemens, Phys. Lett. 126B:9 (1983).
2. L.P. Csernai and J. Kapusta, Phys. Rep. 131:223 (1986) and refs. therein.
3. Peilert et al., Phys. Rev. C39:1402 (1989).
4. Boal and J.N. Glosli, Phys. Rev. C37:91 (1988); C42: R50 (1990) .
5. L.G. Moretto et al., Phys. Rev. Lett. 69:1884 (1992).
6. W. Bauer et el., Phys. Rev. Lett. 69:1888 (1992).
7. Charity et al., Nucl. Phys. A483:371 (1988).
8. Friedman and W.G. Lynch, Phys. Rev. C28:16 (1983); C28:950 (1983).
9. Friedman, Phys. Rev. C42:667 (1990).
10. Bondorf et al., Nucl. Phys. A443:321 (1985); Nucl. Phys. A444:460 (1986); H.W. Barz, et al., Nucl. Phys. A448:753 (1986).
11. M.B. Tsang et al., Phys. Rev. C40:1685. (1989).
12. R. Stock, Phys. Reports. 135:259 (1986).
13. H. Stöcker and W. Greiner, Phys. Reports 137:277 (1986).
14. C. Cavata et al., Phys. Rev. C42:1760 (1990).
15. H.G. Ritter, Nucl. Phys. A488:651c (1988).
16. C.A. Ogilvie et al., Phys. Rev. C40:654 (1989).
17. J. Hubele et al., Z. Phys. A340:263 (1991).
18. L. Phair, Ph.D. thesis, Michigan State University (1993).
19. L. Phair et al., Nucl. Phys. A548:489 (1992).
20. L. Phair et al., Nucl. Phys. (in press).
21. D.R. Bowman et al., Phys. Rev. Lett. 67:1527 (1991).
22. R.T. de Souza et al., Phys. Lett. B268:6 (1991).
23. D.R. Bowman et al., Phys. Rev. C46:1834 (1992).
24. M.B. Tsang et al., Phys. Rev. Lett. 71:1502 (1993).
25. Bauer et al., Phys. Lett. 150B:53 (1985); Nucl. Phys. A452:699 (1986); Phys. Rev. C38:1927 (1988).
26. X. Campi, J. Phys. A19:L917 (1986); Phys. Lett. 208B:351 (1988).
27. H. Ngô et al., Z. Phys. A337:81 (1990).
28. T.S. Biro et al., Nucl. Phys. A459:692 (1986).
29. A.S. Hirsch et al., Phys. Rev. C29:508 (1984).
30. L. Phair et al.,Phys. Lett. B285:10 (1992).

31. L. Phair et al., Phys. Lett. B314:271 (1993).

32. A.D. Panagiotou et al., Phys. Rev. Lett. 52:496 (1984).

33. C.A. Ogilvie et al., Phys. Rev. Lett. 67:1214 (1991).

34. M. Mahi et al., Phys. Rev. Lett. 60:1936 (1988).

35. T. Li et al., Phys. Rev. Lett. 70:1924 (1993).

36. X. Campi and H. Krivine, Z. Phys. A344:81 (1992).

37. L. Phair et al., Phys. Lett. B291:7 (1992).

38. J. Pochodzalla et al., Phys. Rev. C35:1695 (1987).

39. Z. Chen et al., Phys. Rev. C36:2297 (1987) and C38:2630 (1988).

40. G.J. Kunde et al., Phys. Lett. B272:202 (1991).

41. C. Schwarz et al., Phys. Rev. C48:676 (1993).

42. W.A. Friedman, Phys. Rev. Lett. 60:2125 (1988).

43. Y.D. Kim et al., Phys. Rev. Lett. 67:14 (1991); Phys. Rev. C45:338 (1992); Phys. Rev. C45:387 (1992).

44. D. Fox et al., Phys. Rev. C47:R421 (1993).

EQUATION OF STATE OF NEUTRON STAR MATTER, LIMITING ROTATIONAL PERIODS OF FAST PULSARS, AND THE PROPERTIES OF STRANGE STARS

Fridolin Weber

Institute for Theoretical Physics, University of Munich, Theresienstr. 37/III, 80333 Munich, FRG, and Lawrence Berkeley Laboratory, Berkeley, CA 94720, USA

Norman K. Glendenning

Nuclear Science Division, Lawrence Berkeley Laboratory, University of California, Berkeley, CA 94720, USA

INTRODUCTION

One of the most challenging but also complicated problems of modern physics consists in exploring the behavior of matter under extreme conditions of temperature and/or density. Knowledge of its behavior is of key importance for our understanding of the physics of the early universe, its evolution in time to the present day, compact stars, various astrophysical phenomena, and laboratory physics. On earth, relativistic heavy-ion collisions provide the only means to learn about the behavior of dense nuclear matter. What is not widely appreciated is the fact that from the study of the properties of compact stellar objects, e.g. neutron stars, one too gains knowlege of the behavior of dense matter, as illustrated in Fig. 1.

Neutron stars are associated with two classes of astrophysical objects, – pulsars and compact X-ray sources. Matter in their cores possess densities ranging from a few times the density of normal nuclear matter (2.5×10^{14} g/cm^3) to about an order of magnitude higher, depending on mass. They thus contain matter in one of the densest forms found in the universe! The equation of state of the stellar matter decisively links neutron stars with nuclear and particle physics (plus various other branches of physics). It is the basic input quantity whose knowledge over a broad

Hot and Dense Nuclear Matter, Edited by
W. Greiner *et al.*, Plenum Press, New York, 1994

range of densities (ranging from the density of iron at the star's surface up to ~ 15 times the density of normal nuclear matter reached in the cores of massive stars) is necessary when solving the Einstein equations for the properties of neutron stars.

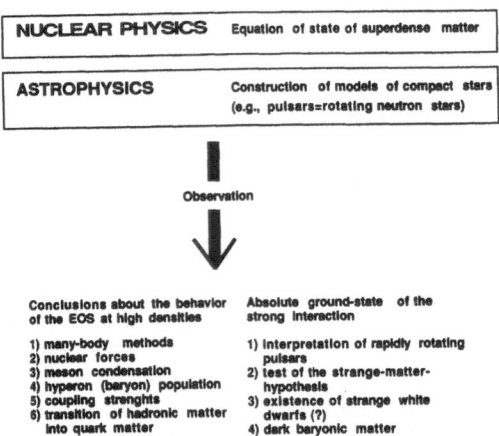

Figure 1. The marriage of nuclear physics with astrophysics enables one to learn about the behavior of the nuclear equation of state at high densities as well as the absolute ground-state of the strong interaction.

The physical behavior of matter under such extreme densities as in the cores of massive stars is rather uncertain and the associated equation of state is only poorly known. The models derived for it differ considerably with respect to the dependence of pressure on density, which has its origin in various sources. To mention several are: (1) the many-body technique used to determine the equation of state, (2) the model for the nucleon-nucleon interaction, (3) description of electrically charge neutral neutron star matter in terms of either (a) only neutrons, (b) neutrons and protons in β-equilibrium with electrons and muons, or (c) nucleons, hyperons and more massive baryon states in β-equilibrium with leptons, (4) inclusion of meson (π, K) condensation, and (5) treatment of the transition of confined hadronic matter into quark matter. By means of marrying nuclear physics with astrophysics the compatibility of these phenomena with observed data on pulsars can be tested, which addresses an open fundamental problem of nuclear physics concerning the behavior of the high-density equation of state.

In these two lectures the following items will be treated:

- The present status of dense nuclear matter calculations and constraints on the behavior of the associated equation of state at high densities from data on rapidly rotating pulsars.

- Recent finding of the likely existence of a mixed phase of baryons and quarks forming a coulomb lattice in the dense cores of neutron stars.

- Review of important findings of recently performed calculations of rapidly rotating compact stars. These are constructed in the framework of general relativity theory for a representative collection of realistic nuclear equations of state which account for items (1)-(5) from above.

Table 1: Nuclear equations of state applied for the construction of models of general relativistic rotating neutron star models. Their tabulated representations, i.e. pressure versus energy and baryon density $P(\epsilon, \varrho)$, are given in Ref. [1].

Label	EOS	Description (see text)	Reference
Relativistic field theoretical equations of state			
1	G_{300}	H, K=300	2
2	HV	H, K=285	3, 4
3	G^{DCM2}_{B180}	Q, K=265, $B^{1/4}=180$	5, 6
4	G^{DCM2}_{265}	H, K=265	7
5	G^{π}_{300}	H, π, K=300	2
6	G^{π}_{200}	H, π, K=200	8
7	Λ^{00}_{Bonn} + HV	H, K=186	9
8	G^{DCM1}_{225}	H, K=225	7
9	G^{DCM1}_{B180}	Q, K=225, $B^{1/4}=180$	5, 6
10	HFV	H, Δ, K=376	4
11	Λ^{00}_{HEA} + HFV	H, Δ, K=115	9
Non-relativistic potential model equations of state			
12	BJ(I)	H, Δ	10
13	WFF(UV_{14}+TNI)	NP, K=261	11
14	FP(V_{14}+TNI)	N, K=240	12
15	WFF(UV_{14}+UVII)	NP, K=202	11
16	WFF(AV_{14}+UVII)	NP, K=209	11

- Establish the minimum-possible rotational periods of gravitationally bound neutron stars and self-bound strange stars. Its knowledge is of fundamental importance for the decision between pulsars that can be understood as rotating neutron stars and those that cannot (signature of hypothetical self-bound matter of which strange stars are the likely stellar candidates).

- Investigate the properties of sequences of strange stars. Specifically, we answer the question whether such objects can give rise to the observed phenomena of pulsar glitches, which is at the present time the only astrophysical test of the strange-quark-matter hypothesis.

COLLECTION OF NUCLEAR EQUATIONS OF STATE

A representative collection of nuclear equations of state, which are determined in the framework of non-relativisitic Schroedinger theory and relativistic nuclear field theory, is listed in Table 1. It is this collection that has been applied for the construction of models of general relativistic rotating neutron star models. The specific properties of these equations of state are described in the third column of Table 1, where the following abbreviations are used: N = pure neutron; NP = n, p, leptons; π = pion condensation; H = composed of n, p, hyperons ($\Sigma^{\pm,0}$, Λ, $\Xi^{0,-}$), and leptons; Δ

$= \Delta_{1232}$-resonance; Q = quark hybrid composition, i.e. n, p, hyperons in equilibrium with u, d, s-quarks, leptons; K = incompressibility (in MeV); $B^{1/4}$ = bag constant (in MeV). Not all equations of state of our collection account for neutron matter in β-equilibrium (i.e. entries 13-16). These models treat neutron star matter as being composed of only neutrons (entry 14), or neutrons and protons in equilibrium with leptons (entries 13, 15, 16), which is however not the ground-state of neutron star matter predicted by theory[3]. The relativistic equations of state *account for all baryon states* that become populated in dense star models constructed from them. An inherent feature of the relativistic equations of state is that they do not violate *causality*, i.e. the velocity of sound is smaller than the velocity of light at all densities, which is not the case for the non-relativistic equations of state. Among the latter only the WFF(UV$_{14}$+TNI) equation of state does not violate causality up to densities relevant for the construction of models of neutron stars from it[13].

POSSIBILITY OF A MIXED PHASE OF BARYONS AND QUARKS IN THE CORES OF NEUTRON STARS

A special feature of the two equations of state denoted G_{B180}^{DCM1} and G_{B180}^{DCM2} is that these, additionally to hyperons, also account for the possible transition of baryon matter to quark matter at high densities. Model G_{B180}^{DCM1} is shown in Fig. 2, where pressure as a function of energy density (in units of the density of normal nuclear matter, $\epsilon_0 = 140$ MeV/fm^3) is exhibited. Most interestingly, the transition of baryon matter to quark matter sets in already at a density $\epsilon = 2.3\,\epsilon_0$[5, 6] (arrow in Fig. 2), which lowers the pressure relative to confined hadronic matter. The mixed phase of baryons and quarks ends, i.e. the pure quark phase begins, at $\epsilon \approx 15\,\epsilon_0$, which is larger than the central density encountered in the maximum-mass star model constructed from this equation of state. Furthermore, one sees that the pressure in the mixed phase *varies with density* (i.e., is not constant!), which implies the existence of a mixed matter phase in neutron stars constructed from it. This finding, which goes back to recent work by Glendenning[5, 6], is in sharp contrast to all other investigations on this topic (for details, see[5, 6]). The reason for this lies in the fact that the transition between confined hadronic matter and quark matter takes place subject to the conservation of baryon charge and electric charge. Correspondingly, there are two chemical potentials, and the transition of baryon matter to quark matter is to be determined in three-space spanned by pressure and the chemical potentials of the electrons and neutrons (rather than two-space). This circumstance has not been realized in the numerous investigations published on this topic earlier. The only existing investigation which accounts for this properly has been performed by Glendenning[5, 6]. For an investigation of the structure of the mixed phase of baryons and quarks predicted by Glendenning, we refer to recent work by Heiselberg, Pethick, and Staubo[14]. This geometric structure of the mixed phase is likely to have dramatic effects on pulsar observables including transport properties and the theory of glitches.

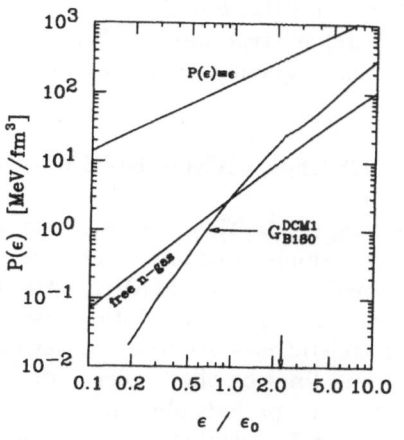

Figure 2. Pressure versus density of equation of state G_{B180}^{DCM1}. For comparison, the free neutron gas equation of state is shown too.

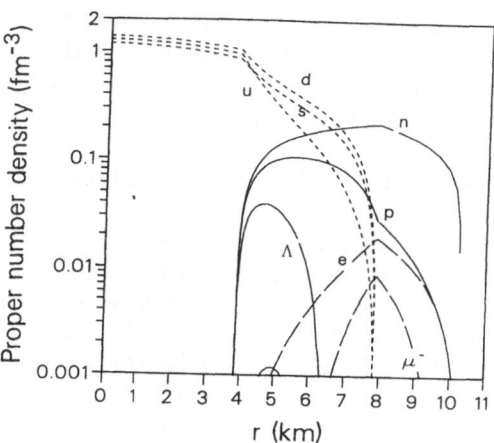

Figure 3. Baryon/Quark composition of a massive neutron star constructed for G_{B180}^{DCM1}.

MODELS OF ROTATING COMPACT STARS IN GENERAL RELATIVITY

Neutron stars are objects of highly compressed matter so that the geometry of space-time is changed considerably from flat space-time. Thus for the construction of realistic models of rapidly rotating pulsars one has to resort to Einstein's theory of general relativity. In the case of a star rotating at its *absolute* limiting rotational periods, i.e. the Kepler (or mass-shedding) frequency, Einstein's equations,

$$\mathcal{R}^{\kappa\lambda} - \frac{1}{2} g^{\kappa\lambda} \mathcal{R} = 8\pi \, \mathcal{T}^{\kappa\lambda}(\epsilon, P(\epsilon)) \; , \tag{1}$$

are to be solved in combination with the general relativistic expression describing the onset of mass-shedding at its equator:[15, 16, 17]

$$\Omega_K = \omega + \frac{\omega'}{2\psi'} + e^{\nu-\psi} \sqrt{\frac{\nu'}{\psi'} + \left(\frac{\omega'}{2\psi'} e^{\psi-\nu}\right)^2} \; . \tag{2}$$

The quantities $\mathcal{R}^{\kappa\lambda}$, $g^{\kappa\lambda}$, and $\mathcal{R}$ denote respectively the Ricci tensor, metric tensor, and Ricci scalar (scalar curvature). The dependence of the energy-momentum tensor $\mathcal{T}^{\kappa\lambda}$ on pressure and energy density, P and ϵ respectively, is indicated in Eq. (1). The quantities ω, ν, and ψ in Eq. (2) denote the frame dragging frequency of local inertial frames and time- and space-like metric functions, respectively. The primes denote derivatives with respect to Schwarzschild radial coordinate, and all functions on the right are evaluated at the star's equator. All the quantities on the right hand side of

149

Eq. (2) depend also on Ω_K, so that it is not an equation for Ω_K, but a transcendental relationship which the solution of the equations of stellar structure, resulting from Eq. (1), must satisfy if the star is rotating at its Kepler frequency. (For more details, we refer to Ref. [16].)

NEUTRON STARS ROTATING AT THE KEPLER FREQUENCY

The computed general relativistic Kepler periods P_K ($\equiv 2\pi/\Omega_K$), defined in Eq. 2, are graphically depicted in Fig. 4 for a representative sample of equations of state of Table 1. The smallest rotational periods are obtained for the WFF(AV_{14} + UVII) equation of state (label 16). We find that the relativistic equations of state lead in general to larger rotational periods than the non-relativistic ones due to the somewhat larger radii of the associated star models[16]. The upper limit on the Kepler period is set by the relativistic HV (label 2) equation of state. The periods obtained from all other equations of state of Table 1 lie between curves "2" and "16". The rectangle in Fig. 4 covers both the approximate range of neutron star masses as determined from observations[18], as well as the measured rotational periods ($P \geq 1.6$ msec). One sees that all pulsar periods so far observed are larger than the absolute limiting Kepler values. The observation of pulsars possessing masses in the observed range but rotational periods that are smaller than say ~ 0.7 msec (depending on the star's mass) would be in contradiction to our collection of equations of state. Consequently the observation of such pulsars cannot be reconciled with the interpretation of such

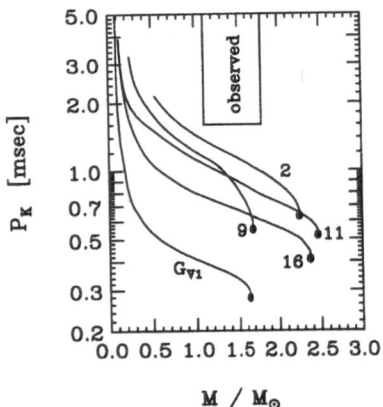

Figure 4. Kepler period versus rotational neutron star mass. The labeling of the curves is explained in Table 1, with the exception of G_{V1} which is explained in the text. Only pulsar periods $P > P_K$ are possible, which is consistent with the pulsar periods known to date.

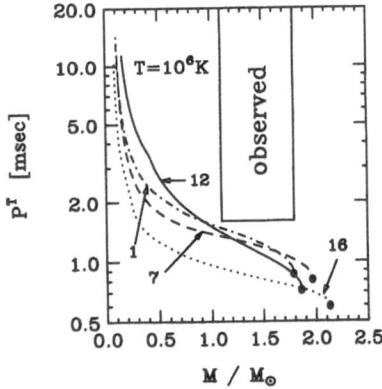

Figure 5. Rotational periods P^T at which instability against emission of gravity waves in cold neutron stars of temperature $T = 10^6$ K sets in versus mass. Stable pulsar periods are restricted to $P > P^T$, which is consistent with the pulsar periods known to date.

objects as rapidly rotating *neutron* stars. Half-millisecond periods, for example, are completely excluded for pulsars made of baryon matter. Therefore, the possible future discovery of a single sub-millisecond pulsar, rotating with a period of say ~ 0.5 msec, would give a strong hint that such an object is a rotating self-bound *strange star*, not a neutron star, and that 3-flavor strange quark matter is the true ground-state of the strong interaction, as pointed out by Glendenning[19, 20].

The fact that any successful model for the nuclear equation of state must accomodate pulsars with rotational periods of (at least) 1.6 msec and masses larger than typically 1.5 $M_\odot$ leads to an overall constraint on its density dependence (double constraint of fast rotation and a large enough neutron star mass): it must behave soft in the vicinity of the density of normal nuclear matter and intermediate nuclear densities in order to lead to small enough rotational pulsar periods, but rather stiff at high nuclear densities to account for large enough masses![9, 21, 22]

An investigation of the limiting rotational Kepler period of neutron stars that is performed *without* taking recourse to any particular models of dense matter but derives the limit only on the general principles that (a) Einstein's equations describe stellar structure, (b) matter is microscopically stable, and (c) causality is not violated has only recently been performed by Glendenning[20]. He establishes a lower bound for the minimum Kepler period for a $M = 1.442\,M_\odot$ neutron star of $P_K = 0.33$ ms (equation of state denoted G_{V1} in Fig. 4). This curve sets an absolute limit on rotation on any star bound by gravity. The equation of state that nature has chosen need not be the one that allows stars to rotate most rapidly, so the above is a strict model independent limit.

GRAVITATIONAL RADIATION-REACTION DRIVEN INSTABILITY IN NEUTRON STARS

Besides the absolute upper limit on rotation set by the Kepler frequency, there is another instability that sets in at a lower rotational frequency, and which therefore sets a more stringent limit on stable rotation[23]. It originates from counter-rotating surface vibrational modes, which at sufficiently high rotational star frequencies are dragged forward. In this case, gravitational radiation which inevitably accompanies the aspherical transport of matter does not damp the modes, but rather drives them[24, 25]. Viscosity plays the important role of damping such gravitational-wave radiation-reaction instabilities at a sufficiently reduced rotational frequency such that the viscous damping rate and power in gravity waves are comparable[26]. The instability modes are taken to have the dependence $\exp[i\omega_m(\Omega)t + im\phi - t/\tau_m(\Omega)]$, where ω_m is the frequency of the surface mode which depends on the angular velocity Ω of the star, ϕ denotes the azimuthal angle, and τ_m is the time scale for the mode which determines its growth or damping according to the sign of τ_m. The rotation frequency Ω at which it changes sign is the critical frequency for the particular mode, m $(=2,3,4,...)$. It is conveniently expressed as the frequency, Ω_m^ν (ν refers to the viscosity dependence, see below), that solves[23]

$$\Omega_m^\nu = \frac{\omega_m(0)}{m}\left[\tilde{\alpha}_m(\Omega_m^\nu) + \tilde{\gamma}_m(\Omega_m^\nu)\left(\frac{\tau_{g,m}}{\tau_{\nu,m}}\right)^{\frac{1}{2m+1}}\right], \tag{3}$$

where

$$\omega_m(0) \equiv \sqrt{\frac{2m(m-1)}{2m+1} \frac{M}{R^3}} \tag{4}$$

is the frequency of the vibrational mode in a non-rotating star. The time scales for gravitational radiation reaction[27], $\tau_{g,m}$, and for viscous damping time[28], $\tau_{\nu,m}$, are given respectively by

$$\tau_{g,m} = \frac{2}{3} \frac{(m-1)[(2m+1)!!]^2}{(m+1)(m+2)} \left(\frac{2m+1}{2m(m-1)}\right)^m \left(\frac{R}{M}\right)^{m+1} R, \tag{5}$$

$$\tau_{\nu,m} = \frac{R^2}{(2m+1)(m-1)} \frac{1}{\nu}. \tag{6}$$

The shear viscosity is denoted by ν. It depends on the temperature, T, of the star ($\nu(T) \propto T^{-2}$). It is small in very hot ($T \approx 10^{10}$ K) and therefore young stars and larger in cold ones. A characteristic feature of the above equations (3) - (6) is that Ω_m^{ν} merely depends on radius and mass (R and M) of the spherical star model and its assumed temperature (respectively viscosity).

The functions $\tilde{\alpha}_m$ and $\tilde{\gamma}_m$ contain information about the pulsation of the rotating star models and are difficult to determine[23, 29]. A reasonable first step is to replace them by their corresponding Maclaurin spheroid functions α_m and γ_m[23, 29]. We therefore take $\alpha_m(\Omega_m)$ and $\gamma_m(\Omega_m)$ as calculated in Refs. [30, 31] for the oscillations of rapidly rotating inhomogeneous Newtonian stellar models (polytropic index $n=1$), and Ref. [23] for uniform-density Maclaurin spheroids (i.e. $n=0$), respectively. Managan has shown that Ω_m^{ν} depends much more strongly on the equation of state and the mass of the neutron star model (through $\omega_m(0)$ and $\tau_{g,m}$, see Eqs. (4) and (5)) than on the polytropic index assumed in calculating α_m[32].

Figure 5 shows the limiting rotational neutron star periods P^T ($\equiv 2\pi/\Omega_m^{\nu}$), i.e. the solution of Eq. (3), at which the gravitational radiation-reaction instability in stars set in. Considered are old (and therefore cold) neutron stars of temperature $T = 10^6$ K.[33] For such stars we find that the $m = 2$ mode is largest and thus is exited first. It therefore sets the limit on stable rotation for neutron stars of such temperatures. As in the case of rotation at the Kepler period, equation of state "16" here too leads to the smallest stable periods obtained for all equations of state of our collection. No neutron star can possess a stable rotational period lying below this curve. In other words, by taking the gravity wave instability into account, the minimum-possible stable rotational periods of cold pulsars are restricted to values larger than $\sim 0.8 - 1.4$ msec, depending on mass and equation of state. This is still in agreement with observation. The latter value, however, is already rather close to the rotational period of the fastest pulsars observed to date. For young and therefore hot pulsars, which are not considered here, the corresponding range of stable periods has been established somewhere else to be $\sim 1 - 1.6$ msec for $M \sim 1.5 \, M_\odot$[13, 16, 33].

PROPERTIES OF STRANGE STARS

THE STRANGE-MATTER HYPOTHESIS

The hypothesis that strange quark matter may be the absolute ground state of the strong interaction (not ^{56}Fe) has been raised by Witten in 1984[34]. If the hypothesis is true, then a *separate* class of compact stars could exist, which are called strange stars. They form a distinct and disconnected branch of compact stars and are not part of the continuum of equilibrium configurations that include white dwarfs and neutron stars. In principle both strange and neutron stars could exist. However if strange stars exist, the galaxy is likely to be contaminated by strange quark nuggets which would convert all neutron stars that they come into contact with to strange stars[19, 35, 36]. This in turn means that the objects known to astronomers as pulsars are probably rotating strange matter stars, *not* neutron matter stars as is usually assumed. Unfortunately the bulk properties of models of neutron and strange quark stars of masses that are typical for neutron stars, $1.1 \lesssim M/M_\odot \lesssim 1.8$, are relatively similar (cf. Fig. 7) and therefore do not allow the distiction between the two possible

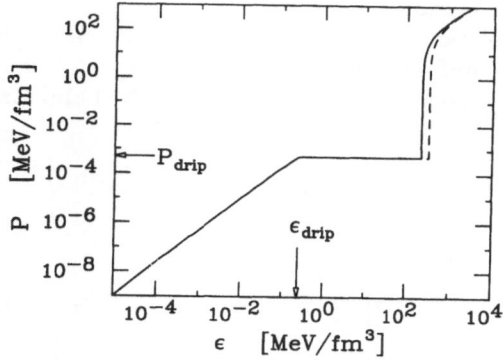

Figure 6. Equation of state of a strange star surrounded by a nuclear crust with inner density below neutron drip (see text). The sybmols P_{drip} and ϵ_{drip} refer to drip pressure and drip energy density. The quark matter equation of state is calculated for $B^{1/4} = 145$ MeV (solid line) and 160 MeV (dashed).

pictures. The situation changes however as regards the possibility of *fast rotation* of strange stars. This has its origin in the different mass-radius relationships of neutron stars and strange quark stars (see Fig. 7)[37]. As a consequence of this the entire familiy of strange stars can rotate rapidly, not just those near the limit of gravitational collapse to a black hole as is the case for neutron stars. As an example, above the minimum possible rotational periods of maximum-mass neutron stars have been determined to be larger than ~ 0.8 msec; this is to be compared with $\approx (0.4-0.6)$ msec calculated for strange stars[38].

HADRONIC CRUST ON STRANGE STARS

At the present time there appears to be only one crucial *astrophysical test* of the strange-quark-matter hypothesis, and that is whether strange quark stars can give rise to the observed phenomena of pulsar glitches. In the crust quake model an oblate solid nuclear crust in its present shape slowly comes out of equilibrium with the forces acting on it as the rotational period changes, and fractures when the built up stress exceeds the sheer strength of the crust material. The period and rate of change of period slowly heal to the trend preceding the glitch as the coupling between crust and core re-establish their co-rotation. The existence of glitches may have a decisive impact on the question of whether strange matter is the ground state of the strong interaction.

The only existing investigation which deal with the calculation of the thickness, mass, and moment of inertia of the nuclear solid crust that can exist on the surface of a rotating strange quark star has only recently been performed by Glendenning and Weber[38]. Their calculations account for the fact that strange stars can possess a solid nuclear crust, which is suspended out of contact with the strange star via the strong electric field on its surface. The maximum density of the inner crust is strictly limited by the neutron drip density (about 4.3×10^{11} g/cm^3), since free neutrons, being electrically neutral particles, cannot exist in the star. These would be dissolved into quark matter as they gravitate into the core. The equation of state of such a strange star with crust is shown in Fig. 6.

The mass-radius relationship of strange stars with crust, which follows from the equation of state exhibited in Fig. 6, is shown in Fig. 7[38, 39]. Here the solid dots denote the maximum-mass star of the neutron star (NS) and strange quark star (SS)

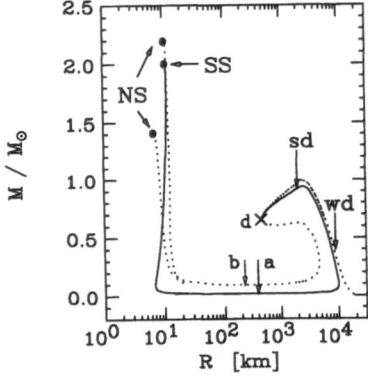

Figure 7. Mass–radius relationship of strange matter stars with nuclear crust (strange stars and strange dwarfs, solid curve), and neutron stars and ordinary white dwarfs (dotted curve).

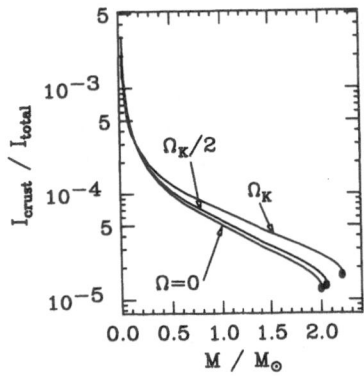

Figure 8. The ratio $I_{\text{crust}}/I_{\text{total}}$ as a function of star mass. Rotational frequencies are shown as a fraction of the Kepler frequency, Ω_K[5].

sequences. The arrows indicate the minimum-mass star of each sequence ('a': strange star, 'b': neutron star). White-dwarf-like strange star configurations ('sd': strange dwarfs) terminate at the crossed point labeled 'd'. The symbol 'wd' indicates the region of ordinary white dwarfs. A value for the bag constant of $B^{1/4} = 145$ MeV for which 3-flavor strange matter is stable has been chosen. This choice represents strongly bound strange matter with an energy per baryon ~ 830 MeV, and thus corresponds to strange quark matter being absolutely bound with respect to ^{56}Fe.

Since the crust is bound by the gravitational interaction (and not by confinement, which is the case for the strange matter core), the mass-radius relationship of strange stars is qualitatively similar to the one for neutron stars. The radius being largest for the lightest and smallest for the heaviest stars in the sequence. Just as for neutron stars the relationship is not necessarily monotonic at intermediate masses. The radius of the strange quark core is proportional to $M^{1/3}$ which is typical for self-bound objects. This proportionality is only appreciably modified near the mass where gravity terminates the stable sequence. The sequence of strange stars has a minimum mass of $\sim 0.015\, M_\odot$ (radius of ~ 400 km) or about 15 Jupiter masses (label 'a'), which is smaller than that of neutron star sequences, about $0.1\, M_\odot$ (label 'b')[40]. These low-mass strange stars may be of considerable importance since they may be difficult to detect and therefore may effectively *hide baryonic matter*. Furthermore, of interest to the subject of cooling of strange stars is the crust thickness of strange stars[41]. It ranges from ~ 400 km for stars at the lower mass limit to ~ 12 km for stars of mass $\sim 0.02\, M_\odot$, and is a fraction of a kilometer for the star at the maximum mass[38]. Those strange stars which result from solving the Oppenheimer-Volkoff equations for central star densities that are smaller than the corresponding central density of the minimum-mass star, but larger than the smallest possible one (determined by $\epsilon = 3\, P_{drip} + 4\, B$, P_{drip} denotes the drip pressure) are shown too in Fig. 7. The cross refers to that particular star whose strange-matter-core radius has shrunken to zero, thus possessing mass and radius values of a white dwarf star[42].

The moment of inertia of the hadronic crust, I_{crust}, that can be carried by a strange star as a functin of star mass for a sample of rotational frequencies of $\Omega = \Omega_K, \Omega_K/2$ and 0 is shown in Fig. 8. Because of the relatively small crust mass of the maximum-mass models of each sequence, the ratio I_{crust}/I_{total} is smallest for them (solid dots in Fig. 8). The less massive the strange star the larger its radius (Fig. 7) and therefore the larger both I_{crust} as well as I_{total}. The dependence of I_{crust} and I_{total} on M is such that their ratio I_{crust}/I_{total} is a monotonically decreasing function of M. One sees that there is only a slight difference between I_{crust} for $\Omega = 0$ and $\Omega = \Omega_K/2$.

Of considerable relevance for the question of whether strange stars can exhibit glitches in rotation frequency, one sees that I_{crust}/I_{total} varies between 10^{-3} and $\sim 10^{-5}$ at the maximum mass. If the angular momentum of the pulsar is conserved in the quake then the relative frequency change and moment of inertia change are equal, and one arrives at[38]

$$\frac{\Delta\Omega}{\Omega} = \frac{|\Delta I|}{I_0} > \frac{|\Delta I|}{I} \equiv f\, \frac{I_{crust}}{I} \sim (10^{-5} - 10^{-3})\, f \ , \ \text{with} \quad 0 < f < 1 \ . \ (7)$$

Here I_0 denotes the moment of inertia of that part of the star whose frequency is changed in the quake. It might be that of the crust only, or some fraction, or all of the star. The factor f in Eq. (7) represents the fraction of the crustal moment of inertia that is altered in the quake, i.e. $|\Delta I| = f\, I_{crust}$. Since the observed glitches have relative frequency changes $\Delta\Omega/\Omega = (10^{-9} - 10^{-6})$, a change in the crustal moment of inertia by less than 10% would cause a giant glitch even in the least favorable case (for more details, see[38]). Finally we find that the observed range of the fractional change in $\dot{\Omega}$ is consistent with the crust having the small moment of inertia calculated and the quake involving only a small fraction f of that, just as in Eq. (7). For this purpose we write[38]

$$\frac{\Delta\dot{\Omega}}{\dot{\Omega}} = \frac{\Delta\dot{\Omega}/\dot{\Omega}}{\Delta\Omega/\Omega} \frac{|\Delta I|}{I_0} = \frac{\Delta\dot{\Omega}/\dot{\Omega}}{\Delta\Omega/\Omega} f \frac{I_{\text{crust}}}{I_0} > (10^{-1} \text{ to } 10) f , \tag{8}$$

which yields a small f value as before: $f < (10^{-4} \text{ to } 10^{-1})$. Here measured values of the ratio $(\Delta\Omega/\Omega)/(\Delta\dot{\Omega}/\dot{\Omega}) \sim 10^{-6}$ to 10^{-4} for the Crab and Vela pulsars, respectively, have been used.

SUMMARY

This work begins with an introduction of a representative collection of realistic models for the nuclear equation of state of neutron star matter. Specifically, the recent finding of the possible existence of a mixed phase of baryon matter and quark matter in dense neutron star matter is stressed. These equations of state are then applied for the construction of models of rapidly rotating neutron stars, whose masses and rotational periods are compared with observational data on pulsars. The indication of this work is that the gravitational radiation-reaction instability sets a lower limit on stable rotation for massive *neutron stars* of $P \sim 0.8$ msec. Lighter ones having typical pulsar masses of $\sim 1.45\, M_\odot$ are predicted to have mimimum rotational periods of ~ 1 msec. This finding may have very important implications for the nature of any pulsar that is found to have a shorter period, say below ~ 0.5 msec. At least for the studied broad collection of nuclear equations of state, rotation at such small periods is not allowed, and thus the interpretation of such objects as rapidly rotating neutron stars fails. Such objects, however, can be understood as rapidly rotating self-bound *strange stars*. The plausible ground-state state in that event is the deconfined phase of (3-flavor) strange-quark-matter. From the QCD energy scale this is as likely a ground-state as the confined phase. At the present time there appears to be only one crucial astrophysical test of the strange-quark-matter hypothesis, and that is whether strange quark stars can give rise to the observed phenomena of pulsar glitches. We demonstrate that the nuclear solid crust that can exist on the surface of a strange star can have a moment of inertia sufficiently large that a fractional change can account for the magnitude of pulsar glitches. Furthermore low-mass strange stars can have enormously large nuclear crusts (up to ~ 400 km) which might considerably alter the cooling rate of strange stars and enables such objects to be possible hiding places of baryonic matter.

If strange-quark-matter is the ground-state of baryonic matter at zero pressure then the conclusion that the confined hadronic phase of nucleons and nuclei is only metastable would be almost inescapable, which would have far-reaching consequences for laboratory nuclear physics, the early universe, and astrophysical compact objects.

Acknowledgement: This work was supported by the Director, Office of Energy Research, Office of High Energy and Nuclear Physics, Division of Nuclear Physics, of the U.S. Department of Energy under Contract DE-AC03-76SF00098.

REFERENCES

1. F. Weber, Habilitation Thesis, January 1992, University of Munich, Munich, F.R.G.

2. N. K. Glendenning, Nucl. Phys. **A493** (1989) 521.

3. N. K. Glendenning, Astrophys. J. **293** (1985) 470.

4. F. Weber and M. K. Weigel, Nucl. Phys. **A505** (1989) 779, and references contained therein.

5. N. K. Glendenning, Nucl. Phys. B (Proc. Suppl.) **24B** (1991) 110.

6. N. K. Glendenning, Phys. Rev. D **46** (1992) 1274.

7. N. K. Glendenning, F. Weber, and S. A. Moszkowski, Phys. Rev. C **45** (1992) 844, (LBL-30296).

8. N. K. Glendenning, Phys. Rev. Lett. **57** (1986) 1120.

9. F. Weber, N. K. Glendenning, and M. K. Weigel, Astrophys. J. **373** (1991) 579.

10. H. A. Bethe and M. Johnson, Nucl. Phys. **A230** (1974) 1.

11. R. B. Wiringa, V. Fiks, and A. Fabrocini, Phys. Rev. C **38** (1988) 1010.

12. B. Friedman and V. R. Pandharipande, Nucl. Phys. **A361** (1981) 502.

13. F. Weber and N. K. Glendenning, *Neutron Stars, Strange Stars, and the Nuclear Equation of State*, in: Proceedings of the First Symposium on Nuclear Physics in the Universe, September 24-26, 1992, Oak Ridge, Tennessee, U.S.A., Ed. M. R. Strayer, IOP Publishing Ltd, Bristol, UK, 1993.

14. H. Heiselberg, C. J. Pethick, and E. F. Staubo, Phys. Rev. Lett. **70** (1993) 1355.

15. J. L. Friedman, J. R. Ipser, and L. Parker, Astrophys. J. **304** (1986) 115.

16. F. Weber and N. K. Glendenning, *Hadronic Matter and Rotating Relativistic Neutron Stars*, Proceedings of the Nankai Summer School "Astrophysics and Neutrino Physics", p. 64 - 183, Tianjin, China, 17-27 June 1991, Eds. D. H. Feng, G. Z. He, and X. Q. Li, World Scientific, 1993.

17. N. K. Glendenning and F. Weber, *Impact of Frame Dragging on the Kepler Frequency of Relativistic Stars*, (LBL-33401), 1993.

18. F. Nagase, Publ. Astron. Soc. Japan **41** (1989) 1.

19. N. K. Glendenning, Mod. Phys. Lett. **A5** (1990) 2197.

20. N. K. Glendenning, Phys. Rev. D **46** (1992) 4161.

21. N. K. Glendenning, *Strange-Quark-Matter Stars*, in: Proc. Int. Workshop on Relativ. Aspects of Nucl. Phys., ed. by T. Kodama, K. C. Chung, S. J. B. Duarte, and M. C. Nemes, World Scientific, Singapore, 1990.

22. J. M. Lattimer, M. Prakash, D. Masak and A. Yahil, Astrophys. J. **355** (1990) 241.

23. L. Lindblom, Astrophys. J. **303** (1986) 146.

24. S. Chandrasekhar, Phys. Rev. Lett. **24** (1970) 611.

25. J. L. Friedman, Phys. Rev. Lett. **51** (1983) 11.

26. L. Lindblom and S. L. Detweiler, Astrophys. J. **211** (1977) 565.

27. S. L. Detweiler, Astrophys. J. **197** (1975) 203.

28. H. Lamb, Proc. London Math. Soc. **13** (1881) 51.

29. C. Cutler and L. Lindblom, Astrophys. J. **314** (1987) 234.

30. J. R. Ipser and L Lindbolm, Phys. Rev. Lett. **62** (1989) 2777.

31. J. R. Ipser and L. Lindblom, Astrophys. J. **355** (1990) 226.

32. R. A. Managan, Astrophys. J. **309** (1986) 598.

33. F. Weber and N. K. Glendenning, Z. Phys. **A339** (1991) 211.

34. E. Witten, Phys. Rev. D **30** (1984) 272.

35. J. Madsen and M. L. Olesen, Phys. Rev. D **43** (1991) 1069, ibid., **44**, 4150 (erratum).

36. R. R. Caldwell and J. L. Friedman, Phys. Lett. **264B** (1991) 143.

37. N. K. Glendenning, *Supernovae, Compact Stars and Nuclear Physics*, invited paper in Proc. of 1989 Int. Nucl. Phys. Conf., Sao Paulo, Brasil, Vol. 2, ed. by M. S. Hussein et al., World Scientific, Singapore, 1990.

38. N. K. Glendenning and F. Weber, Astrophys. J. **400** (1992) 647.

39. F. Weber and N. K. Glendenning, *Fast Pulsars, Compact Stars, and the Strange Matter Hypothesis*, in: Proceedings of the 2nd International Conference on Physics and Astrophysics of Quark-Gluon Plasma, January 19-23, 1993, Calcutta, India, World Scientific.

40. G. Baym, C. Pethick, and P. Sutherland, Astrophys. J. **170** (1971) 299.

41. P. Pizzochero, Phys. Rev. Lett. **66** (1991) 2425.

42. N. K. Glendenning, Ch. Kettner, and F. Weber, *Unusual Configurations of Strange Stars*, in preparation (1993).

STRANGELETS AT THE AGS

Bruce Libby

Department of Physics and Astronomy
Iowa State University

INTRODUCTION

Strangelets can be considered to be a low baryon number (A) example of strange quark matter (SQM), in which up, down, and strange quarks are confined in a single bag rather than as individual protons, neutrons, lambdas, or other baryons[1-4]. It has been hypothesized that SQM may in fact be the true ground state of nuclear matter[5], that neutron stars may be strange stars[5], and that cosmic rays may have a SQM component[6]. Even if SQM is only metastable, with an energy per baryon number between that of nuclei and the mass of the lambda , it is possible for SQM to form by the decay of hypernuclei or multi-hypernuclei[7,8] or even to be formed directly in heavy-ion induced reactions[9-12]. Only if the energy per baryon number of SQM is greater than the mass of the lambda does SQM's hypothesized existence make no sense.

Although SQM may have been a constituent of the early universe, its production on earth might be accomplished only by the use of heavy-ion accelerators, such as the Alternating Gradient Synchrotron (AGS) at Brookhaven National Laboratory in the United States. Production of strangelets has been hypothesized to occur by different mechanisms, including the distillation from droplets of quark-gluon plasma (QGP)[9-12], or, more conservatively, as the product of a coalescence mechanism[8].

The distillation of strangelets from the QGP can be visualized as the end product of the cooling of droplets of QGP formed in a heavy-ion induced reaction. *If* a droplet of QGP is formed in a central collision at the AGS, then free valence quarks (u and d) encounter s and $\bar{s}$ (and u $\bar{u}$ or d $\bar{d}$) quarks formed in the plasma. The QGP will then expand and cool by the emission of pions and kaons. $K^+(u\bar{s})$ and $K^o(d\bar{s})$ will be preferentially emitted from the plasma, because it is more likely for an $\bar{s}$ quark to encounter a free valence quark than for an s quark to encounter an $\bar{u}$ or $\bar{d}$

quark. This leads to a strange enhancement of the plasma. As the plasma cools and hadronizes, the end product can either be a stranglet or a hypernucleus[11,13]. If the energy per baryon number of the hypernucleus is greater than that of the stranglet, then hypernucleus production can serve as a doorway state to the production of stranglets[8]. At RHIC energies, it may be less likely for stranglets to form in the QGP predicted to occur. This is because at high energies the plasma will not be formed in a baryon-rich region, so the valence quarks (who carry away antistrangeness as K^+ and K^o) will not be present in excess as they are at AGS energies.

The probability of forming a stranglet in the QGP scenario P_{prod} can be estimated as the product of various probabilities[10]:

$$P_{prod} = P_{QGP} P_{sp} P_{A,S} P_{cool} \tag{1}$$

in which P_{QGP} is the probability of forming a droplet of quark-gluon plasma, P_{sp} is a spatial factor, $P_{A,S}$ is the probability of forming a stranglet of baryon number A and strangeness S (or -S, depending on notation), and P_{cool} is the probability of the stranglet cooling to its ground state.

A more conservative scenario for the production of stranglets is the coalescence mechanism, which has been shown to be valid at Bevalac and AGS energies for the production of light nuclei (d, t, ^3He, etc)[14]. In coalescence, baryons that overlap in phase space can bind. For example, a proton and neutron can overlap to form a deuteron. Heavier nuclei are built up by the further coalescence of deuterons and neutrons or protons, in a stepwise manner.

The number of stranglets of baryon number A and strangeness S N(A,S) produced per collision can be estimated by:

$$N(A,S) = \frac{N(A,S)}{N(A,0)} \frac{N(A,0)}{N_\alpha} N_\alpha \tag{2}$$

in which N(A,0) is the number of particles of baryon number A with no strangeness and N_α is the number of α particles produced per collision. The ratio N(A,0)/Nα involves a penalty factor P-i.e. to add additional nucleons to the cluster is increasingly difficult as more nucleons are added. The ratio N(A,S)/N(A,0) involves a suppression factor λ, which is the probability of converting a proton to a Λ^o(conversion of an up quark to a strange quark). This leads to:

$$N(A,S) = N_\alpha P^{(A-4)} \lambda^{|S|} \tag{3}$$

The penalty factor P has been estimated to have a value ranging from about 1/10 to 1/20[8,14,15], while the suppression factor λ has be estimated to be between 0.1 and 0.3[8,15,16]:

$$N(A,S) = N_\alpha (0.05 - 0.1)^{(A-4)} (0.1 - 0.3)^{|S|} \tag{4}$$

In this way, the number of stranglets N(A,S) formed in each collision is normalized to an easily measured parameter, the number of α particles produced in each reaction. The number of α particles produced in a central Au-Au collision has been calculated to be about 0.015[8]. Thus, the production rate of an A=7, S=3 stranglet at the AGS is around 5×10^{-8}. This rate should be easily measured by high sensitivity

SEARCHES FOR STRANGELETS AT THE AGS

No matter what the mechanism that produced a strangelet in a heavy-ion collision is, the important fact to note is that strangelets are a new form of matter, not a new type of particle[13]. Thus, strangelets can be produced with a range of sizes, spanning a large range of mass, charge, and strangeness. The most important physical character-istic of strangelets is their unusual charge to mass ratio, which will be close to zero or even negative, depending on the strangeness content of the strangelet[11,13,17–19]. The nuclear charge as a function of mass of non-strange nuclei is shown in Figure 1. It is

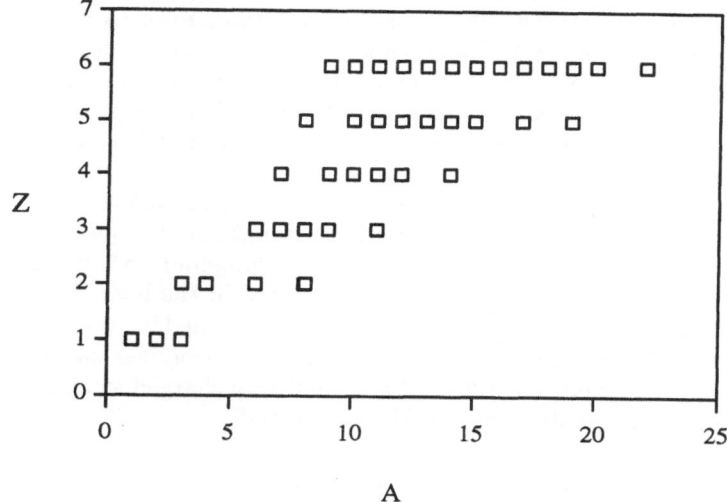

Figure 1: Nuclear charge as a function of mass of known nuclei.

clear from this figure that any particle detected with a charge to mass ratio of less than 0.25 (^8He) is a potential strangelet candidate. Radiochemical and mass spec-trometer based searches for particles with unusual charge to mass ratios, which could include strangelets, have failed to detect them at various levels of sensitivity[20–23]. For example, an accelerator mass spectrometer based search for low Z (less than 9) isotopes with masses between 100 and 10,000 proved to be negative[20]. However, strangelets produced at the AGS at rates large enough to be detected by present and future experiments are expected to have masses less than 30[10,12].

A necessary condition for the production of strangelets at the AGS is the for-mation of a large number of strange quarks. Experiment 802 measured the rapidity distributions of both K$^+$ and K$^-$ produced in the reaction Si + Au at 14.6 GeV/c/A[24]. Also, the K° and $\bar{K}$° rapidity distributions were estimated to be equal to the K$^+$ and

K$^-$ distributions, respectively. Rapidity distribution dN/dy were determined to be 2.9 for K$^+$ and 0.53 for K$^-$ [24]. This could mean that more $\bar{s}$ quarks are emitted than s quarks, leading to a strange enhancement of the participant zone formed in this reaction. Of course, other strange particles, such as the Λ^o (uds) could be emitted and not detected, providing a sink for s quarks. Even inclusion of this point does not alter the conclusion of the study, which is the large production of s and $\bar{s}$ quarks at AGS energies, a necessary condition for the production of strangelets and other strange particles.

One further point on the experimental searches for strangelets at the AGS should be made. None of the experiments discussed below measure the strangeness content of particles detected. All these experiments are attempting to measure is particles with unusual, even negative, charge to mass ratios.

Experiment 814

Experiment 814 attempted to measure positively charged strangelets in the reaction Si + Cu at 14.6 GeV/c/A[18,19]. A total of 2.8 million events were taken with two different trigger conditions, which focused the search on either Z = 1 particles or Z $\geq$ 2 particles. Particles of interest were detected in scintillators and a calorimeter, with a multiplicity trigger to focus on central collisions and either of the two secondary triggers in effect. Certain conditions were applied to the detected events off-line. These conditions included[18,19]:

1) The mass of the particles had to be greater than 10 GeV/c^2.
2) The particles had a consistent position throughout the spectrometer.
3) The particles showed a consistent velocity throughout the spectrometer.
4) There was only a single beam particle detected in the beam counter.

After application of these conditions, no events remained for further analysis. The sensitivity for the production of strangelets in this reaction was calculated to be 8.3 $\times 10^{-4}$ for Z=1 strangelets and 1.2 $\times 10^{-4}$ for multiply charged strangelets, providing an upper limit on the production of SQM at the AGS.

Focussing Spectrometer Experiments

As opposed to open spectrometer experiments, such as E814 (above) and E864 (below), two current AGS experiments, E878 and E886, are focussing spectrometer experiments, in which a series of bending and focussing magnets transport a few "interesting" particles far away from the unreacted beam. Open spectrometers, on the other hand, accept a large number of particles into the spectrometer and depend on large event rates to increase the sensitivity of the spectrometer to the interesting particles.

Experiment 878 is a continuation of E858, an experiment that searched for particles with a negative Z/A ratio and detected two anti-deuterons[17]. The E878 spectrometer consists of scintillators, Cerenkov counters and drift chambers. Triggering is accomplished by use of Cerenkov radiators to eliminate light particles at high velocity[25]. The sensitivity of this experiment was determined by use of a quark-gluon plasma production model[10,12,17], which may or may not be appropriate at AGS energies. This sensitivity approaches 1 $\times 10^{-10}$, but is highly model dependent[25].

While E878 concentrated on searching for negatively charged particles, E886 searched for positively charged particles produced in the reactions Si + Pt at 14.6

GeV/c/A and Au + Pt at 11.7 GeV/c/A[26]. The E886 spectrometer consists of drift chambers, Cerenkov counters and timing scintillators. The trigger requires that the "interesting" particles meet four conditions[26]:

1) The time-of-flight (TOF) is greater than that of a triton.
2) There is no signal in a lucite Cerenkov counter.
3) The nuclear charge is greater than some threshold.
4) The TOF is less than the TOF of "fast" particles.

Preliminary results have indicated the production of the isotopes ^{6}He and ^{8}He and also the presence of 30 "unidentified" particles with Z=1 and a mass greater than 3.5 GeV in the Si-induced reaction. Preliminary analysis indicates the presence of no unidentified particles in the Au-induced reaction[26].

Experiment 864

Experiment 864 is scheduled to come on-line with the AGS heavy-ion beam in 1994[15]. E864 can be best described as a broad-ranged, open spectrometer to detect coalescence nuclei, anti-particles and anti-nuclei and strangelets or other highly

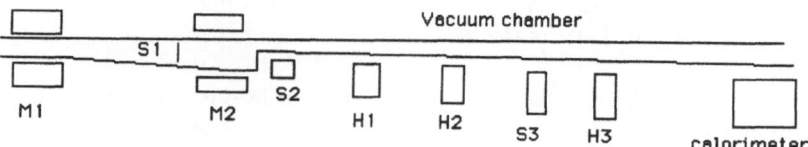

Figure 2: Schematic diagram of the E864 spectrometer.

strange particles. A schematic diagram of the spectrometer is shown in Figure 2, with M1 and M2 dipole magnets, S1-S3 being straw tube stations and H1-H3 being plastic scintillator hodoscopes.

A key feature of E864 is redundancy of measurements. Each hodoscope slat is read out by 2 photomultiplier tubes, giving the vertical position by the time difference between the signals and the time of the particle by the mean of the two signals. The nuclear charge of each particle is calculated by dE/dx in each hodoscope station; the times-of-flight and position are also repeatedly measured. Additionally, excellent time resolution of the calorimeter has been achieved, on the order of 500 ps[27]. This allows the calorimeter to confirm both time-of-flight and position measurements of particles in the spectrometer. The use of a calorimeter also allows for the identification of neutral particles, such as the hypothesized H dibaryon[28-31].

Another important feature of E864 is the use of a large vacuum chamber to contain non-interacting beam particles (and forward-focussed fragments from peripheral reactions). The vacuum chamber will increase the sensitivity of the experiment by reducing the scattering of beam particles and fragments into the spectrometer by interactions not occurring in the target.

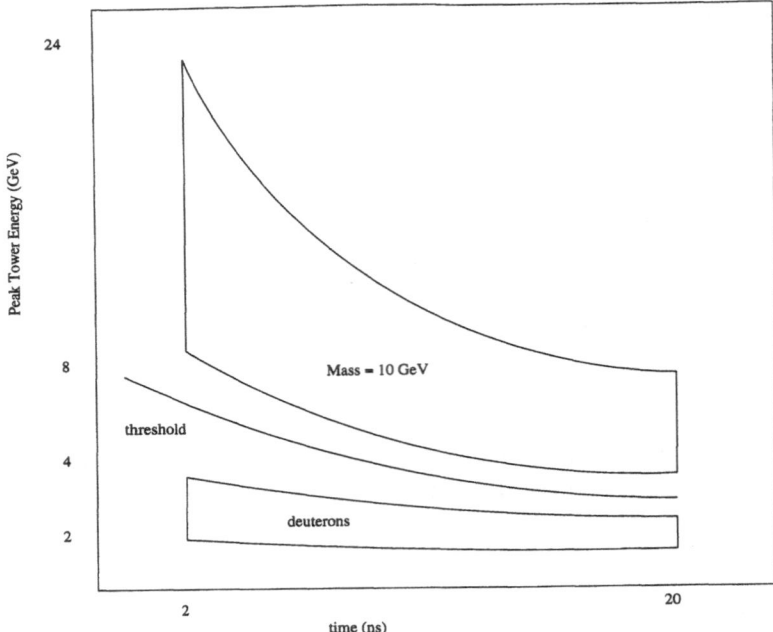

Figure 3: Peak tower energy in the calorimeter as a function of time-of flight for deuterons and a mass 10 GeV particle.

A third feature of the spectrometer is the use of a late energy trigger in addition to a multiplicity trigger. This trigger will attempt to focus on "interesting" events in the calorimeter. This can be illustrated by examination of Figure 3[15], in which the peak tower energy in the shower of a particle detected in the calorimeter is plotted as a function of the time-of-flight of the particle (with respect to a v=c particle). This figure is for deuterons and for a mass 10 GeV/c^2 particle. It is clear that elimination of the deuteron region should be readily attainable.

As previously stated, E864 can be considered to be a spectrometer with which both "bread-and-butter" physics and exotic physics can be performed, and the run plan reflects this dichotomy. In 1994, only 30 % of the calorimeter and the multiplicity trigger will be available, so the focus will be on the coalescence production of nuclei with a Z/A $\approx$ 0.5. This will allow a calibration point for the coalescence production of strangelets, as previously discussed. By changing the setting of the spectrometer magnets, one can focus on strangelets with very small charge to mass ratios (both positive and negative) and other rare composites that might be produced in these reactions.

CONCLUSIONS

Past, present, and future experiments at the AGS have been designed to detect particles with an unusual charge to baryon number ratio. A key feature has been the increasing sensitivity of the experiments, providing better parameters for the production of SQM. If any of the experiments detect particles with unusual Z/A ratios, either positive or negative, then further characterization of the strangeness content will have to be done. None of the current or future AGS experiments are capable of this, requiring new experiments or modification of existing spectrometers to perform this function.

Acknowledgements

The author would like to thank J. Sandweiss and G. Diebold for their fruitful discussions. This work was supported by the United States Department of Energy and the National Science Foundation.

REFERENCES

1. S.A. Chin and A.L. Kerman, Phys. Rev. Lett. 43(1979) 1292.
2. E. Farhi and R.L. Jaffe, Phys. Rev. D30 (1984) 2379.
3. M.S. Berger and R.L. Jaffe, Phys. Rev. C35 (1986)213.
4. E. Blackman and R.L. Jaffe, Nucl. Phys. 324B (1989) 205.
5. E. Witten, Phys. Rev. D30 (1984) 272.
6. E. Farhi, Nucl. Phys B (Proc. Suppl.) 24B (1991) 3.
7. R. Mattiello, et al., Nucl. Phys. B (Proc. Suppl.) 24B (1991) 221.
8. C.B. Dover, et al., Invited Talk Presented at HIPAGS '93.
9. C. Greiner, et al., Phys. Rev. Lett. 58 (1987) 1825.
10. H.J. Crawford, et al., Phys. Rev. D45 (1992) 857.
11. C. Greiner, et al., Z. Phys.C 38 (1988) 283.
12. H.J. Crawford, et al., Nucl. Phys. B (Proc. Suppl.) 24B (1991)199.
13. C. Greiner, et al., Invited Talk Presented at Quark Matter '93 and to be published in the proceedings (1993).
14. S. Nagamiya, et al., Phys. Rev. C24 (1981) 971
15. C. Dover, et al., BNL Proposal 864, October, 1990.
16. J.L. Bailly, et al., Phys. Let. B195 (1987) 609.
17. H.J. Crawford, et al., Nucl. Phys. B (Proc. Suppl.) 24B (1991) 251.
18. F.S. Rotondo, Nucl. Phys. B (Proc. Suppl.) 24B (1991) 251.
19. J. Barrette, et al., Phys. Lett. B252 (1990) 550.
20. T.K. Hemmick, et al., Phys. Rev. D41 (1990) 2074.
21. W.J. Dick, et al., Phys. Rev. Lett. 53 (1984) 431.
22. E.B. Norman, et al., Phys. Rev. Lett 58 (1987) 1403.
23. R.N. Boyd, et al., Phys. Let. B72 (1978) 484.
24. T. Abbott, et al., Phys. Rev. Lett. 64 (1990) 847.
25. H.J. Crawford, Invited talk present at RHIC School '93, Brookhaven National Laboratory, Upton, NY (1993).
26. G. Diebold, private communication.
27. T. Cormier, private communication.
28. C.B. Dover, et al., Phys. Rev. C40 (1989) 234.
29. A.T.M. Aerts and C.B. Dover, Phys. Rev. D28 (1983) 450.
30. A.S. Carroll, et al., Phys. Rev. Lett. 41 (1978) 777,
31. B.A. Shahbazian, et al., Phys. Lett. B316 (1993) 593.

THE RELEVANCE OF THE FRAGMENTS IN THE STUDY OF HOT AND DENSE NUCLEAR MATTER

J.P. Coffin for the FOPI Collaboration

IN2P3-CNRS/Université Louis Pasteur
B.P. 20, F-67037 Strasbourg Cedex 2, France

INTRODUCTION

A great deal of work has been devoted over the past two decades to the study of the properties of the nuclear matter formed in Relativistic Heavy-Ion reactions. A solid background about the main achievements may be found for example in references 1-3. The most salient characteristics of the collisions at medium and low impact parameters (b) are the formation of a hot and dense medium in the region of contact and a pronounced collective motion of all the particles and clusters involved while a rapid thermalization of this zone occurs. The net result is the emission of a large variety of products from light particles ($Z = 1$, 2) to intermediate mass fragments (IMF, typically $3 \leq Z \leq 16$), the so-called multifragmentation process.

The analysis of these phenomena addresses a few fundamental questions whose answers should shed some light on the properties of the highly excited nuclear matter produced. Among the most challenging expectations are the hope to derive the equation of state of the nuclear matter by studying the behaviour of this latter under various regimes of temperature and pressure, the full understanding of the IMF formation possibly through a phase transition or due to large fluctuations in a regime of structural instability, the thorough description of the collective flow of nuclear matter and its implication on the compressibility and the viscosity of this latter, and the determination of the partition of the thermal and the collective energies stored in the system. Of crucial interest is the duality statistics-dynamics since each approach seems to be able to partially reproduce some of the measured properties. Of capital importance are the promptitude and the degree of thermalization of the system. They determine the condition of applicability of a statistical treatment and of a possible thermodynamic approach to extract the entropy and the temperature of a system. However,

Hot and Dense Nuclear Matter, Edited by
W. Greiner *et al.*, Plenum Press, New York, 1994

specific and accentuated homogeneous bulk motion - particularly in violent collisions - may feature an explosion-like type for the reactions and contradicts the idea of a very quick equilibration. It leads to temperature values incompatible with the former suggesting that there might be two temperatures relevant to this problem.

These appealing questions may have elements of an answer only in quite elaborate experiments requiring very exclusive measurements in a close 4π geometry. Attempts of this nature have taken place in works conducted mostly at the Bevalac,[1] Diogène,[4] MSU,[5] and Ganil.[6] Except in some specific cases however the measurements were focused on light products ($Z \leq 3$). Moreover, the criteria requested to select the violent collisions did not always eliminate contributions from semi-peripheral reactions which may consequently distort the data, hence their interpretation.

As shown earlier in this conference[7] the *FOPI* facility has been built at GSI within a European collaboration with the aim of measuring simultaneously all the products - from mesons π and K to IMF 's - emitted in the reactions of identical nuclei with an elaborate level of constraint and versatility in the triggering. We present here two specific aspects of the Au +Au reaction studied at 150, 250, 400, 600 and 800 A MeV incident energy with the *FOPI* apparatus used in its Phase I,[8] namely restricted to the detection of the outgoing products in nearly the forward hemisphere of the centre-of-mass (c.m.). This experimental context offers the best conditions for selecting a large fraction of one of the most massive participant zone (cf. the participant-spectator Model) producible and in which a large amount of incident kinetic energy has been stored. In connection with the duality mentioned above, one aspect concerns the baryonic entropy per nucleon (S/A) created in the system, thus probing its degree of disorder and of equilibration. The other aspect relates to the flow - nearly azimuthally balanced in the highly central collisions since directed emissions in and perpendicularly to the reaction plane merge - about which very little is presently known. In both cases temperature values have been extracted and are compared.

THE PARTICIPANT SOURCE - MULTIPLICITY AND ANGULAR DISTRIBUTIONS - GLOBAL VARIABLES

It is essential in the two topics discussed here to isolate a single source as massive as possible to insure the high centrality of the collisions and to guarantee a large number of participant nucleons and a large amount of deposited energy, i.e. to create the hottest and most compressed possible participant zone as well as the best chance for thermal equilibrium. Different manners to prepare a participant source formed in reactions at various impact parameters and particularly in highly central collisions have been discussed previously.[7] In order to select the lower b reactions we have used the so-called ERAT5 or (PM5-D ≤ 0.2) criteria.[7, 9] In the second case however the additional condition $(0 \leq y / y_p \leq 0.6)_{c.m.}$ on the rapidity y - rescaled to the projectile y_p rapidity - was superimposed. This extra condition simply eliminates some products resulting from residual projectile-like decay.[10]

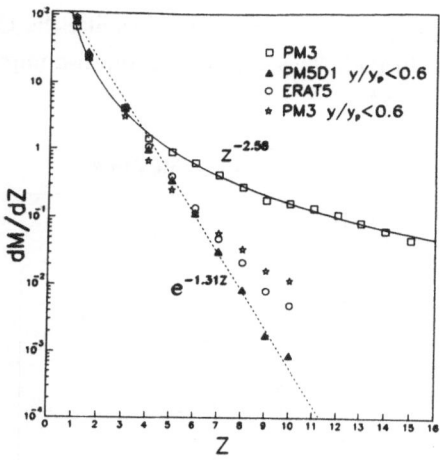

Figure 1. dM/dZ multiplicity distributions extrapolated to 4π measured for the Au + Au reaction at 400 A MeV. The different distributions correspond to semi-central (PM3) and violent (PM5-D1, i.e. D < 0.2 and ERAT5) collisions.[7] The effect of the extra-cut $y/y_p < 0.6$ is shown for PM3. The solid curve and the dashed line are, respectively, a power law $Z^{-2.58}$ and an exponential $e^{-1.31Z}$ fit to the data.

An example of multiplicity distributions dM/dZ measured at 400 A MeV is presented in Fig. 1. They are multiplicities extrapolated to the complete Au + Au system by multiplying them by the ratio of the total number of protons (158) to the measured total charge Z_{tot}. It shows that the distribution trend changes considerably from a power law $Z^{-\alpha}$ to an exponential $e^{-\sigma Z}$ when b decreases from around 10 fm to about 2 fm and less (PM3 to PM5-D $\leq$ 0.2). One also notices the remarkable fact that when the additional cut on the rapidity is applied (e.g. on PM3) only a slight reduction of the multiplicities is observed for the lower Z's whereas the higher Z component originating from the projectile-spectator is eradicated. Thus all the distributions exhibit comparable fall-offs at least up to $Z \approx 7$. This demonstrates that if the participant zone is correctly determined the size of this latter is irrelevant to the determination of quantities like the multiplicity of all the emitted products M_{ap}, that of the IMF's M_f, the average charge of the fragments $\langle Z_f \rangle$ and the slope parameter σ. This is clearly illustrated in Fig. 2. where these variables plus Z_{tot} (charge of the participant zone) are displayed as a function of the centrality. One sees that Z_{tot} grows logically when b increases while M_{ap}, M_f and $\langle Z_f \rangle$ vary substantially (dots) in an expected way. However these latter, plus σ, remain nearly independent of the violence of the collision (open symbols) when the participant zone is properly selected by use of the rapidity cut. Let us observe that the dM/dZ distributions measured at 400 A MeV extend typically up to $Z \approx$ 10 while they reach $Z \approx$ 15 at 150 A MeV where the IMF multiplicity represents almost one fifth of the total particle multiplicity and the charge carried by the IMF's is nearly 30% of the total charge measured. This underlines the relevance of the IMF's in measurements relying upon yield distributions. The exponential fall-off characterising the low b distributions agrees well with the trend predicted by statistical decay models like the Quantum Statistical Model (QSM[11]) as will be shown further. This suggests a quasi-complete thermalization of the participant zone.

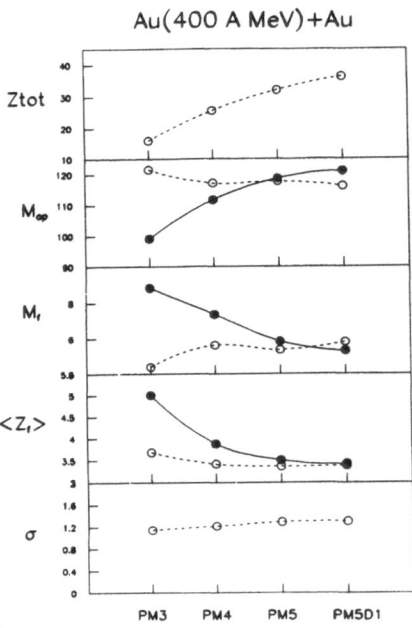

Au(400 A MeV)+Au

Figure 2. Trends of different variables as a function of increasing centrality (from PM3 to PM5D1, i.e. D < 0.2) with the extra condition $y/y_p < 0.6$ (open circles) and without it (dots). The variables are the total charge of the participant zone Z_{tot}, the multiplicity of all the emitted products M_{ap}, that of the Z > 2 fragments M_f, the average charge of the fragments $\langle Z_f \rangle$, and the slope σ of the exponential fall-off of the distributions.

Let us look now at the angular distributions measured for the various ejectiles. An example is presented in Fig. 3 showing the c.m. polar angle $dN/(\sin\theta\, d\theta)$ distributions. These distributions are for the case where the kinetic energy $E_{c.m.}$ of the fragments is less than 13.2 A MeV. Such a limitation allows a selection somewhat comparable to $(0 \le y/y_p \le 0.6)_{c.m.}$ and avoids the acceptance cut beyond $\theta_{lab} = 30°$ imposed by the *FOPI*-detector boundary.[7-10] However the distributions remain partly affected by acceptance effects. The most marked is due to shadowing around $\theta_{c.m.} = 20°$ and 65° and appears as minima in the figure. Nonetheless, even after correction the trends observed here remain. One sees that while the lighter products (Z = 1, 2) feature a grossly isotropic behaviour the heavier Z's exhibit a broad 90° peaked distribution. This tendency is particularly marked for $Z \ge 6$. The effect gets accentuated when $E_{c.m.}$ is reduced and begins to show up for the lighter Z's. Such a comportment expresses the presence of a collective motion in the azimuthal plane and reveals the possible existence of a "pancake" shaped flowing-zone shortly after the two nuclei have collided at very low b's.

One deals here with the dual approach dynamics versus statistics since this interpretation of collective flow is *a priori* in conflict with that of the equilibrated source observed before. In the first case the multifragmentation results from the mechanical rupture of the system under strong compression, in the second, the IMF's are produced out of an equilibrated source according to thermodynamics laws. One way to solve such an apparent contradiction is to conceive the participant zone as a global promptly-achieved statistic equilibrium while its expansion is governed by a homogeneous bulk-motion, i.e. a flow with kinetic energy equally distributed among all flowing nucleons. We shall analyse this possible description through studies of the entropy per nucleon and transversal flow.

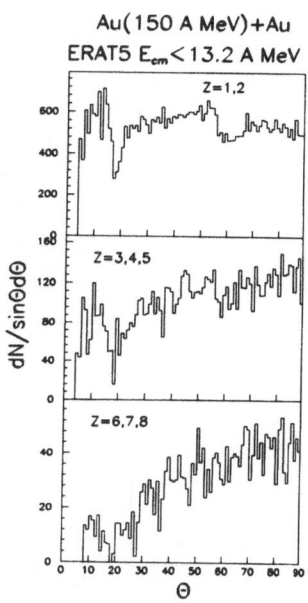

Figure 3. Centre-of-mass polar-angle dN/sinθdθ distributions of various fragments Z for the Au (150 A MeV) + Au central reaction (ERAT5). The kinetic energy of the fragments is limited to $E_{c.m.}$ < 13.2 A MeV.

THE BARYONIC ENTROPY PER NUCLEON

The entropy production has been discussed many times since the first suggestion of Siemens and Kapusta[12] to deduce S/A from the deuteron to proton yield ratio. Works have also shown that entropy is almost entirely created shortly after the maximal compression of the system has been reached.[13] It remains nearly unaffected over the subsequent expansion phase and for this reason it is viewed as a global variable, measurable late in the collision, but carrying the signature of the early state of the reaction. Here only the baryonic entropy has been measured whereas meson production also contributes to entropy. This latter component however is low at the energies considered here, a point which has been discussed elsewhere.[11, 14]

The S/A values were extracted only for the participant zone for which a statistical treatment is applicable ; we have used QSM.[11] In principle, any statistical analysis should be appropriate and a fair number of these models has been made available. However, this particular model makes explicit and extensive predictions about S/A and has been much used in the past. Consequently, it is worth using it to analyse the effect of the IMF inclusion by comparison with earlier works. Furthermore, since the experimental dM/dZ distributions are the main input in this study, as a pre-requisite the models which are used should reproduce correctly the data. This is the case for QSM but not for some other models. The main drawback is that Monte Carlo calculations cannot be performed with the available form of QSM, thus filtered simulations cannot be done and the experimental data have to be extrapolated to the 4π geometry.

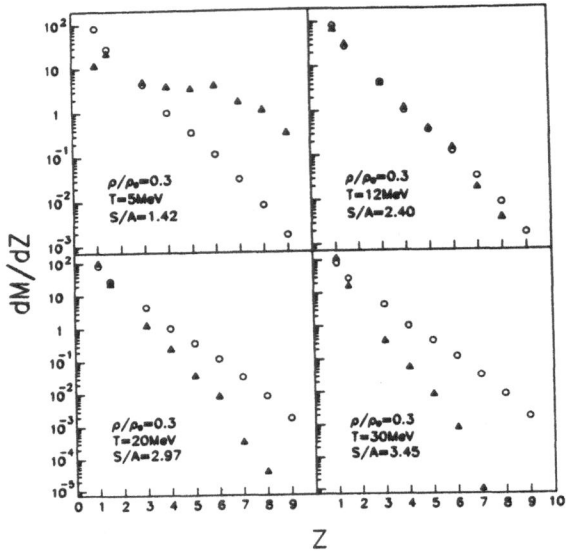

Figure 4. Comparisons between experimental (open circles) and calculated (full triangles) multiplicity distributions for the Au + Au reaction at $400\,A$ MeV. The data are for the participant zone selected with (PM5-D < 0.2 plus y/y_p < 0.6) and extrapolated to 4π. The calculated dM/dZ's are from QSM for a break-up density $\rho/\rho_0 = 0.3$. The corresponding entropy S/A and temperature T values are reported.

An extensive description of QSM[11] and of the exact procedure which is presently applied[14] to extract S/A may be found in the literature. Let us simply recall that the entropy can be calculated from the yields of the various species (i.e. dM/dZ distributions) formed in the reaction. Thus S/A may be determined from various parameters characterising the distributions like M_{ap}, M_f - both extrapolated to 4π -, Z_f and the slope σ of the distributions. The consistency of the values obtained from different methods implies that the whole experimental distributions fully agree with that calculated with QSM. In this analysis the temperature T and the break-up density ρ - re-scaled to the saturation density ρ_0 - are parameters. For each choice of ρ or T a set of values (ρ, T and S/A) may be determined from the best agreement between the calculated yield distributions and the experimental dM/dZ, hence a unique value of S/A cannot be found.

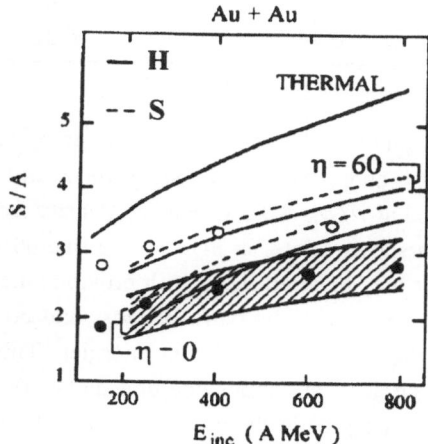

Figure 5. Baryonic entropy per nucleon S/A excitation function determined (dots) for the Au + Au reaction by averaging values obtained with a break-up density ρ/ρ_0 0.1 and 0.9. The hatched region represents the overall limits of S/A when the break-up density is equal to these lower and upper values. Results of Doss et al[23] are also reported (open circles). The curve labelled THERMAL is from a Fireball type calculation. The solid and the dashed curves are from Schmidt et al[13] for a hard (H) and a soft (S) equation of state, respectively, and for two different values of the shear viscosity coefficient η (expressed in MeV/fm^{2}c) indicated in the figure.

Examples of comparison between experimental and calculated dM/dZ distributions are given in Fig. 4 for the case, at 400 A MeV incident energy, where $\rho/\rho_0 = 0.3$ and T is varied between 5 and 30 MeV ; the corresponding S/A values are also reported. One sees that quite a satisfactory agreement is obtained for $T = 12$ MeV. From the comparison of the different cases shown in Fig. 4 one may infer the accuracy of the method to determine the temperature and the entropy once the density is fixed. Furthermore, by comparing the upper-right and the middle-left frames of Fig. 4 one concludes that limiting the analysis to light Z's (i.e. restricting the good agreement between calculations and data only to Z = 1 and 2) may lead to incorrect values of T and S/A. This underlines again the relevance of the IMF's in the baryonic entropy determination.

As proposed by QSM the entropy has been extracted from the M_{ap}, M_f, $\langle Z_f \rangle$ and σ experimental values by varying in the calculations the density ρ/ρ_0 by steps of 0.1 unit. The S/A results, summarised in Table I, show the mean values $\langle S/A \rangle$ obtained by averaging the values deduced at $\rho/\rho_0 = 0.1$ and at 0.9 - these values being themselves the average values of those determined from the four different variables - at the various bombarding energies. The $\Delta(\langle S/A \rangle)$ values correspond to the overall limits of $\langle S/A \rangle$ when

Table 1. Average entropy per nucleon $\langle S/A \rangle$ in the Au + Au reaction and its overall limits $\Delta(\langle S/A \rangle)$ when the break-up density ρ/ρ_0 is varied between 0.1 and 0.9 at the different bombarding energies reported.

Energy (A MeV)	150	250	400	600	800
$\langle S/A \rangle$	1.85	2.21	2.41	2.60	2.69
$\Delta(\langle S/A \rangle)$	0.35	0.34	0.36	0.36	0.36

ρ/ρ_0 is varied between 0.1 and 0.9. The S/A values change by less than 0.1 unit when the density remains between 0.1 and $0.5\rho_0$, and $\langle S/A \rangle$ turns out to be very close to the particular entropy determined at $\rho/\rho_0 = 0.3$ which is a density frequently speculated.[15] The $\langle S/A \rangle$ excitation function is shown (dots) in Fig. 5. The overall limits of $\langle S/A \rangle$ define the hatched area in the figure. It may be confronted with previous measurements,[16-23] the most straightforward comparison being with the work of Doss and co-workers[23] devoted to the Au + Au system at almost the same set of incident energies. These particular values - also indicated in Fig. 5 as open circles - are about one unit higher that those measured here. In fact, the entropies found here are practically lower than those found in any earlier studies. A thorough explanation is difficult to work out because, even when the same theoretical model has been used, the experimental approaches are generally different.

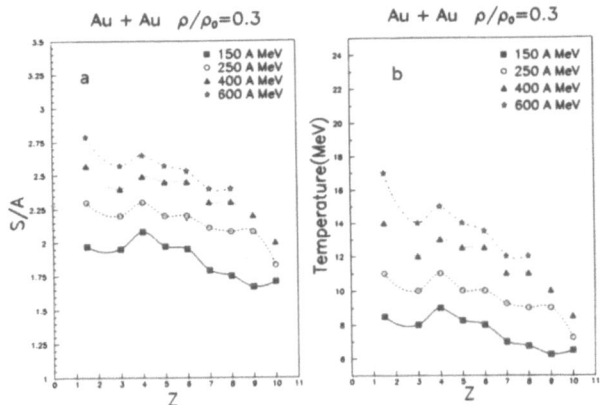

Figure 6 - a) Baryonic entropy per nucleon S/A determined for individual Z for a break-up density $\rho/\rho_0 = 0.3$ for the Au + Au reaction at various bombarding energies reported in the figure. A unique average value has been reported for $Z \leq 2$. - b) Temperatures extracted for individual Z for a break-up density $\rho/\rho_0 = 0.3$ for the Au + Au reaction at various bombarding energies reported in the figure.

In the comparison two inputs are particularly relevant : the "quality" of the participant-zone selection and the specific particle yields used to determine S/A. Not to speak of the inclusive measurements where the b averaging leads evidently to include ejectiles from different origins, it is clear that the way of selecting the participant zone does matter and changes the extracted entropy.[23] In the present work we have shown, with the help of Fig. 2, that the variables used to determine S/A remain nearly independent of the centrality, therefore the *same entropy values are found irrespective of the participant zone size*. The influence of the IMF inclusion has already been discussed before,[21, 22] and was recognised as lowering the extracted entropy. The S/A values measured here for the individual fragment Z for $\rho/\rho_0 = 0.3$ is plotted in Fig. 6-a. The entropy seems to decrease slightly as Z rises but in view of the relative uncertainties attached to the Z yields S/A may be regarded as nearly constant. If indeed a decrease exists it is much less marked than observed before[22] : There, the departure between the entropy found with Z = 1 and 2 and with Z > 2 was estimated - from an analysis quite similar to the present one in the C (100 A MeV) + Au reaction - to be of the order of one unit. Furthermore, this work showed also that the entropy deduced from IMF's stays constant over the 30-100 A MeV domain. Thus it was concluded that the IMF entropy may be a kind of "universal" value possibly connected to invariant temperature and break-up density for clusters. In opposition to this, Fig. 6-a shows that the S/A dependence on Z remains grossly the same at all the energies and that the entropy deduced only from the IMF's increases from about 2 up to 2.6 between 150 and 600 A MeV. Such a result nuances somewhat the earlier thinking that IMF's being formed late after the light particle emission the disorder is spread over a larger fraction of the system, hence a smaller entropy. One should keep in mind however that these earlier conclusions were for incident energies where the compression effects are inexistent.

Coming back to Fig. 5, one may now compare the excitation function measured here with some theoretical calculations. One sees that the thermal Fireball-type calculation fails totally to reproduce the experimental results by overestimating them. This clearly shows that a fraction of the excitation energy retained in the participant zone is in the form of collective energy. The results may be successfully compared with fluid-dynamic calculations[13] incorporating viscous nuclear forces. An ensemble of curves corresponding to hard (H) and soft (S) equations of state with and without shear viscosity (η) is set out in Fig. 5. The comparison shows a low dependence of S/A on the incompressibility coefficient K - although a hard equation of state seems more appropriate - and suggests a low viscosity.

The temperature values obtained from this analysis for each individual Z are presented in Fig. 6-b. On the average they increase from 8 MeV up to about 14 MeV over the energy domain studied. Of course, they exhibit the same type of relationship with Z as the entropy. These temperature values are consistent with a system holding between one fourth and one third of the total kinetic energy put into the system as thermal energy.

TRANSVERSAL FLOW - THERMAL AND COLLECTIVE TRANSVERSE ENERGY PARTITION

Let us examine now the collective motion governing the participant zone. This is one of the most striking phenomenon susceptible to yield information about the stiffness and the viscosity of nuclear matter. It has been extensively studied at intermediate impact parameter reactions and is shown to be accentuated in the reaction plane and perpendicular to it.[1-3] The difficulty for selecting only very low b collisions and the fact that most of the methods to measure the flow requires the use of the reaction plane[24] explain perhaps why very little is known today about the transversal flow : Indeed to study it the condition of high centrality is mandatory and the transversal flow being azimuthally balanced no specific plane may be defined. Probably for these reasons, extented studies of this process have started only very recently.[25-32]

The $dN(V_t)/(V_t\,dV_t)$ transverse velocity distributions, scaled to light velocity (c), measured at $150\,A$ MeV with the cuts $(PM5\text{-}D \leq 0.2) + (0 \leq y/y_p \leq 0.6)_{c.m.}$ are shown in Fig. 7. These distributions feature a broad structure whose width decreases as Z increases. They do not present a Boltzmann behaviour as one would expect from a pure thermal source emission. Indeed corresponding to a functional

$$f\left(V_t^{th}, V_l^{th}\right) = \left(\frac{\eta}{\pi}\right)^{3/2} V_t \int_0^{2\pi} e^{-\eta\left(\left(V_t^{th}\right)^2 + \left(V_l^{th}\right)^2\right)} d\varphi \qquad (1)$$

- where V_t^{th} ans V_l^{th} are the transverse and longitudinal velocity of the thermal source, $\eta = M/(2kT)$, M is the mass of the particle, k is the Boltzmann constant and T the temperature - the higher the $Z = M/2$ the narrower should be their width. This seems to be actually the case, but they should also all be peaking at $V_t = 0$. In fact, even though these distributions are partly distorted by lower energy threshold and acceptance effects of the FOPI detector, they depart markedly from a Boltzmann distribution by a shift of their maximum to higher velocity as Z increases. Such a trend may be explained by the presence of an extra velocity added to V_t^{th}, hence the idea of a collective motion superimposed.

In order to explore this approach - since in $b \approx 0$ reactions the flow may be considered to be nearly oriented into the azimuthal plane where it should be homogeneously distributed - we have calculated[28] the transverse velocity distributions by assuming a flow restricted to the azimuthal plane $(V_l^{flow} = 0)$ and of constant module transverse velocity $|V_t^{flow}|$ for each given Z. The distributions may then be expressed as

$$f\left(V_t, V_l\right) = \left(\frac{\eta}{\pi}\right)^{3/2} V_t\, e^{-\eta\left(V_l^2 + V_t^2 + \left(V_t^{flow}\right)^2\right)} 2\pi I_0\left(2\eta V_t\, V_t^{flow}\right) \qquad (2)$$

with $\vec{V}_t = \vec{V}_t^{th} + \vec{V}_t^{flow}$ and where I_0 is a Bessel function.

A total of 5.10^4 filtered emissions has been simulated with Monte Carlo calculations for each given Z. The direction of $\vec{V}_t^{th}$ was generated randomly as well as that of $\vec{V}_t^{flow}$ with respect to $\vec{V}_t^{th}$ and the experimental biases characterising the FOPI detector have been introduced. These calculated distributions, in which T and $\vec{V}_t^{flow}$ are parameters, have been fitted to the data. The best fits determined from the minima of the χ^2 distributions correspond to the solid curves in Fig. 7, and yield quite a satisfactory agreement.

The analysis is summarised in Fig. 8 which shows the $|V_t^{flow}|$ values found for the different Z. It evidences that the velocity modules exhibit only a weak decrease when Z rises. Only lighter Z feature somewhat larger velocities, but it should be noticed however that they correspond to the cases where $|V_t^{flow}|$ is the most difficult to determine due to the least difference between the simulated velocity spectra with and without $\vec{V}_t^{flow}$. Furthermore $Z = A/2$ is an oversimplified assumption for $Z = 1$ and 2. The results suggests that all fragments move with nearly the same velocity close to $0.12c$ and $0.23c$ at 150 and 400 A MeV, respectively. This substantiates the concept of a flow of particles and fragments all having the same velocity, thus generating a prompt expansion of the participant zone. This corroborates other analyses [30], [31] considering however an expansion not restricted to the transverse direction. In this work the Coulomb repulsion has been ignored. If included, it would generate a repulsive velocity at the expense of compression effect. However, this effect shoud be moderate and stronger for lighter Z's than for heavier.[31]

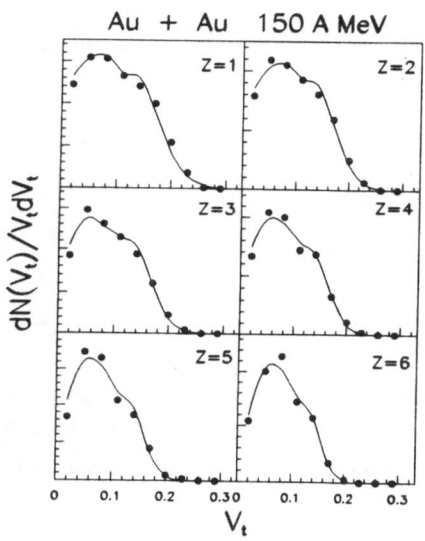

Figure 7. Invariant transverse velocity distributions $dN(V_t) / (V_t dV_t)$ expressed in c unit for various fragments measured (dots) in the Au + Au reaction at 150 A MeV. The curves are the best fits to the data obtained with filtered Monte Carlo calculations as explained in the text.

The total E_t/A and flow E_t^{flow}/A transverse-kinetic energies per nucleon, deduced from the velocities, are presented in Fig. 9-a for the two bombarding energies. They show a moderate decrease with rising Z which saturates from Z ~ 3 onwards. The E_t^{flow}/A values amount to about 75% of E_t/A irrespective of Z. This demonstrates the importance of the collective motion in the azimutal plane thus creating a kind of blast of nuclear matter in this plane. The present results are compatible with the possible existence of a sidewards boosted source or a transversally expanding source, both scenarios leading to the same transverse velocities.[25, 29]

The slope parameter T values are shown in Fig. 9-b, they look fairly independent of Z. If they may be considered truly as temperature, they are much higher than the values previously extracted from the entropy analysis. The difficulty to fit accurately the fall-off of the distributions on which T depends directly must by itself lead us to take them with great prudence. On the other hand finding high temperatures is not so surprising since one is looking at a regime not implying a complete thermalisation. Similar discrepancies have been observed before, for example when measuring the temperature from the excitation and from the kinetic energies of the fragments.[33] The description of the flow presently used may be also too restrictive and a radial flow instead of a transversally oriented flow should perhaps be considered [31, 32], which would lead to lower T values as would the accurate account

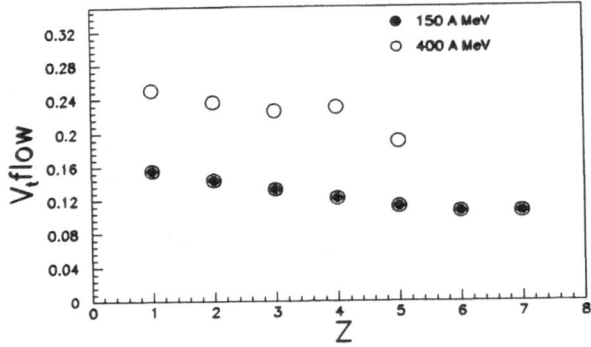

Figure 8. Transverse flow velocity V_t^{flow} deduced from the analysis described in the text plotted as a function of Z for the Au + Au reaction measured at 150 and 400 A MeV. The V_t^{flow} values show a weak dependance on Z suggesting that all products experience the same collective velocity.

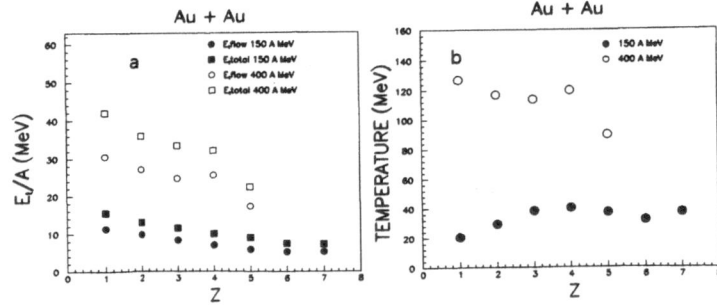

Figure 9 - a. Total- (squares) and collective- (circles) transverse energies, deduced from the analysis described in the text, plotted as a function of Z for the Au + Au reaction measured at 150 and 400 A MeV. **- b** Temperatures deduced from the same experimental conditions as a function of Z.

of Coulomb effects. Several elucidations are required about the applicability of thermodynamics to finite and relatively small nuclear systems, and about the magnitude of the temperature fluctuations before reliable temperature determination become possible. An interesting regime[29,] to explore would be to consider the participant zone as a large ensemble of thermal sources, each characterised with its own chemical potential and temperature and moving with different velocities, which would destroy the concept of a unique boost velocity but would lead to a velocity distribution from which the extracted global temperature is different from that of a fully thermalised system.

CONCLUSION

We have attempted to show via the analysis of the Au + Au reaction between 150 and 800 A MeV that the full account of all the particles and fragments emitted in the reaction is highly recommended for studying the properties of hot and dense nuclear matter. When this is realised it is possible to select in an effective way the participant zone formed in highly central collisions, corresponding to the largest region of excited and compressed nuclear matter. On this basis we have examined two phenomena reflecting opposite behaviour of the reaction. On the one hand the equilibrated character was probed by measuring the entropy created in the system. The inclusion of IMF's leads to values of S/A lower than those previously found - from essentially lighter Z - without showing two radical "regimes" of entropy for lighter and heavier ejectiles as advanced before. The entropy excitation function may be described by hydrodynamic calculations and suggests a low viscosity of nuclear matter. The corresponding temperatures are in agreement with an equilibrated system in which about only one third of the stored energy is thermal. On the other hand a dynamic aspect has been analysed in the case of nearly $b = 0$ reactions by fitting the transverse velocity distributions with filtered Monte Carlo calculations supposing that a boost transverse-velocity is randomly combined with the thermal transverse-velocity. The results show that all particles and fragments are experiencing nearly the same extra transverse velocity. They are compatible with an explosion-like type reaction in which the ejected nuclear matter is transversally propelled before promptly reaching equilibrium.

REFERENCES

1. High Energy Nuclear Collisions, The GSI-LBL Collaboration at the Bevalac Papers 1975-1987, H.H. Gutbrod, A. Sandoval, and R. Stock, ed., GSI Report 87-10 (1987)
2. Proceedings of a NATO Advanced Study Institute on the Nuclear Equation of State, Peniscola, Spain, ed. W. Greiner and H. Stöcker, NATO ASI Series B : *Physics* 216 A and B (1989).
3. Relativistic Heavy Ion Physics, Int. Rev. of Nucl. Phys., L.P. Csernai, and D.D. Strottman, ed., Publisher, World Scientific (1991).
4. D. L'Hote et al., *Nucl. Phys.* A519, 331c (1990) and references therein.

5. G. Westfall in this Institute.

6. J. Peter in this Institute.

7. K.D. Hildenbrand in this Institute.

8. A. Gobbi and the *FOPI* Collaboration at GSI, *Nucl. Inst. Meth.* A324, 156 (1993).

9. J.P. Alard and the *FOPI* Collaboration at GSI, *Phys. Rev. Lett.* 69, 889 (1992).

10. C. Kuhn, PhD thesis, Strasbourg (1993), reprint CRN 93-20.

11. D. Hahn and H. Stöcker, *Nucl. Phys.* A476, 718 (1998).

12. P.J. Siemens and J.I. Kapusta, *Phys. Rev. Lett.* 43, 1486 (1979).

13. W. Schmidt et al, *Phys. Rev.* C47, 2782 (1993).

14. C. Kuhn and the *FOPI* Collaboration at GSI, *Phys. Rev.* C (in press).

15. See for example J.P. Bondorf et al., *Nucl. Phys.* A443, 321 (1985) and *Nucl. Phys.* A444, 460 (1985)

16. S. Nagamiya et al., *Phys. Rev.* C24, 971 (1981).

17. H.H. Gutbrod et al., *Phys. Lett.* 127B, 317 (1983).

18. K.G.R. Doss et al., *Phys. Rev.* C32, 116 (1985).

19. B.V. Jacak et al., *Phys. Rev. Lett.* 51, 1846 (1983), *Phys. Rev.* C29, 1744 (1984) and *Phys. Rev.* C35, 1751 (1987).

20. R. Wada et al., *Phys. Rev. Lett.* 58, 1829 (1987).

21. L.P. Csernai et al., *Phys. Rev.* C38, 2681 (1988).

22. R. Trockel et al., Phys. Rev. C38, 576 (1988).

23. K.G.R. Doss et al., Phys. Rev. C37, 163 (1988).

24. P. Danielewicz and G. Odyniec, *Phys. Lett.* 157B, 146 (1985).

25. P. Danielewicz and Q. Pan, *Phys. Rev.* C46, 2002 (1992)

26. N. Herrmann for the *FOPI* Collaboration, *Nucl. Phys.* A553, 739c (1993).

27. T. Wienold et al., GSI Scientific Report 1992, GSI 93-1, p. 35.

28. C. Kuhn and the *FOPI* Collaboration, Proc. of the XXXI Int. Winter Meeting on Nucl. Phys., Bormio (Italy), 1993, p. 59.

29. J. Konopka, H. Graf, H. Stöcker and W. Greiner, Proc. of the XXXI Int. Winter Meeting on Nucl. Phys., Bormio (Italy), 1993, p. 192.

30. B. Kämpfer and the *FOPI* Collaboration, *Phys. Rev.* C48, R955 (1993)

31. S.C. Jeong et al., GSI Scientific Report 1992, GSI 93-1, p. 34 ; and *Phys. Rev. Lett.* (submitted).

32. W. Bauer et al., *Phys. Rev.* C47, R1838 (1993).

33. For a discussion see A.R. Deangelis and A.Z. Mekjian, Int. Rev. of Nucl. Phys., L.P. Csernai, and D.D. Strottman, ed., Publisher, World Scientific (1991), Chap. 6, p. 363.

RELATIVISTIC NUCLEAR FLUID DYNAMICS

L.P. Csernai, L.B. Bravina, T. Csörgő[1], P. Lévai[1],
D.D. Strottman[2] and E.E. Zabrodin

Centre for Theoretical Physics, University of Bergen
Allegaten 55, N - 5007 Bergen, Norway

1 INTRODUCTION

This lecture is the third at this NATO Advanced Study Institute on the subject. After the previous lectures of Joachim Maruhn and Mark Gorenstein in this lecture we will concentrate on two subjects, strongly connected with the formation of Quark Gluon Plasma (QGP).

The first subject is the study of fluid dynamical solutions in the presence of a first order phase transition to QGP, while the second is the description of the final hadronization and freeze out if QGP was present in earlier stages of the heavy ion collision.

The introductory part as well as the details of the QGP equation of state is based on the textbook "Introduction to Relativistic Heavy Ion Collisions" by L.P. Csernai (J. Wiley and Sons, 1994, [1]). The discussion of measurable consequences of QGP formation and hadronization is based on recent and present research achievements and progress.

We will discuss local equilibrium situations unlike in microscopic string and parton cascade models and unlike in multifluid dynamical approaches. If QGP is formed, the equilibration is extremely fast and the assumption of local equilibrium is applicable except the very initial stages of the reaction. In some cases we will also use the Quark Gluon String Model (QGSM) for comparison and to test the observability of some typical fluid dynamical effects in a microscopic model.

1.1 Flow Characteristics from Numerical Solutions

Let us first see some examples for the solutions of the relativistic Euler equations. In the collaboration between Bergen, Budapest and Los Alamos recently we performed calculations for AGS and planned CERN experiments, Au+Au at 14 A·GeV and Pb+Pb at 160 A·GeV beam energy. The PIC method, developed in Los Alamos, was used in these calculations. This method takes advantage of both the Eulerian and Lagrangian

---

[1] KFKI-Res. Inst. for Particle and Nuclear Physics, Budapest, Hungary
[2] Los Alamos National Laboratory, Los Alamos, New Mexico, USA

Hot and Dense Nuclear Matter, Edited by
W. Greiner *et al.*, Plenum Press, New York, 1994

methods of the solutions of the fluid dynamical equations. The fluid is divided to Lagrangian fluid elements, which are then represented by so-called marker particles which move in a calculational grid. This ensures very accurate energy, momentum and baryon number conservation. The transport among the grid cells is quantized according to the motion of the marker particles.

Recent, developments were, however, necessary in the method. At ultra-relativistic energies the initial compression is rather large, $\approx 2\gamma_{c.m.}n_0$, and so a relatively fine grid with $\Delta z <$ 1fm should be introduced. At late expansion stages, however, the matter becomes dilute and the small grid cells will contain very few marker particles if any. Consequently at late stages the method becomes inaccurate. This was recently remedied by smoothing out the transport carried by the marker particles and introducing a regridding for the accurate description of large changes in the marker particle density or matter density which is necessary at late, dilute stages of the collision.

1.1.1 Pb+Pb Collisions at 160 A·GeV Energy

In a 160 A·GeV Pb + Pb reaction the density increases in relatively sharp fronts of widths of a few fermi. The reason is that the speed of the incoming projectile and consequently the flow velocity is *supersonic*. The sound speed in (ground state) nuclear matter is about $v_{sound} \approx 0.2c$. The average increase of the proper baryon density may reach $< n >_{Vol.} /n_0 \approx 20$, and the lifetime of the hot dense matter is about 4-6 fm/c.

If a phase transition is included then the phase transition takes place in these fronts also. About 50 fm^3 of QGP is formed in a reaction. The collision time and baryon density increase is about 20-30% larger if the phase transition is present. At the final expansion in such cases deflagration fronts may develop where the final rehadronization takes place. The dynamics of the rehadronization will be discussed later. It is interesting to mention that in the Bjorken scenario, applicable at RHIC and LHC energies, the reaction time is much longer, in the order of 50 fm/c if phase transition is present due to the one dimensional nature of the expansion.

1.1.2 Au+Au Collisions at 11.6 A·GeV Energy

In these reactions the average baryon density increases to 10 (7) n_0 during a central collision with a QGP (Hadronic) Equation of State (EOS), and the break-up time is around 9-11 (7-8) fm/c respectively. *I.e.*, the phase transition leads to an increase of the collision time by about 30-40%!

The baryon rapidity distributions indicate strong stopping for both EOS's, but their dependence on the break-up time and on the EOS is about the same. Thus, the baryon rapidity distribution is not an obvious signal for the phase transition. The average transverse momentum for baryons, $< p_t/A >$, is about 1 GeV/c at all rapidities. It does not depend strongly on the EOS. The pion rapidity distribution is even less of a phase transition signal, because in the fluid dynamical model we consider thermal pions at their break-up only. This is because of the extremely strong temperature dependence of the thermal pion spectra.

In the QGSM the rapidity distributions of protons and negatives indicate similarly strong stopping. The spectra are even more peaked that in the fluid dynamical model.

It should not surprise us that the collective fluid dynamical effects are exhibited much more clearly at finite impact parameters. We have studied Au+Au reactions at b=5 fm. The average density increase is smaller now due to the spectators: 5.5 (4) n_0

for QGP (Hadronic) EOS, see Fig. 1 in ref. [2]. The amount of pure QGP formed is negligible (1 fm^3).

We can characterize the collective flow in the x (the reaction reaction plane is $[x, z]$), y and z (beam) directions by $F_j = \sum_{cells} p_j p_j$, where $j = x, y, z$. The longitudinal flow initially decreases strongly, at the maximum compression it is only about 10% of the initial value, then it rises back to about 30-40% depending on the EOS. In the reaction plane the flow F_x is about 25% larger for the hadronic EOS at its maximum.

The strongest and most clearly observable effect is in the Squeeze-out (y-) direction where the flow, F_y is **twice as large** for the hadronic EOS than for QGP EOS! (see ref. [2]).

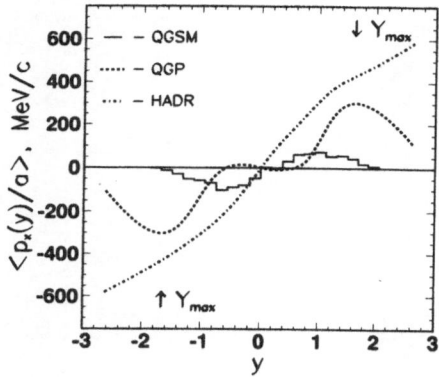

Figure 1: Average transverse flow, $< p_x/A >$ versus Y_{cm} for $Au + Au$ collisions at 11.6 A·GeV obtained in QGSM with $b = 3$ fm (histogram), and in FDM for $b = 5$ fm with Hadronic (dashed line) and QGP EOS (dashed-dotted line).

The rapidity distribution is again not strongly dependent on the EOS, although it indicates strong stopping. A double hump structure representative of the slowed down spectators is more apparent for the hadronic EOS.

The transverse flow analysis, $< p_x/A >$, is also a sensitive indicator of the EOS. $< p_x/A >$ is much smaller than $< p_t/A > \approx 1$GeV/c. For the Au+Au, 11.6 A·GeV reaction at b=5 fm $< p_x/A > \approx 500$, (350)MeV/c for Hadronic, (QGP) EOS at the target or projectile rapidity. In the QGSM the transverse flow is smaller, but still a large and observable effect.

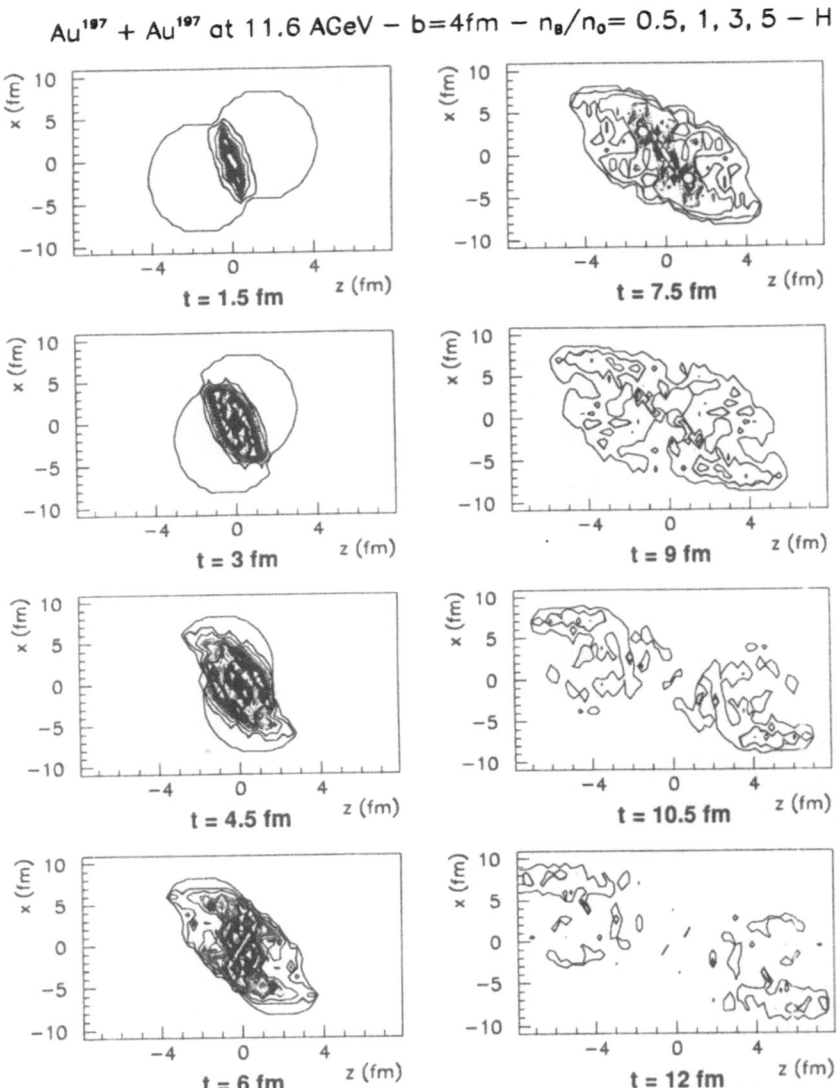

Au^{197} + Au^{197} at 11.6 AGeV – b=4fm – n_B/n_0= 0.5, 1, 3, 5 – H

Figure 2: Contour plots of the density distribution in the reaction plane calculated in the fluid dynamical model with hadronic EOS for Au+Au collision at 11.6 A·GeV. The impact parameter is b=4fm.

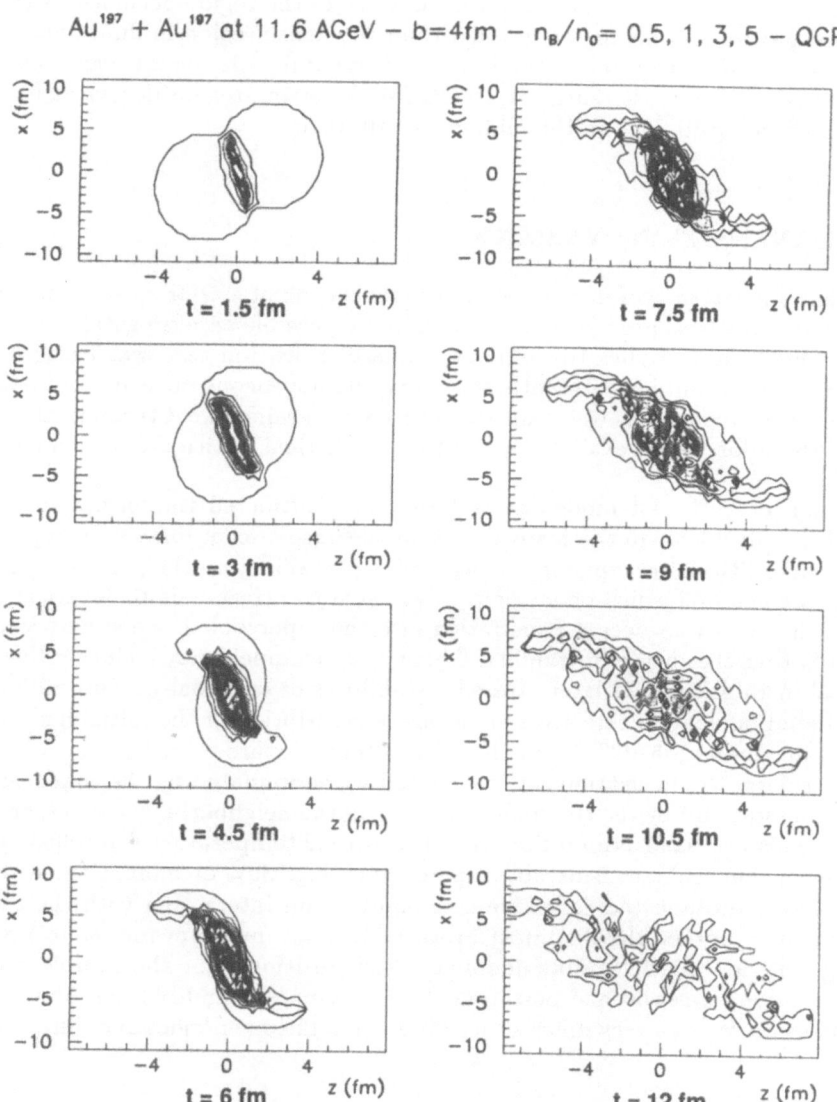

Figure 3: The same as the previous figure with QGP Equation of State. The higher initial compression and the slower final expansion both in longitudinal and in transverse directions are clearly observable.

The space-time pattern of a collision is different in the two models, and in the fluid dynamical model it strongly depends on the EOS. Due to the higher compression with QGP the area of the most dense region is almost half of this area with the hadronic EOS. On the other hand the lifetime of the dense matter is about 60-80% longer (see Figs. 2 and 3).

In the QGSM we studied the space-time pattern of the region occupied by strings or non-formed particles. This region is wider than the dense region in fluid dynamics and it decays in a different pattern. While in fluid dynamics the dense region disappears along t=const. or τ=const. surfaces, in QGSM the string-region decays at its outside surface surviving quite long in the middle, ca. 10 fm/c.

2 PHASE TRANSITION DYNAMICS

If the phase transition is of first order the hadronization of a QGP is an interesting and complicated dynamical process in itself, with many possible experimental consequences. In phenomenological studies the estimated phase transition rate was based on crude dimensional considerations. Recently the rate of homogeneous nucleation of relativistic first order phase transitions was evaluated in a coarse grained field theoretical approach, where all the information available so far from numerical lattice QCD calculations was utilized.[3]

As usual, only the 1st mode of instability was considered the formation of spherical bubbles or droplets. In the many dimensional space of coordinates corresponding to possible instabilities after crossing the borderline of stability on the phase diagram there is always one channel which opens first. This usually corresponds to spherical configurations of instability. A deeper penetration into the supercooled region may lead to the opening of other channels of instability. These other channels may include string-like, or cylindrical instabilities and later layered instabilities or spinodal decomposition. Thus the calculated nucleation rate gives an accurate description of the initial hadronization at small supercooling, 5-10% below the critical temperature.

Furthermore, the nucleation rate calculated was dominated by thermal near - equilibrium processes and by the thermal interaction of the neighboring particles, or thermal damping. This is a valid assumption when the critical temperature is reached, but after further expansion and a considerable supercooling, e.g. 30% or more, the matter is not so dense any more and the collective near-equilibrium interaction with the surrounding matter may not be the dominant process. Instead quantum mechanical processes including very few particles may dominate the transition. For the complete study of the reaction dynamics and hadronization all these aspects should be considered and the final conclusion may be very different for different collision energies and different nuclei.

2.1 Nucleation Rate

The rate for the nucleation of the hadronic phase out of the plasma phase can be written as [3]

$$I = I_0 \, e^{-\Delta F_*/T}, \tag{1}$$

where ΔF_* is the change of the free energy of the system with the formation of a critical size hadronic bubble and I_0 is the prefactor. In general, statistical fluctuations at $T < T_c$

will produce bubbles with associated free energy

$$\Delta F = \frac{4\pi}{3}[P_q(T) - P_h(T)]R^3 + 4\pi R^2\sigma. \tag{2}$$

Here P is the pressure of the quark or hadron phase at temperature T, and σ is the surface free energy of the quark-gluon/hadron interface. Since $P_q - P_h < 0$ it follows as usual that there is a bubble of critical radius $R_*(T) = \frac{2\sigma}{P_h(T)-P_q(T)}$. Larger bubbles will grow, while smaller bubbles tend to shrink because the surface energy is too large relative to volume energy. The free energy of the critical size bubble is therefore $\Delta F_* = \frac{4}{3}\pi\sigma R_*^2$. If we define the free energy density at a given temperature for all values of the energy density not only for the stable and metastable equilibrium energy density values, we can discuss many nonequilibrium phenomena. One can determine the free energy of the critical droplet as a functional of the energy density profile of the droplet. This already enables us to estimate the probability of having such a critical instability in a unit volume of supercooled QGP due to statistical fluctuations in supercooled QGP.

Then, we can study the dynamics of such a critical droplet, determined by dissipative processes. This will give us a time-scale for the formation of such a critical droplet. The statistical fluctuations and the time-scale together determine the prefactor in the rate of nucleation.

2.1.1 Baryon-free Plasma

This prefactor has very recently been computed in a course-grained effective field theory approximation to baryon-free QCD[3]

$$I_0 = \frac{4}{\pi}\left(\frac{\sigma}{3T}\right)^{3/2}\frac{\sigma(\zeta_q + 4\eta_q/3)R_*}{\xi_q^4(\Delta w)^2}, \tag{3}$$

where η_q is the shear viscosity in the plasma phase, ξ_q is a correlation length in the plasma phase, and Δw is the difference in the enthalpy densities of the two phases. The nucleation rate is limited by the ability of dissipative processes to carry latent heat away from the bubble's surface, as indicated by the dependence on the viscosity. At the critical temperature, $R_* \to \infty$, and the rate vanishes.

2.1.2 Baryon-rich Plasma

Venugopalan and Vischer very recently extended this work for baryon-rich QGP.[4] They assumed that the correlation length, the droplet profile and the free energy of the droplet are not influenced considerably by the baryon content. By assuming finite baryon charge density they could allow for a temperature difference within the droplet compared to its surrounding.

Approximating the external temperature profile of the droplet by $T(r) = \Theta_0 R/r$ for $r > R$, they arrived at a prefactor

$$I_0 = \frac{4}{\pi}\left(\frac{\sigma}{3T}\right)^{3/2}\frac{\sigma(\lambda_q T/2 + \zeta_q + 4\eta_q/3)R_*}{\xi_q^4(\Delta w)^2}, \tag{4}$$

where λ_q is the heat conductivity in the plasma phase, and T is the average temperature $T = (T(0) + T(\infty))/2$.

2.2 Hadronization Dynamics in Heavy Ion Collisions

Now let us consider this hadronization in the presence of the fluid dynamical expansion. A standard picture [5] of a central collision at 100 A·GeV or above is that the two nuclei pass through each other, creating a hot plasma of quarks and gluons. This plasma subsequently cools by expanding hydrodynamically, mainly along the beam axis. Eventually the energy density becomes low enough that the quarks and gluons hadronize. If there is a first order thermodynamic phase transition the associated latent heat must somehow be gotten rid of before the hadronization can be completed. Usually an idealized Maxwell construction for two-phase equilibrium is invoked as a model of the hadronization process in fluid dynamical approaches. However, it is by no means clear that the QCD nucleation rate is large enough for this idealization to be anywhere near reality.

Given the nucleation rate one would like to know the (volume) fraction of space $h(t)$ which has been converted from QCD plasma to hadronic gas at the proper time t, which is the time as measured in the local comoving frame of an expanding system. This requires kinetic equations which use I as an input. In [6] the dynamics of hadronization is described by the following dynamical equations. If the system cools to T_c at time t_c then at some later time t the fraction of space which has been converted to hadronic gas is

$$h(t) = \int_{t_c}^{t} dt' I(T(t'))[1 - h(t')]V(t', t).$$

Here $V(t', t)$ is the volume of a bubble at time t which had been nucleated at the earlier time t'; this takes into account bubble growth. The factor $1 - h(t')$ takes into account the fact that new bubbles can only be nucleated in the fraction of space not already occupied by hadronic gas. This conservative approach does not take into account collisions and fusion of bubbles, which would tend to decrease the time needed to complete the transition.

The growth of hadronic bubbles has been studied numerically with relativistic hydrodynamics by Miller and Pantano [7]. After applying a perturbation, the bubble begins to grow. As the radius increases, the surface curvature decreases, and an asymptotic interfacial velocity is approached. The asymptotic radial growth velocity was determined numerically. Their results are consistent with the growth law $v(T) = v_0[1 - T/T_c]^{3/2}$, where v_0 is a model-dependent constant. For numerical purposes we shall use $v_0 = 3c$, which corresponds to their parameter $\alpha = 1$. This expression is intended to apply only when $T > \frac{2}{3}T_c$ so that the growth velocity stays below the speed of sound of a massless gas, $c/\sqrt{3}$. Actually, the interior of the bubble is at a slightly higher temperature than the exterior; we neglect this small temperature difference. Our simple illustrative model for bubble growth then is

$$V(t', t) = \frac{4\pi}{3} \left(R_*(T(t')) + \int_{t'}^{t} dt'' v(T(t'')) \right)^3. \tag{5}$$

The closer T is to T_c the slower the bubbles grow.

For simplicity let us model the hadronic phase by a massless gas of pions, and the plasma phase by a gas of gluons and massless quarks of two flavors, with a bag constant B to simulate confinement dynamics, and use the parameters $\sigma = 50$ MeV/fm^2, $B^{1/4} = 235$ MeV, $\xi_q = 0.7$ fm, and $\eta_q = 14.4T^3$. (The corresponding $T_c = 169$ MeV).

One must distinguish between the nucleation time and the time it actually takes to complete the transition. If the plasma is first equilibrated at a temperature $T_0 = 2T_c$ at time $t_0 = 3/8$ fm/c as suggested by the uncertainty principle and by detailed simulations [8], then the plasma will cool according to the Bjorken model $T(t) = T_0(t_0/t)^{1/3}$ until the time $t_c = 8t_0 = 3$ fm/c. The matter continues to cool below T_c until T falls to about $0.95 T_c$, when noticeable nucleation begins. When the temperature has fallen to about $0.8 T_c$, bubble formation and growth is sufficient to begin reheating the system, due to the release of latent heat. The bubble growth had two components in this model calculation one arising from the Bjorken expansion ad the other from the phase transition dynamics. When the temperature exceeds about $0.95 T_c$ nucleation of new bubbles shuts off. We remark that during this stage of the transition the radius of critical sized bubbles is on the order of 1 fm [6]; this is a nontrivial result since bubbles much larger than several fermi would not be contained within the nuclear diameter. The transition continues only because of the growth of previously created bubbles. However, the temperature must remain somewhat below T_c in order that these bubbles do grow. Compared to the idealized adiabatic Maxwell construction of phase equilibrium at T_c the finite transition rate delays completion of the transition by about 11 fm/c. In the Bjorken hydrodynamics the proper volume of the system increases linearly with time, $V(t) = V(t_c)t/t_c$. Since completion of the phase transition is delayed from 37 fm/c to 48 fm/c, and the entropy density at completion is the same, 30% extra entropy is generated.

Recently in a simplified approach[9] the above results were reproduced within 1 %. In this approach droplet fusion, rapid spherical expansion and other effects were also studied. The growth and completion is somewhat faster this way. The completion happens at a temperature 5% below T_c. The more rapid spherical expansion, allows the possibility of a sizeable supercooling by about 30%! Here the assumptions of nucleations of spherical droplets break down already and other approaches to the problem should be studied.

It is possible to study the space-time evolution of the phase transition with hadron interferometry [10] and/or correlations [11]. It would be quite exciting to decide the issue of the existence of a QCD phase transition experimentally.

2.3 Sudden Freeze-out

If we reach 30% supercooling in the pure QGP phase, which may happen quite early in the reaction (5-15 fm/c) the parton density becomes quite low. The possibility to continue the fluid dynamical expansion assuming local thermal equilibrium may break down.

From trends in interferometry data there are indications that the *freeze-out* time scale is short enough to prevent reheating and the completion of the *rehadronization* of the quark-gluon plasma through bubble formation in the supercooled state. This in turn implies that other mechanisms must dominate the final stages of the hadronization.

The WA85 collaboration found large production rates of strange antibaryons at CERN SPS sulphur + tungsten interactions [12, 13]. The ratio for Ξ^-/Λ observed by WA85 was found to be compatible with those from other interactions. However, the ratio $\overline{\Xi}^-/\overline{\Lambda}$ was found to be about five times greater than those obtained by the AFS collaboration, corresponding to a two standard deviation effect. Rafelski was able to reproduce this enhancement only by assuming sudden rehadronization from QGP near equilibrium, which would not change the strangeness abundance [14]. Really, the long time-scale of the nucleation compared to the short time-scales of the pion freeze-out times at CERN SPS energies support the coincidence of the maximal supercooling of the QGP with freeze-out time of 4.5-6.5 fm/c, leaving very short time for the strange antibaryons for reinteraction in the hadronic gas already at CERN SPS energies.

The latent heat during such a sudden breakup might be released as high kinetic energy of the hadrons, in qualitative agreement with the observation that the multi-strange antibaryons observed by the WA85 collaboration are all at transverse momenta above 1.2 GeV/c, and show an effective $T_{slope} > 200$ MeV.

The freeze-out time turns out to be very close to the time when the supercooling reaches the minimum temperature, when bubbles are formed and are growing at the maximum rate. Using the above parameters of nucleation the pressure of the supercooled QGP vanishes at $T = 0.98T_c$ already. According to the above considerations and ref. [15] the temperature of the system in the supercooled phase reaches $T = 0.7 - 0.9T_c = 135$ MeV, which is in the vicinity of the usually assumed freeze-out temperatures (120-140 MeV). This implies that the hadronic phase freezes out when it is formed, i.e. the width of the proper-time distribution becomes very short.

Let us consider the sudden freeze out from supercooled QGP [16, 17, 18]. The baryon-free case was discussed in [17] in detail including the possibility of converting latent heat to final kinetic energy locally and instantly [in timelike deflagration], while [18] did not include this possibility. Although in ref. [17] it is argued that a superheated hadronic state is not realizable as final state [because one has to pass the mixed state on the way], this restriction does not apply for sudden freeze-out which we consider as a discontinuity across a hypersurface with normal Λ^μ ($\Lambda^\mu \Lambda_\mu = +1$). We can satisfy the energy and momentum conservation across this discontinuity expressed via the energy momentum tensors of the two phases, $T^{\mu\nu}$,

$$(T_H^{\mu\nu} - T_Q^{\mu\nu})\Lambda_\nu = 0,$$

with entropy production in the $Q \to H$ process, if $p_Q^{(init)} \leq \frac{\frac{4}{3}-r^{1/3}}{r^{1/3}-1}B$ or

$$T_Q^{(init)} \leq \left[\frac{B}{3a_Q(r^{1/3}-1)}\right]^{1/4} \approx 118 \text{MeV} = 0.7T_c. \tag{6}$$

(see eq. (30) of ref. [17], $r = \frac{37}{3}$, $a_Q = 37\pi^2/90$, $a_H = 3\pi^2/90$). From this state in a sudden [timelike deflagration] process we reach a superheated hadronic state of $p_H^{(final)}$ $= p_Q^{(init)} + \frac{4}{3}B$ (eq. (24) of ref. [17]) or

$$T_H^{(final)} = \left[\frac{Br^{1/3}}{3a_H(r^{1/3}-1)}\right]^{1/4} = 270 \text{MeV} = 1.6T_c. \tag{7}$$

This strong superheating is a consequence of the large latent heat of the EOS used in [3]. With somewhat more than 30% supercooling in the QGP entropy can be produced during the freeze-out and lower final temperatures may also result even with the same EOS.

In terms of particle composition, the idea that the quark-gluon plasma has to hadronize suddenly in a deeply supercooled state has the consequence that the production rate of strange antibaryons as suggested by [14, 12, 13] could become a clean signature of the quark-gluon plasma formation at RHIC and LHC energies as well as at the present CERN SPS energy.

The mechanism described in [18] could not give a complete account of sudden freeze-out because the suddenly released latent heat reheated the hadronic phase, and caused a high particle density, so that the hadronic matter was still not frozen out. Our present consideration of four additional processes makes such a sudden freeze-out feasible: i) Primarily the process of timelike deflagration allows us to convert large part of the latent heat to the collective kinetic energy of expansion instead and not into the internal thermal energy of the hadronic phase. ii) If we start the freeze-out deflagration from a 30% supercooled state the initial state is a mixture including already 15-25% hadronic phase, so the released latent heat is less. iii) We can take into account the time needed for the completion of the microscopic quantummechanical processes that lead to the sudden timelike deflagration. Thus the deflagration front will have a timelike thickness of about 1-2 fm/c, which leads to a further dilution of the matter after the deflagration is over. iv) the sudden freeze-out will lead to a baryon excess and particularly to a strange baryon excess compared to thermal and chemical equilibrium in the hadronic phase. This reduces the number of light mesons with high thermal velocities and thus, advances freeze-out in the hadronic phase also.

3 Conclusions

Further work is necessary to investigate microscopic mechanisms of rapid freeze-out from a supercooled plasma state. The mechanism reported by L. Wilets at this NATO ASI may be one candidate for such mechanism.

In summary, we considered the time-scales of rehadronization for a baryon-free QGP at RHIC and LHC energies. Pion freeze out times are estimated based on the analysis and extrapolation of present high energy HBT data. We found that the time-scale for reaching the bottom of the temperature curve during the cooling process via homogeneous nucleation [3, 6] is surprisingly close to the time-scale of the freeze out. We argue that the QGP has to complete rehadronization in a 10 - 30 % supercooled phase quite suddenly, and we have shown that such a sudden process is possible and satisfies energy and momentum conservation with non-decreasing entropy. This rehadronization mechanism should also be visible in a clean strangeness signal of the QGP. Detailed calculations have to be performed for making more quantitative predictions.

4 ACKNOWLEDGEMENTS

This work was supported in part by the Norwegian Research Council under Grant No's. 422.93/011, & /008, by the NATO Collaborative Research Grant under Grant No. 920322-644, by the Hungarian Science Foundation under Grant No. OTKA–F4019, and by the US National Science Foundation under Grant No. PHY89-04035.

REFERENCES

[1] L.P. Csernai. "Introduction to relativistic heavy ion collisions" J. Wiley and Sons, (1994) in press.

[2] L. Bravina, L.P. Csernai, P. Levai and D. Strottman, in "Proc. of Heavy Ion Physics at the AGS, HIPAGS '93", ed. by G.F.S. Stephans, S.G. Stedman, and W.L. Kehoe (Massachusetts Institute of Technology, MITLNS-2158, 1993) p. 329.

[3] L.P. Csernai and J.I. Kapusta, Phys. Rev. **D46**:1379 (1992).

[4] R. Venugopalan and A.P. Vischer, University of Minnesota, Minneapolis preprint, No. TPI-MINN-93-34/T and NUC-MINN-93-17/T (1993).

[5] J.D. Bjorken, Phys. Rev. **D27**:140 (1983).

[6] L.P. Csernai and J.I. Kapusta, Phys. Rev. Lett. **69**:737 (1992).

[7] J. C. Miller and O. Pantano, Phys. Rev. **D40**:1789 (1989); ibid. **D42**:3334 (1990).

[8] K. Geiger and B. Müller, Nucl. Phys. **B369**:600 (1992); K. Geiger, Phys. Rev. **D** **46**:4965,4986 (1992).

[9] L.P. Csernai, J.I. Kapusta, Gy. Kluge and E.E. Zabrodin, in "Book of Abstracts, 7th Nordic Meeting on Nuclear Physics", Vigsø , Denmark, Aug. 17-21, Univ. of Aarhus (1992) p. 76-78.

[10] S. Pratt, Phys. Rev. **D33**:1314 (1986); S. Pratt, P. J. Siemens and A. P. Vischer, Phys. Rev. Lett. **68**:1109 (1992). For a review see D. H. Boal, C.-K. Gelbke and B. K. Jennings, Rev. Mod. Phys. **62**:533 (1990).

[11] D. Seibert, Phys. Rev. Lett. **63**:136 (1989).

[12] S. Abatzis, et al., WA85 Collaboration, Phys. Lett. **B259**:508 (1991).

[13] S. Abatzis, et al., WA85 Collaboration, Phys. Lett. **B270**:123 (1991).

[14] J. Rafelski, Phys. Lett. **B262**:333 (1991); Nucl. Phys. **A544**:279c (1992).

[15] L. P. Csernai, J. I. Kapusta, Gy. Kluge and E. E. Zabrodin, Z. Phys. **C 58**:453 (1993).

[16] A. K. Holme, et al., Phys. Rev. **D40**:(1989) 3735.

[17] L.P. Csernai and M. Gong, Phys. Rev. **D37**:(1988) 3231.

[18] J. Kapusta and A. Mekjian, Phys. Rev. **D33**:(1986) 1304.

NUCLEAR CLUSTER EQUATION OF STATE

Jens Konopka, Harald Graf, Horst Stöcker, and Walter Greiner

Institut für Theoretische Physik
Johann Wolfgang Goethe-Universität
D–60054 Frankfurt am Main

INTRODUCTION

The nuclear equation of state has been investigated for a long period. Beside neutron stars and other astrophysical objects, heavy nuclei and reactions between them serve as an excellent tool to study the properties of hadronic matter [1,2], because they allow for a variation of several parameters like the mass of the system the bombarding energy and, in a more qualitative way, the impactparameter of the reaction.

There is an additional parameter to be controlled in heavy ion experiments, namely the initial isospin or equivalently the neutron to proton ratio N/Z of the combined system. In particular, this article adresses the question what we can learn from systematic variation of this parameter.

In the following we consider only symmetric projectile-target combinations. Such a reaction can be viewed as follows: The two heavy nuclei impinging on each other will form a hot and dense region of nuclear matter, which is centered around the center of mass rapidity. Due to the strong compression and the rapid sequence of hard nucleon–nucleon scattering, nucleons will be stopped and then escape from the central reaction zone. During the phase of high pressure projectile and target matter is still streaming in. This hinders the escape of the nucleons from the participant zone in the longitudinal direction. Therefore matter is squeezed out of the reaction plane and projectile and target material bounces off in the reaction plane. These collective flow phenomena have been predicted by nuclear hydrodynamics [1,3] and VUU/QMD [4,5,6] models and have been discovered experimentally [7,8,9,10,11]. In more quantitative theoretical analyses it has been demonstrated that the strengths of these effects can provide some information on the nuclear equation of state, in particular about the pressure in the central reaction zone. From an artist's point of view the system develops as depicted in Figure 1.

Hot and Dense Nuclear Matter, Edited by
W. Greiner *et al.*, Plenum Press, New York, 1994

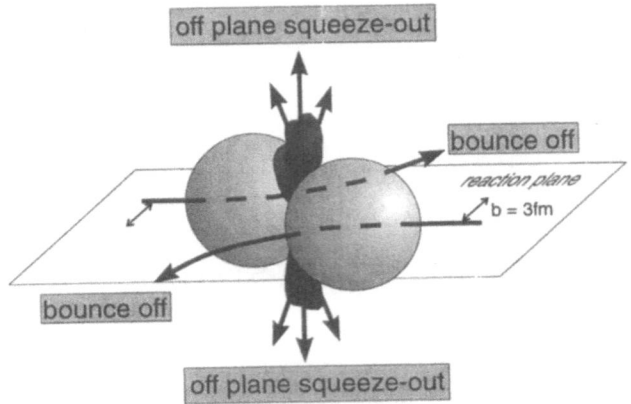

Figure 1. Schematic sketch of a symmetric semi-central heavy ion reaction. Projectile- and target-like particles are bounced in the reaction plane. A large fraction of the particles from the innermost reaction zone escapes towards directions perpendicular to the reaction plane leading to a preferential emission pattern which is peaked around 90° and −90° and which is dubbed squeeze-out.

However, beside these collective single-particle phenomena, the correlated emission of nucleons is another basic feature of nucleus-nucleus reations. Because a large fraction of the total mass of the system in the final state is contained in complex fragments and clusters, one has also to account for these ingredients when considering the underlying equation of state which rules the reaction dynamics.

At least in most central collisions and in the innermost reaction zone, the rapid sequence of nucleon-nucleon collisions an equilibrated source may be established. After the direct reaction has ceased, the system expands again and the density drops below the saturation density, ρ_0. This expansion can be viewed almost isentropic [12]. From what has been stated, one concludes that also the low density behaviour of the nuclear equation of state has an impact on the final state of these reactions.

CONCEPTS FOR NUCLEAR MATTER AT LOW DENSITIES

There are several possibilities for a phenomenological description of the thermodynamical properties of nuclear matter which are footed on different assumptions.

Interacting Fermi-gas

A widely used concept to describe nuclear matter is the assumption of homogenuity of the matter. One of the most straightforward ansätze for infinite nuclear matter is a homogenous Fermi-gas which interacts by a density dependent potential only. A representative equation of state of this type is shown in Figure 2. It exhibits a region of mechanical instability, i.e. where

$$\left.\frac{\partial p}{\partial \rho}\right|_{S=\text{const.}} < 0 \tag{1}$$

Although such an ansatz is rather useful for dynamical calculations, in particular as far as compressional processes are concerned, its applicability at densities below the saturation density is rather questionable since an overall homogenity of the matter is considered. Especially at very low densities ($\rho/\rho_0 \approx 0.1$) and low temperatures (hence low entropies) it predicts low internal energy, contradictary to the situation favoured, when also a non homogenous distribution of matter is allowed in the calculation.

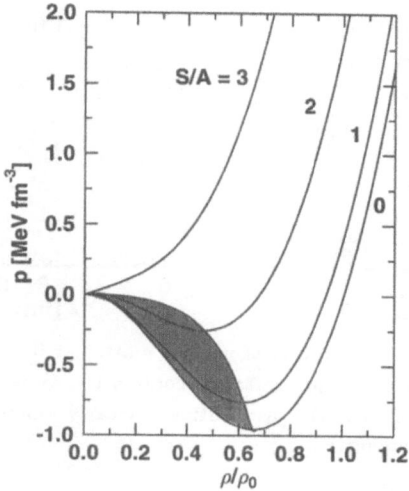

Figure 2. Pressure as a function of density for various isentropes. The equation of state displayed is a standard hard equation of state which is built in various microscopic models. It explicitly obeys the assumption of an overall homogenuity of the matter. The grey area corresponds to the mechanical instability region.

Microscopic modelling by phenomenological nucleon-nucleon forces

Alternatively nuclear matter as well as finite nuclei can be described as an ensemble of classical molecules interacting with each other by phenomenological potentials. Such a concept is widely used in Classical Molecular Dynamic and Quantum Molecular Dynamic calculations. Such a method, combined with the Metropolis procedure is a powerful tool to study thermodynamical properties of many body systems.

This has been worked for the nuclear case out in more detail in [13]. The authors calculated the total energy of infinite nuclear matter in terms of a Metropolis procedure. The Fermionic character of the nucleons, which is essential for providing non-vanishing kinetic energies of the particles at zero temperature, was mimicked by means of a so called Pauli potential. Such a potential has been proposed very early [14] but in this particular approach the parametrisation of [15] has been used. This potential prevents two identical nucleons from coming too close together in configuration as well as in momentum space. This potential was supplemented by Skyrme, Coulomb and Yukawa potentials in order to model the nuclear equation of state on a microscopic footing.

As depicted in Figure 3 the ground state configuration of nuclear matter at its saturation density ρ_0 is homogenously distributed in configuration space, wheras lowering the density leads to the effect that at low thermal excitations the system tends to cluster in order to lower its interal energy. In general there are two effects interplaying with each other when regarding nuclear matter at densities below the saturation point. On the one hand the system favours to be in a rather low energetic state, which is especially true for low temperatures, on the other hand there is the counterbalancing effect of gaining entropy when breaking the system in several smaller pieces than to bind all nucleons in heavy clusters.

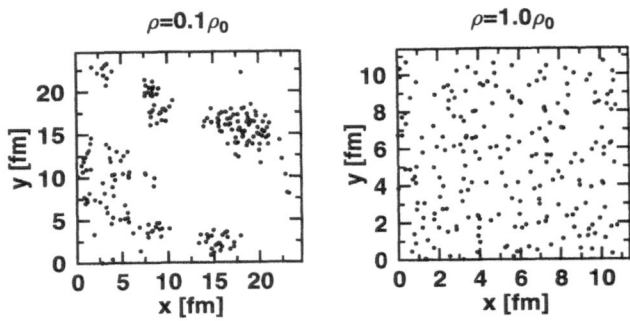

Figure 3. Spatial ground-state configuration of nuclear matter at $0.1\rho_0$ (left) and ρ_0 (right) as a projection onto the x-y-plane. Both representations contain the same number of particles (216), but in a factor of 10 different volume. The distribution is clearly inhomogenous at low densities, evidencing the importance of clusterisation.

Nuclear matter at $T \approx 0$ and at densites below ρ_0 prefers a configuration, where the nucleons are glued together to form heavier clusters [13, 16]. Spatial homogenuity is not established on the nucleon level but on a larger scale, where the combined clusters can be regarded as the elementary objects. This interpretation is somewhat density dependent, at the higher densities in particular the lighter clusters immerse in the continuum – the Mott transition in nuclear matter [17].

Chemical equilibrium of different fragment species

The Quantum Statistical Model (QSM) [18] treats excited nuclear systems as a chemically equilibrated mixture of a large variety of nuclear quantum states. Presently more than 1000 stable and instable states with their proper distribution functions (Bose-Einstein and Fermi-Dirac) and statistical weights are considered. QSM operates in the grandcanonical ensemble, therefore the system is completely characterized by a temperature a density and its N/Z ratio plus the condition of chemical equilibrium, which reads for fragment specie i

$$\mu_i = Z_i \mu_{\mathrm{p}} + N_i \mu_{\mathrm{n}} + E_i , \tag{2}$$

where Z_i, N_i, and E_i denote the charge, the neutron number, and the total binding energy of this particular state. The proton- and neutron chemical potentials are μ_{p} and μ_{n}, respectively. The available volume is reduced by the eigenvolume of the nucleons and clusters.

This approach completely neglects the interfragment interactions, whereas the intrafragment interactions are excatly treated since measured binding energies are put in the calculation. On the other hand it explicitly accounts for a clustered configuration of nuclear matter.

QSM calculates two different fragment distributions: The equilibrium configuration, frequently labelled primordial and the final distribution. The former also includes excited states with half life times, which are too short in order to detect the isotope directly. After the treatment of the decay of these instable states one is left only with those fragments, which can be considered stable in this context. The importance of the secondary decay effects will be discussed in one of the following sections.

PARTICLE AND CLUSTER EMISSION IN HEAVY ION REACTIONS

To understand the isospin dependence in detail, we recall that the chemical potential for each individual fragment species i is (2). The chemical potential can be decomposed into a contribution which is proportional to the mass-number of the isotope under consideration and a term which grows linearly with the isospin component

$$T_i^3 = Z_i - \frac{1}{2}A_i = \frac{Z_i - N_i}{2}. \tag{3}$$

If the binding energy is neglected, which is a good approximation especially at higher temperatures and for heavier clusters, such a representation can be established when the chemical potentials μ_p and μ_n are decomposed into a nucleo- and isochemical potential:

$$\mu_{\text{nucleo}} = \frac{\mu_{\text{p}} + \mu_{\text{n}}}{2}, \qquad \mu_{\text{iso}} = \mu_{\text{p}} - \mu_{\text{n}}. \tag{4}$$

Thus the fragment chemical potential reads

$$\begin{aligned}
\mu_i &= (Z_i + N_i)\frac{\mu_{\text{p}} + \mu_{\text{n}}}{2} + \frac{Z_i - N_i}{2}(\mu_{\text{p}} - \mu_{\text{n}}) + E_i \\
&= A_i \mu_{\text{nucleo}} + T_i^3 \mu_{\text{iso}} + E_i.
\end{aligned} \tag{5}$$

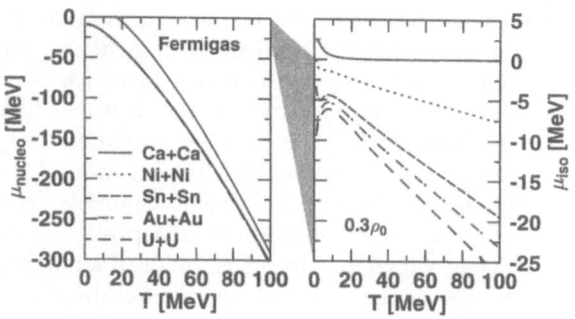

Figure 4. Nucleo- and isochemical potential as a function of temperature. Various system distinguished by their N/Z ratio are displayed. In the $T \to 0$ limit μ_{nucleo} approaches ≈ -8 MeV, the average binding energy of heavy nuclei, rather than $+16.5$ MeV as for a 4-fold degenerated Fermi-gas at the same density, indicated in the left portion of the figure. The break-up density has been kept constant at $0.3\rho_0$. The influence of the isospin becomes less important with increasing temperature – note the different ranges on the μ axes.

As T tends to zero, μ_{nucleo} converges to ≈ -8 MeV. This value is due to the clusterization of complex fragments in low entropy nuclear matter at subsaturation densities as shown in the previous section. In absence of clusterization the chemical potential of an equivalent system at the same density consisting out of free protons and neutrons only, the (nucleo)chemical potential is positive, $+16.5$ MeV.

When the temperature is rised, the system more and more behaves like an ideal Fermigas of protons and neutrons. The influence of the isochemical potential becomes less important (note the different scales on the axes in Figure 4).

In order to compare systems of different size without making use of an intensive thermodynamical variable like T or S/A we introduce the reduced total fragment multiplicity, M_{red} [19], which is the number of all charged reaction products devided by the total charge of the system. Note, that for an evaluation of this quantity all fragments must be known.

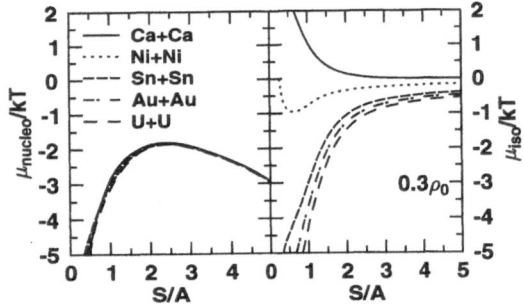

Figure 5. Logarithm of the fugacities corresponding to the nucleo- and isochemical potentials of various systems with different isospins.

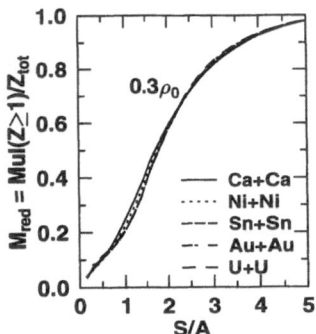

Figure 6. Reduced total fragment multiplicity as a function of the baryonic entropy per nucleon for Ca+Ca, Ni+Ni, Sn+Sn, Au+Au, and U+U systems.

Different systems are normalized with respect to their total charge rather than to any other possible number like the total mass or neutron number of the system. Frequently used heavy ion systems have a neutron excess, while almost all produced clusters have masses below 20, where the most stable isotopes have more or less equal proton and neutron numbers. Hence the primary fragment distribution is basically controlled by charge conservation, whereas the conservation of the neutron number is responsible for the abundances in specific isotopic channels. However, the situation is not as clear as that, because many of the primarily produced fragments are particle instable and undergo decay, which may feed the yields of a particular species to a high extend, which is not in accordance with a chemical equilibration of the final fragment distribution.

Figure 6 shows the reduced total fragment multiplicity M_{red} versus the baryonic entropy per nucleon S/A for various systems at a break-up density of $0.3\rho_0$. It is remarkable that the entropy dependence of M_{red} does itself not depend on the N/Z ratio of the initial system. Nevertheless the relation between incident energy and M_{red}, or equivalently the relation between S/A and E_{lab} may be isospindependent. The former relation, $M_{\text{red}}(E_{\text{lab}})$ can easily be obtained from systematic experimental studies or dynamical calculations which have incorporated a proper treatment of fragments.

It is apparent from Figure 5+6 that the gross features of the fragment distribution like the integrated mass and the charge yields are essentially controlled by the nucleochemical potential. The influence of the initial isospin of the system becomes important when considering the yields of specific isotopes: Figure 7 shows the ratio of the free neutrons over free protons as a function of M_{red} for Ca+Ca, Ni+Ni, Sn+Sn, Au+Au, and U+U. In particular at the lower values of M_{red}, corresponding to low bombarding energy and entropy, the emission of free nucleons is substantially affected by the N/Z ratio of the entire system. Unfortunatly, it is not a simple experimental task to detect free neutrons and protons simultanously.

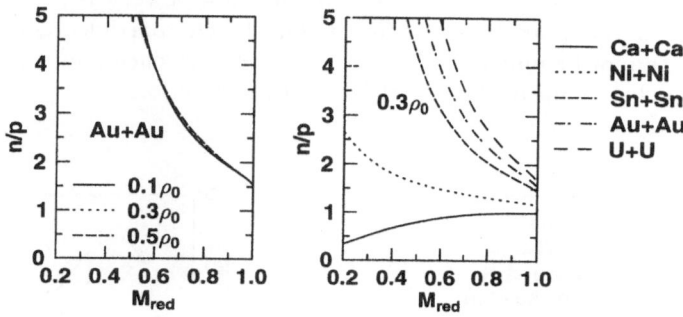

Figure 7. Number of free neutrons over free protons. In the left portion of the figure the results for Au+Au are displayed for various densities. The right part shows the dependence on the isospin.

Nevertheless, the predictions made can more easily be tested in other channels, because the strong isospin dependence is preserved when considering the t over ^{3}He ratio. Similarily to the n/p case, here the nucleochemical potentials of both species equal, but their isospin differs by one unit. A small disturbance enters through the different binding energies of t and ^{3}He, which will become relevant at the lowest temperatures only. The t/^{3}He (see Figure 8) ratio shows a behaviour similar to n/p, compare to Figure 7. Large deviations are observed around $M_{red} < 0.7$ from the expectation when the proton–neutron composition of the the initial system is assumed. For systems with a high neutron excess t/^{3}He exceeds the N/Z ratio of the heavy ions by a factor of 3 and more at the lower energies. This is in agreement with the findings in α-induced reactions at 43 MeV/nucleon [20].

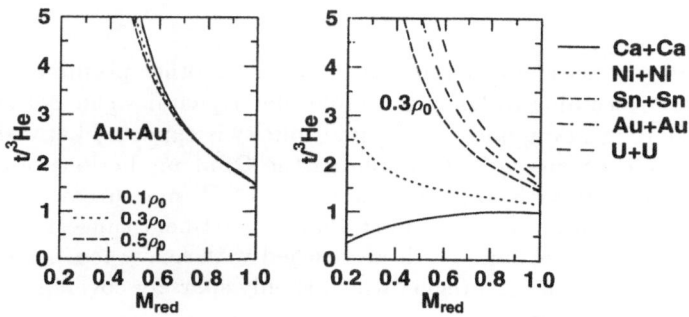

Figure 8. Same as Figure 7 but for the triton over ^{3}He ratio.

As the system's isospin goes towards zero (Ni+Ni and Ca+Ca), the inicident energy- c.f. M_{red} dependence becomes much weaker, the n/p and the t/^{3}He ratios reveal the N/Z ratio of the entire system over a wider range of multiplicities. The predicted large differences emerging when considering e.g. Ca+Ca and U+U could easily be tested in experiments. In addition our predictions do only weakly depend on the break-up density used in the calculation.

Using the entropy values obtained with a shock calculation for a equation of state with a compressibility of 270 MeV [21], QSM can predict also the evolution of ensemble averaged observables with bombarding energy. The most drastic trends are expected for systems with a large N/Z ratio and incident energies below 200 MeV/nucleon, see Figure 9. Neutrons (or equivalently tritons) may be emitted 10 times more frequently than protons (^{3}He) in the case of very heavy systems. This large number demonstrates that the formation of light particles and fragments at these energies cannot be understood in terms of a coalescence picture and an overall homogenous nucleon soup in the initial stage of the reaction.

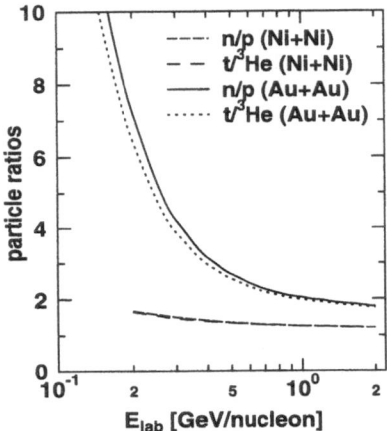

Figure 9. Excitation function of n/p and t/^{3}He for the Ni+Ni and Au+Au system. Shock calculation of ref. [21] has been used to extract the entropy as a function of the bombarding energy.

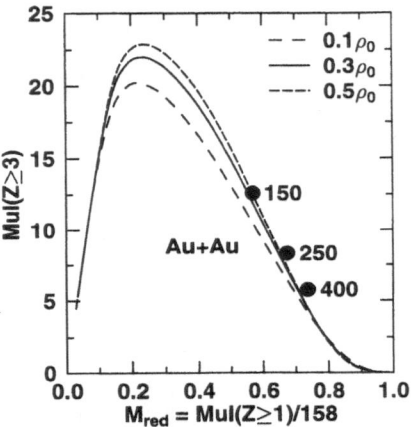

Figure 10. Intermediate mass fragment ($Z \geq 3$) multiplicity for Au+Au. The lines correspond to QSM calculations at 0.1, 0.3 and 0.5ρ_0, whereas the dots represent the measured Au+Au data at 150, 250 and 400 MeV/nucleon of ref. [22].

Even more so, at these low bombarding energies another prominent effect is predicted to occur: According to Quantum Molecular Dynamics calculations the maximum intermediate mass fragment (IMF) multiplicity is expected between 50 and 150 MeV/nucleon incident energy [6]. Considering the QSM predictions of the evolution of the IMF multiplicity with M_{red} and the recent FOPI measurements [22], as can be seen from Figure 10, also the quantum statistical treatment suggests that the maximum IMF production probability will be reached with heavy ion beams at energies lower than 150 MeV/nucleon, a region which is only sparsely covered up to now with experiments in full 4π geometry.

PARTICLE INSTABLE STATES – DISTORTION OF THE FRAGMENT DISTRIBUTION

One should be cautious, though, about interpreting measured particle yields directly in terms of a possible chemical equilibration. In the equilibrium calculation a large number of particle instable states have to be taken into account, because their life-times are longer than the time scales involved for the equilibration processes. On the contrary, the life-times of these isotopes are far too short in order to be detected directly. However, they can be measured by two particle correlations. A wide variety of particle instable states (and γ instable states) has been observed and their abundancies are in accordance with the QSM predictions for final temperatures about $T \approx 5\,\mathrm{MeV}$ [23, 24, 25]. The final fragment distribution including feeding from instable states does not exhibit the shape of an equilibrium distribution: several components, which participated in the equilibrium configuration have vanished completely. This is a large effect as shown in Figure 11. Here the relative yields of d, t, α are plotted in the equilibrium (labelled primordial) and in the final stage. Secondary decays, as mentioned, feed particular channels with multiples of the original yields of this isotope. Thus, it cannot be expected that the initial isospin trivially will have an impact on the final fragment distribution as it has been demonstrated in the previous section. Such a dependence, although present in the primordial (i.e. equilibrium) stage, does not neccessarily need to survive the distortions due to the strong secondary decay effects.

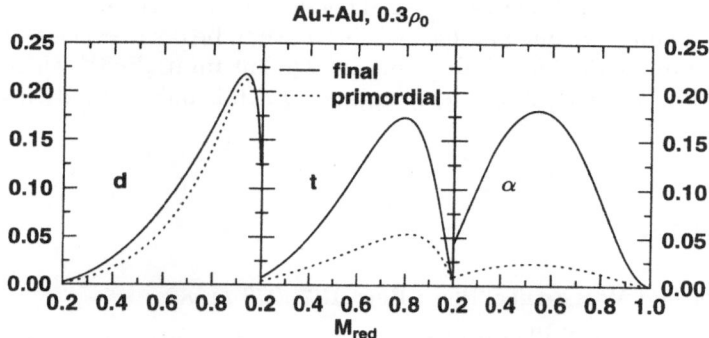

Figure 11. Primordial and final relative yields of d, t and α for the Au+Au system. The yields are normalized to the total charge of the system.

From Figure 12 it is obvious that this effect is present for nucleons, too. In the interesting range $M_{\mathrm{red}} = 0.6 - 0.8$ only one third of the free nucleons in the final stage directly originate from the initial hot and dense region. Therefore the measurement of free nucleons, in particular of free neutrons, gives only partially a direct view into the interesting zone. A large portion of the free nucleon yields is affected by the motion of (excited) fragments plus their nuclear spectroscopy.

In the nuclear soup emerging at intermediate energies most of the nucleons belong to excited nuclear clusters rather than being free nucleons or bound in ground state nuclei. This finding is expressed in Figure 13, where the fraction of matter which is in an excited state is displayed as a function of the baryonic entropy created. The increase at the low entropy values, corresponding to low excitations of the system is partly due to the fact that above mass 20 only stable isotopes had been considered in the present calculation.

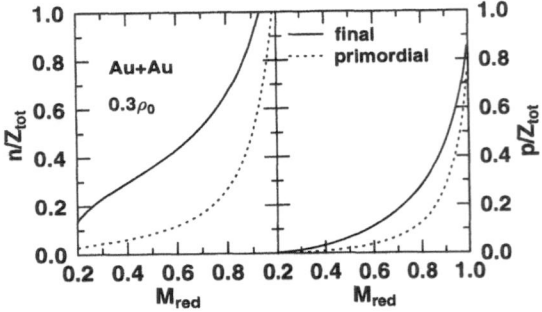

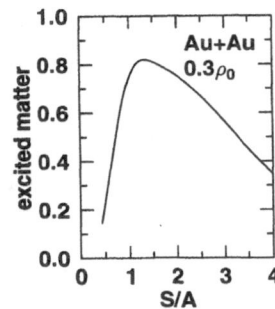

Figure 12. Neutron- and proton number over total charge as function of M_{red} for Au+Au. A substantial amount of free nucleons stems from particle emission from heavier excited clusters.

Figure 13. Fraction of nucleons which are contained in excited nuclear states as function of baryonic entopy.

We have demonstrated that (excited) clusters are important ingredients of the nuclear (thermo)dynamics. They can help to determine thermostatic properties of excited nuclear matter, like temperature and entropy. Furthermore, heavy particles' spectra are much less affected by secondary decay and show the collective effects more pronounced than single particles. Therefore further studies of the cluster and multifragment emission in heavy ion collisions have to be done, especially with respect to the maximum of the intermediate mass fragment multiplicity, which is supposed to be reached in a bombarding energy regime, which is only sparsely investigated up to now.

REFERENCES

1. H. Stöcker and W. Greiner, Phys. Rep. **137**, 277 (1986).

2. Proceedings on the NATO Advanced Study Institute Programme on The Nuclear Equation of State, eds. W. Greiner and H. Stöcker, NATO ASI Series B, Plenum, N.Y. (1990).

3. W. Scheid, H. Müller, and W. Greiner, Phys. Rev. Lett. **32**, 741 (1974).

4. J. J. Molitoris and H. Stöcker, Phys. Rev. C **32**, 346 (1985).

5. J. Aichelin, G. Peilert, A. Bohnet, A. Rosenhauer, H. Stöcker, and W. Greiner, Phys. Rev. C **37**, 2451 (1988).

6. G. Peilert, H. Stöcker, W. Greiner, A. Rosenhauer, A. Bohnet, and J. Aichelin, Phys. Rev. C **39**, 1402 (1989).

7. D. Beavis, S. Y. Chu, S.Y. Fun, W. Gorn, A. Huie, D. Keane, J. J. Liu, R. T. Poe, B.C. Shen, and G. vanDalen, Phys. Rev. C. **27**, 2443 (1983).

8. H. A. Gustafsson, H. H. Gutbrod, B. Kolb, H. Löhner, B. Ludewigt, A. M. Poskanzer, T. Renner, H. Riedesel, H. G. Ritter, A. Warwick, F. Weick, and H. Weimann, Phys. Rev. Lett. **52**, 1590 (1984).

9. R. E. Renfordt, D. Schall, R. Bock, R. Brockmann, J. W. Harris, A. Sandoval, R. Stock, H. Ströbele, D. Bangert, W. Rauch, G. Odyniec, H. G. Pugh, and L. S. Schröder, Phys. Rev. Lett. **53**, 763 (1984).

10. K. H. Kampert, J. Phys. G **15**, 691 (1989).

11. H. H. Gutbrod, A. Sandoval, and H. G. Ritter, Rep. Prog. Phys. **52**, 1267 (1989).

12. W. Schmidt, U. Katscher, B. Waldhauser, J.A. Maruhn, H. Stöcker, W. Greiner, Phys. Rev. C **47**, 2782 (1993).

13. G. Peilert, J. Randrup, H. Stöcker, and W. Greiner, Phys. Lett. **260B**, 271 (1991).

14. L. Wilets, E.M. Henley, M. Kraft, and A.D. McKellar, Nucl. Phys. **A282**, 341 (1977).

15. C. Dorso, S. Duarte, and J. Randrup, Phys. Lett. **188B**, 287 (1987).

16. O. Civitarese, A. Plastino, and A. Faessler, Z. Phys. A **291**, 239 (1979).

17. G. Röpke, L. Münchow, and H. Schulz, Nucl. Phys. **A379**, 536 (1982) and Nucl. Phys. **A399**, 587 (1983).

18. D. Hahn and H. Stöcker, Nucl. Phys. **A476**, 718 (1988).

19. J. Konopka, H. Graf, H. Stöcker, and W. Greiner, to be published.

20. B. Ludewigt, G. Gaul, R. Glasow, H. Löhner, and R. Santo, Phys. Lett. **108B**, 15 (1982).

21. H. Stöcker, M. Gyulassy, and J. Boguta, Phys. Lett. **103B**, 269 (1981).

22. C. Kuhn, J. Konopka, J. P. Coffin, C. Cerruti, P. Fintz, G. Guillaume, A. Houari, F. Jundt, C. F. Maguire, F. Rami, R. Tezkratt, P. Wagner, Z. Basrak, R. Čaplar, N. Cindro, S. Hölbling, J. P. Alard, N. Bastid, L. Berger, S. Boussange, I. M. Belayev, T. Blaich, A. Buta, R. Donà, P. Dupieux, J. Erö, Z. G. Fan, Z. Fodor, R. Freifelder, L. Fraysse, S. Frolov, A. Gobbi, Y. Grigorian, N. Herrmann, K. D. Hildenbrand, S. C. Jeong, M. Jorio, J. Kecskemeti, P. Koncz, Y. Korchagin, R. Kotte, M. Krämer, I. Legrand, A. Lebedev, V. Manko, T. Matulewicz, G. Mgebrishvili, J. Mösner, D. Moisa, G. Montarou, I. Montbel, W. Neubert, D. Pelte, M. Petrovici, S. Ramillien, W. Reisdorf, A. Sadchikov, D. Schüll, Z. Seres, B. Sikora, V. Simion, S. Smolyankin, U. Sodan, K. M. Teh, M. Trzaska, M.A. Vasiliev, P. Wessels, T. Wienold, Z. Wilhelmi, D. Wolfarth, and A. V. Zhilin, Phys. Rev. C **48**, 1232 (1993).

23. J. Pochodzalla, W. A. Friedman, C. K. Gelbke, W. G. Lynch, M. Maier, D. Ardouin, H. Delagrange, H. Doubre, C. Grégoire, A. Kyanowski, W. Mittig, A. Péghaire, J. Péter, F. Saint-Laurent, Y. P. Viyogi, B. Zwieglinski, G. Bizard, F. Lefèbvres, F. Tamain, and J. Québert, Phys. Lett. **161B**, 275 (1985).

24. M. A. Bernstein, W. A. Friedman, W. G. Lynch, C. B. Chitwood, D. J. Fields, C. K. Gelbke, M. B. Tsang, T. C. Awes, R. L. Ferguson, F. E. Obenshain, F. Plasil, R. L. Robinson, and G. R. Young, Phys. Rev. Lett. **54**, 402 (1985).

25. C. Schwarz, W. G. Gong, N. Carlin, C. K. Gelbke, Y. D. Kim, W. G. Lynch, T. Murakami, G. Poggi, R. T. de Souza, M. B. Tsang, H. M. Xu, D. E. Fields, K. Kwiatkowski, V. E. Viola, Jr., and S. E. Yennello, Phys. Rev. C **48**, 676 (1993).

THE DISAPPEARANCE OF FLOW AND CRITICAL BEHAVIOR

IN NUCLEUS-NUCLEUS COLLISIONS

G.D. Westfall, A.M. Vander Molen, W.J. Llope,
S.J. Yennello, T. Li, J. Yee, E. Gualtieri, S. Hannuschke,
R. Pak, D. Craig, W. Bauer, and D. Klakow
National Superconducting Cyclotron Laboratory and
Department of Physics and Astronomy
Michigan State University
East Lansing, Michigan 48824-1321 USA

R.A. Lacey, J. Lauret, and A. Elmaani
Department of Chemistry
State University of New York
Stony Brook, New York 11794-3400 USA

A. Nadasen
Department of Physics
University of Michigan at Dearborn
Dearborn, Michigan 48128 USA

E. Norbeck
Department of Physics, University of Iowa
Iowa City, Iowa 52242 USA

INTRODUCTION

The equation of state of nuclear matter (EOS) is basic information that can be extracted from detailed studies of nucleus-nucleus collisions. Using the MSU 4π Array, we have carried out a series of studies concerning the EOS including the disappearance of collective flow, critical behavior of nuclear matter, the onset of multifragmentation. In the disappearance of flow work, we attempt to extract information about the EOS by making systematic measurements of collective flow and determining the incident energy at which flow disappears. We studied collective flow from the systems C+C, Ne+Al, Ar+Sc, and Kr+Nb at a variety of incident energies. Comparing these results

Hot and Dense Nuclear Matter, Edited by
W. Greiner *et al.*, Plenum Press, New York, 1994

with Boltzmann-Uehling-Uehlenbeck (BUU) calculations, we can extract information about the EOS using an observable that is less sensitive to details of the theoretical calculation. In the critical behavior work, we studied central collisions of Ar+Sc at 15 to 115 MeV/nucleon from which we extracted Z distributions. We related these results to critical phenomena in nuclear matter using a percolation model which we fit to the observed Z distributions and then extrapolated to infinite nuclear matter.

The measurements described below were carried out with the MSU 4π Array[1] at the National Superconducting Cyclotron Laboratory (NSCL) using beams from the K1200 cyclotron. The 4π Array consists of a main ball of 170 phoswich counters (arranged in 20 hexagonal and 10 pentagonal subarrays) covering angles from 23° to 157° and a forward array of 45 phoswich counters covering angles from 7° to 18°. The 30 Bragg curve counters installed in front of the hexagonal and pentagonal sub-arrays were operated (in three different experiments) with pressures of 500, 250, and 100 torr of P5 (95% argon, 5% methane), CF_4, and C_2F_6 gases respectively[2]. The hexagonal anodes of the five most forward Bragg curve counters are segmented creating a total of 55 separate ΔE gas counters. To detect fission fragments, 30 low pressure multi-wire proportional counters (LPMWPCs) were used. The LPMWPCs have an angular resolution of around 1° in θ and ϕ.

The Bragg curve counters served as ΔE detectors for charged particles that stopped in the fast plastic scintillator of the main ball. Consequently, the Array was capable of detecting charged fragments from Z=1 to Z=12. Low energy thresholds (in each experiment) were ≈ 17 MeV/nucleon, ≈ 3 MeV/nucleon and ≈ 5 Mev/nucleon for fragments of Z=1,3, and 12 respectively. In addition fragments with Z > 12 were detected in the LPMWPCs giving angular information.

MASS DEPENDENCE OF THE DISAPPEARANCE OF FLOW

The study of global collective variables in nucleus-nucleus collisions can provide information about the nuclear equation of state (EOS)[3,4]. In recent years much emphasis has been placed on the study of collective flow and several groups have performed experiments to study its disappearance[5,6,7,8,9,10]. The disappearance of collective flow is predicted to occur at an incident energy (termed the balance energy, E_{bal}[5]) corresponding to the point where the attractive scattering, dominant at incident energies around 10 MeV/nucleon, balances the repulsive interactions observed at energies around 400 MeV/nucleon[5,11,12]. Because the magnitude of parameters of the EOS can be related to the dominance of repulsive or attractive scattering, there is great value in the determination of E_{bal}. We report here results from the first systematic study of the disappearance of flow in nucleus-nucleus collisions. We present data for the C+C, Ne+Al, Ar+Sc, and Kr+Nb systems which show that E_{bal} scales as $A^{-1/3}$, where A is the mass of the combined projectile-target system. The general trend of this result is reproduced by Boltzmann Uehling Uehlenbeck (BUU) model calculations confirming the interpretation of E_{bal}.

We interpret the scaling of E_{bal} with mass to be the result of a competition between the dominantly attractive mean field interaction which scales as the surface area, $A^{2/3}$, and the dominantly repulsive nucleon-nucleon scattering which scales as the volume of the two nuclei, A.

The targets used in the flow measurements consisted of 1 mg/cm^2 natural carbon, 1 mg/cm^2 natural aluminum, 1.6 mg/cm^2 natural scandium, and 1 mg/cm^2 natural niobium. Beam intensities were approximately 100 particle pA. The beam energies

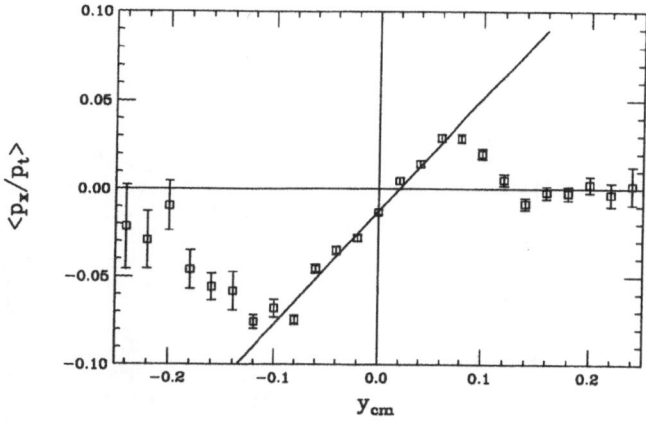

Figure 1: Average transverse momentum in the reaction plane divided by the total transverse momentum versus the center of mass rapidity for He fragments from 35 MeV/nucleon Kr+Nb. The solid line is fit over the region of -0.08≤y_{cm} ≤0.08.

used were 55 to 155 MeV/nucleon for ^{12}C, 55 to 140 MeV/nucleon for ^{20}Ne, 35 to 115 MeV/nucleon for ^{40}Ar, and 35 to 75 MeV/nucleon for ^{86}Kr. Data were taken with a minimum bias trigger (charged particle multiplicity, $m \geq 2$) and a more central trigger ($m \geq 5$).

To extract collective flow in nucleus-nucleus collisions one must determine the impact parameter, b, and the reaction plane. We determined the centrality of each event by way of cuts on the total transverse momentum. "Central" collisions corresponding to an average impact parameter ≈ 0.40 b_{max} were used for each system. Here, b_{max} is the maximum estimated impact parameter. The reaction plane was determined using the method of azimuthal correlations[13].

Having selected central collisions and determined the reaction plane for each event, the average fraction of the total transverse momentum in the reaction plane, $< p_x/p_t >$, was evaluated and plotted as a function of the center of mass rapidity, y_{cm}. Fig. 1 shows an example of such a plot for He fragments from collisions of 35 MeV/nucleon Kr+Nb. The characteristic "S" shape commonly associated with flow is quite evident in this figure. The errors shown are statistical, and the solid line corresponds to a linear least square fits for the region -0.08≤y_{cm} ≤0.08. The slope of this line is defined as the reduced flow.

The reduced flow is shown as a function of beam energy for protons from collisions of Ar+Sc in Fig. 2. The top panel of Fig. 2a clearly shows that the reduced flow goes through a minimum. Our measurements are unable to distinguish the sign (+ or -) of collective flow which implies that such a minimum is indicative of the disappearance of flow. In order to determine E_{bal}, the extracted values of the reduced flow for beam energies below 95 MeV/nucleon are plotted as negative values in Fig. 2b. This procedure is in accordance with expectations from BUU calculations. E_{bal} is determined from this plot (Fig. 2b) by performing a linear least square fit to the reduced flow values. The solid line shown in Fig. 2b represents such a fit. The value of E_{bal} is fixed by the point or intersection of this line with the dashed line (zero reduced flow). In a similar manner, E_{bal} was extracted for the C+C, Ne+Al, Ar+Sc, and Kr+Nb data. Varying the point at which the values for the reduced flow changes

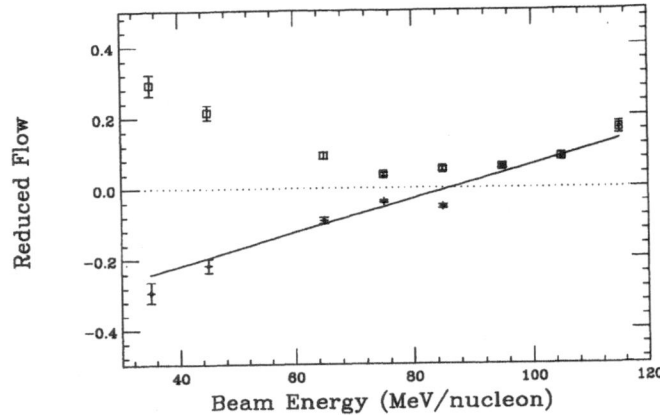

Figure 2: Reduced flow versus incident energy for protons from Ar+Sc. Squares represent reduced flow values as extracted while crosses stand for inverted reduced flow values for incident energies below 95 MeV/nucleon. Solid line is a linear fit.

sign makes little difference in the resulting value for E_{bal}. Nevertheless, this effect is included in the estimated errors for E_{bal}.

We have plotted E_{bal} (solid squares) as a function of the mass of the entrance-channel in Fig. 3. The solid line in this figure corresponds to a power law fit of the form $A^{-1/3}$ where A is the mass of the combined system. The results of BUU model calculations[14] (solid circles) are also included in Fig. 3. A power law fit to the calculated E_{bal} values is represented by a dotted line in Fig. 3. For these calculations, the in-medium nucleon-nucleon cross sections (σ_{nn}) were taken to be the free nucleon-nucleon cross sections and the incompressibility constant K was fixed at a value of 200 MeV.

The calculated dependence of E_{bal} on entrance-channel mass (dotted line in Fig. 3) is strikingly similar to that of the experimental data (solid line). On the other hand, the calculated E_{bal} values are systematically 10 to 20 MeV/nucleon below the experimental values. Also shown in Fig. 3 are BUU calculations with K=200 MeV in which the in-medium cross sections are reduced. In previous studies the free nucleon-nucleon cross section was multiplied with an overall constant scaling factor[5,12,14]. However, this approach fails when one has collisions in low-density nuclear matter, where the in-medium cross section should approach its free-space value. A more realistic approach uses a Taylor expansion of the in-medium cross section in the density variable.

We introduce the dimensionless parameter α_1 to describe the in-medium cross sections in terms of the free cross sections, $\sigma_{nn}^{in-medium} = (1 + \alpha_1 \frac{\rho}{\rho_0})\sigma_{nn}^{free}$. In principle, α_1 is dependent on energy, but we have here, as a first approximation, taken α_1 as energy independent. The BUU calculations with $\alpha_1 = -0.1$ predict a higher E_{bal} than the calculation incorporating the free cross section and are given by the diamonds in Fig. 3. The calculations with $\alpha_1 = -0.2$ roughly reproduce the observed values for E_{bal} as can be seen by the triangles in Fig. 3. The dotted lines correspond to power law fits. The BUU calculations with different α_1 clearly have a different slope in Fig. 3 none of which completely reproduced the quantitative behavior of E_{bal} although the qualitative behavior is correct.

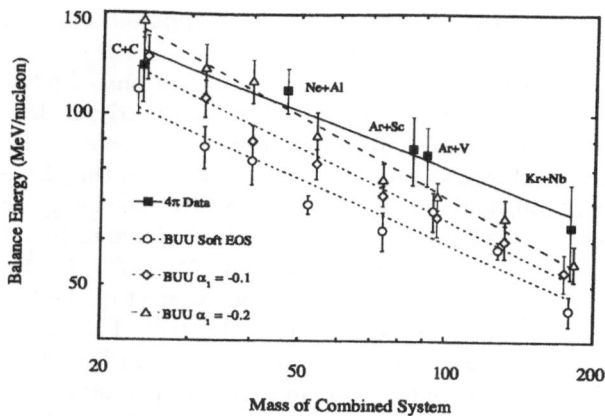

Figure 3: Extracted values of the balance energy (squares) for C+C, Ne+Al, Ar+Sc, and Kr+Nb compared with the predictions of the BUU model for a soft equation of state (circles). Also shown are BUU calculations corresponding to a 10% density-dependent reduction in the nucleon-nucleon cross sections (diamonds) and a 20% density-dependent reduction (triangles). The value for Ar+V is from previous work[9]. The lines are power law fits.

The dependence of E_{bal} on the mass of the combined system can be interpreted as a signature of the competition between the attractive mean field and the repulsive nucleon-nucleon interactions. The mean field can be associated with the surface of the two interacting nuclei and hence should scale as $A^{2/3}$. Similarly, the repulsive nucleon-nucleon interaction should scale as A. This being the case, one expects E_{bal} to show the mass dependence illustrated in Fig. 3. The fact that BUU calculations reproduce this dependence serves to confirm the interpretation of the disappearance of flow as a balancing of the attractive scattering (dominant at low energies) by the repulsive nucleon-nucleon scattering (dominant at high energies).

CRITICAL BEHAVIOR OF NUCLEAR MATTER

Central collisions of heavy ions at intermediate beam energies from 10 MeV/nucleon to several hundred MeV/nucleon go through two stages: an initial compression stage and a final expansion stage. In these reactions the compression can create nuclear matter density as high as twice normal nuclear density and equilibrated matter with mean excitation energy of several tens of MeV per nucleon. Therefore, intermediate energy heavy ion reactions provide information on the thermodynamic properties of nuclear matter, i.e. the nuclear matter equation of state (EOS), if finite size effects of the reaction system are taken into account. From consideration of the long range attractive mean field interaction and short range repulsive nucleon-nucleon interactions in nuclear reactions, a liquid-gas phase transition has been predicted by comparing the nuclear matter EOS with the Van der Waals EOS [15,16,17,18]. This phase transition is of first order, terminating in a second order phase transition at the critical point. The cluster size distribution at the critical point is predicted to be given by $\sigma(A) \propto A^{-\tau}$, where τ is a critical exponent with a value characteristic of the universality class of the phase transition and A is the cluster size. It has been shown [19,20,21] that for small-size systems in the vicinity of the critical point, the clusters size distribution can be fitted by a power law with apparent exponent λ which has a minimum, $\lambda_{min} = \tau$, at the critical point.

The previously measured inclusive cluster distributions of proton induced reaction [22,23] do exhibit power law features and indicate a critical behavior. However, these measurements summed over different impact parameters which have different excitation energies. To obtain an unambiguous signature of the phase transition, well-characterized central collisions with well-defined excitation energy deposition into the system need to be measured and the finite size effects have to be addressed.

Beam energies of 15, 25, 35, 45, 65, 75, 85, 95, 105, 115 MeV/nucleon were used to cover the excitation energy region for which most of the theories predict the occurrence of the second order phase transition [15,17,20,24]. The correction for detector acceptance and centrality selection have been discussed in Ref. [25].

To estimate the finite size effect, a bond breaking percolation calculation is performed. The bond breaking percolation model assumes that each nucleon is "linked" with its nearest neighbors by potential bonds. Each bond can absorb a maximum energy called the bond breaking energy, E_b, and has a probability, P_b, to break. Such simulations have allowed the fitting of P_b to experimental data for Z-distributions from heavy ion reactions [24].

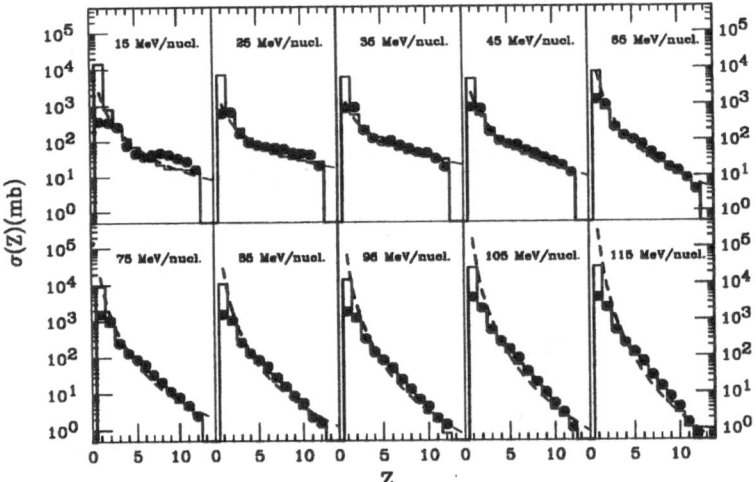

Figure 4: Z-distributions of both experimental data and percolation calculations of ^{40}Ar + ^{45}Sc at 15 to 115 MeV/nucleon. The solid circles are Z-distributions corrected for detector acceptance from central collisions of ^{40}Ar + ^{45}Sc and the histograms are percolation calculations with an initial lattice size of 68 and a binding energy of 7.8 MeV/nucleon. The percolation calculations are normalized to experimental data for $3 \leq Z \leq 12$. The dashed curves are power law fits to the percolation calculations.

We assume that the energy distributed into each bond, ϵ_b, can be described by a Boltzmann distribution with a mean energy $\langle \epsilon_b \rangle$. Each site of the lattice has an average of α bonds. The average excitation energy deposited per site is $\langle E_s \rangle = \alpha \langle \epsilon_b \rangle$; and the binding energy per nucleon of the initial nuclear system is $B = \alpha E_b$. When the system expands, any bond which has an energy greater than E_b will break. Therefore the bond breaking probability is,

$$P_b = \int_{E_b}^{\infty} \sqrt{\epsilon_b} e^{-\epsilon_b/t_b} d\epsilon_b \bigg/ \int_0^{\infty} \sqrt{\epsilon_b} e^{-\epsilon_b/t_b} d\epsilon_b \qquad (1)$$

$$= \int_{B}^{\infty} \sqrt{E_s} e^{-E_s/T_s} dE_s \bigg/ \int_0^{\infty} \sqrt{E_s} e^{-E_s/T_s} dE_s. \qquad (2)$$

Where $t_b = \frac{2}{3}\langle\epsilon_b\rangle$ and $T_s = \alpha t_b = \frac{2\alpha}{3}\langle\epsilon_b\rangle = \frac{2}{3}\langle E_s\rangle$ are slope parameters. We note that the bond breaking probability P_b calculated by equation (1) is independent of α, therefore the calculation is independent of the lattice structure. By fitting the proton kinetic energy spectra with a single moving Boltzmann source at midrapidity[26,27,28], we obtain the slope parameters T_s for each beam energy. The initial size of the lattice is assumed to be given by the fireball geometry for an overlap region of projectile and target with impact parameter of 0.25 b_{max} where b_{max} is the sum of the radii of the projectile and the target nuclei. Therefore we used an initial cubic lattice of 68 sites with the bond breaking probabilities calculated by equation (1) using the slope parameters of protons and a binding energy of 7.8 MeV/nucleon. The binding energy was used as a fitting parameter. We also compare this calculation to fragmentation data of 600 MeV/nucleon Au + C, Al, C in reference [29]. We convert the excitation energies calculated by reference [29] to beam energies of a symmetric system (projectile and target have equal masses) assuming total inelastic collisions. Then the same beam energy as ^{40}Ar + ^{45}Sc and proton slope parameters are used with an initial lattice of 150 sites to reproduce the Au + C, Al, Cu data. For Au + C, Al, Cu a 7.0 MeV/nucleon binding energy was found.

Fig. 4 shows the experimental Z-distributions corrected for detector acceptance from central collisions of ^{40}Ar + ^{45}Sc at beam energies from 15 to 115 MeV/nucleon (solid circle) compared to our percolation calculation (histogram). The dashed curve is the percolation calculation fitted to a power law distribution, $\sigma(Z) \propto Z^{-\lambda}$. The percolation results are normalized to the experimental data for $3 \leq Z \leq 12$.

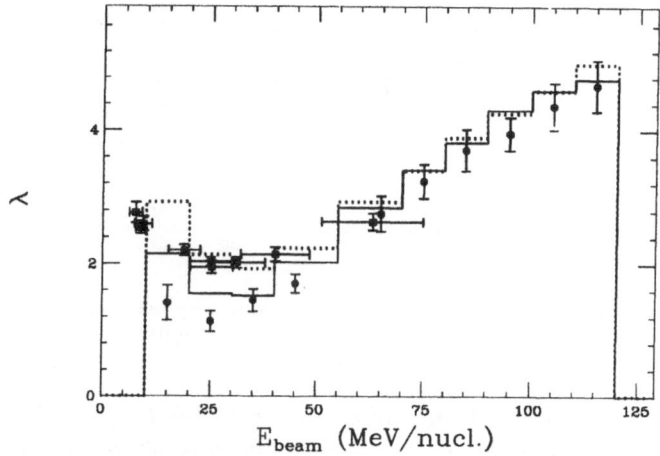

Figure 5: Comparison of power law fitting parameters, λ, to the percolation calculations with experimental results. The solid circles are the power law ($\sigma(Z) \sim Z^{-\lambda}$) fit to the experimental data of Ar + Sc from 15 to 115 MeV/nucleon and the open squares are GSI data of Au + C, Al, Cu at 600 MeV/nucleon [29]. The solid histogram is the power law fit to the percolation calculations with a lattice size of 68 calculated from fireball geometry and a binding energy of 7.8 MeV/nucleon. The dashed histogram is the percolation calculation for a lattice size of 150 with a binding energy of 7.0 MeV/nucleon.

The apparent exponent of the power law, λ, vs. beam energy is shown in Fig. 5. The solid circles are the power law fits to the experimental data of ^{40}Ar + ^{45}Sc and the solid histogram is the percolation calculation with 68 sites and a binding energy of 7.8 MeV/nucleon. The open squares are GSI data of Au + C, Al, Cu at 600 MeV/nucleon [29] and the dotted histogram is the percolation calculation with 150 sites and 7.0 MeV/nucleon binding energy. The equivalent beam energy on the plot for the GSI data is obtained by converting the calculated excitation energy[29] to a symmetric system assuming a total inelastic collision. To obtain the critical exponent, τ, we fit the λ vs. E_{beam} with a four term polynomial. For ^{40}Ar + ^{45}Sc we get $\tau = 1.21 \pm 0.01$ at a beam energy of 23.9 ± 0.7 MeV/nucleon. The percolation calculation with 68 sites and a binding energy of 7.8 MeV/nucleon gives $\tau = 1.5 \pm 0.1$ at a beam energy of 28 ± 0.4 MeV/nucleon. For GSI data of Au + C, Al, Cu we get $\tau = 2.0 \pm 0.01$ at a beam energy of 29 ± 0.2 MeV/nucleon. The percolation calculation with 150 sites and a binding energy of 7.0 MeV/nucleon gives $\tau = 1.98 \pm 0.03$ at a beam energy of 32.7 ± 0.1 MeV/nucleon. All errors are statistical.

In order to estimate the finite size effects and to obtain the critical excitation energy for infinite nuclear matter, we performed percolation calculations using a binding energy of 8 MeV/nucleon for different lattice sizes, ranging from 50 sites to 800 sites, and for slope parameters T_s ranging from 5 MeV to 19 MeV. The critical excitation energy increases when the lattice size increases. Above 400 sites, the critical value for the slope parameter converges to 13.1 ± 0.6 MeV. This value can be compared with the theoretical calculation of 15.3 MeV[15].

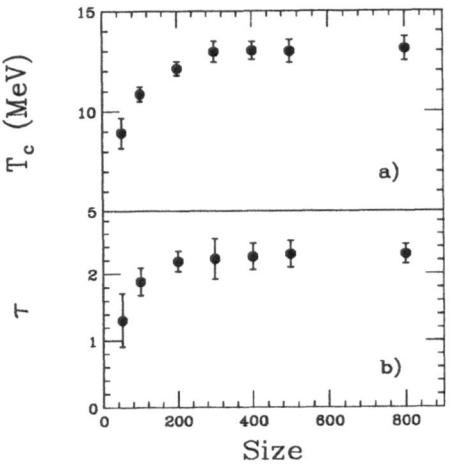

Figure 6: The size dependence of the critical value of slope parameter $T_c = T_s(\tau)$ and the critical exponent τ. a) The critical slope parameter T_c with different initial lattice size. b) The critical power law exponent τ as function of initial lattice size.

Fig. 6 a) shows the critical value of slope parameter $T_c = T_s(\tau)$, extracted from the polynomial fits, vs. the size of the lattice. Fig. 6 b) shows the critical exponent τ as a function of the lattice size. It approaches a limit of 2.3 ± 0.2 at a large size.

CONCLUSIONS

The disappearance of collective flow can provide information concerning the in-medium cross sections necessary for transport models to describe nucleus-nucleus collisions. We find that a reduction of about 20% in the scattering cross sections used in BUU is necessary to obtain qualitative agreement with observed balance energies. We related Z distributions from central collisions to critical behavior in nuclear matter using a percolation model. Using this model we extrapolated our results for the finite system, Ar+Sc, to infinite nuclear matter.

REFERENCES

[1] G.D. Westfall, J.E. Yurkon, J. van der Plicht, Z.M. Koenig, B.V. Jacak, R.Fox, G.M. Crawley, M.R. Maier, B.E. Hasselquist, R.S. Tickle, and D. Horn, Nucl. Inst. Meth. **A238**, 347 (1985).

[2] D. Cebra, S. Howden, J. Karn, D. Kataria, M. Maier, A. Nadasen, C. A. Ogilvie, N. Stone, D. Swan, A. M. Vander Molen, W. K. Wilson, J. S. Winfield, J. Yurkon, G. D. Westfall, E. Norbeck, Nucl. Instr. and Meth. **A300**, 518 (1991).

[3] H.H. Gutbrod, A.M. Poskanzer, and H.G. Ritter, Rep. Prog. Phys. **52**, 1267 (1989).

[4] P. Danielewicz, H. Ströbele, G. Odyniec, D. Bangert, R. Bock, R. Brockmann, J.W. Harris, H.G. Pugh, W. Rauch, R.E. Renfordt, A. Sandoval, D. Schall, L.S. Schroeder and R. Stock, Phys. Rev. **C38**, 120 (1988).

[5] C.A. Ogilvie, W. Bauer, D.A. Cebra, J. Clayton, S. Howden, J. Karn, A. Nadasen, A. Vander Molen, G.D. Westfall, W.K. Wilson, and J.S. Winfield, Phys. Rev. **C42**, R10 (1990).

[6] D. Krofcheck, W. Bauer, G.M. Crawley, C. Djalali, S. Howden, C.A. Ogilvie, A. Vander Molen, G.D. Westfall, and W.K. Wilson, R.S. Tickle, and C. Gale, Phys. Rev. Lett. **63**, 2028 (1989).

[7] J.P. Sullivan, J. Péter, D. Cussol, G. Bizard, R. Brou, M. Louvel, J.P. Patry, R. Regimbart, J.C. Steckmeyer, B. Tamain, E. Crema, H. Doubre, K. Hagel, G.M. Jin, A. Péghaire, F. Saint-Laurent, Y. Cassagnou, R. Lebrun, E. Rosato, R. Macgrath, S.C. Jeong, S.M. Lee, Y. Nagashima, T. Nakagawa, M. Ogihara, J. Kasagi, and T. Motobayashi, Phys. Lett. **B249**, 8 (1990).

[8] W.M. Zhang, R. Madey, M. Elaasar, J. Schambach, D. Keane, B.D. Anderson, A.R. Baldwin, J. Cogar, J.W. Watson, G.D. Westfall, G. Krebs, and H. Wieman, Phys. Rev. **C42**, R491 (1990).

[9] D. Krofcheck, D.A. Cebra, M. Cronqvist, R. Lacey, T. Li, C.A. Ogilvie, A. Vander Molen, K. Tyson, G.D. Westfall, W.K. Wilson, J.S. Winfield, A. Nadasen, and E. Norbeck, Phys. Rev. **C43**, 350 (1991).

[10] D. Krofcheck, W. Bauer, G.M. Crawley, S. Howden, C.A. Ogilvie, A. Vander Molen, G.D. Westfall, W.K. Wilson, R.S. Tickle, C. Djalali, and C. Gale, Phys. Rev. **C46**, 1416 (1992).

[11] J. Molitoris and Horst H. Stöcker, Phys. Lett. **162B**, 47 (1985).

[12] G.F. Bertsch, W.G. Lynch and M.B. Tsang, Phys. Lett. **189B**, 384 (1987).

[13] W.K. Wilson, R. Lacey, C.A. Ogilvie, and G.D. Westfall, Phys. Rev. **C45**, 768 (1992).

[14] W. Bauer, Phys. Rev. Lett. **61**, 2534 (1988) and W. Bauer, C.K. Gelbke, and S. Pratt, Ann. Rev. Nucl. Part. Sci. **42**, 77 (1992).

[15] David H. Boal, Alan L. Goodman, Phys. Rev. **C33**, 1690 (1986).

[16] Alan L. Goodman, Joseph I. Kapusta, Aram Z. Mekjian, Phys. Rev. **C30** 851 (1984).

[17] Joseph Kapusta, Phys. Rev. **C29** 1735 (1984).

[18] H. R. Jaqaman, A. Z. Mekjian, and L. Zamick, Phys. Rev. **C29** 2067 (1984).

[19] W. Bauer, D. R. Dean, U. Mosel and U. Post, Phys. Lett. **150B** 53 (1985).

[20] A. D. Panagiotou, M. W. Curtin, H. Toki, D. K. Scott, and P. J. Siemens, Phys. Rev. Lett. **52**, 496 (1984).

[21] M. E. Fisher, Physics **3**, 255 (1967).

[22] N. T. Porile, A. J. Bujak, D. D. Carmony, Y. H. Chung, L. J. Gutay, A. S. Hirsch, M. Mahi, G. L. Paderewski, T. C. Sangster, R. P. Scharenberg, and B. C. Stringfellow, Phys. Rev. **C39**, 1914 (1989).

[23] A. S. Hirsch, A. Bujak, J. E. Finn, L. J. Gutay, R. W. Minich, N. T. Porile, R. P. Scharenberg, and B. C. Stringfellow, Phys. Rev. **C29**, 508 (1984).

[24] W. Bauer, Phys. Rev. **C38**, 1297 (1988).

[25] T. Li, W. Bauer, D. Craig, M. Cronqvist, E. Gualtieri, S. Hannuschke, R. Lacey, W. J. Llope, T. Reposeur, A. M. Vander Molen, G. D. Westfall, W. K. Wilson, J. S. Winfield, J. Yee, S. J. Yennello, A. Nadasen, R. S. Tickle and E. Norbeck, Phys. Rev. Lett. **70** 1924 (1993).

[26] G. D. Westfall, J. Gosset, P. J. Johansen, A. M. Poskanzer, W. G. Meyer, H. H. Gutbrod, A. Sandoval and R. Stock, Phys. Rev. Lett. **37**, 1202 (1976).

[27] G. D. Westfall, B. V. Jacak, N. Anantaraman, M. W. Curtin, G. M. Crawley, C. K. Gelbke, B. Hasselquist, W. G. Lynch, D. K. Scott, B. M. Tsang, M. J. Murphy, T. J. M. Symons, R. Legrain, and T. J. Majors, Phys. Lett. **116B**, 118 (1982).

[28] B. V. Jacak, G. D. Westfall, G. M. Crawley, D. Fox, C. K. Gelbke, L. H. Harwood, B. E. Hasselquist, W. G. Lynch, D. K. Scott, H. Stöcker, M. B. Tsang, and G. Buchwald, Phys. Rev. **C35**, 1751 (1987).

[29] C. A. Ogilvie, J. C. Adloff, M. Begemann-Blaich, P. Bouissou, J. Hubele, G. Imme, I. Iori, P. Kreutz, G. J. Kunde, S. Leray, V. Lindenstruth, Z. Liu, U. Lynen, R. J. Meijer, U. Milkau, W. F. J. Müller, C. Ngô, J. Pochodzalla, G. Raciti, G. Rudolf, H. Sann, A. Schüttauf, W. Seidel, L. Stuttge, W. Trautmann, and A. Tucholski, Phys. Rev. Lett. **67**, 1214 (1991).

ANTISYMMETRIZED MOLECULAR DYNAMICS

H. Horiuchi

Department of Physics, Kyoto University, Kyoto 606-01, Japan

INTRODUCTION

The purpose of this lecture is to explain the framework of the new microscopic simulation theory of heavy ion collisions named antisymmetrized molecular dynamics (AMD)[1~7] and to discuss the results of some analyses of experimental data with this theory which is to elucidate characteristic features of this theory.

The AMD can be regarded as being an antisymmetrized version of the quantum molecular dynamics (QMD)[8,9] or the quasi-particle dynamics (QPD)[10]. The construction of the AMD is made by incorporating the stochastic two-nucleon collision process into the fermionic molecular dynamics (FMD)[11]. The FMD is a special case of the theory of the time-dependent cluster model (TDCM)[12] where every cluster in the system is composed of a single nucleon. Hence the physical nucleon coordinate of the AMD which enables the incorporation of the two-nucleon collision process into the FMD is made by mimicking the theory[13] of the canonical coordinate of the TDCM.

The AMD is a quantum mechanical theory, since it describes the time development of the system wave function. It is to be stressed that the initialization problem is now solved satisfactorily because we initiate the collision calculation by boosting the quantum mechanical ground state wave functions of colliding nuclei. Furthermore, as will be shown later explicitly, we are able to treat the shell effect in the fragment production which is important for the investigation of the dynamical stage of the collision process through the analysis of the fragment data.

FORMULATION OF AMD

Equation of Motion of AMD

In AMD, the wave function of the A-nucleon total system is described by a Slater determinant,

$$
\Phi(\{\vec{Z}_j\}) = \frac{1}{\sqrt{A!}} \det\left[\varphi_j(\vec{r}_k)\chi_{\alpha_j}(\xi_k)\right], \quad \vec{Z}_j = \sqrt{\nu}\vec{D}_j + \frac{i}{2\hbar\sqrt{\nu}}\vec{K}_j,
$$

$$
\varphi_j(\vec{r}) = (2\nu/\pi)^{3/4} \exp[-\nu(\vec{r} - \frac{\vec{Z}_j}{\sqrt{\nu}})^2 + \frac{1}{2}\vec{Z}_j^2] \propto \exp[-\nu(\vec{r} - \vec{D}_j)^2 + i\frac{\vec{K}_j}{\hbar} \cdot \vec{r}],
$$

(1)

where $\chi_{\alpha_j}(\xi)$ stands for the spin-isospin function and α_j represents the spin and isospin label of the j-th single particle state, $\alpha_j = $ p $\uparrow$, p $\downarrow$, n $\uparrow$, or n $\downarrow$. The

Hot and Dense Nuclear Matter, Edited by
W. Greiner *et al.*, Plenum Press, New York, 1994

time developments of the coordinate parameters $\{\vec{Z}_j\}$ are determined by the time-dependent variational principle;

$$\delta \int_{t_1}^{t_2} dt \, \frac{<\Phi(\{\vec{Z}_j\})|(i\hbar\frac{d}{dt}-H)|\Phi(\{\vec{Z}_j\})>}{<\Phi(\{\vec{Z}_j\})|\Phi(\{\vec{Z}_j\})>} = 0, \tag{2}$$

which leads to the equation of motion for $\{\vec{Z}_j\}$;

$$i\hbar \sum_{j\tau} C_{k\sigma,j\tau} \dot{Z}_{j\tau} = \frac{\partial}{\partial Z_{k\sigma}^*} \frac{<\Phi(\{\vec{Z}_j\})|H|\Phi(\{\vec{Z}_j\})>}{<\Phi(\{\vec{Z}_j\})|\Phi(\{\vec{Z}_j\})>},$$

$$C_{k\sigma,j\tau} \equiv \frac{\partial^2}{\partial Z_{k\sigma}^* \partial Z_{j\tau}} \log <\Phi(\{\vec{Z}_j\})|\Phi(\{\vec{Z}_j\})>, \tag{3}$$

where $\sigma, \tau = x, y, z$. We can prove that the Hermitian matrix $\{C_{k\sigma,j\tau}\}$ is positive definite.

In order to make AMD applicable to heavy ion collisions, the following two problems should be further treated. One is the *initialization problem*, namely the construction of the ground state wave functions of colliding nuclei, and the other is the *incorporation of two-nucleon collisions*. The construction of the ground state wave function means to determine the values of all the coordinate parameters $\{\vec{Z}_j\}$ which minimize $<\Phi(\{\vec{Z}_j\})|H|\Phi(\{\vec{Z}_j\})> / <\Phi(\{\vec{Z}_j\})|\Phi(\{\vec{Z}_j\}) >$. It is a multi-dimensional variational problem. The incorporation of two-nucleon collisions is not a simple problem because the position parameters $\{\vec{D}_j\}$ do not represent nucleon positions and also the momentum parameters $\{\vec{K}_j\}$ do not represent nucleon momenta. This fact can be easily understood when we consider the following example. Let all $\vec{D}_j$ and $\vec{K}_j$ be vanishingly small in ^{16}O system. The resulting total wave function Φ is almost equal to the harmonic oscillator shell model wave function $\Phi_{c.s.}$ of the double closed shell configuration of $0s$ and $0p$ shells, $(0s)^4(0p)^{12}$. Nucleons of the state $\Phi_{c.s.}$ are neither all located at the space-coordinate origin nor all condensed to zero momentum state in spite of $\vec{D}_j \approx 0$, $\vec{K}_j \approx 0$ for all j.

The initialization problem is solved by the use of the frictional cooling method, which is discussed in the next section. There we will see that the AMD wave function together with the frictional cooling technique presents us a novel and powerful method for the the nuclear structure study. The incorporation of two-nucleon collisions is solved by the introduction of the physical nucleon coordinate parameters $\{\vec{W}_j\}$, which we discuss below in this section.

Time-Dependent Cluster Model

As we mentioned before, the FMD, namely the AMD without two-nucleon collisions, is nothing but a special case of the TDCM (time-dependent cluster model). Therefore we can utilize the ideas and techniques of the TDCM in order to incorporate the two-nucleon collision into the FMD. The wave function of the TDCM has the form

$$\hat{\Phi}(\{\vec{Z}_j\}) = \det\left[\Psi_I(\vec{D}_I, \vec{K}_I)\Psi_{II}(\vec{D}_{II}, \vec{K}_{II})\cdots \right], \quad \vec{Z}_j \equiv \sqrt{A_j\nu}\vec{D}_j + \frac{i}{2\hbar\sqrt{A_j\nu}}\vec{K}_j,$$

$$\Psi_j(\vec{D}_j, \vec{K}_j) = Slater\ determinant \propto \exp\left[-A_j\nu\left(\vec{r}_{Gj} - \vec{D}_j\right)^2 + i\frac{\vec{K}_j}{\hbar}\vec{r}_{Gj} \right]\Phi(j),$$

$$\propto \quad \exp\left[-A_j\nu\left(\vec{r}_{Gj} - \frac{\vec{Z}_j}{\sqrt{A_j\nu}}\right)^2 + \frac{1}{2}\vec{Z}_j^2\right]\Phi(j),$$

$$j = I, II, III, \cdots, \quad A_j = mass \ number \ of \ cluster \ j,$$

$$\vec{r}_{Gj} = center \ of \ mass \ coordinate \ of \ cluster \ j,$$

$$\Phi(j) = internal \ wave \ function \ of \ cluster \ j. \tag{4}$$

The time development of the coordinate parameters $\{\vec{D}_j\}$ and $\{\vec{K}_j\}$ (or $\{\vec{Z}_j\}$ is determined by the time-dependent variational principle;

$$\delta \int_{t_1}^{t_2} dt \frac{< \hat{\Phi}(\{\vec{Z}_j\})|(i\hbar\frac{d}{dt} - H)|\hat{\Phi}(\{\vec{Z}_j\}) >}{< \hat{\Phi}(\{\vec{Z}_j\})|\hat{\Phi}(\{\vec{Z}_j\}) >} = 0, \tag{5}$$

which leads to the equation of motion ;

$$i\hbar \sum_{j\tau} C_{k\sigma,j\tau} \dot{Z}_{j\tau} = \frac{\partial}{\partial Z_{k\sigma}^*} \frac{< \hat{\Phi}(\{\vec{Z}_j\})|H|\hat{\Phi}(\{\vec{Z}_j\}) >}{< \hat{\Phi}(\{\vec{Z}_j\})|\hat{\Phi}(\{\vec{Z}_j\}) >}, \quad and \quad c.c.,$$

$$C_{k\sigma,j\tau} \equiv \frac{\partial^2}{\partial Z_{k\sigma}^* \partial Z_{j\tau}} \log < \hat{\Phi}(\{\vec{Z}_j\})|\hat{\Phi}(\{\vec{Z}_j\}) >, \tag{6}$$

where $\sigma, \tau = x, y, z$. We can clearly recognize that the FMD is a special case of the TDCM where every cluster is composed of single nucleon.

In TDCM, it is well known that the center-of-mass position parameter $\vec{D}_j$ and the center-of-mass momentum parameter $\vec{K}_j$ of the j-th cluster do not necessarily represent the real position and momentum of the j-th cluster center of mass on account of antisymmetrization. An important progress made in the theoretical framework of the TDCM was the proposal of the theory of the canonical coordinates of Ref.[13]. In the case of the two-cluster system, the authors of Ref.[13] showed how to construct explicitly the exact canonical position and momentum coordinates from the original coordinates $\{\vec{D}_j\}$ and $\{\vec{K}_j\}$. The canonical relative position and momentum coordinates, $\vec{R}_r$ and $\vec{P}_r$, are given as

$$\vec{W}_r \equiv \sqrt{Q}\vec{Z}_r, \quad Q \equiv \frac{\partial}{\partial(\vec{Z}_r^* \vec{Z}_r)} \log < \hat{\Phi}(\{\vec{Z}_j\})|\hat{\Phi}(\{\vec{Z}_j\}) >,$$

$$\vec{Z}_r = \sqrt{\mu\nu}\vec{D}_r + \frac{i}{2\hbar\sqrt{\mu\nu}}\vec{K}_r, \quad \vec{W}_r = \sqrt{\mu\nu}\vec{R}_r + \frac{i}{2\hbar\sqrt{\mu\nu}}\vec{P}_r,$$

$$\vec{D}_r \equiv \vec{D}_1 - \vec{D}_2, \quad \vec{K}_r \equiv \mu(\vec{K}_1/A_1 - \vec{K}_2/A_2), \quad \mu \equiv A_1 A_2/(A_1 + A_2). \tag{7}$$

An important property of the canonical coordinate $\vec{W}_r$ is that there exists a *Pauli-forbidden region*[13] in the phase space of $\vec{W}_r$. We can easily prove that $\vec{W}_r$ satisfies

$$\vec{W}_r^* \vec{W}_r = \frac{1}{\hbar\omega}\left(\frac{\vec{P}_r^2}{2\mu} + \frac{\mu\omega^2}{2}\vec{R}_r^2\right) \geq (N_{\text{osc}})_{\text{min}},, \tag{8}$$

where $(N_{\text{osc}})_{\text{min}}$ stands for the lowest number of the oscillator quanta which the wave function of the relative motion can have[13,14,15]. Hence in other words, $\vec{W}_r$ cannot enter the Pauli-forbidden region defined by $|\vec{W}_r| < (N_{\text{osc}})_{\text{min}}$. The Pauli-forbidden region expresses, in the language of the phase space, the functional space spanned by the Pauli-forbidden states of the inter-cluster relative motion[14,15].

Two-Nucleon Collisions in AMD

We mimick the theory of the canonical coordinates of the TDCM in constructing physical nucleon coordinates. The physical nucleon coordinates $\{\vec{R}_j, \vec{P}_j, j = 1 \sim A\}$ we use are defined as follows,

$$\vec{W}_j = \sqrt{\nu}\vec{R}_j + \frac{i}{2\hbar\sqrt{\nu}}\vec{P}_j \equiv \sum_{k=1}^{A} \left(\sqrt{Q}\right)_{jk} \vec{Z}_k,$$

$$Q_{jk} \equiv \frac{\partial}{\partial(\vec{Z}_j^* \cdot \vec{Z}_k)} \log < \Phi(\{\vec{Z}_j\})|\Phi(\{\vec{Z}_j\}) > . \tag{9}$$

In the absence of antisymmetrization we of course have $\vec{W}_j = \vec{Z}_j$. We can easily prove the following relations:

$$N_{\text{osc}} = < \Phi(\{\vec{Z}_j\})| \sum_j \vec{a}_j^\dagger \vec{a}_j |\Phi(\{\vec{Z}_j\}) > / < \Phi(\{\vec{Z}_j\})|\Phi(\{\vec{Z}_j\}) >= \sum_{jk} \vec{Z}_j^* \vec{Z}_k Q_{jk}$$

$$= \sum_j \vec{W}_j^* \vec{W}_j = \frac{1}{\hbar\omega}\sum_j (\frac{1}{2m}\vec{P}_j^2 + \frac{m\omega^2}{2}\vec{R}_j^2),$$

$$< \Phi(\{\vec{Z}_j\})|\vec{L}|\Phi(\{\vec{Z}_j\}) > / < \Phi(\{\vec{Z}_j\})|\Phi(\{\vec{Z}_j\}) >= (-i\hbar) \sum_{jk} \vec{Z}_j^* \times \vec{Z}_k Q_{jk}$$

$$= (-i\hbar) \sum_j \vec{W}_j^* \times \vec{W}_j = \sum_j \vec{R}_j \times \vec{P}_j, \tag{10}$$

where $\vec{a}_j$ stands for the destruction operator of harmonic oscillator quanta of j-th nucleon, $\vec{a}_j = \sqrt{\nu}\vec{r}_j + (i/2\hbar\sqrt{\nu})\vec{p}_j$, and $\vec{L}$ is the total orbital angular momentum operator. Furthermore as for the center-of-mass coordinate we can show $\sum_j \vec{Z}_j = \sum_j \vec{W}_j$. Also there holds the following relation; namely if $\vec{Z}_j$ are simultaneously displaced by $\vec{c}$, $\vec{Z}_j \rightarrow \vec{Z}_j + \vec{c}$, then $\vec{W}_j$ are simultaneously displaced by the same amount $\vec{c}$, $\vec{W}_j \rightarrow \vec{W}_j + \vec{c}$. From these properties of $\{\vec{W}_j\}$, we regard that $\{\vec{W}_j\}$ are physical coordinates.

In the case of two-nucleon system, these physical coordinates are nothing but the canonical coordinates introduced by Saraceno et al. in the two-cluster system in the TDCM.[13] However, in general systems containing more than two nucleons, our physical coordinates are not exact canonical coordinates.

Once we have physical nucleon coordinates, we can treat two-nucleon collisions just as the two-nucleon collisions in QMD. When $\vec{R}_j$ and $\vec{R}_k$ come near each other, two nucleons j and k are made to scatter. Namely initial $\vec{P}_j$ and $\vec{P}_k$ are changed into final $\vec{P}_j'$ and $\vec{P}_k'$ keeping initial $\vec{R}_j$ and $\vec{R}_k$ unchanged. Speaking in terms of complex coordinates $\vec{W}$, initial $\vec{W}_j$ and $\vec{W}_k$ are changed into final $\vec{W}_j'$ and $\vec{W}_k'$. In order to continue the calculation of time development of the system wave function after this two-nucleon collision, we need to back-transform $\{\vec{W}_1, \cdots, \vec{W}_j', \cdots, \vec{W}_k', \cdots, \vec{W}_A\}$ into $\{\vec{Z}_1', \vec{Z}_2', \cdots, \vec{Z}_A'\}$. However, in general, the back-transformation from $\{\vec{W}_j, j = 1 \sim A\}$ to $\{\vec{Z}_j, j = 1 \sim A\}$ does not always exist. When the back-transformation does not exist, we regard that the two-nucleon collision is Pauli-blocked.

218

The existence of $\{\vec{W}_j, j = 1 \sim A\}$ for which we have no corresponding $\{\vec{Z}_j, j = 1 \sim A\}$ is easily verified by using the relation $N_{\text{osc}} = \sum_j \vec{W}_j^* \vec{W}_j$ of Eq.(9). This immediately shows that the coordinate origin $\{\vec{W}_j = 0, j = 1 \sim A\}$ cannot be constructed from any $\{\vec{Z}_j, j = 1 \sim A\}$, because N_{osc} can not be zero if $A > 4$. When we choose the center-of-mass at the coordinate origin, $\sum_j \vec{W}_j = 0$, we have

$$(N_{\text{osc}})_{\min} \leq \frac{1}{A} \sum_{i<j} |\vec{W}_i - \vec{W}_j|^2, \tag{11}$$

from which we get $\sqrt{< |\vec{W}_i - \vec{W}_j|^2 >_{\text{average}}} = \sqrt{2(N_{\text{osc}})_{\min}/(A-1)} \geq \sqrt{2}$. If the relative distances $|\vec{W}_i - \vec{W}_j|$ are too small as to violate the relation of Eq.(11), such $\{\vec{W}_j, j = 1 \sim A\}$ cannot be back-transformed to any $\{\vec{Z}_j, j = 1 \sim A\}$. $\{\vec{W}_j, j = 1 \sim A\}$ is defined to be in Pauli-forbidden region if it cannot be back-transformed to any $\{\vec{Z}_j, j = 1 \sim A\}$.

The notion of the Pauli-forbidden region defined above is obviously an extension of that of the TDCM explained in the previous subsection.

INITIALIZATION
—Construction of Ground State Wave Functions—

Frictional Cooling Method

The frictional cooling method which we use for the construction of ground state wave functions in AMD[5,6] is originally due to the cooling technique of Wilet et al.[16] in classical molecular dynamics approach which was used for example also in Ref.[10]. The AMD wave function to be constructed by the frictional cooling method can be extended into more general form than in Eq.(1) as follows:

$$\Phi^\pm = \Phi^\pm(\{\vec{Z}_j\}) + C\Phi^\pm(\{\vec{Z}_j'\}) + C'\Phi^\pm(\{\vec{Z}_j''\}) + \ldots\ldots,$$
$$\Phi^\pm(\{\vec{Z}_j\}) \equiv (1 \pm P)\Phi(\{\vec{Z}_j\}), \tag{12}$$

where P is the parity operator.

In order to construct the wave function Φ for the ground state we must determine the values of all the parameters, $(\{\vec{Z}_j\}, \{\vec{Z}_j'\}, \{\vec{Z}_j''\}, \ldots, C, C', \ldots)$, which minimize $< \Phi^\pm|H|\Phi^\pm > / < \Phi^\pm|\Phi^\pm >$. First we choose randomly parameters $\{u_k\} \equiv (\{\vec{Z}_j\}, \{\vec{Z}_j'\}, \{\vec{Z}_j''\}, \ldots, C, C', \ldots)$. The wave function $\Phi^\pm$ (or Φ) with these parameters expresses a highly excited state with gaseous configuration of nucleons. This initial state is then cooled down to the energy minimum state by applying the equation of frictional cooling which is given as

$$i\hbar \frac{d}{dt} u_k = (\lambda + i\mu) \frac{\partial}{\partial u_k^*} \frac{< \Phi^\pm|H|\Phi^\pm >}{< \Phi^\pm|\Phi^\pm >}. \tag{13}$$

We can easily show the relation

$$\frac{d}{dt} \frac{< \Phi^\pm|H|\Phi^\pm >}{< \Phi^\pm|\Phi^\pm >} < 0, \quad for \ \mu < 0, \quad \lambda = arbitrary. \tag{14}$$

Actual calculations have shown that the convergence to the minimum energy state is very fast (especially of course when a single AMD Slater determinant is adopted).

The most important feature of the construction of the ground state with the frictional cooling method is that it is without prejudice i.e. free from model assumptions. It does not use any assumptions such as axial symmetry, existence of clusters, and so on.

Converged ground state configurations of the self-conjugate 4N nuclei have proved to be as follows[5,6]: They have negligibly small values for all K_j in all the cases of nuclei. In ^{8}Be, ^{12}C, and ^{16}O, ground state wave functions are almost identical to the Brink-type[17] two-alpha intrinsic wave function, the intrinsic state wave function of the Elliott SU(3) model[18], $(0s)^4(0p)^8(\lambda, \mu) = (0, 4)$, and the double closed shell wave function, $(0s)^4(0p)^{12}$, respectively. These results are due to the calculations when a single AMD Slater determinant is adopted. But it was checked that when the linear combination of two AMD Slater dterminants is adopted, the main AMD Slater determinant is almost the same as the AMD Slater determinant obtained by the calculation using a single AMD Slater determinant and the minor AMD Slater determinant mixes in with very small coefficient C. The above results show two things: First, the $\alpha - \alpha$ structure of ^{8}Be has been proved theoretically without assuming the existence of any cluster structure. Second, the ground state wave functions expressed by the AMD theory are good and realistic.

The method of the frictional cooling can be extended so as to be used under given constraints[6,7]. An example of the application of the constrained frictional cooling method is the study of the structure-change of the ^{20}Ne nucleus along its yrast line for both positive and negative parities. The constraint in this case is that on the expectation value of the total angular momentum operator $\vec{J}$. The structure change from $\alpha + ^{16}$O clustering strucure to oblately deformed shape and then to ^{12}C + 2α was obtained.

Application to the Structure Study of Li, Be, and B Isotopes

In order to demonstrate that the AMD wave function together with the frictional cooling technique presents us a novel and powerful method for the nuclear structure study, we briefly report here some of the results of the structure study of Li, Be, and B isotopes with AMD[19]. Since the AMD wave function constructed with the frictional cooling method represents the intrinsic state, we make the angular momentum projection by performing the three dimensional numerical integration. The two-nucleon force in the Hamiltonian H consists of the density-dependent nuclear force MV1-case 3[20], the two-nucleon spin-orbit force G3RS[21], and the Coulomb force. The oscillator constant ν was chosen so as to get the optimum binding energy for each instrinsic state with definite parity.

Calculations with linearly combined two AMD Slater determinants were compared to those with single AMD Slater determinant. In most cases it was found that the main AMD Slater determinant is almost the same as the AMD Slater determinant obtained by the calculation using a single AMD Slater determinant and the minor AMD Slater determinant mixes in with very small coefficient C. When we discuss the state with neutron halo, however, the linear combination of only two AMD Slater determinants is not sufficient.

The energy level spectra are reproduced well like the results reported in Ref.[6]. Fig.1 shows good reproduction of magnetic moments which are close to the Schmidt value for nuclei with $N \approx 8$. This good reproduction is due to the change of the structure which has pronounced clustering for $N \approx Z$ but is shell-model-like for $N \approx 8$.

Since the single-particle wave function of AMD is a Gaussian wave packet with common oscillator constant, the long tail of neutron wave function of the neutron halo nucleus can not be sufficiently expressed by the linear combination of two AMD Slater determinants. Hence in the case of the positive parity state of ^{11}Be, we have superposed several AMD Slater determinants. The resulting density distributions of neutron and proton are shown in Fig.2.

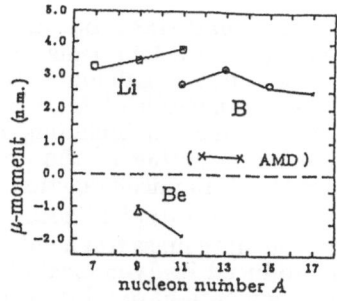

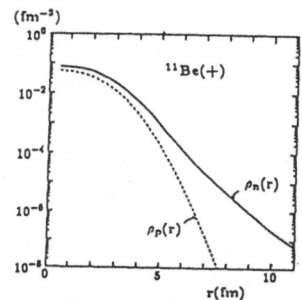

Fig.1 Magnetic moments. Crosses connected with lines are calculated values with AMD, while squares, circles and triangles represent data.

Fig.2 Density distribution of the lowest $1/2^+$ state of ^{11}Be calculated with AMD by superposing Slater determinants.

SHELL EFFECTS IN FRAGMENT FORMATION

We analysed fragment production in ^{12}C + ^{12}C reaction at 28.7 MeV/u by our new microscopic simulation framework AMD[1,2]. Since in this Fermi energy region the compression effect is not large, we used the Volkov No.1 force[22] which is a density-independent force. We checked that the ground state energies of various nuclides lighter than ^{12}C are reproduced very well in our AMD framework. Needless to say, the ground state wave functions of these nuclides are constructed with the frictional cooling method. The r.m.s. radius of ^{12}C is calculated to be 2.49 fm while its observed value is 2.48 fm.

Fig.3 shows the comparison of the calculated production cross sections of fragments with the data[23]. We see gross feature of the data is reproduced by theory. Especially the observed large cross section of α particle is reproduced rather well by the theory. The enhanced production cross section of the α particle by the AMD implies that the AMD can treat the shell effect in the fragment production. This ability of the AMD is due to the quantum mechanical nature of the AMD theory. It is to be noted that the previous simulation frameworks such as QMD and VUU / BUU (Vlasov / Boltzmann Uehling Uhlenbeck) can not describe this kind of shell effect during the dynamical reaction process since they are of classical nature.

We note in this figure that the calculation underestimates the observed cross section of the intermediate mass fragments ($A_f \sim 6, 7$). Produced fragments which

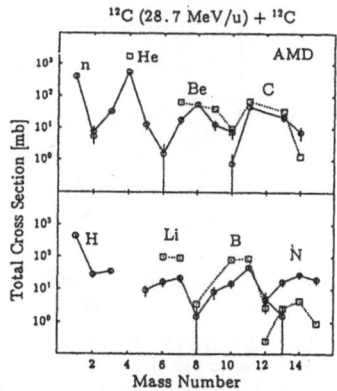

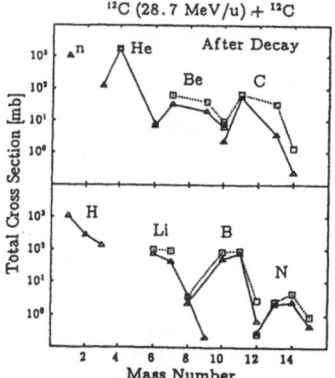

Fig.3 Isotope distribution. Circles with error bars are the results of AMD calculation truncated at 200 fm/c, while boxes represent observed values. Lines connect isotopes.

Fig.4 Isotope distribution after the statistical cascade decays. Triangles are the results of the cascade calculation while boxes represent observed values. Lines connect isotopes.

are analysed at around 200 fm/c are not in their ground states but are fairly excited. These fragments will decay by evaporating particles in a long time scale. Therefore in order to compare theory with experiments, we need to calculate statistical decay processes of produced fragments. Fig.4 is the comparison of fragment production cross section between the data and the AMD calculation including the statistical cascade decay calculation. We see very nice reproduction of the data by theory. About 1/3 of the α particle yield comes from the dynamical reaction process while large part of the remaining 2/3 of α yield is due to the statistical decay of excited ^{12}C produced in dynamical stage. The yield of intermediate mass fragments is supplied by statistical decays of various heavier clusters produced in dynamical stage, while the the dynamically produced intermediate mass fragments have all decayed into lighter clusters.

FRAGMENT FLOW AND NUCLEON FLOW

The collective flow in heavy ion reactions in the intermediate and high energy regions has been studied extensively under the expectation that one can extract the nuclear equation of state (EOS) from these studies. It has turned out, however, that the extraction of the EOS is not so easy because the collective flow depends not only on the mean field but also on the nucleon-nucleon cross section which has theoretical ambiguity in the nuclear medium of the colliding system.

Recently the collective flow has been measured with the identification of charges and/or mass numbers of fragments[24,25]. The large flow of fragments (or clusters) is observed compared to the flow of nucleons. It suggests that the flow of composite fragments carries the direct information of the EOS (or the mean field), while most nucleons are emitted by the hard stochastic collisions which erase the effect of the mean field. Therefore we can expect that the study of the fragment flow together with the nucleon flow may give us an important information on the in-medium cross section and hence on EOS.

For the theoretical analysis of the fragment flow, the theory should be able to describe the dynamical formation of of fragments. Moreover since the flow of α particles is important among the flows of various fragments, the theory is required to have the ability of describing the shell effect in the fragment formation. Thus we see that the AMD is very advantageous for the study of the fragment flow in comparison of the nucleon flow.

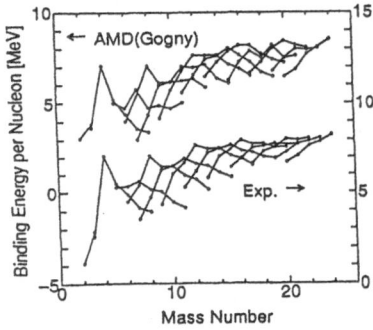

Fig.5 Binding energies per nucleon of nuclei calculated with AMD with the Gogny force. Isotopes are connected with lines. Experimental data are shown by shifting by -5 MeV.

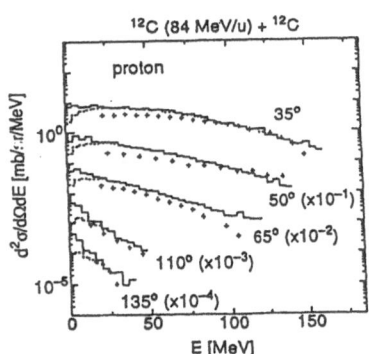

Fig.6 Energy spectra of protons. Dashed histograms are the proton spectra calculated before the statistical cascade decay, and the solid histograms are the final calculated results including the contributions from statistical decays. Experimental data are shown by crosses. Energy E and angles are in the laboratory system.

Under the above motivation, we studied the fragment flow and the nucleon flow in the system of $^{12}C + ^{12}C^4$. Below we report some results of the study. The study of heavier systems such as $^{40}Ar + ^{27}Al$ is in progress and will be reported elsewhere.

The balance energy of the $^{12}C + ^{12}C$ system was reported in Ref.[25] to be 122 ± 12 MeV/nucleon and hence our study was made in the energy region 45 MeV/nucleon $\le E \le 150$ MeV/nucleon. As the effective nuclear force we used the Gogny force[26] which is density-dependent and of finite range. The momentum dependence of the mean field given by the Gogny force is reliable below 200 MeV, since the energy dependence of the real part of the nucleon optical potential reproduced in this energy range. The Gogny force can describe the saturation property of the nuclear matter and gives the soft EOS with the incompressibility $K = 228$ MeV. Fig.5 shows binding energies of nuclei lighter than ^{24}Mg calculated with the AMD with the Gogny force, which are in good accordance with the data.

In order to show the reliability of the AMD in the incident energy region around 100 MeV/nucleon, we display in Fig.6 the good reproduction of the proton energy spectra in the 84 MeV/nucleon $^{12}C + ^{12}C$ reaction. The dotted histograms are the proton energy spectra at 150 fm/c, while the solid histograms are the final results containing the contributions from the statistical decay of excited frafments after 150 fm/c.

Here we define the flow of fragments with mass number A_F as the mean in-plane transverse momentum, $< wP_x > /A_F$, where $w = 1$ for $P_x > 0$ and $w = -1$ for $P_x < 0$. P_x and P_z are the components of the in-plane momentum of the fragment in the center-of-mass system which are perpendicular to and along with the beam direction, respectively. This definition of the flow was already used in Ref.[27]. We checked that the qualitative features of the calculated results did not change even if we adopt the usual definition of the flow as the slope of $< P_x > -P_z$ curve.

Figure 7 shows the energy dependence of the calculated flows of nucleons and α particles, where as well as in all the results hereafter the impact parameter is fixed to be 2 fm. We see that the calculated balance energy E_{bal} is 100 ± 20 MeV/nucleon which is close to the observed value. A remarkable feature is that the absolute value of the α-particle flow is much larger than the nucleon flow in the energy region $E < E_{bal}$. In order to clarify the reason for this large flow of fragments, we distinguished the flow of dynamically produced particles from the flow of the particles produced by the statistical decays after the dynamical stage of the reaction. Figure 8 shows the dynamical flows calculated at 150 fm/c and the evaporation flows due to the statistical decays after 150 fm/c. We see a clear difference between dynamical flow of

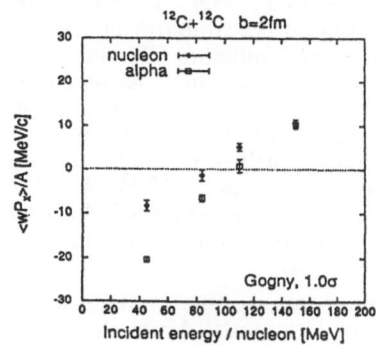

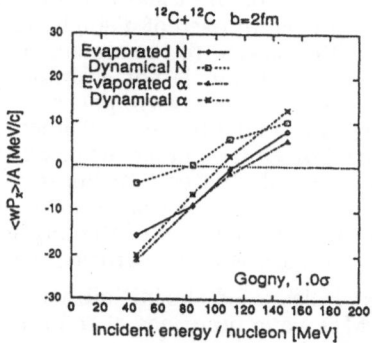

Fig.7 Calculated flows of nucleons and α particles for various incident energies for the reaction $^{12}C + ^{12}C$. The impact parameter is fixed to be 2 fm. The Gogny force is used. The definition of flow is $< wP_x > /A$ (see text).

Fig.8 The dynamical flows of nucleons and α particles which are calculated before the statistical decay, and the flows of evaporated particles which are calculated only from the decay products of the the statistical decay.

223

nucleons and evaporation flow of nucleons while there is almost no difference between dynamical flow of α particles and the evaporation flow of α particles. The evaporation nucleon flow is much larger than the dynamical nucleon flow. Furthermore we see that the evaporation nucleon flow is close to the α particle flow. Since the number of dynamically emitted nucleons is larger than that of the evaporated nucleons, the net flow of nucleons is closer to the dynamical nucleon flow.

These results lead us to the following interpretation. We can separate the collective flow into two components. The first one is the dynamical nucleon flow, namely the flow of nucleons generated by the hard stochastic collisions in the dynamical stage of the reaction. Since stochastic collisions erase the memory of the mean field, the absolute value of the negative flow is small. The shadowing effect of the projectile and the target may be another reason for this small flow. The second component of the collective flow is the dynamical fragment flow generated in the dynamical stage of the reaction. This component is directly affected by the mean field and has large absolute value. The flows of evaporated nucleons and fragments inherit the flow of the parent fragments and hence they have the large value similar to the original dynamical fragment flow. This interpretation is just in accordance with what one expected as we discuissed at the biginning of this section.

Since the nucleon flow and fragment flow depend differently on the stochastic collisions, the study of these different flows will give us the way to determine the in-medium cross section σ. Thus we studied the σ dependence of the flows. We found that the dynamical nucleon flow is almost independent on σ while the σ dependence of the fragment flow is large. We analysed the reason of the large σ dependence of the fragment flow and found that it is due to the large σ dependence of the dissipated component in the momentum distribution of fragments.

SUMMARY

In this lecture I explained the the framework of the novel microscopic simulation theory of heavy ion collisions named antisymmetrized molecular dynamics (AMD) and discussed the results of some analyses of experimental data with this theory which is for the sake of elucidating characteristic features of this theory.

First I discussed the formulation of the AMD. The AMD is a quantum mechanical theory, since it describes the time development of the system wave function. The time development consists of two parts: One is the time development due to the equation of motion derived from the time-dependent variational principle. This part of AMD is the same as the Feldmeier's fermionic molecular dynamics (FMD), and the FMD is a special case of the time-dependent cluster model (TDCM) where every cluster of the system is composed of a single nucleon. The other time development of the AMD wave function is due to the stochastic two-nucleon collision. The incorporation of the two-nucleon collision is not a simple problem because the position and momentum parameters of nucleon wave packets of the AMD Slater determinant do not represent real positions and momenta of nucleons. The two-nucleon collision in the AMD is formulated by constructing the physical nucleon coordinate which is the generalization of the canonical cluster coordinate of the theory of the TDCM. The Pauli-blocking of the two-nucleon collision is naturally formulated in terms of the Pauli-forbidden region of the phase space of the physical nucleon coordinate. The notion of the Pauli-forbidden region is also borrowed from the TDCM.

Next I discussed the initialization problem, namaly the construction of the ground state AMD wave functions of colliding nuclei. This multi-dimensional variation problem is solved efficiently by the frictional cooling method. It has been shown that the resulting ground state wave functions of the AMD are very realistic and are in very good accordance with our knowledge on the ground state structure of nuclei. The AMD wave function together with the frictional cooling technique presents us with a novel and powerful method for the nuclear structure study, because we can construct very good wave functions without any prejudice, namely without any model assumptions such as axial symmetric deformation, existence of clusters, and so on. To demonstrate this point, we showed some results of successful studies of light neutron-rich nuclei.

Thirdly we discussed the AMD study of the fragment production reaction in the incident energy region around the Fermi energy in the system of $^{12}C + ^{12}C$. We saw that the observed production cross sections of many fragments were all reproduced very well by the AMD. What is very important in this study is the fact that the AMD gives very large production cross section of α particles in the dynamical reaction stage. It means that the AMD can describe the shell effect in the dynamical production of fragments, which is impossible for the classical simulation frameworks such as QMD and VUU/BUU.

Finally we discussed the collective flows of α particles and nucleons in the system of $^{12}C + ^{12}C$. The AMD is advantageous for this problem since it can describe large cross sections of the dynamical production of α particles. We found that the separate measurements of fragment flow and nucleon flow are very important because they carry different information on the mean field and in-medium cross sections. Since dynamical nucleons are produced by the stochastic collisions, the dynamical nucleon flow is smaller than the dynamical fragment flow and is only weekly dependent on the magnitude of in-medium cross sections. On the other hand the dynamical fragment flow is directly affected by the mean field resulting in larger magnitude and its large magnitude is inherited by the evaporated nucleons and fragments. The dependence of the fragment flow on the magnitude of in-medium cross sections was found to be large in contrast to the dynamical nucleon flow.

REFERENCES

1. A. Ono, H. Horiuchi, T. Maruyama, and A. Ohnishi, Phys. Rev. Letters **68**, 2898 (1992).

2. A. Ono, H. Horiuchi, T. Maruyama, and A. Ohnishi, Prog. Theor. Phys. **87**, 1185 (1992).

3. A. Ono, H. Horiuchi, T. Maruyama, and A. Ohnishi, Phys. Rev. **C47**, 2652 (1993).

4. A. Ono, H. Horiuchi, and T. Maruyama, Phys. Rev. **C48**, No.6 (1993); preprint KUNS 1213 (Kyoto University)

5. H. Horiuchi, Nucl. Phys. **A522**, 257c (1991).

6. H. Horiuchi, T. Maruyama, A. Ohnishi, and S. Yamaguchi, *Proc. Int. Conf. on Nuclear and Atomic Clusters*, Turku (1991), eds. M. Brenner, T. Lönnroth, and F. B. Malik, (Springer, 1992), p.512; *Proc. Int. Symp. on Structure and Reactions of Unstable Nuclei*, Niigata (1991), eds. K. Ikeda and Y. Suzuki, (World Scientific, 1992), p.108.

7. H. Horiuchi, A. Ono, Y. Kanada, T. Maruyama, and A. Ohnishi, *Proc. First Joint Italian-Japanese Meeting within the INFN-RIKEN Agreement on Perspectives in Heavy Ion Physics*, Catania (1992), eds. M. Di Toro and E. Migneco, *Italian Physical Society Conference Proceedings* **Vol. 38** (1993), p.223.

8. J. Aichelin and H. Stöcker, Phys. Lett. **176B**, 14 (1986); J. Aichelin, Phys. Reports **202**, 233 (1991).

9. G. Peilert, H. Stöcker, W. Greiner, A. Rosenhauer, A. Bohnet and J. Aichelin, Phys. Rev. **C39**, 1402 (1989).

10. D. H. Boal and J. N. Glosli, Phys. Rev. **C38**, 1870 (1988).

11. H. Feldmeier, Nucl. Phys. **A515**, 147 (1990); *Proc. NATO Advanced Study Institute on the Nuclear Equation of State*, Peñiscola(1989), eds. W. Greiner and H. Stöcker, (Plenum, 1989), p.375.

12. S. Drożdż, J. Okolowcz, and M. Ploszajczak, Phys. Lett. **109B**, 145 (1982); E. Caurier, B. Grammaticos and T. Sami, Phys. Lett. **109B**, 150 (1982); W. Bauhoff, E. Caurier, B. Grammaticos and M. Ploszajczak, Phys. Rev. **C32**, 1915 (1985).

13. M. Saraceno, P. Kramer, and F. Fernandez, Nucl. Phys. **A405**, 88 (1983).

14. S. Saito, Prog. Theor. Phys. **40**, 893(1968); **41**, 705 (1969).

15. H. Horiuchi, Prog. Theor. Phys. Supplement **62**, 90 (1977).

16. L. Wilet, E. M. Henley, M. Kraft and A. D. MacKellar, Nucl. Phys. **A282**, 341 (1977).

17. D. M. Brink, *Proc. Int. School of Phys. "Enrico Fermi"*, **course 36** (1965), ed. C. Bloch, p.247.

18. J. P. Elliott, Proc. Roy. Soc. (London) **A245**, 128, 562 (1958).

19. H. Horiuchi, Y. Kanada-En'yo, and A. Ono, to appear in *Proc. Second Int. Conf. on Atomic and Nuclear Clusters*, Santorini (1993); preprint KUNS 1214 (Kyoto University).

20. T. Ando, K.Ikeda, and A. Tohsaki, Prog. Theor. Phys. **64**, 1608 (1980).

21. H. Furutani et al., Prog. Theor. Phys. Supplement No.**68**, Chapt.III (1980).

22. A. B. Volkov, Nucl. Phys. **74**, 33 (1965).

23. J. Czudek, L. Jarczyk, B. Kamys, A. Magiera, R. Siudak, A. Strzałkowski, B. Styczen, J. Hebenstreit, W. Oelert, P. von Rossen, H. Seyfarth, A. Budzanowski, and A. Szczurek, Phys. Rev. **C43**, 1248 (1991).

24. J. Péter, *Proc. Int. Symp. on Heavy Ion Physics and Its Application*, Lanzhou (1990), eds. W. Q. Shen, Y. X. Luo, and J. Y. Liu, (World Scientific, 1991), p.191.

25. G. D. Westfall, W. Bauer, D. Craig, M. Cronqvist, E. Gualtieri, S. Hannuschke, D. Klakow, T. Li, T. Reposeur, A. M. Vander Molen, W. K. Wilson, J. S. Winfield, J. Yee, and S. J. Yennello, Phys. Rev. Letters **71**, 1986 (1993).

26. D. Gogny, *Proc. Int. Conf. on Nuclear Self-Consistent Field*, Trieste (1975), eds. G. Ripka and M. Porneuf, (North Holland, 1975), p.176, p.209, p.265, p.266; J. Dechargé and D. Gogny, Phys. Rev. **C43**, 1568 (1980).

27. D. Klakow, G. Welke, and W. Bauer, *Proc. 8th Winter Workshop on Nuclear Dynamics*, Jackson Hole (1992), eds. W. Bauer and B. Back, (World Scientific, 1992), p.190; preprint MSUCL-884, WSU-NP-93-1.

THERMAL PHOTON RADIATION FROM A PLASMA WITH FINITE BARYON DENSITY

A. Dumitru, H. Stöcker, W. Greiner

Institut für Theoretische Physik der J.W.Goethe Universität
Postfach 111932, D-60054 Frankfurt am Main 11
Germany

INTRODUCTION

One of the main goals for the study of ultrarelativistic heavy-ion collisions is the search for the so-called quark-gluon plasma (QGP). If the plasma is created in a heavy-ion collision, it will emit lots of particles which in principle represent 'probes' of the QGP. Especially electromagnetic probes, like real photons or dileptons, are of great interest, since they do not interact strongly and thus their mean free path is large enough to leave the plasma volume without further interactions [1, 2, 3, 4, 5]. Therefore, several experiments were proposed in order to measure these particles [6]. In the following we will study the production of real photons from a plasma with arbitrary net baryon number. Up to now, most calculations of photon production from a QGP [1, 2, 3, 4, 5] followed Bjorken's scenario [7], where zero baryon density (and thus $\mu = 0$) in the plasma is assumed. Then the photon production rate depends only on the plasma temperature which is related to the energy density by the Stefan-Boltzmann law $\epsilon \sim T^4$. The energy density ϵ created in the collision can be estimated from the measured transverse energy and geometrical considerations [1, 7, 8, 9] or HBT-measurements [10] of the plasma volume.

We assume (local) thermodynamic equilibrium in the QGP and consider only u- and d-quarks with vanishing mass. Then the quark-chemical potential μ of the plasma is related to the baryon-chemical potential by $\mu = \mu_B/3$. The photon production rate $R \equiv dN/d^4X$ (i.e. the number of photons produced per space-time volume) is a function of both temperature T and quark-chemical potential μ of the plasma.

Hot and Dense Nuclear Matter, Edited by
W. Greiner *et al.*, Plenum Press, New York, 1994

ESTIMATE OF μ/T

In the absence of a self-consistent EOS derived from first principles (i.e. QCD), we will, in the following, use the phenomenological MIT bag model EOS [11] given by

$$\epsilon = AT^4 + CT^2\mu^2 + D\mu^4 + B \tag{1}$$

where

$$A = \frac{37\pi^2}{30} - \frac{11\pi\alpha_s}{3}, \quad C = 3(1 - 2\alpha_s/\pi), \quad D = \frac{C}{2\pi^2} \ .$$

We fix $B = (200 \, MeV)^4$ and $\alpha_S = 0.3$. Using this EOS we relate the ratio μ/T to the number of mesons and baryons produced during the hadronization of the QGP. The derivation of this relation will not require any model assumptions for the dynamical evolution of the plasma !

The entropy density s and the net quark density n may be computed from (1): $s(\mu,T) = \partial\epsilon(\mu,T)/3\partial T$, $n(\mu,T) = \partial\epsilon(\mu,T)/3\partial\mu$. We now assume that entropy is conserved also during the hadronization phase transition [12, 13]. Most of the produced particles are mesons and thus the entropy will be approximately proportional to the number of mesons N_π yielding [1] $n_\pi = 45\zeta(3)s/2\pi^4$.

The baryon density n_B in the QGP is given by $n_B = n/3$ and therefore, dividing n_B by n_π, we find

$$\left(\frac{\mu}{T}\right)^3 + \pi^2 \frac{\mu}{T} = \frac{3}{2} \frac{45\zeta(3)}{\pi^2} \left(2\frac{A}{C} + \left(\frac{\mu}{T}\right)^2\right) \frac{n_B}{n_\pi} \ , \tag{2}$$

which is a cubic equation for μ/T. This equation allows to calculate the ratio μ/T from the baryon and meson densities in a heavy-ion collision. In the region were $\mu \leq T$ it simplifies to $\mu/T \approx 6\, n_B/n_\pi$.

If we assume that μ/T does not depend on the transverse coordinates (i.e. that s and n are independent of $\vec{x}_\perp$), and that the longitudinal coordinate z is a differentiable function of the space-time rapidity[2] η, the relation

$$\frac{n_B}{n_\pi} = \frac{dN_B/d\eta}{dN_\pi/d\eta} \tag{3}$$

holds, i.e. eq. (2) allows one to compute μ/T as a function of space-time rapidity from the known baryon and meson rapidity distributions. We will, however, directly use eq. (2) in the following.

For RHIC-energies, the baryon and meson densities may be calculated in a variety of microscopic models, as e.g. in the RQMD approach [14]. We use the events of Ref. 14 and proceed in the following way: we first boost into some inertial frame with momentum-rapidity Δy relative to the center-of-mass system (CMS) and consider the evolution of the collision in this frame. We then determine the meson and net

---

[1]The factor $2\pi^4/45\zeta(3)$ is the entropy of any massless meson at vanishing meson-chemical potential. We therefore denote by N_π the number of *all* produced mesons, like π, ρ, ω etc., neglecting the masses.

[2]The space-time rapidity η is defined as $\tanh(\eta) = v_\parallel(z)$, were $v_\parallel(z)$ is the longitudinal component of the local *average* 3-velocity $\vec{v}(z)$ (e.g. the fluid velocity in a hydrodynamical picture). A specific dynamical model yields the explicit dependence of z on η, for instance, in Bjorken's scaling solution $v_\parallel = z/t$ (t is the CMS-time).

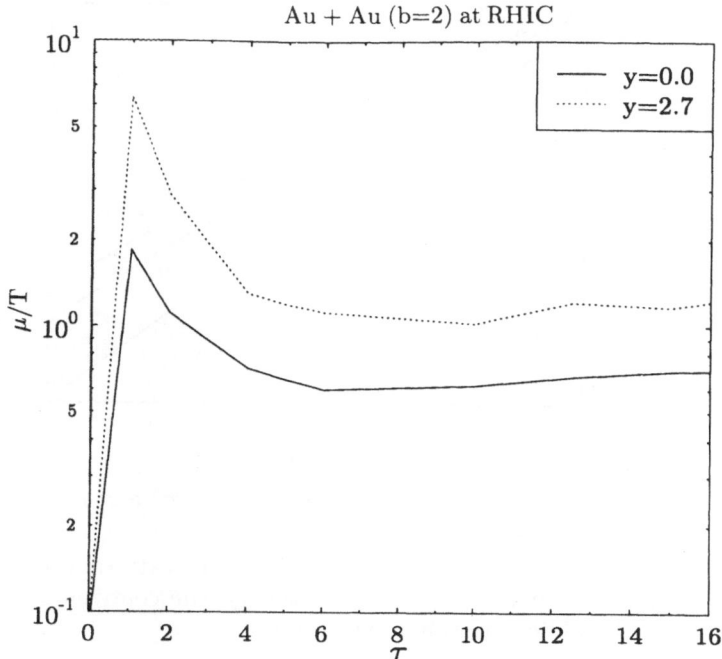

Figure 1: μ/T-values as calculated from eq. (2) as a function of eigentime for a Au+Au collision at RHIC. Input data taken from Ref. 14.

baryon local rest frame densities in each space-time cell. In order to avoid purely kinematic effects, we consider only particles in a rapidity interval $\pm \delta y$ around that of the respective inertial frame (we use $\delta y = 1.35$). Finally, the obtained local rest frame densities are averaged over all cells which exceed a critical energy density of $\epsilon_{crit} = 1 \ GeV/fm^3$ (our results do not change essentially if we vary ϵ_{crit} by factors $2-3$) and μ/T is calculated using eq. (2). We thus obtain μ/T as a function of the eigentime τ in the respective inertial frame. In the following we consider the inertial frames with $\Delta y = 0$ (i.e., the CMS itself) and $\Delta y = 2.7$, where the maximum net baryon density is found [14] and which at the same time is in the middle between CMS- and target/projectile-rapidity. Also, the RHIC dilepton spectrometer acceptance covers this rapidity region [15].

With this procedure, using the data of Ref. 14 as input for eq. (2), we find μ/T-values as shown in Fig. 1. Observe that μ is of the same order of magnitude as T. At CERN-SPS bombarding energies, a similar analysis [16] yields μ/T-values of the order of $1-5$, depending on y.

THERMAL PHOTON PRODUCTION AT FINITE μ/T

Let us now turn to the computation of thermal photon production from a plasma with a finite μ/T-ratio. The lowest order contributions to the photon rate are the

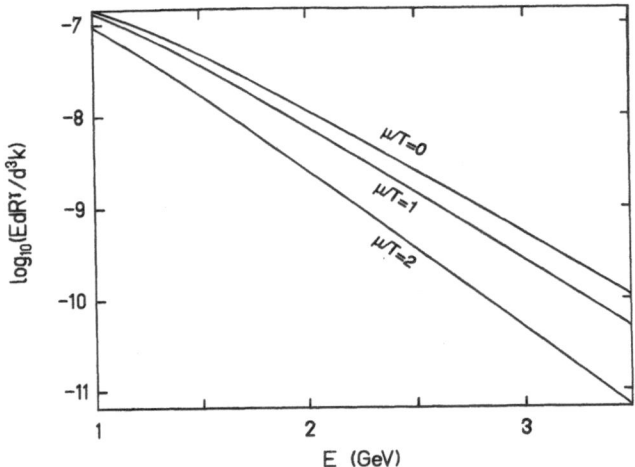

Figure 2: Photon production rate for $\epsilon = 8\ GeV/fm^3$ and $\mu/T = 2, 1, 0$.

QCD annihilation reaction ($q\bar{q} \to \gamma g$) and the Compton scattering ($qg \to q\gamma$, $\bar{q}g \to \bar{q}\gamma$) of a quark or an antiquark. Using kinetic theory, the contribution from each of these reactions to the (differential) photon rate is

$$2E\frac{\mathrm{d}R_i^\gamma}{\mathrm{d}^3k} = \frac{1}{(2\pi)^8} \int \frac{\mathrm{d}^3p_1}{2E_1} \frac{\mathrm{d}^3p_2}{2E_2} \frac{\mathrm{d}^3p_3}{2E_3}\ n_1 n_2 \left(1 \pm n_3\right) \delta^{(4)}(p_1 + p_2 - p_3 - k)|\mathcal{M}_i|^2 \quad (4)$$

where n_i are the particle distributions (functions of E_i, μ, T), the sign in front of n_3 is chosen according to Bose enhancement or Pauli suppression for the strongly interacting final state particles and $|\mathcal{M}_i|^2$ is the amplitude for the corresponding reaction [5]. For high energy photons with $E \pm \mu \gg T$ the integrals in eq. (4) can be performed [17], leading to Fig. 2. Obviously, the photon rate decreases appreciably with increasing μ: for a photon energy of $E = 3\ GeV$ there is a suppression by one order of magnitude if $\mu/T = 2$ as compared to the case of $\mu/T = 0$. This effect does also occur in the case of dileptons, cf. Ref. 19. Finite net baryon densities ($\mu/T > 0$) thus result in a decrease of electromagnetic radiation but, on the other hand, are necessary for the destillation of strangelets [20].

REFERENCES

1. See for instance:
   L. McLerran, T. Toimela: Phys. Rev. D31 (1985) 545

2. B. Sinha: Phys. Lett. B128 (1983) 91; B157 (1985) 221
   P.V. Ruuskanen: Univ. of Jyväskylä preprint 28/1991
   and references therein

3. R. Baier, H. Nakkagawa, A. Niégawa, K. Redlich: Z. Phys. C53 (1992) 433

4. M. Neubert: Z. Phys. C42 (1989) 231

5. J. Kapusta, P. Lichard, D. Seibert: Phys. Rev. D44 (1991) 2774

6. See for example:

   G. Roche: contributed paper, Proc. of the NASI on The Nuclear Equation of State, Peñiscola, Spain, May 22 – June 3, 1989, Plenum Press, N.Y., p.161
   P. Braun-Munzinger and G. David: contributed paper, Proc. of the Int. Workshop on Gross Properties of Nuclei and Nuclear Excitations XX, Hirschegg, Austria, Jan. 20–25 (1992), H. Feldmeier, ed.
   R.S. Hayano et al. (TALES coll.): Two-Arm Electron/Photon/Hadron Spectrometer, Letter of Intent for RHIC Experiment, Sep. 1990
   S.H. Aronson et al.: A Lepton/Photon Spectrometer for RHIC, Presentation to the BNL Advisory Commitee, 28 Aug. 1991
   H. Gutbrod et al. (WA98-coll.): CERN-proposal SPLC 91-17 P260
   J. Schukraft and the HELIOS-coll.: Nucl. Phys. A498 (1989) 79c

7. J.D. Bjorken: Phys. Rev. D27 (1983) 140

8. W. Heck and the NA35-coll.: Z. Phys. C38 (1988) 19

9. J. W. Harris and the NA35-coll.: Nucl. Phys. A498 (1989) 133c

10. T. Humanic: Z. Phys. C38 (1988) 79

11. A. Chodos, R.L. Jaffe, K. Johnson, C.B. Thorn, V. Weisskopf:
    Phys. Rev. D9 (1974) 3471
    J. Kapusta: Nucl. Phys. B148 (1979) 461

12. R.C. Hwa, K. Kajantie: Phys. Rev. D32 (1985) 1109

13. J. Kapusta, L. McLerran, D. Srivastava: Phys. Lett. B283 (1992) 145

14. Th. Schönfeld, H. Sorge, H. Stöcker, W. Greiner: UFTP preprint 314/1992, to be published in Mod. Phys. Lett. A
    Th. Schönfeld: PhD-thesis, Frankfurt 1993

15. The PHENIX proposal, S. Nagamiya: talk given at GSI, Sept. 1992

16. M. Hofmann et al: Quark Matter 1993, proceedings

17. A. Dumitru, D.H. Rischke, H. Stöcker, W. Greiner: Mod. Phys. Lett. A 8 (1993) 1291

18. K. Kajantie, P.V. Ruuskaanen: Phys. Lett. B121 (1983) 352

19. A. Dumitru, D.H. Rischke, Th. Schönfeld, L.A. Winckelmann, H. Stöcker, W. Greiner: Phys. Rev. Lett. 70 (1993) 2860

20. C. Greiner, P. Koch, H. Stöcker: Phys. Rev. Lett. 58 (1987) 1825
    C. Greiner, D.H. Rischke, H. Stöcker, P. Koch: Phys. Rev. D38 (1988) 2797

FRAGMENT–FRAGMENT CORRELATIONS

G. Peilert, T.C. Sangster, M.N. Namboodiri, and H.C. Britt

Lawrence Livermore National Laboratory
Livermore, CA 94550
USA

A topic of great interest in nuclear physics is the multifragmentation of heavy nuclei at moderate excitation energies. In the experiments from the Purdue group [1, 2] first attempts have been made to relate the results in these reactions to the critical exponents of a phase transition, analogous to the liquid gas transition in condensed matter physics. Within the last years it became possible to study such reactions in highly exclusive experiments and to vary the order parameter of the proposed phase transition in order to extract the critical properties. This can be achieved by investigating the moments of the mass distributions as it was proposed by Campi [3]. Investigations utilizing the percolation model [4] show indeed that this infinite matter phase transition is observable even in small finite systems of the size of nuclei [5].

Such a procedure, however will not provide the physical quantities that drive the transition from a nuclear liquid to the vapor phase. If one wants to extract quantities like a critical temperature and density one has to rely on models that describe the whole dynamical process of a heavy ion collision. Unfortunately there is presently no complete model available that describes the process of the thermally driven multifragmentation in heavy ion collisions in a single, consistent approach. Current modeling involves a pre equilibrium stage described in the earliest approaches by an Intranuclear Cascade Model, INC, [6, 7] and more recently by molecular dynamics pictures such as the Quantum Molecular Dynamics Model, QMD, [8, 9] or by using single particle models from the VUU/BUU type [10]. Following this stage there remains a distribution of nucleons and complex fragments which can themselves be highly excited. These excited fragments then undergo further statistical decay. This statistical decay process has been modeled in various codes as a sequential evaporation [11, 12, 13] or as an explosive simultaneous multifragmentation [14, 15, 16, 17, 18]. The simultaneous multifragmentation models assume a statistical break–up of a hot source into many intermediate mass fragments and then calculate the most probable fragment distribution by exploiting the available phase space. On the other side the sequential models assume that the source decays by the successive, binary emission of nucleons

Hot and Dense Nuclear Matter, Edited by
W. Greiner *et al.*, Plenum Press, New York, 1994

and heavy fragments. None of these models, however, includes the dynamics of the fragmentation.

Recent investigations with the QMD model have shown, that for central collisions of heavy nuclei at high energies a rapid compression–decompression mechanism emerges in the initial stage of the reaction and leads to an direct multifragmentation process, where a highly excited heavy residue is no longer formed with a high probability [9, 19]. These investigations suggest that the region to search for the thermally driven multifragmentation in heavy ion reactions is in central collisions at low bombarding energies (E/A $\approx$ 100 A MeV) or in peripheral (or highly asymmetric) reactions at high energies (E/A $\approx$ 1 A GeV) . In this type of reaction the direct reaction leads to a highly excited source that then breaks up into many IMFs.

Many experiments [19, 20, 21] have unambiguously proven that nuclear multifragmentation is indeed a new decay mechanism for highly excited nuclei, but up to now there is still little understanding of the physical process that drives this multifragmentation. One of the basic questions that still is not answered is whether the fragmentation takes place by sequential emission of fragments or by a simultaneous "explosion". Since for both scenarios most of the final state observables like multiplicities, fragment yields, energy spectra etc. are identical, one has to find new observables that are sensitive to the details of break–up mechanism, e.g., the emission time scale and the Freeze–Out volume.

Recently attempts have been made to extract the emission time scale out of different experimental data sets [21, 22, 23]. Complementary to these investigations we will show in this letter, that the same data can be explained by assuming a simultaneous multifragmentation and we will extract the corresponding Freeze–Out volume.

For the following comparison to experimental data we will use the QMD+SMM approach to model the dynamical reaction with the following statistical decay of the excited pre fragments. The QMD model provides a microscopic dynamical calculation of a heavy ion reaction. The basic version of the QMD model uses a local two and three–body Skyrme interaction and a Coulomb and Yukawa interaction. The parameters of the interactions are adjusted to reproduce the properties of infinite nuclear matter, as well as the binding energy and radii of nuclei in the mass regime from A=2 - 200. The numerical simulation of the collision takes place in three steps. First the projectile and target are initialized in their rest frames. Successfully initialized nuclei are then boosted towards each other with the proper center of mass velocities using relativistic kinematics. Each nucleon is then propagated corresponding to the Hamilton equations with a second order Runge-Kutta integration algorithm. After each integration step the hard N-N collisions are treated via a stochastic scattering term. Two nucleons can scatter if the spatial distance of the centroids of their Gaussians is smaller than $\sqrt{\sigma/\pi}$. For the present calculation we used an isotropic, isospin independent scattering cross section of 41 mb. Whenever a collision occurs, the resultant phase-space distribution around the final states of the scattering partners is checked whether the resultant configuration violates the Pauli principle. Whenever a collision is blocked, the momenta of the scattering partners are replaced by the values they had prior to the scattering.

In order to simulate the Fermi motion we employ a so called Pauli potential [25]. The implementation of the Pauli potential into the dynamical QMD model yields two major improvements. First the ground states are now well defined and the Monte Carlo procedure can be used to initialize projectile and target in their real ground state. Secondly the excitation energy of the resulting fragments can then be determined with respect to their true ground state.

Previous investigations with the QMD model have shown that there are two different mechanisms leading to the multifragmentation process. One is related to the mechanical rupture of the system, whenever compression effects are important. The other mechanism produces fragments thermally from an equilibrated source. This thermal multifragmentation has so far not been described in a microscopic model like QMD [9]. In the first comparison to inclusive IMF data it was shown that a two step model was necessary to reproduce the experimental angular distributions [24]. This two step model involved the calculation of initial kinematics and excitation energy of all pre fragments with QMD model and a subsequent de excitation calculation utilizing the Statistical Multifragmentation Model (SMM) of Botvina et al. [16]. The input for the SMM stage of the reaction is the mass and the excitation energy of the fragments produced in QMD. These values are consistently determined within the QMD approach.

This model describes the multifragmentation of highly excited nuclei based on the statistical approach and a liquid–drop description of hot fragments. It is assumed that the excited primordial fragments, characterized by their excitation energy, mass and charge, break up into an assembly of nucleons and fragments. All these decay products are described as Boltzmann particles in a Freeze–Out volume $V = V_0(1+\kappa)$, where κ is a model parameter and V_0 is the volume of the system corresponding to normal nuclear matter density. Since all the produced fragments are excited (only particles with A ≤ 4 are considered as elementary particles) the final de excitation of those fragments is treated as a Fermi break up for light fragments (A < 16) and via an evaporation of nucleons and clusters up to ^{18}O for heavier fragments (for details see Ref. [16]).

In the following we will vary the volume parameter κ of the SMM model in order to extract the Freeze–Out volume out of the two fragment correlation data.

Figure Figure 1 shows a comparison between the experimental data [24] and the QMD+SMM calculations using $\kappa = 2, 5$ and 10 for two different projections of the double differential cross section $d^2\sigma/(dZ d\Omega)$. Both the data and the calculations are for central collisions only (for details see Ref. [24]). The upper part of figure Figure 1 shows the charge yield distribution at a laboratory angle of 72° (±13°), while the lower part shows the angular distribution of fragments with Z=10. In both cases it can be seen that the variation of the volume parameter κ does not influence those semi exclusive observables and all calculations agree reasonably well with the data. The calculated angular distributions, however, are still slightly too steep and under predict the data at backward angles.

Recently it has been shown from Kim et al., [21] that the technique of intensity interferometry can be applied to light fragments, in order to extract the time scale of the reaction. For this purpose they used a final state Coulomb interaction model based on the Koonin–Pratt formalism [26]. This formalism is, however, restricted to the case, where the fragment-fragment correlation function is governed by the final state Coulomb interaction of the fragment pair under consideration only. Whether this assumption can hold in the case where one has many (up to 10) fragments which may have been produced in a small volume at the same time, is at least questionable.

A different approach to extract the emission time scales has been used by Sangster et al. [22] and Lacey et al. [23]. They used a the classical three-body trajectory code *MENEKA* [27] that simulates an isotropic surface emission of IMFs from a spherical source characterized by a unique radius parameter with angular momenta chosen randomly from a triangular distribution. The emission energies are selected to reproduce the experimentally measured distributions and the distribution of time delays between fragment emission is given by a Gaussian. Those studies showed that the correlation data are consistent with a emission time between 50 and 500 fm/c [22] and that this time scale saturates with decreasing bombarding energy [23].

All the limitations that restrict the studies using the Koonin Pratt formalism or trajectory calculation do not apply to the QMD calculations. In this case the

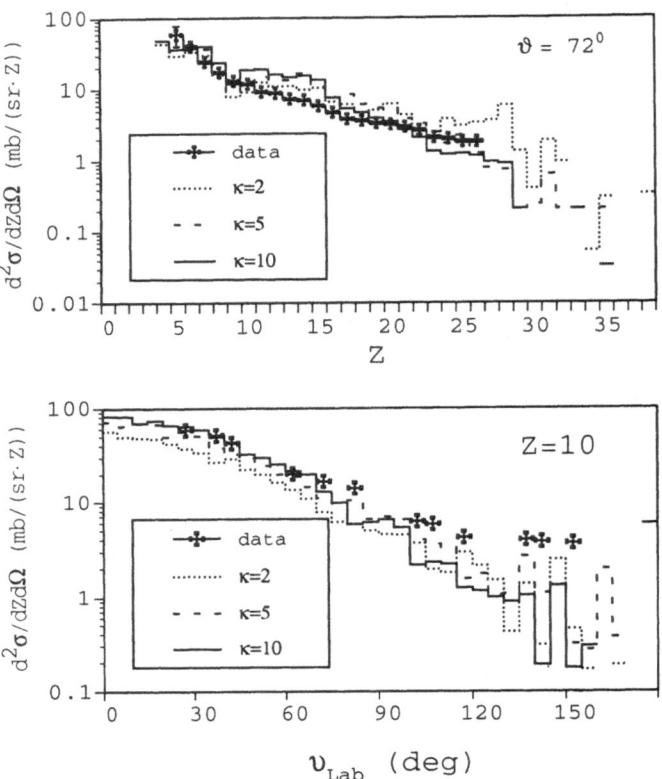

Figure 1. Projections of the semi exclusive triple differential cross section $d^2\sigma/(dZ\,d\Omega)$ for the reaction Fe (100 A MeV, b = 0 - 6 fm) + Au. The symbols show the data [24], for the charge distributions for fragments at $\vartheta_{Lab} = 72°$ (upper panel) and the angular distributions for fragments with $Z = 10$ (lower panel). The histograms show the calculations with the QMD+SMM model for different volume parameters κ as indicated.

Hamilton equations of all nucleons are integrated, which implies, that correlations between all the existing nucleons and fragments are treated in all orders. This is especially important in the case when one deals not only with two IMF's, but with many. In the following we present the fragment- fragment correlation function for the

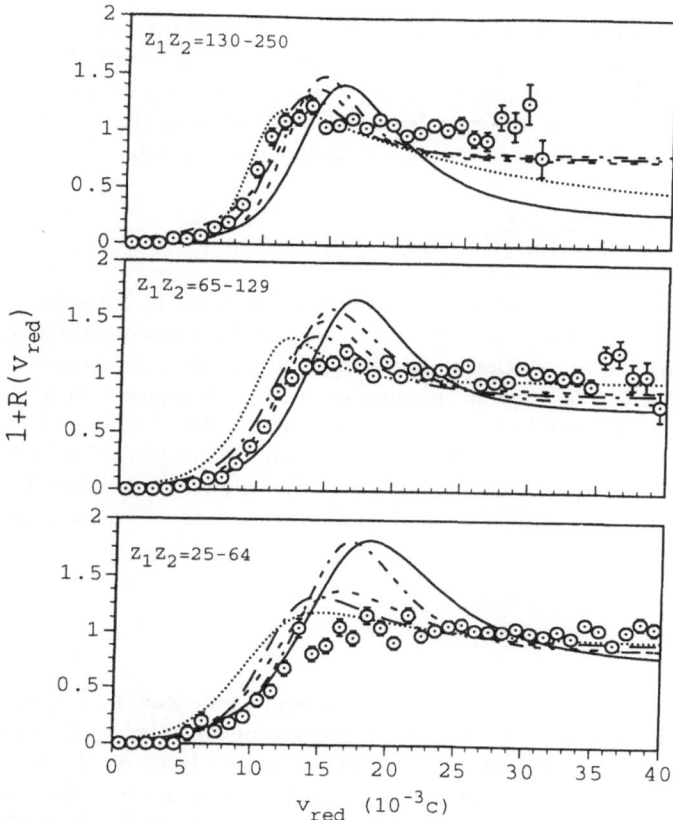

Figure 2. Mixed fragment reduced velocity correlation functions for three different constraints on the Coulomb product $Z_1 \cdot Z_2$. The symbols show the data of ref. [22], while the curves are fits to equation 2 obtained with the QMD+SMM model. The different lines show calculations done with the volume parameter $\kappa = 1$ (solid line), $\kappa = 2$ (variable length dashed line), $\kappa = 5$ (dashed line), $\kappa = 10$ (dash–dotted line) and $\kappa = 15$ (dotted line),

reaction Fe (100 A MeV) + Au for central collisions (b = 0 - 6 fm) in comparison to the data of Ref. [24]. After the break-up of the hot target remnant within the SMM model (this takes place after 300 fm/c reaction time) we neglect the nuclear forces and follow the Coulomb trajectories of the charged particles. This propagation is stopped when the total Coulomb interaction is sufficiently small.

The reduced velocity correlation functions shown in figure Figure 2 have been obtained by taking ratios between correlated IMFs in a single event divided by the same quantities where the quantity of interest was obtained from two IMFs taken from different physical events, thereby eliminating the final state correlations. In all cases the true ($Y_{true}(v_{red})$) and background distributions ($Y_{back}(v_{red})$) were developed separately for two classes of events: 1) two IMFs detected on opposite sides of the beam and 2) two IMFs detected on the same side of the beam. In the areas where these two correlations overlap in relative velocity it was found that the correlations obtained from the two acceptance ranges were effectively identical and so in this paper the results have in all cases been combined into single distributions. The two fragment correlation function is then calculated according to

$$1 + R(v_{red}) = \frac{Y_{true}(v_{red})}{Y_{back}(v_{red})}. \tag{1}$$

Note that there is no additional normalization factor in equation (1). If both the true yields and the background are normalized and there are no additional correlations (and only then) the correlation function approaches unity for large values of v_{red}.

Fig. Figure 2 shows the experimental correlation function (symbols) for the three different charge product bins $Z_1 \cdot Z_2 = 25 - 64$, $65 - 129$ and $130 - 250$ (the smallest detected charge is Z=5 in all cases) compared to the calculated results (lines) for different volume parameters κ in the SMM model (note that $V = V_0(1 + \kappa)$). The theoretical curves show fits to the actual correlation functions using the fitting function

$$1 + R(v_{red}) = a \frac{1 + e^{\frac{d - v_{red}}{e}}}{1 + e^{\frac{b - v_{red}}{c}}}. \tag{2}$$

It can be seen that the size of the Coulomb hole can be explained within the QMD+SMM approach if a volume parameter between $\kappa = 2$ and 10 is used; both $\kappa = 1$ and $\kappa = 15$ clearly disagree with the data for the heavier fragments ($Z_1 \cdot Z_2 = 65 - 129$ and $130 - 250$). For the lightest fragments under consideration the smallest volume parameter seems to describe the data best, but here the sensitivity to the Freeze-Out volume is less pronounced. This may indicate that the fragments do not come from one Freeze-Out volume, but the smaller fragments are emitted earlier from a smaller volume.

This results are in good agreement with previous investigations of the same data using the sequential MENEKA model [22]. In this case a emission time in the order between 50 and 500 fm/c was found. These time scales are comparable to the typical time a fragment needs to traverse a Freeze-Out volume of roughly 5 times the nuclear volume.

Another feature that can be observed in the calculated correlation functions in figure Figure 2 is the strongly pronounced peak at $v_{red} \approx 15 - 20$. This kind of additional correlation is absent in the present data and it is also not seen in the data of the MSU group [21], except for very peripheral collisions [28]. Our calculations show indeed that this effect is more pronounced for the more peripheral collisions and it also increases (see figure Figure 2) with decreasing volume parameter in the SMM model (a similar behavior has recently been found in the Berlin model [29] when the excitation energy is decreased). This effect is due to an additional correlation in the fragment–fragment correlation function because of the Coulomb repulsion due to a heavy third body. In figure Figure 3 we show the average charges of the three largest fragments as well as the average charge asymmetry $< \Delta Z > = \sqrt{((Z_1 - Z_2)^2 + (Z_1 - Z_3)^2 + (Z_2 - Z_3)^2)/3}$. and the average multiplicity of IMF s with Z=4-20. All the results in this figure have not been filtered according to the geometry of the PAGODA array, only the lower energy threshold has been taken into account. The upper panel of figure Figure 3 shows these values for the reaction Fe (100 A MeV) + Au for a fixed volume parameter $\kappa = 2$ versus the impact parameter b, while in the lower panel the impact parameter interval is kept fixed (b = 0 - 6 fm) and the volume parameter κ is varied. It can be seen that both for peripheral collisions and for impact parameter averaged collisions with a small volume parameter the charge distribution is very asymmetric, while the average IMF

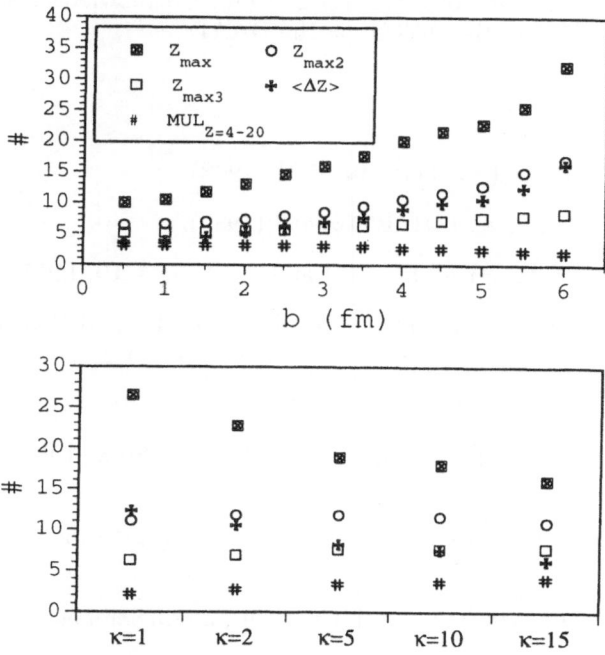

Figure 3. The three largest fragments as well as the average charge asymmetry $< \Delta Z > = \sqrt{((Z_1 - Z_2)^2 + (Z_1 - Z_3)^2 + (Z_2 - Z_3)^2)/3}$. and the average multiplicity of IMFs with Z=4-20 is are shown for the reaction Fe (100 A MeV) + Au for a fixed volume parameter $\kappa = 2$ versus the impact parameter b (upper figure). In the lower figure the impact parameter interval is kept fixed $(b = 0 - 6 \ fm)$ and the volume parameter κ is varied.

multiplicity stays almost constant. From this we conclude that the additional bump in the calculated correlation functions in figure Figure 2 results from a strong charge asymmetry in the models that is not observed in the data.

It also has been found that this bump is , especially for the lightest fragments, almost absent in the unfiltered correlation function for the reaction Fe (100 A MeV, b=0-6fm) + Au. This is easily understood if one considers the geometry of the PA-GADA detector array (for details see Ref. [24]). Since this setup consist of 8 detector modules at the polar angles ±36° , ±72° , ±108° and ±104°, the small relative momenta between IMFs are strongly suppressed in forward direction. An IMF pair with small relative angle (velocity) can therefore only be detected at large laboratory angles. The main mechanism that leads to fragments at backward angles, however, is the Coulomb repulsion from a heavy third body at rest in the c.m. frame. Because of this geometry the filtering of the calculated results triggers predominantly on fragment pairs from a asymmetric decay and therefore over pronounces the bump in the correlation function.

Acknowledgement

This work was supported by the US Department of Energy by LLNL under Contract W-7405-ENG-48. One of us (G.P.) gratefully acknowledges support from the Wissenschaftsausschuss of the NATO via the DAAD.

REFERENCES

1. J.E. Finn et al., Phys. Rev. Lett. **49**, 1321 (1982).

2. R.W. Minich et al., Phys. Lett. **B118**, 458 (1982).

3. X. Campi, Phys. Lett. **B124**, 8 (1984) and J. Phys. A **19**, L917 (1986).

4. D. Stauffer, Introduction to percolation theory, Taylor and Francis, London, 1985 and Phys. Rep. **54**, 1 (1979).

5. X. Campi, Phys. Lett. B **208**, 351 (1988).

6. Y. Yariv and Z. Fraenkel, Phys. Rev. C **20**, 2227 (1979) and Phys. Rev. C **24**, 488 (1981).

7. K.K. Gudima and V.D. Tonnev, Yad. Fiz. **27**, 67 (1978) [Sov. J. Nucl. Phys. **27**, 67 (1978)].

8. G. Peilert et al., Phys. Rev. C **39**, 1402 (1989); J. Aichelin et al., Phys. Rev. C **37**, 2451 (1988).

9. G. Peilert et al., Phys. Rev. C **46**, 1457 (1992).

10. J.J. Molitoris, H. Stöcker, and B.L. Winer, Phys. Rev. C **36**, 220 (1987); G. Bertsch and P.J. Siemens, Phys. Lett. B **126**, 9 (1983); C. Gregoire et al., Nucl. Phys. A **465**, 317 (1987).

11. M. Blann, Phys. Rev. Lett. **54**, 2215 (1985) and Phys. Rev. C **32**, 1231 (1985); M. Blann, T. Komoto, and I. Tservaya, Phys. Rev. C **40**, 2498 (1989); M. Blann et al, Phys. Rev. C **44**, 431 (1991).

12. R.J. Charity et al., Nucl. Phys. A **453**, 371 (1988); D.R. Bowman et al., ibid. A **523**, 386 (1991).

13. W.A. Friedman, Phys. Rev. Lett. **60**, 2125 (1988).

14. G. Fai and J. Randrup, Nucl. Phys. A **381**, 557 (1982) and Nucl. Phys. A **404**, 551 (1983).

15. J.P. Bondorf et al., Nucl. Phys. A **443**, 321 (1985); J.P. Bondorf et al., Nucl. Phys. A **444**, 460 (1985).

16. A.S. Botvina et al., Nucl. Phys. A **475**, 663 (1987); A.S. Botvina, A.S. Iljinov, and I.N. Mishustin, Nucl. Phys. A **507**, 649 (1990).

17. D.H.E. Gross and X.Z Zhang, Phys. Lett. B**161**, 47 (1985); D.H.E. Gross, X.Z. Zhang, and S.Y. Xu, Phys. Rev. Lett. **56**, 1544 (1986); D.H.E. Gross, Rep. Prog. Phys. **53**, 605 (1990).

18. D. Hahn and H. Stöcker, Nucl. Phys. A **476**, 718 (1988).

19. B. Tsang et al., Phys. Rev. Lett. **71**, 1502 (1993).

20. J. Hubele et al., Z. Phys. A **340**, 263 (1991); C.A. Ogilvie et al., Phys. Rev. Lett. **67**, 1214 (1991).

21. Y.D. Kim et al., Phys. Rev. C **45**, 338 (1992); Y.D. Kim et al., Phys. Rev. C **45**, 387 (1992).

22. T.C. Sangster et al., Prog. Part. Nucl. Phys. Vol. 30, 189 (1993) and Phys. Rev. C **47**, R2457 (1993) .

23. R. Lacey et al., Phys. Rev. Lett. **24**, 3705 (1993).

24. T.C. Sangster et al., Phys. Rev. C**46**, 1404 (1992).

25. G. Peilert et al., Phys. Lett. B **260**, 271 (1991).

26. S.E. Koonin, Phys. Lett. B**70**, 43 (1977); W.G. Gong et al., Phys. Rev. C43, 781 (1991).

27. A. Elmaani et al., Nucl. Instrum. Methods A**313**, 401 (1992).

28. D.R. Bowman et al., Phys. Rev. Lett. **23**, 3534 (1993).

29. O. Schapiro, A.R. DeAngelis, and D. Gross, Preprint HMI 1993/P1-Schap 1.

Two Pion Correlations in A-A Collisions at CERN Energies

H. Ströbele

Universität Frankfurt/Main

1   Introduction

Nucleus-Nucleus collisions at high energies produce hadronic systems at high energy and particle number densities. Single particle observables are used to study the reaction dynamics[1]. Information about the space and time configuration of the zone, in which the particles are produced, can be obtained from particle correlations. In this contribution we discuss in very simple terms how this information is extracted from the experimental data on two-pion correlations and show that different correlations are expected for static and (longitudinally) expanding sources. For a comprehensive description of the techniques of source size measurements the reader is deferred to references [2, 3].

The Experiment NA35 at CERN detects charged particles by means of a Streamer Chamber and a Time Projection Chamber (TPC) in conjunction with a strong magnetic field. Deflection and curvature allow to determine momentum and charge. Negative particles are interpreted as pions and used to construct two pion correlation functions. Most of the longitudinal and transverse phase space is covered. Sulphur and Oxygen projectiles at an energy of 200 GeV/nucleon have been used to produce and record central collisions with target nuclei ranging from Suphur to Gold. The experimental results presented here are preliminary. Final results are expected to appear in Zeitschrift f. Physik in late 1994.

Hot and Dense Nuclear Matter, Edited by
W. Greiner *et al.*, Plenum Press, New York, 1994

2 Pion Correlations

Correlations between two or more particles can be due to forces between the particles or the result of quantum statistical effects. An example for the former is the Coulomb force. The latter are also known as the Pauli Principle and the Bose Einstein enhancement. These are relevant only for identical particles which have a reduced, respectively enhanced, probability to be found in the same 6-dimensional phase space cell. In general, particle correlations are measured in momentum space for an ensemble of equivalent but independent events or particle pairs. It is intuitively clear that the correlation observed in momentum space will be weak, if the particles are far away in configuration space. Let us specify the meaning of "weak" for the case of the Coulomb force: two particles with small relative momenta will experience a small (large) change in relative momentum, if they are at large (small) distance in configuration space; in the limit of infinite distance no change in relative momentum will occur. Thus "weak" here means a small change in momentum difference.

As a practical example for the construction of a correlation we consider the distribution of the momentum differences between all combinations of two particles. The gross features of its shape are given by the square of the single particle density distribution in momentum space. We are not interested in this trivial type of correlation. Dynamical and statistical effects can be extracted by taking the ratio of the true distribution of momentum differences to a distribution which contains only the correlation due to the single particle phase space distribution. The latter is obtained, e.g., by pairing only particles from different events and the former by pairing particles from the same event. This ratio as a function of the momentum difference represents the relevant correlation function. Its shape is affected by the average distance between the two particles which form the pair. To be more precise: the distribution of distances between the particles of all pairs in the ensemble considered is reflected in the shape of the correlation function with large (small) distances affecting small (large) momentum differences.

We turn now to the correlation between pairs of identical pions. The laws of (Bose-Einstein) statistics tell us that two such pions have an enhanced probability to be found in the same 6-dimensional phase space cell given by $\Delta \vec{p} \cdot \Delta \vec{r} \sim \hbar$. Comparing two ensembles of pairs coming from sources with $\Delta \vec{r}$ small and large (on the average) will result in correlation functions showing an enhancement in a large and small range of $\Delta \vec{p}$ (momentum differences) respectively. This inverse correspondence holds true also for each component of the momentum difference and distance. Fig. 1 shows an example of a correlation function as obtained from central S+Au collisions at 200 GeV/nucleon. The variable is the Lorentz invariant momentum difference $Q_{inv}=(p_1-p_2)$ with p_i being the four-vector of particle i. The Bose Einstein enhancement is clearly seen at low Q_{inv}. For a more detailed view of what can be measured we consider all six degrees of freedom spanned by the 3-momenta of two pions $(\vec{p}_1, \vec{p}_2)$. Three momentum differences are chosen in the following way: $\Delta p_{long}=Q_{long}$, in the direction of the incident beam; $\Delta p_{T(out)}=Q_{out}$, the component of the transverse momentum difference vector projected onto the sum vector $(\vec{p}_{1T} + \vec{p}_{2T})$; $\Delta p_{T(side)}=Q_{side}$, the remaining orthogonal component of the transverse momentum difference. The corresponding size parameters are R_{long}, R_{side}, and R_{out}. The special role of R_{long} will be discussed below.

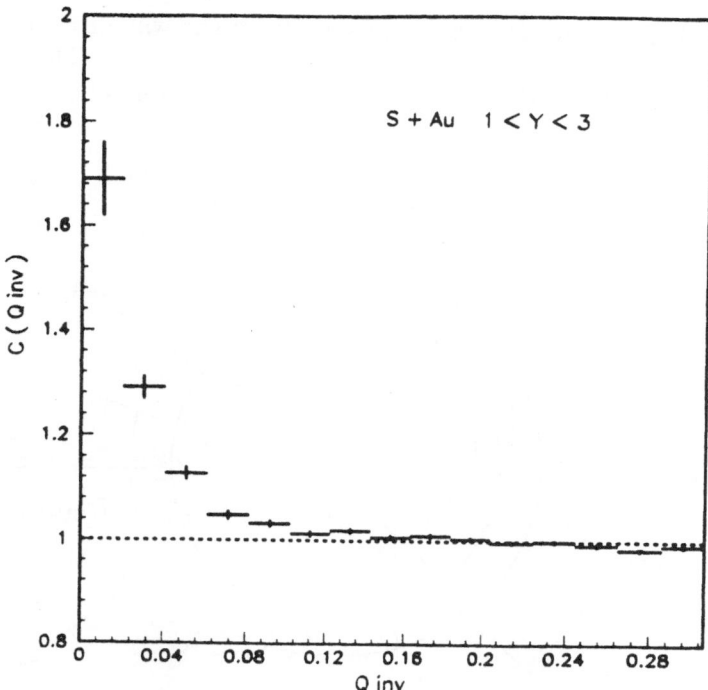

Fig. 1. Two-pion correlation as a function of the invariant momentum difference.

The meanings of R_{out} and R_{side} are illustrated in Fig. 2 . They stand for a measure of the depth of the source and its transverse size, respectively. Details of the evaluation of the radius parameters and their interpretation can be found in reference [2]. With 3 out of 6 degrees exhausted we consider the remaining three, which are the components of the vector sum $\vec{p}_1 + \vec{p}_2$. One degree of freedom, the azimuthal angle, should be integrated over because of the rotational symmetry with respect to the beam axis. The two remaining degrees of freedom are chosen to be the rapidity (y) and the transverse momentum (p_T) of the pair. With these categories it is natural to study R_{long}, R_{side}, and R_{out} as function of rapidity and p_T. An experimental difficulty should be mentioned here: the influence of non genuine pion pairs, in which at least one particle does not originate from the main interaction point but instead from a decay or a secondary interaction, is not corrected for in the results presented below. The corresponding distortion will cause only minor systematic errors on the radius parameters, but will reduce the λ-parameter, which is introduced explicitly for this type of effects[4].

Different scenarios for the shape and time evolution of the pion emitting source will be discussed next. The simplest one is a static spherical source which emits all pions instantly and with a momentum distribution which could be thermal or of any other simple shape. In this case an observer will see the same shape of the correlation function in Q_{out}, Q_{side} and Q_{long}, if the momentum measurements are done in the Lorentz frame of the source. In the case of relative motion (e.g. in the longitudinal direction) the source will appear Lorentz contracted (i.e. R_{long} will be smaller than R_{out} and R_{side}). The correlation functions will all be independent of where in phase space the momentum measurements are done. Thus the shape of the correlation functions will look alike in all intervals of rapidity and p_T.

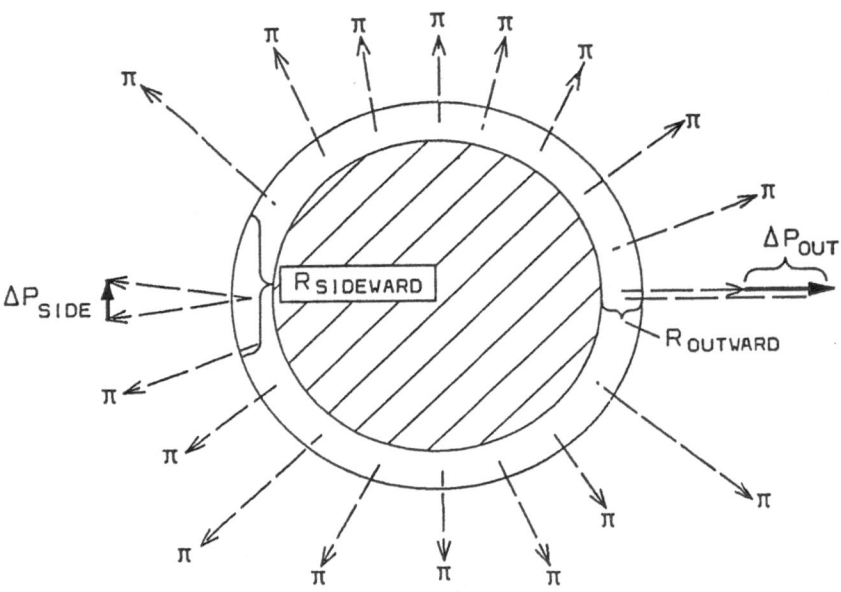

Fig. 2. Illustration of the meaning of the outward (R_{out}) and sideward (R_{side}) radii.

Another extreme scenario is a static source in the same configuration but with continuous emission of pions. In this case (and if in addition the measurement time is short enough) R_{out} will be larger than R_{side}, because pions emitted radially and at different times may lead to pairs of pions in phase cells which are outside of the original source volume. (In fact both scenarios could be described more appropriately by a spherical set of radius parameters.) We will not pursue the picture of a static source any further, because it is not consistent with the rapidity distributions of produced particles (see [1]). They are roughly two times broader than expected for thermal emission, whereas in the transverse directions the momentum distributions resemble closely a thermal distribution. There seems to be a preference for longitudinal motion, for which several mechanisms have been proposed, one being Bjorken's longitudinal scaling expansion[5] and, another, two or more sources having longitudinal velocities, which depend on their relative distance in configuration space. The longitudinal motion of the observed pions would then be the superposition of a thermal component and the motion of several different pion sources (or one expanding pion source).

In both pictures a strong correlation between rapidity and the longitudinal co-ordinate in configuration space is established. In the expansion regime particles emitted from the centre of the source will be observed at midrapidity (y=0) and particles emitted from a certain distance from the centre will be shifted by a corre-sponding Δ y. (The thermal motion will wash out this 1-to-1 relation somewhat.) In the scenario of a longitudinal expanding source the correlation between R_{long} and Q_{long} is lost, because particle pairs emitted from distant locations in the source will never be near in rapidity and thus never at small Q_{long}! The correlation in Q_{long} measures instead the duration of expansion from the formation of the source until the particles become free (freezout!). This can be qualitatively understood from the original 'definition': the probability to observe two identical bosons in the same phase space cell is enhanced. The corresponding longitudinal phase space cell dimen-sion is given by $\hbar = \Delta x \cdot Q_{long}$. In the longitudinal expansion scheme Δx (e.g. the distance to the origin) will increase with time according to $\Delta x = \tau \cdot v_{expansion}$. On the other hand $v_{expansion}$ is correlated with the longitudinal momentum difference, thus with Q_{long}. With this correlation the uncertainty relation reads $\hbar = \tau \cdot f(Q_{long}) \cdot Q_{long}$. With Δx eliminated from the expression, the measurement of the Bose enhancement as a function of Q_{long} will give information on τ only. These arguments hold, if the observer measures pairs which have small longitudinal momenta (rapidities) in his c.m. frame. As soon as one looks at parts of the source which are far away they move away too, and thus the scale in configuration space is Lorentz contracted, the same way as in the expanding universe the light from distant stars is blue shifted. The clear prediction is that $\tau = \Delta x. / v_{expansion}$ decreases with increasing rapidity.

These qualitative arguments can be easily checked by studying the dependence of R_{long} on rapidity. The complete theory ([6, 7, 8, 9]) predicts:

$$R_{long} = \sqrt{2T/m_T} \cdot \tau \cdot cosh^{-1}(y - y_0)$$

Figs. 3a+3b show the experimental results obtained from central S+S and S+Ag collisions. The data points represented by squares are derived from the TPC de-tector; they nicely extend the rapidity coverage beyond midrapidity and show good agreement with the corresponding value below midrapidity in the symmetric S+S system.

For R_{long} the $cosh^{-1}$ dependence is obvious. Using the experimental values T=180 MeV and m_T =200 MeV and R_{long} at y=y_0 yields $\tau \simeq 5$ fm. Before turning to the results on R_{side} and R_{out} a word of caution is neccessary: the experimental evidence for a longitudinal expanding source is quite solid and in accordance with the shape of the single particle rapidity distribution. The quantitative result on τ, however, is subject to many uncertainties and, thus, has a large (50%) systematical error.

We turn now to the other radius parameters which are presented in Fig.3 disre-garding, however, the λ-parameter (see above). No significant rapidity dependence of R_{out} and R_{side} are observed. The difference between R_{out} and R_{side} being small we don't observe an effect which was predicted in references [10, 11] for the emission of pions from a long lived source (like the Quark-Gluon-Plasma).

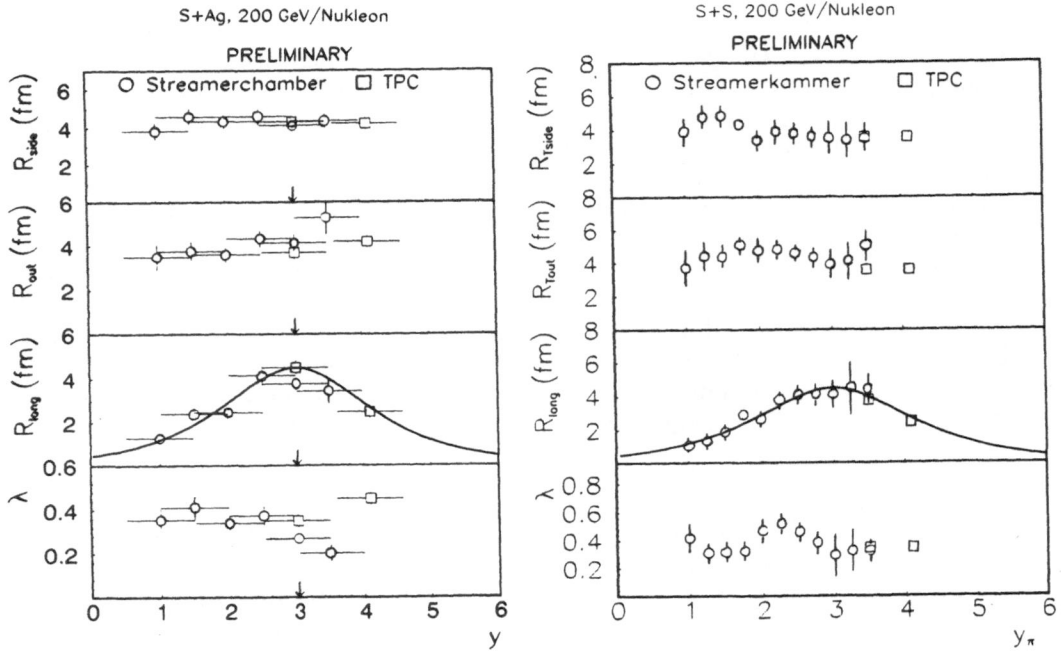

Fig. 3. Radius parameters R_{side}, R_{out} and R_{long} as function of rapidity in central $^{32}S+^{32}S$ (a) and $^{32}S+Ag$ (b) collisions. All pions in each rapidity interval are combined to form the pairs which are used to calculate the correlation functions.

A systematic overview of transverse source sizes of different collision systems is presented in Fig. 4 [12]. On the abszissa various projectile/target combinations are grouped at arbitrary distances; they are ordered, however, according to the geometrical transverse size of of the interaction volume (see below). R_{side} seems to increase monotonically with the transverse size of the system. Any statement about absolute magnitudes or comparisons with sizes of nuclei are dangerous.

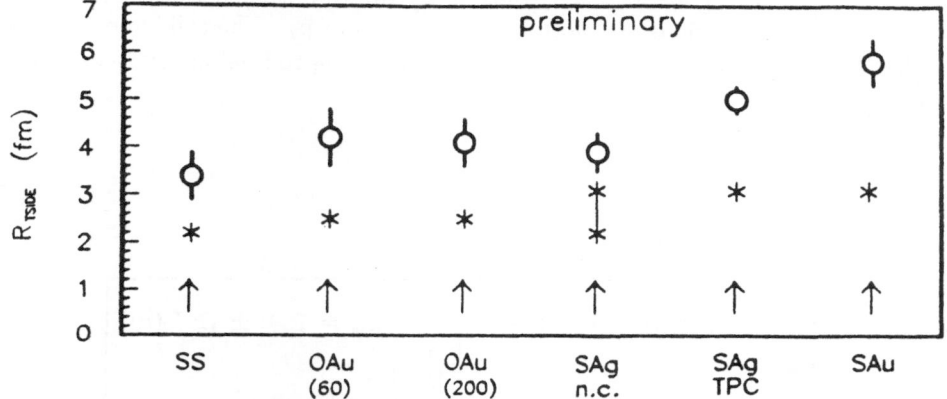

Fig. 4. Radius parameters R_{side} as obtained in various nucleus-nucleus collisions at 60 and 200 GeV/nucleon. The '*' points represent effective transverse sizes of the incident nuclear densities.

The transverse source size parameter (R_{side} which is computed as the rms-width of a Gaussian density distribution) should be compared to the mean transverse radius of the interaction volume as defined by the initial conditions. More precisely, the relevant size has to be deduced from the distribution of the pion production points which may be identified with the loci of the nucleon-nucleon interaction points. The density distribution of these points is given by the product of the nucleon density distributions in the overlapping parts of the two colliding nuclei. The projection of the resulting density distributions onto a plane perpendicular to the beam direction, or rather their rms, provides the quantity to which R_{side} must be compared. It turns out (see Fig.4) that the radius parameters obtained from the two-pion correlation analysis is consistently a factor of 1.5 - 2 larger than expected from the initial distribution of the pion production points. If this first impression is confirmed by experiments with better statistics and smaller systematical errors and/or larger projectiles and higher energies one could consider the scaling factor a sort of normalization of the method of determining the radius parameter. An increase of this parameter with energy or pion density could be an indication of transverse expansion before the pions decouple.

So far the rapidity dependence of the source size parameters have been looked at. What do we expect for their variation with transverse momentum? In the case of the longitudinal dimension in the longitudinal expanding scenario the R_{long} dependence on m_T as given by the formula presented earlier provides a concrete prediction. R_{long} should decrease with inreasing p_T as $\sqrt{m_T^{-1}}$. We demonstrate this sort of dependence by showing the correlation functions of Q_{long} for low and high transverse momentum intervals (Fig. 5) as seen in central S+Au collisions. The difference between the correlation functions is striking. Clearly the high p_T pairs give a smaller R_{long} and thus a smaller apparent decoupling time. More detail of the p_T dependence is given in Fig.6. Here R_{long} values obtained from central S+S, S+Ag, and S+Au collisions are plotted as a function of p_T. The solid lines indicate the $\sqrt{m_T^{-1}}$ function. Again good agreement with the longitudinal expansion scenario is apparent.

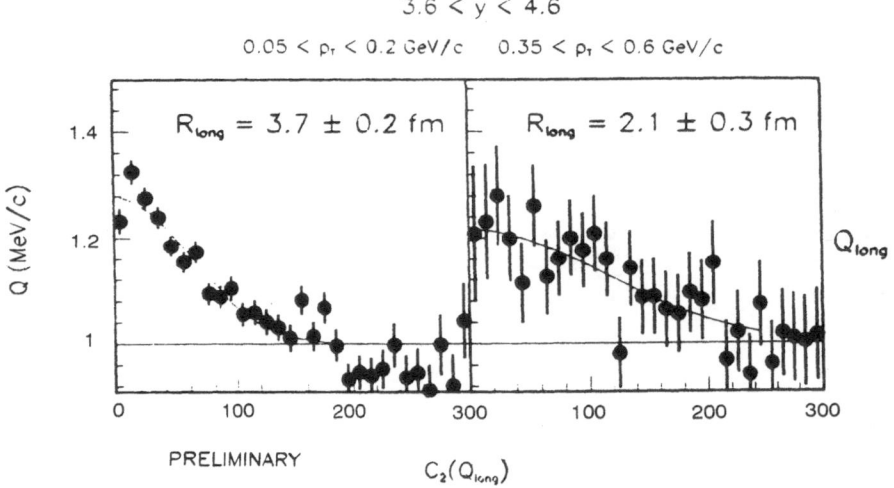

Fig. 5. Two pion correlation functions as obtained in two different transverse momentum intervals in ^{32}S+Au collisions at 200 GeV/nucleon.

Going back to Fig.5 we consider the p_T dependence of R_{side} and R_{out}. Here the statistical errors are too large to draw any conclusions. The fits to the correlation functions yield smaller radii at large transverse momenta, but a simple overlay of the measured correlations in the two momentum intervals confirms that the differences are not significant. A decrease of the transverse radii with transverse momentum would be indicative of transverse expansion. More high statistics data are needed before the existence of transverse expansion can be claimed. To this end NA35 has embarked on the analysis of 15000 Streamer Chamber events. Preliminary results should be available early 1994.

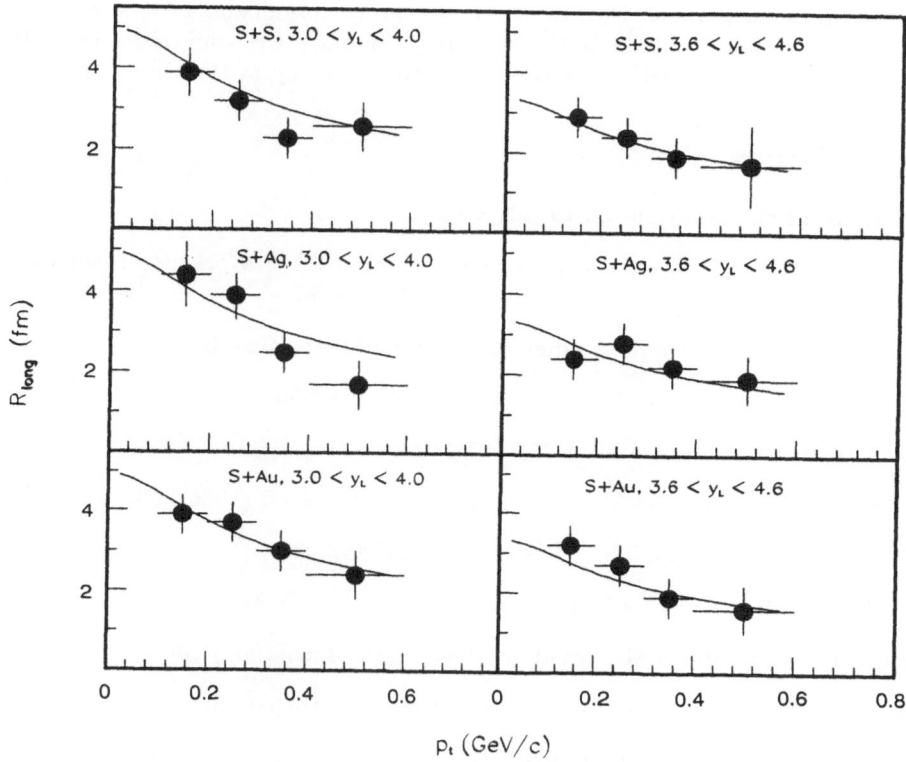

Fig. 6. Radius parameters R_{long} as function of transverse momentum in central $^{32}S+^{32}S$, $^{32}S+Ag$ and $^{32}S+Au$ collisions.

3   Summary

Two pion correlation studies provide information on the dimension of the particle emitting source. The size parameters are determined as function of rapidity and transverse momentum of the pairs. Subject to the uncertainties of various model assumptions the results strongly suggest that the pion source is expanding longitudinally and that the decoupling time is of the order of 5 fm/c. The transverse dimensions seem to be systematically larger than the effective projectile-target overlap region. The corresponding factor is approximately constant (1.7).

4 Acknowledgements

Without the contribution and engagement of all members of the NA35 collaboration [14] the results presented here would not be at hand. They are based especially on the work of D. Ferenc[3] and G. Roland[13]. The author acknowledges financial support from the German Government and from the organizers of this workshop. It was a pleasure to participate in the workshop, to follow the scientifique programme, and to enjoy the stimulating atmosphere during the stay in Bodrum.

References

[1] D. Röhrich; Contribution to this Volume

[2] M. Plümer, S. Raha and R.M. Weiner, editors CAMP Workshop Marburg 1990, World Scientific Publ. Co. 1991, ISBN 981-02-0331-4

[3] D.Ferenc, PhD Thesis, Rudjer Bošković Institute, Zagreb, 1992

[4] M. Deutschmann et al.; Nucl. Phys. B204(1982)333

[5] J.D. Bjorken; Phys. Rev. D27(1983)140

[6] K. Kolehmainen and M. Gyulassy; Phys. Lett. B180 (1986)203

[7] Y. Hama and S. Padula; Phys.Rev. D37(1988)3237

[8] A. Makhlin, and Yu.M. Sinyukov; Z. Phys. C39(1988)69

[9] B. Lorstad and Yu.M. Sinyukov; Phys. Lett. B265(1991)159

[10] G.F. Bertsch and G.E Brown; Phys. Rev. C40(1989)1830

[11] S. Pratt; Phys. Rev. D33(1986)1314

[12] D. Ferenc et al., Nucl. Phys. A544(1992)531c

[13] G. Roland, PhD Thesis, Univ. Frankfurt, 1992

[14] The NA35 Collaboration J. Bächler[7], J. Bartke[4], H. Bialkowska[12], M.A. Bloomer[3], R. Bock[5], R. Brockmann[5], P. Bunčić[13], S.I. Chase[3], J.G. Cramer[10], I. Derado[9], V. Eckardt[9], J. Eschke[6], C. Favuzzi[2], D. Ferenc[6,13], P. Foka[5], M. Fuchs[5], M. Gaździcki[6],E. Gladysz[4], J.W. Harris[3], M. Hoffmann[7], P. Jacobs[3], S. Kabana[6], K. Kadija[13], J. Kosiec[11], M. Kowalski[9], A. Kühmichel[6], J.Y. Lee[6], A. Ljubičić jr.[13], S. Margetis[3], J.T. Mitchell[3], R. Morse[3], E. Nappi[2], G. Odyniec[3], G. Paić[5,13], A.D. Panagiotou[1], A. Petridis[1], A. Piper[8], F. Posa[2], A.M. Poskanzer[3], F. Pühlhofer[8], G. Rai[3], W. Rauch[9], R. Renfordt[6], W. Retyk[11], D. Röhrich[6], G. Roland[6], H. Rothard[6], K. Runge[7], A. Sandoval[5], J. Schambach[3], N. Schmitz[9], E. Schmoetten[7], P. Seyboth[9], J. Seyerlein[9], E. Skrzypczak[11], P. Spinelli[2], P. Stefansky[4], R. Stock[6], H. Ströbele[6], L. Teitelbaum[3], S. Tonse[3], T. Trainor[10], G. Vasileiadis[1], M. Vassiliou[1], D. Vranic[13], S. Wenig[6], B. Wosiek[9]

[1]Dept. of Physics, Univ. of Athens, Athens, Greece, [2]INFN Bari, Bari, Italy, [3]LBL, Berkeley, CA, USA, [4]Inst. of Nucl. Physics, Cracow, Poland, [5]GSI,

Darmstadt, Germany, [6]Fachbereich Physik der Univ., Frankfurt, Germany, [7]Fakultät für Physik der Universität, Freiburg, Germany, [8]Fachbereich Physik der Univ., Marburg, Germany, [9]Max–Planck–Inst. für Physik, München, Germany, [10]Univ. of Washington, Seattle, USA, [11]Inst. for Experimental Physics, Univ. of Warsaw, Poland, [12]Inst. for Nuclear Studies, Warsaw, Poland, [13]Rudjer Bošković Institute, Zagreb, Croatia.

METASTABLE EXOTIC MULTIHYPERNU-CLEAR OBJECTS

Jürgen Schaffner,[1] Carl B. Dover,[2] Avraham Gal,[3] Carsten Greiner,[4] Horst Stöcker[1]

[1]Institut für Theoretische Physik, J.W. Goethe Universität, D-60054 Frankfurt am Main, Germany

[2]Physics Dept., Brookhaven National Lab., Upton, NY 11973

[3]Racah Institute of Physics, The Hebrew University, Jerusalem 91904, Israel

[4]Physics Dept., Duke University, Durham, NC 27706

INTRODUCTION

When one is looking at the area of normal nuclei one knows for a very long time that there exists a valley of stability and that the binding energy can be well described by the Bethe-Weizsäcker formula. Nowadays, it is assumed that QCD is the underlying basic theory. If one studies another degree of freedom of QCD, the strangeness, one realizes that this opens a new dimension of finite nuclear systems. Hence, one has to deal with a three dimensional space of nuclear composites of nucleons and hyperons.

There are several calculations and experimental data for one to two units of strangeness [1] but a nuclear desert opens for larger numbers. Speculations about small nuggets of strange quark matter have been proposed as the true ground state of matter for large strangeness [2]. But the more natural way is to extend the known strange nuclear systems, hypernuclei, to this unknown domain. Here we want to examine the properties of strange hadronic matter in a controlled fashion extending the Bethe-Weizsäcker formula.

Let us consider possible combinations of hyperons and nucleons taking care of the conserved quantum numbers of QCD, charge, strangeness, and baryon number. First,

Hot and Dense Nuclear Matter, Edited by
W. Greiner *et al.*, Plenum Press, New York, 1994

Table 1: Metastable combinations of nucleons and hyperons.

$-S\backslash Z$	-2	-1	0	+1	+2
0			nn	np	pp
1		$\Sigma^- n$	$\Lambda\, n$	$\Lambda\, p$	$\Sigma^+ p$
2	$\Sigma^-\Sigma^-$	$\Xi^- n$	$\Lambda\Lambda$	$\Xi^0 p$	$\Sigma^+\Sigma^+$
3	$\Xi^-\Sigma^-$	$\Xi^-\Lambda$	$\Xi^0\Lambda$	$\Xi^0\Sigma^+$	
4	$\Xi^-\Xi^-$	$\Xi^0\Xi^-$	$\Xi^0\Xi^0$		
5	$\Xi^-\Omega^-$	$\Xi^0\Omega^-$			
6	$\Omega^-\Omega^-$				

there exist several strong nonmesonic reactions inside a hypernucleus:

$$\Sigma^- p \to \Lambda n \quad , \qquad \Sigma^+ n \to \Lambda p \quad , \tag{1}$$

$$\Sigma^-\Lambda \to \Xi^- n \quad , \qquad \Sigma^+\Lambda \to \Xi^0 p \quad , \qquad \Sigma^-\Sigma^+ \to \Lambda\Lambda \quad , \tag{2}$$

$$\Xi^0 n \to \Lambda\Lambda \quad , \qquad \Xi^- p \to \Lambda\Lambda \quad . \tag{3}$$

with a large energy release of 50 MeV or more except for the last two reactions (23 MeV and 28 MeV respectively). So are there only Λ hypernuclei metastable? This is not the case at all, as can be seen in the table.

Here the baryon pairs are ordered according to their strangeness S and charge Z. Note that these are all metastable combinations because they occupy the lowest lying state for given strangeness and charge. For three different baryons one finds the following metastable combinations: npΛ (hypernuclei), $\Xi^-\Sigma^- n$, $\Xi^-\Lambda n$, $\Xi^0\Lambda p$, $\Xi^0\Sigma^+ p$, $\Xi^-\Xi^0\Lambda$, and $\Xi^-\Xi^0\Omega^-$. A composite of more than three different species of baryons is not metastable in free space. This might be different when in-medium effects are taken into account. We will come to this point later.

What do we know about strangeness in nuclei experimentally?

As we have seen that there exist metastable combinations of strange baryons from first principles. The experimentally known nuclei with strangeness are Λ, $\Lambda\Lambda$, Σ, and Ξ hypernuclei [1]. The Λ hypernuclei were first seen 1953 by Danysz and Pniewski in cosmic radiation [3]. Over the next decades one has found Λ hypernuclei from $^3\text{He}_\Lambda$ to $^{15}\text{N}_\Lambda$ [4], seven events of Ξ hypernuclei [5], and three events of $\Lambda\Lambda$ hypernuclei [6]. The situation for Σ hypernuclei is still controversal due to the bad statistics and the strong reaction decay with an energy release of about 80 MeV [7]. In the mid seventies spectroscopy of Λ hypernuclei started using meson factories enabling to measure even deep lying single particle states of the Λ [1]. The main feature extracted out of these hypernuclear data from $^9\text{Be}_\Lambda$ to $^{209}\text{Bi}_\Lambda$ is a potential depth of the Λ in nuclear matter of about 30 MeV and a very small spin orbit splitting. The scarce data about Ξ hypernuclei indicates a potential depth of about 20 to 25 MeV [8]. Therefore the YN interaction is weaker than the NN interaction (the nucleons feel a potential depth of about 50 to 60 MeV in nuclear matter). Interestingly the $\Lambda\Lambda$ interaction extracted from the three available $\Lambda\Lambda$ hypernuclei is strong, about 3/4 of the NN interaction [6]. In summary for this section experimentalists can only tell us something about nuclei with one or two units of strangeness.

What are the properties of multiply strange nuclear systems?

To learn something about nuclei with arbitrary strangeness content we have to go back to theory. Our starting point for a nuclear model to describe nuclei and hypernuclei is the relativistic mean field model [9]. The nucleons interact through mesons assuming that only low spin and isospin is needed. This is motivated from the succes of one-boson exchange models. In the mean field approximation the pion field vanishes due to its pseudoscalar character. The scalar meson effectively describes the 2π exchange and is attractive while the vector meson exchange is repulsive. The ρ meson takes care of isotopic trends. In addition we have nonlinearities of the scalar field to get a correct compression constant of nuclear matter and a static Coulomb field. We start from the following Lagrangian for arbitrary baryons (model 1):

$$
\begin{aligned}
\mathcal{L} &= \mathcal{L}_{Mesons}^{free} + \mathcal{L}_{Baryons}^{free} + \mathcal{L}_{coupling} \\
\mathcal{L}_{Mesons}^{free} &= +\frac{1}{2}\partial_\nu\sigma\partial^\nu\sigma - U(\sigma) - \frac{1}{4}G_{\mu\nu}G^{\mu\nu} + \frac{1}{2}m_\omega^2 V_\mu V^\mu - \frac{1}{4}\vec{B}_{\mu\nu}\vec{B}^{\mu\nu} \\
&\quad + \frac{1}{2}m_\rho^2\vec{R}_\mu\vec{R}^\mu - \frac{1}{4}F_{\mu\nu}F^{\mu\nu} \\
\mathcal{L}_{Baryons}^{free} &= \sum_B \overline{\psi}_B\left(i\gamma^\nu\partial_\nu - m_B\right)\psi_B \\
\mathcal{L}_{coupling} &= -\sum_B g_{sB}\overline{\psi}_B\psi_B \cdot \sigma - \sum_B g_{vB}\overline{\psi}_B\gamma_\mu\psi_B \cdot V^\mu \\
&\quad - \sum_B g_{rB}\overline{\psi}_B\gamma_\mu\vec{\tau}_B\psi_B \cdot \vec{R}^\mu - \sum_B e \cdot q_B\overline{\psi}_B\gamma_\mu\psi_B \cdot A^\mu
\end{aligned}
\tag{4}
$$

In the mean field and no sea approximation one gets Klein-Gordon equation for the meson fields and Dirac equations for the baryons which are solved iteratively and selfconsistently [10]. The spherical relativistic mean field theory has been proven to give an excellent description of the properties of nuclei [11] and hypernuclei [12]. One gets a rather good agreement even for deformed and light hypernuclei. Now we can extrapolate to multi hypernuclei [13] within this approach taking special care of metastability.

Systems of $\{n,p,\Lambda,\Xi^0,\Xi^-\}$ baryons are most likely to be studied in the following due to the following arguments. Ξ's can be Pauli-blocked against strong reactions inside a nuclear system because of their rather low energy release [14]:

$$\Lambda\Lambda \leftrightarrow \Xi^- p - 28 \text{ MeV} \quad , \qquad \Lambda\Lambda \leftrightarrow \Xi^0 n - 23 \text{ MeV} \quad .$$

Multi Λ-hypernuclei collapses therefore to systems of $\{n,p,\Lambda,\Xi^0,\Xi^-\}$ baryons for a certain number of Λ's [15]. This breakdown is visualized in fig. 1 yielding objects with binding energies of about 13 MeV/A. Note that the Pauli-blocking mechanism makes them metastable yielding lifetimes of the order of 10^{-10} sec. or more. These objects we call metastable exotic multihypernuclear objects (MEMOs).

On the other hand Σ's are generally not stable due to

$$\Sigma N \to \Lambda N + 75 \text{ MeV} \quad , \qquad \Sigma\Xi \to \Lambda\Xi + 80 \text{ MeV} \quad .$$

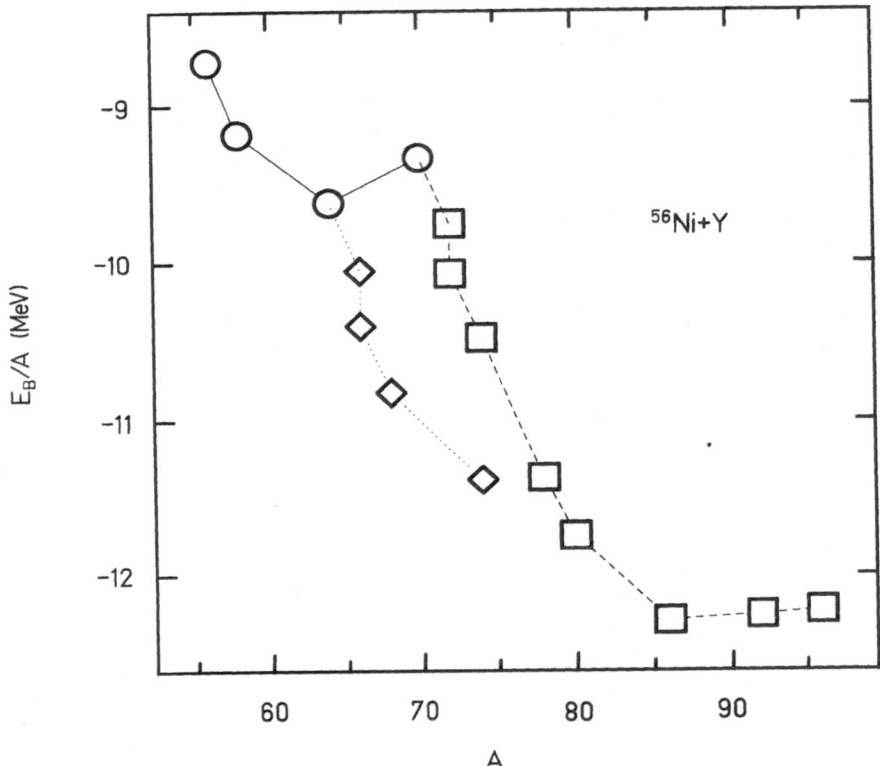

Figure 1. The breakdown of multi Λ hypernuclei exemplified by a ^{56}Ni core (circles stand for $n_\Lambda = 0, 2, 8, 14$ added to the core, diamonds for $n_\Lambda = 8$ plus Ξs, and squares for $n_\Lambda = 14$ plus Ξs added).

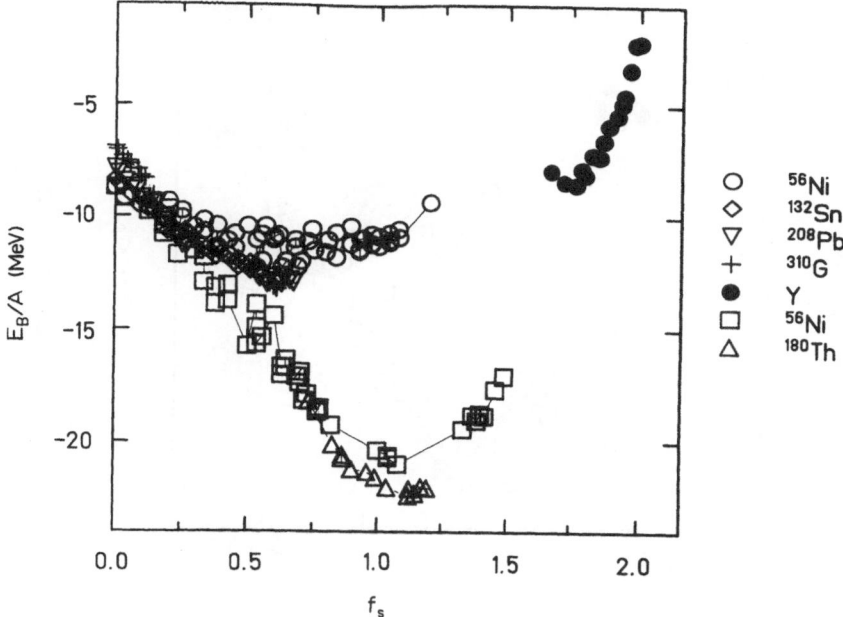

Figure 2. The binding energy of several MEMOs versus the strangeness fraction f_s.

Next we have to implement the strong YY interaction in the mean field model. This is motivated by the few available $\Lambda\Lambda$ hypernuclei and by a one boson exchange model consistent with NN, NΛ, NΣ scattering data. The $\sigma + \omega$ model generally underestimates the $\Lambda\Lambda$ interaction. But introducing a strange meson exchange ($\sigma^* + \phi$) improves the situation considerably [16]. The σ^* can be identified with the f_0 meson. This gives an additional Lagrangean of the form

$$\mathcal{L}_{YY} = -\frac{1}{4}S_{\mu\nu}S^{\mu\nu} + \frac{1}{2}m_\phi^2\phi_\mu\phi^\mu + \frac{1}{2}\partial_\nu\sigma^*\partial^\nu\sigma^* - \frac{1}{2}m_{\sigma^*}^2\sigma^{*2}$$
$$- \sum_B g_{s*B}\overline{\psi}_B\psi_B \cdot \sigma^* - \sum_B g_{\phi B}\overline{\psi}_B\gamma_\mu\psi_B\phi^\mu \tag{5}$$

defining model 2. The coupling constants to the strange vector meson are taken to be

$$\frac{g_{\phi\Lambda}}{g_{vN}} = -\frac{\sqrt{2}}{3} \quad , \qquad \frac{g_{\phi\Xi}}{g_{vN}} = -\frac{2\sqrt{2}}{3} \tag{6}$$

using SU(6) relations.

The binding energy of several branches of MEMOs, starting with different normal nucleonic cores, is plotted in fig. 2 versus the strangeness fraction $f_s = |S|/A$. There is a sizable increase to $E_B/A \approx -21$ MeV for model 2. The minimum lies at large strangeness contents of 0.6 (model 1) and 1.1 (model 2) respectively which is in the range of hypothetical strange quark matter nuggets (strangelets) [2].

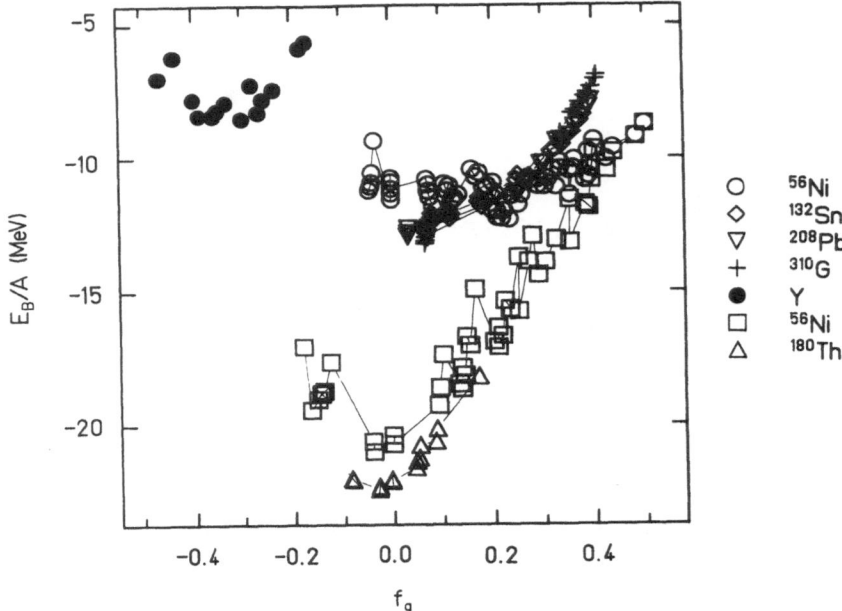

Figure 3. The binding energy of several MEMOs versus the charge fraction $f_q = Z/A$.

The rather exotic and unusual charge to mass ratios can be read off from fig. 3. Negatively charged composites with a positive baryon number are possible. The minimum is located around zero charge for both models striking similar to strangelets. Another new form of hadronic matter, purely hyperonic nuclei exist in model 2 for $f_s \approx 1.7$ and $Z/A \approx -0.3 \pm 0.1$. This negative charge makes MEMOs well distinguishable from normal nuclei.

How can we find MEMOs?

Nevertheless one has to find a source of strangeness to generate these multiply strange nuclear systems predicted in the last section. The lifetime of Λ hypernuclei has been measured to be around the lifetime of the free Λ, i.e. 10^{-10} sec. Therefore they are not visible in the 'normal' world. Mesonic and electromagnetic production of hypernuclei will be not able to yield $|S| \geq 3$ composites. The early universe or the interior of neutron stars will be enriched with abundant strangeness but these is out of reach. Relativistic heavy ion collisions constitute a terrestrial prolific source of strangeness. Here one can possibly find strange hadronic matter which is build during the evaporation process. Predictions of RQMD presented at this conference gives about 30 hyperons for central collisions of Au(11.7 AGeV)Au and about 100 (!) hyperons for Pb(160 AGeV)Pb per one event.

The phase space distribution of the baryons of the RQMD can be used as input for a coalescence model where two baryons are sitting together for a certain momentum sphere ($\Delta p \leq p_0$) [17]. Fig. 4 shows the number of produced nuclear and hyper-

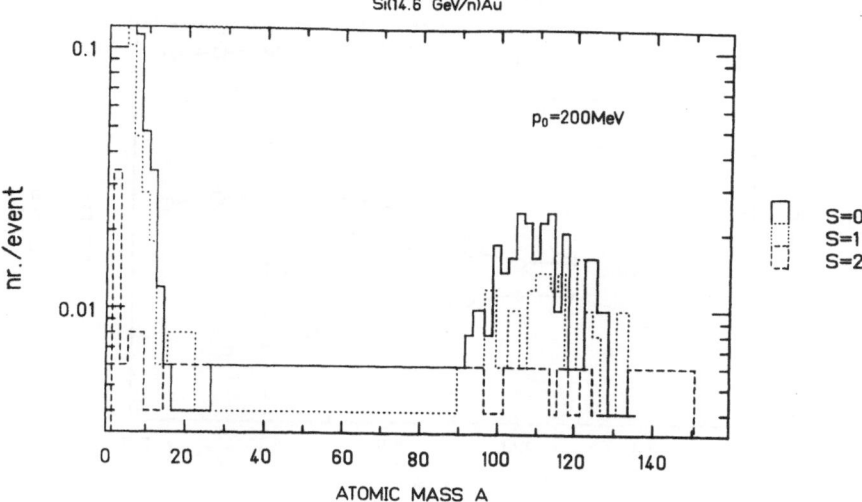

Figure 4. The production of hyperfragments for different mass numbers.

nuclear fragments for the reaction Si(14.6 AGeV)Au. There is a peak around the mass number $A \approx 100$ which are fragments of the spectator matter of the Au target. Here the production of heavy double hypernuclei is possible. Many light multistrange composites $(A > 10)$ spread out of the central region of the collision.

Fig. 5 shows the number of produced dibaryons in central collisions of Pb(160 AGeV)Pb which will be soon available at CERN. The number of dihyperons is by an order of magnitude larger than the number of antideuterons. Therefore negatively charged tracks of fragments can be mainly attributed to strange hadronic matter. This is insensitive to the coalescence parameter p_0. Note that we only take into account pairs which have no other baryons inside their momentum sphere.

Using the coalescence model one can deduce a simple 'empirical' formula for the production of strange composites in heavy ion collisions. The possibility for a composite with baryon number A and strangeness S is

$$P(A, S) = \frac{P(A,S)}{P(A,0)} \cdot \frac{P(A,0)}{N_{p,n}} \cdot N_{p,n} \approx p^{|S|} \cdot q^{A-1} \cdot N_{p,n} \approx 10^{3-A-|S|}$$

where $p \approx 0.1$ (at AGS) and $q \approx 0.1 - 0.4$ are penalty factors. For a detection sensitivity of 10^{-r} of the experiment MEMOs can be detected up to

$$A + |S| < r + 3 \quad .$$

High sensitivity experiments are under way at AGS presently with $r \approx 7 - 12$. For all these experiments the lifetime of MEMOs will be crucial for their detection (up to now they are only sensitive to $\tau > 10^{-9}$ sec.).

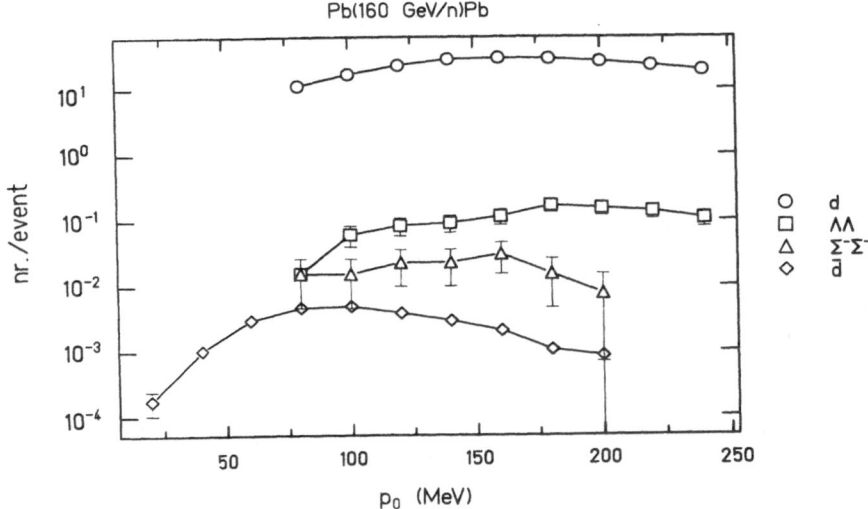

Figure 5. The production of dibaryons for different coalescence parameters p_0.

Conclusion

In summary the strange meson exchange introduced here gives a correct strenght of the YY interaction. We find stable objects composed of $\{n,p,\Lambda,\Xi^0,\Xi^-\}$ baryons with $f_s > 1.0$ and $f_q < 0$. Purely hyperonic matter of $\{\Lambda,\Xi^0,\Xi^-\}$ baryons exist for $f_s \approx 1.7$ and $f_q \approx -0.3 \pm 0.1$. Heavy ion collisions seems to be a promising tool to generate this objects with reasonable production possibilities. Nevertheless the sensitivity to a proposed lifetime of $\tau \approx 10^{-10}$ sec. gives a new challenge for experimentalists.

REFERENCES

1. R.E. Chrien and C.B. Dover: Ann. Rev. Nucl. Part. Sci. **39**, 113 (1989)
   H. Bandō, T. Motoba, and J. Žofka: Int. J. Mod. Phys. **A5**, 4021 (1990)

2. S.A. Chin and A.K. Kerman: Phys. Rev. Lett. **43**, 1292 (1979)
   E. Witten: Phys. Rev. **D 30**, 272 (1984)
   Carsten Greiner, D.H. Rischke , P. Koch and H. Stöcker: Phys. Rev. **D 38**, 2797 (1988)

3. M. Danysz and J. Pniewski: Phil. Mag. **44**, 348 (1953)

4. M. Juric et al.: Nucl. Phys. **B52**, 1 (1973)

5. D.H. Wilkinson, S.J.St. Lorant, D.K. Robinson and S. Lokanathan: Phys. Rev. Lett. **3**, 397 (1959)

A. Bechdolff, G. Baumann, J.P. Gerber and P. Cüer: Phys. Lett. **26B**, 174 (1967)
J. Catala, F. Senet, A.F. Tejerina and E. Villar: Proceedings of the International Conference on Hypernuclear Physics, Vol. 2 (Argonne, Illinois, 1969), p. 758
A.S. Mondal, A.K. Basak, M.M. Kasim and A. Husain: Il Nuovo Cimento **54**, 333 (1979)

6. M. Danysz et al.: Phys. Rev. Lett. **11**, 29 (1963)
   D.J. Prowse: Phys. Rev. Lett. **17**, 782 (1966)
   S. Aoki et al.: Prog. Theor. Phys. **85**, 1287 (1991)

7. C.B. Dover, D.J. Millener, and A. Gal: Phys. Rep. **184**, 1 (1989)

8. C.B. Dover and A. Gal: Ann. of Phys. **147**, 309 (1983)

9. B.D. Serot and J.D. Walecka: Adv. in Nucl. Phys. **16**, 1 (1986)

10. P.-G. Reinhard: Rep. Prog. Phys. **52**, 439 (1989)

11. P.-G. Reinhard, M. Rufa, J. Maruhn, W. Greiner and J. Friedrich: Z. Phys. **A323**, 13 (1986)
    M. Rufa, J. Maruhn, W. Greiner, P.-G. Reinhard, M. Strayer: Phys. Rev. **C 5**, 390 (1988)

12. R. Brockmann and W. Weise: Phys. Lett. **69B**, 167 8 (1977)
    A. Boussy: Nucl. Phys. **A290**, 159 (1977)

13. M. Rufa, H. Stöcker, J. Maruhn, P.-G. Reinhard and W. Greiner: J. Phys. **G 13**, 143 (1987)
    J. Mares and J. Žofka: Z. Phys. **A333**, 209 (1989)
    M. Rufa, J. Schaffner, J. A. Maruhn, H. Stöcker, W. Greiner and P.-G. Reinhard: Phys. Rev. **C 42**, 2469 (1990)

14. J. Schaffner, Carsten Greiner and H. Stöcker: Phys. Rev. **C 46**, 322 (1992)

15. J. Schaffner, C.B. Dover, A. Gal, Carsten Greiner, and H. Stöcker: Phys. Rev. Lett. **71**, 1328 (1993)

16. J. Schaffner, C.B. Dover, A. Gal, Carsten Greiner, D.J. Millener, and H. Stöcker: Annals of Physics in press

17. R. Mattiello, C. Hartnack, A. v. Keitz, J. Schaffner, H. Sorge, H. Stöcker, and Carsten Greiner: Nucl. Phys. **B24B**, 221 (1991)

WAVELET CORRELATIONS
IN SELFSIMILIAR CASCADES

Martin Greiner[1], Peter Lipa[2] and Peter Carruthers[3]

[1]Institut für Theoretische Physik, Justus Liebig Universität, Heinrich-Buff-Ring 16, 35392 Giessen, Germany
[2]Institut für Hochenergiephysik der Österreichischen Akademie der Wissenschaften, Nikolsdorfergasse 18, A-1050 Wien, Austria
[3]Department of Physics, University of Arizona, Tucson AZ 85721, USA

INTRODUCTION

It is always a good thing to look for new methods and technologies, which have been developed even outside of physics; in some cases this is extremely beneficial. Here we report on such a new method: the wavelet transformation [1, 2]. It has found widespread application in engineering for signal processing and data compression [3].

We will give a very basic introduction of the wavelet transformation. Because of its selfsimiliar structure, we suspect that wavelets represent a very convenient tool to describe selfsimiliar processes; therefore we introduce the wavelet transformation to correlation studies of selfsimiliar cascades.

SOME BASICS OF WAVELETS

We will summarize only some very basic concepts of wavelets and the related idea of a multiresolution analysis. For a more profound introduction we refer the reader to some excellent reviews [1-4].

Given an arbitrary and for simplicity one-dimensional function $\epsilon(x)$ supported in the interval $[0,1]$, we seek to approximate it in terms of a histogram with 2^J bins. A histogram is a collection of individual bins, represented by the set of box functions

$$\phi_{Jk}^H(x) = \phi^H(2^J x - k) = \begin{cases} 1 & \text{for } k2^{-J} \leq x \leq (k+1)2^{-J} \\ 0 & \text{otherwise} , \end{cases} \tag{1}$$

which are all constructed from the unit box function $\phi_{00}^H(x)$ by a discrete dilation factor 2^J and a translation governed by an integer k. Within a given scale the box functions $\phi_{Jk}^H(x)$ are orthogonal with respect to the shift index k.

The approximation of the function $\epsilon(x)$ at the finest scale J is then written as a histogram:

$$\epsilon(x) \rightarrow \epsilon^{(J)}(x) = \sum_k \epsilon_k^{(J)} \phi_{Jk}^H(x) ; \tag{2}$$

compare with the upper left histogram of fig. 1. If we were to approximate $\epsilon(x)$ with the box functions $\phi_{J-1,k}^H$ belonging to the rougher resolution scale $J-1$, which are again orthogonal with respect to the shift index k, but are not orthogonal to the box functions ϕ_{Jk}^H of the finer 'resolution' scale J, evidently some detail is lost compared to the approximation (2). This detail is the difference between approximations (2) with 'resolution' scales J and $J-1$. It can be fully expressed in terms of the difference functions $\psi_{J-1,k}^H(x) = \psi^H(2^{J-1}x - k)$ with

$$\psi_{00}^H(x) = \psi^H(x) = \begin{cases} 1 & \text{for } 0 \leq x \leq 1/2 \\ -1 & \text{for } 1/2 \leq x \leq 1 \\ 0 & \text{otherwise} . \end{cases} \tag{3}$$

Again the functions $\psi_{J-1,k}^H(x)$ are orthogonal with respect to the shift index k within the given 'resolution' scale J-1; they are also orthogonal to the box functions $\phi_{J-1,k}^H(x)$ at scale J-1, but are not orthogonal to the box functions $\phi_{Jk}^H(x)$ at scale J. On the other hand the $\phi_{Jk}^H(x)$ are expressible in terms of the $\phi_{J-1,k}^H(x)$ and $\psi_{J-1,k}^H(x)$, together; we get for example

$$\phi_{10}^H(x) = \frac{1}{2}(\phi_{00}^H(x) + \psi_{00}^H(x)) \quad \text{and} \quad \phi_{11}^H(x) = \frac{1}{2}(\phi_{00}^H(x) - \psi_{00}^H(x)) . \tag{4}$$

This defines a multiresolution analysis of the function $\epsilon(x)$: At first $\epsilon(x)$ is approximated with box functions $\phi_{Jk}^H(x)$ of scale J according to eq.(2). Then according to eq.(4) the $\phi_{Jk}^H(x)$ are expressed in terms of box functions $\phi_{J-1,k}^H(x)$ and the difference functions $\psi_{J-1,k}^H(x)$, both of lower scale J-1. Thereafter the $\phi_{J-1,k}^H(x)$ are rewritten in terms of $\phi_{J-2,k}^H(x)$ and $\psi_{J-2,k}^H(x)$ and so on. Going from one scale j to the next lower scale j-1, only the difference between the two resolutions is memorized and expressed in terms of the difference functions $\psi_{j-1,k}^H(x)$. This procedure is depicted in figure 1. As a consequence of this multiresolution analysis the amplitudes $\epsilon_k^{(J)}$ of the box functions at the finest scale J are transformed into the amplitudes $\tilde{\epsilon}_{jk}^{(J)}$ of the difference functions with scales in the range $0 \leq j \leq J - 1$ together with the amplitude $\epsilon_0^{(0)}$ of the box function at the roughest scale j=0:

$$\epsilon^{(J)}(x) = \sum_k \epsilon_k^{(J)} \phi_{Jk}^H(x)$$

$$= \epsilon^{(0)}(x) + \sum_{j=0}^{J-1} \tilde{\epsilon}^{(j)}(x) = \epsilon_0^{(0)} \phi_{00}^H(x) + \sum_{j=0}^{J-1} \sum_{k=0}^{2^j-1} \tilde{\epsilon}_{jk}^{(J)} \psi_{jk}^H(x) \ ; \tag{5}$$

confer again fig. 1.

In general the difference function $\psi_{jk}(x)$ is called a wavelet and the approximation function $\phi_{jk}(x)$ the corresponding scaling function. More formally they are defined via the dilation equation

$$\phi(x) = \sum_m c_m \, \phi(2x - m) \tag{6}$$

and

$$\psi(x) = \sum_m (-1)^m c_{1-m} \, \phi(2x - m) \ , \tag{7}$$

where the finite number of nonzero coefficients c_m have to satisfy various conditions stated by Daubechies [1]. Once a finite set of admissible c_m is chosen, the solutions ϕ and ψ can be found by (numerical) iteration of eqs.(6) and (7).

The scaling function $\phi^H(x)$ of eq.(1) and the Haar wavelet $\psi^H(x)$ of eq.(3) represent the simplest example; for them the only nonvanishing coefficients are $c_0 = c_1 = 1$. Another choice,

$$c_0 = \frac{1}{4}(1 + \sqrt{3}) \ , \quad c_1 = \frac{1}{4}(3 + \sqrt{3}) \ , \quad c_2 = \frac{1}{4}(3 - \sqrt{3}) \ , \quad c_3 = \frac{1}{4}(1 - \sqrt{3}) \ , \tag{8}$$

leads to the smoother orthogonal Daubechies D4-wavelet [1]. These wavelets and their corresponding scaling functions are exemplified in fig. 2.

As a generalization of the multiresolution analysis presented for the Haar wavelet before, the equations (5), (6) and (7) define a multiresolution analysis for any specific choice of wavelets. The resulting "histograms" at the various scales then acquire the smoothness of the underlying scaling function and wavelet.

WAVELET CORRELATION DENSITIES
OF THE SELFSIMILIAR P-MODEL CASCADE

We will briefly introduce the p-model, which successfully describes the multifractal spectrum of the energy dissipation in turbulent flow [5]. We determine its conventional correlation densities as well as its wavelet correlation densities [6].

p-model

Without any loss of generality we consider the interval $[0,1]$, and normalize the 'energy' E to unity in this interval. We then split this interval into two equal parts with energies $E_1 = pE$ and $E_2 = (1-p)E$, where E_1 goes randomly, with equal probability, to the left or right subinterval (bin). Let us pick the left subinterval and, say, it goes with the 'energy' E_1; we then again split this subinterval into two parts with corresponding 'energies' $E_1' = pE_1$ and $E_2' = (1-p)E_1$, where again E_1' goes randomly to the new left or right subinterval. For the right subinterval we proceed in the same way. The whole procedure is repeated over and over again, which

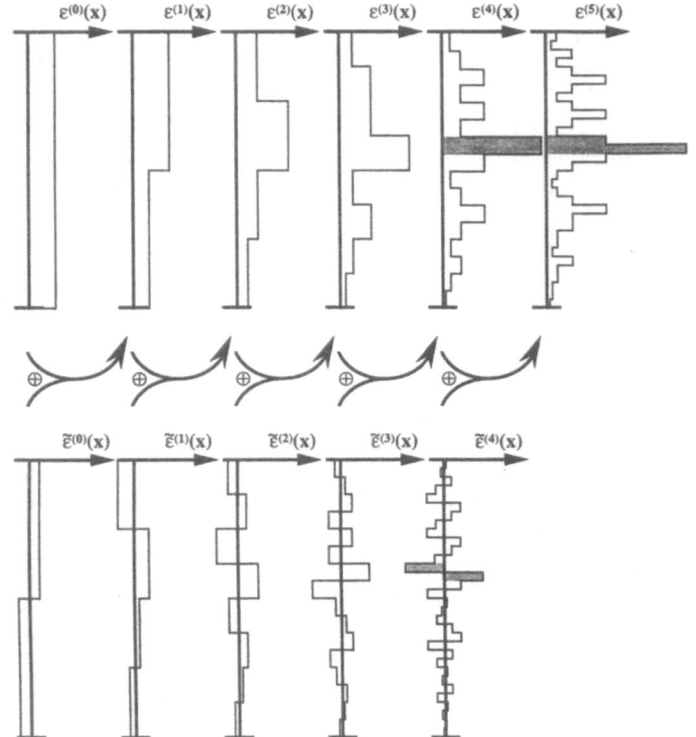

Figure 1: A multiresolution decomposition of a random function (upper left corner) at initial resolution scale $J = 5$ with respect to the Haar-wavelet basis. The left column shows a sequence of approximations to the original signal, while the right column represents the Haar-wavelet transform of $\epsilon^{(5)}(x)$.

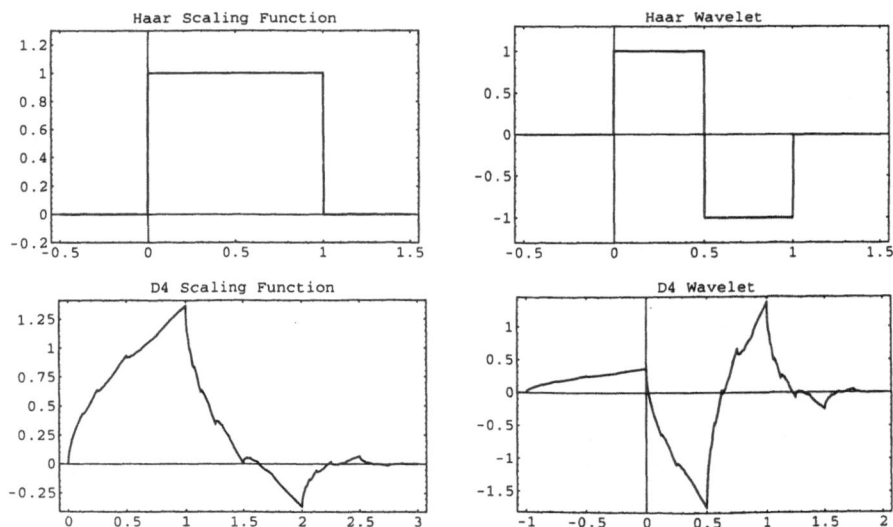

Figure 2: Scaling function and wavelet for the Haar and Daubechies D4 case.

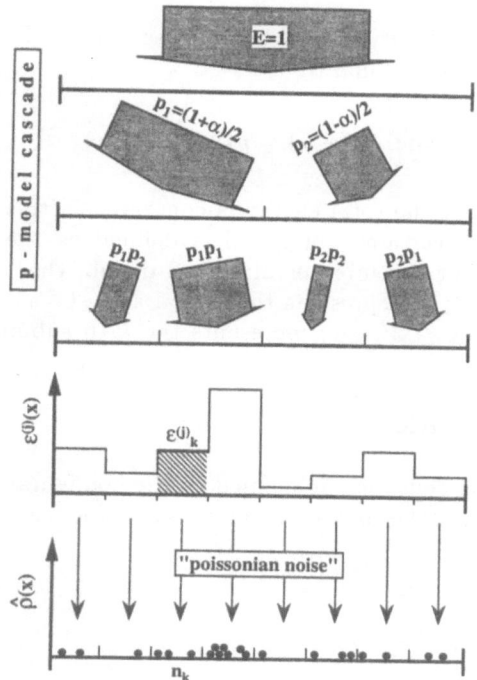

Figure 3: The p-model cascade, where towards smaller scales the length of the interval is divided into two halves and the flux of energy is transferred in nonequal fractions $p_1 = (1 + \alpha)/2$ and $p_2 = (1 - \alpha)/2$. At the last step a number of particles is tossed in each bin according to a Poisson distribution with mean $\bar{n}\epsilon_k^{(J)}$.

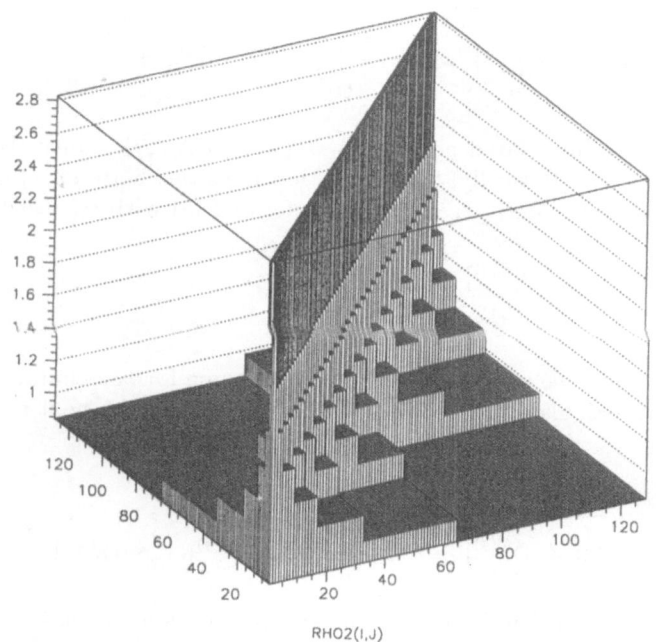

RHO2(I,J)

Figure 4: Two-bin correlation density $\rho_{k_1 k_2}$ after $J=7$ p-model cascade steps ($\alpha = 0.4$).

characterizes the selfsimiliarity of this cascade; see fig. 3. For later convenience we set the 'splitting'-parameter equal to

$$p = p_1 = \frac{1}{2}(1 + \alpha) \quad \text{and} \quad (1 - p) = p_2 = \frac{1}{2}(1 - \alpha) \tag{9}$$

with $0 \leq \alpha \leq 1$. Clearly the total energy is conserved in this cascade. The 'energy' density contained in a certain subinterval is defined as the corresponding energy divided by the length of the subinterval; let us denote the 'energy' density as $\epsilon_k^{(J)}$, where the index J $(J \geq 0)$ represents the actual scale (= number of cascade steps) and the index k $(0 \leq k \leq 2^J - 1)$ represents the k-th subinterval belonging to the scale J.

Bin Correlation Densities

The bin correlation densities are defined as the configuration average $\langle \rangle$ of products of energy densities of the corresponding bins at some scale j (= number of cascade steps):

$$
\begin{aligned}
\rho_{k_1}^{(j)} &= \langle \epsilon_{k_1}^{(j)} \rangle \ , \\
\rho_{k_1,k_2}^{(j)} &= \langle \epsilon_{k_1}^{(j)} \epsilon_{k_2}^{(j)} \rangle \ , \dots \quad .
\end{aligned}
\tag{10}
$$

For convenience we consider the characteristic function for the correlation densities:

$$
\begin{aligned}
Z^{(j)}[\vec{\lambda}^{(j)}] &= \left\langle \exp\left(i \sum_{k=0}^{2^j-1} \lambda_k^{(j)} \epsilon_k^{(j)} \right) \right\rangle \\
&= 1 + i \sum_{k_1} \lambda_{k_1}^{(j)} \rho_{k_1}^{(j)} + \frac{i^2}{2!} \sum_{k_1,k_2} \lambda_{k_1}^{(j)} \lambda_{k_2}^{(j)} \rho_{k_1,k_2}^{(j)} + \cdots ,
\end{aligned}
\tag{11}
$$

where we use an abbreviated notation $\vec{\lambda} = (\lambda_0, \dots, \lambda_{2^j-1})$. Once the characteristic function is known analytically, the correlation densities between $2, 3, \dots, q$ bins can be obtained by partial derivatives of q-th order of $Z[\vec{\lambda}]$ with respect to $\lambda_k^{(j)}$:

$$
\rho_{k_1,\dots,k_q}^{(j)} = \frac{1}{i^q} \frac{\partial^q Z^{(j)}[\vec{\lambda}^{(j)}]}{\partial \lambda_{k_1}^{(j)} \cdots \partial \lambda_{k_q}^{(j)}} \Bigg|_{\vec{\lambda}^{(j)}=0} .
\tag{12}
$$

For the p-model we introduce an iterative scheme for the characteristic function, which allows to calculate it at one scale recursively from the next rougher scale. We find:

$$
\begin{aligned}
Z^{(j+1)}[\vec{\lambda}^{(j+1)} &= (\vec{\lambda}_L^{(j+1)}, \vec{\lambda}_R^{(j+1)})] \\
&= \frac{1}{2} \left(Z^{(j)}[(1+\alpha)\vec{\lambda}_L^{(j)}] \, Z^{(j)}[(1-\alpha)\vec{\lambda}_R^{(j)}] + Z^{(j)}[(1-\alpha)\vec{\lambda}_L^{(j)}] \, Z^{(j)}[(1+\alpha)\vec{\lambda}_R^{(j)}] \right) .
\end{aligned}
\tag{13}
$$

This iterative solution for the characteristic function can now be used to derive

270

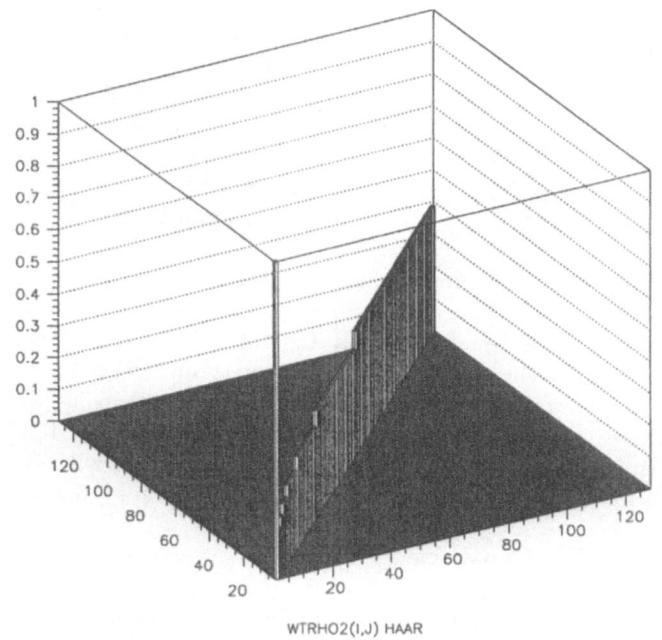

WTRHO2(I,J) HAAR

Figure 5: Haar-wavelet transformed two-bin correlation density $\tilde{\rho}_{(j_1 k_1),(j_2 k_2)}$ for the p-model cascade.

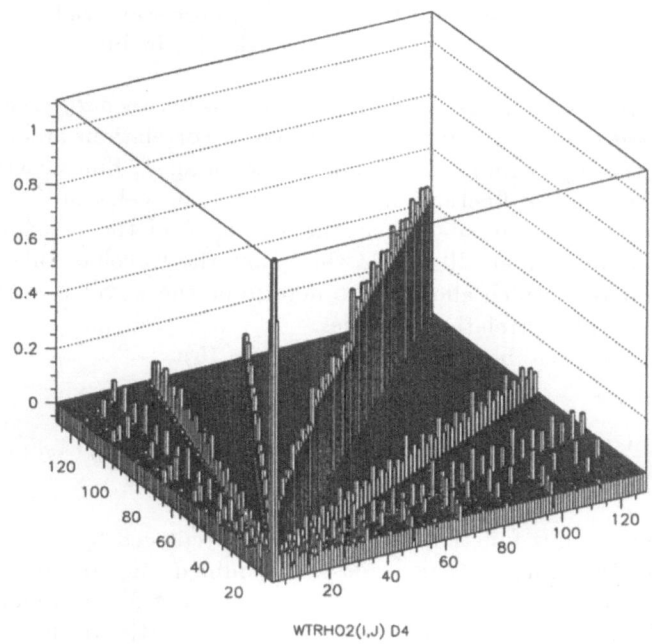

WTRHO2(I,J) D4

Figure 6: D4-wavelet transformed two-bin correlation density $\tilde{\rho}_{(j_1 k_1),(j_2 k_2)}$ for the p-model cascade.

271

recursion relations for the correlation densities (12). We get:

$$\rho_{k_1}^{(j)} = 1 \quad , \tag{14}$$

$$\rho_{k_1,k_2}^{(j+1)} = \begin{cases} (1+\alpha^2)\rho_{k_1,k_2}^{(j)} & , \text{if } k_1, k_2 \in \{L\} \\ (1+\alpha^2)\rho_{k_1-2^j,k_2-2^j}^{(j)} & , \text{if } k_1, k_2 \in \{R\} \\ (1-\alpha^2) & , \text{if } k_1 \in \{L\} \text{ and } k_2 \in \{R\} \text{ or vice versa;} \end{cases} \tag{15}$$

$\{L\}$ and $\{R\}$ stand for all the bins belonging to the left and right branch of the dyadic cascade.

The one-bin density is equal to one as it should be, because 'energy' is conserved in the p-model, so that the average 'energy' density on every scale j is equal to the initial density. The two-bin correlation density is depicted in fig. 4; it shows clearly the multiplicative structure (15) as the scale becomes finer. The closer the bins are together, the stronger they are correlated; a scaling law with decreasing bin-bin distance shows up as a result of the selfsimiliarity of the p-model-cascade.

Wavelet Correlation Densities

So far we have determined the correlation densities between the various bins at the last step of the p-model cascade. If we would first average over several adjacent bins and then study the correlations of the local averages, we could extract information about earlier stages of the cascade; referring to the left column in fig. 1 this would correspond to the analysis of correlations among the amplitudes within each averaged histogram $\epsilon^{(j)}(x)$ separately. Obviously this procedure yields highly redundant information since the correlations at a given scale also include the correlations at all rougher scales.

As an alternative we suggest to study correlations between *differences of bin amplitudes at adjacent scales* ; in other words, we look at correlations among the difference-amplitudes of the right column histograms $\tilde{\epsilon}^{(j)}(x)$ of fig. 1. Since each such histogram reflects the difference of fluctuations at two adjacent scales only and thus is independent of the fluctuations at earlier stages (scales) of the cascade, this approach contains no redundancy at all! — At this point the wavelets enter the game. We develop a formalism, which allows us to determine the wavelet correlation densities directly from the bin correlation densities.

We start with a suitable generalization of eqs. (5):

$$\epsilon(x) \longrightarrow \epsilon^{(J)}(x) = \sum_{k=0}^{2^J-1} \epsilon_k \phi_{Jk}(x) = \epsilon_0^{(0)}\phi_{00}(x) + \sum_{j=0}^{J-1}\sum_{k=0}^{2^j-1} \tilde{\epsilon}_{jk}\psi_{jk}(x) \, , \tag{16}$$

where the Haar scaling function $\phi_{Jk}^H(x)$ has been replaced by an arbitrary orthogonal scaling function $\phi_{Jk}(x)$. This leads to a modified multiresolution analysis. The resulting "histograms" at the various scales are now not discontinuous step functions as in fig. 1, but acquire the smoothness and other properties of the underlying scaling function and wavelet.

Eq. (16) can be viewed as a linear transformation of the amplitudes ϵ_k, written compactly as

$$\vec{\epsilon} = (\epsilon_0, \epsilon_1, \epsilon_2, \ldots, \epsilon_{2^J-1}) \, , \tag{17}$$

to the wavelet amplitudes $\tilde{\epsilon}_{jk}$

$$
\begin{aligned}
\vec{\tilde{\epsilon}} &= (\epsilon_0^{(0)}, \tilde{\epsilon}_{00}, \tilde{\epsilon}_{10}, \tilde{\epsilon}_{11}, \tilde{\epsilon}_{20}, \ldots, \tilde{\epsilon}_{23}, \tilde{\epsilon}_{31}, \ldots, \tilde{\epsilon}_{J-1,2^{J-1}-1}) \\
&=: (\tilde{\epsilon}_0 \ , \tilde{\epsilon}_1 \ , \tilde{\epsilon}_2 \ , \tilde{\epsilon}_3 \ , \tilde{\epsilon}_4 \ , \ldots, \tilde{\epsilon}_7 \ , \tilde{\epsilon}_8 \ , \ldots, \tilde{\epsilon}_{2^J-1} \qquad) \ ,
\end{aligned} \tag{18}
$$

so that

$$
\vec{\tilde{\epsilon}} = \mathbf{W}\vec{\epsilon} \ . \tag{19}
$$

Note that the transformation matrix $\mathbf{W}$ depends only on the coefficients c_m of eqs. (6) and (7). For the characteristic function (11) we derive:

$$
\begin{aligned}
Z[\vec{\lambda}] &= \left\langle \exp\left(i\vec{\lambda} \cdot \vec{\epsilon}\right) \right\rangle = \left\langle \exp\left(i\vec{\lambda}\mathbf{W}^{-1} \cdot \mathbf{W}\vec{\epsilon}\right) \right\rangle \\
&= \left\langle \exp\left(i\left(\left(\mathbf{W}^{-1}\right)^T \vec{\lambda}\right) \cdot \vec{\tilde{\epsilon}}\right) \right\rangle = \left\langle \exp\left(i\vec{\eta} \cdot \vec{\tilde{\epsilon}}\right) \right\rangle \\
&= Z[\vec{\eta}] \ .
\end{aligned} \tag{20}
$$

The new set of variation parameters

$$
\vec{\eta} = (\eta_0, \eta_{00}, \eta_{10}, \eta_{11}, \ldots, \eta_{J-1,2^{J-1}-1}) \tag{21}
$$

are the wavelet transformed $\vec{\lambda}$:

$$
\vec{\eta} = \left(\mathbf{W}^{-1}\right)^T \vec{\lambda} \ . \tag{22}
$$

The wavelet correlation densities are obtained by appropriate derivatives of the characteristic function with respect to the η-variables. For example, the first order wavelet correlation density reads

$$
\begin{aligned}
\tilde{\rho}_{k_1} &= \left.\frac{1}{i} \frac{\partial Z[\vec{\eta}]}{\partial \eta_{k_1}}\right|_{\vec{\eta}=0} = \sum_{k_2=0}^{2^J-1} \frac{\partial \lambda_{k_2}}{\partial \eta_{k_1}} \frac{1}{i} \left.\frac{\partial Z[\vec{\lambda}]}{\partial \lambda_{k_2}}\right|_{\vec{\lambda}=0} \ , \\
&= \sum_{k_2=0}^{2^J-1} (\mathbf{W}^T)_{k_2 k_1} \rho_{k_2} = \sum_{k_2=0}^{2^J-1} (\mathbf{W})_{k_1 k_2} \rho_{k_2} \ .
\end{aligned} \tag{23}
$$

For the second order wavelet correlation densities we obtain analogously:

$$
\tilde{\rho}_{k_1 k_3} = \sum_{k_2,k_4} (\mathbf{W})_{k_1 k_2} (\mathbf{W})_{k_3 k_4} \rho_{k_2 k_4} \ . \tag{24}
$$

This approach is model independent and applicable for any compact wavelet. The only prerequisite is that the bin correlation densities are known as is the case for the here studied p-model (eqs. (14)–(15)).

For the first order Haar-wavelet correlation density we obtain:

$$
\begin{aligned}
\tilde{\rho}_0 &= 1 \ , \\
\tilde{\rho}_{(j_1 k_1)} &= 0 \quad \text{for } 0 \le j_1 \le J-1, \ 0 \le k_1 \le 2^{j_1}-1 \ .
\end{aligned} \tag{25}
$$

That outcome is easy to understand, since all bins in the p-model cascade have an average 'energy'-density equal to one and the average fluctuation between 'energy'-densities is equal to zero.

The results for the second order Haar-wavelet correlation density are more striking:

$$
\begin{aligned}
\tilde{\rho}_{00} &= 1 \,, \\
\tilde{\rho}_{(j_1 k_1),(j_1 k_1)} &= \alpha^2 (1 + \alpha^2)^{j_1} \quad \text{with } 0 \leq j_1 \leq J - 1 \,, \ 0 \leq k \leq 2^{j_1} - 1 \,, \\
\tilde{\rho}_{(j_1 k_1),(j_2 k_2)} &= 0 \quad\quad\quad\quad\quad \text{for } (j_1, k_1) \neq (j_2, k_2) \,;
\end{aligned}
\tag{26}
$$

see fig. 5. This correlation density matrix is diagonal! So to say, the Haar wavelet basis represents the adequate normal coordinates for the p-model. All off-diagonal contributions vanish simply because the average of the product of two differences belonging to different scales is zero; as a consequence there are no second order correlations between different scales of the p-model cascade! As the resolution j_1 increases, the diagonal contributions reflect a power-law (scaling law), which is a clear evidence of the selfsimilarity of the p-model cascade. Compared to the two-bin correlation density (15), the Haar-wavelet correlation density has a much simpler and clearer structure; compare fig. 4 with fig. 5.

Fig. 6 shows the result for the second order D4-wavelet correlation density $\tilde{\rho}_{k_1 k_2} = \tilde{\rho}_{(j_1 k_1),(j_2 k_2)}$. The dominating contributions are on the diagonal and also indicate a scaling, which besides fluctuations is approximately equal to the scaling occurring in the Haar-wavelet case of fig. 5. Off-diagonal contributions arise in bands, which reflect correlations between different scales belonging to the same evolution branch of the p-model cascade tree. Although, as expected, the D4-wavelet basis does not lead to a complete diagonalisation of the covariance matrix, it gives rise to a "quasi" diagonalisation since off-diagonal contributions are strongly suppressed with respect to the diagonal contributions. This "quasi" diagonalisation of the covariance matrix for selfsimilar stochastic processes seems to be a general feature of the wavelet basis!

CONCLUSIONS AND OUTLOOK

The main goal of this study is to explore the wavelet transform applied to correlation studies. Orthogonal wavelets define a multiresolution representation of a (random) signal, which like a microscope dissects the latter into contributions from different scales. We expect, that data arising from hierarchically organized random processes exhibit a uniquely simple correlation structure once they are represented in an appropriate wavelet basis.

These expectations are supported by the present analysis of a simple cascade model. For the selfsimilar p-model we have shown, that the simple Haar wavelet diagonalizes the covariance matrix exactly and, moreover, the diagonal contributions belonging to different scales exhibit a scaling law. More general wavelets, as for example the Daubechies D4-wavelets, are not ideal normal coordinates for the p-model and lead only to a "quasi" diagonalisation of the covariance matrix. The power-law scaling along the diagonal is still recovered, but now minor off-diagonal contributions arise in band structures, which reflect small correlations between "D4–fluctuations" at different scales.

The results of this study cannot be more than a very first study; there are still long ways to go. The above observations let us hope, that the successive "local smoothing" and "differentiation" operations of an appropriate multiresolution decomposition might facilitate comparisons of experimental cluster-correlation studies at hadron level with theoretical calculations of parton shower models. We further believe that wavelet correlations serve as a useful tool to study the selfsimilarity aspect of QCD parton cascades, which occur in e^+e^- and hh collisions. Of course, these hopes and expectations remain to be verified by further analyses.

REFERENCES

1. I. Daubechies, Comm. Pure Appl. Math., 41:909 (1988); "Ten Lectures on Wavelets", Society for Industrial and Applied Mathematics (SIAM), Philadelphia (1992).

2. Y. Meyer, "Wavelets and Operators", Cambridge University Press, New York (1992); "Wavelets: Algorithms and Applications", Society for Industrial and Applied Mathematics (SIAM), Philadelphia (1993).

3. S. Mallat, IEEE Trans. Pattern Anal. and Machine Intell. 11:674 (1989).

4. M.B. Ruskai et. al., "Wavelets and Their Application", Jones and Bartlett, Boston (1992).

5. C. Meneveau and K.R. Sreenivasan, Phys. Rev. Lett. 59:1424 (1987).

6. M. Greiner, P. Lipa and P. Carruthers, preprint HEPHY–PUB 586/93.

NON-PERTURBATIVE ASPECTS OF QCD AT HIGH TEMPERATURE

Janos Polonyi

Laboratory of Theoretical Physics, Department of Physics, Louis Pasteur University, 67087 Strasbourg, and CRN, 67037 Strasbourg, Cedex France, On leave from L. Eötvös University and CRIP, Budapest, Hungary

I. INTRODUCTION

The studies of QCD at finite temperature have been motivated by the observation that the renormalized coupling constant becomes small at high temperature. This circumstance raised the hope of testing QCD by means of the perturbative expansion and revealing its basic constituents. Soon Monte Carlo calculations indicated a phase transition at high temperature, $T = T_c$, and the possibility of isolating a single quark in the high temperature phase. There have been a series of papers and models build on the weakly interacting quark-gluon gas and the dynamics of the high temperature phase of QCD seems to be a well understood and closed subjects for a number of researchers. The goal of this contribution is to emphasis that we are far from the understanding of the high temperature phase which is not at all 'more perturbative' than the hadronic vacuum of QCD.

What can be wrong with the perturbation expansion when the renormalized coupling constant is small ? There are two ways to see the difficulties. One possibility is to start with the observation that the insertion of additional vertices into the finite temperature Feynman graphs multiplies the contribution by a power of $g^2 T$ instead of g^2. Thus the perturbative result for an observable, $\mathcal{O}$, is

$$\mathcal{O} = \sum_n (g^2 T)^n I_n, \tag{1}$$

where I_n is a loop integral. Its dimension,

$$[I_n] = [\mathcal{O}] - n, \tag{2}$$

is negative for sufficiently high order. Since the theory has no perturbative mass gap the higher order graphs will be increasingly infrared singular and the perturbation expansion is ill defined. This is very similar to the situation of the superrenormalizable theories which contain coupling constants with positive dimension. There a partial resummation of an infinite subset of diagrams is needed to render the perturbation expansion reliable [1]. A more careful analysis reveals the presence of a dimensionless expansion parameter in high temperature QCD whose value can only be obtained by a partial resummation of the perturbative contributions [2]. It is not known what resummation is needed for the evaluation of this expansion parameter.

Another qualitative way of seeing the the failure of the perturbation expansion in high temperature QCD is to note that the time dependent modes are suppressed at high temperature [3, 4]. In fact, the time dependent modes have non-vanishing Matsubara frequency which is proportional to the temperature. This frequency acts as an infrared mass gap and suppresses the infrared divergences which would spoil the perturbation expansion. The size of the system in the Euclidean time direction is around the confinement radius at the phase transition. It is by no means surprising that the non-perturbative time dependent modes which are responsible for the confining features of the theory are suppressed in the high temperature phase. In other words, the system has a short extension along the Euclidean time at high temperature and all non-constant periodic modes have short wavelength and high energy. But the high temperature, i.e. short extension in time does not influence the static modes. Static modes can have arbitrarily long wavelength and small energy and thus create strong effective coupling strength at large distances just like in the vacuum. In fact, one finds that the non-perturbative correlations appear at the same length scale at $T = 2T_c$ as at $T = 0$ [5].

One of the reason one thinks about the high temperature phase as a plasma of weakly interacting quarks and gluons is the surprising agreement between the internal energy computed in lattice QCD and the Stefan-Boltzmann law. One has to be cautious at this point. First, the internal energy is saturated by the light modes. It verifies the multiplicity, mass and the weakness of the interactions of the light quasiparticles. The (long range) structure of these quasiparticles or the existence of other, heavier local excitations play little role in forming this observable. But the more careful analysis [6] reveals the presence of some massive collective modes in the simulations. Another source of the troubles is a finite size effect. The temperature is usually changed in the simulations by modifying the lattice spacing. It is true that the temperature, i.e. the inverse size of the system in the Euclidean time direction can be controlled in this manner. But one changes the spatial extent of the system, too. At high temperature when the lattice spacing is chosen to be small the spacelike size becomes small as well. This in turn eliminates the non-perturbative effects and the agreement between the simulation and the Stefan-Boltzmann law is a trivial consequence of asymptotic freedom. The deviation from the Stefan-Boltzman result should be investigated by keeping the spatial size fixed.

The relevance of the perturbation expansion for long range issues in high temperature QCD would be negligible even in the absence of the problems mentioned above. At the temperatures $T \approx T_c$ the non-perturbative phenomena driving the phase transition are essential. So one may hope that the perturbation expansion will ultimately be useful at higher temperatures. But this is not to happen in our world [7]. Due to the plasmon effects which show up at $O(g^3)$ the naive expansion parameter at finite temperature is not $\frac{g^2}{4\pi}$ but g. The actual value of Λ_{QCD} is so small that the running coupling constant $g(T) \approx \frac{1}{2}$ for temperatures around the Planck mass. At lower temperatures where QCD is unified by the other interactions the coupling constant is larger and renders the perturbative approaches useless. Naturally asymptotic freedom keeps the short distance phenomena perturbative as in the vacuum. What is stressed here is that the long range structure of QCD is not 'more perturbative' at high temperature than in the vacuum.

I believe that in the view of these difficulties more effort should be devoted to the search of clean signatures of the high temperature phase. Instead of some quantitative changes of the branching ratios or spectrum lines which are based on the weakly interacting quark-gluon plasma scenario more genuine and qualitatively new signatures are needed. Originally the search of the 'new phase of matter' was compared to the discovery of the superconducting state. The superconducting nature of the QED ground state could not have been found out from the few particle reactions, its discovery needed experiments on multi-particle systems. Similarly, the new physics we are after in the relativistic heavy ion reactions should come from the collective, non-perturbative aspects and is expected to be as surprising as the superconductivity.

It is well know that the non-perturbative methods to investigate many-particle systems are non-systematical and under-developed. Nevertheless there are some interesting results about the high temperature phase of QCD what one can support by the numerical simulations in lattice QCD. Some of them are mentioned briefly in this paper to raise the interest of the search of qualitatively new signatures of the quark-gluon plasma. The signatures can be divided into two classes depending whether they show up only in the vicinity of the phase transition or for the entire high temperature phase. In Sections II and III our concern is the high temperature behaviour. In Section IV a non-perturbative effect in the vicinity of the phase transition is mentioned. The conlusion is briefly summarised in Section V.

II. HILBERT SPACE

In order to understand the quasiparticle content of the high temperature phase we start with a more careful characterization of the Hilbert space for gluons. One expects some unusual phenomena concerning the Hilbert space due to the following reason. They used to compare the Mott-type insulator-conductor phase transition with the deconfining transition. The lower energy level are localized in a solid state crystal and the conducting band is situated at higher energy. The effect of the temperature is roughly to select those states which contribute to the physical processes. Thus at low temperature when the low lying, localized states are important the conductance is negligible for certain materials. At higher temperature the energy-entropy balance selects higher energy region to be relevant and the conducting, extended states start to contribute and the conductance is finite. Such a sudden increase of the conductance is a Mott-type phase transition.

The analogy with the deconfining phase transition seems reasonably at the first sight. Indeed, the 'quark conductance', the component of the baryon current which is carried by isolated quarks, shows similar behavior as the function of the temperature. But the essential difference is that there are no finite energy eigenstates of the QCD hamiltonian which would be the analogs of the conducting, extended states of QED. But what do we mean by an isolated quark in the deconfined phase ? The Monte Carlo results suggest the possibility of separating a quark from a hadron in the high temperature phase. Thus one is in principle in the position to create a state with fractional baryon number. But there is definitely no such state with finite energy for the confining QCD hamiltonian.

The resolution of the problem is rather simple: The Hilbert space over which the partition function is computed changes at the phase transition. For $T < T_c$ it includes states which can be observed in the vacuum but for $T > T_c$ its dimension becomes three times larger and new states, gluons with the color quantum number of a quark or anti-quark contribute, too. The support of this unusual behavior comes directly from the simulations of lattice QCD [8].

In order to minimize the technical details the structure of the Hilbert space for gluons will be sketched by analogy with the many-body systems. In Quantum Mechanics the coordinates of the N-body system are x_k^i, where i is the vector index, $i = 1, 2, 3$ and k labels the particles, $k = 1, \cdots, N$. We use the coordinates $A_j^a(\mathbf{x})$ in temporal gauge QCD and treat the color index, a, like a vector index and $j = 1, 2, 3$ and $\mathbf{x}$ as the analogs of k, the index which distinguishes different degrees of freedoms. The dimension of the (internal) space of the SU(n) gauge theory is $n^2 - 1$. The canonical momentum is

$$p_k^i = \frac{1}{i} \frac{\partial}{\partial x_k^i}, \tag{3}$$

and

$$E_j^a(\mathbf{x}) = \frac{1}{i} \frac{\delta}{\delta A_j^a(\mathbf{x})}, \tag{4}$$

in Quantum Mechanics and QCD, respectively. The states are described by wave functions or wave functionals, $\psi(x_k^i)$ and $\Psi[A_j^a(\mathbf{x})]$ and the hamiltonians are

$$H = \sum_k \frac{\mathbf{p}_k^2}{2m_k} + \sum_{kk'} V_i(|x_k^i - x_{k'}^i|) + \sum_k V_e(|x_k^i|), \tag{5}$$

and

$$H = \frac{1}{2} \int d^3x [\mathbf{E}^2(\mathbf{x}) + \mathbf{B}(\mathbf{x})]. \tag{6}$$

The states such kind of hamiltonians are acting upon can be classified by the help of the symmetries. In Quantum Mechanics we use the rigid rotations,

$$x_k^i \to \mathcal{M}^{ii'} x_k^{i'}, \tag{7}$$

where $\mathcal{M}$ is an SO(3) matrix. Another parametrization which is based on spinors the fundamental quantity is a linear superposition of the Pauli matrices $\{\sigma^i\}$,

$$x_k = x_k^i \sigma^i, \tag{8}$$

and the rotation 7 is written as

$$x_k \to \omega x_k \omega^\dagger, \tag{9}$$

in terms of the SU(2) matrix, ω. $\mathcal{M}$ is the adjoint representation of ω.

The rigid (gauge) rotations for QCD are of the form

$$A_j^a(\mathbf{x}) \to \mathcal{M}^{aa'} A_j^{a'}(\mathbf{x}),$$ (10)

or

$$A_j(\mathbf{x}) \to \omega A_j(\mathbf{x})\omega^\dagger,$$ (11)

where the matrix

$$A_j(\mathbf{x}) = A_j^a(\mathbf{x})\tau^a,$$ (12)

has been introduced and $\{\tau^a\}$ are the generators of the gauge group, i.e. the Pauli or the Gell-Mann matrices for SU(2) or SU(3) gauge theories.

The quantum number provided by rigid rotations is the angular momentum. It can be made explicit by the help of the collective variables $\mathbf{X}$, $\alpha\beta\gamma$ and $\{r_{kk'}\}$. The first one is the center of mass, the second is the Euler angles of the SO(3) rotation connecting the laboratory and a suitable defined body fixed frame and finally the relative coordinates are $r_{kk'} = |x_k^i - x_{k'}^i|$. It is useful to introduce the product base in the Hilbert space,

$$\psi(x_k^i) = e^{-i\mathbf{P}\mathbf{X}} \mathcal{D}_{mm'}^j(\alpha\gamma\beta)\chi(r_{kk'}).$$ (13)

The index j of the irreducible representation matrix of the SO(3) group is the total angular momentum of the state. The Hilbert space is the sum of two subspaces,

$$\mathcal{H} = \mathcal{H}_0 \oplus \mathcal{H}_1,$$ (14)

$\mathcal{H}_0$ and $\mathcal{H}_1$ contains the states with integer and the half integer values of j, respectively. The two superselection classes, $\mathcal{H}_0$ and $\mathcal{H}_1$, are not mixed with physical, single valued i.e. integer spin operators, such as the coordinates and the momenta. The hamiltonian 5 has finite matrix elements in $\mathcal{H}_0$. The hamiltonian naturally has different matrix elements in the other superselection class. It is not difficult to see that the hamiltonian in $\mathcal{H}_1$ which has similar form in polar coordinate system than 5 in $\mathcal{H}_0$ can be obtained by introducing an external gauge potential [9]. Naturally there is no physical way of verifying the form of the hamiltonian in $\mathcal{H}_1$ because we have no possibility to mix states in $\mathcal{H}_0$ and $\mathcal{H}_1$ by the help of the operators available for measurements. The existence of the half-integer angular momentum subspace with double-valued wave functions remains an intriguing mathematical possibility in Quantum Mechanics whose physical relevance is unclear.

The key quantity in the previous steps is the center of the rotation group. The center of the group G is its subgroup C which commutes with the whole group, $[G, C] = 0$. For $G = SU(n)$ the center consists of the identity matrix multiplied by an n-th root of one. In the case of the rotation group the only nontrivial center element, z, is a rotation by 2π around an arbitrary axis. Since $[x_k, z] = 0$ the center transformation, $x_k \to z x_k z^\dagger$, keeps the coordinates invariant. But such invariance of the coordinates does not imply the invariance of the wave functions. It is the doubly-connected topology of the rotation group $SO(3)$ which actually makes possible the construction of wave functions which realize a nontrivial representation of the center, i.e. which are not center invariant. These are just the states which form $\mathcal{H}_1$. The index 0 and 1 of the Hilbert space labels the representation of the center.

Similar construction is possible for gauge theories. The center of the gauge group SU(3) has two nontrivial elements, $z = \exp(i\frac{2\pi}{3})$ and z^2. Global gauge transformation by z leaves the gauge field invariant but there are wave functionals which are not invariant. The wave functionals which transform as

$$\Psi \to e^{it_a\frac{2\pi}{3}}\Psi, \tag{15}$$

under z belong to the superselection class $\mathcal{H}_{t_a}$, where the quantum number $t_a = 0, 1, 2$ is called algebraic triality [10]. The wave functionals with non-vanishing algebraic triality can be constructed along the lines mentioned above without any difficulties. An example of a state with $t_a = 1$ is

$$|A(\mathbf{x})_j >_1 = \int_{SU(3)} d\omega \omega_{jk}^* |\omega A(\mathbf{x})_j \omega^\dagger >, \tag{16}$$

where ω_{jk} is an arbitrarily chosen matrix element of the SU(3) matrix ω. The form of the QCD hamiltonian, 6, is known only for $\mathcal{H}_0$. As in Quantum Mechanics the hamiltonian has different matrix elements in the other superselection classes of the complete Hilbert space,

$$\mathcal{H} = \mathcal{H}_0 \oplus \mathcal{H}_1 \oplus \mathcal{H}_2. \tag{17}$$

We did not find physical relevance for the multi-valued sector of the many-body system. What is the situation for QCD ? The multi-valued sector decouples from the conventional, single valued subspace for many-body systems because our operators, the coordinates and the momenta have integer spin and are single-valued. But we do have coordinates which transform like the fundamental representation of the global gauge rotations in QCD. They are just the quark fields !

Imagine a state of an infinitely heavy quark test and some gluons. This state, being constructed only by gluonic operators, is element of the gluonic Hilbert space 17. For example, the free energy of a static test charge, F_q, can be obtained by the thermal average of the Polyakov line,

$$\Omega(\mathbf{x}) = T[e^{\frac{i}{2}\int_0^\beta dt A_0(\mathbf{x},t)}], \tag{18}$$

as

$$e^{-\beta F_q} = << \Omega(\mathbf{x}_0) >>_{\text{thermal}}. \tag{19}$$

This state is actually in $\mathcal{H}_1$ since its color charge is that of a single quark. In the confined phase it decouples from the partition function and has infinitely high free energy. But it contributes to the thermal averages in the high temperature phase where the Polyakov line expectation value is non-vanishing. Thus we see the appearance of the superselection sector $\mathcal{H}_1$ in the partition function of the high temperature phase [8].

The argument outlined above should leave the careful reader unsatisfied. The more detailed argument to support the presence of the multi-valued states in the deconfined phase proceeds along the following line. Consider the expectation values

$$< \Psi' | T[\psi(x)\bar{\psi}(y)] | \Psi >, \tag{20}$$

and

$$< \Psi' | \Omega(\mathbf{x}_0) | \Psi >, \tag{21}$$

where $|\Psi'>$ and $|\Psi>$ are states from $\mathcal{H}_0$, $\psi(x)$ is the quark field operator, and T stands for time ordering. The expectation values 20 and 21 represent contributions to the quark propagator in full QCD and to the 'order parameter of confinement', 19, in pure gluon system, respectively. It is easy to prove that the states

$$T[\psi(x)\bar{\psi}(y)] | \Psi >, \tag{22}$$

and

$$\Omega(\mathbf{x}_0) | \Psi >, \tag{23}$$

belong to $\mathcal{H}_1$ [11]. Thus the expectation values 20 and 21 are the overlaps between states in $\mathcal{H}_0$ and $\mathcal{H}_1$ and test the orthogonality of these superselection classes. As long as they are manifestly orthogonal the quark propagator and the Polyakov line expectation value are vanishing, i.e. the system confines. Thus the multi-valued states may occur in the computation of the usual expectation values in gauge theories if fields which transform according to the fundamental representation of the gauge group are present. But these multi-valued subspaces are explicitly orthogonal to the single-valued one and do not contribute in the confining phase of the theory.

The physical role of the multi-valued states can be elucidated by pointing out their relation to the quark triality charge which was introduced in the Eightfold Way. The quark triality of the multi-quark state with N_q quarks and $N_{\bar{q}}$ anti-quarks is

$$t_q = N_q - N_{\bar{q}}(\text{mod}3). \tag{24}$$

The confinement of quarks is equivalent with the claim that only the states with vanishing quark triality, $t_q = 0$ are observed. In the low temperature phase where the superselection classes, $\mathcal{H}_{t_a}$, are kept orthogonal and only $\mathcal{H}_0$ is present then the two triality charges agree,

$$t_q = t_a. \tag{25}$$

In the high temperature phase the thermal average, 19, is non-vanishing and the manifest orthogonality of states from $\mathcal{H}_1$ and $\mathcal{H}_2$ is lost. The multi-valued states appear and they make the two triality charges different. One can show that the correct statement which holds for arbitrary temperature is that only states with vanishing algebraic charges, $t_a = 0$, are observable [8]. The deconfined quark, a possible asymptotic state of the high temperature phase must have vanishing algebraic triality. In other words, the gluon cloud of the deconfined quark is a state from $\mathcal{H}_2$ with color charge which screens the quark charge completely. The appearance of the multi-valued gluon states provides the mechanism to free a quark without ever seeing a naked color charge in isolation!

III. Quasi particles

The short distance structure of the correlation functions of the high temperature phase is described by the perturbation expansion according to asymptotic freedom. Thus the partons are important elements of the high energy reactions in the quark-gluon plasma. The interesting deviations from the hadronic phase should show up beyond the confinement radius where the correlation functions are governed by non-perturbative phenomena. The massless gluons and current quarks give rise to an excitation spectrum with a gap at long distances in the vacuum. Similarly one expects a major deformation of the structure of the excitations and their spectrum as the observation length scale becomes non-perturbative at high temperature. The structure of the large extended quasi-particles is not accessible either by the present resummation techniques of Feynman graphs or the numerical simulations so we have to be contented by some speculations as of the nature of the quasi-parrticles is concerned.

The absence of coloured asymptotic states in the high temperature phase is supported by the numerical studies, as well, [12]. Thus the deconfined quark is actually a rather complicated composite object since it contains the gluon cloud from $\mathcal{H}_2$ which screens the color charge of the quark. The hamiltonian has certainly finite eigenvalues in all of the superselection classes since the Polyakov line average would be vanishing otherwise. It is the orthogonality of the superselection classes rather than the infinity of the quark self energy which keeps the isolated quarks hidden from observations.

We now face the question whether there is a gap in the spectrum of the multi-valued states [13]. If a gap is formed above the ground state then the first excited state must be stable. This state has a finite extent since its size, the characteristic length of the correlation function $<< \Omega(\mathbf{x})\Omega^*(\mathbf{y}) >>_{\text{thermal}}$ is the Debey length which is finite. The screening longitudinal photon states of the QED plasma are certainly not stable in the absence of the charge to be screened. But the non-linearity of the gluon field which generates the gap in $\mathcal{H}_0$ may lead to similar effects in the other sectors. The absence of colored asymptotic states in the high temperature phase, [12], indicates the absence of soft gluons which could fill up a gap. The chromomagnetic monopoles are possible candidates for gluonic states which are stabilized by the non-linearity of the gluon system [14].

Suppose that the spectrum indeed has a gap, an assumption to be verified by strong coupling expansion and Monte Carlo simulation. Then we have particle like gluonic states carrying the color charges of a quark or anti-quark in the high temperature phase ! The high temperature phase is then similar to the following toy-model: Consider a variant of our world, where the top and the bottom quarks do not contribute to the electro-weak currents. A lepton-hadron deep-inelastic experiment performed in this hypothetical world on a hadron with net top or bottom flavor quantum number is interpreted as the finding of isolated quark(s). In fact, the top or bottom quark which is needed to make up the vanishing quark triality of the hadron is not seen and the remaining states which are well identified by the experiment carry the color charge of a quark or anti-quark. The analogs of the 'gluonic quarks' of the high temperature phase are the top or bottom quarks in this hypothetical world.

Thus the assumption of the gap in all superselection classes, a claim to be verified, has far reaching consequences. It introduces new quark 'flavors' and suggests the use of this extended quark model to classify the quasiparticles of the high temperature phase as normal 'hadrons'. The high temperature phase should be more appropriately called screened rather than deconfined.

The chromomagnetic monopoles [14], being described by hedgehog-like configurations, have half-integer spin when their algebraic triality is non-vanishing. This mechanism which is the analogy of the spin-half quantization of the Skyrmions creates fermionic *and* bosonic components for the gluonic cloud which screens the deconfined quark. Thus the new gluonic quark can be a fermion *or* boson. The deconfined quark is then boson or fermion depending on the statistics of the screening cloud.

IV. Quark temperature at the phase transition

The unphysical degrees of freedom of gauge theories corresponding to $A_0(x)$ may lead to observables effects in the vicinity of the phase transition. The Euclidean $A_0(x)$ influences the quark propagator $< T[\psi(x)\bar{\psi}(y)] >$ after analytic continuation,

$$\text{tr} < T[\psi(\mathbf{x}, t + i\beta)\bar{\psi}(\mathbf{x}, t)] >, \tag{26}$$

where the trace is taken over the color and the spin indices. For heavy quark this quantity is the Polyakov line, [18]. For arbitrary quark mass 26 is an order parameter for the same center symmetry as the Polyakov line, its thermal average is non-vanishing only in the high temperature phase. Note that the center transformation z changes the phase of the operator in 26, $\Phi(x)$, by $\frac{2\pi}{3}$.

The center symmetry is broken explicitly by dynamical quarks in the grand canonical ensemble. This limits the importance of the quantities like 26. But it turns out that the canonical and the grand canonical ensembles disagree in the confined phase due to the long range confining forces and one should use the canonical ensemble only [10]. In this ensemble the center symmetry is formally respected by the dynamical quarks and is broken only dynamically.

Introduce the effective integer valued field, $k(x)$, by requiring that $k(x)\frac{2\pi}{3}$ be the closest integer multiple of $\frac{2\pi}{3}$ to $\Phi(x)$. In the low temperature phase $k(x)$ fluctuates rapidly in space and time. At high temperature when the center symmetry is broken dynamically $k(x)$ is constant. Thus one expects domains of finite extent where $k(x)$ is constant in the vicinity of the critical point. Slightly above the critical point these domains should be large and stable enough that the thermodynamical equilibrium is reached for each domain. For such temperatures the long time thermal averages are the averages of the domain averages. This is similar to the quenched averaging of solid state physics.

The dynamics of the quarks in a domain can be approximated in the zeroth order by neglecting all interactions and keeping only a constant homogeneous Euclidean background A_0 field which reproduces the desired value of $k(x)$. The quark occupation number density for a domain characterized by the integer k is found in this approximation to be

$$n_k(\epsilon, \mu) = \frac{1}{1 + e^{ik\frac{2\pi}{3}}e^{\beta(\epsilon-\mu)}} + \frac{1}{1 + e^{-ik\frac{2\pi}{3}}e^{\beta(\epsilon+\mu)}}, \tag{27}$$

where ϵ and μ are the single particle energy and the chemical potential, respectively. Averaging over the domains yields

$$n(\epsilon, \mu) = \frac{1}{3}\sum_k n_k(\epsilon, \mu) = \frac{1}{1 + e^{3\beta(\epsilon-\mu)}} + \frac{1}{1 + e^{3\beta(\epsilon+\mu)}}. \tag{28}$$

The comparison of 28 with the usual Fermi-Dirac distribution indicates that the quark component has the temperature

$$T_q = \frac{T}{3}.$$ (29)

The interactions which were neglected in this simple argument should modify the result slightly without leading to the equality of the temperature of the quark and the gluon components.

The deviation of the temperature of different components of the plasma shows up only for long term observables. If short time measurements are used to extract the temperature one naturally must find the same value for each components of a system in thermal equilibrium. We may call the temperature which is read off by fitting the measured results by the help of some thermal distribution the 'running temperature'. Its value refers to the time scale which characterizes the measurement. Thus the short time temperature must be equilibrated for the plasma. But as the observational time exceeds the life time of the center domains then the distribution functions reflect a lower effective temperature due to the interference between the domains. Thus the quark temperature is reduced by a factor 3 in the infrared, long time limit [15].

V. CONCLUSION

Some non-perturbative phenomena of finite temperature QCD are outlined in this paper. It is stressed that QCD does not become more perturbative by increasing the temperature and one has to be prepared to find challenging and highly nontrivial problems in trying to settle the long range stability and structure of the quark-gluon plasma. The difficulties we face in this process are comparable to those we find in the vacuum. In this situation the search for qualitatively new and clean signatures of the high temperature phase is encouraged. As possible candidates the structure, in particular the statistics of the deconfined quark and the temperature of the quark component are mentioned. Naturally it requires much more work to judge whether the observation of these signatures is realistic enough in the given experimental situation.

Finally let me draw the attention of the theoretical colleagues of this community to lattice gauge theory. The technique of the numerical investigation of lattice QCD is rather stable by now. Though it is far from being satisfactory in providing a general purpose scheme for ab initio calculations, it can answer certain appropriately posed questions. The questions and problems which occupy the genuine lattice gauge theorists are usually different from those which are relevant for this community. Thus it seems desirable to initiate a thoroughly planned and exhausting project for the investigation of the properties of the quark-gluon plasma in the thermal equilibrium by means of lattice QCD.

REFERENCES

1. R. Jackiw and S. Templeton, *Phys. Rev.* **D23**,2291,(1981).

2. A. Linde, *Phys. Lett.* **93B** (1980) 327.

3. D. Gross, R. Pisarski and A. Yaffe, *Rev. Mod. Phys.* **53** (1981) 53.

4. S. Nadkarni, *Phys. Rev.* **D33** (1986) 3738.

5. E. Manousakis and J. Polonyi, *Phys. Rev. Lett.* **58** (1987) 847.

6. M.I. Gorenshtein, D.H. Rischke, H. Stoecker, W. Greiner and K.A. Bugaev, *J. Phys.* **G19** (1993) L69-L75.

7. I thank M. Gyulassy for pointing out this problem.

8. J. Polonyi, *Phys. Lett.* **B213** (1988) 340.

9. J. Polonyi, in preparation.

10. M. Oleszczuk and J. Polonyi, submitted to *Ann. Phys.*

11. J. Polonyi, in Quark-Gluon Plasma, ed. R. Hwa, World Scientific, 1989.

12. C. DeTar, *Phys. Rev.* **D37**, 2328, (1988).

13. I thank Prof. Shimon Levit for drawing my attention to this problem.

14. J. Polonyi, *Nucl. Phys.* **A461** 279c, (1987).

15. M. Oleszczuk and J. Polonyi, *Ann. of Phys.* **226** (1993).

DIRECTED FLOW IN AU ON AU COLLISIONS AT INTERMEDIATE ENERGIES

Thomas Wienold

FOPI - Collaboration
Gesellschaft für Schwerionenforschung
64220 Darmstadt, Germany

ABSTRACT

Reactions of Au on Au from 100 to 800MeV/u were studied using the Phase 1 setup of the FOPI - detector. Charge and velocity of the reaction products were measured with an azimuthally symmetric acceptance in the angular region from 1 to 30 degrees in the laboratory. The directed flow is analyzed using azimuthal many - particle - correlations. The strength of this correlation is shown as a function of 'centrality' (stopping) and the beam energy. A first interpretation of the data is given using the Quantum Molecular Dynamics Model. It is shown, that the combined analysis of i) the strength of the correlation, ii) the degree of stopping and iii) the integrated cross section where the strength of the correlation reaches its maximum is very powerful to solve ambiguities in the interpretation. In the current parametrization of the model a hard equation of state clearly yields the best agreement with the data although a fully consistent description seems not yet possible.

INTRODUCTION

In this contribution we want to address the directed sidewards flow, i.e. the flow of particles in the direction of the reaction plane, which is produced in Au on Au collisions. The study of this phenomenon is of high interest since it is believed to be correlated with the nuclear mean field and thus with the equation of state (EOS) [1,2].

Hot and Dense Nuclear Matter, Edited by
W. Greiner *et al.*, Plenum Press, New York, 1994

On the other hand it was shown that the directed sidewards flow could be influenced by the momentum dependent interaction [3,4,5], the (in medium) nucleon nucleon cross section [6] or the viscosity [7] as well. It remains therefore a complicated task to extract some information about the EOS, specially because the latter effects are not negligible.

One of the most popular methods to analyse the directed sidewards flow is the transverse momentum analysis introduced by P.Danielewicz and G.Odyniec [8]. It consists in reconstructing an apparent reaction plane and projecting the transverse momenta onto that plane. The slope of the resulting $< P_x/A >$ at midrapidity is then taken as a 'flow signal' [9]. As was shown by the Plastic Ball Group this 'flow signal' increases with the mass of the system and the fragment charge [9,10]. Recently this finding was qualitatively reproduced by our data for the Au on Au system [11,12].

Here we follow a slightly different line by looking to the directed flow in a more global way. The contribution is separated in two parts:

The first part describes the method and the experimental results. In the second part we compare the data to the result of the Quantum Molecular Dynamics Model.

EXPERIMENTAL METHOD AND RESULTS

The setup used in the experiment on Au on Au from 100 to 800MeV/u allowed the measurement of nuclear charge and velocity of the reaction products from 1 to 30 degree in the laboratory. Charge identification up to about $Z = 15$ was achieved by a $\triangle E$ versus time-of-flight (TOF) method. TOF and $\triangle E$ was measured with a highly granular and azimuthally symmetric Plastic Scintillator Wall. The $\triangle E$ of slow and heavy particles which were stopped in the Scintillator Wall was detected in an ionization chamber and a thin plastic scintillator array. For technical details we refer to [13,14].

At each incident energy about 10^6 events have been recorded with a multiplicity trigger which corresponds roughly to an impact parameter regime less than 9 fm in a clean-cut model. The results we present have been analysed under this trigger condition. Thus peripheral collisions are strongly suppressed. In the following we sort the events according to a quantity ERAT [15] which is defined as the ratio of the transversal- to longitudinal kinetic energy in the forward hemisphere:

$$ERAT = \frac{\sum p_t^2/2m}{\sum p_z^2/2m}$$

Large values of ERAT correspond to events with a large transfer of the initial longitudinal momentum into transverse momemtum indicating more central collisions, whereas small values of ERAT can be asigned to more peripheral collisions. In figure 1 the accuracy of our reactionplane determination at 150MeV/u is shown qualitatively using the method proposed in ref. [8]: for each event two reaction planes Q1, Q2 are calculated by dividing the event randomly in two subevents. The strength of the azimuthal correlation of $\vec{Q1}$ and $\vec{Q2}$ gives an information how well the reactionplane can be calculated. We observe a strong correlation for low and moderate ERAT's which vanishes for larger ERAT's. Thus the reactionplane has no meaning for events with a high degree of 'stopping'. To be more quantitative we calculate the strength

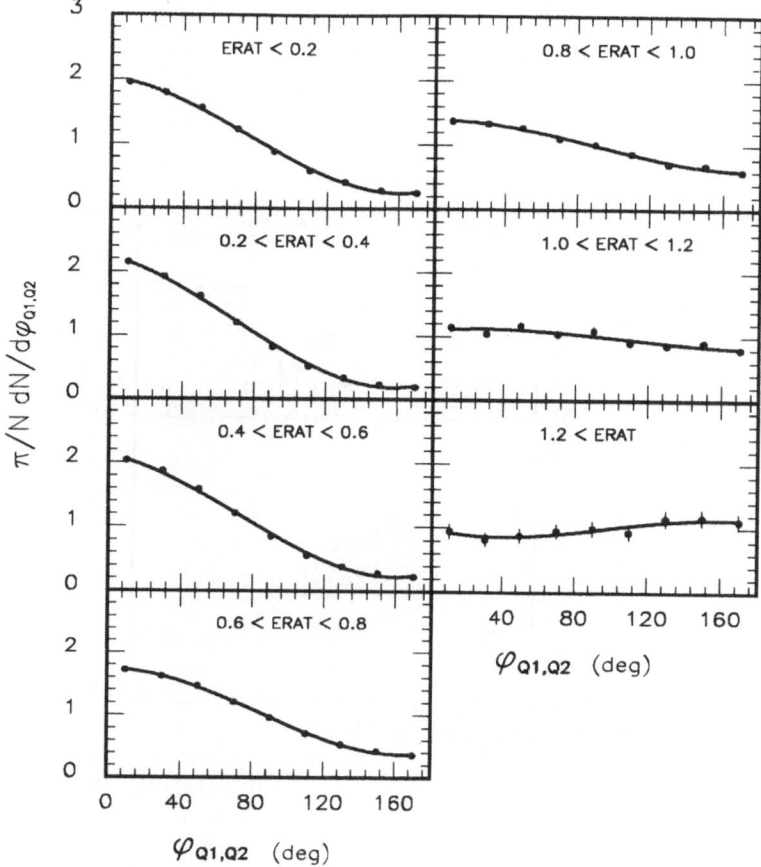

Figure 1: Azimuthal correlation of the two reactionplanes calculated from the randomly subdivided event.

of the azimuthal correlation with a quantity F_s (compare to [8,16,17]) defined as:

$$F_s = \frac{N_c}{N_c - 1} * \frac{\sum_i \sum_{j \neq i} |\vec{p}_{t(i)}| |\vec{p}_{t(j)}| \cos\varphi_{ij}}{(\frac{P_{proj}}{A_{proj}} * \sum_i A_i)^2}$$

The indices i,j are running over the number of charged particles N_c measured in the forward hemishere. For the momenta we use the assumption $A = 2 * Z$. The normalization was chosen in a way that F_s is dimensionless.

The correlation of this observable with the degree of stopping (fig.2) is consistent with the result from fig.1. The mean of F_s has a well pronounced maximum at moderate ERAT's and vanishes for large ERAT's where the cross - section becomes small (see lower part of fig.2). Further information is given by the energy dependence of the azimuthal correlation. In figure 3 (left panel) we display the maximum of $< F_s >$ as a function of the incident energy. Since this observable was normalized to the projectile momentum (in the cm - system), we conclude that there is no scaling at least below 400MeV/u. Also above 400MeV/u a slight decrease is indicated. In the right panel $< F_s >$ was rescaled to illustrate the change in absolute units of

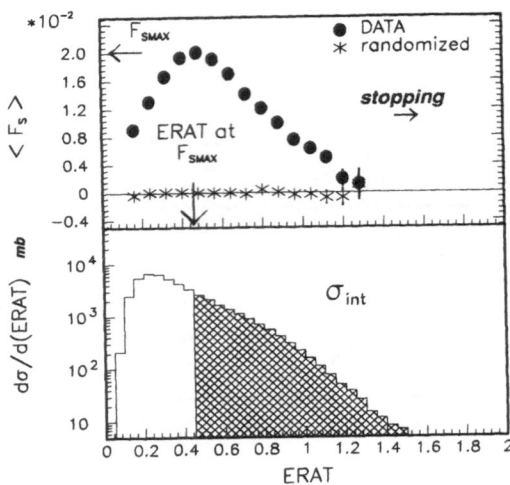

Figure 2: Correlation of the quantity F_s (see text) with the degree of 'stopping' and the corresponding 'stopping' - distribution at 150 MeV/u.

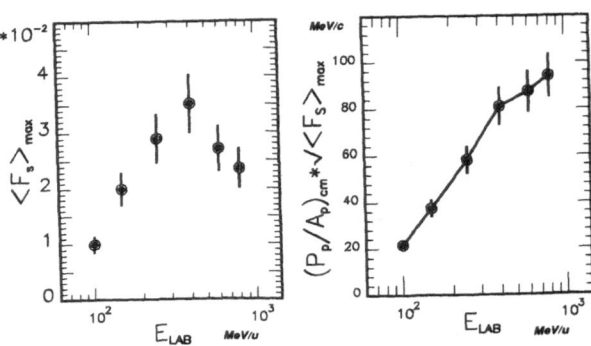

Figure 3: Maximum strength of the azimuthal many-particle correlation ($< F_s >$) as a function of the incident energy.

momentum per nucleon. A rise with the incident energy is observed.

COMPARISON TO QMD CALCULATIONS

For the comparison we choose the Quantum Molecular Dynamics Model in the version from C.Hartnack [18]. Recently it has been shown that the QMD model reproduces more or less the charge distributions at least around 200MeV/u in central Au on Au collisions [19,20,21]. In addition many flow effects are not too different from those seen in the data [2,22,18]. Therefore an experimental filter can be applied to the model calculations without severe problems.

With the QMD model (parametrizations H, HM, SM) we have calculated about 5000 events each for incident energies of 150, 400, 600 and 1000MeV/u using a flat impact parameter distribution between 0 and 14fm. These events have then been filtered and weighted according to their impact parameter. We have not applied any cut on the impact parameter directly and treat the simulation in the same way as the data. Figure 4 shows as an example the distributions of the charged particle multiplicity ($7.5^0 < \Theta_{lab} < 30^0$), of ERAT and the correlation of $< F_s >$ with ERAT at 400MeV/u. Whereas the multiplicity distribution is described fairly well (the difference at low multiplicities is partially due to the trigger condition), the other two observables seem to be more sensitive to the momentum dependence and the incompressibility.

Since our acceptance is limited to laboratory angles below 30^0 it is important to study the influence of our filter to the observables of interest. From figure 5 we conclude that $< F_s >$ is only at the percent level distorted by the filter at 250MeV/u. At the higher bombarding energies the effect of multiple hits (below 10% at 800MeV/u central Au on Au collision) has to be corrected. In figure 6 we compare the measured data with the corrected QMD results. All three parametrizations reproduce the trend seen in the data. The version HM gives the best result, but fails at the lower energies. Very recently it was reported that the new parametrization of the momemtum dependence using the optical potential from the p + U data increases the sidewards flow specially at the lower incident energies compared to the old parametrization from the p + Ca data used here [23]. However, the change from the soft to the hard EOS (SM, HM) yields only 20% difference (at 800MeV/u) and the result from H and SM is almost the same! To solve this ambiguity additional observables are needed. As one such observable we suggest the associated degree of stopping at the point where $< F_s >$ has its maximum (see fig. 2). This is illustrated in figure 7. Now we observe a clear separation between H and SM which is even larger at low energies where the directed flow effects are relatively small! Obviously the nucleons are less stopped at the point of $max(< F_s >)$ when comparing SM with H. As before the best agreement with the data is given by the hard EOS plus momentum dependent interaction (HM)!

A further observable could be the associated cross - section σ_{int} (see fig. 2). In figure 8 (left panel) we show this quantity. Also here a clear separation between H and SM can be observed. Contrasting the previous results HM does no longer reproduce the data. The integrated cross - section differs significantly from the calculation with HM. In the right panel we show (for the model calculation) the impact parameter at $max(< F_s >)$. It was extracted directly from the correlation of $< F_s >$ with b using a gaussian fit near the maximum. The result is very similar to the one of the associated cross - section supporting the interpretation of this cross - section in terms of a geometrical cutoff impact parameter b_{geo}. This can also be concluded from the

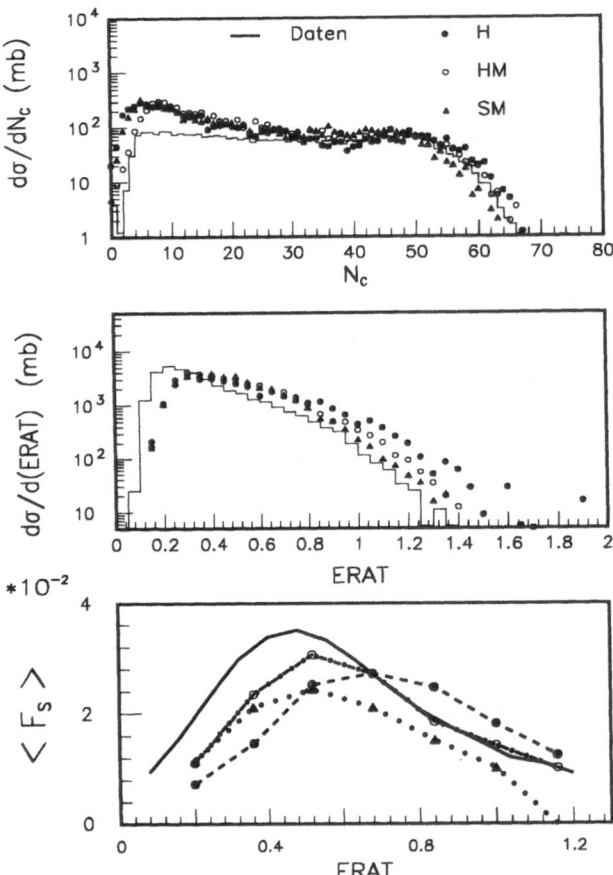

Figure 4: Charged particle multiplicity, 'stopping' distribution and the correlation of $< F_s >$ with ERAT at 400 MeV/u in comparison to the (I)QMD calculation.

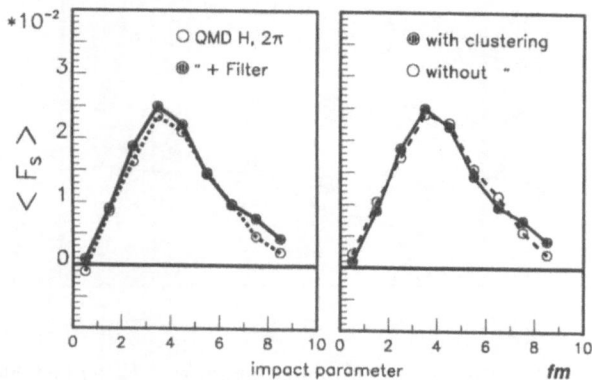

Figure 5: Influence of the experimental filter to the quantity $< F_s >$ at 250 MeV/u.

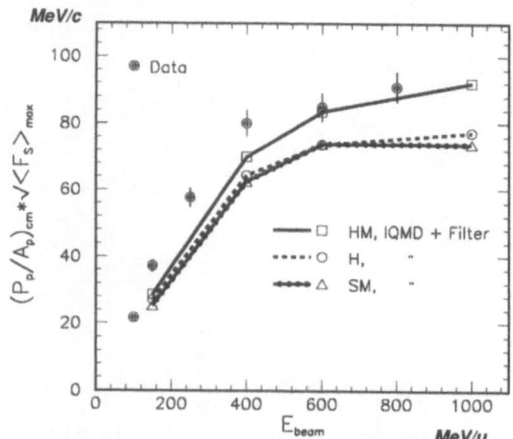

Figure 6: Rescaled quantity $\max(< F_s >)$ in comparison to (I)QMD calculations.

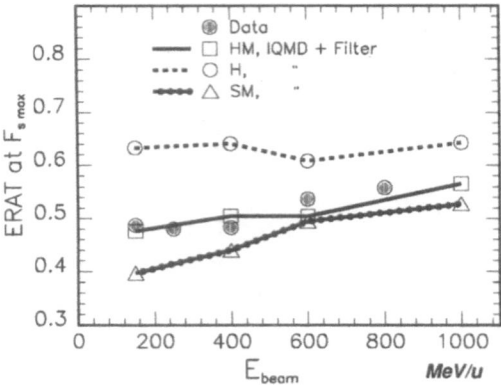

Figure 7: Associated degree of 'stopping' at $\max(< F_s >)$ (see also fig.2).

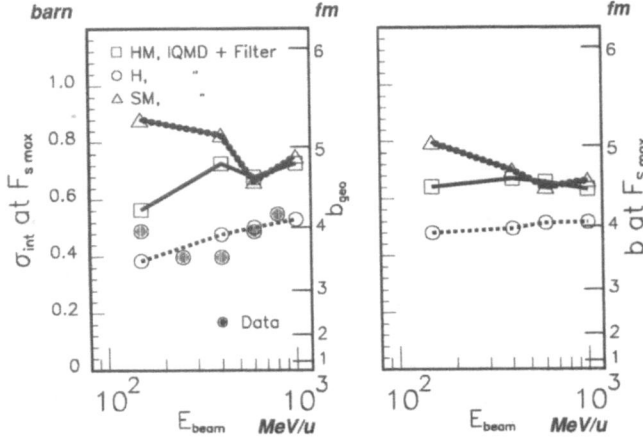

Figure 8: Associated cross section at $\max(< F_s >)$ (see also fig.2).

quite sharp correlation of the impact parameter with ERAT [15,24,25]. Neverthe-less we don't want to overstress the interpretation at the moment since the fluctuations might be larger in the data than in the model calculations.

SUMMARY AND CONCLUSIONS

We have presented recent data on azimuthal many-particle correlations and stop-ping for the system Au on Au at 100 to 800 MeV/u obtained with the phase 1 setup of the 4π - detector. The strength of these correlations depends strongly on the degree of stopping achieved in the collisions. A rise of the strength with the beam energy is observed.

A detailed comparison to (I)QMD calculations was performed. It shows that the combined analysis of several observables helps to solve ambiguities in the interpreta-tion. Remarkably, the model reproduces two of the presented observables quite well using a hard EOS with momentum dependent interaction but fails completely in the third. At the moment we have not studied the influence of the in-medium nucleon-nucleon cross section to our observables which is also an important quantity in the parameter space of the model. In this sense the analysis is not yet fully complete. Nevertheless we hope that the free parameter space of the model was reduced by our data and we are one step further on the way to extract the EOS with heavy ion collisions.

Acknowledgement: We would like to thank Dr.C.Hartnack, Prof.J.Aichelin and Prof.H.Stöcker for fruitful discussions and Dr.N.Herrmann for generating more than 100000 QMD events on the GSI - machines.

REFERENCES

[1] H.Stöcker and W.Greiner, Phys.Rep. 137 (1986) 277
[2] G.Peilert et al., Phys.Rev. C39 (1990) 1402
[3] K.Weber et al., Nucl.Phys. A515 (1990) 747
[4] J.Jaenicke et al., Nucl.Phys. A536 (1992) 201
[5] T.Maruyama et al., Phys.Lett. B297 (1992) 228
[6] A.Bohnet et al., Nucl.Phys. A494 (1989) 349
[7] W.Schmidt et al., Phys.Rev. C47 (1993) 2782
[8] P.Danielewicz et al., Phys.Lett. B157 (1985) 146
[9] K.G.R.Doss et al., Phys.Rev.Lett. 57 (1986) 302
[10] K.G.R.Doss et al., Phys.Rev.Lett. 59 (1987) 2720
[11] F.Rami et al., GSI Report 93 - 1 (1992) 40
[12] J.P.Alard et al., GSI Report 93 - 1 (1992) 39
[13] A.Gobbi et al., Nucl.Inst.Meth. A324 (1993) 156
[14] K.D.Hildenbrand, contribution to this conference
[15] W.Reisdorf, Proceedings, Hirschegg 1993
[16] P.Danielewicz et al., Phys.Rev. C38 (1988) 120
[17] W.M.Zhang et al., Phys.Rev. C42 (1990) R491
[18] C.Hartnack, GSI Report 93 - 05 (1993)

[19] G.Peilert, GSI Report 92 - 13 (1992) 178

[20] N.Herrmann, Nucl.Phys. A553 (1993) 793c

[21] T.Wienold, GSI Report 93 - 28 (1993)

[22] J.Aichelin, Phys.Rep. 202 (1991) 233

[23] C.Hartnack et al., Rapport Interne LPN 93 - 13, Universite de Nantes

[24] Y.I.Grigorian et al., GSI Report 92 - 1 (1991) 23

[25] T.Wienold et al., to be published in Phys.Rev.C

REALISTIC FORCES
AND HEAVY ION COLLISIONS

Amand Faessler

Institut fuer Theoretische Physik
Universitaet Tuebingen
D-72076 Tuebingen, Germany

Abstract: The goal of heavy ion collisions with 100 MeV up to 2000 MeV per nucleon is often claimed to be the determination of the equation of state of nuclear matter, which one needs for example for neutron stars, supernova explosions and the early universe. But the situation in heavy ion collisions is quite different from thermal equilibrium with a spherical momentum distribution and a fixed temperature. The effective nucleon-nucleon interaction as determined by the solution of the Bethe-Goldstone equation depends through the Pauli operator and through the single particle energies on the surrounding nuclear matter. Here results are presented using for the description of heavy ion reactions at intermediate energies Quantum Molecular Dynamics (QMD) in a non-relativistic and in a completely covariant (RQMD) form. The production of gamma-rays, pions and the inclusive spectra of nucleons and light nuclei are not sensitive to the equation of states. An observable sensitive to the equation of state is the directed transversal momentum in heavy ion collisions. Also sensitive is the production of heavier particles like K^+ below the NN threshold. These collisions can have enough energy to compress and heat up nuclear matter and since they are below the NN threshold several nucleons must cooperate. It is also shown that for a quantitative description of antiproton production one needs to take into account the effective mass of the nucleons and antiprotons depending on density and temperature of the surrounding nuclear matter.

1. Introduction

Infinite nuclear matter has been studied in the past at zero temperature and for the saturation density. Heavy ion collisions at intermediate energies from about 100 to 2000 MeV per nucleon allow to get information on nuclei and nuclear matter at finite temperature and also at densities other than the saturation density of about 0.17 nucleons per fm^{-3}. Thus it has often been claimed that one of the main aims of heavy ion collisions at intermediate energies is the determination of the equation of state (EOS) of nuclear matter. The equation of state is not only an interesting relation for nuclear physics, but is also needed in astrophysics to describe supernova explosions, neutron stars and phases in the early universe. But at least in the earlier phase of heavy ion collisions one is far away from the thermal equilibrium, which is assumed to be present in nuclear matter for the EOS. The momentum distribution is unisotropical and this is influencing the effective nucleon-nucleon interaction during the collision. The bare nucleon-nucleon interaction as determined by the scattering between two nucleons in vacuum is highly momentum dependent (see qualitative sketch of this dependence in fig. 1).

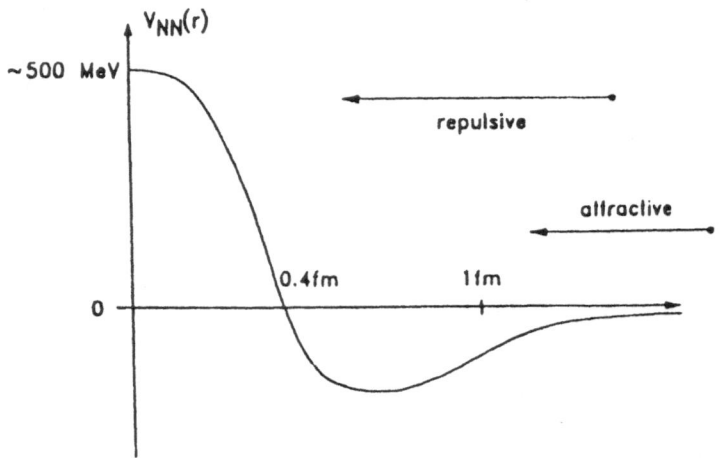

Figure 1. *Qualitative sketch of the radial dependence of the nucleon-nucleon interaction in a S wave. The interaction is highly momentum dependent. At high relative momentum it is strongly repulsive and at low momenta it is attractive.*

The strong momentum dependence of the nucleon-nucleon interaction does not allow to use perturbation theory to calculate the nucleon-nucleon matrix elements in nuclear matter, but the distortion of the nucleon-nucleon wave function due to the interaction has to be taken into account. This is done with the help of the Brueckner-Theory in solving the Bethe-Goldstone equation. In a heavy ion collision the Pauli operator of the Bethe-Goldstone equation is not spherical in momentum space, but reflects the bombarding energy of the incoming heavy ion.

$$G(\rho_t, \rho_p, E/A, \epsilon_1 + \epsilon_2) = V + V \frac{Q}{\epsilon_1 + \epsilon_2 - H_o + i\eta} G \qquad (1)$$

The Brueckner-reaction matrix depends as the free nucleon-nucleon interaction on the momenta of the two incoming and the two outgoing nucleons. But in addition it depends also on the distribution of the surrounding nucleons in orbital and momentum space for the early stage of the nucleus-nucleus collision that means on the densities of the target ρ_T and the projectile ρ_P at the interaction point, on the bombarding energy E/A and on the starting energy $W = \epsilon_1 + \epsilon_2$. The dependence on the densities and the bombarding energy enters the Bethe-Goldstone equation through the Pauli operator Q. A further dependence on the medium comes through the single particle energies ϵ_1 and ϵ_2, which are modified from the free kinetic energies due to the interacting with the neighbouring nucleons [4,5,6].

Opposite to the situation in heavy ion collisions one needs for the equation of state of nuclear matter a spherical Fermi distribution in momentum space, where only the surface can be smeared out due to a finite temperature. Figure 2 shows the situation for the EOS and a heavy ion collision in momentum space.

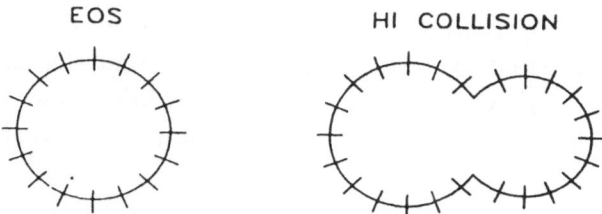

EOS   HI COLLISION

Figure 2. *Local distributions in momentum space for the equation of state (EOS) and for the initial states of a heavy ion collision (right- hand side). The slashes perpendicular to the Fermi surface indicate that the surface is smeared out due to a finite temperature. The volume inside the Fermi sphere is proportional to the density of nuclear matter at the spatial point considered. The momentum distribution in a heavy ion collision (right-hand side) is not spherical.*

To extract from a heavy ion collision (right-hand side of figure 2) information on the EOS (left-hand side of figure 2) one needs a detailed knowledge of the effective medium dependent interaction in heavy ion collisions. In the present contribution we will discuss the following points:

(i) By solving the Bethe-Goldstone equation for the momentum distribution shown on the right-hand side of figure 2 one can calculate the local energy density, which is complex. By subtracting the energy of two heavy ions at large distance one can obtain the optical potential between the two nuclei. This optical potential allows to describe in a quantitative way elastic and inelastic scattering of two heavy ions.

(ii) At intermediate energies we describe heavy ion collisions with the help of Quantum Molecular Dynamics (QMD), which is distinct from the Vlasov-Uehling-Uhlenbeck approach by the fact that in each collision one does not average over an ensemble and thus includes fluctuations and correlations. The collision term is derived from the Brueckner reaction matrix G obtained from solving equation (1).

$$d\sigma_{NN}/d\Omega \propto |G|^2$$
$$U(i) = \sum_j < i,j|G|i,j > \rho_j \tag{2}$$

The selfconsistent potential U(i) for particle i is also determined by the Brueckner reaction matrix solved for the density and momentum distribution derived from the Brueckner reaction matrix G determined including the special Pauli operator and the selfconsistent single particle energies in the surrounding nuclear medium.

(iii) We extend QMD to a fully covariant version into Relativistic Quantum Molecular Dynamics (RQMD). We especially will see that the directed transversal flow is increased by relativistic effects. Each particle has its own time. Putting all times equal, as done in some "relativistic" BUU codes, reduces this essentially to the non-relativistic result.

(iv) Finally we look into the sensitivity of particle (γ, π, K^+, antiprotons) production to the EOS. The highest sensitivity is found for K^+, where one can compress and heat up nuclear matter and still is below the production threshold in NN collisions. The long mean free path of the K^+ in nuclear matter also guaranties that the K^+ spectrum still contains information about the dense and hot production area. The situation is quite different for the production of antiprotons which have in nuclear matter a mean free path of less than 1.5 fm. Only an inclusion of the dependence of the effective mass of the antiprotons, the nucleons and the deltas involved in the production on density and temperature give good results. This dependence is described here in the Nambu-Jona-Lasinio model using Thermo Field Dynamics. The scattering and the absorption of the antiprotons is treated dynamically by describing the antiprotons on the same level as the nucleons moving in a selfconsistent field where the vector part did change sign.

2. Elastic and inelastic heavy ion collisions

In this chapter we calculate the optical potential for the collision of two nuclei using a local density approximation. We solve the Bethe-Goldstone equation (1) for each spatial point in a heavy ion collision, where the momentum distribution is characterized as indicated on the right-hand side of figure 2. The radii of the two spheres are given by the densities of the target and the projectile at the spatial point considered. The distance of the two centres of the two momentum space spheres are given by the bombarding energy and indicate the average relative momenta of the nucleons in the targets relative to the nucleons in the projectile. The Brueckner reaction matrix obtained in this way by solving equation (1) is complex, because the two interacting nucleons can be scattered into states on shell allowed by the Pauli operator Q. A complex energy density is then calculated at each spatial point using the Hartree-Fock approach. By integrating over the two heavy ions at a given distance and by subtracting the energy of the two heavy ions at large distance one obtains a good approximation for the real and the imaginary part of the optical potential. The optical potential and the scattering cross section obtained in this way for ^{12}C on ^{12}C at 1016 MeV bombarding energy in the lab system is shown in figures 3 and 4.

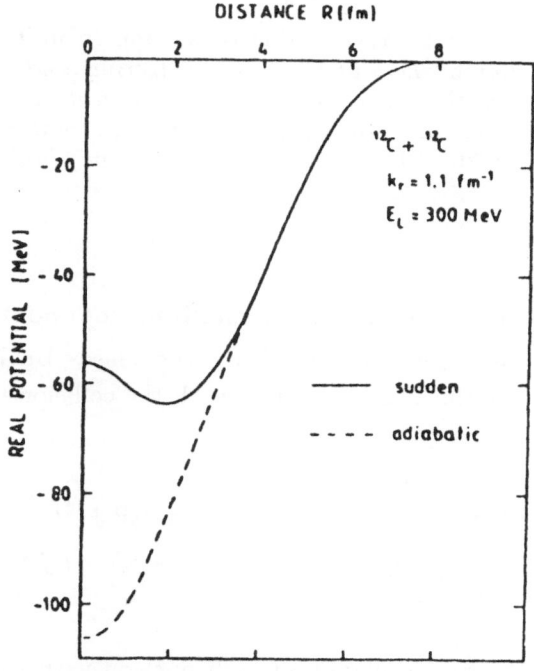

Figure 3. *Microscopically calculated real part of the optical potential for ^{12}C at a bombarding energy of 1016 MeV. The solid line is the microscopically calculated potential using the Reid-Soft core interaction. The dashed line is a fit to the data by Bunere et al* [7].

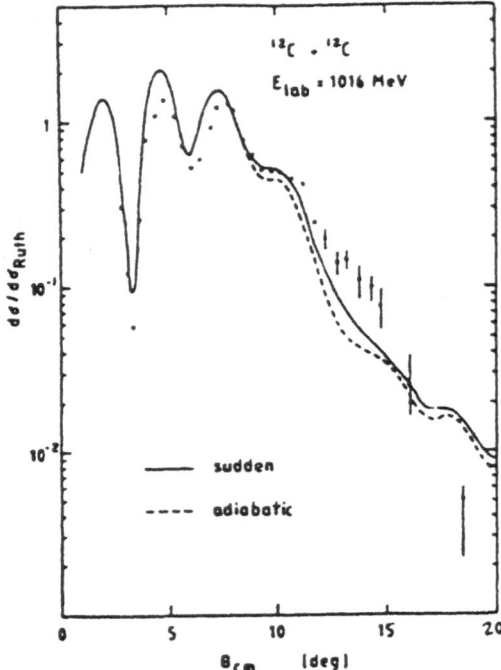

Figure 4. *Elastic $^{12}C - ^{12}C$ scattering cross section in units of the Rutherford cross section as a function of the centre of mass scattering angle. The data are from reference 7, the two theoretical curves assume for the density distribution of the two ^{12}C nuclei a sudden and an adiabatic approximation. The total reaction cross section is experimentally 996 (+50 -250) mb and theoretically one obtains 993 mb.*

3. Quantum molecular dynamics with medium dependence

In quantum molecular dynamics like also in the Vlasov-Uehling- Uhlenbeck approach one is not solving the Vlasov equation with the collision term directly

$$\delta_t f_i(\mathbf{p}, \mathbf{r}, t) - \nabla_{\mathbf{r}_i} U_i(\mathbf{p}, \mathbf{r}, t) \cdot \nabla_{\mathbf{p}_i} f_i(\mathbf{p}, \mathbf{r}, t) + \mathbf{r}_i \nabla_{\mathbf{r}_i} f_i(\mathbf{p}, \mathbf{r}, t) =$$
$$\sum_2 \sum_3 \sum_4 \int d2 \cdot d3 \cdot d4 \ \sigma_{i2;34} \ |\mathbf{v}_{rel}|[f_3 f_4 (1 - f_i)(1 - f_2) - f_i f_2 (1 - f_3)(1 - f_4)] \quad (3)$$

but one follows the time evolution of each nucleon with the help of Hamilton equations [2,3].

$$\dot{\mathbf{r}}_i(t) = \nabla_{p_i} H(1, ...A, t)$$
$$\dot{\mathbf{p}}_i(t) = -\nabla_{r_i} H(1, ...A, t) \quad (4)$$

$f_i(\mathbf{p}, \mathbf{r}, t)$ is the Wigner transform giving the momentum and spatial distribution of nucleon i. The collision term on the right-hand side of equation (3) contains a gain and a loss term, describing scattering into the state i and out of the state i. The factors in the square brackets take into account the Pauli principle. For the Hamiltonian in equation (4) we use the Hartree- Fock approximation.

$$H = \sum_{n=1}^{A} \left[\frac{P_n^2}{2M_N} + U_n(t) \right] + V_{SU} + V_{SY} + V_{coul}$$

$$U_n(t) = \sum_2 \oint d2 \, G(n,2) f_2(\mathbf{r}_2, \mathbf{p}_2, t) \qquad (a)$$

$$U_n(t) = \alpha \rho(n,t) + \beta \rho^\gamma(n,t) + [U_p(\mathbf{p}_n)] \qquad (b) \qquad \qquad (5)$$

$$V_{SU} = \frac{\eta}{8M_N} (\nabla \rho)^2$$

$$V_{SY} = C(\rho_n - \rho_p)^2 / \rho^{1/3}$$

$$V_{coul} = Coulomb$$

$$\sigma_{i2;34} \propto |<3,4|G|i,2>|^2$$

For the selfconsistent potential (a) $U_n(t)$ for the nucleon n we use a microscopic expression calculated as the Hartree-Fock potential from the Brueckner reaction matrix $G(n,2) = <n,2\,|G|n,2>$. Sometimes we use also a phenomenological expression from a Skyrme force (b) with the parameters α and β adjusted to give the right binding energy per nucleon and the right saturation density in nuclear matter. In this expression we sometimes also take into account a momentum dependence $U_p(\mathbf{p}_n)$. The potential contains also corrections for the surface (V_{SU} = Weizsäcker surface correction term) and for the symmetry energy depending on the neutron ρ_n and the proton ρ_p density.

4. Photo and K^+ production

Figure 5 shows the photo production cross section of ^{12}C on ^{12}C with 84 MeV per nucleon. The left-hand side shows the calculation using the Skyrme potential and the nucleon-nucleon cross section of Cugnon [8]. On the right-hand side we show the gamma ray production cross section as a function of the energy of the produced gammas using the microscopic collision term from the Bethe-Goldstone equation (1) and (2). The selfconsistent potential is here again calculated with a Skyrme force (5). The four different curves have been calculated with and without momentum dependence in the self-consistent potential $U_n(t)$ and with a different photo production cross sections.

Photo and pion production turn out to be not sensitive to the nuclear equation of state (EOS). On the one side one needs a bombarding energy below the NN threshold to ensure cooperative effects from many nucleons to be sensitive to the EOS. But one needs also to compress and to heat up nuclear matter. Both conditions can not

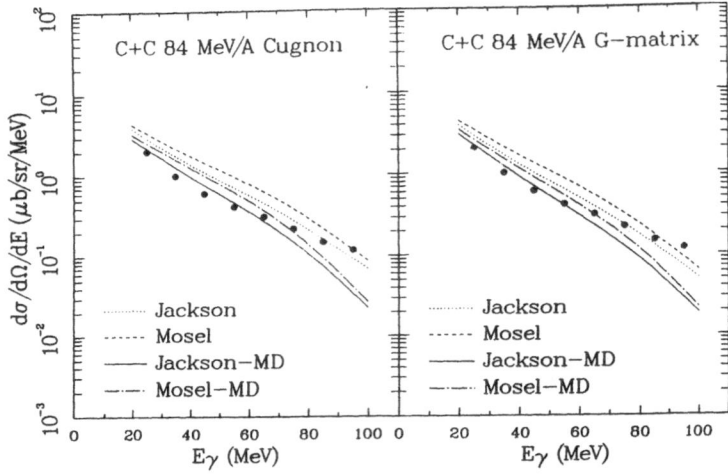

Figure 5. *Photon production cross section calculated with the Soft EOS (incompressibility $K = 200$ MeV) at the emission angle of 90° with respect to the beam axes in $^{12}C + ^{12}C$ collisions at $E_{lab} = 84$ MeV/A. Results are obtained with the Cugnon [8] (left-hand side) and the G-matrix (right-hand side) in-medium NN cross section. Dotted (solid) and dashed (dashed-dotted) curves are obtained using the momentum independent (dependent) interaction with the Jackson formula and the microscopic expression of reference 10 for the $pn\gamma$ cross section, respectively. The experimental data are taken from reference 11. The two lowest curves contain the momentum dependence in the selfconsistent field (5). At higher energies the production cross section is smaller including the momentum dependence due to the fact that the momentum dependence is repulsive and reduces the probability for nucleon-nucleon collisions and by that also the production cross section for gamma rays.*

be fulfilled for the production of light particles like photons ($m_\gamma = 0$) and π mesons ($m_\pi = 140 MeV$). This situation is better for K meson production ($m_K \approx 500 MeV$). Especially the K^+ production seems promising, since this meson has also a large mean free path ($\lambda_{K^+} = 7 - 9 fm$) in nuclear matter, so that it can come out without rescattering or absorption.

The Lorentz-invariant double-differential K^+ multiplicity in a HI reaction at a given impact parameter, b, is calculated from its multiplicity in BB collisions ($B_1 + B_2 \rightarrow B + Y + K^+$) via:

$$\frac{E}{p^2} \frac{d^2 N(b)}{dp d\Omega} = \sum_{BB_{coll}} \int \frac{E'}{p'^2} \frac{d^2 \sigma_{BB}(\sqrt{s})}{dp' d\Omega'} / \sigma_{tot}(\sqrt{s})[1 - f(\mathbf{r}, \mathbf{p}, t)] \frac{d\Omega''}{4\pi}, \qquad (6)$$

where the primed and unprimed quantities are in the center-of-mass (c.m.) systems of the two interacting baryons and the two nuclei, respectively. $\sigma_{tot}(\sqrt{s})$ is the total cross section of two baryons with invariant mass, $\sqrt{s}$, taken from the parametrization by Cugnon [17]. The term $[1 - f(\mathbf{r}, \mathbf{p}, t)]$ takes care of the Pauli blocking for the final nucleon.

The Lorentz-invariant double-differential K^+-production cross section in the HI collision is then given by:

$$\frac{E}{p^2}\frac{d^2\sigma}{dpd\Omega} = 2\pi \int bdb \frac{E}{p^2}\frac{d^2N(b)}{dpd\Omega}.$$ (7)

The elementary K^+-production cross section in BB collisions in eq.(2) is parametrized as:

$$\frac{E}{p^2}\frac{d^2\sigma_{BB}(\sqrt{s})}{dpd\Omega} = \sigma_{K^+}(\sqrt{s})\frac{12}{4\pi p^2}\frac{E}{p_{max}}(1-\frac{p}{p_{max}})(\frac{p}{p_{max}})^2,$$ (8)

with the maximum possible momentum p_{max} of K^+ given by:

$$p_{max}^2 = \frac{1}{4s}[s-(m_B+m_Y+m_K)^2][s-(m_B+m_Y-m_K)^2],$$

and m_B, m_Y and m_K are the masses of the non-strange baryon, the hyperon, and the kaon, respectively. For the total cross section $\sigma_{NN\to K^+}(\sqrt{s})$, there are several parametrizations [18-20]. We use in this work the parametrization proposed in Ref. 18:

$$\sigma_{NN\to K^+}(\sqrt{s}) = 72(p_{max}/m_K) \ \mu b.$$ (9a)

The contributions to the K^+ cross section from the different baryonic channels are related to the NN channel:

$$\sigma_{B_1 B_2 \to K} \approx \begin{cases} \sigma_{NN\to K} & \text{for N N}; \\ \frac{3}{4}\sigma_{NN\to K} & \text{for N } \Delta; \\ \frac{1}{2}\sigma_{NN\to K} & \text{for } \Delta \Delta \end{cases}$$ (9b)

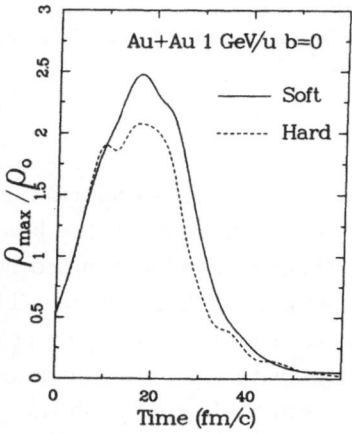

Figure 6a. *Time evolution of the maximum central density, defined as the maximum value reached anywhere inside a sphere of radius of 2 fm with its centre located at the central point of the whole HI system, for head-on Au-Au collision at 1 GeV/u.*

In order to study the time evolution of the hot and dense nuclear matter formed in the central HI collision, we present here the results for the maximum central density, ρ_{max}, defined as the maximum value reached anywhere inside a sphere of radius of 2 fm with its centre located at the central point of the whole HI system, and also for the averaged density ρ_{ave} inside this sphere. Figs. 6(a) and 6(b) show the time evolution of ρ_{max} and ρ_{ave} in the head-on ^{197}Au+^{197}Au collision at 1 GeV/u, respectively. It is found that nuclear matter is compressed to a high density of about 2 times normal nuclear density, ρ_o, at zero impact parameter. We note that the maximum density does not differ much from the averaged density, and this indicates that a high density region is formed. The number of two-body collisions also increases at this compression stage.

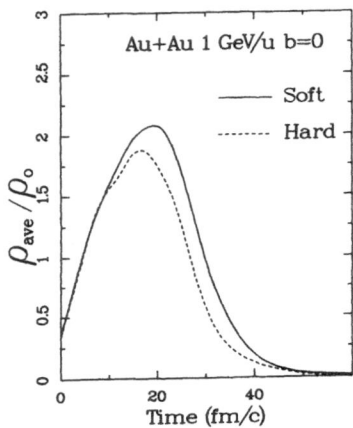

Figure 6b. *Same as Fig. 6(a) but for the averaged central density, defined as the averaged density in a sphere of radius of 2 fm with its centre located at the dentral point of the whole HI system.*

In Fig. 7 theoretical results for the K^+-production cross section in this collision are shown as a function of kaon momentum in the laboratory system. The kaons are observed at $\theta_{lab} = 44^0$. Different results obtained with different kinds of mean fields, corresponding to soft and hard EOS, are shown. The recent experimental data [16] obtained from SIS at GSI are shown in the figure as open squares. One can see that the theoretical predictions with soft mean fields are in reasonable agreement with experimental data. The results obtained with the soft and hard EOS differ by approximately a factor of between 2 and 3.

To study the subthreshold particle production mechanism in HI collisions, we study the contributions to the total cross section from multiple BB collisions. We

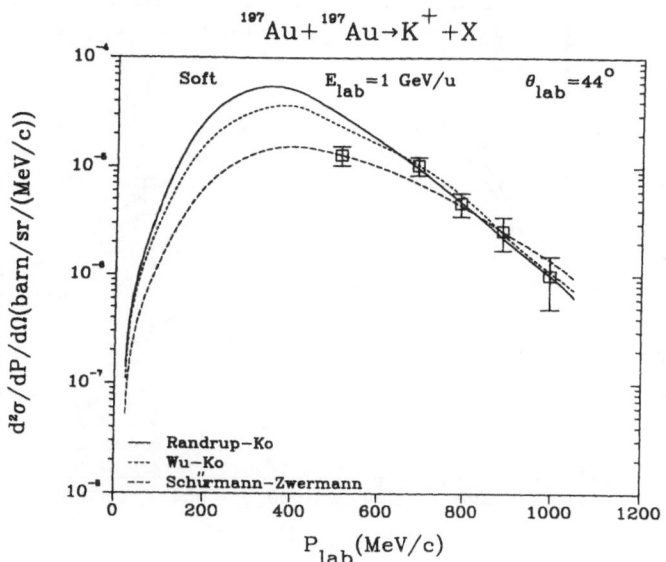

Figure 7. *The differential K^+ production cross section for $Au + Au$ at 1 GeV/u lab energy for the soft (incompressibility $K = 200$ MeV) Skyrme EOS for different elementary cross sections (instead of eq. (9)) for the K^+ production in NN collisions [18,19,23] as a function of the antiproton momentum measured at $\vartheta_{lab} = 44^o$.*

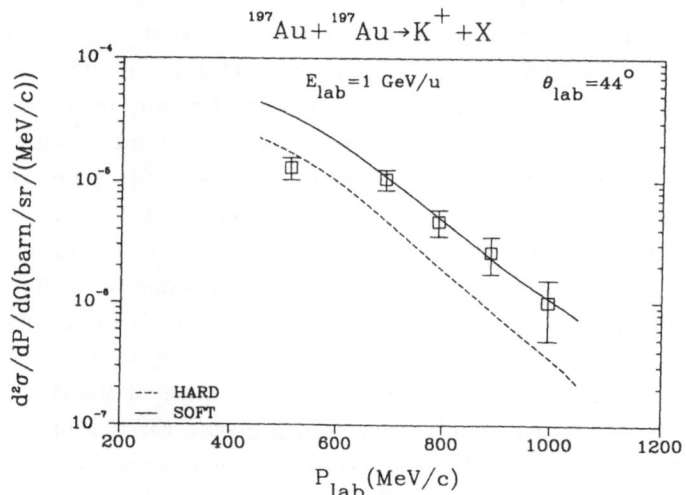

Figure 8. *The differential K^+ production cross sections of $Au+Au$ at 1 GeV/u obtained with soft and hard mean fields. The experimental data (open squares with error bars) are taken from ref. 16.*

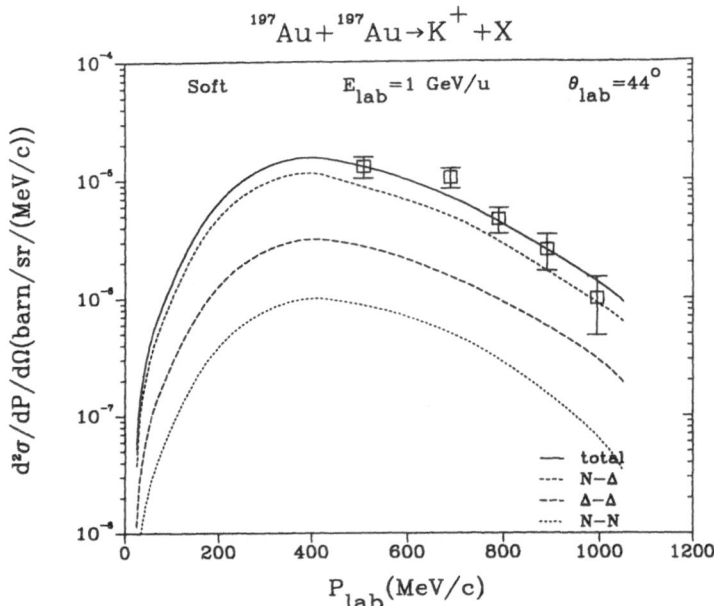

Figure 9. *Decomposition of the differential K^+ production cross section into contributions from different incoming channels.*

decompose the total cross section into different parts coresponding to different types of BB collisions. It is found that the first-chance and second-chance BB collisions both give a negligible contribution to the total cross section, while the later collisions give the main contribution (figure not shown). This is in contrast to the case of photon production, where the first-chance and second-chance collisions give the main contribution to the total cross section, which has been found in our earlier work [21]. This means that, for subthreshold K^+ production, very few primary nucleons can provide sufficient energy from the Fermi motion to produce a K^+ in the first- chance and second-chance NN collisions. To see this clearly ,we decompses, as shown in Fig. 9 , the total cross section into different contributions from different incoming channels. We found that the $\Delta+N$ and $\Delta+\Delta$ channels give main contributions to the total cross section. Since the relative momentum required to produce a K^+ in $\Delta+N$ and $\Delta+\Delta$ collisions is lower due to the large mass of the delta, the majority of the kaons are produced through a two step process. First, a baryonic resonance, such as Δ (1232), is produced from a first-chance or a second-chance NN collision. Next a nucleon or the produced Δ gain enough energy through multiple collisions and the mean field to produce a K^+. This is the reason why the K^+-mesons are mainly produced from $\Delta+N$ and $\Delta+\Delta$ collisions, and these processes occur only after a first-chance collision.

In the present calculation, we neglect the rescattering of the produced kaons with surrounding nucleons due to the small kaon-nucleon cross section. This is a good aproximation for small angles as in our case here, which is supported by a schematic study of Randrup [22]. However, in the general case, especially for large angles, one must take into consideration the rescattering effect.

It should be noted that, in this calculation at a subthreshold energy of 1 GeV/u, we use the free kaon mass in the evaluation of the kaon- production cross section. The reduction of the kaon mass in the nuclear medium has been studied recently by C. M. Ko *et al.*[23]. They showed that the cross section for the process $\pi\pi \to K\overline{K}$ and $\rho\rho \to K\overline{K}$ increases significantly in hot and dense hadronic matter as a result of the restoration of chiral symmetry. The dependence of the process $B_1 + B_2 \to B + Y + K^+$ on density and temperature is, however, less significant. In our calculation we consider only the latter reaction channel. The process $\pi\pi \to K\overline{K}$ and $\rho\rho \to K\overline{K}$ become important at higher relativistic energies. For example, in heavy ion collisions at Brookhaven, these production mechanisms are important for the explanation of the enhanced K^+/π^+ ratio observed. In ^{197}Au$+^{197}$Au collisions at 1 GeV/u, which are considered here, the $\pi\pi \to K\overline{K}$ process contributes only 2.5% to the total K^+ cross section [21].

5. Antiproton production

Figure 10 shows the antiproton production cross section in the ^{12}C on Cu collision with a laboratory energy $E/A = 3.65 \, GeV$ measured at Dubna. The elementary

$$B + B \to B + B + B + \bar{p} \tag{10}$$

cross section is taken from ref. 29. The antiprotons have in fig. 10 the free mass of 938 MeV. The experimental cross section can in this approach only be reproduced if the antiprotons are assumed to have an infinite mean free path, which contradicts to the data. To get a satisfactory description we include the reduction of the mass of the antiproton and the other baryons (N and Δ) involved in the reaction (10) by the interaction with the surrounding nuclear and Δ matter.

Figure 11 shows the dependence of the effective mass as a function of the density for different temperatures [25]. We calculated in each time step and at each location the density from the QMD simulation but used the temperature as a parameter. The antiprotons are moving through the selfconsistent field with inverted vector part of the nucleon-nucleus selfconsistent potential. The scattering and the absorption of the antiproton is treated time step by time step with Monte Carlo techniques. The antiproton- baryon cross sections for scattering and absorption are taken from experimental data parameterized by Cugnon [26]. Figure 12 shows the results for Ni

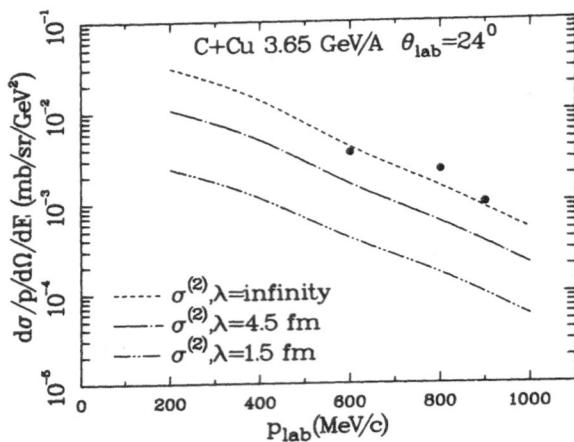

Figure 10. *Antiproton production measured at Dubna[28] in ^{12}C on Cu with $E_{Lab}/A = 3.65\ GeV$. In the theoretical calculation the free mass of the antiproton is assumed and the absorption of the antiproton is calculated by following a straight line with the mean free path $exp(-x/\lambda)$. The data can not be reproduced with a realistic mean free path of the antiprotons.*

on Ni with 1.93 GeV/A (GSI data[27] and C on Cu with 3.65 GeV/A (Dubna data [28]). One gets agreement with the data for a temperature around $T = 140\ MeV$. For $C + Cu$ we show also (dotted line) the result with the bare antiproton mass $m_p = 938\ MeV$.

6. Perpendicular momentum flow

Figure 13 shows the perpendicular momentum flow as a function of the time for Nb on Nb at 400 MeV/A for an impact parameter of b=3 fm. The three curves shown are calculated with the soft (circles) equation of state (incompressibility K=200 MeV), with the hard equation of state (crosses; K=380 MeV) and with the collision term and the selfconsistent potential calculated from the Brueckner reaction matrix (Reid-soft core potential; squares). The directed transverse momentum is defined by

$$< p_x^{dir} >= \frac{1}{N} \sum_{i=1}^{N} sign\ (y_i)p_x(i) \tag{11}$$

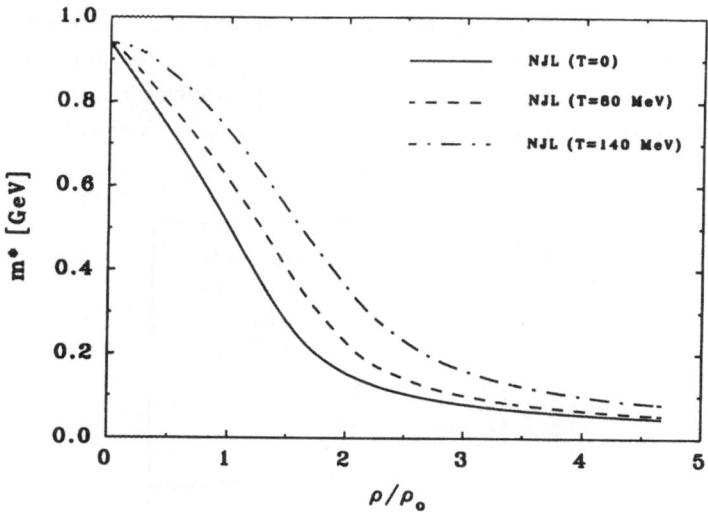

Figure 11. *Dependence of the effective mass of a nucleon or an antiproton on the density and the temperature of the surrounding nuclear and delta matter calculated in the Nambu-Jona-Lasinio model using Thermo Field Dynamics*[25].

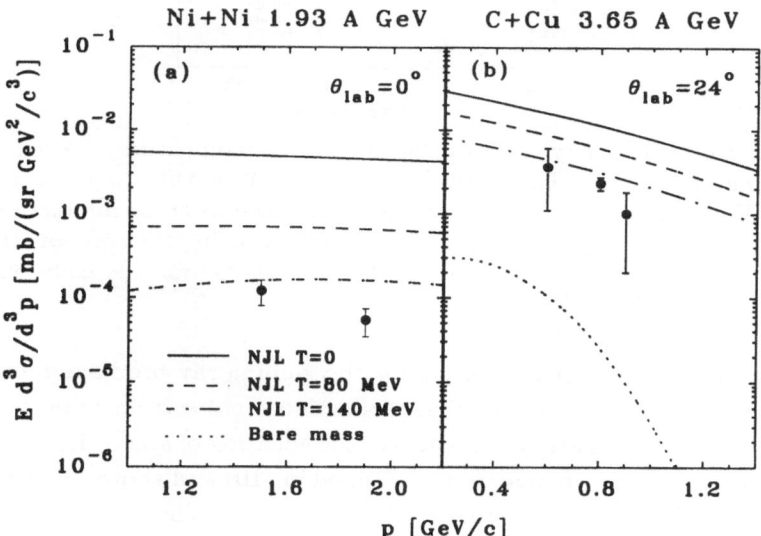

Figure 12. *Production cross section for antiprotons as a function of the antiproton momentum in the lab system. The production cross section in Ni + Ni at 1.93 GeV/A has been measured at GSI* [27] *at zero degrees, while for C + Cu with 3.65 GeV/A it has been measured at Dubna* [28] *at $\vartheta_{lab} = 24°$.*

as a function of time. The signum of the rapidity y_i of the different emitted particles in a heavy ion collision guarantees that the directed transverse momentum is different from zero. The transverse momenta of the target particles scattered to one side and of the projectile particles scattered to the other side are counted by expression (11) all with the same sign.

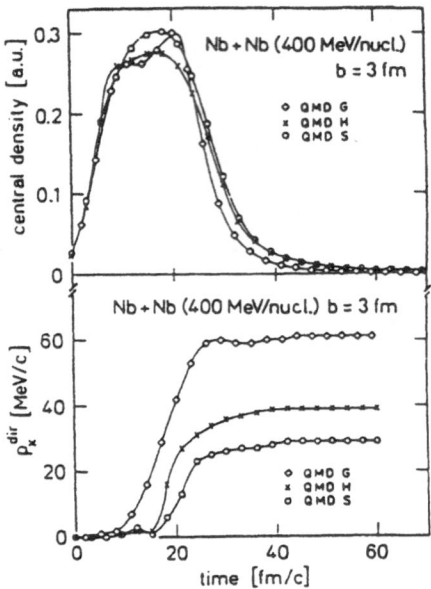

Figure 13. *Central density and directed transverse momentum as a function of the reaction time for the reaction Nb on Nb at 400 MeV/A with an impact parameter b=3 fm. We compare the results obtained using the Skyrme interaction with the hard equation of state (QMD H) with those obtained using the soft equation of state (QMD S) and the Brueckner reaction matrix (QMD G) for the collision term and the selfconsistent potential.*

The results shown here in figure 5 for the gamma ray production cross section and corresponding results for the pion production cross section indicate that these production cross sections are not sensitive to the equation of state. The most sensitive quantity is the transverse momentum as defined in (10) and shown in figure 13. Also sensitive is the K^+ production cross section discussed in chapter 4.

7. Relativistic quantum molecular dynamics

A classical covariant treatment of the many body problem was first attempted by Dirac [12]. But his approach cannot include interactions between the nucleons. An extension which can take interactions fully into account was given by Samuel [13] and first used by Sorge, Stöcker and Greiner [14] to describe nuclear collisions. We follow here the approach of Samuel [13]. The relativistic phase space has 8 dimension for each nucleon. We reduce this to a six dimensional phase for each nucleon by constraining the momenta and energies of each nucleon on shell and by fixing the different times of each nucleon in a covariant fashion[15]. The time evolution parameter is defined as the average time of all the nucleons in the total centre of mass system.

$$\tau = Q \cdot P/|P| \Rightarrow \frac{1}{A} \sum_{i=1}^{A} t;$$

$$Q = \frac{1}{A} \sum_{i=1}^{A} (t_i, \mathbf{r}_i) \qquad (12)$$

$$P = \sum_{i=1}^{A} (E_i, \mathbf{p}_i)$$

The interaction between the nucleons is assumed to be a Lorentz scalar and is so defined that their non-relativistic reduction is the Skyrme force.

$$V_{NN}(1,2) = t_1 \delta(\mathbf{r}_{12}) + t_3 \rho^{\gamma-1} \delta(r_{12})$$
$$U(i) = \alpha \rho(i, \tau) + \beta \rho^{\gamma}(i, \tau) \qquad (13)$$

Figure 14 shows the collision of ^{40}Ca on ^{40}Ca for E/A=1000 MeV for an impact parameter b=3 fm. The directed transverse momentum defined in equation (11) is given as a function of time for the hard and the soft relativistic (RQMD) and non-relativistic (QMD) treatment. One sees an appreciable difference between the relativistic and the non-relativistic treatment. RQMD gives a smaller transverse momentum than QMD. As expected the transverse momentum is also larger for the hard equation of state (K=380 MeV) than for the soft one (K=200 MeV).

Figure 14 shows for the hard EOS (K = 380 MeV) again the directed transversal flow in the non-relativistic (solid line) and the relativistic (short dashed) QMD. Each nucleon has in RQMD its own time determined by Lorentz transformations. If one puts all times equal to τ defined in eq. (12) as done in some "relativistic" BUU codes, the modification of the directed transversal flow by the relativistic contraction disappears again. Thus this "equal time" approximation is not allowed if one wants to include relativistic effect.

We have also calculated relativistically the production of gamma rays and pions and found no dependence on the equation of state and also practically no dependence on a relativistic and a non-relativistic treatment. Thus the strongest differences

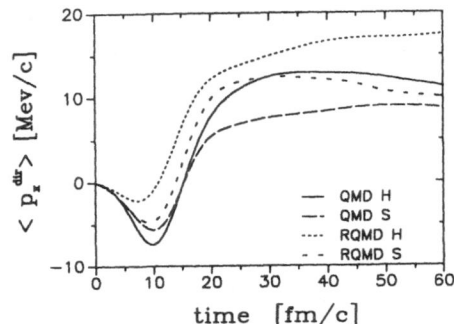

Figure 14. *Directed transverse momentum as defined in equation (10) for the collision ^{40}Ca on ^{40}Ca with $E/A=1000$ MeV and an impact parameter $b=3$ fm as a function of time. The different curves give the non-relativistic (QMD) and the relativistic results (RQMD) for the soft ($K=200$ MeV) and the hard equation of state ($K=380$ MeV).*

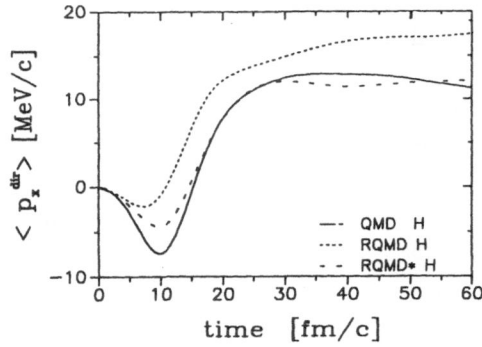

Figure 15. *Directed transversal flow for ^{40}Ca on ^{40}Ca with $E/A = 1000$ MeV and $b = 3$ fm and the hard Skyrme EOS ($K = 380$ MeV). The solid line is the result of QMD, the short dashed line of RQMD and the long dashed line of RQMD with all times equal to τ (eq. (12)).*

depending on the EOS and on QMD and RQMD can be seen in the directed transverse momentum (10).

8. Summary

The free nucleon-nucleon interaction in vacuum is strongly momentum dependent. This strong momentum dependence of the free nucleon-nucleon interaction translates in a strong medium dependence if the nucleon-nucleon collision is considered within a nuclear medium as in a collision between two heavy ions. If one wants to extract the equation of state (EOS) from data of heavy ion collisions one has to take into account this medium dependence. In the present work we have solved the Bethe-Goldstone equation for the local density and momentum distribution and obtained in this way a medium dependent nucleon-nucleon interaction the G-matrix. With this Brueckner reaction matrix we calculate the collision term and the self-consistent potential for a quantum molecular dynamics (QMD) calculation to describe heavy ion collisions.

Due to the strong repulsion at high relative momenta one needs in this more realistic and microscopic approach no high densities as for Skyrme forces to get a repulsive potential between two nuclei. It is enough to have high relative momenta.

Presently we calculated the collision term using the G-matrix calculated by solving the Bethe-Goldstone equation approximating the local momentum and density distribution by two spheres in momentum space for each spatial collision point. But we have not yet taken into account the smearing of the Fermi surface due to temperature effects. The self-consistent potential has been calculated in the Hartree-Fock approach using Brueckner reaction matrix elements calculated again for temperature zero. Since in the potential one averages over the second nucleon we simplified here further by taking for the Pauli operator only a spherical momentum distribution. In addition we compared with results calculated with Skyrme forces.

We also extended QMD to a fully covariant treatment (Relativistic Quantum Molecular Dynamics =RQMD).

The production cross section for gamma rays and pions turned out to be not sensitive to the EOS or to relativistic and non-relativistic treatments. The most sensitive quantity turned out to be the directed transverse momentum defined in equation (10) and the production of K^+ mesons. To obtain a good description of the antiproton production in heavy ion collisions one has to include the medium dependence of the effective mass. This dependence of the nucleon and antinucleon mass on density and temperature is here calculated within the Thermo Field Dynamics using the Nambu Jona Lasinio model.

ACKNOWLEDGEMENTS

I would like to thank Drs. Lehmann, Puri, Huang and Batko, with whom this work has been performed.

REFERENCES

1. G. F. Bertsch, H. Kruse, S. D. Gupta, Phys. Rev. **29** (1984) 673
2. H. Stöcker, W. Greiner, Phys. Rep. **137** (1986) 277
3. J. Aichelin, H. Stöcker, Phys. Lett. **163B** (1985) 59
4. M. Trefz, A. Faessler, W. H. Dickhoff, Nucl. Phys. **A428** (1985) 499 and S. Krewald, A. Faessler, Nucl. Phys. **A341** (1980) 319
5. N. Ohtsuka, R. Linden, A. Faessler, F. B. Malik, Nucl. Phys. **A465** (1987) 550
6. A. Bohnet, N. Ohtsuka, J. Aichelin, R. Linden, A. Faessler, Nucl. Phys. **A494** (1989) 349
7. M. Buenerd et al, Phys. Rev. **C26** (1982) 1299
8. J. Cugnon, T. Mitzutani, J. Vandermeulen, Nucl. Phys. **A352** (1981) 505 and Phys. Rev. **C22** (1980) 1885
9. D. T. Khoa, N. Ohtsuka, S. W. Huang, M. Ismail, A. Faessler, M. El Shabshiry, J. Aichelin, to be published in Nucl. Phys. (1991)
10. M. Schäfer, T. S. Biro, W. Cassing, U. Mosel, UGI- preprint 89-1
11. E. Grosse, P. Grimm, H. Heckwolf, W. F. J. Müller, H. Noll, A. Oskarsson, H. Stelzer, Europhys. Lett **2** (1986) 555
12. P. A. M. Dirac, Proc. Roy. Soc. **A246** (1958) 326
13. J. Samuel, Phys. Rev. **D26** (1982) 3475 and 3482
14. H. Sorge, H. Stöcker, W. Greiner, Ann. Phys. **5** (1989) 266
15. T. Maruyama, S. W. Huang, N. Ohtsuka, A. Faessler, J. Aichelin, submitted to Nucl. Phys. A
16. Kaos-Collaboration GSI Darmstadt (private communication E. Grosse) and E. Grosse et al in: Progress in Part. and Nucl. Phys. **30** (1993); ed. A. Faessler
17. J. Cugnon, T. Mitzutani and J. Vandermeulen, Nucl. Phys. **A352** (1981) 505
18. J. Aichelin, C. M. Ko, Phys. Rev. Lett. (1985) 2661
    J. Randrup, C. M. Ko, Nucl. Phys. **A343** (1980) 519
    J. Randrup, C. M. Ko, Nucl. Phys. **A411** (1983) 537
19. W. Zwermann, B. Schürmann, Nucl. Phys. **A423** (1984) 525
    B. Schürmann, W. Zwermann, Mod. Phys. Lett. **A3** (1988) 441
20 W. Cassing, W. Metag, U. Mosel, K. Niita, Phys. Rep. **188** (1990) 363
21 N. Ohtsuka, M. Shabshiry, M. Ismail, A. Faessler, J. Aichelin, J. Phys. **G16** (1990) L155.

D. T. Khoa, N. Ohtsuka, S. W. Huang, M. Ismail, A. Faessler, J. Aichelin, M. Shabshiry, Nucl. Phys. **A529** (1991) 363.

G. Q. Li, D. T. Khoa, T. Maruyama, S. W. Huang, N. Ohtuska, A. Faessler, J. Aichelin, Nucl. Phys. **A534** 81991) 697

22. J. Randrup, Phys. Lett. **B99** (1981) 9

23. C. M. Ko, Z. G. Wu, L. H. Xia, G. E. Brown, Phys. Rev. Lett. **66** (1991) 2577

    G. E. Brown, C. M. Ko, K. Kubodera, Z. Phys. **A341** (1992) 301

24. V. N. Russhikh, Yu. B. Ivanov, Nucl. Phys. **A543** (1992) 751

25. K. Tsushima, T. Maruyama, A. Faessler, Nucl. Phys. **A535** (1991) 497

    T. Maruyama, K. Tsushima, A. Faessler, Nucl. Phys. **A537** (1992) 303

26. J. Cugnon, J. Vandermeulen, Ann. de Physique **14** (1989) 49

27. A. Schrter et al. Nucl. Phys. **A553** (1993) 775c

28. A. A. Baldin et al. Nucl. Phys. **A519** (1990) 407c

29. G. Batko, W. Cassing, U. Mosel, K. Niita, Gy Wolf, Phys. Lett. **256 B** (1991) 331

PHOTON-INDUCED PROCESSES IN
ULTRARELATIVISTIC HEAVY-ION COLLISIONS

G. Soff,[1] M. Vidović,[1] M. Greiner,[2] S. M. Schneider,[3] D. Hilberg[3]

[1]Gesellschaft für Schwerionenforschung (GSI)
Planckstraße 1, Postfach 110 552
D-64220 Darmstadt, Germany

[2]Institut für Theoretische Physik
Justus Liebig Universität
Heinrich-Buff-Ring 16
D-35392 Gießen, Germany

[3]Institut für Theoretische Physik
Johann Wolfgang Goethe-Universität
Postfach 111 932
D-60054 Frankfurt am Main, Germany

Introduction

Traditionally, the investigation of ultrarelativistic heavy-ion collisions is motivated by the search for a phase transition of nuclear matter into a quark-gluon plasma [1]. However, the strong photon field [2] with photon energies up to $\hbar\omega = 100$ GeV accompanying relativistic nuclei at LHC energies provides additional opportunities for fundamental studies. Associated two-photon processes in peripheral collisions may lead to the formation of new particles [2-17] such as Higgs bosons and top quarks. Particle production by photon-gluon fusion is rather sensitive on the gluon distribution of a bound nucleon inside a nucleus [18]. Furthermore, the electromagnetic dissociation of nuclei represents one of the dominant contributions for the luminosity loss of a relativistic heavy-ion beam and thus requires a closer inspection [19,20]. Even the hadronic content of the photon [21] can be explored.

After this brief introduction we examine the electromagnetic dissociation of Au nuclei at RHIC energies of $E_{\text{ion}} = 100$ GeV/u. The next section deals with the electromag-

Hot and Dense Nuclear Matter, Edited by
W. Greiner *et al.*, Plenum Press, New York, 1994

netic creation of mesons [4]. In particular, we present the rapidity distribution of generated boson pairs. Then we turn the discussion to central ultrarelativistic heavy-ion collisions [5,9,22]. After introducing the general working scheme for the evaluation of particle yields we concentrate our considerations on the Z^0-boson production [9].

Electromagnetic dissociation of Au nuclei in peripheral relativistic heavy–ion collisions

The sudden electromagnetic pulse accompanying a fast impinging nucleus may lead to the dissociation of the collision partner [19,20,23]. We calculate the total electromagnetic dissociation cross section in peripheral heavy–ion collisions ^{197}Au + ^{197}Au for the RHIC collider. We employ an impact-parameter dependent version of the equivalent photon method [3] with experimental photon-nucleus dissociation cross sections [24-26]. Because of the high collider energy it is justified to utilize the equivalent photon method [3,20]. The electromagnetic dissociation of nucleus A is determined by the equivalent photon spectrum $n_B(\omega, b)$ of nucleus B, multiplied with the photon–nucleus dissociation cross section $\sigma_{\gamma A}(\omega)$. For the computation of the total cross section we have to integrate over all photon energies ω and impact parameters b

$$\sigma_{\text{dis}} = \int\limits_{b \geq R_{12}} \mathrm{d}b \, 2\pi \, b \, P(b) \tag{1}$$

with

$$P(b) = \int \mathrm{d}\omega \, n_B(\omega, b) \, \sigma_{\gamma A}(\omega) \quad . \tag{2}$$

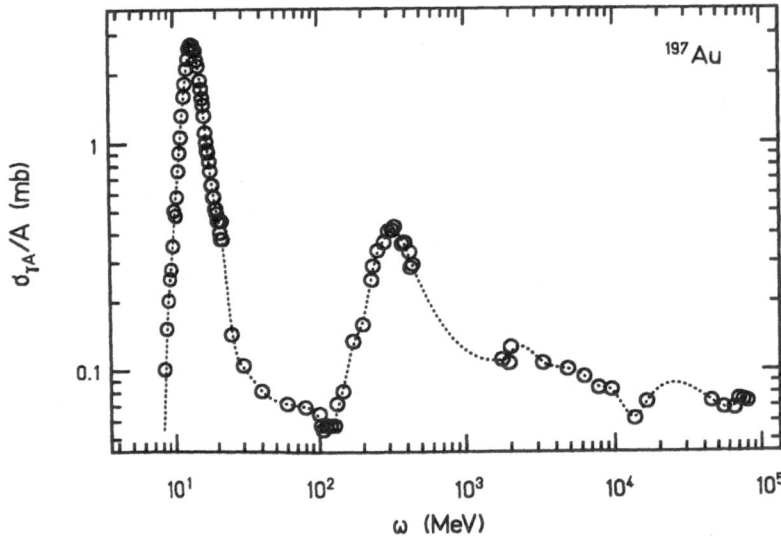

Figure 1: Measured and extrapolated total photon absorption cross section of ^{197}Au devided by the nuclear mass number $A = 197$ in dependence on the photon energy ω. The dotted line depicts the employed interpolation.

Since we only deal with peripheral collisions in order to avoid any direct hadronic interactions we perform the impact-parameter integration from the sum of the nuclear radii $R_{12} = R_A + R_B$ up to infinity. The photon spectrum is given by [3]

$$n(\omega, b) = \frac{Z^2 \alpha}{\pi^2} \frac{1}{\omega} \left| \int_0^\infty dk_\perp \, k_\perp^2 \frac{F(k_\perp^2 + \omega^2/\gamma^2)}{k_\perp^2 + \omega^2/\gamma^2} J_1(bk_\perp) \right|^2 \quad , \tag{3}$$

which indicates the number of photons with energy ω for nuclear trajectories at an impact parameter b. $\gamma = (1 - v^2/c^2)^{-1/2}$ is the Lorentz contraction factor and F denotes the nuclear charge form factor. J_1 is the Bessel function of order 1. The Lorentz factor γ has to be taken in the rest frame of the collision partner, so that a corresponding Lorentz transformation of the c.m. system of the collider with γ_{coll} to the rest frame yields $\gamma = 2\gamma_{\text{coll}}^2 - 1$. For the photon–nucleus cross section $\sigma_{\gamma A}(\omega)$ of ^{197}Au we adopt measured data as far as they are available [24-26]. Otherwise we scaled the data for the photodissociation of ^{208}Pb.

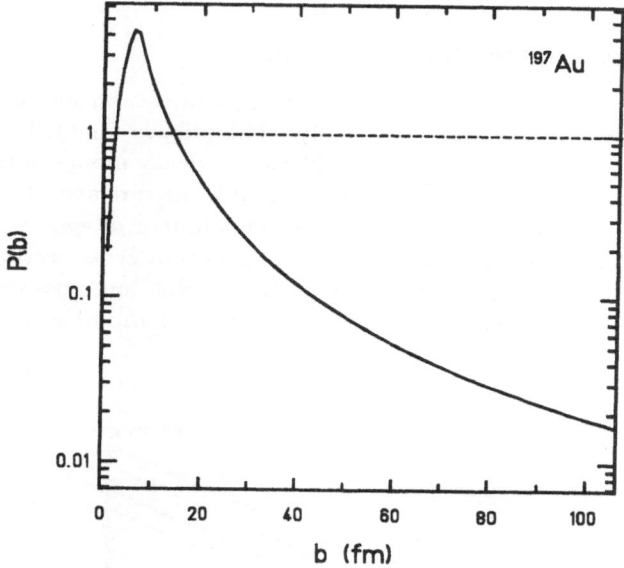

Figure 2: Nuclear dissociation probability $P(b)$ in dependence on the impact parameter b for Au+Au collisions at RHIC energies ($\gamma = 108$). Also indicated is the maximum value 1 for the probability $P(b)$.

The largest contribution emerges from the excitation of the giant dipole resonance (8 MeV $\leq \omega \leq$ 40 MeV), which predominantly leads to neutron emission. With increasing photon energy it follows the quasi–deuteron region and the Δ–resonance region (40 MeV $\leq \omega \leq$ 2 GeV). High-energy photon–nucleus cross sections are measured up to $\omega = 9.5$ GeV for ^{197}Au. In figure 1 we plot the employed data for $\sigma_{\gamma A}(\omega)$ for ^{197}Au divided by the nuclear mass number A. – In table 1 we present the calculated contributions to the total cross section σ_{dis} of eq. (1) from the three different photon energy regions, 8 - 40 MeV, 40 - 2000 MeV, and 2 - 80 GeV. As expected, the dominant portion originates from the giant resonance region.

In eq. (1) we applied lowest-order perturbation theory which may violate unitarity conservation, because the probability $P(b)$ formally may exceed its maximum value 1.

Unitarity conservation would be restored by taking into account higher order corrections. Figure 2 depicts the probability $P(b)$ in dependence on the impact parameter b for Au+Au collisions at RHIC energies. Unitarity violation may occur for small impact parameters up to about 15 fm, which are irrelevant for the integration domain in eq. (1). Even very large impact parameters contribute to the total cross section. The approximate $1/b^2$ decline of $P(b)$ is partly compensated by the area element $2\pi b\,db$.

Table 1: Contributions from various photon energy ranges to the dissociation cross section $\sigma_{\gamma\text{Au}}$ in a peripheral Au+Au collision. Σ indicates the sum from the different ranges including an extrapolation for $\omega > 80$ GeV.

γ_{coll}	σ_{dis} (barn)			
	(8 - 40) MeV	(40 - 2000) MeV	(2 - 80) GeV	Σ
108	65.4	22.9	5.6	94.3

Electromagnetic production of mesons

In peripheral heavy-ion collisions also scalar and pseudoscalar mesons as for example π^0, η, η' can be created electromagnetically. This allows for QCD studies; electromagnetic formfactors, two-photon decay widths and decay modes of the mesons could be deduced. Since the photons become quasi real in ultrarelativistic heavy-ion collisions, the generated mesons can only be either of spin 0 or of spin 2. The two-photon decay width entering the expression for the production cross section is taken from experimental data. In ref. [4] we summarized the production cross sections of various mesons. Roughly spoken, they are in the range from 0.1 mb to several tens of mb for the mesons.

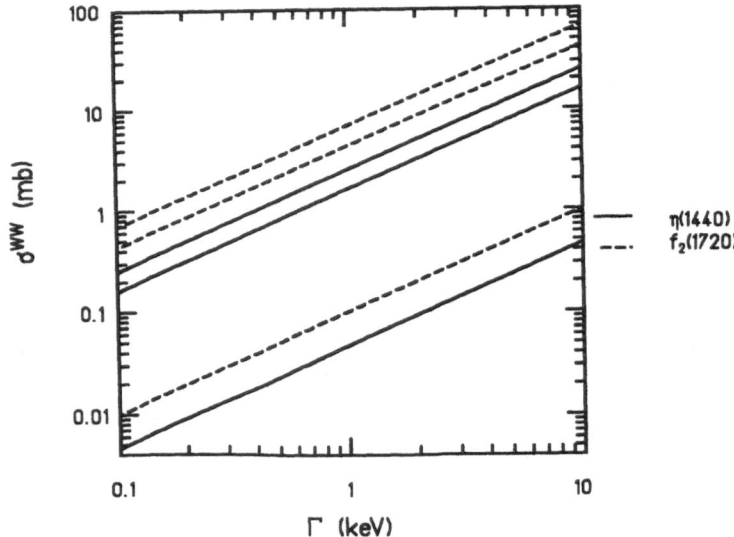

Figure 3: Weizsäcker-Williams cross section for the production of the two glueball candidates $\eta(1440)$ (solid lines) and $f_2(1720)$ (dashed lines) in a Pb+Pb collision in dependence on the two-photon decay width for SSC, LHC and RHIC energies.

One prediction of QCD is the existence of exotic bound states containing several gluons and quarks. Bound states, which predominantly contain gluons compared to quarks, are called glueballs. A definite experimental evidence for their existence has not been provided. One problem of searching for an evidence is the possible mixture between gluonic and quarkonic states. The electromagnetic production of glueballs should also be studied, because this could contribute a strong background to a possible signal for a quark-gluon plasma. We consider here the two glueball candidates $\eta(1440)$ with mass $m = 1440$ MeV and quantum number J^{PC} 0^{-+} and $f_2(1720)$ with mass $m = 1713$ MeV and quantum number 2^{++}, which eventually mix with other mesons. Again the physical important input for the electromagnetic production is the two-photon decay width. For our considerations we will vary the two-photon decay widths of the two canditates from 0.1 keV to 10 keV.

In figure 3 we show the calculated Weizsäcker-Williams cross sections of the two candidates in dependence on their two-photon decay widths $\Gamma_{X \to \gamma\gamma}$ for the three collider: SSC ($\gamma = 8000$), which yields the largest cross sections, LHC ($\gamma = 3500$) and RHIC ($\gamma = 100$). In general the production cross sections for the glueball canditates are expected to be inbetween 0.01 mb and 100 mb. With a RHIC luminosity of 10^{26} cm^{-2}sec^{-1} and a running time of 10^7 sec about 10^5 glueball mesons per year could be produced ($\Gamma_{\gamma\gamma} = 1$ keV).

The electromagnetic production of a pair of charged mesons in an ultrarelativistic heavy-ion collision would allow to study their properties as for example electromagnetic formfactors, decay modes and decay widths. We summarize the computed production cross sections of various charged mesons in table 2.

Table 2: Cross sections of various charged mesons

particle	mass (MeV)	σ (mb)	
		$\gamma = 100$	$\gamma = 3500$
$\pi^{\pm}$	139.57	12.8	194
$K^{\pm}$	493.65	0.14	7.54
$D^{\pm}$	1869	$1.8 \cdot 10^{-3}$	0.19
$D_s^{\pm}$	1968.8	$0.13 \cdot 10^{-3}$	0.17
$B^{\pm}$	5278.6	$0.03 \cdot 10^{-6}$	$8.12 \cdot 10^{-3}$

Next we discuss the rapidity distribution of the produced particles. It results for the differential cross section

$$\frac{\mathrm{d}\sigma^{WW}_{A_1 A_2 \to A_1 A_2 X}}{\mathrm{d}P_{\parallel}} = \int \mathrm{d}\omega \, n_{A_1}\left(\omega - \frac{P_{\parallel}}{2}\right) n_{A_2}\left(\omega + \frac{P_{\parallel}}{2}\right) \sigma_{\gamma\gamma \to X}\left(\omega - \frac{P_{\parallel}}{2}, \omega + \frac{P_{\parallel}}{2}\right) \quad (4)$$

with the longitudinal momentum $P_{\parallel}$ of the produced particle system with the invariant mass $M = 4\omega^2$. It is advantageous to introduce the rapidity

$$y = \frac{1}{2} \ln \left(\frac{E + P_{\parallel}}{E - P_{\parallel}}\right) = \tanh^{-1}\left(\frac{P_{\parallel}}{E}\right) \quad . \quad (5)$$

With $E = \omega_1 + \omega_2$ and $P_{\parallel} = \omega_1 - \omega_2$ the rapidity simplifies to ($\omega = \omega_1$)

$$y = \frac{1}{2} \ln \left(\frac{\omega_1}{\omega_2}\right) = \frac{1}{2} \ln \left(\frac{\omega}{\omega - P_{\parallel}}\right) \quad (6)$$

and we obtain the rapidity dependent differential cross section

$$\frac{\mathrm{d}\sigma^{WW}_{A_1 A_2 \to A_1 A_2 X}}{\mathrm{d}y} = \int \mathrm{d}\omega\; 2\omega\; n_{A_1}(\omega\, \mathrm{e}^{+y})\, n_{A_2}(\omega\, \mathrm{e}^{-y})\, \sigma_{\gamma\gamma \to X}(\omega\, \mathrm{e}^{+y}, \omega\, \mathrm{e}^{-y}) \quad . \tag{7}$$

For the double differential cross section with respect to rapidity and to the impact parameter we find

$$\begin{aligned}\frac{\mathrm{d}^3\sigma^{WW}}{\mathrm{d}y\,\mathrm{d}b^2}(y,\vec{b}) = \int \mathrm{d}\omega\; 2\omega\; \Big[& n_{\mathrm{s}}(\omega\, \mathrm{e}^{+y}, \omega\, \mathrm{e}^{-y}; \vec{b})\, \sigma_{\mathrm{s}}(\omega\, \mathrm{e}^{+y}, \omega\, \mathrm{e}^{-y}) \\ & + n_{\mathrm{ps}}(\omega\, \mathrm{e}^{+y}, \omega\, \mathrm{e}^{-y}; \vec{b})\, \sigma_{\mathrm{ps}}(\omega\, \mathrm{e}^{+y}, \omega\, \mathrm{e}^{-y}) \Big] \quad . \end{aligned} \tag{8}$$

The restriction to peripheral collisions implies, that we have to integrate over the impact parameter space from twice the nuclear radius, $2R$, up to infinity. In figure 4 we display the rapidity distribution of a charged boson pair which is produced electromagnetically in Pb + Pb collisions at LHC energies. The differential cross section $\mathrm{d}\sigma/\mathrm{d}y$ with respect to the rapidity y is plotted normalized to the total Weizsäcker-Williams cross section σ^{WW} [4]. The considered boson masses are indicated. Heavy bosons are produced predominantly at central rapidity ($y = 0$) while lighter boson pairs may be distributed over the full rapidity range.

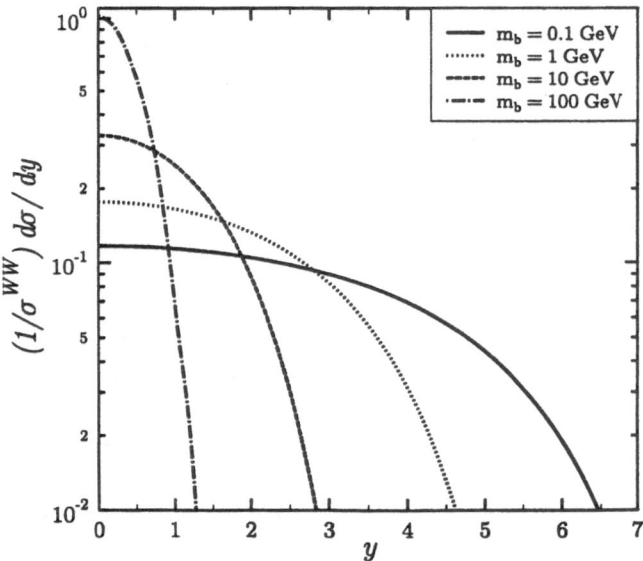

Figure 4: Normalized rapidity distribution of a charged boson pair being produced in Pb + Pb collisions at LHC energies. Different masses of the produced bosons are considered.

The electromagnetic creation of various particles open further possibilities for fundamental studies. E.g., the production of $W^{\pm}$-bosons [2] may serve as a test of the gauge structure of the standard model. It has been demonstrated [28,29] that the $W^{\pm}$-boson production in heavy-ion colliders can be employed to determine the magnetic dipole moment μ_W and the electric quadrupole moment Q_W of the W-boson.

In addition, the large production rates for τ leptons in relativistic nuclear collisions allow for measurements of the anomalous magnetic moment and a possible electric dipole moment of the τ lepton [30].

Z^0-boson production

The general scheme for the evaluation of particle production rates incorporating central collisions is elucidated in figure 5. As fundamental processes we first calculate the cross section for particle production on the parton level. As example we indicate the collision of two partons with masses m_a and m_b and associated four-momenta p_a and p_b. The dynamical dependences are expressed via the Mandelstam variables s, t and u. The produced particles with masses m_1 and m_2 carry the momenta p_1 and p_2. On the next higher level we consider the collisions of two hadrons h_1 and h_2 with momenta P_1 and P_2, respectively. We employ the parton model and the parton distribution functions yielding the momentum fractions x_1 and x_2 to determine the particle formation cross sections. In the ultrarelativistic domain we treat the nucleus-nucleus collision system as an ensemble of individual nucleon-nucleon collisions. The corresponding number of individual hadron collisions has been provided by the RQMD model [27]. As a new feature compared with elementary particle collisions we have to take into account strong contributions emerging from the coherent photon field around a nucleus. Again as a particular example we depict the Compton scattering of a photon from a nucleus at a quark belonging to another nucleus resulting finally in the formation of a Z^0 boson. Here we utilize the Weizsäcker-Williams method in conjunction with the parton model to investigate the generation of new particles.

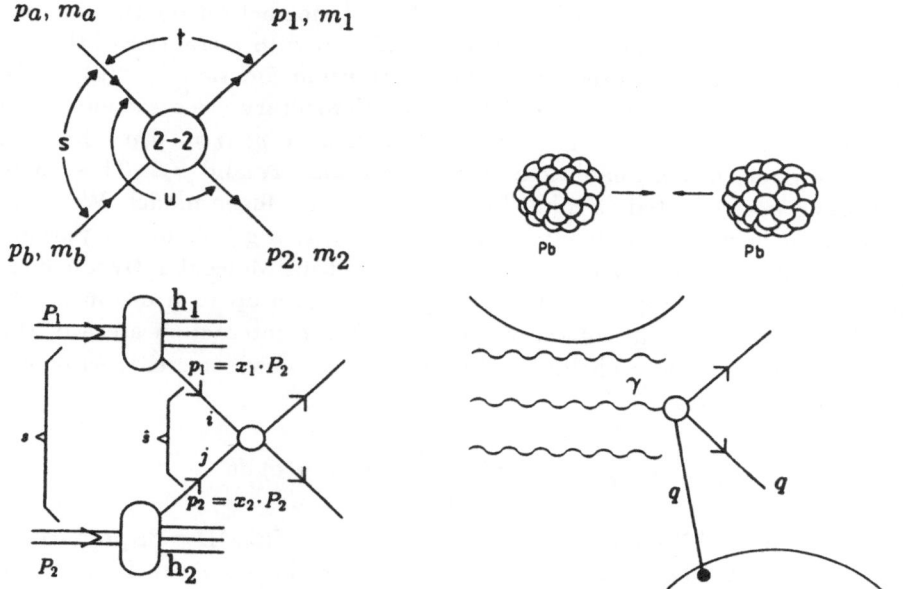

Figure 5: The general working scheme for the evaluation of particle production cross sections in ultrarelativistic heavy-ion collisions.

We investigate the production of Z^0 bosons in heavy–ion collisions at LHC energies.

Different elementary production mechanisms incorporating quarks, gluons, and photons are considered and implemented into the employed parton model. The detection of the produced Z^0 bosons should proceed via their decay into dileptons. Our evaluation incorporates processes such as quark-antiquark annihilations and gluon-gluon fusion for central heavy-ion collisions as well as quark-photon fusion for peripheral collisions. According to the standard model of electroweak interactions we have to take into account all possible subprocesses $a + b \rightarrow Z^0 + X$, which yield a Z^0 boson in the final state. For a and b being quarks and antiquarks, respectively, the processes $q + \bar{q} \rightarrow Z^0 \rightarrow l^+l^-$, $q + \bar{q} \rightarrow Z^0 + Z^0$, $q + \bar{q} \rightarrow Z^0 + \gamma$, $q + \bar{q} \rightarrow Z^0 + g$ contribute in first-order perturbation theory. In the resonant process the l stands for any kind of lepton such as e, μ, τ. For only one incoming quark or antiquark and additionally one photon or gluon the processes $(\bar{q})q + \gamma \rightarrow q + Z^0$, $(\bar{q})q + g \rightarrow q + Z^0$ have to be taken into account. The loop diagrams from gluon fusion and photon fusion may be significant, although they are of higher order. Gluon fusion processes occur only via quark loops, while photon fusion subprocesses occur through quark loops, lepton loops and W-boson loops. Here the cross sections of $pp \rightarrow ZZ + X$ and $pp \rightarrow Z\gamma + X$ from $gg \rightarrow ZZ$ and $gg \rightarrow Z\gamma$ were evaluated with a top quark mass of 140 GeV and a Higgs mass of 500 GeV.

The hadronic cross sections were determined using the parton model formalism. Thus the total cross section σ_{NN} for Z^0-boson formation in nucleon-nucleon collisions reads

$$\sigma_{NN}(s) = \int\int dx_1\, dx_2 \sum_{i,j} \left\{ f_i^{N_1}(x_1, Q^2) \cdot f_j^{N_2}(x_2, Q^2) \right.$$
$$\left. + f_j^{N_1}(x_1, Q^2) \cdot f_i^{N_2}(x_2, Q^2) \right\} \cdot \sigma_{\text{el.}}(\hat{s}) . \qquad (9)$$

Here $\hat{s} = x_1 \cdot x_2 \cdot s$ denotes the c.m.-energy squared in the parton-parton frame, where the partons carry a fraction $x_{1/2}$ of the momentum of the nucleon and the possibility to find such a parton in the nucleon N_k when a collision with a transfer of the squared momentum Q^2 occurs is expressed by the distribution functions $f_{i/j}^{N_k}(x_{1/2}, Q^2)$. To get the hadronic cross section $\sigma_{NN}(s)$ from the elementary cross section $\sigma_{\text{el.}}(\hat{s})$ one has to sum over all contributing partons ($i = u, d, s, c, g$; $j = \bar{u}, \bar{d}, \bar{c}, \bar{s}, g$) and to integrate over all their available momentum fractions x_1 and x_2. These hadronic cross sections were adopted as input in our heavy-ion collison model. We estimated the number of hard nucleon-nucleon collisions $n(b)$ at a given impact parameter b using a simplified version of the Relativistic Quantum Molecular Dynamics model (RQMD). After averaging over all impact parameters b up to a distance $a > 2R$, where R denotes the radius of the nuclei so that the interaction areas of the two nuclei are distinct and normalizing to the total nucleon-nucleon cross section σ_{NN}^{tot} we obtained the expression

$$\sigma_{\text{Pb}-\text{Pb}}(s) = 2\pi \frac{\sigma_{NN}(s)}{\sigma_{NN}^{tot}(s)} \int_0^{a>2R} b \cdot n(b)\, db \qquad (10)$$

for the total ^{208}Pb $-$ ^{208}Pb cross section. The runs of the RQMD code produced values up to $n(b = 0) \approx 1450$ for LHC energies according to this scenario. Furthermore, we calculated the number of produced Z^0 bosons using the Weizsäcker-Williams method in conjunction with the parton model. It follows

$$\sigma_{\text{Pb}-\text{Pb}}(s) = p_{\text{scr}} \int_0^{\omega_{\max}} n(\omega) \int_0^1 \sum_{\substack{\text{nucleons} \\ h}} \sum_{\substack{\text{flavour} \\ i}} f_i^h(x_q, Q^2) \cdot \sigma_{q\gamma}(\hat{s})\, dx_q\, d\omega . \qquad (11)$$

Here, massless quarks possess the energy $E_q = x_q\sqrt{s}/2$, x_q denotes the fraction of the momentum of the nucleon, and the photon energy is $E_\gamma = \hbar\omega$, so that the squared c.m.-energy in the quark-photon frame amounts to $\hat{s} = 4E_q E_\gamma = 2x_q\sqrt{s}\,\hbar\omega$. The factor p_{scr}, which describes the photon screening, was set to 0.65.

The detection of the produced Z^0 bosons can proceed via their decay into dileptons, which can leave the collision zone without participating in strong interactions and which will provide a clear detectable signal. Following the procedures described above one computes the production rates during a "one years" run ($= 10^6$ sec) of the modified LHC with an estimated luminosity of $\mathcal{L} = 10^{27}$ cm^{-2}sec^{-1} for a ^{208}Pb beam as provided in table 3. The first column indicates the considered processes. Results for the gluon-gluon fusion $gg \to Z^0 Z^0$ are splitted according to the incorporation or neglection of a triangle Higgs-graph. The second column contains the computed hadronic cross sections $\bar{\sigma}_{NN}$, which were used in the geometrical collision model to compute the total heavy-ion cross sections $\bar{\sigma}_{\mathrm{Pb-Pb}}$ in the third column. Multiplying this value with the projected luminosities and the beam time leads to the total number of produced Z^0 bosons N_{Z^0} after one "year" beam time. Counting only the Z^0 bosons which decay into one dilepton species results in the values $N_{l^+l^-}$ in the last column. These values have to be doubled if the experimental setup allows to count for electrons and muons at the same time.

Table 3: Displayed are the calculated cross sections and the numbers of particles which are to be expected after one "year" of colliding ^{208}Pb on ^{208}Pb at LHC energies. The first column indicates the considered processes. For the processes $q\bar{q} \to Z^0\gamma$ and $gg \to Z^0\gamma$ we employed the cuts $|y| < 3.0$ and $p_t > 10$ GeV, whereas for $q\bar{q} \to Z^0 g$ we had to use $|y| < 3.0$ and $p_t > 40$ GeV. The averaged nucleon-nucleon and Pb-Pb cross sections are presented in the second and third columns, respectively. Multiplying the Pb-Pb cross section with the luminosity and the estimated beam time leads to the annual number of produced Z^0 bosons in the fourth column. The number of detectable dilepton pairs are contained in the last column.

Z^0 production (LHC, $\sqrt{s}/2 = 3.5$ TeV/u, $\mathcal{L} = 10^{27}$ cm^{-2}s^{-1})				
Process	$\bar{\sigma}_{NN}$ [pb]	$\bar{\sigma}_{\mathrm{Pb-Pb}}$ [pb]	N_{Z^0}	$N_{l^+l^-}$
$q\bar{q} \to Z^0 \to l^+l^-$	$1.3 \cdot 10^{+3}$	$2.2 \cdot 10^{+6}$	≈ 66300	≈ 22100.0
$q\bar{q} \to Z^0 Z^0$	$3.5 \cdot 10^{+0}$	$5.8 \cdot 10^{+3}$	58	1.9
$q\bar{q} \to Z^0\gamma$	$3.3 \cdot 10^{+1}$	$5.4 \cdot 10^{+4}$	545	18.3
$q\bar{q} \to Z^0 g$	$2.8 \cdot 10^{+2}$	$4.7 \cdot 10^{+5}$	4690	157.1
$q\gamma \to qZ^0$	—	$8.8 \cdot 10^{+5}$	8790	294.5
$gg \to Z^0 Z^0$				
with Higgs graph	$5.1 \cdot 10^{-1}$	$8.4 \cdot 10^{+2}$	8	0.3
without Higgs graph	$3.9 \cdot 10^{-1}$	$6.4 \cdot 10^{+2}$	6	0.2
$gg \to Z^0\gamma$	$4.3 \cdot 10^{-1}$	$7.0 \cdot 10^{+2}$	7	0.2

The largest contribution of detectable Z^0 bosons is due to the resonance scattering $q\bar{q} \to Z^0 \to l^+l^-$, as expected. Quark-photon fusion $q\gamma \to qZ^0$ results in the second highest yield which arises from the fact that in the Weizsäcker-Williams model for peripheral collisions a photon from one nuclei is allowed to react with any quark of the other nuclei, whereas the number of quark-quark reactions is restricted due to the hadron collision counting method in the applied central collision model. There will be a sufficiently high production rate even if one detects the Z^0 bosons via their decay into dileptons.

Finally we point out that also supersymmetric particles [31] may be generated effectively in peripheral [8] as well as in central nuclear collisions. Supersymmetric particles may be bound as constituents in so-called R-mesons. In collisions of very heavy ions the presence of hundreds of hadrons and the quark environment in the available phase-space volume enhances considerably the formation of R-mesons compared with elementary particle collisions.

Summary

Effective photon energies up to $\hbar\omega = 100$ GeV will be available in the Lorentz-contracted Coulomb field of heavy ions at LHC energies of about $E_{ion} = 3.5$ TeV/u [2]. For the electromagnetic dissociation cross section of lead nuclei we obtained $\sigma_{dis} \simeq 220$ b for Pb + Pb collisons with $\gamma = E_{ion}/M = 3100$ [19]. The corresponding value for Au nuclei at RHIC energies ($\gamma = 108$) amounts to $\sigma_{dis} = 94$ b. We have calculated the cross sections for the production of exotic particles such as Higgs bosons, top quarks, SUSY particles, vector bosons $W^{\pm}$, Z^0, and glueballs in ultrarelativistic heavy-ion collisions [4]. They are relatively large compared with collisions of single hadrons. Furthermore, we propose to utilize the photon-gluon fusion into $c\bar{c}(b\bar{b})$ pairs in order to determine the gluon distribution of a nucleon inside a nucleus [18].

As future developments we have to perform a more rigorous background analysis for the production rates of new particles. In addition, the electron-positron pair creation with simultaneous capture of the electron in atomic bound states deserves a closer inspection.

References

1. H. R. Schmidt, J. Schuhkraft, J. Phys. G11 (1993) 1705
2. M. Grabiak, B. Müller, W. Greiner, G. Soff, P. Koch, J. Phys. G15 (1989) L25
3. M. Vidović, M. Greiner, C. Best, G. Soff, Phys. Rev. C47 (1993) 2308
4. M. Greiner, M. Vidović, G. Soff, Phys. Rev. C47 (1993) 2288
5. S. M. Schneider, W. Greiner, G. Soff, Phys. Rev. D46 (1992) 2930
6. G. Soff, D. Hilberg, Ch. Hofmann, Ch. Best, S. Schneider, M. Vidović, M. Greiner, Rev. Mex. Fis. 38, Supl. 2 (1992) 196
7. M. Greiner, M. Vidović, J. Rau, G. Soff, J. Phys. G17 (1991) L45
8. J. Rau, B. Müller, W. Greiner, G. Soff, J. Phys. G16 (1990) 211
9. D. Hilberg, W. Greiner, C. Kao, G. Soff, Phys. Lett. B318 (1993) 231
10. E. Papageorgiu, Phys. Lett. 250B (1990) 155
11. M. Drees, J. Ellis, D. Zeppenfeld, Phys. Lett. 223B (1989) 454
12. G. Baur, L.G. Ferreira Filho, Nucl. Phys. A518 (1990) 786
13. R. N. Cahn, J. D. Jackson, Phys. Rev. D42 (1990) 3690

14. C. Bottcher, A. K. Kerman, M. R. Strayer, J. S. Wu, Part. World 1 (1990) 174
15. B. Müller, A. J. Schramm, Phys. Rev. D42 (1990) 3699
16. K. J. Abraham, R. Laterveer, J. A. M. Vermaseren, D. Zeppenfeld, Phys. Lett 251B (1990) 186
17. J. W. Norbury, Phys. Rev. D42 (1990) 3696
18. Ch. Hofmann, G. Soff, A. Schäfer, W. Greiner, Phys. Lett. B262 (1991) 210
19. M. Vidović, M. Greiner, G. Soff, Phys. Rev. C48 (1993) 2011
20. C. A. Bertulani, G. Baur, Phys. Rep. 161 (1988) 299
21. S. M. Schneider, W. Greiner, G. Soff, J. Phys. G19 (1993) L39
22. M. Greiner, D. Hilberg, Ch. Hofmann, G. Soff, J. Phys. G19 (1993) 261
23. J. W. Norbury, Phys. Rev. C43 (1991) R368
24. A. Veyssiere, H. Beil, R. Bergere, P. Carlos, L. Lepretre, Nucl. Phys. A159 (1970) 561
25. A. Lepretre, H. Beil, R. Bergere, P. Carlos, J. Fagot, A. De Miniac, and A. Veyssiere, Nucl. Phys. A367, 237 (1981).
26. S. Michalowski, D. Andrews, J. Eickmeyer, T. Gentile, N. Mistry, R. Talman, K. Ueno, Phys. Rev. Lett. 39 (1977) 737
27. H. Sorge, H. Stöcker, W. Greiner, Ann. Phys. 192 (1989) 266
28. F. Cornet, J. I. Illana, Phys. Rev. Lett. 67 (1991) 1705
29. G. Couture, Phys. Rev. D44 (1991) 2755
30. F. del Aguila, F. Cornet, J. I. Illana, Phys. Lett. B271 (1991) 256
31. H. P. Nilles, Phys. Rep. 110 (1984) 1

TRANSPORT EQUATIONS WITH PARTICLE PRODUCTION AND BACK-REACTION

J.M. Eisenberg

TRIUMF, 4004 Wesbrook Mall
Vancouver, B.C., V6T 2A3, Canada

and

School of Physics and Astronomy
Raymond and Beverly Sackler Faculty of Exact Sciences
Tel Aviv University, 69978 Tel Aviv, Israel*

ABSTRACT

A summary is presented of studies of systems in which pairs are produced in a strong electric field, are accelerated by it, and then react back on it through the counter-field produced by their current. Our motivation is eventually to apply our results to the construction of a phenomenological formalism for use in the analysis of the quark-gluon plasma (QGP). The field formulation of the problem is sketched for charged bosons, and a phenomenological transport equation is presented that embodies physics essentially identical to that of the field theory. Numerical results are shown to support this. Extensions are noted for fermions, and for boost-invariant variables (needed for the QGP). The formal derivation of the transport equation as an approximation to the original field theory is reviewed, and work in progress on the study of more realistic conditions—namely, nonclassical (electric) fields, non-Abelian fields, and spatial dependence—is noted.

INTRODUCTION

The usefulness of transport-equation formalisms for studies of the physics of the quark-gluon plasma (QGP) has long been recognized and exploited[1]. A crucial part of the dynamics of the QGP involves the production of partons, and the transport-equation approaches have usually incorporated this by the somewhat artificial use of a source term. More recently[2-6], a collaboration involving groups at Los Alamos and at Tel Aviv have been attempting to base this approach on a proper field-theory study

---

* Permanent address.

Hot and Dense Nuclear Matter, Edited by
W. Greiner *et al.*, Plenum Press, New York, 1994

of the physics involved. Here I shall give a brief summary of the current status of this problem, and sketch how work is proceeding to provide a practical calculational scheme for the QGP, based on transport equations founded in field theory. The hope is that this will eventually yield a tool for reliable phenomenological studies of the QGP.

In a common picture of the first stage of the production of the QGP at ultra-relativistic energies, two highly contracted nuclei collide, generate color charges, and pass through each other. They leave behind a chromoelectric field produced by their color charges. At the first level of the treatment here, this chromoelectric field is taken as Abelian; it is further approximated as a classical field[4], and is regarded as filling all space homogeneously. These highly unrealistic assumptions insofar as QGP application is concerned we hope eventually to relax as we develop a better understanding of how the various physical features translate into transport equations.

Out of the (chromo)electric field of the above scenario there will tunnel pairs of partons that comprise the plasma. The tunneling mechanism in question is very well known[7], and an exact solution[8] for the pair production rate for a fixed, external electric field has also long been available. What is happening[9] in the tunneling process is envisioned by imagining a fictitious potential that binds the latent pair at the combined rest-mass energy $2mc^2$. The electric field provides an additional $-eEx$, which lowers the potential at large distances, and allows the pair to tunnel out. It is well known that the production rate is given[7-9] roughly by

$$\text{rate} \sim \exp[-\pi m^2 / eE], \tag{1}$$

The tunneling process has, of course, no perturbative expansion about $E = 0$.

The application[1] of pair tunneling to QGP production has been made through the use of a Boltzmann equation in which the pair rate serves as a source term,

$$\frac{\partial f}{\partial t} + \frac{\vec{p}}{(\vec{p}^2 + m^2)^{1/2}} \cdot \frac{\partial f}{\partial \vec{x}} + e\vec{E}(t) \cdot \frac{\partial f}{\partial \vec{p}} = \cdots \exp\left[-\frac{\pi m^2}{eE(t)}\right], \tag{2}$$

where $f(\vec{x}, \vec{p}, t)$ is the density of particles at position $\vec{x}$, with momentum $\vec{p}$ and at time t. The electric field in eq. (2) is time-dependent because of the inevitable appearance of back-reaction: The charged pairs make their appearance, are accelerated by the electric field, and produce a current which in turn makes an electric field opposing the original field direction. Eventually field and plasma oscillations are set up. This back-reaction enters through Maxwell's law,

$$\dot{\vec{E}}(t) = -\vec{j}(t) = -e \int d\vec{p} \frac{\vec{p}}{(\vec{p}^2 + m^2)^{1/2}} f(\vec{x}, \vec{p}, t) - \vec{j}_{\text{pol}}(t), \tag{3}$$

where $\vec{j}_{\text{pol}}$ is the polarization current induced by the pair source term of eq. (2). (For the case of a homogeneous system filling all of space, only a constant magnetic field can arise, which we ignore.) Equations (2) and (3) now form a coupled system which must be solved to incorporate back-reaction. Note that eq. (2) has not in fact been derived from basic field equations, but instead has been put together by physical intuition. This immediately raises the questions, (i) How does back-reaction emerge in a description of this same physical system based on field theory, and (ii) Can one derive a transport equation resembling eq. (2) directly from a field-theory formulation?

BACK-REACTION AND RENORMALIZATION

I here sketch back-reaction in its simplest terms; a far more complete discussion is given in refs. 4-6. Consider a system of charged bosons of mass m satisfying a Klein-Gordon equation

$$[(\partial + ieA)^\mu (\partial + ieA)_\mu + m^2]\phi = 0, \tag{4}$$

where for the homogeneous electric field filling all space we take a vector potential (in a particular gauge) $A_\mu = (0, \vec{A}(t))$ which satisfies the Maxwell equation

$$\ddot{\vec{A}}(t) = \vec{j} = -ie[\phi^\dagger \vec{\nabla}\phi - (\vec{\nabla}\phi^\dagger)\phi] - 2e^2 \vec{A}\phi^\dagger \phi. \tag{5}$$

After second quantization

$$\phi(\vec{x}, t) = \int \frac{d\vec{k}}{(2\pi)^3} \left[f_{\vec{k}}(t)\, a_{\vec{k}}\, \exp[i\vec{k}\cdot\vec{x}] + f^*_{-\vec{k}}(t)\, b^\dagger_{\vec{k}}\, \exp[-i\vec{k}\cdot\vec{x}] \right], \tag{6}$$

where $a_{\vec{k}}$ and $b^\dagger_{-\vec{k}}$ are the particle annihilation and antiparticle creation operators, respectively. The $f_{\vec{k}}$ are mode amplitudes for bosons with momentum $\vec{k}$; upon substituting eq. (6) into eq. (4), they satisfy

$$\ddot{f}_{\vec{k}}(t) + \omega^2_{\vec{k}}(t) f_{\vec{k}}(t) = 0, \tag{7}$$

with $\omega^2_{\vec{k}}(t) = [\vec{k} - e\vec{A}(t)]^2 + m^2$. The Maxwell equation now becomes

$$\ddot{\vec{A}}(t) = \langle\vec{j}(t)\rangle = e \int \frac{d\vec{k}}{(2\pi)^3} [\vec{k} - e\vec{A}(t)]\, 2|f_{\vec{k}}(t)|^2, \tag{8}$$

where the brackets on $\vec{j}(t)$ are intended to yield a classical electric field, as we have chosen to restrict ourselves. They here imply an expectation value in the initial vacuum.

There arises the question as to whether the integral in eq. (8) converges. In fact, renormalization is required for back-reaction[2-6]. This is usually accomplished in the context of so-called adiabatic regularization[4]. The basic notion is that one can identify the logarithmically divergent term in the expression for the current in Maxwell's equation (8) by using a WKBJ-like method to study the high-momentum behavior of $f_{\vec{k}}(t)$. This term can then be subtracted off, yielding, in place of eq. (8),

$$e\ddot{\vec{A}}(t) = e^2_R \int \frac{d\vec{k}}{(2\pi)^3} \left[2|f_{\vec{k}}(t)|^2 [\vec{k} - e\vec{A}(t)] + \frac{k^2 e\ddot{\vec{A}}(t)}{12\omega^5_{\vec{k}}(0)} \right], \tag{9}$$

which is now convergent. The renormalized charge in eq. (9) is $e^2_R = Ze^2$, where the infinite renormalization constant is

$$Z = \left[1 + e^2 \int \frac{d\vec{k}}{(2\pi)^3} \frac{k^2}{12\omega^5_{\vec{k}}(0)} \right]^{-1}, \tag{10}$$

and the renormalized electromagnetic field is $\vec{A}_R = \vec{A}/Z^{1/2}$. (The combination eA is of course unchanged, and we generally drop the subscript R.) It is important to note

335

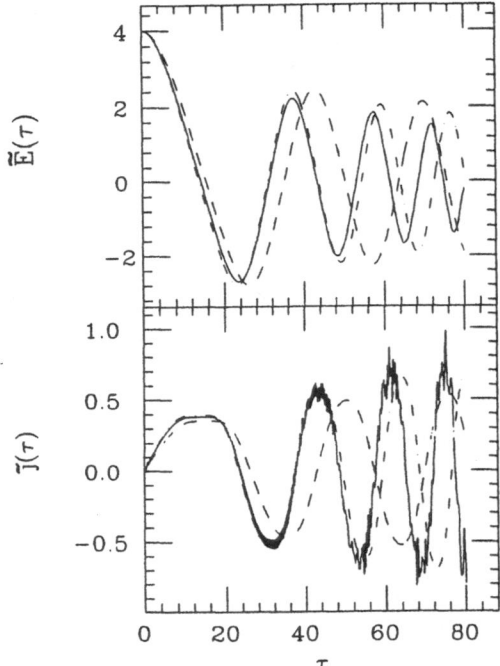

Figure 1. Time evolution of scaled field $\tilde{E}(\tilde{t})$ and $\tilde{j}(\tilde{t})$ for $\tilde{E}(0)=4$ and $e^2/m^2=0.1$. Units are such that $\tilde{E}=eE/m^2$ and $\tilde{t}=mt$. Solid line is field-theory calculation, dashed curve is Boltzmann result without Bose enhancement, and dot-dash curve includes that enhancement.

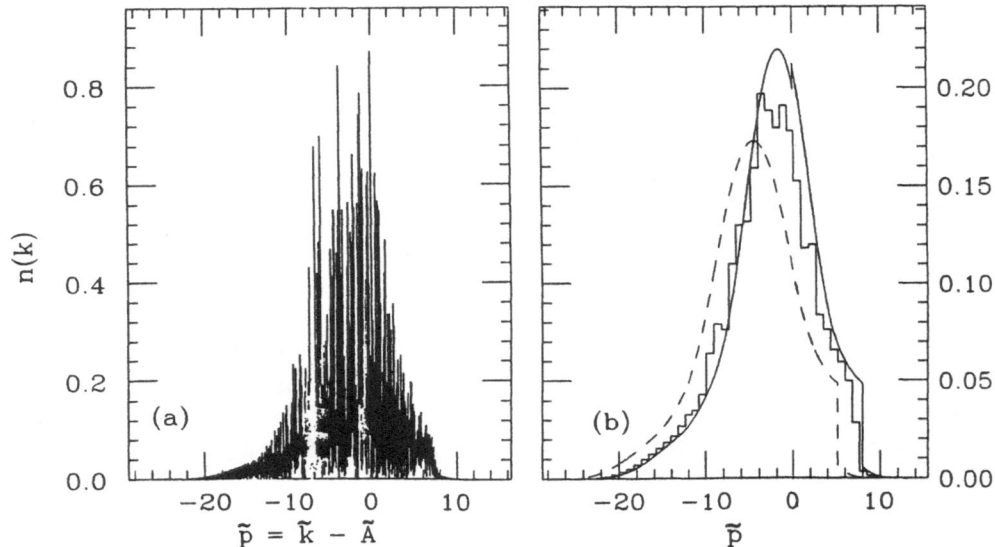

Figure 2. (a) Momentum distribution of produced pairs at $\tilde{t}=550$ for $\tilde{E}(0)=1$. Abscissa is scaled kinetic momentum $p=k-eA$, in units of m. (b) Results after smoothing (histogram) compared with Boltzmann model with (smooth curve) and without (dashed curve) Bose enhancement.

336

that e_R^2 involves the usual charge renormalization of QED, a result that is necessary if our scheme is to be consistent. This also hints at the possibility of renormalizing in the back-reaction problem without using the adiabatic procedure by merely replacing the electric charge with its renormalized value $e_R^2(\Lambda)$ for some momentum cutoff Λ, calculating the current $\vec{j}_\Lambda(t)$ in terms of Λ, and verifying the independence of the result for sufficiently large Λ. This may prove to be an important scheme when there is mode mixing (e.g., if spatial variation is present) and adiabatic regularization does not work.

A CLASSICAL TRANSPORT EQUATION AND NUMERICAL RESULTS

We now consider a transport equation[2] close to that of (2) and (3), and test it against the field solutions. This equation is

$$\frac{\partial f}{\partial t} + eE(t)\frac{\partial f}{\partial p} = \frac{|eE(t)|}{2\pi} \log\left[1 + \exp[-\frac{\pi m^2}{|eE(t)|}]\right]\delta(p), \tag{11}$$

where the right-hand side is the pair-production rate of the tunneling mechanism, shown here for one spatial dimension. The δ-function in momentum space is suggested by microscopic arguments on pair tunneling[9]. Equation (11) is now easily solved together with the Maxwell equation (3). The description of back-reaction is vastly improved[2] if a factor embodying Bose enhancement $(1 + 2f)$ is inserted on the right-hand side of eq. (11). Numerical comparisons between the field-theory results and those of the transport equation then show that the interpretation of the Boltzmann problem can in a major degree replace that of field theory, thus justifying the applications[1] to the quark-gluon plasma.

A few numerical results will help to establish this main physical conclusion: the equivalence between the Boltzmann model and field theory. We restrict ourselves to one spatial dimension (see, e.g., ref. 6 for $3 + 1$ dimensions). Figure 1 shows $\tilde{E}(t)$, and its derivative $\tilde{j}(t)$, as functions of time for an initial value $\tilde{E}(0) = 4$ and $e^2/m^2 = 0.1$. The quantities are scaled according to $\tilde{E} \equiv eE/m^2$, $\tilde{j} \equiv j/m^3$, and $\tilde{t} \equiv mt$, and adiabatic initial values are taken for $f_{\tilde{k}}(t)$. These quantities show plasma oscillations having slightly increasing frequency with time, which corresponds to the additional production of pairs in the electric field, mainly at its peak values. The current $j(t)$ shows a plateau where its first (and, in some cases, second) oscillatory peak is expected, because the initial field strength is large, and the acceleration of the particles brings them to the speed of light, saturating the current. When the solution of eq. (11), shown by the dashed line in fig. 1, is compared with that of the field equations the initial oscillatory and plateau behavior are reproduced quite remarkably. Later, the oscillations drift out of phase. The agreement is made even more striking—shown in fig. 1 by the dot-dash line—when the factor[2] $(1 + 2f)$ is introduced to include boson enhancement.

The highly oscillatory numerical results for the distribution of produced pairs, shown as $n(k)$ in fig. 2, bear little resemblance to the Boltzmann function $f(p, t)$. But if we smooth the former, say by grouping every 75 or 100 momentum points into one bin, we obtain the curve shown in that figure on the right-hand side; this again shows a very strong resemblance to the result for $f(p, t)$ given by the dashed line (no Bose enhancement) and the solid curve (with enhancement).

In fig. 3 the same comparison is shown for fermions, because of the dramatic result: The Boltzmann function—without Pauli blocking—can of course yield several

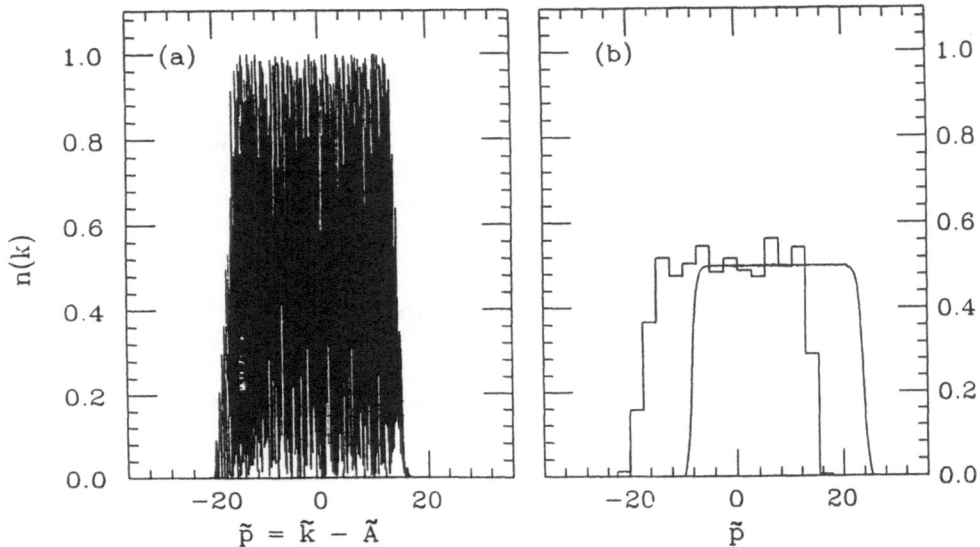

Figure 3. Results analogous to those of fig. 2, but for fermions (with $\tilde{E}(0)=4$ and $\tilde{t}=200$). The smooth curve includes the effects of Pauli blocking through a factor $(1-2f)$ on the right-hand side of the Boltzmann equation for fermions; without this factor, the Boltzmann distribution can and does put five or six fermions in the central momenta bins.

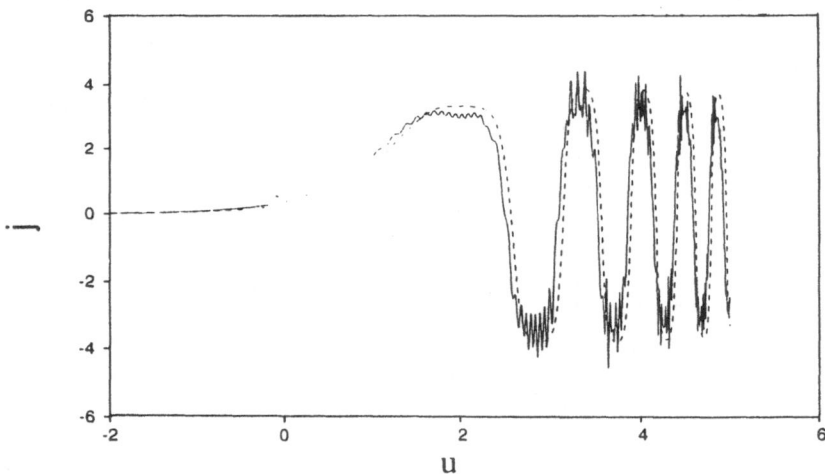

Figure 4. The current in the case of boost-invariant variables for $\tilde{E}(u=-2)=4$, where $u \equiv \log m\tau$. The solid line is the field-theory result, and the dashed line is that of the transport equation.

fermions in a momentum bin. This is corrected by a factor $(1-2f)$ multiplying the fermion pair-production rate, which immediately leads to a stable nonzero solution around $f = 1/2$. Last, in fig. 4 a small sample of results for boost-invariant coordinates is given, illustrating the same general physical features of the cartesian coordinates. This is the case that eventually will be important for studies of the QGP.

FORMAL DERIVATION OF THE TRANSPORT EQUATION

It has proved possible to derive[10] the transport equation (11) in an approximate form starting with the single-time Wigner-operator formalism[11] (for the split-time formalism, see ref. 12). In a certain sense, this shows explicitly how to divide the consequences of the electric field between "virtual" acceleration of the pairs (i.e., their physical production) and the subsequent real acceleration of the existing charged particles.

We define the Wigner function as

$$W(\vec{x}, \vec{p}, t) = \int d\vec{y} \, \langle 0 | \{ \Phi_\alpha (\vec{x} - \tfrac{1}{2}\vec{y}, t), \Phi_\beta^\dagger (\vec{x} + \tfrac{1}{2}\vec{y}, t) \} | 0 \rangle$$

$$\times \exp \left[i\vec{p} \cdot \vec{y} + ie \int_{\vec{x} - \frac{1}{2}\vec{y}}^{\vec{x} + \frac{1}{2}\vec{y}} \vec{A}(\vec{\xi}) \cdot d\vec{\xi} \right]. \tag{12}$$

In order to have first derivatives in time for the transport equation for W, we require first derivatives in time for our field equations as well. This we achieve by using the Feshbach-Villars representation for the Klein-Gordon solution ϕ, taking Φ to be a two-component wave function satisfying

$$\left[\frac{(\vec{p} - e\vec{A})^2}{2m} (\sigma_3 + i\sigma_2) + m\sigma_3 + eA^0 \right] \Phi = i \frac{\partial \Phi}{\partial t}, \tag{13}$$

where $\vec{\sigma}$ are the usual Pauli matrices.

We expand W according to

$$W(\vec{x}, \vec{p}, t) = \sum_{i=1}^{4} P_i \, f_i, \tag{14a}$$

where

$$P_1 = \frac{1}{4} \begin{pmatrix} m/E + E/m & m/E - E/m \\ m/E - E/m & m/E + E/m \end{pmatrix}, \tag{14b}$$

$$P_2 = \frac{1}{2} \begin{pmatrix} 1 & 0 \\ 0 & 1 \end{pmatrix}, \tag{14c}$$

and

$$P_{3/4} = \frac{1}{4} \begin{pmatrix} m/E - E/m & m/E + E/m \mp \frac{1}{2} \\ m/E - E/m \pm \frac{1}{2} & m/E - E/m \end{pmatrix}, \tag{14d}$$

and $E = \sqrt{\vec{p}^2 + m^2}$. For the homogeneous case the f_i satisfy a form suggested by the method of characteristics (or even by a method of initial test particles!), namely,

$$f_i(\vec{p}, t) = \int d\vec{p}_0 \, g_i(\vec{p}_0, t) \, \delta(\vec{p} - \vec{p}(\vec{p}_0, t)), \tag{15a}$$

with

$$\vec{p}(\vec{p}_0, t) = \vec{p}_0 + \int_{-\infty}^{t} dt' \, e\vec{\mathcal{E}}(t'), \tag{15b}$$

for initial momenta at $\vec{p}_0$ and an electric field $\vec{\mathcal{E}}(t)$. In turn the g_i fulfill

$$\frac{dg_1}{dt} = \frac{e\vec{\mathcal{E}} \cdot \vec{p}}{E^2} (g_3 + g_4), \qquad \frac{dg_2}{dt} = 0, \tag{16a, b}$$

and

$$\frac{dg_{3/4}}{dt} = \frac{e\vec{\mathcal{E}} \cdot \vec{p}}{E^2} \frac{g_1}{2} \pm 2iEg_{3/4}. \tag{16c, d}$$

These can be expressed as the solutions to a variable-frequnecy oscillator,

$$\ddot{\xi}(t) + E^2(t)\xi(t) = 0, \tag{17}$$

through

$$g_1 = \frac{E}{2E(0)}\left[|\xi|^2 + \frac{1}{E^2}|\dot{\xi}|^2\right], \tag{18a}$$

$$g_3 + g_4 = \frac{E}{2E(0)}\left[|\xi|^2 - \frac{1}{E^2}|\dot{\xi}|^2\right], \tag{18b}$$

and

$$i(g_3 - g_4) = \frac{1}{2E(0)}[\xi^*\dot{\xi} + \dot{\xi}^*\xi]. \tag{18c}$$

The solutions for $\xi(t)$ then exhibit[10] the usual pair-production rate at large times; the oscillatory term $g_3 + g_4$ can be dropped from the final expression for the current, which is then

$$\vec{j}(\vec{x}, \vec{p}, t) = \frac{e\vec{p}}{E}(g_1 + g_3 + g_4) \sim \frac{e\vec{p}}{E}g_1. \tag{19}$$

The particle density that is then read off from this satisfies, approximately, the transport equation (11) in the presence of a boson-enhanced pair-production source term. The rather roundabout generation of this result points up the artificiality of the separation in the classical description of the two roles of the electric field—particle production and particle acceleration—which appear together so naturally in the field theory.

EXTENSIONS

We list here with various extensions beyond the classical, Abelian, boson case:

a. *Fermions*[3]:- It turns out that fermions satisfy the classical Boltzmann formulation along the same lines as bosons. In ref. 3 the fermion case was set up to avoid the occurrence of a current starting with a nonzero value, which jeopardizes adiabaticity, but in fact this complication now appears to be unnecessary[13].

b. *Boost-invariant coordinates*[3]:- For relativistic heavy-ion collisions one uses boost-invariant coordinates, and there too one might at first have doubts as to whether the Boltzmann description can handle the introduction of the spatial dependence incorporated intrinsically in those variables. The singularity introduced into the d'Alembertian by the proper time coordinate also causes some technical difficulties, but there again proves to be no real problem with this, so that applications to QGP boost-invariant coordinates are straightforward.

c. *Nonclassical electric fields*[13]:- The group at Los Alamos has formulated the next order in the $1/N$-approximation (which, to lowest order, gives[4] the classical electric field) in the Schwinger-Keldysh formalism in $1 + 1$ dimensions. This includes the effects of scattering between the charged particles. The resulting computation is extremely tedious, but the hope is that, if its results are well represented by the transport equation, the latter may be used to simplify subsequent calculations.

d. *Non-Abelian fields*[13]:- Also at Los Alamos, the issue of non-Abelian fields is being addressed, with a similar expectation and motivation to that involved in examining the next order in $1/N$.

e. *Spatial variation*[14] :- At Tel Aviv we are engaged in the calculation of pair production in a finite volume. This may allow for eventual handling of a system that is boost-invariant but has limited transverse extent as in the QGP. Since this introduces mode-mixing, so that adiabatic renormalization is no longer applicable, success depends in large measure on the ability to carry out charge renormalization along the lines hinted at in section 2.

In summary, we do not expect to be able to incorporate all these features into a *usable* field-theory formulation for the QGP, but we do hope to show that each of them is faithfully represented in a transport-equation version of the problem, which can then be used for phenomenological treatments of the QGP behavior.

It is a pleasure to thank my various collaborators in the projects reported on here for their many illuminating contributions: C. Best, F. Cooper, Y. Kluger, E. Mottola, and B. Svetitsky. Part of the work reported on here was carried out while I was a visitor at the Institute for Theoretical Physics at the University of Frankfurt (and indeed the references show a number of collaborators from there); it is a pleasure to express my gratitude for the hospitality I met there. This work was partially supported by the U.S.–Israel Binational Science Foundation and by the Ne'eman Chair in Theoretical Nuclear Physics at Tel Aviv University.

REFERENCES

1. A. Białas and W. Czyż, Acta Phys. Pol. B **17** (1986) 635, and references therein; G. Gatoff, A.K. Kerman, and T. Matsui, Phys. Rev. D **36** (1987) 114.

2. Y. Kluger, J.M. Eisenberg, B. Svetitsky, F. Cooper, and E. Mottola, Phys. Rev. Lett. **67** (1991) 2427.

3. Y. Kluger, J.M. Eisenberg, B. Svetitsky, F. Cooper, and E. Mottola, Phys. Rev. D **45** (1992) 4659; F. Cooper, J.M. Eisenberg, Y. Kluger, E. Mottola, and B. Svetitsky, Phys. Rev. D **48** (1993) 190.

4. For a more complete formulation of the problem in this regard, as well as material that pertains to renormalization, see F. Cooper and E. Mottola, Phys. Rev. D **40** (1989) 456.

5. Y. Kluger, J.M. Eisenberg, and B. Svetitsky, Acta Phys. Polonica, B **23** (1992) 577.

6. Y. Kluger, J.M. Eisenberg, and B. Svetitsky, Int. J. Mod. Phys. E, **2** (1993) no. 2.

7. F. Sauter, Z. Phys. **69** (1931) 742.

8. J. Schwinger, Phys. Rev. **82** (1951) 664.

9. This view is presented in A. Casher, H. Neuberger, and S. Nussinov, Phys. Rev. D **20** (1979) 179 and in C. Itzykson and J.-B. Zuber, *Quantum field theory*, (McGraw–Hill, New York, 1985), See also W. Greiner, B. Müller, and J. Rafelski, *Quantum electrodynamics of strong fields*, (Springer, Berlin, 1985).

10. C. Best and J.M. Eisenberg, Phys. Rev. D **47** (1993) 4639.

11. I. Bialynicki-Birula, P. Górnicki, and J. Rafelski, Phys. Rev. D **44** (1991) 1825.

12. J.M. Eisenberg and G. Kälbermann, Phys. Rev. D **37** (1988) 1197.

13. F. Cooper, E. Mottola, and Y. Kluger, private communications (June, 1993) and papers in the Workshop on Pre-equilibrium Parton Dynamics in Heavy-Ion Collisions, M. Gyulassy, B. Mueller, and X.-N. Wang, eds., (Lawrence Berkeley Laboratory, August, 1993).

14. G. Dror, S. Graf, and J.M. Eisenberg, work in progress.

CHIRAL DYNAMICS IN DENSE MATTER*

W. Weise

Institute of Theoretical Physics
University of Regensburg
D-93040 Regensburg, Germany

Abstract

We discuss the chiral dynamics of hadrons and implications for changes of hadron structure in a dense nuclear medium. This includes issues such as the density dependence of the chiral (quark) condensate, the related appearance of strong scalar mean fields in nuclei and the stability of the pion mass against compression in dense matter.

INTRODUCTION

How does the intrinsic structure of hadrons change in dense nuclear matter? This is one of the important issues in nuclear and heavy-ion physics. In this presentation I discuss this topic from the point of view of chiral dynamics, with special emphasis on pions and scalar fields in matter. Our main theme is that possible changes of hadron properties in a nuclear medium are controlled to a large extent by symmetries and symmetry breaking patterns of Quantum Chromodynamics (QCD).

An important aspect of QCD with light quarks is Chiral Symmetry and the concept of spontaneous symmetry breaking. It governs a large variety of low energy, long-wavelength hadronic and nuclear phenomena. Spontaneous chiral symmetry breaking manifests itself in the non-trivial structure of the QCD ground state. The vacuum hosts a condensate $< \bar{q}q >$ of (scalar) quark-antiquark pairs. Hadrons are excitations of this vacuum, and hadronic properties, especially those of the low-mass mesons and baryons, are closely related to vacuum properties. With this in mind, we elaborate a question of central interest from a nuclear physics point of view: how does the chiral condensate $< \bar{q}q >$ change with increasing nuclear density $\rho = < N^+N >$? We show that a strong attractive scalar mean field in nuclei arises as the magnitude of the quark condensate decreases with increasing baryon density, and we discuss its significance in comparison with the large scalar potential commonly found in nuclear Dirac phenomenology. Then we focus on pions in dense matter. The pion has an exceptional status. Its mass is nearly one order of magnitude smaller than the typical hadronic mass scale of 1 GeV/c^2. It is the "soft mode" of hadron physics. The role of the pion as a Goldstone boson of spontaneously broken chiral symmetry implies that

---

*Work supported in part by BMFT grant 06 OR 735, GSI grant OR Wei T and DFG grant We 655/11-1

its mass, unlike other hadron masses, remains remarkably stable in dense matter. We summarize our present understanding of this important feature and confront it with the quest for "s-wave pion condensation" raised in the recent literature.

FOUNDATIONS: QCD AND CHIRAL SYMMETRY

Our starting point is QCD with three light quark fields $q = (u, d, s)$. The QCD Lagrangian

$$\mathcal{L}_{QCD} = \mathcal{L}_{QCD}^{(o)} + \Delta\mathcal{L}_{mass} \tag{1}$$

has a massless part,

$$\mathcal{L}_{QCD}^{(o)} = \bar{q}i\gamma_\mu(\partial^\mu - igG^\mu)q \quad + \text{"glue"} \tag{2}$$

where G_μ is the gluon field and the "glue" term involves the squared gluonic field strength tensor. The mass term

$$\Delta\mathcal{L}_{mass} = -m_u\bar{u}u - m_d\bar{d}d - m_s\bar{s}s \tag{3}$$

introduces the current quark masses m_i. To the extent that these masses are small compared to typical hadronic mass scales, $\Delta\mathcal{L}_{mass}$ can be treated as a perturbation. Apart from local $SU(3)_{color}$ gauge symmetry, QCD has approximate global symmetries which govern much of the low energy dynamics of hadrons and their nuclear interactions. Consider massless QCD with $\mathcal{L}_{QCD}^{(o)}$ only. The massless quarks can be separated into right- and left-handed fields, $q_{R,L} = \frac{1}{2}(1 \pm \gamma_5)q$. The quark current couplings to gluons G_μ in $\mathcal{L}_{QCD}^{(o)}$ do not change quark chirality, so that right- and lefthanded quarks remain decoupled as long as they are massless. In other words, $\mathcal{L}_{QCD}^{(o)}$ is invariant under the chiral flavour group $SU(3)_R \otimes SU(3)_L$, i.e. the group of separate global transformations

$$q_{R,L} \rightarrow exp\left[i\alpha_{R,L}^a\frac{\lambda^a}{2}\right]q_{R,L} \tag{4}$$

of the three right- and lefthanded quark fields. Here $\lambda^a (a = 1, \ldots, 8)$ are the standard $SU(3)$ flavour matrices. As a consequence of this symmetry there are eight vector currents $V_\mu^a = \bar{q}\gamma_\mu(\lambda^a/2)q$ and eight axial vector currents $A_\mu^a = \bar{q}\gamma_\mu\gamma_5(\lambda^a/2)q$ which are rigorously conserved $(\partial^\mu V_\mu^a = \partial^\mu A_\mu^a = 0)$ in the chiral limit with $m_u = m_d = m_s = 0$.

Independent of the quark masses, the QCD Lagrangian is also invariant under $U(1)_V$ which generates a global phase transformation $q \rightarrow e^{i\alpha}q$. The corresponding conserved current is the baryon current $\bar{q}\gamma_\mu q$. Moreover, $\mathcal{L}_{QCD}^{(o)}$ with massless quarks is formally invariant with respect to the axial $U(1)_A$-symmetry, i.e. under transformations $q \rightarrow e^{i\gamma_5\alpha}q$. This symmetry is broken in nature, a phenomenon referred to as the axial anomaly in QCD which is reflected in the large mass of the η' meson. The point of importance for hadron and nuclear physics is that the underlying $SU(3)_R \otimes SU(3)_L \otimes U(1)_V$ symmetries of QCD and their corresponding conserved currents, establish rules which govern the low energy dynamics of strongly interacting systems, irrespective of whether their relevant degrees of freedom are elementary quarks and gluons or composite hadrons.

The quark mass term $\Delta\mathcal{L}_{mass}$ of eq.(3) breaks chiral $SU(3)_R \otimes SU(3)_L$ symmetry *explicitly*. It involves the scalar quark density $\bar{q}q = \bar{q}_L q_R + \bar{q}_R q_L$ which mixes left- and right-handed quarks. The generally accepted values of the bare (current) quark masses (at a renormalization point $\Lambda \sim 1\ GeV$) are[1] :

$$m_u = (5 \pm 2)\ MeV, \quad m_d = (9 \pm 3)\ MeV, \quad m_s = (175 \pm 55)\ MeV. \tag{5}$$

The smallness of $m_{u,d}$ as compared to typical hadron masses, e.g. the nucleon mass $M_N \sim 1\ GeV$ suggests that $m_u \bar{u}u + m_d \bar{d}d$ can be treated perturbatively. This is the starting point of Chiral Perturbation Theory. For the larger strange quark mass m_s this is less obvious.

There is evidence both from low-energy hadron phenomenology and from lattice QCD that chiral symmetry is *spontaneously* broken. The QCD ground state, or vacuum, is in the Nambu-Goldstone realization of chiral symmetry. It is characterized by a non-vanishing vacuum expectation value $< \bar{q}q >$ of the scalar quark density (the chiral quark pair condensate). Typical "empirical" values given in the literature[2] are

$$< \bar{u}u > \ \simeq\ < \bar{d}d > \ \simeq -(230 \pm 25\ MeV)^3, \tag{6}$$

and results for $< \bar{s}s >$ are of similar magnitude. For later discussions, one should note that the magnitude of the condensate (6) is a large number on nuclear scales:

$$|< \bar{u}u > + < \bar{d}d >| \sim 3 fm^{-3}. \tag{7}$$

Recall that the baryon density in nuclear matter is only $\rho \simeq 0.17 fm^{-3}$. The number of valence u- plus d-quarks in nuclear matter is three times larger, but still small compared to the scalar vacuum density (7).

Lattice QCD results[3] confirm that the chiral order parameter $< \bar{q}q >$ is non-zero in the strong coupling limit. With decreasing coupling strength g^2 (or increasing temperature), the condensate decreases in magnitude, indicating a tendency towards restoration of chiral symmetry above a critical temperature $T_c \sim (150 - 200)\ MeV$. Explicit symmetry breaking by the non-zero bare quark masses has a significant influence, in that there is no sharp chiral phase transition, but only a smooth crossover in the region where one commonly expects the transition from hadronic to quark matter.

According to Goldstone's theorem a spontaneously broken global symmetry implies the existence of massless bosons with spin zero. For spontaneously broken chiral $SU(3)_R \otimes SU(3)_L$ the whole pseudoscalar meson octet would be massless in the chiral limit with $m_u = m_d = m_s = 0$. Weak decays of the pseudoscalar mesons are described by their axial current matrix elements. For the pion, the matrix element $< 0|A_\mu^a(o)|\pi^b(q) >= if_\pi q_\mu \delta^{ab}$ introduces the pion decay constant f_π as a new scale parameter, with the empirical value $f_\pi = 93.3 MeV$. Roughly speaking, this f_π is related to the normalization of the intrinsic quark-antiquark wave function of the pion at the origin, i.e. at zero distance. It is a measure of spontaneous chiral symmetry breaking just like the $< \bar{q}q >$ condensate. It has in fact been suggested[4] that the characteristic 1 GeV scale that governs chiral dynamics is just $4\pi f_\pi$.

Explicit chiral symmetry breaking by the non-zero quark masses (5) shifts the Goldstone bosons to the observed pseudoscalar mesons. This can be seen by studying a sumrule related to the axial charge $Q_A^a = \int d^3 x q^+ \gamma_5 \frac{\lambda^a}{2} q(x)$. Consider the vacuum expectation value of the double commutator $[Q_A^a, [\mathcal{H}(o), Q_A^b]]$ with the QCD Hamiltonian density $\mathcal{H}$. In the chiral limit, $\mathcal{H}$ commutes with the axial charge. Non-vanishing contributions come only from the quark mass term (3). Making use of the fact that the Goldstone bosons exhaust the spectrum of Q_A^a one finds, e.g. for the pion (with $a=1,2$):

$$< 0|[Q_A^a,[\mathcal{H}(o),Q_A^a]|0> \quad = -m_\pi^2 f_\pi^2 = \frac{1}{2}(m_u + m_d) < \bar{u}u + \bar{d}d > +\mathcal{O}(m_{u,d}^2), \quad (8)$$

the Gell-Mann, Oakes, Renner (GOR) relation[5]. Sum rules of the form (8) are familiar to nuclear physicists. The GOR relation is in fact reminiscent of RPA sum rules. The pion can be viewed as a collective particle-hole (quark-antiquark) excitation of the QCD ground state which exhausts the pseudoscalar sum rule.

A Schematic Model. In fermionic many-body systems, and in the nuclear many-body problem in particular, much insight is often gained already at the level of the Hartree-Fock and RPA approximations. An analogous approach in hadron physics is provided by the Nambu and Jona-Lasinio (NJL) model[6]. It offers a simple way to illustrate the basic mechanism that drives spontaneous chiral symmetry breaking. Its modern flavour SU(3) version[7-8] has become quite popular also among nuclear theorists in recent years.

The NJL model is an effective Lagrangian of relativistic fermions (quarks) which interact through local current-current couplings. It approximates important features of QCD in the low energy, long wavelength limit, assuming that gluonic degrees of freedom can be frozen into pointlike effective interactions between quarks. The symmetry breaking scenario in the NJL model involves two basic steps. First, quarks develop large dynamical effective masses due to their strong interactions. In the mean field (or Hartree-Fock) approximation, these masses are determined by self-consistent gap equations. For example, the u-quark mass equation has the form

$$M_u = m_u - G < \bar{u}u >, \qquad (9)$$

where m_u is the small bare mass of the non-interacting quark and G is a coupling strength of dimension (length)^2. The condensates

$$< \bar{u}u >= -\frac{3}{2\pi^3} \int_{|\vec{p}|\leq\Lambda} d^3p \frac{M_u}{\sqrt{\vec{p}\,^2 + M_u^2}}, \quad etc., \qquad (10)$$

represent the scalar densities of the negative energy Dirac sea integrated up to a characteristic cutoff Λ in momentum space. In essence, a quark gets its large dynamical mass by polarizing the Dirac sea. There is an "active" part of that Dirac sea, with low momenta $|\vec{p}| \leq \Lambda$, which participates in the strong interactions, whereas the QCD forces are weak for high momenta, and the NJL model simplifies this by effectively turning off the interactions for $|\vec{p}| > \Lambda$.

The solution of the gap equation (9) gives dynamical masses as shown in Fig.1. Typical values of the cutoff Λ turn out to be around 1 GeV. Together with a dimensionless scalar coupling constant $g_S^2 = G\Lambda^2$ of about 15 this gives the "empirical" values of the condensates and constituent quark masses $M_u = M_d \simeq 350\ MeV$, $M_s \simeq 500\ MeV$. The effective potential U_{eff} derived from the Lagrangian of this model, plotted as a function of the condensate $| < \bar{q}q > |$, shows the characteristic feature of spontaneous symmetry breaking: once the coupling strength exceeds a critical value G_{crit} (to be specific: $G_{crit}\Lambda^2 = 4\pi^2/3$), U_{eff} develops a minimum at a non-zero value of the order parameter $<\bar{q}q>$ (see Fig.2). In this picture, spontaneous chiral symmetry breaking and dynamical fermion mass generation have the same origin.

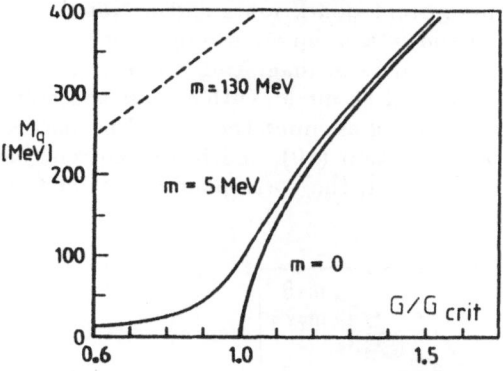

Figure 1: Dynamical quark mass M_q as a function of $g^2 = G\Lambda^2$ for different current quark masses m_q. The critical coupling strength is $g^2_{crit} = 4\pi^2/3$ (from ref. 9)

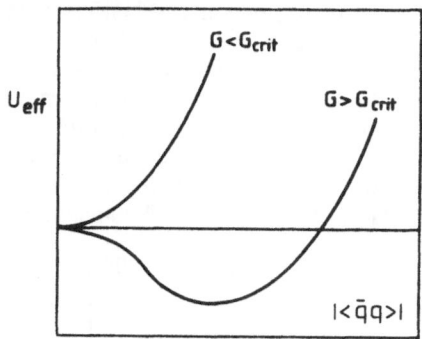

Figure 2: Illustration of the effective potential of the NJL model, showing spontaneous chiral symmetry for $G > G_{crit}$

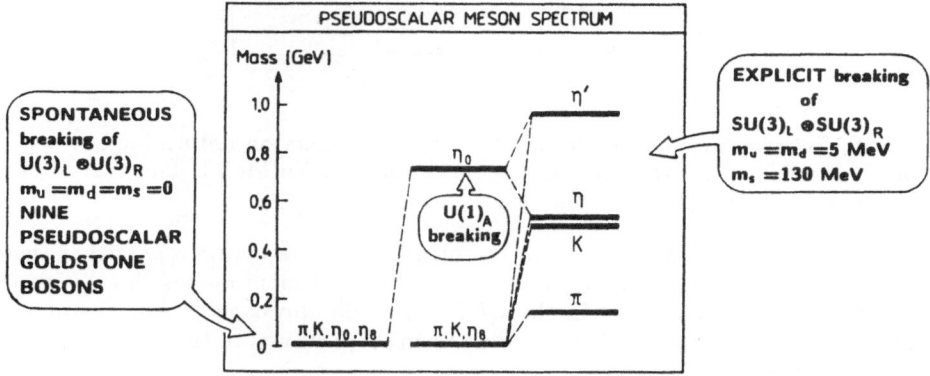

Figure 3: Pseudoscalar meson spectrum calculated in the NJL model[9], showing the chiral and flavour symmetry breaking pattern.

The second step in the NJL approach is to solve the Bethe-Salpeter equation using the two-body interaction kernel derived from the chiral effective Lagrangian, in order to study quark-antiquark bound states. The pseudoscalar meson nonet is of particular interest since it directly reflects the chiral and flavour symmetry breaking pattern (see Fig.3). In the limit of exact chiral $U(3) \otimes U(3)$ symmetry there are nine massless pseudoscalar Goldstone bosons (the pions, kaons, the η_0 and the η_8). The axial $U(1)_A$ anomaly shifts the singlet η_0 mass up. When the non-zero current quark masses are introduced, the pseudoscalar spectrum falls right into place with its observed pattern, including $\eta - \eta'$ mixing. Together with other important chiral low energy theorems, the GOR relation (8) is now a natural outcome of the model.

The Pion on the Lattice. The very special features of the pion are also evident in recent QCD lattice calculations[11]. Consider the pseudoscalar Euclidean correlation function $\Pi_P(x) = \langle TJ(x)J(o)\rangle$ with $J = \bar{u}\gamma_5 d$. It describes the propagation of a pionic quark-antiquark pair over a distance x in Euclidean space. This correlation function, normalized to the one for a free, uncorrelated quark-antiquark pair, has been determined on the lattice, varying the current quark mass and extrapolating down to $m_q = 8\ MeV$. The result (Fig.4) shows a uniquely strong enhancement of Π_P with increasing distance, by almost two orders of magnitude for $x > 1.5 fm$. This behaviour is characteristic of a strongly bound $\bar{q}q$ mode, with a mass spectrum completely dominated by a single small mass. Using as input the special properties of the pion as they are manifest in the GOR relation (10), the lattice correlation function is indeed well reproduced[11]. In comparison, the corresponding correlation function of the ρ meson[10] looks fairly "normal".

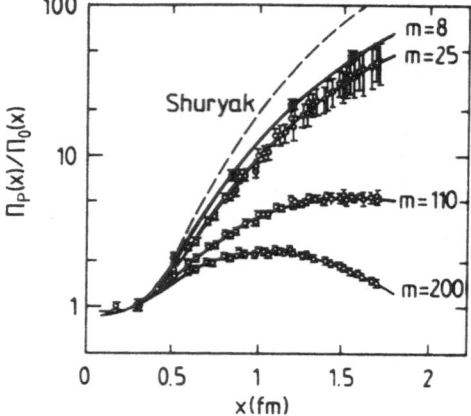

Figure 4: Pseudoscalar quark-antiquark correlation function determined in a QCD lattice calculation[10] for varying current quark mass m (in MeV). The dashed curve is the analysis of ref. (11).

Intermediate Summary To close this introductory survey, we repeat that there is generally a close connection between hadron masses and spontaneous chiral symmetry breaking. This is also evident in the QCD sum rule approach[12]. For example, the nucleon mass in this approach has a leading term proportional to the quark condensate $\langle \bar{q}q \rangle = \langle \bar{u}u + \bar{d}d \rangle$:

$$M_{Nucleon} \simeq -\frac{4\pi^2}{\mathcal{M}_B^2}\langle \bar{q}q \rangle, \qquad (11)$$

where the characteristic scale is the Borel mass $\mathcal{M}_B \sim 1\ GeV$. Eq.(11) is quite reminiscent of the gap equation in the NJL model.

However, pions (and kaons) have a distinctly different behaviour because of their Goldstone boson nature. They are massless in the limit $m_u = m_d = m_s = 0$. These special features will be important later in the discussion of pion properties in dense matter.

VACUUM STRUCTURE

We now come to several points of central interest from a nuclear physics point of view. We describe how the vacuum structure of QCD, expressed in terms of the quark condensate, changes when nucleons are added. As a consequence this implies a strong scalar mean field in nuclei.

The Chiral Condensate in Matter. How does the chiral condensate $< \bar{q}q >= < \bar{u}u + \bar{d}d >$ change in a nuclear environment with increasing baryon density $\rho = < N^+N >$? At $\rho = 0$ we have $< \bar{q}q >_{\rho=0} = -m_\pi^2 f_\pi^2/m_0$ following the GOR relation (8), with the average current quark mass $m_0 = \frac{1}{2}(m_u + m_d) = (7 \pm 2) \ MeV$. Adding nucleons to the vacuum implies a change of the average scalar density. To leading order in ρ, this change is determined by the nucleon matrix element $< N|\bar{q}q|N >$ of the quark scalar density:

$$< \bar{q}q >_\rho \ = \ < \bar{q}q >_{\rho=0} + < N|\bar{q}q|N > \rho + \text{higher orders in } \rho. \qquad (12)$$

Here we use the normalization $< N(p')|N(p) >= (2\pi)^3\delta^3(\vec{p}\,' - \vec{p})$ so that the matrix element $< N|\bar{q}q|N >$ is dimensionless. This matrix element is related to the pion-nucleon sigma term

$$\Sigma_N = \frac{1}{2}(m_u + m_d) < N|\bar{u}u + \bar{d}d|N > \equiv m_0 < N|\bar{q}q|N > = m_0 \frac{\partial M_N}{\partial m_0}. \qquad (13)$$

It measures the effect of explicit chiral symmetry breaking in the nucleon mass M_N. An extrapolation from low energy pion-nucleon scattering data gives[13]:

$$\Sigma_N = (45 \pm 8) \ MeV. \qquad (14)$$

Using the GOR relation (8) one finds the relative change of the chiral condensate to leading order in the density:

$$\frac{< \bar{q}q >_\rho}{< \bar{q}q >_0} = 1 - \frac{\Sigma_N}{m_\pi^2 f_\pi^2}\rho + \cdots \qquad (15)$$

This important relation was first written down by Drukarev and Levin[14] and has subsequently been derived and discussed in a variety of ways[15,16].

The consequences implied by eq. (15) are quite remarkable. At normal nuclear matter density $\rho = \rho_0 = 0.17 fm^{-3}$, using $\Sigma_N = 45 \ MeV$, the condensate decreases in magnitude by about 35 % from its vacuum value, indicating a strong tendency towards chiral symmetry restoration already in the bulk of ordinary heavy nuclei. In order to draw firm conclusions, however, one must have a reliable estimate of the higher order terms in the expansion (17). A more general relation based on the application of the Hellmann-Feynman theorem[17] is:

$$< \bar{q}q >_\rho \ = \ < \bar{q}q >_{\rho=0} + \frac{\partial}{\partial m_0}\mathcal{E}(\rho), \qquad (16)$$

where

$$\mathcal{E}(\rho) = M_N \ \rho + \mathcal{E}_{kin}(\rho) + \mathcal{E}_{int}(\rho) \qquad (17)$$

is the energy density of nuclear matter. The mass term $M_N\rho$ alone gives the lowest order result (12,15). The $\rho^{5/3}$ correction from the kinetic energy is small, a few percent at $\rho = \rho_0$. Terms of order ρ^2 from the interaction part $\mathcal{E}_{int}$ are not so easy to get. Their evaluation requires not only a detailed knowledge of correlations in the nuclear many-body problem, but also a clear picture of how the NN interaction varies with changes of the current quark mass m_0. A simple estimate can be obtained using nuclear Dirac phenomenology with scalar and vector mean fields. In this approach[18] the nucleons experience a large attractive scalar potential $U_S \simeq -(g_S^2/m_S^2)\rho \simeq -350\ MeV(\rho/\rho_0)$ and an almost equally large repulsive vector potential $U_V = (g_V^2/m_V^2)\rho \simeq 300\ MeV(\rho/\rho_0)$. Assuming that U_S changes with m_0 (or, equivalently, with m_π^2) roughly according to the rules of the linear sigma model[19] and that the vector boson mass scales with the nucleon mass, we find the following approximate expression to order ρ^2:

$$\frac{<\bar{q}q>_\rho}{<\bar{q}q>_0} \simeq 1 - \frac{\Sigma_N}{m_\pi^2 f_\pi^2}\rho \left[1 - \frac{3}{10}\left(\frac{p_F}{M_N}\right)^2 - \frac{3U_S}{2M_N} - \frac{U_V}{M_N}\right], \qquad (18)$$

with the Fermi momentum $p_F = (3\pi^2\rho/2)^{\frac{1}{3}}$. Short range $NN-$ correlations will reduce both the U_S and U_V terms further, so that one expects the correction to the leading model independent term (15) to be 10 % or less at $\rho = \rho_0$. A similar result is found in ref. (15). However, one should be aware of the fact that this relatively small correction arises from a cancellation between two large numbers, so that this estimate is not very well under control. So far, different authors reach different conclusions[19,20] about the significance of $\mathcal{O}(\rho^2)$ terms, and more work needs to be done to sort this out.

Scalar Mean Field in Nuclei. The change of the chiral condensate with baryon density (18) can be translated into an effective scalar mean field experienced by nucleons in a nucleus, as follows. Assume that the nucleon mass M_N scales with $<\bar{q}q>$, as it is commonly the case in spontaneous chiral symmetry breaking scenarios. This implies that $M_N^*(\rho)/M_N =<\bar{q}q>_\rho / <\bar{q}q>_0$. One can introduce a Dirac-scalar mean field $U_S = M_N^* - M_N$ and finds

$$U_S(\rho) = M_N\left(\frac{<\bar{q}q>_\rho}{<\bar{q}q>_0} - 1\right) = -\frac{M_N \Sigma_N}{m_\pi^2 f_\pi^2}\rho \simeq -7.2\ \Sigma_N\left(\frac{\rho}{\rho_0}\right). \qquad (19)$$

Using $\Sigma_N \simeq 45\ MeV$ this gives $U_S(\rho = \rho_0) \simeq -325\ MeV$, in remarkably close correspondence with values $U_S(\rho = \rho_0) = -(300 \div 400)\ MeV$ usually used in the nuclear Dirac phenomenology. QCD sum rule considerations[21] arrive at similar values for U_S. In addition they predict $U_V/U_S = -(8m_0/\Sigma_N) \sim -1$. One should note that the scalar mean field experienced by an antiproton in matter, through the mechanism (19), is identical to the one for the nucleon, while U_V changes sign. This source of substantial attraction may influence the rate for subthreshold antiproton production in heavy ion collisions.

PIONS IN DENSE MATTER

Studies of pion propagation in a nuclear environment have a long history[22]. Given the systematics of data from pion-nucleus scattering and pionic atoms, the parameters of the pion-nuclear interaction are quite well determined, at least at low densities. The spectrum of pionic modes with energy ω and momentum $\vec{q}$ in nuclear matter is determined by the dispersion equation

$$\omega^2 - \vec{q}\,^2 - m_\pi^2 - \Pi(\omega, \vec{q}\,; \rho) = 0, \tag{20}$$

where Π is the pion self-energy. Much of pion-nuclear physics at non-zero momentum $\vec{q}$ is governed by the propagation of the p-wave $\Delta(1232)$ resonance. Its description in terms of Δ-hole excitations is by now a standard part of nuclear many-body theory.

More recently there has been a revival of interest in the role played by s-wave pion interactions in a nuclear medium. These s-wave interactions are relevant for pions "at rest" relative to the surrounding matter. They determine the in-medium pion mass:

$$m_\pi^{*2}(\rho) \equiv \omega^2(\vec{q} = 0; \rho) = m_\pi^2 + \Pi(\omega = m_\pi^*, \vec{q} = 0; \rho). \tag{21}$$

For nuclear physics, the question about a possible density dependence of the pion mass is certainly of crucial importance, given the fact that the long-range NN-interaction is determined by one-pion exchange.

Brown et al.[23] have discussed the possibility that the pion effective mass drops with increasing density at the same rate as the chiral condensate (see eq.(15)), leading to s-wave pion condensation at a density not far above twice that of normal nuclear matter. Their starting point is the explicit symmetry breaking term of the $SU(2) \otimes SU(2)$ chiral effective Lagrangian,

$$\Delta \mathcal{L} \simeq -\Sigma_N (1 - \frac{\pi^2}{2f_\pi^2})(\bar{N}N), \tag{22}$$

written in terms of the pion field π and the nucleon scalar density $\bar{N}N$. In the mean field approximation, with $< \bar{N}N > \simeq \rho$ for non-relativistic nucleons, and (22) translates into a correction of the pion mass,

$$\Delta m_\pi^2 \simeq -\frac{\Sigma_N}{f_\pi^2} \rho, \tag{23}$$

which suggests a substantial decrease of m_π with increasing density. Using $\Sigma_N \simeq 45 \; MeV$ one observes a tendency towards s-wave pion condensation at a critical density $\rho_{crit} \sim m_\pi^2 f_\pi^2 / \Sigma_N$, less than three times the density of nuclear matter.

Such large effects evidently deserve a closer look. They are in fact not observed in nature. At low density the s-wave pion self-energy, or optical potential, for symmetric nuclear matter is given in terms of an effective isospin averaged scattering length b_{eff}:

$$\Pi(\vec{q} = 0; \rho) = 2m_\pi U_{opt}(\rho) = -4\pi(1 + \frac{m_\pi}{M_N})b_{eff} \; \rho, \tag{24}$$

which is constrained by low energy pion-nucleus scattering and pionic atom data. Systematic analysis of these data[24] gives best fits for b_{eff} ranging between zero and about $-0.07fm$ (This range includes the uncertainty due to poorly known dispersive terms from pion absorption). In any case, this analysis suggests a weakly repulsive s-wave pion-nuclear interaction, whereas eq.(23) would imply $b_{eff} = (\Sigma_N / 4\pi f_\pi^2)(1 + m_\pi/M_N)^{-1} \simeq +0.07fm$, i.e. substantial attraction instead.

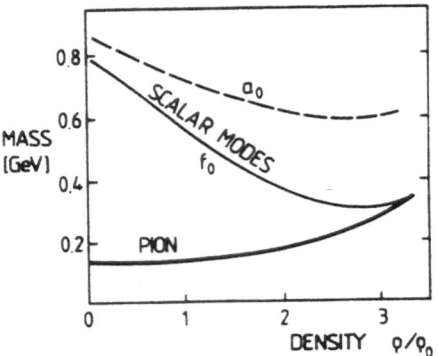

Figure 5: Pion mass and masses of scalar mesons as a function of baryon density ρ, calculated in a SU(3)-NJL model[16].

Recent detailed studies arrive at the common conclusion that the pion mass stays remarkably constant in nuclear matter. For example, in calculations using the SU(3)-NJL model[16] it is found that, whereas the masses of scalar mesons (the a_0 and f_0) drop with increasing density, the pion mass is almost constant, at least up to $\rho = \rho_0$ (see Fig.5). The main point demonstrated in detail in ref.16) is that the GOR relation (8) holds not only in vacuum, but also at finite densities and temperatures. Any changes of pion properties in dense and hot matter are subject to constraints from the GOR relation. Hence

$$\frac{\Delta m_\pi^2(\rho)}{m_\pi^2} = \frac{\Delta < \bar{q}q >_\rho}{< \bar{q}q >_0} - \frac{\Delta f_\pi^2(\rho)}{f_\pi^2}. \tag{25}$$

The variation of the pion mass in matter involves two effects which enter in eq. (25) with opposite signs: the change of the chiral condensate and the change of the pion decay constant with density. To leading order in density we can write

$$\Delta m_\pi^2(\rho) = -\frac{\Sigma_N \, \rho}{f_\pi^2}(1 - \delta), \tag{26}$$

where the parameter δ describes the variation of f_π with density. This δ represents loop corrections which result from medium effects on the intrinsic pion wave function. These corrections turn out to be large: any chiral model must give $\delta \sim 1$. In the NJL model, for example, one finds $\delta = 1 + (4/3) \, M_q^2 < r_S^2 >_\pi$ where M_q is the constituent quark mass and $< r_S^2 >_\pi$ is the mean square scalar radius of the pion. Consequently, the pion mass in medium tends to increase very weakly (rather than decrease substantially) with increasing density. There is no tendency towards s-wave pion condensation, and this is consistent with low-energy pion-nuclear data. Chiral symmetry protects the pion mass against strong in-medium variations.

Delorme et al.[25], and more recently Yabu et al.[26], come to similar conclusions. They discuss the same issue from the point of view of the off-shell energy dependence of the πN s-wave amplitude, with constraints imposed by chiral symmetry. We show in ref. 27) that the result, eq. (26), derives indeed from a systematic expansion of the s-wave pion self-energy in powers of the energy ω. Consider for example a π^- with zero momentum $\vec{q} = 0$ in matter with proton and neutron densities ρ_p and ρ_n. We find the following off-shell π^- self-energy to order ω^2:

$$\Pi_{\pi^-}(\omega, \vec{q} = 0; \rho) = \frac{\omega}{2f_\pi^2}(\rho_n - \rho_p) + \frac{\Sigma_N}{f_\pi^2}\left[1 - \frac{2\omega^2}{m_\pi^2}\left(1 - \frac{\delta}{2}\right)\right](\rho_n + \rho_p). \qquad (27)$$

The first term on the r.h.s. is the model independent Weinberg-Tomozawa term. It vanishes for symmetric nuclear matter with $\rho_p = \rho_n = \rho/2$. The second term has a model independent "soft" limit at $\omega = 0$ determined by yet another chiral low energy theorem (the Adler-Weisberger theorem). Note that at this point the effective πN s-wave interaction is actually repulsive! The interpolation between the soft point and the physical threshold at $\omega = m_\pi$ is subject to a strong energy dependence which involves the model dependent parameter δ, just the one that enters in eq. (26). Chiral models typically give values around $\delta \sim 1.2$. The change of the squared pion mass Δm_π^2 then results from $\Pi = -4\pi b^{(+)}\rho$ taken at $\omega = m_\pi$ and $\rho = \rho_p + \rho_n$, where the isospin-symmetric πN scattering length $b^{(+)} = (\Sigma_N/4\pi f_\pi^2)(1 - \rho)$ has a small negative value consistent with data. A naive identification of Δm_π^2 directly from (22) misses the Adler soft point. On the other hand a consistent construction of the pion propagator with inclusion of the symmetry breaking term does give the proper result[28].

SUMMARY

Changes of hadron properties in a nuclear medium are governed to a large extent by symmetries and symmetry breaking patterns of QCD, with chiral dynamics playing a prominent role. Hadrons are excitations of the QCD vacuum. The properties of low-mass hadrons are therefore closely connected with vacuum structure. Hence changes of hadron properties in medium carry signals of the way in which the vacuum changes in a nuclear environment.

An important part of this discussion is the change of the quark condensate $< \bar{q}q >$ in matter. To leading order in the baryon density, the rate of change of this condensate is determined model-independently by the pion-nucleon sigma term. Corrections of higher order in density have been only crudely estimated so far and need to be worked out in detail.

The change of the chiral condensate in medium implies a strong scalar mean field in nuclei. Its possible connection with Dirac phenomenology and the relativistic nuclear many-body problem is of great interest.

The very special nature of the pion as a Goldstone boson of spontaneously broken chiral symmetry protects its mass against variations in matter. The pion mass stays almost constant up to normal nuclear matter density and even beyond. This important feature can be traced to the detailed chiral symmetry properties of the πN s-wave amplitude. The Gell-Mann, Oakes, Renner relation remains valid in medium. It follows that the pion decay constant f_π decreases with density like the square root of the chiral condensate. An interesting nuclear physics issue is to find observables which might carry signatures of a dropping f_π.

REFERENCES

1. J. Gasser and H. Leutwyler, Phys. Reports 87 (1982) 77; J.F. Donoghue, Ann. Rev. Nucl. Sci. 39 (1989) 1.

2. M. Shifman, A. Vainshtein and V. Zakharov, Nucl. Phys. B 147 (1979) 385, 448; L. Reinders, H. Rubinstein and S. Yazaki, Phys. Reports 127 (1985) 2.

3. J.B. Kogut, D.K. Sinclair and K.C. Wang, Phys. Lett. B 263 (1991) 101.

4. A. Manohar and H. Georgi, Nucl. Phys. B 234 (1984) 189.

5. M. Gell-Mann, R. Oakes and B. Renner, Phys. Rev. 175 (1968) 2195.

6. Y. Nambu and G. Jona-Lasinio, Phys. Rev. 122 (1961) 345.

7. T. Hatsuda and T. Kunihiro, Phys. Lett. B 185 (1987) 304.

8. V. Bernard, R.L. Jaffe and U.-G. Meißner, Nucl. Phys. B 308 (1988) 753.

9. S. Klimt, M. Lutz, U. Vogl and W. Weise, Nucl. Phys. A 516 (1990) 429, 469; U. Vogl and W. Weise, Prog. Part. Nucl. Phys. 27 (1991) 195, and refs. therein.

10. M.-C. Chu, J.-M. Grandy, S. Huang and J.W. Negele, Phys. Rev. Lett. 70 (1993) 255.

11. E.V. Shuryak, Rev.Mod. Phys. 65 (1993) 1.

12. B.L. Ioffe, Nucl. Phys. B 188 (1981) 317.

13. J. Gasser, H. Leutwyler and M. Sainio, Phys. Lett. B 253 (1991) 252.

14. E.G. Drukarev and E.M. Levin, Nucl. Phys. A 511 (1990) 679; Prog. Part. Nucl. Phys. 27 (1991) 77.

15. T.D. Cohen, R.J. Furnstahl and D.K. Griegel, Phys. Rev. C 45 (1992) 1881.

16. M. Lutz, S. Klimt and W. Weise, Nucl. Phys. A 542 (1992) 521.

17. R.P. Feynman, Phys. Rev. 56 (1939) 340.

18. B.D. Serot and J.D. Walecka, Adv. Nucl. Phys. 16 (1986) 1.

19. M.C. Birse and J.A. McGovern, Phys. Lett. B 309 (1993) 231.

20. M. Ericson, Phys. Lett. B 301 (1993) 11.

21. T.D. Cohen, R.J. Furnstahl and D.K. Griegel, Phys. Rev. Lett. 67 (1991) 961.

22. T.E.O. Ericson and W. Weise, Pions and Nuclei, Oxford Press (1988).

23. G.E. Brown, V. Koch and M. Rho, Nucl. Phys. A 535 (1991) 701; C. Adami and G.E. Brown, Phys. Reports (1993), (in print).

24. R. Seki and K. Masutani, Phys. Rev. C 27 (1983) 2799; C. de Laat et al., Nucl. Phys. A 523 (1991) 453.

25. J. Delorme, M. Ericson and T.E.O. Ericson, Phys. Lett. B 291 (1992) 355.

26. H. Yabu, S. Nakamura and K. Kubodera, preprint USC(NT)-93-3.

27. M. Lutz, A. Steiner and W. Weise, preprint TPR-93-19, subm. to Nucl. Phys. A.

28. A. Wirzba, private communication.

$\mid S \mid= 2$ AND $\mid S \mid= 3$ HYPERON PRODUCTION
IN HEAVY ION EXPERIMENTS AT CERN

Domenico Di Bari

Universitá di Bari and Sezione INFN, I-70126 Bari, Italy

INTRODUCTION

The principal aim of heavy ion experiments is to find evidence for a new state of matter, the Quark Gluon Plasma (QGP). It has been suggested that strange baryon and, in particular, strange antibaryon production in such collisions provides a useful probe in the search for a QGP formation.

Up to now, interesting results have been obtained by measurements of $\mid S \mid= 2$ hyperon production in WA85 and NA36 experiments at CERN. Moreover the first signal of $\mid S \mid= 3$ hyperon production has been detected by WA85 experiment.

Strangeness and QGP

Heavy ion collisions actually provide the only way to study, in laboratory, the hadronic matter under extreme conditions. In fact a new state of matter, the Quark Gluon Plasma (QGP), is expected to be created in central heavy ion collisions at ultrarelativistic energy. To investigate the possibility of a QGP formation in such a scenario, many signals have been suggested as useful probes. In particular, in the hypotesis of a phase transition from the normal hadronic matter to deconfined quarks and gluons, an enhancement of strange particles in the final state with respect to the tipical level of the normal hadronic collisions is predicted [1].

Experimentally, enhancement of various strange particle species have been observed in heavy ion collisions when compared to nucleon-nucleon and nucleon-nucleus interactions [2]. However, such enhancement could be explained even in absence of an initial QGP phase: secondary interactions in a very hot hadronic gas could produce, given enough time to achieve chemical equilibrium, a similar straneness enhancement [3,4]. Unlike the kaon and hyperon case, the rescattering mechanism itself could not influence significantly the enhancement of the antihyperons, expecially those carrying

Hot and Dense Nuclear Matter, Edited by
W. Greiner *et al.*, Plenum Press, New York, 1994

more than one units of strangeness [5]. The antihyperon production is in fact too slow a process to play an important role over the short lifetime of the hadronic fireball. On the contrary, in a QGP scenario, both strange baryons and antibaryons could reach easily the chemical equilibrium, due to the low $s\bar{s}$ pair production threshold [6].

Therefore, a systematic study of the hyperon and antihyperon production, for different projectiles masses and energies, is expected to be crucial in understanding a possible QGP formation.

This kind of study is the aim of some heavy ion experiments; in particular, in the following section, I will show some results about the production of Ξ^- and $\overline{\Xi^-}$ ($| S |= 2$) obtained up to now by two experiments at CERN, WA85 and NA36.

Finally, I will present also the first results about the Ω^- and $\overline{\Omega^-}$ ($| S |= 3$) signals in heavy ion interactions, obtained by the WA85 collaboration.

Results about $| S |= 2$ particle production

The detection of multistrange hyperon decays (Ξ^-, Ω^-) in heavy ion interactions is rather difficult because of the large event multeplicities; while many experiments have succeded in measuring $| S |= 1$ strange particle spectra, up to now the Ξ^- spectra has been obtained only by WA85 and NA36 experiments.

WA85 has been designed to study strange particle production, by measuring their charged decay particle, in central S + W collisions at 200 GeV/c per nucleon at $p_T > 1$ GeV/c and central rapidity [7]. New results about Ξ^- and $\overline{\Xi^-}$ production obtained by WA85 collaboration [8] come from the analysis of the total amount of data collected during the last run (1990), which represents a sixfold increase in statistics with respect the previous one [9]. Fig.1 shows the transverse mass distributions for (a)

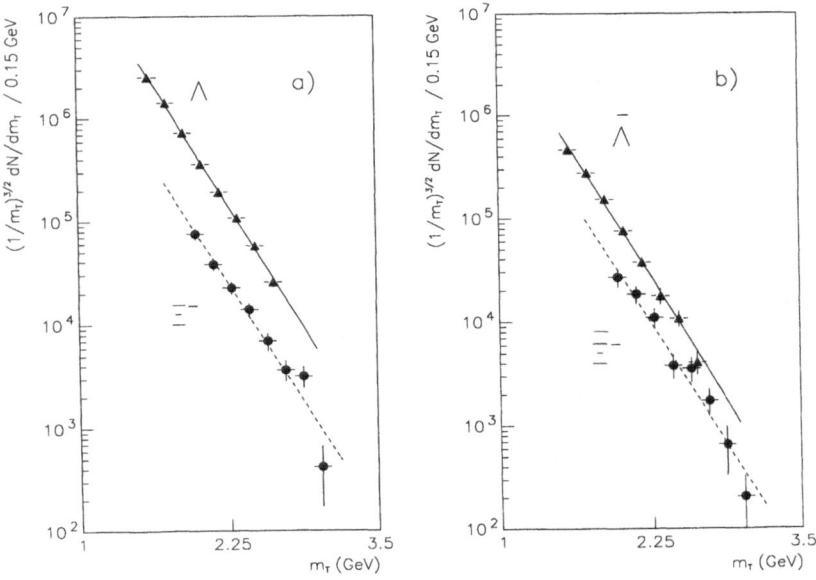

Figure 1: m_T distributions for a) Λ and Ξ^-, and b) $\overline{\Lambda}$ and $\overline{\Xi^-}$.

Λ and Ξ^-, and (b) $\overline{\Lambda}$ and $\overline{\Xi^-}$ hyperons in the the Λ ($\overline{\Lambda}$) distribution being corrected for contamination from Ξ^- ($\overline{\Xi^-}$) decays. The distributions have been fitted to the exponential law $\exp(-\beta m_T)$, as provided by thermal model. The inverse slope $(1/\beta)$ are given in table 1 (the errors are statistical only); they are all compatible with each other and around 230 MeV. From these distributions, the relative hyperon ratios can

Table 1: Inverse slopes of hyperons in S + W interactions (WA85).

Particle	inverse slope (MeV)
Λ	232±3
$\overline{\Lambda}$	230±6
Ξ^-	239±11
$\overline{\Xi^-}$	234±15

be evaluated. Table 2a shows their values in $1 < p_T < 2$ GeV/c and $2.3 < y_{lab} < 2.8$.

The same ratios have been calculated by NA36 collaboration [10], and are shown in table 2b. Nevertheless the comparison between the two sets of results is difficult,

Table 2: Strange hyperon ratios calculated by (a) WA85 and (b) NA36.

Ratio	WA85 (a)	y-p_T	NA36 (b)	y-p_T
$\overline{\Lambda}/\Lambda$	0.20±0.01	2.3 <y< 2.8 1.0<p_T<2.0	0.207±0.014	1.5 <y< 3.0 0.6<p_T<1.6
$\overline{\Xi^-}/\Xi^-$	0.41±0.05	same	0.276±0.108	2.0 <y< 2.5 0.8<p_T<1.8
Ξ^-/Λ	0.09±0.01	same	0.066±0.013	1.5 <y< 2.5 0.8<p_T<1.8
$\overline{\Xi^-}/\overline{\Lambda}$	0.20±0.03	same	0.127±0.022	2.0 <y< 3.0 0.6<p_T<1.6

due to the different phase space windows, which only partially overlap, and different target (S+W WA85, S+Pb NA36); the different y-p_T phase space windows where the ratios Ξ^-/Λ and $\overline{\Xi^-}/\overline{\Lambda}$ are calculated by WA85 and NA36 are schematically shown in Fig.2. It is interesting to compare Ξ^-/Λ and $\overline{\Xi^-}/\overline{\Lambda}$ ratios with those measured by other experiments in pp, $p\bar{p}$ and e^+e^- interactions. As can be seen in Fig.3, while the

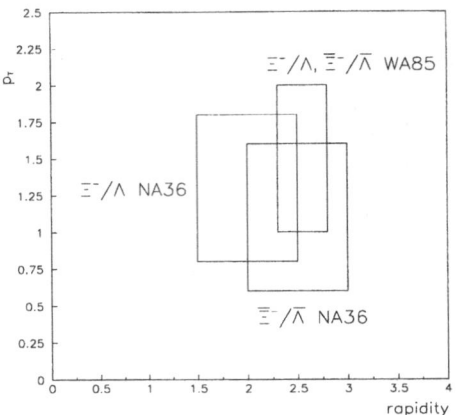

Figure 2: different phase space windows chosen by WA85 and NA36 to calculate hyperon ratios.

Ξ^-/Λ ratio measured by WA85 is compatible with those from other experiments, the $\overline{\Xi^-}/\overline{\Lambda}$ ratio is three times greater than the value obtained by the AFS collaboration (pp interactions) in the same p_T interval, with a statistical significance of four standard deviations. An enhancement of a factor two is seen by NA36, as can be obtained from the table 2b, even if in a different phase space window.

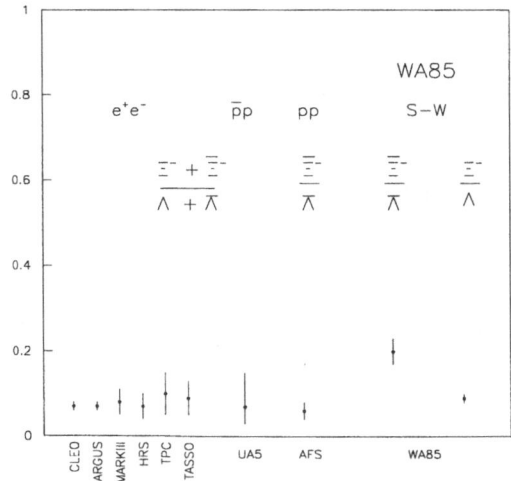

Figure 3: Ξ^-/Λ and $\overline{\Xi^-}/\overline{\Lambda}$ ratios for different experiments

Ω^- signal in heavy ion collisions

The observation of Ω^- and $\overline{\Omega^-}$ in nucleus-nucleus collisions comes from the WA85 experiment, which has also measured the $\overline{\Omega^-}/\Omega^-$ production ratio at central rapidity ($2.5 < y_{lab} < 3.0$) and medium transverse momentum ($p_T > 1.6$ GeV/c) [11].

The Ω^- particles are identified through the two-step decay $\Omega^- \to \Lambda + K^-$, where the Λ also is selected through its charged decay particles.

The selection criteria for the Ω^- decays are described in [11]. However, it should be pointed out that the Ω^- signal is rare (three order of magnitude weaker than Λ signal), and than it is crucial to take under control all known sources of background. In WA85 the main sources of background are summarized as follows:

- uncorrelated Λ-negative track combinations, mainly coming from the target region;

- secondary interactions in air;

- reflection of the Ξ^- peak.

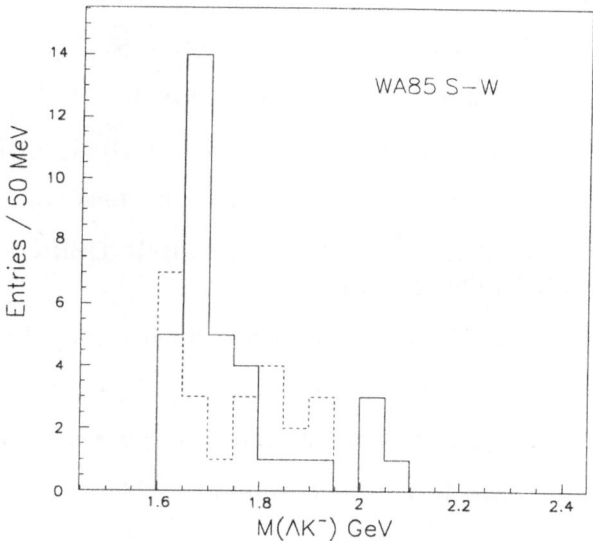

Figure 4: $\Lambda K^- + \overline{\Lambda} K^+$ invariant mass spectra. Solid line = data, dashed line = combinatorial background.

The Fig.4 shows the (ΛK^-) effective mass distribution, unambigous with the Ξ^- ($\overline{\Xi^-}$) reflection, from the full 1990 S+W statistics. A clear peak in the Ω^- mass region can be seen; the dotted line shows the normalized combinatorial background extimate. After subtraction, 7.0 ± 3.6 Ω^-'s and 4.0 ± 2.0 $\overline{\Omega^-}$'s have been left, giving the ratio $\overline{\Omega^-}/\Omega^- = 0.57 \pm 0.41$, in the phase space region $2.5 < y_{lab} < 3.0$ and $p_T > 1.6$

GeV/c. This ratio is, however, uncorrected for possible differences in acceptance and efficiency for Ω^-'s and $\overline{\Omega^-}$'s, even if a big difference between them is not expected.

CONCLUSIONS

In this paper it has been presented a review of some recent results about multi-strange baryon and antibaryon production. In particular $\overline{\Xi^-}/\overline{\Lambda}$ have been calculated by WA85 and NA36, showing a significant enhancement with respect to nucleon-nucleon interactions. A first signal of Ω^- and $\overline{\Omega^-}$ has been detected in heavy ion collisions at high energies by WA85 collaboration. The ratio $\overline{\Omega^-}/\Omega^- = 0.57 \pm 0.41$ has been calculated in the phase space region $2.5 < y_{lab} < 3.0$ and $p_T > 1.6$ GeV/c.

REFERENCES

[1] J. Rafelski and B. Müller, Phys. Rev. Lett. 48 (1982) 1066

[2] E. Quercigh, in "Particle Production from Highly Excited Matter", H.H. Gutbrod and J. Rafelski (eds.), Plenum Press, New York, 1993, pag. 499.

[3] T. Matsui, B. Svetisky and L.D. McLerran, Phys. Rev. D 34 (1986) 2047.

[4] K.S. Lee, M. Rhoades-Browne and U. Heinz, Phys. Rev. C 37 (1988) 1452.

[5] H. C. Eggers and J. Rafelski, Int. J. Mod. Phys. A6 (1991) 1067.

[6] P. Koch, B. Müller and J. Rafelski, Phys. Rep. 142 (1986) 167.

[7] WA85 Proposal, CERN/SPSC/84-76 P206 (1984), CERN/SPSC/87-18 P206 Add. (1987), CERN/SPSC/88-20 P206 (1988).

[8] D. Evans et al., Quark Matter '93 , Borlänge, Sweden, June 1993.

[9] S. Abatzis et al., Phys. Lett. 270B (1991) 123

[10] J.M. Nelson et al., Quark Matter '93 , Borlänge, Sweden, June 1993.

[11] S. Abatzis et al., Phys. Lett. B316 (1993) 615.

DIELECTRON PRODUCTION IN HEAVY ION COLLISIONS BETWEEN 1 AND 13AGeV

L. A. Winckelmann [†*], H. Sorge [†] H. Stöcker [†], W. Greiner [†]

[†] Institut für Theoretische Physik JWG Universität
60054 Frankfurt, Germany
*Gesellschaft für Schwerionenforschung
64291 Darmstadt, Germany

INTRODUCTION

The measurement of electromagnetic probes in todays heavy ion experiments are an unique tool to investigate the properties of hot, compressed nuclear matter such as in medium modifications of hadrons or the nuclear equations of state (EOS) which would yield a deeper understanding of astrophysical processes like supernovae explosions and the physics of neutron stars. The observation of the modification of the ϱ mass in a dielectron experiment at SIS or AGS energies considerably constrains currently available low density approximations to QCD and would serve as a probe of quark and gluon condensates at finite baryon density[1, 2, 3, 4, 5, 6].

Dilepton production in proton and heavy ion induced reactions at low dielectron masses and at moderately high incident energies has been studied experimentally by the DLS-collaboration at the BEVALAC[7, 8, 9, 10] and in theory[12, 13, 14, 15, 16, 17, 18, 19]. Many theoretical works about dielectron production in heavy and light ion induced reactions concentrate on the determination of different dielectron sources by comparison to experimental data. This is a necessary task since dielectron and also diphoton production in heavy ion collisions yields a distinctive insight into the composition of resonance matter, because their sources like ϱ's, η's and ω's have different couplings to baryon resonances: while ϱ mesons couple to nearly all baryon resonances up to a mass of 2GeV, η's couple only to N^*_{1535} and N^*_{1710}. Unfortunately, the coupling of ω mesons to baryon resonances and therefore their production mechanism even in $p+p$ reactions is not clear up to now. The experiments at $E_{kin} = 1\,A$GeV for light systems like p+Be and Ca+Ca seem to be explained by pn Bremsstrahlung, Δ and η Dalitz decays and pion annihilation[16, 17].

Relativistic Quantum Molecular Dynamics (RQMD) has been used to predict the differential dielectron cross section for the interaction of proton on proton and proton on deuteron at 4.9GeV incident energy[18]. Dalitz decays of η's dominate the dilepton mass spectra above the π^0 mass up to 0.5GeV while decays of vector mesons such as ω's and ϱ's produce a peak around the ϱ mass. pn Bremsstrahlung is strongly reduced at these masses. This is due to the forward peak in the differential elastic cross section $\mathrm{d}\sigma/\mathrm{d}t$ at energies above $E_{kin} \approx 1$GeV, as compared to previous calculations. The results for p+Be at 4.9GeV compare well with available DLS-data[18].

The scope of this contribution is to explore the present understanding of dielectron production by the application of a microscopic phase space model (RQMD) in comparison to experimental data for neutral mesons in $p + p$ collisions and dielectron production in light and heavy ion induced reactions, and the introduction of new dielectron sources such as vector meson production by resonance decays and electromagnetic Bremsstrahlung in elastic and inelastic collisions or decays of short lived resonances at BEVALAC and AGS energies. Further we will give an example of the influence of a medium dependent ϱ meson mass on a dielectron mass spectrum in Au on Au at AGS energies.

ELECTROMAGNETIC BREMSSTRAHLUNG OF HADRONS

Since in proton and heavy ion induced reactions they are much more inelastic collisions between nucleons, resonances and/or mesons than elastic pn collisions, previous models for pn Bremsstrahlung[18] are extended in the following to inelastic collisions between arbitrary hadrons and decays of resonances. The transition amplitude for photon production $a + b \rightarrow 1, .., n, \gamma$ is obtained by the so called "soft photon approximation" in the leading order and the limit for $q^0 \rightarrow 0$ by factorisation[20]

$$\mathcal{M}^\gamma(p_a, p_b; p_1..p_n, q) = e\,\epsilon \cdot J \mathcal{M}(p_a, p_b; p_1..p_n) \ . \tag{1}$$

with the current of the in- and outgoing particles

$$J^\nu = \sum_{i=1}^n Q_i \frac{p_i^\nu}{p_i \cdot q} - Q_a \frac{p_a^\nu}{p_a \cdot q} - Q_b \frac{p_b^\nu}{p_b \cdot q} \ . \tag{2}$$

Were Q_i denotes the charge of particle i in units of the elementary charge e. Therefore the cross section for soft photon production can be written in terms of the associated process without electromagnetic radiation as

$$\mathrm{d}\sigma^\gamma = \frac{\alpha}{4\pi^2}|\epsilon \cdot J|^2 \frac{\mathrm{d}^3 q}{q^0}\mathrm{d}\sigma \ . \tag{3}$$

Here $\mathrm{d}\sigma^\gamma$ denotes the cross section for $a + b \rightarrow 1, ..., n, \gamma$ and $\mathrm{d}\sigma$ corresponds to $a + b \rightarrow 1, ..., n$. An extrapolation of $q^0 \mathrm{d}\sigma^\gamma/\mathrm{d}^3 q$ to hard and massive photons can be made by multiplying a fraction of two phase space integrals to eq.(3).

$$\frac{R_2(\bar{s})}{R_2(s)} \quad \text{with} \quad R_2(s) = \frac{1}{16\pi}\sqrt{\left(1 - \frac{(m+m')^2}{s}\right)\left(1 - \frac{(m-m')^2}{s}\right)} \ . \tag{4}$$

This Ansatz has been made first by Gale and Kapusta[12] to evaluate dielectron production by pn-Bremsstrahlung in p+Be collisions at 4.9GeV. Unfortunately the approximations for the t dependence of the differential elastic pn cross section made there leads to a much higher yield as compared to a calculation were the t dependence is fitted to experimental data[18, 19]. $\bar{s}$ has the physical interpretation of an effective (squared) CMS-Energy

$$\bar{s} = (p + p' - q)^2 = s^2 - 2\sqrt{s}q_0 + M^2 \,, \tag{5}$$

were p, p', q are the four momenta of the two colliding hadrons and the radiated photon respectively. We obtain the differential cross section for dilepton production by Bremsstrahlung in terms of the cross section for the process $ab \to X$ were X denotes an arbitrary n-particle final state

$$d\sigma_{ab}^{Xe^+e^-} = \frac{\alpha^2}{12\pi^3}|\epsilon \cdot J|^2 \left(1 + \frac{2m_\ell^2}{M^2}\right)\sqrt{1 - \frac{4m_\ell^2}{M^2}}\frac{R_2(\bar{s})}{R_2(s)}\frac{d^3q}{q^0}dM^2d\sigma_{ab}^X \,. \tag{6}$$

m_ℓ and M denotes the lepton mass and the invariant mass of the pair respectively. Analog we obtain the differential decay rate for dilepton production by Bremsstrahlung in terms of the decay rate for the process $a \to X$ by the replacement $d\sigma_{ab}^{\cdots} \to d\Gamma_a^{\cdots}$ and $\sqrt{s} \to m_a$ and by time reversal symmetry the cross section for electromagnetic Bremsstrahlung with only one hadron in the final state in terms of the cross section for the production of a resonance $ab \to c$ by $m_a, m_b \to m, m'$ in eq.(4). Note that we do not approximate the squared current $|\epsilon \cdot J|^2$ but keep its exact relativistic dipol structure by using the four momenta generated by RQMD.

NEUTRAL MESON PRODUCTION IN THE RQMD-MODEL

The RQMD-model describes particle dynamics in a covariant way, including collisions between the ingoing nucleons, decays of particles and secondary rescattering [23]. Stable mesons and baryons are included as well as their well established higher resonances. Mesons and baryons can be produced according to experimental branching ratios by meson and baryon resonance decays. Resonances are excited in nucleon-nucleon, meson-nucleon, meson-meson collisions as well as in collisions between mesons/baryons and/or resonances themselves[24].

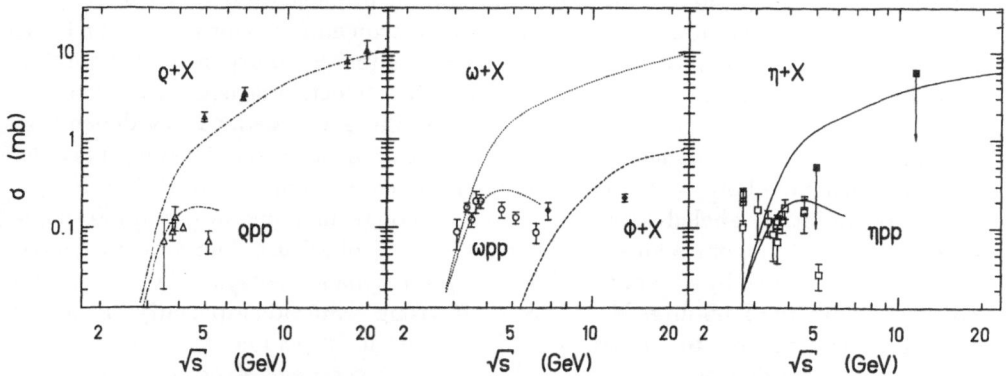

Figure 1. Cross sections for inclusive and exclusive meson production in pp collisions as implemented in the RQMD model via the resonance excitation mechanism described in[22, 24]. An

additional branch of the N^*_{1990} resonance into ωN of 45% has been included to fit the ωpp cross section. For the ϱ meson the experimental data[25] are compared to the symmetric Lorentz component of the ϱ meson mass spectrum from RQMD at each energy. η's are emitted via decays of N^*_{1535} and N^*_{1710} resonances.

Figure 1 shows cross sections for inclusive and exclusive meson production in pp collisions as implemented in the RQMD model. Here an additional branch of the N^*_{1990} resonance into ωN of 45% has been included to fit the exclusive cross section for $pp \mapsto \omega pp$. At higher energies most mesons stem from string decays. The experimental data[25] for the ϱ meson are compared to the symmetric Breit-Wigner component of the ϱ meson mass spectrum from RQMD at each energy. This is necessary since these data are fitted to the symmetric Breit-Wigner component in the di-pion mass spectrum, while especially at lower energies a lot of ϱ's are produced "off-shell", i.e. there mass spectrum is strongly phase space modified, which leads to an enhancement at masses below $m_\varrho = 770 \mathrm{MeV}$. η's are emitted via decays of the N^*_{1535} and N^*_{1710} resonance at low energies. Note that at energies regarded here the inclusive channel is much higher then the — measured — inclusive channel.

DIELECTRON PRODUCTION IN LIGHT SYSTEMS

Dielectron production in light systems at $4.9 \mathrm{GeV}$ is strongly linked to the production of dielectrons in moderate and heavy systems at energies around 1 and $2A\mathrm{GeV}$: while in the latter case heavy baryon resonances are excited via multi step processes, they are directly excited in light systems at $4.9 \mathrm{GeV}$. This baryon resonances can radiate mesons such as η's, ϱ's and ω's which can decay into a dilepton pair. For example in the present model in $p + p$ at $E_L = 4.9 \mathrm{GeV}$ 15% of al η's are radiated by decays of the N^*_{1710} resonance while in central Au+Au reactions at $1.6 \mathrm{GeV}$ this value is 10%. Therefore the comparison of any microscopic model model to data for neutral mesons and especially dielectrons at $4.9 \mathrm{GeV}$ is an important proof also for dielectron production in heavy ion reactions at SIS/BEVALAC energies since in both cases a very similar phase space for particle production is reached.

ϱ and ω meson masses are distributed randomly according to their width and may be absorbed by resonance formation. This differs from Wolf et al.[16] were ϱ mesons can only be created by pion annihilation and are not propagated dynamically.

In Figure 2 the differential cross section for dielectron emission for $p+p$ and $p+d$ at $4.9 \mathrm{GeV}$ versus invariant mass of the dielectron is compared to experimental data[7]. Dalitz decays of the Δ_{1232}[12] are included as well as electromagnetic decays of the following particles: N^*_{1520}, η, ω[17, 18, 21]. The differential decay rates depend on the invariant mass of the dielectron and on the actual mass of the decaying particle. ω mesons contribute both by a direct vector meson conversion and by Dalitz decays.

Pion annihilation labeled separately by $\triangleright$ — though included in our microscopic monte carlo simulation for $p+d$ — yields only $\sim 1.5\%$ of all ϱ's. The ϱ decay contribution is overwhelmed by ϱ's emitted from baryon resonance decays. The calculation is corrected with the DLS-filter V2.0. Note the strong reduction especially in the low mass region as compared to our prediction[18] were the DLS-filter V1.6 was used. In the higher mass region our calculation shows a stronger peak structure due to the decays ϱ's ($+$) and ω's ($\triangleleft$). This is caused by the DLS-resolution of $15\% * M$ which is not yet included in the present DLS-filter V2.0. Pion annihilation labeled separately by $\triangleright$ — though included in our microscopic monte carlo simulation for $p+d$ — yields only $\sim 1.5\%$ of all ϱ's. The ϱ decay contribution is overwhelmed by ϱ's emitted from baryon resonance decays.

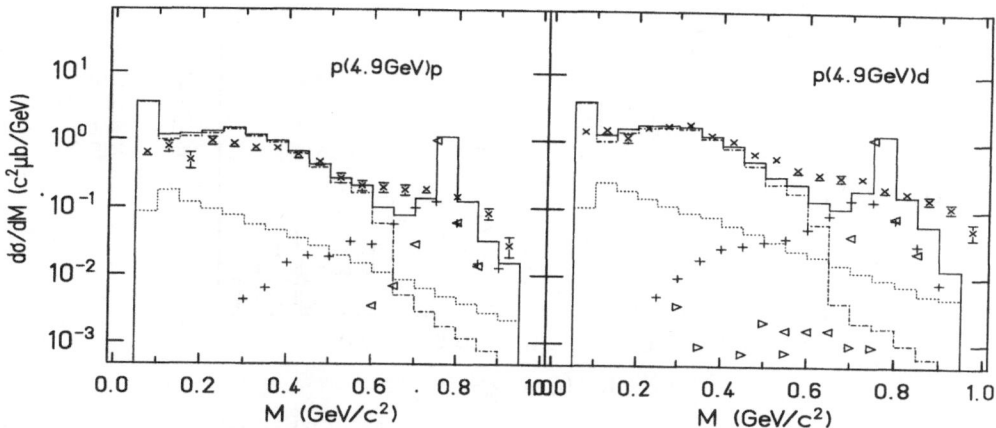

Figure 2. The differential cross section for dielectron emission for $p + p$ and $p+d$ at $4.9\,\mathrm{GeV}$ versus invariant mass of the dielectron is compared to experimental data[7] ($\times$). The dashed-dotted line refers to the sum of all Dalitz decays. The crosses $+$ refer to decays of ϱ mesons. The dotted lines gives the sum of all contributions from the electromagnetic Bremsstrahlung. The solid line is the sum of all. The calculation is corrected with the DLS-filter V2.0. Note the strong reduction especially in the low mass region as compared to our prediction[18] made with the DLS-filter V1.6. Our calculation shows a slightly stronger peak structure in the ϱ/ω mass region. In $p+d$ only 1.5% of all ϱ mesons are formed via dynamical annihilation of co-moving pions ($\triangleright$).

This can be compared with calculations[13, 19] were a source that yields the experimental pion spectra is found to be the major contribution to the dilepton mass spectrum for this system in a mass region around $\sim 0.5 - 1\,\mathrm{GeV}$. However, since we use a microscopic model we can distinguish between a pion annihilation that proceeds via the dynamical formation of the ϱ meson and a ϱ meson radiated by a baryon resonance decay. Thus our ϱ meson direct decay contribution is compatible with the "pion annihilation" signal of refs.[13, 19].

Our result deviates strongly from the contribution of "hadronic decays" shown in ref.[9]. The reasons are that there only the measured exclusive cross sections ($pp \to$ pp+meson) are used and further in case of the ϱ meson the "off shell" production as mentioned in the previous section is not included.

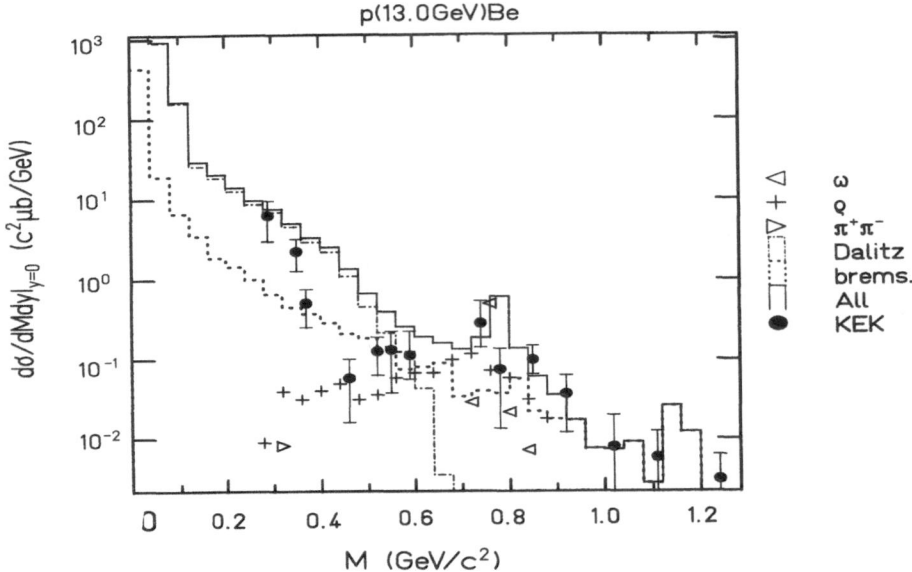

Figure 3. The differential cross section for dielectron emission at lab rapidity $y = 0$ for p+Be at $13\,GeV$ versus invariant mass of the dielectron is compared to experimental data[11]. The crosses + refer to decays of ϱ mesons. ω mesons contribute both by a direct vector meson conversion and by Dalitz decays. The dotted lines gives the sum of all contributions from electromagnetic Bremsstrahlung. The solid line is the sum of all.

In Figure 3 the differential dielectron cross section at target rapidity $y = 0$ versus invariant dielectron mass for p+Be at $13A\,GeV/c$ is compared to experimental data[11]. In the low mass region up to 400MeV the Dalitz decay of the η and ω gives the dominant contribution. In contrast to light and heavy systems at SIS and BEVALAC energies electromagnetic Bremsstrahlung of hadrons is the strongest background source for dielectron masses higher than ~ 400MeV. The comparison to the KEK data seems to indicate that the background contributions from Dalitz decays and electromagnetic Bremsstrahlung in small systems are well understood within our model.

RHO MESON PRODUCTION IN MODERATE AND HEAVY SYSTEMS: PION ANNIHILATION VERSUS RESONANCE DECAYS

In the previous sections we have seen that ϱ meson production via resonance decays in pp collisions is in accordance with existing experimental data and necessary to understand dielectron mass spectra for light systems up to KEK energies. In heavy systems ϱ mesons are not only emitted in resonance decays but also created by pion annihilation. The latter source was assumed to yield the dominant contribution for moderate and heavy systems[16]. However, since the decays of ϱ mesons radiated in resonance decays determine completely the dielectron mass spectrum in the region around $\sim 500MeV$ for light systems, we also consider them in moderate and heavy systems and extract the corrections as compared to a calculation including only pion annihilation.

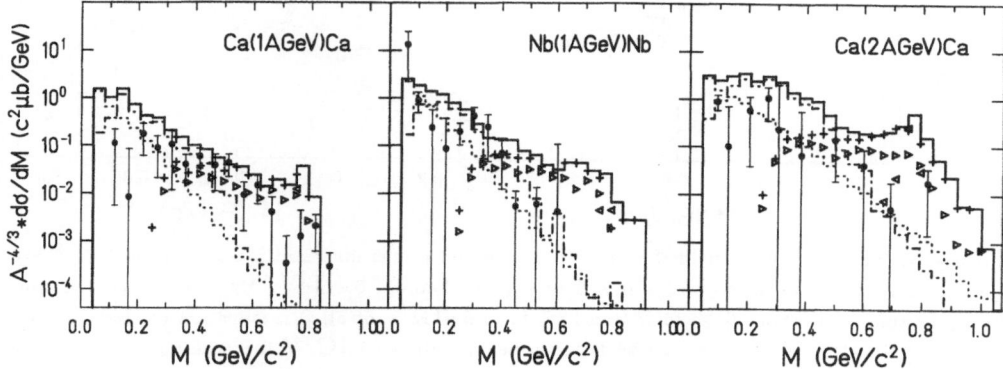

Figure 4. The differential cross section for dielectron emission for Ca+Ca at 1 and $2AGeV$ and Nb+Nb at $1AGeV$ versus invariant mass of the dielectron is compared to experimental data[7, 10]. The calculation is corrected with the DLS-filter V1.6. The dashed-dotted line refers to the sum of all Dalitz decays. The crosses + refer to decays of ϱ mesons. The contribution to the decays of ϱ mesons from pion annihilation is shown separately, labeled by $\triangleright$. The dotted lines give the sum of all contributions from the electromagnetic Bremsstrahlung. The solid line is the sum of all. Only about 64% at $1AGeV$ and 50% at $2AGeV$ of the decaying ϱ mesons stem from pion annihilation. The inclusion of new dielectron sources such as ϱ mesons emitted by baryon resonances and electromagnetic Bremsstrahlung in various hadronic reactions leads to an incompatibility to the experimental data in the ϱ mass range.

In Figure 4 the differential dielectron cross section versus invariant dielectron mass for Ca+Ca at 1 and $2AGeV$ and Nb+Nb at $1AGeV$. is compared to experimental data[7]. In the low mass region up to $500MeV$ Dalitz decays of $\eta's$ and $\Delta's$ and electromagnetic Bremsstrahlung of hadrons give the dominant contributions. The consideration of reasonable structures in the differential cross section according to eq.(3) leads to a strong reduction of bremsstrahlung as compared to ref.[16]. At Ca($2AGeV$)Ca the slope of the mass distribution is much harder as compared to the same system at $1AGeV$, were only a "ϱ shoulder" instead of a "peak" as for $2AGeV$ develops. Further, a strong correction of the contribution of ϱ meson decays from resonance formation is predicted from our calculation as compared to previous results from microscopic models[16], i.e. only 55% at $1AGeV$ and 47% at $2AGeV$ for Ca+Ca and 57% for Nb+Nb of the decaying ϱ mesons stem from pion annihilation. The inclusion of these new dielectron sources in our calculation leads to an incompatibility at the ϱ/ω mass: thus one might expect that in medium effects of the ϱ meson are already achieved at the BEVALAC but not discovered by the DLS due to insufficient statistics. Therefore new high statistic dielectron experiments at these energies are strongly required.

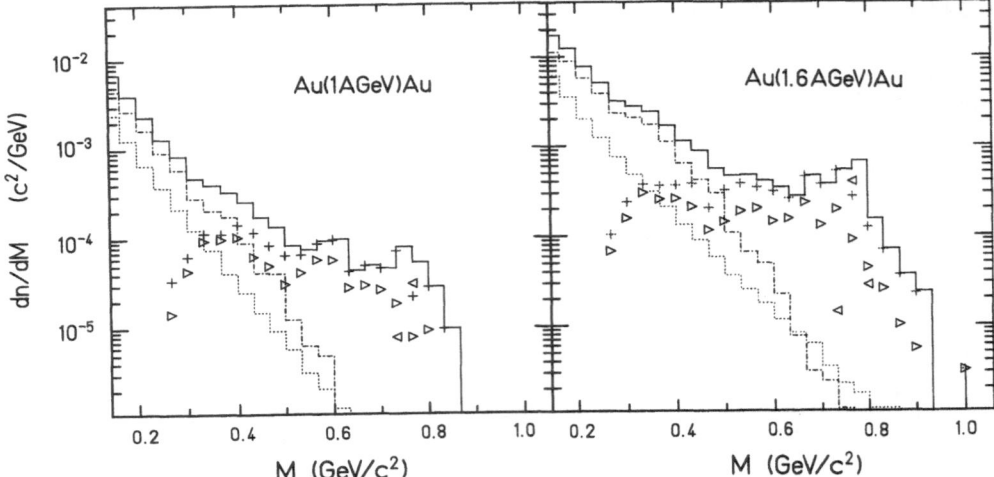

Figure 5. The differential yields for dielectron emission versus invariant mass of the dielectron are shown for central collisions ($b < 1$fm) of Au+Au at 1 and $1.6AGeV$. Even in this central and heavy system only about 64% for 1 and 53% for $1.6AGeV$ of all ϱ mesons are created by pion annihilation. At $1.6AGeV$ the ω exeeds the ϱ peak, while at $1AGeV$ it is still hidden under the ϱ shoulder within our resolusion.

Figure 5 shows differential dielectron yields versus invariant dielectron mass for Au+Au in central collisions at 1 and $1.6AGeV$ incident kinetic energy. These systems will possibly be measured with a future high acceptance dielectron spectrometer (HADES) at GSI's SIS[26]. Again the yields below $M \sim 500MeV$ are dominated by Dalitz decays of the η and Δ as well as electromagnetic Bremsstrahlung from hadronic reactions. Here only about 64% at $1AGeV$ and 53% at $1.6AGeV$ of all decaying ϱ mesons stem from pion annihilation. This indicates that at these energies the extraction of the contribution from pion annihilation can not be made by increasing the mass of the system or triggering on central collisions.

Figure 6 shows background contributions for the differential dielectron cross sections versus invariant dielectron mass for Au+Au at $10.7AGeV$ incident kinetic energy. The yields below $M \sim 500MeV$ are dominated by the Dalitz decays of the η. The pn Bremsstrahlung component is suppressed by more than two orders of magnitude as compared to contributions from Dalitz decays and Bremsstrahlung from baryon-baryon collisions including inelasticities.

DIELECTRONS AS A PROBE OF CHIRAL SYMMETRY RESTORATION

Vector meson properties, i.e. mass and width — easily accessible in dielectron mass spectra — are expected to be strongly modified in a dense and hot baryonic environment. While boson exchange models[3, 4, 5] yield only a small change in mass but a substantial increase of the width, QCD-sum rule (QSR) calculations[2] predict a linear dependence of the pole position of the ϱ meson propagator on the nuclear density:

$$m_\varrho(\rho/\rho_0) = m_\varrho(0)(1 - \lambda(\rho/\rho_0)) \ . \tag{7}$$

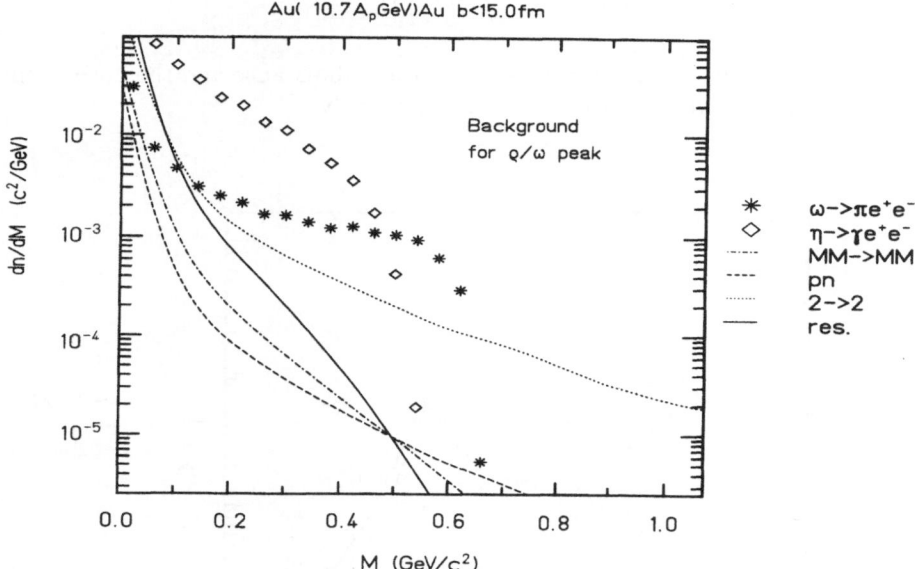

Figure 6. The background contributions for the differential cross section of dielectron emission versus invariant mass of the dielectron is shown for Au+Au at $10.7\,AGeV$. The dashed-dotted line refers to the sum of all Dalitz decays. Various Bremsstrahlung contributions are shown separately: $MM \rightarrow MM$: meson meson Bremsstrahlung, pn: pn Bremsstrahlung, $2 \rightarrow 2$: sum of all Bremsstrahlung sources involving two in and outgoing hadrons. This channel includes also the so called "inelastic Bremsstrahlung"(i.e. for example $NN \rightarrow \Delta Ne^{\pm}$), res: Bremsstrahlung by decays or creation of resonances (i.e. for example $N^*_{1990} \rightarrow N\pi e^{\pm}$ or $KN \rightarrow \Sigma^* e^{\pm}$). The most important Dalitz decay contributions: $\diamond$: $\eta \rightarrow \gamma e^{\pm}$, $*$: $\omega \rightarrow \pi^0 e^{\pm}$.

Here ρ_0 denotes the ground state density of nuclear matter, and $\lambda = 0.18[1, 3]$. Recently, the scattering length of the ϱ meson has been calculated[6]. However, the result obtained there leads to a mass shift of $+60 MeV$ which yields $\lambda = -0.078$ in contrast to refs.[1, 2, 3]. Both these results are obtained in the dilute gas approximation for a ϱ meson at rest, their extension to large densities and non vanishing momentum is not clear. A separate question is whether — a sizable density dependence assumed — a rho mass shift would show up over the background of other dielectron sources. This will be studied in the following.

In the RQMD-model the baryon density ρ is given as the sum over the one particle densities represented as gaussians boosted into the rest frame of the decaying ϱ meson:

$$\rho = (\pi L)^{-3/2} \sum_{\text{baryons}} \gamma \exp\left[-\left(x_\perp^2 + \gamma^2 x_\parallel^2\right)/L\right] , \qquad (8)$$

were $x_\perp$, $x_\parallel$ and γ denote the transverse, longitudinal distance and the inverse dilatation between the baryon and the ϱ meson respectively.

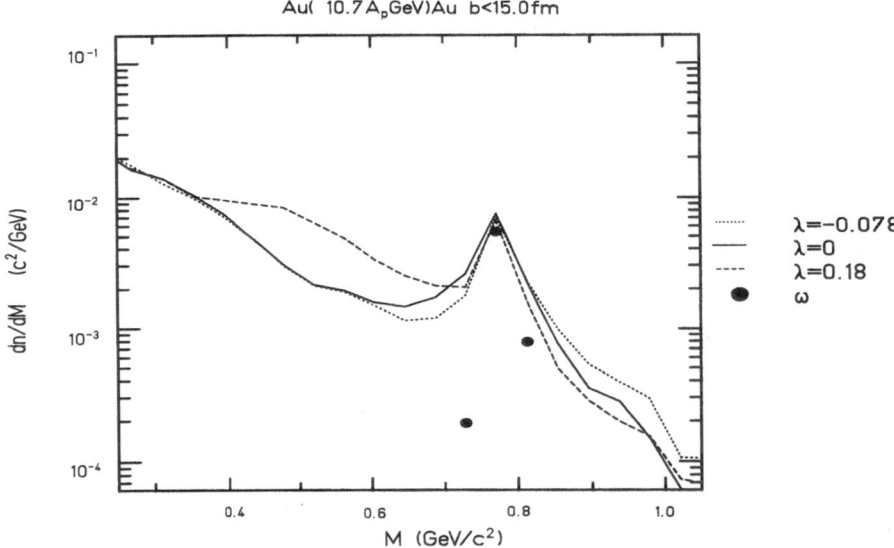

Figure 7. Dielectron mass spectrum in Au($10.7 AGeV$)Au for decreasing, constant and increasing density dependence of the ϱ meson mass: $\lambda = 0.18$ dashed line, 0 solid line, -0.078 dotted line. The ω meson is not affected since its lifetime is much larger as compared to the ϱ meson: only 1% of all ω mesons decay at a density larger than $0.2\rho_0$.

In Figure 7 the dielectron mass spectrum for decreasing, constant and increasing ϱ meson masses is displayed ($\lambda = 0.18, 0, -0.078$ in eq.(7)). However, a similar equation holds also for the ω meson, but since its lifetime is much larger as compared to the ϱ meson only 1% of all ω mesons decay at a density larger than $0.2\rho_0$. The inclusion of QSR motivated parametrisations of the ϱ meson mass in our calculations leads to a change in the peak structure: the peak at the ω meson mass gets sharper as compared to the calculation without a density dependent ϱ mass.

In conclusion, we have shown that ϱ mesons from resonance decays leads to a correction by approximately a factor of two in heavy ion collisions with respect to previous applications of microscopic models at BEVALAC energies. For central collisions at $1.6 A GeV$ and peripheral collisions at $2 A GeV$ the ω exeeds the moderately pronounced ρ peak, while at $1 A GeV$ it is still hidden under the broad ϱ shoulder. A measurement of the modification of the ϱ mass in a dielectron experiment at AGS energies considerably constrains currently available low energy approximations to QCD and would serve as a probe of quark and gluon condensates or the creation of large baryon densities.

REFERENCES

[1] G. E. Brown and M. Rho, *Phys. Rev. Lett.* **66**(1991)2720.

[2] T. Hatsuda and S. H. Lee, *Phys. Rev.* **C46**(1992)R34.

[3] M. Asakawa, C. M. Ko, P. L'evai and X. J. Qui, *Phys. Rev.* **C46**(1992)R1159.

[4] G. Chanfray and P. Schuck, Nucl. Phys. A 489(1992)271c.

[5] M. Herrmann, B. L. Frimann and W. Nörenberg, *Z. Phys.* **A343**(1992)119.

[6] Y. Koike, Michigan State University preprint no. MSUCL-898.

[7] G. Roche et al., *Phys. Lett.* **B226**(1989)228; C. Naudet, *Phys. Rev. Lett.* **62**(1989)2652; P. A. Seidl, *Nucl. Phys.* **A525**(1991)299c; J. Carrol, H. Huang, L. S. Schroeder, K. Wilson and the DLS-collaboration, private communication.

[8] A. Letessier-Selvon et al., *Phys. Rev.* **C40**(1989)1513.

[9] H. Z. Huang et al., *Phys. Lett.* **B297**(1992)233.

[10] S. Beedoe et al., *Phys. Rev.* **C47**(1993)2840

[11] S. Mikamo et al., *Phys. Lett.* **B106**(1981)428.

[12] C. Gale and J. Kapusta, *Phys. Rev.* **C35**(1987)2107; C. Gale and J. Kapusta, *Phys. Rev.* **C40**(1989)2397.

[13] J. Kapusta and P. Lichard, *Phys. Rev.* **C40**(1989)R1574.

[14] M. Schäfer, T. S. Biró, W. Cassing and U. Mosel, *Phys. Lett.* **B221**(1989)2397.

[15] K. Haglin, J. Kapusta and C. Gale, *Phys. Lett.* **B224**(1989)433; K. L. Haglin, *Ann. Phys.* **212**(1991)84.

[16] L. Xiong, J. Q. Wu, Z. G. Wu, C. M. Ko, and J. H. Shi, *Phys. Rev.* **C41**(1990)R1355; L. Xiong, Z. G. Wu, C. M. Ko, and J. Q. Wu, *Nucl. Phys.* **A512**(1990)772; G. Wolf, G. Batko, W. Cassing, U. Mosel, K. Niita, M. Schäfer, *Nucl. Phys.* **A517**(1990)615.

[17] L. A. Winckelmann, A. Jahns, A. v. Keitz, Th. Schönfeld, R. Mattiello, H. Sorge, H. Stöcker, W. Greiner, Proceedings of the Symposium " Physics with the PS Collider ", KEK, Japan (1990); A. Jahns, L. A. Winckelmann, R. Mattiello, A. v. Keitz, Th. Schönfeld, H. Sorge, H. Stöcker, W. Greiner, Proceedings of the XXIX International Winter Meeting on Nuclear Physics, ed. I. Iori, Bormio (Italy), p. 474 (1991).

[18] L. A. Winckelmann, H. Stöcker, W. Greiner, and H. Sorge, *Phys. Lett.* **B298**(1993)22.

[19] K. Haglin and C. Gale, *Phys. Rev.* **C49**(1994)401

[20] N. S. Craigie, H. N. Thompson, *Nucl. Phys.* **B141**(1978)121.

[21] L. G. Landsberg, *Phys. Rep.* **128**(1985)301.

[22] H. Sorge, H. Stöcker, W. Greiner, *Ann. Phys.* **192**(1989)266.

[23] H. Sorge, A. v. Keitz, R. Mattiello, H. Stöcker and W. Greiner, *Z. Phys.* **C47**(1990)629; A. von Keitz, L. A. Winckelmann, A. Jahns, H. Sorge, H. Stöcker, W. Greiner, *Phys. Lett.* **B263**(1991)353.

[24] H. Sorge, L. A. Winckelmann, H. Stöcker, W. Greiner, *Z. Phys.* **C59**(1993)85.

[25] Flaminino et al., CERN-HERA report-84-1 and references therein.

[26] H. Neumann for the HADES coll., Proceedings of the International Workshop on Gross Properties of Nuclei and Nuclear Excitations XXI, ISSN 0720-8715.

STRING-PARTON MODEL DESCRIPTION OF RELATIVISTIC HEAVY-ION COLLISIONS

A. S. Umar[2,1], D. J. Dean[3], and M. R. Strayer[1]

[1]Center for Computationally Intensive Physics
Physics Division, Oak Ridge National Laboratory, Oak Ridge,
TN 37831-6373
[2]Vanderbilt University, Department of Physics & Astronomy,
Nashville, TN 37235
[3] W. K. Kellogg Radiation Laboratory, California Institute of
Technology, Pasadena, CA 91125

INTRODUCTION

Various models have been developed to address the ordinary hadronic physics that occurs in relativistic heavy-ion collisions. These include string-based fragmentation models such as the LUND model[1], and its extensions in FRITIOF[2], which assume that excited hadrons behave as a chain of color dipoles that move like one-dimensional relativistic strings. Interactions are introduced via multiple small momentum exchanges between the color dipoles of two overlapping strings. Other nondynamical models are the dual-parton model[3], in which the strings are formed by soft gluon exchange between the valence partons of the colliding hadrons. The quark-gluon string model[4] (QGSM), also based on the dual parton model, has been developed to study soft parton collisions, and includes rescattering. The *strings* in the above models are in fact one-dimensional constructions in momentum space, and string evolution is carried out in this space. They are sometimes referred to as the longitudinal phase space models. Any coordinate space quantities that these models may study come from transformations from momentum space one-dimensional string coordinates to configuration space. Relativistic quantum molecular dynamics (RQMD) calculations have also been performed to study relativistic collision phenomena[5]. This approach combines resonance formation and decay of light hadronic states, and one-dimensional string fragmentation (LUND model) for very heavy resonances. RQMD follows the full space-time evolution of the light hadronic states, and uses one-dimensional momentum space evolution for the heavy states via the LUND string description.

We have developed a real-time dynamical model for studying the inclusive properties of hadronic collisions in three-dimensions[6, 7, 8, 9]. The model is based on the Nambu-Gotō string description of hadrons supplemented by extensions to incorporate the basic features of the parton model, together with a hadronization mechanism. The advantage of having a fully dynamical model is the possibility to explore the detailed time evolution of the hadronic matter. Furthermore, since the model is fully three dimensional, all of the transverse degrees of freedom are available to the evolving system. The model has been successfully applied to the study of high-energy e^+e^-, pp[6], μp, and μA[7, 8] collisions. These calculations have been used to fix all of the parameters present in the model. For all of the calculations presented here, the model parameters have been held fixed at their predetermined values.

STRING-PARTON MODEL ESSENTIALS

In building the phenomenology of the dynamical string-parton model description of relativistic heavy-ion collisions, it is desirable to start from a description which entails many of the features observed for elementary high-energy processes. One of the most important properties of hadrons is their substructure observed mainly via deep-inelastic charged lepton-hadron collisions. In the parton model the scaling behavior is explained in terms of the presence of point-like charged constituents generically called partons. It can be shown in the infinite momentum frame that the scaling variable x is the fraction of the momentum of the nucleon carried by the struck parton. This relation is true only in this frame; however, it is approximately valid in other frames, if the partons are assumed to be massless. Corrections arising from the finite parton mass are usually neglected, as well as the small differences between neutron and proton distribution functions. The string-parton model reproduces the observed structure functions by utilizing the endpoint dynamics of strings; at any given instant the energy and momentum is shared by the string segment (string shrinks or stretches) and the endpoints. This allows the natural identification of these endpoints as partons belonging to the hadron described by the entire string. This identification was first used in the string-parton model to establish a connection between strings and parton structure functions. For the description of baryons, one end represents a single quark whereas the other end a diquark. Each quark carries a baryon number of 1/3 thus giving $B = 1$ for baryons. The description of mesons involves a quark at one end, and an antiquark at the other. For relativistic strings at rest, it is natural to define a fractional momentum variable associated with the string endpoints. Assuming collinear motion along the z-direction, which will be the boost axis, we define the string longitudinal momentum fraction in terms of the ratio of the light-cone variables,

$$x_s = \frac{k_0 + k_3}{P_0 + P_3} \, , \tag{1}$$

where k is the endpoint four-momentum and P is the total string four-momentum. The variable x_s is Lorentz invariant for boosts in the longitudinal direction. We have shown[6] that different dynamical states of motion of the strings give rise to different fractional momentum distributions of the string endpoints, which are identified as massless partons. An ensemble average of different dynamical states of the strings accurately reproduces the valence quark structure function of the proton[6].

The initial setup of a nuclear collision is done as follows: nuclei are randomly generated within an impact parameter range by employing a Fermi-density distribution. The nuclear volume is then populated with nucleons, simulated by strings, such that no two nucleons overlap in the initial state. Each nucleon is initialized according to a Monte-Carlo sampling of the strings so as to reproduce the correct parton distribution functions, as described above. The nuclear impact parameter, b, may be fixed at a particular value, or given a range, and is distributed as $2\pi b db$. The two nuclei are then boosted with the collision γ, and the time evolution proceeds via the string equations of motion.

In order to simulate hadron-hadron collisions, the strings must interact with each other. This interaction mechanism should lead to excited strings which must then decay via a suitable hadronization mechanism. An assumption is made in the string-parton model, namely that the final state interactions, which confine the partons, act at large space-time distances of the order of the hadron size, much larger than the parton size and the time scale of the current parton interactions. Then, during the time of interaction the parton can be regarded as quasi-free, and the cross section calculated. In the string-parton model all string-string interactions are specified via effective parton-parton scattering and exchange. The probability that an interaction of two strings takes place has the form

$$W_{AB} \sim \rho_{i/A}(x_A)\rho_{j/B}(x_B)F_{ij}(\vec{b})\mathcal{P}_{ij} ,$$

where $\rho_{i/A}(x_A)$ denotes the probability density of finding a parton of type i with a given parton momentum fraction x_A in nucleon A. F_{ij} is the parton impact parameter dependence. $\mathcal{P}_{ij}(\theta)$ denotes the probability for scattering two partons with incoming momentum states p_i and p_j, and outgoing momentum states p_1 and p_2, where θ is the angle between p_i and p_1 measured in the center-of-momentum frame of the partons. For a single parton-parton scattering, this probability has the t- and u- channel form as given by first-order perturbative QCD calculations[12], except here it is treated phenomenologically as

$$\mathcal{P}_{ij}(\theta) = \mathcal{A} \int_0^\theta d\theta' \left[\frac{s^2 + u^2}{(t - m_t^2)^2} + \frac{s^2 + t^2}{(u - m_u^2)^2} \right] , \tag{2}$$

where $\mathcal{A}$ is the normalization such that $\mathcal{P}_{ij}(\pi) = 1$. We sample this distribution with Monte-Carlo techniques. We use Monte Carlo techniques to obtain the distribution by randomly choosing $\mathcal{P}_{ij}$ and inverting the equation to find θ. In practice we choose gluon masses $m_t = m_u = 0.25$ GeV which give a range of the interaction corresponding to the 30 mb pp inelastic cross section. The effective scattering results in a very different behavior in comparison to the simple Born scattering with zero gluon mass. The effective scattering probability (2) includes the minijet terms in an average fashion. However, minijet phenomena are not expected to be significant at CERN energies. In order to preserve color neutrality, the gluon exchange is followed by a quark exchange[13]. The above interaction mechanism results in excitations of the strings due to the energy-momentum transfer. In nucleus-nucleus collisions all strings are allowed to interact with each other, including the strings produced by fragmentation as described below. In this sense multiple scattering effects are included in the dynamics of the evolution.

The real-time dynamics of interacting strings must be supplemented by a hadronization mechanism. Here, the string-parton model utilizes the pair-creation followed by a string breakup method which is similar to flux-tube breaking of the strong-coupled QCD calculations[14]. Some experimental evidence is also provided by the studies of jets in e^+e^- and pp collisions. These experiments demonstrate that jets originate from hard quarks and gluons and provide support that fragmentation takes place within color neutral systems, and not from isolated partons[15].

In our simulation of high-energy collisions, string-string interaction mechanisms lead to excited strings which stretch and decay by breaking until they reach a pre-defined minimum mass. Each string may only decay into segments whose masses are above this cutoff mass. The minimum masses are $M_q = 0.28$ GeV for the mesonic strings, and $M_{qq} = 0.94$ GeV for baryon strings. The cutoff masses are determined by reproducing the correct hadron multiplicities and charge distributions in e^+e^- collisions[6]. We also note that the cutoff masses are just the minimum and, in practice, a spectrum of final string masses is produced (simulating excited states). The mass spectra obtained in the string-parton model correctly reproduce the observed distributions. The choice for the spatial decay point along the string is based on the invariant area decay law[16, 17, 6], in which the probability, $\mathcal{P}$, for a small segment of string to decay is a function of the invariant area it sweeps as it propagates, ΔA, and is given by $\mathcal{P} = 1 - \exp(-\Lambda \Delta A)$. The decay constant, Λ, could also be expressed in terms of a proper time interval for decay, $\Lambda = 1/\tau_0^2$, where $\tau_0 = 0.5$ fm/c which was fitted from μp collision data[7].

The pair-creation process is expected, not only to reproduce the longitudinal distributions observed in high-energy collisions, but also to contribute to the transverse momentum distributions. The quark and the antiquark of the created pair could carry equal and opposite nonvanishing transverse momenta. This source of transverse momenta will primarily contribute to the low momentum (approximately $p_T \leq 1.0\ GeV$) part of the total transverse momentum distribution. In the absence of any fundamental calculations, we choose to parameterize the transverse momentum assignment with a simple exponential distribution function

$$f(p_T)p_T dp_T \propto e^{-\alpha p_T} p_T dp_T\ . \tag{3}$$

In practice we have used $\alpha = 3.88\ GeV^{-1}$ which accurately reproduces transverse momentum distributions in high-energy pp collisions[6]. The created quarks are initially virtual and become on-shell by absorbing energy from the string. The virtual quarks may not interact until they become real; however, other quarks on the string are allowed to scatter as the string propagates. If the source of the transverse momentum is that acquired by the created virtual quarks, this directly influences the time taken by these particles to come on-shell. This can simply be viewed as a manifestation of the Landau-Pomeranchuk effect.

COMPARISON WITH DATA

We have applied the string-parton model to collision data for S+S, O+C, S+Al, and O+Au measured at NA35[11] and WA80[10] experiments at CERN. The collisions are followed to 100 fm/c in the collision frame, which is a sufficient amount of time for all particle production to have taken place.

The NA35 data which we have addressed is for S+S collisions in the impact parameter range $0 < b < 2.5$ fm at 200 A GeV[11]. In Fig. 1 the rapidity distributions

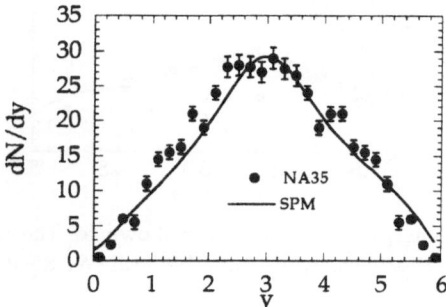

Figure 1: Calculations of the charged pion rapidity distribution for S+S collisions (solid line). Experimental data are from NA35[11] (dots). Details of the calculation are given in the text.

are shown for all charged pions, and in Fig. 2 for the charged baryons. The important aspect of this data is that there has been some filling of the central rapidity region, but for the most part the baryons have not stopped. We have shown by calculations that multiple scattering of the baryons and mesons during the collision process is the principal cause for the Gaussian nature of the meson rapidity distributions, and the primary reason for the observed mid-rapidity nucleons.

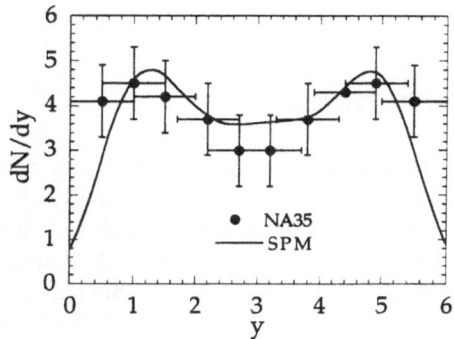

Figure 2: Calculations of the charged baryon rapidity distribution for S+S collisions at 200A GeV (solid line) compared to experimental data from NA35[11] (dots). Details of the calculation are given in the text.

We have also performed calculations to compare with data from WA80[10] experiments. These experiments consider the transverse energy per unit of pseudo-rapidity produced in asymmetric collisions. We have investigated collisions of O+C and S+Al. We show for the O+C and S+Al systems the transverse energy spectra in Fig. 3.

All experimental cuts have been imposed on these calculations. In Fig. 4 we show the cross section $d\sigma/dE_T$ as a function of E_T for the O+C and S+Al systems.

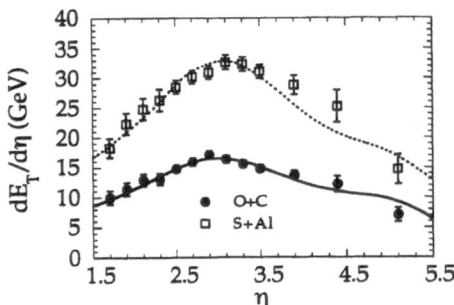

Figure 3: The transverse energy of all hadrons is shown for the O+C and S+Al systems. Calculations are given by the lines, while experimental WA80 data[10] are shown by the symbols listed on the figure.

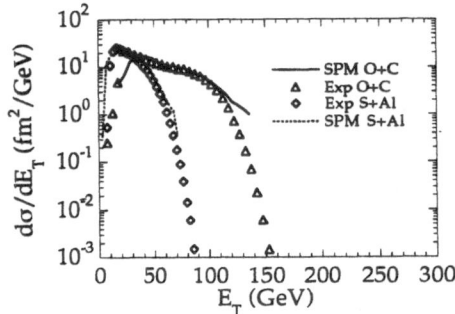

Figure 4: The differential cross section $d\sigma/dE_T$ is plotted as a function of E_T for the O+C and S+Al systems.

DYNAMICAL ENERGY DENSITIES

As a fully dynamical theory, the string-parton model can be used to study the time evolution of the meson energy density produced in relativistic nucleus-nucleus collisions. Several procedures may be used to calculate the central energy density. For example, the quark-gluon string model counts all hadrons, including those not yet formed, in energy density calculations[4]. Thus, this calculation obtains significant energy densities for 160 A GeV Pb+Pb collisions (20 GeV/fm^3 at a time of t=1.05 fm/c). The VENUS model calculates the energy density with on-shell hadrons, using the "rather arbitrary definition of the hadronization point which is defined as the point where two corresponding [produced] partons meet" for the first time[18]. VENUS calculations obtain $\varepsilon_{max} \sim 4$ GeV/fm^3 for 200 A GeV O+Au collisions. The time scale to this maximum density was $\tau \sim 1.0$ fm/c, which is expected since $\tau \propto \kappa$ in this calculation.

In our calculations of the central energy-densities we will consider only produced on-shell mesons. Although this is in the spirit of the calculations of Ref.[18], several differences are explained by the different timescales involved in the problem, and of the dynamics of our decay mechanism, as discussed below. Calculations are performed in the center-of-momentum frame of the produced mesons. In this frame the maximal

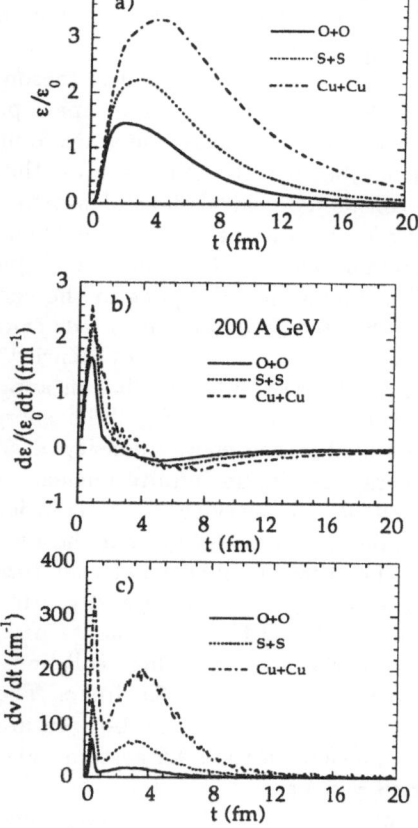

Figure 5: (a) The central meson energy density, defined in the text as a function of time, is plotted for O+O, S+S, and Cu+Cu collisions at 200A GeV. (b) The time derivative of the energy density is shown for the two systems. (c) The collision rate $d\nu/dt$ is shown for the two systems. In (b) and (c) the derivatives are numerical.

longitudinal extent of the volume, ℓ, is determined by the separation of the two leading nuclei from the time of their initial contact. The energy and number densities for a radial shell of area $2\pi r dr$ are then given by $\epsilon(r) = E(r)/(2\pi r dr \ell)$ and $n(r) = N(r)/(2\pi r dr \ell)$, where $E(r)$ and $N(r)$ are the energy and number of mesons contained in the shell. The transverse extent of the volume is obtained by sampling the radial meson distribution, $N(r)$. The meson number and energy densities at a given time are then

$$\langle \epsilon \rangle = \frac{\int N(r)\epsilon(r)d^2r}{\int N(r)d^2r}, \quad \langle n \rangle = \frac{\int N(r)n(r)d^2r}{\int N(r)d^2r} . \tag{4}$$

The calculation of the energy density is primarily influenced by various time scales involved in hadroproduction. These time scales include the interaction time, i.e. the time taken for the two nuclei to cross each other, the decay or excitation time, and the time required for the created virtual mesons to become on-shell. The model correctly reproduces the experimental transverse energy spectra, shown in the previous section, which is the quantity used in the Bjørken formula for predicting the central energy density. For S+S collisions the ratio of baryons to mesons in the central rapidity region is on the order of 5% [11]. Thus they do not contribute significantly to the measured transverse energy spectrum, $dE_T/d\eta$.

In Fig. 5a we show the time dependence of the meson energy density for 200 A GeV O+O, S+S, and Cu+Cu collisions at zero-impact parameter. We note that the maximum energy density in the central region scales roughly as $A^{0.62}$. Statistical fluctuations in the radial number density, $N(r)$ are on the order of 40%. For the Cu+Cu collisions we also considered central slabs of widths 1.0, 2.0, and 3.0 fm, and found insignificant differences with the maximum central energy density obtained as described above. We have also checked that slices in rapidity about $y = 0$ alter the results by only 10-20%. Inclusion of baryons in the central region, ($|y| < 0.5$), increases the energy density by 10%. We have also made calculations for determining the beam energy dependence of the meson energy density. We find a dependence which roughly scales as $\ln \gamma$. These results predict an approximately 2.9 GeV/fm^3 meson energy density for U+U at 100 A GeV collider energy. In Fig. 5b we show the time derivative of the scaled meson energy density, $d(\varepsilon/\varepsilon_0)/dt$, where $\varepsilon_0 = 0.16$ GeV/fm^3, which is the energy density in infinite nuclear matter. We also show in Fig. 5c the collision rate, $d\nu/dt$. An overview of the collision becomes quite clear as we view Fig. 5. For example, for the S+S system the two nuclei begin to overlap at 0 fm/c, as indicated by the onset of collisions at that time. By approximately 0.5 fm/c, the two nuclei are completely overlapping and particle production has begun, as shown by the dm/dt curve. At 1.0 fm/c the energy production rate has reached its maximum, as indicated in Fig 5b. Secondary collisions of baryons and mesons are indicated by the broad peak of $d\nu/dt$ at 3.0 fm/c. The maximum meson mass density appears at roughly the same time as can be seen in Fig. 5a. Note that there are many more secondary collisions in the S+S system. These collisions are actually occurring in the baryon-rich regions.

String-parton model calculations of the maximum energy density for mesons formed during nuclear collisions are smaller than the densities given by estimates obtained from the Bjørken formula[10] by factors of 3 to 4, depending on the nuclear system involved. This formula is based on the assumption of longitudinal rapidity scaling which is not supported by the data at SPS energies[10] or by the NA35 data presented above. The Bjørken formula does not take into account the finite size of the nucleus nor the nuclear collision times. Furthermore, the time scale used in the formula is taken to be the strong interaction time, 1 fm/c, which does not account for the time required for the virtual particles to become on-shell.

This research was sponsored in part by the U.S. Department of Energy under contract No. DE-AC05-84OR21400 managed by Martin Marietta Energy Systems, Inc., and under contract No. DE-FG05-87ER40376 with Vanderbilt University.

REFERENCES

1. B. Andersson and G. Gustafson, Z. Phys. C **3**, 223 (1980); B. Andersson, G. Gustafson, G. Ingelman, and T. Sjostrand, Phys. Rep. **97**, 33 (1983).

2. B. Andersson, G. Gustafson, and B. Nilsson-Almqvist, Nucl. Phys. **B281**, 289 (1987).

3. K. Werner, Z. Phys. C **42**, 85 (1989).

4. N. S. Amelin, E. F. Staubo, L. P. Csernai, V. D. Toneev, K. K. Gudima, and D. Strottman, Phys. Lett. **B261**, 352 (1991).

5. H. Sorge, A. v. Keitz, R. Mattiello, H. Stöcker, and W. Greiner, Z. Phys. C. **47** 629 (1990).

6. D. J. Dean, A. S. Umar, J. -S. Wu, and M. R. Strayer, Phys. Rev. C **45**, 400 (1992).

7. A. S. Umar, D. J. Dean, and M. R. Strayer, in Proceedings of *Quark Matter '91*, edited by T. C. Awes, F. E. Obenshain, F. Plasil, M. R. Strayer, and C. Y. Wong, Nucl. Phys. **A544**, 475c (1992).

8. D. J. Dean, M. Gyulassy, B. Müller, E. A. Remler, M. R. Strayer, A. S. Umar, and J.-S. Wu, Phys. Rev. C **46**, 2066 (1992).

9. D. J. Dean, A. S. Umar, and M. R. Strayer, Intl. Jour. of Mod. Phys. E.

10. R. Albrecht, et al., WA80 Collaboration, Phys. Rev. C **44**, 2736 (1991).

11. H. Ströbele et al., NA35 Collaboration, Nucl. Phys. **A525**, 59c (1991).

12. R. Cutler and D. Sivers, Phys. Rev. D **17**, 196 (1978).

13. N. Isgur, Nucl. Phys. **A497**, 91c (1989); K. Maltman and N. Isgur, Phys. Rev. D **29**, 952 (1984).

14. R. Kokoski and N. Isgur, Phys. Rev. D **35**, 907 (1987); G. A. Miller, Phys. Rev. D **37**, 2431 (1988); P. Geiger and N. Isgur, Phys. Rev. D **41**, 1595 (1990).

15. P. Mättig, Phys. Rep. **177**, 141 (1989).

16. K. Sailer, B. Müller, and W. Greiner, J. Mod. Phys. **A4**, 437 (1989); K. Sailer, B. Müller, and W. Greiner, Proc. of *The Nuclear Equation of State*, ed. W. Greiner and H. Stöcker, (Plenum, New York, 1990) p.531.

17. E. A. Remler, Proc. of *Gross Properties of Nuclei and Nuclear Excitations*, Hirschegg, 1987, p.24.

18. K. Werner, Phys. Lett. **B219**, 111 (1989).

STRANGENESS IN NUCLEAR COLLISIONS
AT JINR

Edgar Okonov

Laboratory of High Energies
Joint Institute for Nuclear Research,
141980, Dubna, Moscow region, Russia

Strangeness produced in nucleus-nuclleus (A_P A_T) collisions is argued to be an useful tool to study highly excited hadron matter.

It is predicted as an effective probe for quark-gluon plasma (QGP) formation in the stopping ("baryon-rich") regime[1] which could be realized at rather low energies of projectile nuclei ($E_P \simeq 2 \div 10$ A GeV)[1÷4].

The production of Λ-hyperons and K_S^o-mesons has been investigated at JINR using the two-meter streamer spectrometer and propane bubble chamber with various targets inside feducial volumes (A_T=^{6}Li, ^{12}C, ^{20}Ne, Cu, Zr, Ta, Pb) exposed to nuclear beams (A_P = d, ^{4}He, ^{12}C, ^{16}O, ^{20}Ne, ^{24}Mg) of the Dubna Synchrophasotron at energies of 3.3÷3.7 A GeV[6÷12]. One might consider it to be a Nature's favour that the degree of thermalization (randomization) of hadron matter in AA-collisions could be easely estimated looking at the Λ-hyperon peculiarities in their angular distributions which are known to be forward-backward peaked in the initial reaction NN→ ΛNK due to the leading effect of baryonic diquark.

As can be seen from Fig.1 the "centrally" produced Λ(K_S^o) particles are emitted near isotropically in contrast to the forward (backward) peaked emission from noncentral CC-collisions which reproduce the particular feature of initial NN-interactions.

Very similar regularities have been observed in angular distributions of Λ(K_S^o) particle energies also in the CM-system ($dE^*_{\Lambda,K}/dCos\Theta^*$). These effects, obtained first from our early Λ data and confirmed later by our K_S^o ones, suggest a full stopping

---

[1]This is not likely the case in baryon-free regime [5] predicted to be realized at much higher energies.

with formation of a single thermalized source (fireball) in midrapidities of very central AA-collisions.

The study of Λ-hyperon polarization appears to be another profitable tool for examination of excited hadron matter. The polarization $\wp_\Lambda$ which is likely also due to the leading diquark effect,has been found to be rather large in pA-interactions for a high P_T-region. This parameter $\wp_\Lambda$ is expected to vanish for Λ's from central AA-collisions with a formation of a thermalized fireball.

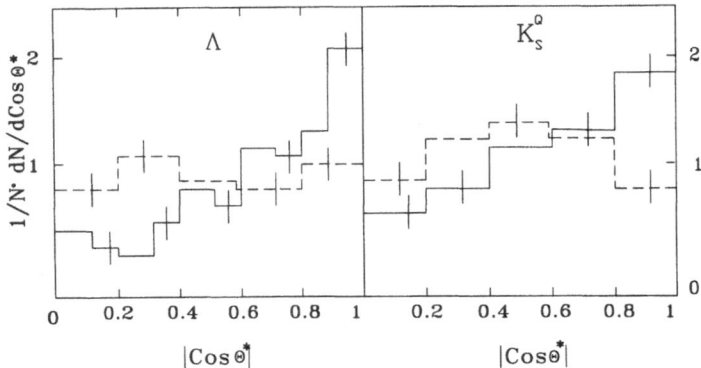

Figure 1: Folded (in $Cos\Theta^*$) angular distributions of Λ hyperons and K_S^0-meons produced: in noncentral CC-collisions — solid lines, and in central ones — dashed lines.

We have seen some increase of $|\wp_\Lambda|$ when increasing P_T of Λ's from noncentral AA-collisions. As for centrally produced Λ's there is no polarization observed, within rather large errors though: $\Delta(\alpha\wp_\Lambda) \simeq 0.2$. Statstically richer data are needed for more significant results. Anyhow the obtained data support the above suggestion derived from the analysis of angular distributions.

The dependence of hadron matter excitation upon a collision centrality has been studied by estimating parameters $<P_T>_{\Lambda,K}$ and temperatures T_B extracted from Boltzmann-like spectra (or an inverse slope of invariant cross sections spectra T_0, treated often wrongfully as temperature). Our early analysis have been revealed a considerable rise of T_B with degree of centrality: from $T_\Lambda = (75 \pm 8)$MeV up to $T_\Lambda = (158 \pm 11)$MeV which corresponds to $T_0 \simeq 200$MeV. The same increase from $T_K = (73 \pm 11)$MeV up to $T_K = (162 \pm 8)$MeV has been observed when K_S^0 mesons have been used as "thermometer"[13]. This signifies a collective effect of the heating of hadronic matter(the created fireball) up to temperatures being near critical ones predicted for a phase transition into QGP.

Such a fireball appeared to be not only very hot but also rather dense. We have observed in central AA-collisions a considerable portion of Λ's with anomalously large P_T, emitted (rescattered) from midrapidities (above 12% compared with $\sim$ 1% in noncentral ones). Taking into account this effect some model dependent estimation could be obtained which gives for the baryonic density $\rho = (4 \pm 1)\rho_o$.

The search for a possible strangeness enhancement has been performed looking at the measured rellative yields ($< n_\Lambda > / < \pi^- >$) of Λ_K-hyperons with $P_T > 1$ GeV/c being beyond kinematical limits of reaction NN→ ΛNK at 3.7 AGeV. This cut, used to eliminate the background of Λ's from NN-interactions, has been supported by theoretical considerations which have argued in favour of the study of strange particles with anomalously high $P_T(E_T)$ in order to search for QGP[14]. We have found that for such a set of Λ's, which is free of background Λ's from NN-interactions, the ratio $< n_\Lambda > / < \pi^- >$ increase by factor 9±2 when going from peripherical to central AA-collisions.

Most of effects found in Dubna experiments are summarized in the Table and Fig.2 which illustrate dependences of main characteristics of produced Λ-hyperons upon the degree of nuclear collision centrality (upon the number of projectile nucleons-participants $<Q>$).

Table 1

Effects observed with increasing of degree of collision centrality	Predicted as signals of...
– flattening of angular distributions $dN_{\Lambda,K}/dCos\Theta^*$ and $dE^*_{\Lambda,K}/dCos\Theta^*$ from strongly forward-backward peaked to nearly isotropic ones; – Boltzmann-like Λ and K^0_S spectra; – decrease of Λ polarization to $\alpha\wp_\Lambda \simeq 0 \pm 0.2^{*)}$	stopping, randomization, thermalization (at least local)
– anomalous increase of tranverse momenta $P_T(\Lambda)$ in midrapidities	increase of baryonic density to $\rho = (4 \pm 1)\rho_0$
– increase of relative yield of Λ's $< n_\Lambda > / < n_{\pi^-} >^{*)}$ (beyond an background from NN interactions) by factor 9±2	QGP formation(?)
– raise of Boltzmann temperatures of Λ's and K0's from $T_B \simeq 75$ Mev up to $\sim$ 160 MeV (to $T_0 \simeq 200$ MeV) with a cessation of further raise, approaching of T_B to a plateau$^{*)}$	heating with possible 1th order phase transition and QGP+hadr.gas mixed phase formation(?)

*) Supported by recent BNL and CERN data[16,17].

To examine a further dependence of hadron matter excitation upon the the total released energy, a study has been performed with an analysis P_T specrum of Λ's from very central MgMg collisions[15] which involve a twofold number of nucleons (with twice as great released energy) than central CC collisions. The value of $T_B=137\pm9$ MeV has been found from the mentioned analysis which does not differ within errors from $T_B=158\pm11$ MeV obtained for central CC collisions.

This gives an indication that the temperature stops to raise and seems to go to a plateau.

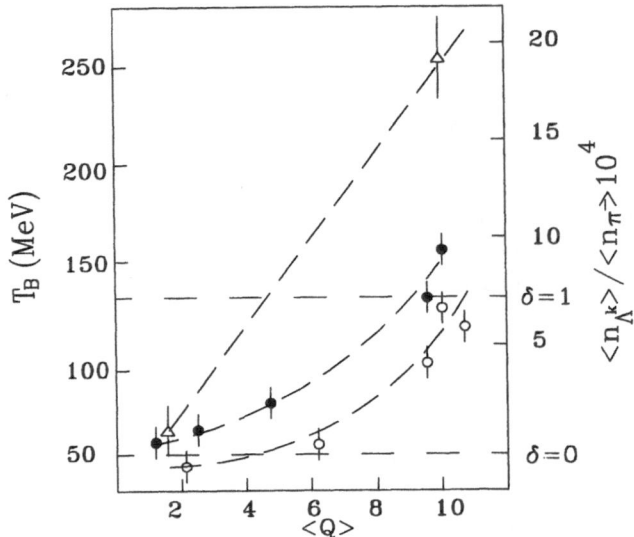

Figure 2: Dependenses upon collision centrality (i.e. upon $<Q>$ — average number of projectile nucleons-participants) of the following parameters: – the degree of flattening δ which is $\delta=0$ for the peaked distribution from pp→ ΛK^+p and $\delta=1$ for the isotropic distribution (open circles); – the Boltzman temperature T_B (black circles); – the relative yield $<n_\Lambda>$ / $<n_{\pi^-}>$ of Λ hyperons with $P_T >1$GeV/c (triangles).

The recent data of the experiments at AGS BNL[16] and SPS CERN[17] have suggested the evidence for such a plateau extending to much higher energies as can be seen from Fig.3.

Moreover in these experiments strangeness enhancement has been also observed in central AA-collisions,and not only for a relative yield of Λ's but for those of $K^\pm$ and $\bar\Lambda$ (with different cuts $P_T >0.4$-0.5 GeV/c).

This chain of the revealed effects, mentioned above is predicted as signals of a stopping, thermalization and heating of hadronic matter with a formation of a dense strangeness abundant fireball(mixed phase) via first order transition. Possibilities of such a transition under conditions (stopping, density, temperature, energy) similar to our ones have been considered in many theoretical papers. Nevertheless, even being confirmed by data of other groups, these results need more detailed comparative analyses and looking for possible alternative interpretations (beside QGP) to make final conclusions.

Anyway our data and other similar ones could be treated as a strong evidence for a creation of a hot and dense fireball (possibly mixed phase) in violent AA-collisions. This provides favourable conditions to search for Metastable Exotic Multihypernuclear Objects(MEMO's)[18] and Strange Quark Matter (SQM) states which are predicted to be considerably enhanced in such a fireball due to expected strangeness enrichment[19]. Such an investigation being of great importance itself might give a proof of the QGP (mixed phase) formation[20].

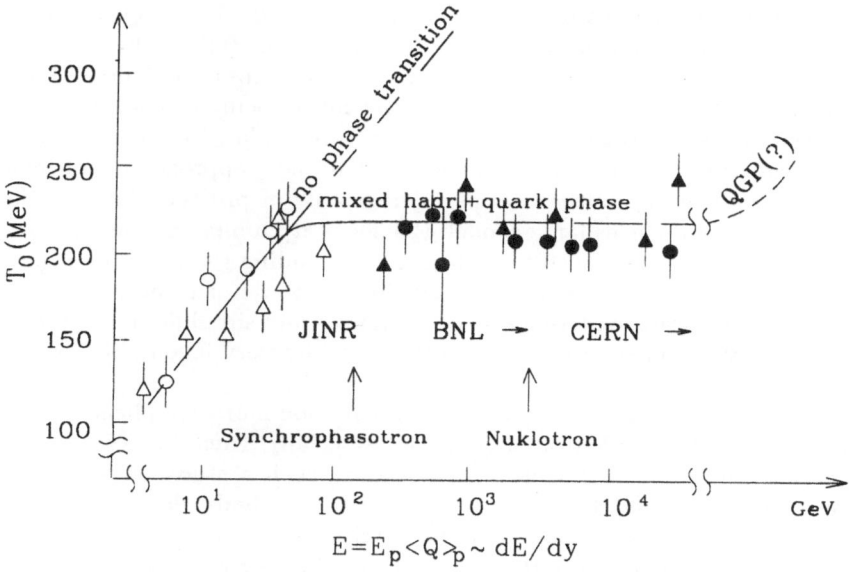

Figure 3: Inverse slope parameters T_0 versus $E=E_P <Q>$: open circles and open triangles — K_S^0 and Λ JINR data; black circles and black triangles — neutral (charged) kaons and Λ data of BNL [16] and CERN [17] (To make an adeqate comparision, the T_0 parameters (not the Boltzmann temperatures T_B) are shown on this figure, because T_B values are not given in CERN papers.)

At the first stage of this investigation we plan to look for H-dihyperon (the ground SQM state) and few baryon MEMO's by re-analyzing anomalous events which have been detected in an open (4π) geometry from central AA-collisions in streamer chambers and recorded in DST but failed to be fitted as decays of "usual" strange particles. The requirement of a coexistence with 3 double hypernuclei, observed by now, provides the most probable properties of H-particle [21]: M_H=2.22 GeV and $\tau \sim 1 \div 10$ ns with the main decay mode $H \rightarrow \Sigma^-$ p [22] followed by $\Sigma^- \rightarrow n\ \pi^-$.

Some computer program have been elaborated for analyses of data:

- the program of the kinematical reconstruction of H-decays to determine masses and to identify H-particles by fitting;

- the code for the simulations of H production from a hot midrapidity fireball with an subsequent H decay in feducial volumes of chambers to obtain detection efficiencies and H yields (their upper limits), which depend on the parameters T_B, τ, A_P/A_T and E_P.

The preliminary estimations show that the search for H-particles and MEMO's could be performed at sensitivity levels being near (above) of predicted yields.

Additional scannings of the available data will be likely required to look for MEMO's (after theoretical consideration of their main decay modes).

Our further plans in this research field are connected with a development of the new approach which has been proposed[23,24] and successfully realized at JINR[25].

The previous ("usual") methods in hypernuclear physics have to do with hypernuclei which are produced as fragments of a target nuclei being at rest. Therefore they have low energies and short ranges which makes impossible to observe directly decays of hypernuclei with A>14 even in emulsions. The new approach offers relativistic hypernuclei produced in nuclear beams as fragments of projectile nuclei. Such hypernuclei have decay ranges large enough (20-30cm at Dubna energies) to use rather thick targets and various detectors in their path including them in an trigger.

Velocities of these hypernuclei are very close to those of projectile nuclei which makes possible to measure lifetimes by decay ranges and simplifies fairly their identification and analyses of their decays. All this provides very favourable conditions to hypernuclear studies

Similar conditions could be obtained for a detection more complicated metastable strange objects formed in midrapidity fireball especially when $A_P > A_T$.

The streamer spectrometer used previously in such a study, is able to be triggered once per pulse and to detect only mesonic two body decays of the lightest hypernuclei because of the impossibility to discriminate background triggers from interactions of projectile nuclei in the gas filled the chamber when many body mesonic and non-mesonic hypernucleus decays occur. To remove these drawbacks the triggering system has been constructed with a vacuum cavity (V) as decay volume and trigger detectors inside it. This system is designed to be incorporated in the wide aperture spectrometer (with fast coordinate and charge detectors), which has been used to study a fragmentation of relativistic nuclei from Synchrophasotron with hundreds triggers per pulse.

Fig.4 exhibits a lay-out of the designed experiment (with a production of $^{14}_{\Lambda}$N in the target M by ^{19}F projectile and subsequent decay in V as an example). Two multywire proportional chambers of three planes (PC$_{1,2}$), Cherenkov (Č$_{1-4}$) and scintillation counters (C$_{1,2}$) will be used to measure the charge (Z) of A_P and coordinates of its track, eight MWPC's (PC$_{3-10}$) – to obtain the A/Z ratio for fragments-products of interaction/decay (by their rigidities in magnet CP-40), the set of 30 Cherenkov detectors (Čr) – to determine Z of fragments.

Three levels of triggering will be used with the following logic:
a decrease of the charge, increase of multyplicity detected by two sets of Si-microstrip

dE/dx detectors S_{1-3} and S_{4-6} (as a result of a hypernuclear decay in vacuum volume V) and reconstruction of a vertex within V using fast processors.

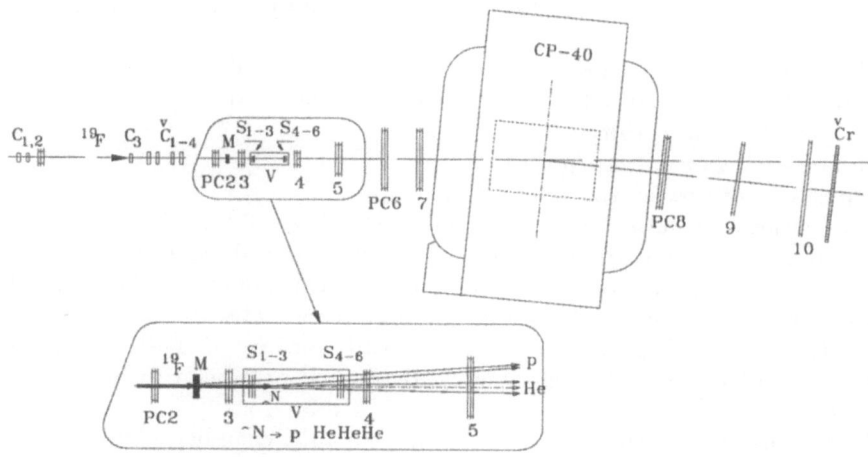

Figure 4: A lay-out of the designed experiment.

Such a system will eliminate a background from any possible imitating processes. A final identification will be done by off-line analyses to reject any accident events.

It has been estimated that this approach will increase data taking rate by factor $10^2 \div 10^3$ and make possible to detect heavier hypernuclei and double ones which are predicted [26] to be formed at E_P=5 AGeV (Nuclotron) about 10 times more frequent than at E_P=14.5 AGeV (AGS) due to a good overlap between momentum distributions of Λ's and nuclear fragments.

The triggering system for a detection of relatively light metastable SQM states and MEMO's is designed to be somewhat different taking into account peculiarities of their decays (e.g.two section decay volume will be created). This system (with approriate trigger detectors and electronics) will be constructed as rather detached one to make possible using it not only with JINR spectrometers at Nuclotron but also with other proper ones in nuclear beams at AGS (BNL), SPS (CERN) and possibly at KEK.

It should be emphasized that the consdered approach aims to detect metastable strange objects in (nearly) open geometry and to identify them revealing strangeness by particular decays, whereas few other experiments, proposed thus far[27] (except P-864 to AGS) intend to search for SQM by unusual A/Z ratio which is not necessary a peculiar feature of SQM but also of other possible anomalous forms of matter.

In any case the experiments mentioned in [19] are unable to distinguish between MEMO's and SQM states and can not detect objects with τ <10ns.

The above considered results were obtained in 1982-1989 in nuclear beams of JINR Synchrophasotron which is able by now to accelerate nuclei up to ^{32}S to an energy

$E_P \simeq V4$ A GeV, but the most part of the proposed research plans to study the strangeness in AA-collisions will be performed at the JINR Nuclotron.

The Nuclotron, a strong focusing synchrotron with a superconducting magnetic system, was successfully tested in 1993 and will gain in the near future following main designed parameters:

- maximum energy: for protons – 12 Gev , for nuclei – 6 A GeV;
- duration of beam pulse from slow extraction – up to 10 s.;
- duty cycle – up to 75%;
- momentum spread - $\Delta p/p = 10^{-3}$;
- effectivity of slow extraction – 96% ;
- intensities (per pulse): $p \sim 10^{11}$, $^{24}Mg \sim 10^{8}$, $^{84}Kr \sim 10^{7}$, $^{238}U \sim 10^{5}$.

Hereafter main setups are listed which are appropriate/designed to detect strange particles/objects expected to be formed in nuclear beams of the Nuclotron (including fragments with an unusual A/Z ratio as potential candidates to SQM states):

-SPHERE- a multipurpose 4π detector with forward magnetic spectrometer, MWPC's, MWDC's, Pb glass calorimeter, dE/dx and TOF scintillation hodoscopes, Cherenkov counters;

– GIBS-a hybrid spectrometer with 2m.long streamer chamber, dE/dx scintillation counter, MWPC's;

– ANOMALON-a multipurpose spectrometer, MWPC's, Cherenkov hodoscope, scintillation counters, TOF;

– PAMIR-a magnetic spectrometer with Cherenkov dE/dx detectors as an active target, Si strips as a vertex detector, MWPC's, Pb glass hodoscope (calorimeter), TOF, liquid argon calorimeter;

– INESS ALPHA- a magnetic two-arm spectrometer with MWPC's, scintillation telescopes, TOF, threshold Cherenkov counters;

– SYAO- a nuclear recoil spectrometer with telescopes of dE/dx Si-detectors, arrays of plastic scintillators and NaI (Tl) cristals;

– DISC a magnetic two-arm spectrometer with ΔE-E detectors, TOF, threshold Cherenkov counter;

–KASPIY-a magnetic spectrometer with MWPC's and MWDC's, ΔE-E detectors, scintillation telescopes, TOF, Cherenkov plexiglass and lead glass counters.

All these setups being used in experiments at Synchrophasotron, provide promissing possibilities for widespread studies of hot and dense matter with strangeness as a effective probe in nuclear beams of the Nuclotron.

ACKNOWLEDGEMENTS

I am very grateful to all my colleagues participated in the mentioned Dubna investigations. This work was supported by the Russian Foundation for Fundametal Investigations(RFFI), grant No 93-02-15583

REFERENCES

1. T.Biro, J.Zymanyi, Phys.Lett.**B113**,6(1982).

2. H.Stoecker, Nucl.Phys.**A418**,587(1984).

3. M.Gyulassy, LBL-16895,Berkeley(1984).

4. N.Glendening, Nucl.Phys.**A512**,737,(1990).

5. T.Matsui et al.,Phys.Rev.**D34**,2047(1986).

6. M.Anikina et al.,Phys.Rev.Lett.50,1971(1982); Z.Phys.**C33**,(1986).

7. E.Okonov, JINR D2-82-568,Dubna(1982); JINR P1-86-312,Dubna(1986).

8. M.Gazdzicki et al.,JINR E1-85-989, Dubna(1985); Z.Phys.**C33**,895(1986).

9. E.Okonov, in: Modern Developments in Nuclear Physics, World Scient. Ed.,166(1988).

10. M.Anikina et al.,JINR E1-85-578,Dubna(1985).

11. D.Armutlijski et al.,Sov.Nucl.Phys.**43**,366(1986).

12. V.Boldea et al.,Sov.Nucl.Phys.**47**(1988).

13. K.Iovchev, E.Kladnitskaya and E.Okonov, JINR Rap. Com. 7, 27, Dubna(1990).

14. M.Danos and J.Rafelski, Phys.Lett.**B192**,492(1987).

15. S.Avramenko et al.,JINR P1-91-235,Dubna(1991).

16. T.Abbot et al.,Phys.Rev.Lett.**64**,847(1990).

17. J.Bartke et al., Z.Phys. **C48**, 191(1990). P.Seyboth et al., Nucl. Phys. **A544**, 293(1992); D.Greiner et al., ibid., 309; J.Kinson et al., ibid., 321.

18. J.Schaffner,C.Greiner and H.Stoecker, Phys.Rev.**C46**,322(1992).

19. C.Greiner,A.Diener,J.Schaffner and H.Stoecker, to be published in Proc.of Quark Matter-93 Conference(1993).

20. C.Greiner,D.Rischke,H.Stoecker and P.Koch,Phys.Rev.**D38**,2797(1989).

21. R.Dalitz et al.,Proc.Royal Soc.Lon.**A426**,1(1989); S.Aoki et al.,DFNU-91-07(1991).

22. J.Donoghue, E.Golowich and B.Holstein, Phys.Rev.**D34**,3434(1987).

23. M.Podgoretski, JINR-8309,81(1974).

24. E.Okonov, JINR-8309,104(1974); see also JINR B1-7113,Dubna(1973), JINR p1-87-191,Dubna(1987), JINR E1-90-591,Dubna(1991).

25. S.Avramenko et al.,JINR D1-88-691,Dubna(1988); Nuov.Cim.**A102**,95 (1989).

26. M.Sano and M.Wakai, RIKEN-NP-105,246(1991).

27. J.Sandweiss, Nucl.Phys.**24B**,234(1991).

THE MESOZOIC ERA OF RELATIVISTIC HEAVY ION PHYSICS AND BEYOND

John W. Harris

Lawrence Berkeley Laboratory
University of California
Berkeley, CA 94720

INTRODUCTION

In order to understand how matter 15 billion years ago in the form of quarks, gluons and leptons at a temperature of 2×10^{12} °K evolved to become today's Universe, the goal of relativistic and ultra-relativistic heavy ion physics is to understand the equation of state of nuclear, hadronic and partonic matter. This quest is of cross-disciplinary interest. The phase transition from partonic matter to hadronic matter tens of micro-seconds after the beginning of the universe is of interest to cosmology. Fluctuations during this phase transition would influence nucleosynthesis and the understanding of baryonic inhomogeneities in the universe. The nuclear matter equation of state, which describes the incompressibility of nuclear matter, governs neutron star stability. It determines the possible existence of strange quark matter stars and the dynamics of supernova expansion in astrophysics. The existence of collective nuclear phenomena in nuclear physics is also determined by the nuclear equation of state. In relativistic heavy ion collisions collective nuclear flow has been observed and is being studied extensively to obtain a better understanding of the incompressibility of nuclear matter. In high energy nuclear and particle physics, production and excitations of hadronic final states have been studied in detail and are important to an overall understanding of the equation of state of nuclear matter at finite temperature. The possibility in ultra-relativistic heavy ion collisions to create and study highly excited hadronic and partonic degrees of freedom provides a unique opportunity for understanding the behavior of nuclear, hadronic and partonic matter. Study of the QCD vacuum, of particular interest in particle physics, would provide a better understanding of symmetry-breaking mechanisms and the origins of the masses of the various quarks and particles. Creation and study of the quark-gluon plasma, an excitation of the QCD vacuum, is the goal of physicists in the new field of ultra-relativistic heavy ion collisions.

In the pre-Mesozoic days, there were herbivores and carnivores.
In the early days, there were experimentalists and theoreticians.

Hot and Dense Nuclear Matter, Edited by
W. Greiner *et al.*, Plenum Press, New York, 1994

EVOLUTION OF RELATIVISTIC HEAVY ION PHYSICS - MESOZOIC ERA

The Mesozoic Era spanned the 165 million years from 230 million years ago until 65 million years ago. It consisted of the Triassic, Jurassic and Cretaceous Periods. During this Era the dinosaur species developed, evolved, flourished and vanished. At the same time plant species multiplied. Birds began to develop near the end of this Era and the mammals developed soon after the extinction of the dinosaurs at the end of this Era.

Today we are viewing the "Evolution of Relativistic Heavy Ion Physics." The field recently experienced a transformation which was unexpected and unforeseen in its early days. In this lecture I present the evolution of the field of relativistic heavy ion physics in analogy with the evolution of the dinosaurs in the Mesozoic Era. At the same time, I note that the evolution of high p_t physics and the parton model developed coincidentally in time with that of relativistic heavy ion physics. These developments and the understanding of high p_t (transverse momentum) physics in terms of perturbative quantum chromodynamics are important ingredients helping to end the "Mesozoic Era of Relativistic Heavy Ion Physics" and are the basis of current theoretical developments in the field of ultra-relativistic heavy ion physics.

Triassic Period

The Mesozoic Era started with the Triassic Period which ended approximately 190 million years ago. In this period the early dinosaurs developed. By the end of the period, there were two genetic lines of dinosaurs - the plant-eating ornithiscians, with bird-like hip-bone structure, and the carnivorous saurischians, with lizard-like hip structure. Near the end of this period, there was global mass extinction of most of the large creatures and only the smaller creatures survived into the next period. *

The earliest interests in the possible existence of dense nuclear matter involved primarily its astrophysical implications.[1,2] These investigations concentrated on the possible existence of collapsed nuclei (with radii considerably smaller and densities considerably larger than those of normal nuclei), their properties, their interactions with normal nuclear matter, and the astrophysical consequences of their possible existence. Possible formation and existence during the early universe and in the cores of dense celestial bodies were hypothesized and discussed in terms of the quark model and a composite hadron model.

Jurassic Period

After the Triassic Period came warmer habitats where new dinosaur populations flourished. The first known bird, the Archaeopteryx, emerged and has been linked to the dinosaur. During the same time super-giant dinosaurs, called sauropods, developed. †
Near the end of the Jurassic Period, the climate on earth became arid and the sauropods began to decrease in numbers. The Jurassic Period ended 136 million years ago.

Experimentalists and theoreticians in the field flourished during this period. With the advent of the first relativistic heavy ion accelerators at the Bevalac in Berkeley and the Synchophasotron in Dubna, searches could be made for abnormal states of dense nuclear matter. Although none were found, many new theoretical ideas came to light. Early speculations of possible exotic states of matter continued to focus on the astrophysical

---

* Originally, paleontologists hypothesized that the impact of an extraterrestrial object which formed the one hundred kilometer diameter Manicouagan Impact Crater in Canada was responsible for this catastrophic event. The massive, dust cloud following such an impact would have blocked sunlight, killing most plants, and would have led to starvation of at least the larger dinosaurs. However, recent dating techniques have placed the formation time of the crater several million years before the death of these dinosaurs. Climatic changes may have been responsible, since at this time the large super-continent which spanned the earth was just beginning to break apart. Widespread volcanic activity, geological activity and the appearance of new seas as the super-continent was pulled apart had drastic effects on weather patterns and the dinosaurs as well.
† The sauropods were herbivores, as were the large Diplodocus, Brachiosaurus and Seismosaurus species.

implications of abnormal states of dense nuclear matter.[3] Subsequent field theoretical calculations, assuming chiral symmetry in the σ model, resulted in predictions of abnormal nuclear states and excitation of the vacuum.[4] This generated considerable interest in transforming the state of the vacuum by using relativistic nucleus-nucleus collisions.[5,6] A deconfinement phase transition to quark matter or a quark-gluon plasma[7,8,9] was predicted. At the same time there were also predictions of phase transitions resulting from pion condensation in nuclear matter[10] with possible formation of the condensate in relativistic nucleus-nucleus collisions.[11]

Cretaceous Period

The beginning of this period was marked by a climatic change caused by the drifting apart of the continents. The climate became hot and humid, and shallow seas arose. As the sauropods decreased in numbers, the ornithiscians (bird-hipped herbivores) became dominant. By the late Cretaceous Period, most dinosaurs had developed warm-blooded metabolisms and social tendencies, * *which aided in their survival. At the same time, plant species that reproduced by flowering and seed distribution grew rapidly. Again with good food sources, the carnivores (e.g. the Tyrannosaurs) grew larger in size.*

At the Bevalac collective nuclear flow[12,13] was discovered, initiating a new "industry" of nuclear flow studies at lower energy accelerators around the world. With predictions of possible chiral symmetry restoration and quark-gluon plasma formation at high energy densities, the focus of the field moved to higher energies. Initial experiments were undertaken with 14.5 GeV/n Si projectiles at the BNL AGS and 200 GeV/n S at the CERN SPS incident on various nuclear targets. The Dual Parton[14] and Lund/FRITIOF[15] string models were developed, generalized and used to describe relativistic heavy ion collisions at CERN fixed target energies. Although no quark-gluon plasma was observed, the experiments suggested that high baryon densities could be produced in the AGS experiments and high energy densities in the CERN experiments.[16] In addition, new models were becoming available. The Lorentz invariant molecular dynamics approach used in the RQMD Model[17] and the Regge-Gribov-Veneziano approach used in the VENUS model[18] predicted substantial baryon stopping at midrapidity, even for the CERN experiments with relatively light ion (A=32) beams.

There were new species of experiments[19,20] operating at the Bevalac, as it was shut down, producing new data on collective nuclear flow and the role of medium effects.[†] The quest was and still is to extract the nuclear equation of state by separating out compression effects from medium effects from these experiments and new experiments at the SIS accelerator at the GSI. Furthermore, new data is anticipated to investigate nuclear matter in the high baryon density regime using 11.5 GeV/n Au + Au at the BNL AGS and 170 GeV/n Pb + Pb at the CERN SPS. New plans have been made to investigate the high energy density regime using higher energies in heavy ion experiments at the Relativistic Heavy Ion Collider (RHIC) and the Large Hadron Collider (LHC).

The Cretaceous Period ended with the extinction of the dinosaurs 65 million years ago. ** *After extinction of the dinosaurs, new species would develop, including seasonal primates and somewhat later the primates.*

---

* Social development along the lines of herding instincts and parental and group care.

† such as the pion dispersion relation, medium effects on pion and delta propagators, resonance widths in-medium, inelastic scattering cross sections in the nuclear medium, and many others.

** One hypothesis is that the impact of an extraterrestrial object with earth 65 million years ago created dust and darkness, upsetting the entire ecosystem of the earth and extinguishing the dinosaurs. This hypothesis originated from the discovery on earth of 65 million year-old deposits of irridium, found rarely on earth but common in asteroids. It was further supported by the recent finding of a 65 million year old crater, approximately 180 kilometers in diameter, on the Yucatan peninsula of Mexico. However, the dinosaurs may have been in decline by the end of the Cretaceous Period and could have vanished before such an impact.

EVOLUTION OF HIGH P_T PARTON PHYSICS - MESOZOIC ERA

Triassic Period

The early quark model[21] was able to describe the static properties of hadrons: hadron spectroscopy, electromagnetic properties and weak decays. However, due to the complexity of the experimentally observed phenomena, hadron dynamics were yet to be successfully described in a theoretically-consistent framework. The observation of scaling behavior of the nucleon structure functions in deep inelastic electron scattering[22,23] led to a description of the hadron, in a frame where the hadron's momentum is infinite, as a system of independent point-like particles.[24,25] This was the basis and origin of the parton model.[26,27]. It immediately led to discussion and predictions of deep inelastic hadronic processes.[28]

Jurassic Period

A connection between the quark and parton models was made and the parton model extended[29] to describe inclusive hadron production at high transverse momentum. The production of highly collimated groups of particles resulting from fragmentation of scattered quarks, called jets, was hypothesized. If a photon were exchanged between incident quarks and the scattered quarks hadronize, then jets of particles would emerge. Furthermore, new elementary parton processes were possible, mediated by the exchange of a spin 1 gluon. The cross section for an exchange of such a gluon would exceed the cross section for photon exchange, which is electromagnetic in nature, by approximately four orders of magnitude. It was then hypothesized that hard-scattered quarks would be emitted from the interaction "back-to-back". The remaining non-interacting "spectator" quarks would continue forward and backward as "beam-jets" resulting in four-jet processes. Proposals were made to measure parton-parton cross-sections directly. However, the early detectors at Fermilab and at the CERN ISR were not designed with the study of these processes in mind. Bjorken[30] pointed out that the use of calorimetry to measure the energy of the hard-scattered partons would for the most part eliminate the problems of parton fragmentation confusing the measurements. Furthermore, Ellis and Kisslinger[31] pointed out that the leading fragment of a jet, which contains the scattered quark as a valence quark, provided a high p_t trigger particle to trigger experiments measuring these high p_t and relatively low cross section processes. They also noted that to be able to interpret the inclusive p_t spectra, better information on structure and fragmentation functions was necessary.

Cretaceous Period

With such good advice, fixed target experiments with specialized jet triggers started accumulating data at Fermilab and CERN. The incident energies of a few hundred GeV, e.g. $\sqrt{s}$ = 20 GeV at the CERN SPS, were too low to successfully study hard parton-parton cross sections. No jets were observed. High multiplicity events dominated over jets.

Subsequently, the Intersecting Storage Ring (ISR) experiments at CERN, $\sqrt{s}$ = 63 GeV, were able to measure large angle jets. Those experiments found that the hard-scattering parton-parton cross sections increased more rapidly (as $\sqrt{s}$) than the high multiplicity events, where $dn/d\eta \sim \ln \sqrt{s}$. This was especially true for p_t > 2 GeV/c, where a power law behavior was observed in p_t. Extensive jet studies were then undertaken at the $Sp\bar{p}S$ and Tevatron Colliders. Hard parton-parton scattering was measured and found to be described well using perturbative QCD. Furthermore, the structure of nucleons could be described using structure functions in terms of the quark densities.

QCD is able to describe the basic structure and interactions of hadrons in terms of partons. The existence of gluons is postulated in the parton model. The parton evolution is derived from the simple branching transitions

$$q \rightarrow q + g \qquad\qquad g \rightarrow g + g \qquad\qquad g \rightarrow q + \bar{q}$$

and the Altarelli-Parisi evolution equations.[32] The connection between perturbative QCD and experimental observations is made using a phenomenological approach to describe how partons fragment into hadrons, known as fragmentation functions. (Note that a jet is a collection of hadrons resulting from the fragmentation of a parton.)

THE DYNAMICS OF ULTRA-RELATIVISTIC HEAVY ION COLLISIONS

A new field of theoretical studies has developed using perturbative quantum chromodynamics to calculate the large momentum transfer processes expected in future ultra-relativistic heavy ion collider experiments. This new theoretical approach is a product of the high p_t physics (hard scattering) studied in the "Mesozoic Era of High P_t Parton Physics" and the understanding of low p_t (soft) processes gained from previous relativistic heavy ion studies in the "Mesozoic Era of Relativistic Heavy Ion Physics." A comprehensive description of ultra-relativistic heavy ion physics must necessarily incorporate perturbative QCD to describe the high p_t processes and approaches which describe the low p_t (soft) processes in the ultra-relativistic heavy ion environment.

High P_t Processes

The initial parton distributions in nuclei, the nuclear structure functions, are essential to be able to specify the initial conditions of these collisions. In turn, specific knowledge of the gluon, valence quark and sea quark distributions in the nucleon is required, prior to understanding their distributions in nuclei. There is evidence for the presence of semi-hard processes with $p_t \geq 2$ GeV/c,[33] often referred to as mini-jets:[34,35] the inelastic cross sections are observed to increase with $\sqrt{s}$, the tail of the charged particle p_t distribution (dN_{ch}/dp_t at $p_t \geq 2$ GeV/c) increases with $\sqrt{s}$, the pseudorapidity density of charged particles ($dN_{ch}/d\eta$) increases with $\ln (\sqrt{s})$ and fluctuations in N_{ch} increase with $\sqrt{s}$. Extremely hard processes lead to jet production, which has been studied extensively at the colliders. There is systematic data on jet production as seen in Fig. 1 and the production processes are well understood using perturbative QCD. Notice that the jet cross section at $\sqrt{s_{nn}} = 200$ GeV (RHIC energy) in Fig. 1 is significant and that at $\sqrt{s_{nn}} = 6300$ GeV (LHC energy) jets will become even more prevalent. These semi-hard and hard scattering processes will influence the evolution of ultra-relativistic heavy ion collisions and must be considered for a complete description.

Nucleon and Nuclear Structure Functions

In order to develop a description of ultra-relativistic heavy ion collisions from the partonic degrees of freedom, the structure functions (parton distributions) of the nucleon and nucleus must be measured over the range of x and Q^2 relevant to future RHIC and LHC heavy ion collisions to be studied. Here $Q^2 = 4EE'\sin^2(\Theta_L/2)$ is the four-momentum transferred to the struck quark in the nucleon, where E is the incident energy, E' and Θ_L the energy and scattering angle of the scattered particle. The variable $x = Q^2/2M(E-E')$ is the ratio of the four-momentum transferred to the energy transferred. It is a measure of the fraction of the total momentum of the nucleon that is carried by the struck quark.

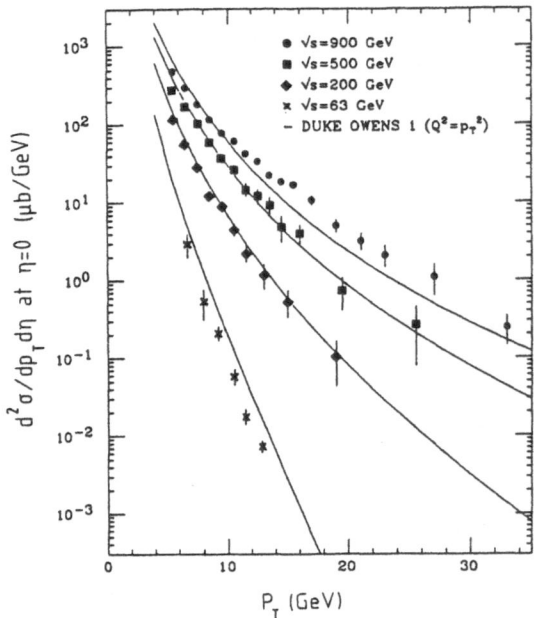

Fig. 1. Inclusive jet production cross section data at $\eta = 0$ for various incident energies.[36]

The regions of x probed by semi-hard and hard collisions will be significantly different for RHIC and LHC interactions. For example, back-to-back gluon jets at midrapidity corresponding to semi-hard (p_t = 2 GeV/c) and hard collisions (p_t = 10 GeV/c) at the SPS, RHIC and LHC will have the following values of x:

p_t	x at SPS	x at RHIC	x at LHC
2 GeV	2×10^{-1}	2×10^{-2}	6×10^{-4}
10 GeV	1.0	1×10^{-1}	3×10^{-3}

The parton distributions of the nucleon have been studied at fixed target facilities for the x and Q^2 regimes relevant to the SPS and at slightly larger x-values (and lower Q^2 values) than anticipated for semi-hard and hard collisions at RHIC. Only recently in the electron-proton collider at HERA have structure functions of the proton been investigated[37] in the region $10^{-4} < x < 10^{-2}$ at Q^2 relevant to hard collisions at the LHC. Future information from HERA, RHIC and LHC will be necessary to understand the parton distributions of the proton in the x and Q^2 regimes relevant to RHIC and LHC.

The parton distributions of nucleons in the nucleus were first observed to be different from those of free nucleons in the well-known EMC experiments.[38] These initial studies covered the region $0.05 < x < 0.7$. Since then, extensive studies of parton distributions of nucleons in nuclei have been made over the range $6 \times 10^{-4} < x < 1$ at Q^2 values relevant to RHIC and LHC.[39] A schematic curve representing the ratio of cross sections for deep inelastic lepton scattering from heavy nuclei relative to deuterium is displayed in Fig.2. Notice that the ratio rises above 1 near x = 1 due to Fermi motion.[40] The EMC effect is observed as a dip at approximately $0.2 < x < 0.6$. Here the valence quark distribution in nucleons of the nucleus is depleted relative to that of a free nucleon. Several explanations have been proposed for this effect, ranging from an excess of virtual pions in the nucleus to enlarged bags of freely-interacting valence quarks. At lower values of x, i.e. $0.07 < x <$

0.2, the ratio rises above 1. This region is known as the anti-shadowing region. For $x <$ 0.05 the ratio again shows a depletion and shadowing occurs. Explanations of the shadowing and anti-shadowing regions are intimately related. Shadowing has long been observed in hadron-nucleus interactions.[41] Due to the strong interaction, scattering occurs in these interactions near the surface of the target nucleus and cross sections are proportional to the surface area, i.e. $\sim A_T^{2/3}$. The front surface of the target nucleus shadows the rear target nucleons. This was not expected, a priori, in lepton scattering experiments, where the interactions occur through the transfer of a virtual photon. However, shadowing via real and virtual photo-absorption can be described in terms of the Vector Dominance Model.[42] This can be understood simply from the fact that the photon is sometimes a superposition of hadronic states which have the same quantum

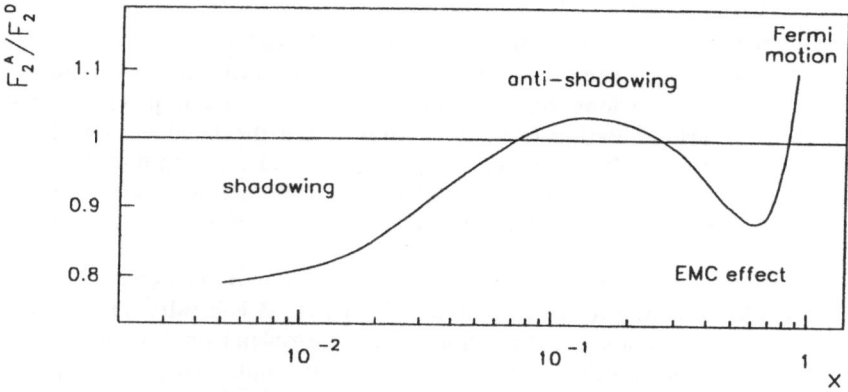

Fig. 2. Schematic curve representing the ratio of cross sections for deep-inelastic lepton scattering from heavy nuclei relative to deuterium.[39] See text for description.

numbers as the photon, i.e. the vector mesons. It is not surprising that photon-nucleus cross sections are also shadowed similar to those of hadron-nucleus interactions. Recently, partonic models based on QCD have been developed to describe the shadowing and anti-shadowing observed in deep-inelastic lepton scattering. In these models, the low momentum partons are spread out over a large distance, due to the Uncertainty Principle. This distance is comparable to the separation of nucleons in the nucleus. Thus partons from different nucleons overlap, fuse and increase the number of high momentum partons at the expense of the lower momentum ones. The anti-shadowing and shadowing result naturally from these models. New data from the Fermilab E665 experiment at lower values of x and the CERN NMC experiment are expected to tightly constrain these models in the low x shadowing region. Nuclear shadowing of gluons[43] could have a large effect on the parton evolution and dynamics of ultra-relativistic heavy ion collisions. It will be important to have more information on the gluon shadowing and to measure the effects of nuclear shadowing by systematic studies of pp and pA interactions at RHIC.

Rapid Gluon Dominance

In ultra-relativistic heavy ion collisions a rapid dominance of gluons in the early stages of the collisions has been hypothesized.[44] This can easily be understood by considering the following. Compare the expected behavior of the elementary parton-parton cross sections

$$\sigma(g + g \rightarrow g + g) \;>\; \sigma(g + q \rightarrow g + q) \;>\; \sigma(q + q \rightarrow q + q) \;>\; \sigma(g + g \rightarrow q + g)$$

where q = quark and g = gluon. The resulting parton mean-free paths are such that $\Lambda(\text{quark}) \sim 9/4 \times \Lambda(\text{gluon})$. Thus, on the average gluons interact more frequently than quarks when traversing partonic matter. If we now consider that the branchings $g \rightarrow g + g$ and $q \rightarrow q + g$ occur more frequently than $g \rightarrow q + q$, then in AA collisions if the elementary cross sections are sufficiently large, we expect that gluons multiply more quickly than quarks. In fact, the cross section $\sigma(g + g \rightarrow g + g)$ is expected to be large, and thus gluons rapidly dominate the system. This is exacerbated by the possibility of multiple-gluon branching processes $g \rightarrow ng$, where $n > 1$.

Parton Models for Ultra-Relativistic Heavy Ion Collisions

Several models have been developed recently to describe the evolution of ultra-relativistic heavy ion collisions from the partonic degrees of freedom. These include the HIJING Monte Carlo,[45] Parton Cascade Model,[46] and the Dual Parton Model with cascading and minijets.[47] The general results of these codes are similar in many ways. The primary differences can be attributed to the cut-off momentum used in each code to separate the soft and hard processes, and the different methods of treating the soft processes.

I will outline the approach of one of these models, the Parton Cascade Model (PCM), to give the reader a feeling for what is involved. The PCM is a fully relativistic, space-time approach. It assumes that the short-range parton-parton interactions and the hadronization process factorize. A relativistically co-variant transport equation with a collision term is used. The parton substructure is derived from nucleon structure functions. Parton-parton interactions are calculated using perturbative QCD and the hadronization is governed by the parton fragmentation functions. Displayed in Fig. 3 are the results of the PCM for the time evolution of the pressure, number density, energy density, entropy and temperature in central Au + Au collisions at $\sqrt{s_{nn}} = 200$ GeV. There is a consensus among the HIJING, PCM and lowest order QCD[48] calculations. Initially there are large energy and parton (primarily gluon) densities, a high "temperature" $T \equiv 4/3(\varepsilon/s)$, rapid entropy production per particle within the first 0.3 - 0.5 fm/c, and thermal equilibration of gluons within the first 1 - 2 fm/c. A hot, gluon (minijet) plasma is formed just after the onset of interactions. It appears to be locally thermally equilibrated. However, an outstanding question is whether rapid chemical equilibration occurs. The model calculations suggest not, primarily due to the small number of quarks. However, better information on the gluon shadowing in nuclei is necessary to be able to make accurate predictions about the parton evolution. The hot, gluon plasma expands becoming a parton gas, which later hadronizes into a hadron gas. Since the models and some of the input distributions are still in an adolescent stage, improvements are expected and predictions may change. In particular, work is needed to bring together in one model a more realistic treatment of the soft interactions coupled to the perturbative QCD treatment of the hard-scattering interactions. An example would be an approach integrating a perturbative QCD description for the hard-scattering processes and a string model description for the soft processes.

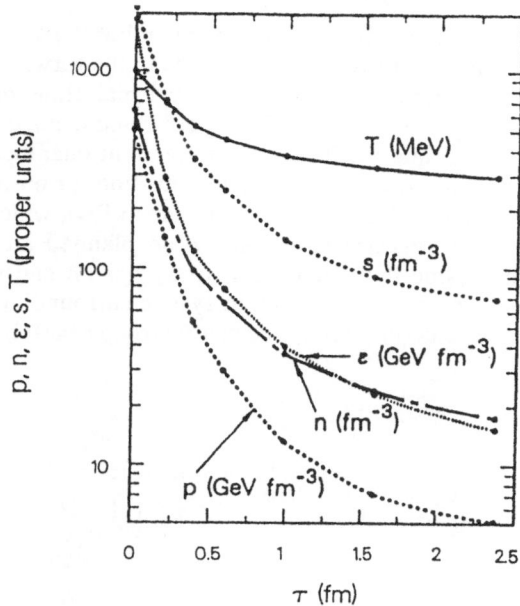

Fig. 3. Parton Cascade Model predictions for the pressure, number density, energy density, entropy and temperature in the local rest frame as a function of time for central Au + Au collisions at $\sqrt{s_{nn}} = 200$ GeV.[49]

FUTURE ULTRA-RELATIVISTIC HEAVY ION EXPERIMENTS

With RHIC and LHC on the horizon, heavy ion collider experiments are presently being designed and constructed to be able to address a range of physics from the soft-scattering processes to the hard-scattering ones. Systematic studies of pp, pA and various AA interactions will be made to understand the underlying structure of the incident nucleons and nuclei. Likewise, systematic studies will be made to obtain a detailed understanding of the space-time evolution, i.e. dynamics, of the AA collision processes. Unless nature is extremely generous and quark-gluon plasma signatures[50] standout visibly, it is most likely that the quark-gluon plasma may be studied only after systematic studies determine the structure of the incident nuclei and the collision dynamics. The physics goals and status of the heavy ion experiments in various stages of development for RHIC and LHC have recently been presented. These include two large experiments at RHIC (STAR[51] and PHENIX[52]), two smaller ones at RHIC (BRAHMS[53] and PHOBOS[54]) and the ALICE[55] experiment at the LHC. I will briefly discuss the STAR experiment and refer the reader to the above references for details on the future ultra-relativistic heavy ion experiments that are presently being planned.

The Solenoidal Tracker At RHIC (STAR)

The detector systems of the STAR experiment at RHIC are shown in Fig. 4. Measurements will be made at midrapidity over a large pseudo-rapidity range ($|\eta| < 4.5$) with full azimuthal coverage ($\Delta\phi = 2\pi$) and azimuthal symmetry. This experiment is designed around a large solenoid magnet with a 0.5 T field for precision tracking. Momentum measurements and identification via dE/dx at low p_t for charged-particles will be made using a silicon vertex tracker (SVT, covering pseudorapidities $-1 \leq \eta \leq 1$)

near the intersecting beams and a time projection chamber (TPC, covering $-2 \le \eta \le 2$). Additional tracking of charged-particles will be made in the forward regions (XTPC, $2 \le |\eta| \le 4.5$) downstream from the solenoid in external time projection chambers. Triggering on the centrality of collisions will be performed using a central trigger barrel ($-1 \le \eta \le 1$) of scintillators and the TPC endcap readout chambers ($1 \le |\eta| \le 2$) to trigger on charge multiplicity. Approximate vertex position for on-line triggering will be determined from timing in the vertex position detectors (VPD), which consist of arrays of Cherenkov counters in the forward directions. Also planned are an electromagnetic calorimeter (EMC) just inside the magnet coil to trigger on and measure jets and the transverse energy of events, and a time-of-flight system surrounding the TPC for particle identification at high momenta (in place of the central trigger barrel).

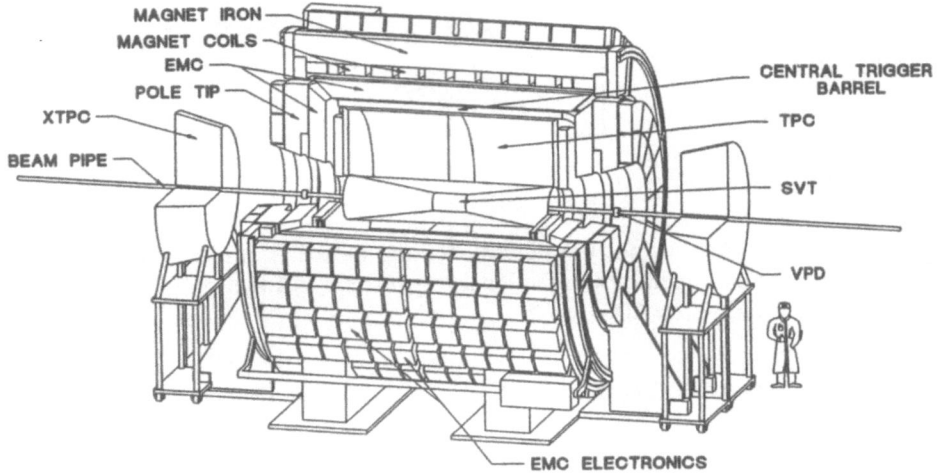

Fig. 4. Layout of the STAR Experiment. See text for description.

Event-by-event measurement of global observables - such as temperature, flavor composition, collision geometry, reaction dynamics, and energy or entropy density fluctuations - will be possible in STAR because of its large solid angle coverage and the very high charged particle densities, $dn_{ch}/d\eta \approx 1000$ expected in nucleus-nucleus collisions at RHIC. This will allow novel determination of the thermodynamic properties of single events. Correlations between observables made on an event-by-event basis may isolate potentially interesting events. These studies will focus on identifying special events with thermodynamic properties characteristic of a quark-gluon plasma as well as determining the collision dynamics.

Measurable jet yields at RHIC will allow investigations of hard QCD processes via both highly segmented calorimetry and high p_t single particle measurements in a tracking system. A systematic study of particle and jet production will be carried out over a range of colliding nuclei from pp through p-nucleus up to Au-Au, over a range of impact parameters from peripheral to central, and over the range of energies available at RHIC. The pp interactions will help establish the gluon structure functions, the p-nucleus interactions will be used to study the nuclear gluon distributions and thus the extent of shadowing of gluons in the nucleus, while the nucleus-nucleus interactions are essential to determine the degree of quenching[56] of hard-scattered partons in the surrounding nuclear, hadronic, and partonic matter. Measurements of the remnants of hard-scattered

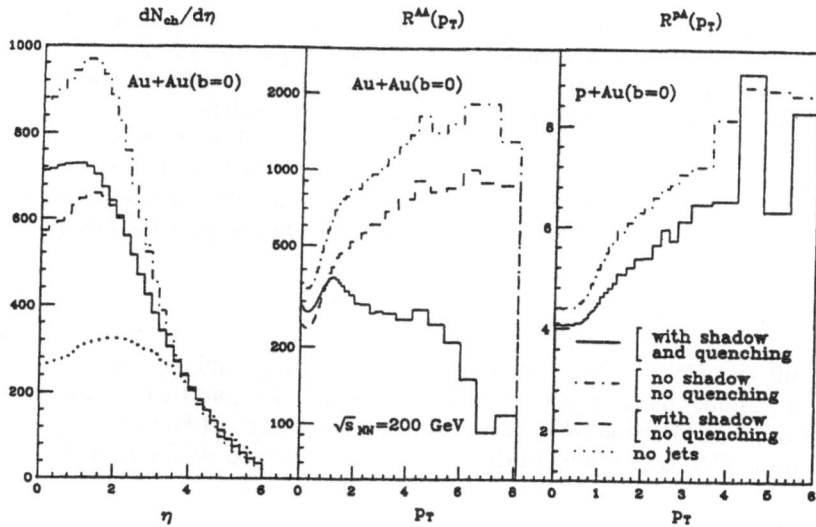

Fig. 5. Predictions of the HIJING Model for the pseudo-rapidity and transverse momentum distributions of charged hadrons in central Au + Au and p + Au collisions at RHIC. Displayed are the pseudo-rapidity distribution for central Au + Au (left panel), the ratio $R^{AB}(p_t)$ of the inclusive p_t spectrum of charged hadrons in central Au + Au collisions relative to that of pp (central panel), and the ratio $R^{PA}(p_t)$ of the inclusive p_t spectrum of charged hadrons in central p + Au collisions relative to that of pp (right panel).[56]

partons will be used as a penetrating probe of the QGP, and will provide new information on the nucleon and nuclear structure functions and parton shadowing in nuclei. An example of the effects of parton shadowing and quenching on the single particle distributions at midrapidity and at high transverse momentum can be seen in Fig. 5.

SUMMARY

Understanding the early stages of ultra-relativistic heavy ion collisions is fundamental to understanding the subsequent evolution of the system and determining the existence and type of phase transition between partonic and hadronic matter. The early dynamics of these collisions involving hard parton-parton interactions can be calculated using perturbative QCD. Various theoretical approaches result in predictions that highly excited ($T_{effective} \sim 500$ MeV), predominantly gluonic matter will be formed rapidly, within the first 0.3 fm/c of the collision process. The system then evolves through subsequent interactions and reinteractions, effectively cooling and expanding. The most important question affecting potential phase transitions between partonic and hadronic matter is whether chemical and/or thermal equilibrium are reached among the partonic degrees of freedom. Answering such questions requires a detailed understanding of the partonic content of the incident nuclei. Obtaining this information is of interest to both nuclear and particle physicists and will require measurements at various accelerators, including proton-proton and proton-nucleus measurements at high energy colliders. The Relativistic Heavy Ion Collider (RHIC) is being constructed and the Large Hadron Collider is being considered for construction and injection with heavy ions in order to investigate in the laboratory these new and fundamental properties of matter.

I have attempted to place in perspective the evolution of relativistic heavy ion physics and that of high transverse momentum parton physics. Both evolved simultaneously and their intersection has led to new theoretical and experimental approaches to the study and understanding of ultra-relativistic heavy ion collisions. A combination of what has been learned from both fields will be necessary to understand ultra-relativistic heavy ion collisions; the behavior of nuclear, hadronic and partonic matter at high energy densities; and how the Universe evolved from a hot plasma of quarks, gluons and leptons 10 micro-seconds after its beginning to what we now know.

ACKNOWLEDGMENTS

I thank Tim Hallman for comments on the manuscript and Joy Lofdahl for technical assistance with the manuscript. This work was supported in part by the Director, Office of Energy Research, Division of Nuclear Physics of the Office of High Energy and Nuclear Physics of the U.S. Department of Energy under contract DE-AC03-76SF00098.

REFERENCES

[1] E. Feenberg and H. Primakoff, Phys. Rev. 70, 980 (1946).

[2] A.R. Bodmer, Phys. Rev. D4, 1601 (1971).

[3] G. Baym and S.A. Chin, Phys. Lett. 62B, 241 (1976).

[4] T.D. Lee and G.C. Wick, Phys. Rev. D9, 2291 (1974).

[5] T.D. Lee, Rev. Mod. Phys. 47, 267 (1975).

[6] In the article above, T.D. Lee points out that "in high-energy physics we have concentrated on experiments in which we distribute a higher and higher amount of energy into a region with smaller and smaller dimensions. In order to study the question of 'vacuum', we must turn to a different direction; we should investigate some 'bulk' phenomena by distributing high energy over a relatively large volume."

[7] J.C. Collins and M.J. Perry, Phys. Rev. Lett. 34, 1353 (1975).

[8] G. Chapline and M. Nauenberg, Phys. Rev. D16, 450 (1977).

[9] L. Susskind, Phys. Rev. D20, 2610 (1979).

[10] A.B. Migdal, Rev. Mod. Phys. 50, 107 (1978) and references therein.

[11] V. Ruck, M. Gyulassy and W. Greiner, Z. Phys. A277, 391 (1976).

[12] H.-A. Gustafsson et al., Phys. Rev. Lett. 52, 1590 (1984).

[13] R.E. Renfordt et al., Phys. Rev. Lett. 53, 763 (1984).

[14] A. Capella and J. Tran Thanh Van, Z. Phys. C10, 249 (1981) and Phys. Lett. 93B, 146 (1980).

[15] B. Andersson et al., Phys. Rep. 97, 31 (1983); B. Andersson, G. Gustafson and B. Nilsson-Almqvist, Nucl. Phys. B281, 289 (1987); B. Nilsson-Almqvist and E. Stenlund, Comp. Phys. Comm. 43, 387 (1987).

[16] see for example Proceedings of the Sixth International Conference on Ultra-Relativistic Nucleus-Nucleus Collisions - Quark Matter 1987, Z. Phys. C38 (1988) and Proceedings of the Seventh International Conference on Ultra-Relativistic Nucleus-Nucleus Collisions, Nucl. Phys. A 498 (1989).

[17] H. Sorge, H. Stöcker and W. Greiner, Ann. Phys. 192, 266 (1989).

[18] K. Werner, Z. Phys. C42, 85 (1989).

[19] see H.G. Ritter, Proceedings of this Summer School and D. Keane, Proceedings of this Summer School.

[20] see G. Roche, Proceedings of this Summer School.

[21] J.J. Kokkedee, The Quark Model, W.A. Benjamin Pub., 1969; see reprints within.

[22] W.K.H. Panofsky, Proc. 14th International Conference on High Energy Physics, Vienna 1968, pub. CERN, Geneva, 23 (1968).

[23] M. Breidenbach et al., Phys. Rev. Lett. 23, 935 (1969).

[24] R.P. Feynman, in High Energy Collisions, Third International Conference held at SUNY, Stony Brook, 1969, Gordon and Breach Pub., New York, 1969; R.P. Feynman, Phys. Rev. Lett. 23, 1415 (1969).

[25] J.D. Bjorken, Phys. Rev. 179, 1547 (1969).

[26] J.D. Bjorken and E.A. Paschos, Phys. Rev. 185, 1975 (1969).

[27] R.P. Feynmann, Photon-Hadron Interactions, Benjamin 1972.

[28] S.M. Berman and M. Jacob, Phys. Rev. Lett. 25, 1683 (1970).

[29] S.M. Berman, J.D. Bjorken and J.B. Kogut, Phys. Rev. D4, 3388 (1971).

[30] J.D. Bjorken, Phys. Rev. D8, 4098 (1973).

[31] S.D. Ellis and M.B. Kislinger, Phys. Rev. D9, 2027 (1974).

[32] G. Altarelli and G. Parisi, Nucl. Phys. B26, 298 (1977).

[33] C. Albajar et al. (UA1 Collaboration), Nucl. Phys. B335, 261 (1990).

[34] P.V. Landshoff, Nucl. Phys. A498, 217c (1989).

[35] K. Kajantie, P.V. Landshoff and J. Lindfors, Phys. Rev. Lett. 59, 2517 (1987); K.J. Eskola, K. Kajantie and J. Lindfors, Nucl. Phys. B323, 37 (1989).

[36] A.R. Norton, *Multiparticle Production*, ed. R. Hwa and X. Qu-Bing, World Scientific Pub., p. 87 (1988).

[37] L. Jönsson, Nucl. Phys. A566, 5c (1994).

[38] J.J. Aubert et al. (EMC Collaboration), Phys. Lett. B123, 275 (1983).

[39] see M. Arneodo, CERN Preprint No. CERN-PPE/92-113 (1992) to be published in Physics Reports.

[40] K. Saito and T. Uchiyama, Z. Phys. A322, 299 (1985).

[41] W. Kittel, Act. Phys. Pol. B12, 1093 (1981).

[42] see D. Perkins, *Introduction to High Energy Physics*, Addison-Wesley Pub. (1982).

[43] see K.J. Eskola, Nucl. Phys. B400, 240 (1993).

[44] E. Shuryak, Phys. Rev. Lett. 68, 3270 (1992) and Nucl. Phys. A566, 559c (1994).

[45] X.N. Wang and M. Gyulassy, Phys. Rev. D44, 3501 (1991), Phys. Rev. D45, 844 (1992).

[46] K. Geiger and B. Müller, Nucl. Phys. B369, 600 (1992), Phys. Rev. D47, 133 (1993).

[47] I. Kawrakov, H.-J. Möhring and J. Ranft, Nucl. Phys. A544, 471c (1992).

[48] E. Shuryak, Nucl. Phys. A566, 559c (1994).

[49] K. Geiger, Nucl. Phys. A566, 257c (1994).

[50] for current status of signatures see Proceedings of the Tenth International Conference on Ultra-Relativistic Nucleus-Nucleus Collisions, Nucl. Phys. A566 (1994).

[51] J. Harris et al. (STAR Collaboration), Nucl. Phys. A566, 277c (1994).

[52] S. Nagamiya, Nucl. Phys. A566, 287c (1994).

[53] F. Videbaek, et al. (BRAHMS Collaboration), Nucl. Phys. A566, 299c (1994).

[54] B. Wyslouch, Nucl. Phys. A566, 305c (1994).

[55] J. Schukraft et al., (ALICE Collaboration), Nucl. Phys. A566, 311c (1994).

[56] X.N. Wang and M. Gyulassy, Nucl. Phys. A544, 559c (1992) and references within.

CORRELATIONS IN THE QUARK VACUUM

S. Schramm

Nuclear Theory Center, Indiana University
Bloomington, IN 47408, USA

INTRODUCTION

During the last two decades $SU_C(3)$ gauge theory has been firmly established as the underlying theory of strong interactions. Many experimental signals support the concept of QCD. For instance, 2-jet [1] and 3-jet [2] events in high-energy e^+ e^- annihilation revealed the hadronic substructure as consisting of spin $\frac{1}{2}$ and spin 1 particles. These can be identified with the states of the fundamental and adjoint SU(3) multiplets, i.e. the quarks and gluons, respectively. At large momenta QCD is increasingly well understood at a qualitative and quantitative level. The situtation is much worse, however, in the regime of low-enery QCD. The primary difficulty for this can be understood by considering the momentum dependence of the QCD running coupling constant $\alpha_{QCD}(q^2)$, which in one-loop summation reads

$$\alpha_{QCD}(q^2) = \frac{4\pi}{(11 - 2N_f/3)\ln(-q^2/\Lambda^2)} \ ,\tag{1}$$

where N_f is the number of quark flavors. The parameter $\Lambda \sim 100 - 500$ MeV sets the overall scale in QCD. As can be seen from Eq. (1), α is small for high momenta $q^2 \gg \Lambda^2$. This guarantees the validity of perturbative calculations for large $q^2 \gg 1\,\text{GeV}^2$ (asymptotic freedom) and is the reason for the quantitative success of perturbative QCD. For low momenta $q^2 \sim \Lambda^2$, however, the coupling becomes large, $\alpha \gtrsim 1$, excluding the application of standard perturbation theory with respect to the coupling constant, since higher-order diagrams are not suppressed any more. Therefore non-perturbative methods have to be applied in order to calculate low-energy QCD processes and states like baryons, mesons or glueballs.

One approach is to solve QCD exactly on the lattice. With the development of faster computers and more efficient numerical techniques considerable progress has been made in this field [3]. There is still much more work to be done, especially in including dynamic fermions in the calculation and in treating systems with non-zero chemical potential.

Another alternative approach to treating non-perturbative effects in field theories is based on variational model calculations, which will be discussed in some more detail in the following. One advantage of calculations of this kind is that it is possible (and necessary) to incorporate some physical idea in the approximation of the QCD hamiltonian or in the restriction of the relevant Fock space. Furthermore it is straight forward to extend the calculation to non-zero temperature and chemical potential.

The outline of the article is as follows. First, in a general discussion some concepts of treating non-trivial vacuum structures are introduced. Then an approximation of the QCD hamiltonian is derived and a variational calculation of the non-perturbative vacuum state is performed which is followed by an analogous discussion using the Nambu-Jona-Lasino model [4]. The paper concludes with a few remarks on the possibility of four-quark clusters in the vacuum.

Particle Condensation

The field operator $\psi(x)$ of a spin-$\frac{1}{2}$ field can be expanded in Fock space in the form

$$\psi(x) = \sum_n a_n \phi_n(x) \ . \tag{2}$$

a_n denote the annihilation operators for a particle in state n. The summation (integration) is performed over a complete set of states n. If one uses a set of wavefunctions ϕ_n which are solutions of the non-interaction Hamiltonian of the system the corresponding annihilation and creation operators define the perturbative vacuum of the system $|0>$ in the following way:

$$a_n|0> = 0 \quad , \quad \forall \, n \ . \tag{3}$$

That is, the perturbative vacuum does not contain particles defined in the way described above. This is generally not true if one introduces a different expansion in Eq. (2). For Eq. (3) to be valid even in the non-interacting case, the spectrum of the one-body Hamiltonian should be positive definite. This is not the case in a relativistic system where the energy spectrum of free particles with mass m is $E = \pm\sqrt{p^2 + m^2}$. Of course, this problem can be solved by introducing antiparticle operators b_n so that the new operators read

$$a_n \equiv a_n^{(+)} \quad , \quad b_n \equiv (a_n^{(-)})^\dagger \tag{4}$$

where $\pm$ refers to the sign of the energy of the state n. As can be seen later the redefinition (4) is a special case of introducing a non-trivial groundstate. In the case of an external field it can also happen that the single-particle spectrum exhibits negative energy eigenvalues. One famous example is the case of electrons in the supercritical external electric field of a hypothetical nucleus with charge larger than Z=173. It has been shown [5] that in this case spontaneous electron-positron pair production occurs, which again is a special case of introducing particle-antiparticle pairs in the groundstate via a transformation of the vacuum as explained in the following.

Generally one can introduce a new vacuum state through a unitary Fock-space transformation U of the original state:

$$|\tilde{0}> \equiv U|0> \quad , \quad U^{\dagger} = U^{-1} \quad , \tag{5}$$

where U should preserve the known symmetries of the groundstate. U is a functional of the particle operators a_n and can contain general particle-antiparticle correlations so that the new state $|\tilde{0}>$ has a complicated clustered structure. It is straightforward to define new particle operators corresponding to the new groundstate (5). Starting from the definition of $|0>$, Eq. (3), one gets

$$
\begin{aligned}
U a_n |0> &= 0 \quad , \\
U a_n U^{\dagger} \, U|0> &= 0 \quad ; \\
\tilde{a}_n |\tilde{0}> &= 0 \quad .
\end{aligned}
\tag{6}
$$

Eq. (6) defines the new quasiparticle operator $\tilde{a}_n$

$$\tilde{a}_n \equiv U a_n U^{\dagger} \quad . \tag{7}$$

Since U is unitary the quasiparticle operators obey the same anti-commutation rules as the particle operators:

$$\{\tilde{a}_n, \tilde{a}_m^{\dagger}\} = \{a_n, a_m^{\dagger}\} \quad , \quad \{\tilde{a}_n, \tilde{a}_m\} = \{a_n, a_m\} \quad . \tag{8}$$

A simple example is a BCS-type transformation which introduces irreducible two-particle correlations in the groundstate. In this case U reads

$$U = \exp\{\sum_k \alpha_k \left(a_n^{\dagger} a_{-n}^{\dagger} - a_{-n} a_n\right)\} \quad , \tag{9}$$

where $-n$ denotes the time-reversed state of n. The vacuum $|\tilde{0}>$ can be determined easily as

$$|\tilde{0}> = \prod_k (\cos \alpha_k + \sin \alpha_k a_k^{\dagger} a_{-k}^{\dagger})|0> \quad . \tag{10}$$

The quasiparticle operator follows from Eq. (7) as

$$\tilde{a}_k = U a_k U^{\dagger} = \cos\alpha_k \, a_k - \sin\alpha_k a_{-k}^{\dagger} \tag{11}$$

which is the well-known Bogoljubov transformation. Finally, the parameters of the trial transformation can be determined by minimization of the energy, i.e. by solving

$$\frac{\delta}{\delta \alpha_n} < \tilde{0}|H|\tilde{0}> = 0 \quad , \forall \, n \quad . \tag{12}$$

In principle, the structure of the transformation (5) is not restricted to simple two-particle or particle-antiparticle correlations, but one can introduce also higher-order terms. For illustration consider the example of a nucleus with protons (a_n)

and neutrons (b_n). One can add alpha-particle type correlations by considering the unitary transformation [6]

$$U_\alpha = \exp\{\sum_n \beta_n \left(a_n^\dagger b_n^\dagger a_{-n}^\dagger b_{-n}^\dagger - a_n b_n a_{-n} b_{-n} \right)\} \tag{13}$$

The corresponding vacuum and particle operators can be constructed in the same way as described above. The results are

$$U_\alpha |0> = \prod \left[\cos(\beta_n) + \sin(\beta_n) a_n^\dagger b_n^\dagger a_{-n}^\dagger b_{-n}^\dagger \right] |0> \tag{14}$$

and

$$\begin{aligned}\tilde{a}_n &= \cos\beta_n a_n - \sin(\beta_n) b_{-n}^\dagger a_{-n}^\dagger b_n^\dagger \\ &+ [1 - \cos(\beta_n)]\left[N_{b+} + N_{b-} + N_{a-} - N_{a-}N_{b+} - N_{b-}N_{b+} - N_{b-}N_{a-} \right] a_n \end{aligned} \tag{15}$$

with the number operators $N_{a\pm} = a_{\pm n}^\dagger a_{\pm n}$ and $N_{b\pm} = b_{\pm n}^\dagger b_{\pm n}$. Note that although the quasiparticle operators are still anti-commuting, they are not merely rotations in the single-particle sector but contain clusters of "perturbative" particles. These higher clusters might be of importance in QCD, too, as will be discussed later. So far, however, most field theoretical calculations are restricted to correlations of the type (9).

Simple Quark Dynamics from QCD

Non-Local Interactions

One can derive a simple quark hamiltonian starting from the original QCD Lagrangian by neglecting gluonic degrees of freedom [7]. The QCD Lagrange density reads

$$\mathcal{L} = \bar{q}_{\alpha,f}(x) \left(i(D_\mu)^{\alpha\beta}\gamma^\mu - m_{0,f}\delta^{\alpha,\beta} \right) q_{f,\beta}(x) - \frac{1}{4} F_{\mu\nu}^a(x) F^{\mu\nu,a}(x) \ . \tag{16}$$

$m_{0,f}$ is the current quark mass of a quark with flavor f which is a few MeV in the case of the up and down quark and about 150 MeV for the strange quark. The covariant derivative is given by

$$D_\mu^{\alpha\beta} = \partial_\mu \delta^{\alpha\beta} - ig\frac{1}{2}\lambda^{\alpha\beta,a} A_\mu^a \tag{17}$$

and the gluonic field tensor reads

$$-ig\frac{\lambda^a}{2}F_{\mu\nu}^a = \left[D_\mu^a, D_\nu^a \right] \ . \tag{18}$$

Note that with the exception of the small mass term the Lagrangian (16) is chirally symmetric, i.e. left- and right-handed quarks are decoupled. Performing a Legendre transformation in the usual way, and introducing Coulomb gauge for the gluon fields, i.e.

$$\vec{\nabla} \cdot \vec{A}^a = 0 \quad , \quad \forall a \tag{19}$$

one can derive the (classical) QCD-Hamiltonian

$$\begin{aligned}
H \;=\; & \int d^3x \, q_f^\dagger(x)(\vec{\alpha}\vec{p} + \beta m_{0,f}) q_f(x) \,- \\
& -\frac{g^2}{2} \int d^3x \, d^3x' \, j^{0,a}(x) \left[(\vec{\nabla}\cdot\vec{D})^{-1}(\vec{\nabla}\cdot\vec{\nabla})(\vec{\nabla}\cdot\vec{D})^{-1} \right]_{(x-x')} j^{0,a}(x') \,+ \\
& + \frac{1}{2} \int d^3x \left(\vec{E}^a \cdot \vec{E}^a + \vec{B}^a \cdot \vec{B}^a \right)
\end{aligned} \tag{20}$$

where the color density operator $j^{0,a}(x)$ is given by

$$j^{0,a}(x) = \bar{q}(x)\gamma^0 \frac{\lambda^a}{2} q(x) + f^{abc} \vec{A}^b \cdot \vec{E}^c \quad . \tag{21}$$

Quantization introduces additional factors arising from the curvilinear metric in the case of Coulomb gauge. These complications and others concerning the Gribov ambiguity [8] will not be considered here, since we neglect physical gluons in this discussion. With this condition, i.e.

$$A_i^a \, |F> \;=\; 0 \tag{22}$$

the Hamiltonian reads

$$\begin{aligned}
H = & \int d^3x \, q_f^\dagger(x)(\vec{\alpha}\vec{p} + \beta m_{0,f}) q_f(x) \,- \\
& -\frac{g^2}{2} \int d^3x \, d^3x' \, \bar{q}(x)\gamma^0 \frac{\lambda^a}{2} q(x) \left(\frac{1}{\vec{\nabla}\cdot\vec{\nabla}} \right)_{(x-x')} \bar{q}(x')\gamma^0 \frac{\lambda^a}{2} q(x') \quad .
\end{aligned} \tag{23}$$

In a similar spirit one can introduce effective interactions between the quarks by considering general quark currents $\bar{q}\Gamma q$, where Γ is a Dirac matrix, and by modifying the propagator of the exchanged particle in Eq. (23). Care has to be taken no to violate the basic symmetries of the original QCD Lagrangian.

Expanding the quark operator in momentum space yields

$$q_{\alpha,f}(x) = \int \frac{d^3p}{(2\pi)^{3/2}} \sqrt{\frac{m_{0,f}}{p_0}} \left[u(p,s) b_{\alpha,f}(p,s) + v(-p,s) d_{\alpha,f}^\dagger(-p,s) \right] e^{ip\cdot x} \tag{24}$$

with $u(p,s), v(p,s)$ being the spinor of a quark and antiquark, respectively, with momentum p and helicity s. The anti-commutation rules for the quark creation and annihilation operators are:

$$\begin{aligned}
& \{ b_{\alpha,f}(\vec{p},s), b_{\alpha',f'}(\vec{p}',s') \} = \{ b_{\alpha,f}^\dagger(\vec{p},s), b_{\alpha',f'}^\dagger(\vec{p}',s') \} = 0 \quad , \\
& \{ b_{\alpha,f}(\vec{p},s), b_{\alpha',f'}^\dagger(\vec{p}',s') \} = \delta(\vec{p}-\vec{p}')\delta_{ff'}\delta_{\alpha\alpha'}\delta_{ss'} \quad .
\end{aligned} \tag{25}$$

411

Following the procedure outlined in Section 2 one can introduce quark correlations in the vacuum by applying a unitary transformation to the perturbative vacuum $|0>$ [9]:

$$U|0> \equiv |\tilde{0}>$$
(26)

with

$$U(\{\zeta_p\}) = e^{i\sum_{s,\alpha}\int d^3p\,\zeta_p\left[sb_\alpha^\dagger(\vec{p},s)d_\alpha^\dagger(-\vec{p},s)-sd_\alpha(-\vec{p},s)b_\alpha(\vec{p},s)\right]}$$
(27)

where ζ_p is assumed to be real, depending on the absolute value $|\vec{p}|$ only. The quasi-particle operators $\tilde{b}, \tilde{d}$ are defined by

$$\tilde{b}_\alpha(\vec{p},s) = U(\{\zeta_p\})b_\alpha(\vec{p},s)U^\dagger(\{\zeta_p\}) \quad ,$$

$$\tilde{d}_\alpha(\vec{p},s) = U(\{\zeta_p\})d_\alpha(\vec{p},s)U^\dagger(\{\zeta_p\}) \quad .$$
(28)

Note that one can write the transformation U in terms of the particle operators $b_\alpha(\vec{p},s)$, $d_\alpha(\vec{p},s)$ as well as in terms of the quasiparticles $\tilde{b}_\alpha(\vec{p},s)$, $\tilde{d}_\alpha(\vec{p},s)$:

$$
\begin{aligned}
& e^{i\sum_{s,\alpha}\int d^3p\,\zeta_p\left[sb_\alpha^\dagger(\vec{p},s)d_\alpha^\dagger(-\vec{p},s)-sd_\alpha(-\vec{p},s)b_\alpha(\vec{p},s)\right]} \\
&= U(\{\zeta_p\})e^{i\sum_{s,\alpha}\int d^3p\,\zeta_p\left[sb_\alpha^\dagger(\vec{p},s)d_\alpha^\dagger(-\vec{p},s)-sd_\alpha(-\vec{p},s)b_\alpha(\vec{p},s)\right]}U^\dagger(\{\zeta_p\}) \\
&= e^{i\sum_{s,\alpha}\int d^3p\,\zeta_p U(\{\zeta_p\})\left[sb_\alpha^\dagger(\vec{p},s)d_\alpha^\dagger(-\vec{p},s)-sd_\alpha(-\vec{p},s)b_\alpha(\vec{p},s)\right]U^\dagger(\{\zeta_p\})}
\end{aligned}
$$
(29)

which, according to Eq. (28), is equal to

$$= e^{i\sum_{s,\alpha}\int d^3p\,\zeta_p\left[s\tilde{b}_\alpha^\dagger(\vec{p},s)\tilde{d}_\alpha^\dagger(-\vec{p},s)-s\tilde{d}_\alpha(-\vec{p},s)\tilde{b}_\alpha(\vec{p},s)\right]} \quad .$$
(30)

Eq. (18) yields the Bogoljubov transformation

$$\tilde{b}_\alpha(\vec{p},s) = \cos\zeta_p\,b_\alpha(\vec{p},s) + s\sin\zeta_p\,d_\alpha^\dagger(-\vec{p},s)$$
(31)

$$\tilde{d}_\alpha(\vec{p},s) = \cos\zeta_p\,d_\alpha(\vec{p},s) - s\sin\zeta_p\,b_\alpha^\dagger(-\vec{p},s)$$

and the inverse relation

$$b_\alpha(\vec{p},s) = \cos\zeta_p\,\tilde{b}_\alpha(\vec{p},s) - s\sin\zeta_p\,\tilde{d}_\alpha^\dagger(-\vec{p},s)$$
(32)

$$d_\alpha(\vec{p},s) = \cos\zeta_p\,\tilde{d}_\alpha(\vec{p},s) + s\sin\zeta_p\,\tilde{b}_\alpha^\dagger(-\vec{p},s) \quad ,$$

respectively. By inserting Eq.(32) into (23) and normal-ordering the expression one obtains a structure of the hamiltonian of the form

$$H = E_{\tilde{0}} + H_2(\tilde{b},\tilde{d}) + H_4(\tilde{b},\tilde{d})$$
(33)

where H_2 contains the energy operator for the quasiquarks and H_4 includes the residual interaction between the quasiparticles. Minimization of the vacuum energy E_0 with respect to ζ_p completely fixes the quasiquark hamiltonian and one can determine quantities like the mass of the quasiquarks and masses of mesons as bound states of a quasi-quark and -antiquark.

By introducing scalar $q\bar{q}$ pairs in the QCD trial vacuum (26) one has spontaneously broken the chiral symmetry of the system. This is the so-called Nambu-Goldstone mode of symmetry breaking which is induced through the state vector rather than through explicit symmetry breaking terms in the Lagrangian. The fact that chiral is broken is illustrated in Fig. 1. A left-handed quark which interacts with one of the

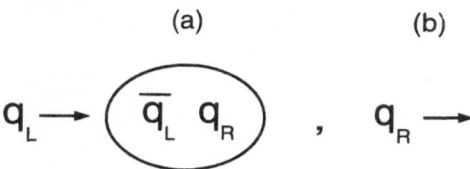

Figure 1. (a) A left-handed quark scatters from a scalar $q\bar{q}$ pair. (b) A right-handed quark is emitted.

vacuum pairs annihilates with the corresponding antiquark emitting a right-handed quark. This coupling of left- and right-handed particles corresponds to a mass term in the Lagrangian:

$$m\bar{q}q = m\left(\bar{q}_{\mathrm{L}}q_{\mathrm{R}} + \bar{q}_{\mathrm{R}}q_{\mathrm{L}}\right) \quad . \tag{34}$$

The vacuum correlations generate a dynamic mass of the particle, which, with the right choice of parameters, lies in the range of ~ 350 MeV in the case of the light quarks. Thus one might identify the quasiparticle with the heavy quark of the constituent quark model. Another important aspect of this method of symmetry breaking comes from the following argument. The vacuum Eq. (26) breaks chiral symmetry, i.e. a chiral rotation U_5 applied to (26) generates a new state $|\tilde{0}>$ as U_5 does not commute with U. The chirally invariant Hamiltonian commutes with U_5. This means that $|\tilde{0}>$ is a groundstate of the system degenerate with $|\tilde{0}>$. That is, as long as the hamiltonian does not break chiral symmetry explicitly the vacuum expectation value of the energy remains unchanged under a chiral rotation of the groundstate. The quantal excitations corresponding to these chiral rotations are therefore (in the limit of vanishing current masses) massless Goldstone bosons which can be identified with the pions. They aquire mass through the small chiral symmetry breaking current mass term in the Lagrangian. This property of the dynamic symmetry breaking yields a natural explanation how one can have heavy constituent quarks and light pseudoscalar mesons in a consistent picture. The equations for determining the pion decay constant f_π and pion mass m_π were derived in [10] by considering the Bethe-Salpeter equation for the pion vertex. In the chiral limit one gets the famous PCAC

Table 1: Results for a quark model calculation using different coupling strengths α, and cut-offs Λ (see [9]). The constituent quark mass M_Q, the scalar condensate $< \bar{q}q >$, pion mass m_π and pion decay constant f_π, and the vacuum energy density $B^{1/4}$ are shown (units in MeV). The last column lists the phenomenological or experimental value.

α	0.5	0.75	1.	1.25	1.5	
Λ	1.008	0.432	0.268	0.191	0.146	
M_Q	0.276	0.345	0.370	0.3866	0.399	0.340
$| < \bar{q}q > |^{1/3}$	0.306	0.255	0.230	0.212	0.195	0.235
m_π	0.135	0.135	0.135	0.135	0.135	0.135
f_π	0.083	0.079	0.074	0.069	0.065	0.095
$B^{1/4}$	0.101	0.130	0.132	0.131	0.127	0.145

relation:

$$m_\pi = \sqrt{2\bar{m}_0| < \bar{q}q > |}/f_\pi \tag{35}$$

$\bar{m}_0$ is the average current quark mass

$$\bar{m}_0 = \frac{1}{2}(m_{0,u} + m_{0,d}) \quad , \tag{36}$$

and $< \bar{q}q >$ is the scalar condensate which is the measure of the admixture of $q\bar{q}$ pairs to the vacuum.

Table 1 shows some numerical results for a calculation based on Eq. (23) (for details see [9]).

The NJL model

A simple realization of a relativistic quark Lagrangian is given by the Nambu-Jona-Lasino model [4], which has been quite successfully used to study the QCD chiral symmetry breaking. NJL calculations reproduce a number of low-energy QCD quantities to good accuracy. This suggests that the simple form of the Lagrangian incorporates basic features of the QCD chiral symmetry breaking in the quark sector. To demonstrate the method we adopt an extended version of the Nambu-Jona-Lasinio model [4] for the quark dynamics. Extensions of the original NJL model have been discussed in a number of papers [11].

A standard NJL Lagrangian has the following structure:

$$\mathcal{L} = \mathcal{L}_0 + \mathcal{L}_4 + \mathcal{L}_6 \tag{37}$$

with the Lagrangian for free quarks

$$\mathcal{L}_0 = \bar{q}(x)(\not{p} - \hat{m}_0)q(x) \tag{38}$$

$\hat{m}_0$ is the diagonal mass matrix of the current quarks. The quark field $q(x)$ is a vector in color $SU_C(3)$ and flavor $SU_F(3)$, respectively. As in the discussion above, with the exception of explicit current quark mass terms the QCD lagrangian is chirally invariant. Hadron masses are assumed to be generated through dynamical chiral symmetry breaking. Analogously, in the NJL approach one introduces a chirally symmetric interaction. $\mathcal{L}_4$ is a flavour-symmetric $U_{F,L}(3) \times U_{F,R}(3)$ invariant quark-quark point-like interaction [12]:

$$
\mathcal{L}_4 = c \left\{ \left[\bar{q}(x) \frac{\lambda^a}{2} q(x) \right]^2 - \left[\bar{q}(x) \frac{\lambda^a}{2} \gamma_5 q(x) \right]^2 \right\} \tag{39}
$$

where $\lambda^a (a = 0, ..., 8)$ are the generators of the flavor U(3) group with the normalisation $\text{Tr}[(\lambda^a)^2] = 2$. $\mathcal{L}_6$ is a $U_A(1)$ breaking three-quark interaction which has been introduced by t'Hooft [15]:

$$
\mathcal{L}_6 = \epsilon_{abc} \epsilon_{def} \left[\kappa (\bar{q}_{a,L} q_{d,R})(\bar{q}_{b,L} q_{e,R})(\bar{q}_{c,L} q_{f,R}) \right] + (L \leftrightarrow R) \tag{40}
$$

with $q_{\genfrac{}{}{0pt}{}{R}{L}} \equiv \frac{1}{2}(1 \pm \gamma_5)$. η_{abc} is the Levi-Civita tensor in flavor space. This term breaks the $U_A(1)$ which is phenomenologically required for achieving the correct $\eta - \eta'$ splitting [14]. As derived in [15] instanton configurations of the gluon gauge fields induce quark interactions of the structure Eq. (40). Furthermore there is increasing evidence from QCD lattice calculations that instantons play an important role in the chiral symmetry breaking [16].

For a sufficiently strong coupling constant c the interaction terms Eqs. (39,40) generate a condensation of scalar quark-antiquark pairs, which in turn induce a non-perturbative mass δm_f of the quasiparticle. The resulting integral equation reads:

$$
\delta m_f \equiv m_f - m_{0,f} = -8c < \bar{q}_f q_f > -\frac{7}{3} \kappa \prod_{f' \neq f} < \bar{q}_{f'} q_{f'} > \tag{41}
$$

with the scalar quark condensate

$$
< \bar{q}_f q_f > = -3i \int \frac{d^4 p}{(2\pi)^4} \text{Tr} S_{ff}(p) \quad . \tag{42}
$$

The quark with mass m_f can be identified with a constituent quark. The quark propagator is diagonal in flavor space and has the simple structure in the NJL approach:

$$
S_{ff'}(p) = \frac{\not{p} + m_f}{p^2 - m_f^2 + i\epsilon} \delta_{f,f'} \tag{43}
$$

The integral in Eq. (42) is divergent. One has to apply a cut-off procedure to regularize the integral. The most common method is to introduce a cut-off in the momentum integration performed in euclidean space-time [4].

To lowest order mesons can be described as poles in the quasiparticle - anti-quasiparticle scattering neglecting two or more quark-antiquark pair contributions to the mesonic state. The T-matrix of $q\bar{q}$ scattering reads [14, 13]:

$$
T_{\Gamma_i \Gamma_j}^{aa'} \left(\Gamma_i \lambda^a \otimes \Gamma_j \lambda^{a'} \right) = \left(\Gamma_i \lambda^a \otimes \Gamma_j \lambda^{a'} \right) \left\{ C_{\Gamma_i \Gamma_j}^{aa'} \right.
$$
$$
\left. - \sum \int \frac{d^4 q}{(2\pi)^4} \text{Tr} \left[C_{\Gamma_i \Gamma_k}^{ab} \Gamma_k \lambda^b S(p + q/2) T_{\Gamma_i \Gamma_j}^{ba'} \Gamma_i \lambda^b S(p - q/2) \right] \right\} \tag{44}
$$

415

Table 2: NJL calculation of the mass spectrum of the pseudoscalar mesons (π, K, η, η') compared to experiment (in units of MeV).

	m_π	m_K	m_η	$m_{\eta'}$
NJL	136	496	485	941
Exp.	135	498	549	958

$\hat{T}$ and the couplings $\hat{C}$, which are determined by the interactions (39,40) are matrices in flavor and Dirac space [14, 13, 17]. The interaction term in Eq. (40) generates a coupling in the $q\bar{q}$ scattering channel by contracting two of the six quark fields [17]. Γ_i denote the tensors $\{\mathbb{1}, \gamma_5, \gamma^\mu, \gamma_5\gamma^\mu, \sigma^{\mu\nu}\}$ for the different quantum numbers of $q\bar{q}$ channels. Poles in the solution of the self-consistent set of equations (44) define the mesonic states of the theory.

Table 2 shows the result of a model calculation for the pseudoscalar mesons. It shows that the model can reproduce meson spectra reasonably well on the level of simple quark-antiquark correlations in the vacuum.

Conclusions, Outlook

Basic variational methods treating non-perturbative vacuum states were discussed. In the case of QCD it was shown that simple models based on quark-antiquark correlations in the groundstate are already able to reproduce many low-energy QCD properties. However, recent calculations using finite-density QCD sum rules might hint to a more complex structure of the vacuum [18]. There the usual assumption of the factorization of the local four-quark condensate

$$O_4 \equiv < \bar{q}(x)q(x)\bar{q}(x)q(x) > \tag{45}$$

might not be valid anymore. The factorization hypothesis says that the O_4 can be approximated by the product of the two-quark condensate

$$O_4 \sim < \bar{q}(x)q(x) >^2 \tag{46}$$

One can understand the approximation by inserting a complete set of states in (45)

$$O_4 = \sum_n < \bar{q}(x)q(x)|n >< n|\bar{q}(x)q(x) > \tag{47}$$

If one assumes that the scalar operator $\bar{q}(x)q(x)$ couples mainly to the vacuum state $|n > = |0 >$ then Eq. (46) follows automatically. In the case of a BCS-type groundstate there are by construction only irreducible two-particle correlations present in the vacuum by construction. Therefore, the four-quark condensate reduces to a product of two-quark condensates as in (46). A violation of the factorization hypothesis would imply that the simple vacuum structure (26) is not sufficient. Introducing four-quark correlations in the spirit of Eq. (13), however, would produce non-factorizable contributions which might explain a possible failure of the factorization assumption[17]. Clearly, more work has to be done along this line.

416

REFERENCES

1. G. Hanson et al., Phys. Rev. Lett **35** (1975) 1609.

2. R. Brandelik et al. (TASSO collab.), Phys. Lett. **B86** (1979) 243;
   D.P. Barber et al. (Mark-J collab.), Phys. Rev. Lett. **43** (1979) 830;
   C. Berger et al. (PLUTO collab.), Phys. Lett. **B86** (1979) 418;
   W. Bartel et al. (JADE collab.), Phys. Lett. **B91** (1980) 142.

3. see, e.g., D. Touissant, Nucl. Phys. **B26** (1992) 3.

4. Y. Nambu and G. Jona-Lasino, Phys. Rev. **122**, 345 (1961); Phys. Rev. **124**, 246 (1961).

5. W. Pieper and W. Greiner, Z. Phys. **218** (1969) 327.

6. S. Schramm, B. Müller and W. Greiner, J. Phys. **G13** (1987) L69.

7. J.R. Finger and J.E. Mandula, Nucl. Phys. **B199** (1982) 168.

8. V.N. Gribov, Nucl. Phys. **B139** (1978) 1.

9. S. Schramm and W. Greiner, Int. J. Mod. Phys. **E1** (1992) 73.

10. J. Govaerts, J.E. Mandula, and J. Weyers, Nucl. Phys. **B237** (1984) 59.

11. For a recent review see: S. P. Klevansky, Rev. Mod. Phys. **64**, 649 (1992).

12. D. Ebert and H. Reinhardt, Nucl. Phys. B271, 188 (1986).

13. S. Klimt, M. Lutz, U. Vogl and W. Weise, Nucl Phys. **A516**, 429 (1990); U. Vogel, M. Lutz, S. Klimt and W. Weise, Nucl. Phys. A516, 469 (1990).

14. M. Takizawa, K. Tsushima, Y. Kohyama and K. Kubodera, Nucl. Phys. **A507**, 611 (1989).

15. G. t'Hooft, Phys. Rev. **D14** (1976) 3432, Phys. Rev. **D18** (1978) 2199.

16. M-C. Chu and S. Huang, Phys. Rev. D **45**, 2446 (1992).

17. S. Schramm, in preparation.

18. R. J. Furnstahl, D. K. Griegel, T. D. Cohen, Phys. Rev. D **46**, 1507 (1992).

COMPRESSION, EXPANSION, AND FREEZE-OUT IN NUCLEUS-NUCLEUS COLLISIONS AT THE AGS

Peter Braun-Munzinger

Department of Physics
SUNY at Stony Brook
Stony Brook, NY 11794-3800, USA

E814/E877 Collaboration:
J. Barrette[3], R. Bellwied[8], S. Bennett[8], P. Braun-Munzinger[6], W.E. Cleland[5], T.M. Cormier[8], G. David[6], J. Dee[6], G.E. Diebold[9], O. Dietzsch[7], J.V. Germani[9], S. Gilbert[3], S.V. Greene[9], J.R. Hall[8], T.K. Hemmick[6], N. Herrmann[2], B. Hong[6], K. Jayananda[5], D. Kraus[5], B. Shiva Kumar[9], R. Lacasse[3], Q. Li[8], A. Lukaszew[8], W.J. Llope[6], T.W. Ludlam[1], S. McCorkle[1], R. Majka[9], S.K. Mark[3], R. Matheus[8], J.T. Mitchell[9], M. Muthuswamy[6], E. O'Brien[1], S. Panitkin[6], C. Pruneau[8], M.N. Rao[6], M. Rosati[3], F. Rotondo[9], N.C. daSilva[7], S. Sedykh[6], U. Sonnadara[5], J. Stachel[6], H. Takai[1], E.M. Takagui[7], J. Wessels[6], C. Winter[9], G. Wang[3], D. Wolfe[4], C.L. Woody[1], N. Xu[6], Y. Zhang[6], Z. Zhang[5], C. Zou[6]

[1]BNL – [2]GSI – [3]McGill Univ.– [4]Univ. of New Mexico – [5]Univ. of Pittsburgh – [6]SUNY Stony Brook – [7]Univ. of São Paulo – [8]Wayne State Univ. – [9]Yale Univ.

INTRODUCTION

In the following we will present selected aspects of recent data on central nucleus-nucleus collisions taken by the E814/E877 collaboration at the BNL-AGS. We will first demonstrate that stopping is nearly complete in these collisions, and discuss the initial energy and baryon densities inferred from model comparisons with the data. Next we will discuss the amount of expansion the system undergoes from the initial phase until freeze-out by presenting recent results on pion-pion correlations. The (baryonic) resonance composition of the system at freeze-out is obtained by analyzing pion spectra at low transverse momentum p_t and by inspection of direct measurements of the $\Delta(1232)$ resonance. Based on analysis of these measurements we will then discuss a "local" freeze-out scenario, where the freeze-out volume is connected with the temperature averaged pion-nucleon cross section. A summary of the current results and of future plans for Au-Au collisions with the E877 apparatus concludes the paper.

The data presented were obtained for central 14.6 A GeV/c Si + Al and Si + Pb collisions and 11.4 A GeV/c Au+Au collisions. The E814/E877 experimental setup has been described previously (see e.g. [1, 2]); all data presented here were taken in runs in spring 1991 and 1992 in the 'open spectrometer configuration' described in [3]. The forward spectrometer covers in one fixed setting an angular range in the magnet bend

Hot and Dense Nuclear Matter, Edited by
W. Greiner *et al.*, Plenum Press, New York, 1994

plane of $-115 < \theta_x < 14$ mr and $-21 < \theta_y < 21$ mr perpendicular to it. Centrality is selected by measuring associated charged particle multiplicity and/or transverse energy. The degree of centrality selection is quantified by the ratio of the measured cross section σ to the geometric cross section $\sigma_{geo} \approx 3.6$ b (1.6 b) for Si+ Pb (Al) and ≈ 6 b for Au + Au.

BARYON RAPIDITY AND TRANSVERSE ENERGY DISTRIBUTIONS
2.1

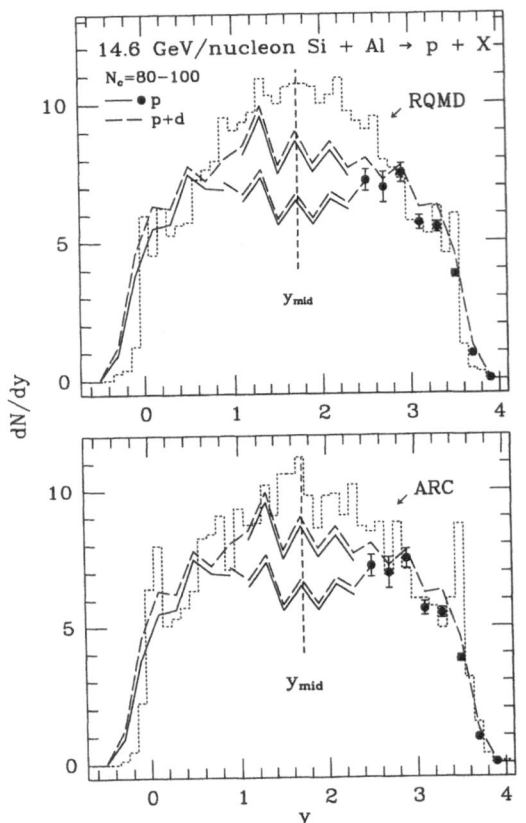

Figure 1: Proton rapidity distribution [3] (solid dots and lines) and sum of proton and deuteron distributions (long dashed lines) together with predictions from the event generators RQMD and ARC.

The measured proton rapidity distribution for very central ($\sigma/\sigma_{geo} = 0.2$ %) Si + Al collisions is presented in Figure 1. The data are published in [3], where details about the setup, the analysis and the determination of systematic errors are described.

The absence of a peak at beam rapidity indicates that a large degree of stopping is achieved in the system. However, the distribution is much wider than expected for a fully stopped isotropic proton source where $dN/dy \propto \exp[-(y - y_f)^2/2 \cdot 0.34^2]$. A quantitative analysis of the degree of stopping for this system has been made in [4, 5]. There it is shown that either the amount of stopping is about 65 % or that the protons were initially fully stopped and expanded longitudinally with expansion velocity $\beta_{\parallel} \approx 0.6$.

To get the full baryon distribution we add, from a recent E814 analysis of deuteron distributions, the number of protons bound in deuterons to the proton distributions (see the dashed lines in Fig. 1). The data in Fig. 1 are compared to predictions from the event generators RQMD [6] and ARC [7], both based on cascading and resonance production. The overall degree of stopping is reasonably described by both models. The slight discrepancy at midrapidity could indicate that the (inverse) slope constant of the proton m_t spectra is actually larger than we assumed for the evaluation of dN/dy.

Full stopping and very high baryon densities are also inferred from our recent measurements [2] of transverse energy production in Au+Au collisions at 11.4 GeV/c per nucleon. Results for the angular distribution of the produced transverse energy are shown in Fig. 2. Values of 200 GeV/unit of pseudorapidity η are observed for the Au+Au system, a factor of 8.7 larger than what is observed for Si+Al. Accounting for the different beam energy this implies a 48 % increase in the central rapidity density of E_t compared to what is expected from the available energy in the center-of-mass. In the ARC [7] and RQMD [6] models, which describe the transverse energy production rather well (see [2]), initial energy and baryon densities of $> 1.5\ \mathrm{GeV/fm^3}$ and > 8 times normal nuclear matter density are achieved over significant time periods ($> 5 fm/c$) for these heavy systems.

PION CORRELATIONS AND SOURCE SIZE AT FREEZE-OUT

The results of $\pi^+\pi^+$ and $\pi^-\pi^-$ correlation function measurements for central Si + Pb collisions (top 10 %) [8] are presented in Figure 3. Because of the correlation between momentum and space for pions and its strong influence on the correlation function measured within a finite acceptance we do not attempt to extract source radii directly from a fit to the data. Rather we compare our measured correlation functions with predictions from the event generator RQMD [6], using the Koonin/ Pratt formalism to compute the correlation function within the acceptance of our spectrometer. More details are given in [8].

Fig. 3 shows this comparison. Excellent agreement is found for both $\pi^+\pi^+$ and $\pi^-\pi^-$ correlation functions. The shape of the RQMD source for pions at freeze-out, defined by the distribution of space-time points corresponding to the last (strong) interaction of the pions, was determined from events corresponding to roughly

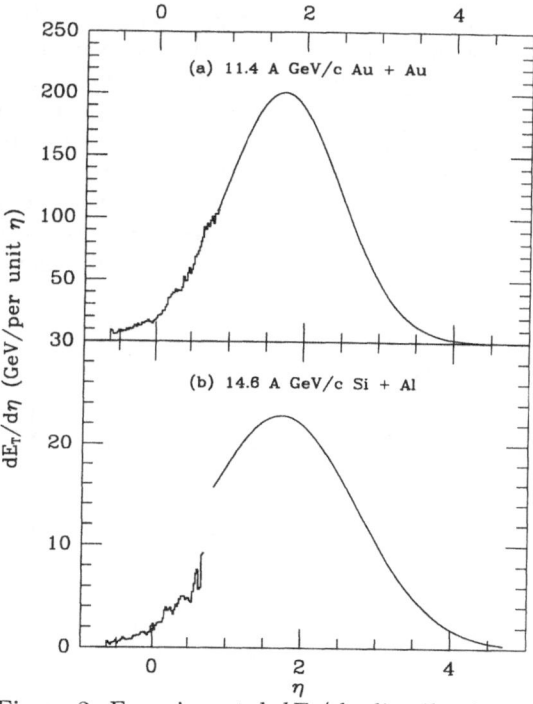

Figure 2: Experimental $dE_t/d\eta$ distributions for central Si+Al and Au+Au collisions.

the same centrality as in the data. This analysis yielded a nearly spherical source size of transverse radius $R_T = 6.7$ fm and longitudinal radius (in the c.m. frame) $R_L = 5.0$ fm, resulting in a rms radius of R = 8.3 fm. The agreement between data and RQMD calculations shown in Fig. 3 implies that our measured correlations are consistent with such a large source.

The rms radius of the source should be compared to the rms radius of Si of 3.04 fm [9]. This corresponds to a transverse radius of $R_T(\mathrm{Si}) = 2.5$ fm yielding a transverse expansion of the sytem of about a factor of 3. Because of the resonance decays (see below) it is difficult to get a unique result on transverse expansion velocities. Preliminary results [5] indicate $\beta_\perp \approx 0.5$, leading to an estimate for the collective expansion time of about 10 fm/c. We also note that very similar (large) source radii are obtained from recent E814 measurements of the deuteron to proton ratio in central Si + Pb collisions and its interpretation in a thermal model [10].

PARTICLE SPECTRA AND RESONANCE MATTER

The shape of pion spectra as function of transverse momentum or transverse mass contains information about the collision dynamics. For example, pion emission from an equilibrized system with temperature T_B leads to a Boltzmann distribution for the invariant cross section, i.e. $\sigma_{inv} \propto m_t \cdot \exp(-m_t/T_B)$, with the transverse mass $m_t = \sqrt{m^2 + p_t^2}$. The Boltzmann shape can be modified, particularly at low values of transverse momentum, by pions resulting from the decay of resonances at freeze-out (see below for a definition of freeze-out), even if the system stays in thermal equilibrium throughout [11]. Analysis of this low p_t component then yields information about the population of, in particular, nucleon excited states.

Figure 4 shows the π^- spectrum plotted versus transverse mass m_t divided by a Boltzmann spectrum $\sigma_{inv} \propto m_t \cdot \exp(-m_t/T_B)$ fitted to the range $p_t \geq 0.3$ GeV/c. The data are for central (top 2 %) Si + Pb collisions. More details and results for other systems and pion charges can be found in [12]. For $m_t - m_\pi$ values larger than 0.2 all spectra very closely follow a thermal shape. At smaller transverse mass, they exhibit a significant enhancement over a thermal distribution. We find very similar enhancement in both π^+ and π^- spectra, and for Si+Al collisions.

To show the sensitivity of the transverse momentum spectra to the resonance content at freeze-out, we also plot, in Fig. 4, the predictions for the spectral shape assuming various amounts of $\Delta(1232)$ resonance excitation. Assuming a ratio of pions from Δ decay to direct (thermal) pions of π_Δ/π_T between 0.4 and 0.6 nicely brackets all the data.

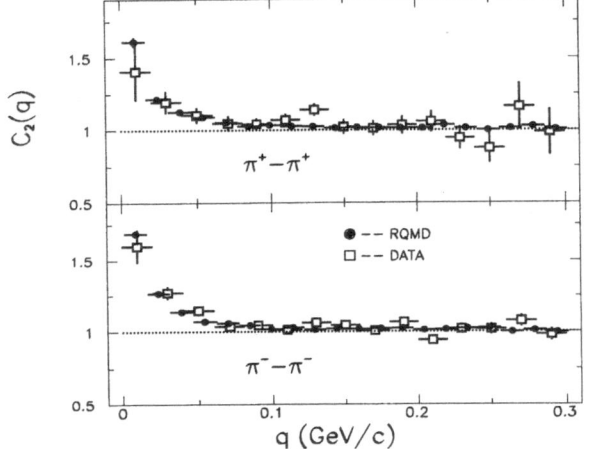

Figure 3: Experimental 2 pion correlation function together with the corresponding correlation function constructed from RQMD events (see [8]).

From the number of observed pions and nucleons (see, e.g., [13, 3, 12]) we get $(\pi/N)_{exp} = (\pi_\Delta + \pi_T)/(N + \Delta) \approx 1.1$. This then implies that $\frac{\Delta}{\Delta+N} = \frac{\pi_\Delta}{\pi_\Delta+\pi_T} \frac{\pi_\Delta+\pi_T}{\Delta+N} = 0.37 \pm 0.05$. As discussed in [12], RQMD reproduces our pion spectra extremely well and there the rapidity averaged fraction of nucleons in the $\Delta(1232)$ resonance at freezeout is 35 % [14], in good agreement with the above number.

In Figure 5 it is shown that the $\Delta(1232)$ resonance is also seen directly in E814 by reconstructing the pπ^+ invariant mass in central (top 2 %) Si + Pb collisions. From the observed number of Δ resonance states one can determine their multiplicity. Taking into account the acceptance of the E814 spectrometer for Δ^{++} reconstruction of $1.5 \cdot 10^{-3}$ for the rapidity interval y = 1.9 - 3.1 leads to a Δ^{++} multiplicity of 2.1±0.7. For comparison, we measured [3] 5.9 protons in the same rapidity interval. RQMD predicts 1.8 Δ^{++} for this interval in good agreement with our preliminary experimental number.

The measured Δ resonance population can also be understood in the context of a purely thermal model. To illustrate this, and to determine the temperature T of the equilibrized system from the measured Δ to nucleon ratio we present in Fig. 6 as a function of the temperature T, the population probabilities of all non-strange nucleon resonances with masses less than 2 GeV. In this calculation the widths of all states and, of course, all statistical factors have been included.

Clearly the Δ resonance is the dominant excited state. However, because of the many heavier resonances, the population of the Δ never exceeds about 45 %. From our measured Δ to nucleon ratio of 0.37 ± 0.05 we conclude that $T = 0.14 \pm 0.02 GeV$. This is consistent with the temperature determination from the analysis of pion spectra, where $T \approx 0.15$ GeV [4].

While the enhancement at low p_t in pion spectra is well understood in terms of baryon resonance decays, recent preliminary E814 data imply [4] that there is also a strong enhancement at low p_t in kaon spectra. The source of this enhancement is presently not understood: resonance decays are too weak to explain the size of the effect. An exciting possibility is that the kaon enhancement is due to a drop of the kaon mass in the hot and dense medium formed during the collision. Data recently taken with Au beams should shed light on this situation.

FREEZE-OUT SCENARIO

At freeze-out we assume that the system consists of a gas of pions and nucleons. The dominant cross section is then the pion-nucleon cross section $\sigma_{\pi N}$. We define freeze-out of the expanding system by requiring that at freeze out $< d_{\pi N} > = \sqrt{< \sigma_{\pi N} > / \pi}$. Here, $< d_{\pi N} >$ is the mean distance of a pion and the nearest nucleon in the gas, and $< \sigma_{\pi N} >$ is the pion-nucleon cross section averaged over spin, isospin, and the Boltzmann distribution of pions and nucleons at the freeze-out temperature T_f. From the previous arguments about spectral shapes and resonance population we deduce $T_f \approx 0.14 GeV$. At this temperature, $< \sigma_{\pi N} > \approx 70$ mb. Using a simple Monte Carlo procedure we have determined that $< d_{\pi N} > \approx \frac{0.75}{(n_\pi + n_N)^{1/3}}$, where n_π and n_N are the pion and nucleon number densities at freeze-out. Note that this implies that the freeze-out radius $R_f \propto dN/dy^{1/3}$. For the freeze-out volume V_f we obtain:

$$V_f = (\frac{< \sigma_{\pi N} >}{\pi})^{3/2} \frac{N_\pi + N_N}{0.75^3}.$$

Here, N_π and N_N are the number of pions and nucleons in the fireball. A freeze-out volume directly proportional to the total particle multiplicity, *i.e.* freeze-out at constant density, was already discussed long ago by Pomeranchuk [15].

For the system Si+Pb and very central collisions we obtain [13, 3, 12] $N_N \approx 138$ and $N_\pi \approx 150$, depending somewhat on what rapidity range is included. This implies a freeze-out volume of $V_f = 2270 fm^3$ corresponding to $R_f = 8.1 fm$. The equivalent rms radius is then 6.3 fm, somewhat smaller than that

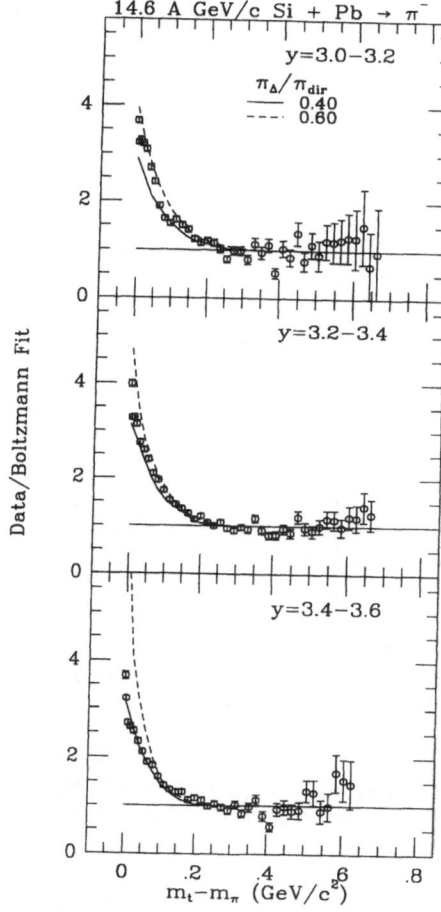

Figure 4: Experimental π^- spectra normalized to a Boltzmann distribution fitted to the data for $p_t \geq 0.3$ GeV/c. Solid and dashed lines: thermal model with different fractions of Δ decay vs. direct pions.

determined from the correlation analysis as discussed above. The resulting pion and nucleon densities at freeze-out are $n_\pi = 0.066/fm^3$ and $n_N = 0.061/fm^3$. This nucleon density is less than half of the density in the center of nuclei in their ground state. Freeze-out at such low density arises principally because of the very large pion-nucleon cross section and because, at AGS energies, the fireball contains roughly equal numbers of pions and nucleons. Extrapolating this picture to CERN energies we would expect a smaller (by about a factor of $\sqrt{3}$) freeze-out radius because of the dominance of pions in the fireball. This seems indeed to be the case, as demonstrated by measurements of the Na44 collaboration [16].

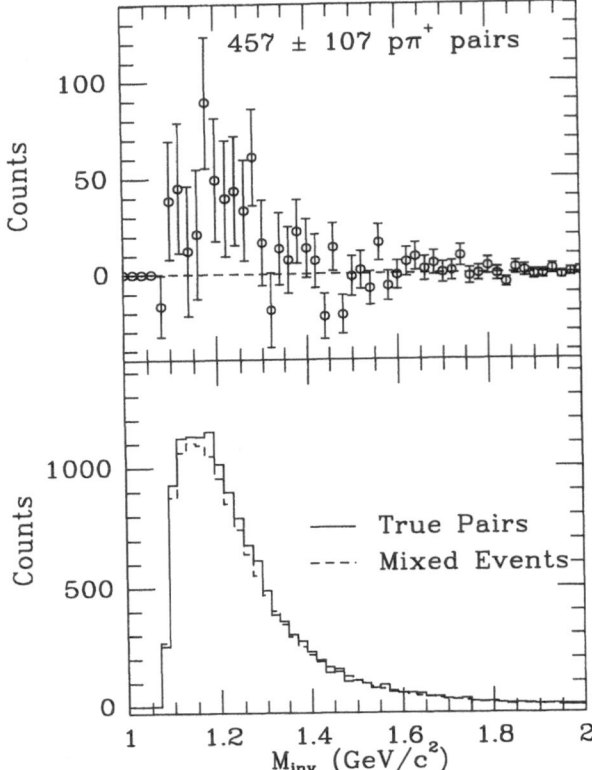

Figure 5: Reconstruction of the Δ^{++} resonance for central Si-Pb collisions.

From T_f and the particle densities n_N and n_π we can determine both the pion and nucleon chemical potentials via

$$ n = \frac{g}{2\pi^2} \int \frac{p^2 dp}{exp((E - \mu)/T_f) \pm 1}, $$

with $g = 4$ for nucleons and $g = 1$ for each pion charge. A (small) excluded volume correction [17] is also applied. This yields for the nucleon chemical potential (at $T_f = 0.14$ GeV) $\mu_B = 0.51$ GeV. For the pion chemical potential we get $\mu_\pi = -44$ MeV, (-15 MeV for $T_f = 0.13$ GeV), i.e. fairly close to zero and not in agreement with the large positive values used recently [18] to explain the low p_t enhancement in pion spectra. From the measured K^+/K^- ratio [19] one can also determine the strangeness chemical potential $\mu_S = 0.105$ GeV.

If the system is in thermodynamic and chemical equilibrium at freeze-out, then the above obtained values for the baryon and strangeness chemical potentials determine the baryon production ratios $\bar{p}/p = 7 \cdot 10^{-4}$ and $\bar{\Lambda}/\Lambda = 3.7 \cdot 10^{-3}$. Both ratios are close to what is observed experimentally at AGS energies, namely $(\bar{p}/p)_{exp} = (6 \pm 3) \cdot 10^{-4}$ [20], and $(\bar{\Lambda}/\Lambda)_{exp} = (2 \pm 1) \cdot 10^{-3}$ [21], lending further support to the equilibrium scenario.

In addition to the chemical potentials, the particle densities and freeze-out temperature also completely determine the total entropy of the system via [15]

$$ S_f = \frac{V_f}{T_f} \sum_{i=\pi,N} g_i \int \frac{p^2 dp}{2\pi^2} \frac{p^2/3E_i - \mu_i + E_i}{\exp \frac{E_i - \mu_i}{T_f} \pm 1}. $$

Numerical evaluation leads to an entropy per baryon of $S_f/N_N = 15$ for $T_f = 0.14$ GeV. Alternatively, one can also deduce the entropy/ baryon of the system from the measured ratios [10] of deuterons/ protons and pions/ baryons. This uses the formula [22] of Siemens and Kapusta, with a result of $S_f/N_N = 13$, in beautiful agreement with the entropy from the freeze-out scenario.

We note that the freeze-out scenario changes somewhat if one uses the RQMD freeze-out radius of 8.3 fm rather than the 6.3 fm deduced from our freeze-out considerations above. This reduces all densities by a factor of 0.44 and would imply that the freeze-out temperature is closer to 0.12 GeV.

SUMMARY AND FUTURE PLANS

We have demonstrated, using recent data from the E814/E877 collaboration and from the AGS heavy ion program in general, that strong evidence exists for the formation of a hot and dense fireball at AGS energies. Comparison to models successfully describing our data implies that $\epsilon \approx 1.5$ GeV/fm^3 and $\rho/\rho_0 \approx 8$ are achieved for significant time scales (≈ 5 fm/c), especially for the Au+Au system. This would imply that the fireball reaches densities at least close to that expected for a mixed phase of hadrons and quarks and gluons.

A considerable transverse expansion (by about a factor three) ensues, leading to a freeze-out radius of 8.3 fm (rms) for Si+Pb. Analysis of the Δ resonance population shows that about 37 % of the baryons are excited to the Δ resonance. This and the inverse slope constants of pion m_t spectra imply a freeze-out temperature of $T_f \approx$ 0.14GeV.

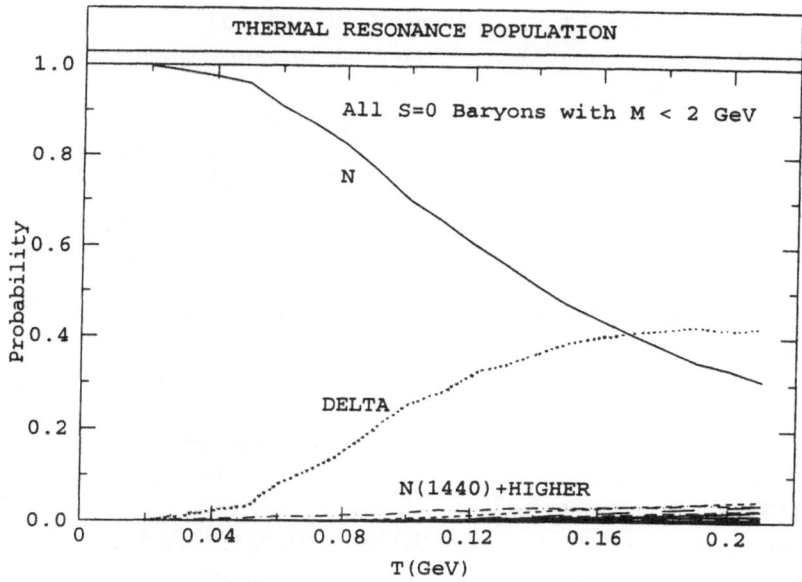

Figure 6: Population of Nucleon Resonances as function of the temperature. The widths of the resonances is included in the calculation.

Using the criterion of freeze-out at a constant density and assuming full thermodynamical equilibrium we developed a consistent freeze-out scenario. In this scenario freeze-out happens at rather low density because of the very large pion-nucleon cross section and the large baryon content of the fireball at AGS energies. The pion and baryon chemical potentials and the entropy at freeze-out are then completely determined. The deduced values are nicely consistent with observed particle production ratios. A large positive pion chemical potential is not in agreement with our results.

The thermodynamical equilibrium also insures that event generators based on nucleon-nucleon collisions reproduce many aspects of the data, even if their input parameters differ significantly.

The future plans of our collaboration (E877) are centered on the analysis of the recent run with Au projectiles at 11 GeV/nucleon. These data will be analyzed to provide, e.g., three-dimensional information on pion source sizes, a high statistics measurement of the low p_t part of the kaon transverse momentum spectra to hopefully confirm our intriguing Si+Pb results, and a first look at inclusive photon spectra following central Au+Au collisions. In addition, preliminary analysis of transverse energy distributions has yielded a first indication of collective transverse flow and this direction will also be vigorously pursued.

Financial support by the US DoE, the NSF, the Canadian NSERC, and CNPq Brazil is gratefully acknowledged.

References

[1] J. Barrette et al., E814 Coll., Phys. Rev. Lett. **64** (1990) 1219; Phys. Rev. **C45** (1992) 819; Phys. Rev. **C46** (1992) 312.

[2] J. Barrette et al., E814/E877 Coll., Phys. Rev. Lett. **70** (1993) 2996.

[3] J. Barrette et al., E814 Coll., Z. Physik **C59** (1993) 211.

[4] J. Stachel, Invited Paper, Quark Matter '93, Borlänge, Sweden, June 1993, Nucl. Phys. A (in print)

[5] P. Braun-Munzinger, J. Stachel, N. Xu, unpublished

[6] H. Sorge, H. Stöcker, W. Greiner, Ann. Phys. (NY) **192** (1989) 266; A. Jahns, H. Sorge, H. Stöcker, W. Greiner, Z. Physik **A341** (1992) 243.

[7] T. J. Schlagel, S. H. Kahana and Y. Pang, Phys. Rev. Lett. **69** (1992) 3290.

[8] N. Xu, Contributed Paper Quark Matter '93, Borlänge, Sweden, June 1993, Nucl. Phys. A (in print)

[9] H. Überall, Electron Scattering from Complex Nuclei, Academic Press, N.Y. 1971, p. 210.

[10] J. Germani, Ph.D. thesis, Yale Univ., July 1993, and J. Barrette et al., E814 Coll., preprint Dec. 1993

[11] G.E. Brown, J. Stachel, and G.M. Welke, Phys. Lett. **B253** (1991) 19.

[12] T.K. Hemmick, Contributed Paper, Quark Matter '93, Borlänge, Sweden, June 1993, Nucl. Phys. A (in print)

[13] C. Parsons, proceedings Workshop HIPAGS '93, MITLNS-2158.

[14] M. Hofmann and R. Mattiello, private communication.

[15] E. Shuryak, "The QCD Vacuum, Hadrons and the Superdense Matter", World Scientific, Singapore 1988, p. 333

[16] J.P. Sullivan *et al.*, Phys. Rev. Lett **70**(1993)3000

[17] J. Cleymans and H. Satz, Z. Physik **C57**(1993)135

[18] J.L. Goity and M. Leutwyler, Phys. Lett **228B**(1989)517; M. Kataja and V. Ruuskanen, *ibid* **243B**(1990)181

[19] T. Sung, proceedings Workshop HIPAGS '93, MITLNS-2158.

[20] see e.g. overview by H.J. Crawford, proceedings Workshop HIPAGS '93, MITLNS-2158 and refs. there.

[21] G.S.F. Stephans, Invited Paper, Quark Matter '93, Borlänge, Sweden, June 1993, Nucl. Phys. A (in print)

[22] P. Siemens and J. Kapusta, Phys. Rev. Lett. **43**(1978)1486, 1690E

ANTIPROTON PRODUCTION IN HEAVY ION COLLISIONS: MULTI-STEP PROCESSES VS ANNIHILATION

A. Jahns, C. Spieles, H. Sorge, H. Stöcker, W. Greiner

Institut für Theoretische Physik, Frankfurt/Main, Germany

INTRODUCTION

Antibaryons have a large annihilation cross section in baryon-rich enviroments formed in heavy ion reactions. By studying their production and absorption mechanisms we hope to get informations about the time evolution of the baryon density in the reaction region [1]-[4]

In previous calculations [5] experimental antiproton data at the AGS energy regime have been explained on a microscopic level. The final particle yields have been interpreted in terms of two counter balancing effects: On the one hand, $\bar{B}$ *production* is enhanced due to collective effects, on the other hand $\bar{B}$ *annihilation* becomes stronger as nuclear stopping and therefore the baryon density increases. The strength of these competing processes depend strongly on the incident energy and the reaction volume. Recent measurements of inclusive antiproton spectra with proton, silicon and gold beams at the AGS (10-15 AGeV) [6]-[10] do not give clues about the strength of such opposite effects. These results - with an uncertainty of a factor 2 in the $\bar{p}$ yields as extrapolated from pp collisions - are compatible with RQMD calculations as well as with the first collision model: Antibaryons are produced similar to the first collision yields if the absorption is neglected. There are other theoretical calculations which predict that antibaryons are also enhanced in a quark gluon plasma event [11], by chiral symmetry restoration [12], by in medium-effects [13] or by string-string interactions [14].

In this article we demonstrate that strong annihilation distorts considerably the momentum distribution of antibaryons. Due to the built-up of a high density- and pressure zone, the nucleons stemming from the projectile bounce-off in the reaction plane just into the opposite direction of the target nucleons in noncentral collisions. Antibaryons - and also pions and negatively charged kaons - show as a result of rescattering and absorption a strong "anti-flow", i.e. anticorrelations to the nucleons in the reaction plane. This leads to an observable asymmetry in the azimuthal angular distributions $(dN_{\bar{p}}/d\phi)$ or in the mean directed transverse momentum of antiprotons in the reaction plane as a function of the rapidity $\vec{p}_x(y)$. This can be experimentally tested once the reaction plane is determined, e.g. by measuring the azimuthal distribution of foreward and backward going baryons or transverse energy in the same event as which detect the antiprotons. First experimental evidence for the collective flow [15] has recently been reported by the E877 group [16].

Hot and Dense Nuclear Matter, Edited by
W. Greiner *et al.*, Plenum Press, New York, 1994

THE RQMD MODEL

The calculations presented here are based on a microscopic phase space approach, the relativistic quantum molecular dynamics model (RQMD 1.07). Let us briefly describe the relevant features of this model [17][18]. The fate of all particles – the original as well as the newly produced ones – is followed in RQMD by propagating those particles on classical trajectories and allowing collisions between all of them until the cascade process is finished. The interaction probabilities are given by geometry, i.e. the minimum two-body CMS distance has to be below $\sqrt{\sigma/\pi}$ with σ the binary reaction cross section. The dominant reaction mechanism in the first stage of a reaction when projectile and target interpenetrate each other is the excitation of both collision partners to resonances or strings. Strings are a 1+1 dimensional idealization of the chromoelectric flux tube with constituent quarks moving at their ends. The decay of the excited states depends on its mass. If the mass is below some threshold, e.g. for nonstrange baryon resonances $2\,\text{GeV}/c^2$, they are projected onto the experimentally given resonance spectrum. Higher mass states fragment as colour strings. The string fragmentation parameters are universal for soft hadronic multiparticle production and extracted from e^+e^- and lh collisions. The free $N\bar{N}$ annihilation cross section enters into the RQMD calculations. $B\bar{B}$ annihilation is described as the annihilation of two quarks from each collision partner, $qq\text{-}\bar{q}\bar{q}$, leading to a mesonic $q-\bar{q}$-string.

RESONANCE MATTER AND MULTI-STEP EXCITATION

Schematically there are three different energy regions in hadron-hadron interactions: the (quasi-)elastic scattering, the region of resonance production and formation and the high energy region chacterized by abundant production of particles. In the resonance region the kinetic energy of the reaction partners is sufficient to form excited states of the ingoing hadrons wich subsequently decay again into the stable hadrons. The most prominent examples are

$$
\begin{aligned}
\pi N &\longrightarrow \Delta(1232) \\
\pi\pi &\longrightarrow \varrho(770) \\
N N &\longrightarrow N\Delta(1232)
\end{aligned}
\tag{1}
$$

This reactions are contained in RQMD. They are calculated with the help of the Breit-Wigner formula for s channel resonance formation or via one-meson-exchange in the t channel.

The formation of resonances influences the particle production rates and their momentum distributions, e.g. the p_t spectra. The particle chemistry may change, too, destorting signatures from earlier reaction stages, e.g. due to reactions like

$$
\begin{aligned}
\pi Y &\longleftrightarrow Y^\star &\longleftrightarrow \bar{K}N \\
\pi\pi &\longleftrightarrow S^\star(970) &\longleftrightarrow K\bar{K}
\end{aligned}
\tag{2}
$$

A second source for excitation of baryons which is of utmost importance at AGS energy as discussed in [5][18] are secondary interactions, the dominant process being the annihilation of produced mesons on baryons which leads to the formation of s channel resonances or strings. The formation time of newly produced secondaries is given by the finite lifetime of a resonance or by the time which a string needs to break into parts which is given by the momenta of the fragments. Those resonances formed are not only responsible for strangeness enrichment [18], but they may be further excited in a subsequent collision to a mass larger than $3m_N$ which allows for $\bar{N}N$ creation.

Provided the projectile nucleon gets excited in the first collision, but it does not loose momentum itself for target excitation (in a subset of cases). A further excitation in a subsequent collision then may help to bring the projectile beyond the $\bar{B}$ creation threshold. The strength of this effect results from the nearby production threshold which makes the $\bar{p}$ yield around AGS energies extremely energy dependent. There is the subtlety involved in these multi-step processes that the "optimum mass" of the intermediate resonance state is influenced by two counteracting factors.

This process is schematically demonstrated in Fig. 1. Shown is a proton nucleus collision in which the projectile nucleon hits the target several times. In each of these collision it is possible that the nucleon gets excited. The excited masses are labeled by B^* (first excitation), B^{**} (second excitation) and τ denotes the formation time.

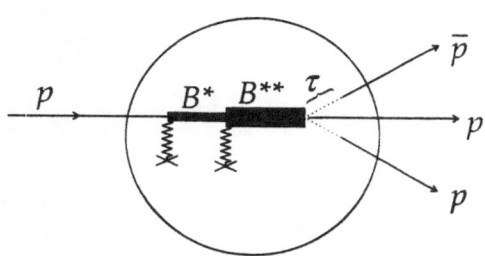

Figure 1: Multi step process in a proton nucleus collision (see text for explanation).

On the one hand, the larger the mass the better the chance to overcome the $\bar{B}$ creation threshold in a subsequent collision, because the second step of excitation can be smaller. On the other hand, the life time goes down with increasing mass reducing the chance to hit another target nucleon. Therefore the multi-step process of $\bar{p}$ production has to be viewed as a complicated interplay between excitation and decay processes. One may get a feeling for the importance of this finite resonance lifetime aspect by comparing with a calculation in which the resonance is forbidden to decay during its passage through the target. This is realized for instance in the model FRITIOF in which the strings are excited very similarly to RQMD, but are kept stable and decay only afterwards into hadrons [19]. The difference between RQMD and FRITIOF – leaving out annihilation and with a choice of parameters giving the same $\bar{p}$ yields in pp interactions – can amount for large targets to approximately 15 percents.

RESULTS

First measurements, simultaneously exploring the collision geometry and target mass dependence of antiproton production, have been presented by the E814 group [9]. Figure 2 shows the antiproton yields per event plotted as functions of N_{int}. N_{int} denotes the number of interacting projectile nucleons and is therefore a direct observable for the centrality of the colliding system.

The data are compared with RQMD results by reconstructing the experimental trigger conditions. This has been done using GEANT, to determine the fraction of antiprotons that will be detected in the momentum and rapidity cuts of the E814 experiment [20]. RQMD ingredients for a comparison with the data are the inverse slopes B, the width of the dN/dy distribution σ_y and the rapidity value where y_0 the distribution peaked (see values of the table).

Both, the data and the RQMD calculations give yields that are rather independent of the target mass. The antiproton yields are roughly linearly rising with the number of interacting projectile nucleons (increasing centrality).

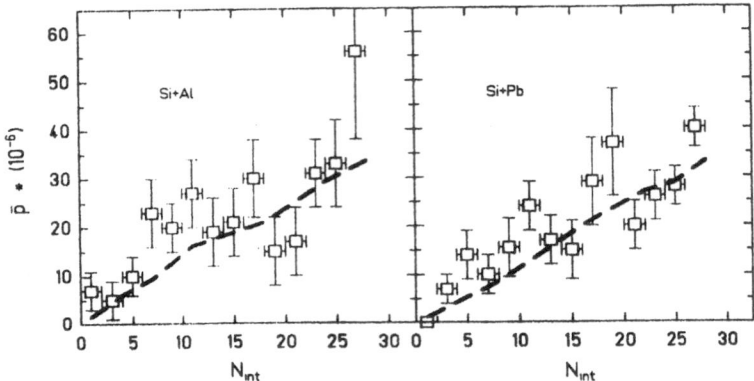

Figure 2: The yield of antiprotons per event measured by E814 in $Si + Al$ and $Si + Au$ interactions plotted as a function of the number of nucleons in the projectile that have interacted. The data are comared with RQMD calculations (dashed lines).

As pointed out in [9] this rules out an explanation in terms of first chance collision production followed by absorption. Concerning the RQMD model these behaviour reflects obviously the interplay between increased antiproton production and increased annihilation at higher centrality.

system	trigger	B	σ_y	y_0
$p + p$	-	115 ± 15	0.6 ± 0.1	1.72 ± 0.1
$Si + Al$	min bias	145 ± 20	0.7 ± 0.1	1.72 ± 0.1
$Si + Al$	central	165 ± 25	0.8 ± 0.1	1.72 ± 0.1
$Si + Au$	min bias	165 ± 25	0.8 ± 0.1	1.74 ± 0.1
$Si + Au$	central	195 ± 30	0.8 ± 0.1	2.1 ± 0.2

We have studied $Au + Au$ reactions both at 10.7 AGeV and also at 2 AGeV and 160 AGeV. In accord with recent antiproton measurements without centrality trigger [21] we focus here on minimum bias calculations. We want to point out, however, that much more detailed informations can be obtained if impact parameter selection criteria are applied as was clearly demonstrated in the $E \sim 1$ AGeV energy domain [22]-[24].
It is expected that multi-step processes might become even more important for large mass target-projectile combinations due to the larger reaction volume and lifetime. Note that the Au beam energy is closer to the $\bar{p}$ production threshold than the Si beam energy at the AGS. Therefore, additional production processes - apart from first collisions - should become even more important. On the other hand, high nuclear density – caused by strong nuclear stopping – should strengthen the $\bar{p}$ suppression. In contrast, a microscopic calculation with the cascade model ARC [25] was presented in which such multi-step processes are less important, because the excitation spectrum is restricted to the Δ mass only. Instead, in a high density enviroument the annihilation of antiprotons is suppressed.
Fig.3 shows the calculated rapidity distribution of the antiprotons, $dN_{\bar{p}}/dy$, for minimum bias $Au(10.7 \text{ AGeV})+Au$ reations with (solid histogram) and without annihilation

(dotted histogram). RQMD calculations are also given for the $\bar{p}$ distributions in pp reactions. This distribution (solid curve) is multiplied by the calculated number of first
nucleon nucleon collisions (~ 12) in $Au + Au$.

Let us investigate the first collision model by comparing the first collision yield (fermi
momentum considered) of antiprotons with the total (initially produced) antiproton
yield: The initially produced antiproton yield in the full RQMD calculation is a factor
of *twenty* higher than in the first collision model! Even after annihilation is included,
this factor is still *two* (i.e. the model predicts $\approx 90\%$ annihilation). Hence, we infer
that in heavy systems like gold on gold the enhanced production of antibaryons is *not*
completely counter balanced by the annihilation.

The absorption is strongest around midrapidity. The final $dN_{\bar{p}}/dy$ distribution is
broader than expected for antiprotons which originate from first pp reactions, because
$\bar{B}$ annihilation is correlated with the rapidity density of baryons which is highest at
midrapidity. This goes along with large baryon stopping.

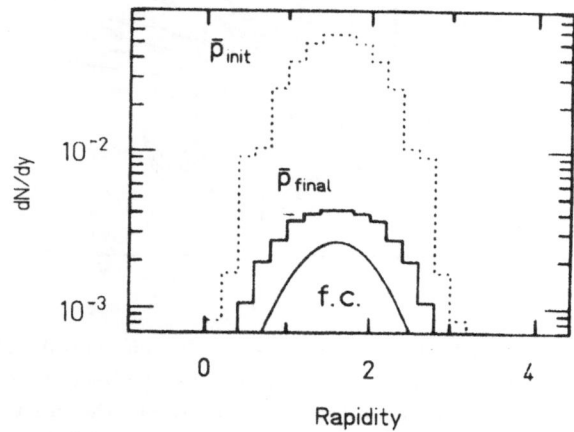

Figure 3: The antiproton rapidity distribution $dN_{\bar{p}}/dy$ is shown for minimum bias $Au + Au$
collisions at 10.7 AGeV. The two histograms represent the RQMD results with (solid) and
without (dotted) annihilation. The solid curve corresponds to the antiprotons calculated in
pp collisions multiplied by 12 (number of first collisions in minimum bias $Au + Au$ reactions).
The fermi momentum is also considered.

Preliminary data measured at $p_t = 0$ [26] seem to exhibit this broadening of the $dN_{\bar{p}}/dy$
distribution. This support the scenario presented here. The antiproton yield should
also be measured in small systems (pp, $Si + Si$, etc.) - in addition to $Au + Au$ - at the
same energy. A systematic study of the mass and centrality dependence of antiproton
production can help to disentangle the competing effects of multi-step production and
enhanced absorption predicted in this work.

One can look at observables which exhibit collective behaviour more clearly. For instance, baryon stopping and creation of a baryon dense region may lead to collective
baryon flow predicted long time ago on the basis of the fluid-dynamical model [22].
The observation of flow in turn is of vital importance for diagnostic purposes: The
predicted bounce-off and squeeze-out effects can be used as barometers to measure the

pressure built up in the hot and dense participant matter [27]. The bounce-off can be quantified e.g. via the measurement of the directed in-plane transverse momentum transfer which has been widely used at BEVALAC/SIS energies [23][24]. While the nucleonic flow at target and projectile rapidity respectively can easily be pictured as repelled and deflected matter, the antibaryons will show an anticorrelated behaviour due to annihilation and rescattering.

Fig. 4 shows this behaviour for a medium impact parameter (b=6 fm): A snapshot of the baryon density contours in the reaction plane after 16 fm/c (cms). The antiproton momenta are represented by the arrows. Obviously "anti-matter" goes in the direction opposite to the "matter". Note that the snapshot shows antiprotons from nearly 10^4 (overlayed) events.

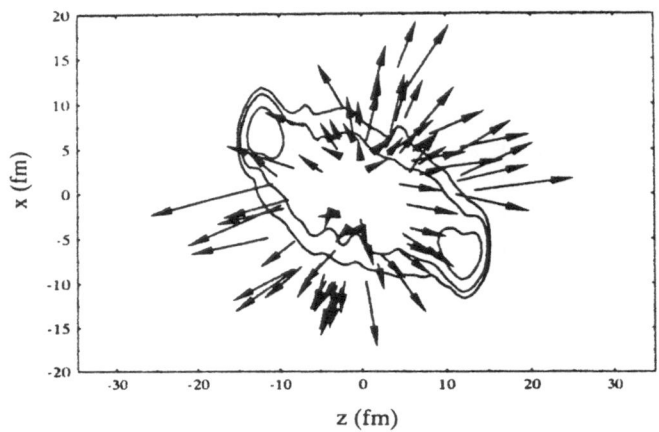

Figure 4: Snapshot (after 16 fm/c) in the reaction plane for semi peripheral (b=6 fm) $Au + Au$ collisions at 10.7 AGeV. Shown are RQMD calculations for the baryon density (contour plots) and antiprotons (arrows). The lengths of the arrows represent the momenta of the particles, projected on the reaction plane. Antiprotons are obtained from 10^4 (overlayed) events.

Which observables are relevant to quantify this effect ? One possibility is to look at the azimuthal (ϕ) distribution of antiprotons. ϕ is the angle between $\vec{p}_T$ and the x- axis ($\tan \phi = |p_y|/p_x$). Thus, a vector with $\phi = 0$ degrees points into the direction of the x-axis. Strong antibaryon annihilation leads to a decreasing $dN/d\phi$ distribution when going from 0° to 180°. If annihilation is turned off, this distribution is flat: The antiprotons are -in the present model- produced azimuthally isotropic. The antiproton yield in a given rapidity bin (in percent) which goes to the upper hemisphere ($\phi < 90°, p_x > 0$) is shown in Fig. 5 as a function of rapidity for $Si + Au$(b=3.5 fm) and $Au + Au$ (minimum bias).

Up to 70% of the antiprotons exhibit $\phi < 90°$ at target rapidity for both systems. This means more antiparticles survive if they are emitted in opposite direction than the matter flow. This effect disappears at projectile rapidity for $Si + Au$ due to the stopping of the Si projectile.

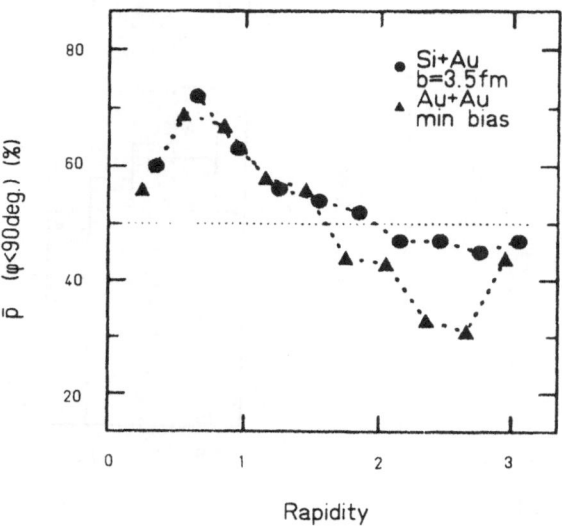

Figure 5: Percentage of final antiprotons with an azimuthal angle $< 90°$. Analysed are the systems $Si + Au$ (b=3.5 fm) and $Au + Au$ (minimum bias) at AGS energies with RQMD. At target rapidity both systems show a azimuthal asymmetry up to 70 %. At projectile rapidity this effect disapears for $Si + Au$ due to the stopping of the Si projectile.

Fig. 6 shows the in-plane directed transverse momentum $\vec{p_x}(y)$ as a function of rapidity for protons (including potentials) and all produced negatively charged particles (devided by the particles mass). Minimum bias events are calculated. Rescattering modifies also the π^- and K^- distributions from the primary nucleon-nucleon collisions. At lower energies this anticorrelation of pions to the nucleons can be explained by pion absorption [28] or by multiple πN scattering[29] - the observed pions in the target-projectile rapidity frame are reflected from the spectator pieces. The latter argument is not relevant for antibaryons.

Nucleon flow is predicted for SIS, AGS and SPS energies [15]. "Anti-flow" of antinucleons is also obtained for SIS- energies ($E_{Lab.} \sim 2 AGeV$) and for the CERN - Pb beam - energies ($E_{Lab.} \sim 160$ AGeV) as demonstrated in Fig. 7.

The directed transverse momentum of the antiprotons at projectile and target rapidity is most prominent for low energies. Around midrapidity the $\bar{p}$ statistics is best in our calculations as well as in the experiment. Here, the details of the baryon dynamics, the initial transverse momenta of the antiprotons and the $\bar{p}$ absorption lead to a characteristic maximum for the flow effect near 11 AGeV. At the AGS energy regime - considering the ammount of flow and the statistic - a confirmation of our predictions seems favourable at $0.4 \leq y \leq 1.2$ (resp. at $2.0 \leq y \leq 2.8$). This rapidity region is accessible by recent measurements.

The simultaneous determination of the reaction plane and the detection of the antibaryons is mandatory on an event by event basis. Correlations between particle production and transverse momenta of projectile fragments have been demonstrated by the E802 collaboration [30]. Azimuthal asymmetries for the $\pi^\pm/p$ and $(d + t)/p$ ratios have been found at the target rapidity region. Measurements of antiproton correlations are in progress [21][31].

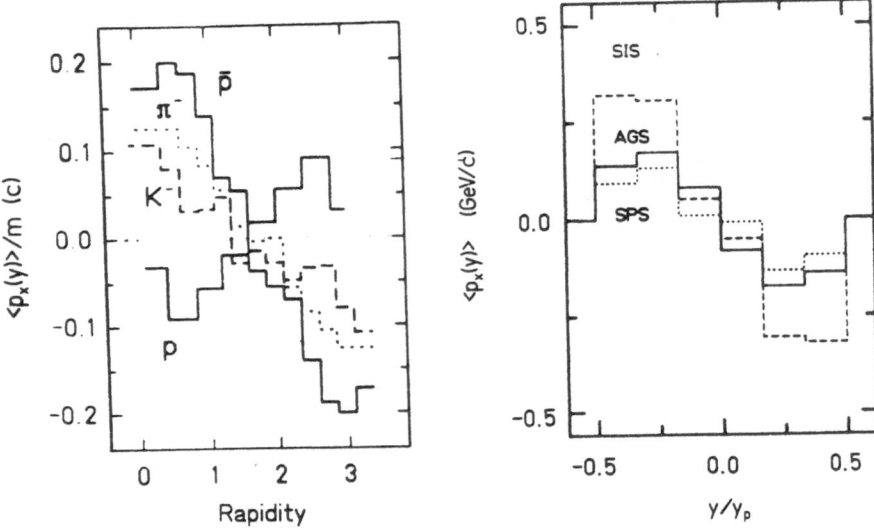

Figure 6 (left): Directed in-plane transverse momentum for different particles (p, $\bar{p}, \pi^-, K^-$) in the reaction $Au + Au$ (minimum bias) at 10.6 AGeV devided by the particle masses as a function of rapidity $< p_x(y)/m >$. The flow effect of antiprotons is in the mean seven times larger than for π^- or K^-.

Figure 7 (right): Directed in-plane transverse momentum for antiprotons in minimmum bias reactions at SIS- ($Au + Au$ at $E_{\text{Lab.}} \sim 2$ AGeV), AGS- ($Au + Au$ at $E_{\text{Lab.}} \sim 11$ AGeV) and CERN- ($Pb + Pb$ at $E_{\text{Lab.}} \sim 160$ AGeV) energies as a function of the scaled rapidity (y/y_P).

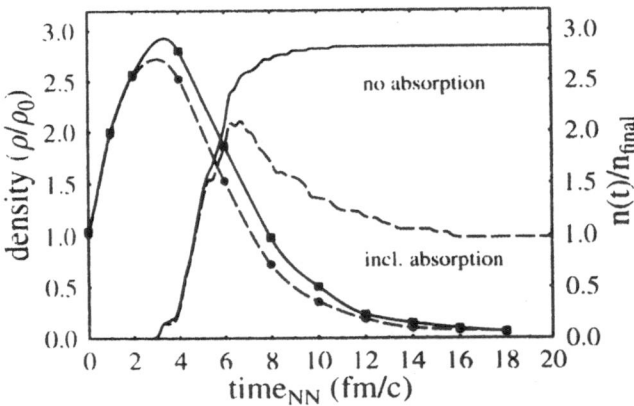

Figure 8: Maximum (time dependent) baryon density ρ_{max}/ρ_0 in central reactions of $Ne + NaF$ at 2.0 AGeV. Shown are RQMD calculations with (circles) and without potentials (squares). Also shown are the (time dependent) number of antiprotons in minimum bias reactions of $Ne + NaF$ at 2.0 AGeV with and without absorption.

Subthreshold production of antiprotons has first been observed experimentally for $Si + Si$ at BEVALAC [32]. Experimental studies have been done recently with a 2.0 AGeV and 1.8 AGeV Ne beam from the heavy ion synchrotron (SIS) at GSI and will be continued [33].

Systematic studies of antiproton production and annihilation at this energy regime [34] show a) that the antiproton production takes place dominantly in central collisions, b) that the annihilation probabilities are higher compared to AGS energies but in the same order of magnitude, and c) that the antiprotons start to interact with the nuclear medium not necessarily at highest densities (where they are produced via multi-step excitation). This situation is demonstrated in Fig. 8.

CONCLUSIONS

The cancelation of two large numbers - multi-step processes and annihilation - hinders the direct extraction of information on the intermediate baryon dense system as suggested by Gavin *et al* [3]. It is very difficult to disentangle $\bar{p}$ production and absorption from inclusive data. All systems exhibit nearly the same balance of these opposing processes. In RQMD this is at the origin for the lack of target dependence in pA as well as in AA collisions. RQMD predicts that this is not longer true for massive $Au + Au$ collisions: The enhanced production of antibaryons is *not* completely counter balanced by the annihilation.

The rapidity dependence of the in-plane directed transverse momentum $\vec{p_x}(y)$ of antiprotons shows the opposite sign of the nucleon flow, which has indeed recently been discovered at 10.7 AGeV by the E877 group [16]. The "anti-flow" of antiprotons is also predicted at 2 AGeV and at 160 AGeV and appears at all energies also for pions and negatively charged kaons. The energy dependence of "anti-flow" of antiprotons involves many elementary and collective properties of hadronic matter. Therefore, it can serve as a stringent test for different models. These predicted $\bar{p}$-anticorrelations are a direct proof of strong antibaryon annihilation in massive heavy ion reactions. The experimental discovery of this effect would rule out the first chance collision model with screening of antibaryon annihilation, which has recently been put forward to explain experimental data.

ACKNOWLEDGEMENTS

This work was supported by the Gesellschaft für Schwerionenforschung (GSI) and the Bundesministerium für Forschung und Technik (BMFT). One of us (A.J.) likes to thank the experimentalists P. Braun-Munzinger, G. Diebold, W. Koenig, B. Shiva Kumar and J. Stachel for fruitful and stimulating discussions.

REFERENCES

[1] P. Koch, C. Dover: Phys. Rev. C40, 195 (1989).

[2] H. Sorge, A.v. Keitz, R. Mattiello, H. Stöcker, W. Greiner: Phys. Lett. B243, 7 (1990)

[3] S. Gavin, M. Gyulassy, M. Plümer, R. Venugopalan: Phys. Lett B234, 175 (1990)

[4] A. Jahns, C. Spieles, R. Mattiello, H. Sorge, H. Stöcker, W. Greiner: Proceedings of Heavy Ion Physics at the AGS (HIPAGS '93), Cambridge USA, edited by G.S.F Stephans, S.G. Steadman and W.L. Kehoe (MIT-Press) p.120, and Contribution to the Proceedings of the 11th International Conference on Ultrarelativistic Nucleus-Nucleus Collisions "Quark Matter '93" ,Borlänge (Sweden) 1993, to appear in Nucl. Phys. A.

[5] A. Jahns, H. Stöcker, W.Greiner, H. Sorge: Z. f. Phys. A341, 243 (1992); Phys. Rev. Lett. 68, A. Jahns, C. Spieles, R. Mattiello, H. Stöcker, W.Greiner, H. Sorge: Phys. Lett B308, 11 (1993);

[6] E814, J. Barette et al.: Phys. Rev. Lett 70, 1763 (1993);

[7] E858, A. Aoki et al.: Phys. Rev. Lett. 69, 2345 (1992);

[8] E802, T. Abbot et al.: Phys. Lett. B271, 447 (1991);

[9] E802, T. Abbott et al.: Phys. Rev. C47 R1351 (1993);

[10] E886, G. Diebold et al.: Phys. Rev. C48, 2984 (1993);

[11] U. Heinz, P.R. Subramanian, H. Stöcker, W. Greiner: J. Phys. G: Nucl. Phys. 12, 1237 (1986)

[12] J. Ellis, U. Heinz, H. Kowalski: Phys. Lett. B233, 233 (1989)

[13] J. Schaffner, I.N. Mishustin, L.M. Satarov, H. Stöcker, W. Greiner: Z. Phys. A341, 47 (1991)

[14] H. Sorge, M. Berenguer, H. Stöcker, W. Greiner: Phys. Lett. B289, 6 (1992)

[15] H. Sorge, A.v. Keitz, R. Mattiello, H. Stöcker, W. Greiner: Nucl. Phys. A525, 95c (1991), A. v. Keitz, L. Winckelmann, A. Jahns, H. Stöcker, W. Greiner: Phys. Lett. B263, 353 (1991).

[16] P. Braun-Munzinger et al. (E877 collab.): This volume.

[17] H. Sorge, H. Stöcker, W. Greiner: Ann. Phys. (N.Y.) 192 (1989) 266.

[18] R. Mattiello, H. Sorge, H. Stöcker, W. Greiner: Phys. Rev. Lett. 63, 1459 (1989); H. Sorge, R. Mattiello, A. Jahns, H. Stöcker, W. Greiner: Phys. Lett. B271, 37 (1991); H. Sorge, R. Mattiello, H. Stöcker, W. Greiner: Phys. Rev. Lett 68, 286 (1992).

[19] B. Anderson, G. Gustafson, G. Ingelman, T. Sjoestrand: Phys. Rep. 97, 31 (1983); T. Sjoestrand: Comp. Phys. Comm. 39, 347 (1986). B. Nilsson-Almquist and E. Stenlund: Comp. Phys. Comm. 43 387 (1987); B. Anderson, G. Gustafson and B. Nilsson-Almquist: Nucl. Phys. B242, 251 (1987).

[20] B. Shiva Kumar, P. Braun-Munzinger (E814 collab.): Private communication

[21] G. Diebold and the E886 collab.: Private communication

[22] H. Stöcker and W. Greiner: Phys. Rep. 137, 277 (1986)

[23] H.A. Gustafson, H.H. Gutbrod, B. Kolb, H. Loehner, B. Ludewigt, A.M. Poskanzer, T. Renner, H. Riedesel, H.G. Ritter, A. Warwick, F. Weik and H. Wieman: Phys. Rev. Lett. 52. 1590 (1984).

[24] R.E. Renfordt, D. Schall, R. Bock, R. Brockmann, J.W. Harris, A. Sandoval, R. Stock, H. Stroebele, D. Bangert, W. Rauch, G. Odinec, H.G. Pugh and L.S. Schroeder: Phys. Rev. Lett. 53, 763 (1984).

[25] S. H. Kahana, Y. Pang, T. Schlagel and C.B. Dover: Phys. Rev. C47, R1356 (1993)

[26] B. Shiva Kumar (for the E858/878 collab.): Contribution to the Proceedings of the 11th International Conference on Ultrarelativistic Nucleus-Nucleus Collisions "Quark Matter '93" ,Borlänge (Sweden) 1993, to appear in Nucl. Phys. A.

[27] R. Mattiello *et al* and references therein: This volume

[28] B.A. Li, W. Bauer, G.F. Bertsch: Phys. Rev. C44, 2095 (1991)

[29] S.A. Bass, C. Hartnack, R. Mattiello, H. Stöcker, W. Greiner: Phys. Lett. B302, 381 (1993).

[30] E802, T. Abbott *et al.*: Phys. Rev. Lett70, 1393 (1993)

[31] P. Rotshild, S. Nagamiya: Private communications

[32] J.B. Carroll *et al.*: Phys. Rev. Lett. 62 (1989) 1829.

[33] A. Schröter , E. Berdermann, A. Gillitzer, J. Homolka, P. Kienle, W. Koenig, B. Povh, F. Schumacher, H. Ströher: Physica Scripta Vol. 48 (1993) 184; W. Koenig: This volume.

[34] C. Spieles, A. Jahns, H. Sorge, H. Stöcker, W. Greiner: Mod. Phys. Lett A27, 2547 (1993); C. Spieles: diploma thesis Uni. Frankfurt 1994

RECENT PROGRESS IN UNDERSTANDING QUARK AND GLUON DISTRIBUTION FUNCTIONS FOR LARGE NUCLEI AT SMALL x

Larry McLerran

Theoretical Physics Institute
School of Physics and Astronomy
University of Minnesota
Minneapolis, MN 55445

INTRODUCTION

The theoretical question which will be addressed in this talk is the computation of quark and gluon distribution functions at small Bjorken x. The reason why such computations may be possible is because at small values of Bjorken x, which correspond to central values of rapidity, the density of partons per unit area satisfies

$$\frac{1}{\pi R^2}\frac{dN}{dy} >> \Lambda^2_{QCD} \tag{1}$$

This density of partons per unit area is the only local parameter with dimensions of an energy squared which we can construct, and the coupling therefore should be evaluated at this scale. Defining

$$\Lambda^2 = \frac{1}{\pi R^2}\frac{dN}{dy} \tag{2}$$

the strong coupling constant at this scale must be

$$\alpha_s(\Lambda) << 1 \tag{3}$$

so that it may be possible to formulate the problems in weak coupling.

There are several problems which might be solved if one can compute the quark and gluon distribution functions at small Bjorken x. At small x, it is expected that the gluon distribution function for a single proton behaves as [1]

$$\frac{dN}{dx} \sim \frac{1}{x^{1+C\alpha_s}} \tag{4}$$

This implies that the distribution of partons for a proton diverges as $x \to 0$. If this is the case, then this problem should be possible to analyze using weak coupling methods. The distribution functions at small x have recently been measured at HERA, and do seem to be singular at small x.

In addition to understanding the Lipatov enhancement, it would also be useful to compute quantities such as the ratio of sea quark to gluon distributions. For example, if the typical energy scale has been increased due to a high density of partons, we should expect that heavier quarks will become of increasing importance. Charm quark production might for example become substantial.

Another interesting physical problem is deep inelastic scattering and di-lepton production using nuclei with $A \gg 1$. If the value of Bjorken $x \ll A^{-1/3}$, then the nucleus as seen by a parton moving with that value of x is Lorentz contracted to a scale size which is much smaller than the wavelength of the parton in a frame comoving with its longitudinal momentum. It is expected that in this kinematic region, there should be non-trivial effects which might screen the effects of the valence nuclear matter distribution. On the other hand, we expect that the distribution functions for partons should become large for large nuclei, and if the effects due to screening can be ignored (as we will see they can be when we do the computation) then

$$\frac{1}{\pi R^2} \frac{dN}{dy} \sim A^{1/3} \gg \Lambda^2_{QCD} \tag{5}$$

as $A \to \infty$. In addition to the questions about the Lipatov enhancement and the ratio of sea quarks to glue, one can also ask about the A dependence of the distribution functions.

Finally, there is the problem of determining the initial conditions for quarks and gluons in heavy ion collisions. If one can understand the distribution functions, then these may provide information about the boundary conditions for the evolution of the matter into a quark-gluon plasma. Recall that since the density of partons per unit area is the only scale in the problem, and it goes like $A^{1/3}$, the energy density will have to scale as $E/V \sim A^{2/3}$, where this is the energy density scale at a time which corresponds to the dimensional scale constructed from the density of partons per unit area. In order for these considerations to be valid, the Lorentz contracted size of the nucleus must be smaller than the size scale constructed from the density of partons per unit area. This is $E_{CM}/A \gg \sqrt{A}$ which for large nuclei requires that $E_{CM}/A \gg 50 \; GeV$ which is within the range accessible at RHIC.

In what follows, I will describe how to compute the quark and gluon distribution functions for very large nuclei. We have not yet succeeded in being able to compute for a single hadron. It may also be true that our weak coupling analysis may be marginal at best for realistic values of A, and therefore our results may only be useful as a theoretical laboratory which will give us some insight into the structure which we might expect for realistic nuclei. If it is true that weak coupling techniques cannot be used for the earliest stages of a heavy ion collision, then it also will imply that such techniques are probably never applicable in the subsequent evolution of the matter produced in such a collision.

SUMMARY OF RESULTS

Before going into a discussion of how to analyze the problem of computing distribution functions for very large nuclei, I will first summarize our results to date.[2] First I define

$$\mu^2 = \frac{4}{3} \frac{N_Q^{valence}}{\pi R^2} \sim 1.1 A^{1/3} \, Fm^{-2} \tag{6}$$

which is the the density of valence quark color charge squared per unit area. I will show that there is a many body theory which describes the quark and gluon distribution functions so long as we restrict our attention to parton transverse momenta which satisfy

$$q_T^2 << \mu^2 \tag{7}$$

and we require that we are at small values of x

$$x << A^{-1/3} \tag{8}$$

There are two expansion parameters for this many body theory $\alpha_s(\mu)$ and $\alpha_s(\mu)\mu/q_T$

In the range $\alpha_s(\mu)\mu << q_T << \mu$ the gluon distribution function may be evaluated to lowest order in $\alpha_s(\mu)$ as

$$\frac{1}{\pi R^2} \frac{dN}{dx d^2 q_T} = \frac{\alpha_s(N_c^2 - 1)}{\pi^2} \frac{\mu^2}{x q_T^2} \tag{9}$$

This is just the Weizsacker-Williams distribution function for a Lorentz boosted distribution of Coulombic gluons, scaled by the factor μ^2 which is the average value of the charge squared per unit area.

The result above is however only a formal result if it is true that there is a Lipatov enhancement. The Lipatov enhancement is of the form $x^{-C\alpha_s}$ If we formally expand this as a series in α_s, we find the expansion parameter is $\alpha_s(\mu)ln(1/x)$. Since we have assumed that $x << A^{-1/3}$ and $\mu^2 \sim A^{1/3}$, this expansion is not well behaved. To correct for this behavior if it occurs, it will be necessary to go beyond the naive weak coupling expansion. It should however still be true that weak coupling methods are still applicable.

If we sum to all orders in $\alpha_s\mu$ but to first order in α_s, we have shown that the gluon distribution function is of the form

$$\frac{1}{\pi R^2} \frac{dN}{dx d^2 q_T} = \frac{N_c^2 - 1}{\pi^2} \frac{1}{x} \frac{1}{\alpha_s} H(q_T^2/\alpha_s^2\mu^2) \tag{10}$$

where the function $lim_{y\to\infty} H(y) \to 1/y$ The function H is a correlation function for an ultraviolet finite two dimensional Euclidean quantum field theory. The strong coupling limit of this field theory is the small q_T limit. In this limit, we expect disorder since the theory is strongly coupled. This implies exponentially falling correlations in coordinate space which implies that $H(0)$ should be finite.

SETTING THE PROBLEM UP

To begin computing the distribution functions, we will assume we are working in a frame where the nucleus is moving close to the speed of light. In this frame the nucleus appears as a Lorentz contracted pancake. The value of Bjorken x in this frame is essentially the ratio of longitudinal momentum of the parton to that of the projectile nucleus per nucleon.

Since the variation in the nuclear valence quark distribution function with respect to transverse coordinate is

$$\frac{1}{N}\frac{dN}{dr} \sim \frac{1}{R_{nuc}} \tag{11}$$

and since the average scale of transverse momenta is $q_T >> 1/R_{nuc}$, we see that locally, the variation in the transverse nuclear matter distribution can be ignored. Therefore to compute the local properties of the gluon distributions function as a function of transverse coordinate,we need only consider a problem where the transverse matter distribution is assumed to be uniform and of infinite extent in the transverse direction. The problem is therefore of a sheet of valence quarks uniformly distributed on a thin sheet of infinite extent in the transverse direction. We will take the density of valence quarks to be $N_{quark}/\pi R^2 \sim A^{1/3}$.

The natural variable with which to analyze the dynamics are light cone variables,

$$a^{\pm} = \frac{a^0 \pm a^z}{\sqrt{2}} \tag{12}$$

If we let k^+ be a parton light cone momentum and p^+ that of a projectile nucleon, then $x = k^+/p^+$ In these variables, instead of constructing eigenstates of the Hamiltonian, it is simpler to construct the eigenstates of the generator of x^+ transformations,

$$P^- = \frac{1}{\sqrt{2}}(H - P^z) \tag{13}$$

Finally, when constructing the light cone Hamiltonian and action for the theory in terms of these variables, it is simplest to work in light cone gauge,

$$A_- = -A^+ = 0 \tag{14}$$

In this gauge, the light cone Hamiltonian has the form

$$
\begin{aligned}
P^- &= \int d^3x \, \frac{1}{4}F_T^2 + \frac{1}{2}(\rho_F + D_T \cdot E_T)\frac{1}{P^{+2}}(\rho_F + D_T \cdot E_T) \\
&\quad + \frac{1}{2}\psi^\dagger(M - \gamma \cdot P_T)\frac{1}{P^+}(M + \gamma \cdot P_T)\psi
\end{aligned}
\tag{15}
$$

In this equation,

$$P^\mu = \frac{1}{i}\partial^\mu - g\tau \cdot A^\mu \tag{16}$$

and $E_k = -\partial_- A_k$. The Lagrangean is generated in the usual way by adding in $-i\psi^\dagger \partial_+ \psi + E_k \partial_+ A_k$ The fermion charge density is

$$\rho_F^a = \overline{\psi}\gamma^+ \tau^a \psi \tag{17}$$

The quantized fields are

$$
\begin{aligned}
\psi_\alpha(x) &= \int_{k^+>0} \frac{d^3k}{(2\pi)^3}\left(b_\alpha(k)e^{ikx} + d_\alpha^\dagger(k)e^{-ikx}\right) \\
A_i^a(x) &= \int_{k^+>0} \frac{d^3k}{(2\pi)^3\sqrt{2k^+}}\left(a_i^a(k)e^{ikx} + a_i^{a\dagger}(k)e^{-ikx}\right)
\end{aligned}
\tag{18}
$$

The last ingredient we need to set the problem up is how to describe the valence quarks. These quarks are traveling close to the speed of light. For small values of α_s, these quarks occasionally emit a small x gluon. This emission does not change the path of the valence quark. It should therefore be a good approximation to treat the trajectories for the valence quarks as straight line propagation at light velocity, that is

$$J_a^+ = \rho_a(x^+, x_T)\delta(x^-) \tag{19}$$

Unlike the case for QED, in general in QCD the charge density will have to depend on the time x^+ since the extended current conservation condition requires that

$$\partial_+ Q^a + f^{abc}A_b Q_c = 0 \tag{20}$$

This forces the charge to rotate as

$$\tau \cdot Q(x^+) = U(x^+)\tau \cdot Q(0)U^\dagger(x^+) \tag{21}$$

where

$$U(x^+) = T \exp\left(\int_0^{x^+} dy^+ \, \tau \cdot A_+(y^+)\right) \tag{22}$$

To summarize, we have shown the problem which must be solved is to compute the ground state expectation values for a system with the valence quarks traveling with light velocity localized along an infinite sheet in the transverse space. The density of the valence quarks is uniform. This problem is well posed since the constraint on the valence quarks is equivalent to specifying the space-time coordinates of the electromagnetic charge and baryon number. These operators commute with the QCD Hamiltonian.

THE EXAMPLE OF QED

In QED, the problem we want to solve is the photon structure function generated by a fast moving electron. We will ignore pair production of electron-positron pairs. The source for the electron is x^+ independent and is

$$\rho_e = e\delta(x^-)\delta^{(2)}(x_T) \tag{23}$$

The light cone Hamiltonian is

$$P^- = \int d^3x \ \left(\frac{1}{4}F_T^2 + \frac{1}{2}(\rho_e + \nabla_T \cdot E_T)\frac{1}{P^{+2}}(\rho_e + \nabla_T \cdot E_T)\right) \tag{24}$$

For this Hamiltonian, the ground state is a coherent state

$$\mid \Psi >= C exp \left(i \int d^3x A^{op}(x) E^{cl}(x)\right) \mid 0 > \tag{25}$$

Letting P^- operate on this state, we see that the ground state has $P^- = 0$ so that the classical field is purely longitudinal and

$$\begin{aligned} B_T &= 0 \\ \nabla_T \cdot E_T &= -e\rho_e \end{aligned} \tag{26}$$

The solution for these equations are that

$$\vec{A_T} = e \ \frac{1}{k^+} \frac{\vec{k_T}}{k_T^2} \tag{27}$$

In space-time, the vector potential is $\theta(x^-)\vec{\nabla}\lambda$ which is for $x^- < 0$ vanishing and a pure gauge for $x^- > 0$

The field above is precisely the Weizsacker-Williams field for the Lorentz boosted Coulomb field. The distribution function for the photons can be computed and is

$$\begin{aligned} F(k^+, k_T) &= \frac{1}{(2\pi)^3} < a^\dagger(k)a(k) > \\ &= \frac{2e^2}{(2\pi)^3} \frac{1}{k^+k_T^2} \end{aligned} \tag{28}$$

or

$$F(x, k_t) = \frac{\alpha}{\pi^2} \frac{1}{xk_T^2} \tag{29}$$

THE DISTRIBUTION FUNCTIONS FOR QED

For QCD, we have not been successful in constructing the ground state wavefunction in the presence of the external source corresponding to the valence quarks. We have however concentrated on computing ground state expectation values. To do this, we first note that

$$Z = lim_{T \to \infty} \sum_N < N \mid e^{iTP^-} \mid N >$$ (30)

will project onto the ground state. The sum over N here includes a sum over the color labels of the external source of color charge generated by the valence quarks.

To do the sum over N, we break our transverse space into a grid of squares with size $d^2x >> \pi R^2 / N_{quark} \sim A^{-1/3} F m^2$ Our approximation will therefore only be good when we look at transverse momentum scales where $q_t^2 << \mu^2$. In these limits, the number of valence quarks in each square is much larger than 1. If this is the case, then typically the charge in each square will be much larger than 1. If Q is this charge, then $Q^2 >> Q \sim [Q, Q]$ so that the charge may be treated classically.

If the total charge of interest is also much less than the maximum possible charge in the square, then the density of states for charge Q is $e^{-Q^2/2\mu^2}$. Summing over the states in the definition of Z is therefore equivalent to inserting into the path integral the integration

$$exp\left(-\frac{1}{2\mu^2} \int d^2x_T \, \rho^2(x)\right)$$ (31)

where ρ is the charge density per unit area.

For such a Gaussian charge distribution we have

$$< \rho^a(x)\rho^b(y) > = \mu^2 \delta^{ab} \delta^{(2)}(x_T - y_T)$$ (32)

where it is straightforward to estimate $\mu^2 = 1.1 A^{1/3} \, F m^2$

The problem we must solve is therefore that described by the theory in the presence of an arbitrary external source of surface charge on the light cone and then integrating over all possible values of the charge. The fields generated correspond to a stochastic source of charge. The problem has therefore been reduced to a many body theory. It is possible to integrate out the sources entirely and get an action in terms of the quark and gluon fields with a term involving μ^2 which is associated with the valence quark charge density. The value of μ^2 sets the scale for the coupling constant. For large μ^2 the coupling is small and weak coupling methods should be reliable

We can now compute the gluon distribution function to lowest order in α_s and to lowest order in μ^2. We first compute the change in the propagator induced by the sources. This is

$$\delta < A^\mu A^\nu > = = \int [d\rho] \, g^2 \delta^{\mu i} \delta^{\nu j} < \left(\frac{\nabla_T^i}{\partial + \nabla_T^2}\right) \rho^a(x) \left(\frac{\nabla_T^i}{\partial + \nabla_T^2}\right) \rho^b(x) >$$ (33)

or

$$\delta D^{\mu\nu}(k, q) = g^2 \mu^2 \delta^{ab} (2\pi)^4 \delta(k^-)\delta(q^-)\delta^{(2)}(k_T - q_T)\delta^{\mu i}\delta^{\nu j} \frac{k_T^i q_T^j}{k^+ q^+ k_T^2 q_T^2}$$ (34)

Using this form of the propagator, it is now easy to show that

$$\frac{1}{\pi R^2} \frac{dN}{d^3k} = \frac{\alpha_s \mu^2 (N_c^2 - 1)}{\pi R^2} \frac{1}{k^+ k_T^2}$$ (35)

445

This is just the Weizsacker-Williams distribution function weighted by the average charge squared per unit area. This result reflect the RMS fluctuations in the stochastic background field induced by the source associated with the valence quarks.

We will soon see that this result is valid only in the range of momentum where $\alpha_s^2 \mu^2 << k_T^2 << \mu^2$ The last term in the previous limit is just the region of validity of our derivation of the many body theory. The first term is the limit of validity of assuming that μ^2 is small and that one can expand to first order in this quantity.

We can generalize our results to all orders in $\alpha_s \mu$ and first order in α_s To do this, we solve the classical problem of computing the fields in the background of an arbitrary source and then integrate over the source. This classical problem will be accurate to first order in α_s

We must solve the equations of motion

$$D_\mu F^{\mu\nu} = g \delta^{\nu+} \delta(x^-) \rho(x^+, x_T) \tag{36}$$

There is a solution of these equations of motions with $A_+ = A_- = 0$ which also has $F_T = 0$. This solution is of the form

$$A_j(x^+ x_T) = \theta(x^-) \alpha_j \tag{37}$$

The condition that $F_T = 0$ is equivalent to the condition that the field α_j is a gauge transformation of the vacuum configuration of a two dimensional gauge theory. The condition that $D_T \cdot E_t = -g\rho$ is equivalent to the two dimensional gauge condition

$$\nabla \cdot \alpha = -g\rho \tag{38}$$

The above field configuration may be written as

$$\tau \cdot \alpha = -\frac{1}{ig} U(x_T) \nabla_i U^\dagger(x_T) \tag{39}$$

and the gauge condition is

$$\vec{\nabla}(\cdot U \vec{\nabla} U^\dagger) = ig^2 \rho \tag{40}$$

The integration over the sources may be written as

$$\int [dU] exp\left(-\frac{1}{\mu^2 g^4} tr\left(\vec{\nabla}_T \cdot U \frac{1}{i} \vec{\nabla}_T U^\dagger\right)^2\right) \tag{41}$$

There is of course a Fadeev-Popov determinant for this measure which comes from restricting to Feynman gauge, but we will not be concerned with this measure here. (The determinant does not affect the arguments we present for the validity of perturbative expansions nor the scaling behavior in k_T.)

Note that the coupling constant for this theory is $g^2 \mu$ so that the expansion parameter is $g^2 \mu / q_T$. The fluctuations over the external charge generate an ultraviolet finite theory. The correlation function of $U \nabla U^\dagger / g$ generates the modifications due to the sources of the gluon propagator. This correlation function should die exponentially at long distances corresponding to small coupling , that is, in momentum space, the correlation function should be finite at zero momentum.

The step function $\theta(x^-)$ in the solution of the classical field equations guarantees that the distribution of gluons is proportional to $1/k^+$ when the theory is solved to first order in α_s The gluon distribution function is therefore of the form

$$\frac{1}{\pi R^2}\frac{dN}{dx\,d^2k_T} = \frac{N_c^2-1}{\pi^2}\frac{1}{x}\frac{1}{\alpha_s}H(k_T^2/\alpha_s^2\mu^2) \tag{42}$$

where $lim_{y\to\infty}H(y) = 1/y$

SUMMARY

There are of course many problems which must be solved before this approach can provide a realistic theory of the distribution functions at small x. The first order corrections in α_s must at least be understood. This will generate the induced contribution of sea quarks. It will also presumably lead to the Lipatov enhancement. This must be nontrivial since for this theory where the valence quarks are localized to a delta function on the light cone, there is no scale of P^+, the momentum per nucleon of the nucleon in the nucleus. This can only arise from the cutoff dependence of a regularized delta function. Such a dependance will not affect the physics to leading order in α_s nevertheless, since if we change the cutoff by a finite amount $(\Lambda/k^+)^{C\alpha_s}$ changes by only an amount proportional to α_s

In addition to the above problems, the issue of how large a transverse momentum one can use to define a sensible theory remains open. Even though our approximations were only valid for $q_T << \mu$ the validity of assuming straight line trajectories along the light cone should be valid in a much broader region. The extent to which the theory can be solved in the large momentum transfer region and the extent of the region of validity of the external source approximation remains open.

Finally, there is the issue of actually computing structure functions for deep inelastic scattering or Drell-Yan particle production. We have here only discussed computing the expectation values of quark distributions in the ground state wavefunction. When a probe of momentum q^2 is introduced, in general there will be two parameters of interest q^2 and μ^2. For $q^2 >> \mu^2$, presumably one can analyze the distribution functions using Altarelli-Parisi equations. For $q^2 \leq \mu^2$, the situation must be more complicated since here the momentum of the quarks inside the hadron wavefunction are important and one is not in the scaling region.

Acknowledgements

I thank Stan Brodsky, Miklos Gyulassy, Al Mueller and Janos Polonyi for useful comments and insights. This work was stimulated in part by conversations with Klaus Kinder-Geiger. I thank Raju Venugopalan with whom the results described above were derived. This work was supported under DOE High Energy DE-AC02-83ER40105 and DOE Nuclear DE-FG02-87ER-40328.

REFERENCES

1. E. A. Kuraev, L. N. Lipatov and V. S. Fadin, *Sov. Phys. JETP*, **45**, 2 (1977)

2. L. McLerran and R. Venugopalan, University of Minnesota Preprints TPI-Minn-93-44/T and TPI-MINN-93-52/T

FINITE MEMORY IN THE COLLISION PROCESSES OF A FERMIONIC SYSTEM AND ITS EFFECT ON RELATIVISTIC HEAVY ION COLLISIONS

Carsten Greiner
Department of Physics, Duke University, Durham, NC 27708

Klaus Wagner and Paul-Gerhard Reinhard
Institut für Theoretische Physik, Universität Erlangen
D-8520 Erlangen

Abstract

We consider equilibration in relativistic nuclear dynamics starting from a nonequilibrium Greens-functions approach. The widely used Boltzmann-Uehling-Uhlenbeck equation is obtained only as the Markovian limit (i.e. negligible memory time). The actual memory time in energetic nuclear collisions turns out to be $\sim$ 2-3 $\frac{fm}{c}$, which interferes substantially with the time-scale of the relaxation process. The memory kernels of the two-particle collisions and their feedback on the complete dynamics (by an explicite comparison with the standard Boltzmann treatment) will be presented.

MOTIVATION

The present work is concerned with quantum transport equations for central energetic heavy ion collisions. Thereby we have in mind a regime of incident bombarding energies about $200\,MeV/A - 2\,GeV/A$ where two-body collisions play an important role and where relativistic kinematics is needed. Most present microscopic simulations adopt a Markovian (i.e. kinetic) approach in that they assume as the basic mechanism *instantanous* binary collisions of two quasi-free, i.e. on-shell nucleons. This is handled in practice by a Boltzmann-Uehling-Uhlenbeck collision term[1]. Conclusions on extracting for example the equation of state or more direct observables like the collective flow[2] depend sensitively on a delicate interplay between the collision term and the mean field. Much has been done to understand medium effects on the cross section in the collision term, e.g. in a G-matrix approach[3] to equilibrated or momentum deformed, yet still *quasistatic* nuclear matter. There remains, however, the question of memory effects beyond the Markovian limit!

Starting from the fundamental equations of motion of quantum mechanics like the N-body Liouville and Schrödinder equation one can construct a hierarchy of N coupled equations for the reduced density operators. Any attempt to truncate this hierarchy through expressing correlations in terms of lower order density operators inevitable leads to at least principally non Markovian integral equations, reflecting the finite time the system needs to build them up. Memory effects, as were originally studied at rather low excitation energies[4, 5, 6], come into play if the collision time interferes with the typical timescale over which the mean field changes[4, 5, 6], either due to coherent motion or due to relaxation of occupation numbers. A finite collision, or correlation time is generated by a destructive interference of the various scattering channels (including off-shell) building up for times going more and more towards the past. This goes as far that one sees large memory effects at low energies, e.g. for the damping of giant resonances, which oscillate at a rate of about $60 \, \text{fm}/c$ [5].

For heavy ion collisions at much higher energies, a rather rough (and entirely classical) estimate of the duration of a collision is the time which a nucleon moves through the potentials range. The Boltzmann equation is strictly valid if one could separate between two time scales $\tau_{Cor} \ll \tau_{Rel}$, distinguishing between rapidly changing ('irrelevant') intrinsic variables and smoothly behaving ('relevant') observables. But either the mean field potentials may very well change at the same scale in energetic heavy-ion collisions[7] as well as relaxation processes can be very fast there, as we will see.

SCHEMATIC VIEW OF (POSSIBLE) MEMORY EFFECTS

To further illustrate the potential consequences consider a pure relaxation process which obeys the integral equation

$$\frac{\partial}{\partial T}\rho(T) = \int_0^\infty d\bar{t} \, M^{in}(T,\bar{t}) \, (1 - \rho(T - \bar{t})) - \int_0^\infty d\bar{t} M^{out}(T,\bar{t})\rho(T - \bar{t}) \qquad (1)$$

where $M^{in/out}$ define two memory kernels which should vanish in the remote past $\bar{t} \to \infty$. The first term on the right hand side specifies a back-scattering ('in')-rate, the other one a direct-scattering ('out')-rate. The Markovian (or Boltzmann) limit of equation (1) is obtained by assuming that the distribution function ρ varies sufficiently slow compared to the range in time of the kernels:

$$\frac{\partial}{\partial T}\rho(T)\bigg|_B = \left[\Gamma^{in}(T)(1 - \rho(T)) - \Gamma^{out}(T)\rho(T)\right] \qquad (2)$$

where the rates are defined as the corresponding integrals

$$\Gamma^{in/out}(T) := \int_0^\infty d\bar{t}\, M^{in/out}(T,\bar{t}) \geq 0 \; .$$

To see the possible interference of the finite extension in time of the kernels we can perform a Taylor expansion of ρ in the past up to first order in equation (1). This yields

$$\left.\frac{\partial}{\partial T}\rho(T)\right|^{(1)} = (1 + \frac{\tau_{Mem}(T)}{\tau_{Rel}(T)})\left.\frac{\partial}{\partial T}\rho(T)\right|_B + \delta\Gamma^{in}(T)(1-\rho(T)) - \delta\Gamma^{out}(T)\rho(T)\; , (3)$$

where the following abbrevations are introduced:

$$\tau_{Mem}(T) \quad := \quad \frac{\int_0^\infty d\bar{t}\,\bar{t}\,(M^{in}(T,\bar{t}) + M^{out}(T,\bar{t}))}{\Gamma^{in}(T) + \Gamma^{out}(T)} \; , \qquad (4)$$

$$\tau_{Rel}(T) \quad := \quad \frac{1}{\Gamma_{in}(T) + \Gamma^{out}(T)} \; .$$

The first correction factor $\mathcal{Z}(T) := 1 + \tau_{Mem}(T)/\tau_{Rel}(T)$ stems from the zeroth and first order expansion of the explicit ρ and $(1-\rho)$ dependence of eq.(1). If one reminds that the Boltzmann collision term contains additional three distribution functions to be integrated over phase space, the memory kernels $M^{in/out}$ are also merely complex functionals in $\rho(t')$ $(t' < T)$, so that in addition they also should be subject of an Taylor expansion in the historical evoluion of ρ. The thus occuring corrections of first order we have indicated in (3) by $\delta\Gamma^{in/out}(T)$ being proportional to the functional derivative $\frac{\delta M^{in/out}(\rho(t'))}{\delta\rho(t)}$. At the moment, however, in this schematic view, we do not intend to further stress the possible manifestation of these additional terms. The Markov assumption clearly states that $\rho(T-\bar{t}) \approx \rho(T)$ over the range of the kernels, i.e. $|\tau_{Mem}| \ll \tau_{Rel}$. At this stage of approximation we obtain $\mathcal{Z}(T) \approx 1$ and regain the Boltzmann–Uehling–Uhlenbeck equation for $\left.\frac{\partial}{\partial T}\rho(T)\right|_B$. The relaxation rates $\frac{\partial}{\partial T}\rho(T)$ are renormalized by the 'memory factor' $\mathcal{Z}$ as soon as the here defined memory time τ_{Mem} starts to interfere with the relaxation time τ_{Rel}. Note that strong oscillations in the kernels $M^{in/out}$ may allow *both signs* for τ_{Mem}, refering to *our* definition given in eq. (4). According to equation (3), one thus observes enhanced relaxation for $\tau_{Mem} > 0$, as well as reduced relaxation for $\tau_{Mem} < 0$ in comparison to the pure Markovian limit of the underlying collision process equation (1). The Taylor (or gradient) expansion breaks down altogether if $\left|\frac{\tau_{Mem}}{\tau_{Rel}}\right| \gtrsim 1$ and the full structure of $M^{in/out}$ needs to be considered.

TRANSPORT EQUATION

In this section we only want to give a brief description of the transport equation for a relativistic fermionic system[8]. For our study we choose the quantum hadrodynamical (QHD) Lagrangian with σ- and ω-mesons[9]. We formulate the emerging equations of motion with the technique of real time Green functions as introduced by Schwinger and Keldysh[10]. Here time arguments are defined and ordered on a special time contour path C running from an initial time t_0 to $+\infty$ and back to t_0. The equation of motion for the 1-particle nucleonic Green function $G(1,1')$ looks similar to the nonrelativistic case

$$
\left(i\gamma_\mu \partial_1^\mu - m_N - g_s\langle\hat\phi(1)\rangle - g_v(\gamma_\mu)\langle\hat V^\mu(1)\rangle \right) G(1,1') =
$$

$$
\delta_C^4(1,1')\mathbf{1} + \int_{t_0}^{t_0} d2\,\Sigma_B(1,2)G(2,1') \quad , \tag{5}
$$

where $\Sigma_B(1,2)$ is the irreducible Born collision self-energy operator. The integral in eq. (5) runs over space and time along C. In the following we restrict our considerations only to *spatially* homogenous systems (which means that the effect of spatial nonlocalities in the collisions will not be investigated), and we assume spin and isospin saturation. The equation of motion for the one-particle distribution function, $\rho(\mathbf{p},T) = -i\frac{1}{4}\mathrm{Tr}\{\gamma_0 G^<(\mathbf{p};T,T)\}$, is obtained by taking the difference of eq. (5) and its adjoint, i.e.

$$
\partial_T\,\rho(\mathbf{p},T) = \frac{1}{2}\Re\left(\int_{t_0}^{T} d\bar t\,\mathrm{Tr}\,\{\Sigma^<(\mathbf{p};T,\bar t)G^>(\mathbf{p};\bar t,T) - \Sigma^>(\mathbf{p};T,\bar t)G^<(\mathbf{p};\bar t,T)\} \right)
$$

$$
\equiv \int_{t_0}^{T} d\bar t\,M^{in}(\mathbf{p},T;\bar t)(1-\rho(\mathbf{p},\bar t)) - \int_{t_0}^{T} d\bar t\,M^{out}(\mathbf{p},T;\bar t)\rho(\mathbf{p},\bar t)\,. \tag{6}
$$

In neglecting the influence of damping on $G(\mathbf{p};t_1,t_2)$ inside the collision term and parametrizing it purely in terms of a mean-field propagation[8], the memory kernels are given by the expression

$$
M^{in/out}(\mathbf{p},T;\bar t) = \frac{1}{2}\Re\left(\mathrm{Tr}\left\{ \frac{\not p + m^*}{\pm 2iE_p^*}\Sigma^{\lessgtr}(\mathbf{p};T,\bar t) \right\} e^{-i(E_p^* + g_v V^0)(\bar t - T)} \right)\,. \tag{7}
$$

Besides of a trivial substitution $\bar t \to T - \bar t$, equation (6) becomes identicel to eq (1). For the direct term of the Born collision self-energy operator the kernels take the explicit form (after some involved disentangling of the time structure of the meson propagators on the contour path C)

$$
M^{in}(\mathbf{p},T;\bar t) =
$$

$$
\Re\left(\sum_{\alpha,\beta=1}^{5} g_\alpha^2 g_\beta^2 \int_{t_0\to-\infty}^{\infty} dt_3 \int_{t_0\to-\infty}^{\infty} dt_4 \int \frac{d^3p_2\,d^3p_3\,d^3p_4}{(2\pi)^6}\,\delta^3(\mathbf{p}-\mathbf{p_2}-\mathbf{p_3}+\mathbf{p_4}) \right.
$$

$$
e^{-i(E_p^*-E_{p_2}^*)(\bar t - T)}\,e^{-i(E_{p_4}^*-E_{p_3}^*)(t_4-t_3)}\,D_\alpha^{0\,ret}(\mathbf{p}-\mathbf{p_2};T,t_3)\,D_\beta^{0\,adv}(\mathbf{p}-\mathbf{p_2};t_4,\bar t)
$$

$$
\mathrm{Tr}\left\{ \Gamma_\beta\frac{\not p + m^*}{2E_p^*}\Gamma^\alpha\frac{\not p_2 + m^*}{2E_{p_2}^*} \right\}\,\mathrm{Tr}\left\{ \Gamma_\beta\frac{\not p_4 + m^*}{2E_{p_4}^*}\Gamma^\alpha\frac{\not p_3 + m^*}{2E_{p_3}^*} \right\}
$$

$$
\rho(\mathbf{p_2},\bar t)\rho(\mathbf{p_3},\min(t_3,t_4))\,(1 - \rho(\mathbf{p_4},\min(t_3,t_4)))) \tag{8}
$$

452

where in compact notation $g_\mu = g_v$; $\Gamma_\mu = \gamma_\mu$ for $\mu = 1, \dots, 4$ and $g_5 = i g_s$; $\Gamma_5 = \mathbb{1}$. The kernel M^{out} is generated by the replacement $\rho \leftrightarrow (1 - \rho)$.

Considering (6) and (8), the Markovian approximation is performed by neglecting the historical evolution of the internal quantities, which means to shift all time arguments of the densities, the effective masses and the single particle energies to the actual time T, i.e. replacing $\rho(\mathbf{p}, \bar{t}) \to \rho(\mathbf{p}, T)$. From the explicit expression (8) it is then obvious that the semi-infinite time integration in (6) restricts the contributing processes to *on-shell* scattering, thus leading to the usual energy conserving δ-function. One ends up with the Boltzmann equation including the Pauli blocking and the spin- and isospin averaged Born cross sections of the QHD-Lagrangian. The δ–Function in the four quasi–particle energies leads to the usual feature of the Boltzmann collision term that it will vanish only for an equilibrated and stationary (hot) Fermi distribution, which, in return, will also be approached by the more general eq. (6). It can be shown[11, 13] that the quasi-particle ansatz together with the *Markov* assumption immensely simplifies to disentangle even more involved Feynman diagramms. The assumption of *quasistationarity* (and quasihomogenity) is the general idea for all derivations of quantum *kinetic* equations[11, 12]. Yet quasi–stationarity is not a priori justified for such a violent reaction expected in a relativistic heavy ion collision.

Instead, from eq. (8) one sees that the time structure of the memory kernels is governed by the *off-shell* behaviour of the two-particle scattering amplitude. In case, if the interference of the extension in time of the kernels play a significant role on the dynamics of the system, i.e. $|\tau_{Mem}| \approx \tau_{Rel}$, the single particle energies alone are no longer conserved. Indeed, it can be shown[13] that the more general transport equation also obeys energy (and particle number) conservation, as it should. However, in addition to the contribution of the one-particle states, a correlation energy has to be properly introduced and incorporated.

MEMORY KERNELS – RESULTS AND IMPLICATIONS

To simulate a central heavy ion collision we model the equilibration by considering an infinitely extended, spatially homogenous system of two interpenetrating and counterstreaming Fermi fluids:

$$\rho(\mathbf{p}, t_0) = \frac{1}{\exp\left\{\gamma\beta_0(E_p^* - v|p_z|) - \gamma\beta_0\mu\right\} + 1} \leq 1 . \tag{9}$$

The velocity v provides the anisotropy in momentum space and represents the mean velocity of each streaming fluid in the center of mass frame. The incident bombarding energy in the laboratory frame is given by $2\left(\frac{(E_{tot,cm}/A)^2}{m_N} - m_N\right)$. The chemical potential and the effective mass $m^* = m_N + g_s\langle\phi\rangle$ are determined in fixing the baryon density $\rho_B = 4\int d^3p\, \rho(\mathbf{p}, t_0)$ to $2\gamma\rho_0$, where ρ_0 is the nuclear ground state density and γ accounts for the Lorentz contraction of the two oppositely boosted nuclei. For zero temperature the shape of the generalized Fermi surface represents two ellipsoides shifted in longitudinal momentum.

First, we want to investigate the time-structure of the memory kernels $M^{in/out}$ as such. To this end, we calculate them for a fixed mean-field and distribution function, i.e. for fixed $\rho(\mathbf{p})$, m^*/m and E_p^* accordingly. For this we have employed the coupling parameters of the *linear σ-ω – model*, being aware of overestimating (somewhat) the total scattering cross-sections (a more careful adjustment will be carried out in the next section).

We show in Figure 1 the results for a small center of mass velocity $v = 0.4\,c$ and zero temperature $\beta^{-1} = 0$. This corresponds to a bombarding energy of $E_{kin,Lab} \simeq$ 200 MeV/A. In the left upper part (1a) the density distribution $\rho(p_r, p_z = 0)$ and

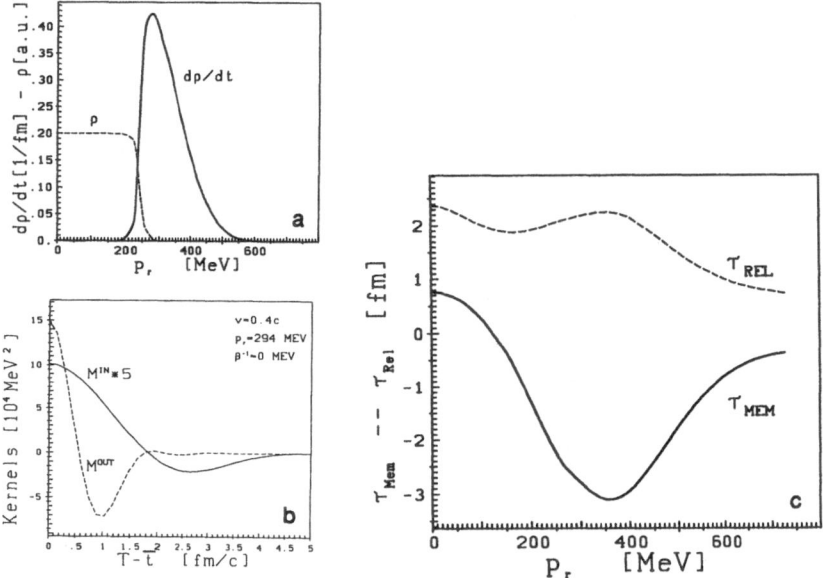

Figure 1. (a): The distribution function $\rho(p_r, p_z=0)$ (in arbitrary units, going from 1 to 0) and the Boltzmann–Uehling–Uhlenbeck scattering rate $\partial_T \rho(p_r, p_z=0)|_B$ are shown along the radial axis of momentum space for the situation of two overlaping Fermi ellipsoides with a a velocity parameter of $\frac{v}{c} = 0.4$.

(b): Memory kernels $M^{in/out}$ as function of time in the past.

(c): The memory time $\tau_{Mem}(p_r, p_z=0)$ (full line) and the relaxation time $\tau_{Rel}(p_r, p_z=0)$ (dashed line) as function of radial momentum.

the Boltzmann–Uehling–Uhlenbeck rate $\dot\rho|_B$ along the radial axis p_r are depicted. As expected, the reaction is strongest at $p_r \sim 290$ MeV, slightly above the Fermi momentum where the on-shell in-scattering is possible and unblocked. The right part of Fig. 1 shows the two memory kernels at that momentum. M^{out} exhibits a pronounced oscillation through zero, becoming negative after $\sim 1\,\text{fm}/c$ in the past (compare Fig. 1b). Visually they stretch up to $\sim 2-4\,\text{fm}/c$. Because of the oscillation in M^{out}, the memory time defined in (4) turns out to be $\tau_{Mem} \sim -3\,\text{fm}/c$ and is hence negative. The relaxation time is evaluated to be $2\,\text{fm}/c$. In the left lower part (1c), both times are plotted along the whole radial momentum axis p_r, showing that the

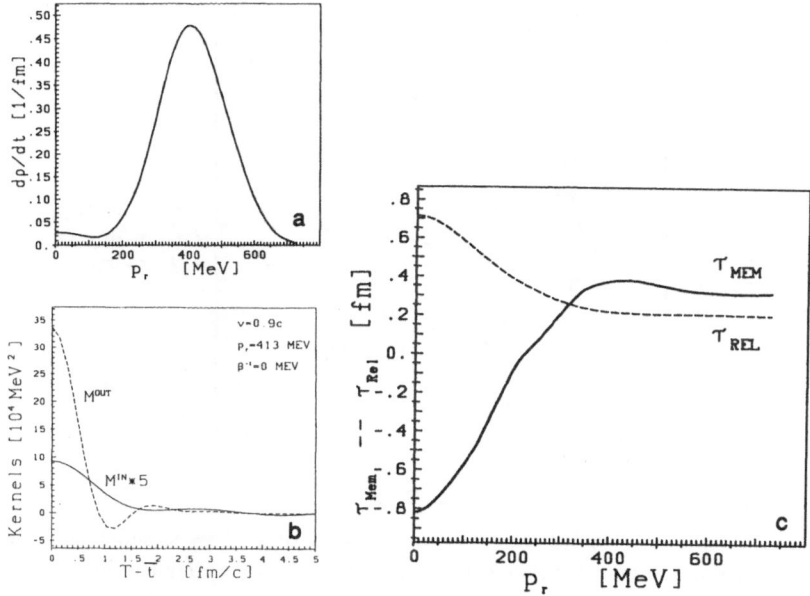

Figure 2. (a) The scattering rate $\partial_T \rho|_B$ is shown for the situation of separated Fermi ellipsoides with $\frac{v}{c} = 0.9$.

(b): Memory kernels $M^{in/out}$ as function of time in the past.

(c): The memory time $\tau_{Mem}(p_r, p_z=0)$ (full line) and the relaxation time $\tau_{Rel}(p_r, p_z=0)$ (dashed line).

Markovian condition $\left|\frac{\tau_{Mem}}{\tau_{Rel}}\right| \ll 1$ is clearly violated nearly everywhere, particularly at the most reactive (deformed) Fermi surface. In general, the pattern of each kernel change with the position (p_r, p_z) in momentum space. For large momenta well above the Fermi surface, M^{in} displays several oscillations, whereas M^{out} turns out to become more smooth. This behaviour is almost completely reverted around the Fermi level and even more in the interior of the density distribution. But the deep interiour or the far outer space are physically not relevant, because the total scattering rates $\dot{\rho}|_B$ are vanishingly small (see Fig. 1a).

One recognizes that a negative renormalization in eq. (3) physically does not make sense, so the process clearly *must* be non Markovian. Nonetheless trying to give arguments along the first order expansion one is tempted to reason that the actual relaxation in momentum space should be considerably slowed down compared to the dynamics treated in the Markovian limit.

As a further (and maybe extreme) example we consider in Fig. 2 a very violent reaction of $v = 0.9$ c, where the two ellipsoides are separated. This would correspond to a bombarding energy of $\simeq 4$ GeV/A which goes beyond the limits of our approach in that inelastic channels are not included. The reaction rate $\dot{\rho}|_B$ has a maximum of about 0.45 c/fm at $p_r \sim 410$ MeV (Fig. 2a). The characteristic oscillations of M^{out} almost disappear (Fig. 2b) and the memory time $\tau_{Mem} \sim 0.4$ fm/c turns out to be small, but positive. Now, although $\dot{\rho}|_B$ is moderate, the corresponding relaxation rate Γ_B reaches 5 c/fm, and thus the critical ratio becomes $\frac{\tau_{Mem}}{\tau_{Rel}} = 1.7$ (Fig. 2c). The

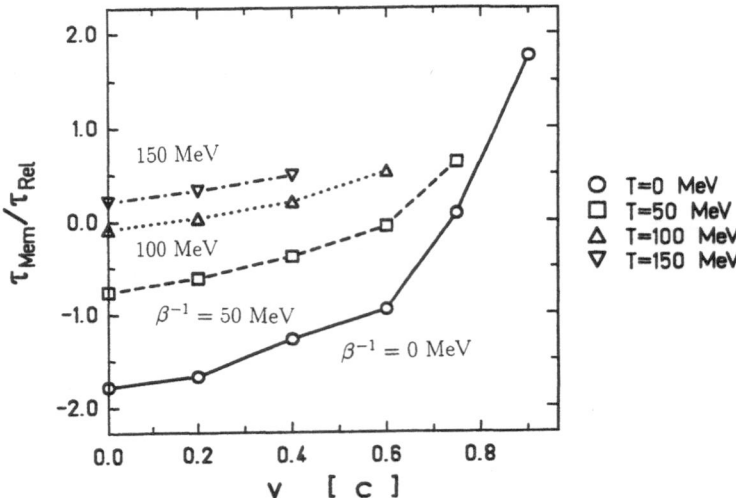

Figure 3. The ratio of memory time versus relaxation time $\frac{\tau_{Mem}}{\tau_{Rel}}$ is summarized for different velocities and temperatures at the momentum grid point p_r with the highest Boltzmann scattering rate. Exclusively, the couplings of the linear QHD-model were employed. Note that the Markov assumption is only guaranteed for some special configurations.

Boltzmann equation would be invalidated again.

The relaxation in the process will reduce the velocity $\pm v$ and thermalize the distribution. In order to investigate the typical memory kernels at later stages, we also consider cases with effective temperatures $\beta^{-1} = 50 - 150 \, \text{MeV}$ in the parametrization (9). The results are summarized in Fig. 3 by plotting the critical ratio $\frac{\tau_{Mem}}{\tau_{rel}}$ for several velocities v and different temperatures at the gridpoint p_r ($p_z = 0$) with largest scattering rate $\dot{\rho}|_B$ of each case. A ratio $\frac{\tau_{Mem}}{\tau_{rel}} \geq 0.5$ is encountered for high temperatures and/or velocities, both related to high incident energies; here the relaxation should be enhanced. The Markov assumption is not valid and even the gradient expansion may fail in some cases. An interesting aspect is the possibility to obtain more stopping power in heavy collisions arising from the occurence of *constructive* dynamical interference during the collision process. For intermediate energies $\leq 1 \, \text{GeV/A}$ one may observe a reduction of the relaxation in the first stages of the evolution because $\frac{\tau_{Mem}}{\tau_{Rel}} \leq -0.5$ at small incident temperatures $\beta^{-1} \approx 0$. But soon larger temperatures are reached and the process could behave Markovian, because the oscillations in the still extended kernels cancel to a small memory time τ_{Mem}. In a nonrelativistic energy domain of $\lesssim 400 \, \text{MeV/A}$ ($v \lesssim 0.5 \, \text{c}$) we observe $\frac{\tau_{Mem}}{\tau_{rel}} \leq -1.0$. Due to the presence of the oscillations the interference of the relaxation and the off-shell part of the kernels happens mainly in a *destructive* way. Thus the reaction as described in our model will be significantly slowed down due to memory effects and does not behave Markovian.

At this point of argumentation one has to remember that a renormalization of the 'naked' scattering processes is not the complete story when dealing with first order corrections. Due to the functional dependence of the memory kernels on the distribution function over the past time, additional corrections will show up, which, however, have not as a simple form and interpretation. Casting eq. (3) as

$$\frac{\partial}{\partial T}\rho(T)\bigg|^{(1)} = \mathcal{Z}^{in}(T)\Gamma^{in}(T)(1-\rho(T)) - \mathcal{Z}^{out}(T)\Gamma^{out}(T)\,\rho(T)\,, \qquad (10)$$

indeed two different renormalization should show up[8, 13] containing both the common contribution discussed so far. An explicit calculation shows that they still might differ quite sizeably, nearly as large as their common ratio τ_{Mem}/τ_{Rel}. Of course, a different renormalization scheme of the two individual scattering parts could have pronounced effects too! It is clear by now that in any case a Markovian treatment is valid only if the ratio between the two scales is sufficiently small. For obtaining decisive predictions for this complicated nonlinear and nonlocal transport process only a comparison between the solution of the full equation (6) and the equivalent Boltzmann equation can support our temptative conclusions.

Before doing so let us once more come back to the shape of the memory kernels. Insight of their structure can be obtained by a Fourier transformation[13] which shows that $M^{in/out}(\mathbf{p}, \omega)$ may either be rather Gaussian centered around $\omega = 0$ or shifted around some finite frequency. This resembles the importance of the off-shell structure of the kernels, which is basically determined by the possible off-shell scattering contributions due to the phase space occupancy. (The on-shell Boltzmann rates are obtained for $\omega = 0$.) For higher temperatures or energies it always tends to be more Gaussian, giving raise to nearly no oscilations when transformed back to physical time. Such an analysis should in principle allow to at least qualitatively predict the form of the kernels considered at different regimes.

Another interesting issue deals with the question of classical ($\hbar = 0$) and quantum mechanical contributions to the kernels[13]. In principle the kernels will contain both parts. For example, assuming a dilute and equilibrated 'Maxwell-Boltzmann' gas at some finite temperature T and also assuming a nonrelativistic Gaussian potential $V(r) = V_0 \exp(-r^2/r_0^2)$, the kernels may be worked out rather analytically and the typical expansion in time is given by $\sim \sqrt{r_0^2 m\beta + (\hbar\beta)^2}$. The first part reflects the intuitive expectation, the time a particle passes through the range of the potential, the second part merely reflects the average temporal extent associated via the time-energy uncertainty relation with the *characteristic (off-shell) energy scale* of the system. For our presented modeling of the situation inside a heavy-ion collision, where a dilute gas is not appropriate, we only want to state that here the phase space distributions and thus the quantum mechanical effects seems to determine the structure of the kernels (the range of the potential is given by the inverse of the mass of the heavy mesons and is here thus negligible small – for pion exchange potentials, hovever, the classical contribution should be quite sizeable and might be of importance).

REALISTIC SIMULATION

Finally, we investigate a *complete* time evolution according to eqs. (6,8).

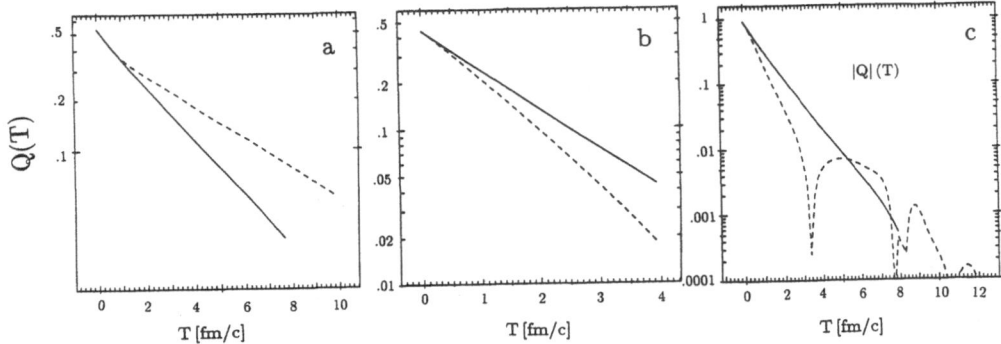

Figure 4. The relaxation in time of the quadrupole moment $Q(T)$ for various momentum anisotropical systems is depicted: (a) $\frac{v}{c} = 0.4$, $\beta^{-1} = 0$ MeV; (b) $\frac{v}{c} = 0.4$, $\beta^{-1} = 50$ MeV; (c) $\frac{v}{c} = 0.8$, $\beta^{-1} = 0$ MeV. The Boltzmann dynamics is represented by the full line, whereas the dynamics *including memory* is given by the dashed line.

It is to be remarked that in the last section we have used coupling parameters g_s and g_v which are adjusted to describe the mean-field in the linear QHD-model. But the absolute strength of the collision term depends also sensitively on these couplings, $\sim g^4$. Reducing the effective cross section will enhance the relaxation time τ_{Rel} while the memory time τ_{Mem} remains nearly unaffected. Thus the whole dynamics may come a bit closer to the Markovian limit. Therefore we decided to use in the following calculations different sets of coupling constants for the mean-field (nonlinear version of the QHD-model) and the collision term (with additional formfactor corrections). Here the coupling constants for the collision term were selected to reproduce the second moment, i.e. $\sigma^{(2)} = \int d\vartheta \frac{d\sigma}{d\Omega}(\vartheta) \sin^2 \vartheta$, of the *free* differential crossection (at the momentum of the in-medium particles), which is crucial for the stopping power and relaxation in a two-fluid scenario[14].

The proper initialisation requires that one runs the dynamics with fixed mean-field and distribution some 'preparation time' from $\bar{t} = t_0 \approx -10$ fm/c to $\bar{t} = T_0 = 0$ in order to build up properly the memory kernels and to describe complete collisions as in a Boltzmann–Uehling–Uhlenbeck equation. At $T = T_0 = 0$, the full dynamics is released and relaxation can go its way.

As an indicator for the overall relaxation we depict in Fig. 4 the time evolution of the quadrupole moment of the distribution function, $Q(T) := \int d^3 p \rho(\mathbf{p}, T) \frac{1}{2}(\frac{3p_z^2}{p^2} - 1)$, which is a measure for the systems deviation from equilibrium.

In Fig. 4a the time evolution of the moment is shown for a reaction with an initial velocity of $v = 0.4\,c$. Here the two Fermi-ellipsoides are still connected and the situation corresponds roughly to an initial bombarding energy of 200 MeV/A. As we can see the equilibration within the non Markovian process is significantly slower than the corresponding equilibration within the Boltzmann treatment. This behaviour reflects the implications emphasized in the last section. During the whole evolution the memory times at the relevant momenta are negative. According to the slope of the relaxation of $Q(T)$ the 'equilibration' time in the Boltzmann case is about 2.6 fm/c, whereas in the full treatment it increases to 4.2 fm/c. Visually the phase space distribution has reached full equilibrium, i.e. a Fermi distribution, after ~ 5 fm/c in the Boltzmann case and after ~ 10 fm/c in the non Markovian case. This suggests that the effects due to the dynamical interference will lead to large corrections (in this case the deviations are about 60%).

In Fig. 4b the initial temperature was set to 50 MeV, which should resemble an intermediate step of a reaction at higher energies. In this case the equilibration went faster in the full approach. Such an behaviour can also be understood by our discussion given so far.

As a last example we show the result for an initial velocity $v = 0.8\,c$, where the ellipsoides are separated. Such a reaction models the situation for a central heavy ion collision at bombarding energies of 2 GeV/A. The equilibration in the non Markovian treatment seems now to be obviously faster compared to the Botzmann process. Note that we have plotted here the absolute value of the moment $Q(T)$, because it turns negative after approximately 2.5 fm/c in the full dynamics. The reason is that a lot of particles initially are scattered more outwardly and after some short time do populate more the transversal states. The system then finally relaxes back more lately towards full equilbrium (~ 10 fm/c) than in the Boltzmann treatment (~ 4 fm/c). These 'overshootings' demonstrate once more the to be expected difference between both 'solutions'. The reaction according to the full dynamics is much *more violently* in the beginning, populating states which are not reached in the Boltzmann dynamics. The evolutions of the phase space distribution are by far not similar. Overall, the system then needs more time to approach (very smoothly) full equilibrium in the end.

It is fair to say that non Markovian effects should be present in heavy ion collisions. How they will manifest and what observables are affected needs further investigation.

SUMMARY

The intention of our work was to investigate the possible manifestation of memory effects in transport theories for energetic heavy ion collisions and its consequences for the equilibration times. Equations of motion containing memory integrals are the inevitable result of a reduction of the multi-particle dynamics on e.g. the one-body level which induces phase correlations into the history of the system[15]. Their interference back in the past time gives raise to a characteristic correlation or collision time. Dynamics modeled by Boltzmann-type equations leads to the 'impression' that the collisions itselves are rather instanous processes, so that in modeling transport dynamics the collisions are treated as independent and individual from one timestep to the next. Large phase spaces for the intermediate correlations should lead to large destructive interference and thus sufficiently short correlation times are expected compared to the scale on which observables change in time. This believe is the general idea of kinetic equations.

For the theoretical description we employed the techniques of real-time Greens functions for non-equilibrium dynamics leading to a transport equation posessing an in- and out scattering contribution. The Boltzmann description is obtained as the Markovian limit of this equation. However, the Markov assumption is only valid if the characteristic time scale of the relaxation in phase space is large compared to the correlation time. As a quick and easy measure for the memory effects we introduced the (more precise and well-defined) memory time as first integral moment of the memory kernel in the collision term. The ratio of memory time to relaxation time serves as a criterion for the validity of the Markovian approximation. We have calculated the memory kernel for a relativistic σ-ω-model where the heavy-ion collision is simulated in a model of two counterstreaming homogenous nuclear fluids. We find that the memory times depend sensitively on the energy in the system. They are large for low energies and ranges around a few fm/c for the typical relativistic collisions. Thus they still interfere with the relaxation times which are of the same order. Moreover, it is an interesting feature that the memory times can have both signs, positive or negative, due to the oscillatory pattern of the memory kernels. As has been shown explicitly in the numerical solutions of the equation of motion, the corrections due to memory effects can act in both directions, reduced dissipation for negative memory times and enhanced dissipation for positive memory times.

It was found that the shapes of the memory kernels follow very general pattern in spite of the complicated microscopic expressions which have been used to generate them. It is an important issue for future investigations to relate these pattern to simple phase space arguments. This may allow eventually much simpler and nonetheless microsopic estimates of the memory effects which could turn out to be useful also in other areas where cascade-like descriptions are used (e.g. parton cascades, quark-gluon dynamics, hadron dynamics).

To conclude, we find that a full treatment of a relativistic quantum collision term in the interiour of a typical central heavy ion collision can produce relaxation rates which differ significantly from the standard Boltzmann–Uehling–Uhlenbeck treatment. The changes go in different directions depending on the kinematical regime: an enhanced collision rate can be expected at very high bombarding energies whereas a reduced rate is found at lower energies.

REFERENCES

1. G. Bertsch and S. das Gupta, Phys. Rep. **160**, 189 (1988);
   W. Cassing and U. Mosel, Prog. Part. Nucl. Phys. **25**, 235 (1990)

2. G. Peilert, A. Rosenhauer, J. Aichelin, H. Stöcker and W. Greiner, Phys. Rev. C**39**, 1402 (1989);
   J. Jaenicke, J. Aichelin, N. Ohtsuka, R. Linden and A. Fäßler, Nucl. Phys. A**536**, 201 (1989)

3. A. Bohnet, N. Ohtsuka, J. Aichelin, R. Linden and A. Fäßler, Nucl. Phys. A**494**, 349 (1989);
   B. terHaar and R. Malfliet, Phys. Lett. B**172**, 10 (1986); Phys. Rev. C**36**, 1611 (1987)

4. P. Grangé, H.A. Weidenmüller and G. Wolschin, Ann. Phys. **136**, 190 (1981)

5. H.L. Yadav, P.G. Reinhard and C. Toepffer, Nucl. Phys. A**458**, 301 (1986)

6. M. Tohyama, Phys. Lett. B**163**, 14 (1985); Phys. Rev. C**36**, 187 (1987)

7. R.Y. Cusson, P.-G. Reinhard, H. Stöcker, J.J. Molitoris, M.R. Strayer and W. Greiner, Phys. Rev. Lett **55**, 2786 (1985);
   J.J. Bai, R.Y. Cusson, J. Wu, P.-G. Reinhard, H. Stöcker and M.R. Strayer, Z. Phys. A**326**, 269 (1987)

8. C. Greiner, K. Wagner and P.-G. Reinhard, 'Memory effects in relativistic heavy ion collisions', to be publ. in Phys. Rev. C

9. J.D. Walecka, Ann. Phys. **83**, 491 (1974)

10. J. Schwinger, J. Math. Phys. **2**, 407 (1961);
    L.V. Keldysh, Sov. Phys. JETP **20**, 1018 (1965)

11. S. Mrowczynski and U. Heinz, University of Regensburg preprint TPR-92-37 (1992),
    subm. to Ann. Phys.

12. S.R. DeGroot, W.A. Van Leeuwen and Ch.G. Van Weert, 'Relativistic Kinetic Theory',
    North-Holland-Publishing-Company (1980)

13. C. Greiner, K. Wagner and P.-G. Reinhard, publication in preparation;
    C. Greiner, Thesis, Erlangen (1992);
    K. Wagner, Thesis, Erlangen (1993)

14. A. R. Bodmer and C. N. Panos, Phys. Rev. C **15**, 1342 (1977)

15. E. Fick und G. Sauermann, 'The Quantum Statistics of Dynamic Processes', Vol. 86,
    Springer Series in Solid-State Sciences (1990)

"COLLECTIVE EFFECTS" OF MESONS AT SIS ENERGIES

S. A. Bass[a,b], C. Hartnack[b,c], R. Mattiello[a],
H. Stöcker[a] and W. Greiner[a]

[a]Institut für Theoretische Physik der Universität
 Postfach 11 19 32, 60054 Frankfurt am Main, Germany
[b]GSI Darmstadt, Postfach 11 05 52, 64220 Darmstadt, Germany
[c]Laboratoire de Physique Nucléaire, 44072 Nantes, France

Introduction

New experimental facilities at Darmstadt (GSI) and Berkeley (LBL) allow for the first time the experimental investigation of correlations of secondary particles – pions and other mesons – with the outgoing baryon resonace matter. This is important to probe the properties of hot and dense baryon rich matter in heavy ion collisions [1, 2, 3]. It has been thought that the pion–multiplicity reflects the thermal energy per nucleon in addition to the compressional energy of high nuclear density. The large cross section for pion nucleon interactions in the middle and late phases of heavy ion collisions has severely hampered the usefulness of pion spectra in the investigation of nuclear properties and reaction dynamics. The new experimental setups KaoS and FOPI (together with TAPS and LAND) at GSI and EOS/TPC at LBL enable us to investigate the emission pattern and properties of secondary particles in a far more detailed manner than ever before.

The IQMD-Model

For our investigation we use an extension of the **Q**uantum **M**olecular **D**ynamics model (QMD) [4, 5, 6, 7] which expicitely incorporates isospin and pion production via the delta resonance (IQMD) [8, 9, 10]. In the QMD model the nucleons are represented by Gaussian shaped density distributions. They are initialized in a sphere of a radius $R = 1.14A^{1/3}$ fm, according to the liquid drop model. Each nucleon is supposed to occupy a volume of h^3, so that the phasespace is uniformly filled. The initial momenta are randomly choosen between 0 and the local Thomas-Fermi-momentum. The A_P and A_T nucleons interact via two- and three-body skyrme forces, a Yukawa potential, momentum dependent interactions, a symmetry potential (to achieve a correct distribution of protons and neutrons in the nucleus) and explicit Coulomb forces between the Z_P and Z_T protons. They are propagated according to Hamiltons equations of motion. Hard N-N-collisions are included by employing the collision term of the well known VUU/BUU equation [3, 11, 12, 13, 14, 23]. The collisions are done stochastically, in a similar way as in the cascade models [16, 17]. In addition, the Pauli blocking (for the final state) is taken into account by regarding the phase space densities in the final states of a two body collision.

Pions are treated in the IQMD model via the delta resonance. The following inelastic reactions are explicitly taken into account:

a) $\quad N N \quad \rightarrow \quad \Delta N \quad$ (*hard–delta–*production)
b) $\quad\quad \Delta \quad \rightarrow \quad N \pi \quad$ (Δ–decay)
c) $\quad \Delta N \quad \rightarrow \quad N N \quad$ (Δ–absorption)
d) $\quad N \pi \quad \rightarrow \quad \Delta \quad$ (*soft–delta–*production)

Experimental cross sections are used for processes a) and d) [18], for the delta absorption, process c), we use a modified detailed balance formula [19] which takes the finite width of the delta resonance into account. A mass-dependent decay width is used for the Δ-decay. In between these inelastic reactions pions are propagated on curved trajectories with Coulomb forces acting upon them. The different isospin channels are taken into account using the respective Clebsch–Gordan–coefficients:

$$\Delta^{++} \rightarrow 1(p + \pi^+) \qquad\qquad \Delta^+ \rightarrow \tfrac{2}{3}(p + \pi^0) + \tfrac{1}{3}(n + \pi^+)$$
$$\Delta^0 \rightarrow \tfrac{2}{3}(n + \pi^0) + \tfrac{1}{3}(p + \pi^-) \qquad \Delta^- \rightarrow 1(n + \pi^-)$$

After a pion is produced (be it free or **bound** in a delta), its fate is governed by two distinct processes:

1. absorption $\quad \pi N N \rightarrow \Delta N \rightarrow N N$

2. scattering (resorption) $\quad \pi N \rightarrow \Delta \rightarrow \pi N$

By suppressing first the *soft-delta*-production and then the delta-absorption (while allowing the *soft-delta*-production) we are able to distinguish between effects caused by pion absorption and by pion scattering.

Angular correlations in the reaction plane

Figure 1 shows the $\langle p_x \rangle(y)$ distribution for π^+ and protons in Au(1AGeV)Au collisions with a minimum bias impact parameter distribution and a ϑ_{lab}-angular cut according to the Phase-II setup of the FOPI-spectrometer at GSI. The protons show the expected collective flow [20, 21, 22]. The $\langle p_x \rangle$ of the pions is anticorrelated to that

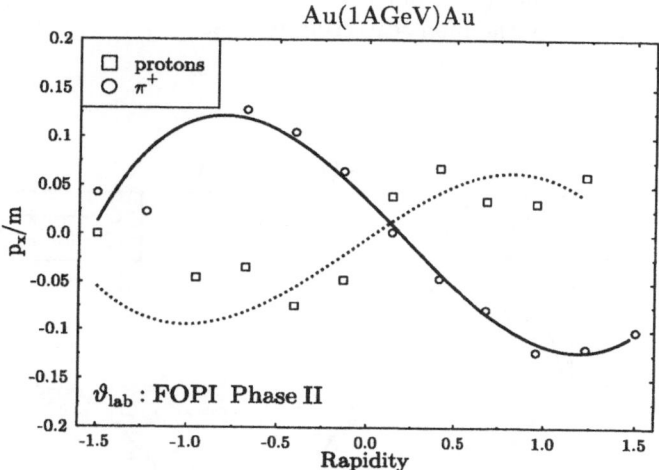

Figure 1. Rapidity y vs. $\langle p_x \rangle/m$ for π^+ and protons in a Au(1AGeV)Au collision with minimum bias impact parameter distribution. The protons show the expected bounce-off. The $\langle p_x \rangle$ of the pions is always directed oppositely to that of the protons.

of the protons. Before investigating the cause of this anticorrelation we study the impact parameter dependence of the pionic $\langle p_x \rangle$ distribution. To accomplish this we use the observable p_x^{dir} which for nucleons is defined as

$$p_x^{dir} = \frac{\sum\limits_{i=1}^{A_P+A_T} p_x^i \cdot sgn(y_i - y_{CM})}{A_T + A_P} \quad .$$

For positive values of p_x^{dir} the pionic $\langle p_x \rangle$ vs. rapidity distribution is correlated to the nucleonic one. For negative values an anticorrelation is observed. Figure 2 shows the respective calculation: For small impact parameters nucleon flow and pion flow are correlated – this is a strong hint towards the existance of delta-matter flow because the pions are produced via the delta resonance. At semiperipheral impact parameters we observe a sign reversal. The shift between π^+ and π^- is caused by Coulomb interaction.

We have studied the origin of the particular shape of the pion angular distribution and the $\langle p_x \rangle$ spectrum by sequentially suppressing first the *soft-delta*-production and then the delta-absorption (while allowing the *soft-delta*-production). If we deactivate the *soft-delta*-production, $\pi N \rightarrow \Delta$, pions are neither scattered nor absorbed after the initial production. No $\langle p_x \rangle$ for pions is observed. In order to decide whether the $\langle p_x \rangle$ spectrum is caused by absorption or by scattering we now deactivate the reaction $\Delta N \rightarrow N N$. We thus suppress pion absorption but allow scattering – the anticorrelation between pions and protons in the $\langle p_x \rangle$ returns. In contrast to previous publications, which investigated the asymmetric system Ne(800AMeV)Pb and suggested the anticorrelation of pionic and nucleonic $\langle p_x \rangle$ at target rapidities to be caused by pion absorption [23], our investigation reveals the $\langle p_x \rangle$ spectrum of the pions to be dominated by the pion scattering process.

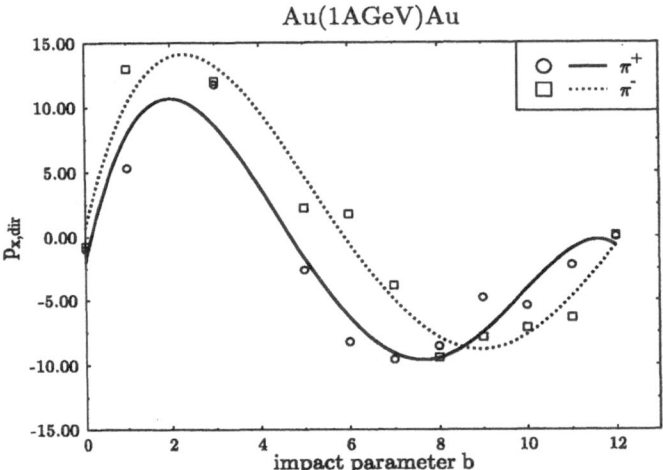

Figure 2. p_x^{dir} versus impact parameter for π^+ and π^-. The shift between π^+ and π^- is due to Coulomb forces.

The following simplified picture can explain the origin of the observed phase space distribution [24]: The Δ decays isotropically in its rest–frame, therefore 50 % of the pions are emitted with a positive p_x and 50 % with a negative p_x. At target rapidity those pions which obtain a positive p_x–value usually do not have the chance to rescatter: Most of the target nucleons are located in the **negative** p_x **area!** Those pions which **do** rescatter at target rapidity are the ones with an initially **negative** p_x: Every time a Δ decays (isotropically) there is a 50% chance that this pion is emitted **upward**, i.e. into an azimuthal angle between $-90° \leq \phi \leq 90°$. These ϕ–values characterize the hemisphere of positive p_x, by definition. This leads for$\approx$ 50% of the pions with – originally – negative p_x to a shift towards a positive p_x. This remains true even after transforming back into the laboratory frame. The same consideration applies vice versa for projectile rapidity: Most projectile nucleons are located in the **positive p_x area**. The pions are rescattered in this area which results in a negative $\langle p_x \rangle$ and a maximum in the azimuthal angular distribution in the $90° \leq \phi \leq 270°$ interval.

It is important to note, however, that the $\langle p_x \rangle$ spectrum of the pions is caused by totally different physics than the *flow* of the nucleons.

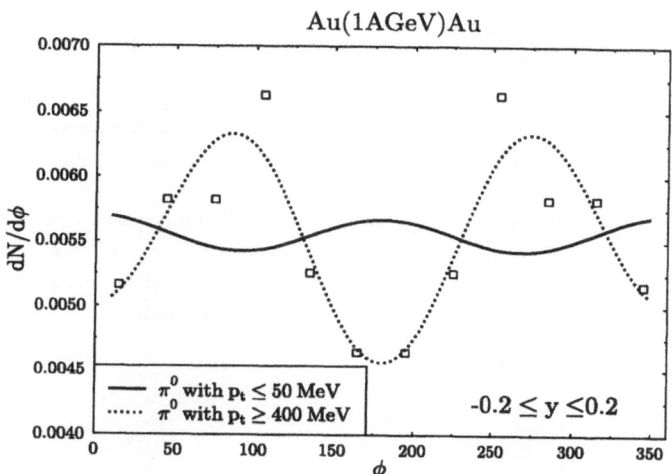

Figure 3. Azimuthal angular distribution $dN/d\varphi$ for π^0 with low and high transverse momentum p_t at mid-rapidity in the reaction Au(1AGeV)Au and minimum bias impact parameter distribution. The maximum at $\varphi = 90°$ corresponds to a preferential emission of high p_t pions perpendicular to the event-plane which is due to pion absorption by large pieces of spectator matter predominantly located in the reaction plane. Low p_t pions have rescattered more often which is only possible in the reaction plane.

Angular correlations perpendicular to the reaction plane

In order to investigate the emission of pions perpendicular to the event-plane we plot the azimuthal (φ) distribution of the pions. φ is the angle between the transverse momentum vector $\vec{p_t}$ and the x-axis (which lies in the reaction plane and is perpendicular to the beam axis). Thus $\varphi = 0$ degrees denotes the projectile hemisphere and $\varphi = 180$ degrees corresponds to the target hemisphere.

Figure 3 shows the respective distributions for neutral pions in the transverse momentum bins $p_t \leq 50$ MeV and $p_t \geq 400$ MeV at a minimum bias impact parameter distribution. The azimuthal angular distribution for π^0 with low p_t shows maxima at $\varphi = 0°$ and $\varphi = 180°$ corresponding to a preferential emission in the reaction plane. The high p_t π^0, however, show a maximum at $\varphi = 90°$ which is associated

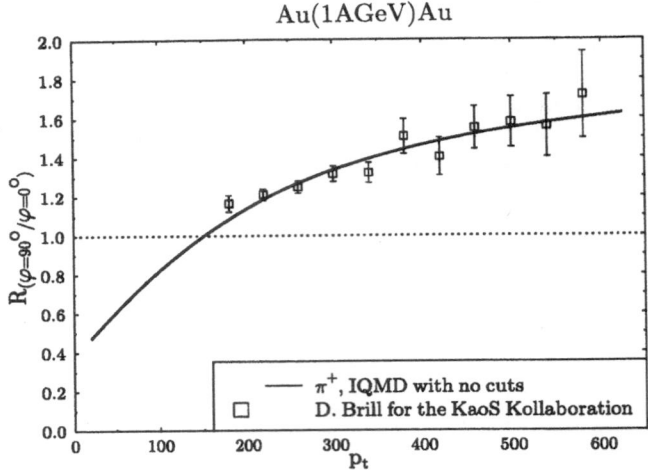

Figure 4. $R_{out/in}$ versus transverse momentum p_t for π^+. Only pions with high p_t are preferentially emitted perpendicular to the reaction plane. Pions with $p_t \leq 100$ MeV are rather emitted in the reaction plane because they have undergone frequent rescattering which can only happen in the reaction plane. The data points stem from the KaoS collaboration.

with preferential emission perpendicular to the reaction plane. The magnitude of the observed anisotropy and its dependence on impact parameter and transverse momentum is best studied by using the following ratio:

$$R_{out/in} = \left. \frac{\frac{dN}{d\varphi}(\varphi = 90°) + \frac{dN}{d\varphi}(\varphi = 270°)}{\frac{dN}{d\varphi}(\varphi = 0°) + \frac{dN}{d\varphi}(\varphi = 180°)} \right|_{y=y_{CM}}$$

For positive values pions are emitted preferentially perpendicular to the event-plane.

Figure 4 shows the transverse momentum dependence of $R_{out/in}$ for Au+Au collisions with an impact parameter of b=6 fm: In contrast to pions with low transverse momentum, which are emitted preferentially in the reaction plane, high p_t pions are preferentially emitted perpendicular to the reaction plane. The data points stem from the KaoS collaboration [25] but serve only as a qualitative comparison. Due to our lack of statistics we were unable to employ the correct angular cuts of the KaoS spectrometer needed for a full comparison. Nevertheless the mid-rapidity pion-acceptance of the KaoS-spectrometer compares well with the IQMD calculations. Simultaneously with the KaoS collaboration the TAPS collaboration has reported an azimuthal asymmetry for the emission of π^0 [26].

The cause of the preferential emission of high p_t pions perpendicular to the event-plane is studied in the same manner as the in-plane anticorrelation between pions and protons: By deactivating the reaction $\pi N \rightarrow \Delta$, pion absorption as well as scattering are eliminated and no preferential emission perpendicular to the event-

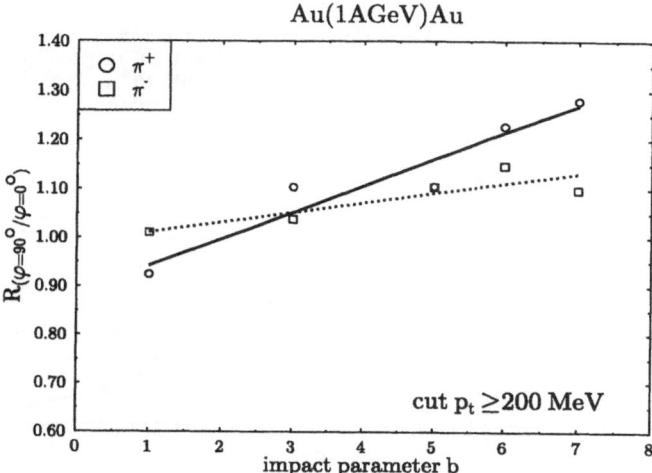

Figure 5. $R_{out/in}$ versus impact parameter for π^+ and π^- with a transverse momentum cut $p_t \geq 200$ MeV. The anisotropy is more pronounced at large impact parameters. This is due to the anisotropy caused by pion absorption by large pieces of baryonic spectator matter which do not exist for small impact parameters. The difference between π^+ and π^- is caused by the different $\pi N \rightarrow \Delta$ production cross section for π^+ and π^- and by Coulomb forces.

plane is observed. In order to decide whether the anisotropy is caused by absorption or by scattering we now deactivate the reaction $\Delta N \rightarrow N N$. We thus suppress pion absorption but allow scattering – no real anisotropy is observed. Therefore we assume the anisotropy to be dominated by the pion absorption process. This conjecture is supported by the impact parameter dependence of $R_{out/in}$ which is shown in Figure 5: No preferential emission is observed for central collisions, the anisotropy increases with the impact parameter. This behaviour stresses the importance of spectator matter for the observed effect. The difference between π^+ and π^- is due to the different $\pi N \rightarrow \Delta$ production cross section for π^+ and π^- and Coulomb effects: The π^+ are pushed away from the spectator protons whereas the π^- are attracted by them. We conclude that the preferential emission of pions perpendicular to the reaction-plane stems from pion absorption by large pieces of nucleonic spectator matter which are located in the reaction-plane. Perpendicular to the plane there is no such spectator matter and pions with high p_t can leave the reaction zone without further interaction [27]. In the reaction plane pions are likely to scatter several times. This causes a loss of transverse momentum. Therefore we observe an excess of low p_t pions in the reaction plane.

Figure 6 shows the distribution of the number of delta resonances n_Δ a pion goes through before its freeze out. Here n_Δ is shown for π^+ emitted both in the reaction plane as well as perpendicular to it. n_Δ represents the number of delta resonances a pion goes through before its freeze out. $(n_\Delta - 1)$ is therefore the number of times a

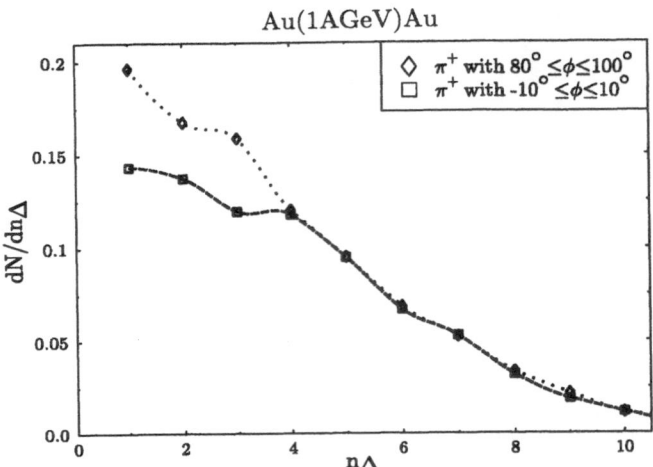

Figure 6. Distribution of the number of delta resonances n_Δ a pion goes through before its freeze out for π^+ emitted in the reaction plane and perpendicular to it. 90% of the produced pions scatter at least once before leaving the reaction zone. The observed preferential emission perpendicular to the reaction plane is due to an excess of pions which on the average have undergone fewer collisions (≤ 2) than the pions in plane.

pion scatters before freeze out. We observe that 90% of the produced pions scatter at least once before leaving the reaction zone. A large number of pions scatters even more often, some up to 10 times! The observed preferential emission perpendicular to the event-plane is due to an excess of pions which either have not rescattered or have rescattered only once or twice. These pions therefore carry information about the hot early phase of the heavy ion reaction which has not been distorted by multiple rescattering. The correlation between high p_t and early freeze-out time perpendicular to the reaction plane can be seen in Figure 7. A contour-plot is displayed in order to show the width of the distribution: High p_t pions are only emitted in the early stages of the reaction until approximately 20 fm/c.

Unfortunately the background of uniformly emitted pions is very high. An analysis of the inclusive pion spectrum reveals that high energy pions, stemming from heavy delta resonances, are produced mostly in the early phases of the reaction. Therefore high p_t pions emitted perpendicular to the event-plane should be the most sensitive pionic probes for the investigation of the hot and early reaction zone. They might even be used as experimental probes to investigate the behaviour of the Δ resonance in hot and dense nuclear matter.

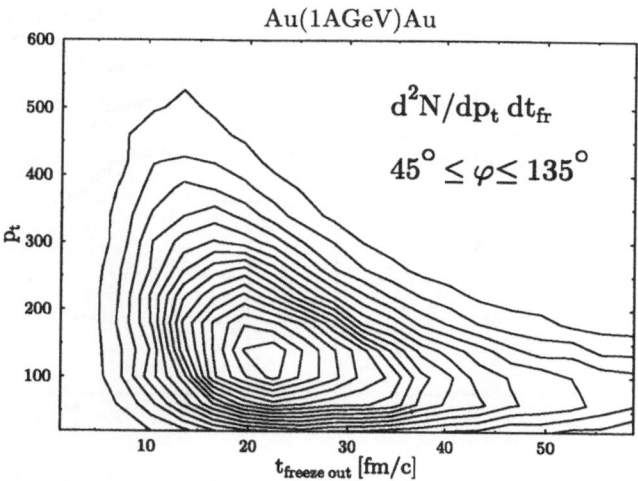

Figure 7. Contour plot of p_t vs. freeze out time t_{fr} for pions emitted perpendicular to the reaction plane. Pions with high transverse momentum are mostly emitted in the early reaction stages at times up to 20 fm/c.

Kaon production

Kaon production at SIS energies is a subthreshold process. This means that the $\sqrt{s}$ of a collision of a nucleon at beam energy with a target nucleon at rest is too small to produce the desired kaon. In order to produce the kaon either one of the colliding nucleons must have gained energy in a preceding collision or the intrinsic fermi momenta of the nucleons must add up to increase the available energy above the threshold for kaon production. Therefore kaons may be sensitive to the reaction dynamics and (especially when produced in a multistep process) the nuclear equation of state [28]. In the IQMD model there are three relevant processes for kaon production:

$$
\begin{array}{rcl}
N + N & \to & N + Y + K^+ \\
N + \Delta & \to & N + Y + K^+ \qquad Y = \{\Lambda, \Sigma\} \\
\Delta + \Delta & \to & N + Y + K^+
\end{array}
$$

The cross section parametrization which is used has been developed by Randrup and Ko [29].

Apart from their small production cross section kaons have also a very low reabsorption cross section due to strangeness conservation. Their production can therefore be calculated in a perturbative way regarding only the production probabilities

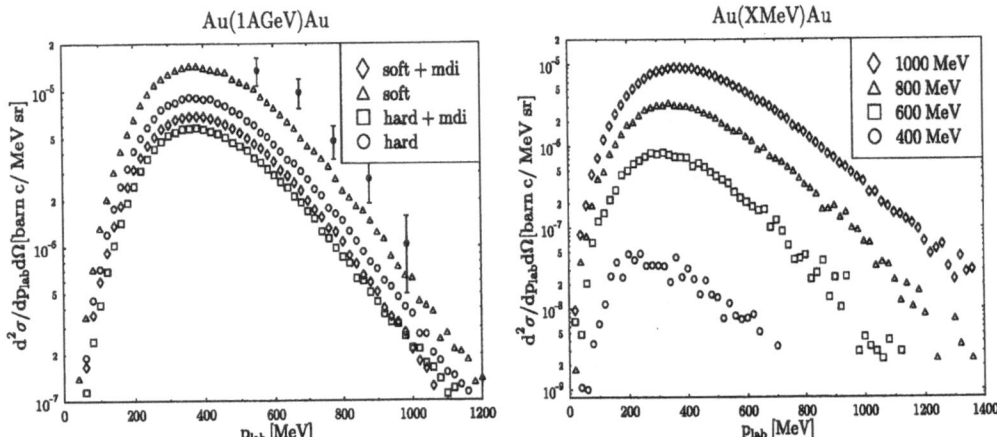

Figure 8. Equation of state dependence of kaon spectra for Au+Au at 1 GeV/nucleon and Kaon spectra for Au+Au collisions at 400, 600, 800 and 1000 MeV per nucleon (calculated with a hard eos) as calculated by the IQMD model. Ratios between spectra at different incident energies might help to dissolve ambiguities concerning the eos-dependence of the spectra with regard to the elementary production cross sections and the in-medium nucleon-nucleon cross section.

in each collision. Figure 8 shows the equation of state dependence of kaon spectra at 1 GeV/nucleon incident energy (left frame) as well as spectra for 400, 600 and 800 MeV beam energy (right frame, calculated with a hard eos) as calculated by the IQMD model. The data points stem from the KaoS collaboration [30]. Apart from the equation of state the kaon yield unfortunately depends strongly on the in medium nucleon nucleon scattering cross section which is also unknown. Furthermore different parametrizations for the elementary production processes can amount to a change in the kaon yield by orders of magnitude. To dissolve these ambiguities it might help to investigate ratios between kaon spectra at different energies.

The differences between the Kaon spectra shown in Figure 8 and the respective QMD calculations performed by Hartnack et. al. [31, 32] are caused by the use of the frozen delta approximation in the QMD model and a different treatment of the Pauli-blocking in the collision term.

REFERENCES

1. S. Nagamiya, M.C. Lemaire, E. Moeller, S. Schnetzer, G. Shapiro, H. Steiner, and I. Tanihata. Phys. Rev. **C24**, 971 (1981).

2. J. Harris , R. Bock, R. Brockmann, A. Sandoval, R. Stock, H. Stroebele, G. Odyniec, L. Schroeder, R. E. Renfordt, D. Schall, D. Bangert, W. Rauch and K. L. Wolf. Phys. Lett. **B153**, 377 (1985).

3. H. Stöcker and W. Greiner. Phys. Reports **137**, 277 (1986).

4. J. Aichelin and H. Stöcker. Phys. Lett. **B176**, 14 (1986).

5. J. Aichelin, A. Rosenhauer, G. Peilert, H. Stöcker, and W. Greiner. Phys. Rev. Lett. **58**, 1926 (1987).

6. G. Peilert, A. Rosenhauer, J. Aichelin, , H. Stöcker, and W. Greiner. Phys. Rev. **C39**, 1402 (1989).

7. J. Aichelin. Phys. Reports **202**, 233 (1991).

8. Ch. Hartnack, H. Stöcker, and W. Greiner. In H. Feldmeier, editor, *Proc. of the International Workshop on Gross Properties of Nuclei and Nuclear Excitation, XVI, Hirschegg, Kleinwalsertal, Austria* (1988).

9. C. Hartnack, L. Zhuxia, L. Neise, G. Peilert, A. Rosenhauer, H. Sorge, J. Aichelin, H. Stöcker, and W. Greiner. Nucl. Phys. **A495**, 303 (1989).

10. Ch. Hartnack. PhD thesis, GSI-Report 93-5 (1993).

11. H. Kruse, B. V. Jacak, and H. Stöcker. Phys. Rev. Lett. **54**, 289 (1985).

12. J. J. Molitoris and H. Stöcker. Phys. Rev. **C32**, 346 (1985).

13. J. Aichelin and G. Bertsch. Phys. Rev. **C31**, 1730 (1985).

14. G. Wolf, G. Batko, W. Cassing, U. Mosel, K. Niita, and M. Schäfer. Nucl. Phys. **A517**, 615 (1990).

15. G. Bertsch, S. Das Gupta, and H. Kruse. Phys. Rev. **C29**, 673 (1984).

16. Y. Yariv and Z. Frankel. Phys. Rev. **C20**, 2227 (1979).

17. J. Cugnon. Phys. Rev. **C22**, 1885 (1980).

18. B. J. VerWest and R. A. Arndt. Phys. Rev. **C25**, 1979 (1982).

19. P. Danielewicz and G. F. Bertsch. Nucl. Phys. **A533**, 712 (1991).

20. J. J. Molitoris and H. Stöcker. Phys. Lett. **B162**, 47 (1985).

21. H.-A. Gustafsson, H. H. Gutbrod, B. Kolb, H. Löhner, B. Ludewigt, A. M. Poskanzer, T. Renner, H. Riedesel, H. G. Ritter, A. Warwick, F. Weik, and H. Wieman. Phys. Rev. Lett. **52**, 1590 (1984).

22. J. J. Molitoris, H. Stöcker, and B. L. Winer. Phys. Rev. **C36**, 220 (1986).

23. B. A. Li, W. Bauer, and G. F. Bertsch. Phys. Rev. **C44**, 2095 (1991).

24. S. A. Bass, C. Hartnack, R. Mattiello, H. Stöcker, and W. Greiner. Phys. Lett. **B302**, 381 (1993).

25. D. Brill and the Kaos Collaboration. Phys. Rev. Lett. **71**, 336 (1993).

26. L. Venema and the TAPS-Collaboration. Phys. Rev. Lett. **71**, 835 (1993).

27. S. A. Bass, C. Hartnack, H. Stöcker, and W. Greiner. Phys. Rev. Lett. **71**, 1144 (1993).

28. A. Shor et. al. Phys. Rev. Lett. **48**, 1597 (1982).

29. J. Randrup and C. M. Ko. Nucl. Phys. **A343**, 519 (1980) and **A411**, 537 (1983).

30. P. Senger and the KaoS collaboration. Nucl. Phys. **A553**, 757c (1993).

31. C. Hartnack, S. A. Bass, J. Aichelin, H. Stöcker and W. Greiner Proceedings of the *International Workshop on Gross Properties of Nuclei and Nuclear Excitations XXI*, Hirschegg, Kleinwalsertal, Austria 1993.

32. C. Hartnack, J. Jaenicke and J. Aichelin. Rapport Interne LPN 93-11 and submitted Nucl. Phys. **A**

QUANTUM DYNAMICS IN PHASE SPACE

A.Smerzi[1,3], *V.Kondratyev*[1,2], *A.Bonasera*[1]

1) INFN-Laboratorio Nazionale del Sud, viale A.Doria, Catania,
Italy
2) Institute for Nuclear Research, 47, Pr.Nauki, Kiev, 252022
Ukraine
3) Dipartimento di Fisica dell' Universita' di Catania, 57, Corso
Italia, 95129 Catania, Italy

INTRODUCTION

Semiclassical kinetic equations, like for example the Boltzmann-Nordheim-Vlasov
(BNV) [1,2] and classical hydrodynamics [3], are used extensively to describe heavy
ions collisions at intermediate and high energy [1,4]. The processes as the deep-
inelastic or collective excitations are tentatively described in the Vlasov approximation [5,8] that represent the lowest order term in $\hbar$-series expansion of the time
dependent Hartree Fock equation (TDHF) [9,10]. This approximation, suitable in
order to study average quantum observables or system at high entropy, washes out
shell effects and neglects other typical quantum effects as tunneling and coherence.

In this contributions we will study the quantum corrections to the Vlasov approach
in the context of the Wigner representation of quantum dynamics, considering the
higher order corrections in $\hbar$ expansion of the phase-space projection of TDHF [11,13].
Dipole collective motions and heavy ions collisions just near the Coulomb barrier will
be analysed as well as the ground state of nuclei.

In the next section we consider a general statement of the problem and describe
the numerical method. Realistic applications and discussion of the results will follow
in the third section. Our conclusions will be drawn in section 4.

Hot and Dense Nuclear Matter, Edited by
W. Greiner *et al.*, Plenum Press, New York, 1994

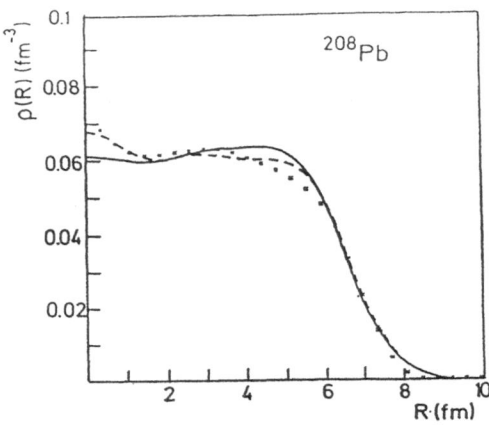

Figure 1. Proton density for Pb. Full line represents the data, the dashed line are H.F. results [19] and our results are given by the circles

GENERAL BACKGROUND AND NUMERICAL METHOD

The Wigner representation of the quantum dynamics represent an useful tool to obtain semiclassical approximations in the phase space. The central quantity in the theory is the Wigner distribution function given as a shifted Fourier transform of the one-body density matrix $\rho(\vec{r}, \vec{r}_1, t)$:

$$f(\vec{r}, \vec{p}, t) = (2\pi\hbar)^{-3} \int exp(-i\vec{s}\vec{p}) \; \rho(\vec{r} - \vec{s}/2, \vec{r} + \vec{s}/2, t) \; d\vec{s} \tag{1}$$

The equation for the one-body density matrix is:

$$i\hbar \; \partial\rho/\partial t = [H, \rho] \tag{2}$$

that in the Wigner formalism is given by:

$$\partial f/\partial t + \vec{p}/m \; \overrightarrow{\nabla} f - (\overrightarrow{\nabla} U)(\overrightarrow{\nabla}_p f) = F_q[f] \tag{3}$$

with:

$$F_q[f] = \sum_{n=1}^{\infty} (-1)^n/(2n + 1)! \; (\hbar/2)^{2n} \; U[\rho](\overleftarrow{\nabla} \overrightarrow{\nabla}_p)^{2n+1} f \tag{4}$$

The direction of the arrows over the gradients indicates on which quantities the operators act and $H = T + U$ is the single particle Hamiltonian. We suppose here that the self-consistent mean field is a local quantity For practical calculations we will use a Skyrme mean field with 200 MeV compressibility including symmetry term and Coulomb interaction. Our method breaks down for models (e.g. with sharp edges) where the power series expansion of potential does not exist or is divergent about any finite point. We assume here the convergence of the $\hbar$-expansion (4). This condition is met in the most practical cases of many-body problem (e.g., systems with high entropy [13]), when the quantum effects can be considered in perturbative way.

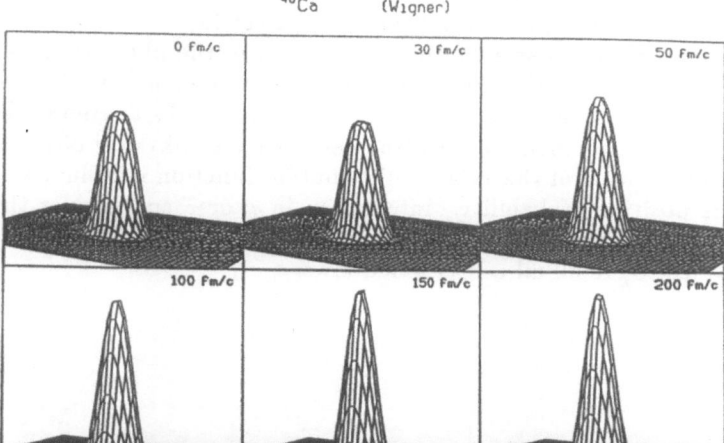

Figure 2. Time evolution of the density for Ca

The effect of quantum term is to perturb about the classical motion.

We call the quantum equation (3) Quantum Wigner Equation (QWE). Putting $F_q[f] = 0$ we get the Vlasov equation (VE).

In our calculation we include only the term $O(\hbar^2)$ in the series $F[q]$:

$$F_q = -\hbar^2/24 \ U(\vec{r})(\overleftarrow{\nabla}\overrightarrow{\nabla}_p)^3 \ f(\vec{r},\vec{p}) \tag{5}$$

The algorithm of solution of eq.s(3,5) is based on the generalization of the test particle method (TP) [1,4,18], that will be referred here as a quantum test particle method (QTP) [11,12].

We parametrize the Wigner distribution function as a collection of NA weighted δ-functions:

$$f(\vec{r},\vec{p},t) = \frac{1}{N}\sum_{i=1}^{NA} \ \sigma_i(t) \ \delta(\vec{r} - \vec{r}_i(t))\delta(\vec{p} - \vec{p}_i(t)) \tag{6}$$

where A is the total mass number of the system and N is an integer large enough to ensure numerical convergency.

To fulfill the Vlasov equation each test particle moves along the classical trajectory in the $U(\rho)$ potential:

$$\delta\vec{r}_i = \delta t\vec{p}_i/m; \qquad \delta\vec{p}_i = -\delta t\vec{\nabla}_r U(r_i); \qquad i = 1,....N_A \tag{7}$$

The quantum step is simulated by dividing the phase space into cells and increasing or decreasing the distribution function in that cell according to:

$$\delta f(\vec{r},\vec{p},t + \delta t) = f(\vec{r},\vec{p},t) + F_q[f(t)] \ \delta t \tag{8}$$

An alternative method, but more CPU time expensive, is to create new TP in each cell, randomly distributed, with a weight $\sigma_i = \pm 1$ [11,12].

We can consider eq.(3) as a continuity equation in the phase space with a source or well term. The distribution function can be positive or negative in some point of the phase space and cannot be considered as a probability anymore. However this does not represent a problem because any calculation of physical observables implies phase space integration of the Wigner distribution function in volumes $\Delta r \Delta p \geq \hbar/2$ that gives us positive probability. Integration in $\vec{p}$ or $\vec{r}$ space gives the density in coordinate or momentum space. Moreover it is very simple to verify that eq.(3) with the condition eq.(5) fulfill all conservation laws.

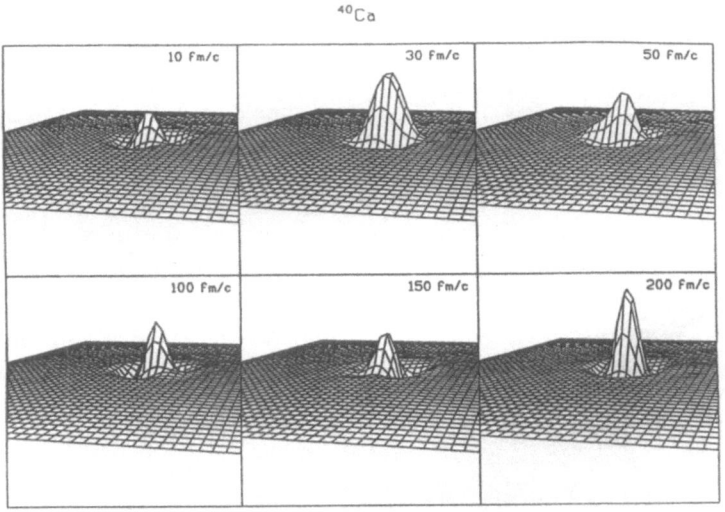

Figure 3. Difference between quantum and classical weight eq.(9) for Ca. The density is increased by a factor 30 compared to fig. 2

GROUND STATE OF THE NUCLEI

Self-consistent ground states of nuclei can be generated by the QTP method. Starting by randomly distributing TP with $\sigma_i(t=0) = 1$ in a Fermi sphere in phase space, we let our algorithm to proceed until solution stabilized.

In fig.(1) we plot the proton density for Pb versus the radial distance r. Our results agree rather well with H.F. calculations (dashed-line) [19]. Some discrepancy appears at the surface which could be due to the lack of a surface term in our Skyrme force. Nevertheless, this resonable agreement to H.F. calculations may indicate that the neglected term in eq.(4) are not important.

In fig.(2) it is plotted the density for the Ca until the ground state is estabilithed. We can see a more compact structure for asymptotic times ($t \geq 100 fm/c$). Until this time the quantum term, eq.(5) destroys part of the distribution function at the surface and creates new matter in the inner part of the nucleus. This can be see in fig.(3), where the difference between quantum and classical weights in a point $\vec{r}$ at

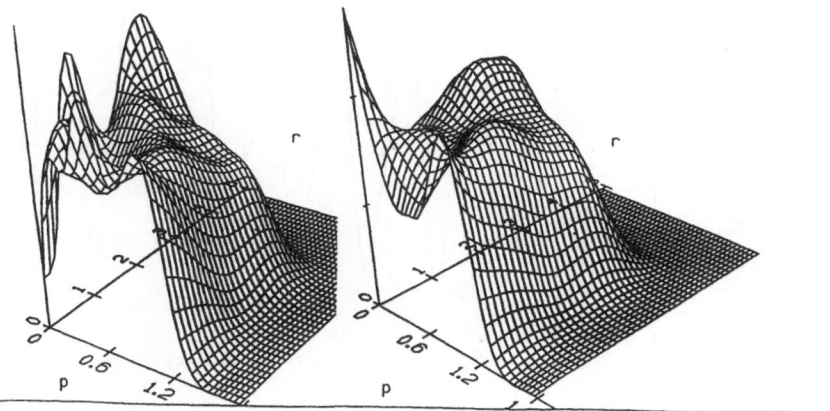

Figure 4. The semiclassical distribution function averaged over all directions of momentum p and distance

time t is drawn:

$$\Delta\Sigma(\vec{r}, t) = \frac{1}{N}\sum_{i=1}^{N_j}[\sigma_i(t) - \sigma_{cl}] = \frac{1}{N}\sum_{i=1}^{N_j}[\sigma_i(t) - 1] \tag{9}$$

N_j is the number of test particles in the cell j of the coordinate space.

In fig.(4) we show the distribution function averaged over angle Ω_r and Ω_p for protons and neutrons of Ca. Because the averaging of the Wigner distribution over a phase space volume greater then $\hbar/2$ makes it positive, the angle averaged distribution also turned out positive and allowed it to be compared directly to the classical Thomas-Fermi results. The Wigner distribution shows strong quantum oscillation which can be identified with shell effects absent in the classical results. The maxima of protons densities correspond to the minim of neutrons and viceversa due to the conflicting role of the Coulomb energy and the symmetry term in the self-consistent field.

The actual shape of $f(\vec{r}, \vec{p}, t)$ is of course very sensitive to the choice of the mean field. In the surface region we have a completely smeared-out behaviour of the Wigner function, similar to ref.[20].

COLLECTIVE MOTIONS AND HEAVY-IONS REACTIONS

Damping of small amplitude collective motions, as giant resonances, is a very interesting and not yet fully understood subject in nuclear physics. In the BNV semiclassical approach, the total width of the giant dipole resonance is obtained by an interplay between one-body and two-body dissipation [21], while the explanation of the width of giant monopole resonances is still an open problem [22].

We solve the WVE and the VE to study the one-body dissipation (or Landau-damping) for the GDR. This damping is related with absorption of collective vibra-

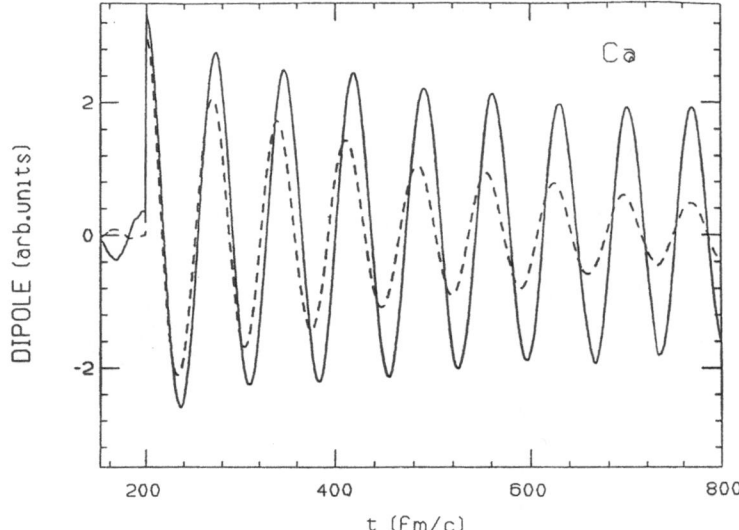

Figure 5. Time evolution of dipole oscillation for Ca. Full line is the solution of WVE, the dashed line of VE

tions energy by particles which move in resonance with collective field [7,23,24].

A single nucleus is prepared in its self-consistent ground state until the time t=100 fm/c, after that the neutrons and protons distributions in momentum space are shifted of a value Δp_z along the z axis, where Δp_z is calculated according to GDR systematics [7,20]. In fig.(5) we draw the time evolution of the expectation value of the dipole operator obtained in the WVE calculation (solid line) and VE one (dashed line). We can see that the energy of oscillation is unchanged and in agreement with the experimental data but the quantum correction leads to coherence that decreases the Landau-Damping as compared to the Vlasov approach. This effect is due to the more compact structure for the ground state of the nucleus fig.(3) and to the dynamical role of $\hbar^2$ correction.

Another interesting application of the QTP method is a nucleus-nucleus collision at beam energy just above the Coulomb barrier. We will consider here a Ru+Pb reaction at the beam energy E=8 MeV/A in the Lab. frame at impact parameter b=0. In fig.(6) we plot the time evolution of the density in the reaction plane. While in the Vlasov dynamics the two nuclei touch, exchange few nucleons but, after some time, separate due to the strong Coulomb field, fig.(6,a), in the Wigner distribution we finally get a fused system fig.(6,b).

The difference is due, principally, to two effects:

1) More important transfer of protons from the projectile to the target, due to quantum tunneling, absent in Vlasov, which decrease the Coulomb barrier.

2) Annihilation of nucleons at the center of the two nuclei and creation in the neck with consequent strong attractive mean field.

It is possible to see the two effects in fig.(7) where it is plotted eq.(9) for protons fig.(7,a) and neutrons fig.(7,b).

We can see, for protons, that a t=500 fm/c there is a considerable transfer of protons from Ru to Pb, while the neutrons, at the same time, are destroyed at the center of the two nuclei and create a pronounced neck. The combination of this two

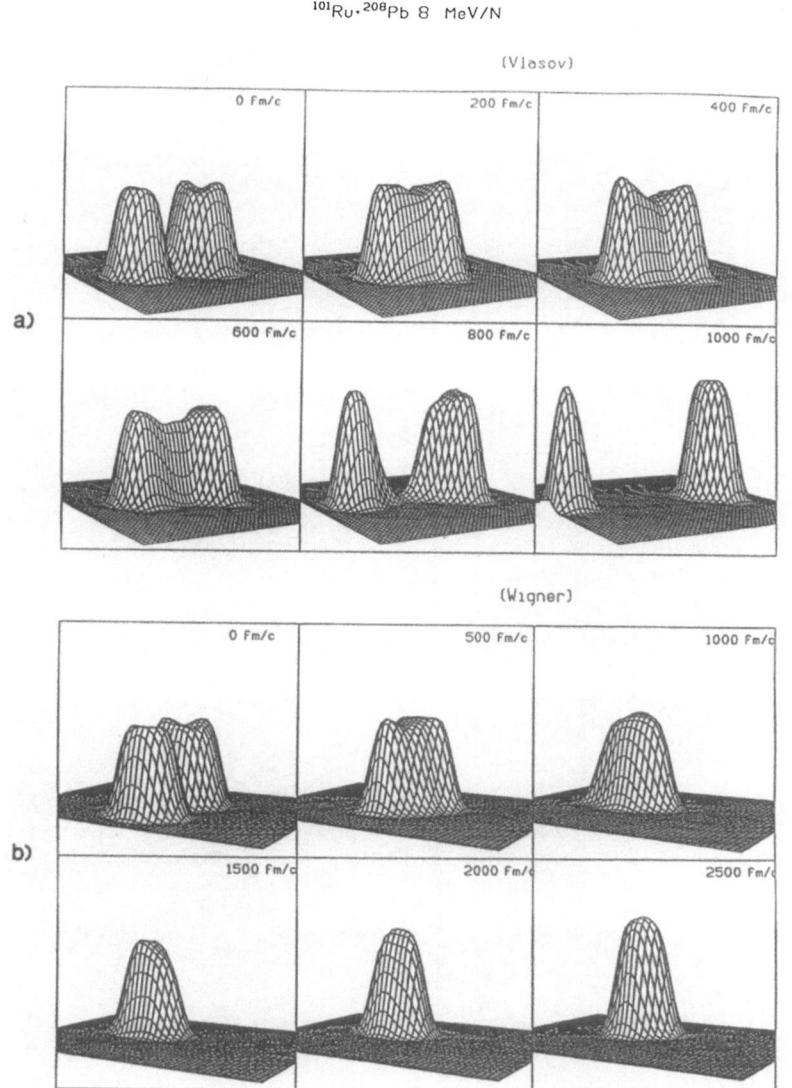

^{101}Ru·^{208}Pb 8 MeV/N

Figure 6. Time evolution of the density for Ru+Pb collision a) Vlasov solution b) solution of WVE

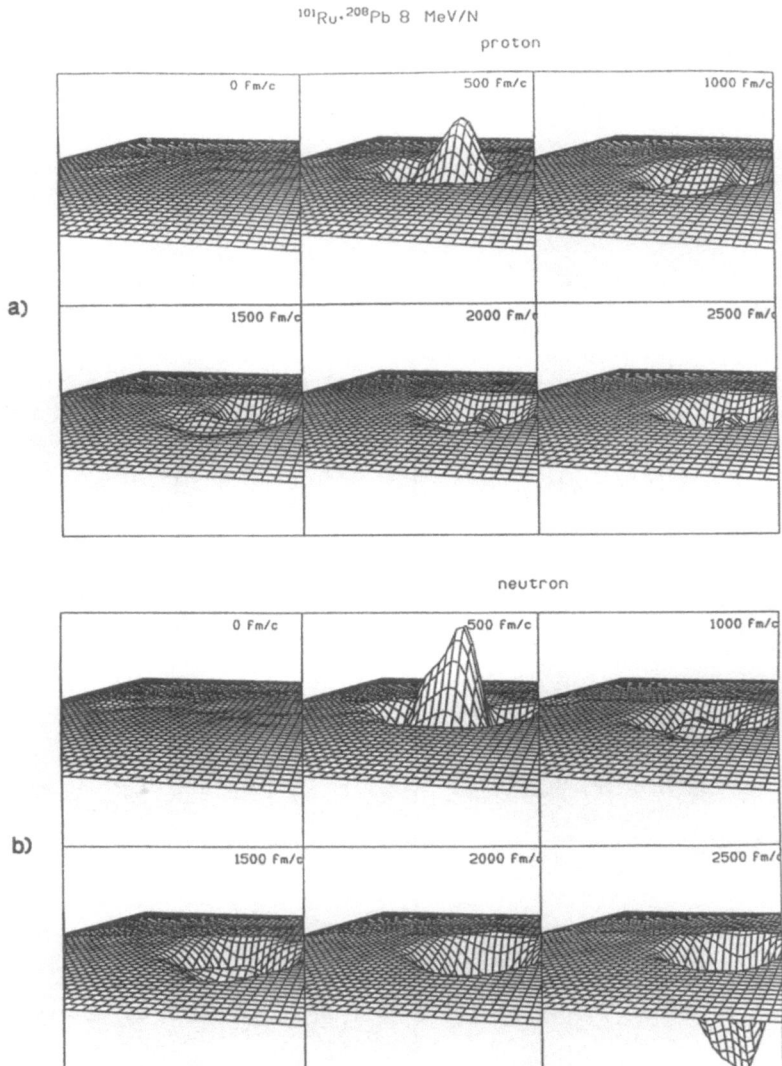

Figure 7. The same as fig.(3) for the reaction Ru+Pb a) proton b) neutron The density is increased by a factor 30 compared to fig. 6

effects gives a very different dynamical behaviour of WVE compared to VE. It is clear that this effect can be very important to study the possibility to get super-heavy nuclei. In our case the fused nucleus fig.(6,b) is evaporating because of the large excitation energy, with the final fused mass around $A = 200$.

CONCLUSIONS

We have presented here a general method to include higher order quantum corrections in Vlasov dynamics. The equation of motions for the density operator in Wigner representation is solved by a weighted test particles method. An alternative (but equivalent) method, i.e. using a time dependent number of test particles, is proposed.

Satisfactory description of the ground states of the nuclei can be obtained in this framework. The Wigner distribution function show strong quantum oscillations which can be identified with shell effects. Very compact structure for Ca and Pb are obtained.

The application of our method to giant resonances shows that quantum corrections gives coherence effects that are important to describe the one-body dissipation of the collective motion.

Finally, WVE calculations gives a very different dynamic behaviours for nuclear collisions near the Coulomb barrier in comparison with VE calculations. This is due to the nucleons transfer that is not described in VE calculation.

We note that this approach is completely general and could be used to study dynamics of any finite quantum system (electrons in atoms, molecules or clusters, quarks in bags, etc.) just changing the properties of the mean field.

ACKNOWLEDGEMENTS

Authors are indebted to Prof.s P.Carruthers, W.Greiner, J.Marhun, and Y.Oganessian for constructive discussions.

REFERENCES

1. A.Bonasera, F.Gulminelli and J.Molitoris Phys.Rep. (to be published)

2. A.Bonasera et al. Phys.Lett. B246(1990)337

3. This conference

4. G.F.Bertsch and S.Das Gupta Phys. Rep. 160(1988)189

5. M. Papa et al. submitted for publication

6. D.M. Brink, A.Dellafiore and M.Di Toro Nucl.Phys. A456 (1986)205; G.F.Burgio and M.Di Toro Nucl.Phys. A476(1988)189

7. A.Smerzi, A.Bonasera amd M.Di Toro Phys.Rev. C44 (1991) 1713

8. Chomaz, M. Di Toro, A.Smerzi, Nucl.Phys. A563 (1993) 509

9. E.P.Wigner Phys.Rev. 40 (1932) 749

10. P.Carruthers and F.Zachariesen Rev.Mod.Phys. 55(1983)245

11. A.Bonasera, V.N.Kondratyev, A.Smerzi and E.A.Remler Phys. Rev. Lett. 71 (1993) 505

12. V.N.Kondratiev, A.Smerzi and A.Bonasera,Nucl.Phys.A(1994) (in press)
    A.Smerzi, Ph.D.Thesis, Catania 1994, unpublished

13. S.John and E.A.Remler Ann.Phys.180 (1987) 152

14. E.J.Heller J.Chem.Phys.65(1976)1289

15. G.R.Chin and J.Rafelski Phys.Rev.A48(1993)1869

16. K.Takashi J.Phys.Soc.Japan.55(1986)762; 1443; 57(1988)442

17. Y.Kluger, J.M.Eisenberg, B.Svetitsky, F.Copper and E.Mottola Phys. Rev. Lett. 67(1991) 2427; Phys. Rev. D45(1992) 4659

18. C.Wang Phys.Rev. C25(1982)1460 A398 (1983) 544

19. P.Ring and P.Shuck The Nuclear Many-Body Problem (Springer-Verlag, New York, 1980)

20. M.Durand, V.S.Ramamurty and P.Schuck Phys.Lett. 113B (1982) 116

21. A.Bonasera, M.Di Toro, A.Smerzi and D.Brink, Nucl. Phys. A in press

22. V.Abrosimov, M.Di Toro and A.Smerzi Zeit. Fur Physik, in press

23. L.D.Landau Sov.Phys. JETP 16 (1946)574; E.Lifshits and L.Pitaevskyi, Physical Kinetics (Pergamon, NY, 1981)

24. A.Bonasera, F.Gulminelli and P.Schuck Phys.Rev. 46C(1992)1431

COLLAPSE OF FLUX TUBES

L. Wilets

Department of Physics
University of Washington
Seattle, WA 98195, USA

1 INTRODUCTION

Flux tubes are one of the most elementary systems of quantum chromodynamics. They are the idealized configurations of heavy quark-antiquark pairs at large separation L such that the region between can be assumed to possess axial-cylindrical symmetry. They play a central role in lattice QCD calculations and in models of QCD, as well as in the phenomenology of QCD processes.

From the spectroscopy of heavy quarkonia and Regge trajectories, it has been possible to extract a reliable measurement of the string tension by assuming the two body potential for heavy quarks to be

$$V(L) = -\frac{4\alpha_s}{3L} + \theta L + \text{ constant} \qquad (1)$$

The "experimental" value is[1] $\theta = 913$ MeV/fm $= 0.18$ GeV^2=4.63/fm^2. In fact the linear region of the flux tube potential is not directly explored by quarkonia or by lattice QCD calculations:

The wave functions in quarkonia calculations tends to be centered near the "knee" of the potential although high-lying states do explore more deeply into the linear region. However other forms of the heavy quark potential (including a logarithmic term) have also achieved some success in reproducing quarkonia spectrum.

Lattice QCD calculations on flux tubes are generally limited to the quenched approximation (no massless quarks) and allow for a separation of the heavy quark-antiquark of only about 1 fm.

Static flux tubes are unstable at separations greater than 1 fm, since the energy required to stretch the tube by 1 fm is about 1 GeV and that is about the energy difference between a quarkonium, $Q\overline{Q}$, and a pair of heavy-light mesons, $Q\,\overline{q} + \overline{Q}\,q$. Lattice calculations without light quarks cannot explore this instability.

Recently there have been a number of papers which have explored the creation of light quark pairs as a mechanism for flux tube breaking. They employ a variation of the Schwinger parallel plate capacitor model in which the infinite transverse geometry is replaced by MIT boundary conditions for the light quarks. The electric field is constant over all space in the longitudinal (z) direction. It is found that pair

Hot and Dense Nuclear Matter, Edited by
W. Greiner *et al.*, Plenum Press, New York, 1994

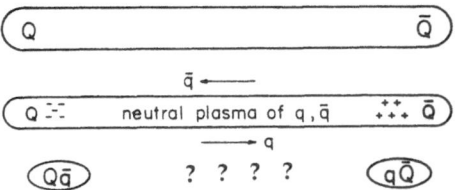

Figure 1. Schematic sequence in the collapse of the flux tubes.

production in the cylinder is suppressed relative to the Schwinger formula due to transverse confinement which give the quarks an effective mass proportional to $1/a$ where a is the radius of the flux tubes.

2 THE MIT MODEL

In the MIT model, the calculation of the flux tube configuration is straightforward since it involves no quarks. The flux through the tube is equal to the color charge $Q = \frac{1}{2}\lambda g_s$; the electric field is $E = Q/A$. Here Q is the charge at one end of the flux tube ($\bar{Q}$ at the other). It is this charge which will be shielded as a function of time due to pair production out of the sea. We distinguish it from the charge q of the light quarks being created in pairs, although for the static model $Q = q$.

The energy per unit length is given by

$$\theta = B\,A + \frac{1}{2}E^2\,A = B\,A + \frac{Q^2}{2A}, \tag{2}$$

where $A = \pi a^2$, with a the cylinder radius, and B is the bag constant. Minimization with respect to A yields $A = Q/\sqrt{2B}$ and

$$\theta = \left[2\,Q^2\,B\right]^{1/2} = 2\,B\,A = \frac{Q^2}{A}. \tag{3}$$

The translation to QCD obtains by the replacement

$$Q^2 \rightarrow =< Q^2 >= \frac{g_s^2}{4} < \lambda^2 >= \frac{16\,\pi\alpha_s}{3}. \tag{4}$$

3 THE SCHWINGER MECHANISM

The rate of pair creation in the Schwinger mechanism with MIT boundary conditions has been calculated by several groups[2,3,4]. The point of departure is the QED Schwinger problem of infinite parallel plate capacitors infinitely far apart.[5] The rate of pair production per
unit time per unit volume is

$$w(\infty) = \frac{\kappa^2}{4\pi^3} \sum_{n=1}^{\infty} \frac{e^{-\pi n m_0^2/\kappa}}{n^2}$$

$$\rightarrow \frac{\kappa^2}{4\pi^3}\frac{\pi^2}{6} = \frac{\kappa^2}{24\pi} \approx 0.25\ c/\text{fm}^4 \qquad \text{for} \quad \kappa = \theta \quad m_0 \rightarrow 0, \tag{5}$$

where m_0 is the Fermion mass and $\kappa = |q\,\mathbf{E}|$. In the MIT model, κ (with $q = Q$) is just the string tension θ. (We note that Pavel and Brink[2] identify $\kappa = 2\theta$, which is *not* the MIT model. Those authors denote the string tension by σ whereas we use θ.)

As we will see below, the rate is strongly suppressed by confinement.

4 CONFINEMENT AND TRANSVERSE MASS

Transverse confinement has the effect of introducing a "transverse mass" in the formula for w: $m_0^2 \rightarrow m_0^2 + x_n^2/a^2$. Using the results of Pavel and Brink[2] [*cf.* their Eq. (4.3)], the pair production rate per unit time per unit length W is

$$W(a) = -\frac{\kappa}{\pi} \sum_{x_n > 0} \ln\left[1 - e^{-\pi x_n^2/a^2 \kappa}\right] \approx \frac{\kappa}{\pi} e^{-\pi\,x_1^2/a^2\kappa} . \tag{6}$$

For reasonable a and $m_0 = 0$, the two lowest transverse modes, $x_{\pm 1}$ dominate, and we consider these only here: $|x_{\pm 1}| = 1.4347$.

5 SIMPLE ESTIMATE OF COLLAPSE TIME

We now consider $Q = Q(t)$, since the charge at the ends is shielded due to the flow of opposite charge into the ends. We will calculate it more carefully in the next section, but it is a good approximation to calculate the current by using the initial radius of the flux tube and the electric field for short times.

Then

$$\frac{dQ(t)}{dt} = -J = -2\,\frac{n_f\,W(0)\,q^2}{2\,A\,m}\,t^2\,Q(0) , \tag{7}$$

where n_f is the number of light flavors. One more time integration gives

$$Q(t) = Q(0)\left[1 - \frac{n_f\,W\,q^2}{3\,A\,m}\,t^3\right] . \tag{8}$$

We now estimate the characteristic lifetime of the tube to be

$$\tau(a) \approx \left[\frac{3\,m\,A}{n_f\,W\,q^2}\right]^{1/3} = \left[\frac{3\,m}{n_f\,W\,\kappa}\right]^{1/3} = \left[\frac{3\,\pi\,x\,e^{\pi\,x_1^2/a^2\kappa}}{n_f\,\kappa^2\,a}\right]^{1/3} , \tag{9}$$

where we have made the identification $m = m_\perp = x_1/a$. Note that there is no dependence on the length of the flux tube!

Taking $n_f = 2$, we find

$\kappa =$	θ	2θ	
$\tau(0.5) =$	7.0	1.7	fm/c,
$\tau(1.0) =$	1.4	0.68	fm/c.

6 TIME-DEPENDENT DIFFERENTIAL EQUATION

We must now distinguish between the elementary quark charge q and the *shielded* heavy quark charge $Q(t)$, where $Q(0) \equiv q$. In the (adiabatic) MIT model, there is no reference to the elementary (light) quark charge. Using $R(0) \equiv a$, we have

$$R^2(t)/a^2 = Q(t)/q, \tag{10}$$

which gives the interesting result that the electric field is independent of time, even as the shielded charge decreases.

The pair production rate depends upon E, the elementary charge q and $R(t)$. The quantity $\kappa = qE$ is independent of time. This gives

$$W(t) \approx \frac{\kappa}{\pi} e^{-\pi x_1^2 q/\kappa a^2 Q(t)}. \tag{11}$$

For the kinetic mass we take

$$m(t) = \frac{x_1}{R(t)} = \frac{x_1}{a} \sqrt{\frac{q}{Q(t)}}. \tag{12}$$

Then

$$J(t) = 2 n_f q^2 E \int_0^t m^{-1}(t') W(t') (t - t') dt', \tag{13}$$

and

$$\frac{dQ(t)}{dt} = -J(t). \tag{14}$$

where J is the sum of the q and $\bar{q}$ currents.

We write the final integro-differential equation in terms of $\tilde{Q} \equiv Q/q$:

$$\tilde{Q}^{-1/2} \frac{d\tilde{Q}(t)}{dt} = 2 \frac{d\tilde{Q}^{1/2}(t)}{dt} = -2n_f \frac{\kappa^2 a}{\pi x_1} \int_0^t e^{-\pi x_1^2/\kappa a^2 \tilde{Q}(t')} (t - t') dt', . \tag{15}$$

We can convert this to a differential equation by expressing

$$\frac{d^2 \tilde{Q}^{1/2}(t)}{dt^2} = -n_f \frac{\kappa^2 a}{\pi x_1} \int_0^t e^{-\pi x_1^2/\kappa a^2 \tilde{Q}(t')} dt', \tag{16}$$

$$\frac{d^3 \tilde{Q}^{1/2}(t)}{dt^3} = -n_f \frac{\kappa^2 a}{\pi x_1} e^{-\pi x_1^2/\kappa a^2 \tilde{Q}(t)}. \tag{17}$$

with the initial conditions $\tilde{Q}(0) = 1$, $\dot{\tilde{Q}}(0) = \ddot{\tilde{Q}}(0) = 0$.

We have solved Eq. (17) for a number of cases and have found that it is reproduced quite well by Eq. (9) for short times, but does lead to a longer collapse time, see Figure 2.

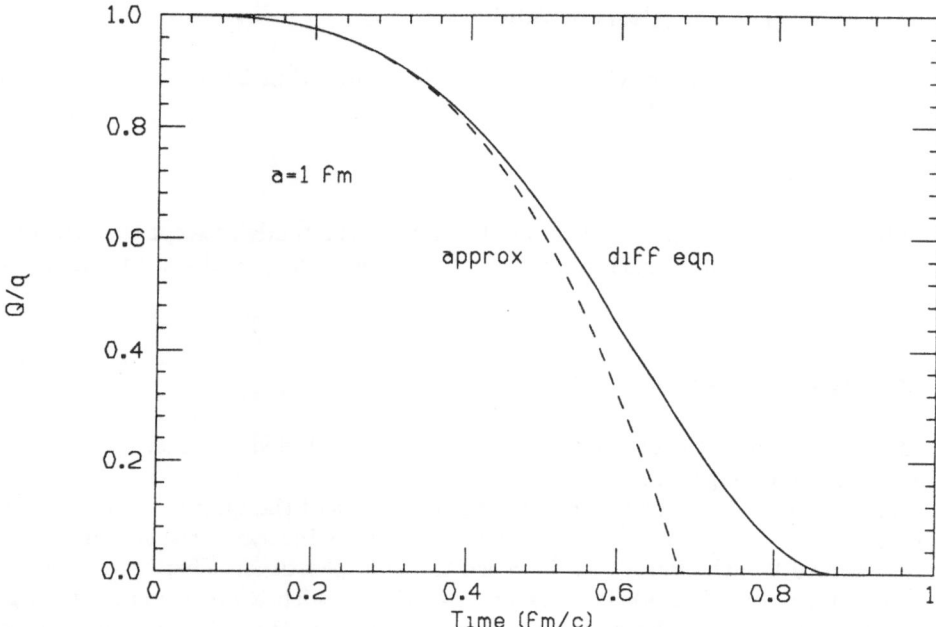

Figure 2. $Q(t)/Q(0)$ calculated according to Eqns. (a) [approx] and (12) [diff. eqn]; a = 1, $\kappa = 2\theta$.

7 MODELS OF FLUX TUBES

Since the collapse time is strongly dependent upon the initial flux tube radius $R(0) = a$, we now address more realistic flux tube calculations to determine a and the relationship between κ and θ.

7.1 The MIT Model

In this model, the field energy $\frac{1}{2}E^2$ is uniform within a radius a which is characteristically 1 fm or larger. The flux tube energy is shared equally between the electric field and volume energies: $\theta = 2\frac{1}{2}E^2A = Q^2/A = q|E|$.

7.2 Lattice Gauge Calculations

Lattice calculations at present are usually restricted to the quenched approximation (no zero-mass dynamical quarks) and to $Q\overline{Q}$ separations L of less than about 1 fm. Nevertheless Sommer[6] obtains a dependence of the (longitudinal) electric field energy which appears to be stable with L. The functional form is roughly exponential, $\frac{1}{2}E^2 \propto e^{-r_\perp^2/b}$ with $< r_\perp^2 >= 6b^2 \approx (0.2 \text{ fm})^2$. The *equivalent* radius (the radius of a square form giving the same $< r_\perp^2 >$) is $r_{eq} = \sqrt{2} < r_\perp^2 >^{1/2} \approx 0.3$ fm. Here $\theta = 2q|E|$.

7.3 Dual Superconductivity Model

The dual superconductivity model[7] yields a somewhat larger radius than lattice calculations, with $r_{eq} = \approx 0.4$ fm.

7.4 The CDM

The chromodielectric model[8] is similar to the MIT model, except that the electric field energy has a diffuse surface and the flux tube energy θ also contains a surface term such that $\theta > q|E|$.

8 FURTHER CORRECTIONS

There are several effects which are yet to be included in order to improve the estimate of the collapse time:

(1) Inclusion of the relativistic longitudinal mass of the quarks. Instead of $m_\perp = x_1/R$ we have $m = \sqrt{(x_1/R)^2 + p^2}$. This larger mass increases the collapse time.

(2) Finite length effects also increase the collapse time. There are at least two such effects to consider: a) The current associated with a given momentum group terminates when all the charge in the group is transported the full length of the tube. b) No pairs can be created unless the change in potential the length of the tube is greater than $m_\perp$: $|qEL| > x_1/R$. These also lengthens the collapse time.

(3) Dynamics of the confinement mechanism and associated energy conservation. One might consider the chromodielectric soliton model to handle this.

(4) Pressure of the neutral $q\,\overline{q}$ plasma on the walls of the flux tube. This could not only delay collapse but also leave a residual tube of pure plasma which decays into mesons (see next item).

(5) Multi-meson production. The present discussion suggests division into two heavy mesons, $Q\,\overline{q} + \overline{Q}\,q$ which are highly excited (for long original tubes) and emit multiple light $q\,\overline{q}$ mesons. Alternatively, one might consider a sausage instability in the flux tube that decays into a linear chain $Q\,\overline{q} + q\,\overline{q} + \cdots + q\,\overline{q} + \overline{Q}\,q$. It is not clear which mode should be favored. This brings us back to item (3).

9 FINAL COMMENTS

It is important to note that the calculations presented here do not depend on the velocity of the heavy quarks in the longitudinal direction. The collapse time depends sensitively on the radius of the flux tube and indeed on detailed structure of the tube.

For quite a difference approach to the problem, see ref. [9].

Acknowledgments

I wish to thank Profs. M. Baker, H. A. Pirner and R. Sommer for useful discussions. This work is supported in part by the U. S. Department of Energy.

REFERENCES

1. A. Casher, H. Neuberger and S. Nussinov, Phys. Rev. D. **20**, 179 (1979).
2. H.-P. Pavel and D. M Brink, Z. Phys. **C 51**, 119 (1991).
3. T. Schenfeld, B. Muller, K. Sailer, J. Reinhardt, W. Greiner, and A. Schafer, Phys. Lett. **B 247**,5 (1990);K. Sailer, Z. Hornyak, A. Schaefer, and W. Greiner, Phys. Lett. **B 287**, 349 (1992).
4. C.S. Warke and R.S. Bhalerao, Pramana J. Phys. **38**, 37 (1992).
5. J. Schwinger, Phys. Rev. **82**, 664 (1951).
6. R. Sommer, Nuc. Phys. **B 306**, 181 (1992)
7. M. Baker, J. S. Ball and F. Zachariesen, Phys. Rev. D **41**, 2612 (1990); **47**, 3021 (1993), and private communication.
8. G. Fai, R. Perry and L. Wilets, Phys. Lett. **B 208**, 1 (1988).
9. Y. Kluger, J. M. Eisenberg and B. Svetitsky, Int. J. Mod. Phys. **E 2**, 333 (1993).

STUDY OF THE DYNAMICS OF THE Au+Au COLLISIONS BY MEANS OF ISOTOPE-RESOLVED EMITTED LIGHT PARTICLES

G. Poggi, M. Bini, A. Olmi, P.R. Maurenzig,
G. Pasquali and N. Taccetti

I.N.F.N. and University of Florence, Italy

P. Danielewicz

Michigan State University, East Lansing - Michigan

and

FOPI Collaboration

Darmstadt - Heidelberg - Dresden - Strasbourg
Clermont-Ferrand - Florence - Bucarest - Budapest
Warsaw - Moscow - Zagreb

INTRODUCTION

In this lecture an apparatus of the so called *"Phase I"* of the 4π Detector ("FOPI") which operates at the SIS accelerator of GSI (Darmstadt) is described and the obtained experimental data are presented. A detailed discussion of the main 4π facility and of the associated experimental results can be found elsewhere in this meeting [1, 2, 3].

During Phase I the main 4π Detector did not provide any information about the mass of detected particles, while this information in Phase II is provided by the drift chambers placed inside a solenoidal magnet around the target position. Our apparatus, which operated in conjunction with the main detector, consists of 8 multiple telescopes which can be moved in the horizontal plane over an angular range from 6° up to 90°. It was designed and built mainly in order to study the intensity and energy distributions of light particles with full mass resolution. The chosen geometry allows to sample data in a quite narrow azimuthal angle range (few degrees), while a relatively large range of polar angle is covered.

In this lecture, we first describe the chacteristics of the apparatus and some of the technical problems isolated and solved during the design and construction. The

Hot and Dense Nuclear Matter, Edited by
W. Greiner *et al.*, Plenum Press, New York, 1994

problem of phase-space coverage will be addressed in particular, also with reference to the models to which comparison is made. The experimental energy spectra of light particles emitted during the reaction Au+Au at 100, 150, 250 $AMeV$ are next confronted with the calculation of Danielewicz and Qiubao Pan [4]. The experimental data obtained with our apparatus allow to assess the presence of a collective expansion of the emitted light fragments. The agreement between experimental results and calculations may permit to determine the amount of stopping in the collision.

DESCRIPTION OF THE DETECTOR

The apparatus [5], which has been designed and built by the Florence subgroup of the FOPI Collaboration, relies on the classical, well known, $\Delta E - E_R$ technique for particle identification and energy measurement. This technique consists in measuring the amount of energy deposited by a given particle in the various elements of a stack of detectors constituting a so-called $\Delta E - E_R$ telescope.

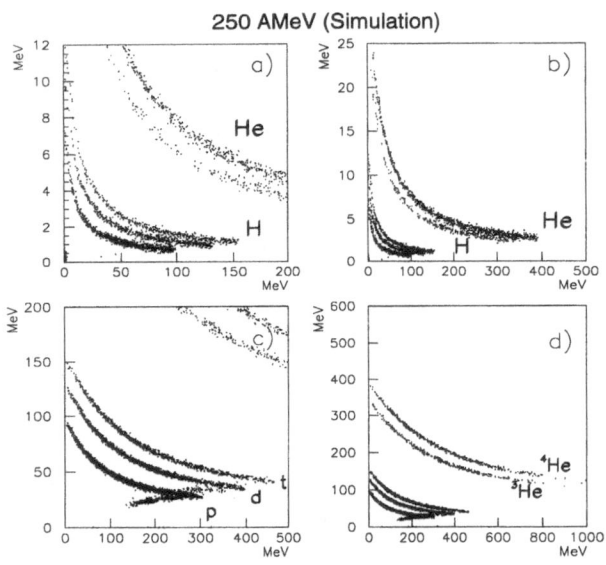

250 AMeV (Simulation)

From the shape of the correlation among the energies deposited by the particle in the various elements of the telescope, it is possible to identify the particle and to measure its kinetic energy, if the particle is stopped inside the telescope. In our case the first detector (ΔE element) is a planar ion-implanted Silicon detector having a nominal thickness of 500 μm; the next detectors are inorganic scintillators (CsI(Tl)) of various thicknesses. In fig.1 a) and b) the results of a simulation are shown, representing an ideal correlation among the energies deposited in one of our telescope by $Z = 1, 2$ particles produced in the Au+Au reaction at 250 $AMeV$ bombarding energy. In part a) and b) only those particles that are fully stopped in the second detector 30 mm thick are shown. In part c) and d) the correlation is shown between the sum of the energies deposited in the first two detectors (500 μm Si + 30 mm CsI(Tl)) and the energy deposited in the next detectors (two CsI(Tl) scintillators having a total thickness of 200 mm).

Figure 1. $\Delta E - E_R$ correlations. The reaction is Au+Au at 250 $AMeV$ and data refer to a simulation. In part a) and b) the energy deposited in Silicon detector is plotted versus the energy deposited in the first scintillator; the horizontal and vertical axes have different ranges in a) and b)). In part c) and d) the energy deposited in the Silicon detector plus the one deposited in the first scintillator is plotted versus the total residual energy. Energy ranges are different. Absence of nuclear interactions and of multiple hits is assumed in the simulation. The finite width of the correlations derives from electronic noise, non-uniformity of detector thickness, non-homogeneity of light collection and energy straggling, which were included in the simulation.

Note the punch-through branches, due to high energy particles that could not be stopped inside the telescope.

The $\Delta E - E_R$ technique was developed many years ago and it has been of fundamental importance for nuclear spectroscopy at low energies. The technique is presently sound and developed, therefore it would be useless to spend a single word on this subject, if not for the unusual application of this technique in our case, where **high energy particles** emitted in **high multiplicity events** were to be identified.

Let us first discuss the problems connected to the energy of the detected particles: the $\Delta E - E_R$ technique in practice requires the use of residual energy (E_R) detectors thick enough to stop the particle to be identified. However, as the energy of the particle increases, the thickness of the detector stack must increase more than linearly. The probability becomes correspondingly higher that this particle during its slowing down process, mainly determined by collisions with atomic electrons, experiences also a collision with the nuclei of the detector material; in case of such a nuclear collision the measurement of the energy deposited by the particle is not reliable anymore; apart from problems associated to the Q-value of the nuclear collision, secondary particles are in fact emitted in the reaction and only a fraction of the initial kinetic energy is deposited inside the telescope volume.

In Table 1 the probability of inelastic nuclear interaction in CsI(Tl) is shown for various particles at various energies [6, 7].

Table 1. Percent non-elastic nuclear interaction probability P for various particles of different energies in CsI(Tl).

	Particle	p	α	^7Li	^{11}B	^{12}C
P (50 $AMeV$)	[%]	2.8	6.5	4.2	2.3	1.6
P (100 $AMeV$)	[%]	8.8	20.0	13.7	8.7	6.7
P (150 $AMeV$)	[%]	16.1	34.4	24.4	16.4	13.0
P (200 $AMeV$)	[%]	23.8	47.9	35.3	24.7	19.9
P (250 $AMeV$)	[%]	31.7	59.8	45.5	28.3	26.9
P (300 $AMeV$)	[%]	39.5	69.5	54.9	40.5	33.8

Note that a proton of 300 MeV kinetic energy has a probability of about 40% for experiencing a nuclear collision during its slowing down process. Obviously if such an energetic particle is detected in a telescope whose ΔE element is a relatively thin detector (as a $500 \mu m$ Si) and the particle undergoes a nuclear collision (most probably in the E_R detector), the pair of the $\Delta E, E_R$ values will fall outside of the correct $\Delta E - E_R$ correlation. The ΔE value will be the same as one would observe in the absence of nuclear collision, while the residual energy value E_R will be smaller than the unperturbed value. Consequently, the efficiency for proper identification of the particle is reduced. The presence of an increased nuclear interaction probability forces the use of different identification techniques when the energy of the particles to detect increases; these techniques usually require the use of magnetic fields and track reconstruction. This is actually the solution adopted in Phase II of the FOPI Collaboration; during Phase I to which our apparatus belongs, we decided to take the risk (at a much reduced cost and solid angle with respect to Phase II) of pushing the $\Delta E - E_R$ technique to its limits; of course we were acquainted that an additional effort would be needed to develop specific analysis procedures (described in the following) for correcting the efficiency losses due to the finite probability of nuclear interactions.

With regard to the solid angle coverage, the high multiplicity of the studied events demands for a high granularity of the device, eventhough it covers less then 1% of 4π.

The required granularity could be simply achieved by providing many independent telescopes small enough to keep the multiple hit probability on a single module low: this trivial, but expensive solution (expensive because of the multiplication of electronic chains) would have reduced the savings associated with the adopted $\Delta E - E_R$ solution. Therefore the solid angle coverage has been obtained by telescopes with a high degree of granularity of the ΔE detectors (each with an area of $30 \times 30 \ mm^2$) followed by three large (and thick) blocks of CsI(Tl). The CsI(Tl) blocks have thicknesses of 30, 100 and 100 mm and cross sections of $70 \times 70 \ mm^2$ and $100 \times 100 \ mm^2$ for the so called "small" and "large" telescopes, which have 4 and 9 Silicon detectors respectively. Each ΔE Si detector has its independent electronic chain, while each CsI scintillator has many (4-8) photodiodes connected to it placed on the sides; all photodiodes belonging to the same crystal are read out together by an associated electronic chain. In this way it is possible, by exploiting the information from the independent Si detectors, to identify multiple hits (of charged particles) and to reject the associated events.

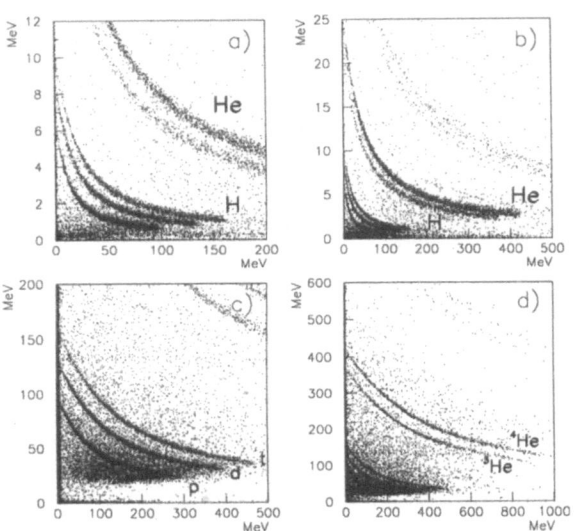

Figure 2. Same as fig.1, but for experimental data.

Note that, for a central Au+Au collision at 150 $AMeV$ bombarding energy, one has for the telescope placed at 32.5° polar angle a probability of about 28% that two or more Silicon detectors are hit; without the aforementioned precaution for identifying and rejecting these events, the $\Delta E - E_R$ correlation among various detectors would have been severely spoiled. The adopted granularity of Si detectors partially solves the problem, but not completely. Obviously there is no way to recognize the case where two particles have hit the same Silicon detector; moreover, particles which penetrate the CsI scintillator without passing through any Silicon detector (small dead areas are present between the Si detectors) are not properly accounted for in the multiple hit reconstruction. Moreover neutrons, which practically do not interact with the thin Si detectors, have a sizeable probability of depositing some energy in the thick scintillator. These effects distort the correct correlation of charged particles, adding some background correlations.

Because of the presence of nuclear interactions and of non-identified multiple hits, the $\Delta E, E_R$ correlation looks like fig.2 which shows actual experimental data taken for the Au+Au collisions at 250 $AMeV$ bombarding energy. Let us now discuss briefly the procedures devised and followed for correcting the effects of nuclear interactions and residual multiple hits.

ANALYSIS PROCEDURES

Since the aforementioned effects (nuclear interactions and non-identified multiple hits) have the same consequences, i.e. of removing particles from their proper correlation branches and to move them to background zones, a good technique to correct the data would be that (A) of considering only those particles not removed from the correct branches and (B) of correcting the so-obtained intensities with the efficiencies accounting for both the nuclear interactions and the multiple hit probabilities. Obviously this technique is applicable only to a finite set of events, and not to every single event.

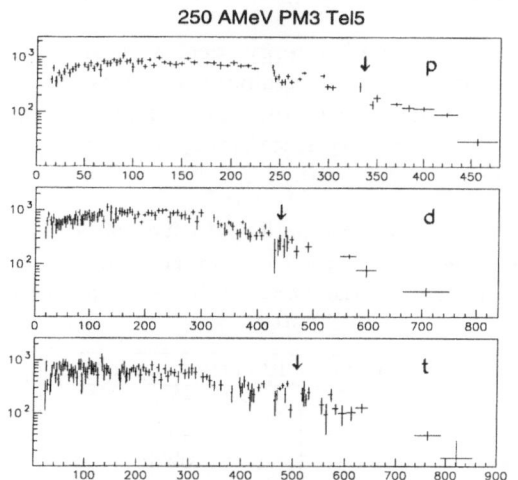

Figure 3. Spectra of p, d and t reconstructed by the method presented in the text. The reaction is Au+Au at 250 $AMeV$ and data refer to the telescope placed at 32.5°. For the various particles the punch-through values are marked with arrows.

Part (A) of the procedure, i.e. the intensity evaluation of the particles belonging to the branches is further complicated, at least for the $Z = 1$ particles, by the presence of the punch-through branches, i.e. of high energy particles not stopped in the detector stack. The procedure that we have devised and successfully applied to our data, allows the evaluation of the intensities of the various particle species as a function of the total kinetic energy, even in presence of background correlations. It consists in a multiple peak fitting procedure applied to adjacent "slices" of the $\Delta E - E_R$ correlation, projected onto one of the two axes. The procedure is fully automatic, both in terms of slice selections, projections and peak fitting. With this technique is virtually possible to reconstruct the intensities of particles as a function of the total kinetic energy also beyond their punch-through energy, once their total kinetic energy has been estimated from the amount of energy deposited on the active volume of the telescope. As an example of this reconstruction technique, fig.3 shows the kinetic energy spectra dN/dT for p,d and t, as obtained from the correlation shown in fig.2. Please note that the spectra of various particles extend well beyond the corresponding punch-through energies.

The so-obtained spectra must now be corrected (part (B) of the procedure) for the efficiency losses; in fact, only "clean" particles which neither experienced nuclear interactions nor constituted a multiple hit, contribute to the reconstructed intensity. The efficiency correction for the nuclear interaction probability is obtained by dividing the dN/dT spectra of a given particle for $(1 - P(T))$, where $P(T)$ represents the probability that the considered particle, of kinetic energy T, undergoes a nuclear collision. This probability has been taken from the literature [6] when available or estimated from the knowledge of the reaction cross sections of various particles in CsI (or Cs or I) [7].

As far as efficiency corrections for multiple hit losses are concerned, they have been obtained from realistic MonteCarlo simulations, which took into account all the relevant detection mechanisms. In fact, the amount of multiple hits depends critically on the geometry of the apparatus as well as on the charged particle multiplicity of the examined event, but it also depends on other parameters, like neutron multiplicity and detailed angular and energy distributions of emitted particles. The latter effects can be safely and correctly accounted for by means of realistic simulations, also exploiting the information from the telescopes and the main detector system. Different simulations have been run to determine the actual dependence of the correction on the charged particle multiplicity detected by the Plastic Wall. The accuracy of the simulation in describing the behaviour of the apparatus has been carefully checked [5].

In conclusion, spectra like those presented in fig.3 (obtained with part (A) of the procedure) are eventually corrected with the efficiencies (part (B) of the procedure) to give the final kinetic energy spectra.

As a final check we compared our corrected kinetic energy spectra for $Z = 1$ and $Z = 2$ particles (i.e. summing over the resolved isotopes) against the results obtained by the main 4π detector, at least for those angular and kinetic energy regions covered by both apparatuses. This comparison showed a very satisfactory agreement, thus confirming the correctness of the whole procedure.

Table 2. In the first column the identification thresholds as determined by the ΔE Silicon detector are reported. Maximum energies of particles stopped in the telescopes are given in the second column, while the third column gives the energy up to which the spectra of the various particle can be reconstructed.

Particle		Lower thr.	Punch-through	Effective upper thr.
p	$[AMeV]$	8	330	≈ 450
d	$[AMeV]$	5.5	215	≈ 350
t	$[AMeV]$	4.5	165	≈ 250
α	$[AMeV]$	8	330	≈ 450

We can conclude that we were able to measure the energy spectra of $Z = 1, 2$ particles with lower thresholds as low as $\approx 10\ AMeV$ and up to an energy significantly higher than the punch-through energy in the telescope which is anyhow a thick detector (see Table 2).

PHASE-SPACE COVERAGE OF THE APPARATUS

As already anticipated the telescopes have been used to study the Au+Au collisions at 100, 150, 250 and 400 $AMeV$ bombarding energy, in the framework of the FOPI Collaboration at SIS. During the experiment 8 telescopes were placed at polar angles around the beam direction between 6° and 60°, covering a total solid angle of about 3×10^{-2} sr. The large Plastic Wall detector covered the polar angle region under 30° with full azimuthal coverage. Since our telescopes operate in coincidence with the "4π" detector (which at these bombarding energies cover the forward solid angle in symmetric collisions), it is possible to study the emission of light particles (in terms of their intensity and energy distributions) in correspondence to various classes of reactions as operatively defined by the 4π detector. Different types of selections criteria have been devised, with the aim of making the best selection as possible (mainly in terms of efficiency) of centrality. The most used centrality selection criteria are the Plawa Multiplicity (PM), the Directivity (D) and the Energy Ratio (ERAT). I will not discuss neither these selection criteria, nor their efficiencies and their possible biases, since a detailed discussion is presented elsewhere in this Conference [1, 3].

In order to describe the performance of a detector it is always important to define its phase-space acceptance: a useful description of this coverage is obtained by plotting the covered region in a $p_\perp$ (or $p_\perp$ scaled with CM momentum) versus rapidity y plane (see fig.4, where the telescopes coverage is shown for $250 AMeV$ bombarding energy). The region around midrapitity is covered, though not in an uniform way. In the adopted representation all particles types cover an identical region (hatched in the figure), apart from lower and upper thresholds (marked with thick lines), which on the contrary change with particle type, as they depend on the stopping in the detectors. When changing bombarding energy, the covered regions also change; consequently if one wants to identify a common covered region at all bombarding energies, this region will be further reduced with respect to the coverage valid for a specific bombarding energy.

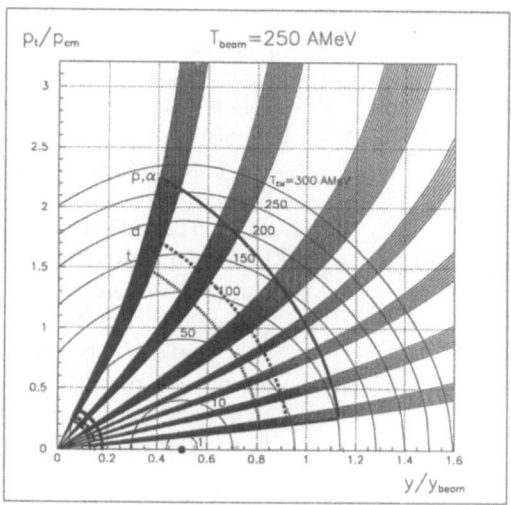

Figure 4. Phase-space coverage of the telescopes for $250 AMeV$ Au+Au collision expressed in the $p_\perp$ (scaled with CM momentum) versus normalized rapidity y plane. Hatched areas refer to geometrically covered regions.

The unavoidable limitation of phase-space coverage induces some problems in the data analysis and -more important- in the comparison with theoretical calculations and with other experimental results. Before entering this discussion, it is to be reminded that all detectors, even those which are claimed to cover 4π solid angle, have a limited phase-space coverage because of both (possibly small) geometrical cuts and (often much more important) lower and upper detector thresholds, either in energy or velocity; in other words all the problems which we were faced with and which are going to be described, are not specific of only our apparatus, but they more or less apply to all particle detectors; however in order to be precise and didascalic, these aspects will be discussed with specific reference to our apparatus, without generalizing the conclusions.

A typical problem is that associated to the trasformation of the reference frame: this trasformation is generally needed for passing from the lab frame to the more priviledged center of mass frame; in our case (symmetric system) the CM frame is particularly priviledged, since for "central" collisions a source at rest in this frame is expected to be responsible of a large fraction of the total intensity of emitted particles. Since the physical quantity almost directly measured by our apparatus is the kinetic energy T of the detected particle, an experimental information that one can immediately get is the yield of various particles represented in the plane θ_{lab}, T_{lab} (the data are already corrected for multiple hits, nuclear losses and so on, as explained before). Fig.5 a) shows such a result for protons measured in the Au+Au reaction at $250 AMeV$. Each stripe in the plot represents the contribution of a telescope, covering a fixed polar angle in the laboratory frame. When moving from the lab reference frame to the CM one, the representation of part a) transforms into that of part b). A selection of a specific CM angular region is very often applied to the experimental data in order to eliminate residual contributions from target- and projectile-like fragment emission; if such a selection is applied to the correlation of fig.5 part b), it is obvious that the associated CM kinetic energy spectra, integrated over such a region, cannot be obtained directly, due to the presence of various "uncovered regions". Although the only physically true information remains that of covered regions, a possible solution for that problem (not the unique solution) can be obtained by "interpolating" through the experimental lab data, and afterwards transforming in the CM (see fig.5 c) and d) respectively).

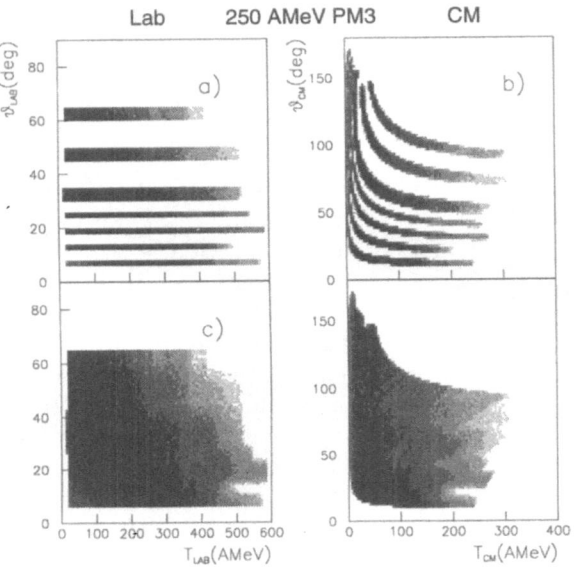

This is actually the procedure that we have followed. This procedure has mainly a cosmetic character, but it is also useful because it produces quantities having physical meaning (in our case: CM kinetic energy spectra in a given angular range) and it allows a simple comparison of the reconstructed energy spectra with theoretical expectations. Of course when these comparisons are made, the uncertainty associated to this data "interpolation" cannot be disregarded and experimental errors must keep memory of the procedure. Besides, before applying this procedure, a careful test of the influence of the "interpolation" in possibly altering the "true" data, is necessary and this has been done by studying the effect of the procedure on simulated data.

Figure 5. Part a): yield of protons for semicentral Au+Au reactions at 250 $AMeV$ represented in the lab polar angle θ_{lab} versus kinetic energy T_{lab} plane. Each stripe is associated to one telescope. Part b): yield of protons for semicentral Au+Au reactions at 250 $AMeV$ represented in the CM polar angle θ_{CM} versus kinetic energy T_{CM} plane, obtained after tranforming from lab to CM the data presented in part a). Part c): same as part a), but for data interpolation. Part d) is obtained tranforming from lab to CM the data presented in part c).

An alternative and well known method exsists, if the theoretical model is an event generator. In this case one can simply apply to the calculated (simulated) events the so called "experimental filter". In this way, calculated quantities are obtained which can be directly compared with experimental results. This method is extremely clean and straightforward; however it has the disadvantage of producing quantities which tend to loose their physical meaning when the cuts associated with the filter are severe. Besides this procedure intrinsically priviledges the confrontation between specific experimental results and theory (or theories), in practice discouraging the comparison between different experimental results, thus excluding consistency checks among different experiments. Apart from the particular solution chosen for these comparison, it is important to note that

500

the geometrical coverage of our apparatus is definetely limited, for instance extremely smaller than that of the Plastic Ball [8], to cite an apparatus which produced a precious and vast amount of information on light particles emitted in relativistic heavy ion collisions; however the "kinetic" coverage is definitely higher, both in terms of low and high energy thresholds. As it will be shown in the next section, the improved energy coverage can be very important, since it allows a more detailed comparison in the high energy regions of the spectra, where the unavoidable contamination from low energy reaction mechanisms is expected to be smaller.

EXPERIMENTAL RESULTS

In fig.6 are presented the CM transverse energy spectra calculated from 60° to 90° CM polar angle) for p, d, t, ^3He and α at 100, 150 and 250 $AMeV$ for the Au+Au collisions. The reported quantity actually is the differential multiplicity (per event) $d^3M/dp^3 = 1/pE \ d^2M/dTd\Omega$ and not the invariant cross section given by $1/p \ d^2M/dTd\Omega = E \ d^3M/dp^3$, where E and T are the total and kinetic energy of the particle respectively.

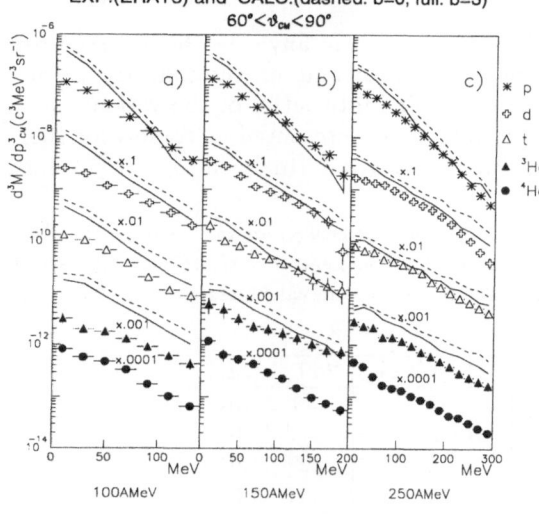

The chosen representation is particularly suggestive since spectra of thermally emitted particles would appear as pure exponentials with slope equal to the temperature τ of the emitting system. The data presented in fig.6 are selected in terms of centrality by ERAT5 cut. The adopted polar angle selection is meant to keep the contamination from spectator emission at a minimum. From fig.6 one observes that at all bombarding energies the proton spectra are the steepest. An overall trend of increasing hardness of the spectra with particle mass is also apparent. To be more quantitative, the mean transverse energy for each particle species has been calculated, using the experimental kinetic energy distributions dN/dT, obtained with the same centrality and angular selections already mentioned.

Figure 6. Experimental Kinetic energy spectra in the CM system for p, d and t for central Au+Au reactions at 100, 150 and 250 $AMeV$. The centrality selection is ERAT5 (see text). Data are integrated over the CM polar angle range 60° 90°. Protons are represented by stars, deuterons by open crosses, triton by open triangles, ^3He by full triangles and ^4He by full dots. Results of the transport model are presented as dashed lines for $b = 0 \ fm$ and with full lines for $b = 3 \ fm$. Stiff EOS has been assumed.

In Table 3 the mean tranverse energy are reported for 100, 150 and 250 $AMeV$ bombarding energies; the mean is calculated over the same energy regions shown in fig.6. Accordingly to the qualitative behaviour of the spectra, protons always present a reduced value of the mean tranverse energy. If the distribution of collective velocities at freeze out is the same for all particles and if it factorizes in a gaussian form [4] one has $< T_{\perp x} >_{90°} \approx 3/2(\tau + AE_{\perp N}^{coll})$

where τ is the maxwellian temperature and $E_{\perp N}^{coll}$ is the collective tranverse energy per nucleon. From the differences of the mean transverse energies of deuterons and protons reported in the table, one estimates a collective transverse energy per nucleon varying from 5 to 15 $AMeV$, depending on the bombarding energy. As anticipated, a definite trend of increasing mean kinetic energy as a function of the fragment mass is indeed observed; however the actual behaviour does not closely follow the naïf formula shown before. That means that a precise value of the transverse collective

Table 3. Experimental mean transverse kinetic energies for various fragments detected in the Au+Au collisions at 100, 150 and 250 $AMeV$. The centrality selection is ERAT5. Data refer to the CM polar angle range 60° - 90°. The mean is calculated over the energy ranges shown in fig.6.

x		100 $AMeV$	150 $AMeV$	250 $AMEV$
p	[MeV]	48 ± 1	65 ± 2	89 ± 1
d	[MeV]	55 ± 1	74 ± 3	112 ± 1
t	[MeV]	51 ± 1	85 ± 4	119 ± 2
^3He	[MeV]	63 ± 2	94 ± 5	124 ± 2
^4He	[MeV]	56 ± 1	80 ± 3	118 ± 2

energy per nucleon cannot presently be assessed, but in any case the observed results rule out a purely thermal description of the emission and, on the contrary, point to a sizeable contribution from collective motion. The intensities of the various particles integrated over the energy ranges shown in fig.6, are presented in Table 4 for 100, 150 and 250 $AMeV$ bombarding energies. Intensities are normalized to proton intensity.

Table 4. Experimental intensities (with respect to proton intensity) for various fragments detected in the Au+Au collisions at 100, 150 and 250 $AMeV$. The centrality selection is ERAT5. Data refer to the CM polar angle range 60° - 90°. The intensities are calculated over the energy ranges shown in fig.6.

x	100 $AMeV$	150 $AMeV$	250 $AMEV$
d	$.84 \pm .03$	$.86 \pm .05$	$.86 \pm .02$
t	$.72 \pm .03$	$.74 \pm .05$	$.70 \pm .02$
^3He	$.21 \pm .01$	$.34 \pm .03$	$.24 \pm .01$
^4He	$.78 \pm .03$	$.74 \pm .05$	$.47 \pm .01$

One could use the information contained in the Table 4 to calculate the entropy produced in the early stage of the collision, using one the various recipes of the QSM [9]. This requires an evaluation of the energy sharing between thermal and collective parts which should take into account all the available experimental information of both light particles and clusters. This task is presently under way.

COMPARISON WITH THE TRANSPORT MODEL

Recently Danielewicz and Qiubao Pan [4] have carried out a dynamical calculation specifically aimed at the determination of the energy spectra of light fragments (up to mass 3) emitted in heavy-ion reactions. In their model, which will be presented also in this meeting [10], the light composite are produced in a few-nucleon processes inverse to composite break-up. In the limit of many interactions, at a low baryon density, the model yields the required law of mass action. For light systems the model reproduces the phenomenological power-law scaling of inclusive spectra.

While for central Au+Au collisions some characteristics typical of thermal equilibrium are indeed found in the final state, these were reached in the course of the collision dynamics and do not follow from explicit assumptions.

In fig.6 the results of the calculations for the three bombarding energies are reported. The dashed lines represent the results obtained assuming an impact parameter $b = 0$ fm. The comparison is absolute, i.e. both experimental and calculation are separately normalized to the number of examined events. As already anticipated, α particles are presently not included in the calculation. The overall agreement is quite good; apart from protons, which in the calculation are softer than in the experiment, the shape of calculated and experimental spectra are almost the same. An overstimate of the experimental cross section is however apparent for the $b = 0$ fm calculation. This is not surprising at all; in fact from the experimentally known cross section associated to ERAT5 centrality selection one estimates, with the sharp cut off approximation, an upper limit between 3 and 4fm for the relevant impact parameter. Accordingly to this observation, the comparison between experiment and calculation with $b = 3$ fm (solid lines in fig.6) shows an improved agreement, mainly as far as absolute cross section is concerned.

Calculations presented in fig.6 were done assuming stiff EOS and free-space cross sections. Tests performed on simulations at 250 $AMeV$ bombarding energy have shown very similar results for soft EOS, while a sensitivity of intensity ratios of fragments has been detected on the collision rate; in particular an impressive agreement of experimental and calculated isotopic ratios for $Z = 1$ has been obtained once the collision rate has been changed increasing by a factor of two the collision cross sections. The associated increasing of stopping produced, as a side effect, a little reduction of the slopes of the spectra.

CONCLUSIONS

Kinetic energy spectra for mass resolved $Z = 1, 2$ fragments have been measured for central Au+Au collisions at 100,150 and 250 $AMeV$, using the $\Delta E - E_R$ technique. The energy range of detected particles extends from few $AMeV$ up to energies significantly higher than the punch-through values in the telescope. Special care has been devoted for correcting the measured spectra for the finite probability of nuclear interactions and for multiple hits.

The agreement between theoretical and experimental CM energy spectra for fragments with $A \leq 3$ detected in central Au+Au collisions at 100, 150 and 250 $AMeV$ suggests that the main collision mechanisms are indeed included in the transport model [4].

Tests are in progress to check the sensitivity of the results on parameters of the calculation, in particular on the collision rate.

REFERENCES

1. K.D. Hildenbrand
   Invited talk at this Conference.

2. J.P. Coffin
   Invited talk at this Conference.

3. T. Wienold
   Invited talk at this Conference.

4. P.Danielewicz and Qiubao Pan,
   Phys. Rev. **C46** (1992) 2002

5. G. Poggi, G. Pasquali, M. Bini, P.R. Maurenzig, A. Olmi and N. Taccetti,
   Nucl. Instr. and Meth. **A324** (1993) 177

6. J.F. Janni
   Atomic Data and Nuclear Data Tables **27** (1982) 147

7. D.E. Measday and R.J. Schneider
   Nucl. Instr. and Meth. **42** (1966) 26

8. H. Gutbrod, A.M. Poskanzer, and H.G. Ritter,
   Rep. Prog. Phys. **52** (1989) 1267

9. D. Hahn and H. Stöcker,
   Nucl. Phys. **A476** (1988) 718

10. P. Danielewicz
    Invited talk at this Conference.

Hadron Production in Nucleus-Nucleus Collisions at SPS- and AGS-Energies

Dieter Röhrich

University of Frankfurt

Experiment NA35

Experiment NA35 at the CERN SPS studies collisions of p– and ^{32}S–projectiles of 200 GeV/nucleon incident energy with nuclear targets. The main detectors are two large–volume tracking devices (see Fig.1), a streamer chamber inside a 1.5 Tesla magnet and a time projection chamber (TPC) positioned downstream of the magnet. Both detectors record the space trajectories of charged particles, from which the particle momenta are derived. The acceptances of the detectors for negatively charged hadrons are complementary to each other and cover full phase space with some overlap (Fig.2). Central events are selected by the absence of projectile spectators in the veto calorimeter.

The objective of the experiment is to measure the final hadronic state in nucleus-nucleus collisions (Fig.3). In order to achieve this aim the phase space for many different particle species was well covered by the two detectors. Besides negatively and positively charged particles the streamer chamber detects decay topologies of charged and neutral weakly decaying particles. The analysis of the streamer chamber pictures has been fully automated, which not only results in increasing the measuring rate by an order of magnitude but also allows for a complete simulation of the detector based on the GEANT-package. The results discussed in the following are mainly based on the automated analysis and traditional measurements have been included where appropriate. The TPC on the other hand identifies charged particles by multiple sampling of their specific ionisation, the dE/dx–resolution of 5% allowing a statistical particle identification. As a result, negative hadrons, net protons $(p - \bar{p})$, Λ, $\overline{\Lambda}$, K_s^0, K^+, K^- and $\bar{p}$ can be detected.

Thus information about almost all baryons and mesons in the final hadronic state can be accessed due to the large phase space coverage. Especially the production of antibaryons like $\overline{\Lambda}$ and $\bar{p}$ near midrapidity may shed light on the reaction dynamics.

Hot and Dense Nuclear Matter, Edited by
W. Greiner *et al.*, Plenum Press, New York, 1994

Simulations based on the measured $\overline{\Lambda}$ production show that typically 25% of all detected $\overline{p}$ stem from $\overline{\Lambda}$-decays. Therefore the measurement of $\overline{\Lambda}$ in the same experiment allows a model-independent correction of this contamination in the $\overline{p}$-spectra.

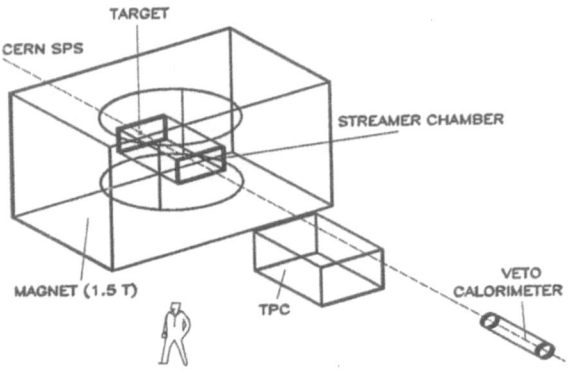

Figure 1: Experimental setup of experiment NA35 at the CERN SPS in a typical heavy ion run configuration.

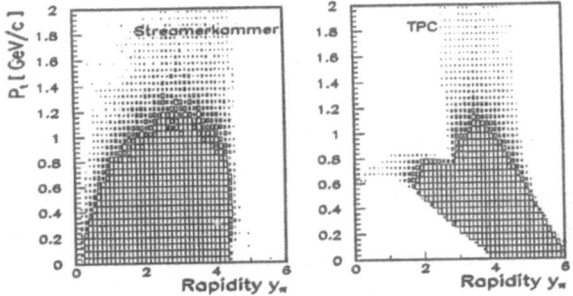

Figure 2: Acceptance of the two large volume tracking detectors Streamer Chamber and TPC, illustrated by the phase space population of negatively charged hadrons in a typical heavy ion run configuration.

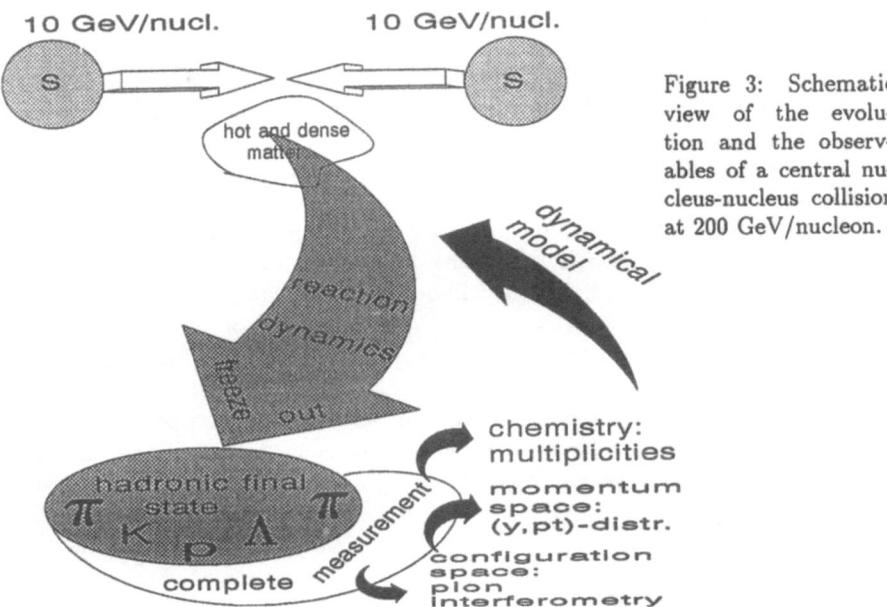

Figure 3: Schematic view of the evolution and the observables of a central nucleus-nucleus collision at 200 GeV/nucleon.

NA35 studied a variety of collision systems, ranging from simple systems such as minimum bias p+S interactions to central S+Au reactions. The central collision fractions of the inelastic cross sections of the nucleus–nucleus collisions on which we are focussing here are 3%, 3.2% and 6% for the S+S, S+Ag and S+Au systems, respectively. The discussion of the reaction dynamics for the various collision systems will be guided by the rapidity distribution of the participating baryons. First of all, participating nucleons are identified and their rapidity loss is measured by deducing the rapidity density of net protons $(p - \bar{p})$. The rapidity density of $(\Lambda - \bar{\Lambda})$ is also discussed because some of the participating nucleons are transformed into Λs. The energy deposited by the participating baryons is to a great extent used up for hadron production. Particle production will be discussed in terms of yields and ratios of the various species, e.g. rapidity distributions of h^-, K_s^0, K^+, K^-, Λ, $\bar{\Lambda}$ and $\bar{p}$. Invariant cross sections of strange particles as a function of m_T are described by exponential slope parameters. If necessary, an extrapolation of particle yields into the m_T regions outside the acceptance is made assuming an exponential m_T distribution which leads to a set of hadron multiplicities in 4π.

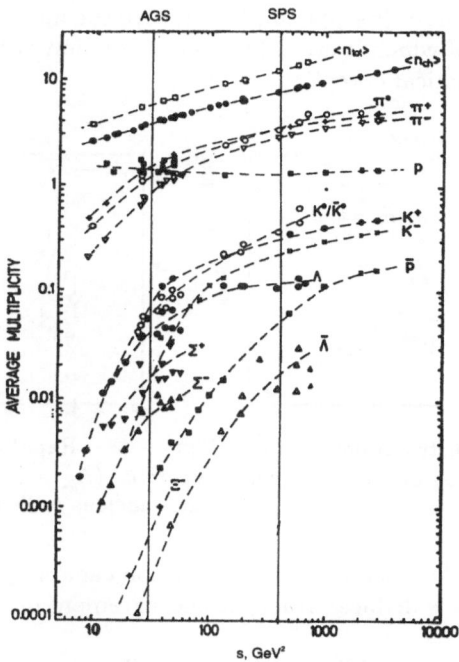

Figure 4: Particle multiplicities in pp-collisions as a function of the CM-energy.

The results at 200 GeV/nucleon are compared to data obtained in Si+nucleus collisions at 14.6 GeV/nucleon by the experiments E802 and E810 at the AGS.[1, 2] At the lower energy the multiplicity of produced particles - mainly mesons - is comparable to the number of participating nucleons while at the SPS mesons are typically five times more abundant than nucleons. In order to compare rapidity densities the different rapidity gap and the energy dependence of the inelastic cross sections have to be taken into account. The rapidity densities at 200 GeV/nucleon are normalized by the ratio of the rapidity gap at the SPS of 6 and the one at AGS-energies of 3.44. The particle multiplicites at the SPS are scaled by the energy dependence of the particle production measured in pp-interactions (Fig.4).[3] Both normalizations take out the 'trivial' differences between the rapidity densities at both energies.

Several phenomenological models can be applied to heavy ion reactions at both energies. Both macroscopical thermal and hydrodynamical models (see e.g. [4,5]) or microscopical dynamical models like VENUS[6], RQMD[7] or QGSM[8] are used to describe the reaction dynamics. Since we will focus on antibaryon production, the data is compared only to RQMD, from which predictions on e.g. $\bar{p}$-production have been published.[9]

Participating Baryons

The rapidity distribution of the participating baryons leads to the number of incoming nucleons contributing to the reaction dynamics and allows to quantify the amount of stopping of target and projectile nucleons in a central collision. Some of the nucleons are converted into hyperons still carrying two quarks of the participating nucleon; their contribution concerning stopping has to be taken into consideration. The rapidity distributions of $(p-\bar{p})$ and $(\Lambda - \bar{\Lambda})$ give the net baryon density at midrapidity, the knowledge of which is crucial for the discussion of the particle production ratios via the baryochemical potential.

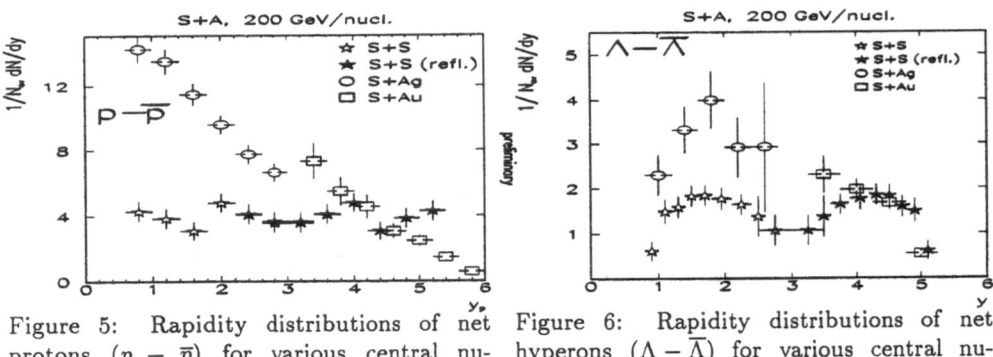

Figure 5: Rapidity distributions of net protons $(p - \bar{p})$ for various central nucleus–nucleus collisions.

Figure 6: Rapidity distributions of net hyperons $(\Lambda - \bar{\Lambda})$ for various central nucleus–nucleus collisions.

Since the streamer chamber has no identification capability for stable particles the net proton distribution is deduced from the measurement of the charge excess

$$\frac{dN_{+,-}}{dy}(AA \to (p-\bar{p})) = \frac{dN}{dy}(AA \to h^+) - \frac{dN}{dy}(AA \to h^-).$$

This is a good approximation for all rapidities concerning collisions between isoscalar nuclei, and even for asymmetric systems deviations are small. The identification of protons with the charge-excess is subject to corrections discussed below. All data are corrected for geometrical acceptance, for e^+ and e^- misidentified as hadrons, for hadrons resulting from the weak decays of K_s^0 and Λ, and for hadrons produced in secondary hadron-nucleus interactions. In addition, when deriving the proton spectra, a correction is applied for the excess of the K^+ over the K^- yield, which may be attributed to associated production-like channels which are open for K^+ but not for K^- production. Since the necessary corrections are based on measurements of the same experiment over a wide acceptance, the corrections are model independent.

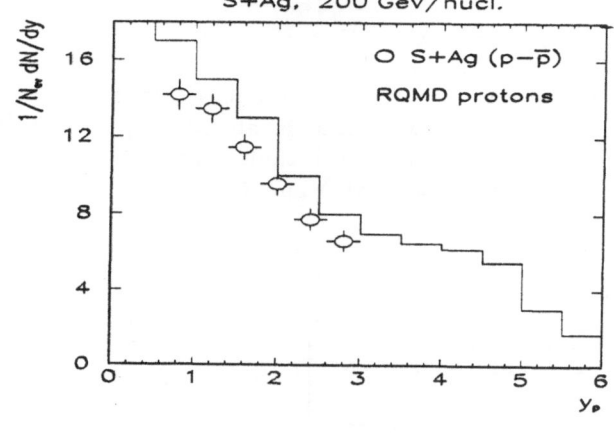

Figure 7: Rapidity distribution of net protons in central S+Ag collisions compared to a RQMD calculation.[10]

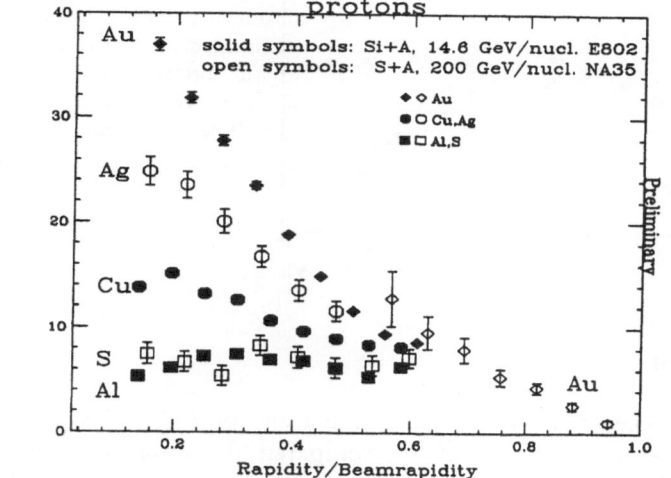

Figure 8: Rapidity distribution of net protons in central S+A collisions at 200 GeV/nucleon compared to proton distributions in central Si+A collisions at 14.6 GeV/nucleon.[1]

The rapidity distribution of $(p - \bar{p})$ in central S+S, S+Ag and S+Au collisions clearly shows a significant net baryon density at midrapidity (Fig.5). With increasing target mass an increasing number of projectile and target nucleons is shifted to midrapidity. The shape of the $(\Lambda - \bar{\Lambda})$ distribution (Fig.6) is rather similar to that of the net protons above the target fragmentation region.

A comparison of the rapidity distribution of the participating protons in central S+Ag collisions at 200 GeV/nucleon to the proton distribution of an RQMD-calculation $(b < 1fm)$[10] shows a good agreement (Fig.7).

By normalizing the rapidity densities at 200 GeV/nucleon to the rapidity gap at the AGS, the net proton distributions can be compared to the protons measured in similar systems at 14.6 GeV/nucleon at the AGS (Fig.8)[1]. The rapidity distributions exhibit almost similar shapes in spite of the slightly different projectiles and of the centrality triggers, e.g. 27 participating protons in S+S collisions as compared to 22 participating protons in Si+Al. This implies that projectile-protons at 14.6 and 200 GeV/nucleon loose the same fraction of rapidity in central nucleus-nucleus collisions.

Hadron Production

Most of the energy deposited in the reaction volume by the participating baryons is available for hadron production. Pions are produced in abundance, negatively charged hadrons are mostly π^-; other produced mesons include charged and neutral kaons. Baryonproduction can be studied by looking at $\overline{\Lambda}$ and $\overline{p}$.

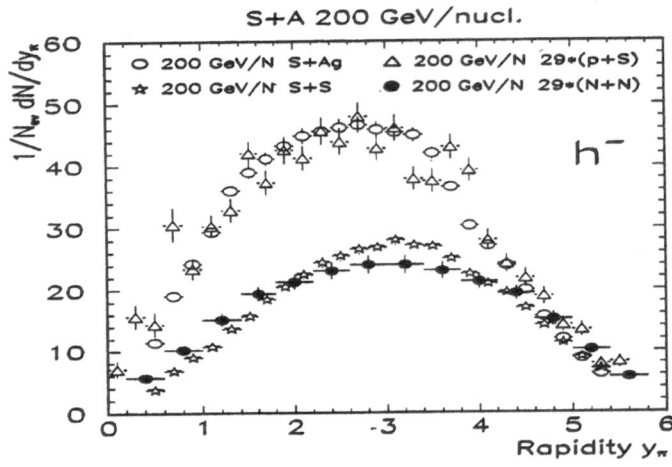

Figure 9: Rapidity distributions of negatively charged hadrons for central nucleus–nucleus collisions compared to hadron induced reactions.

Rapidity Distributions

The rapidity distributions of negatively charged hadrons for S+S and S+Ag collisions at 200 GeV/nucleon are shown in Fig.9 and compared to nucleon–nucleon (NN) for S+S and to p+S for S+Ag. The comparison shows that scaling the NN (p+S) distributions by factors of about 29 can reproduce the heavy systems mentioned above, although pion distributions seem to be slightly narrower in nucleus-nucleus collisions. Based on this observation one may draw the conclusion that, regarding pion production, central S+Ag collisions can be represented as an independent superposition of minimum bias p+S interactions. This is also reflected by the fact that the pion multiplicity per participating baryon is the same in both nucleus–nucleus systems and similar to the one in NN collisions (1.6). Such a scaling behaviour does not hold true for the production of strange particles (see below).

The comparison of the normalized rapidity distributions of negatively charged hadrons for S+S collisions for two different centrality triggers (3% resp. 11% of the inelastic cross section) to the rapidity distribution of positively charged pions in central Si+Al collisions at the AGS (Fig.10) shows a good agreement especially in the case where the number of participating nucleons is roughly the same (a 11% trigger for S+S as compared to a 7% trigger for Si+Al).[11] A scaling of the pion production

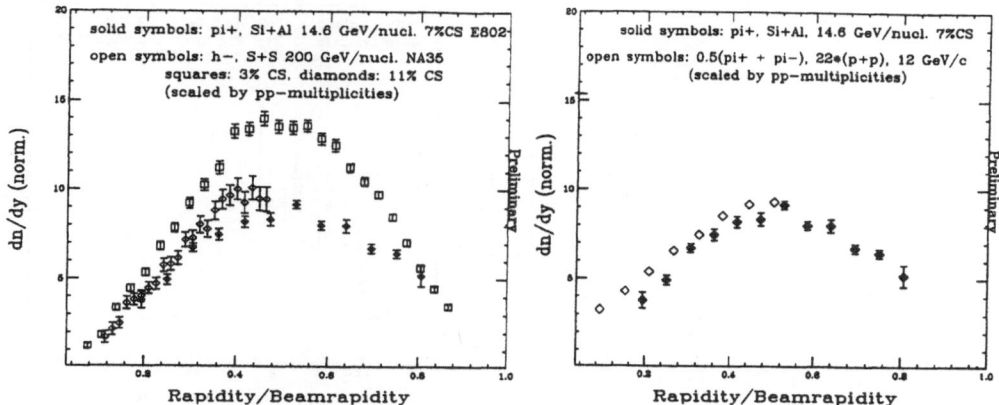

Figure 10: Rapidity distributions of nega- tively charged hadrons for central S+S colli- sions (3% and 11% of the inelastic cross sec- tion) at 200 GeV/nucleon compared to distri- butions of π^+ in central (7%) Si+Al collisions at 14.6 GeV/nucleon.

Figure 11: Rapidity distributions of π^+ in central (7%) Si+Al collisions at 14.6 GeV/nucleon (E802) compared to scaled p+p data.[14]

in NN-collisions interpolated to AGS-energies by roughly a factor of 22 describes the Si+Al data fairly well (Fig.11).

The yields of strange particles relative to the yield of non-strange particles pro- duced in central S+S collisions at 200 GeV/nucleon are higher than the corresponding ratio in nucleon–nucleon (NN) interactions at the same energy.[12] This strangeness enhancement effect is also present in central S+Ag collisions.[13] The production of K_s^0 in S+Ag collisions is enhanced by a factor of 1.7 compared to the scaled yield of p+S collisions (Fig.12). The charged kaons show a similar trend, especially around $y = 1$ there is a large excess of K^+ over the yield in 29·(p+S) (Fig.13).

A similar effect of strangeness enhancement is observed at AGS-energies. A simple scaling of pp-data by a factor of 22 - as for the pions - cannot describe the kaon yields in central Si+Al collisions (Fig.14). Here the rapidity distribution of the average

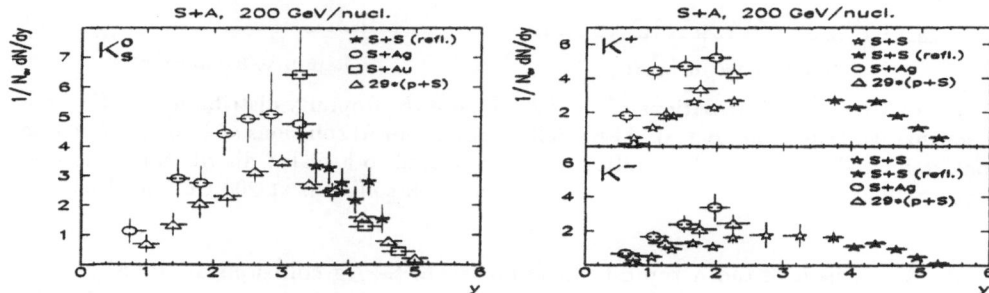

Figure 12: Rapidity distributions of neutral kaons from central nucleus–nucleus collisions compared to a scaled distribution from p+S reactions.

Figure 13: Rapidity distributions of charged kaons from central nucleus–nucleus collisions compared to a scaled distribution from p+S reactions.

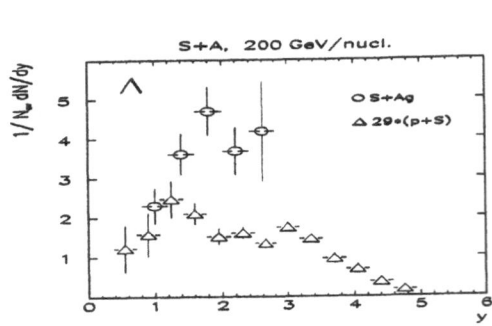

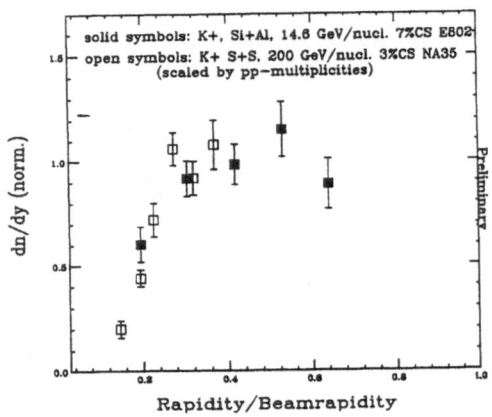

Figure 16: Rapidity distribution of Λ from central S+Ag collisions compared to a scaled distribution from min. bias p+S reactions, both at 200 GeV/nucleon.

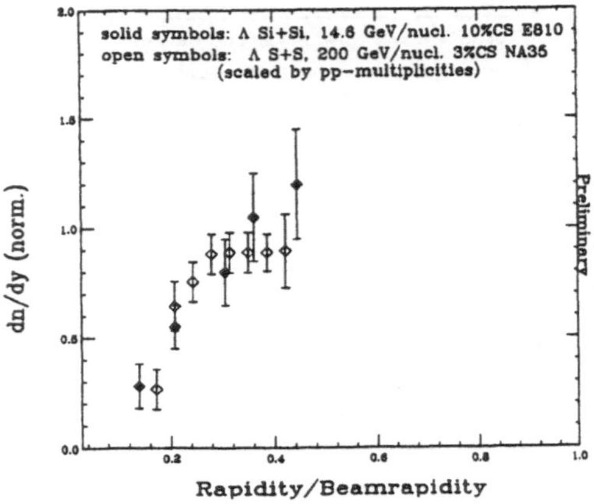

Figure 17: Rapidity distributions of Λ from central S+S collisions at 200 GeV/nucleon compared to a scaled distribution from central Si+Si collisions at 14.6 GeV/nucleon.

charged kaon $(0.5 \cdot (K^+ + K^-))$ is more than a factor of two above the scaled K^0-yield obtained in pp-reactions. [11, 14] As can be seen in Fig.15 the magnitude of the kaon enhancement and the rapidity dependence at both energies seem to be similar.

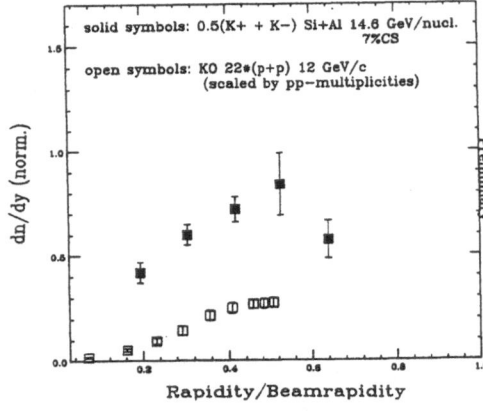

Figure 14: Rapidity distributions of the average charged kaon from central Si+Al collisions compared to a scaled distribution from p+p reactions.

Figure 15: Rapidity distributions of K^+ from central Si+Al collisions at 14.6 GeV/nucleon compared to a scaled distribution from central S+S collisions at 200 GeV/nucleon.

A comparison of the Λ rapidity distribution of S+Ag collisions at the SPS and a scaled distribution of minimum bias p+S interactions (with the scale factor adjusting the multiplicities of negatively charged hadrons) shows that the production is not only enhanced, but that this enhancement is most pronounced around midrapidity (Fig.16). A similar enhancement has been observed in the symmetric systems S+S at 200 GeV/nucleon and Si+Si at 14.6 GeV/nucleon.[2] A comparison of the normalized

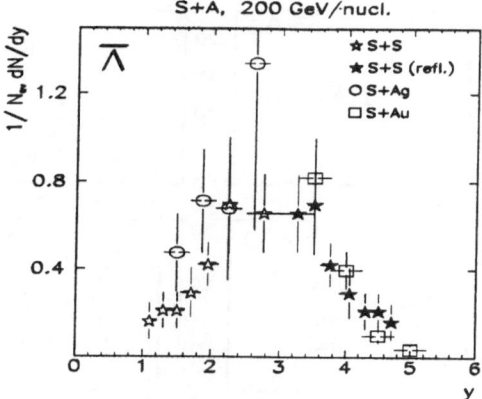

Figure 18: Rapidity distributions of $\overline{\Lambda}$ from various central nucleus–nucleus collisions at 200 GeV/nucleon.

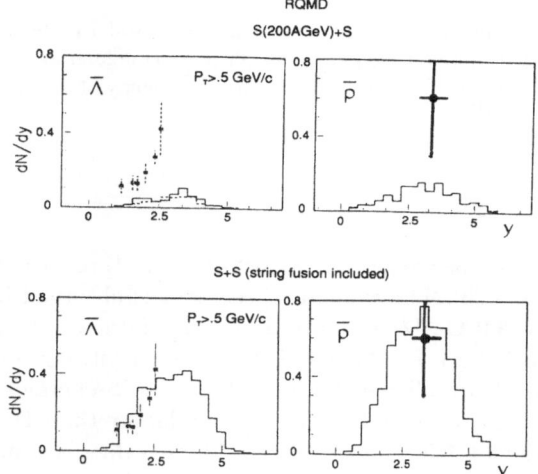

Figure 19: Rapidity distributions of $\overline{\Lambda}$ and $\overline{p}$ calculated by the RQMD-model with and without including color rope formation for central S+S collisions at 200 GeV/nucleon. Preliminary data on $\overline{p}$ production is also shown.[9]

rapidity distributions of the Λ-production in these two systems (Fig.17) shows the similiarity despite the different triggers.

In contrast to the apparent fact that negative hadron production in central nucleus-nucleus collisions can be understood as a superposition of independent NN respectively p+A interactions, strangeness production is clearly enhanced by about a factor of two. This effect is observed at both SPS- and AGS-energies.

$\overline{\Lambda}$-production in nucleus-nucleus collisions at 200 GeV/nucleon is concentrated around midrapidity and seems to be slightly larger in heavy systems (Fig.18). The ratio of the rapidity densities of $\overline{\Lambda}$ to Λ for various collision systems is always similar: it is symmetric around midrapidity, peaks at midrapidity and reaches the value of about 0.4.[17]

$\overline{\Lambda}$-data at AGS-energies is sparse, the only preliminary result is the rapidity density near midrapidity ($1.2 < y < 1.7$) in central Si+Au collisions of $(6 \pm 3) \cdot 10^{-3}$.[15] Here for the first time a simple normalization of the SPS data does not work, the

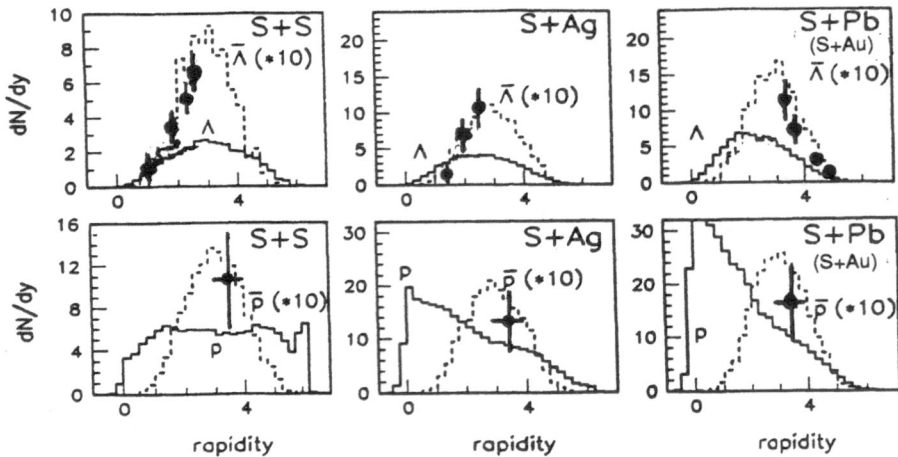

Figure 20: Rapidity distributions of $\overline{\Lambda}$ and $\overline{p}$ calculated by the RQMD-model for various central nucleus–nucleus collisions at 200 GeV/nucleon and the corresponding preliminary measurements. The $\overline{p}$ stemming from $\overline{\Lambda}$-decays have been added to both data and calculations.[10]

value expected for this normalization of SPS-data is three times higher than the measured yield at the AGS. The same disagreement is observed in the $\overline{p}$-production near midrapidity. At 200 GeV/nucleon preliminary data result in rapidity densities ($2.8 < y < 3.8$) of $0.6 \pm 0.3, 0.7 \pm 0.3$ and 1.0 ± 0.4 for central S+S, S+Ag and S+Au collisions respectively. Normalizing these results to AGS-energies the way described above gives values that are at least a factor four larger than the observed $\overline{p}$-yields e.g. in Si+Au reactions $((9.5 \pm 1) \cdot 10^{-3})$.[15] These values are not corrected for $\overline{\Lambda}$-contamination and are therefore an upper limit. On the other hand the acceptance at the AGS is just below midrapitiy, while the SPS-results stem from just above midrapidity.

Due to the lack of solid data of antibaryon production in identical phase space regions at both energies, the discussion of the SPS-results will be guided by RQMD-calculations. The RQMD-model including secondary collisions failed to describe the $\overline{\Lambda}$-production in heavy ion reactions at 200 GeV/nucleon. By adding a new reaction mechanism, namely the fusion of strings in the first moments of the reaction into heavy objects - color ropes -, the model not only reproduced the measured rapidity distributions of $\overline{\Lambda}$, but predicted a $\overline{p}$-yield confirmed by experiments (Fig.19,20, c.f. also [16]). Such a mechanism is not excited at AGS-energies.

Transverse Momentum Spectra

The invariant cross sections for strange particles are compatible with an exponential shape as a function of m_T,

$$\frac{1}{N_{event}} \frac{1}{m_T} \frac{dn}{dm_T} = C \, e^{-\frac{m_T}{T}}, \tag{1}$$

where C is a normalization factor and T is the slope parameter. These two parameters are determined by fitting the experimental data. As an example, Fig.21 shows the transverse mass distribution of neutral and averaged $(0.5(K^+ + K^-))$ charged kaons produced in central S+Ag collisions at 200 GeV/nucleon, the Λ-spectrum for S+Au collisions is shown in Fig.22. The invariant cross section of $\bar{p}$ as a function of m_T for various systems can well be described by an exponential shape with slope parameters around 200 MeV (Fig.23). The slope parameters ("temperatures") of all hadrons except pions are 210 ± 20 MeV, which is surprisingly high for a hadronic system.

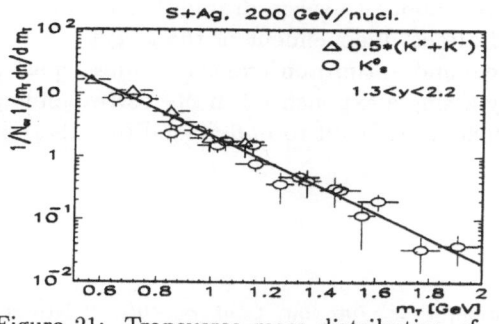

Figure 21: Transverse mass distribution of kaons produced in central S+Ag collisions $(1.3 < y < 2.2)$. The solid line is an exponential fit with a slope parameter of 213 ± 20 MeV.

Figure 22: Transverse mass distribution of lambdas produced in central S+Au collisions $(3 < y < 5)$. The solid line is an exponential fit with a slope parameter of 235 ± 40 MeV.

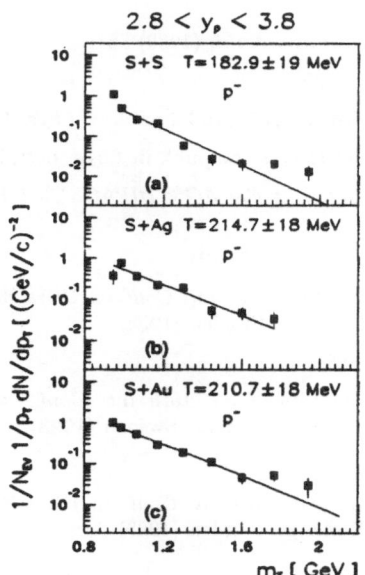

Figure 23: Transverse mass distribution of antiprotons produced in central S+A collisions $(2.8 < y < 3.8)$. The solid lines are exponential fits.

On the other hand, the invariant cross sections of negatively charged hadrons are in general incompatible with a single exponential either in m_T or in p_T over the whole p_T-range.

Summary

NA35 has measured a large number of hadronic observables for various collision systems. The results can be compared to models and to data from collisions of similar systems at the AGS. In central nucleus–nucleus collisions a large net baryon density is observed at midrapidity at both energies. Strangeness production in central S+A collisions at 200 GeV/nucleon is enhanced compared to the yields in NN and pA interactions. This effect seems to occur mainly around midrapidity. A similar enhancement is observed at the AGS, but the enhanced production of antibaryons seen only at SPS-energies suggests a production mechanism beyond rescattering in a hadronic scenario. The measured ratio $\overline{\Lambda}/\overline{p}$ of about 1 near midrapidity is larger than the ratio measured in pp-interactions and almost independent of the target mass in S+A collisions at the SPS. Strange particle and antiproton transverse mass spectra at 200 GeV/nucleon are well described by a single exponential in m_T corresponding to "temperatures" of about 210 MeV, which are difficult to understand in a thermal hadron gas picture.

REFERENCES

1. H. Hamagaki for the E802/E859 Collaboration, *Proc. 10th Int. Conf. on Ultra-Relativistic Nucleus-Nucleus Collisions (Quark Matter 1993)*, Borlänge, Sweden (1993).

2. A. Saulys et al., *Proc. of Heavy Ion Physics at the AGS: HIPAGS'93 MITLNS-2158* (1993) p.196.

3. A. Wróblewski, *Proc. Int. Symp. on Multiparticle Dynamics*, Kayserberg, (1977) p.A1.

4. J. Sollfrank et al., preprint Univ. Regensburg TPR–93–14.

5. U. Katscher et al., *Z. Phys.* **A346** (1993) 209.

6. K. Werner, *Phys. Rev. Lett.* **62** (1989) 2460.

7. H. Sorge, H. Stöcker und W. Greiner, *Nucl. Phys.* **A498** (1989) 567c.

8. N.S. Amelin et al., *Phys. Rev* **C47** (1993) 2299.

9. H. Sorge et al., *Los Alamos Preprint LA-UR- 92-1078*, to be publ. in *Phys. Lett.* **B**.

10. H. Sorge, *Los Alamos Preprint LA-UR- 93-1103*, to be publ. in *Phys. Lett.* **B**.

11. M. Gonin et al., *Proc. of Heavy Ion Physics at the AGS: HIPAGS'93 MITLNS-2158* (1993) p.184.

12. A. Bartke et al. (NA35 Collab), *Z. Phys.* **C48** (1990) 191.

13. M. Gaździcki for the NA35 Collaboration, *Proc. 10th Int. Conf. on Ultra-Relativistic Nucleus-Nucleus Collisions (Quark Matter 1993)*, Borlänge, Sweden (1993).

14. V. Blobel et al., *Nucl. Phys.* **B69** (1974) 454.

15. G. Stephans for the E802/E859 Collaboration, *Proc. 10th Int. Conf. on Ultra-Relativistic Nucleus-Nucleus Collisions (Quark Matter 1993)*, Borlänge, Sweden (1993).

16. B. Jacak, contribution to this conference.

17. D. Röhrich for the NA35 Collaboration, *Proc. 10th Int. Conf. on Ultra-Relativistic Nucleus-Nucleus Collisions (Quark Matter 1993)*, Borlänge, Sweden (1993).

STRING MODEL APPROACH
TO NUCLEAR SCATTERING

K. Werner[1,2]

Institut für Theoretische Physik, Universität Heidelberg
Philosophenweg 19, 69120 Heidelberg, Germany

We discuss the string model of hadronic and nuclear scattering at ultrarelativistic energies. The main purpose is to treat theoretical concepts common to essentially all successful models: strings, Pomerons, and their "marriage" in the string model approach.

1   INTRODUCTION

In this article, we focus on a class of models [1]-[5] of ultrarelativistic scattering being based on Gribov-Regge theory (GRT) and relativistic strings, henceforth referred to as "string models" or "the string model approach". Other successful models not based on GRT [6, 7], use partly similar concepts, at least the string picture.

Why is the string approach so successful? Let us review some basic features of high energy scattering. For all types of reaction, from e^+e^- to nucleus-nucleus scattering, one observes a "longitudinal structure" of particle production: The scattering angles are very small, in the cms one observes two jets of particles with a narrow cone each (see fig. 1). To be more quantitative: Transverse momenta are limited to a few hundred MeV, even at highest energies (TeV). The longitudinal momenta, however, are only limited by the incident energy. On the other hand, limited transverse momenta and correspondingly a one-dimensional character are the fundamental properties of fragmenting relativistic strings. Consider a (so-called yo-yo-) string, consisting of two endpoints acting as energy reservoirs and the interior with constant energy per length (see fig. 2a). The endpoints move longitudinally with the velocity of light either "outwards" (see fig. 2b) by losing energy to the interior, or "inwards" (see fig. 2c) by absorbing energy from the interior. A newly formed massive string will first expand, both endpoints moving outwards, and then successively break into substrings (hadrons finally), as shown in fig. 3. String breaking is a local process, independent of the total energy. Therefore the new endpoints created at a string break acquire a transverse momentum (of a few hundred MeV) which is not affected

---

[1] Heisenberg fellow
[2] Email: werner@dhdmpi5.bitnet,   werner@hobbit.mpi-hd.mpg.de,   28877::werner

Figure 1: Longitudinal structure of produced particles.

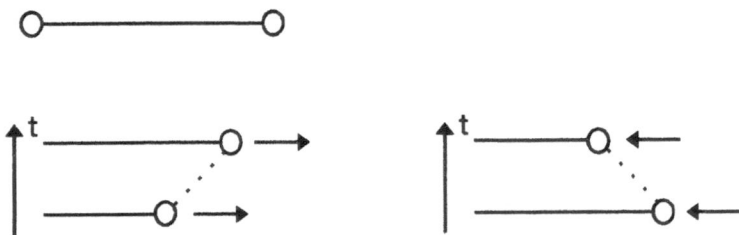

Figure 2: The one–dimensional string.

by the total energy. So, the local nature of string breaking provides limited transverse momenta and correspondingly a longitudinal structure (jets).

Another experimental fact concerns rapidity distributions, the rapidity being

$$y := \frac{1}{2} \ln \frac{1+v}{1-v} \equiv \frac{1}{2} \ln \frac{E+p_l}{E-p_l} \approx \ln \frac{2p_l}{\sqrt{m^2 + p_t^2}}. \tag{1}$$

The rapidity is so important because of its simple behaviour under Lorentz boosts:

$$y' = y + Y, \tag{2}$$

with Y being the relative rapidity of the two frames. At least for elementary interactions (lepton-nucleon) one observes a "rapidity plateau": with increasing energy, the distribution jets wider, not higher (see fig. 4). A rapidity plateau is also a fundamental property of fragmenting relativistic strings, being a consequence of the covariant formulation.

To summarize: the most important features observed in high energy collisions coincide with elementary properties of fragmenting relativistic strings. This is a strong motivation for the use of string models.

2 THE STRING MODEL OF HADRONIC INTERACTIONS

2.1 Hard and Soft Scattering

How should one formulate a model for hadronic interactions? A successful tool in physics is to generalize or modify known methods. So let us start with the simpler case of lepton-nucleon scattering, which is well described by the parton model (see fig. 5): The interaction amounts to an exchange of a photon (γ) between the lepton (l) and a quark (q) of the incident nucleon (N). This lepton-quark interaction is calculated according to the QED Feynman rules (taking the appropriate quark charge). Contributions for different quark momenta are added incoherently.

Can this picture be generalized to nucleon-nucleon scattering? It can: In fig. 6 the interaction amounts to exchanging a gluon (g) between two quarks (q_1 and q_2) of the nucleons, similarly a gluon may be exchanged between two gluons – keeping in mind that a nucleon consists of quarks, antiquarks and gluons. So here we consider elementary QCD diagrams as compared to QED diagrams in lepton scattering. Despite the similarity, there is a fundamental difference. In the QCD-case, the coupling constant α_s is in general not

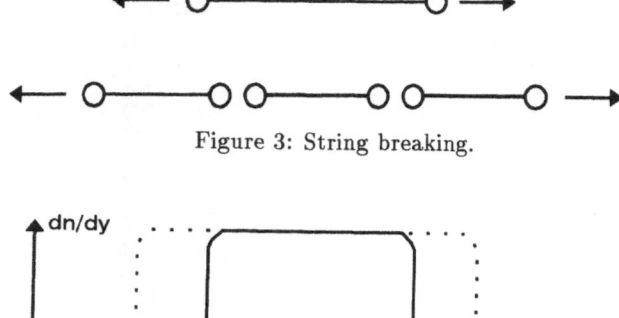

Figure 3: String breaking.

Figure 4: Rapidity plateau: increasing the energy makes the distribution wider.

small: it depends on the transferred momentum q as

$$\alpha_s = \alpha_s(-q^2) = \alpha_s(Q^2) \sim \frac{1}{\ln Q^2/\Lambda^2}. \tag{3}$$

So α_s is only small for sufficiently large transferred momentum Q, otherwise α_s is large and the perturbative QCD treatment as discussed above and shown in fig. 6 breaks down. As a convention, the term "hard scattering" is used for processes with large transferred momentum Q such that α_s is small and perturbation theory applies. For small momentum transfer, with perturbation theory not being applicable, the term "soft scattering" is used.

There are experimental means to separate soft and hard scattering. At high energies, the transferred momentum is purely transverse, so we have $Q = p_t$. Since the transferred transverse momentum is correlated with the transverse momentum of produced particles, one simply needs to study transverse momentum spectra of produced hadrons. One observes qualitatively the behaviour sketched in fig. 7, where dn/dp_t is plotted versus p_t. We find two regions: For small p_t (say $p_t < 2$ GeV/c) the spectra are exponential (straight lines in the log-plot); for large values of p_t, the spectra drop slower than exponential, a power law fit of the form p_t^{-n} represents the data properly. The low p_t region with the exponential behaviour represents soft scattering, the high p_t region with the p_t^{-n} behaviour represents hard scattering. In fact, p_t^{-n} is a typical result of perturbative QCD calculations.

The relative weight of soft and hard scattering is energy dependent: at SPS energies of $\sqrt{s} = 20$ GeV (per nucleon-nucleon) the hard contribution is very small, at most only few percent, we have almost pure soft scattering. With increasing energy, the hard component becomes more important, at LHC energies ($\sqrt{s} = 6.3$ TeV) substantial. The relative weight of hard scattering versus energy is sketched in fig. 8, where the band represents the uncertainty of the statement. For most of the range shown in fig. 8 soft scattering is dominant, a careful treatment of the soft scattering process is crucial in order to realistically model ultrarelativistic hadronic interactions.

2.2 The PQCD Approach of Hard Scattering

Let us be somewhat more explicit about the perturbative QCD approach of hard scattering, before we discuss the theoretical treatment of soft scattering and compare the two schemes.

The perturbative treatment amounts to writing an inclusive cross section as [10, 11,

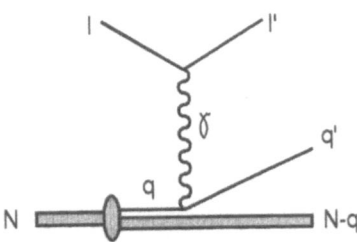

Figure 5: Lepton–nucleon scattering.

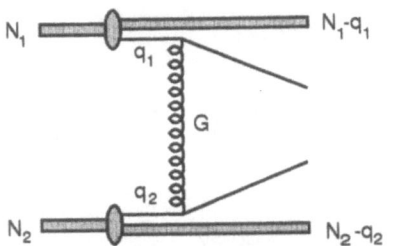

Figure 6: Nucleon–nucleon·scattering.

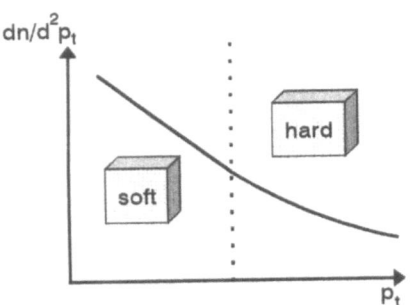

Figure 7: Tranverse momentum distribution separating soft and hard scattering.

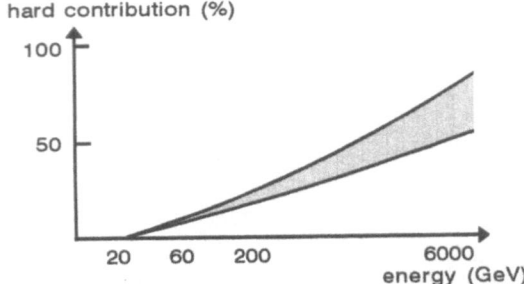

Figure 8: Realative weight of hard scattering versus energy.

12, 8, 9]

$$\sigma_h = \sum_{ij} \int dx_1 \int dx_2 \int dt \frac{d\sigma_{ij}}{dt} f_i(x_1, Q^2) f_j(x_2, Q^2) \tag{4}$$

with the corresponding diagram shown in fig. 9. The indices i and j refer to partons of the nucleons (quarks/antiquarks of a certain flavour or gluons), f_i and f_j are the corresponding momentum distribution functions, with x_1 and x_2 being the longitudinal momentum fractions with respect to the incident momenta, and with Q being the transferred momentum. The quantity $d\sigma_{ij}/dt$ represents the elementary cross section, calculated from summing QCD diagrams of lowest order.

It turned out that the inclusive cross section at high energies exceeds the inelastic cross section,

$$\sigma_h > \sigma_{\text{inel}}, \tag{5}$$

which reflects multiple scattering: The ratio

$$\bar{n} := \frac{\sigma_h}{\sigma_{\text{inel}}} \tag{6}$$

represents the average number of hard scatterings – which may be larger than one. In order to quantitatively reproduce not only p_t spectra but also multiplicity distributions, an impact parameter model has been introduced [11, 8, 9]. Here one introduces an average number of collisions for a given impact parameter as

$$\bar{n}(b) := \sigma_h T(b) \tag{7}$$

with

$$T(b) := \int \rho(b - b') \, \rho(b'), \tag{8}$$

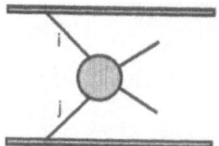

Figure 9: Hard scattering.

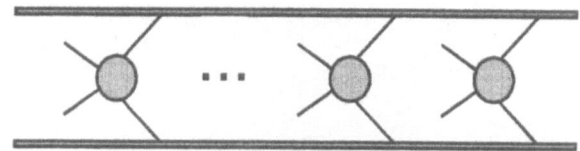

Figure 10: Multiple hard scattering.

where

$$\rho(b) := \int dz\, \rho(b_x, b_y, z) \tag{9}$$

is the transverse matter density of a nucleon. The quantity $T(b)$ is a measure of the overlap of the two nucleons: $T(b)$ is largest for $b = 0$ (representing complete overlap), drops with increasing b, and approaches zero around $b = 2R_{\text{nucleon}}$ (representing two touching nucleons).

Assuming independent multiple scattering, one writes the probability $\text{prob}(j, b)$ of j scatterings at given impact parameter as a Poissonian,

$$\text{prob}(j, b) = \frac{\bar{n}(b)^j}{j!}\, e^{-\bar{n}(b)}, \tag{10}$$

with $\bar{n}(b)$ given in eq. (7) (see fig. 10). Summing over j and integrating over b, one obtains the inelastic cross section,

$$
\begin{aligned}
\sigma_{\text{inel}} &= \int d^2b \sum_{j=1}^{\infty} \text{prob}(j, b) \\
&= \int d^2b \left\{ 1 - \exp[-\sigma_{\text{h}} T(b)] \right\}.
\end{aligned}
\tag{11}
$$

This is the well-known eikonal form, which will be derived later in the Gribov-Regge theory of soft scattering as well.

2.3 The String Model of Soft Scattering

In this section we discuss a general framework ("the string model") for treating soft hadronic interactions, used by a whole class of models, containing the VENUS model [1], the dual parton model (DPM) [2, 3], and the quark gluon string model [4, 5]. The purpose is first of all to introduce the principles; details and differences between the models can be found elsewhere [1].

Let us first introduce the string model in a very qualitative way, before we present the proper theory. To be definite, we consider nucleon-nucleon scattering. The basic interaction is colour exchange as shown in fig. 11: The incident nucleons are considered as "short" diquark-quark $(qq - q)$ strings, short meaning low mass – the nucleon mass $m_N = 0.94$ GeV is very small compared to the longitudinal momenta involved. The interaction amounts to coupling quarks and antiquarks together differently: now a quark and a diquark from different nucleons are connected and form two "long strings", which have in general a large mass.

There is no momentum transfer involved in this process, it is a surely soft process. Nevertheless, due to the colour rearrangement, two low mass objects (nucleons) are transformed into two massive objects (qq-q strings). Energy-momentum conservation is taken care of, just the type of energy changes: Initially the energy is essentially kinetic, finally the relative kinetic energy of the two strings may be quite small, however, the two strings have in general large masses. So, the colour exchange mechanism may be characterized

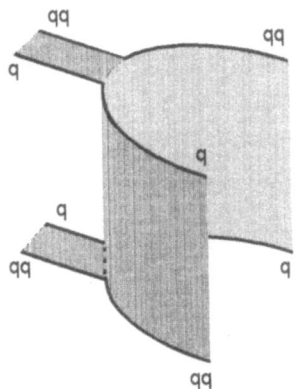

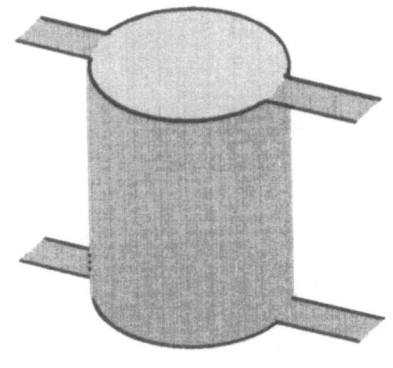

Figure 11: Colour exchange.   Figure 12: Cylinder diagram.

as follows: It is a really soft interaction without momentum transfer, which amounts to transforming the kinetic energy of the incident nucleons into string mass.

The generalization to arbitrary hadrons is obvious. In the case of a meson, the diquark (qq) has to be replaced by an antiquark $(\bar{q})$.

After this introduction let us discuss the theoretical basis of this colour exchange mechanism, being the basis of the string models VENUS, DPM, QGSM. One first considers the elastic amplitude $A(s,t)$ for hadron-hadron scattering, with s and t being the Mandelstam variables $s = (p_1 + p_2)^2$ and $t = (p_1 - p_3)^2$, where p_1, p_2 are the incoming and p_3, p_4 the outgoing momenta. The amplitude $A(s,t)$ is given as a multiple Pomeron exchange series,

$$\blacksquare = \lessgtr + \lessgtr\lessgtr + \ldots$$

with the zigzag lines symbolizing Pomerons. The Pomeron is the elementary exchange object here, without being elementary in terms of quarks and gluons. Originally the Pomeron was thought to be a gluon ladder. According to Veneziano, a Pomeron is the sum of all QCD diagrams of cylindrical topology (see fig. 12). A gluon ladder is, by the way, of cylindrical topology, in some sense the simplest cylinder.

Whatever the precise nature of the Pomeron may be: first of all one simply parametrizes the Pomeron propagator as

$$G(s,t) \sim s^{\alpha(t)} = s^{\alpha(0)+\alpha't}, \tag{12}$$

with this Regge pole form coming from general considerations of amplitudes in the limit $s \to \infty$. Two parameters characterize the Regge trajectory $\alpha(t)$: the "intercept" $\alpha(0)$ and the "slope" α', both being adjusted to fit data. With this simple form for G, the convolutions $G \otimes \ldots \otimes G$ can be worked out and summed over, and one obtains

$$A(s,t) = \frac{i}{4\pi} \int d^2b\, e^{i\vec{k}\vec{b}}\, \gamma(s,b), \tag{13}$$

with

$$\gamma(s,b) = 1 - e^{-\omega(s,b)}, \tag{14}$$

where ω is the Fourier transform of the Pomeron propagator G,

$$\omega(s,b) = \frac{1}{i\pi} \int d^2k\, G(s,t)\, e^{-i\vec{k}\vec{b}}. \tag{15}$$

It should be noted that only two-dimensional integrations occur (d^2b, d^2k). The reason is that at high energies transferred momenta are purely transverse, and therefore longitudinal and transverse degrees of freedom decouple. Longitudinal integrations can be performed first, so the convolutions $G \otimes ... \otimes G$ mentioned earlier are simply two-dimensional integrations referring to transverse momenta. This is the reason why results can be presented in "impact parameter representation", namely in the form $\int d^2b....$

Using the above result for $A(s,t)$, the elastic cross section is given as

$$\sigma_{\text{el}} = \int dt \frac{d\sigma_{\text{el}}}{dt} = \int d^2b \, |\gamma(s,b)|^2 \tag{16}$$

and, using the optical theorem, the total cross section can be written as

$$\sigma_{\text{tot}} = 8\pi \, \text{Im} \, A(s,0) = \int d^2b \, 2\text{Re}\gamma(s,b). \tag{17}$$

As a consequence, the inelastic cross section is given as

$$\sigma_{\text{inel}} = \sigma_{\text{tot}} - \sigma_{\text{el}} = \int d^2b \, \{1 - e^{-2\omega(s,b)}\}. \tag{18}$$

This expression should be compared with the corresponding formula for σ_{inel} in the case of hard scattering (eq.(11)): both expressions are very similar, being of the eikonal form. In the soft case, we have the exponent $-2\omega(s,b)$, representing Pomeron exchange; in the hard case, we have the exponent $-\sigma_{\text{h}}T(b)$, representing parton-parton scattering. The eikonal form is the consequence of the multiple Pomeron exchange in one case and of multiple parton-parton scattering in the other case. So although soft and hard scattering look very different at first sight, they are formally quite close to one another.

After having discussed elastic scattering, we would like to study inelastic amplitudes

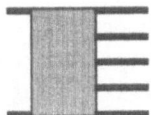

representing particle production. Unitarity relates elastic and inelastic amplitudes via

$$2 \, \text{Im} \quad \blacksquare = \sum \left| \blacksquare \right|^2$$

where the sum extends over a complete set of final states. Since inelastic amplitudes cannot be calculated directly, one first treats the imaginary part of the elastic amplitude and then draws conclusions about inelastic scattering. For this purpose, the AGK technique is employed [13], which provides a method to express $\text{Im}A(s,t)$ in terms of elementary inelastic scatterings, given as $\text{Im}G(s,t)$, with G being the Pomeron propagator. Using this technique and the optical theorem, the total cross section can be expanded as

$$\sigma_{tot} = \sum_{m=0}^{\infty} \sigma_m, \tag{19}$$

where σ_m is the cross section for m elementary inelastic scatterings. The σ_m can be calculated, they are weakly energy dependent, as shown in table 1, where we show $w_m = \sigma_m/\sigma_{\text{inel}}$.

Table 1: Weights w_m for cutting m Pomerons (m colour exchanges).

	19.4 GeV	200 GeV	6.3 TeV
w_1	0.54	0.52	0.47
w_2	0.24	0.23	0.22
w_3	0.12	0.13	0.13
w_4	0.059	0.067	0.081
w_5	0.026	0.033	0.047
w_6	0.011	0.015	0.026
w_7	0.0040	0.0061	0.0130
w_8	0.0013	0.0023	0.0060
w_9	0.00042	0.00080	0.00260
w_{10}	0.00012	0.00025	0.00100

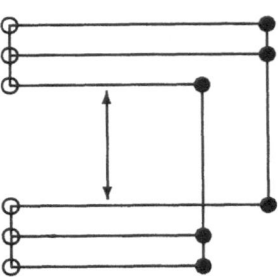

Figure 13: Quark line diagram.

To be more specific about the "elementary inelastic processes", we need to know something about the nature of the Pomeron. We adopt Veneziano's picture of a Pomeron being a cylinder (of gluons and quark loops),

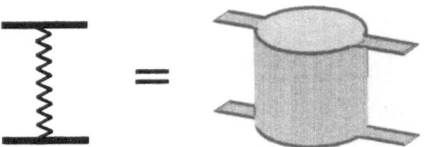

As a consequence, the imaginary part of the Pomeron amplitude is taken to be a squared "cut cylinder",

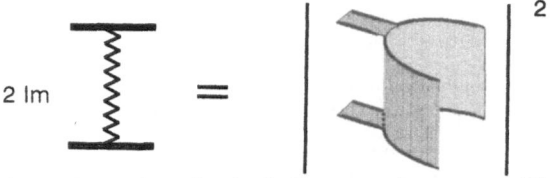

This is somehow the unitarity equation of a single Pomeron (compare with the full unitarity equation). We recognize the "cut cylinder": this is exactly the diagram we used earlier in connection with introducing the colour exchange mechanism. So the elementary inelastic process is just this colour exchange, and correspondingly, multiple inelastic scattering is nothing but multiple colour exchange.

A useful tool to deal with multiple colour exchanges, in particular in case of nucleus-nucleus scattering, are quark line diagrams as shown in fig. 13 for a single colour exchange in the case of the VENUS model. Quarks (and antiquarks) are represented by horizontal lines, the colour exchange is indicated by an arrow, and vertical connection of quark lines represent strings including the incident hadrons. The quarks belonging to a string are highlighted by dots. The generalization to multiple colour exchange is straightforward and obvious. Similar diagrams are also used in the DPM and QGSM.

For a given quark line diagram, strings are determined (in most applications) via Monte Carlo procedures. First of all, the moments of the participation quarks are generated according to distribution functions F, which factorize and scale as

$$F(p_l, p_t) = f\left(\frac{p_l}{p_l^0}\right) g(p_t), \tag{20}$$

with standard distributions f and g. Momentum conservation determines the momenta of the other components like diquarks. Knowing the momenta of the endpoints, strings are unambiguously determined. The strings are fragmented into observable hadrons via standard procedures [1].

3 NUCLEUS–NUCLEUS SCATTERING

Nuclear scattering can be treated as a straightforward generalization of hadron-hadron interactions. There are no new parameters entering apart from those characterizing the nuclear geometry. Applying the concept of multiple Pomeron scattering as in $h - h$, one derives a very simple geometrical picture as discussed in the following. It is important to realize that the classical geometrical picture is the result of a relativistic quantum mechanical treatment, not an assumption. It is just, mainly due to the decoupling of transverse and longitudinal degrees of freedom, that the system behaves (in some way) similar to the classical collision of two bags of hard spheres.

The treatment of the Gribov-Regge theory of nuclear scattering is quite involved. We will only present the major ideas here; details may be found in the review article [1]. In hadron-hadron scattering, the inelastic cross section σ_{in} can be expressed in terms of elementary inelastic processes as

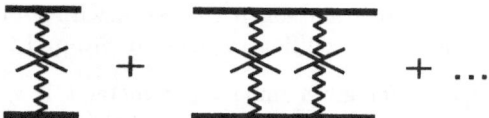

with an elementary process corresponding to colour exchange:

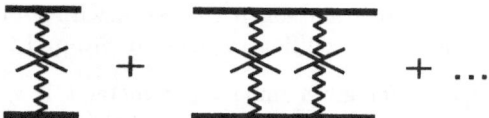

In a similar way, the inelastic cross section for nucleus-nucleus scattering may be expanded. The corresponding expression can be worked out, and one obtains

$$\sigma_{\text{inel}} = \int d^2b \sum_{\alpha_1\beta_1...\alpha_\nu\beta_\nu} \int dT_{AB} \prod_{\mu=1}^{\nu} \xi(b_\mu) \prod_{\mu=\nu+1}^{AB} 1 - \xi(b_\mu). \tag{21}$$

Here, $\alpha_1\beta_1...\alpha_\nu\beta_\nu$ represents collision sequences: α_μ and β_μ are the indices of projectile and target nucleon involved in the μth nucleon-nucleon collision. We further use

$$dT_{AB} = \prod_{\alpha=1}^{A} d^2b_\alpha^A T(b_\alpha^A) \prod_{\beta=1}^{B} d^2b_\beta^B T(b_\beta^B), \tag{22}$$

with b_α^A and b_β^B representing the transverse coordinates of the projectile and target nucleons and with

$$T(b) := \int \rho(b_x, b_y, z) dz \qquad (23)$$

being the transverse nuclear density. We use standard parametrizations of nuclear densities ρ. The quantities

$$b_\mu := b - b_{\alpha_\mu}^A + b_{\beta_\mu}^B \qquad (24)$$

represent the transverse distance between the projectile and target nucleon involved in the μth nucleon-nucleon interaction. The function ξ can be considered as a probability that a nucleon-nucleon interaction occurs at impact parameter b. In fact, ξ is related to the nucleon-nucleon cross section as

$$\sigma_{\text{inel}}^{NN} = \int d^2b \, \xi(b), \qquad (25)$$

with $\xi(b)$ behaving similar to a Fermi function with $\xi(b) = 1$ for $b << 2R_N$ and dropping to zero around $2R_N$, with R_N being the nucleon radius.

So the expression eq. (21) has a very simple structure: dT_{AB} is responsible for projectile and target nucleons being distributed properly (according to nuclear densities); the expression $\prod_{\mu=1}^\nu \cdots \prod_{\mu=\nu+1}^{AB} \cdots$ is the probability of collisions between $\alpha_1\beta_1, \alpha_2\beta_2, ..., \alpha_\nu\beta_\nu$ to occur ($1 - \xi$ is the probability of no collision).

The integrations and summations in eq. (21) are in general very difficult to carry out: for heavy nuclei ($A = B \approx 200$) we have an 800-fold integration. Whereas conventional methods fails, the problem can be easily solved via Monte-Carlo techniques. The following algorithm provides an exact numerical solution of eq. (21):

(A1) Generate A projectile and B target nucleons according to nuclear densities, and then take the transverse projections b_α^A, b_β^B.

(A2) Check, for given b, distances $b_{\alpha\beta} := b + b_\beta^B - b_\alpha^A$. Consider an interaction between projectile nucleon α and target nucleon β successful with probability $\xi(b_{\alpha\beta})$. All successful interactions provide a collision sequence $\alpha_1\beta_1...\alpha_\nu\beta_\nu$.

(A3) The cross section $\tilde{\sigma}_{\text{inel}}(b)$ for given impact parameter b is given as the ratio of successful nucleus-nucleus interactions ($\nu > 0$) and the number of attempts. The cross section is then $\sigma_{\text{inel}} = \int d^2b \, \tilde{\sigma}(b)$.

The Monte Carlo algorithm not only provides a solution, it also shows very clearly the geometrical nature of the interaction. This becomes even more clear, when the profile function $\xi(b)$ is approximated by a step function: $\xi(b) = \theta(2R_N - b)$. In this case, an interaction is successful whenever the transverse distance between two nucleons is less than twice the nucleon radius. The problem is thus identical to the classical collision of two big spheres (nuclei) filled with little hard spheres (nucleons), with an interaction occurring whenever two little spheres touch. It should be noted, however, that this classical picture is not an assumption: it has been derived within the appropriate framework of relativistic quantum theory. It should be noted as well that the system behaves in some way classically, other aspects are quantum mechanical: for example lifetimes and formation times play a crucial role. So our finding is not a general justification for using classical models; it helps, however, to understand some basic experimental results of high-energy nuclear collisions [1].

From somewhat more general considerations as discussed so far, one obtains not only collision sequences $\alpha_1\beta_1...\alpha_\nu\beta_\nu$, but also the numbers of colour exchanges $m_1...m_\nu$ for each individual nucleon-nucleon interaction. Using quark-line diagrams, one can easily construct strings as shown for a simple example in fig. 14. The rules for constructing quark-line diagrams are exactly the same as in NN scattering, and also the hadronization

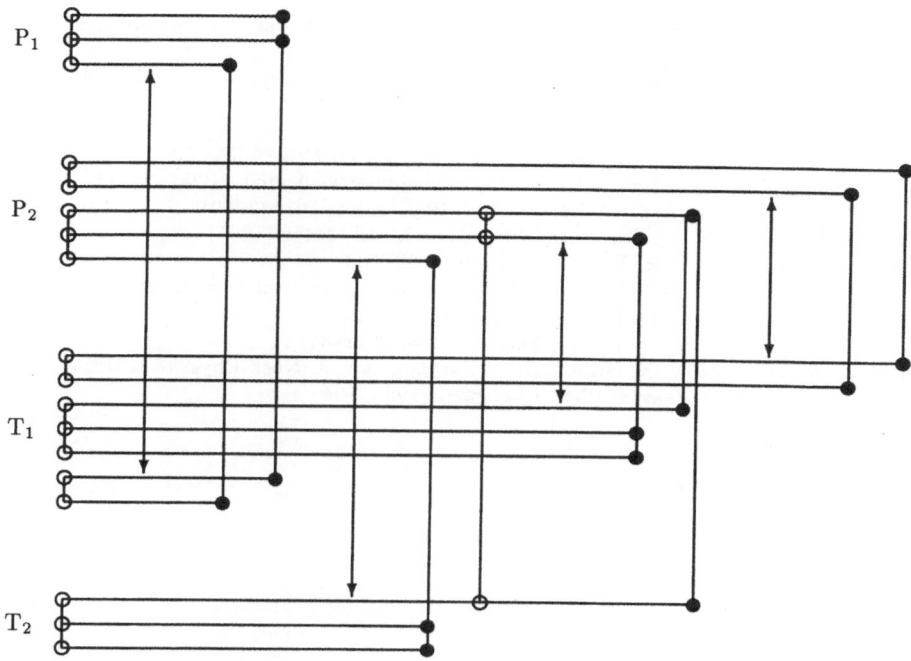

Figure 14: Quark line diagram for a nuclear interaction.

of the strings is done by employing the same (standard) methods. So there is actually little freedom by going from nucleon-nucleon to nucleus-nucleus scattering.

The treatment of nucleus-nucleus scattering as discussed so far ("independent string model") is appropriate as long as the two nuclei are light. For heavy nuclei, many strings are formed which may be quite close to each other. So interactions between strings or between string fragments need to be considered ("secondary interactions"). Several approaches have been proposed: quark cluster formation [1], hadronic rescattering [7, 5], string fusion [7, 14]. The treatment of secondary interactions is a very interesting and important subject, providing the link between independent string models and the formation of a quark gluon plasma, the ultimate aim of all the efforts in the field of ultrarelativistic collisions.

4 SUMMARY

We tried to sketch the theoretical principles and methods underlying essentially all the successful models on nuclear scattering. We have shown that there is a consistent formulation within the framework of relativistic quantum theory, in terms of Pomerons and strings. The fact that the successful models somehow converge to at least a common framework demonstrates that soft scattering is much less arbitrary than some people think. More than 20 years of theoretical and experimental work has been invested on the subject of soft scattering, which must not be ignored, certainly not at SPS energies, but also not at very high energies (RHIC), because soft scattering still plays a very important role.

References

[1] K. Werner, Physics Reports 232 (1993) 87–299

[2] A. Capella, U. Sukhatme, Chung-I Tan and J. Tran Thanh Van, preprint LPTHE 92-38, to appear in Physics Reports

[3] H. J. Möhring, A. Capella, J. Ranft, J. Tran Thanh Van, C. Merino, Nucl. Phys. A525 (1991) 493c

[4] A. Kaidalov, Nucl. Phys. A525 (1991)39c

[5] V. D. Toneev, A. S. Amelin and K. K. Gudima preprint GSI-89-52, 1989

[6] B. Andersen, G. Gustafson and B. Nielsson-Almqvist, Nucl. Phys. B281, 289 (1987)

[7] H. Sorge, H. Stöcker and W. Greiner, Nucl. Phys. A498, 567c (1989)

[8] X. N. Wang and M. Gyulassi, LBL 31036 (1991), LBL 31159 (1991)

[9] K. Geiger and B. Müller, Nucl. Phys. B369 (1992) 600

[10] F. E. Paige, Lecture at "Theoretical Advanced Summer Institute", Boulder, CO, USA, 1989

[11] T. Sjöstrand, M. van Zijl, Phys. Rev. D36 (1987) 2019

[12] G. Marchesini and B. R. Webber, Nucl. Phys. B238 (1984) 1

[13] V. A. Abramovskiĭ, V. N. Gribov, O. V. Kancheli, Sov. J. Nucl. Phys. 18 (1974) 308

[14] M.A.Braun and C. Pajares, Phys. Lett. B287 (1992) 154

COLOR DIFFUSION IN A QUARK-GLUON PLASMA[*]

Miklos Gyulassy and Alexei V. Selikhov[†]

Department of Physics
Columbia University, New York, NY 10027

Abstract

Color diffusion is shown to be an important dissipative property unique to non-abelian quark-gluon plasmas. The characteristic color relaxation time scale is shown to be $t_c \approx (3\alpha_s T \log(m_E/m_M))^{-1}$ revealing its sensitivity to the ratio of the static color electric and magnetic screening masses. Fokker-Planck equations are derived for QCD Wigner distributions taking into this account quantum color dynamics. These equations show that the anomalously small color relaxation time leads to a surprising small color conductivity and possibly overdamping of collective color modes.

INTRODUCTION

The quark-gluon plasma (QGP) phase of QCD matter has been predicted to exhibit collective behavior similar to QED plasmas with Debye screened (color) electric interactions, collective plasmon oscillations, and collective color mode instabilities under a variety of non-equilibrium configurations (see, e.g., reviews in [1, 2, 3]). Such collective behavior of quark-gluon plasmas (QGP) could be observed experimentally via ultra-relativistic heavy-ion reactions if the time scales of dissipative processes which act to damp color collectivity are long compared to the period of oscillation. Most studies of transport properties of QGP have focused thusfar on momentum relaxation processes to estimate viscosity and thermal conductivity coefficients, dE/dx,

---

[*]This work was supported by the Director, Office of Energy Research, Division of Nuclear Physics of the Office of High Energy and Nuclear Physics of the U.S. Department of Energy under Contract No DE-FG-02-93ER-40764.

[†]Supported by the AUI/RHIC Fellowship Fund; Permanent address: *Kurchatov Institute, 123182 Moscow D–182, Russia*

etc. [4, 5, 6]. Relaxation phenomena associated with the color degrees of freedom have received much less attention. In this talk I review our recent work on color diffusion. We derive a new set of Fokker-Planck type equations for QCD which describe the diffusion in both color and momentum space. We calculate not only the momentum relaxation time[5] that controls the friction and diffusion coefficients in momentum space but also the color relaxation time that controls the color diffusion coefficient introduced in [7]. We show that the same (divergent) color diffusion coefficient arises in both quantum and classical treatments of color dynamics. That divergence is caused by unscreened long range color magnetic fluctuations. We regulate that divergence by introducing a non-perturbative magnetic mass as suggested in [8]. The kinetic equations derived here imply that long wavelength collective color modes in a QGP are more strongly damped by very soft color exchange scattering processes (by a factor $1/\alpha_s$) than their abelian counterparts. Non-abelian plasmas are therefore relatively poor (color) conductors. The QCD Fokker-Planck equations show that, unlike in electro-magnetic plasmas, the divergent damping rates of hard partons computed diagrammatically (as in [8]) have considerable physical impact on the collective properties of QGP. It may well be that collective color degrees of freedom may effectively be ignored in the evolution of the quark gluon plasmas, and that a simple billiard ball parton cascade such as developed in ref.[9] may in fact be sufficient to compute the evolution of the partonic system produced in nuclear collisions.

COLOR DIFFUSION: CLASSICAL CONTINUOUS COLOR [7]

To derive color relaxation time, t_c, we follow here ref.[10, 11] and consider the motion of a classical colored particle moving along a world line $x^\mu(\tau)$ with momentum $p^\mu(\tau) = mv^\mu(\tau)$ and $v^\mu = dx^\mu/d\tau$, that couples with an N^2-1 component color charge vector $\Lambda^a(\tau)$ to a background Yang-Mills field, $A^a_\mu(x)$. We follow this approach here because of its intuitive simplicity and because it leads to the same result that we find [12] starting from the kinetic equations for the quark and gluon Wigner density matrices [1] including the Lenard-Balescu type collision terms derived in [13].

Consider the motion of a heavy quark in the d dimensional representation of $SU(N)$ with second order Casimir, C_2, which obeys the equations derived in [10]:

$$\frac{dp^\mu}{d\tau} = gv_\nu F^{\mu\nu}_a \Lambda^a \ , \tag{1}$$

$$\frac{d\Lambda^a}{d\tau} = gf^{abc}v^\mu A^b_\mu \Lambda^c \ , \tag{2}$$

where the fields are evaluated at $x^\mu(\tau)$.

We study the time dependence of the color charge averaging over an ensemble of background field configurations. We assume that the ensemble characterizes a color neutral medium in which the A^a_μ are random fields such that the ensemble averaged potentials vanish in a suitable gauge. The ensemble average of Eq. (2) gives

$$\frac{d}{d\tau}\langle\Lambda^a(\tau)\rangle = gf^{abc}\langle[vA]^b_\tau\Lambda^c(\tau)\rangle \ , \tag{3}$$

where for shorthand we write $[vA]^b_\tau = (p^\mu(\tau)/m)A^b_\mu(x(\tau))$. Integrating (formally) Eq. (2) and substituting the result back into Eq. (3), yields

$$\frac{d}{d\tau}\langle\Lambda^a(\tau)\rangle = g^2 f^{abc} f^{cde} \int_0^\tau d\tau' \langle[vA]^b_\tau[vA]^d_{\tau'}\Lambda^e(\tau')\rangle \ . \tag{4}$$

Our first main physical assumption is that the time scale of field fluctuations is short compared to the time scale of variations of the color orientation. Formally, this is suggested from eq.(4) *if* $g \ll 1$ because the rate of change of the color charge is proportional to g^2. We therefore assume a stochastic (random phase) ansatz[14] for the expectation value of the product

$$\langle[vA]^b_\tau[vA]^d_{\tau'}\Lambda^e(\tau')\rangle \approx \langle[vA]^b_\tau[vA]^d_{\tau'}\rangle\langle\Lambda^e(\tau)\rangle \ . \tag{5}$$

With this ansatz, eq.(4) reduces to

$$\frac{d}{d\tau}\langle\Lambda^a(\tau)\rangle \approx \left\{g^2 f^{abc} f^{cde} \int_0^\tau d\tau' \langle[vA]^b_\tau[vA]^d_{\tau'}\rangle\right\}\langle\Lambda^e(\tau)\rangle \equiv -d^{ae}(\tau)\langle\Lambda^e(\tau)\rangle \ . \tag{6}$$

This defines the color diffusion tensor, $d^{ae}(\tau)$, whose physical interpretation will become clear below. Note that we cannot use the $v^\mu A^b_\mu = 0$ gauge in Eq. (6) because $v^\mu(\tau)$, via eq.(1), is different for every member of the ensemble.

Our second main approximation is to work in the eikonal limit. This is equivalent to the "hard thermal loop" approximation in [15]. We thus assume that our test parton has a high initial four momentum and that energy loss via eq.(1) is small. This also follows formally from eq.(1) in the perturbative $g \ll 1$ limit. In this case, we can factor the four velocity out of the ensemble average and approximate

$$\langle[vA]^b_\tau[vA]^d_{\tau'}\rangle \approx v^\mu v^\nu \langle A^b_\mu(x(\tau))A^d_\nu(x(\tau'))\rangle \equiv v^\mu v^\nu C^{bd}_{\mu\nu}(x(\tau), x(\tau')) \ . \tag{7}$$

Next we note that in this eikonal limit $x^\mu(\tau) - x^\mu(\tau') = (\tau - \tau')v^\mu$, and thus for a homogeneous, color neutral ensemble, the correlation function C can expressed as

$$C^{bd}_{\mu\nu}(x(\tau), x(\tau')) = \delta_{bd} \int (dk)e^{-i(kv)(\tau-\tau')}C_{\mu\nu}(k, u) \ , \tag{8}$$

where $(dk) = d^4k/(2\pi)^4$, u^μ is the four velocity of the ensemble rest frame. The correlation function C measures the spontaneous fluctuations of the background field. In the classical limit those fluctuations can be calculated using kinetic theory[11, 13, 16]. For a system in thermal equilibrium, the $\omega \ll T$ spontaneous fluctuations can also be related to the response function (retarded commutator) via the fluctuation-dissipation theorem[17] as

$$C^{\mu\nu}(k, u) \approx -\frac{2T}{ku}\text{Im}D^{\mu\nu}(k, u, T) \ , \tag{9}$$

where $D^{\mu\nu}(k, u, T)$ is the thermal averaged retarded commutator that arises in linear response theory. Both methods give the same result in the high temperature limit in the 1-loop approximation[11]. The 1-loop result obtained in ref.[18, 15] is

$$D_{\mu\nu}(k, u) = -\frac{Q_{\mu\nu}}{k^2 - \Pi_L} - \frac{P_{\mu\nu}}{k^2 - \Pi_T} + \alpha\frac{k_\mu k_\nu}{k^4} \ , \tag{10}$$

where α is a gauge parameter. This decomposition utilizes the longitudinal and transverse projectors, $Q_{\mu\nu} = \bar{u}_\mu \bar{u}_\nu / \bar{u}^2$ and $P_{\mu\nu} = \bar{g}_{\mu\nu} - Q_{\mu\nu}$, as given in terms of $\bar{g}_{\mu\nu} = g_{\mu\nu} - k_\mu k_\nu / k^2$ and $\bar{u}_\mu = \bar{g}_{\mu\nu} u^\nu$. Furthermore, the longitudinal and transverse polarization functions are related to the gluon self energy $\Pi^{\mu\nu}(k)$ through $\Pi_L = \Pi^{\mu\nu} Q_{\mu\nu}$ and $\Pi_T = \Pi^{\mu\nu} P_{\mu\nu}/2$.

With (9,10), the color diffusion tensor can be expressed via the color diffusion coefficient, d_c, and the corresponding color relaxation time in the plasma rest frame, t_c, via

$$\lim_{\tau \to \infty} d^{ae}(\tau) \equiv \delta_{ae} N d_c \equiv \delta_{ae}(vu)/t_c \ . \tag{11}$$

For ultra-relativistic partons, the color relaxation rate reduces to

$$\frac{1}{t_c} = -2\pi g^2 NT \int \frac{(dk)}{(ku)} \frac{k^2}{(ku)^2 - k^2} \delta\left(\frac{kv}{vu}\right) \mathrm{Im}\left(\frac{1}{k^2 - \Pi_L} - \frac{1}{k^2 - \Pi_T}\right) \ , \tag{12}$$

It is important to note that the dependence on the gauge parameter α dropped out because the delta function constrains $kv = 0$. The gauge invariance of this result is related to the gauge invariance of the damping rate in hard thermal loop calculations in ref.[8] and depends essentially on the eikonal limit. For momenta $\sim 3T$, the eikonal approximation applies perturbatively because the energy loss per interaction, $\sim g^2 T$, is relatively small if $g \ll 1$.

Given t_c, the solution of eq. (6) in the plasma rest frame is

$$\langle \Lambda^a(t) \rangle = e^{-t/t_c} \langle \Lambda^a(0) \rangle \ . \tag{13}$$

Note, that while the ensemble averaged color of the parton vanishes rapidly for $t > t_c$, the equations of motion conserve the magnitude of the color vector, i.e., $\langle \Lambda^a(\tau)\Lambda^a(\tau) \rangle = $ constant.

In fact we find [7] that the color relaxation rate, $1/t_c$, corresponds exactly to the leading log approximation to the gluon damping rate computed in [8, 19]. This coincidence of results based on classical kinetic theory and 1-loop high T QCD is a general result[11] in all problems where the classical $\omega \ll T$ and $k \ll T$ modes of the system are dominant.

Unfortunately, as is well known[8], the transverse contribution to that damping rate is logarithmically divergent. Dynamical screening is sufficient to regulate the Coulomb divergence in the expression for the momentum relaxation rate[5] but not the damping rate. The longitudinal part is finite[6, 8] and can thus be neglected. Several attempts to deal with this problem have been proposed. One is to introduce a nonperturbative magnetic mass, $m_M \sim g^2 T$, as in ref.[8]. Another is to introduce damping self consistently by adding an imaginary part to the bare fermion propagator[20]. However, no satisfactory solution is yet available.

In our kinetic formulation the source of the problem may be traced back to our first assumption that the field fluctuation time is short compared to the color diffusion time. What we learned a postiori was that fluctuations of the transverse magnetic fields are long ranged in time and that quasi-static unscreened magnetic fields lead to the divergent color relaxation rate. This suggests that the change of the color moment with time should taken into account within the integrand in eq.(6). Such a "memory" effect could damp the contribution from early times with $(vu)(\tau - \tau') > t_c$. Inserting a factor $\exp(-(vu)(\tau - \tau')/t_c)$ within the integral over τ' would smear out the delta function, $\delta(kv)$, appearing in eq.(12) into a Lorentzian form $\propto d_c/((kv)^2 + d_c^2)$. The resulting self consistent equation for d_c would lead to a finite result analogous to the method considered by Altherr et al.[20].

In order to regulate that infrared divergence, we follow ref.[8] and introduce formally a non-perturbative color magnetic screening mass, $m_M \sim (g^2 T)$, via $\mathrm{Im}(k^2 - \Pi_T)^{-1} \approx -\mu m_E^2/(\vec{k}^2 + m_M^2)^2 + (\mu m_E^2)^2)$ where $\mu = \omega/|\vec{k}|$. Note that the dynamic screening factor[5] is properly included above. By using this anzatz we obtain [7]

$$t_c^{-1} \approx \alpha_s N T \log(m_E/m_M) \ . \tag{14}$$

To show that d_c actually corresponds to a diffusion coefficient in kinetic theory, we consider next the classical phase space density $Q(x, p, \Lambda)$ of an ensemble of colored particles defined by

$$Q(x, p, \Lambda) = \int d\tau' \sum_i \delta(x - x_i(\tau'))\delta(p - p_i(\tau'))\delta(\Lambda - \Lambda_i(\tau')) \ , \tag{15}$$

where p_i^μ and Λ_i^a obey Wong equations (2) coupled via a self consistent field. The phase space density obeys the (Heinz) transport equation [1, 11]:

$$(p_\mu \partial_x^\mu + g p^\nu F_{\mu\nu}^a \Lambda^a \partial_p^\mu + i g p^\mu A_\mu^a L^a) Q(x, p, \Lambda) = 0 \ , \tag{16}$$

where

$$L^a = -i f^{abc} \Lambda^b \frac{\partial}{\partial \Lambda^c} \ , \tag{17}$$

are generators of $SU(N)/Z^N$ obeying the commutation relations $[L^a, L^b] = i f^{abc} L^c$.

Under the same assumptions made in deriving the color diffusion coefficient previously and with neglect the momentum degradation part to focus exclusively on color diffusion we obtain for the ensemble average $\langle Q(x(\tau), p, \Lambda)\rangle$ the following color diffusion equation

$$\frac{d}{d\tau}\langle Q(x(\tau), p, \Lambda)\rangle = -d_c L^a L^a \langle Q\rangle \ , \tag{18}$$

where color diffusion coefficient d_c is given by eqs.(11,14). Note that Eq. (18) can be written in the form of Fokker-Planck equation

$$\frac{d\langle Q\rangle}{d\tau} = -\frac{\partial}{\partial \Lambda^a}(d^{ab} \frac{\partial}{\partial \Lambda^b}\langle Q\rangle) \ , \tag{19}$$

where $d^{ab} = d_c f^{adc} f^{ceb} \Lambda^d \Lambda^e$. Equation (18) therefore describes diffusion in color space. Note that operator $C_2 = L^a L^a$ just is quadratic Casimir operator of $SU(N)/Z^N$.

To illustrate color diffusion, it is instructive to consider $SU(2)$, where $L^i = -i\epsilon^{ijk}\Lambda^j\partial/\partial\Lambda^k$ is an angular momentum operator in color (iso-spin) space and $C_2 = l(l+1)$. Since the magnitude of $\vec{\Lambda}$ is fixed, any initial distribution in color space can be expanded in the spherical harmonics as

$$\langle Q(\vec{\Lambda}, \tau = 0)\rangle = \sum_{l,m} c_{l,m} Y_l^m(\theta, \phi) \ , \tag{20}$$

where θ, ϕ are polar coordinates in iso-spin space. In this case, the proper time evolution is simply

$$\langle Q(\vec{\Lambda}, \tau)\rangle = \sum_{l,m} c_{l,m} Y_l^m(\theta, \phi) e^{-d_c \tau l(l+1)} \ . \tag{21}$$

¿From Eq. (21) it follows that any non-isotropic distribution in color space evolves to a uniform distribution,

$$\lim_{\tau \to \infty} \langle Q(\vec{\Lambda}, \tau)\rangle = c_0/\sqrt{4\pi} \ , \tag{22}$$

on a proper time scale $\sim 1/2d_c$ (for $N = 2$). The same picture holds qualitatively for the general $SU(N)$ case, where any distribution evolves to uniform distribution which is determined by the zero mode of quadratic Casimir operator. This color equilibrium state is achieved after a proper $\tau_c = 1/Nd_c$. Since the time in the plasma rest frame is $t = (vu)\tau$ and $Nd_c = (vu)/t_c$, color equilibrium in the plasma rest frame is achieved on the time scale t_c.

COLOR DIFFUSION: DISCRETE QUANTUM COLOR [12]

In kinetic theory, non-equilibrium physical system are described by means of one-particle Wigner distribution functions. The gauge covariant quark, $\hat{Q}^+(p,x)$, and antiquark, $\bar{Q}(p,x) = \hat{Q}^-(-p,x)$, Wigner operators for $SU(N)$ are $N \times N$ matrices in color space. They are related to the gauge covariant quark Wigner operator [1, 21] via $\hat{Q}^\pm(p,x) = \theta(\pm p_0)\delta(p^2)\hat{W}(p,x)$, where

$$\hat{W}(p,x) = \int \frac{d^4y}{(2\pi)^4} e^{-ip\cdot y} \bar{\psi}(x) e^{\frac{y}{2}\hat{D}^\dagger} \otimes e^{-\frac{y}{2}\hat{D}} \psi(x) \ , \tag{23}$$

where $\hat{D}_\mu = \partial_\mu + ig\hat{A}_\mu$, $\hat{A}_\mu = \hat{A}_\mu^a t^a$; $\hat{A}_\mu^a$ is the operator of gluon field, t^a are the hermitian generators of $SU(N)$ in fundamental representation, and the field tensor $\hat{F}_{\mu\nu} = [\hat{D}_\mu, \hat{D}_\nu]/(ig)$. The covariant gluon Wigner operator $\hat{G}(p,x)$ is an $(N^2 - 1) \times (N^2 - 1)$ matrix in color space and is defined similarly [22, 13].

Near equilibrium at high temperature $T \gg 200$ MeV, the typical momentum transfers, $k \sim gT$, in the plasma are perturbatively small compared to the average momenta, $p \sim 3T$. In that case, spin effects can be neglected in the first approximation, and the evolution can be treated in the eikonal approximation assuming approximate straight line trajectories. This physical picture forms the physical basis behind the hard thermal loop approximation[15] in high temperature pQCD. With spin effects neglected, the Wigner operator obeys the following dynamical equations in the semi-classical limit [21]:

$$p^\mu \hat{D}_\mu \hat{Q}^\pm(p,x) + gp^\mu \partial_p^\nu \frac{1}{2}\{\hat{F}_{\nu\mu}, \hat{Q}^\pm(p,x)\} = 0 . \tag{24}$$

A similar equation holds for the gluon Wigner operator with generators t^a replaced by those in the adjoint representation [21, 22, 13]. The quark and gluon phase space densities are defined as quantum-statistical averages of the corresponding operators: $Q^\pm(p,x) = \langle \hat{Q}^\pm(p,x) \rangle$, $G(p,x) = \langle \hat{G}(p,x) \rangle$. In equilibrium,

$$
\begin{aligned}
Q_{eq}^\pm &= \frac{2N_f}{(2\pi)^3} 2\theta(\pm p_0)\delta(p^2)(\exp(\pm(p\cdot u)/T)+1)^{-1} , \\
G_{eq} &= \frac{2}{(2\pi)^3} 2\theta(p_0)\delta(p^2)(\exp((p\cdot u)/T)-1)^{-1} .
\end{aligned}
\tag{25}
$$

For small deviations from these color neutral equilibrium distribution, we write

$$Q_{ij}^\pm = Q_{eq}^\pm \delta_{ij} + \Delta Q_{ij}^\pm , \quad G_{ab} = G_{eq}\delta_{ab} + \Delta G_{ab} . \tag{26}$$

The non-equilibrium deviations obey the linearized kinetic equations derived in ref.[13]:

$$p^\mu \partial_\mu \Delta Q^\pm + gp^\mu \partial_p^\nu \Delta F_{\nu\mu} Q_{eq}^\pm = \Delta C_1^\pm(p,x) + \Delta C_2^\pm(p,x) , \tag{27}$$

where the linearized collision term on the right hand side has been decomposed into two parts. In these linearized collision terms quantum statistics is neglected but dynamical polarization effects are included.

The first, which describes diffusion in momentum space, is written in Fokker-Planck form as [12]

$$\Delta C_1^\pm(p,x) = \{t^a, \{t^a, \left(-\partial_p^\mu(a_\mu \Delta Q^\pm(p,x)) + \partial_p^\mu(b_{\mu\nu}\varepsilon(p_0)\partial_p^\nu \Delta Q^\pm(p,x))\right)\}\} + \delta c_1^\pm, \tag{28}$$

where $\{,\{,\}\}$ denotes a double anti-commutator, and the momentum diffusion tensor and friction force vector are given by

$$
\begin{aligned}
b_{\mu\nu} &= \int dp' B_{\mu\nu}(p,p') \mathcal{N}_{eq}(p') , \\
a_\mu &= \int dp' B_{\mu\nu}(p,p')\varepsilon(p_0')\partial_{p'}^\nu \mathcal{N}_{eq}(p') .
\end{aligned}
\tag{29}
$$

The kernel is given by

$$B_{\mu\nu} = \frac{\alpha_s^2}{2}\int dk k_\mu k_\nu \mid p^\alpha D_{\alpha\beta}(k)p'^\beta \mid^2 \delta(pk)\delta(p'k) . \tag{30}$$

In our notation $\varepsilon(p_0) = \theta(p_0) - \theta(-p_0)$, and the effective equilibrium density is $\mathcal{N}_{eq}(p) = \frac{1}{2}(Q_{eq}^+ + Q_{eq}^-) + NG_{eq}$. This density controls the high temperature polarization tensor [11, 2, 13]

$$\Pi^{\mu\nu}(k) = -g^2 \int dp \frac{p^\mu k_\alpha(p^\alpha \partial_p^\nu - p^\nu \partial_p^\alpha)}{pk + \imath\epsilon p_0}\mathcal{N}_{eq}(p) = \Pi_L Q_{\mu\nu} + \Pi_T P_{\mu\nu} , \tag{31}$$

535

where the longitudinal and transverse projectors $Q_{\mu\nu}$, $P_{\mu\nu}$ and polarization functions are defined below eq.(10). As noted in ref.[11], the polarization tensor derived from kinetic theory coincides exactly with the gauge invariant high temperature 1-loop result obtained diagrammatically in ref.[15, 18]. The polarization tensor in turn determines the medium modified (retarded) gluon propagator (10) that appears in the kernel of the above collision integrals. It is important to note that the collision terms are gauge independent because the eikonal mass shell conservation factors, $\delta(pk)$ and $\delta(p'k)$, insures that the convolution of eikonal vertex factors, p^μ and p'^ν, with the gauge fixing term vanishes.

The tensorial structure of $b_{\mu\nu}$ is in general complicated, but the dominant term which leads to a (Rayleigh) friction coefficient proportional to velocity, has the structure

$$b_{\mu\nu} \approx b(p^\mu u^\nu + p^\nu u^\mu - g^{\mu\nu}(p \cdot u)) \ . \tag{32}$$

Neglecting quantum statistics, the friction force and diffusion tensor are related by the analog of the Einstein relation

$$a_\mu = -b_{\mu\nu} u^\nu / T = -b p_\mu / T \ . \tag{33}$$

We now relate $b = -T(au)/(pu)$ to the energy loss per unit length derived in [6]. Note first that in the high T limit, $\varepsilon(p_0)\partial_p^\mu \mathcal{N}_{eq}(p) \approx -(u^\mu/T)\mathcal{N}_{eq}(p)$, and thus from eq.(31) it follows that

$$\mathrm{Im}\,\Pi_{\mu\nu}(k) \approx \frac{\pi g^2 ku}{T} \int dp' p'_\mu p'_\nu \delta(p'k)\mathcal{N}_{eq}(p') \ . \tag{34}$$

Noting next the identity

$$\int dk\ \delta(pk)\ \mathrm{Im}(pD(k)p) = -\int dk\ \delta(pk)\ pD(k)(\mathrm{Im}\,\Pi(k))D^*(k)p \ , \tag{35}$$

we find that

$$\frac{4b}{T} = \frac{1}{C_2}\frac{dE}{dx} = -g^2 2\pi \int \frac{dk}{(2\pi)^4}\frac{ku}{pu}\mathrm{Im}\,(p^\mu D_{\mu\nu}(k)p^\nu)\ \delta(pk) \approx \frac{4\pi}{3}\alpha_s^2 T^2 \log(k^*/m_E), \tag{36}$$

where the color electric screening mass is $m_E = gT\sqrt{(1 + N_f/6)}$ and $k^* \ll 3T$ is a cutoff parameter separating the soft and hard momentum transfer scales. As discussed in [23], the above formula for energy loss is accurate only for low momentum transfers. Physically, this is also clear from our kinetic theory derivation which utilized the eikonal approximation. In practice, however, setting $k^* = 3T$ is adequate to logarithmic accuracy. The effects of hard collisions require of course an extension beyond the Fokker-Planck approximation. The consistency of the relation between b and dE/dx can be verified by multiplying eqs.(27,28) in the plasma rest frame by p_0, taking the trace, and integrating over d^3x. We can also relate b to the gluon momentum relaxation time defined via

$$1/t_p = \langle 1/E\rangle dE/dx \approx (0.7/T)C_A(4b/T) \approx 4\alpha_s^2 T \log(1/\alpha_s) \ , \tag{37}$$

536

which is close to the numerical result obtained in ref.[5].

The second collision term in eq.(27), which has no abelian counterpart and, as shown below, describes diffusion in color space, can be expressed in an analogous Fokker-Planck form [12]

$$\Delta C_2^{\pm}(p, x) = -d^{ab}[t^a, [t^b, \Delta Q^{\pm}(p, x)]]\varepsilon(p_0) + \delta c_2^{\pm} , \qquad (38)$$

The double commutator corresponds to the second order term in the rotation of ΔQ in color space by random angles, θ_a, with $\langle \theta^a \rangle = 0$ but $\langle \theta^a \theta^b \rangle \propto \delta_{ab}$. Hence, the first term in (38) corresponds to diffusion in color space. The color diffusion tensor is a measure of the mean square fluctuations of the rotation angles in color space and is diagonal, $d^{ab} = d_c \delta_{ab}$, with the color diffusion coefficient, d_c, given by

$$\begin{aligned} d_c &= 2\alpha_s^2 \int dp' dk \mid p^\mu D_{\mu\nu} p'^\nu \mid^2 \delta(pk)\delta(p'k)\mathcal{N}_{eq}(p') \\ &= -2\pi g^2 T \int \frac{dk}{(2\pi)^4} \frac{\delta(pk)}{ku} \text{Im}\, (p^\mu D_{\mu\nu}(k)p^\nu) \\ &\approx (pu)\alpha_s T \log(m_E/m_M) , \end{aligned} \qquad (39)$$

Note that we used eqs.(34,35) again to express d_c in a form that is equivalent to the one derived in ref.[7] starting from classical color dynamics[11, 10]. As emphasized in [7] perturbative dynamic screening in a QGP is not enough to make the color diffusion coefficient converge. The momentum diffusion coefficient converges because of an extra two powers of $\omega = ku$ appearing in the integral. The divergence is due to long range unscreened color magnetic interactions.

Note that d_c has dimensions of energy squared unlike that defined in ref.[7] because we treat massless partons here. The color diffusion time, defined in this case by $t_c = (pu)/(Nd_c)$ is, however, identical.

As in abelian plasmas[24], the corrections to the Fokker-Planck terms, $\delta c_1^{\pm}$ and $\delta c_2^{\pm}$, can be generally neglected because they involve integrals over the small non-equilibrium deviations instead of the equilibrium density [12]. The smallness of the correction terms is caused[24, 12] by constraints on the non-equilibrium deviations imposed by conservation laws of particle number, color current, and energy-momentum.

Proceeding analogously with the kinetic equation for gluons[13], we obtain finally the QCD Fokker-Planck equations

$$\begin{aligned} p^\mu \partial_\mu \Delta Q + g p^\mu \partial_p^\nu \Delta F_{\nu\mu}^a t^a Q_{eq} &= -d_c[t^a, [t^a, \Delta Q]] - \partial_p^\mu \left((a_\mu - b_{\mu\nu} \partial_p^\nu)\{t^a, \{t^a, \Delta Q\}\} \right), \\ p^\mu \partial_\mu \Delta \bar{Q} - g p^\mu \partial_p^\nu \Delta F_{\nu\mu}^a t^a \bar{Q}_{eq} &= -d_c[t^a, [t^a, \Delta \bar{Q}]] - \partial_p^\mu \left((a_\mu - b_{\mu\nu} \partial_p^\nu)\{t^a, \{t^a, \Delta \bar{Q}\}\} \right). \\ p^\mu \partial_\mu \Delta G + g p^\mu \partial_p^\nu \Delta F_{\nu\mu}^a T^a G_{eq} &= -d_c[T^a, [T^a, \Delta G]] \\ &\quad - \partial_p^\mu \left((a_\mu - b_{\mu\nu} \partial_p^\nu)\{T^a, \{T^a, \Delta G\}\} \right), \\ \partial^\mu \Delta F_{\mu\nu}^a = \Delta j_\nu^a &= g \int dp p_\nu \left(Sp(t^a(\Delta Q - \Delta \bar{Q})) + Tr(T^a \Delta G) \right). \quad (40) \end{aligned}$$

This system describes the transport of small deviations in a color neutral quark-gluon plasma in terms of the transport coefficients, d_c, a^μ, and $b^{\mu\nu}$, which are controlled by the two time scales, t_p and t_c computed above.

COLOR CONDUCTIVITY:
IS THE QGP A COLOR INSULATOR IN DISGUISE??

One of the interesting consequences of these equations is that they show that the damping of collective modes in a non-abelian plasma is controlled by the (perturbatively divergent) parton damping rates. Non-abelian plasmas are therefore poor color conductors.

To calculate the color conductivity coefficient, we must test the linear response of the system to a weak external field, F_{ex}. To calculate the induced color current, we take the color octet moments of the Fokker-Planck equations (40). In that case the color diffusion terms transform into relaxation terms for the color octet deviations, e.g.

$$p^\mu \partial_\mu Q^a + gp^\mu [F_{ex}]^a_{\nu\mu} \partial^\nu_p Q_{eq} = -(pu)Q^a/t_c \ .$$ (41)

where $Q^a(x,p) = 2Sp(t^a \Delta Q)$. Solving (41) for the Fourier transform, $Q^a(k,p)$, the induced color current is given by

$$j^{\mu a}(k) = \frac{g}{2} \int dp p^\mu Q^a(p,k) \equiv \sigma^{\mu\alpha\beta}(k)[F_{ex}(k)]^a_{\alpha\beta} \ ,$$ (42)

where the conductivity tensor for a homogeneous color neutral plasma of quarks, antiquarks, and gluons is [11, 2]

$$\sigma^{\mu\alpha\beta}(k) = g^2 \int dp \frac{p^\mu p^\alpha}{-i(pk) + (pu)/t_c} \partial^\beta_p \mathcal{N}_{eq}(p) \ ,$$ (43)

where $\mathcal{N}_{eq}(p)$ is defined above eq.(31). For an isotropic plasma this tensor reduces to $\sigma^{\mu\alpha\beta} = \sigma^{\mu\nu}(k)u^\beta$. Finally, from Eq.(43) it follows that [11, 2] the static ($k = 0$) color conductivity is $\sigma^{\mu\nu}(0) = \sigma_c g^{\mu\nu}$, where

$$\sigma_c = t_c \omega_{pl}^2 \ .$$ (44)

Here $\omega_{pl}^2 = 4\pi \alpha_s T^2 (1 + N_f/6)/3$ is the color plasmon frequency[18, 15]. The important difference between our derivation and that in ref.[11, 2] is that the collision rate is derived here directly from the kinetic equation for color fluctuations rather than parameterized via the relaxation time approximation. We see in particular that the color diffusion time rather than momentum degradation time controls the magnitude of the conductivity. In addition, the same color diffusion rate applies for both quarks and gluons unlike in the general ansatz of [2]. Inserting the expression for t_c, we thus find that

$$\sigma_c \approx 2T/\log(m_E/m_M) \ .$$ (45)

We emhasize that in pertubation theory, $m_M = 0$ and hence $\sigma_c = 0$!! Thus the classical QGP is in fact an insulator and the name plasma is a misnomer. Only quantum, nonperturbative effects could salvage some of its plasma properties.

The appearance of the non-perturbative and non-classical magnetic mass in this final expression indicates that the classical ($k \to 0$) or high T limit of a QGP is in fact be very different from QED plasmas because of color diffusion. The long wavelength ($k \ll T$) color plasmon mode, obtained from the dispersion relation $k^2 = \Pi_L(k) = -iQ_{\mu\nu}(\sigma^{\mu\alpha\nu} - \sigma^{\mu\nu\alpha})k_\alpha$, are in fact strongly damped with a rate proportional to $1/t_c$ as can be seen from the general discussion in ref.[11, 2]. A finite damping rate requires going beyond the hard thermal loop approximation and introducing nonperturbative (quantal) effects such as the color magnetic mass or a self consistent approach as discussed in [8]. The nonperturbative damping rate, $\gamma \sim O(g^2 \log(1/g)T)$, of color plasmons is only $O(g \log(1/g))$ times smaller than their natural frequencies, $\omega_{pl} \sim O(gT)$. This is a remarkable difference with respect to abelian plasmas, where the divergent damping rates never enter at the level of semiclassical kinetic equations and collisional damping is controlled instead by the perturbatively smaller momentum relaxation rate, $1/t_p \sim O(g^4 \log(1/g))$. It will be interesting to compute the remnant observable consequences of color dynamics via the non-abelian Fokker-Planck transport equations for nuclear collisions. Possible consequences arise through the color conductive coupling between minijets and beam jets[25] and on the spectrum of soft induced gluon radiation associated with jet quenching[26].

Acknowledgments

We thank H.-Th. Elze, K.J. Eskola, U. Heinz, H. Heiselberg, St. Mrówczyński, R. Pisarski, S. Gavin and T. Ludlam for fruitful discussions.

References

[1] H.-Th. Elze and U. Heinz, Phys. Rep. 183 (1989) 81; U. Heinz, Phys. Rev. Lett., 51 (1983) 351.

[2] St. Mrówczyński, *Quark-Gluon Plasma*, Adv. Ser. on Directions in High En. Phys., Vol 6, ed. R. C.Hwa, (World Scientific, 1990, Singapore); Phys.Rev. D39 (1989) 1940; Phys. Lett. 188B (1987) 129.

[3] Proc. of Quark Matter 91, ed. F. Plasil, Nucl. Phys. A544 (1992) 1c.

[4] P. Danielewicz and M. Gyulassy, Phys. Rev. **D31** (1985) 53; A. Hosoya and K. Kajantie, Nucl. Phys. **B250** (1985) 666.

[5] G. Baym, et al. Phys. Rev. Lett. 64 (1990) 1867; H. Heiselberg and C.J. Pethick, NORDITA NBI-93-19, Les Houches 2-11 Feb.1993, Ed. P. Schuck, in press.

[6] M. Thoma, M. Gyulassy, Nucl. Phys. B351 (1991) 491; M. Gyulassy, et al. Nucl. Phys. A 538 (1992) 37c.

[7] A.V. Selikhov and M. Gyulassy, Phys.Lett. B316 (1993) 373.

[8] R.D. Pisarski, Phys.Rev.Lett. 63 (1989) 1129. R.D. Pisarski, BNL-P-1/92.

[9] K. Geiger and B. Muller, Nucl.Phys. B369 (1992) 600; Phys.Rev.D46 (1992) 4965.

[10] S.K. Wong, Nuovo Cim. 65A (1970) 689.

[11] U. Heinz, Ann. Phys. (N.Y.) 168 (1986) 148; 161 (1985) 48.

[12] A.V. Selikhov and M. Gyulassy, Columbia Univ. Preprint CU-TP-610/93; (Phys.Rev.Lett., submitted).

[13] A.V. Selikhov, Phys. Lett., B268 (1991) 263; (E) B285 (1992) 398; Kurchatov Inst. Preprint IAE-5526/1 (1992).

[14] N.G. van Kampen, Phys. Rep. C24 (1976) 171.

[15] E. Braaten, R. Pisarski, Phys.Rev.Lett. 64 (1990) 1338; Nucl.Phys. B337 (1990) 569.

[16] V.P. Silin, Sov.Phys. J.E.T.P. 11 (1960) 1136.

[17] H.B. Callen and T. A. Welton, Phys.Rev. 83 (1951) 34; W.B. Thompson and J. Hubbard, Rev.Mod.Phys. 32 (1960) 714; J. Hubbard Proc. Roy. Soc. A260 (1961) 114.

[18] H.A. Weldon, Phys. Rev., D26 (1982) 1394; V.V. Klimov, Sov. J. Nucl. Phys. 33 (1981) 934; E.V. Shuryak, JETP 47 (1978) 212.

[19] H. Heiselberg and C.J. Pethick, Phys. Rev. D47 (1993) R769; C.P. Burgess and A.L. Marini, *ibid.* 45 (1992) R17; A. Rebhan, *ibid.* 48 (1992) 482.

[20] V.V. Lebedev and A.V. Smilga, Ann.Phys. (N.Y.) 202 (1990) 229; T. Altherr, E. Petitgirard, T. del Rio Gaztellurutia, Phys. Rev. D47 (1993) 703; R. Kobes, G. Kunstatter, K. Mak, PRD45 (1990) 4632;

[21] H.-Th. Elze, M. Gyulassy and D. Vasak, Nucl. Phys. B276 (1986) 706; Phys. Lett. 177B (1986) 402;

[22] H.-Th. Elze, Z. Phys. 47 (1990) 647.

[23] St. Mrówczyński, Phys. Lett., B269 (1991) 383; E. Braaten and M.H. Thoma, Phys. Rev. D44 (1991) 1298; R2625.

[24] R.L. Liboff, Kinetic Theory (Englewood Cliffs, Prentice Hall, 1990).

[25] K.J. Eskola and M. Gyulassy, Phys.Rev. C47 (1993) 2329.

[26] X.N. Wang and M. Gyulassy, Phys. Rev. Lett.68 (1992) 1480.

COHERENT PION PRODUCTION IN NUCLEUS-NUCLEUS REACTIONS

Barbara Erazmus

Laboratoire de Physique Nucléaire de Nantes
2, rue de la Houssinière
44072 Nantes, France

INTRODUCTION

The coherent pion production process can be considered as the creation of virtual pions followed by their elastic scattering with the nucleus which is left in its ground state. The energy transferred during the collision is converted into one degree of freedom : the real pion.

Virtual pions are created through collective spin-isospin excitations in the projectile and target nuclei.The resonance in the pion-nucleon system is the Δ-isobar. The study of the Δ-isobar in the nucleus is closely related to the behavior of the pion in the nuclear medium. The interaction governing this type of excitations is an important component of the nucleon-nucleon force.

Even if the predicted signatures of the coherent process are very clear, its experimental identification is not easy, namely because of the very small cross section.

The search for coherent pion production in nucleus-nucleus collisions has been made in two energy domains : far below threshold at GANIL, in ^{12}C (^{12}C, ^{12}C) reaction at 95 MeV/nucleon and above threshold in ^{12}C (^{3}He,t) reaction at 2 GeV bombarding energy at SATURNE.

COHERENT SUBTHRESHOLD PION PRODUCTION

Subthreshold pion production in heavy-ion collisions requires a large overlap of the projectile and target density distributions [1-3], pions are therefore created preferably in central collisions. However, to research coherent phenomena, peripheral collisions are certainly more appropriate since the main features of coherent effects are char-

Hot and Dense Nuclear Matter, Edited by
W. Greiner *et al.*, Plenum Press, New York, 1994

acteristic of forward direction and they rapidly disappear with increasing transverse momentum transfer.

Brown and Deutchman [4] suggested that pion production in peripheral heavy-ion collisions can occur because of a coherent spin-isospin excitation through the Δ-hole channel in either projectile or target.

As an example, let us consider the reaction :

$$^{12}C +^{12} C \rightarrow^{12} C(1^+, T = 1; E^* = 15.1 MeV) + \pi^\circ + X. \tag{1}$$

The projectile (target) is excited to the $1^+T=1$ state through a Gamow-Teller transition ($\Delta S = \Delta T = 1$) decaying dominantly by the emission of a M1 photon of 15.1 MeV. Thus, the projectile (target) acts as a source of virtual pions scattering elastically on the target (projectile) nucleus and exciting a Δ-hole state which decays by emitting a pion.[5]

The full identification of the process can be done by a coincidence measurement of the ejectile (^{12}C), the 15.1-MeV photon and the two photons from the π° decay.

An experiment to study this process has been performed at GANIL at 95 MeV/nucleon [6]. The ejectile has been analysed by the SPEG spectrometer which allows the full identification of charge, mass and energy. A set of 30 BaF$_2$ detectors was used for simultaneous measurement of the 15.1-MeV photon and of the photons from the π° decay.

By way of preliminary, only one high-energy gamma, assumed to originate from the π° decay, has been measured giving an upper limit of the process under investigation.

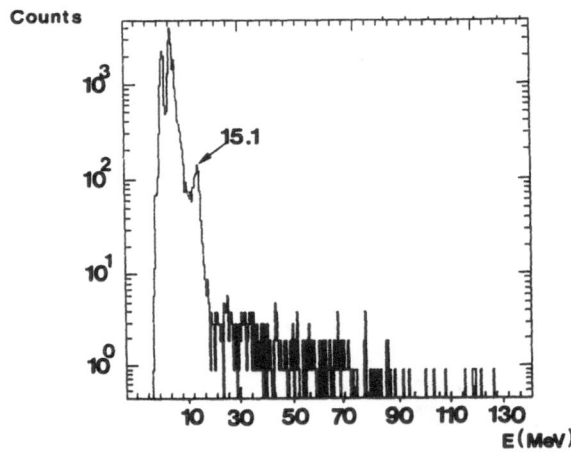

Figure 1. Energy spectrum of photons detected in coincidence with ^{12}C ejectile.

The energy spectrum of photons, triggered by the kinematical conditions imposed to the ^{12}C ejectile, clearly shows a peak around 15.1 MeV and corresponding to the M1 transition in ^{12}C (Fig.1). The estimated upper limit for the cross section for coherent π° production in the ^{12}C[^{12}C, ^{12}C(1^+, T=1 ; E*=15.1MeV)] reaction at 95 MeV/nucleon is 60 nb. This result is consistent with theoretical predictions [7] giving a range of values from 60 to 110 nb depending on the choice of the NN-NΔ interaction.

However, it should be emphasized that at subthreshold energy, where the transferred momenta are as large as 2-4 fm^{-1}, the theoretical predictions may contain large uncertainties. The incident projectile wave function is distorted and attenuated in the nuclear medium apart from the pionic interactions. Since the projectile acts as a source of the virtual pion beam this latter is also attenuated. These distortion effects are difficult to calculate at high momentum where nuclear wave functions are not well known.

COHERENT PIONS IN CHARGE-EXCHANGE REACTIONS ABOVE THRESHOLD

At higher energies, far above threshold, theoretical calculations are more accurate and coherent pion production might be easier to observe. Interesting data come from recent experiments performed at SATURNE using a $^{12}C(^3He, t\,\pi^+)$ reaction at 2 GeV [8].

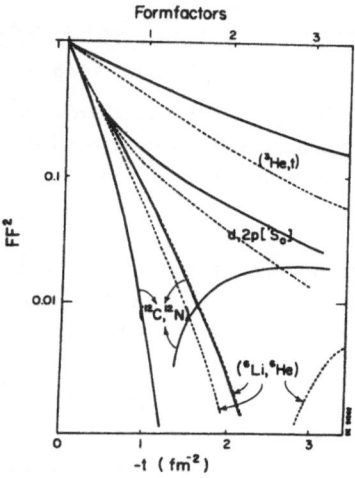

Figure 2. The spin-longitudinal (solid) and spin-transverse (dashed) from factors squared (FF^2) for several projectile-ejectile systems [9].

This reaction is an especially interesting probe since the (^{3}He,t) form factor enhances very much the spin-longitudinal excitations, the pion-like modes [9] (Fig.2). By the way, as we can see in Fig.2, this kind of enhancement is also expected for ($^{12}C,^{12}N$) (or ($^{12}C,^{12}C$)) system as far as large momenta are concerned. Thus, one of the two responses, spin-longitudinal or spin-transverse, can be emphasized by proper choice of projectile-ejectile systems.

The charge-exchange transition of the projectile is a source of virtual pions transferred to the target. The virtual pions propagate through the nucleus exciting Δ-hole states which decay by emitting a real pion.

In the SATURNE experiment the tritons were analysed with a dipole magnet and two sets of drift chambers. The Diogene detector allowed the identification of charged pions and the measurement of their vector momenta.

For events corresponding to the transferred energy $w \sim 250$ MeV, the angular correlation between the transferred momentum $\vec{q}$ and the momentum $\vec{p}_\pi$ of the pion has been examined. In the reaction probing the spin-longitudinal pionic modes with a $\vec{S}.\vec{q}$ spin structure, the angular correlation is, in principle, characterized by a $(\vec{S}^+.\vec{p}_\pi)(\vec{S}.\vec{q})$, i.e. a $(\vec{p}_\pi.\vec{q})$ dependence on the momentum vectors.

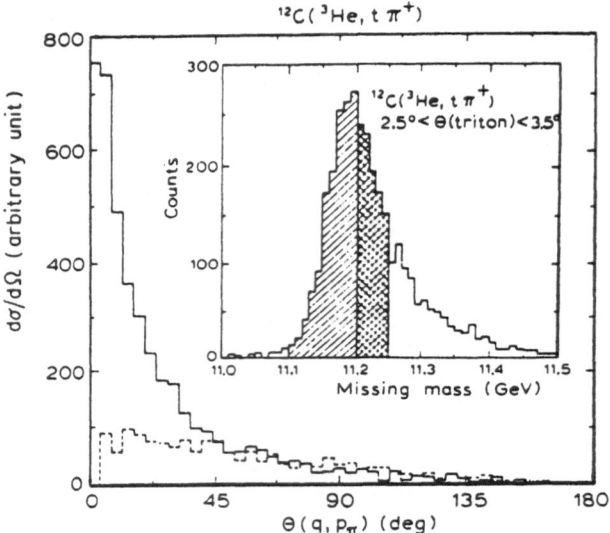

Figure 3. The angular correlation between the transferred momentum $\vec{q}$ and the momentum of pions for different gates in the missing mass spectrum indicated in the insert. The gate around the ground state includes events in which missing mass is less than 11.2 GeV.[8].

In fact, a strong angular correlation has been observed for events in which the ^{12}C target is left in its ground state or low-lying states (Fig.3). It seems that the pion is emitted in the direction of the transferred quantum. Moreover, the virtual pion is not much off-shell and the target recoil energy is of the order of 1 MeV.

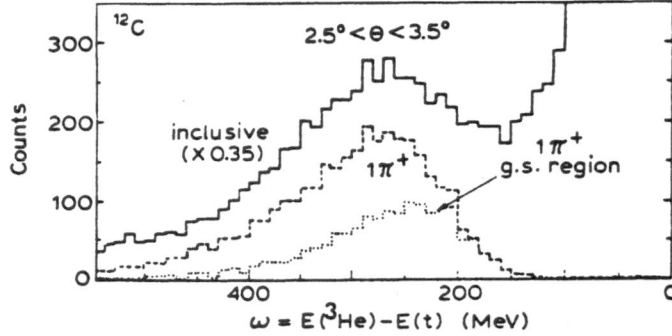

Figure 4. Energy transfer spectra for different channels [8].

The energy-transfer spectra show that, for the events with the target left in its ground state,the Δ-peak is shifted ($w \sim 235$ MeV) in comparison with inclusive data (Fig.4). In general, the downward shift of the Δ-peak for nuclear target is attributed to Δ-hole correlations in the spin-longitudinal channel [9] reflecting the Δ-hole interaction in the medium.

The cross section for coherent pion production has been estimated to be about 7% of the inclusive ^{12}C(He,t) cross section.The width of the peak in the energy-transfer spectra for the coherent pions as well as the observed shift are reasonably well reproduced by theoretical calculations[10].However,the predicted cross section are 2 to 3 times larger than those estimated from the experimental data.

It should be emphasized that the finite resolution in the missing mass (FWHM $\sim$ 25MeV) does not allow the clear selection of the ground state of the target. Furthermore, due to the DIOGENE acceptance, a significant part of the pion events in which the target is left in a bound state is lost.

Therefore, the pion angular correlation cannot be determined for small triton angles ($\theta_t < 2.5°$) and a precise estimation of the coherent pion production cross section cannot yet be done.

COHERENT PIONS PRODUCED NEAR THRESHOLD

Existing data come from two experiments performed far below and above pion production threshold. There is however an interesting intermediate region near threshold where the interpretation of data is easier than at 100 MeV/nucleon and where some of the interesting effects, observed at higher energy, might be enhanced. In fact, Deutchman[11] has recently demonstrated that a coherent addition of Δ-hole states increases the magnitude of the form factor,i.e. of the cross section, particularly at large momentum transfer values (Fig.5). Furthermore, the shift of the Δ-peak down in energy is expected to increase with momentum transfer [9].

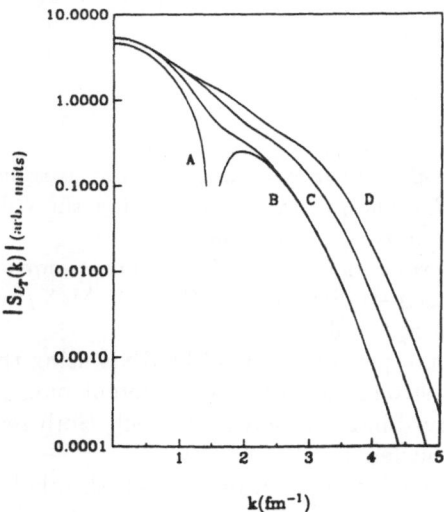

Figure 5. Magnitudes of the coherent form factor as various Δ-hole states in ^{12}C are induced.[11] A corresponds to the $(1p_\Delta) (1p)^{-1}$ state only. B corresponds to the to A plus the $(1s_\Delta) (1s)^{-1}$ state. C corresponds to B plus the $(2p_\Delta) (1p)^{-1}$ and $(2s_\Delta) (1s)^{-1}$ states.D corresponds to C plus the $(3p_\Delta) (1p)^{-1}$ and $(3s_\Delta) (1s)^{-1}$

However, one should be aware of distortion effects which could complicate the analysis. For this reason a new measurement has to be as complete as possible. This implies low thresholds, good accuracy and large solid angles for the detection of the reaction products.

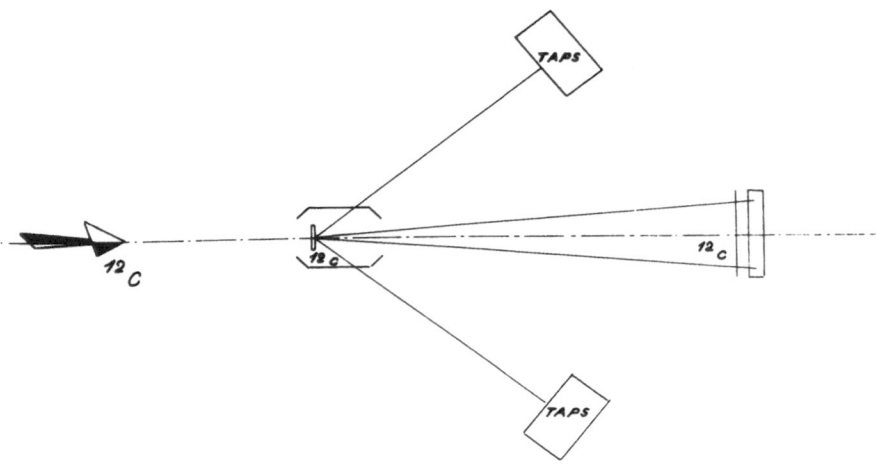

Figure 6. Scheme of the experimental arrangement

As far as the ^{12}C[^{12}C,^{12}C(1^{+},T=1 ; 15.1 MeV), $\pi°$] reaction is concerned, it can be study at SIS at 200-400 MeV/nucleon. TAPS, a large-solid-angle detector of photons is particularly well suited for the registration of the $\pi°$ and of the 15.1-MeV photon. In order to be complete, the measurement should include the identification of the projectile and of the target (Fig.6). The study of the coherence in both nuclei is possible using the additional information given by the Doppler shift of the 15.1-MeV photon.

CONCLUSIONS

Coherent pions are produced through collective spin-isospin excitations in projectile and target nuclei. The energy transferred during the collision is almost totaly converted into one degree of freedom, the pion.

Results from two experiments searching for coherent pions are available. The ^{12}C (^{12}C, ^{12}C) reaction has been studied at GANIL at 95 MeV/nucleon giving an upper limit for the cross section $\sigma_{\pi°}^{coh} \leq 60nb$.

The recent measurement performed at SATURNE using the $^{12}C(^{3}He, t)$ reaction at 2 GeV clearly show a strong indication for coherent pion production and a large shift of the Δ-peak observed in the coherent channel. Both reactions probe the spin-longitudinal response of nuclei.

In the intermediate region, near pion production threshold, the distortion effects are less important than at low energy and moreover, the coherence effects might be enhanced.

Acknowledgments

I would like to express my thanks to C. Guet, T. Hennino, T.Reposeur, C.Lebrun, G.Paic, D.Nouais ,TAPS and KAOS collaborations for numerous discussions on this subject.

I thank E.Gerbaud for the help with preparation of the manuscript, and K.Chawoshi and M.Rio for additional assistance.

REFERENCES

- 1. J. Miller et al, Phys. Rev. Lett. 58, 2408 (1997).

- 2. S. Aiello et al, Europhys. Lett. G, 25 (1988).

- 3. B. Erazmus et al, Nucl. Phys. A 481, 821 (1988).

- 4. G. Brown and P.A. Deutchman, in Proceedings of the Workshop on High Resolution Heavy Ion Physics at 20-100 MeV/A, Saclay, May 31- June 2, 1978, p. 212.

- 5. A. H. Blin, C.Guet and B.Hiller , Nucl. Phys. A 454, 746 (1986).

- 6. B. Erazmus et al, Phys. Rev. C 44, 1212 (1991)

- 7. C. Guet, M. Soyeur, J. Bowlin and G.E. Brown, Nucl. Phys. A 494, 558 (1989)

- 8. T. Hennino et al, Phys. Lett. B 303, 236 (1993)
  M. Roy-Stéphan,Nucl.Phys. A 553,209 (1993)

- 9. C. Gaarde, Annu. Rev. Nucl. Part. Sci. 41, 187 (1991) and references therein

- 10. P. Hernandez de Cordoba, J. Nieves, E. Oset and M. J. Vicente-Vacas, preprint University of Valencia (1992).

- 11. P.A. Deutchman, Phys. Rev. C 47, 2794 (1993).

AZIMUTHALLY ANISOTROPIC EMISSION OF PIONS IN SYMMETRIC HEAVY-ION COLLISIONS

Dieter Brill
for the KaoS-Collaboration[1]

GSI
Postfach 11 05 52
D-64220 Darmstadt

INTRODUCTION

In several experiments the emission pattern of nucleons with respect to the reaction plane has been measured. A sidewards deflection of the spectator fragments ("bounce off") was found as well as directed flow of nucleons from the overlap region between the colliding nuclei (participants) in the reaction plane ("side-splash")[2, 3, 4, 5]. Additionally, an azimuthally anisotropic emission peaked perpendicular to the event plane is observed for protons, neutrons and light nuclei ("squeeze-out")[6, 7]. In addition to the participant and spectator nucleons, newly created hadrons, predominantly pions, are observed in relativistic nucleus-nucleus collisions. In contrast to case of the nucleons, no data were available so far for the emission pattern of pions in symmetric heavy mass systems. We studied therefore the correlation between the momentum vectors of pions and the reaction plane in ^{197}Au+^{197}Au collisions at a beam energy of 1 GeV/u[8, 9].

EXPERIMENT

The experiment was performed at the heavy ion synchrotron (SIS) facility at GSI using the KaoS magnetic spectrometer [10]. This spectrometer is a double-focussing QD configuration with wide momentum acceptance. Particle identification is achieved by measuring momentum and velocity. The momentum is determined from the focal

Hot and Dense Nuclear Matter, Edited by
W. Greiner *et al.*, Plenum Press, New York, 1994

plane position of the particle, whereas the velocity is deduced from time of flight (Fig. 1). The spectrometer can be moved around the target. It was set to a polar angle of $\Theta_{Lab}=44^0$. The horizontal acceptance is about $\Delta\Theta=\pm 4^0$, the vertical acceptance depends on the momentum of the particles and is about $\Delta\phi=\pm(5^0\text{-}9^0)$.

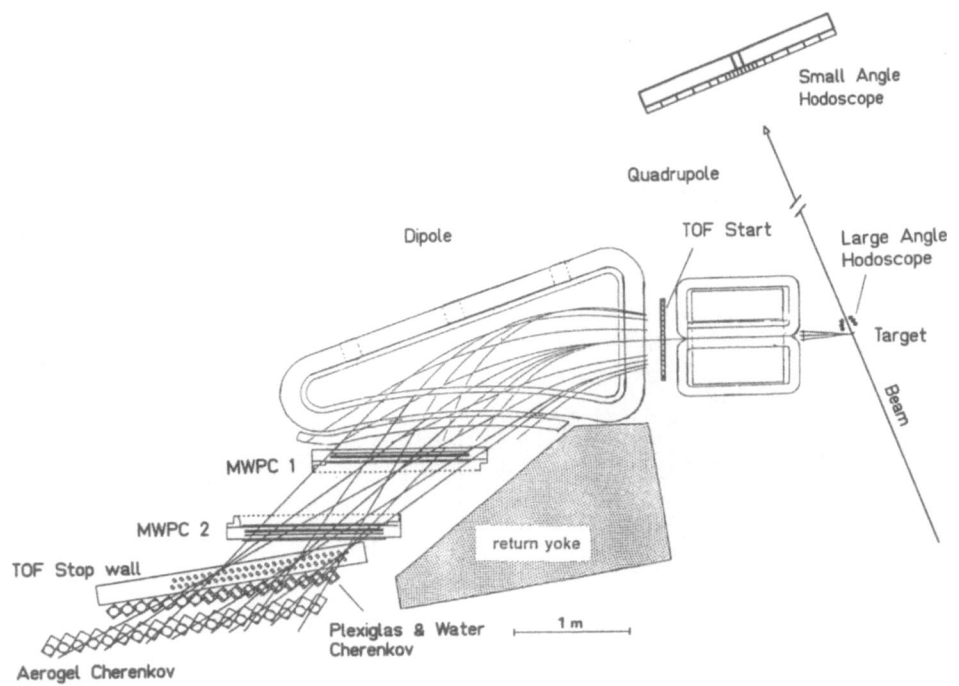

Figure 1. Experimental setup of the magnetic spectrometer KaoS with its detector system: time-of-flight (TOF) start and stop detectors, multiwire proportional chambers (MWPC), Cherenkov detectors with water, lucite and silica aerogel radiators, two hodoscopes for event characterization at target and 7 m downstream (Small Angle Hodoscope).

Two scintillator detector arrays external to the spectrometer provide information to characterize the event by global variables. The Small Angle Hodoscope (Fig. 2), is located 7 meters downstream of the target. This array is composed of 380 modules and covers polar angles $0.5^0 < \Theta_{Lab} < 11^0$. Most of the particles emitted in this angular range are spectator nucleons. This detector system provides information on position, charge and time of flight. Projectile spectators are identified by the ΔE signal in the scintillator modules and by time of flight. The position information of the particles is used to reconstruct the event plane. The second external detector, the Targethodoscope, gives information on the particle multiplicity which is a measure of the impact parameter. This detector consists of 96 modules and accepts charged particles in the polar angular range $12^0 < \Theta_{Lab} < 48^0$. Particles detected in this angular range are predominantly participating protons and charged pions. The distance between the target and the detector modules is about 10 cm. This is too short to measure the velocity of the particles. The time information averaged over the scintillators yielding a signal, is a good approximation of the time when the reaction took place. This information is used as start signal for the time-of-flight measurement from the target to the Small Angle Hodoscope.

The determination of the centrality of the reaction is the first step in a global analysis of heavy-ion collisions. The impact parameter distribution is studied using the correlation of the particle multiplicity in the Targethodoscope and the summed nuclear charge Z_{sum} of spectator fragments in the Small Angle Hodoscope [10, 11]. Fig. 3 shows the multiplicity distribution of charged particles measured with the Targethodoscope. Note that the distribution increases for higher multiplicities since the events are triggered on charged particles in the spectrometer. The peripheral collisions are thereby strongly suppressed by the trigger condition. To classify the reaction centrality we have divided the multiplicity distribution into five bins. MUL1 corresponds to the most peripheral, MUL5 to the most central collisions.

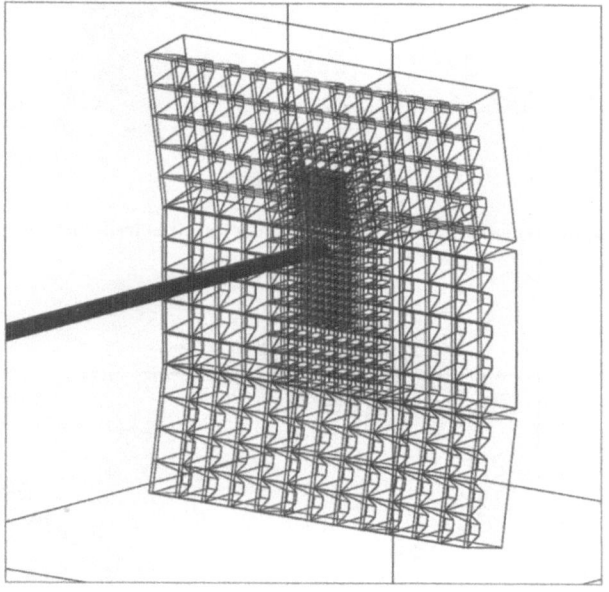

Figure 2. Sketch of the Small Angle Hodoscope. Four modules in the center of the wall are removed for a beam pipe made of carbon fibre tube (diameter 72 mm, thickness 0.85 mm).

For the reconstruction of the event plane the transverse momentum method is used[2]. This method yields a reaction plane for each event defined by a vector $\vec{Q}$ which is the vector sum of the transverse momenta of all spectator particles observed in the Small Angle Hodoscope. The accuracy in this determination of the event plane is estimated by subdividing each event randomly into two subevents and by subsequently evaluating the correlation of these two subevents. The uncertainty obtained by this method as a function of normalized multiplicity is shown in Fig.4. The reconstructed azimuthal angle of the event plane has a dispersion of $\Delta\phi = 20\text{-}25°$ for semi-central collisions. For peripheral collisions the dispersion increases due to the low multiplicity of detected particles in the Small Angle Hodoscope. For central collisions the dispersion increases due to the intrinsic azimuthal symmetry of the reaction geometry.

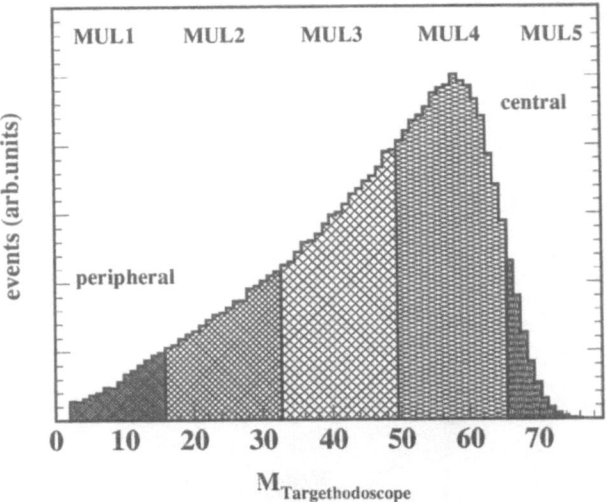

Figure 3. Multiplicity distribution of charged particles measured with the Targethodoscope.

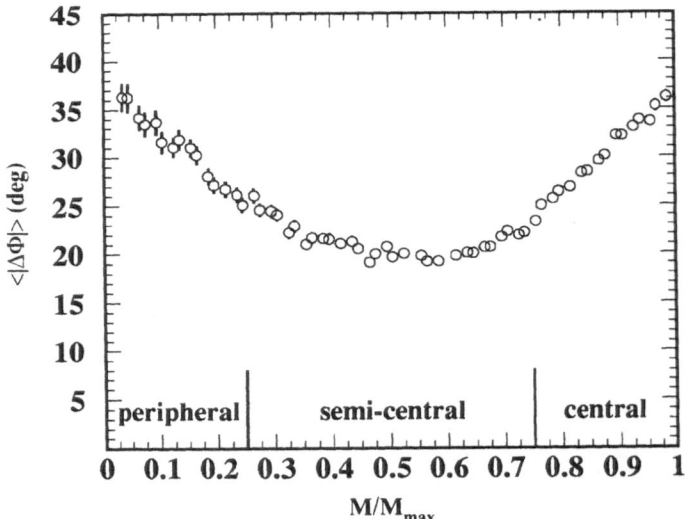

Figure 4. Mean value of the absolute azimuthal angle difference between the measured and the true reaction plane as function of the normalized multiplicity of the Targethodoscope.

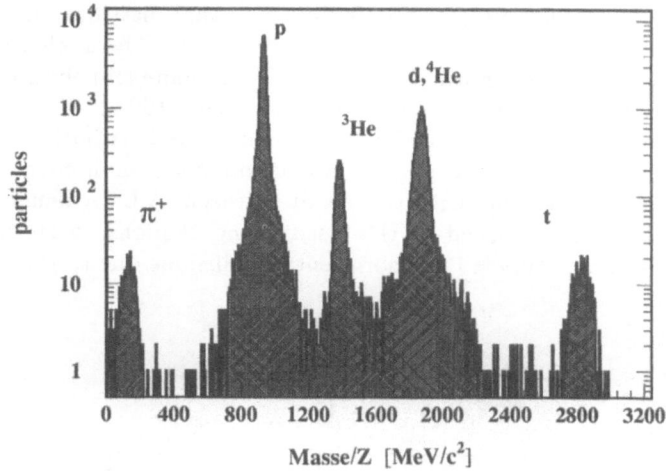

Figure 5. Spectrum of the mass/Z-ratio of particles detected in the spectrometer.

RESULTS

Fig. 5 shows a mass/Z-spectrum of particles detected in the spectrometer. One can identify pions, protons, deuterons, tritons and He-isotopes. The deuterons and ^{4}He can be separated by their different energy loss in the stop detector. The particles shown in Fig. 5 originate in different regions in phase space as can be seen in Fig. 6. The higher the mass of a fixed Z the larger is the shift to target rapidity. Most of the particles measured stem from the backward hemisphere. Only the pions are measured in the forward hemisphere: $0.55 < y/y_{proj} < 0.75$.

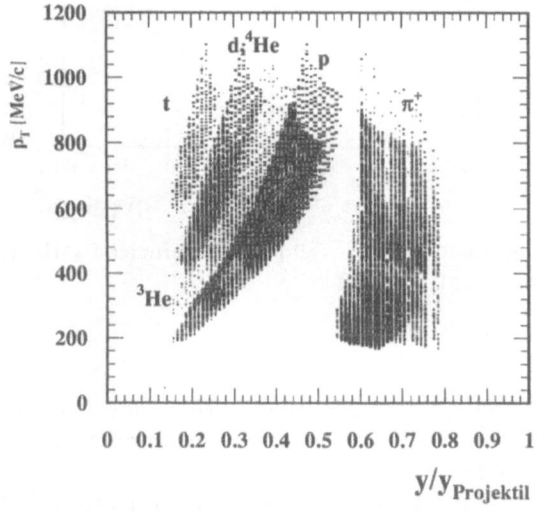

Figure 6. Phase space distribution of particles detected in the spectrometer.

Fig. 7 shows for different centralities the azimuthal distributions of the vector $\vec{Q}$ for events in which a π^+ is observed in the spectrometer. The angle shown is the relative azimuthal angle between the measured reaction plane and the emitted pion. The left hand side covers the transverse momentum range $160 < p_T < 260\ MeV/c$, the right hand side $260 < p_T < 600\ MeV/c$. The azimuthal distribution of the peripheral collisions show a strong anticorrelated flow component for high momentum pions. This behaviour is consistent with the result of recent VUU-calculations predicting anticorrelated pion flow caused by the rescattering of pions on spectator matter[12]. The anticorrelation disappears for more central collisions due to the decreasing number of spectators.

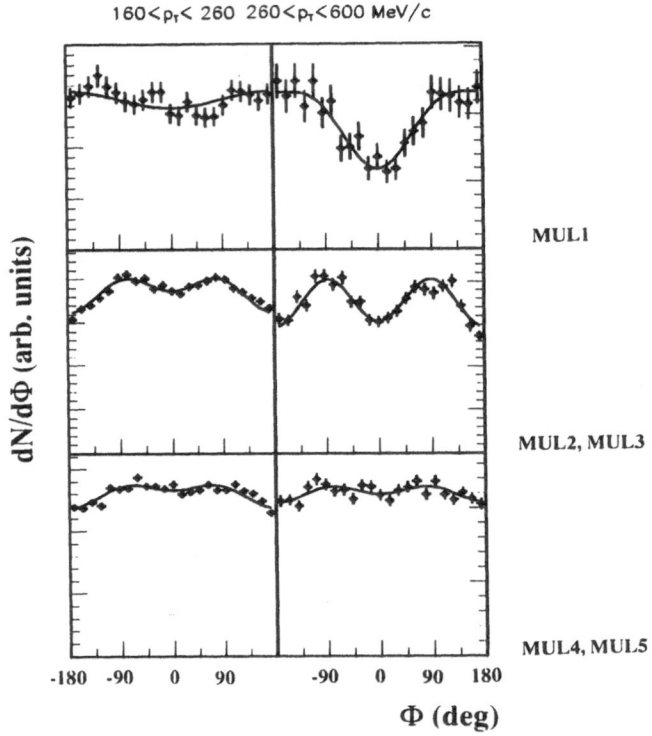

Figure 7. Azimuthal distribution of the vector Q for peripheral (MUL1), semi-central (MUL2, MUL3) and central collisions (MUL4, MUL5).

At semi-central impact parameters the distributions show maxima at $\pm 90^o$ with respect to the event plane. This pattern represents a preferred emission of pions out of the reaction plane (squeeze-out). For central collisions the effect diminishes due to the azimuthal symmetry of the reaction geometry (shown in the bottom part of Fig.7). The distributions can be parametrized by the expression $N(\varphi) \sim 1 + P_1 \cos(\varphi) + P_2 \cos(2\varphi)$. The parameter P_2 reflects the out-of-plane emission and is a measure of the strength of the squeeze-out effect. The value of P_2 is used to quantify the ratio

R of the number of pions emitted perpendicular to the reaction plane to the number of pions emitted in the reaction plane:

$$R = \frac{N(90^\circ) + N(-90^\circ)}{N(0^\circ) + N(180^\circ)} = \frac{1 - P_2}{1 + P_2} \tag{1}$$

A value of R larger than unity implies preferred out-of-plane emission. The dependence of R on the pion transverse momentum is shown in Fig. 8 for semi-central collisions. The value of R is larger than one for all transverse momenta and shows an approximate linear dependence on transverse momentum. The solid line in Fig. 8 represents a linear fit. It is important to note that the fitted values of P_2 are affected by the uncertainty in determining the reaction plane. Fig. 9 shows the corresponding plot for the protons. The dependence on the transverse momentum is very similar to pions.

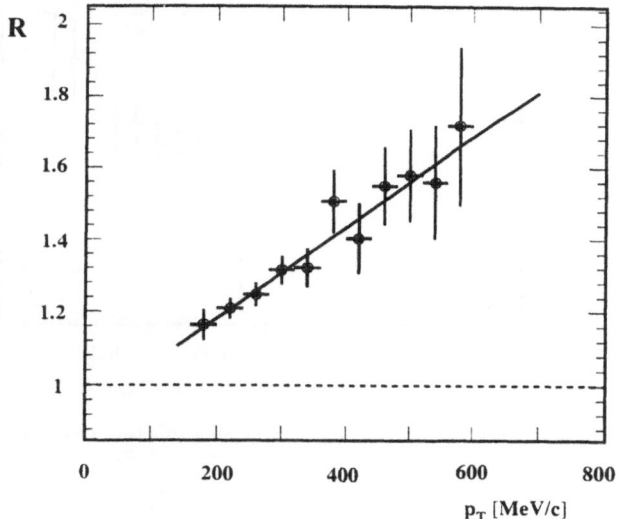

Figure 8. The out-of-plane/in-plane ratio R for π^+ as a function of transverse momentum for semi-central collisions. The dashed line at R=1 indicates no preferred out-of-plane emission.

The anisotropy could be a consequence of the dynamical squeeze-out of the nucleons. In this energy range most of the pions are decay products of the Δ-resonances. These Δ-resonances are expected to flow with the off-plane squeeze-out of the nucleons and therefore similar anisotropy effects could be exhibited. On the other hand one has to take into account the decay kinematics of $\Delta \to N\pi$. This decay kinematics should change the pion emission angle because of the small mass of the pions compared with the Q-value of the decay. Another more obvious interpretation is that the anisotropy is due to shadowing effects caused by absorption and scattering of pions via the channels: $N\pi \rightleftharpoons \Delta$ and $\Delta N \rightleftharpoons NN$. These processes are responsible for the short mean free path of the pions in nuclear matter. The probability of an interaction in the nuclear matter depends on the geometry of the collision. The chance of absorption and scattering is higher in-plane than out-of-plane due to the spectator matter located in the reaction plane. The absorption of pions would imply a reduced total number of pions. On the other hand scattering would lead to similar anisotropy effects. The scattering of pions occurs predominantly in the reaction plane whereas the subsequent reemission is isotropic. In this interpretation the pion yield is just removed from the reaction plane and the total pion multiplicity stays constant. Recent calculations with microscopic modells [13, 14] predict similar asymmetries in the pion emission pattern as observed in our experiment.

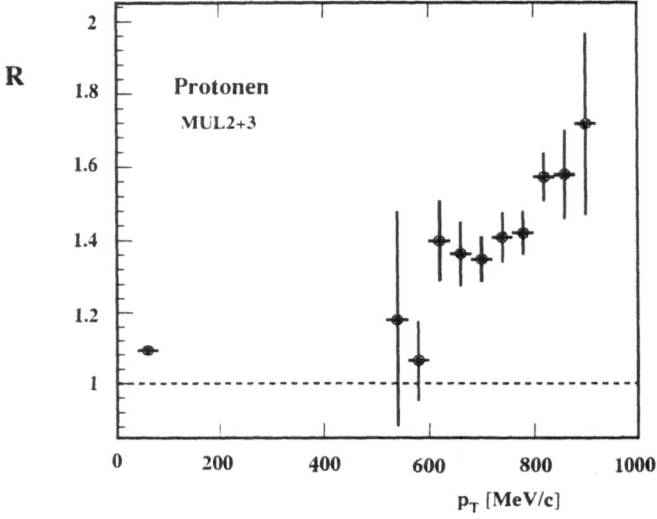

Figure 9. The out-of-plane/in-plane ratio R for protons as a function of transverse momentum for semi-central collisions.

SUMMARY

In summary, we have determined the correlation between the reaction plane and the momentum vectors of charged particles emitted in relativistic heavy ion collisions. An enhanced emission of charged pions perpendicular to the reaction plane is observed. The ratio of pions emitted out-of-plane/in-plane increases very similar to the protons as a function of transverse momentum. The anisotropic emission of the pions is most prominent for the high energy pions. Microscopic models predict that these pions are created in an early state of the collision. The pions emitted perpendicular to the reaction plane can leave the reaction zone nearly undisturbed. Therefore they might serve as a new probe of the hot and dense matter in heavy-ion collisions.

REFERENCES

1. KaoS-Collaboration: W. Ahner, P. Baltes, R. Barth, P. Beckerle, C. Bormann, D. Brill, M. Cieslak, M. Debowski, E. Grosse, W. Henning, P. Koczon, B. Kohlmeyer, D. Miskowiec, C. Müntz, H. Oeschler, H. Pöppl, F. Pühlhofer, R. Schicker, E. Schwab, P. Senger, Y. Shin, J. Speer, J. Stein, R. Stock, H. Ströbele, C. Sturm, K. Völkel, A. Wagner, W. Walus.

2. P. Danielewicz and G. Odyniec, Phys. Lett. **157B**, 147 (1985).

3. H. A. Gustafsson et al., Phys. Rev. Lett. **52**, 1590 (1984).

4. R. E. Renfordt et al., Phys. Rev. Lett. **53**, 763 (1984).

5. D. Beavis et al., Phys. Rev. **C33**, 1113 (1986).
   K. G. R. Doss et al., Phys. Rev. Lett. **59**, 2720 (1987).
   P. Danielewicz, H. Ströbele, G. Odyniec et al., Phys. Rev. **C38**, 120 (1988).
   M. Demoulins et al., Phys. Lett. **B241**, 476 (1990).

6. H. H. Gutbrod, K. H. Kampert, B. W. Kolb, A. M. Poskanzer, H. G. Ritter, R. Schicker, H. R. Schmidt, Phys. Rev. **C42**, 640 (1990).

7. Y. Leifels et al., Phys. Rev. Lett. **71**, (1993).

8. D. Brill, Ph. D. thesis, Uni. Frankfurt/M., GSI Report 1993, No. 93-36, (ISSN 0171-4546)

9. D. Brill et al., Phys. Rev. Lett. **71**, 336 (1993).

10. P. Senger et al., Nucl. Instrum. and Meth. **A327**, 393 (1993).

11. W. Ahner et al., Z. Phys. **A341**, 123 (1991).

12. S. A. Bass, C. Hartnack, R. Mattiello, H. Stöcker, W. Greiner, Phys. Lett. **B302**, 381 (1993).

13. S. A. Bass, C. Hartnack, H. Stöcker, W. Greiner Phys. Rev. Lett. **71**, 1144 (1993).

14. B. A. Li, Phys. Lett. **B**, (1993) in press.
    B. A. Li, Nucl. Phys. **A**, (1993) in press.

RESULTS FROM THE EOS TIME PROJECTION CHAMBER AT THE BEVALAC

D. Keane[2] for the EOS Collaboration:
S. Albergo[6], F. Bieser[1], F. P. Brady[4], Z. Caccia[10], D. A. Cebra[4],
A. D. Chacon[5], J. L. Chance[4], Y. Choi[9], S. Costa[6], J. Elliott[3],
M. Gilkes[3], J. A. Hauger[3], A. Hirsch[3], E. L. Hjort[3], A. Insolia[6],
M. Justice[2], V. Lindenstruth[7], H. S. Matis[1], M. McMahan[1],
C. McParland[1], W. F. J. Mueller[7], D. L. Olson[1], M. Partlan[4],
N. Porile[3], R. Potenza[6], G. Rai[1], J. Rasmussen[1], H. G. Ritter[1],
J. Romanski[6], J. L. Romero[4], G. V. Russo[6], H. Sann[7], R. Scharenberg[3],
A. Scott[2], Y. Shao[2], B. Srivastava[3], T. J. M. Symons[1], M. Tincknell[3],
C. Tuvè[6], S. Wang[8], P. Warren[3], H. H. Wieman[1], and K. L. Wolf[5]

[1]Lawrence Berkeley Laboratory, Berkeley, California 94720
[2]Kent State University, Kent, Ohio 44242
[3]Purdue University, West Lafayette, Indiana 47907
[4]University of California, Davis, California 95616
[5]Texas A&M University, College Station, Texas 77843
[6]Università di Catania and INFN-Sezione di Catania, 95129 Catania, Italy
[7]Gesellschaft für Schwerionenforschung, D-64220 Darmstadt 11, Germany
[8]Harbin Institute of Technology, Harbin 150006, P. R. China
[9]Sung Kwun Kwan University, Suwon 440-746, Republic of Korea
[10]Centro Siciliano di Fisica Nucleare e Struttura della Materia, 95129 Catania, Italy

INTRODUCTION

Experiments at the Bevalac using the Plastic Ball and streamer chamber initiated the study of nucleus-nucleus collisions with close to 4π acceptance for charged particles. [1] Much was learned as a result of these experiments, and by the mid 1980s, there was a widespread belief that it would soon be possible to constrain unambiguously the equation of state of the compressed hadron gas that exists for a few fm/c during the early stage of the collision. During the years since then, we have come to a better understanding of the difficulties involved in inferring properties of the nuclear equation of state, and while these difficulties are surmountable, there are some indications that experimental observables are less strongly influenced by the state variables of interest than was initially assumed. [2, 3] Such considerations motivate new measurements with good statistics and the smallest possible observational biases.

Following the relocation of the Plastic Ball detector to CERN in 1986, it was clear that a modern 4π detector was needed to build on the progress made during the earlier phase of the Bevalac program. The limitations of the previous measurements arose from the fact that the Plastic Ball detector has a complex acceptance that is not easily simulated, and the streamer chamber has limited capabilities for particle identification and provided relatively poor statistics; the EOS Time Projection Chamber, with its simple and seamless acceptance, good particle identification and high statistics, was designed to overcome these limitations.

THE EOS DETECTOR

The EOS TPC has a rectangular geometry, and operates in a 1.3 T dipole field provided by a superconducting magnet at the Bevalac's Heavy Ion Spectrometer System (HISS) facility. Unlike previous TPCs, EOS relies solely on pads for readout. The pad plane covers an area of 1.54×0.96 m^2, with 128 pad rows along the longer dimension, and 120 pads per row. Details about the chamber, the electronics and the data acquisition have been reported previously. [4, 5]

The standard EOS detector configuration, illustrated in Fig. 1, includes the TPC, a multiple sampling ionization chamber (MUSIC II) positioned to intercept projectile spectator fragments, an array of scintillator slats to provide time-of-flight information at small polar angles, and a high efficiency neutron detector (MUFFINS) centered on zero degrees with respect to neutrons emitted from the target. The EOS TPC became operational during the first half of 1992, and the main data-taking phase took place between July and December, 1992; shortly thereafter (February 1993), the Bevalac was permanently removed from service. Fig. 2 depicts the number of events recorded for various combinations of projectile, target and beam energy.

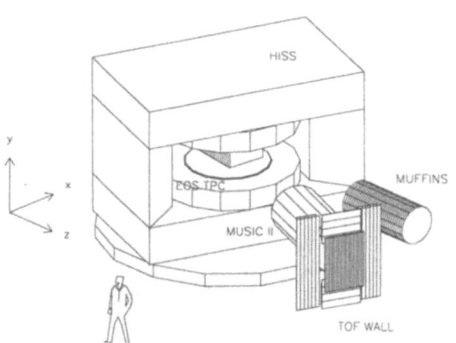

Figure 1: A perspective view showing the relative position of the EOS detector subsystems: the TPC centered in the HISS dipole magnet, the MUSIC ionization chamber, the Time-of-flight wall, and the MUFFINS neutron spectrometer.

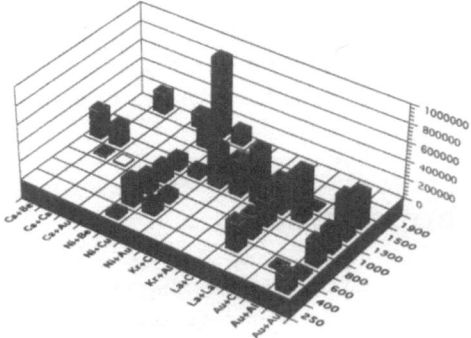

Figure 2: Manhattan plot showing the number of events recorded on tape for various combinations of projectile, target and beam energy.

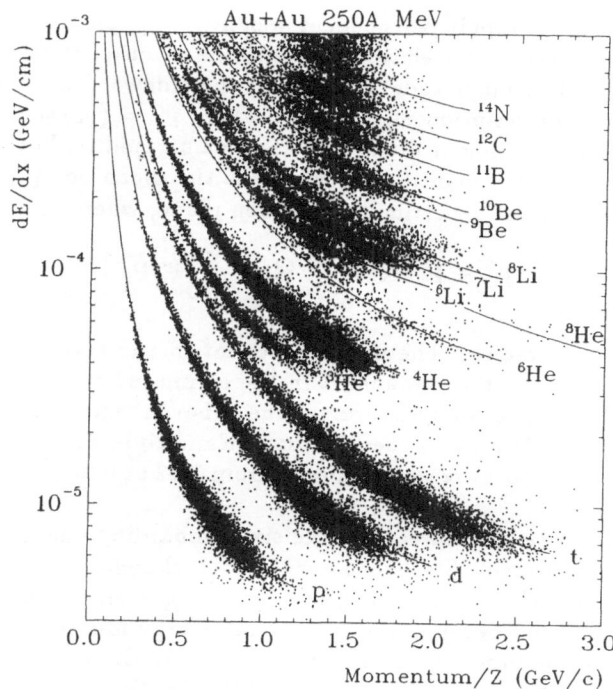

Figure 3: Fragment energy loss versus momentum per charge.

Fig. 3 illustrates the particle identification capabilities of the EOS TPC, based on truncated mean energy loss dE/dx and momentum per charge. This plot includes fragments with 70 or more usable samples per track, emitted from Au + Au collisions at $250A$ MeV. Many isotopes up to ^{10}Be can be identified, while charge resolution up to at least nitrogen $(Z = 7)$ is possible. The MUSIC detector provides good charge resolution for all species above $Z \approx 6$.

Preliminary EOS results on nuclear fragmentation are reported elsewhere in these proceedings. In this article, we report preliminary results on sideward flow in collisions of Au + Au at beam energies of 250, 400, 600, 800, 1000 and $1150A$ MeV. The sample size for each beam energy corresponds to typically one-tenth of the total statistics depicted in Fig. 2. Only data from the TPC have been used in the current analysis.

IN-PLANE TRANSVERSE MOMENTUM

To facilitate comparisons with previous work, our initial investigation of collective effects in Au + Au collisions includes the same in-plane transverse momentum analysis with essentially the same data selection criteria as used by the Plastic Ball group.[6, 1] In particular, all nuclear fragment species up to ^{4}He are included, and we select an interval of multiplicity centered about the value where the flow has its maximum. This multiplicity interval corresponds to baryon multiplicities $0.6M^{max} < M < 0.9M^{max}$, where M^{max} is a value near the upper limit of the M spectrum where the height of the distribution has fallen to half its plateau value. The in-plane transverse momentum method[7] involves estimating the orientation of the reaction plane for each event using the vector $\mathbf{Q} = \sum_{\nu=1}^{M} w(y_\nu)\mathbf{p}_\nu^\perp$, where $\mathbf{p}_\nu^\perp$ is the transverse momentum for the νth baryon track, $w(y_\nu)$ is a rapidity-dependent weighting factor, and M is the observed multiplicity. To optimize the correlation of $\mathbf{Q}$ with the reaction plane, $w(y_\nu)$ should vary according to the relative magnitude and sign of the sideward deflection of fragments at y_ν; we follow the prescription $w(y'_\nu) = y'_\nu$, where $y'_\nu = (y_\nu/y_{beam})_{cm}$ is defined as the rapidity of the νth fragment divided by the beam rapidity, both evaluated in the center of mass frame. The quantity $\langle p^{x'}(y')/A\rangle$ is the mean component of transverse momentum per nucleon in this estimated reaction plane, where

$$p_\nu^{x'} = \mathbf{p}_\nu \cdot \hat{\mathbf{Q}}_\nu\,, \qquad \mathbf{Q}_\nu = \sum_{\mu \neq \nu} w_\mu \mathbf{P}_\mu^\perp\,, \tag{1}$$

and $\hat{\mathbf{Q}}$ denotes a unit vector. The flow component in the true reaction plane, p^x, is systematically larger than the component in the estimated plane, $p^{x'}$, and for consistency with the Plastic Ball analysis, we use the subevent method [7, 6] to correct for this dispersion effect. Because of the high baryon multiplicity in Au + Au collisions, dispersion correction factors are small, ranging from 1.12 at $0.25A$ GeV to 1.07 at $1.15A$ GeV.

Fig. 4 presents $\langle p^x(y')/A\rangle$ at each of the six bombarding energies under investigation. We observe the classic "S"-shaped curve which changes sign at $y' = 0$. Although projectile-target symmetry dictates $p^x(y') = -p^x(-y')$, even an ideal 4π detector cannot satisfy this condition because absorption and energy loss in the target introduces distortion for y' approaching -1. In the EOS detector, the target was located about 14 cm upstream from the active volume of the TPC, leading to optimized performance near mid-rapidity and above, at the expense of a progressive loss of acceptance approaching target rapidity. At each beam energy, we fit the $\langle p^x(y')/A\rangle$ curves over the region indicated by the solid lines in Fig. 4 with a function of the form $my' - m_3 y'^3$; the fitted values of m characterize the overall magnitude of the sideward flow effect among participant fragments, and these slopes are known simply as "flow" in the literature.

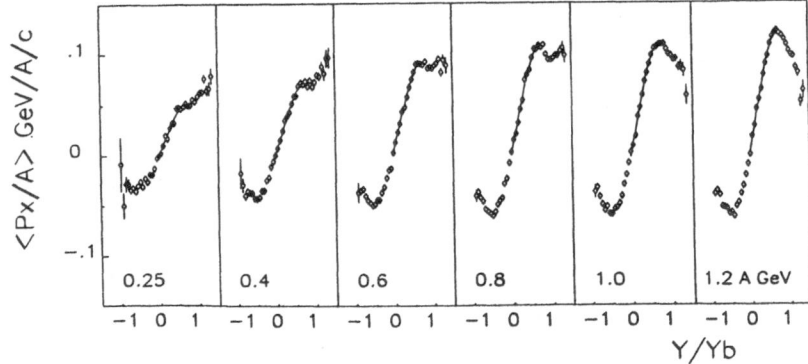

Figure 4: Transverse momentum per nucleon projected on the event reaction plane, as a function of fragment rapidity divided by beam rapidity, for Au + Au collisions at six beam energies between $0.25A$ GeV and $1.2A$ GeV.

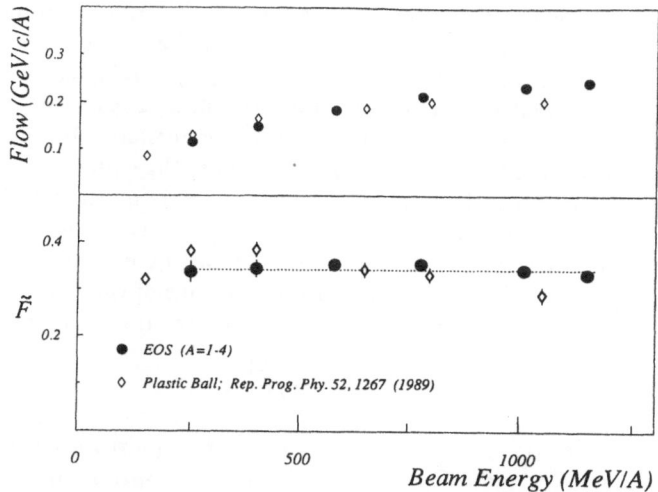

Figure 5: Flow excitation function for Au + Au collisions, and scale-invariant magnitude of sideward flow, $\bar{F}$, ("flow" divided by beam momentum in the cm frame) for Au + Au events measured by EOS and by the Plastic Ball.

The upper panel of Fig. 5 shows the resulting flow excitation function. A noteworthy feature is the steady increase in flow up through the highest Bevalac energy for Au + Au.

The lower panel of Fig. 5 presents the excitation function for the dimensionless flow parameter $\bar{F} = \text{flow}/p_{cm}^{beam}$, where p_{cm}^{beam} denotes the projectile momentum per nucleon in the cm frame. Also plotted in Fig. 5 are the Plastic Ball measurements of $\bar{F}$ for Au + Au [6, 1]. A direct comparison between the EOS and Plastic Ball results in Fig. 5 is only appropriate to the extent that the biases and distortions introduced by the two detectors are both negligible or are the same. Nevertheless, it is seen that agreement is close around the middle of the investigated beam energy range, and discrepancies are no bigger than about two standard errors at either end.

ORTHOGONAL COMPONENTS OF SIDEWARD FLOW

Before considering the interpretation of the measured flow excitation functions, we present results of an alternative analysis which provides additional constraints on the sideward flow effect. A comprehensive description of the phenomenon would allow us to write the observed transverse momentum $\mathbf{p}^\perp$ for a fragment in the form

$$\mathbf{p}^\perp = \mathbf{p}_{uncor}^\perp + F\hat{\mathbf{x}} + G\hat{\mathbf{y}}, \qquad (2)$$

where the unit vector $\hat{\mathbf{x}}$ is aligned with the reaction plane, $\hat{\mathbf{y}}$ is perpendicular to $\hat{\mathbf{x}}$, and $\mathbf{p}_{uncor}^\perp$ comes from an uncorrelated event such as can be generated by event mixing, *i.e.*, randomly selecting M tracks, each from a different observed event with multiplicity M. Existing measurements are inadequate for fully constraining the functions F and G. For example, data on $\langle p^x(y') \rangle$ cannot provide information about the dependence of F on φ, the fragment's azimuth relative to the collision reaction plane. Other flow observables, such as sphericity tensor eigenvalues,[8] and cross sections or yields as a function of azimuth relative to the estimated reaction plane, $dN/d(\phi - \phi_Q)$ [9] are not useful for overcoming this limitation.[1] The alternative analysis presented here involves simultaneous measurements of two orthogonal components of the collective sideward motion: the azimuthal component, associated with rotations of $\mathbf{p}^\perp$ relative to an uncorrelated distribution, and the radial component of sideward flow, associated with changes in the magnitude of $\mathbf{p}^\perp$ relative to an uncorrelated distribution. Measurements of these two components complement each other, and together, they place more complete and stringent constraints on dynamical models. Moreover, this method facilitates study of the fragment mass-dependence of flow, since in this representation, it is easy to test whether the enhanced flow for composite fragments can be explained by coalescence.

The azimuthal pair correlation function [10] makes use of the variable ψ, the smaller angle between the transverse momenta of two fragments, and is defined as

$$C(\psi) = \frac{P_{cor}(\psi)}{P_{uncor}(\psi)}, \tag{3}$$

where $P_{cor}(\psi)$ is the ψ distribution for observed pairs, i.e., pairs in which both fragments belong to the same event, and $P_{uncor}(\psi)$ is the ψ distribution for pairs from mixed events. Sideward flow leads to an enhanced probability for fragments to be emitted with azimuths close to each other, near the reaction plane orientation; thus, if $C(\psi)$ is plotted for a rapidity interval that is not centered on mid-rapidity, we observe $C(\psi) > 1$ at small ψ and $C(\psi) < 1$ at large ψ. If fragments within a given rapidity interval are distributed in φ according to $P(\varphi) \propto 1 + \lambda \cos \varphi$, [10] then $C(\psi) = 1 + 0.5\lambda^2 \cos \psi$. Fitted λ values provide a dimensionless measure of the azimuthal flow component. The azimuthal pair correlation function offers several advantages over previous flow analyses: it circumvents the need for event-by-event estimates of the reaction plane and the need to correct for dispersion in these estimates, it allows flow measurements in different rapidity intervals to be completely independent of each other, and the denominator in $C(\psi)$ automatically corrects for any azimuthal asymmetry introduced by the detector. Detector-related asymmetry in the transverse plane becomes apparent at $y' < 0$; however, only forward rapidities are studied in the current analysis.

To characterize the radial component of sideward flow, we introduce a new quantity which we call the radial pair variance function:

$$\sigma^2(\psi) = \langle p_{sum}^2(\psi) \rangle - \langle p_{sum}(\psi) \rangle^2, \tag{4}$$

where $p_{sum} = p_i^\perp / A_i + p_j^\perp / A_j$ is the sum of the $\mathbf{p}^\perp$ magnitudes per nucleon for the pair. This function offers the same advantages as the azimuthal pair correlation function; in this instance, there is no reason to compute a ratio like P_{cor}/P_{uncor}, because $\sigma^2(\psi)$ is flat for mixed events, even at backward rapidities where detector-related asymmetry in the transverse plane is apparent. The fact that $\mathbf{p}^\perp$ magnitudes tend to be larger when a fragment's azimuth is parallel to the flow direction, and tend to be smaller when antiparallel, leads to an inequality $\sigma^2(\psi \sim 0°) > \sigma^2(\psi \sim 180°)$. If F is independent of φ and is large compared with G, σ^2 decreases linearly with increasing ψ; this linearity can be demonstrated analytically for an idealized example where there are no thermal fluctuation in $\mathbf{p}^\perp$, and simulations with realistic momenta also indicate linear $\sigma^2(\psi)$. To characterize the magnitude of the radial component of sideward flow in momentum units, we define $S = \sqrt{(d\sigma^2/d\psi)}$; to get dimensionless units, we define $S' = S/p_{cm}^{beam}$.

Fig. 6 shows azimuthal pair correlation functions $C(\psi)$ and radial pair variance

---

[1] At rapidities where the G term in Eq.(2) can be neglected, $dN/d(\phi - \phi_Q)$ is observed to have a sinusoidal shape; the amplitude provides information about $F_{\varphi=0} + F_{\varphi=180°}$, but has little or no useful sensitivity to the φ-dependence of F in the interesting region, namely, over quadrants such as $\varphi = 0$ through $\varphi = 90°$.

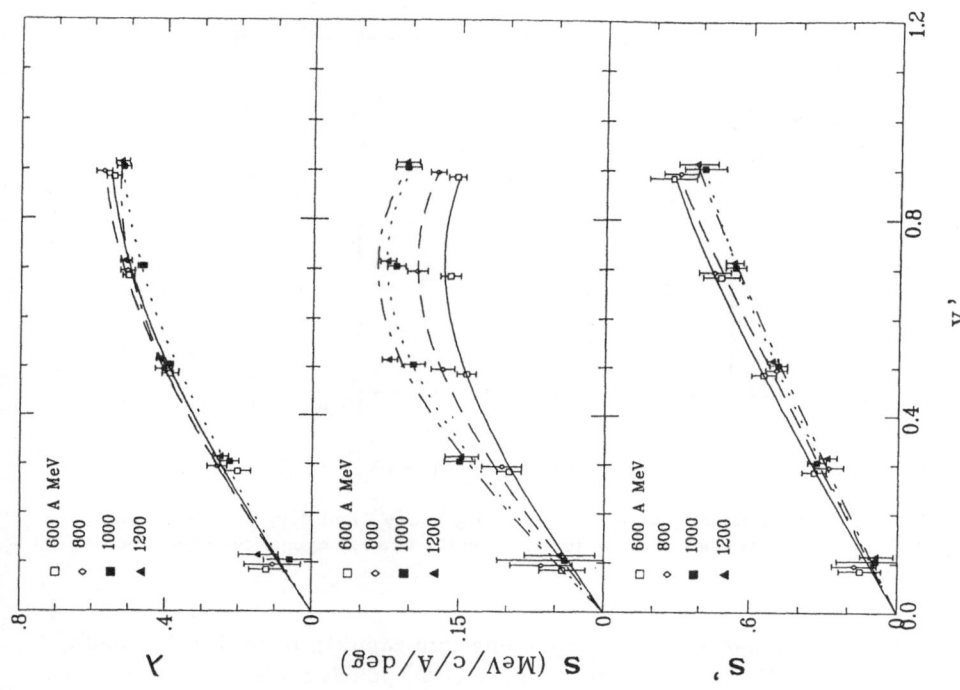

Figure 7: Azimuthal flow component $\lambda(y')$ and radial components of sideward flow $S(y')$ and $S'(y')$ for Au + Au at several beam energies.

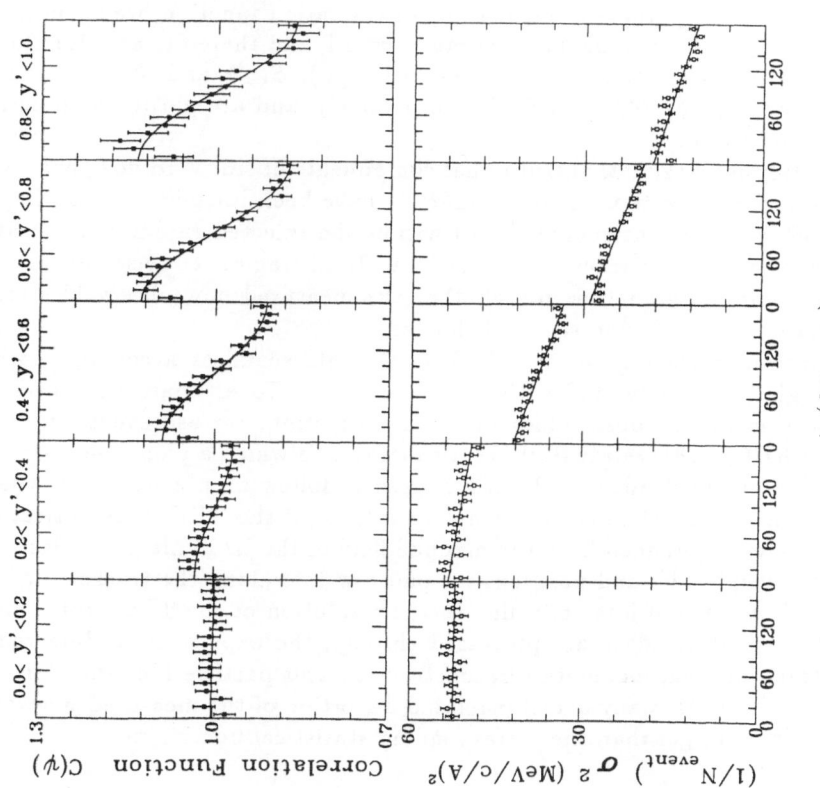

Figure 6: Azimuthal pair correlation functions and radial pair variance functions in five rapidity intervals spanning mid-rapidity to projectile rapidity, for Au + Au at 1.2A GeV.

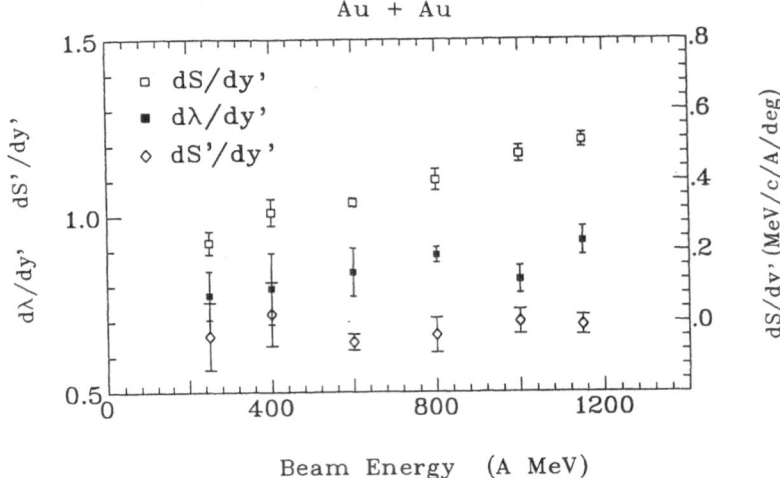

Figure 8: Slopes near midrapidity based on fits to $\lambda(y')$, $S(y')$ and $S'(y')$, versus beam energy. This plot represents the excitation function for the two orthogonal components of sideward flow.

functions $\sigma^2(\psi)$ in five rapidity intervals spanning mid-rapidity to projectile rapidity, for Au + Au at $1.2A$ GeV. The solid curves in the $C(\psi)$ panels represent least-squares fits using the function $1 + 0.5\lambda^2 \cos\psi$, and the $\sigma^2(\psi)$ data in the lower panels are fitted using a straight line. The resulting λ values and slopes S and S' for the 5 rapidity bins are presented in Fig. 7, along with corresponding data for other beam energies under investigation. For the purpose of generating an excitation function, we again fit the y' dependence using a function of the form $my' - m_3 y'^3$, and these fits are also plotted in Fig. 7. The fitted parameters m associated with $\lambda(y')$, $S(y')$, and $S'(y')$ are hereafter denoted by $d\lambda/dy'$, dS/dy', and dS'/dy', respectively, and are plotted as functions of beam energy in Fig. 8.

In this analysis of azimuthal and radial components of sideward flow, only protons and deuterons from events with $M > 0.4 M^{max}$ have been included. These fragments make up 73% of all reconstructed tracks within the selected rapidity range at $1.2A$ GeV, and 47% of all such tracks at $0.25A$ GeV. If all fragment species up to ^{4}He are included, a preliminary analysis reveals that the measured flow observables increase, but show the same trends and scaling behavior.

An important advantage of the EOS TPC is its seamless acceptance, which is simple enough to be simulated with good accuracy. To estimate the effect of detector distortion on the observables under investigation, we use events from a version of the $FREESCO$[11] statistical event generator to which a phenomenological flow correlation[12] has been added; GEANT is used to follow all particles from the interaction point in the target through the active volume of the TPC, with trajectories of charged particles determined from the mapped field of the HISS magnet; a fast simulator generates amplitudes and positions for pad-row hits along the tracks, allowing for smearing and merging of hits according to the resolution of the TPC; from this stage onwards, the simulated data are processed through the experimental data reduction codes for track finding, momentum reconstruction and particle identification. These simulations indicate that any detector-related distortion of the measured sideward flow components is no larger than the corresponding statistical uncertainty.

DISCUSSION AND CONCLUSIONS

It is evident that $\lambda(y')$ is statistically consistent with being independent of beam energy, while $S(y')$ increases with beam energy. Fig. 5 demonstrates that the slope of $\langle p^x(y')/A \rangle$ for essentially the same data increases monotonically with beam energy, and so the current results for $d\lambda/dy'$ and dS/dy' imply that the radial component of sideward flow is solely responsible for the increase in the slope of $\langle p^x(y')/A \rangle$. Moreover, when the radial component $S(y')$ is expressed in a dimensionless scale-invariant form $S'(y')$, it too becomes consistent with being independent of beam energy, as demonstrated in Figs. 7 and 8.

Interesting implications for investigations of flow at higher energies arise from the observation that scale-invariant flow parameters remain constant over an almost 5 to 1 range of beam energy, with no sign of any departure from this scaling at the highest Bevalac energy (equivalent to sideward flow expressed in momentum units increasing steadily with beam energy). Maximum beam momentum for Au at the Brookhaven AGS corresponds to about 5 times the corresponding maximum at the Bevalac. If sideward flow cannot be detected at maximum AGS momentum, that would imply that somewhere above $2A$ GeV/c, there is an unusually precipitous departure from the scale-invariance observed at the Bevalac using EOS. The increasing importance of pion production above Bevalac energies might be expected to result in only a gradual decrease in scale-invariant flow parameters. [14]

The idea of expressing flow magnitudes in scale-invariant form originated with Balazs et al. [13] and Bonasera and Csernai.[14] The simplest class of hydrodynamic model ("ideal" hydrodynamics) has a gaseous equation of state, and dissipative effects such as viscosity are neglected. Such a model predicts that scale-invariant flow parameters at constant y' should be independent of beam energy and independent of the mass of the colliding system. It is clearly not applicable at beam energies of a few tens of A MeV, where the nuclear mean field changes from attractive to repulsive and sideward flow is observed to drop to zero. [15, 16, 17] However, the nuclear mean field is predominantly repulsive at and above $250A$ MeV, and a simple gaseous equation of state becomes more realistic. The EOS data presented in Figs. 5, 7 and 8 show a remarkable adherence to ideal hydrodynamic scaling for Au + Au collisions over the investigated range of y' and bombarding energy.

Previous measurements at these energies reveal a significant increase in $\bar{F}$ as the mass of the colliding system increases,[1] in disagreement with ideal hydrodynamics. The projectile/target mass dependence of $\bar{F}$ has been interpreted as evidence for viscous effects, and based on a calculation of nuclear viscosity by Danielewicz[18], Bonasera and Csernai[14] predicted a modified scaling behavior for flow, in which $\bar{F}$ acquired both a mass dependence and a beam energy dependence. For Au + Au collisions, viscosity was predicted to decrease $\bar{F}$ by 15 to 20% between $0.25A$ GeV and $1.2A$ GeV. EOS measurements based on the current subset of statistics are consistent with no change in $\bar{F}$ within $\pm 10\%$. While viscosity may be needed in hydrodynamic models to explain the smaller flow for lighter systems, a possible conclusion from our preliminary results is that dissipation effects can be neglected to a good approximation in the context of flow for the heaviest systems between $0.25A$ and $1.2A$ MeV.

ACKNOWLEDGEMENTS

This work is supported in part by the US Department of Energy and the US National Science Foundation.

REFERENCES

[1] Recent experimental reviews include: K.-H. Kampert, Nucl. Part. Phys. **15**, 691 (1989); H. H. Gutbrod, A. M. Poskanzer, and H. G. Ritter, Rep. Prog. Phys. **52**, 1267 (1989).

[2] For reviews, see H. Stöcker and W. Greiner, Phys. Rep. **137**, 277 (1986); G. F. Bertsch and S. Das Gupta, Phys. Rep. 160, 189 (1988); *The Nuclear Equation of State*, Vol. 216 of *NATO Advanced Study Institute, Series B: Physics*, edited by W. Greiner and H. Stöcker, (Plenum, New York, 1989), Part A.

[3] J. Jänicke and J. Aichelin, Nucl. Phys. **A547**, 542 (1992).

[4] G. Rai *et al.*, in *The Nuclear Equation of State*, ed. W. Greiner and H. Stöcker, NATO-ASI Vol. B216 part A (Plenum, NY 1989), p. 187.

[5] G. Rai *et al.*, IEEE Trans. Nucl. Sci. **37**, 56 (1990).

[6] K. G. R. Doss, H.-Å. Gustafsson, H. H. Gutbrod, K. H. Kampert, B. Kolb, H. Löhner, B. Ludewigt, A. M. Poskanzer, H. G. Ritter, H. R. Schmidt, and H. Wieman, Phys. Rev. Lett. **57**, 302 (1986).

[7] P. Danielewicz and G. Odyniec, Phys. Lett. **157B**, 146 (1985).

[8] M. Gyulassy, K. A. Frankel, and H. Stöcker, Phys. Lett. **110B**, 185 (1982); P. Danielewicz and M. Gyulassy, Phys. Lett. **129B**, 283 (1983); G. Fai and J. Randrup, Nucl. Phys. **A404**, 551 (1983).

[9] D. Beavis, S. Y. Fung, W. Gorn, D. Keane, Y. M. Liu, R. T. Poe, G. VanDalen, and M. Vient, Phys. Rev. Lett. **54**, 1652 (1985).

[10] S. Wang, Y. Z. Jiang, Y. M. Liu, D. Keane, D. Beavis, S. Y. Chu, S. Y. Fung, M. Vient, C. Hartnack, and H. Stöcker, Phys. Rev. C **44**, 1091 (1991).

[11] G. Fai and J. Randrup, Nucl. Phys. **A404**, 551 (1983); Comp. Phys. Comm. **42**, 385 (1986).

[12] A. F. Barghouty, G. Fai, and D. Keane, Nucl. Phys. **A535**, 715 (1991).

[13] N. Balazs, B. Schürmann, K. Dietrich, and L. P. Csernai, Nucl. Phys. **A424**, 605 (1984).

[14] A. Bonasera and L. P. Csernai, Phys. Rev. Lett. **59**, 630 (1987); A. Bonasera, L. P. Csernai, and B. Schürmann, Nucl. Phys. **A476**, 159 (1988).

[15] D. Krofcheck, W. Bauer, B. M. Crawley, C. Djalali, S. Howden, C. A. Ogilvie, A. VanderMolen, G. D. Westfall, W. K. Wilson, R. S. Tickle, and C. Gale, Phys. Rev. Lett. **63**, 2028 (1989).

[16] C. A. Ogilvie, W. Bauer, D. A. Cebra, S. Howden, J. Karn, A. Nadasen, A. VanderMolen, G. D. Westfall, W. K. Wilson, and J. S. Winfield, Phys. Rev. C **42**, R10 (1990).

[17] W. M. Zhang, R. Madey, M. Elaasar, J. Schambach, D. Keane, B. D. Anderson, A. R. Baldwin, J. Cogar, J. W. Watson, G. D. Westfall, G. Krebs, and H. Wieman, Phys. Rev. C **42**, R491 (1990).

[18] P. Danielewicz, Phys. Lett. **146B**, 168 (1984).

PERTURBATIVE AND NONPERTURBATIVE EM LEPTON PAIR PRODUCTION IN RELATIVISTIC HEAVY-ION COLLISIONS

Volker E. Oberacker[1,3], Jack C. Wells[1,2,3], A. Sait Umar[1,3], and Michael R. Strayer[1,2]

[1]Center for Computationally Intensive Physics, Oak Ridge National Laboratory, Oak Ridge, Tennessee 37831, USA
[2]Physics Division, Oak Ridge National Laboratory, Oak Ridge, Tennessee 37831, USA
[3]Department of Physics & Astronomy, Vanderbilt University, Nashville, Tennessee 37235, USA

INTRODUCTION

In this talk, I will focus on *electromagnetic* dilepton production from the QED-vacuum in relativistic heavy-ion collisions. Heavy ions in relativistic motion generate strong time-dependent EM fields with large Fourier components which give rise to sizable pair production. There are several motivations for our studies: Lepton pair production by *hadronic (Drell-Yan)* processes has been widely discussed as a possible signature of the quark-gluon plasma formation[1]. The dominant background will come from electromagnetic sources and could even mask the signals from the plasma phase[2].

Electromagnetically produced lepton pairs also impose severe constraints on the design of relativistic heavy-ion colliders such as RHIC and LHC. In addition to the *free pair production* discussed above, *pair-production with capture* of the negatively charged lepton into a bound state is also possible, as illustrated in Figure 1. This change of the charge state of the ions is the leading mechanism for beam loss of relativistic colliders. Accurate predictions of the cross section for this process are important because the cross section increases with energy.

A few historical remarks: Free electron-pair production in the collisions of cosmic rays with nuclei was first described in 1937 by Racah[3], based on the "equivalent pho-

Hot and Dense Nuclear Matter, Edited by
W. Greiner *et al.*, Plenum Press, New York, 1994

ton method" developed by Weizsäcker and Williams[4] and by Landau and Lifshitz[5]. In 1971, Brodsky et al.[6] applied this technique to pair creation in particle accelerators.

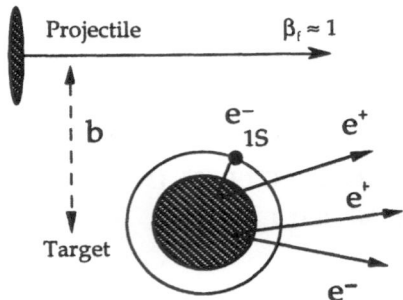

Figure 1. Free electron-pair production and pair production with capture of the electron into a bound state.

One calculates classically the photon flux generated by the particles in motion and folds it with the photoproduction cross section for pairs. In 1977, Soff[7] utilized the Weizsäcker-Williams method for the first time to describe electron-positron production in heavy-ion collisions. For a detailed discussion we refer to the review article by Bertulani and Baur[8].

PERTURBATIVE LEPTON-PAIR PRODUCTION

The equivalent photon method relies on several approximations: it assumes that the photons are on-shell and have zero transverse momenta. This approximation is only valid in the ultra-relativistic limit. Furthermore, the corresponding S-matrix elements are singular and must be regularized with both high- and low-frequency cutoffs. A better approach is to use Feynman perturbation theory and to evaluate the lowest-order two-photon-exchange diagram for e^+e^- pair production which is depicted in Figure 2.

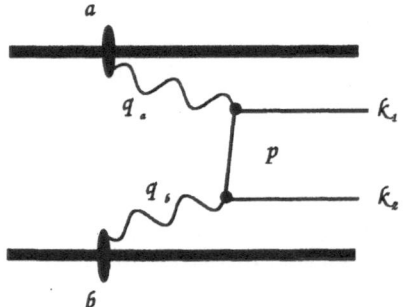

Figure 2. Lowest-order Feynman diagram for free e^+e^- pair production.

The two-photon-exchange diagram can be reduced to an eight-dimensional integral; this integral has been computed exactly by Bottcher and Strayer[9] with a Monte-Carlo technique. The resulting total cross section for electron-pair production is shown in Figure 3 for $^{197}Au +^{197} Au$ as a function of the collider energy (in units of GeV/amu). The cross section is displayed in units of $\sigma_0 = (Z\alpha)^4 \lambda_e^2 = 165$ b; at $E_c = 10$ GeV/amu (corresponding to CERN-SPS energy) the total cross section for free e^+e^- pair production is about 3 kilobarn, and at 100 GeV/amu (i.e. RHIC energy) it is about 30 kilobarn. Figure 3 also gives a comparison between the exact Monte-Carlo evaluation of the lowest-order Feynman diagram and various calculations based on the Weizsäcker-Williams method. It is striking that Racah's result is very close to the exact calculation whereas the result of Bertulani and Baur underestimates the cross section at lower energies.

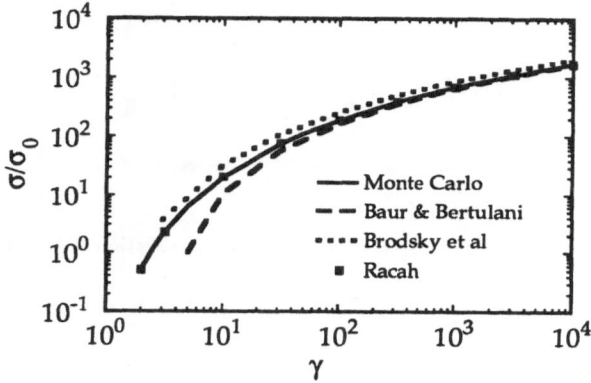

Figure 3. Free electron-pair production cross section vs. collider energy.

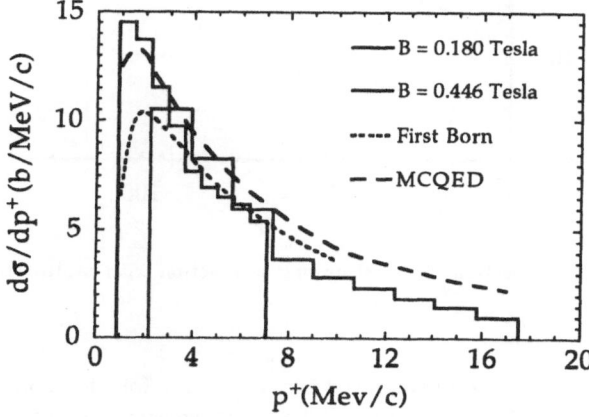

Figure 4. Positron momentum spectrum for $_{16}S +_{79} Au$ at $E_c = 10$ GeV/amu.

In Figure 4 we compare the prediction of the two-photon perturbation theory with data[10] measured at the CERN-SPS for $_{16}S +_{79} Au$; the equivalent collider energy is $E_c = 10$ GeV/amu. Up to positron momenta of about 7 MeV/c, theory and experiment are in excellent agreement; at higher positron momenta, it appears that the lowest-order perturbation theory overestimates the measured differential cross section. The total measured free electron-pair production cross section is (85 ± 22) barn, whereas the Feynman-Monte Carlo calculation yields 98 barn.

Rhoades-Brown, Bottcher and Strayer[11] have also performed perturbative calculations of electron-pair production with electron capture into a bound state of the target atom. The two-photon-exchange Feynman diagram is shown in Figure 5.

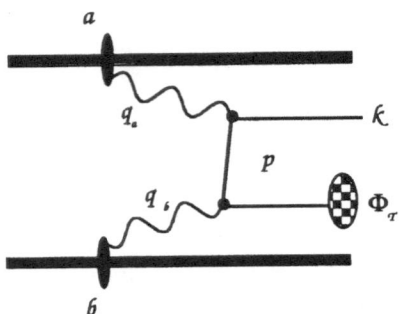

Figure 5. Lowest-order Feynman diagram for e^+e^- pair production with bound state capture of the electron.

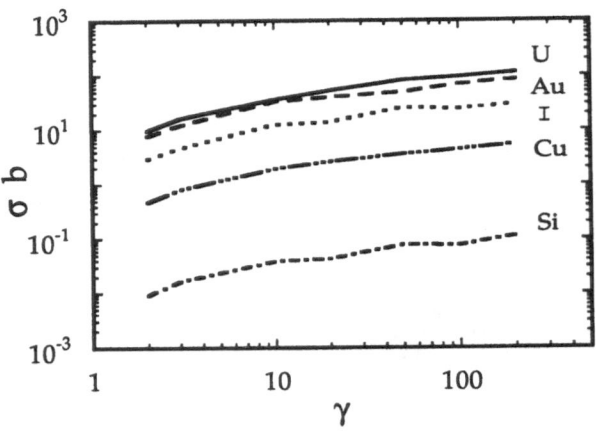

Figure 6. Total cross sections for electron-pair production with capture into the 1s state.

In Figure 6 we display the calculated cross sections for electron capture following pair creation from the QED-vacuum. The cross sections are given for symmetric $_zA +_z A$ collisions as a function of the collider energy (in units of GeV/amu). For the system $Au + Au$ at 100 GeV/amu (RHIC energy) one finds a total cross section of 72 barn; this value corresponds to a 10 hour lifetime of the RHIC beams, enough to carry out meaningful experiments. However, it is doubtful that perturbation theory holds for very heavy systems at high energies: the effective coupling constant is no longer small compared to 1 (for a $Au + Au$ collision we have $Z\alpha \approx 0.6$), and the time-dependent electromagnetic fields generated by the projectile are 20,000 times stronger than the static Coulomb field of the target. First indications that perturbation theory may be inadequate to describe pair creation with capture have recently been reported by experimenters at the BEVALAC[12]. In $_{92}U +_Z A$ collisions at a fixed-target energy of 1 GeV/amu it was found that the cross section dependence on the atomic number of the target is $Z^{(2.8 \pm 0.25)}$. Lowest-order perturbation theory predicts a Z^2 dependence. It is therefore necessary to study lepton-pair production by non-perturbative methods.

NON-PERTURBATIVE LEPTON-PAIR PRODUCTION

We shall now outline our theoretical formalism[13, 14] for nonperturbative lepton-pair production. Bottcher and Strayer have derived the field equations from a semi-classical least-action principle[2]. In the following, we give an alternative derivation which is similar to the approach developed by Reinhardt et al.[15] for nonrelativistic heavy-ion reactions. We start from the standard QED Lagrange density

$$\mathcal{L}_{\text{QED}} = \hat{\psi}^\dagger \gamma^0 [\gamma^\mu i \partial_\mu - 1]\hat{\psi} - \frac{1}{4}\hat{F}_{\mu\nu}\hat{F}^{\mu\nu} - \hat{\jmath}^\mu \hat{A}_\mu \,, \tag{1}$$

where

$$\hat{\jmath}^\mu = \hat{\jmath}^\mu_{\text{lept}} + \hat{\jmath}^\mu_{\text{ext}} = e\,\hat{\psi}^\dagger \gamma^0 \gamma^\mu \hat{\psi} + \hat{\jmath}^\mu_{\text{ext}} \tag{2}$$

denotes the current density operator which describes the lepton current and the conserved, external current of the moving heavy nuclei. The total radiation field $\hat{A}_\mu$ consists of three parts

$$\hat{A}_\mu = \hat{A}^{\text{lept}}_\mu + \hat{A}^{\text{ext}}_\mu + \hat{A}^{\text{free}}_\mu \,, \tag{3}$$

where $\hat{A}^{\text{lept}}_\mu$ is generated by the lepton current $\hat{\jmath}^\mu_{\text{lept}}$ and $\hat{A}^{\text{ext}}_\mu$ by the external current $\hat{\jmath}^\mu_{\text{ext}}$, respectively. The term $\hat{A}^\mu_{\text{free}}$ denotes the free radiation field. By varying the action integral

$$S_{\text{QED}} = \int d^4x \mathcal{L}_{\text{QED}} \tag{4}$$

with respect to the field operators $\hat{\psi}$ and $\hat{A}_\mu$, we obtain the Euler-Lagrange equations of motion for the quantum fields

$$[\gamma^\mu(i\partial_\mu + e\hat{A}_\mu) - 1]\hat{\psi}(x) = 0 \,, \tag{5}$$

$$\partial_\mu \hat{F}^{\mu\nu}(x) = \hat{\jmath}^\nu(x) \,. \tag{6}$$

These coupled quantum field equations are difficult to solve without approximations; hence, we make the following simplifying assumptions: We neglect the leptonic current $\hat{\jmath}_{\text{lept}}^{\mu}$ in Eq. (6) and the associated electromagnetic field $\hat{A}_{\mu}^{\text{lept}}$ in Eq. (5); this is justified because both of of them are much smaller than the strong external heavy-ion current $\hat{\jmath}_{\text{ext}}^{\mu}$ and the external field $\hat{A}_{\mu}^{\text{ext}}$, respectively. With this approximation, the field equations decouple. Furthermore, we treat the external field classically by solving Maxwell's equations, i.e. $\hat{A}_{\mu}^{\text{ext}} \rightarrow \langle \hat{A}_{\mu}^{\text{ext}} \rangle_{\text{class}} = A_{\mu}^{\text{ext}}$. We also neglect QED radiative corrections like vacuum polarization and self energy effects arising from interactions of the lepton field with the free radiation field, i.e. $\hat{A}_{\mu}^{\text{free}} = 0$. The problem is thus reduced to the solution of the time-dependent Dirac equation, Eq. (7), for the lepton field $\hat{\psi}$ interacting with an external, classical four-vector potential A_{μ}^{ext} determined independently by the classical Maxwell equations, Eq. (8), i.e.

$$[\gamma^{\mu}(i\partial_{\mu} + eA_{\mu}^{\text{ext}}) - 1]\hat{\psi}(x) = 0 , \tag{7}$$

$$\partial_{\mu}F_{\text{ext}}^{\mu\nu}(x) = \jmath_{\text{ext}}^{\nu}(x) . \tag{8}$$

We study the electromagnetic production of lepton pairs in a reference frame where one of the nuclei, henceforth referred to as the target, is at rest. The target nucleus and the lepton interact via the static Coulomb field A_{T}^{0}. The only time-dependent interaction $A_{\text{P}}^{\mu}(t)$ arises from the classical motion of the projectile. Thus, it is natural to recast the Dirac equation (7) into the Schrödinger form

$$[H_{\text{F}} + H_{\text{P}}(t)]\hat{\psi}(\vec{r}, t) = i\frac{\partial}{\partial t}\hat{\psi}(\vec{r}, t) . \tag{9}$$

The Furry Hamiltonian

$$H_{\text{F}} = -i\vec{\alpha} \cdot \nabla + \beta - eA_{\text{T}}^{0} \tag{10}$$

describes the leptons in the presence of the static Coulomb field of the target nucleus, and

$$H_{\text{P}}(t) = e\vec{\alpha} \cdot \vec{A}_{\text{P}}(t) - eA_{\text{P}}^{0}(t) \tag{11}$$

is the time-dependent interaction between the lepton field and the projectile. We expand now the lepton-field operator $\hat{\psi}(\vec{r}, t)$ in a complete, orthonormal set of single-particle basis states. In terms of the stationary eigenstates of the Furry Hamiltonian

$$H_{\text{F}}\chi_k(\vec{r}) = E_k\chi_k(\vec{r}) \tag{12}$$

we obtain

$$\hat{\psi}(\vec{r}, t) = \sum_k \hat{a}_k(t)\chi_k(\vec{r}) \exp(-iE_kt) , \tag{13}$$

where the $\hat{a}_k$'s are operator-valued expansion coefficients. The Furry eigenstates are the proper in- and out-states for asymptotic times $|t| \rightarrow \infty$, where the interaction $H_{\text{P}}(t)$ is zero. From the anticommutation relations for the fermion field operators $\hat{\psi}$ and $\hat{\psi}^{\dagger}$, one readily shows that $\hat{a}_k^{\dagger}$ and $\hat{a}_k$ describe the creation and annihilation of leptons in state k in the static Coulomb field of the target nucleus. We also expand the lepton field in terms of the time-dependent basis of solutions to the full Dirac Hamiltonian $H_F + H_P(t)$

$$[H_F + H_P(t)]\phi_j(\vec{r}, t) = i\frac{\partial}{\partial t}\phi_j(\vec{r}, t) \, . \tag{14}$$

This basis expansion results in

$$\hat{\psi}(\vec{r}, t) = \sum_j \hat{\alpha}_j \phi_j(\vec{r}, t) \, , \tag{15}$$

where the $\hat{\alpha}_j$ are quasi-particle destruction operators. Under the influence of the interaction Hamiltonian $H_P(t)$ the single-particle states $\chi_j(\vec{r})$ evolve into the time-dependent states $\phi_j(\vec{r}, t)$ according to the time-dependent Dirac equation (14). Therefore, the index j does not refer to a set of good quantum numbers for $\phi_j(\vec{r}, t)$, but refers to the quantum numbers of the particular Furry state, $\chi_j(\vec{r})$, that satisfies the $t \to -\infty$ boundary condition

$$\lim_{t \to -\infty} \phi_j(\vec{r}, t) \to \chi_j(\vec{r}) \exp(-iE_j t) \, . \tag{16}$$

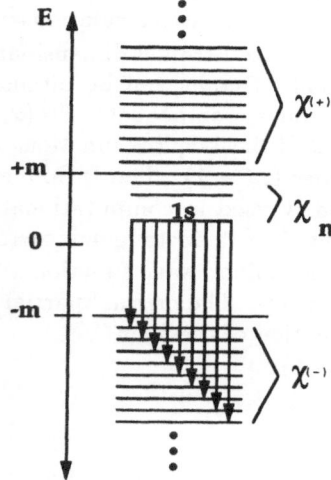

Figure 7. The transitions leading to pair production with capture into the $1s$ state are indicated (see Equation (19)).

Our aim is to treat pair production by completely ionized heavy atoms. In this case, the QED vacuum state in the external Coulomb field of the target nucleus has the structure

$$|\Phi_0\rangle = \prod_{i<F} (\hat{a}_i^\dagger)|0\rangle \, , \tag{17}$$

where $i < F$ denotes states below the Fermi energy $E_F = -m_0 c^2$. The time-evolved QED vacuum state is given by the time-evolution operator $\hat{U}(t, t_0)$

$$|\Phi_0(t)\rangle = \hat{U}(t, t_0 \to -\infty)|\Phi_0\rangle \qquad (18)$$

By making the assumption that $|\Phi_0(t)>$ is at all times a single Slater determinant, one can derive expressions for the inclusive number of leptons and antileptons created after the collision. The total number of leptons captured into the $1s$ state following pair creation from the QED vacuum is given by projecting the time-evolved single-particle state $\phi_{1s}(t \to \infty)$ onto the complete set of Dirac continuum states $\chi_k^{(-)}$ with negative energy $(k < F)$, i.e.

$$N_{capt}^{(1s)}(t \to \infty) = \sum_{k<F} |< \chi_k^{(-)} | \phi_{1s}(t \to \infty) >|^2 . \qquad (19)$$

To compute probabilities for free lepton-pair production is computationally much more involved because one has to time-propagate a large number of negative-energy states and project these onto the Furry continuum states with positive energy.

SOLUTION OF THE TIME-DEPENDENT DIRAC EQUATION

There are essentially two different approaches to the numerical solution of the time-dependent Dirac equation (14): lattice calculations[2, 13, 14, 16] and the coupled-channels formalism[17, 18]. We utilize here a 3-dimensional Cartesian lattice and the Basis-Spline collocation method. Details will be discussed in the presentation by Jack Wells. The four Dirac spinor components $\psi^{(p)}(x, y, z, \tau), p = (1, 2, 3, 4)$ are expanded in terms of a product of Basis-Spline functions $B_i^M(x)$ which are piecewise-continuous polynomials of order $(M - 1)$. The B-Splines are generalizations of the "finite elements" that are widely used in computational physics. The Dirac spinor components $\psi^{(p)}$ are represented on a rectangular cartesian lattice $(x_\alpha, y_\beta, z_\gamma)$ i.e. the lattice representation of the spinor wave function $\psi^{(p)}$ is a vector ψ with $N = 4 \cdot N_x \cdot N_y \cdot N_z$ complex components. The original partial differential equation (14) is transformed into a matrix equation of the form

$$\mathbf{H}\psi(\tau) = i\frac{\partial \psi}{\partial \tau}, \qquad (20)$$

where we use boldface type for matrices and vectors. It is impossible to store $\mathbf{H}$ in computer memory, since this would require the storage of N^2 complex double-precision numbers. Hence, we must resort to iterative methods for the solution of the matrix equation which do not require the storage of $\mathbf{H}$.

We solve the time-dependent Dirac equation in two steps: first, we consider the static Coulomb problem at time $\tau \to -\infty$, i.e. the lepton bound to a heavy nucleus. The stationary Dirac equation for the ground-state spinor is given by

$$\mathbf{H}_0\psi_{gs} = E_{gs}\psi_{gs} \qquad (21)$$

The static problem is solved by an iterative procedure such as the Lanczos algorithm[14] or the damped relaxation method[19].

We solve the time-dependent equation (20) by a Taylor-expansion of the propagator. For an infinitesimal time step $\Delta\tau$ we find

$$\psi\left(\tau+\Delta\tau\right)=\mathbf{U}\left(\tau+\Delta\tau,\tau\right)\psi\left(\tau\right)\approx\left(1+\sum_{n=1}^{N}\frac{(-i\Delta\tau\mathbf{H})^{n}}{n!}\right)\psi\left(\tau\right). \tag{22}$$

We have thus reduced the original problem to a series of (matrix)×(vector) operations which can be executed with high efficiency on vector or parallel supercomputers without explicitly storing the matrix in memory.

RESULTS

In Figure 8 we show the probability of muon-pair production with capture into the K-shell as a function of time. The system studied is $^{197}Au+^{197}Au$ at $E_c=100$ GeV/amu for the grazing impact parameter $b=8.72\lambda_{\mu}=16.3fm$. We notice that the probability increases very steeply as the two nuclei approach each other; after the classical turning point has been passed $P(t)$ reaches a constant value of about 10^{-2}.

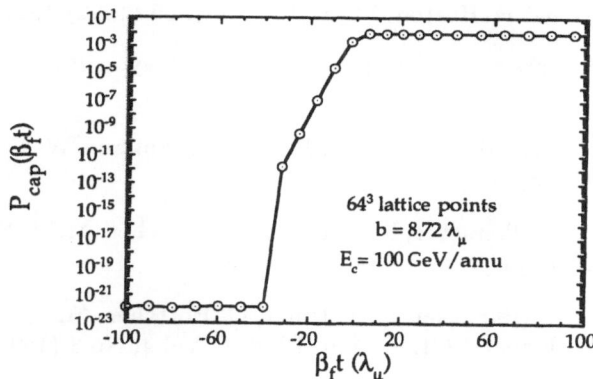

Figure 8. Muon-pair production with capture for a $Au+Au$ collision at RHIC energy.

Further results will be presented by Jack Wells at this conference, in particular the impact parameter dependence of the capture cross section.

ACKNOWLEDGMENTS

This research project was sponsored in part by the U.S. Department of Energy under contract No. DE-FG05-87ER40376 with Vanderbilt University, and under contract No. DE-AC05-84OR21400 with Oak Ridge National Laboratory managed by Martin Marietta Energy Systems. In addition, this research project was partially supported by the U.S. Department of Energy High Performance Computing and Communications Program (HPCC) as the "Quantum Structure of Matter Grand Challenge" project. The numerical calculations were carried out using the Intel iPSC/860 hypercube multicomputer at ORNL and the CRAY-2 at NERSC (Livermore).

REFERENCES

1. K. Kajantie, J. Kapusta, L. McLerran, and A. Mekjian, Phys. Rev. D 34 (1986) 2746

2. C. Bottcher and M. R. Strayer, Ann. Phys. (N.Y.) 175 (1987) 64

3. G. Racah, Nuovo Cim. 70 (1937) 14

4. E. J. Williams, Phys. Rev. 45 (1934) 729

5. L. D. Landau and E. M. Lifshitz, Phys. Zs. Sowj. 6 (1934) 244

6. S. J. Brodsky, T. Kinoshita and H. Terazawa, Phys. Rev. D4 (1971) 1532

7. G. Soff, Ph.D. thesis, Institut für Theoretische Physik, Universität Frankfurt am Main, Germany (1977)

8. C. A. Bertulani and G. Baur, Phys. Rep. 163 (1988) 299

9. C. Bottcher and M. R. Strayer, Phys. Rev. D 39 (1989) 1330

10. C.R. Vane, S. Datz, P.F. Dittner, H.F. Krause, C. Bottcher, M.R. Strayer, R. Schuch, H. Gao and R. Hutton, Phys. Rev. Lett. 69 (1992) 1911

11. M. J. Rhoades-Brown, C. Bottcher and M. R. Strayer, Phys. Rev. A 40 (1989) 2831

12. A. Belkacem, H. Gould, B. Feinberg, R. Bossingham and W. E. Meyerhof, Phys. Rev. Lett. 71 (1993) 1514

13. J. C. Wells, V. E. Oberacker, S. A. Umar, C. Bottcher, M. R. Strayer, J.-S. Wu, and G. Plunien, Phys. Rev. A 45 (1992) 6296

14. J. C. Wells, V. E. Oberacker, S. A. Umar, C. Bottcher, M. R. Strayer, J.-S. Wu, J. Drake and R. Flanery, Int. J. Mod. Phys. C Vol.4, No.3 (1993) 459

15. J. Reinhardt, B. Müller and W. Greiner, Phys. Rev. A 24 (1981) 103

16. J. Thiel, A. Bunker, K. Momberger, N. Grün, and W. Scheid, Phys. Rev. A 46 (1992) 2607

17. K. Rumrich, K. Momberger, G. Soff, W. Greiner, N. Grün, and W. Scheid, Phys. Rev. Lett. 66, (1991) 2613

18. A. J. Baltz, M. J. Rhoades-Brown, and J. Weneser, Phys. Rev. A 44 (1991) 5569

19. C. Bottcher, M.R. Strayer, A.S. Umar and P.G. Reinhard, Phys. Rev. A 40 (1989) 4182

BOLTZMANN-LANGEVIN TRANSPORT MODEL FOR HEAVY-ION COLLISIONS

Sakir Ayik

Tennessee Technological University
Cookeville, TN 38505, USA
and
Joint Institute for Heavy-Ion Research
Oak Ridge, TN 37831, USA

1. INTRODUCTION

Many aspects of heavy-ion collisions can be described by means of the one-body transport models. In these transport models, one deals with a reduced description in terms of the single-particle density, rather than the full many-body information. These models in semi-classical limit with a Boltzmann-Uehling-Uhlenbeck (BUU) form of a collision term has been very successful in describing a large variety of observables associated with heavy-ion collisions at intermediate energies[1-2]. The average description provided by the BUU model is well suited for processes involving small density fluctuations. However, for processes involving large density fluctuations, for example near instabilities and bifurcations, such an average description is inadequate. In these situations the stochastic transport models may provide a more appropriate basis for describing the dynamical evolution. In these stochastic approaches, the one-body transport models are improved beyond the mean-field approximation by incorporating the high order correlations in a statistical approximation, analogous to the treatment of the Brownian motion. The recently developed Boltzmann-Langevin (BL) model constitutes an example of such stochastic transport approaches,[3-5] and it is therefore a promising model for describing catastrophic phenomena, such as phase transitions and nuclear multifragmentations.

In section 2, we briefly describe the BL model. In section 3, we discuss a moment expansion of the BL equation, which is useful for small fluctuations around the average trajectory. In section 4, we describe a projection method for obtaining approximate numerical solutions of the BL equation. In section 5, we present some applications to heavy-ion collisions in which we investigate the dynamics of density fluctuations, and the influence of momentum space fluctuations of kaon production cross-sections at subthreshold energies. Finally, in section 6, we make some remarks on the further developments of the model and give some conclusions.

2. BOLTZMANN-LANGEVIN EQUATION

The BUU model determines the evolution of the average phase-space density $f(\mathbf{r},\mathbf{p},t)$ in the semi-classical limit according to the transport equation,

$$(\frac{\partial}{\partial t} + \mathbf{v}\cdot\nabla_r - \nabla_r U\cdot\nabla_p)\, f(\mathbf{r},\mathbf{p},t) = K(f). \tag{2.1}$$

Hot and Dense Nuclear Matter, Edited by
W. Greiner *et al.*, Plenum Press, New York, 1994

Here the left-hand-side describes the Vlasov propagation determined in terms of the nuclear mean-field U(f). On the right-hand-side, K(f) denotes a binary collision term given by

$$K(f) = \int dp_2\, dp_3\, dp_4\; W(12;34)\; [(1{-}f)\,(1{-}f_2)\,f_3\,f_4$$

$$-f\,f_2\,(1{-}f_3)\,(1{-}f_4)] \qquad (2.2)$$

where $f_j \equiv f(\mathbf{r},\mathbf{p}_j,t)$ and $W(12;34)$ denotes the transition rate between two-particle states, which can be expressed in terms of the in-medium nucleon-nucleon scattering cross-section.[1-2]

In energetic heavy-ion collisions, the nuclear system decays into a great variety of final states. In contrast to the experimental situations, the BUU model yields a unique *(deterministic)* trajectory for a given initial condition, which may be considered as an average over all possible final states *(ensemble averaging)*, and the dynamical branching is not allowed. In particular the BUU description becomes worse when the spreading of the trajectories of the single-particle densities associated with the final states is large. This severe limitation follows from the independent collision approximation employed in the derivation of the BUU model. The derivation of the BUU model involves two levels of approximations. The first one is the truncation of dynamics at a two-body level, which is a good approximation for a sufficiently dilute system. However, this is not sufficient. A more drastic approximation must be introduced by neglecting the correlations between subsequent binary collisions, which is usually referred to as the *molecular chaos assumption*. As a result, the fluctuations are not propagated and the independent binary collisions always drive the system along the average trajectory without breaking the initial symmetries of the system.

In order to describe the fluctuations, i.e., the spreading of trajectories of the single-particle densities, the effects of correlations must be restored into the equation of motion in some ways. The residual interactions, in general, play two different roles: (i) producing dissipation by randomizing the momentum distribution via binary collisions and (ii) inducing fluctuations by propagating correlations in the phase-space. By incorporating these two effects of the residual interactions into the equation of motion, one obtains stochastic transport equations, which can be studied in various representations.[3-5] In the semi-classical limit the equation of motion for the fluctuating phase-space density $\hat{f}(\mathbf{r},\mathbf{p},t)$ takes the form,

$$\left[\frac{\partial}{\partial t} + \mathbf{v}\cdot\nabla_r - \nabla_r\hat{U}\cdot\nabla_p\right]\hat{f}(\mathbf{r},\mathbf{p},t) = K(\hat{f}) + \delta K(\mathbf{r},\mathbf{p},t). \qquad (2.3)$$

which is referred to as the BLE. Here, the collision term $K(\hat{f})$ has the usual BUU form but is expressed in terms of the fluctuating density $\hat{f}(\mathbf{r},\mathbf{p},t)$. On the right-hand-side, in addition to the usual collision term, it appears an additional term $\delta K(\mathbf{r},\mathbf{p},t)$, which is called the *fluctuating collision term*. The fluctuating collision term arises from correlations not accounted for by the collision term. As a matter of fact, such an additional term always arises in transport theory whenever we deal with a reduced description and it describes the coupling to the degrees of freedom, which are not explicitly considered.[6-7] The fluctuating collision term has many properties which are similar to the random force in a typical Langevin equation: (i) It varies rapidly in time with a characteristic time in the order of the duration time of a two-body collision. (ii) It is nearly impossible to calculate the fluctuating collision term explicitly because it is equivalent for exactly solving the many-body problem. (iii) It vanishes on the average with the molecular chaos assumption, hence, does not appear in the average description of the BUU model. The BLE contains full information about dissipation and fluctuation properties of the single-particle density. However, in order to have a tractable model, we need to introduce approximations. In analogy with the Brownian motion, it is assumed that eq. (2.3) describes a stochastic process in which the whole density is a stochastic variable and the fluctuating collision term acts like a random force.[8] In such a stochastic description the fluctuating collision term is characterized by a correlation function,

$$<\delta K(\mathbf{r},\mathbf{p},t) \, \delta K(\mathbf{r}',\mathbf{p}',t')> = C(\mathbf{p},\mathbf{p}') \, \delta(\mathbf{r}-\mathbf{r}') \, \delta(t-t') \tag{2.4}$$

which is assumed to be local in space and time without the *memory effects*. With a specified correlation function, the BLE becomes a well-defined stochastic transport equation for the fluctuating single-particle density. It provides a probabilistic description in contrast to the deterministic description of the BUU model. The BLE has many solutions with a given initial condition. Each solution produces an event and many solutions are needed for describing a collision process.

The BLE was first proposed by Bixon and Zwanzig in order to describe the hydrodynamic fluctuations.[9] They evaluate the correlation function $C(\mathbf{p},\mathbf{p}')$ in equilibrium using the fluctuation-dissipation theorem as an input. As a result, their model is valid only for classical systems near equilibrium. In order to describe non-equilibrium fluctuations in quantal systems, Ayik and Gregoire calculate the correlation function directly in non-equilibrium within a weak-coupling approximation.[3] In the semi-classical limit, the correlation function is given by

$$C(\mathbf{p},\mathbf{p}') = \int d\mathbf{p}_3 \, d\mathbf{p}_4 \, W(11';34) \, [f_1 \, f_1' \, (1-f_3) \, (1-f_4) + (1-f_1) \, (1-f_1') \, f_3 \, f_4]$$

$$-2 \int d\mathbf{p}_2 \, d\mathbf{p}_4 \, W(12;1'4) \, [f_1 \, f_2 \, (1-f_1') \, (1-f_4) + (1-f_1) \, (1-f_2) \, f_1' \, f_4]$$

$$+ \delta(\mathbf{p}-\mathbf{p}') \int d\mathbf{p}_2 \, d\mathbf{p}_3 \, d\mathbf{p}_4 \, W(12;34) \, [f_1 \, f_2 \, (1-f_3) \, (1-f_4) + (1-f_1) \, (1-f_2) \, f_3 \, f_4] \tag{2.5}$$

where $W(12;34)$ is the same transition rate which enters into the collision term and $f_j = f(\mathbf{r},\mathbf{p}_j,t)$ is the locally averaged single-particle density. The correlation function is entirely determined by the one-body properties and is closely related to the collision term. Aside from the mean-field and the nucleon-nucleon cross-section, no other information is needed for describing the fluctuations. The fluctuation and dissipation properties of density, which are described by the collision term and the correlation function, are not independent properties, but must be related to each other (as in any relaxation process) through a fluctuation-dissipation theorem. Therefore, the close relationship between the correlation function and the collision term can be regarded as a fluctuation-dissipation theorem associated with the stochastic evolution of the single-particle density. The BLE satisfies the conservation laws of total energy, total momentum, and total particle number. In contrast to the Brownian motion in which the energy conservation is satisfied on the average, each event of the BLE respects the conservation laws. This property follows from the fact that the fluctuating collision term in the BLE is an *internal noise* and the correlation function satisfies certain sum rules. The correlation function $C(\mathbf{p},\mathbf{p}')$ is valid for large fluctuations in non-equilibrium. For small fluctuations around equilibrium, it reproduces the known result of Bixon and Zwanzig.

3. MOMENT EXPANSION

The BLE generates an ensemble of phase-space densities $\{\hat{f}(\mathbf{r},\mathbf{p},t)\}$ and the proper object of study is therefore the distribution of such densities. Useful information on the deviations $\delta f(\mathbf{r},\mathbf{p},t)$ of the individual densities from the ensemble averaged one $f(\mathbf{r},\mathbf{p},t) = <\hat{f}(\mathbf{r},\mathbf{p},t)>$ is expressed by the correlation function,

$$\sigma(\mathbf{r},\mathbf{p};\mathbf{r}',\mathbf{p}';t) = <\delta f(\mathbf{r},\mathbf{p},t) \, \delta f(\mathbf{r}',\mathbf{p}',t)>. \tag{3.1}$$

The correlation function $\sigma(\mathbf{r},\mathbf{p};\mathbf{r}',\mathbf{p}')$ can be used to calculate the covariances between any two one-body observables $A(\mathbf{r},\mathbf{p})$ and $B(\mathbf{r},\mathbf{p})$,

$$\sigma_{AB}(t) = \int dr\, dp\, dr'\, dp'\, A(r,p)\, \sigma(r,p;r',p';t)\, B(r',p') \tag{3.2}$$

which vanishes if one of the observables is dynamically conserved. By linearizing the BLE around the average trajectory for small fluctuations, it is possible to derive closed equations of motion for the first and second moments. The equation for the first moment, i.e., the average density, is the BUU equation. The equation of motion for the second moment, i.e., the correlation function (3.1), becomes

$$\left[\frac{\partial}{\partial t} + v_1 \cdot \nabla_1 + v_1' \cdot \nabla_1'\right] \sigma(r,p;r',p') = C(r,p;r',p')$$

$$+ I_1 \cdot \sigma(r,p;r',p') + I_1' \cdot \sigma(r,p;r',p') \tag{3.3}$$

where I represents the linearized collision operator evaluated with the average density. On the left, the terms due to the mean-field are dropped by assuming a constant potential, which can always be added to the equation of motion.

The information contained in eqs. (2.1) and (3.3), namely, the ensemble averaged evolution and the character of the fluctuations around this average, is useful in situations when the bundle of trajectories remains reasonably well confined around the average. This is always true for stable modes. When the instabilities are present, the trajectories branch out to very different configurations at large times that the overall average does not make sense. However, for sufficiently short times, the moment approximation provides a useful tool for addressing the early development of the unstable collective modes.[10]

In a particular situation when the system remains close to a local equilibrium, one linearizes the BLE around the local equilibrium, and the linearized equation can be used to extract transport coefficients associated with macroscopic variables.[11] As an example here we consider that a system is initially uniform and in equilibrium at a temperature T with a phase-space density $f_0(\epsilon)$. We can calculate the magnitude of small density fluctuations by linearizing the BLE (2.3) around equilibrium, $\hat{f}(r,p,t) = f_0(\epsilon) + \delta f(r,p,t)$,

$$\frac{\partial}{\partial t}\delta f + v \cdot \nabla_r \delta f + \nabla_r \delta U \cdot \nabla_p f_0 = I_0 \cdot \delta f + \delta K \tag{3.4}$$

where I_0 is the linearized collision operator evaluated with the equilibrium density f_0. This equation is easily solved for the density fluctuations by expanding $\delta f(r,p,t)$ on plane waves,

$$\delta f_p(k,\omega) = \int dr\, dt\, e^{-i(k \cdot r - \omega t)}\, \delta f(r,p,t). \tag{3.5}$$

When integrated over momentum p, this relation gives the Fourier coefficients $\delta n(k,\omega)$ of the local density fluctuations. Then, in weak-damping limit, the density correlation function can be expressed as

$$< \delta n(k,\omega)\, \delta n(k',\omega')\, > = (2\pi)^4\, \delta(k+k')\, \delta(\omega+\omega')\, \sigma(k,\omega) \tag{3.6}$$

with

$$\sigma(k,\omega) = \frac{\Gamma(k,\omega)}{\left[1 + \chi(k,\omega)\frac{\partial}{\partial n}U_k\right]^2 + \left[\Gamma(k,\omega)\frac{\partial}{\partial n}U_k\right]^2} \cdot \frac{T}{\omega}. \tag{3.7}$$

Here U_k denotes the Fourier transform of the mean-field, and the quantities $\chi(k,\omega)$ and $\Gamma(k,\omega)$ are given by

$$\chi(k,\omega) = \int \frac{dp}{(2\pi)^3} \frac{k \cdot \nabla f_0}{\omega - k \cdot v} \tag{3.8}$$

$$\Gamma(\mathbf{k},\omega) = \frac{\omega}{T} \int dp_1\, dp_2\, dp_3\, dp_4\, W(12;34) \left[\frac{\Delta Q}{2}\right]^2 f_1\, f_2\, (1-f_3)\, (1-f_4) \tag{3.9}$$

with $\Delta Q = Q_1+Q_2-Q_3-Q_4$, $Q_j = (1-\mathbf{k}\cdot\mathbf{v_j})^{-1}$. The equilibrium density fluctuations in a given mode $\mathbf{k}$ is determined by integrating $\sigma(\mathbf{k},\omega)$ over frequency,

$$\sigma(\mathbf{k}) = \int \frac{d\omega}{2\pi}\, \sigma(\mathbf{k},\omega) = \frac{n_0}{m} \frac{k^2}{\omega_{\mathbf{k}}} \cdot \frac{T}{\omega_{\mathbf{k}}} \tag{3.10}$$

where n_0 is the equilibrium density and the frequencies $\omega_{\mathbf{k}}$ of the collective modes are determined by the dispersion relation, $1 + \chi(\mathbf{k},\omega)\, \partial U_{\mathbf{k}}/\partial n = 0$. This result is valid in semi-classical limit, $T \gg \omega_{\mathbf{k}}$. At low temperatures, $T \le \omega_{\mathbf{k}}$, the quantal effects become important and should be incorporated into the BLE in the form of memory effects. After this modification, the factor $T/\omega_{\mathbf{k}}$ in eq. (3.10) is replaced by $\coth(\omega_{\mathbf{k}}/2T)$.[11]

4. PROJECTION METHOD FOR NUMERICAL SIMULATION

The stochastic character of nuclear dynamics is especially important when instabilities occur, since different possible trajectories may lead towards configurations that differ drastically from one another. For a quantitative treatment of such catastrophic processes, one must carry out numerical simulations of the BLE. Randrup and coworkers have developed a lattice simulation method by treating the two-body transition rates as random variables characterized by a Poisson distribution.[12] However, a direct simulation of this manner is not very practical and propagates too much detailed information, which is not needed for describing gross properties of the density fluctuations. Therefore one has to develop methods for obtaining approximate solutions of the BLE.

In order to describe the gross properties of density fluctuations, it may be sufficient to propagate the fluctuations associated with a few low order multipole moments of the momentum distribution. As can be seen from the fluid dynamical description, the evolution of density is coupled to the fluctuation-dissipation mechanism through the momentum flow tensor, which is nothing but the local quadrupole moment of the momentum distribution. Based on this observation, Ayik and coworkers have proposed a simulation method by projecting fluctuations on the local multipole moments of the momentum distribution,[13]

$$\hat{Q}_L(\mathbf{r},t) = \int d\mathbf{p}\, Q_L(\mathbf{p})\, \hat{f}(\mathbf{r},\mathbf{p},t). \tag{4.1}$$

Here $Q_L(\mathbf{p})$ is the multipole moment operator of order L in the momentum space which is a $2L+1$ dimensional vector with components $Q_{LM}(\mathbf{p})$. The fluctuations of the multipole moments are characterized by a *diffusion matrix*, which can be deduced from the microscopic correlation matrix $C(\mathbf{p},\mathbf{p}')$ as

$$C_{LL'}(\mathbf{r},t) = \int d\mathbf{p}\, d\mathbf{p}'\, Q_L(\mathbf{p})\, Q_L(\mathbf{p}')\, C(\mathbf{p},\mathbf{p}')$$

$$= \int dp_1\, dp_2\, dp_3\, dp_4\, \Delta Q_L\, \Delta Q_{L'}\, W(12;34)\, f_1\, f_2\, (1-f_3)\, (1-f_4) \tag{4.2}$$

with $\Delta Q_L = Q_L(p_1) + Q_L(p_2) - Q_L(p_3) - Q_L(p_4)$. This quantity determines the early growth rate of the fluctuations of Q_L's, and it can be easily computed at each time step with the pseudo particle simulation. The idea is now to simulate the evolution of Q_L's by performing a multi-dimensional random walk in accordance with the time and position dependent diffusion matrix given above. Then, a single dynamical trajectory can be determined according to the following algorithm: (i)

Starting with a definite density $\hat{f}(r,p,t)$ at time t, its average evolution and the elements of the diffusion matrix are calculated during the time step Δt with the particle simulation, yielding $f(r,p,t + \Delta t)$ and $C_{LL'}(r,t)$. (ii) In the second step, the fluctuations of the multipole moments are determined according to a multi-dimensional Langevin equation,

$$\hat{Q}_L(r,t+\Delta t) = Q_L(r,t+\Delta t) + \sum_{L'} \left[\sqrt{\Delta t\ C(r,t)}\right]_{LL'} \hat{W}_{L'}. \tag{4.3}$$

Here $Q_L(r,t+\Delta t)$ is the multipole moment associated with the locally averaged density $f(r,p,t+\Delta t)$, and the quantity in the second term is the square-root of the diffusion matrix multiplied by the independent Gaussian random numbers $\hat{W}_L$ with unit variance and zero mean for each multipole moment. The square-root of the diffusion matrix is defined in a standard way in terms of the orthogonal transformation which diagonalizes $C_{LL'}(r,t)$.[8] (iii) Finally, the fluctuations are inserted into the phase–space by scaling the local momentum distribution to the new values of $\hat{Q}_L$'s, $f(r,p,t+\Delta t) \rightarrow \hat{f}(r,p,t+\Delta t)$. This procedure is repeated at each time step. In the practical applications of this method, the multipole space must be truncated to a reasonable size. We expect that the propagation of fluctuations by the quadrupole scaling above should provide a good approximation for the gross properties of density fluctuations.

5. APPLICATIONS TO HEAVY-ION COLLISIONS

5.1. Density Fluctuations

The BLE approach provides a useful basis for studying dynamics of density fluctuations in nuclear collisions at intermediate energies and for investigating possible connections between the reaction mechanism of the *multifragmentation* process and the *nuclear matter equation of state*. The multifragmentation process is usually associated with the instabilities in the spinodal region. In the spinodal region the nuclear system becomes unstable with respect to density fluctuations, and that causes the system to break up into clusters. In order to investigate the density fluctuations in heavy-ion collisions, a number of calculations has been carried out.[13] In most of these calculations the BLE events are determined in the lowest order approximation by propagating only the fluctuations associated with the z-component of the quadrupole moment of the momentum distribution, $Q_{20}(p) \equiv Q_2(p) = 2p_z^2 - p_x^2 - p_y^2$. In this case, eq. (4.3) reduces to a one-dimensional Langevin equation involving only a single diffusion coefficient $C_{22}(t) \equiv C_2(t)$. In these calculations a simplified three parameters (t_0, t_3, γ) Skyrme interaction is employed, which gives an incompressibility modulus of 200 MeV, and an energy dependent nucleon-nucleon cross-section together with phenomenological medium effects is used.[14] The computations are performed with 20 pseudo particles per physical nucleon and the collision integral is evaluated by means of, so-called, the full ensemble technique. First, we consider the fluctuations associated with the momentum space and calculate the mean value $<Q_2>$ and the variance $\sigma_2^2 = <Q_2^2> - <Q_2>^2$ of the total quadrupole moment, which provide measures for the energy dissipation and the fluctuations in the momentum space, respectively. As an example, Figure 1 shows the time evolution of the mean value $<Q_2>$, the diffusion coefficient $C_2(t)$, the variance $\sigma_2(t)$ of the total quadrupole moment of the momentum distribution and the collision rate in $^{12}C + ^{12}C$ collisions at various energies. The mean value of the quadrupole moment exhibits a typical relaxation pattern. But the effects of the binary collisions do not show up in Q_2 immediately after touching, which occurs at about 5–10 fm/c in the figure. Due to the effects of the mean-field, there is a time delay of about 15–20

584

fm/c during which the mean value of the quadrupole moment does not show up any damping, but in fact may increase as a result of the initial compression and the diabatic shift of the single-particle energies. On the other hand, the dispersion σ_2 associated with the quadrupole moment starts growing immediately after touching and reaches large values before the mean value of the quadrupole moment exhibits any sizeable damping. The diffusion coefficient C_2 is concentrated during the early stages of the collision and its peak value is much larger than the asymptotic background. As a result, the dispersion σ_2 exhibits a bump during the early stages of the collision, which is a characteristic behavior of a strong transient effect. There are strong correlations between the dissipation rate, the collision rate and the magnitude of fluctuations. The dissipation rate and the magnitude of fluctuations are large when the collision rate is high. We also notice that the magnitude of fluctuations increases for increasing energy. This follows from the fact that with increasing bombarding energy, the Pauli blocking becomes less effective and, consequently, the available phase-space for decay becomes larger, which increases the dissipation rate and the magnitude of fluctuations. Also, calculations are carried out in which the BL events are determined by the quadrupole plus octupole scaling. The results of these calculations for the quadrupole moment are similar to those obtained by the quadrupole scaling alone. However, the situation is different for the octupole moment. The scaling of the quadrupole moment alone does not generate fluctuations in the octupole mode, and as a result its variance remains small. In the case of the quadrupole plus octupole scaling, the fluctuations in the octupole mode is also propagated and its variance exhibits the characteristic bump. These calculations indicate that large dynamical fluctuations are introduced into the momentum space during early non-equilibrium stages of the collision and the amplitude of fluctuations increases with increasing energy.

The fluctuations in the momentum space are subsequently propagated by the mean-field into density. If the system enters into the spinodal region, the fluctuations trigger the instabilities causing the system to break-up into clusters. In a

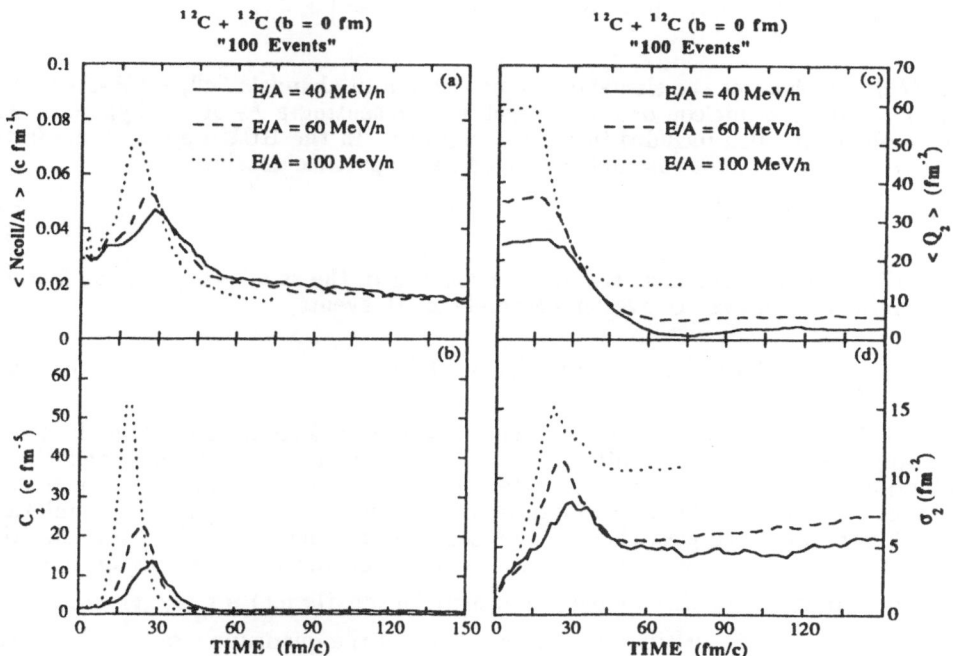

Figure 1. Time evolution of the collision rate (a), the diffusion coefficient (b), the mean value (c) and the variance (d) associated with the total quadrupole moment of the momentum distribution in ^{12}C + ^{12}C collisions at bombarding energies of 40 (solid lines), 60 (dashed lines) and 100 (dotted lines) MeV per nucleon. Taken from ref.[11]

recent work,[15] the BL model has been applied to investigate the multifragmentation processes in ^{40}Ca + ^{40}Ca collisions at bombarding energies of 60 and 90 MeV/n. In these calculations the final state of each event is analyzed using a coalescence model, and the average properties of the intermediate mass fragments are deduced. These calculations indicate that the gross properties of the intermediate mass fragments produced in ^{40}Ca + ^{40}Ca collisions are qualitatively similar to those observed in heavier systems by the MSU group[16] and by the ALADIN collaboration.[17]

5.2. K$^+$ Production at Sub-Threshold Energies

On the basis of the BL approach one can study the rare processes occurring in heavy-ion collisions and investigate the effects of correlations on the sub-threshold particle production and the hard photon emission. The strong transient behavior of the momentum space described in the previous section could make an important effect on the particle production mechanism in heavy-ion collisions below threshold energies. When the fluctuations are large, i.e., during the time interval where the dispersion of the quadrupole moment is going through a maximum, a fraction of the BL events may have sufficient energy to produce mesons below threshold energies, which is otherwise not possible in the average description of the BUU model. In addition, the multiparticle interactions involving more than two nucleons provide an efficient mechanism to produce sub-threshold particles.[18] Both effects should be incorporated into the cross-section calculations. Here, we consider the effects of fluctuations and discuss the calculations carried out for the subthreshold K$^+$ production cross-section in the BL approach and compare the results with those obtained in the average description of the BUU model.[19]

At energies below 1.0 GeV/n, kaons are produced mainly in the elementary baryon-baryon collisions and the invariant K$^+$ production cross-section is evaluated by folding the elementary cross-section with the momentum distribution of the colliding baryons,[20]

$$E \frac{d^3}{dp^3} \sigma_{K^+} = \sum_c \int 2\pi b \, db \int dt \int dr \, dp \, dp' \, |v-v'| \, E \frac{d^3}{dp^3} \sigma_c \cdot F. \qquad (5.1)$$

Here $d^3\sigma_c/dp^3$ denotes the elementary cross-section in channels "c", B+B→B+K$^+$+Y, with B as either a nucleon or a delta and Y representing a Λ- or a Σ-hyperon. The pion channels are also included in the calculations. In the BUU approach the factor F is given in terms of the ensemble averaged single-particle density by

$$F \rightarrow F_{BUU} = f(r,p,t) \, f(r,p',t) \, [1-f(r,p'',t)]. \qquad (5.2)$$

In the BL approach, the cross-section is evaluated in the same way, but the factor F is determined as an average over an ensemble of BL events,

$$F \rightarrow F_{BLE} = <\hat{f}(r,p,t) \, \hat{f}(r,p',t) \, [1-\hat{f}(r,p'',t)]>. \qquad (5.3)$$

An ensemble of BL events is determined using the projection method in the lowest order approximation with quadrupole scaling as discussed in Section 4, and the K$^+$ production cross-section in both the BUU and the BL approaches are calculated with the input data of ref.[21] In order to improve the statistics of the numerical simulations at low bombarding energies, also a model calculation is performed by parameterizing the fluctuating momentum distribution in terms of the z-component of the total quadrupole moment Q of the momentum distribution $\hat{f}(r,p,t) \rightarrow f_Q(r,p,t)$, assuming a Gaussian form for the distribution function P(Q) of the quadrupole moments. In this case, the factor F in eq. (5.1) becomes

$$F \rightarrow F_G = \int dQ \, P(Q) \, f_Q(r,p,t) \, f_Q(r,p',t) \, [1-f_Q(r,p'',t)]. \qquad (5.4)$$

In this Gaussian model, the mean value $<Q>$ and the variance σ_Q of the distribution function $P(Q)$ as well as the momentum distribution $f_Q(r,p,t)$ for each Q-bin, are extracted from a large number of numerical simulations of the BLE, and the cross-section eq. (5.1), is evaluated by numerical integration. Figure 2 shows the total K^+ production cross-section in $^{12}C + ^{12}C$ collisions as a function of bombarding energy. The solid line and the dashed line in Figure 2 represents the result of calculations in the BL and the BUU approaches, respectively. These calculations are performed with a soft mean-field and by taking only the nucleon-nucleon channels into account. The BL model with fluctuations, in particular at low energies, gives much larger cross-sections than those obtained in the BUU description. The dots in Figure 2 represent the calculations performed within the Gaussian model, which agree well with the numerical simulations.

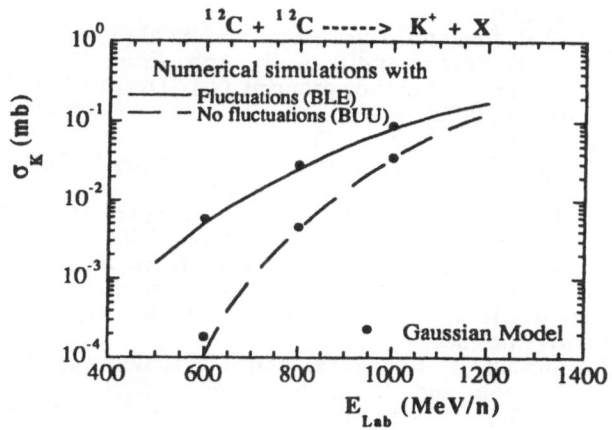

Figure 2. The K^+ production cross-section in $^{12}C + ^{12}C$ collisions as a function of the bombarding energy. Shown are the BLE simulations (solid line), the BUU simulations (dashed line) and the Gaussian model calculations (dots). The calculations are performed with a soft mean-field. Taken from ref.[19]

6. CONCLUSIONS

The development of transport descriptions may lead to novel theoretical tools for understanding the complex reaction mechanism in heavy-ion collisions at intermediate energies. In particular, the recently developed BL model is a leading candidate theory for addressing such catastrophic phenomena as phase transitions and nuclear multifragmentations. Because of numerical complexity, however, so far a limited number of applications have been carried out. The lattice simulation of the BLE, proposed in ref.[12], is so computer intensive that the realistic applications to heavy-ion collisions are difficult to make. The projection method developed in ref.[13] is very efficient for realistic applications, however, one has to truncate the multipole expansion in momentum space rather arbitrarily. Therefore, there is a need for developing approximate methods for realistic applications of the BL model to heavy-ion collisions. One recent effort in this direction may provide a very powerful method for simulating the BLE in realistic problems.[22]

Heavy-ion collisions at bombarding energies of a few GeV/n give rise to highly compressed and excited nuclear matter during the initial stages of the collision. At these energies, the baryonic excitations and the mesonic degrees of freedom become important. In the framework of the Walecka-type field theory, the BL approach has been extended to the relativistic collisions.[23] It will be interesting to investigate the influence of correlations on the particle production mechanism and the collision dynamics at high energies using the relativistic BL model.

ACKNOWLEDGMENTS

I wish to acknowledge my collaborators C. Gregoire, E. Suraud, D. Boilley, M. Belkacem and P. G. Reinhard, with whom I have had the pleasure of collaborating through various periods over the past several years. This work is supported in part by US–DOE grant DE–FG05–89ER40530.

REFERENCES

1. G.F. Bertsch and S. Das Gupta, *Phys. Rep.* 160:190 (1988).
2. W. Cassing and U. Mosel, *Prog. Part. Nucl. Phys.* 25:235 (1990).
3. S. Ayik and C. Gregoire, *Phys. Lett.* B212:269 (1988); and *Nucl. Phys.* A513:187 (1990).
4. J. Randrup and B. Remaud, *Nucl. Phys.* A514:339 (1990).
5. P. G. Reinhard, E. Suraud and S. Ayik, *Ann. Phys.* 213:204 (1992); and P. G. Reinhard and E. Suraud, preprint GANIL–P9107 (1991), *Ann. Phys.* (N.Y.) 216:987 (1992).
6. S. Nakajima, *Prog. Theor. Phys.* 20:948 (1958).
7. R.W. Zwanzig, "Quantum Statistical Mechanics," P.H.E. Meijer, ed., Gordon and Breach, New York, (1966).
8. H. Risken, "The Fokker-Planck Equation," Springer, Berlin, (1984).
9. M. Bixon and R. Zwanzig, *Phys. Rev.* 187:267 (1969).
10. M. Colonna, Ph. Chomaz and J. Randrup, *Nucl. Phys. A* (1993) in press; J. Randrup, Interdisciplinary Workshop on Statistical Description of Transport in Plasma, Astro– and Nuclear Physics, Les Houchers, France (1992), Nova Science.
11. S. Ayik, E. Suraud, J. Stryjewski and M. Belkacem, *Z. Phys.* A337:413 (1990); S. Ayik and D. Boilley, *Phys. Lett.* B276:263 (1992) and 286:482E (1992); S. Ayik, submitted to *Phys. Lett.* (1993).
12. Ph. Chomaz, G. F. Burgio and J. Randrup, *Phys. Lett.* B254:340 (1991) G.F. Burgio, Ph. Chomaz and J. Randrup, *Nucl. Phys* A529:157 (1991); and F. Chapelle, et al., *Nucl. Phys.* A540:227 (1992).
13. E. Suraud, S. Ayik, J. Stryjewski and M. Belkacem, *Nucl. Phys.* A542:141 (1992); S. Ayik, E. Suraud, M. Belkacem and D. Boilley, *Nucl. Phys.* A545:35c (1992); E. Suraud, S. Ayik, M. Belkacem and F.–S. Zhang, preprint GANIL–P9316 and submitted to *Nucl. Phys. A* (1993).
14. C. Gregoire, B. Remaud, F. Sebille, L. Vinet and Y. Raffray, *Nucl. Phys.* A465:317 (1987).
15. F.–S. Zhang and E. Suraud, preprint YITP/K–1009 and submitted to *Phys. Lett. B* (1993).
16. D. R. Bowman, et al., *Phys. Rev. Lett.* 67:1527 (1991).
17. C. A Ogilvie, et al., *Phys. Rev. Lett.* 67:1214 (1991).
18. P. Danielewicz, *Ann. Phys.* 197:154 (1990).
19. M. Belkacem, E. Suraud and S. Ayik, *Phys. Rev.* C47:R16 (1993).
20. W. Cassing, V. Mettag, U. Mosel and K. Niita, *Phys. Rep.* 188:363 (1990).
21. J. Randrup and C. M. Ko, *Nucl. Phys.* A343:519 (1980); and 411:537 (1983).
22. J. Randrup and S. Ayik, preprint LBL–34484, and submitted to *Nucl. Phys. A* (1993).
23. S. Ayik, *Phys. Lett.* B265:47 (1991).

CLUSTERING PHENOMENA IN BOSON- AND FERMION DISTRIBUTIONS FROM HEAVY ION REACTIONS AT RELATIVISTIC ENERGIES

Holger Merlitz[1] and Dietrich Pelte[1,2]

[1]Physikalisches Institut der Universität Heidelberg
[2]Max-Planck-Institut für Kernphysik Heidelberg
D-69120 Heidelberg, Germany

INTRODUCTION

With increasing beam energy (E > 500 AMeV) more and more pions are produced in heavy ion collisions. For example, central S + S collisions at E = 200 AGeV have a π^- multiplicity of the order of $m_{\pi^-} = 100$; but already at E = 1 AGeV, Au + Au collisions produce around $m_{\pi^-} = 15$ negative pions per central event. The counting distribution of these pions in a solid angle $\Omega < 4\pi$ deviates from the Poisson form, i.e. $\sigma^2 > \overline{n}$, where $\sigma^2 = \overline{n^2} - \overline{n}^2$ and $\overline{n}$ is the mean value of the distribution. This deviation is most likely due to the Boson nature of pions, which leads to a clustering of pions in phase space.

Pion clustering in phase space was extensively discussed by Zajc.[1] This short contribution should be considered as an extension of his work, it describes a different method to obtain the phase space distribution and presents exact formulas for the counting distributions of Bosons and Fermions.

THE COUNTING DISTRIBUTIONS FOR BOSONS AND FERMIONS

The starting point is the well known fact that the probability of a Boson to occupy a phase space cell is proportional to n + 1, when this cell is already occupied by n Bosons. For Fermions this same probability is 0 or 1, depending on whether the cell is already occupied or not. The size of the phase space cell δ is given by $\delta x_i \delta p_i = \hbar$, where the index i refers to the three coordinates. This implies that the spacial volume $V = \prod \delta x_i$ of the phase space cell, i.e. the particle source volume, can be determined if $\delta \vec{p}$ can be measured. The measurement requires that the particles remain unperturbed by residual interactions, i.e. $\delta \vec{p} = \vec{p}$, once they have been created. Using spherical coordinates $\vec{p} = (p, \cos \theta_p, \phi_p)$, the absolute value p is often bound by the experiment, i.e. by putting a gate on the particle momenta or by the reaction. The area of the remaining

Hot and Dense Nuclear Matter, Edited by
W. Greiner *et al.*, Plenum Press, New York, 1994

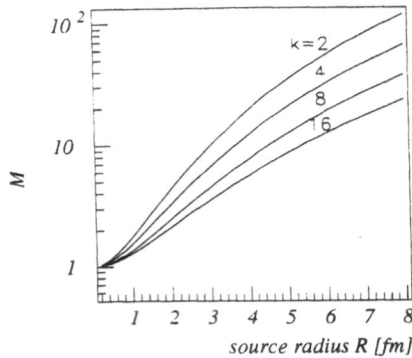

Figure 1: Values of $\mathcal{M}$ for different source radii R and different number k of trial cells.

coordinates $\delta = \delta \cos \theta_p \delta \phi_p$ can be inferred from the counting distribution by placing a trial grid with cell size $\Delta = \Delta \cos \theta_p \Delta \phi_p$ over the available area $-1 \leq \cos \theta_p \leq 1$, $0 \leq \phi_p \leq 2\pi$, i.e. $\Delta = 4\pi/k$ with k the number of trial cells. It is known from the theory of speckle interferometry[2] or it may be derived by using a wave paket expansion of the particles that in case $\Delta \gg \delta$ the ratio Δ/δ is given by

$$\frac{\Delta}{\delta} = \mathcal{M} = \frac{\int dp_1 dp_2 S(p_1) S(p_2)}{\int dp_1 dp_2 S(p_1) S(p_2) \mid g(p_1, p_2) \mid^2} \tag{1}$$

where S(p) represents the momentum distribution of the particles and

$$\mid g(p_1, p_2) \mid^2 = \frac{\int_{\Delta_1} \int_{\Delta_2} \mid \mathcal{F}(\vec{q}, \vec{R}) \mid^2 d\Omega_1 d\Omega_2}{\int_{\Delta_1} \int_{\Delta_2} d\Omega_1 d\Omega_2} \tag{2}$$

with $\vec{q} = \vec{p_1} - \vec{p_2}$ and $d\Omega = d \cos \theta_p d\phi_p$. The Fouriertransform $\mathcal{F}(\vec{q}, \vec{R})$ of the particle source, e.g. for a spherical Gaussian source density with radius R, is given by

$$\mathcal{F}(q, R) = \exp \left\{ -\frac{R^2 q^2}{4} \right\} \tag{3}$$

The values of $\mathcal{M}$ for different source radii R and different cell numbers k are shown in Fig.1. In calculating $\mathcal{M}$ the energy distribution is assumed to have Maxwellian shape with a temperature T = 45 MeV.

For N particles in 4π, the counting distribution of n particles per trial cell is then for Bosons

$$P_B(n) = \binom{n + \mathcal{M} - 1}{n} \frac{(\mathcal{M}k - \mathcal{M} + N - n - 1)! N! (\mathcal{M}k - 1)!}{(N - n)! (\mathcal{M}k - \mathcal{M} - 1)! (\mathcal{M}k + N - 1)!} \tag{4}$$

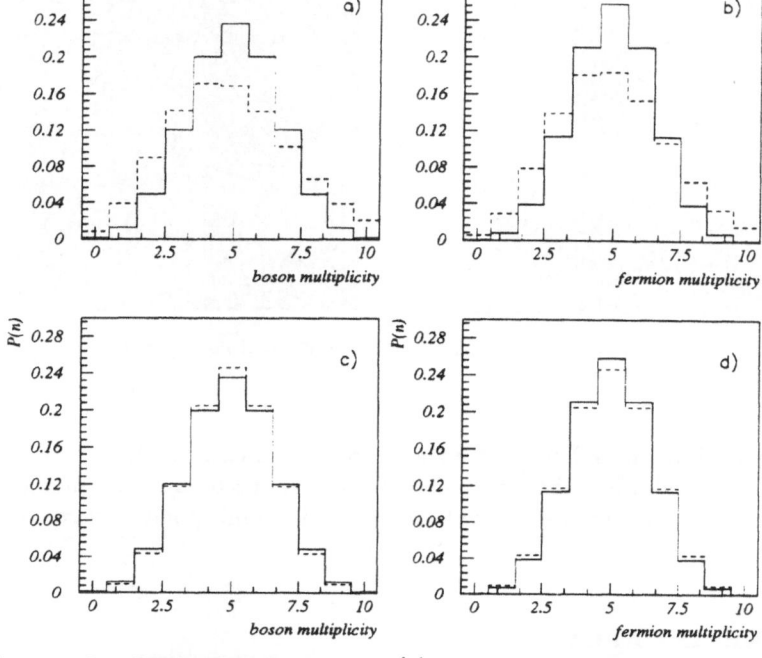

Figure 2: Counting distributions for N = 10, $\mathcal{M}$ = 55 and k = 2. a) Bosons according to eq.(4) (full line) or eq.(6) (dashed line); b) Fermions according to eq.(5) (full line) or eq.(7) (dashed line); c) Bosons according to eq.(4) (full line) or classical binomial (dashed line); d) Fermions according to eq.(5) (full line) or classical binomial (dashed line).

and for Fermions

$$P_F(n) = \left(\begin{array}{c} \mathcal{M} \\ n \end{array} \right) \frac{(\mathcal{M}k - \mathcal{M})!N!(\mathcal{M}k - N)!}{(N-n)!(\mathcal{M}k - \mathcal{M} - N + n)!(\mathcal{M}k)!} \tag{5}$$

These expressions remain valid even when the condition $\Delta \gg \delta$ is not fulfilled.

For a large number of trial cells ($k \rightarrow \infty$) and using $\bar{n} = N/k$, they can be approximated by the more familiar expressions, already quoted in the paper of Zajc[1]

$$P_B(n) \approx \left(\begin{array}{c} n + \mathcal{M} - 1 \\ n \end{array} \right) \left(\frac{\bar{n}}{\mathcal{M} + \bar{n}} \right)^n \left(\frac{\mathcal{M}}{\mathcal{M} + \bar{n}} \right)^{\mathcal{M}} \tag{6}$$

$$P_F(n) \approx \left(\begin{array}{c} \mathcal{M} \\ n \end{array} \right) \left(\frac{\bar{n}}{\mathcal{M} - \bar{n}} \right)^n \left(\frac{\mathcal{M} - \bar{n}}{\mathcal{M}} \right)^{\mathcal{M}} \tag{7}$$

In Fig.2a the exact distribution $P_B(n)$ (full line) is compared to its approximation (dashed line), using the situation found in SIS experiments: total Boson number N = 10, $\mathcal{M}$ = 55 (obtained for a Maxwell energy distribution with temperature T = 45 MeV), R = 6 fm and k = 2. In the Fig.2b the equivalent is shown for Fermions, and the Figs.2c,2d compare the exact quantum distributions to the classical binomial distributions. It is evident that under SIS conditions the influence of the quantum

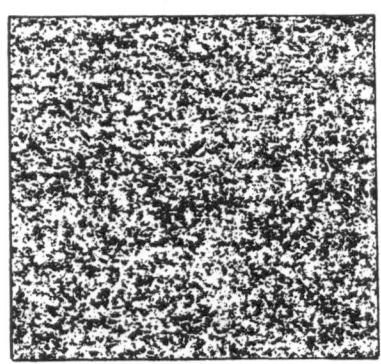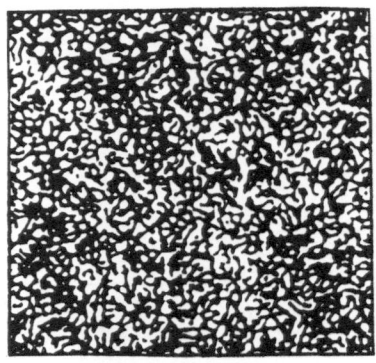

Figure 3: Speckle patterns for pions (left) and light (right).

statistics is weak but nevertheless observable in experiments with a large number of detected pions. In addition the exact distributions have to be used in the analysis. This might change at higher energies with a larger pion multiplicity, where large values of k can be used.

MONTE CARLO CALCULATIONS

The methods described here can be used to implement the clustering phenomenon of Bosons into Monte Carlo simulations. The general idea is to determine the size of the phase space cell by evaluating the integrals of $\mathcal{M}$ for $\Delta = 4\pi$ and for a given source radius R and pion momentum distribution S(p). The probability of a pion to occupy a given phase space cell is proportional to n + 1, where n is the number of pions already in this cell. This method is rather fast and allows one to study cases under extreme conditions. For example, the Fig.3 displays, in the $\cos\theta_p \% \phi_p$ plane, the distribution of 100000 neutral pions produced in one single event. The assumed number of phase space cells is 10000. For comparison the Fig.3 also shows a picture of optical speckles which demonstrates the close resemblance of both phenomena.

In a more realistic situation the Au + Au reaction at E = 1 AGev, as studied by the FOPI collaboration,[3] was simulated. The average total π^- multiplicity was measured to be $m_{\pi^-} = 10$, the low-temperature component of the kinetic energy spectra corresponds to a temperature of 45 MeV, and the source radius was assumed to be R = 6 fm. The pions are isotropically emitted from this source. The phase space distribution of π^- was calculated for 50000 events, it was analyzed in terms of the counting distribution or by the Hanbury-Brown,Twiss (HBT) method.[4] As is well known the HBT method is based on a comparison of the relative momentum distribution P(q) of clustered pions to the one Q(q) of random pions, the latter being obtained by event mixing. The ratio P(q)/Q(q) is given by

$$\frac{P(q)}{Q(q)} = 1 + \mid \mathcal{F}(q, R) \mid^2 \qquad (8)$$

However, the source radii R_m deduced by both methods from the simulated dis-

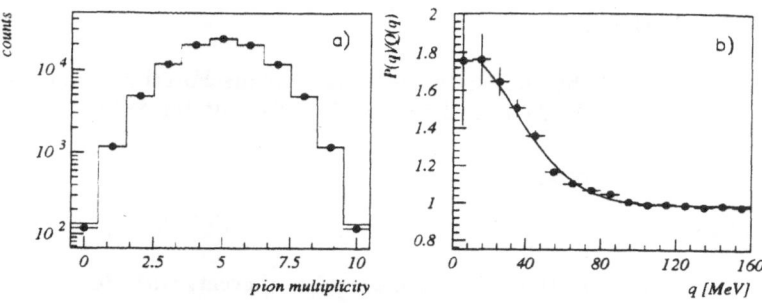

Figure 4: Counting distribution a) and correlation function b) of pions. The dots display the result of the Monte Carlo simulation; the full curves the fit with eq.(4) respectively eq.(8).

tributions, deviated by up to factor 2 from the assumed radius R. It is felt that this disagreement is caused by restricting the cell dimensions to the $\cos\theta_p, \phi_p$ coordinates and neglecting the fact that pions with largely different momenta in this cell should also not be subjected to the Bose statistics. A considerable improvement of the simulation procedure was obtained, when the integration to obtain $\mathcal{M}$ is performed under the restriction $p_1 = p_2$ and when the weight factor in the occupation probability is modified to

$$1 + \sum_{i=1}^{n} \exp\left\{-\frac{R^2 q_i^2}{2}\right\} \qquad (9)$$

where n is the number of pions already in the cell and $q_i = p - p_i$. Certainly this reduces to the usual weight n + 1, when $p = p_i$. It is interesting to note that the method of modifying the weight could probably also be used to simulate other effects, e.g. the pion residual interaction like the Coulomb repulsion of charged pions. First attempts based on a comparison between Coulomb energy and pion temperature have not led to any conclusive results yet.

In the Fig.4 the results of this modified simulation are shown and the corresponding fits to the counting distribution for k = 2 (Fig.4a) and the HBT correlation function (Fig.4b). In the first case the deduced source radius is $R_1 = 6.25 \pm 0.35 fm$, in the second case $R_2 = 5.84 \pm 0.69 fm$. Both results agree with the assumed source radius R = 6 fm. In particular the counting distribution, even in this low-energy situation, is distinctly different from the classical binomial distribution, which yields a minimum χ^2 value 100 times larger than the minimum χ^2 value of the exact distribution.

CONCLUSION

In conclusion, we have developed an algorithm to include Boson clustering into Monte Carlo simulations of heavy ion collisions and have shown that the resulting phase space distributions, when analyzed in terms of counting distributions or by the more familiar HBT correlation function, reproduce reasonably well the assumed source radius. This algorithm is rather fast and can be used also for events with very large Boson multiplicity. It also appears that it is well suited to study the impact of other phenomena like energy flow, onto the Boson clustering.

ACKNOWLEDGMENTS

This work is supported by the Bundesministerium für Forschung und Technologie under Contract No. 06 HD 525 I, and by the Gesellschaft für Schwerionenforschung under Contract No. HD Pel K.

REFERENCES

1. W.A. Zajc, Calculational methods for the generation of events with Bose-Einstein correlations, *Phys. Rev.* D35: 11 (1987).
2. J.C. Dainty, "Laser Speckle and Related Phenomena", Springer-Verlag, Berlin, Heidelberg, New York (1975).
3. M. Gnirs et al., First experiments with the central drift chamber of the 4π-detector, GSI report GSI 93-1: 45 (1993).
4. F.B. Yano and S.E. Koonin, Determining pion source parameters in relativistic heavy-ion collisions, *Phys. Lett:* 78B (1978) 556.

PRODUCTION OF ELECTRON-POSITRON PAIRS IN ATOMIC HEAVY ION COLLISIONS AT RELATIVISTIC ENERGIES

Joachim Thiel, Norbert Grün, and Werner Scheid

Institut für Theoretische Physik der
Justus-Liebig-Universität, Giessen, Germany

INTRODUCTION

The production of electron-positron pairs in relativistic atomic heavy ion collisions has been studied with increasing interest in the last years. With the notation "atomic collisions" we mean collisions with an impact parameter larger than twice the nuclear radius. Theoretical work on the electron-positron creation in collisions of fast charged particles started with the work of Landau and Lifschitz (for references of the former work see e.g. Heitler[1]). With the development of colliders for relativistic heavy ions new elaborate work on electron-positron creation came up[2, 3]. Pair creation with capture of the electron into a bound state of an ion changes the charge of the ion and is one of the main processes for the loss of ions in relativistic heavy ion colliders. The cross section is of the order of 100 barn for RHIC energies ($E_{cm} = 100$ GeV/nucleon). A nonperturbative calculation usually enlarges the cross section for pair creation with capture. Therefore, nonperturbative calculations are urgently needed for high incident energies. Reviews on this field have been given by Bertulani and Baur[4] and Eichler[5].

In the case of projectiles and targets with lower charge numbers and larger impact parameters one can use the perturbation theory of first order to calculate the probabilities for pair creation. However for collisions of ions with very high charge numbers and at small impact parameters it became evident in the last years that the electron-positron creation is a nonperturbative process. A 3-dimensional, fully nonperturbative treatment of lepton-pair production has been carried out by Strayer et al.[6] solving the time-dependent Dirac equation with a basis spline expansion. In this contribution we present three nonperturbative methods, which yield probabilities for electron-positron pair creation enhanced by one or two orders over the results

Hot and Dense Nuclear Matter, Edited by
W. Greiner *et al.*, Plenum Press, New York, 1994

of the perturbation theory. We discuss the coupled channel method worked out by Momberger et al.[7] and Rumrich et al.[8]. This method is also used by Hoffstadt[9] and extended to higher incident energies by Baltz et al.[10]. We present the finite difference method for solving the Dirac equation on a two-dimensional grid and apply it to a collision of U^{92+} (10 GeV/nucleon) on U^{91+} with very small impact parameter[11]. The nonperturbative character of the electron-positron creation can be explained by the effect that the energy of the target K-shell states dives into the negative continuum due to the relativistically enhanced field of the projectile. For CERN energies of $E_{lab} = 200$ GeV/nucleon we present coupled channel calculations for the creation of free pairs, where the electron-positron pairs are treated as bosons.

COUPLED CHANNEL AND FINITE DIFFERENCE CALCULATIONS FOR ELECTRON-POSITRON PAIR PRODUCTION WITH CAPTURE

We make use of the semiclassical approximation. This means that we treat the motion of the nuclei classically assuming the target nucleus as fixed at the origin of the coordinate system and the projectile nucleus moving with constant velocity v_P in the z-direction of the target rest system on a straight line with an impact parameter b. The electron field is quantum mechanically described by the time-dependent Dirac equation ($\hbar = m = c = 1$):

$$(H_T + V_P(t))\Psi(\vec{r}, t) = i\partial\Psi(\vec{r}, t)/\partial t \tag{1}$$

with
$$H_T = -i\vec{\alpha}\vec{\nabla} + \beta - \frac{Z_T e^2}{r}, \tag{2}$$

$$V_P = -Z_P \gamma e^2 (1 - v_P \alpha_z)/ \mid \vec{r}\,' - \vec{R}'(t) \mid, \tag{3}$$

$$\vec{r}\,' = (x, y, \gamma z), \quad \vec{R}'(t) = (b, 0, \gamma v_P t).$$

H_T is the target Hamiltonian, V_P the interaction between electron and projectile and γ the relativistic Lorentz factor: $\gamma = (1 - v_P^2)^{-1/2}$. The time-dependent Dirac equation (1) is solved by expanding Ψ in a complete set of eigenstates ϕ_j of the target Hamiltonian

$$\Psi_i(\vec{r}, t) = \sum_j a_{ji}(t)\phi_j \exp(-iE_j t). \tag{4}$$

We choose the functions ϕ_j as Coulomb-Dirac wave functions which are analytically known. For the positive and negative continuum we discretize these functions by means of wave packets. Insertion of the expansion (4) into the Dirac equation (1) and projection on a particular basis state j yields coupled equations for the expansion coefficients:

$$i\dot{a}_{ji} = \sum_k \langle\phi_j \mid V_P \mid \phi_k\rangle \exp(i(E_j - E_k)t)a_{ki}. \tag{5}$$

These equations have to be solved with the initial conditions $a_{ji}(t \to -\infty) = \delta_{ij}$. In the case of electron-positron pair production we have to treat a system with an infinite number of electrons which initially occupy the negative continuum ($E_F = -mc^2$).

The probability to observe a K-shell electron from pair production is given by

$$P_{1s} = \sum_{E_j < -mc^2} \mid a_{j,1s}(t \to \infty) \mid^2 .$$ (6)

In this formula we used the time reversal symmetry so that only a single time integration of eq.(5) starting with the $1s_{1/2}$-state is necessary to evaluate the probability (6).

First we consider the scattering of U^{92+} on U^{92+} at an incident energy of $E_{lab} = 10$ GeV/nucleon and an impact parameter of $b = 386$ fm (Compton wavelength $\hbar/(mc)$). This impact parameter approximately yields the maximum contribution to the total cross section in first order perturbation theory. The basis set contains 5 bound states ($1s_{1/2}, 2s_{1/2}, 2p_{1/2}, 2p_{3/2}, 3s_{1/2}$) and wave packets at 20 energies for the continua ($E_j = 1.3, ..., 7.3; -1.3, ..., -6.1; \Delta E = 0.6$; unit mc^2) for each of the three angular momenta $s_{1/2}, p_{1/2}, p_{3/2} (\kappa = -1, +1, -2)$. The number of coupled coefficients of this basis set is 172 including all magnetic substates. The restriction in angular momentum summation is presently the most important shortcoming of these calculations. From first order perturbation theory it is known that the summation should be extended up to about $\mid \kappa \mid = 10$, corresponding to 20 different angular momenta. However, coupled channel calculations with basis sets of that size are not feasible for us. Recently, large basis calculations were carried out by Baltz et al.[10] for the Pb on Pb system at ultrarelativistic energies.

In Figure 1 we show results for the differential probability to observe a positron of energy E_{e+} as a function of this energy.[7] The histogram-like curves give the total probability and its decomposition into contributions from pair creation of free pairs and with simultaneous capture. The smooth curves are the results of a calculation in first order perturbation theory, where the same angular momentum quantum numbers are taken into account as in the coupled channel calculation. We find the astonishing result that couplings of higher orders increase the positron spectrum by more than one order of magnitude, especially the capture from vacuum is raised by roughly two orders of magnitude.

For pair production with K-shell capture in a collision of Pb on Pb at an incident energy of E_{lab}=1.2 GeV/nucleon we find an increase of a factor of 50 at zero impact parameter above the result of the first order perturbation theory[8]. With increasing impact parameter the coupled channel results approach the results of the perturbation theory as expected. The total cross section for pair production with K-shell capture is obtained as 0.15 barn and 1.0 barn with perturbation theory and coupled channel calculations, respectively. This increase of the cross section is remarkable. Very recent experiments of Belkacem et al.[12] with U on Au collisions at $E_{lab} = 0.96$ GeV/nucleon yielded a cross section for pair production with K-shell capture of 2.19 ± 0.25 barn, which is in the order of our theoretical nonperturbative result.

Eq. (1) can also be solved by the finite difference method[11]. For reasons of simplicity we assume that the projectile nucleus moves with constant velocity v_P along the z-axis and the target nucleus is fixed at z=0 which means that the impact parameter is set equal to zero. This is a good approximation for atomic collisions with small values of b in the order of two times the nuclear radius. Thus we solve a rotationally symmetric problem with the finite-difference method on a two-dimensional grid. The numerical method and tests of the accuracy were published by Becker et al.[13].

With the finite difference method we calculated the probability for pair creation with K-shell capture and ionization in nearly central collisions of U^{92+} on $U^{92,91+}$

at E_{lab}=10GeV/nucleon. The Dirac equation (1) was solved on a two-dimensional grid by starting with the $1s_{1/2}$ wave function bound at the target ion. The grid had a measure of 600×300 meshes with a size of $\Delta z = \Delta\rho = 5 \cdot 10^{-4}a_o = 26.5$ fm. Figure 2 shows the absolute square of the wave function in the $z\rho$-plane (cylindrical coordinates z and ρ) for the time 11 (unit $\hbar/(mc^2) = 1.288 \cdot 10^{-21}s$) after the collision. For further explanation see the figure caption. In part the wave function follows the projectile indicating large probabilities for ionization and pair production with high linear momenta of the ionized electron and created positron in the direction of the collision axis.

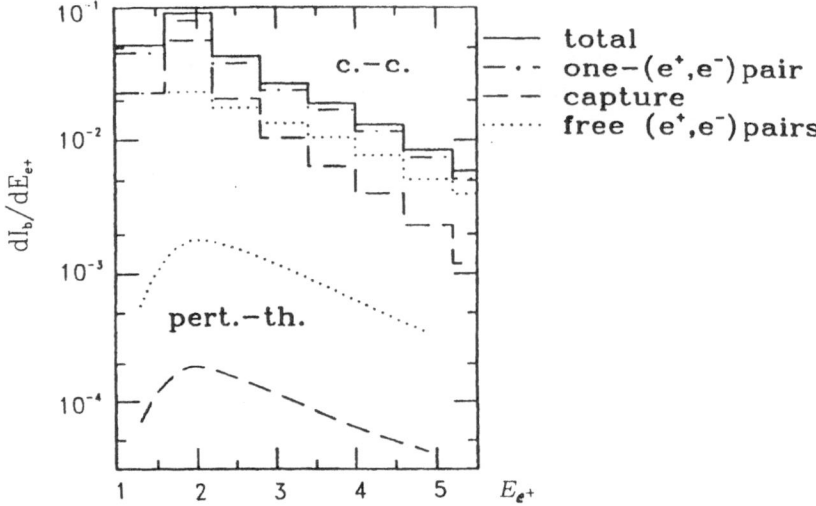

Figure 1. Differential probability for the observation of a positron of energy E_{e+} (in units of mc^2=511 keV) in the collision $U^{92+} + U^{92+}$ at E_{lab}=10 GeV/nucleon and b=386 fm. The histrogram-like curves show the coupled channel results, the continuous curves below are the results of the first order perturbation theory. The solid curve gives the total positron spectrum and the dotted-dashed curve the contribution from the production of a single electron-positron pair during the collision. The dotted curves show the probabilities for the production of free pairs and the dashed curves those for the production of pairs with simultaneous inner shell capture.

The wave function can be used to obtain the amplitudes for the analytical eigenstates of the target Hamiltonian. By projecting on excited bound states we get the probabilities for excitation and by projecting on the positive and negative continuum states the probabilities for ionization and pair production with capture in the $1s_{1/2}$ bound state of the target, respectively. The exact target eigenfunctions with quantum numbers κ (Coulomb-Dirac functions) are not the optimum basis states to describe ionization and pair production with high linear momenta in the forward direction.

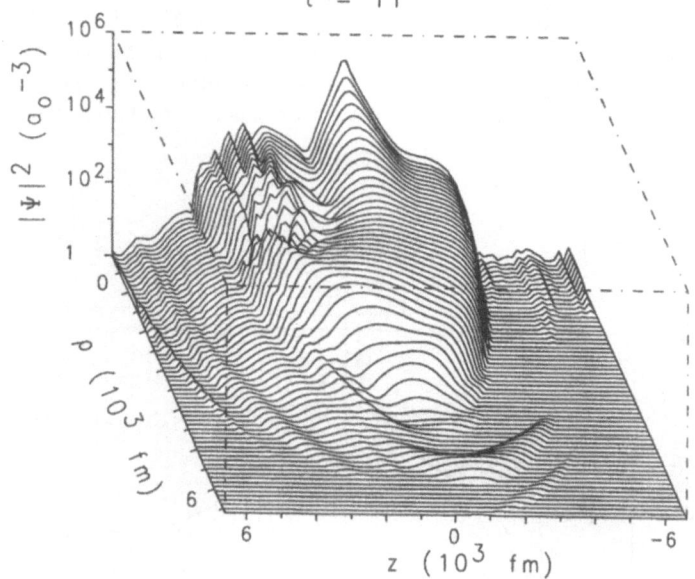

Figure 2. Probability density of the wave function of the electron at time $t= 11$ (unit $1.288 \cdot 10^{-21}$s) for a nearly central collision of U^{92+} on U^{91+} with an incident energy of $E_{lab}=10$GeV/nucleon. The projectile moves from the right to the left hand side and is located at $z= 4.2 \cdot 10^3$fm. The density is given in units of a_o^{-3} ($a_o=5.3 \cdot 10^4$fm).

Therefore, we also projected with Sommerfeld-Maue functions which are approximate solutions of the target Hamiltonian. They contain an incoming spherical wave and a plane wave with linear momentum and helicity, but they are not orthogonal.

Figure 3 shows the probability of pair production with capture for the nearly central collision of U^{92+} (10 GeV/nucleon) on U^{92+} as a function of time. The solid curve represents the values obtained by summing the probability obtained with Coulomb-Dirac wave functions with $\mid \kappa \mid \leq 5$. Values for the probability obtained with Sommerfeld-Maue functions were only calculated for t=0, 6 and 11 because of the limited computer time. They are shown by full dots and are interpolated by a dashed-dotted curve to guide the eyes. The probability is compared with those of a calculation using first order perturbation theory (dashed curve) and of a coupled channel calculation (dotted curve). Again we find the strongly nonperturbative character of the pair production in relativistic collisions of very heavy ions.

The fair agreement of the finite difference method with the coupled channel method seems to be astonishing since only angular momenta with $\kappa = \pm 1, \pm 2$ were used in the coupled channel calculation. We found that continuum states with high angular momenta get occupied in the finite difference calculation at larger times ($t > 0$) via a transfer of the probabilities from smaller to higher angular momenta, whereas for t=0 the finite difference and coupled channel methods yield similar dominant contributions of the $p_{1/2}(\kappa = 1)$ continuum states to the total probability. This may be the reason for the agreement of both methods in the total probability for pair production with capture.

The reason for the failure of the perturbation theory in first order can clearly be seen in Figure 3 where the probability increases up to the order of unity near t=0 although the final probability is very small. Therefore, the perturbation theory becomes invalid around the point of closest approach for relativistic very heavy ions and small impact parameters.

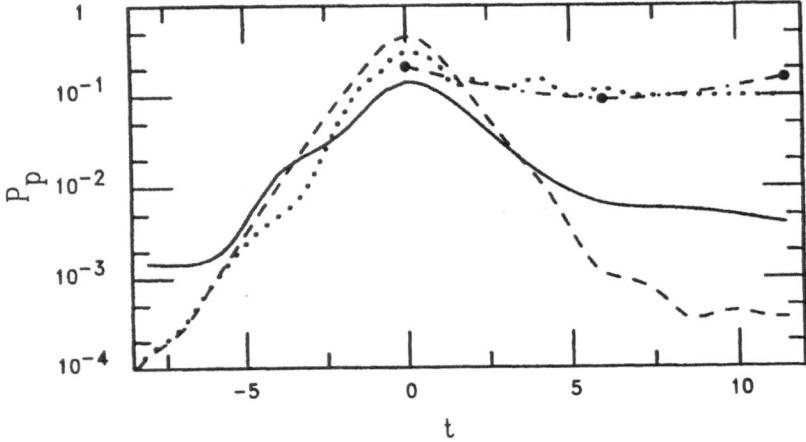

Figure 3. The probability of pair production with capture for nearly central collisions of U^{92+} (10 GeV/nucleon) on U^{92+} is shown as a function of time. The unit of time is $1.288 \cdot 10^{-21}$s. The solid curve is calculated by projecting with Coulomb-Dirac eigenfunctions of the target Hamiltonian up to $|\kappa|=5$. The values calculated by projecting with Sommerfeld-Maue functions are shown by full dots and are interpolated by a dashed-dotted curve to guide the eyes. The result is compared with those obtained by first order perturbation theory (dashed curve) and a coupled channel calculation (dotted curve).

Figure 4 shows the expectation value of the energy of the initial $1s_{1/2}$ state, calculated with the time-dependent numerical wave function[11]. We note that this expectation value immerses into the negative continuum around the point of closest approach because of the relativistically enhanced electromagnetic field of the projectile. Therefore, one expects a large overlap of this state with the states of the negative continuum and, therefore, a large probability to create electron-positron pairs. This situation has strong connections with the effect of the spontaneous production of electron-positron pairs in heavy ion collisions near the Coulomb barrier which has been studied by Greiner and his Frankfurt school[14]. Obviously, this diving of the expectation value of energy into the negative continuum is the reason for the strong nonperturbative behaviour of the pair production process.

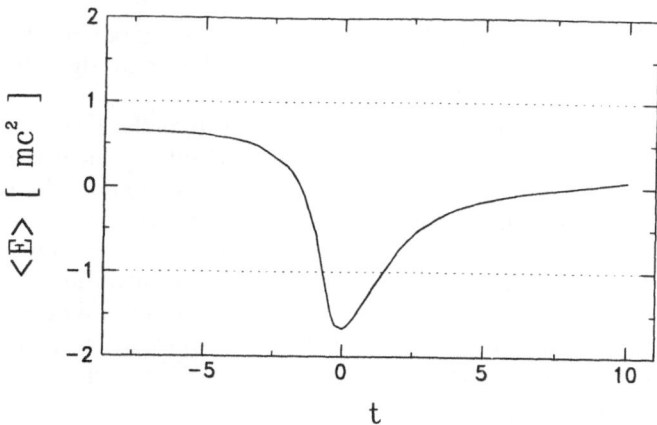

Figure 4. Expectation value of energy calculated with the solution obtained with the finite difference method for U^{92+}(10 GeV/nucleon)+U^{91+}.

COUPLED CHANNEL FORMALISM IN A BOSONIZED SPACE

It was shown[4] that perturbative results at ultrarelativistic incident energies ($E_{lab} \geq$ 500 GeV/nucleon) and impact parameters of $b = 386$ fm violate unitarity. Because of this behaviour, multiple-pair creation is expected to get important for high energies and small impact parameters of the order of the Compton wavelength. In the nonperturbative theories described above multiple-pair creation can only be treated with a very large numerical effort. For example, the number of channels in a coupled channel calculation has to be much larger than the expected number of pairs in each region of the electron and positron energies. The Pauli principle allows only one electron per channel, although each channel includes a large number of physical states. In order to avoid this artificial effect, each electron-positron pair can be treated as a boson and the Pauli principle neglected. If a further approximation is made and rescattering processes of the electrons or positrons of the created pairs are ignored[15, 16, 17], a Poisson distribution is obtained for the multiplicity of the pairs, with the mean number of pairs given by the lowest order perturbation theory. This procedure is actually only consistent up to lowest order perturbation theory.

In this section we present a consistent nonperturbative bosonic theory based on a coupled channel formalism, in order to study nonperturbative effects in the multiple-pair production. As usual we use the semiclassical approximation as described in the second section. The nuclei are assumed as point-like charged particles producing the classical electromagnetic field. The calculations are carried out in the coordinate frame where the nuclei have opposite constant velocities v_0 of the same amount moving on straight lines. The classical electromagnetic fields of the moving ions are given by

$$
\begin{aligned}
A_0^{(T,P)} &= \gamma e Z_{T,P} / \sqrt{(x \pm b/2)^2 + y^2 + \gamma^2(z \pm v_0 t)^2} \\
\vec{A}^{(T,P)} &= \pm v_0 \, A_0^{(T,P)} \, \vec{e}_z,
\end{aligned}
\tag{7}
$$

where $\gamma = 1/\sqrt{1 - v_0^2}$ is the Lorentz factor and $Z_{T,P}$ the charge numbers of the nuclei. The upper signs are valid for the target field and the lower signs for the projectile field.

Standard boson mappings of fermionic theories map the fermionic space one-to-one into a bosonic one, where partly the Pauli principle is neglected[18, 19]. We use the boson mapping in order to reduce the number of channels in the coupled channel calculations. We choose a mapping which maps the physical states of a certain range in the electron and positron momenta to just one bosonic pair state.

Differently to the formalism of the last sections we introduce a field-theoretical treatment and use the interaction picture. The free leptonic field operator is expanded in plane wave solutions $\tilde{\varphi}_{p,s}$ for positive energy eigenvalues and $\tilde{\chi}_{p,s}$ for negative energy eigenvalues of the free Dirac equation $i\partial\Psi/\partial t = \{-i\vec{\alpha}\vec{\nabla} + \beta\}\Psi$, i.e.

$$\Psi(\vec{r}, t) = \sum_{s=-1/2}^{1/2} (2\pi)^{-3/2} \int d^3p \left[b(p, s)\tilde{\varphi}_{p,s}(\vec{r}, t) + d^+(p, s)\tilde{\chi}_{p,s}(\vec{r}, t) \right]. \qquad (8)$$

Here, p is the momentum eigenvalue and s the helicity. The creation and annihilation operators for the electrons are denoted by b^+ and b and those for the positrons by d^+ and d, respectively. The interaction Hamiltonian, which describes the interaction between the lepton field and the classical electromagnetic field $A = A^{(T)} + A^{(P)}$, is given by $H_I = -e \int d^3x :\overline{\Psi}(x)A_\mu\gamma^\mu\Psi(x):$, where $:...:$ denotes normal ordering of the fermionic operators. Obviously the interaction operator consists of four different parts describing creation of pairs, annihilation of pairs, and rescattering of an electron or a positron out of a created pair.

We can neglect the Pauli principle under the condition that the number of created pairs in each bosonic pair state is much smaller than the number of underlying fermionic states. Then are able to map the bi-fermionic pair creation operators b^+d^+ to bosonic operators $A_{\alpha\beta}^+$,

$$\int_{\Delta_\alpha} d^3p \int_{\Delta_\beta} d^3p' \, b^+(ps_\alpha)d^+(p's_\beta) \, / \, \Delta^3P \qquad \longrightarrow \qquad A_{\alpha\beta}^+. \qquad (9)$$

Here we introduced pair channels $\alpha\beta$ with quantum numbers $\vec{p}_\alpha s_\alpha$ for the electron and $\vec{p}_\beta s_\beta$ for the positron, where α and β both represent momentum ranges Δ_α, Δ_β. The momentum spaces for electrons and positrons are divided into cubes of the same size Δ^3P. Each cube Δ_α is labeld by the mean momentum $\vec{p}_\alpha$ and the helicity s_α. The pair creation and annihilation operators $A_{\alpha\beta}^+$ and $A_{\alpha\beta}$ are assumed to fulfil bosonic commutation relations. In a similar way we map the annihilation operators db and the rescattering operators b^+b, d^+d of the interaction Hamiltonian to the bosonic operators $A_{\alpha\beta}$, $R_{\alpha\beta}^{(+)}$, and $R_{\alpha\beta}^{(-)}$, respectively. The transition of the interaction Hamiltonian into the bosonic picture can be illustrated for the creation operator:

$$\int_{\Delta_\alpha} d^3p \int_{\Delta_\beta} d^3p' \int d^3x \, \tilde{\varphi}_{p,s_\alpha}^+ \gamma_\mu A^\mu \tilde{\chi}_{p',s_\beta} b_{p,s_\alpha}^+ d_{p',s_\beta}^+ \longrightarrow \mathcal{M}_{\alpha\beta}^{(A^+)} A_{\alpha\beta}^+.$$

The multiplicity distribution in every channel is a Poisson distribution if the Pauli principle is neglected. Therefore, coherent states are a natural choice for the bosonic

Fock space states:

$$|\Phi(t)\rangle = \exp\left[-\sum_{\alpha\beta}|v_{\alpha\beta}(t)|^2/2\right]\exp\left[\sum_{\alpha\beta}v_{\alpha\beta}(t)A_{\alpha\beta}^+\right]|0\rangle, \tag{10}$$

where $|0\rangle$ is the bosonic vacuum state. The mean number of created pairs in the channel $\alpha\beta$ is given by

$$N_{\alpha\beta} = \lim_{t\to\infty}\,(\,\Phi(t)\,|\,A_{\alpha\beta}^+A_{\alpha\beta}\,|\,\Phi(t)\,) = \lim_{t\to\infty}|v_{\alpha\beta}(t)|^2. \tag{11}$$

Using the variational principle in the boson space with the coherent states (10), we derive coupled channel equations for the coefficients $v_{\alpha\beta}$:

$$i\dot{v}_{\alpha\beta} = \mathcal{M}_{\alpha\beta}^{(A^+)} + \sum_{\delta(E_\delta<0)}\mathcal{M}_{\beta\delta}^{(R^{(-)})}v_{\alpha\delta} + \sum_{\gamma(E_\gamma>0)}\mathcal{M}_{\alpha\gamma}^{(R^{(+)})}v_{\gamma\beta}. \tag{12}$$

We applied the bosonic coupled channel formalism to Pb-Pb collisions at the CERN energy $E_{lab} = 200$ GeV/nucleon. The channels represent states out of cells of the momentum spaces for electrons and positrons with volumes of $(2 \text{ MeV}/c)^3$. The momentum spaces are restricted to ellipsoids with radii of 6 MeV/c in transversal direction and 10 MeV/c in longitudinal direction, which yields 200 channels per helicity value. With this choice we have 400 electron and 400 positron channels. For the bosonic channels we took all combinations into account. This means that we solved 160000 coupled differential equations.

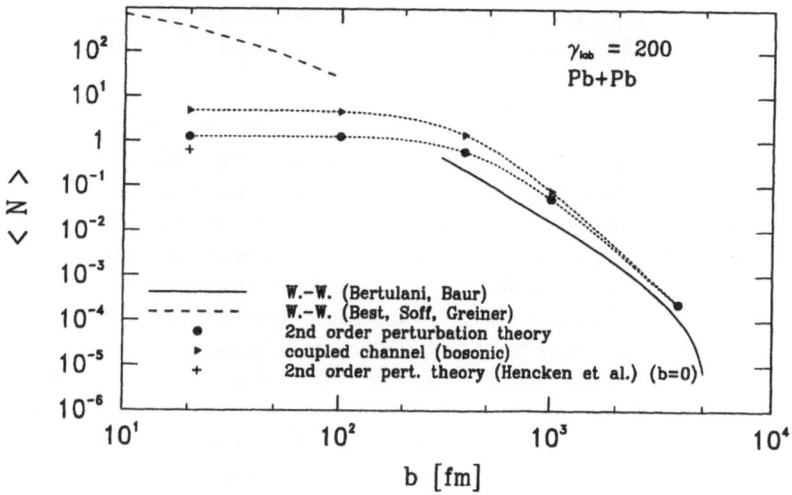

Figure 5. Mean number of created pairs in collsions of Pb^{82+} on Pb^{82+} at an incident energy of $E_{lab} = 200$ Gev/nucleon. The results of this work obtained in second order perturbation theory and by the bosonic coupled channel formalism are shown in comparison other calculations (see legend).

Our results and other calculations are shown as functions of the impact parameter in Figure 5. For impact parameters larger than the Compton wavelength the Weizsäcker-Williams result of Bertulani and Baur[4] agrees well with our results. However the Weizsäcker-Williams approximation is not valid for smaller impact parameters. The results of Best et al.[17] using the Weizsäcker-Williams approach seem to be too large, since the Weizsäcker-Williams procedure is problematic at low impact parameters because of a singularity although it has been regularized by formfactors. At low impact parameters our second order results are in agreement with a similar calculation of Hencken et al.[20] who applied second order perturbation theory with free plane waves at an impact parameter of zero.

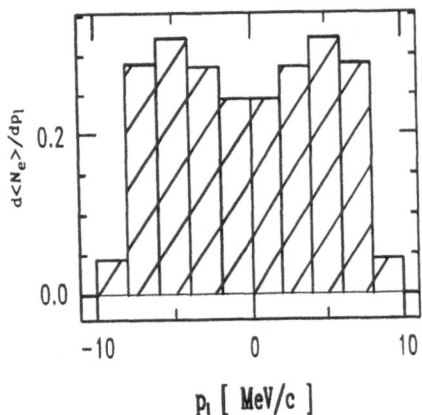 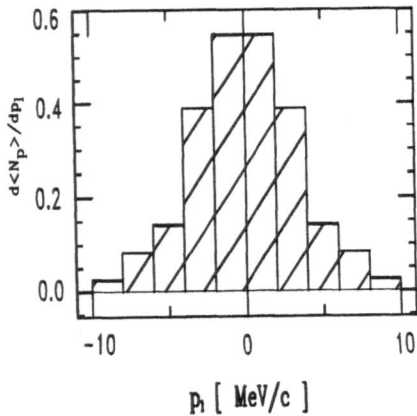

Figure 6. Multiplicity distribution (units: c/MeV) for the longitudinal momenta of the electrons (left part) and positrons (right part) obtained in bosonic coupled channel calculations for collisions of Pb on Pb at $E_{lab} = 200$ GeV/nucleon and an impact parameter of $b = 20$ fm.

The nonperturbative coupled channel results are only enhanced by a factor of about three. The main difference between the nonperturbative and perturbative calculations are found in the momentum distribution of the electrons and positrons. Whereas these particles remain in the perturbative treatments mainly at momenta lower than 2 MeV/c, their distribution is strongly changed in the coupled channel calculations. As shown in Figure 6 the electrons follow the nuclei assuming the same velocity, whereas the positrons tend to have different velocities because of the strong positive charges of the ions. This behaviour shows that the capture of electrons is important at high incident energies, too. Therefore, the incorporation of bound states into our basis should be the next step in our investigations.

Finally we calculated the cross section with the mean numbers of pairs shown in Fig. 5 via

$$\sigma = 2\pi \int_{2R}^{\infty} db \, b \, N(b),$$ (13)

where R is the radius of the Pb-nuclei. The result is $1 \cdot 10^4$ barn in the case of second order perturbation theory and $2.4 \cdot 10^4$ barn in the case of coupled channel calculations.

Our result in perturbation theory can be scaled to the system S on Au yielding 223 barn, which can be compared to the experimental result of 82 ± 22 barn [21]. We want to point out that we agree with the experimental cross section if we restrict the momenta of the electrons and positrons to the measured ones. A large amount of the cross section arises from higher momenta up to about 8 MeV/c in the collider frame.

ACKNOWLEDMENTS

This work was supported by BMFT(06 GI 728), GSI (Darmstadt), HLRZ (Jülich), and the HRZs in Gießen, Darmstadt, and Kassel.
This work is part of the doctoral thesis of Joachim Thiel, Giessen (D26), 1993.

REFERENCES

1. W. Heitler, *The quantum theory of radiation* (Oxford University Press, London) 1957.

2. G. Soff, *Electron-positron pair creation and K-shell ionization in relativistic heavy-ion collisions, Proceedings of the XVIII Winter School, Selected topics in nuclear structure*, eds. A. Balanda and Z. Stachura, (Bielsko-Biala, Poland 1980), p. 201.

3. H. Gould, *Atomic Physics Aspects of a Relativistic Nuclear Collider*, Lawrence Berkeley Laboratory Technical Information, Rep. No. LBL 18593UC-28 (1984).

4. C.A. Bertulani and G. Baur, *Phys. Rep.* **163**, 299 (1988).

5. J. Eichler, *Phys. Rep.* **193**, 167 (1990).

6. M.R. Strayer, C. Bottcher, V. E. Oberacker and A.S. Umar, *Phys. Rev.* **A41**, 1399 (1990); J.C. Wells, V.E. Oberacker, A.S. Umar, C. Bottcher, M.R. Strayer, J.S. Wu, G. Plunien, *Phys. Rev.* **A45**, 6296 (1992).

7. K. Momberger, N. Grün and W. Scheid, *Z. Phys. D-Atoms, Molecules and Clusters* **18**, 133 (1991).

8. K. Rumrich, K. Momberger, G. Soff, W. Greiner, N. Grün and W. Scheid, *Phys. Rev. Lett.* **66**, 2613 (1991).

9. J. Hoffstadt, *diploma thesis, Universität Giessen* (1993).

10. A.J. Baltz, M.J. Rhoades-Brown and J. Weneser, *Phys. Rev.* **A47**, 3444 (1993).

11. J. Thiel, A. Bunker, K. Momberger, N. Grün and W. Scheid, *Phys. Rev.* **A46**, 2607 (1992).

12. A. Belkacem, H. Gould, B. Feinberg, R. Bossingham, W.E. Meyerhof, *Phys. Rev. Lett.* **71**, 1514 (1993).

13. U. Becker, N. Grün and W. Scheid, *J. Phys. B: At. Mol. Phys.* **16**, 1967 (1983).

14. J. Reinhardt and W. Greiner, *Rep. Prog. Phys.* **40**, 219 (1977).

15. G. Baur, *Phys. Rev.* **A42**, 5736 (1990).

16. M.J. Rhoades-Brown, J. Weneser, *Phys. Rev.* **A44**, 330 (1991).

17. C. Best, W. Greiner, G. Soff, *Phys. Rev.* **A46**, 261 (1992).

18. D. Jansen, F. Dönau, S. Frauendorf, R.V. Jolos, *Nucl. Phys.* **172**, 145 (1971).

19. P. Ring, P. Schuck, *The Nuclear Many Body Problem*, Springer Verlag, New York, Heidelberg, Berlin (1980).

20. K. Hencken, D. Trautmann, G. Baur, submitted to *Phys. Rev.* **A** (1993).

21. C.R. Vane, S. Datz, P.F. Dittner, H.F. Krause, C. Bottcher, M. Strayer, R. Schuch, H. Gao, R. Hutton, *Proceedings of the XIII ISIAC, Stockholm* (1993).

ANTIPROTON PRODUCTION IN HEAVY ION COLLISIONS

Barbara V. Jacak
for the NA44 Collaboration

Los Alamos National Laboratory
Los Alamos, New Mexico 87545
USA

INTRODUCTION

In high energy p-p and $\alpha - \alpha$ collisions, baryons are observed predominantly at rapidities near those of target and projectile; the mean rapidity shift of projectile and target nucleons is approximately one unit.[1] In the central rapidity region, the number of baryons is quite small. In fact, the number of baryons and antibaryons is rather similar, indicating that most of these baryons are CREATED particles rather than projectile and target fragments. At the ISR, $\bar{p}/p \approx 0.4 - 0.5$.[2]

Antibaryon production is of interest in heavy ion collisions as enhanced antiquark (and consequently, antibaryon) production has been predicted as a potential signature of quark-gluon plasma formation.[3] Antibaryons also provide a sensitive probe of the hadronic environment, via annihilation[4] and/or mean field effects upon their final distributions[5]. However, the collision dynamics also affect the baryon and antibaryon distributions. Baryons are more shifted toward midrapidity in nucleus-nucleus and p-nucleus collisions than in p-p collisions, [6,7], increasing the probability of annihilating the antibaryons.[4] The interpretation of antibaryon yields is further complicated by collective processes which may take place in the dense hadronic medium formed in nucleus-nucleus collisions. Jahns and coworkers have shown that multistep processes can increase antibaryon production near threshold.[8] At SPS energies, a mechanism allowing fusion of color strings stretched between individual colliding nucleons into color ropes with higher potential has been shown to enhance antibaryon production.[9] To further complicate matters, annihilation of antiprotons by protons may be screened when the hadron density is very high, as suggested by Kahana and coworkers.[10]

Antiproton production is clearly very interesting, but is sensitive to a combination of processes taking place in the collision. The final number of observed antiprotons depends on the balance between mechanisms for extra antiproton production beyond those from the individual nucleon-nucleon collisions and annihilation with surrounding baryons. We can hope to sort out these things by systematic studies, varying the system size and beam energy. I will review what is known about antiproton production at both the AGS and SPS, and look at trends going from p-p to p-nucleus to nucleus-nucleus collisions.

AGS RESULTS

Four experiments at the AGS have measured antiproton distributions in nucleus-nucleus collisions at 14.6 GeV/nucleon: E802/E859[11], E814[12], E858[13], and E886.[14] The measurements were made in somewhat different p_t regions, but all were near y_{nn}. Extrapolating fits to the p_t spectra shows that the measurements all agree within errors.[15]

E859 shows that the rapidity distribution of antiprotons in central Si+Au collisions peaks at y_{nn}, and by reconstruction of $\overline{\Lambda}$ decays, that approximately 50% of the observed $\overline{p}$ in Si + Au collisions arise from $\overline{\Lambda}$. In p-nucleus collisions,[17,16] the antiproton rapidity distribution also peaks at y_{nn}, and the target dependence of the number of $\overline{p}$ produced is minimal. The E859 data indicate that the antiproton transverse mass $(m_t^2 = m^2 + p_t^2)$ inverse slopes are similar to, though perhaps slightly lower than, those of protons in the same experiment. This disagrees somewhat with the lower statistics E802 result, which had different slopes for p and $\overline{p}$, and larger $\overline{p}$ d-N/dy. The E859 results are from a larger data sample, and are within a few standard deviations from E802. The E859 data imply that interpretations based on differing mean field effects[5] for p and $\overline{p}$ may not be valid.

It is interesting to evaluate the importance of first collision production of antiprotons at the AGS, as subsequent collisions of the projectile nucleons are not far from the $\overline{p}$ production threshold. The number of $\overline{p}$ per first collision was estimated at 5.2 x 10^{-6} in the E814 spectrometer for p-p[12,17] and 2.5-3.2 x 10^{-6} for Si + Al, Cu, Pb collisions.[12] This implies that annihilation may be important, however the weak target and centrality dependence is not consistent with only first collision production and annihilation. E858 parameterized their $\overline{p}$ production data with

$$\sigma_{TOT}^{\alpha} = (\pi r^2 [A_P^{1/3} + A_T^{1/3} - 1]^2)^{\alpha}$$

and found $\alpha = 1.3 \pm 0.1$.[13] One may expect that a peripheral process would yield $\alpha = 1/2$, surface emission of $\overline{p}$ result in $\alpha = 1$, and volume emission give $\alpha = 3/2$. The fit to the data suggests that annihilation of the antiprotons may not be extremely important. This analysis does not exclude, however, a balance of annihilation by additional $\overline{p}$ production.

E802 found that $\overline{p}$ production in p-nucleus collisions is rather similar to p-p collisions and the inverse slopes of the distributions are consistent with first collision production.[16] Furthermore, no more than 20% of the $\overline{p}$ need arise from subsequent collisions. E886 found that first collision production can in fact explain ALL the antiproton yield in Si + Pt and Au + Pt collisions.[14] This was interpreted as requiring enhanced production mechanisms in light of the expected $\overline{p}$ annihilation, particularly in Au + Pt. The lack of dependence on the target mass in p-nucleus collisions is not consistent with strong annihilation effects, though of course the number of protons at midrapidity is smaller than in Si or Au on heavy targets.

THEORETICAL WISDOM

Two event generators have been used to successfully reproduce the antiproton production in Si-nucleus collisions at the AGS. The RQMD model, [19] which includes the formation and decay of color strings, produces a large excess of $\bar{p}$ in heavy systems at the AGS.[8,18] The excess arises from multistep excitations of the leading baryons, which become more likely to decay into heavy objects. This excess $\bar{p}$ production is counterbalanced in RQMD by strong final state absorption; approximately 2/3 of all produced antibaryons are annihilated in central Si + Al collisions at 14.6 GeV/nucleon.[18] In contrast, the hadronic cascade model ARC[20] produces most antiprotons in first chance nucleon-nucleon collisions. In ARC, approximately 10% of the $\bar{p}$ in Si + Au come from meson-baryon collisions. The smaller number of produced $\bar{p}$ is balanced by a smaller annihilation probability in ARC, though even in ARC nearly 50% of the $\bar{p}$ in central Si + Au are annihilated.[10] Large annihilation effects in heavy ion collisions were also predicted by Gavin and collaborators.[4] Both RQMD and ARC describe the E802 data rather well, despite their different assumptions. It is clear that quantitative comparisons of these models to the other data sets are crucial to disentangling these effects.

We can also help sort out the different physics issues and address the importance of $\bar{p}$ annihilation by looking at CERN energies. With 200 GeV/nucleon heavy ion beams, threshold effects are no longer important and first collision analyses are not relevant, removing one variable from the interpretation. As $\bar{p}$ may be formed just as easily in subsequent collisions, the ratio $\bar{p}/\pi^-$ may help "calibrate" effects of the varying $\sqrt{s}$ of subsequent nucleon-nucleon collisions on $\bar{p}$ production. At CERN, the baryon density (dN/dy) at midrapidity is considerably lower than at the AGS, so differences in the treatment of the annihilation of $\bar{p}$ should no longer dominate the model results. However, there are still SOME baryons present, so annihilation effects may be investigated and compared to results in the AGS energy regime.

It has also been pointed out that thermal and chemical equilibrium may not be achieved in these collisions. In that case, the abundances of baryons and antibaryons are not controlled by thermodynamic Boltzmann factors, but by the sizes of coherence domains in a low-temperature quark condensate.[21] In this case, the baryon-antibaryon production may be larger than in an equilibrated system, and so the antibaryon production provides a good probe of nonperturbative QCD.

NA44 EXPERIMENT

Experiment NA44 at CERN has measured proton and antiproton spectra with high statistics, and compares p-p (approximated by p-Be), p-nucleus, and nucleus-nucleus collisions in the same experiment. NA44 is a second generation experiment, with a focussing spectrometer optimized for momentum resolution ($\sigma_p/p = 0.2\%$) and identification of charged particles at midrapidity. The trigger level particle identification and high rate capability, along with acceptance for a small number of particles per event, yield high quality single particle spectra, even for rare particles such as antiprotons.

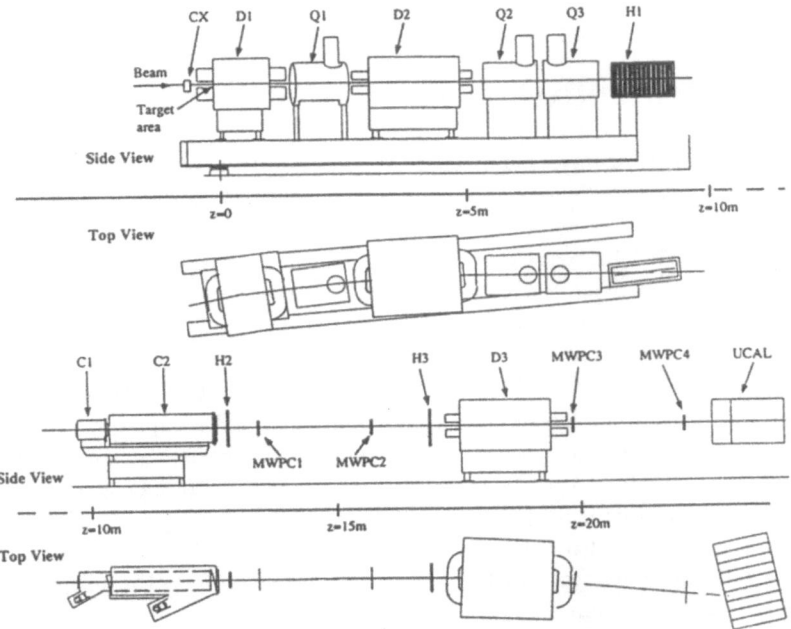

Figure 1: NA44 experiment layout

The layout of NA44 is shown in Fig. 1. There are three dipole magnets (D1, D2, D3) and three quadrupoles (Q1, Q2, Q3); the first two dipoles select the momentum and the last one is used for momentum calibration. The quadrupoles are configured such that the momentum of the particle is strongly correlated with the position on the first hodoscope. Only one charge state can be detected at one time. The momentum range selected by the spectrometer covers a band of $\pm$ 20% around the nominal setting. The angular coverage is approximately -0.3 to 4.5 degrees in the horizontal and $\pm$ 0.3 degrees in the vertical plane. Tracking is done using three highly segmented scintillator hodoscopes (H1, H2, H3), divided into 50, 60, and 50 vertical slats, respectively. Photomultiplier tubes are mounted on either end of the slats to optimize the time-of-flight resolution and allow y-position measurement using the time difference between top and bottom.

The beam rate and time-of-flight start are determined with a Cherenkov beam counter (CX) for heavy ion beams, and a forward scintillator (T0) for proton beams. The forward scintillator is used as a triggering device in both cases, to signal either minimum bias or central collisions via the pulse height. A silicon multiplicity detector is used to measure the charged particle distribution with 2π acceptance in the pseudorapditiy range $1.5 < \eta < 3.3$. The minimum bias data in NA44 do contain a bias toward small impact parameters due to the requirement of finding a particle in the spectrometer acceptance. This bias is stronger for more rarely produced particles. For the sulfur beam, "central" collisions correspond to the top 10% of the geometric cross section. The charged particle spectra have been corrected for the "target out" contribution.

Particles are identified via time-of-flight measured by the hodoscopes ($\sigma_{TOF} \approx$ 100 ps) and pulse height analysis on the two gas Cherenkov counters. The first

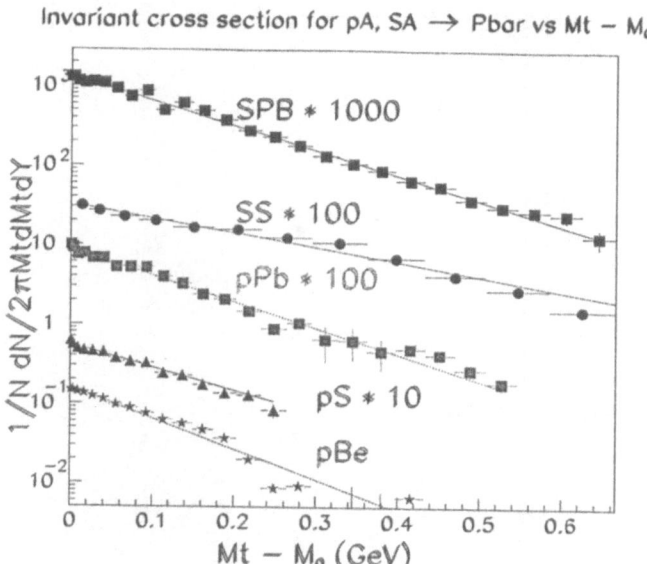

Figure 2: Transverse kinetic energy spectra for antiprotons

Cherenkov (C1) is filled with freon-12 at 1.4 and 2.7 atm. and differentiates between π, K and p. C2 has a nitrogen/neon mixture at 1.0 and 1.3 atm., and rejects electrons and pions. The contamination of any particle sample by particles of another type is less than 4%.

The spectrometer single particle acceptance for protons and antiprotons consists of a rapidity slice of 0.3 units, centered about $y = 2.8$. The p_t range is from 0 - 800 MeV/c. A second spectrometer setting provides p and $\bar{p}$ distributions centered about $y = 2.0$.

Tracks are reconstructed from the recorded positions on the three hodoscopes, with pattern recognition constrained by straight-line trajectories after the magnets. The cross sections have been corrected for geometrical acceptance, multiple scattering, particle identification cuts, and pattern recognition losses using a Monte Carlo simulation of the spectrometer. The simulated events are reconstructed using the full analysis chain, and the input distribution is tuned to the observed particle spectra. The uncertainty on the p_t measurement is 0.15%, arising from multiple scattering and the hodoscope granularity. The systematic uncertainty on the p_t scale is 1.6%.

TRANSVERSE MASS DISTRIBUTIONS

Particle distributions are often shown as a function of m_t, the transverse mass, or k_t, the transverse kinetic energy ($k_t = m_t - m_0$). This is motivated by the possibility of describing the observed distributions with a thermal model; an exponential distribution in m_t, common to all particle species, might be expected for thermal emission.[22] Figure 2 shows the NA44 transverse kinetic energy distributions[23] of antiprotons from 200 GeV/nucleon S + Pb, S + S, and 450 GeV p + Pb, S and Be. The lines are the result of fits with an exponential k_t distribution.

The NA35 collaboration has investigated the decay of $\overline{\Lambda}$, and the contribution of

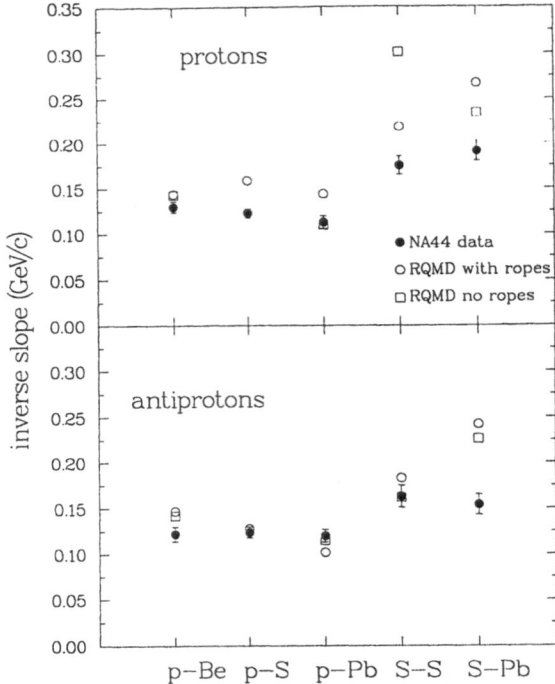

Figure 3: Inverse slopes of proton and antiproton spectra

these decays to the measured $\bar{p}$.[24,25] They find that approximately 30% of $\bar{p}$ in their acceptance come from $\bar{\Lambda}$ decay. We have used RQMD events and our spectrometer acceptance to investigate the effect of $\bar{\Lambda}$ decay upon our antiproton spectra, and find that we reconstruct all $\bar{p}$ arising from $\bar{\Lambda}$. The $\bar{\Lambda}$ contribution is 30% in our data as well, but does not significantly affect shape of the antiproton k_t distribution.

The inverse slopes resulting from the fits in Fig. 2 are shown in Fig. 3 along with the inverse slopes for protons, plotted as a function of the system size. It is clear that both protons and antiprotons have flatter spectra, and larger inverse slopes, in nucleus-nucleus than in p-nucleus collsions. The system size dependence of the p and $\bar{p}$ spectra is considerably stronger than for pions (which show almost no system dependence) or kaons.[22] In S - S and S - Pb collisions, the $\bar{p}$ slope is 1-2 standard deviations lower than the proton slope. NA35 reported at this school on protons and antiprotons identified in the TPC; their $\bar{p}$ inverse slopes agree with NA44 at a similar level. We are currently analyzing additional heavy ion data to reduce the statistical errors on these slopes in order to definitively determine the magnitude of the difference.

Fig. 3 also shows the inverse slopes of protons and antiprotons at midrapidity from the RQMD event generator, as a function of the size of the system. The inverse slopes were determined by fitting an exponential as in the data. Events were generated using RQMD in two modes - with and without the fusion of color strings into color ropes. As expected, the results are the same for p-nucleus collisions which have few nucleon-nucleon collisions and correspondingly few strings which may fuse. In heavy ion

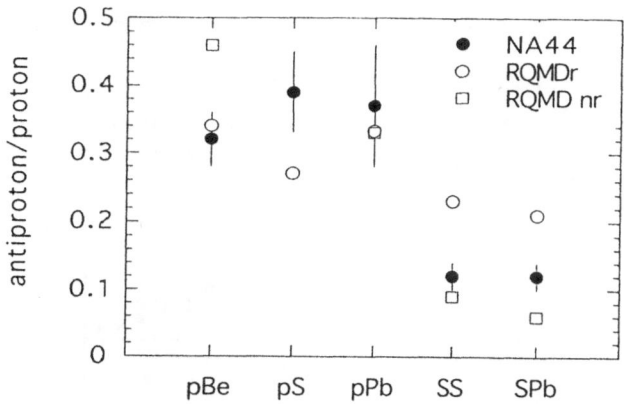

Figure 4: ratio of $\overline{p}/p$ for different systems

collisions, the probability of string fusion is considerably larger,[9] but the effect on the particle momenta is small. Both sets of RQMD calculations successfully reproduce the trends of the slopes with system size. RQMD appears to somewhat overpredict the slopes in S + Pb collsions, but the difference is not very large. In RQMD, the difference between proton and antiproton spectra is small.

PROTON AND ANTIPROTON YIELD

The spectra shown in Fig. 2 were integrated to determine the yield of antiprotons in these collisions; proton distributions were similarly analyzed. The fits were extrapolated to high p_t, beyond the region of measurement, to allow calculation of the rapidity densities. Fig. 4 shows the ratio of dN/dy for $\overline{p}$ and p for the various target-projectile systems. The p + Be ratio agrees reasonably well with the value observed at the ISR.[2] It is rather remarkable that the ratio remains approximately constant for protons on various nuclear targets, then decreases in nucleus-nucleus collisions. Of course, the $\overline{p}/p$ ratio is affected by the proton distributions as well as antiproton production. It is well known that collisions of heavier systems result in a larger number of protons at midrapidity.[6,7] Even proton collisions on heavy targets result in a shift of target protons to larger rapidity.[7] These trends alone may be able to explain the system size dependence observed, though we must also investigate the possible effects of excess $\overline{p}$ production and $\overline{p}$ annihilation. The latter will certainly depend sensitively on p dN/dy, magnifying the effect of midrapidity protons on this ratio.

A powerful tool to address these questions is an event generator containing $\overline{p}$ annihilation at a known level. RQMD provides this feature and can produce excess $\overline{p}$ if string fusion to color ropes is included.[9] The $\overline{p}/p$ ratios from RQMD with and without string fusion for the various systems are compared with the data in Fig. 4. Both sets of RQMD calculations reproduce the trends observed in the data, though $\overline{p}/p$ in the RQMD version without string fusion falls somewhat more steeply. Such behavior may be expected if produced antiprotons are annihilated, and this loss is not balanced by increased production.

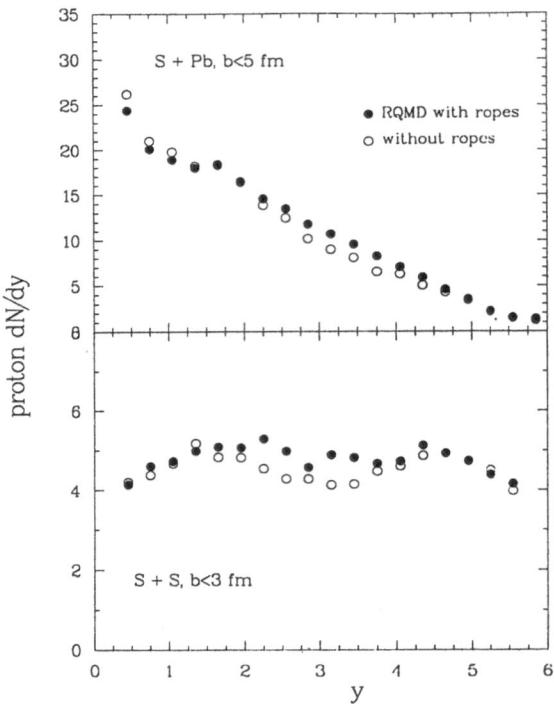

Figure 5: Rapidity distribution of protons from RQMD with and without string fusion to color ropes

THE MODEL VIEW OF ANTIPROTON PRODUCTION AND ANNI-HILATION

Fig. 5 shows the proton rapidity distributions predicted by RQMD for S + Pb and S + S collisions. The impact parameter range was chosen to match the NA44 trigger; the requirement of large pulseheight in the T0 scintillator was simulated using RQMD events as input and including the geometry and response of the trigger scintillator. The closed points are from RQMD events including string fusion to color ropes, and the open points without rope formation. This comparison shows that string fusion does not affect the proton distributions. This may be expected as the proton yield is dominated by protons already existing in the projectile and target, rather than by protons produced in the collision. The predictions in Fig. 5 can be compared with proton production data from HELIOS[7] and with preliminary NA44 results.[23] HELIOS found proton dN/dy = 18 ± 0.9 at y = 1 in central S + W collisions. NA44 sees 16 ± 4 protons at y = 2.1 and 14 ± 4 at y = 2.8. Clearly, the RQMD model successfully reproduces the proton rapidity distribution.

As the proton yields are correct in the model, we may use it to study the competing effects on antiproton production in these systems. The RQMD predictions for the $\bar{p}$ rapidity distributions are shown in Fig. 6. Unlike for protons, the antiproton yield is strongly enhanced (by factors of 2-3) by formation of color ropes. This enhancement is similar to that observed in RQMD for $\bar{\Lambda}$ production,[9] which agrees with data[24] and implies that enhanced antibaryon production signals more collective behavior in

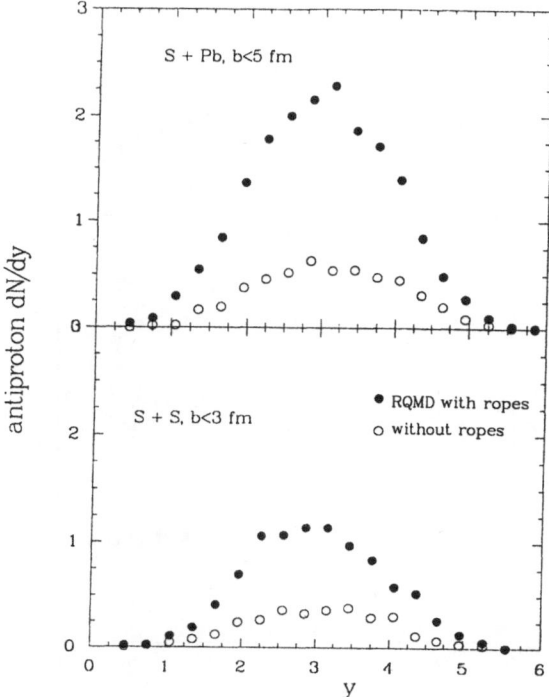

Figure 6: Rapidity distribution of antiprotons from RQMD with and without string fusion to color ropes

heavy ion collisions than just superposition of p - p collisions.[9,25]. Preliminary results from NA44 on $\bar{p}$ dN/dy at $y = 2.8$ are 1.7 ± 0.6 for S + Pb and 0.8 ± 0.2 for S + S collisions. NA35 has reported very similar preliminary values at this conference for antiprotons identified in the TPC.[25] These $\bar{p}$ yields are considerably closer to the predictions of RQMD with color rope formation than without, and support the conclusions drawn from the $\bar{\Lambda}$ results. Of course, the $\bar{p}$ measurement is not totally independent of the $\bar{\Lambda}$ yield, as 30% of the $\bar{p}$ arise from $\bar{\Lambda}$ decays. However, this contribution alone does not remove the need to include color ropes in the model to reproduce the observed $\bar{p}$ yields.

As described above, the RQMD model has a large cross section for annihilation of $\bar{p}$ with protons. At AGS energies, the baryon density at midrapidity is large,[26] and a large fraction of the secondary collisions experienced by produced particles are with baryons. The annihilation of $\bar{p}$ is the dominant process determining the number of $\bar{p}$ produced in the model at those energies. At CERN, the ratio of π/p at midrapidity is much larger,[23,6] and most secondary collisions are with mesons. In this case, we may expect less $\bar{p}$ annihilation and a greater sensitivity to the other processes affecting the antiproton yield.

We try to quantify the annihilation effects at CERN by using RQMD and counting the number of collisions suffered by each proton and antiproton before exiting the system. These are plotted as a function of the final rapidity of the p and $\bar{p}$ for S + Pb, S + S and p + Be in Fig. 7. The events used for this analysis come from

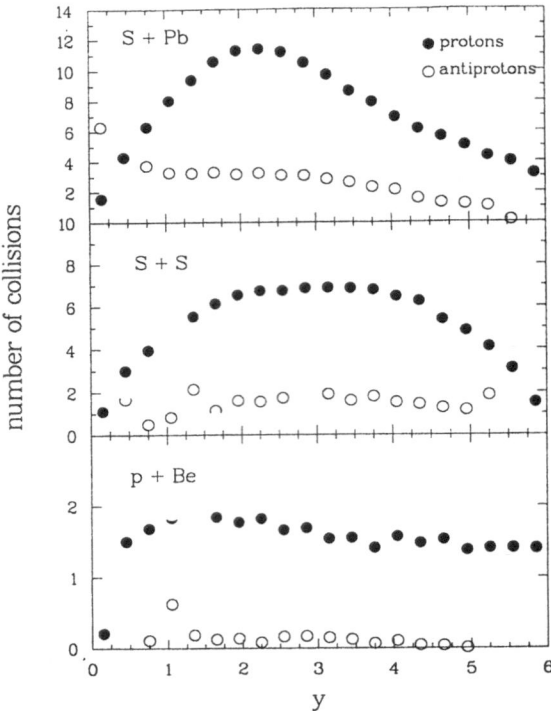

Figure 7: Number of collisions (primary + secondary) of p and $\bar{p}$ in RQMD, as a function of rapidity. S + S and S + Pb are central collisions.

very central collisions ($b \leq 1$ fm) and include string fusion. It is immediately obvious that protons undergo many more collisions than antiprotons in all of the systems. However, the collision count includes PRIMARY collisions of the particles as well as secondary collisions, accounting for at least some of this difference. This may be accounted for by recalling that simple geometrical models which superimpose p + p collisions at differing $\sqrt{s}$ can successfully reproduce the number of pions formed in both p + A and A + A, and consequently do a good job in counting the number of primary nucleon-nucleon collisions.[27] Using such a geometrical picture for central collisions, we estimate that protons undergo an average of 1.5 collisions in p + Be and 2.5 - 3 primary collisions in S + S. In S + Pb, projectile protons suffer approximately 6 primary collisions and target protons about 4. Subtracting these numbers from the collisions experienced by midrapidity protons shown in Fig. 7 shows that there remains a difference in the number of secondary collisions of protons and antiprotons in nucleus-nucleus at CERN. Antiprotons experience an average of 2 - 3 fewer secondary collisions than protons in both S + S and S + Pb. This implies that annihilation of antiprotons does occur at CERN energies.

We have attempted to study the importance of $\bar{p}$ annihilation from the NA44 data by looking at the system dependence of the p/π^+ and $\bar{p}/\pi^-$ ratios. Such an approach uses the pion yields to "count" the number of nucleon-nucleon collisions and normalize for the varying $\sqrt{s}$ of these collisions. This kind of analysis is qualitative rather than quantitative, but may be used to check if the data are consistent with the conclusions drawn from the model; the actual values of the ratios are strongly affected by the spectrometer acceptance and fact that we measure protons around $y = 2.8$ and pions at $y = 3.5 - 4.5$. Nevertheless, the trends are informative. In p-nucleus collisions (which used 450 GeV proton beam), the p/π^+ and $\bar{p}/\pi^-$ ratios do not depend on the target size. p/π^+ exceeds $\bar{p}/\pi^-$, as we would naively expect. In S + Pb and S + S, p/π^+ is larger than in p-nucleus collisions, consistent with the shift of protons toward midrapidity shown in Fig. 5 and supported by data.[6,7] To compare $\bar{p}/\pi^-$ in nucleus-nucleus and p-nucleus collisions, we must first take into account the fact that the $\bar{p}$ production cross section falls by $\approx 25\%$ from 450 GeV to 200 GeV beams, but the π cross section falls by only $\approx 10\%$. The observed $\bar{p}/\pi^-$ ratio in S + Pb and S + S, however, is lower than p-nucleus by more than this trivial drop. Such an additional reduction in $\bar{p}/\pi^-$ can be expected if antiprotons are annihilated. The FRITIOF model,[28] which includes no $\bar{p}$ annihilation or excess production, reproduces the p-nucleus $\bar{p}/\pi^-$ ratio adequately but predicts no change upon going to nucleus-nucleus collisions; it fails miserably on the proton ratios and rapidity distributions, as is well known.[7] RQMD reproduces the general trend in the ratios observed in NA44.

ANTIDEUTERON PRODUCTION

NA44 can also look for antideuterons, using a unique combination of tools for $\bar{d}$ identification and background rejection. The focussing spectrometer gives excellent momentum resolution, which coupled with the very good time-of-flight resolution allows a precise measurement of particle masses. The Cherenkov rejection of pions and kaons greatly reduces the background and makes the main identification problem p-d separation. We also use the uranium/scintillator calorimeter at the end of the NA44 spectrometer to reject kaon and pion decay background and require the $\bar{p}$ or $\bar{d}$ annihilation energy. The combination of these cuts results in rather clean mass spectra; positive tracks show a clear deuteron peak.

Though the statistics of $\bar{d}$ are too small to show a peak, a preliminary analysis shows 4 $\bar{d}$ candidates in S + Pb collisions. From the deuterons observed in the same size sample, we may estimate the number of antideuterons expected using the coalescence model.[29] This yields a prediction of about 4 $\bar{d}$ in our sample. We are currently analyzing a larger data sample to extract a more statistically significant result, but it appears that antideuteron production at CERN energy is range expected from coalescence of antinucleons. This differs from AGS energies, where the $\bar{d}$ cross section is lower than predicted by coalescence by a factor of 10.[13] This difference may serve to again underscore the lesser importance of antinucleon annihilation at CERN than at AGS energy.

CONCLUSIONS

We have discussed the complex tradeoff between threshold effects on the production cross section of antiprotons, and the amount of annihilation in nucleus-nucleus collisions at AGS energies. We use measurements at CERN, where threshold effects are not important, to gain further insight. We find that there is excess antiproton production which cannot be explained by models incorporating no collective behavior of the nucleon-nucleon collisions. The RQMD model including fusion of strings into color ropes agrees well with the observed antiproton yields in S + Pb and S + S collisions. The 30% contribution of $\overline{\Lambda}$ decays to the $\overline{p}$ yield is not sufficient to account for the enhanced $\overline{p}$ production; according to RQMD, $\overline{\Lambda}$ and $\overline{p}$ enhancement are both due to the same mechanism anyway. It appears that some antiproton annihilation occurs at CERN energy, but is less important than at the AGS. The proton and antiproton m_t inverse slopes are fairly similar and do not leave room for large mean field effects on the $\overline{p}$ distibutions.

The NA44 collaboration is currently analyzing a larger data set, extracting antideuteron yields, and working on improving the systematic errors. This along with Pb + Pb results from CERN, and Au + Au at the AGS will help constrain the models and fix production and annihilation cross sections. Armed with these, we may better quantify the magnitude of collective effects on heavy particle production in nucleus-nucleus collisions.

ACKNOWLEGEMENTS

I would like to thank Heinz Sorge, Michael Murray, Jehanne Simon-Gillo, John Sullivan, and Doug Fields for many enlightening discussions about both data and the RQMD model results. Their invaluable aid in compiling and analyzing the model runs for the various systems is most appreciated. This work was supported by the U.S. Department of Energy.

REFERENCES

1. W. Bell, et al., *Z. Phys.* C27, 191 (1985).

2. B. Alper, et al., *Nucl. Phys.* B100, 237 (1975).

3. U. Heinz, P.R. Subramanian and W. Greiner, *Z. Phys.* A318, 247 (1984); P. Koch, B. Mueller, H. Stoecker and W. Greiner, *Mod. Phys. Lett.* A3, 737 (1988).

4. S. Gavin, M. Gyulassy, M. Pluemer and R. Venugopalan, *Phys. Lett.* B234, 175 (1990).

5. V. Koch, G.E. Brown and C.M. Ko, *Phys. Lett.* B265, 29 (1991).

6. H. Stroebele, et al. (NA35 Collaboration), *Nucl. Phys.* A525, 59c (1991).

7. T. Akesson, et al. (HELIOS Collaboration), *Z. Phys.* C53, 183 (1992).

8. A. Jahns, H. Sorge, H. Stoecker and W. Greiner, *Phys. Rev. Lett.* 68, 2895 (1992).

9. H. Sorge, M. Berenguer, H. Stoecker and W. Greiner, *Phys. Lett.* B289, 6 (1992).

10. S.H. Kahana, Y. Pang, T. Schlagel and C.B. Dover, *Phys. Rev.* C47, R1356 (1993).

11. T. Abbott, et al. (E802 collaboration), *Phys. Lett.* B271, 447 (1991) and Quark Matter '93.

12. J. Barette, et al. (E814 collaboration), *Phys. Rev. Lett.* 70, 1763 (1993).

13. A. Aoki, et al. (E858 collaboration), *Phys. Rev. Lett.* 69, 2345 (1992).

14. G. Diebold, et al. (E886 Collaboration), *Phys. Rev.* C48, 2984 (1993).

15. B. Shiva Kumar, Proc. of the XXVI International Conference on High Energy Physics, Dallas TX, August 1992, p. 1006; B. Shiva Kumar, Proc. of HIPAGS '93, Boston MA, Jan. 1993.

16. T. Abbott, et al. (E802 Collaboration), *Phys. Rev.* C47, R1351 (1993).

17. J.V. Allaby, et al., CERN Report No. 70-12, 1970 (unpublished).

18. A. Jahns, H. Sorge, H. Stoecker and W. Greiner, *Z. Phys.* A341, 243 (1992); A. Jahns, C. Spieles, R. Mattiello, H. Stoecker, W. Greiner and H. Sorge, *Phys. Lett.* B308, 11 (1993).

19. H. Sorge, H. Stoecker and W. Greiner, *Ann. Phys.* 192, 266 (1989).

20. Y. Pang, T.J. Schlagel and S.H. Kahana, *Phys. Rev. Lett.* 68, 2743 (1992).

21. J. Ellis, U. Heinz and H. Kowalski, *Phys. Lett.* B233, 223 (1989).

22. B. V. Jacak, Proc. of the NATO School on Particle Production in Highly Excited Matter, Il Ciocco, Italy, July 1992, p.471.

23. M. J. Murray, Proc. of Quark Matter '93, Borlange, Sweden, June 1993.

24. J. Bartke, et al. (NA35 Collaboration), *Z. Phys.* C48, 191 (1990).

25. D. Roerich (NA35 Collaboration), these proceedings.

26. T. Abbott, et al. (E802 Collaboration), *Phys. Rev. Lett.* 66, 1567 (1991).

27. A. D. Jackson and H. Boggild, *Nucl. Phys.* A470, 660 (1987).

28. B. Andersson, G. Gustafson, and B. Nilsson-Almquist, *Nucl. Phys.* B281, 289 (1977).

29. H.H Gutbrod, et al., *Phys. Rev. Lett.* 37, 667 (1976); C. B. Dover, U. Heinz, E. Schnedermann, and J. Zimanyi, *Phys. Rev.* C44, 1636 (1991).

STRANGENESS PRODUCTION IN ULTRARELATIVISTIC NUCLEUS-NUCLEUS COLLISIONS

H. Sorge[1], M. Berenguer[1,2], H. Stöcker[2], W. Greiner[2]

[1]Los Alamos National Laboratory
[2] Institut f. Theoretische Physik d. Universität Frankfurt

INTRODUCTION

Oxygen and sulfur ions have been accelerated at CERN-SPS to a maximum energy of 200 AGeV and collided with different targets. The main goal of heavy ion physics is to study hot, dense strongly interacting matter far beyond the nuclear ground state. Strangeness and antibaryons are strongly suppressed in elementary hadronic reactions. Their enhancement in nucleus-nucleus collisions – as measured for kaons and (anti-)lambda's by the NA35 group [1], for (anti-)cascades by the WA85 group [2] and for ϕ mesons by the NA38 and HELIOS collaborations [3, 4] – indicates therefore the presence of some collective effect.

In this paper collective effects in AA collisions are considered which may have an influence on the hadron chemistry, for instance strangeness, in those reactions: the formation of strong chromoelectric fields (color ropes) and hadronic rescattering. Elementary strings which are pulled out by receding elementary triplet color charges in the initial stage of a nucleus-nucleus collision may combine and form highly charged "rope" fields. The color ropes decay subsequently by quark pair creation screening the initial field. Viewing quark pair creation as a Schwinger type tunneling process, strangeness suppression is weakened due to a stronger rope field as compared to independent string fragmentation. Hadronic rescattering is a second process by which strangeness can be enriched in AA reactions. It drives the system towards chemical equilibration which is not reached in elementary hadron-hadron reactions.

FORMATION AND FRAGMENTATION OF COLOR ROPES

The formation of color ropes in nuclear collisions has been studied back in the eighties after its original proposal in [5]. However, these studies were restricted to an

Hot and Dense Nuclear Matter, Edited by
W. Greiner *et al.*, Plenum Press, New York, 1994

ideal situation of "infinitely" long ropes [6, 7, 8]. Furthermore no attempt had been made to fragment the ropes into a hadronic multi-particle state. The final quark flavor composition – including the effect of stepwise field degradation – had never been calculated. These deficiencies have been recently overcome in the framework of a transport theoretical model [9], the relativistic quantum molecular dynamics approach (RQMD). A complete scenario of rope formation and subsequent decay into hadrons has been implemented into RQMD. RQMD contains – in addition to initial resonance excitation or string formation – rescattering between all hadrons after string and resonance decays [10]. In a unified framework one can study both the initial stage of string, respectively rope dynamics and the importance of final state interactions.

String excitation and rope formation

The initial number of strings is obviously an important ingredient for the rope dynamics, because it determines the colorelectric field strength which can be reached. In the following a short description of the string formation in RQMD is included which shows that the number of created strings goes essentially with the number of elementary collisions. With respect to this question, RQMD is rather similar to dual parton models (DPM) while in the LUND model FRITIOF always the same projectile string gets more and more excited. The latter procedure tends to suppress the achievable string density in nucleus-nucleus collisions.

In RQMD one propagates all degrees of freedom – hadrons and their constituent quarks, strings, ropes – in classical phasespace combined with stochastic interactions, e.g. collisions and decays. In the first stage of a reaction target and projectile nucleus interpenetrate and – at sufficiently high energies – fly through each other due to the onset of transparency. If nucleons pass each other within a certain distance determined by the nucleon-nucleon cross section, longitudinal momentum is exchanged – with an $1/x$ law for the exchanged light cone momentum – from one nucleon to the other. A nucleon gets excited into a resonance or – above some threshold – into a string. Strings are a one-dimensional idealization of color fluxtubes with (anti-) triplet charges as sources and sinks of the color flux at the ends. Due to an interaction, the nucleon wave function is decomposed into two spectator quarks and one interacting quark. In general, the interacting constituent quark will not be accelerated into the reverse direction due to longitudinal momentum transferred from the collision partner. This seem to be incompatible with the dominantly *soft* nature of hadronic interactions. The momentum transferred to the interacting quark polarizes a quark pair from the sea. A backward moving sea quark pulls out a string which connects this quark to the forward moving color charge. Further strings can be formed in subsequent interactions of the interacting constituent quarks, employing the same principle. The cross section of interacting constituent quarks is given from the additive quark model $\sigma_{q-q} = 1/9 \cdot \sigma_{NN}$ which keeps the original projectile nucleon cross section constant while the projectile is traversing the target.

Note that the time scale of soft interactions does not allow to view the constituent quark interaction as a sequence of causally ordered collisions in space-time. Microscopically seen, *different components* of a constituent quark – a complex object – are involved in each interaction. The ingoing constituent quarks get more and more 'undressed' in course of the interactions. The momentum going into string excitations represents the interacting part of the original parton cloud, which can get active in interactions again only after some time has elapsed (Pomeranchuk effect). This effect is simulated in semiclassical string fragmentation schemes like RQMD, because a string has to evolve in space-time before it can break and form secondary hadrons.

The color fields of two or more strings are combining in SU(3)-space if the strings are overlapping, i.e. they share the same transverse area and have some overlap longitudinally. The elementary triplet charges of quarks and antiquarks in one of the receding rope 'condensator' plates are coupled stochastically which determines the total SU_3 color charge of the rope. The charge of the other rope plate has to be opposite to ensure that the whole rope state is white. The statistical combination of nearby located quark charges is a major source of $\bar{3}$-qq and 3-$\bar{q}\bar{q}$ creation. The weight given from SU_3 couplings to form a $\bar{3}$-qq out of two quarks as compared to a $6 - qq$ amounts to 1/3:2/3. A $\bar{3}$-qq will fragment into a baryon if the available phasespace in rope fragmentation allows for this possibility.

The transverse size of a rope is kept unchanged compared to an elementary triplet-antitriplet string (0.8 fm), because the relevant time scales of the decay and the surrounding medium do probably not allow for an adjustment of the radius. Therefore the interaction distance between two strings is (roughly) given by the transverse extension of a color string.

The field strength and the corresponding energy density cannot be calculated for the situation of a QCD condensator with some arbitrary color charge. However, it is inferred from Regge trajectories, hadron structure physics, the strong coupling limit of QCD and calculations on the lattice that a QCD color fluxtube is characterized by a constant energy per unit length, as it is the case in electrodynamics for a condensator with infinite plates. This lends credit to the hypothesis that the chromomagnetic part of QCD is responsible for the vacuum pressure which squeezes the color flux into a tube but can be neglected for the calculation of the average energy density inside a rope. If the chromoelectric approximation is correct, the energy density and the integrated energy per unit length of a rope field are simply proportional to the QCD charge, the Casimir operator $C(p,q) = L(p,q)^\alpha \cdot L(p,q)^\alpha$:

$$\kappa(p,q) = 3/4 \cdot C(p,q) \cdot \kappa_{el} \quad . \tag{1}$$

$\kappa(p,q)$ (κ_{el}) denotes the string tension for a rope (elementary string), (p,q) characterizes the multiplet which the QCD charge belongs to. (Note that in SU_3 one needs two numbers to specify a charge multiplet.) $L(p,q)^\alpha, \alpha = 1, 8$ are the generators of the SU(3) group in the corresponding representation.

Rope fragmentation

The created chromoelectric fields decay via tunneling of quark-antiquark pairs as in the elementary string case. The higher field strength in a rope will affect the time scale of quark pair production, a more rapid creation, and the flavor composition, a weakening of strangeness suppression. In a first approximation the total pair creation rate and the flavor composition can be calculated employing Schwinger's vaccuum persistence rate for a constant chromoelectric field [11, 12]. The screening effect of the created charge pair on the original field should be included, however. This is equivalent to calculating the force which a quark experiences in the rope field as the amount per unit length by which the field lowers its energy due to screening. A field in SU_3 may get screened by a quark-antiquark pair into up to three different color configurations. The tunneling probability per unit time and unit volume can be calulated with the WKB-approximation which gives in addition the transverse momentum spectrum of the created quark pair. If an antiquark is attracted by the outer (p, q) rope charge ($p \geq q$ for definiteness) it turns out as:

$$\frac{d^4 p}{d^4 x} = \frac{\gamma}{4\pi^3} \cdot (\kappa(p,q) - \kappa(p-1,q)) \cdot \sum_{\text{flavors}} \sum_{n=1}^{\infty} \int d^2 p_t \frac{P(p_t)^n}{n}, \tag{2}$$

where $\kappa(p,q)$ and $\kappa(p-1,q)$ are the rope tension before and after charge creation, γ gives the color degrees of freedom ($\gamma = 1$ in this case) and $P(p_t)$ denotes the tunnel probability from a virtual to a real state calculated with the WKB mthod.

What is the creation mechanism for baryon pairs in a string or rope field? An ambiguity arises whether the initial field is solely screened by stepwise production of $3 - \bar{3}$ pairs (quarks). A straightforward generalization to the production of mesons in a fluxtube via $q - \bar{q}$ creation is to assume that baryons are formed by creating a diquark pair in the field. However, one can envision two different mechanisms to produce diquarks, even in elementary strings (see Figure 1). In one scenario one treats the diquark as a unit, i.e. an elementary color charge like an antiquark, but with higher mass. In the other scenario which was employed e.g. in [9] diquarks are produced via a two-step mechanism. This scenario is closer to the fluxtube picture as extracted from strong coupling QCD (see e.g. [13]). A created quark can lower the string or rope charge (p, q) to $(p, q - 1)$ (optimal screening) or to $(p - 1, q + 1)$. This configuration contains a $\bar{3}$-diquark on one side and a diquark with opposite color charge on the other side which normally splits into a baryon-antibaryon pair.

The creation probability of diquark pairs becomes extremely different in those two scenarios if the mass parameters are fixed to give the same diquark suppression in the elementary string decay. In the scenario which treats the diquarks like a unit the probability to produce a diquark becomes for sufficiently strong rope fields larger than to create a quark pair. The suppression caused by the mass becomes negligible and the spin degeneracy favors a diquark by a factor of 2. In contrast, in the two-step scenario the tunneling probabilities are always the smallest, because the corresponding force is weaker than in the optimally screening cases. Some experimental results seem to favor a two-step production model for diquarks [14, 15].

A fragmenting string as well as a decaying resonance neutralizes color charges always locally. This expresses the dominance of confining forces even in situations with high excitation energies. This may be no longer true in the space-time evolution of a fragmenting rope [9] which is schematically shown in Figure 2. Quark and antiquark trajectories cross each other, and they are not forced to form a white state as long as there are other color charges around which can neutralize their charge. There exists no suppression factor acting as a penalty if they do not combine into a color singlet state, but the coalescence probability is given by statistics. Thus rope dynamics is quite different from a hadronic scenario. It may serve as a model for the dynamical evolution of (constituent) quark matter in 1+1 dimensions.

(ANTI-)HYPERON PRODUCTION IN AA COLLISIONS

In the following the effects of hadronic rescattering and rope formation on the final strange (anti-)baryon yields are studied separately. In Figure 3 calculated RQMD rapidity distributions [9] are therefore shown for the system S(200 AGeV)+S *without* rope formation, but with rescattering switched on and off. The corresponding rapidity distributions for neutral strange particles are shown together with the experimental

Figure 1. A fluxtube may split by diquark creation in one step (left side), or in two-steps (right side where only the first step has been indicated and the qq–$\overline{qq}$ configuration will split in the second step). The phenomenological diquark suppression factor in string decays P(qq)/P(q)= $0.08 - 0.12$ can be accounted for in both scenarios. In the single-step scenario diquark creation is suppressed due to the higher diquark mass, in the two-step scenario the quark pair creating field is weaker due to nonoptimal screening.

a)   b)

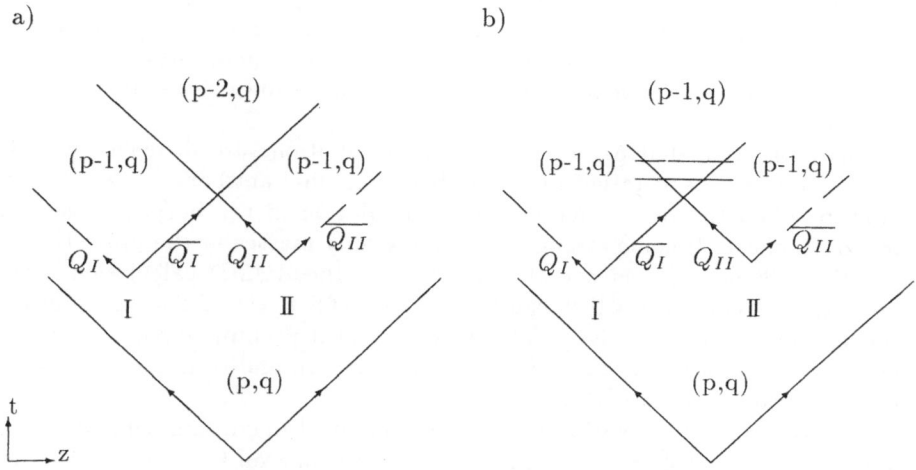

Figure 2. Schematic picture of a rope with with color charge (p,q) (on the right side). While the crossing of a quark trajectory with an antiquark trajectory always leads to hadron formtion in elementary strings ((p,q)=(1,0)) this is not the case in ropes (see case a).

measurements of the NA35 group. The model, including primary collisions only, completely fails to reproduce the observed strangeness enhancement (for K_S, Λ's and $\bar{\Lambda}$'s). It has been demonstrated in [16, 17] that rescattering, dominantly s channel resonance formation in interactions between meson and baryon resonances, is rather effective in creating kaon-hyperon states. Why is the interaction of resonances a major source for enriching strangeness in nuclear collisions? In contrast to πN, other meson-baryon interactions forming s channel resonances in the invariant mass region around $2 \text{ GeV}/c^2$ enrich strangeness, because – in a small percentage of all interactions – they produce a hyperon and kaon, but they do not increase the net number of pions in the system. Note that these processes are speeded up in the *preequilibrium stage* of AA collisions. The mesons which are primarily produced around midrapidity interact with baryon currents which are shifted from the original target or projectile rapidity towards the central rapidity region in course of the reaction. This gives a harder, nonthermal collision spectrum.

The agreement for K_S and Λ's improves noticably, if hadronic rescattering is included, but *not* for the $\bar{\Lambda}$ yields. Hadronic rescattering even decreases the total antibaryon yields, (compare full and dashed line histograms in Figure 3), because produced antibaryons annihilate on surrounding baryons. The $\bar{\Lambda}$ enrichment due to rescattering and annihilation are balancing each other.

The enhancement of single strange hadrons due to rope formation is moderate. While the average strangeness suppression factor in rope fragmentation is strongly enhanced in S+S collisions (to about 0.50), the absolute number of strange particles does not increase that strong, because the total multiplicities from rope decay are roughly 30 percent lower than in the independent string case. A rope decays much faster as an elementary string which results in a smaller amount of created $q\bar{q}$ pairs. In contrast, there is a tremendous effect of rope formation on the final antibaryon yields (see [18]).

In Figure 4 calculated Λ and $\bar{\Lambda}$ rapidity distributions are displayed as a function of the projectile and target mass number. The final antibaryons result from an interplay between the rope dynamics and annihilation in the baryon dense central region. Both effects become stronger with increasing masses of the projectile-target combination. There is good overall agreement of the RQMD calculation with the existing experimental data which can be seen as a strict test of the rope mechanism for antibaryon production. New data which are still preliminary have been taken recently by NA35 for the systems S on Ag and S on Au and compare very well with the RQMD predictions (see [19]).

In a central S collision with a heavy target no local maximum appears any more in the projectile hemisphere. The total baryon source in S on Pb reactions is centered *below* the effective participant centre of mass rapidity $y = 2.5$. There is considerable transport of target baryon number to the central region of this participant fireball, because the baryon number above a rapidity of 2.5 is 20 percent larger than what the S projectile alone provides. Note that the secondary meson rapidity distribution is centered *near* the participant centre of mass rapidity at $y \approx 2.6$.

As can be seen from Figure 4 $\bar{\Lambda}$'s show a rather symmetric behaviour around midrapidity ($y = 3$) in all systems under consideration – even in S+Pb – pointing to yet another source which is present in the calculation of those reactions. The main source of antibaryons are color ropes which are formed by fusing strings. The string overlap in longitudinal direction is the factor which determines how extended the color rope field in space-time will be. This aspect favors the fusion of *long* , i.e. very massive, strings which are preferentially produced in the first collisions between

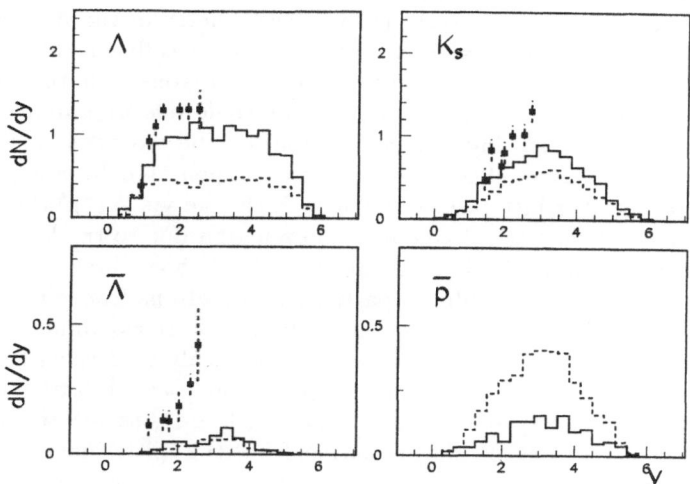

Figure 3. Strange particle – Λ's, $\overline{\Lambda}$'s, and K_S – rapidity distributions in central S(200AGeV)+S collisions: The results of the RQMD calculation *without* the default option rope formation are compared to NA35 data. The histograms represent the calculated rapidity distributions – including secondary rescattering (full line) or primary collisions only (dashed line) – , the symbols the measurement.

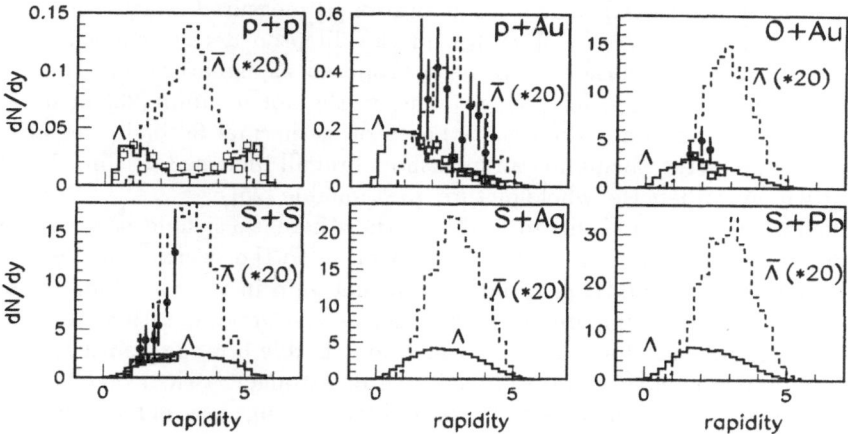

Figure 4. RQMD 1.08 calculation for pp, minimum bias pA and central AA collisions ($b < 1$ fm) with a beam energy of 200 AGeV. The final Λ (full line histogram) and $\overline{\Lambda}$ (dashed line histogram) rapidity distributions are compared to NA35 measurements for p+Au, O+Au and S+S [1]. The NA35 data for O+Au and S+S which have no complete p_t coverage are renormalized by the RQMD ratio of the rapidity density, integrated over all p_t, over the density within the NA35 acceptance.

projectile and target. Thus the relevant centre of mass for interactions which result in rope formation is near to the initial nucleon-nucleon CMS.

It is interesting to note that there is some asymmetry in the $\overline{\Lambda}/\overline{p}$ ratio ($\overline{p}$ without feed-down from hyperon decay) for very asymmetric collisions like S+Pb around midrapidity. The initially symmetric source of antibaryons feels the higher baryon density on the target side. The final $\overline{p}$ distribution (before feeding from $\overline{\Lambda}$ decay) gets distorted by annihilation leaving more antiprotons on the projectile side. In contrast, for the $\overline{\Lambda}$'s the annihilation effect is more than compensated by a gain due to secondary rescattering. For instance, strangeness exchange via $K^{(\star)}\overline{N}^{(\star)}$ interaction is even more important than the charge reversed reaction due to the kaon excess.

The calculated hadron distributions at 200 AGeV show three coupled sources – baryon participants, bulk of produced particles and hadrons from the interior region which are the decay products of fragmenting color ropes. In asymmetric collisions the baryon participants provide an environment of smoothly increasing baryon density towards target rapidity. Measurements of spectra for particle ratios like K^+/K^-, $(\Lambda-\overline{\Lambda})/(p-\overline{p})$ and $\overline{\Lambda}/\overline{p}$ in *asymmetric* collisions which are sensitive to baryon density provide a unique tool to study the transition from the initial nonequilibrium state to the final state [18]. It will be a theoretical as well as an experimental challenge to shed more light on the complicated interplay between additional diquark production mechanisms – like color rope formation – and annihilation due to final state interactions with surrounding baryons. (for ideas pointing into this direction see e.g. [20, 21]).

ϕ MESON PRODUCTION IN AA COLLISIONS

ϕ meson production is strongly suppressed in elementary hadron-hadron collisions because of its quark content, practically solely $s\bar{s}$. Very recently the NA38 collaboration[3] and the HELIOS/3 collaboration[4] measured a factor of 2–3 for the ratio $(\phi/(\omega+\rho^0))_{SU(W)}/(\phi/(\omega+\rho^0))_{pW}$ in the dilepton decay channel. Enhanced ϕ yields in heavy ion collisions have been proposed as signature for a quark-gluon plasma [22], as a sign for a decreasing ϕ mass in the hot medium [23], or in terms of secondary scattering in a hadron gas – based on a thermal fireball calculation [24]. However, in the latter calculation an ultrashort fireball lifetime (≈ 1 fm/c) had to be assumed which renders the whole picture questionable [25].

In the following we are going to study whether there is a sizable extra contribution from rope fragmentation and hadronic rescattering to the ϕ yields in AA collisions. As mentioned above RQMD gives no relevant net gain in the final kaon yields from rope fragmentation as compared to independent string fragmentation. However, this may be no longer true for ϕ mesons which are double strange hadrons. Here the weakening of strangeness suppression is weighted stronger (twice) than the decrease in the absolute yields which cancels the relative strangeness enhancement for the absolute kaon yields from rope fragmentation. Furthermore we reconsider the question of the importance of hadronic rescattering. The reason that the fireball lifetime had to be assumed as ultrashort in [24] is probably that cooling in the expansion stage favors strongly the formation and decay of ρ over ϕ mesons in $\pi\pi$, respectively $K\overline{K}$ annihilations. In contrast, RQMD is a transport model including the preequilibrium interactions in AA collisions. This gives highly energetic interactions which overcome the threshold for ϕ production a much larger weight than in a thermal scenario.

Inspection of the ϕ production mechanisms reveals a large contribution to the

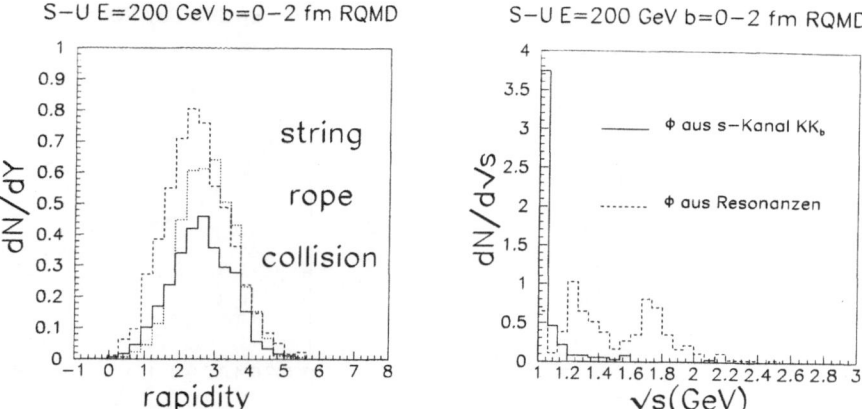

Figure 5. The left side shows the ϕ rapidity distribution for central SU collisions broken up into three contributions, from string (dashed line) and rope (dotted line) decay and from meson-meson collisions. On the right side the center of mass energy distribution of meson-meson collisions with a ϕ in the exit channel is shown for SU collisions: $K\overline{K}$ annihilation dominated by ϕ formation (solid line) and all other channels summed up which are dominated by higher mass meson resonance formation and decay (dotted line).

total final ϕ yields from rope decay (32%) as shown on the the left side of Figure 5. It turns out that ϕ production from ropes in central SU collisions is enhanced by 40 percent in the average if compared to the ϕ yield from independent fragmentation of those strings which fuse into ropes. Most of the enhanced production effect from the initial rope fragmentation survives in the later stages of the reaction. Rope formation alone – i.e. neglecting the gain from rescattering processes – would increase the final ϕ yield by approximately 30 % in central SU reactions.

Meson-meson collisions are the most important contributor to ϕ production in the rescattering sector (24% of the total final yield) dominating over the meson-baryon interactions. Furthermore, the annihilation rate of ϕ mesons in the meson-meson channel turns out to be rather small, approximately 4 percent. The net value of 20 percent for ϕ's produced in the meson-meson sector represents a contribution enhancing the ϕ yield over nonstrange hadrons, because the total multiplicity of produced particles remains practically unchanged by meson-meson interactions (and secondary scattering in general). On the right hand side of Figure 5 the invariant mass spectrum of those meson resonances which decayed into a finally observable ϕ is shown. The most prominent role plays the ϕ formation itself via $K\overline{K}$ and $\pi\rho$ annihilation. $K\overline{K}$ reactions which are favored by comparably large kaon multiplicities (summed up $\approx$ 110) generate – despite the narrowness of the ϕ (Γ=4.4 MeV) – a relevant fraction (8%) of the total final yield. However, formation and subsequent decay of higher mass meson resonances is by far not negligible. These reactions – together with the $\pi\rho$ annihilation into a ϕ – account for 12% of the total ϕ yield. The mass spectrum of meson resonances decaying into a ϕ which is shown in Figure 5 displays two maxima around 1.3 and 1.7 GeV. This reflects the different coupling strengths of the various meson resonances to the ϕ channel which were extracted from the particle data tables.

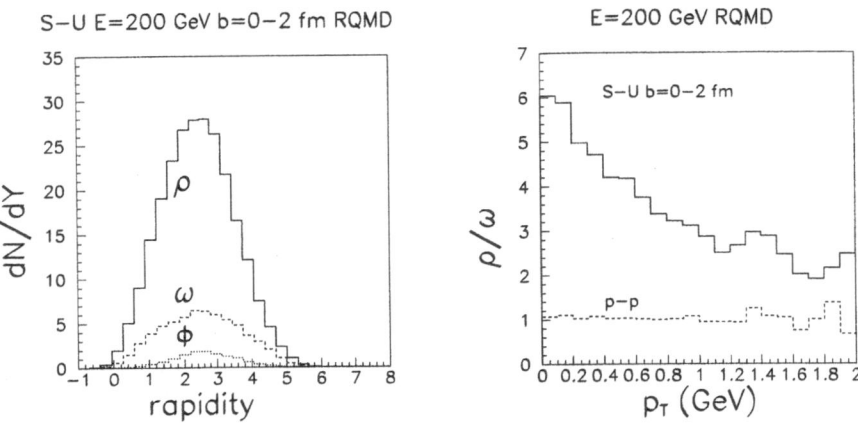

Figure 6. The rapidity distributions of ϕ (dotted line), ω (dashed line) and ρ^0 mesons are shown for central SU collisions at 200 AGeV (left). The ratio ρ^0/ω is shown as a function of the transverse momentum for the same system and for pp collisions. (right).

The following meson resonances with coupling to a decay channel containing a ϕ were considered in the calculation (version RQMD 1.09): $b_1(1235)$, $K_1(1270)$, $K_1(1400)$, $K_2(1430)$, $\rho(1450)$, $\phi_2(1680)$, $\rho_3(1690)$ and $K_2(1770)$. If a partial decay width has not been measured, e.g. for resonances which can decay only off-shell into a ϕ, flavor SU_3 symmetry plus phasespace kinematics was used to relate the corresponding ϕ width to measured branching ratios. In addition, nonresonant meson-meson interactions in the t and u channel were included in the RQMD calculation. A flavor SU_3 symmetric Lagrangian was applied to generate interactions – one meson exchange diagrams in Born approximation – between the pseudoscalar and the vector meson nonets. However, these background processes make a marginal contribution to ϕ production only (approximately 4%) due to their small production cross sections.

We infer from the RQMD calculations that inclusion of rope fragmentation and hadronic rescattering together leads to a doubling of the ϕ yields, but leaves the total particle multiplicity practically constant. Those reaction mechanisms are present in AA collisions but not in elementary pp interactions. Thus the calculated integrated ratio ϕ/π is enhanced by a factor of two in central SU collisions compared to pp. Note, however, that this factor two does not translate directly into a factor of two enhancement for $(\phi/(\omega + \rho^0))_{AA}/(\phi/(\omega + \rho^0))_{pp}$ in the dilepton channel, because the fraction of nonstrange vector mesons (ρ and ω) of all secondaries also changes considerably with target and projectile mass number. This will be discussed in the following.

The rapidity distributions of ϕ, ω and ρ^0 decays are shown for SU collisions at 200 AGeV on the left side of Figure 6. The number of ρ^0 decays outweights the ω yield considerably. In fact, the double ratio $(\rho^0/\pi)_{SU}/(\rho^0/\pi)_{pp}$ is near two, quite similar to the corresponding ratio for the ϕ. In contrast, the yield of ω decays normalized to the total final pion number decreases by 40 %. In elementary pp collisions at high energies both neutral nonstrange vector mesons are produced at equal rates [26], since they have practically the same mass and quark content.

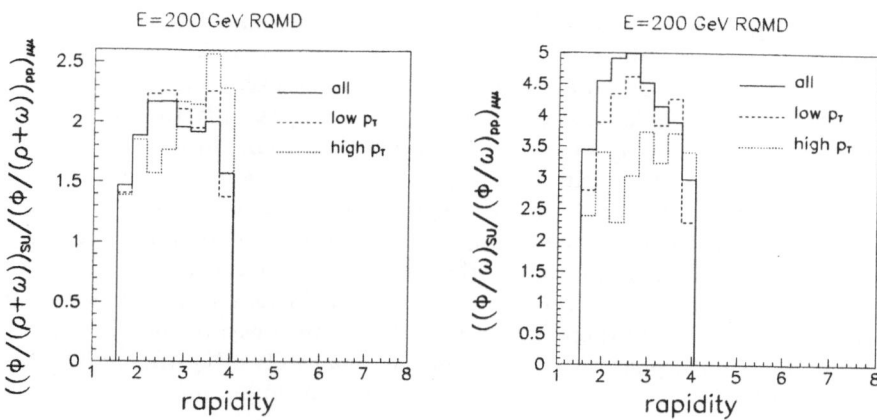

Figure 7. The rate of ϕ decays over ρ and ω decays $\phi/(\omega + \rho^0)$ in the dimuon decay channel for central SU collisions divided by the the corresponding ratio in pp collisions, as calculated with RQMD 1.08 (left), the dimuon decay rate ϕ/ω in central SU collisions to ϕ/ω in pp collisions (right) at 200 GeV integrated over all p_T, for $0.6\,GeV \leq p_T < 1.15\,GeV$ (dashed line) and for $1.15\,GeV \leq p_T < 3.0\,GeV$ (dotted line).

In central SU collisions the ratio of ρ^0/ω decays is strongly p_t dependent, as shown on the right hand side of Figure 6. In the high p_t region, dominated by first collisions, this ratio is not very far from its value 1 in pp reactions. In the low and moderate p_t region, however, where collective effects play an important role, the ρ^0 yield exceeds the ω yield by a factor of up to six. ρ^0 may quite easily be formed in frequent $\pi\pi$ reactions, in contrast to ω which does not couple strongly to a two-particle decay channel. Note that coupling to the two pion channel violates G parity. The differences in the coupling of ρ and ω to possible decay channels are reflected in the short lifetime of ρ^0 ($\tau=1.32\ fm/c$) compared to ω ($\tau=23.4\ fm/c$). Therefore ρ mesons are formed and decay again all the time in course of the AA reaction while most of the ω decays take place after the hadrons have frozen out. Note that pion annihilation is the main but not the sole reason in the RQMD calculation for different yields of ρ and ω. The equality is also broken by formation of s channel baryon and meson resonances in the 1–2 GeV/c^2 mass region in later stages of the reaction. The ρ meson is produced more abundantly in those reactions than the ω.

The double ratio of dimuon production rates in strange and nonstrange vector meson decays in SU and pp collisions $(\phi/(\omega + \rho^0))_{SU}/(\phi/(\omega + \rho^0))_{pp}$ is shown on the left hand side of Figure 7, using the same p_T cuts as the NA38 collaboration [3], The $\mu^+\mu^-$ decay partial widths are chosen according to [27], with the additional assumption that the ω branching ratio into $\mu^+\mu^-$ is the same as into e^+e^-. For the high p_T acceptance a factor of two enhancement is observed in the rapidity region between $3< y <4$, as well as a strong rapidity dependence. The high p_T cut-off and integration over the whole p_T range result in a slightly lower enhancement with no obvious rapidity dependence. From the discussions above it is clear that the calculated enhancement factor in leptonic decays of the strange over nonstrange vector meson decays is an interplay of ϕ and ρ *enhancement* and ω suppression. Enhancement, respectively suppression is defined by the yield normalized to the produced particle multiplicity and compared to pp. Since the vector meson yields scale differently by going from pp to AA, it is more or less accidentally the case that the ϕ enhancement factor in the dilepton channel and the ratio ϕ/π are practically the same.

We conclude that strangeness enrichment due to strong chromoelectric fields, i.e. color ropes, and hadronic rescattering may help to understand the measured factor of 2–3 enhancement of ϕ over $\omega + \rho^0$ in the dilepton decay channel for S on U/W collisions compared to p+W. However, before we can take the good quantitative agreement between the RQMD calculations and the experimental data as a conclusive proof, one has to be careful about drawing too many conclusions from a measurement of a *double ratio*. Obviously there is the urgent need to measure the different neutral vector meson (ϕ,ω,ρ^0) phasespace distributions separately. RQMD predicts different scaling of the vector meson decay yields with atomic mass numbers of the ingoing nuclei. Thus the ratios of the vector meson distributions show peculiar dependencies on transverse momentum and rapidity. First of all it has to be checked that the $\phi/\rho+\omega$ enhancement measured by NA38 and HELIOS/3 reflects an enhancement of ϕ mesons – which is the prediction of RQMD – and not alone the suppression of nonstrange vector mesons as suggested by some other authors. Furthermore, it will be most interesting to learn whether the predicted huge transverse momentum dependent ρ-ω splitting which reflects the lifetime of the system and the importance of low energy processes is indeed realized in nature. The substantial ω suppression predicted by RQMD for AA collisions is in striking contrast to any statistical scenario in which nonstrange hadrons freeze out in chemical equilibrium at high temperatures (on the order of 200 MeV).

REFERENCES

1. J. Bartke et al. (NA35 Collab.): *Z. f. Phys.* **C48** (1990) 191. R. Stock et al. (NA35 Collab.): *Nucl. Phys.* **A525** 221c.

2. S. Abatzis et al. (WA85 Collab.): *Phys. Lett.* **B244** (1990) 130; *Phys. Lett.* **B259** (1990) 508; *Phys. Lett.* **B270** (1991) 123.

3. C. Baglin et al. (NA38 Collab.): *Phys. Lett.* **B272** (1991) 449; R. Ferreira: *Nucl. Phys.* **A544** (1992) 497c.

4. M. A. Mazzoni et al. (HELIOS/3 Collab.): *Nucl. Phys.* **A544** (1992) 623c; contribution to the proc. of "Quark Matter 1993", to appear in *Nucl. Phys.* **A** (1994)

5. T.S. Biro, H.B. Nielsen, and J. Knoll: *Nucl. Phys.* **B245** (1984) 449.

6. A. Bialas, W. Czyz: *Phys. Rev.* **D31** (1985) 198; *Nucl. Phys.* **B267** (1986) 242.

7. G. Gatoff, A.K. Kerman, and T. Matsui: *Phys. Rev.* **D36** (1987) 114; A.K. Kerman, T. Matsui, and B. Svetitsky: *Phys. Rev. Lett.* **56** (1986) 219.

8. B. Andersson, P.A. Henning: *Nucl. Phys.* **B355** (1991) 82.

9. H. Sorge, M. Berenguer, H. Stöcker, and W. Greiner: *Phys. Lett.* **B289** (1992) 6.

10. H. Sorge, H. Stöcker, and W. Greiner: *Ann. Phys.* (USA) **192** (1989) 266; *Nucl. Phys.* **A498** (1989) 567c; H. Sorge, R. Mattiello, A. von Keitz, H. Stöcker, and W. Greiner: *Z. f. Phys.* **C47** (1990) 629.

11. J. Schwinger: *Phys. Rev.* **82** (1951) 664; E. Brezin, C. Itzykson: *Phys. Rev.* **D2** (1970) 1191.

12. A. Casher, H. Neuberger, and S. Nussinov: *Phys. Rev.* **D20** (1979) 179.

13. J. Carlson, J. Kogut, and V.R. Pandharipande: *Phys. Rev.* **D27** (1983) 233.

14. H. Albrecht et al. (ARGUS Collab.): *Z. f. Phys.* **C49** (1991) 349.

15. H. Albrecht et al. (ARGUS Collab.): *Z. f. Phys.* **C9** (1988) 177.

16. H. Sorge, R. Mattiello, H. Stöcker, and W. Greiner: *Phys. Lett.* **B271** (1991) 37.

17. H. Sorge, L. Winckelmann, H. Stöcker, and W. Greiner: *Z. f. Phys.* **C59** (1993) 85.

18. H. Sorge: LA-UR-93-1103 (preprint).

19. D. Roehrich: Contribution to these proceedings.

20. A. Jahns et al: Contribution to these proceedings.

21. S. Mrowczynski: *Phys. Lett.* **B308** (1993) 216.

22. A. Shor: *Phys. Rev. Lett.* **54** (1985) 1122.

23. C. M. Ko, and B. H. Sa: *Phys. Lett.* **B258** (1991) 6.;

24. P. Koch, U. Heinz, and J. Pisùt: *Phys. Lett.* **B243** (1990) 149.; *Z. Phys.* **C47** (1990) 447.

25. U. Heinz, and K. S. Lee: *Phys. Lett.* **B259** (1991) 162.;

26. M. Aguilar–Benitez et al.: *Z. f. Phys.* **C50** (1991) 405.

27. Particle Data Group, K. Hikasa et al.: Review of Particle Properties, *Phys. Rev.* **D45** (1992).

BACK-REACTION BEYOND THE MEAN FIELD APPROXIMATION

Yuval Kluger

Theoretical Division, Los Alamos National Laboratory
Los Alamos, New Mexico 87545 USA

ABSTRACT

A method for solving an initial value problem of a closed system consisting of an electromagnetic mean field and its quantum fluctuations coupled to fermions is presented. By tailoring the large N_f expansion method to the Schwinger-Keldysh closed time path (CTP) formulation of the quantum effective action, causality of the resulting equations of motion is ensured, and a systematic energy conserving and gauge invariant expansion about the electromagnetic mean field in powers of $1/N_f$ is developed. The resulting equations may be used to study the quantum non-equilibrium effects of pair creation in strong electric fields and the scattering and transport processes of a relativistic e^+e^- plasma. Using the Bjorken ansatz of boost invariance initial conditions in which the initial electric mean field depends on the proper time only, we show numerical results for the case in which the N_f expansion is truncated in the lowest order, and compare them with those of a phenomenological transport equation.

INTRODUCTION

In the last few years it has become clear that there are many situations of direct physical interest involving *time-dependent* dynamical evolution, where a perturbative scattering formalism is wholly inadequate. This is the case for strong field electrodynamics in astrophysical plasmas, in nuclear collisions of heavy ions where a phase transition to a quark-gluon plasma is expected to take place and in the hot dense early universe. These problems require detailed knowledge of the dynamical time evolution of the field configurations and energy flow.

An understanding of the dynamics of the quark-gluon plasma following its formation in a heavy ion collision is especially important for the theoretical interpretation of the data that will be produced by RHIC experiments within a few years. One would like to know how the plasma is produced, how efficiently it thermalizes and on what time scale, how it evolves and what are the special signatures (in terms of flow of energy, production of particles, etc.) of this new state of matter.

In the early stages of quark-gluon plasma evolution, phenomenological approaches based on kinetic theory (and subsequently on hydrodynamics) are frequently used[1-5]. The event generators which supply a source term for the kinetic equations in the present models are constructed in the framework of a perturbation theory[5-6]. But, in the range of RHIC energies non perturbative effects like the formation of coherent color-electric fields may play an essential role together with the perturbative quantum fluctuations [see the lecture by Gyulassy at this school]. Moreover, effects like off-shell propagation are not taken into account in these studies [see the lectures by C. Greiner and Neise at this school].

The kinetic approach is an intuitive and tractable one. But, it is not known *a priori* in such a complex system, governed by strong interactions, if it is applicable at all. In order to verify which (if any) of the assumptions of the various kinetic models are valid, it will be useful to investigate whether the inclusion of these various effects in a kinetic model is appropriate for describing the physics. This can be done in three stages: Firstly, by deriving the equations of motion for the mean fields and their quantum fluctuations using the close-time-path method[7-9] and using a systematic approximation (such as the large N_f expansion[10-13]) to the Schwinger-Dyson equations of the full field theory. Secondly, by extracting the transport equations by using approximations like the gradient expansion[14] of Green's functions while keeping consistency with the field theory expansion method in use. Thirdly, by comparing the simulations of the field-theory with those of the corresponding transport equation[15]. If the transport equation does not reproduce similar results to those of the field theory, one must rely on the full field theory approach which is much more complicated.

A practical starting point for this task would be to deal with a simpler theory like QED[16-19], instead of QCD.

I begin with an introduction of the effective action techniques, which are used in the derivation of the effective equations of motion in QED[20] (the first stage). I briefly review the functional method of the effective action in the large N expansion for QED, then the basics of the Schwinger-Keldysh real time CTP formalism and then I bring the two ingredients together and present the effective action and causal equations of motion for electrodynamics.

Finally, I demonstrate the numerical results[21] which were obtained by applying the above methods in the lowest order of the $1/N_f$ expansion for the case when the initial electric mean field is taken to be a function only of the fluid proper time $\tau = \sqrt{t^2 - z^2}$. These results are compared with those obtained by solving a relativistic transport theory with a modified proper time dependent Schwinger source term[22] using boost-invariant coordinates[1-4]. In both cases, these initial conditions kinematically guarantee a flat particle rapidity distribution. The motivation to choose these initial conditions is the fact that in heavy-ion collisions one is dealing with a situation in which particle production in the central rapidity region can be modeled as an inside-outside cascade which is symmetric under longitudinal boosts, and thus produces a plateau in the particle rapidity distributions[23].

THE LARGE N EXPANSION

The large N_f expansion is a perturbation functional method which consist of N_f identical fields. For QED the N_f copies are charged fermions fields (flavors). This method provides a convenient way of parameterizing the separation of the quantum fields into mean fields and their fluctuations and a systematic approximation scheme to the Schwinger-Dyson equations of the full field theory. While the large N_f method is not the only technique possible, it is a gauge and renormalization group invariant expansion which arranges the Feynman perturbation series for scattering diagrams in a way that automatically includes self-energy corrections of scattering to the same order. Only fields which scale like N_f to a positive power for large N_f can be considered strong mean fields in this approach. For such strong fields the leading order approximation is equivalent to the mean field approximation. The Schwinger mechanism for pair production is already realized in the leading order. In the next to leading order in the expansion in powers of $1/N_f$ the effects of collisional and radiation processes back-react on the mean fields and appear for the first time. These processes are essential mechanisms for driving the system to equilibrium.

For QED with N identical charged fermion fields the Lagrangian is

$$L = - \sum_{i=1}^{N} \int d^4x \ \overline{\psi}_i G^{-1}[A]\psi_i - \frac{N}{4e^2} \int d^4x \ F_{\mu\nu}F^{\mu\nu}, \tag{1}$$

where $G^{-1}[A]$ is the inverse fermion propagtor defined as

$$G^{-1}[A] = i\frac{\gamma^\mu}{2}(\vec{\partial}_\mu - \overleftarrow{\partial}_\mu) + \gamma^\mu A_\mu + im \ . \tag{2}$$

Introducing external sources for the gauge potential and Dirac fields, we define the generating functional

$$Z[J,\eta,\overline{\eta}] \equiv e^{iNW[J,\eta,\overline{\eta}]} \equiv \int [\mathcal{D}A_\mu]' \prod_{i=1}^{N} \int [\mathcal{D}\psi_i][\mathcal{D}\overline{\psi}_i] \exp\left\{i \int d^4x \ L[A,\psi,\overline{\psi}]\right\}$$
$$\times \exp\left\{iNJ \circ A + i\overline{\eta} \circ \psi + i\overline{\psi} \circ \eta\right\}, \tag{3}$$

where the prime on the gauge field integration measure denotes that we should integrate only over distinct gauge invariant configurations (or equivalently, fix the gauge), and where the symbol ∘ will denote summation over internal indices and integration over continuous spacetime coordinates in the quantities on either side of it (the de Witt summation convention).

Performing the Gaussian integration over the anti-commuting Dirac fields, and rescaling the Grassman valued sources $\eta \to \sqrt{N}\eta$ so that we can drop the sums over $i = 1, \ldots, N$, we obtain

$$Z[J,\eta,\overline{\eta}] = \int [\mathcal{D}A_\mu]' \exp\left\{iN \int d^4x \ \frac{1}{2e^2} A_\mu \left(g^{\mu\nu}\Box - \partial^\mu\partial^\nu\right) A_\nu\right\}$$
$$\times \exp\left\{N\mathrm{Tr}\ln G^{-1} + iN\overline{\eta} \circ G[A] \circ \eta + iNJ \circ A\right\}, \tag{4}$$

where Tr denotes summation over internal indices and integration over spacetime coordinate. We have defined the sources and coupling with the correct powers of N

to justify performing the remaining functional integration over the electromagnetic potential by the stationary phase method[13]. Varying the argument of the exponent in eq. (4) with respect to A the stationary phase value $A_\mu^s[J, \eta, \overline{\eta}]$ satisfies

$$\frac{1}{e^2}\left(g^{\mu\nu}\Box - \partial^\mu\partial^\nu \right) A_\nu^s(x) = i\mathrm{tr}G(x,x)\gamma^\mu + \overline{\eta}\circ G(\ ,x)\gamma^\mu G(x,\ )\circ\eta - J^a(x), \quad (5)$$

where tr denotes the Dirac matrix trace only (without integration over spacetime coordinates). The second derivative of the exponent in (4) at its stationary point is $-iN\mathrm{Tr}D^{-1}$, where

$$
\begin{aligned}
D^{-1}(x,y)^{\mu\nu} = & -\frac{1}{e^2}\left(g^{\mu\nu}\Box - \partial^\mu\partial^\nu \right)\delta^4(x,y) - i\mathrm{tr}\left[G(y,x)\gamma^\mu G(x,y)\gamma^\nu\right]_{A=A^s} \\
& -2\mathrm{tr}\left[\overline{\eta}\circ G(\ ,x)\gamma^\mu G(x,y)\gamma^\nu G(y,\ )\circ\eta\right]_{A=A^s},
\end{aligned}
\quad (6)
$$

and where symmetrization with respect to interchange of the pair of spacetime labels x, μ with y, ν is understood. Thus, the result of the stationary phase evaluation of (3) and (4) is

$$
\begin{aligned}
W[J, \eta, \overline{\eta}] \equiv\ & W^{(0)} + \frac{1}{N}W^{(1)} + \frac{1}{N^2}W^{(2)} + \cdots \\
=\ & -A^s\circ d^{-1}\circ A^s + J\circ A^s + \overline{\eta}\circ G[A^s]\circ\eta \\
& -i\mathrm{Tr}\ln G^{-1}[A^s] + \frac{i}{2N}\mathrm{Tr}\ln D^{-1}[A^s] + \mathcal{O}(\tfrac{1}{N^2}),
\end{aligned}
\quad (7)
$$

where

$$d^{-1}(x,y)^{\mu\nu} \equiv -\frac{1}{e^2}\left(g^{\mu\nu}\Box - \partial^\mu\partial^\nu \right)\delta^4(x,y) \quad (8)$$

is the differential operator from the classical action.

The c-number fields are given by the variations,

$$A_c(x) \equiv \frac{\delta W}{\delta J(x)} \quad ; \quad \psi_c(x) \equiv \frac{\delta W}{\delta\overline{\eta}(x)} \quad ; \overline{\psi}_c(x) \equiv \frac{\delta W}{\delta\eta(x)} \quad . \quad (9)$$

Note that the c-number field A_c differs from the stationary phase point of the Gaussian integral at order $1/N$, namely

$$A_c = A_s + \frac{1}{N}\frac{\delta W^{(1)}}{\delta A}\circ\frac{\delta A}{\delta J} + \mathcal{O}\left(\tfrac{1}{N^2}\right). \quad (10)$$

When no confusion with the integration variables in (3) exists, we shall drop the subscript "c" from the c-number fields to simplify the notation.

The effective action functional may be defined in terms of the c-number fields (9) by a Legendre transformation in the usual way,

$$\mathcal{S}[A, \psi, \overline{\psi}] \equiv W - J\circ A - \overline{\eta}\circ\psi - \overline{\psi}\circ\eta, \quad (11)$$

where $J, \overline{\eta}$ and η are to be regarded as functionals of the mean fields by inverting eqs. (9). The result of this Legendre transformation is simply

$$\mathcal{S} = -\overline{\psi} \circ G^{-1}[A] \circ \psi - \frac{1}{2} A \circ d^{-1} \circ A - i\operatorname{Tr}\ln G^{-1}[A] + \frac{i}{2N}\operatorname{Tr}\ln D^{-1}[A]. \qquad (12)$$

The photon inverse propagator in the last term is given by

$$D^{-1}[A](x,y)^{\mu\nu} = (d^{-1} + \Pi[A])(x,y)^{\mu\nu}$$
$$\Pi[A](x,y)^{\mu\nu} \equiv -i\operatorname{tr}\{\gamma^{\mu}G[A](x,y)\gamma^{\nu}G[A](y,x)\}, \qquad (13)$$

where Π is the polarization tensor in the presence of the potential A. The inverse propagator cannot be inverted without fixing a gauge, which may be done by a variety of standard methods.

The effective equations for the mean fields $< A > \equiv < out|A|in >$ and $< \psi > \equiv < out|\psi|in >$ are given by

$$\left.\frac{\delta\mathcal{S}}{\delta A(x)}\right|_{A=<A>} = 0 \quad ; \quad \left.\frac{\delta\mathcal{S}}{\delta\psi(x)}\right|_{\psi=<\psi>} = 0. \qquad (14)$$

In the absence of sources the Dirac mean field vanishes. The integro-differential equations for the mean potential read

$$\begin{aligned}
\partial_{\nu}F^{\mu\nu}(x) &= -ie^{2}\operatorname{tr}\{G[A](x,x)\gamma^{\mu}\} + \frac{ie^{2}}{2N}\operatorname{Tr}\left\{D[A] \circ \frac{\delta\Pi[A]}{\delta A_{\mu}(x)}\right\} \\
&= -ie^{2}\operatorname{tr}\{G(x,x)\gamma^{\mu}\} \\
&+ \frac{ie^{2}}{N}\int d^{4}x_{1}\int d^{4}x_{2}\operatorname{tr}\{\gamma^{\mu}G(x,x_{1})\Sigma(x_{1},x_{2})G(x_{2},x)\},
\end{aligned} \qquad (15)$$

where

$$\Sigma(x_{1},x_{2}) \equiv i\gamma^{\mu}G(x_{1},x_{2})\gamma^{\nu}D_{\nu\mu}(x_{2},x_{1}) \qquad (16)$$

is the fermion self-energy. To leading order in $1/N$ these are just the semi-classical Maxwell equations, obtained by replacing the electric current operator of the Dirac field by its expectation value. This leading order semi-classical equation already contains the dynamical reaction of $e^{+}e^{-}$ pairs created by a non-zero electric field (the Schwinger mechanism) back on the electric field itself [see J.M. Eisenberg lecture]. However, at leading order in $1/N$ the created pairs can interact only through the mean field, not directly with each other. The order $1/N$ term with the fermion self-energy Σ contains the quantum Compton scattering, bremmstrahlung and Coulomb interaction effects of these particles on each other and the backreaction of these effects on the self-consistent mean field. Clearly these processes are essential mechanisms for the approach to equilibrium and must be included in any realistic transport theory of a relativistic $e^{+}e^{-}$ plasma. Higher order! processes (such as multiple scat

Having derived the integro-differential equations for the mean fields in electrodynamics to order $1/N$, we turn now to the Schwinger-Keldysh closed time path formulation of the effective action, in order to determine the correct combination of propagator functions needed to obtain a causal (and real) solution to these equations.

THE SCHWINGER-KELDISH CLOSED TIME PATH FORMALISM

The conventional path integral formalism used freely in the preceeding section defines transition elements between states at one time, t (usually taken to be in the infinite past) to states at another time t' (in the distant future). If the class of paths is restricted to be the vacuum configuration at both of its endpoints, then the two states are the $|in\rangle$ and $\langle out|$ vacuum states of scattering theory respectively. The generating functional $Z[J, \eta, \overline{\eta}]$ of Eq. (3) is the transition matrix element

$$Z[J, \eta, \overline{\eta}](t, t') = \langle out, t' | in, t \rangle_{J, \eta, \overline{\eta}} \tag{17}$$

in the presence of the external source J, η and $\overline{\eta}$.

By varying with respect to the external sources we obtain matrix elements of the Heisenberg field operators between the $|in\rangle$ and $\langle out|$ states. For this reason we may refer to the conventional formulation of the generating functional Z as the "in-out" formalism. The time-ordered Green's functions obtained in this way necessarily obey Feynman boundary conditions, and these are the appropriate ones for the calculation of transition probabilities and cross sections between the $|in\rangle$ and $\langle out|$ states. On the other hand the off-diagonal transition matrix elements of the in-out formalism are completely inappropriate if what we wish to consider is the time evolution of physical observables from a given initial condition. As we have remarked the in-out matrix elements are neither real, nor are their equations of motion causal at first order in $1/N$, where direct self interactions between the fields appear for the first time. ! What we require is a generating f

The basic idea of the CTP formalism is to take a diagonal matrix element of the system at a given time $t = 0$ and insert a complete set of states into this matrix element at a different (later) time $t = t'$. In this way one can express the original fixed time matrix element as a product of transition matrix elements from 0 to t' and the time reversed (complex conjugate) matrix element from t' to 0. Since each term in this product is a transition matrix element of the usual (or time reversed) kind, standard path integral representations for each may be introduced. If the same external source operates in the forward evolution as the backward one, then the two matrix elements are precisely complex conjugates of each other, all dependence on the source drops out and nothing has been gained. However, if the forward time evolution takes place in the presence of one source J_+ but the reversed time evolution takes place in the presence of a *different* source J_-, th! en the resulting functional is pr

$$Z[J_+, J_-] \equiv \text{Tr} \left\{ \rho \overline{T} \exp\left[-i \int_0^{t'} dt d^3\vec{x} J_- A \right] T \exp\left[i \int_0^{t'} dt d^3\vec{x} J_+ A \right] \right\} \tag{18}$$

$$= \int [d\Phi][d\Phi'][d\Psi] \langle \Phi | \rho | \Phi' \rangle$$

$$\times \langle \Phi' | \overline{T} \exp\left[-i \int_0^{t'} dt d^3\vec{x} J_- A \right] |\Psi\rangle \langle \Psi | T \exp\left[i \int_0^{t'} dt d^3\vec{x} J_+ A \right] |\Phi\rangle$$

$$= \int [d\phi][d\phi'] \langle \phi | \rho | \phi' \rangle \int [d\Psi] \int_\phi^\Psi [\mathcal{D}A_+]'[\mathcal{D}\psi_+][\mathcal{D}\overline{\psi}_+] \int_{\phi'}^\Psi [\mathcal{D}A_-]'[\mathcal{D}\psi_-][\mathcal{D}\overline{\psi}_-]$$

$$\times \quad \exp\left[i \int_0^\infty dt d^3\vec{x} \left(L[A_+, \psi_+, \overline{\psi}_+] - L[A_-, \psi_-, \overline{\psi}_-] + J_+ A_+ - J_- A_-) \right].$$

where $\rho = |in\rangle\langle in|$ is the initial density matrix (an initial mean electric field corresponds to a coherent initial $|in\rangle$ state), and where in the last equality a path integral representation for each transition element is introduced. Since the time ordering in eq. (19) is forward (denoted by $\mathcal{T}$) along the time path from 0 to t' in the second transition matrix element, but backward (denoted by $\overline{\mathcal{T}}$) along the same path from t' to 0 in the first matrix element, this generating functional receives the name of the closed time path generating functional.

The mean electromagnetic field expectation values are now given by,

$$\left.\frac{\delta W_{in}[J_+, J_-]}{\delta J_+(x)}\right|_{J=0} = -\left.\frac{\delta W_{in}[J_+, J_-]}{\delta J_-(x)}\right|_{J=0} = \langle in|A(x)|in\rangle \tag{19}$$

and the matrix connected two point functions are given by

$$G^{ab}(x,y) = \left.\frac{\delta^2 W}{\delta J_a(x)\delta J_b(y)}\right|_{J=0} ; \qquad a = +, - , \tag{20}$$

more explicitly

$$
\begin{aligned}
G^{-+}(x,y) &\equiv G_>(x,y) = i\mathrm{Tr}\{\rho\,\psi(x)\overline{\psi}(y)\}_{con} , \\
G^{+-}(x,y) &\equiv G_<(x,y) = -i\mathrm{Tr}\{\rho\,\overline{\psi}(y)\psi(x)\}_{con} \\
G^{++}(x,y) &= i\mathrm{Tr}\left\{\rho\,\mathcal{T}[\psi(x)\overline{\psi}(y)]\right\}_{con} = \theta(x,y)G_>(x,y) + \theta(y,x)G_<(x,y) \\
G^{--}(x,y) &= i\mathrm{Tr}\left\{\rho\,\overline{\mathcal{T}}[\psi(x)\overline{\psi}(y)]\right\}_{con} = \theta(y,x)G_>(x,y) + \theta(x,y)G_<(x,y).
\end{aligned}
\tag{21}
$$

We may take over all the results of the previous section on the generating functional, effective action, and equations of motion of QED, provided only we substitute the CTP path ordered Green's function(s) for the ordinary Feynman propagators in internal lines, integrate over the full closed time contour, and satisfy the initial conditions at $t = 0$ corresponding to the given density matrix ρ.

The causal equations of motion for nonequilibrium electrodynamics up to the next to leading order in $1/N$ expansion take the form,

$$
\begin{aligned}
\partial_\nu F^{\mu\nu}(x) = &-\frac{ie^2}{2}\mathrm{tr}\left\{[G_>(x,x) + G_<(x,x)]\gamma^\mu\right\} \\
&+ \frac{2e^2}{N}\mathcal{I}m \int_0^t dt_1 d^3\vec{x}_1 \int_0^{t_1} dt_2 d^3\vec{x}_2\Big\{[G_>(x_1,x) - G_<(x_1,x)]\gamma^\mu \times \\
&[G_>(x,x_2)\Sigma_>(x_1,x_2) - G_<(x,x_2)\Sigma_<(x_1,x_2)]\Big\} .
\end{aligned}
\tag{22}
$$

The Wightman functions for the Dirac field satisfy

$$(i\gamma^\mu\partial_\mu + \gamma^\mu A_\mu + im)\,G_{>,<}(x,y) = 0, \tag{23}$$

together with the initial conditions implied by the first two members of eqs. (21), which satisfy the canonical equal time anticommutator condition,

$$[G_>(x,y) - G_<(x,y)]_{t_x=t_y} = i\mathrm{Tr}\left\{\rho[\psi(x),\overline{\psi}(y)]_+\right\}_{t_x=t_y} = \delta^3(\vec{x} - \vec{y}) , \tag{24}$$

appropriate for Fermi-Dirac statistics.

For a complete initial value problem to order $1/N$, one needs also the two-point function of the Maxwell field obtained by inverting (13) subject to some gauge condition. The simplest way to impose the gauge condition is to split it in the following way

$$D_{\mu\nu}(x,y) = e^2 d_{\mu\nu}(x,y) + \widetilde{D}_{\mu\nu}(x,y) \tag{25}$$

with $d_{\mu\nu}$ the inverse of the differential operator (8) of the *free* action in a definite gauge. The gauge fixing can be performed at the level of the free photon propagator once and for all, independently of the dynamical time evolution problem, and the non-trivial time evolution is contained entirely in $\widetilde{D}_{\mu\nu}$, which then obeys

$$\left(g^{\mu\lambda}\Box - \partial^\mu\partial^\lambda \right) \widetilde{D}_{\lambda\nu}^{>,<}(x,y) = e^2 \int_0^{t_x} d^4x_1 \left[\Pi_>^{\mu\lambda} - \Pi_<^{\mu\lambda} \right](x,x_1) D_{\lambda\nu}^{>,<}(x_1,y)$$
$$-e^2 \int_0^{t_y} d^4x_1 \Pi_{>,<}^{\mu\lambda}(x,x_1) \left[D_{\lambda\nu}^> - D_{\lambda\nu}^< \right](x_1,y). \tag{26}$$

A particularly useful gauge choice for practical implementation of the initial value problem on a computer is the Coulomb gauge, which has the advantage of clearly isolating the physical transverse modes of the photon and allowing the longitudinal and gauge modes to be eliminated from evolution problem, thereby making most efficient use of computer memory.

BACK-REACTION IN BOOST INVARIANT COORDINATES

Equipped with the above formalisms which are suitable for solving an initial value problem, I here present the resulting equations of motion for QED in the case where the initial conditions are symmetric under longitudinal boosts, and where only the leading order of the large N expansion is taken into account. This symmetry is clearly displayed by introducing the light-cone variables τ and η

$$z = \tau \sinh\eta \quad , \quad t = \tau \cosh\eta. \tag{27}$$

and by specifying the initial conditions of the fields at $\tau = \tau_0$, i.e., on a hyperbola of constant proper time.

Covariance of the theory guarantees that once the initial conditions have this symmetry, it will be conserved at any time. This allows us to expand the Dirac field in terms of Fourier modes at a fixed proper time τ:

$$\psi(x) = \int \frac{d\mathbf{k}_\perp dk_\eta}{(2\pi)^3} \sum_{s=1,2} [b_s(\mathbf{k})\psi_{\mathbf{k}s}^+(\tau)e^{ik_\eta\eta + i\mathbf{k}_\perp \cdot \mathbf{x}_\perp} + d_s^\dagger(\mathbf{k})\psi_{-\mathbf{k}s}^-(\tau)e^{-ik_\eta\eta - i\mathbf{k}_\perp \cdot \mathbf{x}_\perp}]. \tag{28}$$

If the electric field is in the z direction and is a function of τ only, upon substituting eq. (28) into eq. (21) and using the boost invariant variables the system of equations

(22)-(24) reduces, in the gauge $(A_\tau = 0, A_x = 0, A_y = 0, A_\eta = A_\eta(\tau))$, to

$$\left[\gamma^0 \left(\frac{d}{d\tau} + \frac{1}{2\tau} \right) + i\gamma_\perp \cdot \mathbf{k}_\perp + i\gamma^3 \pi_\eta + m \right] \psi_{\mathbf{k}s}^\pm(\tau) = 0, \tag{29}$$

$$\frac{1}{\tau} \frac{dE(\tau)}{d\tau} = -\frac{e}{2\tau} \left\langle in \left| \left[\psi^\dagger, \gamma^0 \gamma^3 \psi \right] \right| in \right\rangle, \tag{30}$$

where $\pi_\eta \equiv (k_\eta - eA_\eta)/\tau$.

For an initial coherent state which satisfies $b_s(\mathbf{k})|in\rangle = d_s(-\mathbf{k})|in\rangle = 0$ (an adiabatic vacuum) the Maxwell eq. (30) reads

$$\frac{1}{\tau} \frac{dE(\tau)}{d\tau} = -\frac{e}{2\tau} \int \frac{d\mathbf{k}_\perp dk_\eta}{(2\pi)^3} \sum_{s=1,2} \left\{ -\psi_{\mathbf{k}s}^{+\,\dagger} \gamma^0 \gamma^3 \psi_{\mathbf{k}s}^+ + \psi_{\mathbf{k}s}^{-\,\dagger} \gamma^0 \gamma^3 \psi_{\mathbf{k}s}^- \right\}. \tag{31}$$

The renormalization of the above equation can be done by the adiabatic regularization technique. Unfortunately, it is not applicable for renormalization beyond the leading order of the $1/N$ expansion. However, recently we found a simple prescription for removing the logarithmic divergences by recognizing that renormalization group invariant quantities like eE do not depend on the momentum cutoff Λ, as long as we rescale the coupling constant according to the standard renormalization group flow to the given order in $1/N$. For a fixed renormalized coupling constant e_R and an electric field E_R, the corresponding bare quantities in the lowest order in the large N expansion are defined via

$$e_R^2 \equiv Z_\Lambda e^2 \quad ; \quad E_R \equiv Z_\Lambda^{-1/2} E \quad ; \quad eE = e_R E_R$$
$$Z_\Lambda = [1 + e^2 \delta e^2(\Lambda)]^{-1} = [1 - e_R^2 \delta e^2(\Lambda)]$$
$$\delta e^2(\Lambda) = \frac{1}{4\pi^2} \int_0^\Lambda dk \left[\frac{k^2}{(k^2 + m^2)^{\frac{3}{2}}} - \frac{k^4}{3(k^2 + m^2)^{\frac{5}{2}}} \right]. \tag{32}$$

Using the first line of eq. (32) allows us to rewrite eq. (31) in the form

$$\frac{\epsilon_R}{\tau} \frac{dE_R(\tau)}{d\tau} = -\frac{e_R^2/Z_\Lambda}{2\tau} \int^\Lambda \frac{d\mathbf{k}_\perp dk_\eta}{(2\pi)^3} \sum_{s=1,2} \left\{ -\psi_{\mathbf{k}s}^{+\,\dagger} \gamma^0 \gamma^3 \psi_{\mathbf{k}s}^+ + \psi_{\mathbf{k}s}^{-\,\dagger} \gamma^0 \gamma^3 \psi_{\mathbf{k}s}^- \right\}. \tag{33}$$

which is a renormalization group invariant, and therefore the right hand side is cutoff independent. This cutoff independence has been checked both for scalar QED in flat coordinates and boost invariant coordinates.

Figures 1-2 summarize the results of the numerical simulation for the evolution of the system (29)-(30) in $1 + 1$ dimensions. In Fig. 1(a) the proper time evolution of the electric field is presented, where the initial conditions where fixed at $\tau = 1$. The amplitude falls off as a result of the particle production and the expansion of the system. Having the solution for the fermion mode functions, we calculated the expectation value of the energy-momentum tensor. Fig. 1(b) shows the fermion part of the $T_{\tau\tau}$ component, which can be identified with the energy density ϵ in the comoving frame.

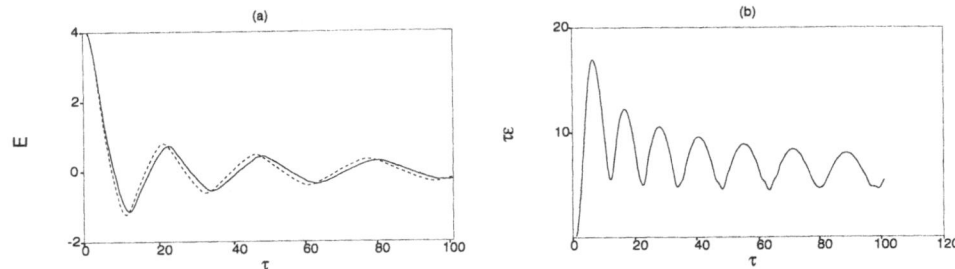

Figure 1. Proper time evolution of (a) the rescaled electric field $E(\tau) \rightarrow eE/m^2$, and (b) the energy density of the fermion field ϵ multiplied by τ, for initial conditions at $\tau = 1$ with a rescaled initial electric field $E(\tau = 1) = 4$ and $e^2/m^2 = 1$.

If we expand the field ψ in terms of the *time-dependent* creation and annihilation operators of the lowest-order adiabatic vacuum

$$\psi(x) = \int \frac{dk_\eta}{2\pi} [a(k_\eta; \tau) y_{k_\eta}^+(\tau) e^{ik_\eta \eta} + c^\dagger(k_\eta; \tau) y_{-k_\eta}^-(\tau) e^{-ik_\eta \eta}], \tag{34}$$

where $y_{k_\eta}^\pm$ are the lowest order adiabatic mode functions, we can relate these operators to those in eq. (28) via a Bogolyubov transformation

$$a(k_\eta; \tau) = \alpha(k_\eta; \tau) b(k_\eta) + \beta^*(k_\eta; \tau) d^\dagger(-k_\eta), \tag{35}$$

where α and β are functions of the known mode functions $\psi_{k_\eta}^\pm$ and $y_{k_\eta}^\pm$. Due to the boost invariant symmetry the rapidity distribution dN/dy is equal to the particle density $dN/d\eta$ which is given by

$$dN/d\eta = \int \frac{dk_\eta}{2\pi} \langle a^\dagger(k_\eta; \tau) a(k_\eta; \tau) + c^\dagger(-k_\eta; \tau) c(-k_\eta; \tau) \rangle. \tag{36}$$

As is shown in Fig. 2(a), most of the production takes place in the early stages. Identifying this particle density with a Boltzmann-like phase density $\tilde{f}$, in Fig. 2(b) we plot the Boltzmann entropy density in the comoving frame, which is defined as

$$s(\tau) = -\frac{1}{\tau} \int \frac{dk_\eta}{2\pi} \{\tilde{f} \ln \tilde{f} + (1 - \tilde{f}) \ln(1 - \tilde{f})\} . \tag{37}$$

By the time particle production has nearly ceased, the quantity τs is roughly a constant, as expected in a hydrodynamics treatment with no source term of particle production[24].

Finally, as is shown in J.M. Eisenberg's lecture at this school, a comparison of the field theory results with those of a transport equation with a Schwinger source term for particle production shows similar results for the case of an homogeneous electric

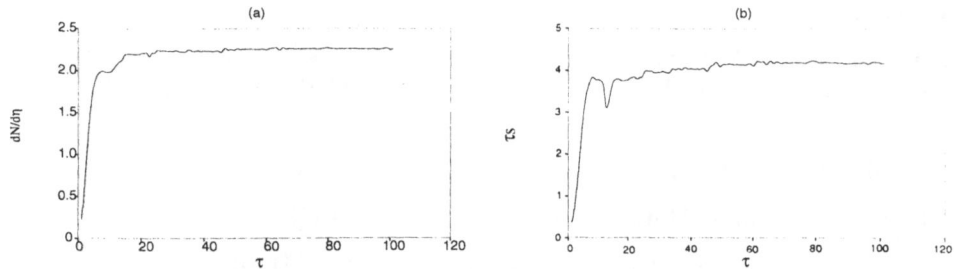

Figure 2. Proper time evolution of (a) $dN/d\eta$, and (b) Boltzmann entropy density multiplied by τ.

field. In the curvilinear coordinates the transport Boltzmann-Vlassov equation reads

$$\frac{\partial f}{\partial \tau} + e\tau E(\tau)\frac{\partial f}{\partial p_\eta} = -[1 - 2f(p_\eta, \tau)]e\tau|E(\tau)| \times \ln\left[1 - \exp\left(-\frac{\pi m^2}{e|E(\tau)|}\right)\right]\delta(p_\eta),$$

$$-\tau\frac{dE}{d\tau} = j_\eta = j_\eta^{cond} + j_\eta^{pol}$$

$$j_\eta^{cond} = 2e\int\frac{dp_\eta}{2\pi\tau p_\tau}p_\eta f(p_\eta, \tau)$$

$$j_\eta^{pol} = -[1 - 2f(p_\eta = 0, \tau)]\frac{me\tau}{\pi}\text{sign}[E(\tau)]\ln\left[1 - \exp\left(-\frac{\pi m^2}{|eE(\tau)|}\right)\right]. \tag{38}$$

In Fig. 1(a) we see that the results of the Boltzmann-Vlassov equation (dashed curve) agree very well with those of the mean field calculation (solid curve) as in the case of flat coordinates with an homogenous electric field.

The agreement of the leading order of the large N expansion and the kinetic approach is encouraging since the latter approach is much simpler for numerical simulations. In the next order, one needs to make a gradient expansion of the field theory equations to obtain a transport theory[25]. It is necessary to see whether the truncation of the gradient expansion reasonably approximates the full field theory results after they are coarse grained in the momentum space. If the next to leading order calculations will reproduce similar results in the homogeneous case, one would then be more confident that the kinetic approach is appropriate for describing more realistic situations (inhomogeneity and higher order collision terms which can be easily derived in the large N expansion). Using the field theory approach, these more difficult problems are beyond the scope of present computing capability.

The CTP project I presented here was done in collaboration with Fred Cooper, Emil Mottola, Juan Pablo Paz, Salman Habib, and Paul Anderson. The last part I presented here was done in collaboration with Fred Cooper, Emil Mottola, Judah Eisenberg, and Ben Svetitsky.

REFERENCES

1. A. Bialas and W. Czyż, Phys. Rev. D **30**, 2371 (1984); *ibid.* **31**, 198 (1985); Z. Phys. **C28**, 255 (1985); Nucl. Phys. **B267**, 242 (1985); Acta Phys. Pol. B **17**, 635 (1986).

2. A. Bialas, W. Czyż, A. Dyrek, and W. Florkowski, Nucl. Phys. **B296**, 611 (1988).

3. K. Kajantie and T. Matsui, Phys. Lett. **164B**, 373 (1985).

4. G. Gatoff, A. K. Kerman, and T. Matsui, Phys. Rev. D **36**, 114 (1987).

5. K. Geiger and B. Muller, Nucl.Phys. **B369**, 600 (1992).

6. X. -N. Wang and M. Gyulassy, Phys. Rev. D **41**, 3501 (1992).

7. J. Schwinger, J. Math. Phys. **2**, 407 (1961).

8. K. T. Mahanthappa, Phys. Rev. **126**, 329 (1962).

9. L. V. Keldysh, JETP **20**, 1018 (1965).

10. K. Wilson, Phys. Rev. D **7**, 2911 (1973).

11. S. Coleman, R. Jackiw, and H. D. Politzer, Phys. Rev. D **10**, 2491 (1974).

12. J. Cornwall, R. Jackiw, and E. Tomboulis, Phys. Rev. D **10**, 2428 (1974).

13. C. M. .Bender, F. Cooper, and G. S. Guralnik, Ann. Phys. **109**, 165 (1977); R. Root, Phys. Rev. D **11**, 831 (1975).

14. H. -Th. Elze, M. Gyulassy and D. Vasak, Nucl. Phys. **B276**, 706 (1986).

15. C. Best and J. M. Eisenberg, Phys. Rev. D **47**,4639 (1993).

16. F. Cooper and E. Mottola, Phys. Rev. D **40**, 456 (1989).

17. Y. Kluger, J. M. Eisenberg, B. Svetitsky, F. Cooper, and E. Mottola, Phys. Rev. Lett. **67**, 2427 (1991).

18. Y. Kluger, J. M. Eisenberg, B. Svetitsky, F. Cooper, and E. Mottola, Phys. Rev. D **45**, 4659 (1992).

19. F. Cooper, in *Particle Production in Highly-Excited Matter, Proc. NATO Advanced Study Institute, Il Ciocco, Italy, July, 1992*, edited by H. Gutbrod and J. Rafelski.

20. P. Anderson, F. Cooper, S. Habib, Y. Kluger, E. Mottola, and J. P. Paz, in preparation.

21. F. Cooper, J. M. Eisenberg, Y. Kluger, E. Mottola, and B. Svetitsky, Phys. Rev. D **48**, 190 (1993).

22. J. Schwinger, Phys. Rev. **82**, 664 (1951).

23. J. D. Bjorken, Phys. Rev. D**27**, 140 (1983).

24. F. Cooper, G. Frye and E. Schonberg, Phys. Rev. **D11**, 192 (1975).

25. E. Calzetta and B. L. Hu, Phys. Rev. D **37**, 2878 (1988).

FROM NUCLEI, VIA HYPERNUCLEI, TO STRANGE HADRONIC MATTER

Avraham Gal

Racah Institute of Physics
The Hebrew University
Jerusalem 91904
Israel

ABSTRACT

The stability of multiply strange baryonic systems is discussed in the context of a generalized Bethe-Weizsäcker mass formula, reproducing the results of mean field calculations which find a broad class of objects composed of neutrons, protons, Λ's and Ξ's, and stable against strong decay, with a strangeness fraction $f_s = |S|/A \approx 1$, density $\rho \approx 2\rho_0$, and charge fraction $q/A \ll 1$, comparable to those of hypothetical stable strange quark matter. The generalized mass formula is used to derive stability regions and weak decay patterns for strange hadronic matter.

INTRODUCTION TO STRANGE HADRONIC MATTER

The possibility of bound strange matter, whether absolutely stable or only against strong interaction decays, has received considerable attention during the last two decades for matter made out of quarks [1-4]. Since the mass difference between the s quark and the (u, d) quarks, about 150 MeV, is less than the Fermi energy for massless nonstrange quarks, about 270 MeV at normal nuclear matter density, the energy of quark matter is lowered by adding up macroscopically strangeness. In particular, two outstanding signatures of strange quark matter (SQM) are (i) a large strangeness fraction $f_s = |S|/A \sim 0.5 - 1.0$ and (ii) a low charge fraction $f_q = q/A < 0.1$. These simply arise from consideration of the basic u, d, s quark triplet for which $f_s = 1, f_q = 0$.

Hot and Dense Nuclear Matter, Edited by
W. Greiner *et al.*, Plenum Press, New York, 1994

In contrast, the possibility of bound strange hadronic matter (SHM) made out of strange baryons ($\Lambda, \Sigma, \Xi, \Omega, ...$) in addition to nonstrange baryons (nucleons) has not attracted much attention simply since the mass differences involved, bounded below by $M_\Lambda - M_N = 177$ MeV, are considerably larger than the Fermi energy of normal nuclear matter ($E_F^{(0)} \sim 40$ MeV). Thus, adding strangeness to nuclear matter increases the total energy, and the stability of SHM holds only with respect to the strong interactions which conserve strangeness. It is often stated that multi strangeness may be realized in nature only through multi Λ hypernuclei since all other hyperons generally convert strongly in nuclear matter via the reactions

$$\Sigma + N \rightarrow \Lambda + N \qquad\qquad (Q \simeq 78 \text{ MeV}) \qquad\qquad (1a)$$

$$\Xi + N \rightarrow \Lambda + \Lambda \qquad\qquad (Q \simeq 26 \text{ MeV}) \qquad\qquad (1b)$$

$$\Omega + N \rightarrow \Lambda + \Xi \qquad\qquad (Q \simeq 178 \text{ MeV}) \qquad\qquad (1c)$$

where the corresponding charge averaged energy releases are given in parentheses.

Relativistic Mean Field (RMF) calculations done recently[5-8] for spherically symmetric configurations have confirmed that multi Λ hypernuclei obtained by adding Λ hyperons to closed shell nuclei indeed are stable against strong interaction decays, with binding energy per baryon $B/A \sim 9$ MeV maximally, and with strangeness fraction $f_s < 0.2$. However, a point missed by all previously published calculations is that, for a sufficiently large number of Λ's, a collapse induced by the reaction $\Lambda + \Lambda \rightarrow \Xi + N$ becomes energetically allowed. Therefore, SHM necessarily involves a consideration of Ξ^- and Ξ^0 hyperons which contribute to its composition inasmuch as the conversion process (1b) is Pauli blocked by bound Λ's. The possibility of such blocking and the implications for binding Ξ's in light systems has been discussed recently by Schaffner et al. [8], and a comprehensive calculation and discussion is offered by refs. 9-11.

In Fig. 1, taken from Ref. 10, we show schematically, based on mass formulae[9] which generalize the Bethe-Weizsäcker formula into strangeness and which mock up the results of RMF calculations, the binding energy per baryon E_B/A for multi Λ hypernuclei obtained by adding n_Λ lambda hyperons to a ^{208}Pb core (solid curve for $n_\Lambda < 41$, dashed curve for $n_\Lambda > 41$). The process $\Lambda\Lambda \rightarrow \Xi^-p$ (Ξ^0n) becomes possible for $n_\Lambda > 41$ (77), and the resulting curve for {p, n, Λ, Ξ^0, Ξ^-} configurations which minimize E_B/A under the constraints of $A = 208 + n_\Lambda$, $Z = 82$ and $S = -n_\Lambda$ is shown in solid. Even if all of the underlying hyperon-hyperon (YY) interactions are switched off, one obtains a curve similar to that of Fig. 1, except that the break induced by $\Lambda\Lambda \rightarrow \Xi^-p$ (Ξ^0n) now occurs for $n_\Lambda = 35$ (59) and E_B/A drops only to -13 MeV instead of -23 MeV for $n_\Lambda > 200$ in the figure. Thus, one must include Ξ's as equal partners to N's and Λ's in any RMF calculation that is designed to incorporate strangeness.

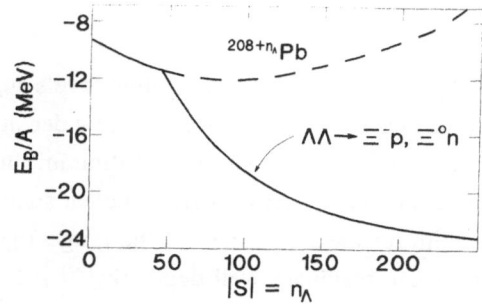

Figure 1. Binding energy per baryon E_B/A for systems consisting of ^{208}Pb core plus hyperons (taken from ref. 10).

This talk is organized as follows: in the next section the phenomenological information on Λ, Ξ and $\Lambda\Lambda$ hypernuclei is summarized. The assumptions made on the YY interactions are motivated in the subsequent section. These are used to constrain the hyperon well depths relating to the latest comprehensive RMF calculations[11]. In the final section a generalization of the Bethe Weizsäcker mass formula for ordinary nuclei is developed for SHM, reproducing the essential features of the RMF calculation. It is used to derive stability regions and weak-decay patterns for strange hadronic matter.

HYPERNUCLEI

Λ Hypernuclei

There is a considerable body of information on the binding energies B_Λ of Λ's bound in various single particle orbitals in hypernuclei with total baryon number from $A = 3$ to $A = 89$. For a review, consult Ref. 12. An analysis of these data in a Skyrme-Hartree-Fock picture[13] led to a well depth

$$V_\Lambda^{(N)} \simeq 27 \text{ - } 28 \text{ MeV} \tag{2}$$

for a single Λ in a nucleus ($-V_\Lambda$ is the potential seen by the Λ in the center of the nucleus, at $\rho(r) \approx \rho_0$). The value of $V_\Lambda^{(N)}$ is roughly 1/3 - 1/2 of the nucleon well depth $V_N^{(N)}$. It is also known that the Λ-nucleus spin-orbit splitting ($\sim \vec{\ell}_\Lambda \cdot \vec{\sigma}_\Lambda$) is at least an order of magnitude smaller than that of the nucleon.

Σ Hypernuclei

The analysis of level shifts and widths of Σ^--atomic levels suggests[14] a well depth of order $V_\Sigma^{(N)} \approx 20 - 30$ MeV. This refers to a Σ^- in a very low density environment. There is some evidence[14] for Σ^--nuclear resonances in the continuum, but the situation remains controversial. Even if these relatively narrow Σ-hypernuclear states are confirmed, they do not imply a large well depth for the Σ^-, perhaps in the range $V_\Sigma^{(N)} \approx 10 - 20$ MeV. The atomic data also determine an imaginary well depth $W_\Sigma^{(N)} \approx 10 - 15$ MeV. As a result, systems involving Σ's together with nucleons or Λ's generally will be unstable with respect to the strong decays $\Sigma N \to \Lambda N$ or $\Sigma \Lambda \to \Xi N$.

Ξ Hypernuclei

In the emulsion experiments with K^- beams, there are a few events attributed to the formation of Ξ^- hypernuclei. Although this evidence is far from being compelling, the data can be interpreted[15] consistently in terms of a potential well for the Ξ^- with a depth

$$V_\Xi^{(N)} \simeq 20 - 25 \text{ MeV} . \tag{3}$$

This range of values is used as a guide for the RMF calculations of systems involving many Ξ's. A value of $V_\Xi^{(N)}$ close to these has been derived[16] from the one boson exchange (OBE) model D of the Nijmegen group[17]. This same model yields an isospin-dependent Ξ-nucleus potential (proportional to $\vec{\tau}_\Xi \cdot \vec{T}_{\text{core}}$) which is an order of magnitude smaller than that observed[16] for the nucleon. In RMF calculations, usually, the isospin potentials are generated by coupling the baryon (N, Σ or Ξ) to the ρ meson.

ΛΛ Hypernuclei

There is a small amount of information[18-20] available on $\Lambda\Lambda$ hypernuclei, namely the ground state binding energies of $^6_{\Lambda\Lambda}\text{He}$, $^{10}_{\Lambda\Lambda}\text{Be}$ and $^{13}_{\Lambda\Lambda}\text{B}$. The analysis of these data yields a $\Lambda\Lambda$ interaction matrix element $\Delta B_{\Lambda\Lambda}$ defined by

$$M\left(^A_{\Lambda\Lambda}Z\right) = M\left(^{A-2}Z\right) + 2m_\Lambda - 2B_\Lambda\left(^{A-1}_\Lambda Z\right) - \Delta B_{\Lambda\Lambda} . \tag{4}$$

One finds

$$\Delta B_{\Lambda\Lambda} \approx 4 - 5 \text{ MeV} \tag{5}$$

corresponding to a rather strong $\Lambda\Lambda$ attraction. In these light systems, the Λ's are in $1s_{1/2}$ orbits, and the relative two-body state is $^1S_0$ ($^3S_1$ is not allowed for $\Lambda\Lambda$ because of the Pauli principle). The value of $\Delta B_{\Lambda\Lambda}$ can be compared to $^1S_0$ matrix elements for light systems of 2-3 MeV for ΛN and 6-7 MeV for NN interactions. Although the $^1S_0$ nn system is known to be unbound, the existence of a weakly bound two-body $^1S_0$ $\Lambda\Lambda$ system cannot be ruled out, owing to the larger Λ mass, and the largely unknown balance between long range attraction and short range repulsion. The existence of such a bound state cannot be established on the basis of the few $\Lambda\Lambda$ hypernuclear events which have been seen. If such a "quasi-nuclear" $\Lambda\Lambda$ state exists, it is distinct from the six-quark H dibaryon proposed by Jaffe [21], which would be an SU(3) singlet with a small $\Lambda\Lambda$ component.

A ratio of $V_{\Lambda\Lambda}/V_{NN} \approx 3/4$, as discussed above for the corresponding $^1S_0$ interaction matrix elements, implies the following estimate for the Λ well depth in "Λ matter" relative to the more familiar nucleon well depth in nuclear matter:

$$V_\Lambda^{(\Lambda)} \bigg/ V_N^{(N)} \approx \frac{3}{4} \frac{1/4}{3/8} \frac{1}{2} = \frac{1}{4} . \tag{6}$$

Here the second ratio in the middle stands for spin-isospin weights appropriate to spatially symmetric two-body configurations, whereas the third ratio of 1/2 approximates the expected scaling of densities ($\rho_p \sim \rho_n \sim \rho_\Lambda$).

HYPERON-HYPERON INTERACTIONS

For the unknown hyperon-hyperon (YY) interactions one relies heavily on the systematics and predictions provided by a one-boson-exchange (OBE) model. For the NN interaction at short range (<0.8 fm) the overall attraction is the result of a sizable cancellation between the spin-independent attractive scalar (σ) exchange and the repulsive vector exchanges. In particular, since the vector meson couplings to the nucleon are dominantly of electric type, ω exchange is repulsive in the NN system for both $^1S_0$ and $^3S_1$ states and ϕ exchange gives a very small (repulsive) contribution ($g_{NN\phi} = 0$ in the quark model). The situation is dramatically different for hyperon interactions, since the f/g ratios can be large for hyperon couplings to vector mesons, particularly the ϕ. The effect of this is to change the sign of the ϕ exchange potential from repulsion to attraction for $^1S_0$ $\Lambda\Lambda$, $\Xi\Xi$ and $\Xi\Lambda$ interactions. In Fig. 2, the $^1S_0$ potentials are plotted for $\{nn, \Lambda\Lambda, \Xi\Xi, \Xi\Lambda\}$ systems, for Model D of the Nijmegen group[17]. With coupling constants satisfying SU(3) symmetry, and phenomenological hard cores at short distances, this model reproduces the NN, ΛN and ΣN scattering data. A key feature of Model D is the SU(3) singlet character of the effective σ meson exchange. With this interaction, Bando[22] has evaluated the $^1S_0$ $\Lambda\Lambda$ matrix element $\Delta B_{\Lambda\Lambda}$ in $^{6}_{\Lambda\Lambda}\mathrm{He}$, obtaining a value of about 4 MeV, comparable to Eq. (5).

In Model F of the Nijmegen group[23], which exhibits a weaker coupling of the σ to strange baryons, $\Delta B_{\Lambda\Lambda}$ is repulsive, in disagreement with the data. Since long range π exchange is absent, ω and ϕ tend to cancel, and η, η' exchange is rather weak, the $\Lambda\Lambda$ interaction is dominated by σ exchange. The large attractive value of $\Delta B_{\Lambda\Lambda}$ thus suggests a strong σ coupling to strange baryons.

The tendency for approximate cancellation of a repulsive ω and an attractive ϕ component persists for other hyperon-hyperon channels ($\Sigma^-\Sigma^-$, $\Xi\Xi$, $\Xi\Lambda$) in Model D, and thus scalar exchange dominates here also. As seen in Fig. 2, the YY $^1S_0$ potentials are all comparable in depth for Model D, reflecting the SU(3) singlet character of the σ. These hyperon potentials are also similar to that for $nn(^1S_0)$, except at distances beyond 2 fm, where the pion exchange term in the nn system finally asserts its dominance.

For $^3S_1$ YY channels, the interaction is weaker, since in this case the η, η', ω, and ϕ exchange potentials are repulsive, and the total vector repulsion is large. The $^3S_1$ $\Xi\Xi$ and $\Xi\Lambda$ potentials in Model D are also shown in Fig. 2. They are seen to be much weaker than the $^1S_0$ potentials, becoming repulsive at shorter distances. Of course, the deuteron is a $^3S_1$ np bound state, but the binding is due to the tensor potential from p exchange; the central np $^3S_1$ potential is shallower than that for $^1S_0$. For $\Xi\Xi$ or ΞN systems, π exchange plays a very minor rôle, since $g_{\Xi\Xi\pi}/g_{NN\pi} = -(1 - 2\alpha_{PS})$ is very small (-1/5 in SU(6), -0.03 in Model D).

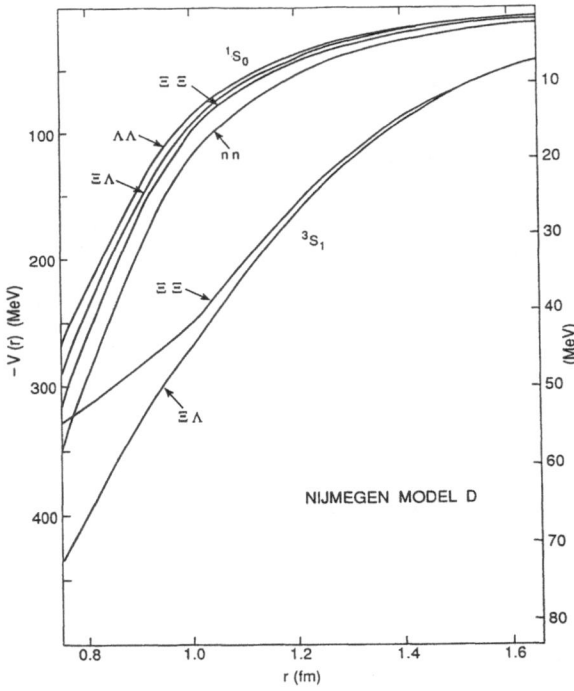

Figure 2. Meson exchange potentials for hyperon-hyperon interactions, from Nijmegen Model D (ref.17). The top four $^1S_0$ curves are assigned values by the left hand scale, the two lower $^3S_1$ curves - by the right hand scale.

By using these YY interaction to construct single-particle potentials, Schaffner *et al.*[11] made a rough estimate:

$$V_{\Xi}^{(\Xi)} \approx V_{\Lambda}^{(\Xi)} = 2V_{\Xi}^{(\Lambda)} \approx 2V_{\Lambda}^{(\Lambda)} \qquad \left(\approx 1/2 \, V_N^{(N)} \right), \qquad (7)$$

where the relationship to $V_N^{(N)}$, in brackets, follows from the estimate given by Eq. (6). Below we describe the results of employing two different sets of interaction parameters. In the first set (set I) all of the YY interactions are neglected, so that $V_Y^{(Y')}$ of Eq. (7) are replaced by zero. This roughly corresponds to multi-Λ hypernuclear RMF calculations[5-8] which use σ and ω bosonic fields only. The second set (set II) assumes the validity of the estimates given by (7), for a typical RMF value of $V_N^{(N)} \approx 80$ MeV, and this roughly corresponds to the RMF calculations[11] which augment the $\sigma + \omega$ model by another scalar-vector meson pair (say f_0 and φ), allowing strangeness to couple more readily to, and thus to boost up, the YY underlying interactions.

When large amounts of strangeness are injected into the system, the mean fields felt by a nucleon and by a hyperon are altered, since YY as well as YN interactions enter the picture. Yet, the well depths considered here are all much smaller than the mass differences $m_Y - m_N$, so the binding energy per particle E_B/A of multi-strange nuclear systems remains small, of order -10 MeV.

THE BETHE-WEIZSÄCKER APPROACH

The Bethe-Weizsäcker (BW) mass formula for nuclei made out of protons and neutrons consists of the following main terms:

$$E_B(\{p,n\}) = -a_V^{(0)} A + a_S^{(0)} A^{2/3} + a_C^{(0)} \frac{Z^2}{A^{1/3}} + a_x^{(0)} \frac{(N-Z)^2}{A}. \qquad (8)$$

Here, $E_B = -B$ is the energy of the nucleus relative to the sum of the rest masses of its constituents. The four terms above represent the contribution of volume energy, surface energy, Coulomb energy and symmetry energy, respectively. Other terms such as pairing are of a lesser importance to the average trend of nuclear masses and to the discussion which follows. The coefficients in Eq. (8) are given approximately by

$$a_V^{(0)} = 16, \quad a_S^{(0)} = 18, \quad a_C^{(0)} = 0.72, \quad a_x^{(0)} = 23. \qquad \text{[in MeV]} \qquad (9)$$

The generalized BW (GBW) mass formula for SHM is constructed[9] in analogy to the ordinary nuclear BW formula, Eq. (8). One assumes that SHM saturates for roughly equal densities of each species and that the Fermi momentum of the underlying strange baryonic Fermi gas is about the same as for ordinary nuclear matter. Whereas a single Coulomb term and a single surface term are retained in the GBW formula, there are now several volume and symmetry terms, e.g.

$$a_V^{(0)} \to a_V - b_V^{(w)} w - b_V^{(y)} y \ , \tag{10a}$$

$$a_x^{(0)} x^2 \to a_x x^2 + a_u u^2 + a_w w^2 + a_y y^2 + a_{wy} wy \ , \tag{10b}$$

where

$$x = (N - Z)/A \ , \quad u = \left(\Xi^0 - \Xi^- \right) \big/ A \ , \quad w = \left(\frac{N + Z}{2} - \frac{\Xi^0 + \Xi^-}{2} \right) \big/ A$$

$$y = \left[\left(N + Z + \Xi^0 + \Xi^- \right) \big/ 4 - \Lambda \right] \big/ A \ . \tag{11}$$

By considering the phenomenological and theoretical input, as discussed in the preceding sections, one generates the potential energy contributions appropriate to the new volume terms (cf. Eqs. (2), (3), (7)). Two parameter sets used in ref. 10 are given in Table 1 and by Eq. (12) below.

Table 1: Parameter sets (in MeV) for use in the generalized BW formula

	a_V	$b_V^{(w)}$	$b_V^{(y)}$	a_x	a_u	a_w	a_y	a_{wy}
set I	10.7	-35.5	-16.75	43	23.7	57.1	45	7.7
set II	28.7	-5.5	-4.75	43	23.7	57.1	45	7.7

$$a_c = a_c^{(0)} = 0.72 \ \text{MeV} \ , \quad a_S^{(0)} \to a_S = 15 \ \text{MeV} \ . \tag{12}$$

In Fig. 3 we show a comparison between using the GBW formula, with parameters defined by Table 1 and Eq. (12), with RMF results[11] for E_B/A as function of A for multi strange nuclei based on given nuclear cores (^{56}Ni, ^{132}Sn, ^{208}Pb and ^{310}G [Z = 126, N = 184] for set I; ^{56}Ni and ^{180}Th for set II). For set I (the upper part of the Figure), E_B/A drops to between -12 and -13 MeV, compared to -9 to -10 MeV in multi Λ hypernuclei for which allowance is not made for Ξ's. Most of the binding energy gain is due to reducing the nuclear Coulomb repulsion by adding Ξ's. The agreement between using the GBW formula and the RMF results is excellent. The GBW formula can then be used trustworthily for determining the appropriate "stability valley"[9, 10]. One finds that the E_B/A curve is monotonically decreasing as function of A, reaching a bulk limit value of -14.1 MeV. Similarly, the strangeness fraction $f_s = |S|/A$ and the charge fraction $f_q = Q/A$ (where Q is the total charge in units of e) also change monotonically, with f_s increasing to its bulk limit value of $f_s = 0.71$ and f_q decreasing to zero.

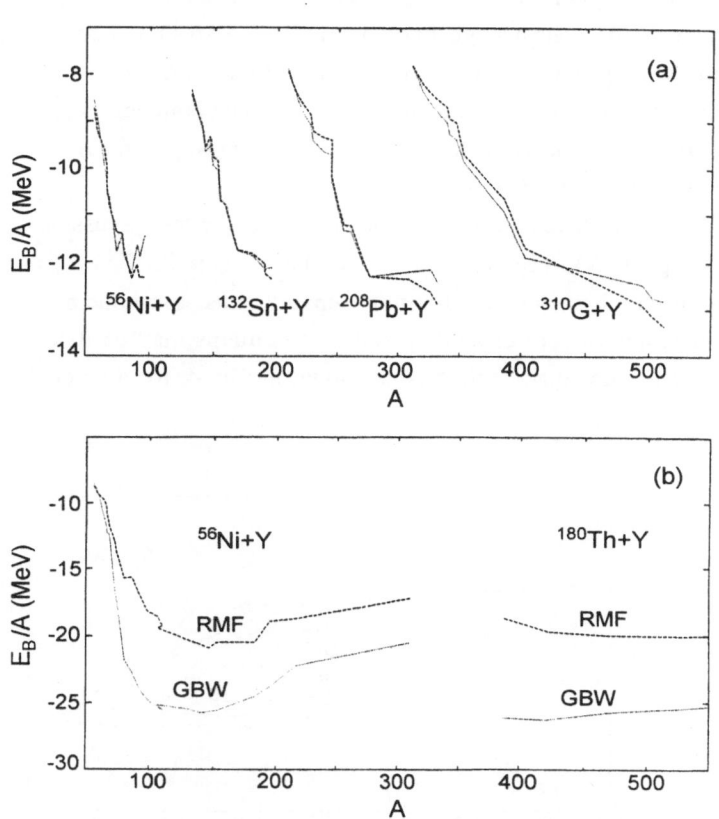

Figure 3. E_B/A as function of A for multi strange nuclei based on nuclear cores as marked. The upper and lower parts, (a) and (b) respectively, correspond to Model 1 and Model 2 respectively of the RMF calculations of ref. 11 (dashed lines), and to set I and set II respectively of the GBW formula parameters, from ref. 10, as given in Table 1 (dotted lines).

For set II (the lower part of the Figure), E_B/A drops to between -25 to -26 MeV by using the GBW formula[10]. This is about 5 MeV lower than what the RMF calculations[11] yield, a difference which is attributed in Ref. 11 to the relativistic nature of the RMF scheme where the scalar attraction gets partially saturated as function of density. The bulk limits for set II are $E_B/A \rightarrow$ -28.3 MeV, $f_s \rightarrow 0.96$, $f_q \rightarrow 0$. For stronger YY interactions f_s exceeds one over a wide range of A values, and f_q is negative throughout most of the stability valley.

The stability of SHM against strong decays is not limited to one configuration per A. For a given A, there will be many ways to distribute the five species, $\{p, n, \Lambda, \Xi^-, \Xi^0\}$, besides the one producing maximum binding, for example by selecting a different strangeness fraction or charge fraction. Once S, q and A are fixed, one can ensure that the $\Xi N \leftrightarrow \Lambda\Lambda$ conversion is at equilibrium. In Fig. 4 we plot the stability domain for SHM with $A = 100$ in the plane spanned by f_s and f_q, for set II. The species within the parallelogram are stable against single baryon emission. Also shown are contours of equal - E_B/A with values (increasing as one moves inward) of 15, 17, 19, 21, 23, 23.5, 24, 24.5, 25 MeV. The dot near the center marks the bulk limit. We comment that as A increases, the stability domain shrinks, mostly with respect to f_q. For example, for $A = 1000$, the charge fraction is bounded by $-0.2 < f_q < 0.2$.

The GBW formula here considered is inappropriate for the discussion of special cases such as stable $\{p, n, \Lambda\}$, $\{\Xi^0, \Xi^-, \Lambda\}$ and perhaps even $\{p, \Xi^0, \Lambda\}$ and $\{n, \Xi^-, \Lambda\}$ systems, none of which possess a bulk limit. Separate mass formulae may be derived[9, 10] for these species, and the corresponding stability domains may be determined[10]. These domains, obviously, are adjacent to the one shown in Fig. 4, for a given A, probably in a discontinuous manner.

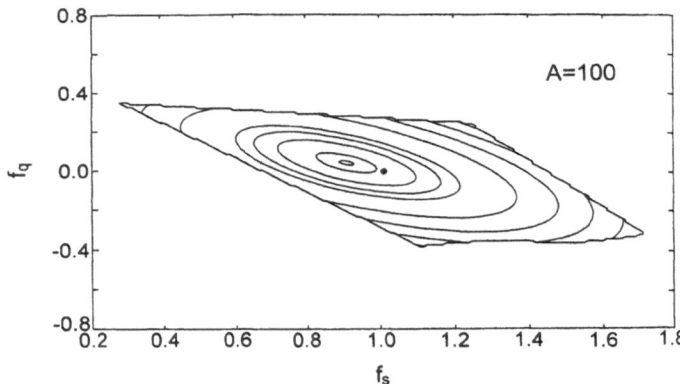

Figure 4. Baryon emission stability domain, calculated in ref. 10 for set II of the GBW formula parameters, for multi strange nuclei with $A = 100$ in the plane spanned by f_s and f_q. The countours of equal E_B/A and the dot are specified in the text.

SHM is always unstable to weak decays, particularly $\Delta S = 1$ nonmesonic transitions, such as $\Lambda + N \rightarrow N + N$, $\Xi + N \rightarrow \Lambda + N$, $\Xi + \Lambda \rightarrow \Lambda + \Lambda$. These will yield lifetimes of order 0.1 ns or less. In Fig. 5 we plot the calculated[10] variation of f_s and f_q, with decreasing A, for sequences of weak decays beginning with $A = 1000$ for set II. The various nonmesonic decay channels are treated incoherently while simplifying assumptions are made for their rates with respect to each other. Three curves, almost coinciding with each other while corresponding to different weak decay sequences evaluated by Monte Carlo, are shown as function of f_s and f_q. In between two successive weak decays, an equilibrium with respect to $\Xi N \leftrightarrow \Lambda\Lambda$ is imposed. The Figure makes it obvious that over a wide range of A values, down to about $A = 300$, the values of f_s and f_q do not significantly change from the initial values $f_s \sim 1, f_q \sim 0$. The transition into ordinary nuclear matter, with a copious increase of the charge fraction into near 0.4, occurs between $A = 200$ to 100. In contrast, for set I (not shown here) the final stages of getting rid of strangeness occur in this example already for $A \sim 300$ so that the outcome is likely to undergo fission.

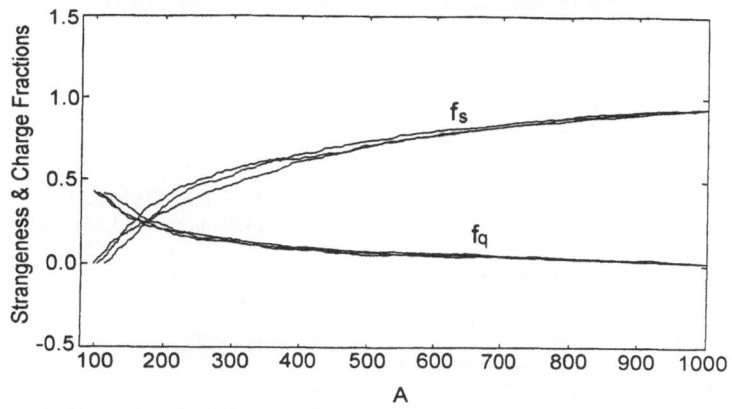

Figure 5. Variation of f_s and f_q throughout weak decay sequences for the most stable multi strange nucleus with $A = 1000$, calculated in ref. 10 for set II of the GBW formula parameters, as function of A -- in decreasing order for temporal correspondence.

Finally, it should be emphasized that the case for stable SHM is made here using phenomenological arguments based on hadronic degrees of freedom at normal nuclear densities and temperature $T = 0$. This phenomenology is trustworthy in its own right, sharing hardly anything in common with the more speculative extrapolation from $B = 1$ quark bags towards strange quark matter (SQM). If SQM existed at densities close to nuclear matter density and at $T = 0$, then SHM would be expected to decay strongly into SQM with a huge energy release in the form of pions, kaons, nucleons and hyperons, and clusters thereof. Of course, the deconfinement process accompanying such decays, from a range of about 1 fermi or less for each of the hadron constituents of SHM to a range of several fermis for SQM, might be heavily suppressed without radical changes in density and/or temperature. Whichever scenario is envisaged for these hypothetical forms of matter and their relationship to each other, undoubtedly SHM poses a challenging opportunity to explore the interplay between hadronic and quark-gluon degrees of freedom for subatomic matter.

ACKNOWLEDGEMENTS

S. Balberg and J. Schaffner produced many of the results reviewed here. I wish to thank them and C.B. Dover for many useful discussions and suggestions. This research was supported by The Basic Research Foundation administered by the Israel Academy of Sciences and Humanities and by the Humboldt Foundation. Last but not least, special thanks are due to W. Greiner and H. Stocker who made it possible for me to participate in the Nato Advanced Study Institute at Bodrum which they organized and ran so skilfully.

REFERENCES

1. S. Chin and A.K. Kerman, Phys. Rev. Lett. **43** (1979) 1292.
2. E. Witten, Phys. Rev. **D30** (1984) 272.
3. E. Farhi and R.L. Jaffe, Phys. Rev. **D30** (1984) 2379; Phys. Rev. **D32** (1985) 2452.
4. M.S. Berger and R.L. Jaffe, Phys. Rev. **C35** (1987) 213; E.P. Gilson and R.L. Jaffe, Phys. Rev. Lett. **71** (1993) 332.
5. M. Rufa et al., J. Phys. **G13** (1987) 143.
6. J. Mares and J. Zofka, Z. Phys. **A333** (1989) 209.
7. M. Rufa et al., Phys. Rev. **C42** (1990) 2469.
8. J. Schaffner, C. Greiner and H. Stöcker, Phys. Rev. **C46** (1992) 322.
9. C.B. Dover and A. Gal, Nucl. Phys. **A560** (1993) 559.
10. S. Balberg, M.Sc. thesis, Hebrew University (1993).
11. J. Schaffner, C.B. Dover, A. Gal, C. Greiner and H. Stocker, Phys. Rev. Lett. **71** (1993) 1328; J. Schaffner, C.B. Dover, A. Gal, C. Greiner, D.J. Millener and H. Stocker, Ann. Phys. (in press).
12. R.E. Chrien and C.B. Dover, Ann. Rev. Nucl. Part. Sci. **39** (1989) 113.
13. D.J. Millener, C.B. Dover and A. Gal, Phys. Rev. **C38** (1988) 2700.
14. C.B. Dover, D.J. Millener and A. Gal, Phys. Rep. **184** (1989) 1.
15. C.B. Dover and A. Gal, Ann. Phys. **146** (1983) 309.
16. C.B. Dover and A. Gal, Prog. in Part. and Nucl. Phys., Ed. D. Wilkinson, Pergamon Press, Oxford (1984), Vol. 12, pp. 171-239.
17. M.M. Nagels, T.A. Rijken and J.J. de Swart, Phys. Rev. **D12** (1975) 744 and **D15** (1977) 2547.
18. M. Danysz et al., Nucl. Phys. **49** (1963) 121; R.H. Dalitz et al., Proc. Royal Soc. London **A426** (1989) 1.
19. D.J. Prowse, Phys. Rev. Lett. **17** (1966) 782.
20. S. Aoki et al., Prog. Theor. Phys. **85** (1991) 1287; C.B. Dover, D.J. Millener, A. Gal and D.H. Davis, Phys. Rev. **C44** (1991) 1905.
21. R.L. Jaffe, Phys. Rev. Lett. **38** (1977) 195, 617(E).
22. H. Bando, Prog. Theor. Phys. **67** (1982) 699.
23. M.M. Nagels, T.A. Rijken and J.J. de Swart, Phys. Rev. **D20** (1979) 1633.

TOWARDS A SELF-CONSISTENT DESCRIPTION OF THE RELATIVISTIC BOLTZMANN-UEHLING-UHLENBECK APPROACH[1]

Zhuxia Li[2], Guangjun Mao, Yizhong Zhuo

Institute of Atomic Energy, Beijing, P.R.China

INTRODUCTION

The heavy ion collisions (HIC) provide a unique opportunity to study the time evolution process of nuclear matter under violent collisions as well as nuclear properties under extreme conditions of high density and temperature. However,the interesting informations on these problems can only be obtained indirectly through certain theoretical models. Thus,one has to choose a good theoretical tool,which should be as general and basic as possible on the one hand and as practicable as possible on the other hand. The most ambitious one is to obtain the relativistic transport equation starting from the time dependent Bruckner G-matrix theory[1]. In this approach the time evolution of the HIC process is described by relativistic kinetic equation and the G-matrix served as a dynamical input of the two body interaction. Therefore, to acquire complete numerical solution will be very difficult and till now there has no real calculation which fulfills this requirement.

It would be more practicable to take the effective interaction rather than the G-matrix as the first covariant transport model for HIC did in ref.[2] where the σ-ω model was used as effective interaction. However, the n-n collisions,which should be medium dependent, are not treated self-consistently.

As the density changes dramatically during the certain period of HIC,the effect of the mean-field and collision term should be added both dynamically and simultaneously. Therefore a self-consistent treatment is highly desirable. Our aim is to develop a model based on the effective Lagrangian,which is practicable but without lossing the self-consistency between mean-field part and in medium n-n collision part.

In the second part,the formalism will be outlined. Then an improved version of the σ-ω model with momentum and density dependent coupling strengths for scalar and vector fields are introduced in the third part. The study for the thermalization process in HIC will be given in the forth part. Finally a short discussion will be given.

---

[1]Supported by National Natural Science Foundation of China
[2]Speaker

THE OUTLINE OF THE DERIVATION OF
THE SELF-CONSISTENT RBUU EQUATION

We start from effective Lagrangian

$$L^{eff} = L_F + L_I. \qquad (1)$$

L_F is the Lagrangian density for free nucleon and σ-ω meson field, L_I is the interaction Lagrangian density of nucleons coupled to σ, ω mesons.

$$L_I = g_\sigma \bar\psi(x)\psi(x)\sigma(x) - g_\omega \bar\psi(x)\gamma_\mu \psi(x)\omega^\mu(x). \qquad (2)$$

The closed-time path Green's function technique is used. For field $\psi, \bar\psi$ the Green's function is a 2×2 matrix with components G^{--}, G^{-+}, G^{+-}, G^{++} according to the time of field ψ and $\bar\psi$ on the time path[3]. The equation of motion for G^{-+}, which we are interested in, has the following form

$$(i\gamma \cdot \partial_1 - M - \Sigma_H^{--}(1))iG^{-+}(1,2) = \int [\Sigma_F^{--}(1,3) + \tilde\Sigma^{--}(1,3)]iG^{-+}(3,2)d3$$

$$+ \int \tilde\Sigma^{-+} iG^{++}(3,2)d3. \qquad (3)$$

where $\Sigma_H^{--}, \Sigma_F^{--}$ and $\tilde\Sigma^{--}(1,2), \tilde\Sigma^{-+}(1,3)$ are the HF term and the high order terms of the corresponding components of the self-energy part. In order to derive the RBUU-like equation the following steps are taken: First, introducing the Wigner transformation, then introducing two approximations, i.e. the semiclassical approximation in which G and Σ are assumed to be peaked around $x' = x_1 - x_2$ and smoothly changing with $x = \frac{1}{2}(x_1 + x_2)$ as well as the quasi-particle approximation we achieve the following kinetic equation,

$$\{[\partial_x^\mu - \Sigma_{HF}^{\mu\nu}(x,p)\partial_\nu^p - \partial_p^\nu\Sigma_{HF}^\mu(x,p)\partial_\nu^x]\frac{P_\mu}{m^*(x,p)}$$

$$+ [\partial_\nu^x\Sigma_{HF}^s(x,p)\partial_p^\nu - \partial_p^\nu\Sigma_{HF}^s(x,p)\partial_\nu^x]\}Tr(iG^{-+}(x,p))$$

$$= Tr[\tilde\Sigma^{+-}(x,p)G^{-+}(x,p) - \tilde\Sigma^{-+}(x,p)G^{-+}(x,p)], \qquad (4)$$

where

$$\Sigma_{HF}^{\mu\nu}(x,p) = \partial_x^\mu\Sigma_{HF}^\nu(x,p) - \partial_x^\nu\Sigma_{HF}^\mu(x,p). \qquad (5)$$

We can further define a distribution function $f(\vec{x}, \vec{p}, \tau)$ as

$$\frac{1}{4}tr[iG^{-+}(x,p)] = -\frac{\pi m^*}{E}\delta(p_0 - E(p))f(\vec{x}, \vec{p}, \tau). \qquad (6)$$

We further introduce the Born approximation when computing the right hand side. Finally, we obtain the following equation

$$\{[\partial_x^\mu - \Sigma_{HF}^{\mu\nu}(x,p)\partial_\nu^p - \partial_p^\nu\Sigma_{HF}^\mu(x,p)\partial_\nu^x]P_\mu$$

$$+ m^*[\partial_\nu^x\Sigma_{HF}^s(x,p)\partial_p^\nu - \partial_p^\nu\Sigma_{HF}^s(x,p)\partial_\nu^x]\}\frac{f(\vec{x}, \vec{p}, \tau)}{E}$$

$$= \frac{1}{2} \int \frac{d\vec{P_2}}{(2\pi)^3}\sigma(s,t)v[F_2^0 - F_1^0]d\Omega, \qquad (7)$$

Here $\Sigma_{HF}^\nu(x,p), \Sigma_{HF}^S(x,p)$ the vector and scalar components of the self-energy part, and $\sigma(s,t)$ the two body scattering cross section, can be given explicitly and simultaneously in our approach. The detailed expression can be found in refs.[4,5]. $(F_2^0 - F_1^0)$ is the

Uehling-Uhlenbeck factor. Differing with the commonly used RBUU equation , in the equation (8) both mean field and two body scattering cross section are calculated self-consistently based on the same Lagrangian and so that the medium effect can be taken into account automatically.

THE MOMENTUM AND DENSITY DEPENDENCE OF THE COUPLING STRENGTHS FOR SCALAR AND VECTOR FIELDS

In the equation (8) the only input is the σ, ω type effective Lagrangian. As an effective interaction the coupling constant g_σ, g_ω are originally adjusted to fit the saturation properties of nuclear matter and the low lying state properties. The parameter sets we used are given in Table 1. But the energy dependence of the optical potential predicted by this model was not in agreement with experimental data. On the other hand in DBHF calculations[7,8] the density dependence of g_σ, g_ω was emphasized. It would be reasonable to phenominalogically introduce both energy and density dependent coupling constants g_σ, g_ω. The experimental data of free n-n cross section and the mean free path in nuclei will provide us a further testing ground in addition to the real part of the optical potential and the saturation properties of nuclear matter. We propose that the dependence of coupling strengths on momentum and density has following form

$$g_\sigma^2(E,\rho) = (g_\sigma^\rho(\rho))^2 (g_\sigma^E(E))^2, \tag{8}$$
$$g_\omega^2(E,\rho) = (g_\omega^\rho(\rho))^2 (g_\omega^E(E))^2, \tag{9}$$

with

$$(g_\sigma^E(E))^2 = g_s^0 + g_s^1/E + g_s^2/E^2 + g_s^3/E^3, \tag{10}$$
$$(g_\omega^E(E))^2 = g_v^0 + g_v^1/E + g_v^2/E^2 + g_v^3/E^3, \tag{11}$$

and

$$(g_\sigma^\rho(\rho))^2 = \frac{1}{0.7941 + 0.2121(\rho/\rho_0) - 0.0062(\rho/\rho_0)^2}, \tag{12}$$
$$(g_\omega^\rho(\rho))^2 = \frac{1}{0.6150 + 0.4347(\rho/\rho_0) - 0.0497(\rho/\rho_0)^2}, \tag{13}$$

which is obtained according to the constraints that the ratios of $\Sigma_H^s(\rho)/\Sigma_H^s(\rho_0)$ and $\Sigma_H^0(\rho)/\Sigma_H^0(\rho_0)$ of DBHF calculations performed by Brockmann et al. [7] have to be reproduced. We parametrize the $g_\sigma^2(E,\rho)$ and $g_\omega^2(E,\rho)$ by fitting the emperical optical potential, the experimental data of free n-n cross section as well as the mean free path, which is related to the in-medium n-n cross section by $\lambda = 1/(\rho \cdot (\sigma^{in} + \sigma^{el}))$, under the constraint that the saturation properties should be guaranteed[9]. The parametrization of $g_\sigma^E(E)$ and $g_\omega^E(E)$ is given in Table 2. Fig.1, Fig.2 show the comparison between the optical potential and mean free path from this work and the experimental data. We find all three nonlinear coupling sets can reproduce the optical potential well, but only set 3 can fit the mean free path well. It implys that the in-medium collision cross section puts forward more strict test to the effective Lagrangian in addition to the optical potential. The Fig. 3 displays the momentum and density dependence of the elastic effective cross section calculated with set 3. The shape of curve at $\rho = 0.25\rho_0$ is close to free n-n cross section. Generally speaking, our results are in agreement with the DBHF calculation of ter Haar and Malfliet within their energy region[7]. Fig. 3 shows that when $E-M \geq$ 400 MeV the density dependence becomes evident and should be taken into account.

From Figs.1, 2, 3 we can conclude that after introducing the momentum and density dependence of the coupling strengths for scalar and vector fields both the energy dependence of the optical potential and n-n cross section can nicely reproduced. It would be more suitable than the constant g_σ and g_ω for applying to the calculation of high energy heavy ion collisions.

THE EQUILIBRATION PROCESS IN RELATIVISTIC HIC

For simplicity in following calculations we use the simple model, i.e. the Cugnon's parametrization is used for n-n collision cross section. We have studied the global and local equilibration process and paid special attention on the interplay between the mean field and two body collisions[4,7]. The head-on collisions for system ^{40}Ca$+^{40}$Ca at 1GeV/u are studied. For the global equilibrium we study the time evolution of the kinetic flow tensor T^{ij} for it constitutes a measure of the degree of relaxation of the system, which is defined as

$$T^{ij} = \int \frac{d^4 p d^3 x}{2m} p^i p^j f(x,p). \tag{14}$$

Fig. 4 shows the time evolution of T_{11}, T_{33} for the cases with linear σ field and nonlinear σ field mean field plus collision part, mean field only and collision part only, respectively. From this figure one can easily find that, firstly, the system never reach complete equilibrium, and secondly, in addition to two body collisions the mean field still play a role in the equilibration process at the energy region ~ 1GeV/u.

We also study the local equilibration process[10]. The quantities we used to measure the local equilibrium are the ratio

$$R(\vec{r}, t) = \frac{2}{\pi} \frac{\langle P_\perp \rangle(\vec{r}, t)}{\langle P_\parallel \rangle(\vec{r}, t)}, \tag{15}$$

and ratio

$$R^*(\vec{r}, t) = \frac{\langle E_{int} \rangle(\vec{r}, t)}{\langle E_{coll} \rangle(\vec{r}, t) + \langle E_{int} \rangle(\vec{r}, t)}, \tag{16}$$

which is taken as a supplemental quantity to $R(\vec{r},t)$ to measure the degree of equilibrium. Fig. 5 and Fig. 6 show the time evolution of R and R^* in the central zone with r$\leq$2.5 fm, the second zone with 2.5 fm $< r \leq$4.5 fm and the third zone with 4.5 fm$< r \leq$7.5 fm in the center of mass system. In the central zone there is no difference between the R values for the case with both mean field and n-n collisions taken into account and with only n-n collisions taken into account. The mean field has a negligible effect in the central zone. So a mild dependence of R on the different mean fields is shown in this zone. In the second zone the effect of the mean field becomes stronger. The particles in this zone feel the Lorentz-force most strongly and it leads to a rapid increase of R value around time 5$-$8 fm/c for smaller effective mass case, in this case the larger R value only reflect the stronger transverse flow but not to larger degree of equilibrium. As concerns the degree of equilibrium reached let us concentrate on the time duration of t=11 fm/c$-$13 fm/c, the late stage of the compression phase. From Fig. 5 and Fig. 6 it can be drawn that only in the central zone the baryons reach a considerable degree of equilibrium of about 80%, while in the outer zone it is not the case.

The mass dependence of the degree of equilibrium is also studied. We have found that the heavier system can reach higher degree of equilibrium. Up to ^{139}La$+^{139}$La, the heaviest system we have investigated, the complete equilibrium is still not reached. So we would conclude that for finite system the concept based on the complete equilibrium is questionable.

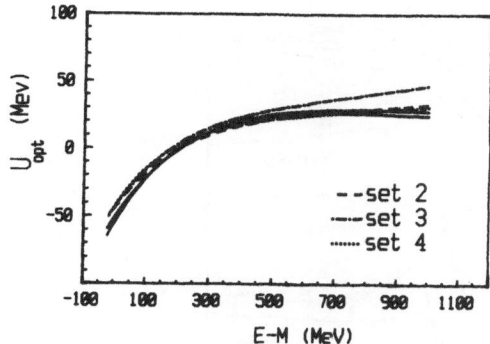

Fig.1 The relativistic optical potential as a function of the kinetic energy. The solid lines are the experimental data.

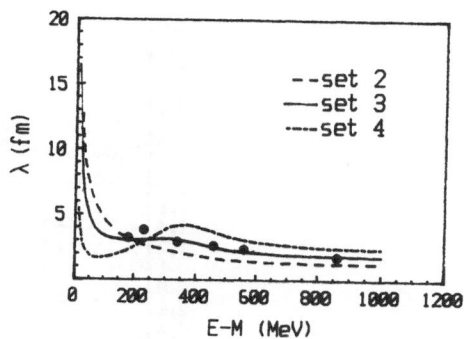

Fig.2 The mean free path in the nuclei. The dots are the experimental data.

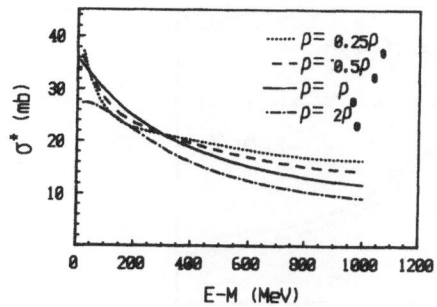

Fig.3 The elastic in-medium cross section as a function of E

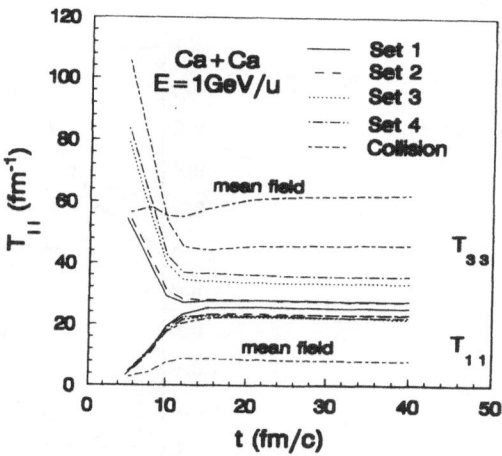

Fig.4 Time evolution of T_{11}, T_{33} for head-on collisions of $^{40}Ca + ^{40}Ca$.

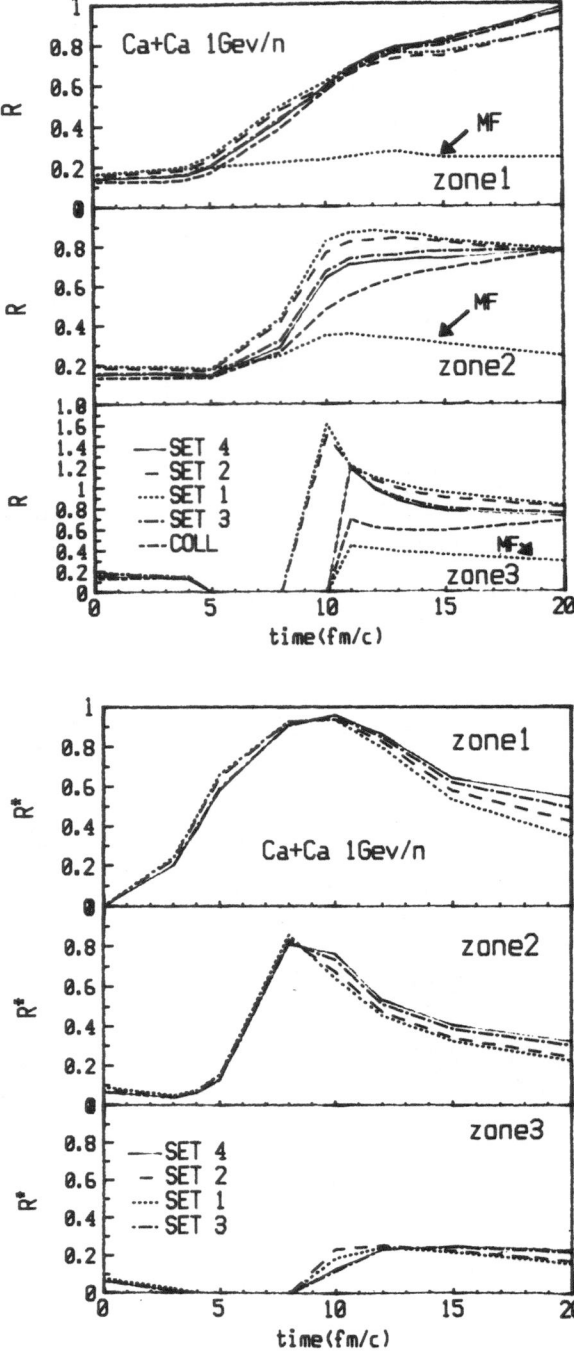

Fig.5,6 The time evolution of R and R* in zone 1, 2, 3 for ^{40}Ca+^{40}Ca at E=1GeV/u.

CONCLUSION AND SUMMARY

We have presented a self-consistent description of the relativistic RBUU equation, in which both mean field and n-n scattering cross section are given explicitly and simultaneously determined by the input Lagrangian. Based on this approach we have presented a improved $\sigma - \omega$ model with momentum and density dependent coupling strengths for σ and ω meson fields, with which both optical potential and mean free path in nuclear medium can be nicely reproduced. The in medium n-n scattering cross section calculated with this approach is in agreement with DBHF calculations. The global and local equilibration process in the HIC is studied for finite system and we have found that the complete equilibrium is not reached for finite system at E~1GeV/u. The role played by the mean field and collision part in the equilibration is investigated.

Table 1

$m_\sigma=550$MeV and $m_\omega=783$MeV are used for all cases

	g_σ	g_ω	bg_σ^3	cg_σ^4	E_{bin}	m^*	K(MeV)	ρ_0
set1	9.57	11.67			-16.00	0.56	540	0.190
set2	9.40	10.95	-0.69	40.44	-15.57	0.70	380	0.145
set3	6.90	7.54	-40.49	383.07	-15.76	0.83	380	0.145
set4	7.937	6.696	42.35	157.55	-16.00	0.85	210	0.153

Table 2

	set 2	set 3	set 4		set 2	set 3	set 4
Λ_σ	704.36	1200.00	1187.41	Λ_ω	950.07	808.29	823.64
g_s^0	206.488	217.231	78.965	g_v^0	164.707	185.157	104.197
g_s^1	-443.308	-526.651	-146.303	g_v^1	-503.419	-572.503	-336.405
g_s^2	589.232	678.793	211.778	g_v^2	939.669	951.254	598.453
g_s^3	-259.114	-311.268	-83.388	g_v^3	-473.722	-491.220	-312.456

REFERENCES

1. W.Botermans and R.Malfliet, Phys. Rep., 198(1990)115.

2. C.M.Ko and Q.Li, Phys. Rev., C37(1988)2270.
   Jin Xuemin, Zhuo Yizhong, Zhang Xizhen, M.Sano, Nucl. Phys. A506(1990)655.

3. Kuang-chao Chuo et al., Phys. Rep., 118(1985). Nucl. Phys., A559(1993)603.

4. Li Zhuxia, Zhuo Yizhong, Gu Yingqi, Sun Zemin, Yu Ziqiang, M.Sano, Nucl. Phys., A559(1993)603.

5. Yu Ziqiang, Mao Guangjun, Zhuo Yizhong, Li Zhuxia and Sun Zheming, High Energy Physics and Nucl. Phys., 16(1992)312.

6. H.Feldmeier and J.Lindner, Z. Phys., A341(1991)83.

7. R.Brockmann and R.Machleidt, Phys. Rev., C42(1990)1965.

8. R.Brockmann, H.Toki, Phys. Rev. Lett., 68(1992)3408.

9. Mao Guangjun, Li Zhuxia, Zhuo Yizhong, et al., Z. Phys. A in press.

10. Li Zhuxia, Zhuo Yizhong, Wu Xizhen, to be published in J.Phys. G: Nucl. Phys.

TESTING THE MEAN-FIELD APPROXIMATION
IN THE SPINODAL REGION

M. Belkacem, V. Latora and A. Bonasera

INFN - Laboratorio Nazionale del Sud
Viale Andrea Doria (ang. via S. Sofia), 95129 Catania (Italy)

INTRODUCTION

The Boltzmann-Nordheim-Vlasov (BNV) equation [1], also known as the Boltz-mann-Uehling-Uhlenbeck equation, has been extensively used in understanding the dynamics of intermediate energy heavy-ion collisions [2]. This kinetic equation provides a good basis for describing the average properties of one-body observables in situations where the fluctuations are small and the effects of correlations are not important. However, when one is interested in processes in which the high-order correlations play a dominant role like, for instance, multifragmentation, such models do not provide a realistic description and are not valid anymore.

In this work, we aim to test the validity of BNV approximation in the spinodal region where the fluctuations grow up quickly leading to symmetry breaking and fragment formation. For that, we compare the predictions of BNV for an expanding classical system with a given excitation energy to the exact classical solution obtained by integrating the coupled Hamilton's equations of motion (Classical Molecular Dynamics) [3, 4]. Outside the critical region, the exact solution and the BNV solution show the same average behaviour which is isentropic, while inside, they differ largely. The exact classical solution enters the spinodal region, reaching very small densities, and remains there. The BNV solution also enters the spinodal region and instead of remaining there, after some time goes back. It cannot reach very small densities as it is the case for the exact solution. For the BNV solution, we test in particular the effects of a surface term added to the mean-field.

The exact (classical) one-body distribution function f_1 satisfies the equation (BB-GKY hierarchy) :

Hot and Dense Nuclear Matter, Edited by
W. Greiner *et al.*, Plenum Press, New York, 1994

$$\partial_t f_1 + \frac{\mathbf{p}}{m}\partial^{\mathbf{r}} f_1 = \int d(2)\partial_{\mathbf{r}} V(\mathbf{r}, \mathbf{r_2})\partial_{\mathbf{p}} f_2 \qquad (1)$$

where $V(\mathbf{r}, \mathbf{r_2})$ is the two-body interaction and f_2 is the two-body distribution function. f_2 in the classical limit, is given by :

$$f_2(\mathbf{r}, \mathbf{r_2}, \mathbf{p}, \mathbf{p_2}) = \sum_{\alpha \neq \beta}^{A} \delta(\mathbf{r} - \mathbf{r}_\alpha)\delta(\mathbf{p} - \mathbf{p}_\alpha)\delta(\mathbf{r_2} - \mathbf{r}_\beta)\delta(\mathbf{p_2} - \mathbf{p}_\beta) \qquad (2)$$

Inserting this equation into eq.(1), one gets :

$$\partial_t f_1 + \frac{\mathbf{p}}{m}\partial_{\mathbf{r}} f_1 - \partial_{\mathbf{r}} U \partial_{\mathbf{p}} f_1 = 0 \qquad (3)$$

where $U = \sum_j V(\mathbf{r}, \mathbf{r_j})$ is the exact potential. Let us now define f_1 and U as sums of an ensemble averaged quantity plus the deviation from this average :

$$
\begin{aligned}
f_1 &= \bar{f}_1 + \delta f_1 \\
U &= \bar{U} + \delta U
\end{aligned}
\qquad (4)
$$

Substituting these equations in eq.(3) and ensemble averaging gives :

$$\partial_t \bar{f}_1 + \frac{\mathbf{p}}{m}\partial_{\mathbf{r}} \bar{f}_1 - \partial_{\mathbf{r}} \bar{U} \partial_{\mathbf{p}} \bar{f}_1 = < \partial_{\mathbf{r}} \delta U \partial_{\mathbf{p}} \delta f_1 > \qquad (5)$$

where one recognizes in lhs the Vlasov term and in rhs the so-called Balescu- Lennard collision term [5]. The mean-field $\bar{U}$ is given by

$$\bar{U}(\mathbf{r}) = \frac{1}{N_{ev}}\sum_{ev}\sum_j V(\mathbf{r}, \mathbf{r_j}) \qquad (6)$$

In practice, it is very difficult to calculate the Balescu-Lennard collision term and we approximate it by the Boltzmann collision term [1] and call the resulting equation classical Boltzmann-Vlasov equation.

So the main objective of this work is to test, in the spinodal region, the validity of the mean-field approximation in the eq.(5) by comparing its solution for a classical system with the exact solution one obtains by solving the eq.(3). For that, we study the expansion of a $A = 100$, $Z = 50$ nucleus. In both calculations, the nucleus is initialized in its groud state by using the frictional cooling method [6,7] and after we give to the nucleons a Maxwell-Boltzmann distribution at a given temperature using the metropolis method [8] and follow the dynamical evolution of the system in the (ρ, T) plane where we plot the average temperature of the biggest fragment versus its average density. The fragments in both calculations are defined as in ref. [9] (percolation models). In this case, it is possible to go from any particle i to another particle j belonging to the same fragment by successive interparticle jumps of a prescribed distance d or less ($d = 4 - 5 fm$ in our calculations), but any path to a particle k belonging to another fragment contains one or more jumps of distance greater than d.

CLASSICAL MOLECULAR DYNAMICS

To solve the eq.(3), we use the Classical Molecular Dynamics [10] (CMD). In the classical limit, the one-body distribution function f_1 is given by :

$$f_1(\mathbf{r}, \mathbf{p}, t) = \sum_{\alpha=1}^{A} \delta(\mathbf{r} - \mathbf{r}_\alpha(t))\delta(\mathbf{p} - \mathbf{p}_\alpha(t)) \tag{7}$$

Inserting in the eq.(3), one gets the classical coupled Hamilton's equations of motion for the A particles :

$$\frac{d\mathbf{r_i}}{dt} = \frac{d\mathbf{p_i}}{m}$$
$$\frac{d\mathbf{p_i}}{dt} = -\nabla \sum_{j \neq i} V(\mathbf{r_i} - \mathbf{r_j}) \tag{8}$$

These equations are solved using the Taylor method at 0(dt**3) [8]. For the two-body interaction V, we use a prescription of Lenk et al [4]. They give two different potentials. One, for identical nucleons interaction, is purely repulsive so no bound state of identical nucleons can exist (to simulate in some sense the Pauli principle). The second, for neutron-proton interaction, is attractive at large r and repulsive at small r. The potentials are assumed to be combinations of Yukawa interactions :

$$\begin{aligned} V_{nn} &= V_{pp} = v_0[exp(-\mu_0 r)/r - exp(-\mu_0 r_c)/r_c] \\ V_{np} &= v_r[exp(-\mu_r r)/r - exp(-\mu_r r_c)/r_c] \\ &- v_a[exp(-\mu_a r)/r - exp(-\mu_a r_c)/r_c] \end{aligned} \tag{9}$$

$r_c = 5.4 fm$ is a cutoff radius. The values of the parameters entering the Yukawa interactions are given in ref.[4] and give a corresponding Equation of State (EOS) of classical nuclear matter having about 250 MeV of compressibility (M model in ref.[4]). This EOS has the same saturation porperties as for infinite nuclear matter, i.e. equilibrium density $\rho_0 = 0.16 fm^{-3}$ and energy $E(\rho_0) = -16 MeV/nucleon$. The average density and temperature are defined as in the ref. [4].

CLASSICAL BOLTZMANN VLASOV EQUATION

To solve the classical Boltzmann-Vlasov equation (eq.(5)), we use the test particle method with a large number of test particles (200 test particles per nucleon in most of our BV calculations) [1,2]. For the mean-field, one can use the eq.(6) or, as done in the ref.[4], fit the EOS of classical matter by a polynomial fit or by a Skyrme fit. In the present BV calculations, we used the Skyrme fit given in the ref.[4] (M model) :

$$U(\rho) = A(\frac{\rho}{\rho_0}) + B(\frac{\rho}{\rho_0})^\sigma \tag{10}$$

which gives a compressibility of $203 MeV$. In these calculations, the average density is defined in the standard test particle way [1] and the temperature is calculated from the transverse momentum of particles in the center of mass of the fragment,

$$T = \frac{1}{N_p} \sum_{i=1}^{N_p} p_{\perp}^2(i)/2m \tag{11}$$

where N_p is the number of test particles in a given fragment. By taking the transverse momentum in the center of mass of the fragment, one gets rid of the collective motion of the expansion.

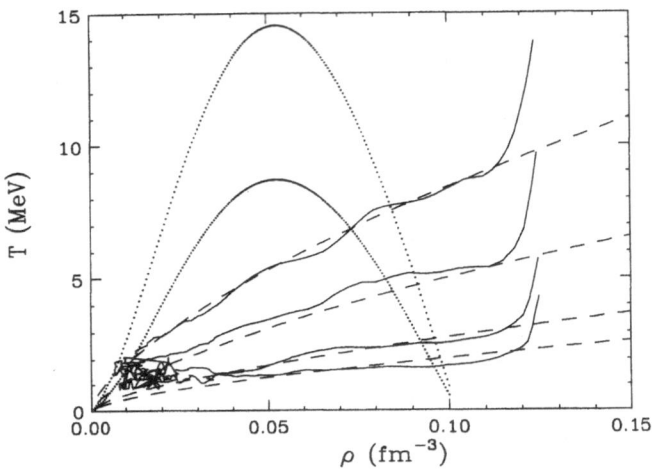

Figure 1. Exact solution. The average temperature of the biggest fragment is plotted versus its average density for various initial conditions. Dotted curves are the isothermal and adiabatic spinodal lines and dashed curves are some of the adiabats of the system.

RESULTS

As indicated, we study the expansion of a classical nucleus having an initial temperature. First, we determine the exact solution by solving the Hamilton's equations of motion and after compare to this exact solution the predictions of the classical Boltzmann-Vlasov equation. In Figure 1, we show the dynamical evolution of the exact solution. We prepare the system at various temperatures and let it evolve according to CMD and follow the evolution, as indicated in the first section, in the (ρ, T) plane. Dotted curves are the isothermal and adiabatic spinodal lines determined by the equations :

670

$$\frac{\partial^2 E(\rho)}{\partial \rho^2}\Big|_T = 0$$

$$\frac{\partial^2 E(\rho)}{\partial \rho^2}\Big|_S = 0 \qquad\qquad (12)$$

using for $E(\rho)$ a polynomial fit of classical matter EOS [4] (which fits very well the EOS). Dashed lines are some of the adiabats of the system. Solid lines show the expansion of the system for different initial conditions. From this figure, one can see that : i) apart the first part of the evolution which is due to a spurious effect of the initialization, the expansion shows an adiabatic behaviour for all the initial conditions, as predicted by Bertsch and Siemens [11, 12] and as seen by Vicentini et al [9] in the fragmentation of hot classical drops where they consider a Lennard-Jonnes two-body interaction between atoms; ii) the system (in this classical case) enters the spinodal region reaching very small densities and remains in this region when developing fragmentation, evaporation or complete vaporisation (depending on the initial condition).

To compare the Boltzmann-Vlasov predictions to the exact (classical) solution, we first start by calculating the Vlasov expansion and after that we add a Boltzmann-type collision term to see the effects of collisions on the expansion and finally we add a surface term to the mean-field. In Figure 2 (a), we show the Vlasov expansion of the system (no collision term) in the (ρ, T) plane as before. We use the same notations as in Figure 1. In this case and for all Boltzmann-Vlasov calculations the isothermal and adiabatic spinodal lines are given by eq.(12) but using for $E(\rho)$ the Skyrme fit of classical matter EOS (eq.(10)). The Vlasov expansion of the system shows outside the spinodal region, an adiabatic behaviour which agrees with the exact solution. However, in the critical region, the Vlasov solution enters this region and after some time (which is about $50-100\,fm/c$) goes back which is in disagreement with the exact solution. The Vlasov expansion does not also reach small densities region as the exact solution. To understand why the Vlasov expansion enters the spinodal region and goes back, we plotted in Figure 2 (b) the potential energy plus the internal energy (defined as $E_{int} = \frac{3}{2}T$) of the biggest fragment versus its average density for various initial conditions. Dotted curves are the isothermal and adiabatic spinodal lines and dashed lines are some of the adiabats of the system. The Vlasov expansion is shown by solid lines. As we saw from Figure 2 (a), the expansion is more or less adiabatic. So if one starts the expansion from a given point on a given adiabatic line and if the expansion is exactly adiabatic, the expansion cannot go more than the other point on the same adiabatic line which has the same energy (due to energy conservation). In the first stage of the expansion, when the energy decreases along the adiabatic line, the internal energy is stored into the collective motion, and after when the energy starts increasing, this collective energy is converted back into internal energy. In some sense, there is an interplay between the internal energy and the collective energy and the system oscillates. But, as the expansion is not completely adiabatic due to evaporation of particles and to Landau damping [5, 13], the system jumps from adiabat to another when emitting a particle and oscillating. This is why the Vlasov expansion (which should be adiabatic because Vlasov equation is time-reversible) goes back in the spinodal region and cannot reach very small densities as it is the case for the exact solution.

In Figure 3, we show the Vlasov plus a Boltzmann-type collision term expansion. For the collision term, we used the Mean Free Path version [1]. In this figure, solid lines show the Vlasov expansion (without collisions) and dashed and dashed-crossed lines the Vlasov plus a collision term. For dashed lines, we used in the collision term a cross-section of $40mb$ while for dashed-crossed lines we used a cross-section of $800mb$. The calculations are done for the same density $\rho = \rho_0$ and two different initial

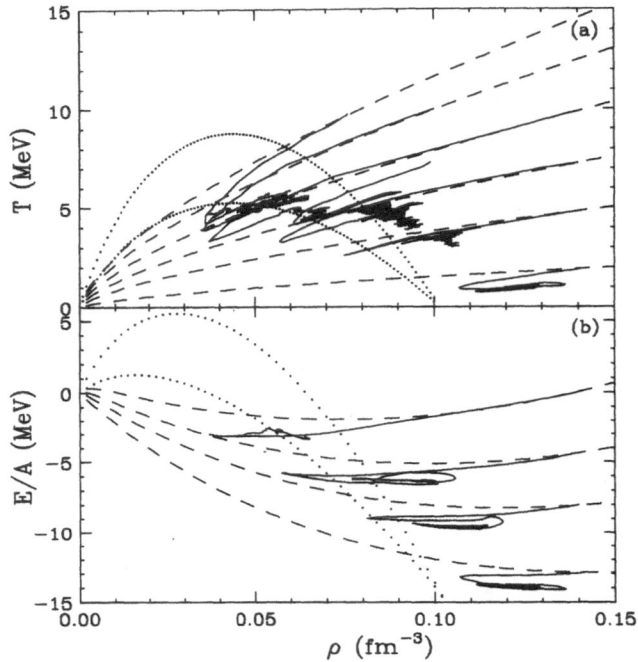

Figure 2. Vlasov expansion. (a) Same as Figure 1 but now for the Vlasov evolution (without the collision term in the eq.(5)). Here we consider different initial densities and temperatures. (b) Potential energy plus internal energy of the biggest fragment is plotted versus its average density for different initial conditions. Dotted curves are the isothermal and adiabatic spinodal lines and dashed curves are some of the adiabats of the system.

temperatures; $T = 10MeV$ and $T = 5MeV$. As one can see from this figure, the collisions seem to have no effect on the expansion of the system even if one increases the cross-section by a factor 20 which means a large increase of the rate of collisions. This could be understood by the fact that, first, in this particular case, the expansion is spherical and in this case, gain and loss terms in the collision integral cancel each other. Second, as the expansion is not fast, the system reaches at each time step equilibrium and in this case also gain and loss terms cancel each other. That's why even when increasing the cross-section, the collision term has always no effect on the expansion.

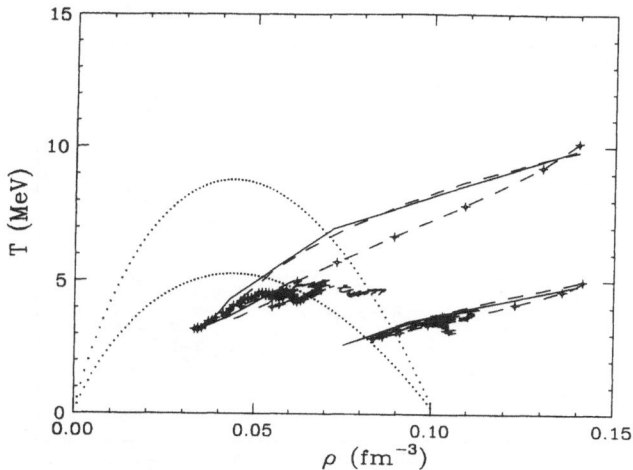

Figure 3. Vlasov plus a Boltzmann-type collision term calculations. Notations are the same as in Figure 1. Solid lines show the Vlasov (without collisions) expansion and dashed and dashed-crossed lines Vlasov plus collision term calculation with a cross-section respectively of 40mb and 800mb.

Finally, we study the effects of a surface term added to the mean-field for the Vlasov expansion (no collisions). We add to the mean-field (eq.(10)) a term given by

$$U(\rho) = A\left(\frac{\rho}{\rho_0}\right) + B\left(\frac{\rho}{\rho_0}\right)^\sigma + c\nabla^2\rho \tag{13}$$

where c is determined so that the $A = 50$ nucleus has the same binding energy with this mean-field as the one given by CMD [4]. In Figure 4, we plot the Vlasov with a surface term in the mean-field expansion (dashed lines) in comparison with the Vlasov (no surface term in the mean-field) expansion (solid lines) for the same initial conditions as in Figure 3. From this figure, one can see that, adding a surface term to the mean-field allows the expansion to go more inside the spinodal region, the system reaching smaller densities than in the case of Vlasov expansion (solid lines). In the test particles simulations of the Vlasov equation, the mean-field is calculated in cells and as we have a first derivative of the mean-field $\partial_r U(\rho(\mathbf{r})) = \partial_\rho U(\rho)\partial_r\rho(\mathbf{r})$, to calculate the effects of the mean-field on a given cell (in the case we don't consider the surface term), one should consider the contribution of the nearest cells to that cell in each direction and this means in some sense that the mean-field has a "range" of the size of the cells. When one introduces a surface term of the form of eq.(13), a third derivative of the density appears and one should consider, to calculate the effects of such mean-field on a given cell, the contribution of the three nearest cells to that cell in each direction which increases somehow the "range" of the mean-field to three times the size of the cells. With these considerations, one could understand why such surface term added to the mean-field allows the system reaching smaller densities than in the case where there is no surface term.

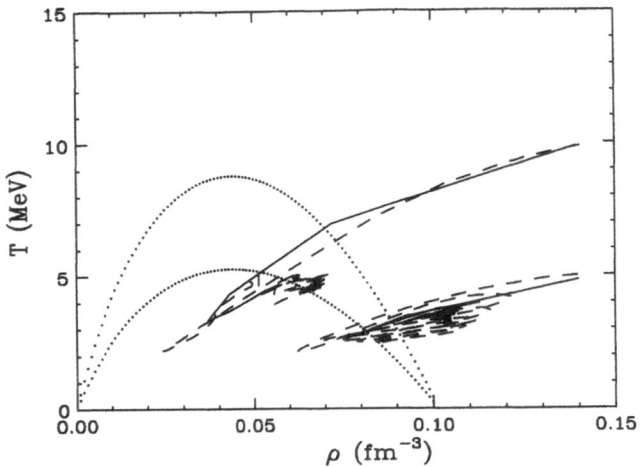

Figure 4. Vlasov (no collisions) with a surface term in the mean-field. Same notations as in Figure 1. Solid lines show the Vlasov expansion (no surface term) and dashed lines Vlasov with a surface term in the mean-field.

CONCLUSION

We have tested the validity of the mean-field approximation in the spinodal region by comparing the predictions of the Boltzmann-Vlasov equation for an expanding classical system to the exact classical solution one obtains by integrating the coupled Hamilton's equations of motion (Classical Molecular Dynamics). These tests are carried out for an expansion of a ($A = 100$, $Z = 50$) nucleus for various initial excitation energies. Both calculations (exact and Boltzmann-Vlasov) show an adiabatic behaviour outside the spinodal region, while inside, they differ largely. The exact expansion enters the spinodal zone, reaching small densities, and remains there, while the Vlasov expansion enters the critical zone and after some time goes back (due to the adiabaticity of the Vlasov equation and to energy conservation). It is also shown that the collisions have no effect on the expansion of the system (probably due to the sphericity of the system and to the low velocity of the expansion).

The surface term added to the mean-field appears to be more important for the expansion allowing the system to enter more inside the critical region and to reach smaller densities than for the Vlasov calculation performed with a mean-field which does not contain a surface term.

Finally, we conclude that, in the classical limit, the Boltzmann-Vlasov equation does not describe the average behaviour of the exact solution in the spinodal region.

REFERENCES

1. A. Bonasera, F. Gulminelli and J. Molitoris, Phys. Rep., in press.

2. G. F. Bertsch and S. Das Gupta, Phys. Rep. **160** (1988) 189.

3. T. J. Schlagel and V. R. Pandharipande, Phys. Rev. **C36** (1987) 162.

4. R. J. Lenk, T. J. Schlagel and V. R. Pandharipande, Phys. Rev. **C42** (1990) 372.

5. E. Lifschitz and L. Pitaevskii, Physical Kinetics, NY:Pergamon, 1981.

6. L. Wilet, E. M. Henley, M. Kraft and A. D. MacKellar, Nucl. Phys. **A282** (1977) 341.

7. H. Horiuchi, Nucl. Phys. **A522** (1991) 257c.

8. S. E. Koonin and D. C. Meredith, Computational Physics, Addison-Wesley, 1990.

9. A. Vicentini, G. Jacucci and V. R. Pandharipande, Phys. Rev. **C31** (1985) 1783.

10. V. Latora, A. Del Zoppo and A. Bonasera, Nucl. Phys. A, in press.

11. G. F. Bertsch and P. J. Siemens, Phys. Lett. **126B** (1983) 9.

12. G. F. Bertsch, Nucl. Phys. **A400** (1983) 221.

13. A. Bonasera, F. Gulminelli and P. Schuck, Phys. Rev. **C46** (1992) 1431.

ENERGY DEPENDENCE OF COLLECTIVE FLOW OF NEUTRONS AND CHARGED PARTICLES IN ^{197}AU + ^{197}AU COLLISIONS

Th. Blaich[5], Th. W. Elze[4], H. Emling[3], H. Freiesleben[1], K. Grimm[4], W. Henning[3,7], R. Holzmann[3], J. G. Keller[1], H. Klingler[4], J. V. Kratz[5], R. Kulessa[2], D. Lambrecht[5], S. Lange[1], Y. Leifels[1], E. Lubkiewicz[2], E. F. Moore[3,6], W. Prokopowicz[2], R. Schmidt[3], C. Schütter[4], H. Spies[4], K. Stelzer[4], J. Stroth[3], E. Wajda[2], W. Waluś[2], M. Zinser[3], E. Zude[3] and the FOPI-Collaboration[+]

[1]   Institut für Experimentalphysik, Ruhr-Universität Bochum,
      44801 Bochum, Germany
[2]   Institute of Physics, Jagellonian University, PL-30-059 Cracow, Poland
[3]   Gesellschaft für Schwerionenforschung, 64291 Darmstadt, Germany
[4]   Institut für Kernphysik, Johann-Wolfgang-Goethe-Universität,
      60325 Frankfurt, Germany
[5]   Institut für Kernchemie, Johannes-Gutenberg-Universität,
      55099 Mainz, Germany
[6]   present address: Dept. of Physics, North Carolina State University,
      P.O.Box 8202, Raleigh, NC 27695-8202
[7]   present address: Physics Division, Argonne National Laboratory,
      Argonne, IL 60439

INTRODUCTION

Collective flow of nuclear matter is one important aspect of the research performed at heavy ion accelerator laboratories. The phenomenon was predicted on the basis of hydrodynamical calculations [1], and experimental evidence was first presented for the systems ^{93}Nb + ^{93}Nb and ^{197}Au + ^{197}Au in the projectile energy range between 150 and 1050 MeV/u [2]. The comparison to microscopic calculations shows that nuclear matter is compressed to about two to three times the ground state density and that a substantial fraction of the kinetic energy in the entrance channel is converted into compressional energy [3]. In these calculations, the relation between density and compressional energy depends on the parameterization of the nucleon-nucleon interaction. The values of its parameters will hopefully be restricted by a comparison of the calculations to experimental data, and thus one can obtain information on the

Hot and Dense Nuclear Matter, Edited by
W. Greiner *et al.*, Plenum Press, New York, 1994

nucleon-nucleon interaction within the nuclear medium which is not accessible from other sources.

This problem has until now mainly been attacked with data for charged particles. Such data are basically incomplete since the information on the neutrons is missing, and it is not a priori clear to what extent collective observables are modified by the long-range Coulomb interaction. It is therefore essential to compare neutrons and protons experimentally. At first sight, this seems to require the construction of a 4π-spectrometer for neutrons which is not feasible within reasonable cost. However, if the global observables of an event – like impact parameter or azimuthal angle of the reaction plane – are provided by a 4π-spectrometer for charged particles, the spectrometer may be combined with a neutron detector of limited solid angle, and the combination will still deliver triple differential cross sections $d^3\sigma/dy\,dp_t d(\triangle\varphi)$ [y: rapidity, p_t: transverse momentum, $\triangle\varphi$: azimuthal angle with respect to the reaction plane]. Thus the full information is obtained which is equivalent to a measurement in the complete solid angle.

Preliminary data from such an experiment have been reported earlier [4, 5]. The neutron data were analyzed in terms of $<p_t>$ in the reaction plane and azimuthal distributions about the beam axis. However, no analogous data for charged particles were taken, and hence the comparison to previous experiments is difficult. That difficulty is overcome by the combination of the FOPI-spectrometer [6] with the **Large Area Neutron Detector LAND** [7]: The FOPI-spectrometer (phase I) covers the forward hemisphere of the cm-system for symmetric systems at energies above ≈ 150 MeV/u and provides the impact parameter vector. LAND determines the momentum vectors of the neutrons and light charged particles in a limited solid angle. They are thus detected in the same region of phase space, and their properties can be compared without any inter- or extrapolation, and without systematic errors due to different acceptances. The combination with FOPI allows for the exclusive investigation with respect to the impact parameter vector.

Our contribution focusses on one particular aspect of collective flow of nuclear matter: the so-called "squeeze-out", i.e. the preferential emission of mid-rapidity particles perpendicular to the reaction plane. The data were taken for the system ^{197}Au + ^{197}Au at 400, 600 and 800 MeV/u. We will cover two topics, the comparison of neutrons and protons, and the bombarding energy dependence of the neutrons' squeeze-out.

EXPERIMENTAL SET-UP AND RESULTS

The experimental set-up is shown schematically in fig. 1, and detailed descriptions of the two detector systems are given in refs. [6, 7]. Results from this experiment at a bombarding energy of 400 MeV/u have been published in [8]. The analysis at two higher energies, 600 and 800 MeV/u, was strictly analogous. Two global observables of each event, the azimuthal angle of the reaction plane and the impact parameter, were obtained from the charged particles in the plastic wall of the FOPI spectrometer: (i) The azimuthal angle φ_R of the reaction plane was taken from the total transverse momentum of the charged particles in the forward hemisphere. A flow tensor analysis was not performed, therefore all azimuthal angles are measured about the beam axis. Contrary to most other experiments, the problem of autocorrelations did not arise, since the particles to be correlated with the reaction plane – those detected in LAND – were not involved in the determination of φ_R. (ii) The impact parameter was determined by the ratio of 'transverse' to 'longitudinal' energy. For each event, $E_\perp = \Sigma\frac{p_\perp^2}{2m}$ and

$E_{\parallel} = \Sigma \frac{p_{\parallel}^2}{2m}$ were calculated. The data were grouped into five bins of the 'energy ratio' $E_{rat} = E_{\perp}/E_{\parallel}$ with bin names E1 through E5, with E5 indicating the highest energy ratio. The transformation of E_{rat} into absolute values of the impact parameter had to be done by means of model calculations. Since the data are to be compared to quantum molecular dynamics calculations (see below), the same model was chosen for the impact parameter transformation. Figure 2 shows the result for one particular version, the so-called 'soft equation of state' with momentum-dependent interactions. As intuitively expected, the highest values of E_{rat} are found to select the most central collisions. The correlation between E_{rat} and impact parameter depends only weakly on the QMD version used.

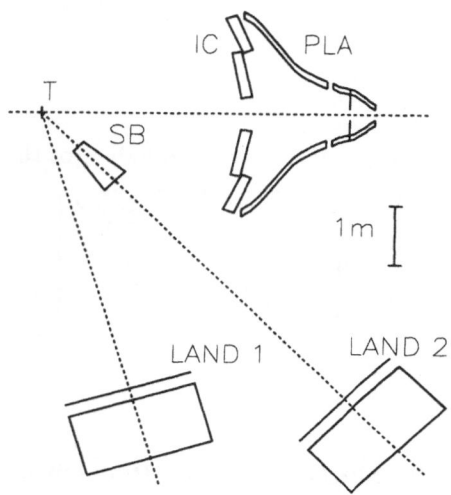

Figure 1: Set-up of the joint experiment of LAND and the FOPI-spectrometer (phase I). The latter detects charged particles at $1° \leq \theta_{lab} \leq 30°$ with full coverage of the azimuthal angle. PLA indicates the internal and external plastic wall of scintillators, IC indicates a shell of ionization chambers for detection of heavy, slow clusters. For details, see ref. 6. The two halves of LAND are positioned at $\theta_{lab} = 45° \pm 8°$ and $\theta_{lab} = 73° \pm 12°$, respectively. The figure gives as example the set-up for a shadow bar (SB) measurement with LAND 2.

Since neutrons have a long range, their spectra contain background which originates from scattering from all massive objects within the experimental area and off the concrete walls. To investigate this effect, iron shadow bars were inserted between the target and each sub-system of LAND separately, in such a way that the neutron detectors were shielded from particles coming directly from the target. During the analysis, all spectra were accumulated twice, once from runs without shadow bars and once from runs with the respective shadow bar in place. The spectra were then subtracted in order to eliminate the background. Table 1 shows the integral values in the two halves of LAND at the three projectile energies investigated. It is important to note that the background is event-correlated, i.e. some of the neutrons originate from the target, but reach the detector indirectly after scattering.

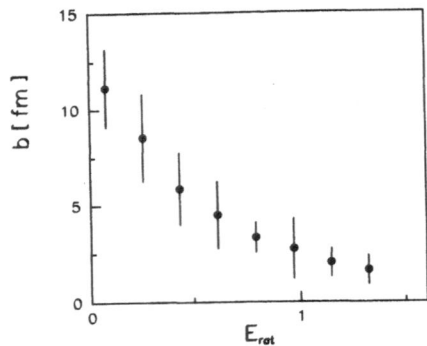

Figure 2: Relation between E_{rat} and impact parameter from QMD calculations

Table 1: Neutron background in the two halves of the LAND detector

E_{proj}	LAND 1	LAND 2
400	10%	30%
600	12%	31%
800	14%	33%

SQUEEZE-OUT OF NEUTRONS AND PROTONS

The squeeze-out effect is depicted in fig. 3 which shows typical data from the system ^{197}Au + ^{197}Au at 600 MeV/u. The figure shows azimuthal distributions about the beam axis of mid-rapidity neutrons (top) and hydrogen ions (bottom) for different impact parameter ranges which were selected by E_{rat}. Only particles with rapidities $0.4 \leq y/y_{proj} < 0.6$ are considered throughout the remainder of this contribution. The maxima of the count rate at $\Delta\varphi = \varphi - \varphi_R = \pm 90°$ indicate the preferential emission perpendicular to the reaction plane. The anisotropy is very large at impact parameters which correspond to the radius of target and projectile, and it decreases with decreasing b as the system becomes more and more azimuthally symmetric.

From a qualitative inspection of fig. 3 it is already obvious that the behaviour of neutrons and hydrogen ions is very similar. However, collective flow effects are known to increase with the mass of the particle considered, and therefore a separation of the hydrogen isotopes is essential. A scintillator wall in front of LAND yields the charge number of a particle from the correlation of energy loss with time of flight, whereas the mass number is obtained from calorimetric information from the neutron detector itself. Fig. 4 shows – for particles with Z = 1 – the ratio of the deposited energy and the kinetic energy per mass, which is given by the Lorentz factor γ-1. One clearly sees a sharp peak at 0.13 GeV, a second peak at about twice that value, and a broad

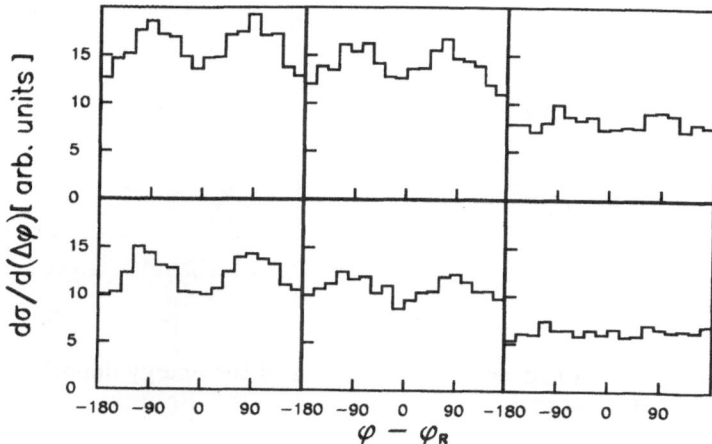

Figure 3: Azimuthal distributions of neutrons (top) and hydrogen ions (bottom) at impact parameters of b=6.9±3.1 fm (left), b=4.2±1.8 fm (middle), and b=2.8±1.3 fm (right).

shoulder around $E_{dep}/(\gamma\text{-}1) \approx 0.4$ GeV. It is obvious that the calorimetric resolution is insufficient to separate the H isotopes event by event. However, we confirmed the correspondence of the peaks to the H isotopes by fitting the spectrum with three asymmetric peaks. Their positions and widths were fixed relative to each other at values of $1 : 2 : 3$ and $1 : \sqrt{2} : \sqrt{3}$, respectively, and the asymmetry was described by only one parameter common to all three components. The resulting fit, which has only six free parameters, provides a good description of the spectrum as shown in the left part of fig. 4. Hence a cut on $E_{dep}/(\gamma\text{-}1)$ can be used to enrich p, d, and t in the azimuthal distributions and to get a feeling for the size of the signal as a function of mass. After applying very conservative cuts (right part of fig. 4), the azimuthal distributions were fitted with Legendre polynomials $\frac{dN}{d\Delta\varphi} = B_0 + B_1 \cdot P_1(x) + B_2 \cdot P_2(x)$ with $x = \cos(\Delta\varphi)$. The coefficient B_2 describes the squeeze-out effect. Its values for mid-rapidity particles are shown in fig. 5. The results for neutrons and 'protons' are within the error bars the same, and the heavier isotopes show a larger anisotropy which is consistent with the earlier finding that collective effects show up more clearly in the data of heavier particles [9]. In addition, our results show that at relativistic energies collective observables are not affected by the Coulomb force despite its long range.

ENERGY DEPENDENCE OF NEUTRON SQUEEZE-OUT

We now turn to the dependence of the squeeze-out signal on the bombarding energy. This investigation is done for the neutrons only, since in that instance we do not need to worry about isotope effects which are known to be substantial (see above). The effect is summarized in fig. 6 which shows the squeeze-out ratio $R_N = \{N(+90^0) + N(-90^0)\}/\{N(0^0) + N(180^0)\}$ as a function of the neutrons' perpendicular momentum. At the highest transverse momenta very high anisotropies are reached, higher than ever observed for baryons before. Fig. 6 (left part) shows that at all bombarding energies

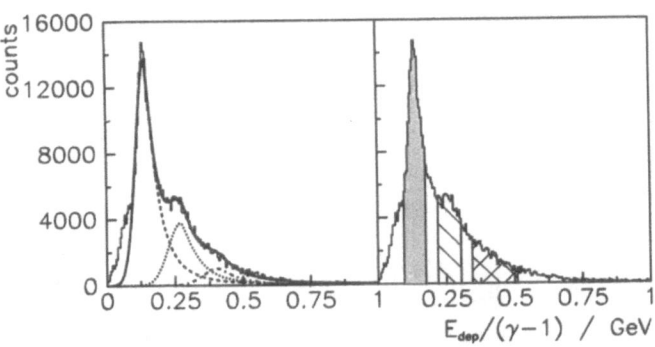

Figure 4: Enrichment of hydrogen ions by means of the energy deposited in the LAND detector. For details, see text.

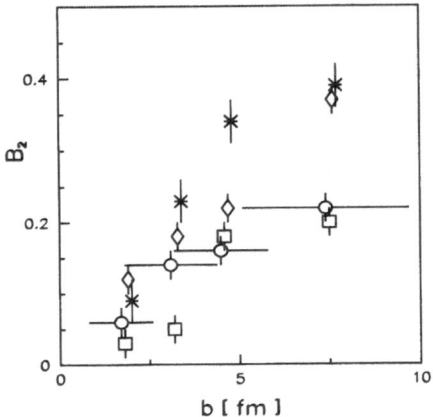

Figure 5: Coefficients of 2nd-order Legendre polynomials for neutrons (circles) and particles in the p-cut of fig. 4 (squares), in the d-cut (diamonds), and in the t-cut(asterisks).

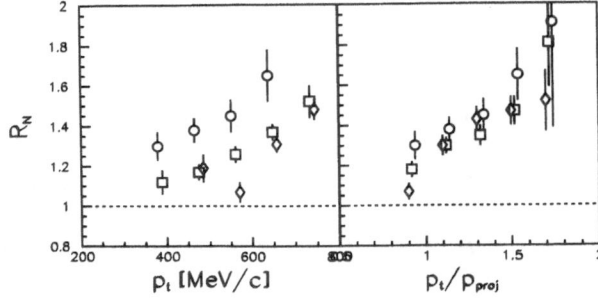

Figure 6: Squeeze-out ratio for neutrons as a function of perpendicular momentum at E_{proj}=400 MeV/u (circles), 600 MeV/u (squares), and 800 MeV/u (diamonds). An E_{rat}-cut was applied which selects impact parameters of about 7±3 fm.

682

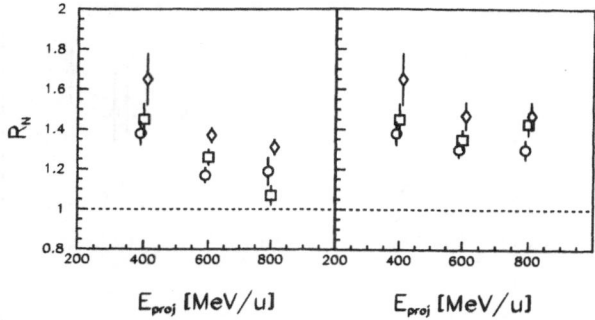

Figure 7: Squeeze-out ratio for neutrons as a function of bombarding energy. An E_{rat}-cut was applied which selects impact parameters of about 7 ± 3 fm. The left part shows R_N for neutrons with $p_t=390$ MeV/c (circles), $p_t=470$ MeV/c (squares), and $p_t=550$ MeV/c (diamonds) in bins of 80 MeV/c. The right part shows R_N for neutrons with $p_t=0.9 \cdot p_{proj}$, $p_t=1.1 \cdot p_{proj}$, and $p_t=1.3 \cdot p_{proj}$, respectively.

the maximum values of R_N are the same, but they are reached at different transverse momenta: With increasing energy, the squeeze-out signal is shifted to increasingly higher p_t. The effect becomes even more striking if the transverse momentum of the neutrons is not measured in absolute numbers, but in fractions of the projectile momentum: The dependence of R_N on p_t leads to a universal curve, independent of bombarding energy, as shown in the right part of fig. 6. We are thus led to a new interpretation of the excitation function: The latter has been shown to decrease with energy, if no cuts on p_t are applied (fig. 12 from [10]). This situation corresponds to an acceptance cut on p_t in absolute values, and the result is reproduced if more detailed cuts are applied (left part of fig. 7). However, if the cut on p_t is implemented in units of projectile momentum, the excitation function of R_N becomes flat (right part of fig. 7).

These findings remind one of the scaling laws of non-viscous fluid dynamics. We need to point out, however, that the effect is clearly observed in a fairly narrow impact parameter range where the squeeze-out signal is large, but the signature gets less clear if we go to smaller b, although indications of the effect remain.

COMPARISON TO THEORY

In an attempt to interpret our data quantitatively, we compared them to microscopic calculations in the framework of quantum molecular dynamics (QMD) [11]. Since we are discussing the behaviour of neutrons, we chose the recently developed version of IQMD [12] which keeps track of isotopic spin. We restricted ourselves to two versions which both contain momentum dependent interactions, but use different compressibilities of infinite nuclear matter ('hard' and 'soft' equation of state). The results of the calculations were filtered event-wise by the acceptance of the LAND- and FOPI-detectors, and then treated exactly like data: E_{rat} was used to cut on the im-

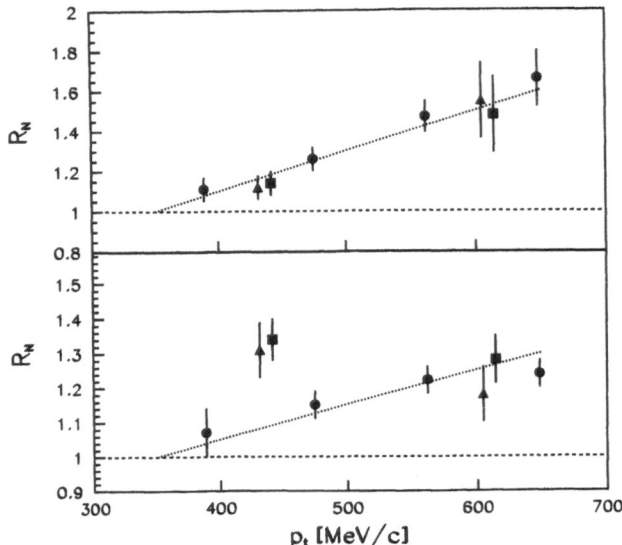

Figure 8: Squeeze-out ratio for neutrons as a function of perpendicular momentum at E_{proj}=400 MeV/u (top) and 600 MeV/u (bottom). An E_{rat}-cut was applied which selects impact parameters of about 7±3 fm. Circles: experimental data, squares: QMD results with hard EOS, triangles: QMD results with soft EOS.

pact parameter, the total transverse momentum was used to re-determine the reaction plane, and azimuthal distributions of neutrons were analyzed strictly analogous to the experimental data. The result of the comparison is shown in fig. 8: At 400 MeV/u, both versions of IQMD fit the data very nicely, however, at 600 MeV, the calculations are not able to reproduce the p_t-dependency of the squeeze-out signal. The experimental values are rising with p_t, whereas the calculated values show the opposite behaviour. As of this writing, the reason is unclear, and detailed investigations of the model are under way.

Until now, we have been dealing with squeeze-out which is related to collective flow perpendicular to the reaction plane. We have shown that in a certain impact parameter range, the squeeze-out signal scales with projectile momentum. Naturally, the question arises whether a similar behaviour can be found in other observables pertaining to collective effects. In particular, it is interesting to look at the 'flow parameter' F, i.e. the slope of the $<p_t>$–y–curve at $y=y_{cm}$. $<p_t>$ is the transverse momentum in the reaction plane, averaged over the particles of the event. At a given bombarding energy, F goes through a maximum if plotted versus impact parameter. A preliminary excitation function of the maximum F_{max} for the system ^{197}Au + ^{197}Au has been presented by the EOS collaboration [13]. It is rising continuously if $<p_t>$ is measured in absolute values, whereas scaling of $<p_t>$ with p_{proj} yields a value of $\approx$30% independent of energy (see fig. 9). Thus the flow of $<p_t>$ *in the* reaction plane exhibits the same scaling behaviour as the flow *out of the* reaction plane. This kind of scaling has been investigated earlier for the in-plane flow ([14]) in the frame work of hydrodynamical models, but our observation of this phenomenon for the out-of-plane flow calls for more quantitative calculations.

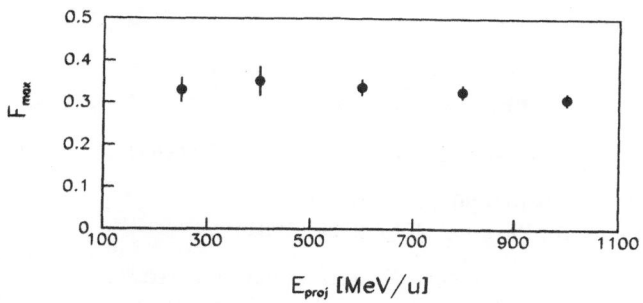

Figure 9: Maximum flow parameter versus bombarding energy for the system ^{197}Au + ^{197}Au. For details, see text.

SUMMARY

In summary, we have presented a detailed investigation of the squeeze-out effect for the ^{197}Au + ^{197}Au system in the energy range 400 MeV/u $\leq E_{proj} \leq$ 800 MeV/u. We have shown that the effect is the same for neutrons and protons, which means that the Coulomb force – despite its long range – does not have a significant effect on the observables of collective flow. The size of the squeeze-out signal depends on the transverse momentum of the particles observed. The p_t-dependence scales with the momentum of the projectile such that a plot of R_N versus p_t/p_{proj} leads to a universal curve, at least for large impact parameters where the signal is large. This scaling behaviour for out-of-plane momentum flow is also found for in-plane momentum flow as measured by the slope of the $<p_t>$–y–curve at y=y_{cm}, and detailed comparisons to microscopic as well as macroscopic calculations seem necessary for an understanding of this effect.

ACKNOWLEDGEMENTS

We thank the accelerator personnel at GSI for providing an excellent ^{197}Au-beam for our experiment. – This work was funded by the German Federal Secretary for Research and Technology (BMFT) under contract numbers 06 BO 103, 06 OF 112, and 06 MZ 106, and by GSI via Hochschulzusammenarbeitsvereinbarungen under contract numbers BO FRE, F ELE, and MZ KRD. This work was partly supported by the Committee for Scientific Research (Poland) under grant No. PB1158/P3/92.

(+) The members of the FOPI-collaboration:
J.P.Alard, Z.Basrak, N.Bastid, I.M.Belayev, M.Bini, R.Bock, A.Buta, R.Čaplar, C.Cerruti, N.Cindro, J.P.Coffin, M.Crouau, P.Dupieux, J.Erö, Z.G.Fan, P.Fintz, Z.Fodor, R.Freifelder, L.Fraysse, S.Frolov, A.Gobbi, Y.Grigorian, G.Guillaume, N.Herrmann, K.D.Hildenbrand, S.Hölbling, O.Houari, S.C.Jeong, M.Jorio, F.Jundt, J.Kecskemeti, P.Koncz, Y.Korchagin, R.Kotte, M.Krämer, C.Kuhn, I.Legrand, A.Lebedev, C.Maguire, V.Manko, T.Matulewicz, G.Mgebrishvili, J.Mösner, D.Moisa, G.Montarou, P.Morel, W.Neubert, A.Olmi, G.Pasquali, D.Pelte, M.Petrovici, G.Poggi, F.Rami, W.Reisdorf, A.Sadchikov, D.Schüll, Z.Seres, B.Sikora, V.Simion, S.Smolyankin, U.Sodan, N.Taccetti, K.Teh, R.Tezkratt, M.Trzaska, M.A.Vasiliev, P.Wagner, J.P.Wessels, T.Wienold, Z.Wilhelmi, D.Wohlfarth, A.V.Zhilin

References

[1] W.Scheid et al., Phys Rev Lett <u>32</u>, 741 (1974)
   H.Stöcker et al., Phys Rev <u>C25</u>, 1873 (1982)

[2] H.A.Gustafsson et al., Phys Rev Lett <u>52</u>, 1590 (1984)

[3] J.Aichelin, Phys Rep <u>202</u>, 233 (1991)
   B.Blättel et al., Phys Rev <u>C43</u>, 2728 (1991)

[4] J.Schambach et al., in: The Nuclear Equation of State,
   Proceedings of a NATO Advanced Study Institute, Peniscola, 1989,
   Vol. 216B, edited by W. Greiner and H. Stöcker, part A, p.115

[5] R.Madey et al., Proceedings of the International Nuclear Physics Conference,
   Wiesbaden 1992, Nucl Phys <u>A553</u>, 779c (1993)

[6] A.Gobbi et al., Nucl Instr Meth <u>A324</u>, 156 (1993)

[7] Th.Blaich et al., Nucl Instr Meth <u>A314</u>, 136 (1992)

[8] Y.Leifels et al., Phys Rev Lett <u>71</u>, 963 (1993)

[9] K.G.R.Doss et al., Phys Rev Lett <u>59</u>, 2720 (1987)

[10] H.H.Gutbrod et al., Phys Rev <u>C42</u>, 640 (1990)

[11] J.Aichelin and H.Stöcker, Phys Lett <u>B176</u>, 14 (1986)

[12] Ch.Hartnack, PhD thesis, University of Frankfurt, 1992

[13] H.G.Ritter, Proceedings of this conference

[14] A.Bonasera et al., Nucl Phys <u>A476</u>, 159 (1988)
    and references therein

TRANSPORT TREATMENT OF AN EXPAND-
ING PION GAS IN ULTRARELATIVISTIC
HEAVY ION COLLISIONS

H.W. Barz[1], G. Bertsch[2,3], P. Danielewicz[2], H. Schulz[4] and G.M. Welke[5]

[1] KAI e.V. and Institut für Kern- und Hadronenphysik
FZ Rossendorf, PF 510119, D-01314 Dresden, Germany
[2] National Superconducting Cyclotron Laboratory and Department of
Physics and Astronomy, Michigan State University, East Lansing,
MI 48824, USA
[3] Department of Physics FM-15,University of Washington, Seattle
WA 98195, USA
[4] The Niels-Bohr Institut, Blegdamsvej 17, DK-2100 Copenhagen
[5] Department of Physics and Astronomy, Wayne State University,
Detroit, MI 48202, USA

INTRODUCTION

In ultrarelativistic collisions of heavy ions a hot central region is formed that subsequently decays mostly into π mesons. One of the main goals of the theory of collisions is to trace the evolution of the hot initial zone until the stage when the freezeout occurs. For this purpose one has to study the dynamics of the pionic gas. One of the problems to be solved is, to what extent the observed pion spectra and, in particular, the enhancement at low momenta have to be associated with the formation of a hot and dense pionic zone right after the hadronization. In other words, one has to find out, whether in the rapidly expanding pion cloud the $\pi\pi$ collisions are effective and frequent enough to thermalize the distribution of the pions. If not, the experimentally observed distribution of the pions might be remainder of the hadronization process itself and could give us, for example, information on whether the pions originate from a hadronizing plasma blob or not.

To answer these questions the evolution of the system has been studied with a Boltzmann equation including Bose statistics [1, 2]. Due to Bose statistics the collision rate is huge and may even lead to condensation in the momentum space. However, it was shown [3] that a consistent description of the $\pi\pi$ scattering in the gas results in an in-medium cross section which is largely reduced compared to the free one. The collision rate turned out to be almost the same as in the ordinary Boltzmann equation containing the unaffected cross section and disregarding the proper Bose occupancy factors.

The goal of the present work is to calculate the evolution of the hot and dense pion gas using the Boltzmann equation with proper statistics and in-medium cross sections. We do not consider the effects of the mean field [4]. Using the phase space distribution we calculate the transverse momentum spectra of pions and analyze the effects on the two-particle correlations.

BOSONIC BOLTZMANN EQUATION WITH IN-MEDIUM $\pi\pi$ SCATTERING

The relativistic Boltzmann equation for the time evolution of the phase space density $f_1(\boldsymbol{x}, \boldsymbol{p})$ of the interacting Bose gas of pions reads

$$
(p^0 \frac{\partial}{\partial t} + \boldsymbol{p}\frac{\partial}{\partial \boldsymbol{x}}) f_1(\boldsymbol{x}, \boldsymbol{p}) = -\frac{1}{4}\sum\int d\omega_2 d\omega_3 d\omega_4 \times
$$
$$
\delta^4(k_1 + k_2 - k_3 - k_4) \mid t \mid^2 \{f_1 f_2 \bar{f}_3 \bar{f}_4 - f_3 f_4 \bar{f}_1 \bar{f}_2\}, \tag{1}
$$

where the abbreviations stand for $d\omega_i = d^3 p_i/(2\pi\hbar)^3 2E_i$, $E_i = \sqrt{m_\pi^2 + p_i^2}$, $\bar{f}_i \equiv \bar{f}_i(\boldsymbol{x}_i, \boldsymbol{p}_i) \equiv 1 + f(\boldsymbol{x}_i, \boldsymbol{p}_i)$, the mass of the pion is m_π and $\boldsymbol{p}_i$ and $\boldsymbol{x}_i$ stand for their momenta and spatial coordinates, respectively.

The in-medium scattering of the pions is described by the t-matrix that may be derived microscopically using the nonequilibrium Green function technique (see refs. [6] and refs. therein). The equation for the t-matrix is then of Bethe-Goldstone type (see also ref. [5]); it describes the interaction of a pair of pions embedded in a pure pionic medium. The influence of the medium is accounted for through the phase space occupancy factors and by the single-particle energies in the propagator that we take equal to the free energies as in (1). In the subsequent calculations we will consider the $\pi\pi$ interaction in the isospin $I = 0$ s-wave and in the $I = 1$ p-wave channels. The former one is referred as to the σ channel and the latter one as to the ρ channel corresponding to the resonance formed in these channels. The square of $\mid t \mid^2$ in (1) is decomposed into the channel contributions α

$$
\mid t \mid^2 = \sum_\alpha \mid t_\alpha \mid^2 P_{l_\alpha}^2(\cos\theta), \tag{2}
$$

where θ is the scattering angle. The solution of the t-matrix equation is greatly simplified when the $\pi\pi$ amplitude is parametrized in the separable form. The potential is in this case (see also Johnstone and Lee [7]) of the form

$$
V_{\pi\pi}^\alpha = <\pi\pi \mid V \mid \alpha> \frac{1}{s - M_\alpha^2 + i\epsilon} <\alpha \mid V \mid \pi\pi>, \tag{3}
$$

where $s = E^2 - \mathbf{P}^2$ is the center of mass energy of the pion pair having a total momentum $\mathbf{P}$. The formfactor is given by

$$
V_{l_\alpha}(k) = 4\pi \, \omega(k) \, g_{l_\alpha} \sqrt{2M_{l_\alpha}} \, (\frac{k}{k_{l_\alpha}})^{l_\alpha} \frac{1}{(1 + \frac{k^2}{k_{l_\alpha}^2})^{l_\alpha+1}} \tag{4}
$$

688

with the parameters: $M_0 = 940$ MeV, $g_0 = 0.60\, m_\pi^{-1/2}$, $k_0 = 2.71\, m_\pi$ in the s-wave channel and $M_1 = 826.7$ MeV, $g_1 = 0.6684\, m_\pi^{-1/2}$ and $k_1 = 3.34\, m_\pi$ in the p-wave channel. The dispersion relation for the pions is $\omega_k^2 = m_\pi^2 + k^2$.

Using eq.(2) the Bethe-Goldstone separates into the different channels

$$t_\alpha = V_{l_\alpha} + V_{l_\alpha} G t_\alpha, \tag{5}$$

where the Green's function reads

$$G_\alpha(s, \boldsymbol{P}^2, k) = \frac{\int d\Omega_{\boldsymbol{k}}\ (1 + f_1(\mathcal{L}\boldsymbol{k}) + f_1(\mathcal{L}(-\boldsymbol{k})))}{\omega(k)(s - 4\omega^2(k) + i\epsilon)}. \tag{6}$$

The symbol $\mathcal{L}$ denotes the Lorentz boost of the momentum $\boldsymbol{k}$ from the $\pi - \pi$ rest frame into the frame with pair momentum $\boldsymbol{P}$. Assuming isotropic momentum distribution the Green function depends only on $\boldsymbol{P}^2$ as a consequence of the integration over the orientation of the pion momentum $\boldsymbol{k}$. The solution of eq.(5) becomes

$$t_\alpha(k, k', s) = \frac{V_\alpha(k)\, V_\alpha(k')}{s - M_\alpha^2 - \frac{1}{4\pi^2} \int_0^\infty dk\ k^2\ \frac{V_\alpha^2(k)\ <\ 1 + f_1(k) + f_1(-k)\ >}{\omega(k)(s - 4\omega^2(k) + i\epsilon)}} \tag{7}$$

In solving the Boltzmann equation (1) we use the test particle method. We mention that in calculating the collision rates the two effects of the medium, namely the enhancement factors $1 + f$ in the Boltzmann equation and the similar factors in the in-medium t-matrix of eq. (7) nearly balance one another [3]. This is illustrated in fig. 1 for the ratio R of the collision rates $< f_1 f_2 v_{12} \sigma_{in-medium} \bar{f}_3 \bar{f}_4 >$ and $< f_1 f_2 v_{12} \sigma_{free} >$ affected and not affected by the medium. In this calculation the distribution function $f(\boldsymbol{p})$ is taken for the case of thermal equilibrium.

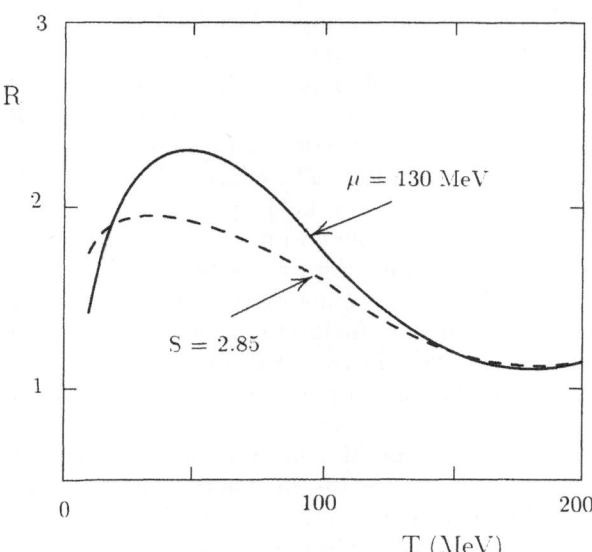

Fig.1. Ratio R of collision rates calculated with and without medium effects as function of the equilibrium density given by the temperature T for constant entropy S or chemical potential μ, respectively.

EXPANSION DYNAMICS

The goal of our study is to find out to what extent the interaction of the pions will drive the pions to occupy preferentially regions with low momenta. To make the situation we want to address clear, let us for a moment assume that the expansion of the pion gas would be purely adiabatic maintaining chemical and thermal equilibrium. In this case the distribution of the pions is given by $f(\mu_\pi, T) = 1/(exp((E - \mu_\pi)/T) - 1)$ where E is the energy, μ_π and T are the chemical potential and the temperature, respectively.

At low energies pion annihilation processes such as $4\pi \rightarrow 2\pi$ are suppressed and entropy and particle number are conserved. Such an assumption would imply that the chemical potential increases during the expansion. The key point is however, whether the system expands slowly enough and there are sufficient collisions to reach such a equilibrium distributions. In other words, we want to find out by means of our realistic calculations to what extent at all a given initial distribution of the pions can be thermalized by the subsequent collisions.

We consider specifically the central collision of ^{16}O on Au at 200 GeV per nucleon measured by the NA35 collaboration [9]. Integration over the experimental rapidity distributions gives about 400 pions. These pions are observed in a rapidity range of $\Delta y = 2.6$. We estimated the initial volume by assuming that the fireball has a cross section given by that of the ^{16}O nucleus and the extension in the longitudinal z-direction is connected to the rapidity as $z = \tau \, sinh \, y$ according to the Bjorken picture. The parameter τ describes the time which is needed for the formation of the pions. Since pions originate from different sources, such as resonance decays or the hypothetical plasma stage, the parameter τ may largely exceed the hadronization time of 1 fm familiar from the string hadronization picture.

A rather large fraction of the final pions will will come from the decay of long lived resonance states such as the ω's and η's while the short lived ρ's decay during the expansion. In fact, invoking the string picture and assuming that strings break by creating pairs of quarks and antiquarks independently, mesons are formed according to their statistical weights. Therefore we expect that they are formed in the ratio $\pi : \rho : \omega : \eta \sim 3 : 9 : 3 : 1$. Phenomenological modelling of quark jets, such as done in the Lund model have statistical weights close to these values. So we start with 38 π, 114 ρ, 38 ω and 13 η mesons and use a momentum distribution in accordance with a temperature of 160 MeV and zero chemical potential. The effective hadronization time τ determines the initial particle density which is roughly proportional to $1/\tau$. The finite life time of the mesons are taken into account which implies some modifications of the Boltzmann equation (1). The in-medium $\pi\pi$ scattering is described by the t-matrix in eq. (7), whereas the elastic scattering of the pions with the ω and η mesons is treated by a Hauser-Feshbach type method outlined in ref. [8]. For the solution od the Boltzmann equation we employ the so-called test particle method, where each particle is replaced by a large number of test particles and the one-particle density is represented by the ensemble of cascading test particles.

In fig. 2 we show the density and velocity profiles in transverse direction at mid-rapidity as a function of the time starting from the moment of hadronization for $\tau = 4.5 fm/c$. At the beginning the pion density increases rapidly due to the hadronization and the decay of the ρ mesons. The maximum density is reached after $2 fm/c$, after this time the density decreases due to the longitudinal and transverse expansion.

The longitudinal expansion is mainly a consequence of the initial conditions and

690

is proportional to $1/(t + \tau)$ (Bjorken scaling). In radial direction a linear velocity profile is generated. The velocity at a fixed radius reaches its maximum value after a time of $4\,fm/c$. The collision rate per pion averaged over the whole volume amounts to $0.5\,c/fm$ which corresponds to a relatively large mean free path with respect to the size of the system. So we compare our results with the expansion of a collisionless mesonic gas (dashed lines in fig.2). One recognizes that there is an overall similarity except in the very central region. In the centre the collisions are frequent enough to generate a pressure which accelerates the expansion. Thus the density drops faster and the maximum velocity is reached earlier than in the case of free streaming.

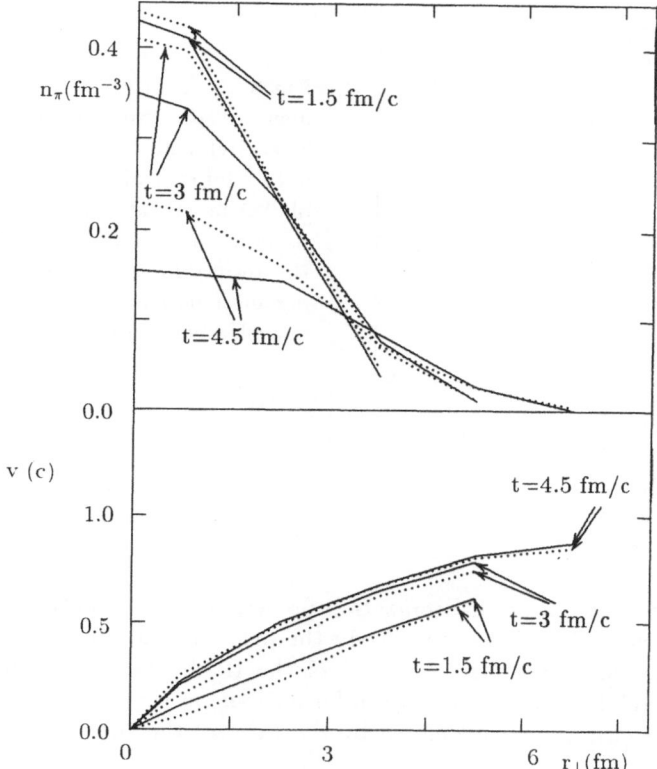

Fig.2. Density and velocity profiles in transverse direction at three different times. The effective hadronization time is $4.5\,fm/c$. For comparison the dashed curves show the corresponding profiles for a collisionless expansion.

Now we discuss the final distribution of the pion momenta which can be compared to experiment. Three calculations are carried out with the hadronization times of $\tau = 1$ fm/c, 4.5 fm/c and 8 fm/c. The obtained transverse momentum distributions of the pions are compared with ^{16}O data in fig. 3. One sees that for the small hadronization time there is a pronounced peak originating from the scattering processes and decaying resonances which concentrate the pions into a narrow region of the phase space with low momentum. Further the spectrum has a concave shape demonstrating that pions with intermediate momentum are redistributed to low momenta by the collisions. Using a large hadronization time the effect is moderate and the initial momentum distribution is not changed significantly.

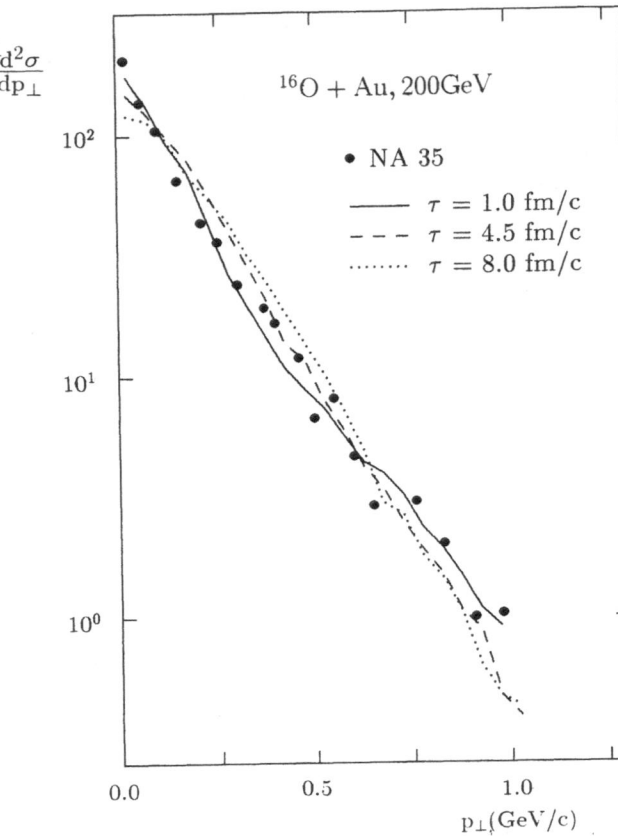

$\dfrac{\mathrm{d}^2\sigma}{\mathrm{dp}_\perp}$

$^{16}\mathrm{O} + \mathrm{Au},\ 200\,\mathrm{GeV}$

- NA 35

——— $\tau = 1.0$ fm/c

– – – $\tau = 4.5$ fm/c

········ $\tau = 8.0$ fm/c

$\mathrm{p}_\perp(\mathrm{GeV/c})$

Fig.3. Transverse momentum spectrum of pions. The circles are the NA35 data [9] for central collisions of $^{16}\mathrm{O}$ on Au at 200 GeV per nucleon compared with the results for an expanding mesonic gas for different effective hadronization times τ.

PION INTERFEROMETRY

To gain an deeper understanding of the hadronization process in ultra-relativistic heavy ion collisions pion interferometry is used to study the properties at the freeze out time (for a recent review see ref.[10]). On the basis of the Hanbury-Brown and Twiss effect the two-pion correlations gives valuable informations on the size of the emitting source. The correlation function is defined by the ratio of the two-pion to the one-pion inclusive yields

$$C_2(\boldsymbol{p}_1, \boldsymbol{p}_2) \quad = \quad \frac{N(\boldsymbol{p}_1, \boldsymbol{p}_2)}{N(\boldsymbol{p}_1)\,N(\boldsymbol{p}_2)} \tag{8}$$

$$\approx \quad 1 + exp(-\frac{1}{2}(\boldsymbol{p}_1 - \boldsymbol{p}_2)^2 R^2) \tag{9}$$

The correlation function is usually fitted to an Gaussian distribution by eq. (9) as a function of the relative momentum and allows the determination of the source radius R.

At ultrarelativistic energies the two-pion interference comes mainly from the fact that two identical pions obey Bose statistics and have a symmetric wave function.

Therefore two pions emitted from space-time points x_1 and x_2 with 4-momenta p_1 and p_2, respectively, are correlated according to $f = 1 + cos((p_1 - p_2)(x_1 - x_2))$. Following the prescription [11] the correlation function can be calculated from the one-particle density matrix of the system via

$$C_2(p_1, p_2) = 1 + \frac{\int d^4x d^4x' \, g(x, P)g(x', P)cos[q(x - x')]}{\int d^4x d^4x' \, g(x, p_1)g(x', p_2)}, \qquad (10)$$

where $q = p_1 - p_2$ and $P = (p_1 + p_2)/2$. The density matrix is connected to the source function $g(x, p)$ which describes the distribution function of the emissions points of the pions.

It is an important task for the various models applied to heavy ion collisions to determine the source function. Using the Boltzmann approach one avoids the use of break-up criteria because the collisions automatically cease as the system expands. The function $g(x, p)$ is given by the last interaction points of the test particles within our Boltzmann simulation:

$$g(x, p) = \sum_i \delta^{(4)}(x - x_i)\delta^{(3)}(\boldsymbol{p} - \boldsymbol{p}_i). \qquad (11)$$

Here, we use the on-shell values for the source function in eq.(10). For a more thorough discussion of this point see ref. [13]. The function (11) is too singular to display interference effects since eq.(10) requires a continuous source function. We therefore smooth the momentum part of eq. (11) by an momentum spread of $30 MeV/c$. The choice of smearing is important since using too large a spread the correlation functions C_2 becomes too narrow.

In the following we consider only momenta $\boldsymbol{q}$ and $\boldsymbol{P}$ in the plane perpendicular to the beam direction. The sideways correlation, is obtained as a function of the relative momentum $\boldsymbol{q}$ which is orthogonal to the total momentum $\boldsymbol{P}$ [12]. Fig. 4 illustrates the obtained correlation function in the central rapidity region $-0.5 < y < 0.5$. Using eq. (9) with R replaced by R_{side} we obtain the values of $R_{side} = 1.9fm$, $3.0fm$, and $3.9fm$ for hadronization times of $\tau = 1fm/c, 4.5fm/c$ and $8fm/c$, respectively. These values have to be compared with the initial transverse radius of ^{16}O nucleus of $2.1fm$. On the first view the result is surprising, because the most strongly interacting pions exhibit the smallest radius. However, this result can be explained by the fact that in calculating the correlation function by eq.(10) those regions have a high weight from which the two pions have roughly the same momentum. In a collective motion of cooled matter this region is small which is reflected by the obtained radius. Therefore the pion interferometry provides us the size of the region in which the momenta are randomized around $\frac{1}{2}\boldsymbol{P}$.

To investigate the behaviour of the correlation function in more detail, we calculate the correlation for pion pairs with different total momentum $P_\perp$. The result is shown in fig. 5. For large momenta $P_\perp$ the values of the extracted radii are nearly independent of the initial density. This means that the fast pions do not collide frequently. The value of the radius is only slightly larger than the initial radius of the source. Obviously the fast pions leave the surface region in an early stage evading the reaction zone. However for small momenta the behaviour is quite different. For small τ, i.e. high density, we expect a highly collective behaviour which results in small radii. On the other hand the relatively independent motion of pions for large τ together with those coming from the ρ decay gives larger radii.

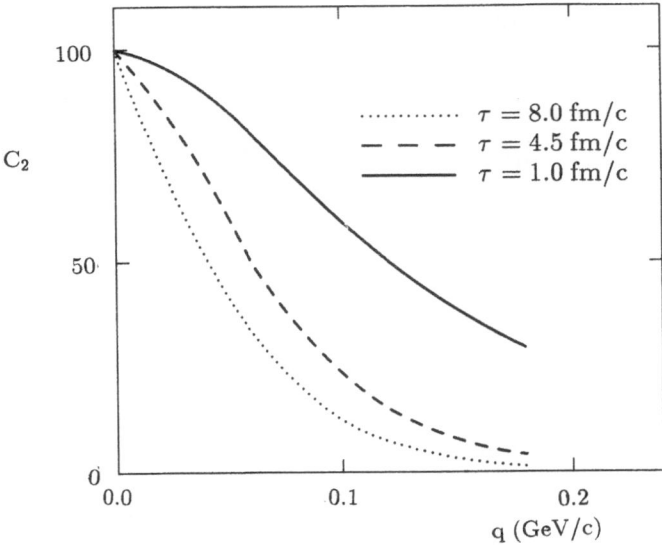

Fig.4. Two-pion correlation functions C_2 as a function of the relative momentum q_{side} which points perpendicularly to the total momentum of the pion pair and the beam direction. The pions are sampled from the central rapidity region. The spectra are shown for different effective hadronization times.

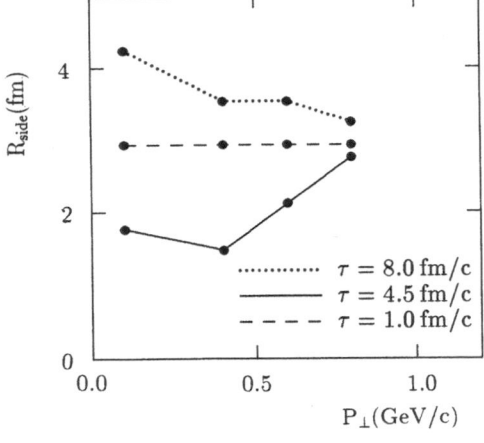

Fig.5. Extracted radii R_{side} from the correlation function using pion pairs with total momentum around $P_{\perp}$.

CONCLUSIONS

We have investigated an ultrarelativistic meson gas within the frame of the Boltzmann approach. Using the Bethe-Goldstone method we have demonstrated that the large Bose enhancement factors in dense matter are nearly compensated by the reduction of the $\pi\pi$ cross sections inside the medium.

The solution of the Bosonic Boltzmann equations provides us with the full phase space distribution of the source. The expanding gas has a complicated space-time structure where the free initial motion of the pions together with the decay of the resonances competes with the thermalization processes via the in-medium interactions.

Our studies have shown that at large initial density particles with low momenta suffer many collisions and show a large degree of collectivity. Effects due to the onset of Bose condensation may occur. This effect would be observable in a low $p_\perp$ enhancement in the transverse pion spectra and by a relatively small source radius extracted from pion interferometry. Such effects should not be seen analyzing pions with large transverse momenta or pions which are created with low density in a late stage of ultrarelativistic heavy ion collisions.

ACKNOWLEDGEMENTS

G.B., P.D. and G.M.W. acknowledge support of the National Science Foundation under Grant Nos. 90-17077 and 89-05933 while H.W.B. acknowledges support of the German ministry BMFT under project No. 06DR107. H.W.B. and H.S. are indebted to the Nuclear Theory Group at MSU-NSCL for the kind hospitality extended to them and for financial support.

REFERENCES

1. G.M. Welke and G.F. Bertsch, Phys. Rev. C45:1403(1992)

2. S. Gavin and P.V. Ruuskanen, Phys. Letters 262B:326(1991)

3. H.W. Barz, P. Danielewicz, H. Schulz and G. Welke, Phys. Lett. 287B:40(1992)

4. E. Shuryak, Phys. Lett. 207B:345(1988); Nucl. Phys. A533:761(1991); J.L. Goity and H. Leutwyler, Phys. Lett. B228:517(1989); A. Schenk, Nucl. Phys. B262:326(1991); V. Koch and G.F. Bertsch, Nucl. Phys. A552:591(1993)

5. M. Schmidt, G. Röpke and H. Schulz, Ann. Phys. (N.Y.) 202:57(1992); Z. Aouissat, G. Chanfray, P. Schuck and G. Welke, Z. Phys. A340:347(1991); H.W. Barz, G.F. Bertsch, P. Danielewicz and H. Schulz, Phys. Lett. 275B:19(1992)

6. P. Danielewicz, Ann. Phys. (N.Y.) 152:1628(1984) 1628

7. J.A. Johnstone and T.-S.H. Lee, Phys. Rev. C34:243(1986)

8. G.F. Bertsch and S. Boggs, Annual Report, 1990, NSCL, Michigan State University

9. J.W. Harris for the NA35 collaboration, Nucl. Phys. A498:133(1989); The NA35 collaboration: H. Ströbele et al., Z. Phys. C38:89(1988)

10. D.H. Boal, B.K. Jennings and C.K. Gelbke, Rev. Mod. Phys. 62:553(1990)

11. S. Pratt, Phys. Rev. Lett. 53:1219(1984)

12. G. Bertsch, M. Gong, and M. Tohyama, Phys. Rev. C37:1896(1988)

13. G. Welke, H.W. Barz, G.F. Bertsch, P. Danielewicz and H. Schulz, to be published

MORE-FLUID MODELS
FOR ULTRARELATIVISTIC
NUCLEAR COLLISIONS

U. Katscher,[1] J.A. Maruhn,[1] W. Greiner,[1] I.N. Mishustin,[2] L.M. Satarov[2]

[1]Institut für Theoretische Physik, 60054 Frankfurt/Main, Germany
[2]The Kurchatov Institute, Moscow 123182, Russia

INTRODUCTION

Since Landau formulated his hydrodynamical model [1], hydrodynamics are regarded as an appropriate tool for the description of heavy-ion collisions [2]. In contrast to microscopic models one has to implement an equation of state explicity and thus is able to study explicitly effects like the phase transition to the quark-gluon-plasma [3]. On the other hand, the applicability of the hydrodynamic model is affected by the question of local thermodynamic equilibrium. Since this assumption is probably not valid for heavy-ion collisions at ultrarelativistic energies, one introduces additional fluids. In a more-fluid model one assumes local thermodynamic equilibrium only for each separate fluid, but not for the sum of the fluids. The first 2-fluid model was developed by Amsden et al. [4] for nonrelativistic energies. Mishustin et al. [5] extended the model to relativistic energies. A nonrelativistic 3-fluid model was constructed by Rosenhauer et al. [6]. In this paper we present a 3-fluid model for ultrarelativistic energies. Besides the baryonic fluids formed by the target and the projectile nucleons, we introduce a third fluid collecting particles produced during the reaction – mainly pions.

After the description of the model, we study the reaction S+S, 200AGeV [7] as well as the excitation function of Au+Au [8] in a simplified version of the model. Finally we discuss spectra resulting from the complete version of the model [9].

FORMULATION OF THE MODEL

Let ρ, e, and p denote the particle number density, energy density, and pressure in the local rest frame of a fluid element and $U^\mu = \gamma(1, \vec{\beta})$ its four-velocity. Then the four-current and the energy-momentum tensor of an ideal fluid are expressed, respectively, as $j^\mu = \rho U^\mu$ and $T^{\mu\nu} = (e + p)U^\mu U^\nu - pg^{\mu\nu}$, where $g^{\mu\nu}$=diag(+1,-1,-1,-1) is the metric tensor.[1] The hydrodynamic equations determine the evolution of the four-currents j_i^μ and the energy-momentum tensors $T_i^{\mu\nu}$ for the three fluids ($i = p, t, f$)

$$\partial_\mu j_i^\mu = S_i, \tag{1}$$

$$\partial_\mu T_i^{\mu\nu} = S_i^\nu. \tag{2}$$

---

[1]Units $\hbar = c = 1$ are used throughout the paper.

Hot and Dense Nuclear Matter, Edited by
W. Greiner *et al.*, Plenum Press, New York, 1994

The baryon numbers of the projectile and the target are conserved ($S_{p,t} \equiv 0$) while the number of hadrons in the fireball increases during the reaction ($S_f > 0$). Energy and momentum can be transmitted between all three fluids subject to the condition

$$\partial_\mu \sum_i T_i^{\mu\nu} = 0 \,. \tag{3}$$

The sources used in Eqs. (1),(2) are very close to the interflow friction force described in [10]. Since [10] includes a detailed study of the interflow friction force, we restrict ourselves to a short sketch of the derivation of the sources.

The four-momentum density loss of the nuclear fluids is assumed to be a superposition of the four-momentum losses occurring in single nucleon-nucleon collisions. Therefore we start with the four-momentum loss of a single nucleon Δp^μ averaged over the cross section

$$\langle \Delta p^\mu \rangle = \int d\sigma_{NN \to xNN}(p^\mu - p'^\mu) \,, \tag{4}$$

where σ_{NN} is the total cross section of the nucleon-nucleon interaction. p^μ and p'^μ denotes the nucleon four-momentum before and after the collision, respectively. In an arbitrary system, this can be written as

$$\langle \Delta p_a^\mu \rangle = \frac{1}{2}\left[(p_a^\mu - p_b^\mu)\sigma_p + (p_a^\mu + p_b^\mu)\sigma_E\right] \tag{5}$$

for a nucleon a colliding with nucleon b.

$$\sigma_p = \int d\sigma_{NN \to xNN}(1 - p'_\parallel/p) \,, \tag{6}$$

$$\sigma_E = \int d\sigma_{NN \to xNN}(1 - E'/E) \tag{7}$$

are two moments of the inclusive cross section proportional to longitudinal momentum loss and energy loss of the nucleon. E and p are defined as $E = \sqrt{s/2}$, $p = \sqrt{s/4 - m_N^2}$, $s = (p_a^\mu + p_b^\mu)^2$, $m_N = 0.939\,\mathrm{GeV}$ the free nucleon mass.

To get the four-momentum loss of a fluid element, the four-momentum loss of a single nucleon has to be multiplied with the collision frequency ν, which is given by

$$\nu = \rho_p \rho_t \sigma_{NN} \sqrt{(U_{p,\mu}U_t^\mu)^2 - 1} \,. \tag{8}$$

$((U_{p,\mu}U_t^\mu)^2 - 1)^{1/2}$ is the invariant relative velocity between projectile and target. We get for the projectile fluid

$$S_p^\mu = \nu\langle \Delta p_p^\mu \rangle = -\frac{1}{2}\rho_p \rho_t \sigma_{NN} \sqrt{(U_{p,\mu}U_t^\mu)^2 - 1}\left[(U_p^\mu - U_t^\mu)\sigma_p + (U_p^\mu + U_t^\mu)\sigma_E\right], \tag{9}$$

which is the same as S_t^μ with indices t and p switched. Eq. (9) is valid only for vanishing temperature of the nuclear fluids $T_{p,t} \equiv 0$. A finite nuclear temperature can be introduced in Eq. (9) via the substitutions $U_p^\mu \to U_p^\mu \phi(T_t)$ and $U_t^\mu \to U_t^\mu \phi(T_p)$ [10], where $\phi(T) = K_1(m_N/T)/K_2(m_N/T)$ is the quotient of two MacDonald functions.

Eq. (3) leads to the fireball source

$$S_f^\mu = -(S_p^\mu + S_t^\mu) = \rho_p \rho_t \sigma_{NN} \sqrt{(U_{p,\mu}U_t^\mu)^2 - 1}(U_p^\mu + U_t^\mu)\sigma_E \,. \tag{10}$$

Note that the part of the nuclear sources proportional to σ_p does not appear in S_f^μ. This underlines the meaning of this part as the energy and momentum exchange between the nuclear fluids, while the part proportional to σ_E describes the energy and momentum loss of the nuclear fluids due to hadron production.

In the present state of the model, the fireball source term S_f contains only the production of pions

$$S_f = \bar{n}\rho_p\rho_t\sigma_{NN}\sqrt{(U_{p,\mu}U_t^\mu)^2 - 1}\,,\tag{11}$$

where $\bar{n}$ is the mean pion multiplicity in a binary nucleon-nucleon collision. The introduction of pion absorption and pion decay into Eq. (11) is in progress.

Furthermore we need equations of state for the fluids. For the fireball fluid we compare two different choices (Fig. 1).

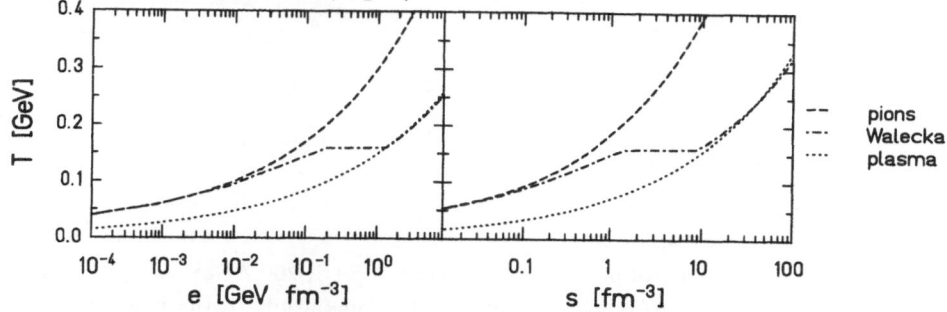

Figure 1: Temperature as a function of energy density (left) and entropy density (right). Dashed line: ideal gas of pions ($m = 140$ MeV, $n = 3$), dotted line: ideal gas of quarks and gluons ($m = 0$, $n = 47.5$), dashed-dotted line: interacting hadron gas obtained in the framework of the Walecka model (see text).

The first belongs to an ideal pion gas, while the second one in addition contains terms of interacting hadrons. The interaction between these hadrons is obtained in the framework of the Walecka model with the condition of vanishing net baryon number [11] according to the principles of the model. The resulting equation of state exhibits a first order phase transition at a transition temperature of 160 MeV [12]. The equation of state is identical with the ideal pion gas equation of state for low densities. At high densities, it approaches the equation of state of an ideal gas of massless quarks and gluons (Fig. 1).

For the baryonic fluids we use an ideal gas equation of state including a compressional part:

$$p_{tot} = p_{th} + p_c = \frac{2}{3}\epsilon + p_c\tag{12}$$

with $\epsilon = e_B - E_c\rho_B$, e_B the baryon energy density and ρ_B the baryon number density. We use the "linear" ansatz for the compression energy E_c [13]

$$E_c = \frac{1}{18}K_c\frac{(\rho_B - \rho_0)^2}{\rho_B\rho_0} + m_N\,,\tag{13}$$

where $K_c = 300$MeV. Of course it would be more consistent to use a mean-field approximation also for the baryons like for the fireball. However, due to the mean-field phase transition [12], we then would be faced with the situation of plasma colliding with plasma, respectively plasma colliding with normal matter. Up to now, no source terms are developed for this situation.

THE (2+1)-FLUID MODEL

On the way to a complete version of the 3-fluid model, we first developed a simplified version for the sake of simplicity, the so-called (2+1)-fluid model [7]. Here we assume that the nuclei retain their initial density and temperature and thus stay spherical throughout the reaction. Pions are emitted from the geometrical overlap region of the nuclei, creating the fireball, which evolves hydrodynamically. The energy stored in the fireball is subtracted from the colliding nuclei and leads to a deceleration of the nuclei.

Fig. 2 shows the time development of the pion density of a central reaction S+S, 200 AGeV (horizontal beam axis), using an ideal pion gas equation of state.

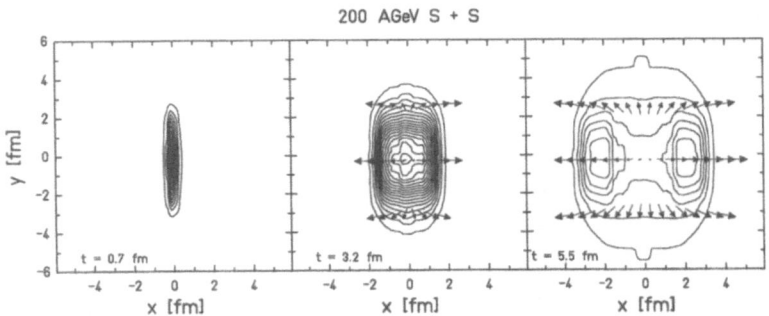

Figure 2: Time development of pion density for central S+S, 200 AGeV using an ideal pion gas equation of state. The arrows show direction and magnitude of the fluid velocity in a given spatial point.

After about 1 fm the nuclei longer have no geometrical overlap and no more pions are produced. The pion cloud is at rest in the rest frame of the nuclei. Afterwards, the cloud expands and cools down. At later stages, a minimum in the middle of the cloud appears, which is a consequence of two overlapping rarefaction waves [14].

The resulting transverse momentum spectra are not sensitive to the equation of state (Fig. 3). The results for both equations of state are within the error bars published by NA35 [15].

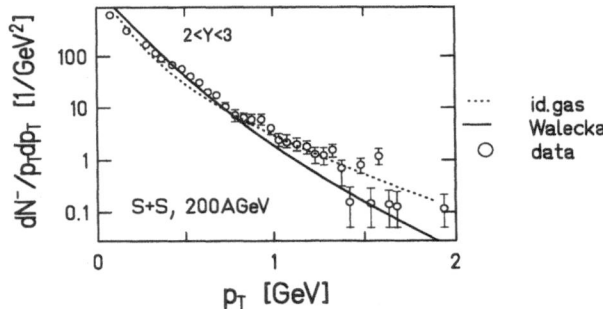

Figure 3: Transverse momentum spectra of negatively charged hadrons in central S+S, 200 AGeV. Pointed line: ideal gas of pions, solid line: interacting hadron gas obtained in the framework of the Walecka model, circles: data from [15].

The situation is different for the rapidity distribution, where a remarkable influence of the equation of state can be seen (Fig. 4). The distribution belonging to the ideal pion gas is – due to the high pressure found in the ideal gas – much broader than the distribution belonging to the interacting hadron gas. The latter one is in good agreement with data. Of course it is not possible to make any conclusions about the phase transition from this result, because one can easily construct an equation of state without phase transition but also yielding the experimental distribution.

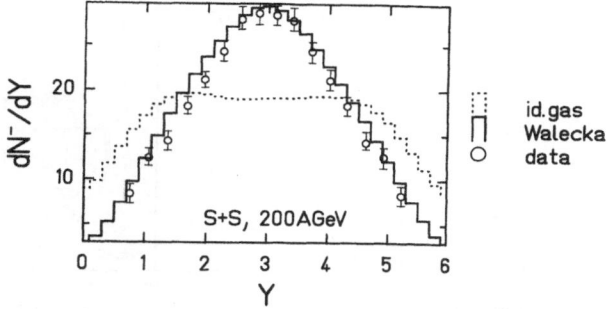

Figure 4: Same as Fig. 3, but for rapidity spectra.

The spectra are normalized by the use of Eqs. (1,11) assuming pion isospin symmetry. Since our equations of state does not depend on the pion number density, the multiplicity and thus the normalization can be achieved also via the temperature determined by Eq. (2). Comparing these methods, it might be possible to draw conclusions about the pion chemical potential [7]. Furthermore, we introduced a factor 1.1 to get the kaon contribution included in the data.

Up to this point, there is no possibility to make any conclusion about the phase transition from the results discussed. To do this, one has to measure the equation of state, e.g., the temperature T as a function of the energy density e. From thermodynamics we know that the mean transverse momentum $\langle p_T \rangle$ reflects the temperature of a fireball. The energy density should be proportional to the bombarding energy E_{LAB}. The idea to connect these variables was initiated by van Hove [16]. The (2+1)-fluid model provides us with a function $\langle p_T \rangle$ vs. E_{LAB} which really reminds of the equation of state (Fig. 5) [8].

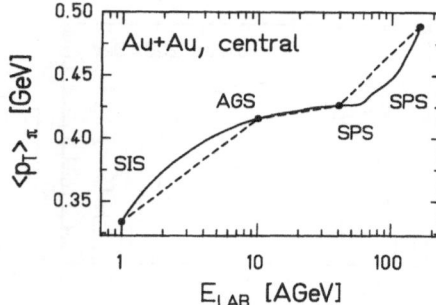

Figure 5: Mean p_T of pions as a function of laboratory energy. The stagnation might be fixed measuring the four indicated points.

$\langle p_T \rangle$ stagnates in an energy region 10AGeV$< E_{\text{LAB}} <$60AGeV, which corresponds to the stagnation of the pressure in the mixed phase of the interacting hadron gas equation of state (Fig. 1). It is rather impossible to vary the incident energy of present heavy-ion accelerators. However, the discussed $\langle p_T \rangle$ stagnation might be fixed by the measurement of four values of $\langle p_T \rangle$: at 1, 10, 40, and 160 AGeV (Fig. 5). The two lower points are under investigation at SIS and AGS, the others are planned for SPS. The effect is of the order of 50 MeV and thus might be measurable.

3-FLUID MODEL

In this chapter we present results of the 3-fluid model [9], i.e. the solution of the complete set of equations (1,2).

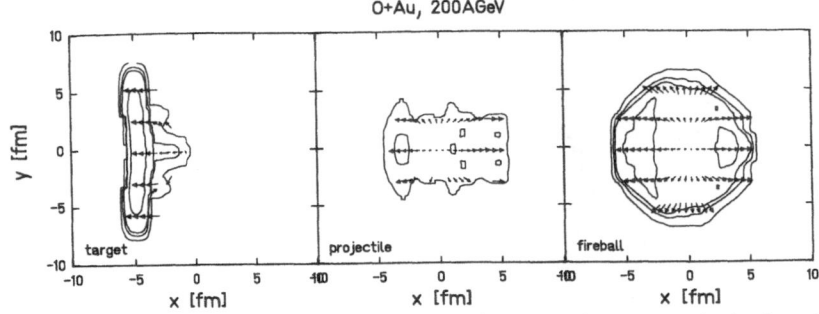

Figure 6: Contour plots of **a)** projectile density, **b)** target density, and **c)** pion density for central O+Au, 200 AGeV collision at time $t = 6$fm. The arrows show direction and magnitude of the fluid velocity in a given spatial point. The contours denote the densities $\rho/\rho_0=0.1$, 0.5, 1.0, 5.0.

Again we start with density contour plots: Fig. 6 shows the reaction O+Au, 200AGeV for $t = 6$ fm. The target is not very strongly deformed (Fig. 6a). We find only a small bump where the projectile came in and some smearing where the projectile went out. In contrast, the projectile is spread out in space (Fig. 6b). Only a small fraction of the oxygen is flying in the initial direction, the majority of the projectile nucleons is dragged by the target. The pion fluid looks very similar to that in the reaction S+S, 200AGeV (Fig. 6c). The main difference to the one-fluid model is the missing shock wave. In contrast to [17] no particular compression of the target can be recognized where the projectile passed through.

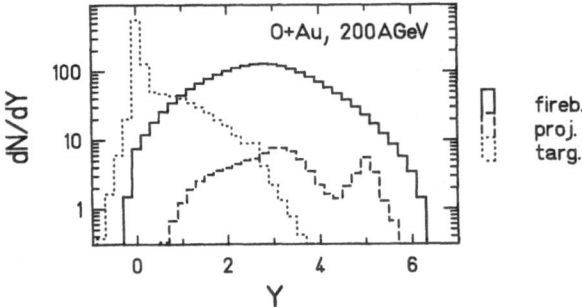

Figure 7: Rapidity distribution of projectile fluid (dashed line), target fluid (dotted line), and fireball fluid (solid line) for the reaction O+Au, 200AGeV.

Fig. 7 shows the rapidity distribution of the reaction O+Au, 200AGeV. A large spectator peak is located at $Y = 0$, the target participants are distributed up to $Y = 3$. A small peak is located at $Y = 5$ containing projectile nucleons which suffered a rapidity loss of one unit. The remaining projectile nucleons populate the whole rapidity region down to $Y = 1$. The pion cloud is centered not exactly at midrapidity $Y = 3$ but somewhat closer to target rapidity due to the asymmetric deceleration of the nuclei.

Fig. 8 demonstrates the difference between 1-fluid, (2+1)-fluid, and 3-fluid model.

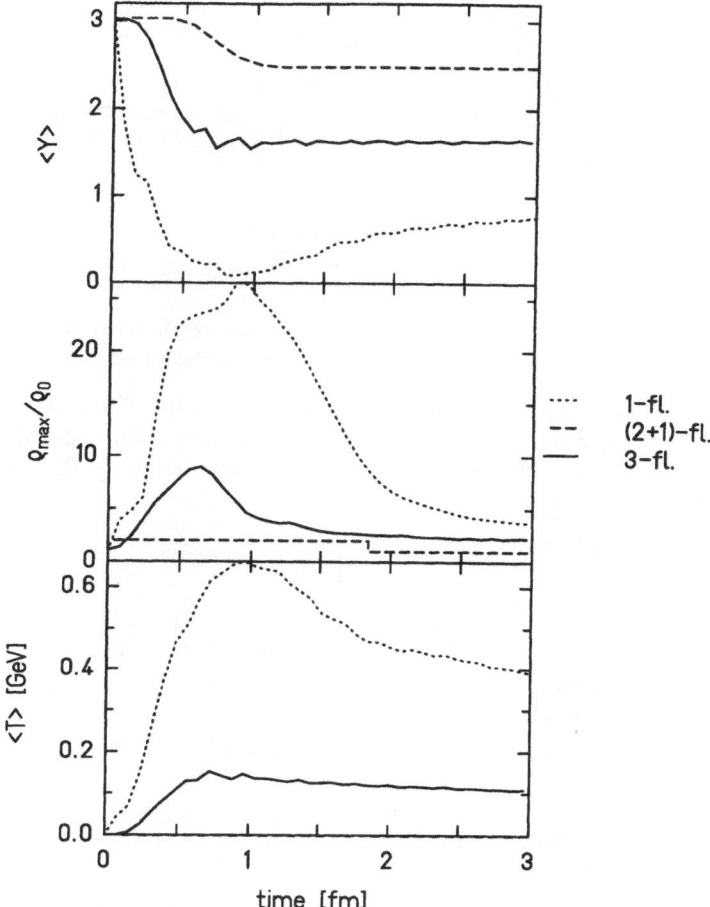

Figure 8: Time development of mean c.m.s-rapidity (above), maximal compression (central), and average temperature (below) for the reaction S+S, 200AGeV. Results of 3-fluid model (solid lines) are inbetween results of (2+1)-fluid model (dashed lines) and 1-fluid model (dotted lines).

We investigated the time dependence of the mean c.m.s-rapidity $\langle |Y| \rangle$, the maximal compression ρ/ρ_0, and the mean temperature $\langle T \rangle$ of the reaction S+S, 200AGeV. In the 1-fluid model, the matter is almost completely stopped at $t = 1$fm. Afterwards, the matter is expanding and accelerating. The final rapidity loss is about two units of rapidity. In the (2+1)-fluid model, we get the smallest rapidity shift which is possible $\Delta Y = 0.55$. The 3-fluid model yields $\Delta Y = 1.3$. Also for the density and the temperature we find realistic values only for the 3-fluid model ($\rho/\rho_0 < 8$, $T < 150$MeV). While the results of the (2+1)-fluid model have only symbolical character ($\rho/\rho_0 < 2$, $T \equiv 0$), the 1-fluid model provides us with rather gigantic values ($\rho/\rho_0 < 26$, $T < 650$MeV).

Another important difference between 1-fluid and 3-fluid model is given by the mass dependence of the reaction. In the 3-fluid model, the reaction Au+Au suffers significantly more stopping then the reaction S+S (Fig. 9). In the 1-fluid model, the two distributions would be on top of each other.

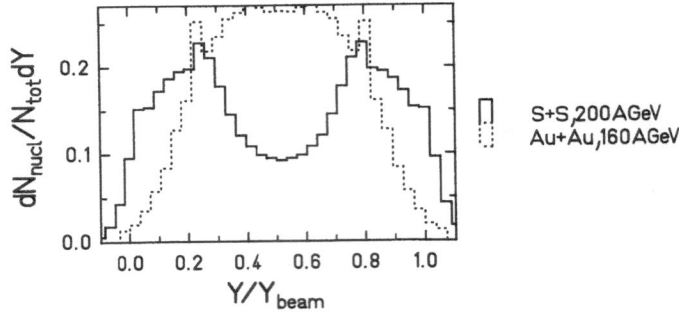

Figure 9: In contrast to 1-fluid model, 3-fluid model yields significantly more stopping power for the reaction Au+Au, 160AGeV (dotted line) than for reaction S+S, 200AGeV (solid line).

Finally we compare results of the 3-fluid model with data. The proton distributions of the reaction S+S, 200AGeV (Figs. 10, 11) demonstrate, that the stopping power obtained in the framework of the model is in the same range as the stopping power observed in the experiment [18].

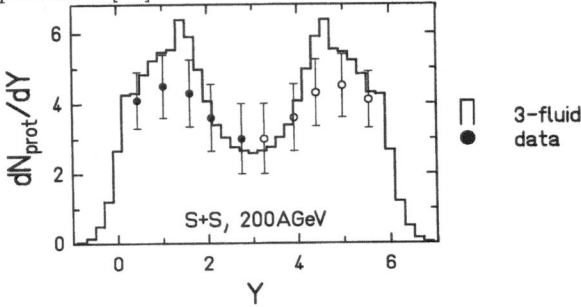

Figure 10: Rapidity distributions of 3-fluid model (lines) and data (circles) [18] for protons of the reaction S+S, 200AGeV.

The double-humped structure of the rapidity distribution is a bit too pronounced compared with data (Fig. 10). Consequently, the transverse motion is slightly underestimated, and the p_T-spectrum resulting from the model is somewhat to steep (Fig. 11). The pion distributions for the reaction S+S, 200AGeV are close to the results of the (2+1)-fluid model [9].

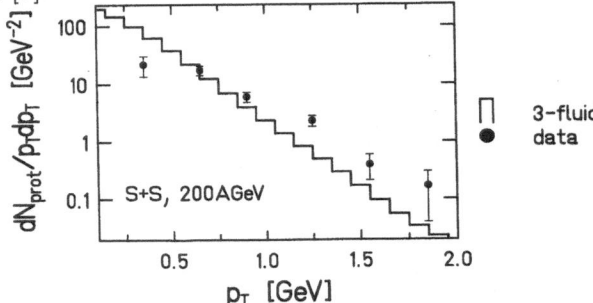

Figure 11: Transverse momentum spectra of 3-fluid model (lines) and data (circles) [18] for protons of the reaction S+S, 200AGeV.

The prediction of the nucleonic rapidity distribution for the reaction O+Au, 200AGeV, and the corresponding prediction of the RQMD-model [19] are almost identical (Fig. 12).

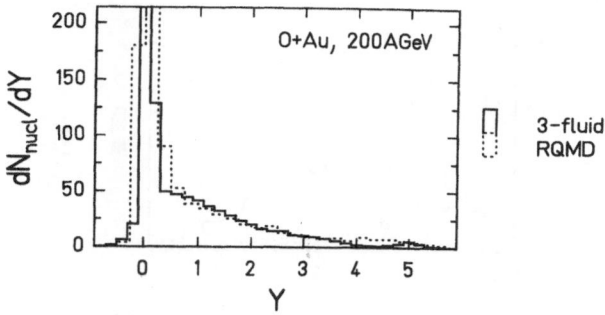

Figure 12: Rapidity distributions of 3-fluid model (solid line) and RQMD (dotted line) [19] for nucleons of the reaction O+Au, 200AGeV.

The rapidity distribution of the pions exhibits again a small lack of stopping power: the shift of the distribution towards target rapidity is not strong enough compared with data [20] (Fig. 13).

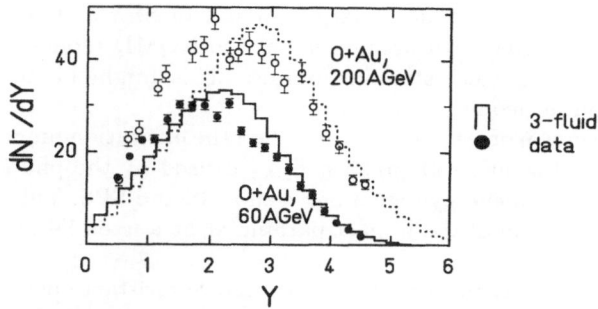

Figure 13: Rapidity distributions of 3-fluid model (lines) and data (circles) [20] for pions of the reactions O+Au, 60 and 200AGeV.

This difference is smaller for lower laboratory energies, e.g. $E_{LAB} = 60$AGeV [20] (Fig. 13). For both energies, the p_T-spectra are again very well reproduced [20] (Fig. 14).

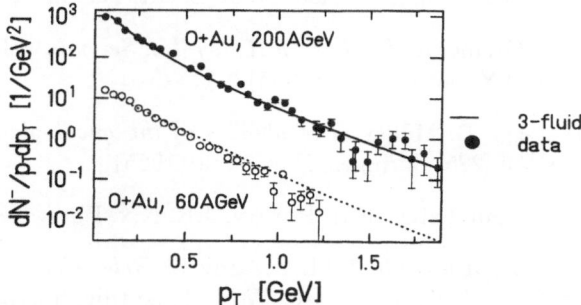

Figure 14: Transverse momentum spectra of 3-fluid model (lines) and data (circles) [20] for pions of the reactions O+Au, 60 and 200AGeV.

A comparison with preliminary data of E866 and E802 [21] proves that the 3-fluid model seems to work not only for CERN-energies but also for AGS-energies (Fig. 15).

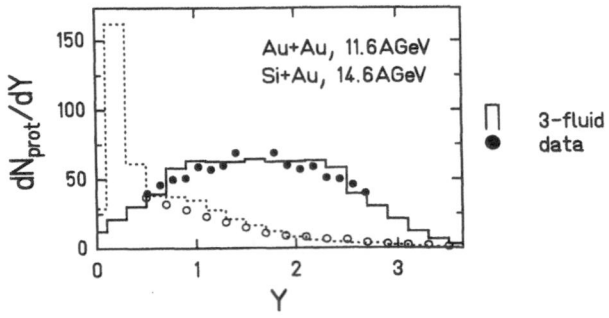

Figure 15: Rapidity distributions of 3-fluid model (lines) and data (circles) [21] for nucleons of the reactions Si+Au, 14.6AGeV and Au+Au, 11.6AGeV.

SUMMARY

We developed a relativistic 3-fluid model of two nucleonic fluids, the target and the projectile fluid, and a third fluid of secondary particles. The coupling terms between the fluids are determined as a superposition of free, inelastic nucleon-nucleon collisions.

The model is able to reproduce a large amount of SPS and AGS data. The predictions of the model agree with predictions of the RQMD model. The model underestimates partially the nuclear stopping power. This might be compensated by the introduction of pion-nucleon rescattering.

In a simplified version of the model, we studied the excitation function of the reaction Au+Au. We find a stagnation of $\langle p_T \rangle$ vs. E_{LAB} caused by the phase transition to the quark-gluon-plasma in an energy region between AGS and SPS. This stagnation cannot be obtained by a variation of the impact parameter at a fixed laboratory energy.

The authors thank J. Schaffner for his most valuable assistance in the application of the mean-field model. The authors are also grateful to V.N. Russkikh for fruitful discussions. This work was supported by the Gesellschaft für Schwerionenforschung (GSI) and the Bundesministerium für Forschung und Technik (BMFT).

References

[1] L.D. Landau, S.Z. Belen'kii, Izv. Akad. Nauk SSSR, Ser. fiz. 17:51 (1953)

[2] J. Maruhn and W. Greiner in *Treatise on Heavy–Ion Science*, vol. 4, ed. D.A. Bromley, Plenum Press, NY and London (1985)

[3] For a review see, e.g., B. Müller, *The physics of the quark-gluon plasma*, Lecture Notes in Physics Vol. 225, Springer, New York (1985)

[4] A.A. Amsden, A.S. Goldhaber, F.H. Harlow, J.R. Nix, Phys. Rev. C17:2080 (1978)

[5] I.N. Mishustin, V.N. Russkikh, L.M. Satarov in *Relativistic Heavy–Ion Physics*, vol. 6, eds. L.P. Csernai, D. Strottman, World Scientific, Singapure (1991)

[6] A. Rosenhauer, J.A. Maruhn, W. Greiner, L.P. Csernai, Z.Phys. A326 (1987) 213

[7] U. Katscher, D.H. Rischke, J.A. Maruhn, W. Greiner, I.N. Mishustin, L.M. Satarov, Z. Phys. A346:209 (1993)

[8] U. Katscher, J.A. Maruhn, W. Greiner, I.N. Mishustin, Z. Phys. A346:251 (1993)

[9] U. Katscher, J.A. Maruhn, W. Greiner, I.N. Mishustin, L.M. Satarov, UFTP preprint Frankfurt/Main (1994)

[10] I.N. Mishustin, V.N. Russkikh, L.M. Satarov, Sov. J. Nucl. Phys. 48:454 (1988)

[11] J. Schaffner, I.N. Mishustin, L.M. Satarov, H. Stöcker, W. Greiner, Z.Phys. A341:47 (1991)

[12] J. Theis, G. Buchwald, J.A. Maruhn, W. Greiner, H. Stöcker, J. Polonyi, Phys. Rev. D28:2286 (1983)

[13] W. Scheid, W. Greiner, Z.Phys. 226:364 (1969)

[14] U. Katscher, B. Waldhauser, J.A. Maruhn, H. Stöcker, W. Greiner, GSI annual report, Darmstadt (1990)

[15] J. Harris et al., Nucl. Phys. A498:133c (1989)

[16] L. van Hove, Phys. Lett. 118B:138 (1982)

[17] B. Waldhauser, D.H. Rischke, U. Katscher, J.A. Maruhn, H. Stöcker, W. Greiner, Z.Phys.C 54:459 (1992)

[18] S. Wenig, GSI-Report, GSI-90-23 (1990)

[19] H. Sorge, A.v.Keitz, R. Mattiello, H. Stöcker, W. Greiner, Z.Phys.C 47:629 (1990)

[20] H. Ströbele et al., Z.Phys.C 38:89 (1988)

[21] F. Videbaek, Proc. of *Heavy-Ion Physics at the AGS (HIPAGS)*, Massachusetts (1993)

THERMODYNAMICAL QUANTITIES
MEASURED IN HEAVY-ION REACTIONS

Martin L. Purschke, GSI

WA80 Collaboration
CERN/PPE, CH-1211 Geneva 23

INTRODUCTION

The search for the phase transition from normal hadronic matter to the quark-gluon-plasma (QGP) can be divided into two different approaches: One is the search for "direct" signals, quantities that are present or absent or change in the presence of a QGP; the other is the measurement of thermodynamical properties of the reaction system that change under the influence of the phase transition. To the first group belong, for example, the J/Ψ suppression[1] and the search for an enhancement of direct photons[2]. The second approach looks for the behaviour of the "temperature" of the system as a function of the system's energy or entropy density.

In this article we will briefly review some theoretical predictions and present data from the WA80 experiment[3] on neutral pions and inclusive photons.

THERMODYNAMICAL PROPERTIES
OF A HEAVY-ION REACTION

The expected behaviour of a heavy-ion collision has some analogy with the behaviour of boiling water. Initially, the temperature rises with increasing energy density until it reaches the 100° C phase transition point. From here, the temperature stays constant since the energy is spent on populating the new degrees of freedom until all the water is vaporized and only then the temperature rises again. A heavy-ion collision is a very dynamic process, and we are looking at the influence of a QGP phase transition on the hadronization phase of particles which have been produced from the available collision energy. For a thermodynamical description to make sense at all, the system must be large enough to be able to be "thermal", and enough collisions between particles must take place to achieve the thermalization. Whether or not these requirements are met is still a subject of active discussion, but the assumption of at least a local if not a global thermalization is supported by the fact that the number of collisions per particle required to achieve a thermalization is model dependent but is invariably small, around 3 to 4. Furthermore, the m_T distributions ($m_T = \sqrt{p_T{}^2 + m^2}$) for hadrons are thermal, $m_T \sim \exp(-m_T/T)$, and m_T-scaling is observed. This means that for distributions of different hadrons the yield is different but not the slope of the spectrum; the slope parameter T is the same and the ratio of the distributions is constant. Though m_T-scaling was already observed in proton-proton collisions and may have a different cause, it can most easily be explained by a thermalized system.

Even if thermalization is achieved, the measurement of the system's temperature is not an easy task. The mean transverse momentum is often taken as a measure for the temperature which is in the order of 160-230 MeV/c, but p_T values below a few hundred MeV/c can usually be determined with large uncertainties only. The same is true for the entropy density S. Here one takes into account that entropy in ultrarelativistic heavy-ion collisions is produced mainly by particle production[4] (rather than by breakup of the nucleus as at BEVALAC energies) and uses the approximation

$$S \sim dN/d\eta \cdot A_{\text{inc}}^{-2/3}, \tag{1}$$

where $dN/d\eta$ is the multiplicity density and A_{inc} is defined as the number of projectile participants.

Early QCD lattice calculations by Celik at al.[5] showed a very significant behaviour of the energy density as a function of the temperature (Figure 1). Blaizot et al.[6] calculate the behaviour of the temperature as a function of the entropy density, taking into account a hydrodynamical expansion of the system. The presence of a QGP phase transition manifests itself in an initial rise followed by a less steep region in the case of a QGP phase transition (Figure 2).

WA80 RESULTS

The WA80 fixed-target experiment uses the heavy-ion beam at the CERN-SPS. Data were taken with a 200 GeV/nucleon ^{16}O (3200 GeV energy) and ^{32}S beam (6400 GeV energy), and various targets (^{12}C to ^{197}Au). One of the goals of the two experiments is the spectroscopy of thermal photons[2] and neutral pions[7]. Photons are measured with the lead glass calorimeter SAPHIR[8]. This detector is located at a distance of 342 cm from the target and covers about 1/6th of the azimuthal angle in the pseudorapidity range $1.5 \leq \eta \leq 2.1$. It consists of 1278 SF5 lead glass modules with 18 radiation lengths and a cross section of 35×35 mm^2. The lead glass calorimeter was later more than tripled in size and moved back to 9 m to better cope with the high particle densities.

A photon hitting the detector induces an electromagnetic shower that spreads over several lead glass modules. The sum of the signals gives the energy of the particle while the lateral profile of the shower allows to distinguish electromagnetic

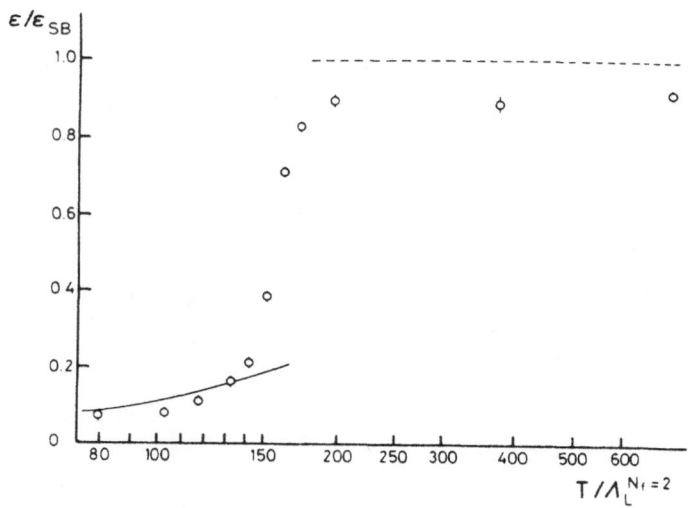

Figure 1. The total energy density ϵ, normalized to the ideal gas limit ϵ_{SB}, calculated on a $8^3 \times 3$ lattice (from Celik et al.[5]).

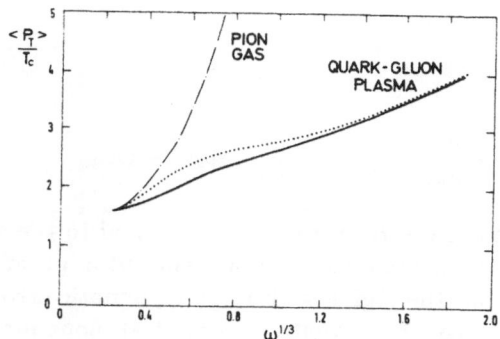

Figure 2. The mean transverse momentum as a function of the dimensionless parameter ω which is proportional to the entropy density (from Blaizot et al.[6]).

from hadronic showers. With decreasing energy this distinction becomes less certain; data points below an energy of about 1500 MeV(which translates to a p_T value around 400 MeV/c) are usually neglected due to the systematic errors.

The spectra are often parameterized with an exponential distribution $dN/dp_T \sim \exp(-p_T/T)$. The slope parameter T is used to describe differences between the distributions from different systems. However, at high p_T values above 2 GeV/c the deviations from a pure exponential distribution become visible (for photons, see Figure 4). In order not to rely on a particular shape of the spectrum, we use the mean value of the truncated p_T distribution,

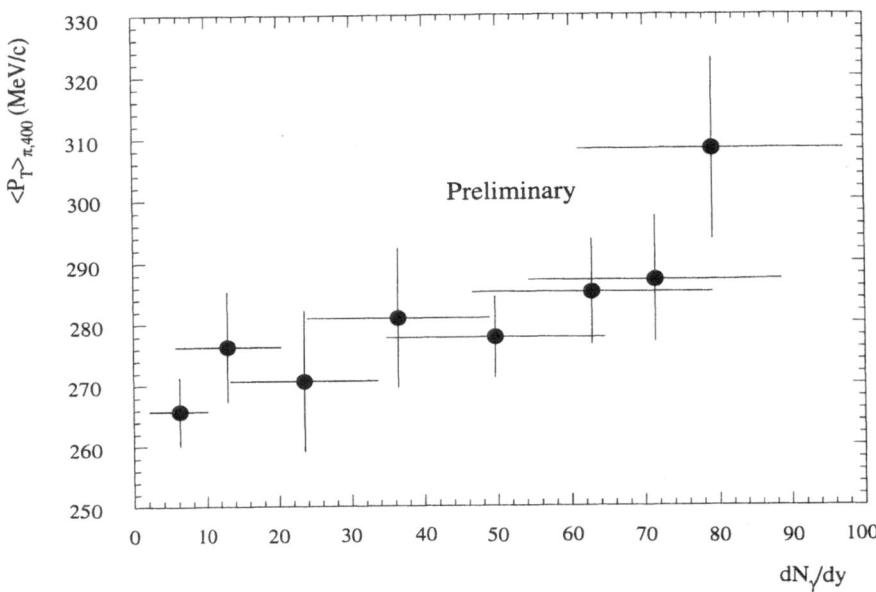

Figure 3. $<p_T>_{\pi^0,400}$ values as a function of dN_γ/dy (which is correlated with the centrality) for neutral pions.

$$<p_T>_C = (\int_C^\infty p_T \frac{dN}{dp_T} \, dp_T \;/\; \int_C^\infty \frac{dN}{dp_T} \, dp_T \;) - C. \tag{2}$$

This value, referred to as the *truncated mean*, is equal to the slope parameter T in the case of an exact exponential distribution. The cutoff parameter C is usually set to $400\,\mathrm{MeV/c}$ to cut out the region with large systematic errors.

Due to the complexity of the analysis of neutral pions measured with the lead glass calorimeter, there is only a preliminary figure available that shows the behaviour of the truncated mean for neutral pions (Figure 3). These data are compatible with an initial steep rise, a plateau, and again a rise of the $<p_T>_{\pi^0,400}$ values with increasing centrality, but the analysis of the data is still at an early stage.

We can reconstruct only those π^0 where both decay photons are actually measured in the detector. Therefore, the inclusive photon statistics exceeds the pion statistics by more than an order of magnitude, allowing for more restrictive cuts. Monte-Carlo simulations show that the properties of a π^0-p_T distribution are to a large extent retained in the distribution of the decay photons. In particular, there is a strong correlation between the slope parameters T of the π^0 and the photon distribution. Also, the truncated mean values from pions and photons are strongly correlated. So we can use the values obtained from photon distributions instead and use the much larger statistics.

If thermalization is achieved, and if the axis values are indeed correlated to the system temperature and the entropy density respectively, there should be an initial steep rise followed by a plateau or at least a less steep region. Furthermore, the

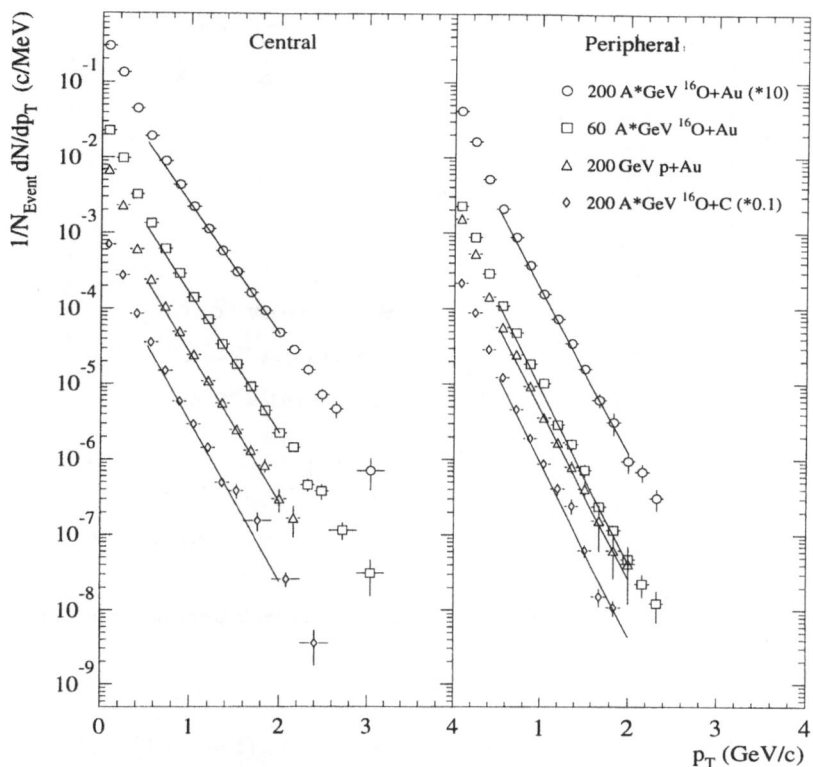

Figure 4. Inclusive photon p_T distributions from central and peripheral events. The lines show the results from a $dN/dp_T \sim \exp(-p_T/T)$ parameterization in the p_T range $0.5\,\mathrm{GeV/c} \leq p_T \leq 2\,\mathrm{GeV/c}$.

data points from different reaction systems should follow the same curve.

Figure 5 shows $<p_T>_{\gamma,400}$ values as a function of the entropy density $S \sim dN/d\eta \cdot A_{\mathrm{inc}}^{-2/3}$. The multiplicity density of charged particles $dN/d\eta$ is measured in the pseudorapidity range $1.2 \leq \eta \leq 4.2$ with the streamer tube detectors, and A_{inc} is defined as the number of projectile participants, which is derived from the forward energy measured in the zero-degree calorimeter; the participant number for a given forward energy is taken from the FRITIOF[9] model. The data do not show the expected behaviour, and there are deviations between data points from different reaction systems.

However, it is possible that kinematic effects like shifts of the center-of-mass rapidity (midrapidity) of the reaction system could have an effect on the values. Since most reaction systems are asymmetric, a different centrality also means a different c.m. rapidity, which varies for the 200 A·GeV ^{16}O + Au system from $\eta = 3$ for the proton-proton-like peripheral reactions to $\eta = 2.5$ for the most central

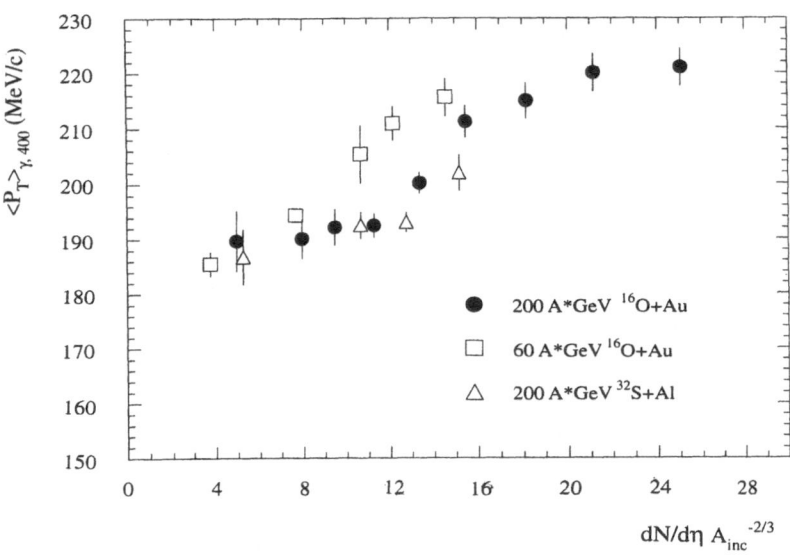

Figure 5. $<p_T>_{\gamma,400}$ values as a function of the entropy density S from the p_T distributions.

collisions with 16 projectile and 50 target participants (Figure 6). This can also be seen in the peak positions of the $dN/d\eta$ distributions from charged particles measured in the pseudorapidity range $1.2 \leq \eta \leq 4.2$[10].

Indeed we observe a dependence of the $<p_T>_{\gamma,400}$ values on the pseudorapidity. For this analysis the SAPHIR detector was divided into 9 pseudorapidity regions, and the $<p_T>_{\gamma,400}$ values determined for each region independently. Figure 7 shows the results. A clear rise is observed for large values of η (small laboratory angle ϑ). Since midrapidity for the 200 A·GeV ^{16}O + Au system is $\eta = 3$, this indicates that the $<p_T>_{\gamma,400}$ values increase for regions closer to midrapidity. The SAPHIR detector ($1.5 \leq \eta \leq 2.1$) measures always backwards from the effective center-of-mass rapidity, but further backwards for peripheral than for central collisions. This means that the detector covers different rapidity regions for different centralities, and this difference in the rapidity region may account for part of the observed differences in the $<p_T>_{\gamma,400}$ values.

This effect can be compensated by selecting the proper η range with respect to the midrapidity value of the particular collision system and the centrality. This is calculated with the projectile and target participant numbers taken from FRITIOF for a given forward energy. For 200 A·GeV ^{16}O + Au, where the midrapidity values vary from 3 to 2.5, and 200 A·GeV ^{32}S + Al which is almost symmetric and has a midrapidity around 3 for all centralities, it is possible to select the η regions always one unit backwards from midrapidity for all centralities and in this way to eliminate the contribution of the c.m. shift.

Figure 8 shows the $<p_T>_{\gamma,400}$ values as a function of S, where the data for each

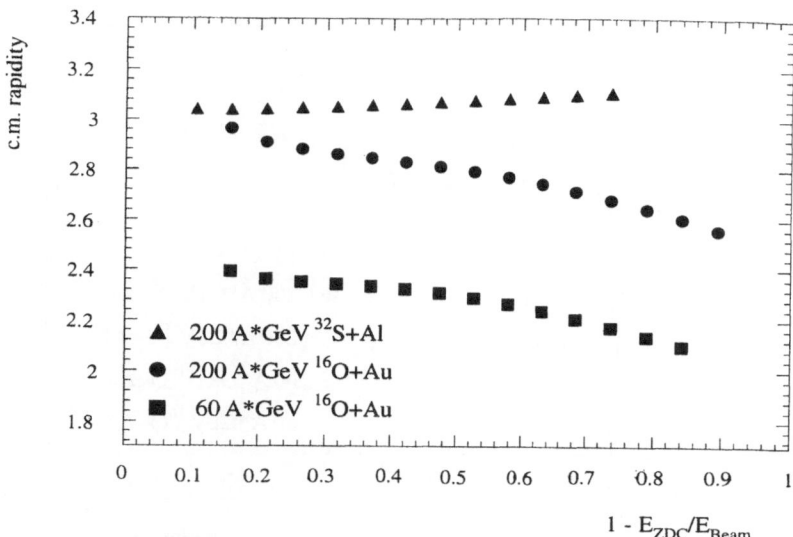

Figure 6. Center-of-mass rapidities for different reaction systems as a function of the forward energy E_{ZDC}. For the asymmetric O + Au systems the c.m. rapidity varies by more than one unit.

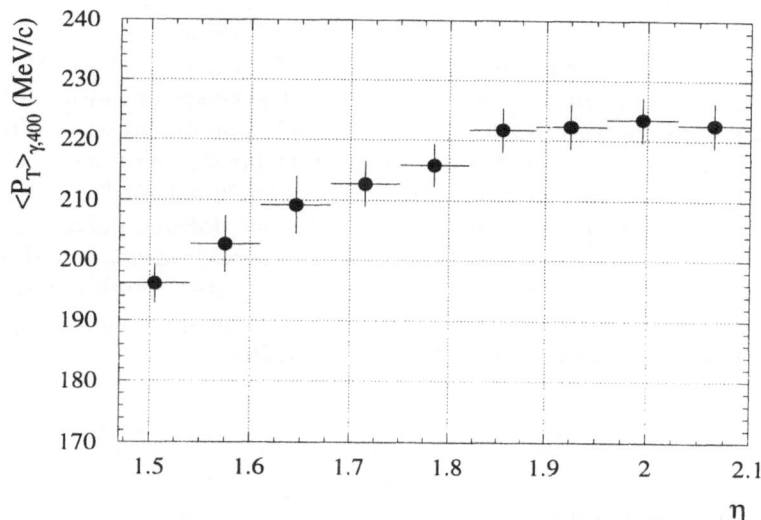

Figure 7. $<p_T>_{\gamma,400}$ values from minimum bias 200 A·GeV ^{16}O + Au data as a function of the pseudorapidity region.

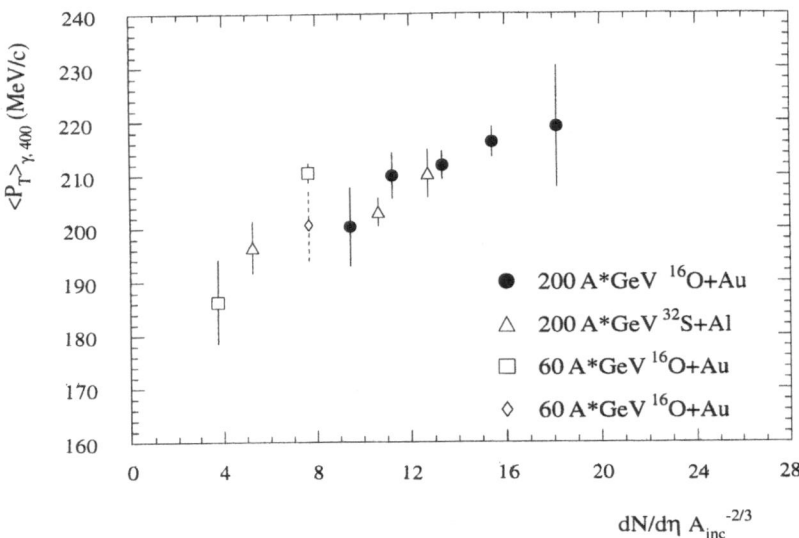

Figure 8. $<p_T>_{\gamma,400}$ values as a function of the entropy density S. To compensate for the midrapidity shift of the asymmetric systems, the detector η region has been selected accordingly. For the 60 A·GeV ^{16}O + Au system the region is only 0.5 units backwards due to the different rapidity range, and for one point (the diamond) 0.7 units backwards.

point are selected from the proper rapidity region for the 200 A·GeV ^{16}O + Au and 200 A·GeV ^{32}S + Al system. For the 60 A·GeV ^{16}O + Au system with a midrapidity range from 2.4 (peripheral) to 2.0 (central) it is only possible to select η regions 0.5 units backwards from midrapidity, placing the detector closer to midrapidity where higher $<p_T>_{\gamma,400}$ values are observed. Indeed the two 60 A·GeV ^{16}O + Au points in Figure 8 are higher than the other systems. However, for one point (the diamond in Figure 8) it is possible select a region 0.7 units backwards, and this point is lower and closer to the overall trend. Though the statistics are not sufficient to give results at the highest values of S, the two systems again show a very similar behaviour, and a steep rise at low values of S is visible.

Conclusion

The WA80 experiment is able to measure the behaviour of the truncated mean p_T values as a function of the entropy density. Taking into account the influence of the shift of the center-of-mass rapidity with the reaction centrality, the data points of different reaction systems follow the same curve. Due to the severe cuts that limit the statistics, this analysis does not yet yield results for the highest centrality values where the plateau is expected. However, the new data taken with the WA93 experiment[11] – which is an upgrade of the WA80 experiment – will enhance the photon statistics significantly, may extend the range of entropy density values, and will also add other reaction systems.

REFERENCES

1. T. Matsui and H. Satz, Phys. Lett. **B178** (1986) 416.

2. R. Albrecht et al., Z. Phys. C – Particles and Fields **51** (1991) 1–10.

3. R. Albrecht et al., Study of relativistic nucleus-nucleus collisions at the CERN SPS, Aug. 1985, preprint CERN/SPSC/85-39 and GSI preprint 85-32.

4. L. van Hove, Nucl. Phys. **A 477** (1985) 443c–454c.

5. T. Celik, J. Engels, and H. Satz, Nucl. Phys. **B 256** (1985) 670–686.

6. J. P. Blaizot and J. Y. Ollitraut, Phys. Lett. **B 191** (1987) 27.

7. R. Albrecht et al., Z. Phys. C – Particles and Fields **47** (1990) 367–376.

8. H. Baumeister et al., Nucl. Instr. and Meth. **A292** (1990) 81–96.

9. B. Nilsson and E. Stenlund, Comp. Phys. Com. **43** (1987) 387 – 397.

10. R. Albrecht et al., Phys. Lett. **B 202** (1988) 596–602.

11. Proposal for a light universal detector for the study of correlations between photons and charged particles, May 1990, CERN/SPSC 90-32.

VECTOR MESON PRODUCTION AT CERN

C. Gerschel

Institut de Physique Nucléaire
F-91406 Orsay Cedex

INTRODUCTION

Since the first oxygen beams at 200 GeV/nucleon which were operated at CERN in 1986, an extensive experimental effort has been devoted to the study of the different proposed signatures of the formation of a quark-gluon plasma (QGP)[1]. At the same time, the obtained results have triggered a considerable amount of theoretical work leading to somewhat controversial interpretations.

Vector meson study gives the opportunity of having insight into two extensively debated signals : charmonium state and strangeness production. In the spirit of this school, I will remind the underlying basic physics, summarize the experimental results and try to discuss the interpretations available on the market at the present time.

CHARMONIUM STATE PRODUCTION

Following the original idea of Matsui and Satz[2], charmonium state production should be suppressed in a QGP because of strong colour screening. This will appear when, at the characteristic formation time τ_0 of the J/ψ, the initial $c\bar{c}$ pair, from which J/ψ is issued, is still in the plasma in space or time. The suppression will thus be stronger at low transverse momentum p_T[3]. It will also be stronger for ψ' than for J/ψ[4]. All this has been widely discussed in the previous years as described in the quoted references.

J/ψ Production

J/ψ production has been studied by the NA38 collaboration . It has been compared to that of dimuons in the mass continuum with an invariant mass $M_{\mu\mu}$ between 1.7 and 2.7 GeV/c^2. The main results obtained are summarized thereafter :
• The ratio of the cross sections $B\sigma_{J/\psi} / \sigma_{cont}$ for J/ψ and continuum formation decreases when increasing the neutral transverse energy E_{0T} of the collision[5]. B is the branching ratio for the decay of a J/ψ into a muon pair. Fig. 1 gives the more recent results of the NA38 experiment in the case of SU collisions[6].
• It is the energy density which is the relevant parameter to relate the ratio $B\sigma_{J/\psi}/\sigma_{cont}$ for

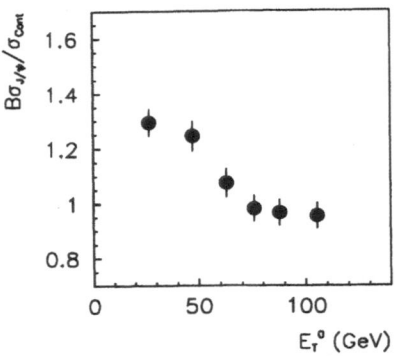

Fig.1 : Ratio of the production cross-sections for J/ψ and continuum dimuons for SU as a function of the neutral transverse energies E_0T.

different target-projectile combinations (fig.2). The energy density ε is defined according to the Björken estimate[7] $\varepsilon = 3 \, E_{0T} / A_\perp \, \tau_0 \, \Delta y$ where $\Delta y = 2.4$ is the rapidity range in which the π^0

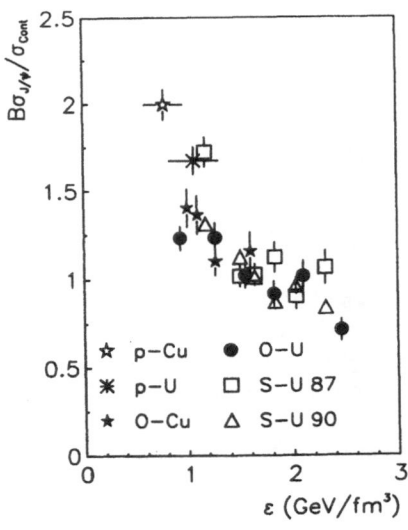

Fig.2 : Ratio of the production cross-sections for J/ψ and continuum dimuons as a function of the energy density.

are measured, $A_\perp$ is the transverse overlap area of the collision at the corresponding E_{0T} value and at the formation time τ_0 which is taken equal to 1 fm/c. $A_\perp$ is deduced from the impact parameter obtained from a modelisation of the E_{0T} distribution[8]. In the case of pA collisions where no selection on the impact parameter is performed, $A_\perp$ is taken equal to $\pi\, r_o^2$ with $r_o = 1.1 \pm 0.1$ fm. The use of the Björken estimate is somewhat incorrect at CERN energies where the transparency regime is not yet reached. However, it provides a reasonable relative scale and, in that sense, is also applied to such a small projectile as the proton.

• The decrease of $B\sigma_{J/\psi}/\sigma_{cont}$ is due to a J/ψ suppression and not a continuum enhancement. The cross-section of the continuum has been measured for different combinations of target A and projectile B. It is found to depend linearly on their atomic numbers : $\sigma_{cont} = \sigma_o\, (AB)^{\alpha_{cont}}$ with $\alpha_{cont} = 1.01 \pm 0.04$ [9]. And it is indeed the J/ψ production which is suppressed. The corresponding α value for J/ψ is $\alpha_\psi = 0.91 \pm 0.04$ [9].

• The J/ψ suppression is stronger at low p_T [10] and more so when the transverse energy increases. The latest results of NA38 [6] are displayed on fig.3a.

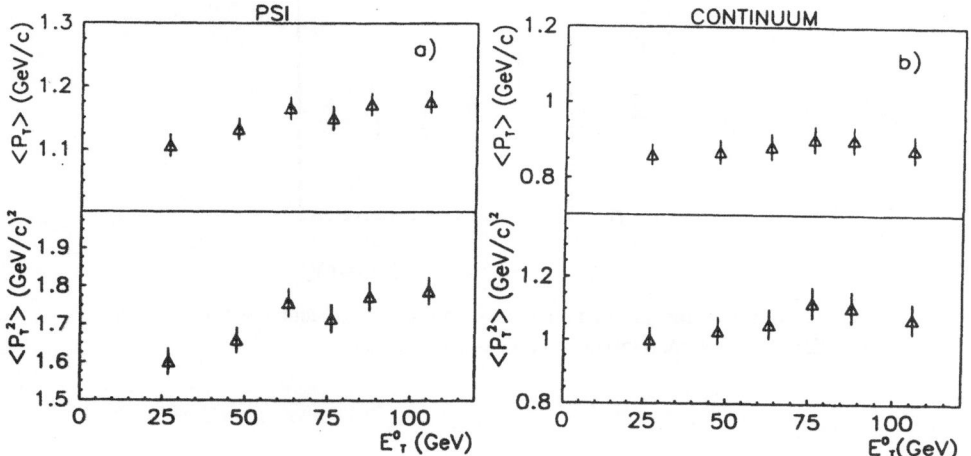

Fig.3 : $< p_T >$ and $< p_T^2 >$ as a function of the neutral transverse energy a) for J/ψ b) for continuum dimuons.

• The effects of nuclear matter on the p_T distribution are stronger for J/ψ than for continuum (fig.3b). One possibility of characterizing the effect of nuclear matter is to define the shift Δp_T^2 of the mean value of p_T^2 for a AB collision at a given E_{0T} as compared to that of a pp collision : $\Delta p_T^2\, (E_{0T}) = < p_T^2 > (E_{0T}) - < p_T^2 >_{pp}$
It is found from fig.2 that $\Delta p_T^2\, (J/\psi) = (2.5 \pm 0.5)\, \Delta p_T^2\, (cont)$ The values of $< p_T^2 >_{pp}$ for J/ψ and for continuum are taken from NA38 [11].

ψ' Production

New results concerning ψ' production have been presented at QM 93 by the NA38 collaboration[12]. They show a stronger suppression for ψ' than for J/ψ. They can be summarized as follows :

• The ratio of the production cross-sections decreases when going from pW and pU to SU collisions (table I).

• This ratio decreases also with the neutral transverse energy of the collision E_{0T} (fig.4) in the case of the SU system.

Table I : Ratio of the production cross-sections for different systems. Only statistical errors are shown.

System	$B\sigma_{\psi'}/B\sigma_{J/\psi}$ (%)
p-W	1.77 ± 0.22
p-U	2.01 ± 0.44
S-U	1.07 ± 0.09

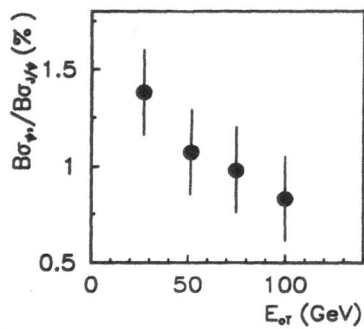

Fig.4 : Ratio of the production cross-sections for ψ' and J/ψ for SU as a function of the neutral transverse energy of the collision.

Interpretation

All these results are expected in the case of a QGP formation. However alternative explanations have been proposed which take care of the complicated environment of a nuclear collision. These effects are present in a pA as compared to pp collisions. They have been extrapolated to BA collisions and should be taken into account before to extract a genuine signal characteristic of a QGP formation in AB collisions.

Informations which are available for J/ψ production deal with the integrated cross-sections for pA, πA $\bar{p}$A and γA collisions, the p_T and x_F - dependences [13,14,15]. The x_F - dependence will not be discussed here and the reader is refered to the paper by E. Quack in these proceedings.

The p_T - dependence in the case of particle-A collisions is usually interpreted in terms of multiple scattering in the initial state of the gluon of the projectile on the nucleons of the target[16,17]. This process has been extrapolated to BA collisions and there is at the present time a general consensus that the p_T - dependence of J/ψ production observed by the NA38 experiment is well reproduced by initial state interactions[11]. It has been stressed[17] that the effect of nuclear matter should be 9/4 times stronger for J/ψ (initial scattering of gluons) than for dimuons in the mass continuum (initial scattering of quarks) and indeed the value of 2.5 ± 0.5 obtained from the new data of NA38 strengthens this hypothesis. Using this interpretation, at a given E_{oT}, the shift Δp_T^2 of $< p_T^2 >$ can provide an "experimental" value of the mean length L_i of nuclear matter available in the initial state at this E_{oT}[15]. This length has to be compared to the geometrical dimension of the nuclei, and any value of L_i larger than the radius of the nuclei for central collisions will sign the appearence of new physics depending on p_T. This has to be checked for the forthcoming Pb-Pb experiments.

The understanding of the cross-sections is much more debated and controversial. In terms of the conventional parametrization $\sigma_{(part\ A)} = \sigma_{(part\ N)}\ A^\alpha$, α is found to be of the order of 0.91 but with large fluctuations from one experiment to another one. It has been shown[15] that, in fact, the length L_f of nuclear matter in the final state is a more adequate parameter to describe the cross-sections and using L_f, no discontinuity is observed between part-A, whatever the nature of the particle, and BA collisions (fig.5). Such a behavior can be

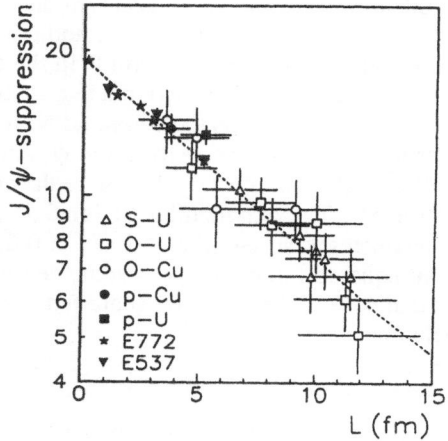

Fig.5 : Suppression of J/ψ as function of the length of nuclear matter in the final state. For BA collisions, it is B $\sigma_{J/\psi}/\sigma_{cont}$ which is plotted. An exponential fit to the data is shown. For the experiments E772 and E537, it is the cross-sections which are plotted. Their values are arbitrarily normalized so that the heavier target agrees with the exponential fit.

interpreted[15] in terms of nuclear absorption leading to the dissociation of J/ψ into a pair of D mesons. Absorption cross-sections are found similar for part-A (6.3 ± 0.3 mb) and BA collisions (6.9± 1.0 mb). However, it is not clear which object interacts with the nucleus. It has been objected[18] that, taking into account the Lorentz boost of J/ψ as compared to nuclear matter, the final resonance is formed far away from the nucleus in the case of particle-A and it should be only a point-like object which crosses the nucleus. Other effects are thus proposed to account for the cross-sections. Final state interactions of J/ψ with comovers[18,19] has the same L_f dependence than nuclear absorption. They depend, of course, on the life-time of the hadron gas. And, as pointed out by Lesniak[20], they are in contradiction with coherent J/ψ production experiments[21]. In such a process, no comovers are produced, and indeed an absorption cross-section of ~ 6 mb is requested to explain the data. This stresses the leading influence of nuclear matter.

A reduction of the production cross-section could also be observed if some shadowing of the gluon densities in the nuclei is present[22]. This initial state effect should exist as observed for quark densities[23] but there is no clear indication about its size. Its proposed parametrization[22] is in contradiction with the observed L_f scaling.

The interpretation of J/ψ suppression is thus rather tricky. However, some recent developments seem promising, thanks to ψ' data. Contrary to J/ψ, ψ' seems to exhibit a different behavior between pA and BA collisions. The ratio R = B$\sigma_{\psi'}$/B$\sigma_{J/\psi}$ looks constant for pA collisions[12]. The Fermilab experiment E772 [13] has also found that $\alpha_{\psi'} \sim \alpha_\psi$ at $\langle x_f \rangle = 0.3$. This similar behavior of J/ψ and ψ' for pA collisions can be explained if it is the same object which crosses the nucleus in both cases.

J/ψ and ψ' productions are considered as a three step process : i) production of a $c\bar{c}$ point-like colour octet state by gluon fusion ii) emission or absorption of a third gluon leading to colour neutrality iii) evolution towards the fully grown resonance state. It has been proposed that the A-dependence of the cross-sections[25] and the x_f-behavior[26] are due to the interaction of the colour octet state with the target nucleus. Of course, the final effect will depend strongly on the time of emission of the third gluon and it is not yet clear whether this can account for the L_f scaling for J/ψ observed at the different center of mass energies. More work needs to be done but this offers an interesting issue. In the case of BA collisions, J/ψ formation times are shorter due to smaller Lorentz γ factor and it is the final size resonances which interact with the matter. The decrease of the ratio R in the case of SU could then be simply explained by the interaction of J/ψ and ψ' with nuclear matter. On fig.6, R is plotted as a function of L_f. The values for pA collisions are corrected both for the $\sqrt{s}$ dependence of the cross-sections[26] and for the feeding of the ψ from the ψ' and the χ, using the values measured for pLi at 300 GeV/c[27]. One observes on the figure the constant value of R for pA collisions ant its decrease for SU. The exponential fit for S-U data gives a ratio of the absorption cross sections between the ψ' and the ψ of 2.1 ± 0.6. If the absorption cross-sections scale like the mean squared radius of the resonances[28], a ratio 2.3 is expected, taking into account the fact that 40% of the observed ψ are fed from direct χ or ψ' [27]. More precise results would be welcomed.

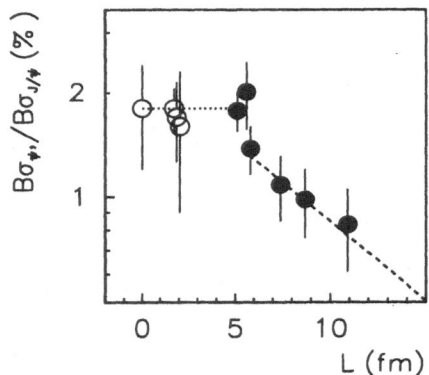

Fig.6 : Ratio of the production cross-sections for ψ' and J/ψ as a function of the length of nuclear matter in the final state. The black dotes correspond to NA38 results, the open dots to other pA experiments[12].

Conclusion

No definite explanation can be given at the present time about ψ and ψ' suppression in heavy ion collision induced by O and S projectiles. J/ψ and ψ' behave differently in comparison with pA collisions. If not unambiguously characteristic of the existence of a QGP, this fact should improve our knowledge of time scales involved in the process. A good reference is now available for any new physics appearing in the forthcoming Pb-Pb experiments.

PRODUCTION OF LOW MASS VECTOR MESONS

The ϕ meson, composed of a strange and anti-strange quark, is the lightest vector meson with a hidden flavor. Its behavior can be compared to that of the ρ and ω which are resonances of the light quark-antiquark systems and thus give access to strangeness content of heavy-ion collisions. It had indeed been proposed[29] that the creation of a QGP would result in an enhanced production of strange quarks. The ϕ could be an excellent probe of QGP[30] as its production and decay in pp collisions is suppressed by the Okubo-Zweig-Iizuka (OZI) rule[31], which requires that transitions take place by means of connected quark diagrams. In case of QGP formation, the lack of the OZI suppression in addition to the abundance of strange quarks would lead to an enhanced production of ϕ meson.

Experimental results

Low mass vector mesons have been studied both by the NA38 and Helios/ 3 collaborations, through their decay channels into two muons. Due to the small mass difference between the ρ and the ω (13 MeV) and to the finite mass resolution of the spectrometers, ρ and ω contributions are not experimentally discriminated and the ϕ production is compared to the sum $\rho + \omega$.

The main results are the following :
• NA38 has shown that the ratio of the cross-sections $B_\phi \sigma_\phi/(B_\rho\sigma_\rho + B_\omega\sigma_\omega)$ increases from pW to SU, and with the neutral transverse energy E_{oT} in the case of SU[32] (fig.7). Similar results have been obtained by Helios/3 for pW and SW[33]. An analogous increase was also observed previously at high p_T by the NA38 collaboration in the case of O-U[34]. The overall increase is of about a factor two.

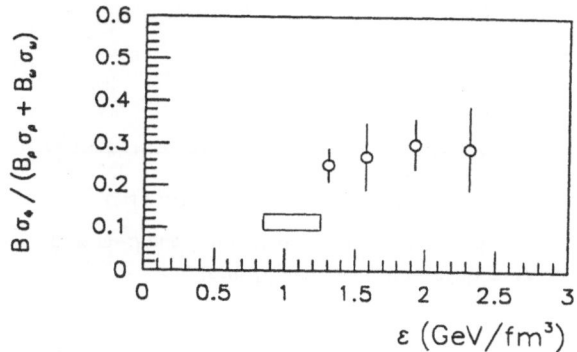

Fig.7 : Ratio of the production cross-sections for ϕ and $\rho + \omega$ from NA38 as a function of the energy density. Open dots : SU. Rectangle : pW.

Table II : Ratio of the production cross-sections for ϕ and $\rho+\omega$ from Helios/3. Comparison between different p_T-bins.

$\phi /(\rho + \omega)$	p + W	S + W
$p_T \le 0.35 / p_T > 0.6$	0.83 ± 0.10	0.95 ± 0.17
$0.35 \le p_T < 0.6 / p_T > 0.6$	0.98 ± 0.11	0.70 ± 0.11

• The increase of this ratio does not seem to depend on p_T [32,33] as observed by NA38 (fig.8) and by Helios/3 (table II).
• As shown by NA38, the increase of this ratio seems to be due more to a ϕ enhancement than to a $\rho + \omega$ suppression[32]. From the A-dependence of the cross-sections between pW and SU, it is possible to deduce the α parameter. It is found that, for $p_T > 0.6$ GeV/c, $\alpha_\phi = 1.24 \pm 0.03 \pm 0.08$ while $\alpha_{\rho+\omega} = 1.01 \pm 0.02 \pm 0.13$.
• There is a discontinuity between pA and BA collisions. From NA38 cross-sections, we get $\alpha_\phi - \alpha_{\rho+\omega} = 0.23 \pm 0.04 \pm 0.09$[32]. This has to be compared to the value obtained for pA collisions $\alpha_\phi - \alpha_{\rho+\omega} = 0.05 \pm 0.05$ [35].

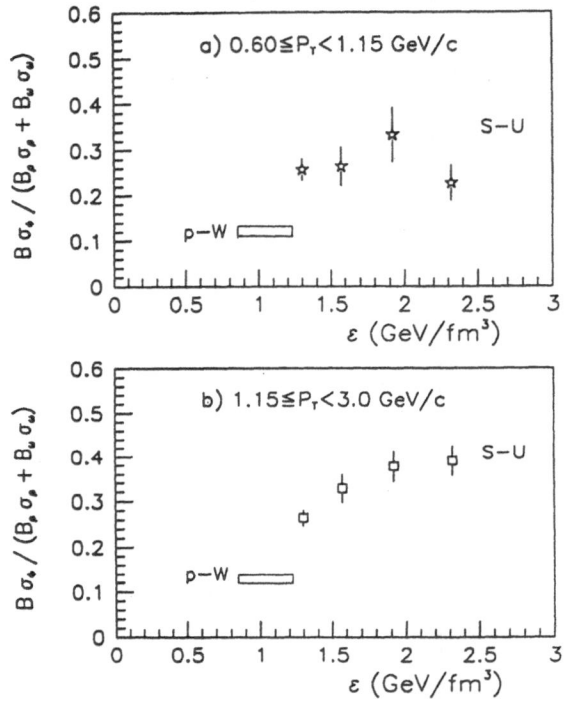

Fig.8 : Ratio of the production cross-sections for ϕ and $\rho+\omega$ from NA38 as a function of energy density, for two different p_T bins.

Interpretation

As in the case of J/ψ, these results are in agreement with what is expected if a QGP is formed[30,36]. However, strange particles can also be produced in hadronic reactions before the nuclear fireball reaches equilibrium [37,38] (see also the paper by H. Sorge in these proceedings). In RQMD calculations, the cascade leading to ϕ production has been studied in detail and shows an important contribution of meson-meson interactions over the whole rapidity range.

The ϕ enhancement must thus be considered in the general framework of strangeness enhancement and do not alone provide a reliable signature for the QGP.

CONCLUSION

Striking experimental effects have been observed both for J/ψ, ψ' and ϕ production in heavy ion collisions. Great theoretical efforts have been made in view of a comparison with avalaible particle-A data. Unfortunately, the initial non ambiguousness of the possible signatures of a QGP has been progressively washed out and it is not clear today wether we deal with a new state of matter or not. As pointed out, in a provocative way, by Blaizot in the summary talk of Quark Matter 93 Conference, are we sensitive, at present CERN energies, to quark and gluon degrees of freedom in the observables we are looking at ? If the answer is yes, we have now a better knowledge of hadronic effects which is a prerequisite to recognize any new physics appearing in the forthcoming Pb-Pb experiments.

REFERENCES

1. Proceedings of Quark Matter Conferences.
2. T. Matsui and H. Satz, Phys. Lett. B178 : 416 (1986).
3. F. Karsch and R. Petronzio, Phys. Lett. B193 : 105 (1987) ; Z. Phys. C37 : 627 (1988).
4. F. Karsch et al., Z. Phys., C37 : 617 (1988).
    C. Gerschel and J. Hüfner, Phys. Lett. B207 : 253 (1988).
5. C. Baglin et al., NA38 Collaboration, Phys. Lett. B220 : 471 (1989) ; B255 : 459 (1991).
    C. Baglin et al., NA38 Collaboration, Phys. Lett. B251 : 465 (1990).
    O. Drapier, NA38 Collaboration, Nucl. Phys. A544 : 209c (1992).
6. R. Mandry, PhD Thesis, Lyon (1993).
7. J.D. Björken, Phys. Rev. D27 : 140 (1983).
8. C. Baglin et al., NA38 Collaboration, Phys. Lett. B251 : 472 (1990).
9. C. Baglin et al., NA38 Collaboration, Phys. Lett. B270 : 105 (1991).
10. C. Baglin et al., NA38 Collaboration, Phys. Lett. B251 : 465 (1990).
    C. Baglin et al., NA38 Collaboration, Phys. Lett. B262 : 362 (1991).
11. C. Baglin et al., NA38 Collaboration, Phys. Lett. B268 : 453 (1991).
12. C. Lourenço, NA38 Collaboration, Proceedings of Quark Matter 93 Conference.
    B. Ronceux, NA38 Collaboration, Proceedings of Quark Matter 93 Conference.
13. D.M. Alde et al., Phys. Rev. Lett. 66 : 133 (1991).
14. S. Katsanevas et al., Phys.Rev. Lett. 60 : 2121 (1988).
15. C. Gerschel and J. Hüfner, Z. Phys. C56 : 171 (1992).
16. S. Gavin and M. Gyulassy, Phys. Lett. B214 : 241 (1988).
    J.P. Blaizot and J.Y. Ollitrault, Phys. Lett. B217 : 392 (1989).
17. J. Huefner, Y. Kurihara and H. Pirner, Phys. Lett. B215 : 218 (1988) .
18. R. Vogt, Nucl. Phys. A544 : 615c (1992).
19. J. Ftanik, P. Lichard and J. Pisut, Phys. Lett. B207 : 194 (1988).
    S. Gavin, M. Gyulassy and A. Jackson, Phys. Lett. B207 : 257 (1988).
    R. Vogt et al., Phys. Lett. B207 : 263 (1988).
    J.P. Blaizot and J.Y. Ollitrault, Phys. Rev. D39 : 232 (1989).
20. L. Lesniak, Phys. Lett. B302 : 140 (1993).
21. M.D. Sokoloff et al., Phys. Rev. Lett. 38 : 263 (1977)
    P. Amaudruz et al., (NMC Coll.), Nucl. Phys. B371 : 553 (1992)
22. S. Gupta and H. Satz, Z. Phys. C51 : 209 (1991).
23. R. Seitz and A. Witzman (NMC Coll.) , Proceedings of the XXVIIIth Rencontre de Moriond (1993).
24. G. Piller , T. Mutsbauer and W. Weise, Nucl. Phys. A560 : 437 (1993).
25. D. Kharzeev and H. Satz, Z. Phys. C60 : 389 (1993).
26. P. Sonderegger, private communication.
27. L. Antoniazzi et al., Phys. Rev. Lett. 70 : 383 (1993).
28. B. Povh and J. Hüfner, Phys. Rev. Lett. 58 : 1612 (1987).
29. J. Rafelski and B. Muller, Phys. Rev. Lett. 48 : 1066 (1982).

30. A. Shor, Phys. Rev. Lett. 54 : 1122 (1985).

31. S. Okubo, Phys. Rev. D16 : 2336 (1977).

32. NA38 Collaboration, R. Ferreira, Nucl. Phys. A544 : 497c (1992).
    R. Ferreira, Ph.D Thesis, Lisbonne (1992).

33. Helios/3 Collab., M.A. Mazzoni, Nucl. Phys. A544 : 623c (1992)
    R.J. Apsimon et al., Z. Phys. C53 : 581 (1992).

34. NA38 Collaboration, C. Baglin et al., Phys. Lett. B272 : 449 (1991).

35. W. Geist, Nucl. Phys. A525 : 149c (1991).

36. H.W. Barz et al., Nucl. Phys. A484 : 661 (1988) ; Nucl. Phys. A525 : 435c (1991).

37. P. Koch et al., Z. Phys. C47 : 477 (1990).

38. M. Berenguer et al., Proc. 9th Winter Workshop on Nuclear Dynamics, Key West, USA (1993).
    M. Berenguer, Ph. D Thesis, Francfort (1993).

KINETIC DESCRIPTION
OF HEAVY ION COLLISIONS WITH
NON-EQUILIBRIUM MEAN FIELDS

C. Fuchs, L. Sehn[+], H.H. Wolter

Universität München, D-85748 Garching, Germany
[+] University of Nantes, France

INTRODUCTION

The investigation of heavy ion collisions at energies of about 1 to 2 GeV/A bombarding energy is largely motivated by the search for the equation of state of nuclear matter, i.e. for the properties of equilibrated nuclear matter at densities away from saturation at non-zero temperatures. In a heavy ion collision the equation of state enters via the mean field in the participant zone of the colliding nuclei. However, through most of the reaction the phase space is far from equilibrium. Thus to determine the equation of state in a heavy ion collision involves the problem of describing realistically the many body system in non-equilibrium situations. Then the same theory can be used to determine the equilibrium equation of state for the ground state.

Transport equations have been derived consistently from non-equilibrium many-body theory by several authors, e.q. by Danielewicz [1] or Botermans and Malfliet [2]. Even with drastic approximations, such as the gradient and quasi particle approximations, they lead to a coupled system of kinetic equations of the Boltzmann-Ueling-Uhlenbeck (BUU) type for the evolution of the phase space distribution function [3] and a Bethe-Salpether equation for the effective interaction (T- or G-matrix) in medium. This system is not solvable in practice at the present. Thus approximate treatments have to be found, which, however, should retain the most important aspects of the non-equilibrium process. In particular, because of the strong momentum dependence of the nuclear interaction, it is important to take the essential features of the momentum distributions into account.

Several approaches have been used for the nuclear input into the kinetic equations, and some of these will be reviewed in more detail later. In early kinetic treatments of heavy ion collisions non-relativistic formulations were employed [3], but with the rise of hadronic relativistic field theories increasingly relativistic treatments are used [4, 5]. These are more appropriate since they incorporate a natural momentum dependence of the optical potential through the Lorentz characters of the various parts of the self energy [6]. With respect to the mean field in most cases a sort of local density approximation has been employed, i.e. the mean field and the cross section are taken as that of ground state or equilibrated nuclear matter at the total local density. With such an approach, however, the difference between the ground state of nuclear matter and the non-equilibrium situation of a heavy ion collision is ignored and dynamic effects are parametrized into the equation of state. Non-equilibrium effects in the form of two-Fermi-sphere momentum space configurations have been incorporated non-relativistically into the collision term [7] in some cases, but usually free NN-cross sections are employed. The mean field is usually taken from empirical parametrizations (Skyrme parametrization [8] or Walecka model [9]), but also fields from microscopic G-matrix calculations have been used [10].

In our approach we attempt to include realistic G-matrix descriptions and non-equilibrium effects approximately in a consistent fashion [11, 12, 13]. We employ a local phase space configuration approximation (called LCA), where the phase space is locally parametrized with two Fermi spheres in momentum space. For this configuration we extrapolate self energies (and in the future cross sections) from Dirac-Brueckner-HF (DB) calculations, and use these in relativistic kinetic calculations. We calculate various observables and try to assess the importance of non-equilibrium effects.

This paper is organized as follows. In section 2 we review the derivation of transport equations, because we feel it is important to keep in mind what is involved in a proper description of complicated processes such as a relativistic heavy ion collision. In section 3 we explain how we determine our non-equilibrium fields and in section 4 we show results of comparative calculations. In this paper we do not discuss very much our procedure to solve the relativistic BUU equation, which has been done in other reports [11].

Non-equilibrium hadronic matter

In this section we briefly review the description of hadronic systems in non-equilibrium using the language of real-time Green functions [14]. Such formulations were given by Kremp et al [15], by Danielewicz [1] and by Botermans and Malfliet [2]. The main purpose of reviewing this material is to set the frame for what is required for a realistic description of nucleus-nucleus collisions and consequently, what is needed to try to learn something from such processes.

One starts from some effective Lagrangian with interactions involving nuclear fields ψ and meson fields π, σ, ω, etc, which we do not need to specify now

$$\mathcal{L} = \mathcal{L}_N^0(\psi, \bar{\psi}) + \mathcal{L}_M^0(\sigma, \omega, \pi, \ldots) + \mathcal{L}^{int} \quad , \tag{1}$$

where the field operators are in the Heisenberg picture. The various 1-particle Green functions that play a role in non-equilibirum can be conveniently formulated on the Schwinger-Keldysh contour [16], which runs along the time axis from $t = -\infty$ to $t = +\infty$ and back

$$G(1, 2) = < T_c(\psi(1)\bar{\psi}(2)) > \quad , \tag{2}$$

where the contour ordering operator T_c orders time arguments along the directed contour. The four possible orderings can be condensed as a 2×2 matrix

$$G = \begin{pmatrix} G^c & G^< \\ G^> & G^a \end{pmatrix} \quad , \tag{3}$$

where $G^{c,a}$ are chronological (antichronological) Green functions and $G^{\lessgtr}$ are correlation functions. One also defines retarded (advanced) Green functions as

$$G^{\pm}_{(1,2)} = \pm\theta(\pm(t - t'))(G^> - G^<) \quad . \tag{4}$$

In a similar way n-particle Green functions G_n are defined which obey a Martin-Schwinger hierarchy. This can be decoupled formally with the introduction of the self energy Σ, which again is a 2×2 matrix.

$$G = G_0 + G_0 \Sigma G \tag{5}$$

$$\Sigma(1, 1') = (-i)(12 \mid U \mid 1'2')G_2(1'2', 1''2)G^{-1}(1'', 1') \tag{6}$$

Here G_0 is the free Green function. In eq.(6) we have the 2-particle Green function G_2 and the elementary 2-particle amplitude U, which contains the meson degrees of freedom and is given as a one-boson exchange potential (OBEP). Eq.(5) is the Dyson equation or - in a different form - the Kadanov-Baym equation [17]. Eqs.(5,6) have to be supplemented by corresponding equations for the meson Green function which involve polarization operators, again depending on G_2.

Approximations to eqs.(5,6) are obtained by approximations for G_2. A well established one is the T-matrix approximation, given by the relations

$$G_2 = GG + i\,GG\,T\,GG + \text{exch.} \tag{7}$$

$$T = U + i\,UGG\,T \quad . \tag{8}$$

With these the self energy is

$$\Sigma = -i\,tr\,TG \quad , \tag{9}$$

where the arguments are implied. The coupled system of eqs.(5,8,9)has to be solved. The first term in eq.(8) is equivalent to Hartree-Fock (HF), the next iterated term is the Born-Approximation and the whole of eq.(8) is a ladder summation of NN-scatterings. The choice of this approximation is motivated by the observation that NN-correlations are the most important ones in nuclear matter at normal density. It should be kept in mind that eq.(5 - 9) are in Keldysh-Notation with 2×2 matrices and involve a complicated coupling between the various components. Some of the relations are more simple using retarded /advanced Green functions.

It is instructive and also important for later applications to first look at the case of equilibrium, i.e. the ground state of nuclear matter with $T = 0$ or an equilibrated state with finite temperature. This leads to the well-known Dirac-Brueckner-HF theory (DB), which e.g. has been described in ref. [18, 19]. In this case only one Green function, namely G^c, appears and Σ is real in the ground state and it is usually a good assumption to assume it to be real generally. It can be decomposed according to its Lorentz character

$$\Sigma(p) = \Sigma_s - \gamma^\mu \Sigma_\mu + \sigma^{\mu\nu} \Sigma_{\mu\nu} \quad , \tag{10}$$

where parity violating terms do not appear and the tensor term is usually small and will be neglected, as is often also true for the space part of the vector term. Then a solution of the Dyson equation (5) is

$$G^\pm = (\gamma^\mu p_\mu^* - m^* \pm i\epsilon)^{-1} \tag{11}$$

with the effective momentum and mass

$$p_\mu^* = p_\mu + \Sigma_\mu \quad , m^* = m + \Sigma_s \quad , \tag{12}$$

which has the form of a free quasi particle Green function with momentum dependent momenta and masses. When G is decomposed into its Dirac and Fermi parts (with $E^*(p^*) = \sqrt{\vec{p}^{*2} + m^{*2}}$)

$$G(p) = \frac{1}{p^* - m^* + i\epsilon} - 2\pi i \Lambda^+(p^*)\delta(p_o^* - E^*)f(p^*) \quad , \tag{13}$$

neglect of the first term eliminates divergences and amounts to a renormalization. The second part takes into account the positive energy poles and $f(p^*)$ is the distribution function. One then introduces effective spinors $\tilde{u}(p^*, m^*)$, whose medium dependence is responsible for saturation in relativistic theories. The self energy is given from eq.(9), using a decomposition of the T-matrix into Lorentz invariants similar as is eq.(10), as

$$\Sigma_s(p) = \frac{1}{8} tr \Sigma(p) = \frac{4}{(2\pi)^3} \int^{k_F} d^3 q \frac{m^*}{E^*(q)} f(q) T_s(q, p) \tag{14}$$

$$\approx \Gamma_s(p)\rho_s(k_F, m^*) \tag{15}$$

for the scalar part, and analogously for the other components. The last line is an approximation, which is valid, if T_s is not to strongly q-dependent. This form is similar to the mean field approximation, where Γ is given by the constant g_s^2/m_s^2 [9]. The approximation of eq.(15) thus parametrizes the self energy in Hartree form with momentum dependent coupling coefficients. It is found to give a reasonable description of realistic results of DB calculations, as shown in ref. [20]. We shall use this parametrization later in the non-equilibrium case.

Generally speaking the DB method for equilibrium has been very successful in describing saturation properties of nuclear matter, properties of finite nuclei, the energy dependence of the optical potential, etc. [19]. However, we now turn to the non-equilibrium situation [1,2]. Since the momentum configuration can now be strongly distorted, open channels are accessible and the self energy has a non-negligible imaginary part. The real part of Σ is now used to define effective spinors $\tilde{u}(p^*, m^*)$. In such a spinor representation the Wigner transform of the Green function $G(p, x)$ is represented by a distribution function F and a spectral function A

$$
\begin{aligned}
G^<(p,x) &= -iA(p,x)F(p,x)2\Theta(p_0^*) \\
G^>(p,x) &= iA(p,x)(1-F(p,x))2\Theta(p_0^*) \quad ,
\end{aligned}
\tag{16}
$$

with the spectral function A is given as

$$
A(p,x) = -\frac{m^*}{E^*}\frac{\Gamma(p)}{(p_0^* - E^*(p^*))^2 - (\Gamma/2)^2}
\tag{17}
$$

$$
\Gamma(p) = 2\,Im(m^*\Sigma_s - p_\mu^*\Sigma^\mu)/E^* \quad .
\tag{18}
$$

It is seen, that the imaginary part of Σ determines the width of the spectral function and thus the amount, to which the momentum arguments of the distribution are on the energy shell. Thus the spectral function takes into account the off-shell propagation of the single particle distribution function. It is not known, how to incorporate these off-shell effects in practical calculations. In order to obtain kinetic equations the so-called quasi particle (QP) approximation has always been introduced, which replaces the spectral function by a delta function

$$
A(p,x) \approx 2\pi\frac{m^*}{E^*}\delta(p_0^* - E(p^*)) \quad ,
\tag{19}
$$

which would be valid for $Im\Sigma \ll Re\Sigma$. The restriction of $F(p,x)$ on the energy shell is then the distribution function $f(x,p)$, which is now on shell. If, in addition, the gradient approximation is made, which sets the Wigner transform of a product equal to the product of Wigner transforms, one obtains a kinetic equation for the distribution function

$$
\left\{p_\mu^*\partial_{(x)}^\mu + \left[p_\nu^* F^{\nu\mu} + m^*(\partial_{(x)}^\mu m^*)\right]\partial_{(p)\mu}\right\}f(x,p) = I^{coll}(x,p) \quad ,
\tag{20}
$$

where p^* and m^* are given as in eq.(12) with $Re\Sigma$, and $F^{\nu\mu}$ is the field tensor $F^{\nu\mu} = \partial^\nu\Sigma^\mu - \partial^\mu\Sigma^\nu$. The collision term is given in terms of the distribution functions and the T-matrix as

$$
\begin{aligned}
I^{coll}(x_1,p_1) = \frac{1}{2(2\pi)^9}\int d^3p_2 d^3p_3 d^3p_4 \left(\prod_{i=1}^{4}\frac{m_i^*}{E(p_i^*)}\right)(2\pi)^4\delta(p_1 + p_2 - p_3 - p_4)\times \\
\times\{<p_1p_2\mid T^+\mid p_3p_4><p_3p_4\mid T^-\mid p_1p_2>\times \\
\times\left(f_3 f_4 \bar{f}_1 \bar{f}_2 - \bar{f}_3 \bar{f}_4 f_1 f_2\right)\} \quad ,
\end{aligned}
\tag{21}
$$

where $\bar{f}(x,p) = (1 - f(x,p))$ are Pauli blocking factors. The T-matrix, from eq.(8), is

$$
\begin{aligned}
<p_1'p_2'\mid T\,(\overset{\pm}{\omega})\mid p_1p_2> &= <p_1'p_2'\mid U\mid p_1p_2> - \int\frac{d^3q}{(2\pi)^3}\frac{m_{1''}^*}{E(p_1''^*)}\frac{m_{2''}^*}{E(p_2''^*)} \\
&\times <p_1'p_2'\mid U\mid p_1''p_2''>\frac{1-f_{1''}-f_{2''}}{\omega - E_{p_1''}^* - E_{p_2''}^* \pm i\epsilon}<p_1''p_2''\mid T^\pm\mid p_1p_2> \quad ,
\end{aligned}
\tag{22}
$$

where the intermediate momenta are $p_{1,2}'' = \frac{p}{2} \pm q$. It is seen, that the T-matrix depends explicitly on the distribution function f. Finally from eq.(9) the self energy is also given by the T-matrix as

$$\Sigma^{\pm} = -itr(T^{\pm}[f]f) \qquad (23)$$

From eq.(20-23) it is seen that the treatment of the non-equilibrium problem involves the solution of a coupled system of a kinetic equation (20,21) with the solution of a non-equilibrium Bethe-Salpether equation for the T-matrix (22). Thus there is a consistency of the self energy, eq.(23) and the effective cross section in the collision term, eq.(21), $\frac{d\sigma}{d\omega} \sim |T[f]|^2$, through the non-equlibrium T-matrix, and also a double consistency of the self energy which explicitly and implicitly, through $T[f]$, depends on the distribution function. This system of equations has not been solved in practical applications.

The approaches taken for the self energies have been the following: In non-relativistic methods the Skyrme parametrization in terms of the total density alone has been used with an empirical momentum dependence [21]. In relativistic applications often the linear or non-linear Walecka parametrization of self energies with constant coupling coefficients were employed, which again is a parametrization for equilibrated nuclear matter [5, 4]. In recent work a non-local parametrization has been used [22], which allows to reproduce better the energy dependence of the optical potential. To take into account non-equilibrium effects, one attempts to calculate the interaction for a standard non-equilibrium situation. For this a configuration of two separated Fermi spheres in momentum space, corresponding to two interpenetrating streams of nuclear matter (colliding nuclear matter), has been considered. In this work we attempt to determine a relativistic expression for the self energies for this situation.

With respect to the cross section in the collision term, very little has been done to achieve consistency with the T-matrix. Usually a parametrization of the free NN-cross section by Cugnon et al.[23] has been used, sometimes with empirical medium modifications. Only by the Tübingen group a cross section derived from the non-relativistic T-matrix of colliding nulcear matter was used, however, not consistently with the mean fields [7]. In this work we also use the Cugnon parametrization. However, we are presently investigating to derive a cross section consistently from the imaginary part of the self energy.

Local phase space configuration approximation

As discussed above we attempt to include non-equilibrium effects in a kinetic calculation by determining the self energy for the configuration of colliding nuclear matter, which we take as a typical non-equilibrium situation not only in the beginning but through a large part of the collision. We term this approach "local phase space configuration approximation (LCA)", since we parametrize the phase space configuration at each space-time cell by this configuration. In contrast to other approaches we consider this method as taking into account not only momentum dependence but momentum *space* dependence realistically.

For this program it would be necessary to have Dirac-Brueckner HF calculations for colliding nuclear matter as a function of the densities of the two colliding streams and the relative velocities. Such calculations exist only non-relativistically [24]. In one line of work we have interpreted these results relativistically [25]. Here we follow a different approach by extrapolating in a consistent way DB calculations for ground state nuclear matter of various densities to colliding nuclear matter [20, 11].

It was discussed in the last section, that the self energy of nuclear matter is given as

$$\Sigma(k, k_F) = \Sigma_s(k, k_F) - \gamma_\mu \Sigma^\mu(k, k_F) . \tag{24}$$

For nuclear matter at rest, eq. (24) takes the form $\Sigma = \Sigma_s - \gamma^0 \Sigma_0 + \boldsymbol{\gamma} \cdot \mathbf{k} \Sigma_v$. The small Σ_v term is taken into account in the calculations but omitted in the following discussion. It was also discussed that the DB results can approximately be represented in a mean field form eq.(15),

$$\Sigma_i(k, k_F) = \Gamma_i(k, k_F) \rho_i(k_F, m^*) , \tag{25}$$

where the index $i = s, o$ is for scalar and vector fields, and $\rho_0 = j_0$ is the baryon density. The DB results taken from ref.[19] are now extrapolated to the configuration of colliding nuclear matter, i.e. to two Fermi ellipsoids. It has to be emphasized that the construction of the configuration is a self-consistent problem, since the positions of the ellipsoids are given as $q = v_{rel} \gamma \sqrt{k_F^2 + m_{12}^{*2}}$ and thus depend on the total self-consistent effective mass m_{12}^*, while v_{rel} is the relative velocity of the two currents. If the Fermi ellipsoids overlap the overlapping volume is removed and the Fermi momenta are increased so that the total density remains constant [25]. The scalar self energy of colliding nuclear matter is now constructed as

$$\Sigma_s^{12}(k, k_F) = \Gamma_s(\Lambda_1^{-1}(k), k_F) \rho_{s1}(m_{12}^*) + \Gamma_s(\Lambda_2^{-1}(k), k_F) \rho_{s2}(m_{12}^*) + \delta \Sigma_p. \tag{26}$$

Thus the coupling coefficients are taken at the properly boosted momenta ($\Lambda_{1,2}$ are Lorentz transformations) and the densities are taken at the self consistent effective mass. $\delta \Sigma_p$ is the contribution from the Pauli correction, if necessary. Σ_0 is constructed in an analogous way. The approximation of eq.(26) neglects the configuration effect in the coupling parameters Γ_i, but does take it into account in the effective masses, the densities and the position of the ellipsoids.

To include the full configuration dependence of the self energy, eq. (26), in a transport calculation is at the moment unfeasable. In the local configuration approximation (LCA) we therefore parametrize the momentum configuration in a cell in coordinate space in the overlap region by two separated Fermi ellipsoids with a set of parameters $\{\alpha\} = \{k_{F_1}, k_{F_1}, v_{rel}\}$, the Fermi momenta and the relative velocity of the subsystems. The set α is extracted from the phase space distribution function, which is decomposed into two parts for projectile and target $f(x, k^*) = f_1(x, k^*) + f_2(x, k^*)$. From the respective currents

$$j_{\mu i}(x) = \int d^3 k k_\mu^* / E^*(\mathbf{k}) \, f_i(x, k^*) \quad , i = 1, 2 \quad , \tag{27}$$

we calculate the Lorentz invariant rest densities, the streaming velocities and the Fermi momenta of the subsystems

$$\rho_{0i}(x) = \sqrt{j_{\mu i} j_i^\mu}, \quad u_{\mu i}(x) = j_{\mu i}(x)/\rho_{0i}(x), \quad k_{F_i}(x) = \left(3/2\pi^2 \rho_{0i}(x)\right)^{\frac{1}{3}} \quad . \tag{28}$$

The relative velocity of the currents $\mathbf{v}_{rel} = (u_{20}\mathbf{u}_1 - u_{10}\mathbf{u}_2) / u_{1\mu} u_2^\mu$ is also an invariant

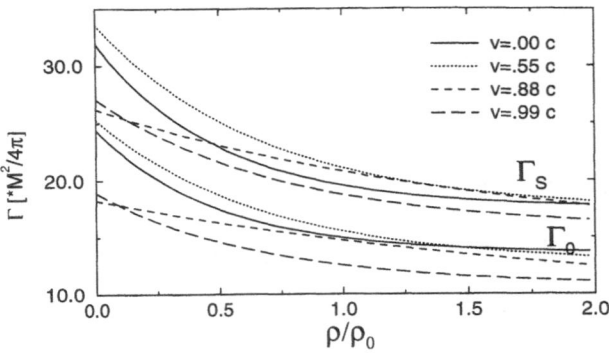

Figure 1: Effective scalar and vector coupling functions $\bar{\Gamma}_S^{12}$ and $\bar{\Gamma}_0^{12}$ in dimensionless units for colliding nuclear matter with different streaming velocities of the subsystems.

quantity. To obtain the selfenergies, eq. (26), in a mean field form the Σ_i^{12} are averaged over the momentum configuration which allows to define averaged coupling functions depending on α

$$\bar{\Sigma}_s^{12} = \bar{\Gamma}_s^{12}(k_{F_1}, k_{F_2}, v_{rel})\rho_s^{12}, \quad \bar{\Sigma}_0^{12} = \bar{\Gamma}_0^{12}(k_{F_1}, k_{F_2}, v_{rel})j_\mu^{12} \quad . \tag{29}$$

In the moment we assume a locally symmetric configuration $k_{F_1} = k_{F_2} = k_F$, but this is no restriction in principle.

The structure of the self energies is now the same as in the mean field approximation and the coupling functions $\bar{\Gamma}_S^{12}$ and $\bar{\Gamma}_0^{12}$ contain in an effective way the correlation effects of the DB calculations as well as the configuration effects as functions of density and relative velocity. A graph of the $\bar{\Gamma}_i^{12}$ is shown in Fig.1, where these are displayed as functions of the subsystems densities and for different relative velocities in units of c. It is seen that they generally decrease with density and relative velocity, but the behavior is more complex due to Pauli effects. The general behavior can be understood as being due to the higher momenta involved, since the basic G-matrix decreases with increasing momentum.

Results and Discussion

The self energies from the last section can now be inserted into the relativistic kinetic equation (20). For the collision term an analogous procedure should be performed with the imaginary part of Σ, but at present we use the Cugnon parametrization [23]. For the solution of the kinetic equation we use the covariant Landau-Vlasov method with covariant Gaussian test particles in an extension from the non-relativistic approach of Grégoire et al. [26]. More details on this are given in refs. [11] and a forthcoming paper [27]. We also included Δ-degrees of freedom since they play an important role in the dynamics and are the source for the pion production. Deltas are created and annihilated through inelastic collisions ($NN \longleftrightarrow N\Delta$). In addition deltas can decay to nucleons and pions ($\Delta \rightarrow N + \pi$) with an exponential decay law $e^{-\Gamma t}$, $\Gamma = 112 MeV$, in the delta rest frame. For the propagation of the deltas we use the same field as for the nucleons, i.e. $m_\Delta^* = M_\Delta - \bar{\Sigma}_S^{12}(x)$. The pions propagate freely with a mean free path $\lambda_\pi = v\rho/4\rho_0$ until they are reabsorbed through the process $NN\pi \rightarrow N\Delta$. In Fig.2 we show a calculation which compares different ways to spec-

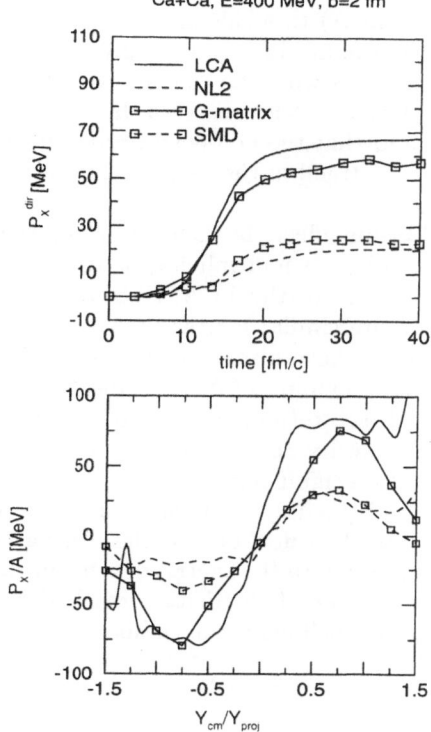

Ca+Ca, E=400 MeV, b=2 fm

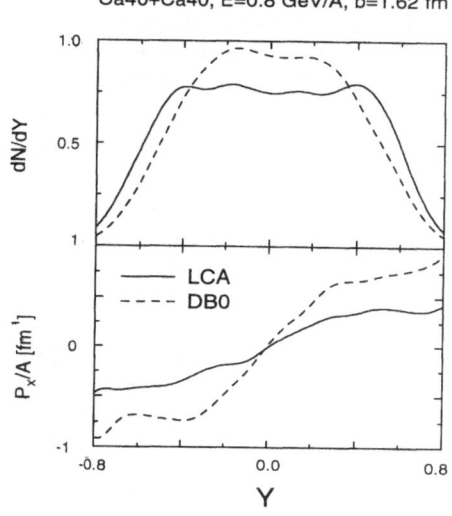

Ca40+Ca40, E=0.8 GeV/A, b=1.62 fm

Fig. 2: Directed flow (above) as a function of time and transverse flow per nucleon (below) for $^{40}Ca +^{40} Ca$, E_{lab}/A=400MeV and b=2.0 fm. Shown are calculations in the present method (LCA) and in the non-linear Walecka model (NL2) [29] and QMD-calculations with a non-relativistic G-matrix and with momentum dependent soft Skyrme parametrisations [28].

Fig. 3: Rapidity distribution (above) and transverse flow per nucleon for $^{40}Ca +^{40} Ca$ at E_{lab}/A=0.8 GeV and b=1.62 fm. Shown are calculations with non-equlibrium effects (LCA) and with ground state mean fields (DB0).

ify the mean fields. Compared are a calculation for $^{40}Ca +^{40} Ca$ at 400 MeV/A lab energy and impact parameter 2 fm. Shown are the directed flow as a function of time and the transverse momentum per nucleon as a function of rapidity. There are two calculations with realistic forces, the curve marked LCA with our method and the curve marked G-matrix, using a non-relativistic G-matrix from the Tübingen group [28]. On the other hand there are two calculations with phenomenological forces with the non-linear Walecka model (parameter at NL2) from the Giessen group [29] and the other a non-relativistic calculation using the Skyrme parametrization (soft equation of state) with momentum dependence [28]. It is seen that the realistic forces produce considerably more flow than the phenomenological ones, where it should be kept in mind, that the latter reproduce experimental data quite well [29]. The origin of this behaviour of realistic forces is not understood yet, but does not seem to depend on the non-equilibrium effects alone.

In Fig.3 we discuss the differences between calculations in our formalism with and without non-equilibrium effects for a $Ca + Ca$ reaction at somewhat higher energies. We compare our non-equilibrium treatment (LCA) to a calculation, where the coupling constants are kept fixed at their saturation point during the entire reaction (DB0). This is equivalent to the linear Walecka model with the coupling constants adjusted to the saturation point of the DB calculations. It is seen that this leads to very strong sidewards flow and strong stopping, and that the non-equilibrium effects soften the repulsion. Referring back to Fig.3 this is a trend in the right direction, but not strong enough.

In Figs. 4 and 5 we investigate the non-equilibrium effects in the pion production rate. In fig. 5 the time evolution of the delta and pion multiplicities is displayed for a central collision of ^{40}Ca on ^{40}Ca at 1.8 GeV/A in the LCA, NL2 and DB0 approaches. The time evolution of the deltas is quite similar for the LCA and NL2 but substantially differs in the DB0, where the deltas have disappeared at an earlier time and consequentely less pions are produced. The amount of deltas is determined by the balance of delta production through inelastic scattering, delta decay, and pion reabsorption. As can be seen in fig. 5 a high Δ-density survives slightly longer in the LCA than in the NL2 approach and gives rise to a significant increase of the pion yield. We see here that the pion production - and possibly particle production in general - is a better probe for the equation of state than flow observables. In fig. 6 the number of deltas (same as in fig. 4) is shown parallel to the central density in the reaction. As can be seen there, not the maximum value of the density reached but the duration of a sufficiently high compression is responsible for the number of deltas in the system and the resulting pion yield.

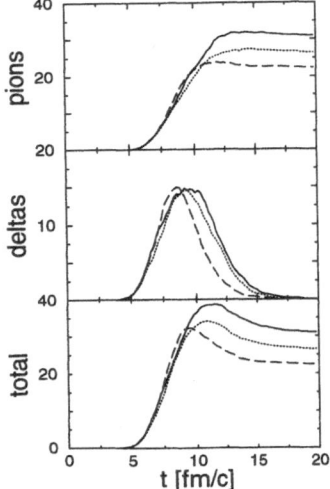

Fig. 4: Pion and delta production for a central collision of ^{40}Ca on ^{40}Ca at E=1.8 GeV/A, LCA (solid), NL2 (dotted) and DB0 (dashed).

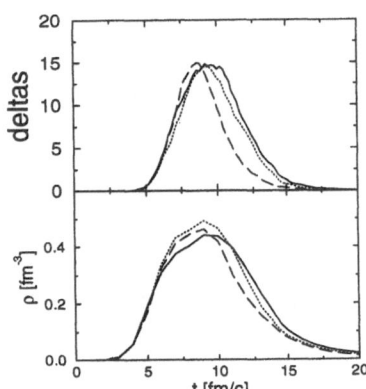

Fig. 5: Delta production (top) and central density (bottom) for the same reaction as fig. 4, LCA (solid), NL2 (dotted) and DB0 (dashed).

To summarize, we have discussed the problems and possible solutions on how to determine the equation of state from heavy ion collisions. From theoretical formulations of such processes it is evident, that the nuclear fields and cross sections in the kinetic equation should be derived from Brueckner-type calculations for non-equilibrium situations. We have attempted an approximation to this difficult problem by deriving self energies for standard 2-Fermi-sphere configurations, i.e. colliding nuclear matter, from Dirac-Brueckner calculations. A similar approach is in progress with respect to the effective NN-cross section in the collision term, derived from the imaginary part of the self energy.

We have incorporated these interactions into a relativistic transport code in this LCA approach. We find that the inclusion of non-equilibrium effects does affect observables considerably, similar as with different equations of state. We found a good agreement of flow data in the LCA approach with other realistic forces but too much flow with respect to non-linear parametrizations of the Walecka model and experiment. Differences between different treatments can also be seen in the pion yields showing that particle production provides as well a sensitive probe for the equation of states as the transverse flow. Finally it is concluded that non-equilibrium effects are important for the description of heavy ion collisions, if the equation of state of nuclear matter is to be reliably extracted from such processes.

REFERENCES

1. P. Danielewicz, Ann. Phys. **152**(1984)239

2. W. Botermans, R. Malfliet, Phys. Rep. **198**(1990)115

3. G.F. Bertsch, S. Das Gupta, Phys. Rep. **160** (1988) 189

4. B. Blättel, V. Koch, W. Cassing, U. Mosel, Phys. Rev. **C38**(1988)1767

5. C.M. Ko, Q. Li, Phys. Rev. **C37**(1988)2270

6. T.L. Ainsworth, G.E. Brown, M. Prakash, H.H. Wolter, Proceedings Hirschegg 1987, ed. H. Feldmeier, Darmstadt 1987

7. A. Bohnet, et al., Nucl. Phys. **A494**(1989)349; J. Jaenicke, et al., Nucl. Phys. **A536**(1992)201

8. H. Stöcker, W. Greiner, Phys. Rep. **137**(1986)277

9. B.D. Serot, J.D. Walecka, Adv. Nucl. Phys. **16**(1986)1

10. D.T. Khoa, et al., Nucl. Phys. **A542** (1992) 671, R.K. Puri, Nucl. Phys. **A**, in print

11. C. Fuchs, L. Sehn, H.H. Wolter, Prog. Part. Nucl. Phys. **30**(1993)247

12. C. Fuchs, L. Sehn, H.H. Wolter, Nucl. Phys. **A545** (1992) 151c

13. C. Fuchs, L. Sehn, H.H. Wolter, Nucl. Phys., in print

14. J. Schwinger, J. Math. Phys. **2**(1961)407

15. D. Kremp, M. Schlanges, T. Bornath, J. Stat. Phys. **41**(1985)661

16. L.V. Keldysh, Sov. Phys. JETR, **20**(1965)1018

17. L.P. Kardanoff, G. Baym, *Quantum Statistical Mechanics*, Benjamin, N.Y. 1962

18. C.J. Horowitz, B.D. Serot, Nucl. Phys. **A464**(1987)613

19. B. terHaar, R. Malfliet, Phys. Rep. **149**(1987)207

20. L. Sehn, H.H. Wolter, Nucl. Phys. **A519**(1990)289c

21. J. Aichelin, A. Rosenbauer, Phys. Rev. Lett. **58**(1987)1926

22. K. Weber, et al., Nucl. Phys. **A539**(1992)713

23. J. Cugnon, et al., Nucl. Phys. **A352**(1981)505

24. N. Othsuka, et al., Nucl. Phys. **A465**(1987)550; N. Othsuka et al., Nucl. Phys. **A490**(1988)715

25. H. Elsenhans, et al., Nucl. Phys. **A536**(1992)750

26. C. Grégoire, et al., Nucl. Phys. **A465** (1987) 317

27. C. Fuchs, H.H. Wolter, Nucl. Phys., submitted

28. D.T. Khoa, et al., Nucl. Phys. **A548**(1992)102

29. B. Blättel, V. Koch, U. Mosel, Rep. Prog. Phys. **56**(1993)1

PHASE STRUCTURE OF $SU(2)$ YANG-MILLS THEORY WITH GLOBAL CENTER SYMMETRY

K. Sailer

Department for Theoretical Physics
Kossuth Lajos University
H-4010 Debrecen, Pf. 5, Hungary

INTRODUCTION

Lattice QCD results indicate that confinement and global center symmetry are strongly correlated. The transition from the confined to the deconfined phase seems to be accompanied by the dynamical breaking of the global center symmetry[1, 2, 3, 4]. In the present article we shall investigate the continuum $SU(2)$ gauge theory with global center symmetry. We shall show that confined phase (with non-vanishing string tension) may exist with global center symmetry, whereas in the deconfined phase (with vanishing string tension) this symmetry is broken. In connection with the phase structure of the theory we also show that only a few of the parameters are of significance.

Our treatment can be outlined as follows. We start with the generating functional of the Green's functions of the lattice regularized $SU(2)$ Yang-Mills theory. In order to ensure global center symmetry the $SU(2)$ invariant Haar measure is used in the path integral[5]. This leads to a cut-off dependent periodic effective potential. For sake of simplicity we neglect the spatial components of the vector potential remaining invariant under global center transformations. The 'skelet' $SU(2)$ Yang-Mills theory obtained in this way contains only the timelike components of the vector potential being sensitive to the global center transformations.

We show that the skelet model is equivalent with the macrocanonical ensemble for a $D = 4$ Coulomb-gas consisting of infinitely heavy point charges. In addition to the strong coupling constant the model contains infinitely many other parameters:

Hot and Dense Nuclear Matter, Edited by
W. Greiner *et al.*, Plenum Press, New York, 1994

the fugacities of charges ± 1, ± 2, etc.

We expand the effective potential at its minima and derive the effective action using mean field approximation and loop expansion[6]. The 1–loop order effective action contains a non-local residual interaction treated in terms of first order perturbation theory.

Finally we derive an expression of the string tension in the effective theory replacing the bare action by the effective one. The string tension is determined through the propagator of the gauge field in the effective theory. According to the value of a particular combination c of the relevant coupling constants the propagator of the effective theory has qualitatively different behaviour. For $c \ll 1$ it exhibits simple poles whereas for $c \gg 1$ double poles occur. In the first case the string tension vanishes, while it may have a non-vanishing value in the second case.

SKELET MODEL

The generating functional $Z[j]$ of the $SU(2)$ Yang-Mills theory is by definition the vacuum to vacuum transition amplitude in the presence of the external source j_μ^a:

$$Z[j] \sim \int \mathcal{D}A_i^a \int \mathcal{D}A_0^a \exp\left\{-\frac{1}{4g^2}\int d^4x\, F_{\mu\nu}^a F_{\mu\nu}^a - S_{gf} - \int d^4x\, j_\mu^a A_\mu^a - S_{gh}\right\}$$

$$(1)$$

with the gauge fixing term S_{gf} and the ghost term S_{gh}. The requirement of color singlet vacuum states leads to the $SU(2)$ invariant integration measure (Haar measure)[4, 5]:

$$\mathcal{D}A_0^a = \prod_x \frac{\sin^2 \frac{a(A_0^a A_0^a)^{1/2}}{2}}{a^2 A_0^a A_0^a} dA_0^a(x),$$

$$(2)$$

with the lattice size a. The integration measure for the spatial components of the vector potential is the usual flat measure $\prod_x dA_i^a(x)$. Due to the use of the group invariant integration measure, invariance of the generating functional under global center transformations can be maintained for vanishing external source. The center of a group consists of the elements commuting with any element of the group. The center of the group $SU(2)$ consists of the elements 1 and -1. Global center transformation means the same center transformation at every point of 3-space at a given time slice. This transformation takes advantage of the fact that we can perform finite gauge transformations at any time slice separately. In the continuum limit this leads to field configurations discontinuous in time.

Under an arbitrary finite $SU(2)$ gauge transformation $h(x)$ the spatial and the timelike components of the vector potential transform as follows[4, 5]:

$$\left(A_i^a \frac{\tau^a}{2}\right)^h = h(x)\left(A_i^a \frac{\tau^a}{2}\right)h^{-1}(x) + h(x)\partial_i h^{-1}(x),$$

$$\left(\exp\left\{ia A_0^a \frac{\tau^a}{2}\right\}\right)^h = \left(\exp\left\{ia A_0^a \frac{\tau^a}{2}\right\}\right)h(x).$$

$$(3)$$

As we see the spatial components of the vector potential remain unaltered under global center transformations. For our goal is to investigate the consequences of the global center symmetry we neglect the spatial components of the vector potential by setting $A_i^a = 0$ in the present work.

Further on we specify the gauge conditions:

$$A_0^1 = A_0^2 = 0, \qquad \partial_0 A_0^3 = 0. \tag{4}$$

The advantage of this gauge fixing is that we end up with a single field component $A_0^3(x) \equiv u(x)$ and the ghosts decouple from it. The disadvantage is that it breaks global center symmetry with except for the particular time independent global center transformation $z(x) \equiv -1$, for which Eq. (3) leads to

$$au^z(x) = au(x) + 2\pi. \tag{5}$$

In this way we obtained the "skelet" model which only contains those degrees of freedom being sensitive to the global center transformations. The corresponding generating functional takes the form ($j_0^3 \equiv q$):

$$Z[q] \sim \int \mathcal{D}u \exp\left\{ \frac{1}{2g^2} \int d^4x\, u \Box u + \int d^4x\, V(u) - \int d^4x\, qu \right\}, \tag{6}$$

where the periodic cut-off dependent potential

$$V(u(x)) = -\frac{1}{a^4} \ln \sin^2 \frac{au(x)}{2} \tag{7}$$

occurs by exponentiating the Haar measure, and the measure $\mathcal{D}u = \prod_x du(x)$ is the flat one. It is the lattice regularized form of the skelet model we consider in the present paper.

COULOMB GAS

Now we show that the generating functional of the skelet model obtained above is equivalent with the macrocanonical partition sum of the $D = 4$ dimensional Coulomb gas with infinitely heavy (static) point charges $\sigma = \pm 1, \pm 2, \ldots$

Let us expand the periodic potential $V(u)$ in Fourier series:

$$V(u) = -\frac{1}{a^4}\left(v_0 + \sum_{\sigma \neq 0} v_\sigma \cos(\sigma a u(x)) \right), \tag{8}$$

with $v_{-\sigma} = v_\sigma$ and $\sigma = \pm 1, \pm 2, \ldots$ Particularly for the potential (7) we obtain: $v_0 = 2 \ln 2$, $v_\sigma = |\sigma|^{-1}$. Expanding now the factor $\exp\left\{ -\int d^4x\, V(u) \right\}$ in Taylor series, we obtain for the generating functional:

$$Z[q] = \sum_{n=0}^{\infty} \frac{(-1)^n}{n!} \left(\prod_{j=1}^{n} \int \frac{d^4x_j}{a^4} \sum_{\sigma(j) \neq 0} v_{\sigma(j)} \right) \int \mathcal{D}u \cdot$$

$$\cdot \exp \left\{ \frac{1}{2g^2} \int d^4 x u \square u - \int d^4 x q u + i \sum_{j=1}^{n} \sigma(j) a \int d^4 x \delta(x - x_j) u(x) \right\}.$$

(9)

Performing the Gaussian integral for the u field we find the macrocanonical partition function of the $D = 4$ dimensional Coulomb gas:

$$Z[0] = \sum_{n=0}^{\infty} \frac{(-1)^n}{n!} \left(\prod_{j=1}^{n} \int \frac{d^4 x_j}{a^4} \sum_{\sigma(j) \neq 0} v_{\sigma(j)} \right) \exp \left\{ -\frac{1}{2} \sum_{j,k} \sigma(j) a^2 D(x_j, x_k) \sigma(k) \right\}.$$

(10)

Here the coupling constants $v_{\sigma(j)}$ are the fugacities and the exponent contains the energy of the Coulomb interaction of the static charges $\sigma(j)$ positioned at x_j since the free u-propagator is given by

$$a^2 D(x_j, x_k) \longrightarrow \frac{a^2 g^2}{4\pi^2 (x_j - x_k)^2} \qquad \text{for} \qquad a \to 0.$$

(11)

In addition to the strong coupling constant g the model contains infinitely many other parameters, the fugacities v_σ.

We conclude that the requirement of global center symmetry leads to the equivalence of the A_0 gluon vacuum with a $D = 4$ dimensional Coulomb gas of static charges.

EFFECTIVE THEORY

Without giving all the details the main steps of obtaining the effective theory are as follows:

1. The effective potential at the tree level has infinitely many minima at $au = \pm (2k + 1)\pi$ $(k = 0, 1, 2, \ldots)$. Our expectation is therefore that the vacuum expectation value $\langle u(x) \rangle \equiv \bar{u}$ does not vanish as a rule, but corresponds to one of the minima of the effective potential. Let us shift the field variable

$$u(x) = \bar{u} + \eta(x).$$

(12)

2. Now we can linearize the exponential of the path integral in terms of $\eta(x)$ by introducing the auxiliary field $h(x)$ by the definition:

$$\exp \left\{ \frac{1}{2g^2} \int d^4 x \eta \square \eta \right\} = \int \mathcal{D} h \exp \left\{ -\frac{1}{2a^8} \int d^4 x d^4 y h(x) a^2 D(x, y) h(y) \right.$$
$$\left. + \int \frac{d^4 x}{a^4} h(x) a \eta(x) \right\}.$$

(13)

In order to clarify the meaning of the auxiliary field let us go back to Minkowski space, which roughly means the replacements $d^4 x \to -i d^4 x$, $\eta \to i\eta$, $q \to iq$,

$h \rightarrow ih$, and forget about lattice regularization. Then the naive integration over $\mathcal{D}\eta$ leads to the Dirac-delta functional:

$$\delta[h(x) - q(x) + \sum_j \sigma(j)a\delta(x - x_j)].\tag{14}$$

According to that the auxiliary field $h(x)$ can be interpreted as the effective charge density being the sum of the external charge density and the polarisation charge density. The latter arises due to the Coulomb charges in the vacuum.

3. Performing the path integral over $\eta(x)$ the generating functional takes the form:

$$Z[q] \;=\; \exp\left\{-a\bar{u}\sum_x q_x\right\} \int \mathcal{D}h \exp\left\{\frac{1}{2}\sum_{xy} h_x a^2 D_{xy} h_y + \ln z[\xi]\right\},\tag{15}$$

with

$$z[\xi] \;=\; \sum_{n=0}^{\infty} \frac{(-1)^n}{n!} \left(\prod_{j=1}^n \sum_{x_j} \sum_{\sigma(j)\neq 0} v_{\sigma(j)}\right) I_\sigma^{(n)}[\xi] \; \exp\left\{ia\bar{u}\sum_{j=1}^n \sigma(j)\right\},\tag{16}$$

$$I_\sigma^{(n)}[\xi] \;=\; \prod_x \frac{\sinh(4\pi\xi_x)}{\xi_x} \prod_{j=1}^n \frac{\xi(x_j)}{\xi(x_j) + i\sigma(j)},\tag{17}$$

and $\xi_x = h_x - q_x$. Here we used the correct discretized expressions of the lattice regularized theory.

4. To the next we perform the path integral over $h(x)$ by the saddle point method taking into account the tree level and the 1–loop contributions. The solution of the saddle point equation

$$\frac{\delta}{\delta h_x}\left\{\frac{1}{2}\sum_{xy} h_x a^2 D_{xy} h_y + \ln z[h - q]\right\} = 0\tag{18}$$

is a local functional $h_{0x}[q]$. The tree level and the 1–loop contributions are given by

$$Z_0[q] \;=\; \exp\left\{\frac{1}{2}\sum_{xy} h_{0x}[q]a^2 D_{xy} h_{0y}[q] + \ln z[h_0[q] - q] - a\bar{u}\sum_x q_x\right\},\tag{19}$$

and

$$Z_1[q] \;=\; \left[\mathrm{Det}(a^2 Q)\right]^{-1/2},\tag{20}$$

resp., with

$$a^2 Q_{xy} \;=\; a^2 D_{xy} + \frac{\delta^2 \ln z[h_0 - q]}{\delta h_{0x}\delta h_{0y}}.\tag{21}$$

5. Let us derive the effective action in 1–loop order defined as the Legendre transform of the generating functional $W[q] = \ln Z[q]$ of the connected Green's functions:

$$\Gamma[\chi] \quad = \quad -\int d^4x \, (\chi + \bar{u}) \, q + W[q], \qquad (22)$$

where the classical field $\chi(x)$ is given by

$$\chi(x) + \bar{u} \quad = \quad \frac{\delta W[q]}{\delta q(x)} \qquad (23)$$

as the functional of the external source $q(x)$. The explicit form of the generating functional $W[q]$ was found by making use of the solution of the saddle point equation (18) including terms up to $\mathcal{O}(q^4)$. Inverting the functional dependence given by Eq. (23) and inserting it in the r.h.s. of Eq. (22) we only retained terms up to the fourth order in the classical field. Thus the following expression was obtained for the effective action:

$$\Gamma[\chi] \quad = \quad \frac{1}{2f_2} \sum_{xyz} \chi_x \left(\delta_{xy} + f_2 D_{xy}^{-1}\right) \left(1 + c Q_0^{-1} D\right)_{yz}^{-1} \chi_z$$

$$+ \text{ quartic terms}, \qquad (24)$$

where $(Q_0)_{xy} = D_{xy} + f_2 \delta_{xy}$. Here we introduced the new coupling constants

$$c \quad = \quad -\frac{f_4}{f_2 g^2}, \qquad (25)$$

and

$$f_2 \quad = \quad -2 \cdot 2! \sum_{\sigma>0} v_\sigma \sigma^{-2} (-1)^\sigma > 0,$$

$$f_4 \quad = \quad 4! \cdot 2 \sum_{\sigma>0} v_\sigma \sigma^{-4} (-1)^\sigma < 0. \qquad (26)$$

These parameters are responsible for the qualitative behaviour of the effective action. The expression (24) is valid in the neighbourhood of any of the minima of the effective potential. It was shown that the position of the minima does not change due to 1–loop corrections: $a\bar{u} = (2k + 1)\pi$ with $k = 0, \pm 1, \pm 2, \ldots$. Therefore the total effective action is periodic:

$$\Gamma_t[\chi] \quad = \quad \sum_k \Gamma \left[(2k + 1)\frac{\pi}{a} + \chi\right] \Theta_k, \qquad (27)$$

where Θ_k is the characteristic function of the interval $2k\pi \leq a\chi \leq (2k + 2)\pi$. Owing to $f_2 > 0$ the expansion is performed at the minima of the effective potential indeed.

The inverse propagator of the u gluons exhibits rather different behaviour in the limiting cases $c \ll 1$ and $c \gg 1$:

$$\tilde{D}_0^{-1}(p) \quad = \quad \begin{cases} \frac{1-c}{f_2} + \tilde{D}^{-1}(p) & \text{for} \quad c \ll 1 \\ \frac{1}{f_2 c} \left(1 + f_2 \tilde{D}^{-1}(p)\right)^2 & \text{for} \quad c \gg 1 \end{cases}$$

$$(28)$$

where $\tilde{D}^{-1}(p)$ is the lattice regularized inverse propagator of the bare theory. Due to the quadratic expression the propagator exihibits double poles for $c \gg 1$, whereas simple poles occur for $c \ll 1$. In both cases a rest mass is generated. This can be read off by comparing the $p_\mu \to 0$ limits of the inverse propagators with the general form $g^{-2}(p^2 + m^2)$. Then we obtain for the rest mass:

$$m^2 = \begin{cases} a^{-2} \frac{-f_4}{f_2^2} \frac{1-c}{c} > 0, & \text{for } c \ll 1 \\[2mm] a^{-2} \frac{-f_4}{2f_2^2} \frac{1}{c} > 0, & \text{for } c \gg 1 \end{cases} \tag{29}$$

Let us note that the cut-off dependent periodic potential of the bare theory, i.e. the usage of the Haar measure lead to the following consequences: (i) the total effective potential is periodic, (ii) rest mass is generated for the A_0 gluons, (iii) non-local self-interaction terms with higher derivatives occurred in the effective action. These higher derivative terms can influence IR physics if the rest mass turns out to be zero.

It turned out that the quartic interaction terms of the effective action can be important in the case $c \gg 1$ for obtaining a non-vanishing string tension. It was observed that these terms are higher order of $1/c$ and, consequently, represent some residual imteraction for $c \gg 1$. Therefore the proper self energy Σ_u of the gluons has been calculated taking the quartic self-interaction into account in first order perturbation theory. The propagator modified by the self-interaction is then given by:

$$\begin{aligned} \tilde{\mathcal{D}}^{-1}(p) &= \tilde{D}_0^{-1}(p) + \tilde{D}_0^{-1}(p)\Sigma_u(p)\tilde{D}_0^{-1}(p) \\ &= -4\frac{(4f_2)^4}{f_4^3}c^2 \left[4 + 3\frac{1}{c}\frac{f_4}{2f_2^2} - \left(1 + \frac{1}{c}\frac{f_4}{2f_2^2}\right) \sum_{\mu=0}^{3} \cos(ap^\mu) \right]^2 . \end{aligned} \tag{30}$$

Considering the limit $p_\mu \to 0$ one recovers the expression of the rest mass obtained before.

STRING TENSION

Let us determine the free energy of a static quark-antiquark pair positioned on the z axis (the quark at $z = 0$, the antiquark at $z = L$). For the effective theory the Wilson-operator can be written in the form:

$$w(L; [\chi]) = 2\cos\left(\frac{1}{2}\sum_x Q_x(\bar{u} + \chi_x)\right) = 2\cos\left(\frac{1}{2}\sum_x Q_x\chi_x\right), \tag{31}$$

with the source

$$Q_x = \delta_{x0}\delta_{y0}(\delta_{zL} - \delta_{z0}). \tag{32}$$

The free energy of the static pair is defined by the vacuum expectation value of the Wilson operator:

$$e^{-\mathcal{F}(L)} = \frac{\int \mathcal{D}\chi\, w(L; [\chi]) e^{-\Gamma[\chi]}}{\int \mathcal{D}\chi\, e^{-\Gamma[\chi]}} \tag{33}$$

with the approximate form of the effective action:

$$\Gamma[\chi] = \frac{1}{2} \sum_{xy} \chi_x (\mathcal{D}^{-1})_{xy} \chi_y. \tag{34}$$

The Gaussian integrals can be performed easily. Then we express the propagator in coordinate space through its Fourier-transform and take the sums over x and y. This results in the following expression for the free energy:

$$\mathcal{F}(L) = -\frac{1}{2} \frac{T}{a} a^3 \int_B \frac{d^3 p}{(2\pi)^3} \frac{\sin^2\left(\frac{1}{2} p_z L\right)}{\tilde{\mathcal{D}}^{-1}(\vec{p}_\perp, p_z; p^0 = 0)} \tag{35}$$

where the 3-momentum integral is taken over the Brillouin zone.

The momentum integral can be performed noticing that the integrand exhibits poles on the complex p_z plane for vanishing transverse momentum $\vec{p}_\perp = 0$. Therefore it is a good approximation to set $\vec{p}_\perp = 0$. The remaining one-dimensional integral has significantly different properties for the limiting cases $c \ll 1$ and $c \gg 1$.

For $c \ll 1$ the integrand has simple poles at $p_z = \pm im$,

$$\int_{-\pi/a}^{\pi/a} dp_z \frac{\sin^2\left(\frac{1}{2} p_z L\right)}{1 + \frac{1}{2} m^2 a^2 - \cos(ap_z)} \approx \frac{2}{a^2} \int_{-\infty}^{\infty} dp_z \frac{\sin^2\left(\frac{1}{2} p_z L\right)}{m^2 + p_z^2}. \tag{36}$$

Rewriting the denominator of the integrand in terms of exponential functions, we can close the integration path in each of the terms through a half circle of infinite radius. Making use of the residuum theorem we obtain:

$$\mathcal{F}(L) = \text{const.} \frac{T}{a} \left[1 + \text{const.} (2 + m^2 a^2)^{-L/a} \right]. \tag{37}$$

Consequently the string tension vanishes:

$$\kappa = \lim_{L \to \infty} \frac{\mathcal{F}(L)}{LT} = 0. \tag{38}$$

For $c \gg 1$ the integrand on the r.h.s. of Eq. (35) has double poles at $p_z = \pm im$:

$$\int_{-\infty}^{\infty} dp_z \frac{\sin^2\left(\frac{1}{2} p_z L\right)}{\left(1 + \frac{1}{2} m^2 a^2 - \cos(ap_z)\right)^2} \approx \frac{4}{a^4} \int_{-\infty}^{\infty} dp_z \frac{\sin^2\left(\frac{1}{2} p_z L\right)}{(m^2 + p_z^2)^2} \sim \frac{L}{a} e^{-mL}. \tag{39}$$

The L dependence of the integral is obtained by expressing the denominator of the integrand in terms of exponential functions, closing the integration path on the complex p_z plane in each term appropriately, and making use of Cauchy's integral formula.

For the string tension we obtain:

$$\kappa = -\lim_{L \to \infty} \frac{1}{2} f_4 m^2 e^{-mL}. \tag{40}$$

This reveals the following possibility. If there is a finite scale $a \to a_*$ at which the mass vanishes, $m^2 \to 0$, but the expression $f_4 m^2$ remains finite, then a finite string tension exists:

$$\kappa = -\lim_{m \to 0} \frac{1}{2} f_4 m^2 \neq 0. \tag{41}$$

CONCLUSIONS AND OUTLOOK

Summarizing we have shown that depending on the coupling constant c the skelet $SU(2)$ Yang-Mills theory with time independent global center symmetry can have a phase structure. Only a finite number of the coupling constants, c, f_2, and f_4 seems to be of significance in this respect.

For $c \ll 1$ the string tension vanishes, and the renormalized rest mass of the A_0 gluons is positive:

$$m^2 = \frac{1}{a^2} \frac{-f_4}{f_2^2} \frac{1-c}{c}. \tag{42}$$

Consequently test $SU(2)$ charges are screened at a distance $1/m$ due to Debye-screening. The effective potential is periodic and its curvature at the minima is defined by the renormalized mass m. The ground state is characterized by a non-vanishing value of the gluon field $\bar{u} \neq 0$ and center symmetry is broken. Thus the theory decribes a deconfined phase for $c \ll 1$.

An extrapolation of the renormalized rest mass to increasing values of the coupling constant $c \to 1$ can be a sign of the possibility of the existence of a phase with vanishing rest mass and restored global center symmetry. The same possibility has been shown investigating the properties of the theory for $c \gg 1$. If the renormalized mass

$$m^2 = \frac{1}{a^2} \frac{-f_4}{2 f_2^2} \frac{1}{c} \tag{43}$$

can vanish with $f_4 m^2$ remaining non-zero, the string tension is non-vanishing. Then the test charges are not screened and the effective potential becomes a constant. This leads to the restoration of global center symmetry in the ground state. These are just the properties of a confined phase.

Finally some remarks on our procedure and the results obtained above:

- Our procedure of deriving the effective theory contains a non-perturbative re-summation of the infinitely many vertices generated by the Haar measure, i.e. by global center symmetry. Due to that we expect our approach to reveal the IR (long distance) features of $SU(2)$ Yang-Mills theory correctly. On the other hand, it is necessary the incorporation of the spatial components of the gauge field for the extension of the model to the UV regime, which is in progress.

- The cut-off dependent effective potential plays a crucial role. Such a potential is absent in the usual perturbative approach. In that case the usage of the flat integration measure in the path integral is sufficient to ensure the invariance of the generating functional under infinitesimal gauge transformations. Even if one would use the Haar measure in the perturbative approach, the effective potential would vanish in dimensional regularization. This is due to the factor a^{-4} of the effective potential corresponding to the vanishing integral over four-momentum[5].

- Our gauge fixing violates global center symmetry except for the symmetry under time independent global center transformation. It is therefore rather probable that the effective action does not include all the excitation modes restoring center symmetry of the vacuum. This can be the reason why we did not really get $m = 0$ for $c \gg 1$. A better choice of the gauge condition is to be think of.

ACKNOWLEDGEMENT

The author is deeply indebted to J. Polónyi and A. Schäfer for the stimulating and useful discussions, and to W. Greiner for his permanent interest and kind hospitality in Frankfurt. This work has been supported by a joint project of the Hungarian Academy of Sciences and the Deutsche Forschungsgemeinschaft, and by the Hungarian Research Fund (OTKA 2192/91).

REFERENCES

1. K.G. Wilson, *Phys. Rev.* D14:2445 (1974).

2. J. Polónyi, *Phys. Lett.* B213:340 (1988).

3. J. Polónyi, *in:* "Quark-Gluon Plasma," R.C. Hwa, ed., World Scientific, Singapore (1990) p. 1.

4. J. Polónyi, F. Csikor, A. Patkós, and K. Szlachányi, "Selected Topics in Quark Confinement," Eötvös University, Budapest (1992)

5. K. Johnson, L. Lellouch, and J. Polónyi, *Nucl. Phys.* B367:675 (1991).

6. J. Zinn-Justin, "Quantum Field Theory and Critical Phenomena," Clarendon Press, Oxford (1989).

DEVELOPING THE DYNAMICAL STRING
MODEL: ROLE OF RESONANCES

B. Iványi[1], Zs. Schram[1], K. Sailer[1], W. Greiner[2]

[1]Department for Theoretical Physics
Kossuth Lajos University
H-4010 Debrecen, Pf. 5, Hungary

[2]Institute for Theoretical Physics
Johann Wolfgang Goethe University
D-60054 Frankfurt am Main 11, Postfach 111932, Germany

INTRODUCTION

We show that the Dynamical String Model[1] improved by incorporating discrete final state resonances provides a rather good quantitative description even for the transverse momentum distribution of the fragments in pp collision, where the earlier version of the model failed.

A Dynamical String Model was worked out in the recent years with the purpose to simulate the time evolution of the hadronization process at high energies[1]. In this model the excited hadrons are described by classical open strings. These hadronic strings exhibiting finite transverse size correspond to the chromoelectric flux tubes of the flux tube model of hadrons. The fragmentation takes place via string decay in which the strings break into smaller ones. It can be considered as a consequence of the quark-antiquark pair production along the string suggested by the flux tube model. The decay constant is determined by the string tension and the string radius[2]. The collision of strings is introduced as a result of classical arm exchange mechanism (rearrangement). The appropriate hadronic cross sections are achieved by fitting the string radius[1]. The radius value $r = 0.5fm$ also explains the observed fraction of strangeness production from hadronic flux tubes[2].

The results for 2-jet events in the $e^+ - e^-$ annihilation obtained by the model were in good qualitative agreement with the experimental data up to $\sqrt{s} \approx 50 GeV$.

However, the transverse momentum distribution was underestimated in the region of $p_T \approx 1.5 GeV/c$. This distribution in pp collision at 28.5 and $200 GeV/c$ laboratory momenta failed even in the low transverse momentum region below $1 GeV/c$. The explanation is that the string decay process and string-string collisions do not provide the string fragments with the transverse momenta sufficient to reproduce the experimental data.

Partially from that reason, it has been suggested in the Dynamical String-Parton Model that one should introduce interaction in which the quarks associated with the string ends are involved[3]. In this way it is easy to make connection between string interaction and partonic processes. In this model the proton structure function is taken into account by an ensemble of strings with appropriate string ends dynamics. The results given by the model fit to the data in $e^+ - e^-$ annihilation and pp collision as well up to $\sqrt{s} \approx 50 GeV$[3].

The descriptions of the string interaction mentioned above represent conceptually different approaches for modelling the collision of highly excited hadrons. In the Dynamical String Model the rearrangement mechanism implies that the interaction is closely related to the gluon field forming the flux tube. Quarks represented by the string ends take part in the interaction used by the other model and they can interact with phenomenological amplitudes. It is probable that the real interaction process is some kind of mixture of both interaction mechanisms.

It is the purpose of this work to investigate the effect of final state hadron resonances emerging during the fragmentation process in the framework of the Dynamical String Model. In the earlier version of the model, no such discrete final states were included: the strings had a continuous mass spectrum. Now we use the more realistic assumption that the hadrons in the low energy region (below a threshold) are strings in the rotating rod mode with a discrete mass spectrum. The rest masses of these resonances and their decay modes are introduced according to the phenomenology[4]. It is the crucial point that the products of resonance decay carry off an appropriate amount of momentum according to phenomenology. We will see that the results obtained by the model with resonances are in good quantitative agreement with the experimental data even for the transverse momentum distribution up to $p_T = 1.5 GeV/c$ in $e^+ - e^-$ annihilation at $\sqrt{s} = 22, 35$ and $50 GeV$ and pp collision at laboratory momentum $p = 29 GeV/c$.

In this way we show that neither the Dynamical String Model with rearrangement like flux tube-flux tube interaction, nor the String-Parton Model with constituent quark interaction via parton scattering can be preferred as far as considering the elementary hadronization processes mentioned above.

RESONANCES IN THE STRING MODEL

The highly excited hadrons represented by classical open strings have a continuous mass spectrum. The assumption, however, that the states of hadrons can be described by classical objects fails in the low energy region. In this region the quantum mechanical effects become important and result in the discrete mass spectrum of hadron resonances. The quantum mechanics of hadronization has not been understood because of its nonperturbative nature. One of the possibilities to take into consideration the discrete final states is to introduce the hadron resonances with their discrete masses and decay properties into the model phenomenologically. One has to choose two mass thresholds, one for baryon resonances and another for meson resonances below which the hadronic strings are considered as discrete resonances.

The resonances are described by strings rotating as a rigid rod, because these objects lie on the leading Regge trajectory. On the other hand this representation lets them interact via string-like interaction. The decay of resonances is governed by their half life taken from particle data table[4]. The decay products are represented also by rotating rods. The magnitude of the momenta distributed among the produced particles in the decay of a resonance are taken to be equal to the average momentum indicated by particle data table. This is a crucial point, because in the string decay process the transverse momenta of the emerging strings are limited as a consequence of the Gaussian distribution of transverse momentum calculated in the flux tube model[5]. Collision and decay process of strings can produce strings having masses below the mass threshold according to the string interaction process. In our simulation code we replace these objects with rotating rods characterized by resonance properties. The decay channel is chosen with the probability proportional to the degeneracy of the emerging resonances. The conservation of energy, momentum and the center of mass is maintained in every elementary interaction process.

RESULTS OF THE SIMULATION

The simulation of hadronization after $e^+ - e^-$ annihilation is the first test for our Dynamical String Model. We have performed calculations for c.m. energies $\sqrt{s} = 22GeV$, $35GeV$ and $50GeV$. In addition to the experimental data we will refer to the results from Dynamical String Model without resonances[1]. The advantage of this comparison is that the effects of hadron resonances can be seen immediately. The simulation provides us with numbers of one-particle distributions and shows the energy dependence of these spectra. 500 two-jet events per energies have been simulated for achieving good statistics. In the process the string collision does not play any role, so it tests our prescriptions for decay.

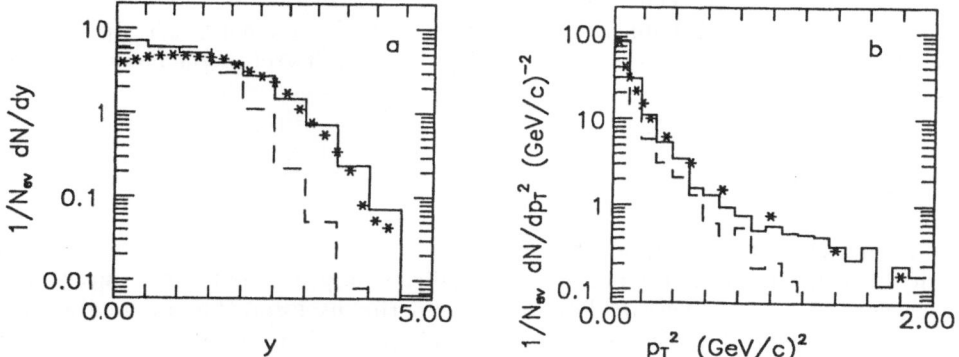

Figure 1. Rapidity (a) and transverse momentum (b) distribution for $e^+ - e^- \rightarrow hadrons$ reaction. The new resuts are at $\sqrt{s} = 22GeV$ (solid line), the results from the model without resonances at $\sqrt{s} = 20GeV$ (dashed line) and experimental data[6] at $\sqrt{s} = 22GeV$ (asterisk).

Comparing to the earlier results without discrete resonances, the rapidity and transverse momentum (Figure 1) distributions are now in good quantitative agreement with experimental data, whereas x_F and KNO distributions remained as good as they were. The charged particle multiplicity is slightly (by the factor 1.2) overestimated. (Rapidity distributions are now calculated using the pion mass, similarly to the evaluation of experimental data.)

We have investigated the pp collision at $29 GeV/c$ beam momentum. The proton initial states are described by rotating rods according to the prescription for resonances. We have performed 800 events at this energy for relatively small statistical error.

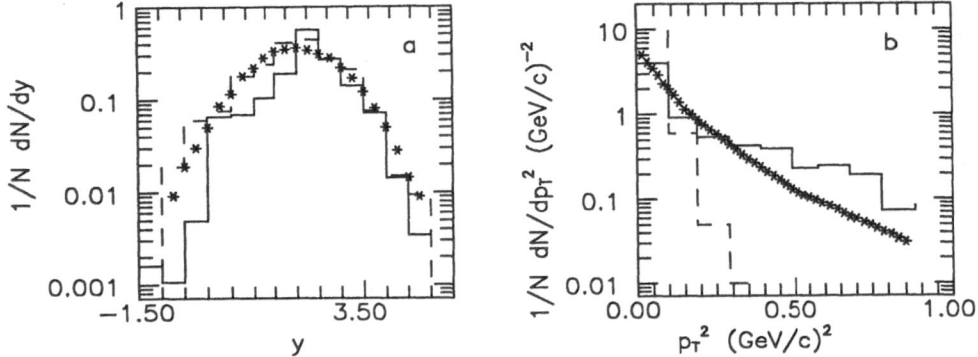

Figure 2. Rapidity (a) and transverse momentum (b) distribution for pp collision. The results are obtained by the model with resonances at $p_{beam} = 29 GeV$ (solid line) and the model without resonances at $p_{beam} = 28.5 GeV$ (dashed line). The experimental data (represented by the line with asterisk) at $p_{beam} = 28.5 GeV$ are taken from Ref.[8] for Fig. a and from Ref.[7] for Fig. b.

For the pp collision multiplicity, rapidity (Figure 2) and x_F distributions give back the experimental data quite well, as before[1]. The transverse momentum distribution of the fragments in pp collision was underestimated by more than an order of magnitude using the version of the model without discrete resonances. Now the transverse momentum distribution (Figure 2) is in good agreement with experimental data. This is due to generating momentum by decay of discrete resonances according to phenomenology.

CONCLUSIONS

Summarizing, the incorporation of discrete final state resonances into the Dynamical String Model results in a rather significant improvement of the quantitative agreement between model calculations and experimental data for fragmentation in $e^+ - e^-$ annihilation (at $\sqrt{s} = 10 - 50 GeV$) and pp collision (at $29 GeV/c$). Therefore it is not unavoidable to turn to the string-parton picture in order to get a reasonable description of these hadronization processes in the energy region considered above.

On this basis neither the Dynamical String Model with rearrangement like flux tube-flux tube interaction nor the String-Parton Model with constituent quark interaction via parton collisions can be preferred.

ACKNOWLEDGEMENT

This work was supported by National Research Fund (2192/91), Foundation for Hungarian Higher Education and Research (57/92) and partly by the Comission of the European Communities (DG XII-B). Computations have been performed on the Microvax 2000 donated by the Humboldt Foundation. Z.S. thanks W. Greiner for his kind hospitality at the Institute for Theoretical Physics, Frankfurt.

REFERENCES

1. K. Sailer, W. Greiner and B. Müller, *in:* "Quark-Gluon Plasma" R. C. Hwa, ed., World Scientific, Singapore (1990) p. 299

2. Th.Schönfeld, A. Schäfer, B. Müller, K. Sailer, J. Reinhardt and W. Greiner, *Phys. Lett.* B247:5 (1990).

3. D. J. Dean, A. S. Umar, J. S. Wu and M. R. Strayer, *Phys. Rev.* C45:400 (1992)

4. Particle Data Group, *Phys. Rev.* D45:I.1 (1992)

5. K. Sailer, Th. Schönfeld, A. Schäfer, B. Müller and W. Greiner, *Phys. Lett.* B240:381 (1990)

6. P. Mättig, *Phys. Rep.* 177:141 (1989).

7. J. Whitmore, *Phys. Rep.* 10:273 (1974)

8. W.H. Sims et. al., *Nucl. Phys.* B41:317 (1972)

ON THE ORIGIN OF MATTER

W.N. Cottingham[1], D. Kalafatis[2], and R. Vinh Mau[3]

[1] Physics Dept., University of Bristol, Bristol BS8 1TH, UK
[2] Dept. of Theoretical Physics, The Schuster Laboratory,
  University of Manchester, Manchester, M13 9PL, UK
[3] Division de Physique Théorique, Institut de Physique
  Nucléaire, 91406 Orsay Cedex and LPTPE, Université P.&M.
  Curie, 4 Place Jussieu 75252 Paris Cedex 05, France

INTRODUCTION

It has been recently realized that the predominance of matter over antimatter apparent in the universe need not be an ad hoc boundary condition set at the beginning of time nor even a feature generated at a very early epoch by an underlying grand unified theory, but a consequence, more or less direct, of the standard models of particle physics and cosmology as they are understood today. There are several important but disparate ingredients that have to be brought together to argue that the standard model of particle physics can by itself generate the baryons and electrons that make up the matter in the universe. At the present time the reasoning is qualitative and imprecise and there are many questions that can only be partially answered. It is the purpose of this paper to present some of our results on these important questions and to bring together these different ideas into a coherent picture.

Within the context of big bang cosmology the universe started very small and very hot and has since been expanding and cooling according to the equations of general relativity and particle physics. The standard model of particle physics is successful in describing physics up to the maximum energies available in our laboratories ≈ 200 GeV, so it is with some confidence that we can try to understand the evolution of the universe after the temperature had fallen to this order of magnitude. Figure 1 illustrates qualitatively how this evolution is thought to have taken place

Hot and Dense Nuclear Matter, Edited by
W. Greiner *et al.*, Plenum Press, New York, 1994

and puts into this context the important epoch for matter genesis. In our scenario, this occured during the electroweak phase transition which, as will be explained in the next section took place when the temperature has fallen to about 100 GeV.

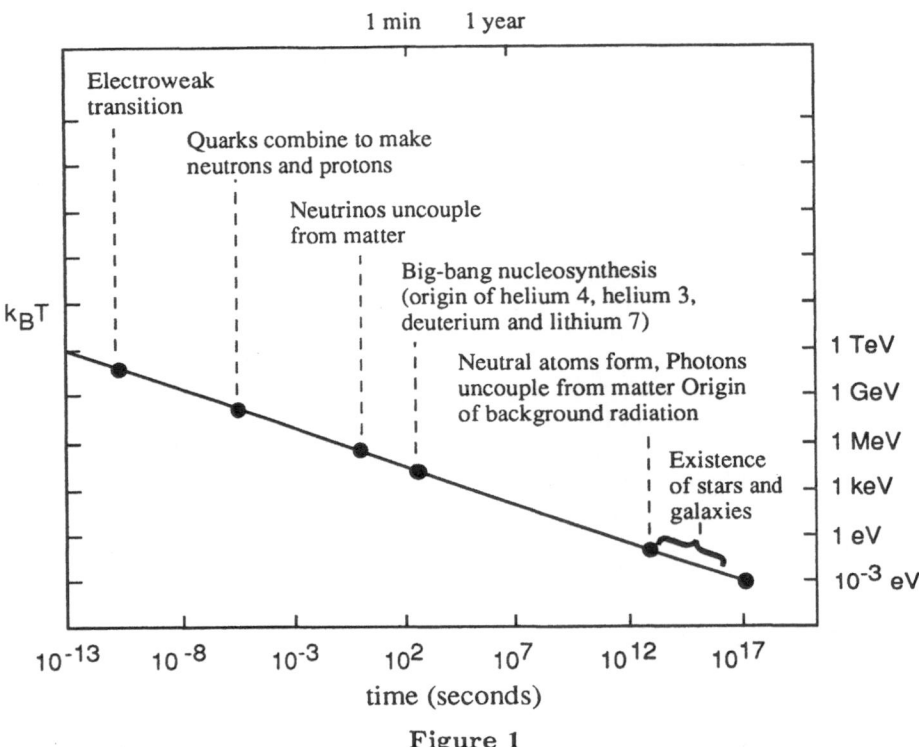

Figure 1

At the semi classical level the standard model has four particle conservation laws, those of baryon number and, separately, electron muon and tau lepton numbers. However, as pointed out by t'Hooft[1] following his analysis of renormalizability, the quantum corrections to the model, coming from single fermion loops, nullify these laws at the absolute level. The currents only obey partial conservation equations[2], for example.

$$\partial_\mu j_B^\mu = \frac{3}{64\pi^2}\epsilon^{\mu\nu\alpha\beta}\left\{\frac{1}{2}g^2\mathrm{Tr}(F_{\mu\nu}F_{\alpha\beta}) + g'^2 f_{\mu\nu}f_{\alpha\beta}\right\}$$

$$F_{\mu\nu} = \partial_\mu A_\nu - \partial_\nu A_\mu + \frac{ig}{2}[A_\mu, A_\nu] \tag{1}$$

$$f_{\mu\nu} = \partial_\mu a_\nu - \partial_\nu a_\mu$$

$A_\mu = A_\mu^a \tau_a$ and a_μ are the $SU(2) \otimes U(1)$ electroweak gauge boson fields.

It is very significant that the right hand side of these equations can also be written as the divergence of a new current, in particular

$$\frac{1}{64\pi^2}\epsilon^{\mu\nu\alpha\beta}\left\{\frac{1}{2}g^2\text{Tr}(F_{\mu\nu}F_{\alpha\beta}) + g'^2 f_{\mu\nu}f_{\alpha\beta}\right\} = \partial_\mu j_T^\mu \tag{2}$$

with $j_T^\mu = \dfrac{1}{32\pi^2}\epsilon^{\mu\nu\alpha\beta}\left\{\dfrac{1}{2}g^2\text{Tr}(A_\nu F_{\alpha\beta} - \dfrac{ig}{3}A_\nu A_\alpha A_\beta) + g'^2 a_\nu f_{\alpha\beta}\right\}.$

This is the current of the topological charge density of the electroweak sector of the standard model. Although the expression of j_T^μ is not gauge invariant the total topological charge

$$N(t) = \int j_T^0(\vec{r}, t)d^3\vec{r} \tag{3}$$

is invariant provided that the integration is over the whole of space and that the fields are only subject to "small" gauge transformations. The partial conservation equations tell us that any change of topological number must be accompanied by a change in baryon and all three lepton numbers since, for example

$$\partial_\mu(j_B^\mu - j_T^\mu) = 0 \tag{4}$$

$N_B(t) - 3N(t)$, and $N_i(t) - N(t)$ are time independent.

The topological charge of a field configuration is changed by an integer if the fields are subject to a large gauge transformation

$$A_\mu \rightarrow A'_\mu = UA_\mu U^\dagger + \frac{2i}{g}\partial_\mu UU^\dagger \tag{5}$$

An example of a large gauge transformation is

$$U = \exp(i\vec{\tau}.\hat{r}f(r)) \tag{6}$$

and f(r) any function of the radial coordinate r with

$f(0) = \pi$ and $f(r) \rightarrow 0$ as $r \rightarrow \infty$.

Although the energy densities e.t.c. in the gauge and Higgs fields are the same in both gauges the topological charge of the A'_μ field is increased by one unit (There is no topological change of charge with gauge transformations of the U(1) field a_μ). Also, there is no simple interpolating field between A'_μ and A_μ i.e. , it is inadmissible, for example, to scale $f(r)$ down ($f(r) \rightarrow \lambda f(r), 0 < \lambda \leq 1$). Such discontinuous transformations are not true gauge transformations. In this example, the singular point at the origin induces an infinite energy density and in fact an infinite energy into the fields. The most simple interpolating field configuration between A'_μ and A_μ involves the sphaleron[3] which presents a large, but finite, energy barrier between the two sectors.

$$E_{\text{barrier}} \approx \frac{4\pi}{g^2}5M_W \tag{7}$$

t'Hooft first investigated quantum tunneling between the different topological sectors of the theory and found that the zero temperature transition rates are suppressed by a factor[4]

$$\exp(-\frac{4\pi}{g^2}\sin^2\theta_W) \approx 10^{-145}$$

thus zero temperature transition rates are negligeably small. However, thermal excitations over the barrier can be expected to increase the transition rate at finite temperature. Also, at temperatures above the electroweak phase transition the Higgs field, and hence all particle masses, have zero expectation value, equation (7) would suggest that the energy barrier is removed and transition rates along with baryo and lepto genesis, can proceed unimpeded by any barrier.

The expectation value of the Higgs field is of paramount importance in all of these considerations. At zero temperature it is the field value that minimizes the potential energy density

$$V(\phi) = \frac{m_H^2}{4\phi_0^2}(\phi^2 - \phi_0^2)^2 \tag{8}$$

m_H is the mass of the Higgs boson and $\phi_0 = 262$ GeV is the VEV of the Higgs field at the minimum. At finite temperature ϕ adopts a mean value that minimizes the free energy density. The next section is devoted to the form of the free energy density as a function of the mean value of ϕ and of the temperature. The electroweak phase transition will be shown, within our approximations, to be of first order. An important aspect of a first order phase transition is the rate at which bubbles of the new phase nucleate in the old. Bubble nucleation rates, as a function of temperature are also estimated in this section. In the following section, we present our calculations of the expanding bubble profiles and their speeds of expansion after nucleation. The expansion rate is of importance for estimating the degree of supercooling attained before the transition is complete and the bubble profiles are needed in the last section which deals with the matter genesis that can take place during this phase transition.

THE ELECTROWEAK PHASE TRANSITION

At any fixed temperature we will consider the Higgs field ϕ to be constrained to some fixed value. The free energy density, which is the negative of the pressure is then a function of ϕ, $f(\phi)$,say. Since regions of high pressure force out regions of low pressure the unconstrained Higgs field adopts the value at which the pressure is maximum and hence the free energy density is minimum. We first compute $f(\phi)$ in the most simple approximation, that of treating the plasma of bosons and fermions as a plasma of free particles and since we are considering temperatures that are far above the quark hadron phase transition (≈ 100 MeV, see Figure 1) the particles and antiparticles in the plasma are the fundamental quarks and leptons. Also, the net quark and lepton number densities are always small and it is appropriate at this stage to neglect their chemical potentials, then, for a fixed magnitude of the Higgs field ϕ the free energy density[5]

$$f(\phi) = V(\phi) \mp k_B T \frac{1}{2\pi^2} \sum_{i=1}^{118} \int_0^\infty dk k^2 \log[1 \pm \exp(-(k^2 + \lambda_i^2 \phi^2)^{1/2})] \qquad (9)$$

leaving aside the Higgs boson the sum is over all particle types i, λ_i is the coupling constant of the particle i to the Higgs field so that $\lambda_i \phi$ is the particle mass. The term -1 is appropriate for fermions, +1 for bosons. Also, in the sum over particle types the W boson for example must be counted six times, three times for spin and twice for its two charge states, similarly one electron (positron) must be counted four times and, including color, each quark twelve times there are a total of 118 terms in the sum. Making a change of integration variable $\beta k = x$ or $k = (k_B T)x$, we can also write

$$f(\phi) = V(\phi) \mp (k_B T)^4 \frac{1}{2\pi^2} \sum_{i=1}^{118} \int_0^\infty dx x^2 \log[1 \pm \exp(-(x^2 + \lambda_i^2 \frac{\phi^2}{k_B T^2})^{1/2})] \qquad (10)$$

All of the couplings λ_i are well known except for that of the top quark t, the top mass is thought to be about 100 GeV and this implies λ_{top}=0.381. We have taken this value and have computed f(ϕ) which is displayed in Figure 2 for selected temperatures in the range 86 GeV $< k_B T <$ 88 GeV. Because of its large coupling the top quark plays an important role in determining these curves but the basic form with the double minimum is not so sensitive to the top mass.

The contributions of Higgs field fluctuations are not included at this stage of the calculations. As we show later in this section it is not difficult to include them in the important regions around the minima of free energy where they correspond to the contributions of the Higgs bosons in the plasma. Anyway, there is only one Higgs boson so these contributions will be small in comparison with the nine gauge bosons and twelve top quarks.

At early times the universe is hot but cooling and expanding. The equation of state of our plasma of free particles with the Einstein equations predict that the temperature decreases with time t according to the formula[6]

$$T = A/t^{1/2}$$

$$A = (\frac{45}{16\pi^3 G N_{eff}})^{1/4} = 0.49 \ 10^{-3} \ GeV(sec)^{1/2} \qquad (11)$$

$N_{eff} = 109$ is the number of relativistic degrees of freedom and the expansion rate $\dot{R}/R = 1/2t$.

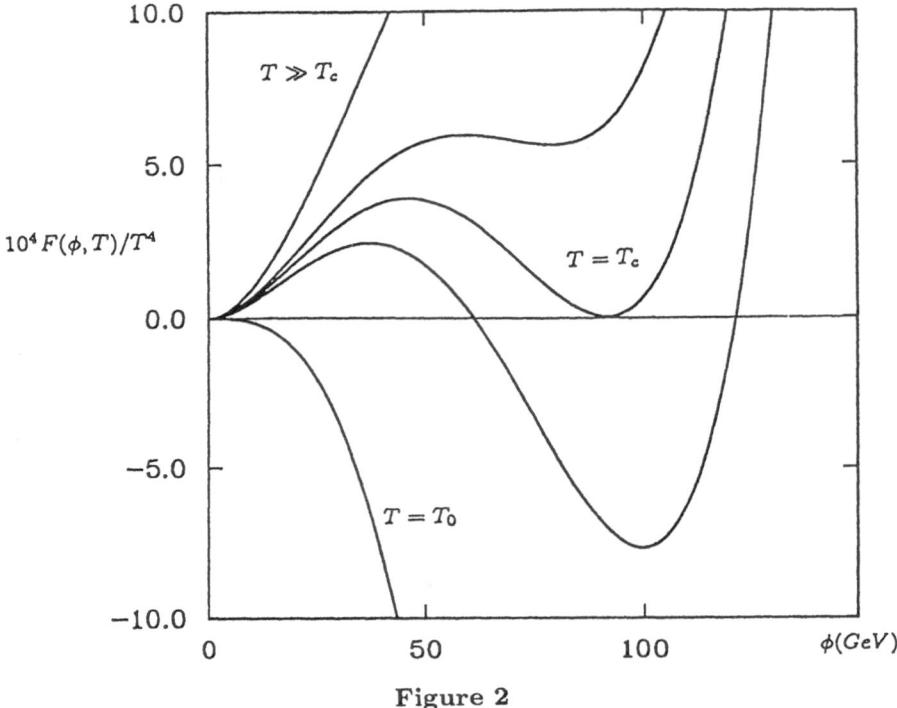

Figure 2

From Figure 2 for $T = T_c = 87.49$ GeV the free energy has two degenerate minima, this is the signature of a first order phase transition. For $T \gg T_c$ ϕ is zero and the particles are massless ; this is the high temperature phase. As the temperature falls below T_c the true minimum is the second minimum with $\phi \approx 100$ GeV. This is the low temperature phase in which the particles acquire mass. The details of how ϕ jumps over the barrier from the first minimum to the second is the subject of the rest of this section. As T falls below T_c the phase transition can be anticipated to proceed first by the formation of bubbles of the stable phase ($\phi \neq 0$) in the unstable phase ($\phi = 0$). If the bubbles are large enough these are driven by their superior pressure to expand and eventually completely replace the $\phi = 0$ phase.

It is at this stage that we consider the fluctuations of the ϕ field. If we work in the high temperature approximation in which the time derivatives of ϕ are neglected then the total free energy F including ϕ field fluctuations is given by

$$e^{-\beta F} = \int \delta \phi e^{-\beta \int [f(\phi) + \frac{1}{2}(\vec{\nabla}\phi)^2] d^3 \vec{r}} \qquad (12)$$

where the functional integral is over ϕ as function of space. To evaluate this integral it is important to find those particular functions $\phi_s(\vec{r})$ that are stationary values of the exponent $\int [f(\phi) + \frac{1}{2}(\vec{\nabla}\phi)^2]d^3\vec{r}$.

For us, two of them are obvious, $\phi=\phi_s=0$ at the first minimum of $f(\phi)$ and ϕ_s = the constant ϕ at the second minimum. The quadratic fluctuations about these stationary values give an estimate of the fluctuating contributions to F. Also, in a large volume, the lower minimum completely dominates exponentially the upper, the lower minimum gives the true free energy.

For $T < T_c$ there is also an important stationary function $\phi_s(\vec{r})$ that describes the critical bubble profile[7]. For a small bubble the forces of surface tension implicit in our expression through the gradient term $\frac{1}{2}(\vec{\nabla}\phi)^2$ dominates the internal pressure and the net force is for contraction. For a large bubble the converse is true and, as we will see in the next section the bubble is driven to expand and expand. The critical bubble profile sits between these two regimes, it is the shape that must be attained, or almost attained for a small bubble, a fluctuation from $\phi(\vec{r}) = 0$, to survive and expand. We can anticipate that the critical bubble will be spherical and if it is centred at the origin of coordinates it will have the form $\phi_s(r)$ and satisfy the Euler Lagrange equation

$$-\frac{1}{r}\frac{d^2}{dr^2}r\phi_s(r) + \frac{\delta f}{\delta \phi}(\phi_s(r)) = 0 \tag{13}$$

with the boundary condition $\phi_s(r) = 0$ as $r \to \infty$ and $\dfrac{d\phi_s(r)}{dr} = 0$ at $r = 0$.

We have solved these equations for various degrees of supercooling. The critical bubble form $\phi_s(r)$ can be used to estimate bubble nucleation rates Γ through the formula[8]

$$\Gamma(T) \approx T^4(\frac{S}{2\pi T})^{3/2}e^{-S/T} \tag{14}$$

with $S = 4\pi \displaystyle\int_0^\infty [\frac{1}{2}(\frac{d\phi_s}{dr})^2 + f(\phi_s(r))]r^2 dr$.

This formula is derived from a consideration of the contributions to the functional integral (eq. (12)) of the fluctuations around $\phi_s(r)$. The prefactor is only a rough estimate of a term coming from the fluctuations[7]. The most important factor $e^{-S/T}$

contains the exponential suppression and it can be reliably calculated. Figure 3 illustrates the evolution of the exponent S/T with temperature.

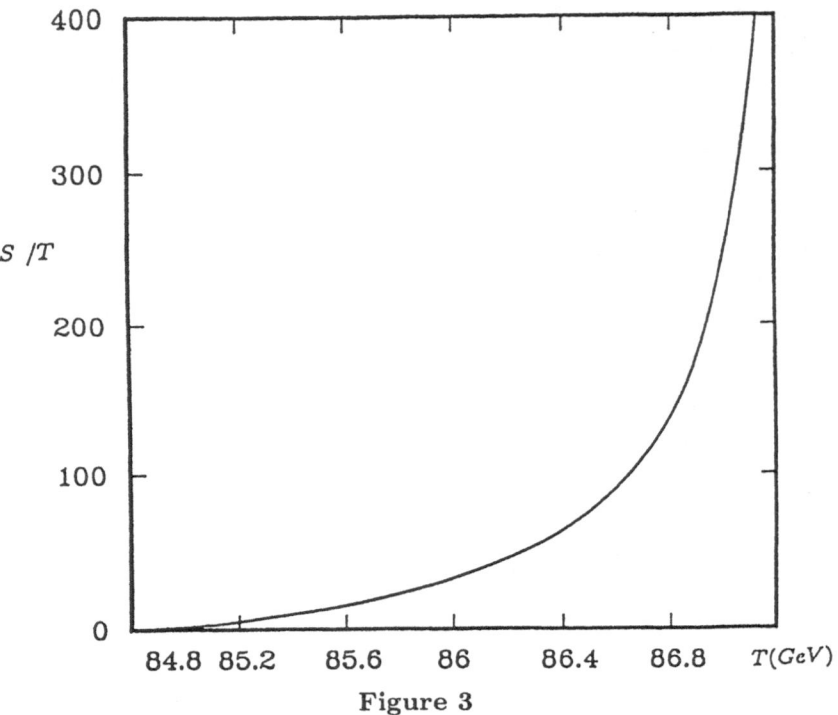

Figure 3

A QUALITATIVE DESCRIPTION OF THE COMPLETION OF THE PHASE TRANSITION

The bubble nucleation rate estimated in the previous section is the formation rate per unit volume of bubbles that are large enough to expand. A simple approximate form for the total free energy which includes now the kinetic energy of expansion is

$$\epsilon(t) = \int \left\{ f(\phi) + \frac{1}{2}(\vec{\nabla}\phi)^2 + \frac{1}{2}(\frac{\partial \phi}{\partial t})^2 \right\} d^3\vec{r} \qquad (15)$$

It will also be that energy is lost into stimulating excitations above the thermodynamic background in those regions where ϕ is out of thermal equilibrium and in particular is changing with time. To understand completely the form of this rate of energy loss would require some detailed field theory calculations but energy loss there should be and we will show later in this section that a plausible form is given by

$$\frac{d\epsilon}{dt} = -\frac{1}{\tau} \int (\frac{\partial \phi}{\partial t})^2 d^3\vec{r} \qquad (16)$$

and τ is a parameter with the dimensions of time.

Substituting this form into eq. (16) gives

$$\int \left\{ \frac{\delta f(\phi)}{\delta \phi} - \vec{\nabla}^2 \phi + \frac{\partial^2 \phi}{\partial t^2} + \frac{1}{\tau} \frac{\partial \phi}{\partial t} \right\} \frac{d\phi}{dt} d^3 \vec{r} = 0 \tag{17}$$

(we have made an integration by parts). A spherically symmetric expanding bubble, $\phi(r,t)$ centred at the origin will satisfy this equation

$$\frac{\delta f(\phi)}{\delta \phi} - \vec{\nabla}^2 \phi + \frac{\partial^2 \phi}{\partial t^2} + \frac{1}{\tau} \frac{\partial \phi}{\partial t} = 0 \tag{18}$$

We have solved this numerically with the starting conditions of perturbations about the critical profiles of Figure 3. We find that in all cases the expanding profile quickly attains an asymptotic form as given by the large r limit

$$\frac{\delta f(\phi)}{\delta \phi} - \frac{\partial^2 \phi}{\partial r^2} + \frac{\partial^2 \phi}{\partial t^2} + \frac{1}{\tau} \frac{\partial \phi}{\partial t} = 0 \tag{19}$$

solutions of this are $\phi(r,t) = \phi(\xi)$ with $\xi = \dfrac{r - vt}{\sqrt{1 - v^2}}$ and

$$-\frac{\delta f(\phi)}{\delta \phi} + \frac{d^2 \phi}{d\xi^2} + +\frac{1}{\tau'} \frac{d\phi}{d\xi} = 0$$
$$\tau' = \frac{\tau (1 - v^2)^{1/2}}{v} \tag{20}$$

The boundary conditions are such that $\frac{df}{d\phi} = 0$ at $\xi = \pm\infty$, $\phi(+\infty) = 0$ and $\phi(-\infty) = \phi_0$, the other minimum of f. Of course the origin ξ is undetermined, we have fixed it by taking $\phi(0) = \frac{1}{2}\phi_0$ and have integrated numerically both forwards and backwards and have adjusted the two parameters $d\phi/d\xi(0)$ and τ' until the two boundary conditions are met. The asymptotic forms that correspond to the critical bubbles of Figure 3 are shown in Figure 4 with the corresponding value of τ'. To estimate the velocity v of bubble wall propagation we need an estimate of the time constant τ. This we do on the assumption that the moving bubble front is equivalent to a beam of Higgs particles with an energy density of twice the kinetic energy density $(\partial \phi / \partial t)^2$ in the front. Suppose the equivalent particles each to have energy q then the number density in the beam is

$$\rho_H = \frac{1}{q} \left(\frac{\partial \phi}{\partial t} \right)^2 \tag{21}$$

The Higgs interact with the particles in the plasma, mostly, presumably, with quarks to produce gluons. If the number density of quark i is ρ_i and the gluon production cross section σ_i then the interaction rate per unit volume is $\sum_i v_i \sigma_i \rho_i \rho_H$ where v_i

is the mean relative velocity of quark i and Higgs. The rate of energy transfer into gluons is then

$$\frac{d\epsilon}{dt} = -q \sum_i v_i \sigma_i \rho_i \rho_H = -(\frac{\partial \phi}{\partial t})^2 \sum_i v_i \sigma_i \rho_i \qquad (22)$$

and this suggests that $\frac{1}{\tau} = \sum_i v_i \sigma_i \rho_i.$

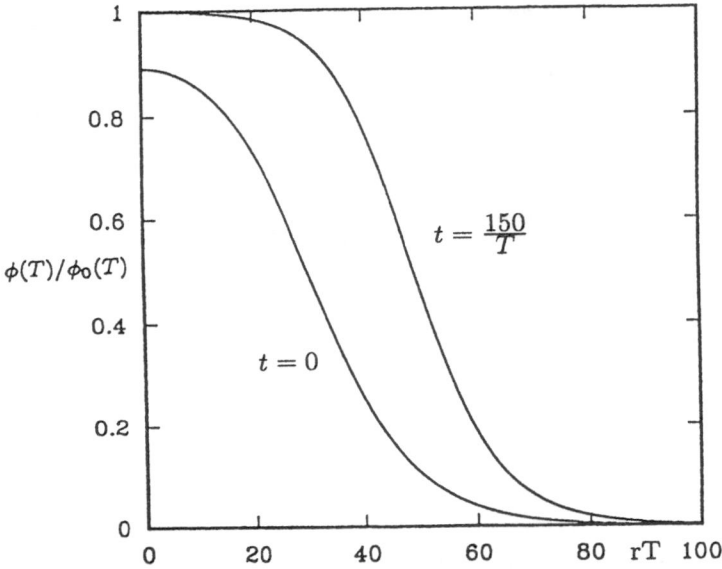

Figure 4. The profile of the Higgs field at T=86.7 GeV.

The cross sections σ_i should contain the dimensionless factors α_S for gluon production, $\lambda_i^2 = m_i^2/\phi_0^2$ for Higgs absorption and dimensionally the cross section should be proportional to $1/m_i^2$. This suggests an equal cross section for every quark $\sigma_i \approx \alpha_S/\phi_0^2$. Since most quarks are light at these temperatures v_i is the velocity of light, and $\rho_i = \frac{T^3}{2\pi^2}\zeta(3)$ (The Riemann zeta function $\zeta(3) = 1.20206$). Hence we take

$$\frac{1}{\tau} = \frac{N_q \alpha_S T^3 \zeta(3)}{2\pi^2 \phi_0^2} \qquad (23)$$

with N_q = the number of quark degrees of freedom N_q = 72. The bubble will propagate with velocity $v = \dfrac{1}{\sqrt{1 + (\tau'/\tau)^2}}$

MATTER GENESIS

It is shown in reference 9 that if the Higgs field is changing so slowly with time that thermal equilibrium can be maintained then one can anticipate that the free energy density will contain a CP violating term proportional to the topological number density ρ_T.

$$f_{CP} = a(t)\rho_T \tag{24}$$

In the standard model with CP violation only in the K.M. matrix, the coefficient function $a(t)$ is too small to account for matter genesis[10]. However it can be much bigger in two Higgs doublet models which have CP violation in the Higgs sector[9] then

$$a(t) = \frac{g^2}{32\pi^2}\frac{d\theta}{dt} \tag{25}$$

with $g = e/\sin\theta w$ the weak coupling constant and θ a CP violating phase difference between the two Higgs fields at the free energy minimum.

Also if ρ_e, ρ_μ and ρ_τ are the net electron, muon and τ lepton number densities and ρ_B the baryon number density, provided that these are small they also contribute to the free energy density a term $(\frac{1}{k_BT})^2[\rho_e^2 + \rho_\mu^2 + \rho_\tau^2 - \frac{3}{4}\rho_B^2]$. In thermal equilibrium the various densities will adopt values that minimize the free energy. However the densities are subject to the constraints that $\rho_B - 3\rho_T$ and $\rho_e - \rho_T$ are constant so that, using the method of Lagrange multipliers the densities are determined by minimizing

$$\begin{aligned} f =&(\frac{1}{k_BT})^2[\rho_e^2 + \rho_\mu^2 + \rho_\tau^2 - \frac{3}{4}\rho_B^2] + a\rho_T - \lambda_e(\rho_T - \rho_e) \\ &+ \lambda_\mu(\rho_T - \rho_\mu) + \lambda_\tau(\rho_T - \rho_\tau) + \lambda_B(3\rho_T - \rho_B) \end{aligned} \tag{26}$$

If the boundary conditions are such that initially $a = 0$, the baryon and lepton densities are zero, as long as thermal equilibrium can be maintained, we will have a common lepton number density ρ_0: $\rho_0 = \rho_e = \rho_\mu = \rho_\tau$ and $\frac{1}{3}\rho_B = \rho_0 = -\frac{2}{39}(k_BT)^2a$. Since a changes with time, the question is, do the processes which maintain thermal equilibrium, proceed at a fast enough rate to keep pace.

In many situations, rates for reactions which are necessary to maintain equilibrium can be equated to the inverse of relaxation times λ and small deviations then approach equilibrium exponentially with time scale λ according to the relaxation equation

$$\frac{d\rho}{dt} = \frac{1}{\lambda}(\rho_0 - \rho) \tag{27}$$

In static situations this has the solution

$$(\rho - \rho_0) = \text{const} \times e^{-t/\lambda} \tag{28}$$

Since in our situation ρ_0 is a function of time and as λ is also a function of time we will modify the relaxation equation to

$$\frac{\partial\rho}{\partial t} = \frac{1}{\lambda(t)}[\frac{2}{39}\beta^2a(t) - \rho] \tag{29}$$

This has the solution

$$\rho(t) = \int_{-\infty}^{t} dt' \frac{2\beta^2}{39} \frac{a(t')}{\lambda(t')} \exp - \int_{t'}^{t} dt'' \frac{1}{\lambda(t'')} \tag{30}$$

The relaxation time λ can be related to the rate of sphaleron transitions per unit volume Γ_s by $\lambda \sim T^3/\Gamma_s$ with

$$\Gamma_s = T^4 \left(\frac{S_s(\phi, T)}{2\pi T} \right)^{3/2} e^{-S_s(\phi,T)/T} \tag{31}$$

This solution exhibits the good properties that outside the bubble $\rho(t)$ is small because the sphaleron rate there is large enough to wash out any net baryon production, but inside the bubble $\rho(t)$ is not suppressed since the sphaleron rate there is exponentially small. Eq. (31) also shows that the baryons are essentially produced in the region where $\frac{\partial\theta}{\partial t}$ is the most significant i.e. near the bubble wall.

References

1.   G. t'Hooft, *Phys. Rev. Lett.* **37**:8 (1976)

2.   S. Deser, R. Jackiw, S. Templeton, *Ann. Phys.* (N.Y.) **140**:372 (1982)
      N.H.Christ, *Phys. Rev.* **D21**:1591 (1980)

3.   F.R. Klinkhammer, N. Manton, *Phys. Rev.* **D30**:2212 (1984)

4.   G. t'Hooft, *Phys. Rev.* **D14**:3432 (1976)

5.   L. Dolan, R. Jackiw, *Phys. Rev.* **D9**:3320 (1974)

6.   S. Weinberg, "Gravitation and Cosmology" John Wiley.

7.   W.N. Cottingham, D. Kalafatis, R. Vinh Mau, *Phys. Rev.* **B48**:6788 (1993)

8.   A.D. Linde, *Nucl. Phys.* **B216**:421 (1983)

9.   N. Turok, J. Zadrozny, *Nucl. Phys.* **B358**:471 (1991)

10.   M.E. Shaposhnikov, *Nucl. Phys.* **B287**:757 (1987) ; *Nucl. Phys.* **B299**:797 (1988).

STOPPING IN THE PRODUCTION OF HIGH
MOMENTUM HADRONS AND DILEPTONS
ON NUCLEI

Engelbert Quack

Institut für Theoretische Physik, Philosophenweg 19
69120 Heidelberg, Germany

INTRODUCTION

On the way towards a thorough understanding of AA collisions, the study of hadron–nucleus reactions constitutes an important step. We will outline in the following, how several nuclear effects can be isolated in inclusive hA reactions by considering specific projectile and final state particles. In particular, we study the energy loss (stopping) of projectiles both on the hadronic and partonic level, and relate it to the observation of final states with high momenta.

For this purpose, the kinematic variable x_F is used, which is defined and measured as $x_F = p_{\parallel}/p_{\parallel}^{max} = [2m_{\perp}/\sqrt{s}]sinh(y)$ with the momenta measured in the c.m. system. Compared to the rapidity y, x_F stretches out the high momentum region.

Nuclear effects in inclusive hA reactions are commonly parametrized by the cross section ratio R *per nucleon* N of the respective process or the related quantity α, defined by

$$\frac{d\sigma}{dx_F}(hA \to h'X) = A^{\alpha(x_F)} \cdot \frac{d\sigma}{dx_F}(hN \to h'X) . \tag{1}$$

Thus, α gives the effectiveness of the nucleus A per nucleon to produce the final state h'; a value of α (or R) < 1 indicates suppression.

Experimentally, a suppression is found and for all final state particles, this is seen to increase with x_F. The suppression is small for Drell–Yan dimuon pairs ($m_{\mu\mu} \geq 4$ GeV), larger for c$\bar{\text{c}}$–pairs in various final states such as J/Ψ, Ψ' and $D\bar{D}$, and most pronounced for light hadrons such as π and K, where α drops from $\alpha(x_F = 0.2) \approx 0.7$ to $\alpha(x_F \to 1) \approx 0.3$ [1, 2, 3].

In this talk we will show that this suppression can be traced back to the energy loss of the initial projectile state.

In this context also the nuclear effects of absorption of final state particles and of the influence of shadowing on the production process need to be considered. For bound states like the J/Ψ, absorption is present, but it results in a constant value of α in the kinematic region of present experiments. Color transparency, on the other hand, would give rise to an *increase* of α with increasing x_F. However, this phenomenon sets in only at lower energies [4]. Since the data do not scale in the Bjorken variable x_2, the effect of shadowing is also not able to account for the observed x_F–dependence [5]. As an alternative mechanism to account for the suppression at least for the $c\bar{c}$–states, the existence of an intrinsic component which is set free in the collison has been proposed, see [6].

Let us now follow the idea of the energy loss of the projectile before producing the final state [7, 8]. It is immediately clear that this effect disables the production of heavy states mainly at high momenta. In order to study such an energy loss on a microscopic level, one must specify to what extent the projectile is resolved in the production process. High Q^2 processes involve partons, and due to the very short reaction time we must consider the propagation of the projectile in terms of its parton constituents. In the next section, we will model the parton energy loss by gluon bremsstrahlung and elaborate its consequences on the production of the heavy final states $\mu^+\mu^-$, J/Ψ and Υ. The production process of light hadrons, on the other hand, does not resolve the projectile partons, and thus the energy loss of the entire projectile hadron has to be considered. This will be outlined in section 4, and it will be shown how the quantity of stopping power can be directly extracted from data on light particle production.

PARTON ENERGY LOSS

Here we first briefly sketch a microscopic treatment of parton energy loss and outline its consequences on the production of heavy final states.

Consider a highly virtual parton traversing nuclear matter which provides color charges upon which the parton can scatter. The lowest order inelastic process that can occur is the bremsstrahlung of gluons. To describe this, we first turn to the corresponding (standard) problem in QED of an electron of mass m traversing a length z of matter which provides a density ρ of charges which are screened within a length l_{scr}. The resulting energy loss of the electron is [9]

$$-\frac{dE}{dz} = \rho \int_0^E \omega d\sigma(\omega) = \rho \sigma_{rad} E, \qquad \text{with} \qquad \sigma_{rad} = \frac{\alpha_e^3}{m^2}[4\ln(l_{scr}m) + \frac{2}{9}] . \quad (2)$$

In integrated form, the electron energy E decreases with the path length z traversed in matter as $E(z) = E(0)exp(-z/\chi_{rad})$ with the radiation length $\chi_{rad} = (\rho\sigma_{rad})^{-1}$.

A limitation of the outlined treatment is given by the possible interference of two emitted charges when the scattering centers are close to each other. In the regime of destructive interference, the cross section saturates to $-dE/dz \propto \sqrt{E}$ as $E \to \infty$. (Landau–Pomeranchuk effect [10, 11]). The situation in QCD is more complicated however. In contrast to QED, there are 8 possible gluonic degrees of freedom, and the borderline between two such regimes is difficult to pin down [12]. In the following we assume that we may stay entirely in the Bethe–Heitler regime.

Taking over the results of QED to the case of strongly interacting quarks, we can write the corresponding energy loss in the form $-dE_s/dz = E/\chi_{rad}$. ¿From eq.(2), $\chi_{rad} \propto Q^2 \approx M^2$ of the final state. The scaling in energy we obtain for the parton energy loss is reflected in the x_F–scaling observed in the data as we will see below.

To get a rough estimate of χ_{rad}, we take as appropriate QCD parameters $\alpha_s(m_\perp^2 = 1\text{GeV}^2) \approx 0.6$, $l_{scr} = 1$ fm and count three color charges within a nucleon. This results in a quark radiation length of $\chi_{rad}^q \approx 200$ fm. Since a gluon scatters with roughly twice the cross section as compared to a quark, an initial state gluon has a lower radiation length of $\chi_{rad}^g \approx \frac{1}{2}\chi_{rad}^q \approx 100$ fm. These estimated values can only serve as a guideline. We will determine them more accurately in the comparison to the experimental data.

We now implement this form for the momentum loss in the standard parton fusion model of heavy fermion pair production in lowest order QCD, which is an adequate approximation for the large masses of interest. For the parton i (q,$\bar{q}$ or gluon) which has travelled a length z in hadronic matter before it fuses with parton j, the cross section is

$$\frac{d^2\sigma}{dx_F dM^2}(ij \to f\bar{f}, z) = \int dx_1 f_i^p(x_1)dx_1'\delta(x_1' - x_1 e^{-\frac{z}{\chi_i}}) \cdot dx_2 f_j^t(x_2)$$

$$\cdot \delta(x_1'x_2 s - M^2)\delta(x_1' - x_2 - x_F) \cdot \hat{\sigma}_{ij}(M^2) \qquad (3)$$

where $f_i^p(x_1)$, $f_j^t(x_2)$ are the parton structure functions of projectile and target, respectively. The symbols x_1 and $f_i^p(x_1)$ refer to the projectile parton momentum and its distribution before it loses energy in the initial state interaction, and σ_{ij} is the elementary parton–parton fusion cross section. The first δ function enforces the energy loss. In the limit $\chi_i \to \infty$, $x_1' \to x_1$ and eq.(3) reduces to the standard factorization form of the cross section.

In this form, the influence of the initial state interaction on the production cross section is rather transparent. The probability that a projectile hadron contributes with a momentum fraction x_1' to the hard fusion process is equal to the probability of a parton being present in the projectile with the larger momentum fraction x_1, and losing a fraction

$$\Delta x_1 = x_1 - x_1' = x_1(1 - e^{-\frac{z}{\chi_{rad}}}) \approx x_1\frac{z}{\chi_{rad}} , \qquad (4)$$

of it in an inelastic process. The parton density is probed at a higher value of x_1 as would be the case for the hard fusion process alone. The decrease of $xf(x)$ with x causes a corresponding decrease of the production cross section which increases with x_F. For illustration purposes, J/Ψ production in a pp collision at 200 GeV is shown in fig. 1 with and without energy loss. The actual value of the energy loss is exaggerated to make the effect visible.

The calculation of the production cross section of a given final state is now straightforward. One has to sum over the different parton–parton fusion contributions, and integrate over the appropriate range of the invariant mass (for Drell–Yan muons, the acceptance range of the corresponding experiment). For bound states like the J/Ψ, one has to account for the absorption. The cross sections so obtained are a function

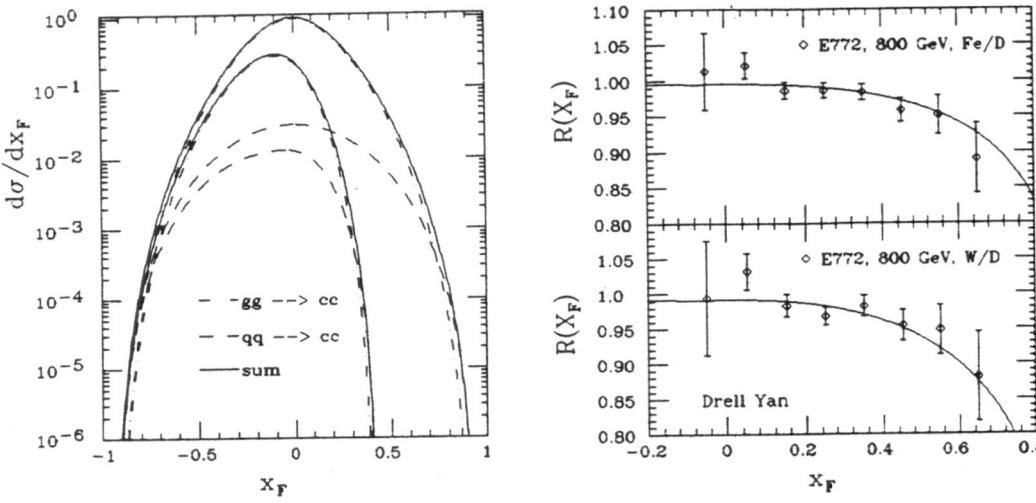

Figure 1: Left side: Differential J/Ψ production cross section in a proton–proton collision. The upper curves are for a 200 GeV incident proton, the lower curves for a proton in the same frame that lost half of its initial energy before producing a J/Ψ. The energy loss causes a depletion of $d\sigma/dx_F$ which increases with x_F. Right side: Result of the energy loss model for Drell–Yan muon pair production compared to E772 data at 800 Gev. Plotted is the ratio $R(x_F)$ for an iron target (up) and for Tungsten (below).

of z and impact parameter $\vec{b}$ and have to be integrated over the appropriate nuclear density.

On the right side of fig. 1, we show the result for the muon pair production. Here, only the $q\bar{q}$ annihilation contributes, and by fitting the results of our model to these data, we get a precise determination of the quark radiation length, with the result $\chi_{rad}^q = (200 \pm 50)$ fm. Note that due to this rather large value, the reduction of the production cross section is only of the order of 1%.

The same procedure can be used for p induced J/Ψ production, which is dominated by gg fusion. The fit to the data yields a value of the gluon radiation length of $\chi_{rad}^g = (100 \pm 25)$ fm, see fig. 2. Both numbers are in good agreement with our estimate. A further test is now provided by considering π and $\bar{p}$ induced J/Ψ production, which constitutes a mixture of gg and $q\bar{q}$ processes. A good agreement with the data is obtained here as well.

In order to check on the Q^2–dependence of χ_{rad}, we mention the production of Υ states. Here, few data are available. These support the results of our calculation, except in the region $x_F \leq 0$. However, as fig. 2 illustrates, the origin of the drop of α at negative x_F is not from a suppression in the heavy nucleus, but is due to an enhancement in the production cross section on deuterium, which remains to be verified experimentally.

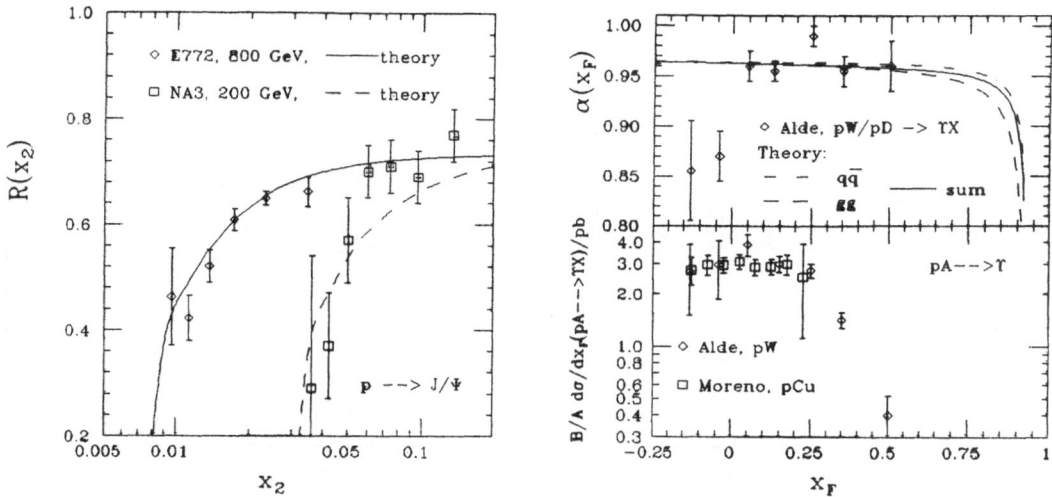

Figure 2: Left side: Proton induced J/Ψ production on heavy targets, plotted in form of the cross section ratio $R(x_2)$. Our results are compared to the measurements of NA3 [1] and E772 [2]. Right side: Results for proton induced Υ production at 800 GeV. The upper part shows $\alpha(x_F)$ for W and D targets, below is plotted $d\sigma/dx_F$ [2, 13]. The data from two different experiments agree well and show no particular asymmetry around $x_F = 0$. The upper data on α at $x_F \leq 0$ therefore do not represent a nuclear effect in the heavy target.

HADRON ENERGY LOSS

In the last part of this talk, we consider final state particles which are so light that their production process does not resolve the internal structure of the projectile hadron, and study the energy loss of this hadron when it enters a nucleus. Only a brief qualitative sketch is given here. For a more detailed treatment see [14].

As before, we concentrate on high momentum states. Particle production is now described in terms of fragmentation functions. Inelastic scattering of the projectile hadron results in a momentum loss with the consequence that the fragmentation function is probed at higher values of x, and thus the production of particles with high momenta is reduced.

We sketch this effect here in a qualitative fashion by considering two limiting cases. First, consider the region of low final state momenta, $x \sim 0.2$. Here, stopping does not severely limit the production of particles, and the A–dependence merely reflects the absorption of the projectile. Due to the large absorption cross sections of hadrons, this results in a surface–like $A^{2/3}$ behavior. Secondly, consider the limit $x \to 1$. Since in this limit the projectile is forced to transfer its momentum entirely to the final state, it cannot undergo inelastic scattering before doing so. Therefore, hadrons with momentum close to the beam momentum are produced in a geometric region where the projectile only scatters once. This corresponds to the spatial region at the edge of the nucleus, and we obtain a value of $\alpha \approx 1/3$ in this limit.

The exact x_F– dependence is determined by nuclear geometry, projectile absorption cross section, fragmentation function and the stopping power (inelasticity) κ.

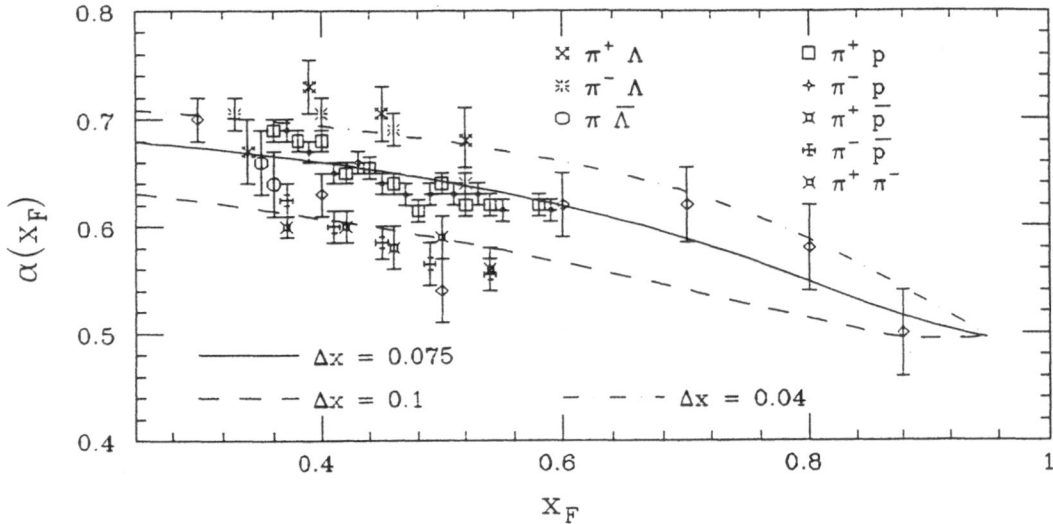

Figure 3: $\alpha(x)$ for different π induced reactions. Indicated is the type of projectile (π^+ or π^-) and of final state hadron. Our results are shown for different values of $\Delta x = \kappa/2$ and are compared to data from [3].

From these quantities, only κ is not well known. It can now be extracted by a fit to the data, as is illustrated in fig. 3.

The result of this analysis yields a stopping power of $\kappa = 0.12 \pm 0.02$ for proton projectiles and $\kappa = 0.15 \pm 0.05$ for pions. Due to the large scattering of the data, a possible energy dependence of the stopping power can not be extracted.

SUMMARY

We have seen that inelastic initial state interaction in nuclei causes an energy loss of the incident projectile before the production of final states.

On the parton level, this is described by a bremsstrahlung process and leads to a suppression of heavy final states that is increasing with x_F. The parameters of quark and gluon radiation lengths can be precisely determined from the data, and a good and systematic description of the x_F–dependence of heavy final states is obtained.

Concerning the production of the heavy states in nucleus–nucleus collisions, the only measurement that has been carried out so far is that of J/Ψ by NA38. Since the acceptance range of this experiment is in the low–x region where the energy loss is negligible, the observed J/Ψ suppression is due to absorption effects.

In the last section, we outlined the consequences of the stopping of a hadronic projectile on the x_F– distribution of light final state particles, and extracted the value of the stopping power κ from the corresponding experimental data.

ACKNOWLEDGEMENTS

I wish to thank the organizers for the invitation to this very inspiring summer school. Financial suppport from NATO and from GSI (project HD Hüf T) is gratefully acknowledged. The meeting gave opportunity to numerous discussions, where in particular I want to thank B.Jacak, C.Gerschel, J.Aichelin, M.Blann and W.Weise.

REFERENCES

1. J.Badier et al., Z. Phys. C 20 (1983) 101

2. E772–Collab., D.M.Alde et al., Phys. Rev. Lett. 64 (1990) 2479, 66 (1991) 133 and 2285

3. W.Busza and M.Zielinski, Phys. Rev. D 31 (1985) 192

4. E. Quack, Nucl. Phys. B 364 (1991) 321

5. P.Hoyer, M.Vänttinen und U.Sukhatme, Phys. Lett. B 246 (1990) 217

6. R. Vogt, S. Brodsky und P. Hoyer, Nucl. Phys. B 360 (1991) 67

7. E.Quack, 'Parton initial state scattering and heavy flavor and lepton pair production in nuclear collisions', preprint HD–TVP–92–2

8. S.Gavin und J.Milana, Phys. Rev. Lett. 68 (1992) 1834

9. W.Heitler, *The Quantum Theory of Radiation*, Oxford Clarendon Press 1970

10. L.Landau und I.Pomeranchuk, Dokl. Akad. Nauk. SSSR 92 (1953) 535

11. S.Brodsky und P.Hoyer, Phys. Lett B 298 (1993) 165

12. M.Gyulassy and X.Wang, 'Multiple collisions and induced gluon bremsstrahlung in QCD', preprint CU-TP-598, June 1993

13. G.Moreno et al., Phys. Rev D 43 (1991) 2815

14. E.Quack and T.Kodama, Phys. Lett. B 302 (1993) 495

RECENT PROGRESS IN NONPERTURBATIVE ELECTROMAGNETIC LEPTON-PAIR PRODUCTION WITH CAPTURE IN RELATIVISTIC HEAVY-ION COLLISIONS

J. C. Wells[1,2], V. E. Oberacker[1,2], M. R. Strayer[1], A. S. Umar[1,2]

[1] Center for Computationally Intensive Physics
Physics Division, Oak Ridge National Laboratory
Oak Ridge, TN 37831-6373
[2] Department of Physics & Astronomy
Vanderbilt University, Nashville, TN 37235

INTRODUCTION

The prospect of new colliding-beam accelerators capable of producing collisions of highly stripped high-Z ions, at fixed-target energies per nucleon up to 20 TeV or more, has motivated much interest in lepton-pair production from the QED vacuum. The time-dependent and essentially classical electromagnetic fields[1] involved in such collisions contain large Fourier components which give rise to sizable lepton-pair production[2] in addition to many other exotic particles[3, 4]. In particular, the process of electron-positron production with electron capture is a principal beam-loss mechanism for highly charged ions in a storage ring, and, thus, plays a central role in the design and operation of these machines[5, 6]. In this process, the electron is created in a bound state of one of the participant heavy ions (most likely the $1s$ state), thus changing the ion's charge state and causing it to be deflected out of the beam. This topic has been discussed in the review articles in Refs. 7 and 8, and recent experimental results are discussed in Refs. 9 and 10.

There is a long and sometimes controversial[5] history concerning the use of perturbative methods in studying electromagnetic lepton-pair production; however, reliable perturbative calculations have been used as input into design models for the Relativistic Heavy-ion Collider (RHIC)[2, 11]. Applying perturbation theory to these processes at high energies and small impact parameters results in probabilities which violate unitarity, and cross sections which violate the Froissart bound[5, 2, 12]. This evidence, along with the initial nonperturbative studies, suggests that higher-order

QED effects will be important for extreme relativistic collisions[13, 5]. Clearly, large nonperturbative effects in electron-pair production with capture would have important implications for RHIC.

In this paper, I will briefly discuss recent progress in nonperturbative studies of the capture problem. In Section II, we state the Dirac equation for a lepton in the time-dependent external field of a heavy ion which must be solved to compute lepton-capture probabilities. The reader is referred to Refs. 15 and 14 for the details of the formalism. Section IV surveys results from recent applications of coupled-channel and lattice techniques to the lepton-capture problem. We draw conclusions and give an outlook in Section V.

SEMICLASSICAL APPROXIMATION

We study the electromagnetic production of lepton pairs with capture in a reference frame in which one of the nuclei is at rest, since recoil may be neglected. The target nucleus and the lepton interact via the static Coulomb field, A_T^0. The only time-dependent interaction, $A_\mathrm{P}^\mu(t)$, arises from the classical motion of the projectile. Splitting the Dirac Hamiltonian into static and time-dependent parts, we write the Dirac equation for a lepton described by a spinor $\phi(\vec{r}, t)$ coupled to an external, time-dependent electromagnetic field

$$[H_F + H_\mathrm{P}(t)]\phi(\vec{r}, t) = i\frac{\partial}{\partial t}\phi(\vec{r}, t), \tag{1}$$

where the static Furry Hamiltonian, H_F, is given by

$$H_\mathrm{F} = -i\vec{\alpha} \cdot \nabla + \beta - eA_\mathrm{T}^0 , \tag{2}$$

and the time-dependent interaction of the lepton with the projectile is

$$H_\mathrm{P}(t) = e\vec{\alpha} \cdot \vec{A}_\mathrm{P}(t) - eA_\mathrm{P}^0(t) . \tag{3}$$

In Refs. 14 and 15, probabilities for vacuum production of leptons with capture into a bound state b are determined by computing the expectation value of the lepton number operator for the state b with respect to the time-evolved vacuum state. The reader is referred to these references for details of the formalism.

The physics of lepton-pair production is defined by the electromagnetic fields of two particles in relative motion, and these fields enter the Hamiltonian via the dimensionless interaction energy, $\bar{A}^\mu \equiv -eA^\mu$, between the lepton and the colliding nuclei in Eqs. (2) and (3). For simplicity of discussion, we assume a point-like charge for both the projectile and the target. However, for heavy-lepton production, finite-size effects are important, and uniform, spherical charge distributions will be used for calculations of these leptons without further discussion. In the target frame, we choose the projectile to move in the z-direction, and the reaction plane to be the y-z plane, i.e. the classical trajectory of the projectile is $x_\mathrm{P}(t) = 0$, $y_\mathrm{P}(t) = b$, $z_\mathrm{P}(t) = \beta_f t$. Since the Dirac equation is covariant under a gauge transformation of the electromagnetic potentials, in principle, the gauge may be chosen for convenience in any problem. The most familiar gauge used in problems with electric sources is the Lorentz gauge, defined by the condition $\partial_\mu A^\mu = 0$. The time-dependent electromagnetic interaction between the projectile and the lepton in the Lorentz gauge can be generated by a Lorentz-boost of the static Coulomb field. This results in

$$\bar{A}_{\mathrm{P}}^0[r'(t)] = \frac{-Z_{\mathrm{P}}\alpha\gamma_f}{\sqrt{x^2 + (y-b)^2 + \gamma_f(z-\beta_f t)}}$$

$$\bar{A}_{\mathrm{P}}^3[r'(t)] = \beta_f \bar{A}_{\mathrm{P}}^0[r'(t)] \tag{4}$$

$$\bar{A}_{\mathrm{P}}^1 = \bar{A}_{\mathrm{P}}^2 = 0 ,$$

where Z_{P} is the atomic number of the projectile, and α is the fine-structure constant. The Lorentz factor in the fixed-target frame is γ_f. The beam energy for a given frame of reference is $E_{\mathrm{kin}} = m_0 c^2(\gamma - 1)$, where γ denotes the Lorentz factor for the frame of interest. The Lorentz factors for the fixed target and collider frames are related by $\gamma_f = 2\gamma_c^2 - 1$.

NONPERTURBATIVE CALCULATIONS

The projectile Hamiltonian represented in the Lorentz gauge is

$$H_{\mathrm{P}}(\vec{r}, t) = (1 - \beta_f\alpha_z)\bar{A}_{\mathrm{P}}^0[r'(t)] , \tag{5}$$

where α_z is the third spin matrix in the Dirac representation. The maximum of the scalar and vector components, which are proportional with proportionality constant $\beta_f \approx 1$, scale as γ_f and their half-width is inversely proportional to γ_f. However, at extreme energies, the probability for capture at a finite impact parameter b is independent of the energy[16, 17]. As can be anticipated from the form of Eq. 5, this independence on the energy implies that large cancellations occur between the scalar and vector amplitudes which are troublesome for most approximate solutions[16]. In addition, the spiked nature of the Lorentz-gauge interaction results in the multipole expansion used in the coupled-channel approach having poor convergence properties[18], and requires extremely fine grid spacing for a faithful representation of the interaction on the lattice. As a result, solving the Dirac equation using the Lorentz-gauge interaction is difficult, and other noncovariant gauge choices have been used in nonperturbative capture calculations. However, good progress has been made by some groups using the Lorentz-gauge interaction. In this section, I will briefly survey recent nonperturbative capture studies, and the choice of gauge when different from the Lorentz gauge.

Coupled-channel Calculations

In the coupled-channel approach, Eq. (1) is solved by expanding $\phi(\vec{r}, t)$ in the complete, orthonormal set of eigenstates $\chi_k(\vec{r})exp(-iE_kt)$ of the Furry Hamiltonian, H_{F}. Insertion of the expansion

$$\phi(\vec{r}, t) = \sum_k a_{jk}(t)\chi_k(\vec{r})exp(-iE_kt) \tag{6}$$

into Eq. (1) and projection leads to the infinite system of first-order coupled differential equations for the occupation amplitudes $a_{jk}(t)$,

$$\dot{a}_{jf}(t) = -i \sum_k a_{jk}(t) \langle \chi_f | H_P | \chi_k \rangle exp(i(E_f - E_k)t) \,, \tag{7}$$

which is equivalent to Eq. (1)[19, 18]. In the coupled-channel approach, Eqs. (7) are integrated numerically after the infinite summation is truncated to include a finite number of states. The amplitudes a_{jf} determine the motion of a single lepton during the collision, and, after substitution into Eq. (6), provide the approximate solution to the time-dependent Dirac equation needed to compute the capture probability. The reader is referred to Refs. 18 and 19 for more details.

Recently, coupled-channels calculations with limited basis sets (334 states) have been performed at moderately relativistic fixed-target energies (1.2 and 2.0 GeV per nucleon) for $U^{92+} + U^{92+}$ which suggest that perturbative approaches underestimate electron-pair production with capture into the K-shell at small impact parameters by a factor of 50 to 100[19]. A strong nonperturbative dependence on the projectile charge is also observed. Reference 19 clearly argues the fact that coupled-channel calculations using a truncated basis are not formally gauge invariant, but fail to discuss the degree to which their calculations depend on the gauge choice.

Reference 20 also contains coupled-channel calculations with the ability to use larger basis sets (1700 and 3400 states) and explores the use of noncovariant gauges. In particular, this work reproduces the exact calculations presented in Ref. 19 and demonstrates that the use of a noncovariant gauge results in a change in the capture probability by a factor of two. This work performs large, coupled-channel calculations to predict an upper limit on the electron-capture probabilities at RHIC energies to be a 10% effect. In related work by this group, the argument is made that the nonperturbative capture cross section should scale with the energy as $\ln \gamma_f$[17]. The significance of this result is that it implies that results of experiments at the lower energies of present-day accelerators may be reasonably extrapolated to the extreme conditions of RHIC.

Lattice Calculations

In lattice solutions to the Dirac equation, one forsakes a continuous representation of the quantum-state vector and coordinate-space operators in favor of a representation only on a discrete set of spatial lattice points, e.g. $\chi(\vec{r}) \rightarrow \chi_{\alpha,\beta,\gamma}$. Therefore, Dirac spinors become discrete vectors of $4N_x N_y N_z$ complex numbers, and the Dirac Hamiltonian becomes a matrix of the same rank. Local operators like the electromagnetic interaction become diagonal matrices with the values along the diagonal being simply the value of the interaction at a given lattice point. Nonlocal operators, like coordinate-space derivatives, are represented by matrices which may be banded, as in the finite-difference method, or full, as in basis-spline and spectral methods. In doing this, one reduces the partial-differential equation with specified boundary conditions to a series of linear equations which may be solved using elimination or iterative techniques[24, 21].

For infinitesimal lattice spacing, lattice solutions reproduce the exact, time-dependent solutions of the analytical Dirac equation, which, as stated earlier, are gauge invariant. To insure that solutions are reliable, one must increase the spatial and temporal resolution of the lattice to test for convergence. However, for finite-lattice spacing, certain gauges may be more easily represented on the lattice, and different gauges may give different results[14].

The Lorentz-gauge interaction has been used in capture calculations during central collisions of $U^{92+} + U^{92+}$ at 10 GeV per nucleon using a two-dimensional finite-difference lattice approach. Using a very fine grid (300×600 points), Ref. 22 reports capture probabilities which are also approximately two-orders of magnitude greater than perturbation theory. The authors state that slight overestimation of the capture probability results from their use of Sommerfeld-Maue analytic continuum states during the calculation of the capture probability, since these states do not form an orthonormal set. Reference 23 extends the coupled-channel calculations, reported in Ref. 19, to $U^{92+} + U^{92+}$ at 10 GeV per nucleon to obtain good agreement with the finite-difference calculations of Ref. 22.

Reference 14 describes our approach to the lepton-capture problem in which we solve Eq. (1) in three-Cartesian dimensions using lattice-basis-spline techniques. These methods have many advantages over finite-difference techniques, which make an unrestricted solution in three spatial dimensions possible, among which are accuracy, stability, and an undoubled energy spectrum[27]. Lattice calculations for the electron-capture process are more interesting and more difficult than the muon-capture problem. The fundamental difficulty is one of having a number of natural length scales involved in the electron capture problem. These length scales are the nuclear radius, the electron's Compton wavelength, and the size of the heavy atom. (For the muon-capture problem, the muon Compton wavelength and the nuclear radius are the same within a factor of 2.) As a consequence, we have presently deferred an attack on the more nonperturbative and interesting electron-capture problem in favor of the muon-capture problem.

To avoid the stated difficulties of the Lorentz-gauge interaction, we have explored the use of noncovariant gauges. One gauge which is easier to represent on the lattice is the axial gauge, defined by requiring the z-component of the Lorentz-gauge interaction to vanish. The axial gauge has several features which result in a more faithful representation of the interaction on the lattice[14].

In computing the capture probability, one needs access to the lattice representation of the Coulomb-Dirac continuum states. However, complete diagonalization of the Hamiltonian matrix exceeds current computational capabilities. As a result, we currently approximate the Coulomb-Dirac continuum states using the lattice representation of Dirac plane wave states modified to be orthogonal to the bound states on the lattice[14]. This approximation may be efficiently implemented and is more accurate for use in electron-capture calculations at high energies where Coulomb distortion is a relatively small effect. We note that numerical methods exist to allow extraction of the probabilities for transition to the exact Coulomb distortion continuum states[28]; however, past implementations of these techniques have proved to entail a prohibitively large computational expense.

PRELIMINARY LATTICE-BASIS-SPLINE RESULTS

In order to gain confidence in our nonperturbative solution of the time-dependent Dirac equation, we seek to reproduce low-order calculations in a perturbative regime. In order to assure that perturbation theory is reasonable, we perform calculations for muon capture at a collider energy per nucleon of 2 GeV for the system $Au^{79+} + Au^{79+}$. At this energy, the maximum frequency in the virtual-photon spectrum of the elec-

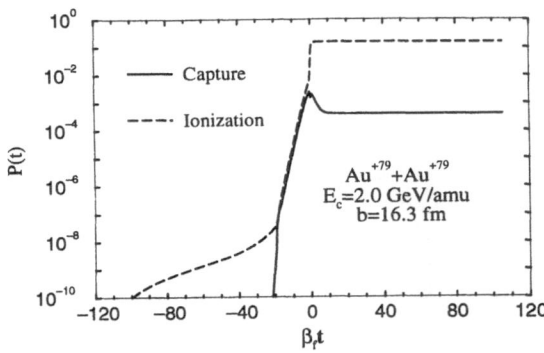

Figure 1: Plotted is the time-dependent probabilities for capture and ionization during collisions of two fully stripped Au nuclei at collider frame energy of 2 GeV per nucleon at an impact parameter of 16.3 fm. The time axis is represented by the position along the z-axis, $\beta_f t$, of the projectile in units of the muon Compton wavelength, λ_μ.

tromagnetic field is approximately $2m_\mu c^2$, the threshold value required to produce a pair. We stress that the calculations presented here should be considered preliminary and no quantitative conclusions should be drawn, as convergence tests with the numerical box size and lattice spacing have not been performed. In addition, as previously mentioned, the approximate continuum states employed for these calculations make quantitative comparison difficult. Nevertheless, calculations with such continuum states are useful, as they can be compared to first-order perturbative calculations performed on the lattice employing the same approximate continuum states. These comparisons provide a stringent test on the dynamics of the nonperturbative solution.

Figure 1 shows the capture and ionization probabilities as a function of time multiplied by the projectile's velocity for a grazing impact parameter of 16.3 fm. The ionization probability is approximately 0.1 and the capture probability is approximately 4×10^{-4}. At no time during the collision is the muon-capture probability greater than 1%. This is consistent with a perturbative process. These numbers highlight another difficulty in performing capture calculations. Sufficient accuracy must be available to extract relatively small capture probabilities out of a time-dependent state dominated by excitation and ionization channels. Figure 2 shows the expectation value of the total Dirac Hamiltonian with respect to the time-dependent Dirac

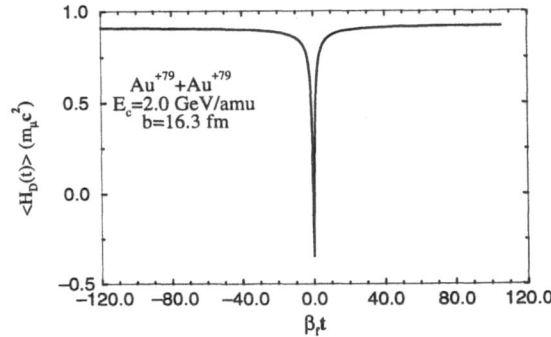

Figure 2: Plotted is the expectation value of the full Dirac Hamiltonian in units of the muon rest mass during the collision described in Fig. 1.

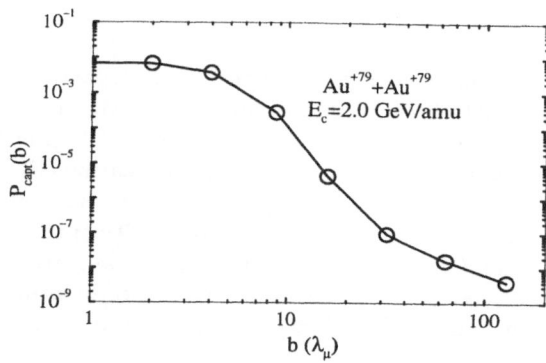

Figure 3: Plotted is the calculated impact-parameter dependence of the muon-capture probability in collisions of $Au^{92+} + Au^{92+}$ at a collider energy of $E_c = 2$ GeV per nucleon.

spinor for the same collision as Fig. 1. One sees that the system is excited near the distance of closest approach, then relaxes as time further increases.

In Fig. 3, we show a preliminary impact parameter dependence for the muon-capture probabilities at collider energies per nucleon of 2 GeV. Notice initially the probabilities fall exponentially with b as expected from perturbation theory for the large impact parameters. However, as the capture probability becomes very small, one observes a change in the b-dependence to a power law. This this power-law dependence is undesirable and results from finite-lattice sizes effecting the small probabilities.

COMMENTS AND OUTLOOK

Recent progress has been made in the theoretical understanding of nonperturbative lepton-pair production with capture as is discussed in this short paper. Most exciting is that pioneering experiments have recently been reported[9, 10], and the second generation of experiments are planned for the near future[25, 26].

As a general comment, all nonperturbative approaches mentioned in this paper sacrifice, to some extent, the gauge invariance of the semiclassical theory used to describe the capture problem. The extent to which gauge invariance is violated in practical calculations gives a measure of the reliability of these calculations and should be reported along with quantitative predictions.

Fundamental improvements have been achieved recently in the application of lattice techniques to the Dirac equation[27], and more progress is anticipated. Specifically, our time-dependent Dirac code currently running on the Intel iPSC/860 hypercube parallel computer will be ported to the Intel's next generation of massively parallel computers, the Paragon XP/S series. This machine will provide the computing power necessary to demonstrate numerical convergence in our large calculations. In addition, a new and efficient algorithm for accurate energy analysis of the time-dependent Dirac state will be implemented and tested[28]. The combination of these two readily accessible improvements will allow us to make accurate predictions for muon-pair production with capture at collision energies accessible to the present-day heavy-ion accelerators.

ACKNOWLEDGEMENTS

This research was sponsored in part by the U.S. Department of Energy (DOE) under contract No. DE-AC05-84OR21400 managed by Martin Marietta Energy Systems, Inc., under contract No. DE-FG05-87ER40376 with Vanderbilt University, and under the High Performance Computing and Communications Program (HPCC), as a Grand Challenge titled the Quantum Structure of Matter. The numerical calculations were carried out on the Intel iPSC/860 hypercube multicomputer at the Oak Ridge National Laboratory, the CRAY-2 supercomputers at the National Energy Research Supercomputer Center (NERSC) at Lawrence Livermore National Laboratory, and the National Center for Supercomputing Applications (NCSA) in Illinois.

REFERENCES

1. J.-S. Wu, C. Bottcher, M. R. Strayer, Phys. Lett. **B252**, 37 (1993).

2. C. Bottcher and M. R. Strayer, Phys. Rev. D **39**, 1330 (1989).

3. J.-S. Wu, C. Bottcher, M. R. Strayer, and A. K. Kerman, Ann. Phys. **210**, 402 (1991).

4. G. Soff, this conference.

5. *Can RHIC be used to test QED?*, Brookhaven National Laboratory Workshop Proceedings, Upton, New York, BNL-52247, April 1990.

6. M. J. Rhoades-Brown, C. Bottcher, and M. R. Strayer, Nucl. Instrum. Meth. **B43** (1989) 301.

7. C. A. Bertulani and G. Baur, Phys. Rep. **163**, 299 (1988).

8. J. Eichler, Phys. Rep. **193**, 167 (1990).

9. A. Belkacem, H. Gould, B. Feinberg, R. Bossingham, and W. E. Meyerhof, Phys. Rev. Lett. **71**, 1514 (1993).

10. A. Westphal and Y. D. He, Phys. Rev. Lett. **71** 1160 (1993).

11. M. J. Rhoades-Brown, T. Ludlam, J.-S. Wu, C. Bottcher, and M. R. Strayer, *Fourth Workshop on Experiments and Detectors for a Relativistic Heavy-ion Collider*, Brookhaven National Laboratory Report, BNL 52262, 325 (July 1990).

12. M. Froissart, Phys. Rev. **123**, 1053 (1961).

13. C. Bottcher and M. R. Strayer, Nucl. Inst. and Meth. **B31**, 122 (1988).

14. J. C. Wells, V. E. Oberacker, A. S. Umar, C. Bottcher, M. R. Strayer, J.-S. Wu, and G. Plunien, *Phys. Rev. A* **45** (1992) 6296.

15. V. E. Oberacker, J. C. Wells, A. S. Umar, and M. R. Strayer, contribution to this school.

16. P. A. Amundsen and K. Aashamar, J. Phys. B. **14**, 4047 (1981).

17. A. J. Baltz, M. J. Rhoades-Brown, and J. Weneser, Phys. Rev. A **44**, 5569 (1991).

18. W. Scheid, contribution to this school.

19. K. Rumrich, G. Soff, and W. Greiner, Phys. Rev. A **47**, 215 (1993); K. Rumrich, K. Momberger, G. Soff, W. Greiner, N. Grün, and W. Scheid, Phys. Rev. Lett. **66**, 2613 (1991).

20. A. J. Baltz, M. J. Rhoades-Brown, and J. Weneser, Phys. Rev. A **47**, 3444 (1993).

21. J. C. Wells, V. E. Oberacker, A. S. Umar, C. Bottcher, M. R. Strayer, J.-S. Wu, J. Drake, and R. Flanery, *Int. J. Mod. Phys. C* **4** (1993) 459.

22. J. Thiel, A. Bunker, K. Momberger, N. Grün, and W. Scheid, Phys. Rev. A **46**, 2607 (1992).

23. J. Thiel, K. Momberger, N. Grün, and W. Scheid, *Proceedings of the International Symposium on Nuclear Physics of our Times*, November 17-21, 1992, Sanibel Island, Florida.

24. U. Becker, N. Grün, and W. Scheid, J. Phys. B **16**, 1967 (1983).

25. Experiment WA99, approved by CERN SPS Program Committee, March, 1993. *Measurements of Pair Production and Electron Capture from the Continuum in Heavy Particle Collisions*, Oak Rdige National Laboratory, Manne Siegbahn Institute of Physics, Univ. of Aarhus, and Univ. of Lund.

26. H. Gould, private communication.

27. J. C. Wells, V. E. Oberacker, M. R. Strayer, and A. S. Umar, Oak Ridge National Laboratory Preprint, ORNL/CCIP/93/12, August, 1993.

28. C. Bottcher, M. R. Strayer, A. S. Umar, and V. E. Oberacker, Phys. Rev. C **37**, 2487 (1988).

CLUSTER-FLOW AND RESONANCE-MATTER FORMATION IN ULTRARELATIVISTIC HEAVY-ION REACTIONS

R. Mattiello, M. Hofmann, Chr. Spieles, A. Jahns,
H. Sorge, H. Stöcker, and W. Greiner

Institiut für Theoretische Physik
Johann Wolfgang Goethe-Universität
Frankfurt am Main

One of the challenges of modern heavy ion physics is the extraction of the equation of state for extremely excited nuclear matter. In particular, the creation and study of hadron matter at high net baryon density has received much attention recently. Not much is known about the strenght of the mean fields at large baryon densities. It is expected that the momentum dependence of the nuclear forces[1], the excitation into resonance matter [2] and the phase transition into the QGP[3] will play a crucial role for the created mean fields. In particular, QCD – as the accepted theory of strong interaction – contains chiral symmetry (in the limit of massless quarks) which is spontaneously broken in its groundstate, the QCD vacuum (see e.g. recent lattice calculations[4]). A rapid restoration of this symmetry with increasing baryon density is predicted by all approaches which embody this fundamental aspect of QCD[5, 6]. Therefore, nucleus-nucleus collisions in the bombarding energy region of baryon stopping could be favorable in order to study such medium effects as compared to ultrahigh energies, for which the two colliding nuclei may become transparent to each other. Transport calculations based on hadronic excitations and rescattering like the RQMD approach (strings, resonances)[7] or the ARC model (resonances)[8] predict that beam energies between 10 to 15 AGeV – as studied experimentally at the BNL-AGS [9, 10, 11] – are well suited to create the desired high stopping and baryon densities.

An observable consequence of the formation of dense nuclear matter – far beyond the groundstate – is the predicted emergence of collective flow driven by compression-induced pressure[12, 13, 14]. Mean fields[12] may give an important contribution to the pressure and could therefore – via the flow effect – be accessible to experimental observation just as in the 1GeV region[12, 15]. In this contribution we particularly focus on:

- Nuclear stopping and collective flow, visible in final proton, deuteron and nucleon spectra.

- Formation of resonance matter and it's signatures in the spectra of pions, antiprotons and strange hadrons.

• Sensitivities of the collective flow to the strength of nuclear mean fields.

We present calculations with the relativistic quantum molecular dynamics model (RQMD 1.07). RQMD is based on the propagation of all hadrons on classical trajectories in the framework of Hamiltonian constraint dynamics. Binary stochastic scattering and decay (with Pauli blocking corrections) simulate quantum features. The model includes in-medium effects like interacting strings and resonances, rescattering of secondaries, a finite formation time for produced particles and modified dispersion relations for baryons[7].

Resonance-matter

In order to formulate hadron interaction at high energies it is necessary to take into account particle production. This is done in RQMD by the excitation of ingoing hadrons into excited states which decay consecutively during the heavy-ion reaction. Fig. 1 shows the energy excitation function of different resonances in central Kr+Kr reactions. The number of excited states corresponds to the maximum value achieved during the evolution. At SIS energies around 1 GeV we observe only little excitation but with increasing energy – in particular around 10 AGeV– the number of resonances even exceed the number of bare nucleons in the system.

The time evolution of the average local densities (upper diagram) and volumes (lower diagram) for baryons and Δs, respectively, is shown in Fig. 2 for the system Au(11.6AGeV)Au. The volume includes all regions with $\rho > 0.1\rho_0$. The rather large volume indicates that the considered beam energies are most promising for the investigation of dense and strongly *resonance*-enriched excited *matter*. Note that the local baryon densities achieve values higher than the simple kinematic overlap[31]. However, a direct proof for the presence of such a state is still missing although strong indications have been discovered already: A considerable amount of delta decays in the final state has recently been extracted via invariant-mass analyses of experimental data in quantitative agreement with RQMD predictions[11].

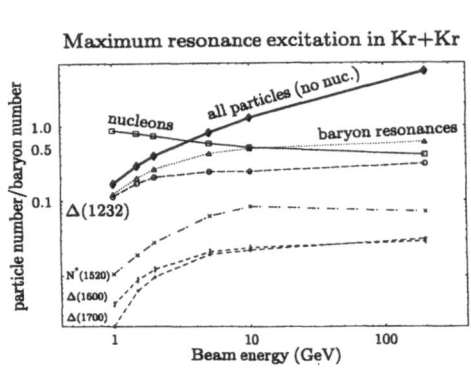

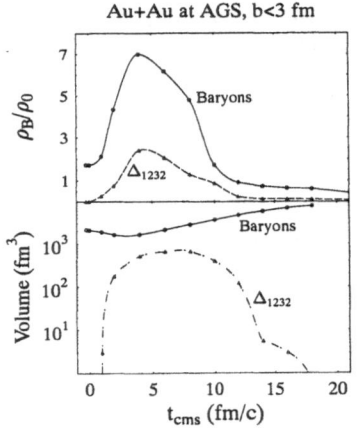

Fig. 1 Energy excitation function for the multiplicities of various resonances in central Kr+Kr collisions calculated with the RQMD. The values correspond to the maximum values achieved during the evolution.

Fig. 2 Lower diagram: Time evolution for averaged *local* baryon densities of all baryons and Δ-resonances, respectively. Upper diagram: Time evolution of the total volume with $\rho_B, \rho_\Delta > 0.1\rho_0$ which contains baryons and Δs, respectively.

In RQMD resonance formation and decay leads to characteristic modifications of the chemical content and the reaction kinematics visible in various final results. Resonance mass acts as an energy storage via the transformation of incident beam energy into the excitation of resonances and strings with high masses but rather small velocities. Resonance formation and also the formation time for produced particles contributes essentially to the baryonic stopping. Characteristic interaction properties of resonances may also contribute to mean-fields. They could modify flow features which are indeed sensitive to in-medium potentials (Sect. 3) even at ultrarelativistic energies.

Resonance formation and (collective) multistep processes change the chemical content markedly (strangeness[21, 16, 17], antimatter[20]). Fig. 3 shows the kaon production rate in Au+Au, divided into the production channels $BB \rightarrow K + X$ (filled), $MB \rightarrow K + X$ (solid) and $MM \rightarrow K + X$ (dashed). As already pointed out for Si+Au contributions[21]. The non-thermal range of the spectrum deviating from single thermal rates. As in Si+Au the yield from the BB-part is similar to the naive superposition of initial NN-collisions. Additional production is provided by secondary MB and MM scattering. The relative enhancement compared to pions is due to the interaction of resonances, particularly the (non-equilibrium) formation of s-channel resonances via $MB \rightarrow B^*$[16]. On the one hand these reactions lead to the excitation of baryons above the production threshold for strange particles and consecutive strangeness production throughout the reaction. On the other hand pions are frequently reabsorbed in resonances and do not increase as much as kaons. Moreover, they even *convert* by such meson absorption and emission channels into *kaons*.

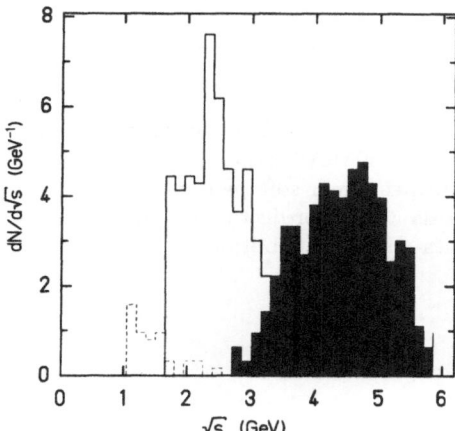

Fig. 3 Kaon production sources in Au(11.6AGeV)Au. The contributions from MM,MB and BB collisions are shown.

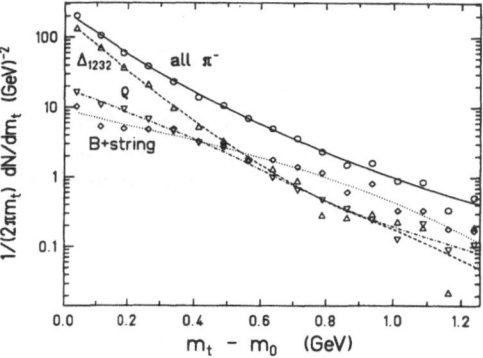

Fig. 4 Transverse mass spectra for π^- in central Au(11.6AGeV)Au collisions. Different contributions from final state resonance decays are depicted which add up to the observed multislope behaviour.

We shall focus in the following on the presence of final-state Δ-resonance decays, i.e. rather direct signatures for resonance production. As mentioned before, invariant mass analyses show evidences for large numbers of delta decays in the final reaction stage. RQMD calculations demonstrate that pions from final decays are preferentially emitted in certain phase space regions: Intergrated over rapidity they contribute dominantly to pions in the low-momentum part of the transverse momentum spectra ($p_t < 300\text{MeV}$). Heavy resonance and string decays dominate the high-p_t region (see Fig. 4). The different contributions add up to a multislope behaviour giving a characteristic concave curvature. The shape of the total π^- distribution agrees very well with experimental data. For π^+, however, RQMD also predicts the spectrum to be bent, while preliminary E866 data claim only *one* slope[29].

One more hint for the presence of the Δ-resonance is given by the analysis of supersoft pions ($p_t < 30MeV$). If deltas dominate, such pions will be supressed always in the rest frame of the decaying Δ. Therefore, the yields at very small p_t-values should increase markedly with increasing distance to the Δ-matter source-rapidity[22]. In Fig. 5 we show the rapidity distribution of delta resonances in Au(11.6 AGeV)Au reactions before their final decay which is obviously centered at midrapidity. Following our argumentation, the relative amount of supersoft pions should increase from central towards target or projectile rapidity. The quantitative result in shape of the ratio $C_y = \sigma(p_t = 30MeV)/\sigma(p_t = 150MeV)$ indeed shows this feature for Au+Au (Fig. 6). Note that this particular behaviour is not achieved with the usually applied concepts to explain low-p_t pion enhancement yields like bose-condensation[23].

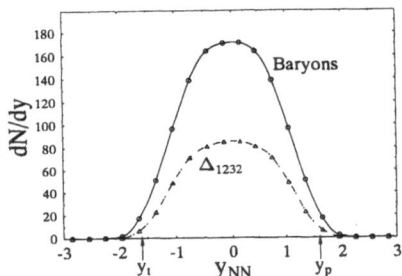

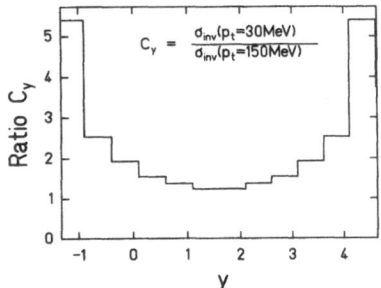

Fig. 5 Rapidity distribution of Δ-resonances before their final freeze-out decay (after which the produced pion moves into the detector).

Fig. 6 Rapidity distribution of the ratio $\sigma(p_t = 30MeV)/\sigma(p_t = 150MeV)$ which indicates super-soft pion enhancement towards target/projectile rapidities due to delta-decay contributions.

Nuclear Stopping and Cluster-Flow

The RQMD calculation does not contain quantum mechanical cluster states (e.g. deuterons) dynamically. The formation of deuterons can be calculated by projecting the generated classical neutron-proton phasespace distribution on the deuteron wave function via the Wigner-function method[24]. This method was applied to bombarding energies around 1AGeV in combination with the intranuclear cascade model[25] and the QMD[1]. The obvious advantage as compared to the usual coalescence model in momentum or phase space[26, 27] is that it avoids the introduction of additional free parameters, namely the cut-off distance in configuration and momentum space [13]. Higher mass fragments are by construction contained in the calculated number of deuterons[13, 25, 28], but they are negligible near midrapidity.

Fig. 7 shows the deuteron yields for Si+Si, Si+Cu, and Si+Au reactions at 14.5

AGeV predicted by RQMD. Fig. 8 shows the rapidity distribution of deuterons and protons in Au(11.6AGeV)Au collisions (b<3fm). The preliminary data points are from[29] and contain approx. 15% systematic error. Deuterons and protons in Fig. 8 show a peak at midrapidity which proofs strong nuclear stopping. Whether stopping or nuclear transparency is actually observed in the AGS experiments has been unclear for quite some time.

However, the ambiguous experimental results have been corrected and confirm the predicted large baryon stopping in central collisions[9, 10, 11, 31, 30].

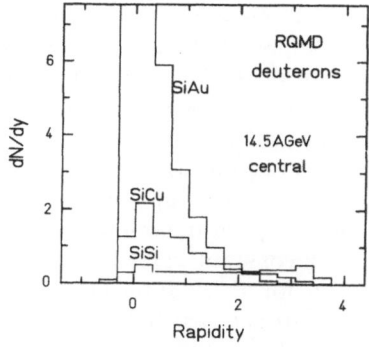

Fig. 7 RQMD prediction for deuterons produced in central Si+Si, Si+Cu and Si+Au collisions at 14.5 AGeV.

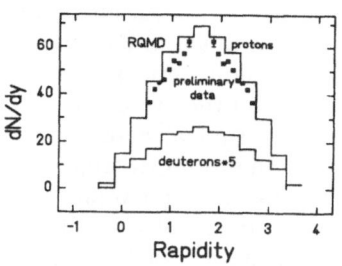

Fig. 8 Rapidity distribution for protons and deuterons in central Au (11.6 AGeV) Au collisions. The data symbols are from [29] and contain 15% systematic uncertainty. The proton distribution is already corrected for the coalesced protons.

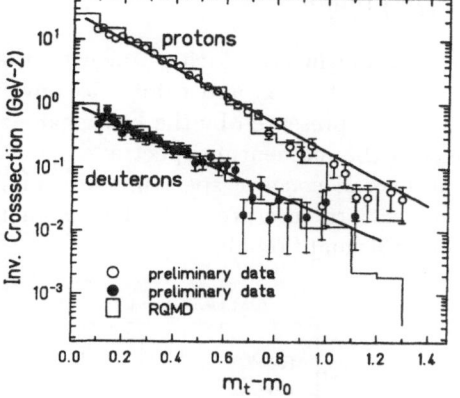

Fig. 9 Transverse mass spectra for protons and deuterons around $y \approx 0.6$ in central Au(11.6AGeV)Au (calculations including potentials).

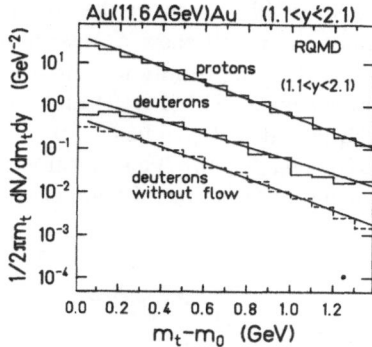

Fig. 10 Transverse mass spectra for protons and deuterons around midrapidity in central Au(11.6AGeV)Au collisions (calculations including potentials).

The comparison of transverse-mass spectra between preliminary E866-data and RQMD (Fig. 9) reveals agreement for both deuterons and protons.

The data have been taken between $0.5 < y < 0.7$ while the calculations were performed in the larger window $0.35 < y < 0.85$. Different apparent temperatures for protons ($\simeq 200\text{MeV}$) and deuterons ($\simeq 250\text{MeV}$) and the strong shoulder-arm shape at midrapidities (Fig. 10) are inconsistent with pure momentum coalescence and the thermal fireball picture.

The slope splitting and the shoulder-arm curvature are due to collective transverse motion. For lower incident energies a similar behaviour was aleady pointed out within the hydrodynamical picture[14]. The flow itself is shown in Fig. 11 with the final transversal velocity of nucleons as function of the freeze-out distance to the central beam axis. The freeze-out is defined by the last strong collision or decay after which the particle is emitted.

Removing this flow with randomly redistributed momenta, i.e. artificially destroyed position-momentum correlations, leads to an apparent temperature similar to protons and a much weaker shoulder-arm curvature for deuterons (dashed histogram in Fig. 10). Furthermore, the absolute value for deuterons decreases considerably because of – on the average – larger distances between neutrons and protons in phasespace. Note that at midrapidity the d/p-ratio increases from 5% in Si+Si to 7.5% in Au+Au due to larger stopping and stronger collective flow components. The scaling with total mass (or volume) in the fireball model[32] –given freeze-out density and temperature in chemical equilibrium– would predict a constant value.

The correlation in Fig. 11 can be parametrized for $r_t < 10\text{fm}$ effectively via $\beta_t = Ar_t^B$ (here: $B \approx 0.44$). At higher r_t values it saturates at $\beta_t \approx 0.7c$. The averaged flow velocity resulting from the integrated profile in Fig. 11 –weighted with the corresponding freeze-out probabilities– is $< \beta > \approx 0.48c$. Such a flow correlation could even lead to a peak in the transverse mass spectra. The effective parametrization $\beta = Ar_t^B$ leads to

$$\frac{\mathrm{d}N}{m_t \mathrm{d}m_t} = \frac{\mathrm{d}N}{p_t \mathrm{d}p_t} = \frac{m_0^2}{BA^{2/B}} \frac{(m_t^2 - m_0^2)^{(1-B)/B}}{(m_t^2)^{(1+B)/B}} \rho(r_t)$$

$\rho = 1/r_t \mathrm{d}N/\mathrm{d}r_t$ denotes the freeze-out density-profile in r_t. With $\rho = $const., the flow-spectrum will show a maximum at finite m_t for ($B < 1$) and exhibit a convex shape. Note that results for clusters up to $Z = 7$ were presented by the European 4π collaboration at SIS (FOPI) and show this peak in the momentum spectra[33].

A quadratic dependence ($B = 2$) yields an overall concave spectrum diverging for $m_t - m_0 \to 0$. This parameter had been used to interprete the low p_t pion enhancement in terms of a spherically expanding thermal fireball[34].

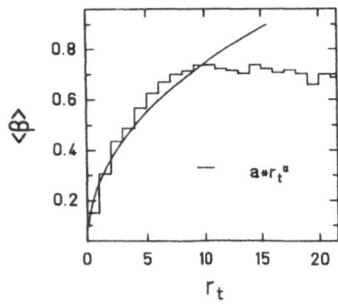

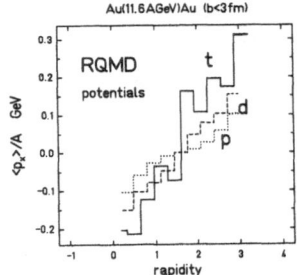

Fig. 11 Averaged final transverse flow velocities as function of the freeze-out distance to the beam axis r_t for nucleons with $1.1 < y < 2.1$ in central Au(11.6AGeV)Au.

Fig. 12 Calculated $p_x(y)$-correlations for protons, deuterons and tritons in Au (11.6AGeV) Au collisions (b<3fm). The calculation includes potential interaction.

In contradiction, the RQMD-calculation shows that the low-p_t behaviour of pions is dominated by the decay properties of the Δ-resonance (Section 1). Therefore the combined analysis of pion and nucleon (cluster) spectra could shed light on the role of (resonance-)flow.

The difference between longitudinal and transverse momentum distributions (the width of the rapidity distribution) indicates *longitudinal* flow expansion. Both together, longitudinal and transversal flow offer the possibility to observe the transverse "bounce-off" of nuclear matter via the $< p_x > (y)$-distribution (see Fig. 12) which was discovered already in the 1GeV/n energy domain[15]. The recently reported azimuthally asymmetric particle correlations indicate sensitivities to the particular collision geometry (event plane) at 10AGeV[35], just as known from the 1 GeV/n energy range[15]. Also regarding the bounce-off effect, the first experimental indications at 10 GeV/n have just been announced[36].

The RQMD calculation shows approximately 1.5-2 times larger p_x-values for deuterons than for protons. Preliminary calculations for three-nucleon clusters – approximated by a 3-body harmonic-oscillator wave-function – predict even higher values.

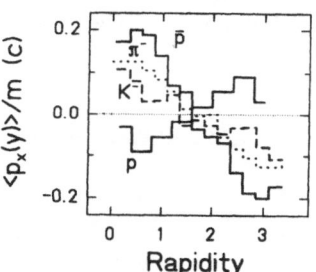

Pions, kaons and antiprotons are preferentially emitted into the opposite direction and thus show an "inverse" directed flow compared to nucleons (Fig. 13). This holds also for SPS energies: Angular correlations in the target fragmentation region measured with the 4π-detector at CERN indicate similar results[19]. The reason for the anticorrelation is different for $\bar{p}$ and pions/kaons: Antiprotons are absorbed in the bouncing nuclear matter[20].

At SIS energies it has been suggested that pions in nuclear spectator matter are basically scattered via $\pi N \rightarrow \Delta \rightarrow \pi N$ and not dominantly absorbed[18].

Fig. 13 Calculated $p_x(y)$-correlations for pions, kaons, protons and antiprotons in Au(11.6AGeV)Au collisions (minimum bias, no potentials). Note the *anti*correlation for pions, kaons and $\bar{p}$.

Those results show that "flow"-correlations can be used to study the reaction-(and flow-) geometry. They could even help to disentangle production and absorption contributions to the final particle yields like $\bar{p}$ which are presently under discussion[20].

Density Dependent Potentials and the Collective Flow

Not much is known about the strength of the mean fields at large baryon densities. At groundstate density the mean field may be decomposed into two large pieces: an attractive scalar field provided by the quark condensate and/or correlated two-pion exchange (the σ field), and a repulsive vector potential (the ω field)[37] which is in accord with Dirac phenomenology for optical potential calculations[38, 39, 40, 41] in p+A and QCD sum-rule estimates[5].

In this contribution we compare two extreme scenarios in order to explore the effect of mean fields on flow observables. In one scenario the acting mean fields are simulated via potential-type interactions along the line presented in[7, 42], i.e. a quasi-potential giving saturation at groundstate density and repulsion at higher baryon densities of the Skyrme-type form $V_i = - \sum_j \alpha_{ij}\rho_{ij} + (\sum_j \beta_{ij}\rho_{ij})^{7/6}$. ρ_{ij} is a Gauss function of the CMS distance vector and expresses the contribution of the baryon with index j ($j \neq i$) to the potential acting on baryon i. In the other scenario the potentials are switched off (cascade mode).

Fig. 14 shows the difference of both scenarios in the final proton and deuteron rapidity distributions for central Au(11.6 AGeV)Au reactions. In the calculation with potentials the spectra exhibit a flater shape indicating stronger *longitudinal* flow expansion. Also the average p_t-values at midrapidity increase by 10-15% indicating stronger *transverse* flow. Without potentials the deuteron and proton yields in the low m_t region ($m_t < 100$ MeV) shows smaller deviation from a thermal shape as indicated in Fig.10.

The larger final momenta and the changes in the shape of the distributions are due to larger flow velocities caused by the pressure which is built up by the repulsive mean fields at high baryon density (up to $8\rho_0$ is achieved[31]). Note that the region of high baryon-densities ($\rho/\rho_0 > 3$) is quite large ($V \simeq$ several hundred fm^3). Up to 2/3 of the baryons in this zone reside in excited states. Therefore, selected features of interacting resonances, e.g. increased effective masses for the Δ-resonances in the nuclear surrounding[2], could contribute essentially to the mean fields and modify the collective behaviour of the system.

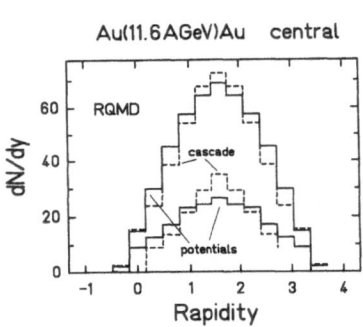

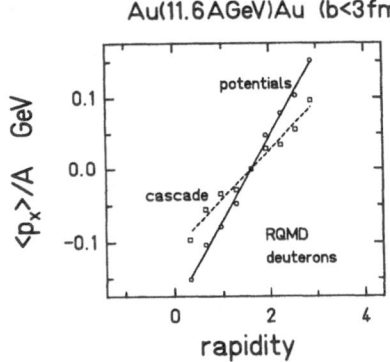

Fig. 14 Rapidity distributions for protons and deuterons in Au(11.6AGeV)Au for two different szenarios: with (solid histograms) and without (dashed histograms) potential interaction.

Fig. 15 $p_x(y)$-correlations for deuterons in Au(11.6AGeV)Au (b<3fm) for calculations with (solid line) and without (dashed line) potential interaction.

The $p_x(y)$-correlations show the strongest sensitivity to the potential interaction employed (Fig. 15). The sideward push due to the mean field – in addition to the repulsion due to the collision term – yields roughly a factor of two higher $p_x(y)$-values. In cascade calculations for low SIS- and high SPS-energies the shapes of the p_x distributions over y/y_p reveal energy independence and even the maximum values are comparable. Hydrodynamics and microscopic models show a similar shape of the directed nuclear flow although the absolute values are higher in hydrodynamic calculations[30], probably due to the unrealistic assumption of instantaneous thermalization. In hydrodynamics the directed flow is also sensitive on the choice of the equation of state. A purely hadronic EOS is stiffer and shows larger flow than the EOS with a phase transition into the quark-gluon plasma[30].

Conclusions

- RQMD predicts a midrapidity peak in rapidity distributions for protons and nuclear clusters in Au(11.6AGeV)Au, i.e. proves **strong stopping** for such heavy systems. This holds from SIS (1 GeV) over AGS ($\approx$ 11 GeV) up to SPS (160 AGeV) energies with the same shape in $dN/d(y/y_p)$. Experiments have confirmed this at 1 and 11 GeV.

- In this dense stopped baryon matter a large number of excited baryon and meson resonances are produced. They store the incident energy in the system. While the Δ_{1232} is the dominant resonance at SIS, for higher energies we obtain much larger contribution of higher resonances, N^*, Δ^* and mesons with densities $\rho_{N^*}, \rho_\Delta \approx \rho_0$. Circumstantial evidence for **resonance matter** is provided by the experimental observation of:

    - concave pion spectra showing "two-slope" behaviour,

    - reconstruction of the Δ^{++} in πN invariant-mass spectra,

    - strangeness enhancement, visible in enhanced K/π and Λ/π-ratios,

- The events show a **nuclear bounce-off** and correlated in-plane flow p_x for baryons, but **anti-correlated** flow for **pions**, **kaons** and **anti-protons**, the latter due to $p\bar{p}$-annihilation, for pions due to scattering off spectator matter.

- **Deuterons** show an enhanced bounce as compared to protons. Different apparent temperatures for protons and deuterons and a strong shoulder-arm shape in their transverse mass spectra at midrapidities can be understood by collective **transverse flow** of nuclear matter ($< \beta > \approx 0.5c$). Heavy clusters exhibit even higher flow-components in p_x as well as in p_t.

- The flow-observables are **sensitive to the mean-field interaction** included in the microscopic simulations and to the **equation of state** in the hydrodynamic approach.

REFERENCES

1. J. Aichelin, A. Rosenhauer, G. Peilert, H. Stöcker, and W. Greiner: Phys. Rev. Lett. (1987)1926; J. Aichelin: Phys. Rep. 202(1991)233

2. F. de Jong and R. Malfliet: Phys. Rev. C46(1992)2567

3. J. Ellis, J. Kapusta, K. Olive: Phys. Lett. B273(1991)122

4. J.B. Kogut, D.K. Sinclair and K.C. Wang: Phys. Lett. B163(1991)101

5. X. Jin, T.D. Cohen, R.J. Furnstahl and D.K. Kriegel: Phys. Rev. C47(1993)2882 and refs. therein

6. U. Vogl and W.Weise: Prog. Part. Nucl. Phys. 27(1991)195; W. Weise: Nucl. Phys. A553(1993)59

7. H.Sorge, H. Stöcker and W. Greiner: Ann. Phys. (NY) 192(1989)266; Nucl. Phys. A498(1989)567; Z. Phys. C47 (1990)629;

8. S.H. Kahana, Y. Pang and T.J. Schlagel: Proc. of the HIPACS Conference 1993 in Cambridge, edited by G.S.F. Stephans, S.G. Steadman and W.L.Kehoe, p. 263

9. Hideki Hamagaki for the E-802/866 collaboration, B. Moskowitz for the E-802/866 collaboration; G.S.F. Stephans for the E-859 collaboration: Proceedings of the Quark Matter Conference 1993, to be published in Nucl. Phys.

10. K.J. Foley for the E810-Collab.: Nucl. Phys. A544(1992)335c

11. J. Stachel for the E-814/877 collaboration; M. Rosati for the E-814/877 collaboration; T. Hemmick for the E-814/877 collaboration; J. Barette for the E-814/877 collaboration; Nu Xu for the E-814/877 collaboration: Proceedings of the Quark Matter Conference 1993, to be published in Nucl. Phys.

12. H.Stöcker and W. Greiner: Phys. Rep. 137(1986)278; H. Kruse, B.V. Jacak, and H. Stöcker:Phys. Rev. Lett. (1985)289; J.J. Molitoris, J.B. Hoffer, H. Kruse and H. Stöcker: Phys. Rev. Lett. (1984)899; G. Buchwald, G. Graebner, J. Theis, J. Maruhn, and W. Greiner: Phys. Rev. Lett. (1984)1594; Ch. Hartnack, M. Berenguer, A. Jahns, A. v. Keitz, R. Mattiello, A. Rosenhauer, J. Schaffner, T. Schönfeldt, H. Sorge, L. Winckelmann, H. Stöcker, W. Greiner: Nucl. Phys. A538(1992)53c

13. P. Danielewicz and Q. Pan: Phys. Rev. C46(1992)2002; Q. Pan and P. Danielewicz: Phys. Rev. Lett. (1993)2062,3523

14. H. Stöcker, A. Ogloblin, W. Greiner: Z. Phys. A303(1981)259; S. Nagamiya, M.-C. Lemaire, E. Moeller, S. Schnetzer, G. Shapiro, H. Steiner, and I. Tanihata: Phys. Rev. C24(1981)971

15. K.-H. Kampert: J.Phys. G15(1989)691; H.H. Gutbrod, K.H. Kampert, B.W. Kolb, A.M. Poskanzer, H.G. Ritter and H.R. Schmidt: Phys. Lett. B216(1989),267

16. H. Sorge, R. Mattiello, H. Stöcker and W. Greiner: Phys. Lett. B271(1991)37

17. H. Sorge, M. Berenguer, H. Stöcker, W. Greiner: Phys. Lett. B289(1992)6

18. S. A. Bass, R. Mattiello, H. Stöcker, W. Greiner: Phys. Lett. B 302 (1993) 381

19. H.R. Schmidt et al. (WA80-Collab.): Nucl. Phys. A544(1992)449c

20. A. Jahns, C. Spieles, H. Sorge, H. Stöcker, W. Greiner: contribution to this volume

21. R. Mattiello, H. Sorge, H. Stöcker, W. Greiner: Phys. Rev. Lett. 63 (1989) 1459

22. H. Sorge: Phys. Rev. C, 1993, in print

23. M. Kataja, P.V. Ruuskanen: Phys. Lett. B143(1990)181

24. E.A. Remler: Ann. of Phys. (NY) 136(1981)293 and refs. therein

25. M. Gyulassy, K. Frankel and E.A. Remler: Nucl. Phys. A402(1083)596

26. H. Gutbrodt, A. Sandoval, P. Johanssen, A. Poskanzer, O. Gosset, W. Meyer, G. Westfall, R. Stock: Phys. Rev. Lett. 37 (1976)667

27. C.B. Dover, U. Heinz, E. Schnedermann, J. Zimanyi: Phys. Rev. C44 (1991)1636

28. R. Mattiello, A. Jahns, H. Sorge, H. Stöcker, W. Greiner: submitted to Phys. Rev. Lett. (1993)

29. M. Gonin for the E802/E859/E866 Collab.: ref. 8, p. 184

30. N.S. Amelin, E.F. Staubo, L.P. Csernai, V.D. Toneev, K.K. Gudima, D. Strottman: Phys. Rev. Lett. 67(1991)1523; R. B. Clare, D. Strottmann, Phys. Rep. 141 (1986) 177;
D. J. Dean et al., submitted to Phys. Rev. E (1993);
B. M. Waldhauser et al., Z. Phys. C43 (1989) 411

31. H. Sorge, A.v. Keitz, R. Mattiello, H. Stöcker, W. Greiner: Phys. Lett. B 243(1990)7

32. A. Mekijan: Phys. Rev. Lett. 38(1977)640; Phys. Rev. C17(1978)1051

33. C. Kuhn et al., Proceedings of the XXXI international winter meeting on nuclear physics, Bormio (1993)

34. K.S. Lee and U. Heinz: Z. Phys. C48(1990)525

35. T. Abott and the E802 Collab.: Phys. Rev. Lett. 70(1993)1393

36. P. Braun-Munzinger for the E877 collaboration: Talk given at the the Nato Advanced Studies on Nuclear Physics Summer School 1993, Bodrum/Turkey

37. B.D. Serot and J.D. Walecka: Adv. Nucl. Phys. 15 (1986); J. Theis et al.: Phys. Rev. D28 (1983)2286

38. E.D. Cooper, B.C. Clark, R. Kozack, S. Shim, S. Hama, J.I. Johansson, H.S. Sherif, R.L. Mercer, B.D. Serot: Phys. Rev. C36(1987)2170

39. M. Jaminon, C. Mahaux, P. Rochus: Nucl. Phys. A365(1981)371

40. B. Ter Haar and R. Malfliet: Phys. Lett. B172(1986)10

41. T.L. Ainsworth, E. Baron, G.E. Brown, J. Cooperstein, M. Prakash: Nucl. Phys. A464(1987)740

42. H. Sorge, R. Mattiello, H. Stöcker and W. Greiner: Phys. Rev. Lett. 68(1992)286.

43. U. Heinz and K. S. Lee: Phys. Lett. B259 (1991) 162.;
M. Aguilar–Benitez et al.: Z. Phys. C50 (1991) 405.
M. Berenguer et al.: manuscript in preparation.

THE RELATIVISTIC TREATMENT OF SYMMETRIC AND ASYMMETRIC NUCLEAR MATTER

Hans Huber and Manfred K. Weigel

Ludwig-Maximilians-Universität München, Sektion Physik, Am Coulombwall 1, D-85748 Garching, FRG

Fridolin Weber

Nuclear Science Division, Lawrence Berkeley Laboratory, University of California, Berkeley, CA 94720, U.S.A.

I. INTRODUCTION

In the last decades it was realized that certain features of the nuclear problem can only be described by going beyond the nonrelativistic approach. Relativistic treatments of the nuclear many-body problem have advantages in several respects, for instance[1,2]: An extremely useful Dirac phenomenology in the description of nucleon-nucleus scattering[3]; the shift of the equilibrium density from the so-called "Coester band" towards the "experimental" value via a new saturation mechanism[4,5]; the natural incorporation of the spin-orbit force[1,2] and the successful description of finite nuclei[6,7], etc.

Naturally, there is a great desire to explore the relativistic many-body quantum field approach in many respects. Among them is the fundamental challenge to understand the properties of nuclei in terms of the interactions between its constituents. A reliable microscopic calculation of the equation of state would be a great benefit for many branches of physics, as for instance, in the physics of supernova explosions[8], neutron stars[9] and heavy-ion scattering[10]. One of the basic attempts in this direction is the relativistic treatment of symmetric and asymmetric nuclear matter in many-body approximations with dynamical two-body correlations with modern one-boson-exchange potentials adjusted to the two-nucleon problem. In the next sections we are going to address this problem in the frame of the Green's function approach and discuss some problems as, for instance, consistence questions, predictive power, limitations etc.

II. GENERAL THEORY

In the model the forces between the nucleons are mediated by the exchange of mesons, hence the dynamics of the particle is governed by a Lagrangian density of the following form[1, 2, 4, 5, 11–15]:

$$\mathcal{L} = L_N + \Sigma_M(L_M + L_M').$$

$\mathcal{L}_N$ denotes the Lagrangian density of noninteracting nucleons; similary, $\mathcal{L}_M$ describes the different free meson fields, which interact via $\mathcal{L}_M'$ with the nucleons.

A suitable tool for the treatment of many-body systems is the Green's function scheme. The formulation of the problem in ladder-type approximation is an established procedure and resembles in its formal structure closely to the nonrelativistic treatment[16, 17]. One obtains a coupled system of the Dyson equations for the G-function and the effective scattering matrix T in matter

$$\{(G^0)^{-1}(1,2) - \Sigma(1,2)\}G(2,1') = \delta(1,1'),$$

$$< 12|T|1'2' > = < 12|V|1'2' - 2'1' > + < 12|V|34 > \Lambda(34,56) < 56|T|1'2' > .$$

We employ the convention to sum or to integrate over all doubly occuring variables. Here V denotes the OBE-potential

$$< 12|V|1'2' > = \sum_{M=(\sigma,\omega,\dots)} < 12|V_M|1'2' >,$$

and the self-energy is given by

$$\Sigma(1,2) = -i < 14|T|52 > G(5,4).$$

The H- or HF-approximation are defined by $T = V(V_{AS})$, respectively.

For the intermediate p-p-propagator a standard choice is the Brueckner propagator (cf. Refs.[4, 5, 11–13, 15]); but one also use the so-called Λ-approximations[13, 14], defined as (G^0 denotes the free propagator; $ij = 00, 01, 11$):

$$-i\Lambda^{ij}(12,34) = \begin{cases} G^0(1,3)G^0(2,4) \\ \frac{1}{2}(G^0(1,3)G(2,4) + G(1,3)G^0(2,4)) \\ G(1,3)G(2,4) \end{cases}$$

which are obtained from the Martin-Schwinger approximation scheme[18] by taking dynamical correlations into account, which are connected with the potential (for instance, $< 12|V|34 > G(34,1'2')$ is included but $< 12|V|34 > G(1'4,32')$ is replaced by $< 12|V|34 > (G(1',3)G(4,2') - G(1',2')G(4,3))$, for more details and a comparison between the different approximations, see Refs.[13, 17, 18, 19, 20]. It turns out that all relativistic approximations give a shift towards the semiempirical values and the RBHF-results are located between the Λ^{00}- and the Λ^{01}-results (Λ^{00} gives the lowest values for E/A[13]; for the treatment of the full ladder approximation see Ref.[14]).

A useful simplification can be achieved by utilizing the spectral representation of G, i.e.

$$G(p) = \int dw \ \frac{A(\vec{p}, \omega)}{(p_0 - \mu)(1 + i\eta) - \omega},$$

since all desired quantities can be determined by the self-energy Σ and the spectral function A alone[13, 17].

III. DISCUSSION

Despite the formal similiarity to the nonrelativistic case the relativistic situation is much more complicated; due to, for instance:

1. Energy-dependent meson-potentials (retardation)

2. Bethe-Salpeter equation in four dimensions

3. Dirac algebra (T-matrix has in principle 256 elements with respect to spin); Σ (and A) has scalar, vector and time-like contributions with the following structure:

$$\Sigma(p) = \Sigma_s(p) + \Sigma_v(p)(\hat{p} \cdot \vec{\gamma}) + \gamma^0 \Sigma_0(p)$$

4. Self-consistent single-particle basis (spinors) are not a priori known; therefore the solution in the self-consistent basis in the full Dirac space is rather complicated[13, 15].

First of all one needs self-consistent spinors, since the relativistic saturation mechanism depends strongly on the lower parts of the spinors, and give a decreasing (increasing) contribution for the $\sigma - (\omega-)$ mesons with increasing density. This feature leads to a non-monotonic behaviour of the kinetic and potential part of the energy. If one uses free spinors one gets back the nonrelativistic features (see Fig. 1).

The consistency problem whether a quasi-particle picture is applicable depends on the energy-dependence of the self-energy. For obtaining a single-particle energy-momentum relation it is necessary, that $|\frac{\partial \Sigma}{\partial \omega}|$ is smaller than 0.5. This question is treated in more detail in Refs.[13, 17]. For instance, the momentum baryon distribution in the relativistic case is given by:

$$\varrho_B(\vec{p}) = \left\{ \frac{W(p)}{|W - [m^*\frac{\partial \Sigma_s}{\partial \omega} + \tilde{k}\frac{\partial \Sigma_v}{\partial \omega} + W\frac{\partial \Sigma_0}{\partial \omega}]|} \right\}_{p_0 = \omega(\vec{p})},$$

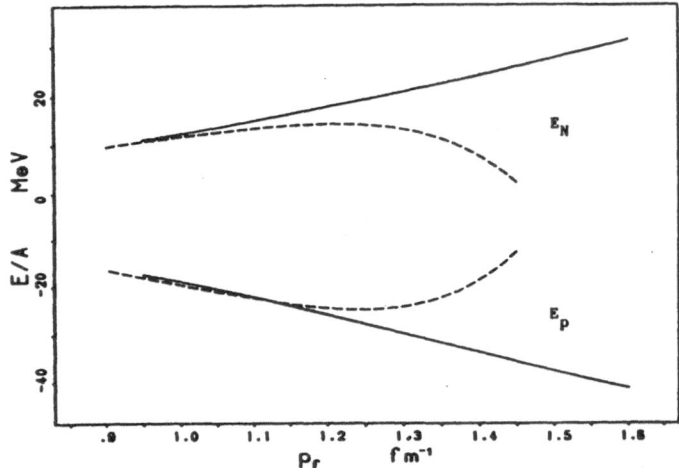

Figure 1. Illustration of the influence of the self-consistent basis (relativistic saturation mechanism): Kinetic (Dirac) and potential energy in RBHF-approximation for the OBE-potential Ho2[13]. The dotted curves correspond to the self-consistent basis (i.e. non-monotonic behaviour); the solid curves give the outcome with free spinors (i.e. similiar to the non-relativistic treatment).

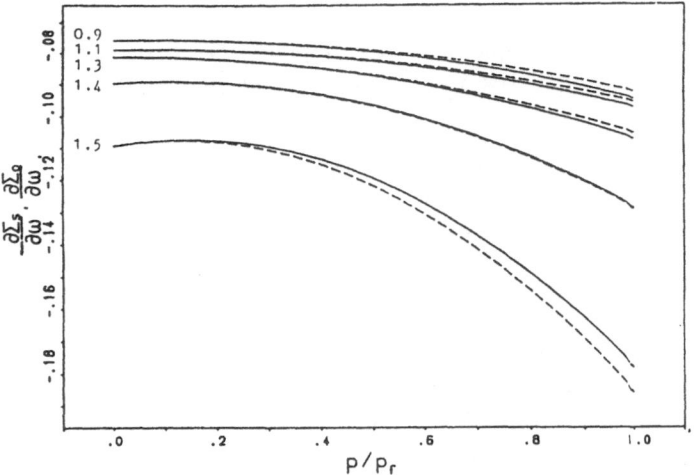

Figure 2. Energy derivatives $\frac{\partial \Sigma_S}{\partial \omega}$ (solid curves) and $-\frac{\partial \Sigma_0}{\partial \omega}$ (dashed curves) vs. p/p_F for different Fermi momenta in Λ^{00} - approximation (OBE-potential Ho2[13]).

Table 1: Saturation properties of nuclear matter in different approximations (Brockmann potential B[5]): RBHF[(1)] (full basis; momentum dependent self-energy), RBHF[(2)] (full basis; momentum averaged self-energy), RBHF[(3)] (positive spinors only; momentum independent self-energy[5]); Λ^{00} ($\Lambda^{00}-$ approximation). For comparison we also give the outcome of the relativistic HF–approximation, where ρ and E/A are adjusted[22]: RHF[(1)] ($\sigma-, \omega-$ mesons only), RHF[(2)] ($\sigma-, \omega-, \pi-$ and $\rho-$ mesons; $f_\rho/g_\rho = 6.6$), RHF[(3)] ($\sigma-, \omega-, \pi-$ and $\rho-$ mesons; $f_\rho/g_\rho = 3.7$).

Method	E/A [MeV]	ρ [fm^{-3}]	K [MeV]	a_4 [MeV]
RBHF[(1)]	-14.8	0.170	263.7	33.9
RBHF[(2)]	-15.7	0.172	248.9	32.8
RBHF[(3)]	-13.6	0.174	249.0	
Λ^{00}	-21.9	0.210	259.6	33.8
RHF(1)	-15.75	0.148	610.0	28.9
RHF[(2)]	-15.75	0.148	360.0	43.3
RHF[(3)]	-15.75	0.148	460.0	38.6

which reduces to the step function for $\frac{\partial \Sigma}{\partial \omega} = 0$. It turns out that the energy-dependence is sufficiently weak for the applicability of the single-particle description (see Fig. 2).

In the pioneering work of the Brooklyn group[11] the problem was treated in the full Dirac space but the relativistic effect was only included in first-order perturbation theory, so avoiding the complicated self-consistency problem. Therefore a comparison with other treatments is rather difficult.

Due to the complexity of the problem one has tried in most treatments to avoid the solution in the full Dirac space.

The standard method, applied in Refs.[4, 5, 12, 14], makes a non-unique ansatz for the T-matrix in terms of five independent Fermi invariants (in Refs.[4, 14] the pseudoscalar invariant is replaced by the pseudovector invariant)

$$T = T^S I^{(1)} I^{(2)} + T^v \gamma_\mu^{(1)} \gamma^{(2)\mu} + T^T \tau_{\mu\nu}^{(1)} \tau^{(2)\mu\nu} + T^{PS} \gamma_5^{(1)} \gamma_5^{(2)} + T^A \gamma_5^{(1)} \gamma_\mu^{(1)} \gamma_5^{(2)} \gamma^{(2)\mu}$$

and obtains the solution in the c.v. frame for positive spinors only. Afterwards they transform the T-matrix into the nuclear matter frame. Once a specific value of m^* is chosen and Lorentz boosting mixes only positive-energy helicity spinors themselves one determines only the positive-energy matrix elements. For that reason the full matrix structure of T, and hence of Σ , is not uniquely determined. It was shown that the results for Σ depend on the chosen decomposition[20].

The other method, used by Brockmann and Machleidt[5], avoids this procedure by the assumption that the scalar and time-like parts of Σ are momentum independent and approximate the positive-energy matrix elements of Σ, obtained from the RBH-solution via

$$< \phi|\Sigma|\phi > = \frac{m^*}{E^*}\Sigma_S + \Sigma_0.$$

Both approaches have been discussed and critized in more detail in Ref.[21]. Despite the weak momentum dependence of Σ_0 for the whole range of $m^*(0.5-1.0\,m_N)$ the absolute values for Σ_0 differ strongly (393–(–117) MeV)[21] and there is no a priori reason to prefer a m^*-value with the smallest deviation from a constant; or otherwise expressed a direct calculation of $m^*(p_F)$ is necessary.

IV. RESULTS

In order to clarify the situation we have calculated for a modern OBE-potential (Brockmann B)[5] the properties of nuclear and asymmetric matter in the full Dirac space. The results with comparison between the different approaches are shown in Table 1. It turns out that, at least for the chosen potential, the differences between the full Dirac space results and the Brockmann treatment are not very large. The Λ^{00}-approximation gives, as expected, a higher density and energy, respectively. Also

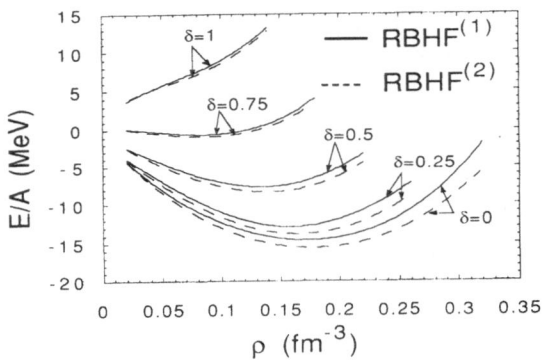

Figure 3. Binding energy per nucleon versus density for different asymmetries in the RBHF-approximation (Brockmann potential B[5]). The solid (dashed) curves correspond to the treatment with (or averaged) momentum dependency of the self-energies.

the bulk symmetry energy a_4 agrees reasonable with the semiempirical value. The phenomenological HF-treatment, it seems, is not capable to reproduce the incompressibility K and a_4[22]. (A further increase in the ρ-meson tensor coupling f_ρ/g_ρ would decrease K). In Figs. 3 and 4 we show the EOS of state in the RBHF- and Λ^{00}-approximation for different aysmmetry parameters $\delta = (\rho_n - \rho_p)/\rho$. Furthermore we tested the validity of the quadratic dependence of the energy upon the asymmetry parameter δ. Our results confirm this empirical law also for higher values of δ (see Fig. 5).

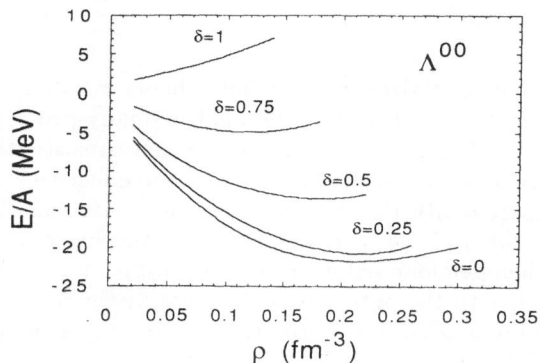

Figure 4. Binding energy per nucleon versus density for different asymmetries in the relativistic Λ^{00}-approximation.

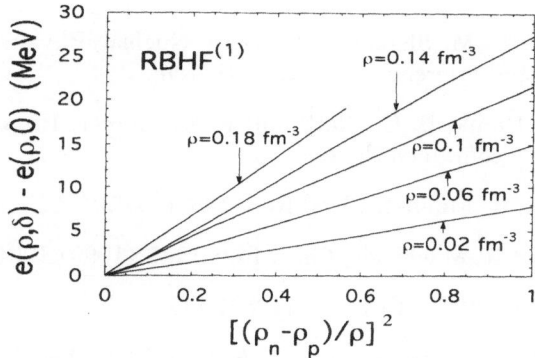

Figure 5. Asymmetry energy per nucleon in the RBHF$^{(1)}$- approximation in the range $0 \le \delta^2 \le 1$ at five densities. The slope of each curve gives the corresponding symmetry energy.

In conclusion, we have investigated the properties of symmetric and antisymmetric nuclear matter by solving self-consistently the relativistic problem with dynamical two-body correlations for a modern OBE-potential in the full Dirac space. It seems that the case of momentum averaged self-energies is applicable for densities equal or below the nuclear matter equilibrium density but for higher densities one should include the momentum-dependence of the self-energy.

V. SUMMARY

In the framework of relativistic nuclear field theory we discuss and compare the different approaches in the treatment of the nuclear-many-problem with inclusion of two-body correlations. The equations are solved self-consistently in the full Dirac space, so avoiding the ambiguities in the choice of the effective scattering amplitude. The results are compared with the standard method, where one only determines the scattering amplitude for positive energy spinors. Furthermore we tested the assumption of momentum independent self-energy. The results for asymmetric matter are in the structure similiar to the outcome of the relativistic Hartree–Fock approximation, but differ from the nonrelativistic treatment. The agreement with the empirical values is quite satisfactory.

REFERENCES

1. B. D. Serot and J. D. Walecka, Adv. Nucl. Phys. **16** (1986) 1, and references contained therein.

2. L. S. Celenza and C. M. Shakin, Relativistic Nuclear Physics (World Scientific, Singapore 1986), and references contained therein.

3. E. D. Cooper, S. Hama, B. C. Clark and R. L. Mercer, Phys. Rev. **C47** (1993) 297, and references contained therein.

4. J. B. ter Haar and R. Malfliet, Phys. Rep. **149** (1987) 209.

5. R. Brockmann and R. Machleidt, Phys. Rev. **C42** (1990) 1964.

6. P. G. Reinhard, Rep. Prog. Phys. **52** (1989) 439.

7. Y. K. Gambhir, P. Ring and A. Thimet, Ann. of Phys. **189** (1990) 132.

8. G. E. Brown, Phys. Rep. **163** (1988) 1, and references contained therein.

9. F. Weber and N. K. Glendenning, Hadronic Matter and Rotating Relativistic Neutron Stars, Proceedings of the Nankai Summer School "Astrophysics and Neutrino Physics", p. 64 - 183, Tianjin, China, 17-27 June 1991, Eds. D. H. Feng, G. Z. He, and X. Q. Li, World Scientific, 1993, and references contained therein.

10. P. Stock, Phys. Rep. **135** (1986) 259, and references contained therein.

11. M. R. Anastasio, L. S. Celenza and C. M. Shakin, Phys. Rev. **C23** (1981) 2258 and 2273.

12. C. J. Horowitz and B. D. Serot, Nucl. Phys. **A 464** (1987) 613.

13. P. Poschenrieder and M. K. Weigel, Phys. Lett. **B 200** (1988) 231; Phys. Rev. **C38** (1988) 471.

14. F. de Jong and R. Malfliet, Phys. Rev. **C44** (1991) 998.

15. A. Amorin and J. A. Tjon, Phys. Rev. Lett. **68** (1992) 772.

16. K. A. Brueckner and C. A. Levinson, Phys. Rev. **97** (1955) 1344.

17. M. Weigel and G. Wegmann, Fortschritte der Physik **19** (1971) 451, and references contained therein.

18. P. C. Martin and J. Schwinger, Phys. Rev. **115** (1959) 1342.

19. R. Puff, A. S. Reiner and L. Wilets, Phys. Rev. **149** (1968) 778.

20. R. Machleidt, K. Holinde and Ch. Elster, Phys. Rep. **149** (1987) 1.

21. C. Nuppenau, Y. J. Lee and A. D. MacKellar, Nuc. Phys. **A504** (1989) 839.

22. M. Lopez-Quelle et al., Nucl. Phys. **A 483** (1988) 479.

MONTE–CARLO MODEL FOR PARTICLE PRODUCTION AT ULTRA–RELATIVISTIC ENERGIES

N. S. Amelin[#], H. Stöcker and W. Greiner
Institut für Theoretische Physik, Universität Frankfurt
Postfach 111932, D–6054 Frankfurt am Main 11, Germany

N. Armesto and C. Pajares
Departamento de Física de Partículas
Universidade de Santiago de Compostela
15706–Santiago de Compostela, Spain

[#]Alexander von Humboldt Research Fellow; on leave of absence from
the JINR, Dubna, Russia

INTRODUCTION

The search of the Quark–Gluon Plasma (QGP) is the goal of the experimental heavy–ion program[1]. Existing data are limited to relatively light ions, and energies in the range $E_{lab} = 10 - 200 \ AGeV$, and most of them can be described by models[2, 3, 4, 5, 6] based on the production of colour strings, which decay independently into the observed hadrons. However, some new phenomena have been discovered, indicating that new theoretical suggestions are required. Particularly, the strangeness enhancement observed by several experiments[7, 8] is not easy to explain by such models. The introduction of additional mechanisms, like colour ropes[3] or string fusion[9], is necessary. It is planned to study the heaviest nucleus collisions, when massive ion beams are available up to collider energies: $E_{cm} = 100 \ AGeV$ at Brookhaven–$RHIC$ and $E_{cm} = 3000 \ AGeV$ at CERN–LHC. A change in the hadron production mechanism is expected: at lower energies, hadrons are produced essentially in soft collisions with small transferred momenta; at collider energies, the so–called semi–hard processes[10] will dominate. The momenta transferred in such collisions are large enough to use Perturbative Quantum Chromodynamics (PQCD) calculations.

Here, the Monte–Carlo Parton String Model predictions for multiparticle production in hadron and nuclear collisions, at $E_{cm} = 19.4$, 200 and 1800 $AGeV$, are presented. The first description of this numerical model can be found in [9], where the influence of string interaction on multiple production was studied. This approach is based on the parton picture of strong interactions, and also on its properties following from the Regge formalism[11]. In this formalism, hadron and nucleus collisions are reduced to interactions of partons with given distributions in the projectile and target. We distinguish "soft" and "hard" parton interactions. Soft parton collisions, when the transferred momenta are neglected, lead to the creation of longitudinal strings. Additionally, we introduce hard perturbative parton–parton collisions and parton bremsstrahlung, whose inclusion leads to the creation of kinky strings.

MODEL PARAMETERS AND COMPARISON TO EXPERIMENTAL DATA

Parton collision probability: string formation

At high energy, a fast moving hadron or nucleus can be considered as the superposition of parton chains. A hadron collision is assumed to be an interaction between slow partons (chain tails), with the additional condition that each parton interacts only once. The distribution in the number of interacting partons is directly connected[11] with the values of the multipomeron vertices in the reggeon theory. In the eikonal approximation, it takes a Poissonian form

$$w_N^{(1)} = \exp(-g(s))g^N(s)/N! \ . \tag{1}$$

The mean number of strongly interacting partons, $g(s) = g_0 s^\Delta$, is a function of the initial energy $\sqrt{s}$. We use $g_0 = 3.0$ and $\Delta = 0.09$. The parton–parton cross section has been assumed energy independent, $\sigma_p = 3.5\ mb$.

Due to diffusion, the parton distribution in impact parameter (relative to the center of the corresponding hadron) is taken to be Gaussian:

$$F(b_p) = (4\pi\lambda)^{-1}\exp(-b_p^2/4\lambda), \tag{2}$$

with the radius depending on the initial hadron energy. For the projectile or target hadron, $\lambda = R^2 + \alpha' \ln\sqrt{s}$, where $\alpha' = 0.01\ fm^2$ and $R^2 = 0.15\ fm^2$. Parton–parton interactions destroy the coherence of the parton chains, which lead to hadronization. As in the standard Dual Parton Model[12, 13] (DPM), where each cut Pomeron is substituted by two strings, we assume that the parton hadronization goes through the decay of strings. Each inelastic parton–parton collision creates two colour strings. The momenta $p_\pm$ of the partons at the ends of the strings are given by the nucleon structure functions in $x_\pm = p_\pm/P_\pm$, where $P_\pm$ is the momentum of the nucleon to which they belong. The nucleon structure function is assumed factorized, except for energy–momentum conservation:

$$u(x_1, x_2, ..., x_n) = \delta(1 - \sum_{i=1}^{n} x_i)\prod_{i=1}^{n} u_i(x_i). \tag{3}$$

For the single parton distributions $u_i(x_i)$, the ones obtained from the Regge theory[13] are used:

$$u_v(x) = u_s(x) = x^{-0.5}, \ u_{vv}(x) = x^{1.5} \ , \tag{4}$$

except for strange sea quarks, for which $u_s(x) = 1/x$, with a cut–off in x ($x > x_{min} = m_t/P_+$), is used. v, s and vv refer to valence and sea quarks and diquarks respectively. We use the strangeness suppression parameter $\gamma_s^h = 0.3$, as in string decay (see below), to find a strange sea quark pair inside the nucleon. From Equation (2), a Gaussian form is also taken for the p_t distribution of partons sitting on the string ends:

$$f(p_t^2)p_t dp_t \sim \exp(-bp_t^2)p_t dp_t \ , \tag{5}$$

with $b = 4 \ GeV^{-2}$.

For nuclear collisions, the nuclear parton wave function is taken as a convolution of the parton distribution within a nucleon, with the distribution of nucleons inside the nucleus[9].

String decay

The modelling of the decay of a string with a given mass, momentum and quark content, is carried out by the Field–Feynman algorithm[14]. At each string break–up, the strangeness suppression parameter is $\gamma_s = 0.3$, and the ratio of the diquark–antidiquark pair production to the quark–antiquark one has been set equal to: $P_{qq,\bar{q}\bar{q}}$: $P_{q\bar{q}} = \gamma_{BB} = 0.09$. Equal probabilities of producing pseudoscalar and vector mesons from quarks and antiquarks, and baryons with spin 1/2 and 3/2 from quarks and diquarks, are assumed. At every string break–up, the $q\bar{q}$ (or $qq\bar{q}\bar{q}$) pair has zero transverse momentum, but, nevertheless, the momenta of the separate quark ($\vec{p}_t$) and the corresponding antiquark ($-\vec{p}_t$) is distributed according to Equation (5), with $b = 8.2 \ (GeV/c)^{-2}$. For a produced hadron, its transverse momentum consists on the transverse momenta of its quarks, and the longitudinal momentum p_z^h and energy E^h are determined through the variable $z = (E^h + p_z^h)/(E^q + p_z^q)$ (E^q energy and p_z^q longitudinal momentum of the fragmenting quark), following the distribution:

$$f_h^q(z) \sim (1 - z)^{\alpha_q^h(p_t)} \ . \tag{6}$$

At $z \to 1$, this function coincides with the fragmentation function $D_q^h(z)$ of the leading quark (antiquark) or diquark (antidiquark) into a hadron. $\alpha_q^h(p_t)$ depends[13] on the flavour of the constituent quark, and on the type of hadron it is transformed into and its transverse momentum. The requirement that the fragmentation function $D_q^h(z) \to 1/z$ for $z \to 0$, is ensured by iterating the string break–ups. If the mass of a string, M_s, is less than $M_c = M_R + \Delta M$, where $\Delta M = 0.35 \ GeV$, and M_R is the mass of the resonance with the same quark composition as the string, the last break is generated. Its kinematics is determined by the isotropy of the emission of two hadrons. The new produced resonances are assumed to be unpolarized, and hence, they decay isotropically, for which the experimentally known branching ratios are used.

Comparison to experimental data at $E_{cm} = 19.4\ AGeV$

In Figure 1 and Figure 2, the model results at $E_{cm} = 19.4\ AGeV$ (CERN–$Sp\bar{p}S$) in proton–proton, proton–nucleus and nucleus–nucleus collisions are shown, together with the experimental data. Here and further on, in the model central nucleus–nucleus collision means impact parameter $b = 0\ fm$. As one can see from Figure 1, the model reproduces successfuly the experimental rapidity and multiplicity distributions for negative particles. It also describes satisfactory strange hadron production in pp and pA collisions, except the shape of the rapidity distribution of Λ's measured by the NA35 Collaboration[9]. However, no low p_t enhancement (which is visible in the NA35 data for central SS collisions, and more pronounced for heavier colliding systems[16]), is predicted. As it was demonstrated earlier[9], our model (without additional mechanisms) fails to reproduce the mean number of strange baryons and their rapidity distributions in central SS collisions. Also, since nuclear cascading of secondaries

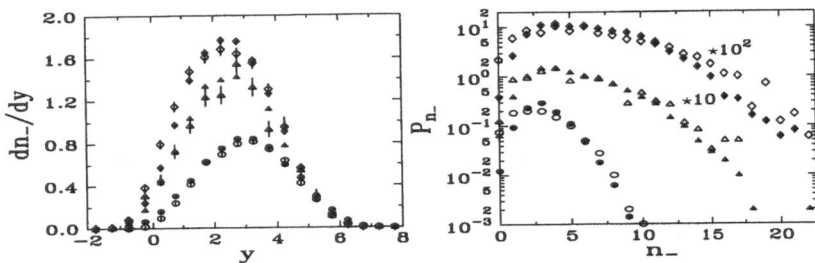

Figure 1. Rapidity and multiplicity distributions of negative particles in pp (circles), pAr (triangles) and pXe (diamonds) collisions. The open points are data[15], the black ones are model predictions.

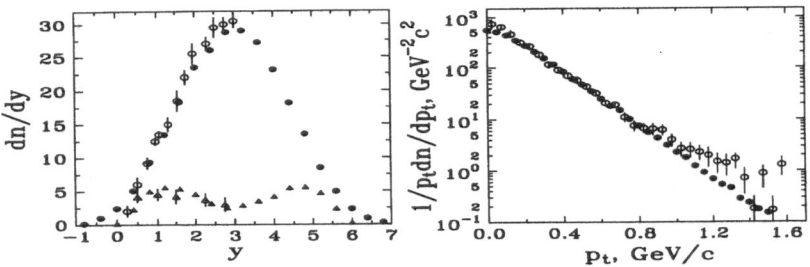

Figure 2. Rapidity distributions of negative particles (circles) and protons (triangles) and transverse momentum distributions of negative particles in central SS collisions. Transverse momentum distributions are calculated and measured in the rapidity interval $0.8 < y < 2.0$. The open points are data[7], the black ones are model predictions.

is not taken into account, it is not able to reproduce the positive particle (protons) rapidity distributions in the nucleus fragmentation region. The secondary cascading could improve the description of the proton rapidity distributions, change essentially the transverse momentum distributions (particularly, the low momentum pion distribution as the result of creation and decay of Δ resonances), and could enhance strangeness production on nuclear targets, due to resonance–nucleon interactions[3, 6].

Charm production: comparison to experimental data

As it was shown in [17, 18], the analytical Quark–Gluon String Model[13] (QGSM) gives a reasonable description of the existing data on charm production: it reproduces satisfactory the cross sections and the spectra of charmed particles produced on nucleon and nuclear targets. Usually, heavy flavour production processes are considered in the framework of PQCD, but these calculations are very sensitive to many factors, particularly, to the low x extrapolation of the gluon structure function. The predictions of QGSM and PQCD (taken into account absorptive corrections) are shown[18] to coincide through a large energy range.

In our Monte–Carlo approach, the fragmentation functions in Equation (6), taken from [13], are used. The exponent $\alpha_q^h(p_t)$ is defined by the quantity α_ψ – the intercept of the Regge trajectory for the ψ family, not well known. Our choice is $\alpha_\psi = -2.2$. Suppression parameters for charmed quark pairs, as for strange quark pairs, are used: from the comparison to experimental data[19], $\gamma_{c\bar{c}} = 0.0025$. The cross sections for different charmed particles are compared with experimental data in Table 1, for 100000 simulated events. This statistics is too small to draw x distributions for the different particles, but enough to show the model prediction for all D mesons (Figure 3).

Table 1. Model prediction for charmed meson production in pp interactions at $E_{cm} = 27.4\ GeV$, compared to experimental data[19].

Meson	Exp. cross section (μb)	Mod. cross section (μb)	Mean multiplicity
D^+	5.7 ± 1.0	10.35	0.00035
D^-	6.2 ± 1.0	12.48	0.00041
D^0	10.5 ± 1.7	10.35	0.00034
D^0	7.9 ± 1.5	19.17	0.00063

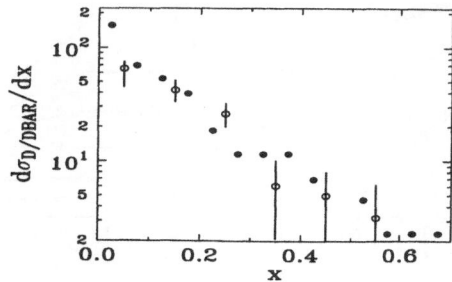

Figure 3. Inclusive spectrum of all D mesons produced in pp interactions at $E_{cm} = 27.4\ GeV$. ($\bullet$) – model predictions, ($\circ$) – data taken from [19].

Comparison to experimental data at $E_{cm} = 200\ AGeV$

In Table 2, we also compare the mean numbers of the produced particles to the experimental data. Production of strange particles and antibaryons is fully determined by the fragmentation parameters: strangeness and baryon–antibaryon suppression factors (described above). In Figure 4, some experimental data are plotted together with the corresponding predictions. With the exception of the low multiplicity event underprediction, the model reproduces well the experimental data. The reason for these discrepancies might be connected with the fragmentation procedure, or with the neccesity of including some additional mechanism, as double diffraction.

Table 2. Experimental data[23] and model predictions on the average numbers of particles in $p\bar{p}$ interactions at $E_{cm} = 200\ GeV$.

Hadron multiplicity	Exp.	Model
$n_{ch}(-3.5 < \eta < 3.5)$	17.6 ± 0.2	16.95
$n_{K_s^0}(-3.5 < y < 3.5)$	0.71 ± 0.08	0.78
$n_{K^\pm}(-3.5 < \eta < 3.5)$	1.15 ± 0.13	1.50
$n_{n+\bar{n}}(-0.5 < y < 0.5)$	0.07 ± 0.03	0.12
$n_{p+\bar{p}}(-3.5 < \eta < 3.5)$	0.6 ± 0.3	0.55
$n_{\Lambda+\bar{\Lambda}+\Sigma^0+\bar{\Sigma}^0}(-2.0 < y < 2.0)$	0.26 ± 0.08	0.25
$n_{\Lambda+\bar{\Lambda}+\Sigma^0+\bar{\Sigma}^0}(-3.5 < \eta < 3.5)$	0.31 ± 0.09	0.29
$n_{\Sigma^\pm+\bar{\Sigma}^\pm}(-3.5 < \eta < 3.5)$	0.16 ± 0.06	0.096
$n_{\Xi^-+\bar{\Xi}^-}(-3.0 < y < 3.0)$	0.03 ± 0.03	0.028
$n_{\Xi^-+\bar{\Xi}^-}(-3.5 < \eta < 3.5)$	0.03 ± 0.03	0.024
$n_{\pi^\pm}(-3.5 < y < 3.5)$	15.9 ± 0.4	15.27

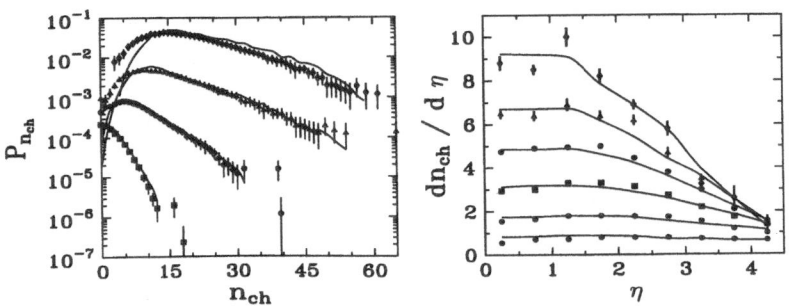

Figure 4. Multiplicity distributions for charged particles in different pseudorapidity intervals ($|\eta| < 0.5$, $|\eta| < 1.5$, $|\eta| < 3.0$, $|\eta| < 5.0$ and full phase space), and semi–inclusive pseudorapidity distributions for different multiplicity bins ($2 \leq n \leq 10$, $12 \leq n \leq 20$, $22 \leq n \leq 30$, $32 \leq n \leq 40$, $42 \leq n \leq 50$ and $52 \leq n$), in non–single diffractive $p\bar{p}$ events, at $E_{cm} = 200\ GeV$, together with experimental data[20, 21]. Full lines are model predictions.

Inclusion of hard parton scattering

So far, only soft interactions between partons have been considered: no momentum, only colour charge, can be transferred during a parton collision. The picture of the interaction is an even number of longitudinal colour strings, spanned between colour triplet quarks or diquarks, moving into opposite directions in the string rest frame. The necessity of the inclusion of hard parton scatterings is demonstrated in Figure 5. At $E_{cm} = 200\ GeV$, the model is not be able to describe the transverse momentum distribution in the whole p_t region, the agreement exists only for $p_t \leq 1.5 - 2$ GeV/c.

In PQCD[10], the fast moving hadrons can be considered as the superposition of parton (mostly gluon) chains, which have increasing transverse momenta along themselves. Small x partons of the chains from the projectile will interact hardly with small x partons from the target, if they are close in impact parameter. The final states obtained in hard processes can be described in term of strings spanned between the quarks, while the gluons are treated as internal kink excitations on the strings[24]. Hence, hard perturbative processes will lead to kinky string states.

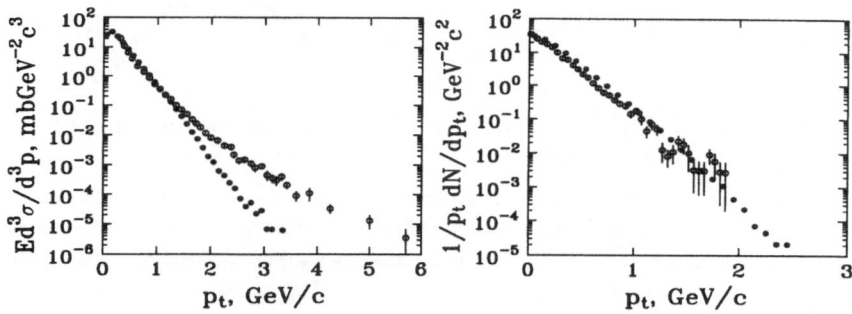

Figure 5. Transverse momentum distributions of charged particles, in the rapidity interval $2 < y < 4$, in pp collisions at $E_{cm} = 19.4\ GeV$ (right figure). Invariant inclusive cross sections of charged particles in $p\bar{p}$ collisions at $E_{cm} = 200\ GeV$ (left figure); both the calculations (o) and the experimental data (•) are in the pseudorapidity region $|\eta| \leq 2.5$.

Up to now, an unified treatment of high energy hadronic collisions, in the framework of QCD, does not exist. There are different phenomenological approaches, which try to combine the "soft" and "hard" parton interactions. In the Regge theory language, it means combinations of exchanges of "soft" and "hard" Pomerons. The most popular one is the eikonal approach[25, 26, 27, 28], in which the probability to have a hard parton interaction (a "hard" cut pomeron) is obtained from the jet production cross section, by using an eikonal unitarization procedure.

Recently, the Lund approach[2] was developed, in which a smooth transition from soft parton to hard parton interactions is provided by the so-called soft gluon radiation. To simulate the gluon radiation, the Lund soft dipole model[31] is used, implemented in the ARIADNE code. This model treats gluon bremsstrahlung in term of colour dipoles, i. e., a gluon g emitted from e. g. a $q\bar{q}$ pair is considered as radiated from the color dipole between q and $\bar{q}$, then a softer gluon is considered as radiated from two independent dipoles, between q and g and between g and $\bar{q}$, and so on. In deep inelastic scattering, the hadron remnants cannot be treated as point–like. Only a fraction $a = \mu/p_t$ of the hadron remnant is allowed to take part in the emission (μ describes the inverse size of the remnant, and p_t the emitted gluon transverse momentum). This reduces the available phase space for gluon radiation in the target fragmentation region. In hadron collisions, the hard gluons (kinks on strings) are the point–like sources of the associated gluon radiation. Quarks or di-quarks sitting at the ends of the strings are considered as extended sources of gluon radiation, leading to its suppression in the fragmentation regions of both ends. The momentum $p_{t,hard}$ transferred during hard gluon–gluon scatterings should be larger than any $p_{t,rad}$ momentum transferred during gluon radiation. This condition is used in the Lund approach[2] as an effective cut–off for hard parton scattering, i. e., hard parton–parton scatterings are accepted only if $p_{t,hard} > p_{t,rad}$.

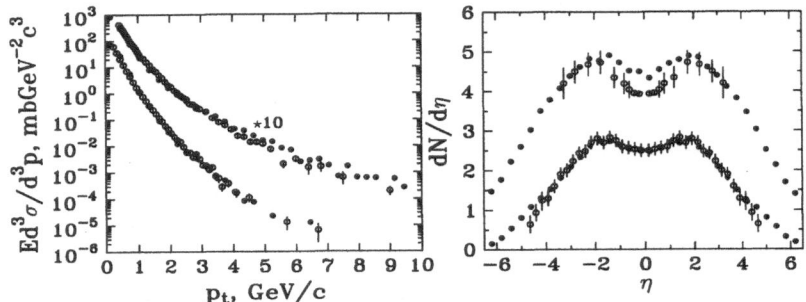

Figure 6. Invariant inclusive cross sections and pseudorapidity distributions of charged particles in $p\bar{p}$ collisions, at $E_{cm} = 200\ GeV$ and $E_{cm} = 1800\ GeV$ respectively. ($\bullet$) – model predictions and (o) – experimental data[21, 22]. For the left figure the calculations and the experimental data[22] are obtained in the pseudorapidity region $\mid \eta \mid \leq 2.5$ at $E_{cm} = 200\ GeV$ and $\mid \eta \mid \leq 1$ $E_{cm} = 1800\ GeV$ respectively.

In the model, we assume that each parton–parton collision can be a hard one with probability $w(s)$, which is a function of the initial hadron energy $\sqrt{s}$. By taking into account both soft and hard processes, we have two mechanisms for string excitation: through longitudinal quark motion, and through hard parton–parton scattering. In the case of hard parton–parton scattering, a kinky string is produced, and its energy–momentum is the sum of the energy–momenta of the hard scattered partons and the longitudinally moving quarks sitting on the string ends.

Our sampling of hard parton–parton interactions is based on the two–jet differential cross section[29], written as a sum of terms, each of them representing a contribution to the cross section due to a particular combination of incoming partons i, j, $(i, j = g, q)$, with momenta p_1 and p_2, and outgoing virtual partons k, l, with momenta p_3 and p_4:

$$\frac{d\sigma}{dy_3 dy_4 dp_t^2} = \frac{\pi}{\hat{s}} \sum_{ij \to kl} f_i(x_1, Q^2)\, f_j(x_2, Q^2)\, \frac{1}{x_1 x_2}\, \frac{d\hat{\sigma}_{ij \to kl}}{d\hat{t}}\, \frac{1}{1 + \delta_{ij}}\,, \qquad (7)$$

where $d\hat{\sigma}_{ij \to kl}/d\hat{t}$ are the differential cross sections of the subprocesses of quarks and/or gluons for massless quarks, δ_{ij} is Kronecker's delta, and $f_{i,j}(x_{1,2}, Q^2)$ are the structure functions evaluated at a momentum scale Q^2, which give the probability of finding partons i, j carrying fractions $x_{1,2}$ of the initial energy and longitudinal momentum of the incoming hadrons. $d\hat{\sigma}_{ij \to kl}/d\hat{t}$ depends on the Mandelstam variables $\hat{s} = (p_1 + p_2)^2$ and $\hat{t} = (p_3 - p_1)^2$. Only gluon–gluon ($gg \to gg$) hard scatterings are considered so far, using the PYTHIA code[30] to obtain the rapidities y_3 and y_4 and the transverse momenta p_t of the outgoing gluons. Since the parton cross sections diverge for $p_t \to 0$, a cut–off $p_t^{min} = 2.3 \; GeV/c$ is introduced. The EHLQ set 1 structure functions[29], inserted in the PYTHIA program as default parameters, are used in these calculations. In the case of n hard collisions, for gluons belonging to the same hadron, the joint structure function should be used; here, each collision is considered independent, except for $\sum_{i=1}^{n} x_i \leq 1$. Some comparison to the experimental data can be seen in Figure 6. To calculate the distributions shown in this figure, we used $w(s) = 1 - \exp\left[-0.025(s - 376.4)^{0.03}\right]$. The associated gluon radiation was simulated by the ARIADNE code[31]. We applied here the Lund string fragmentation model[30, 32] to simulate the decay of the kinky strings. The default parameters of the LUND programs were used for the simulations.

As follows from our calculations, it is possible to predict the shape of the pseudorapidity distributions, and the increase of the central pseudorapidity density with energy, for both versions of the model: the pure "soft" version, in which neither hard gluon scattering nor gluon radiation are included, and the "hard" version, with both effects included. But only the "hard" version gives reasonable predictions for the invariant inclusive cross sections as a function of p_t.

MODEL PREDICTIONS FOR MASSIVE ION BEAMS

The model has given a reasonable description of the existing experimental data at both energies: $E_{cm} = 19.4 \; AGeV$ and $E_{cm} = 200 \; AGeV$. In this section, some predictions for massive ion collisions at these energies are presented. So far, all calculations discussed below were done without hard gluon scatterings and gluon radiation.

Light flavor and charmed particle production

Some results are presented in Table 3. Our model did not take into account neither nucleon elastic scattering nor diffraction dissociation nor secondary rescatterings of the produced hadrons in the nuclear medium; therefore, the number of nucleons should be strongly underestimated. Two factors are essential to produce strange hadrons, baryon–antibaryon pairs and charmed hadrons: suppression parameters for string decay, and string masses. String masses depend on the structure functions, the x cuts, the initial energy and the colliding system. In central $AuAu$ collisions at $E_{cm} = 19.4 \; AGeV$, the initial energy is shared among about one thousand strings. This leads to a kinematical suppression of heavy hadrons. The influence of this kinematical factor is much less at $RHIC$ energy, as can be seen in Table 3. The model fails to reproduce the enhancement of strange baryons in central nucleus–nucleus collisions, observed in the NA35 experiment[7]. For this reason, the predictions for the strange baryon production in central massive ion collisions should be much lower than the experimental data. As shown above, the model gives a reasonable agreement with the existing data on charm production, in pp collision at $E_{cm} = 27.4 \; GeV$. Predictions for charm production on nuclear targets are presented in Table

4, although good quantitative accuracy, specially for central $AuAu$ collisions, is not intended. At $E_{cm} = 200\ AGeV$, the model predicts ~ 5 charmed particles in each central event, and the central rapidity density of charmed particles can reach 1. This large amount of charmed particles can open interesting perspectives for the creation of high energy neutrino beams[33], particularly for the yet unobserved ν_τ.

Table 3. Number of events (N_{evt}), particle production cross sections (σ_{prod}), and mean multiplicities of charged particles (n_{ch}), negative particles (n_{neg}), charged pions ($n_{\pi\pm}$), charged kaons ($n_{K\pm}$), protons (n_p), neutrons (n_n), antiprotons ($n_{\bar{p}}$) and lambdas (n_Λ), and maximum rapidity densities of charged particles (dn_{ch}/dy), negative particles (dn_{neg}/dy), and the difference between the initial nucleon rapidity and the position of the maxima in rapidity net proton distributions ($y_0 - y_{max}$), in pp, pAu, $AuAu$ and central $AuAu$ collisions at $E_{cm} = 19.4\ AGeV$ (left data in each column) and $E_{cm} = 200\ AGeV$ (right data).

Reaction	pp	pAu	AuAu	$AuAu_c$
N_{evt}	40000; 20000	14000; 7000	2000; 1000	200; 100
$\sigma_{prod}\ (mb)$	28.58; 43.56	1642.7; 1776.8	6573.2; 6810.9	
n_{ch}	8.75; 19.62	17.35; 53.11	535.66; 1719.7	2263.3; 7575.7
dn_{ch}/dy	1.74; 3.02	5.26; 8.85	188.0; 334.63	807.5; 1460.8
n_{neg}	3.26; 8.83	7.84; 25.66	252.2; 843.4	1066.4; 3717.3
dn_{neg}/dy	0.81; 1.48	2.39; 4.33	89.45; 165.23	384.0; 713.5
$n_{\pi\pm}$	6.67; 15.65	14.34; 44.82	454.1; 1470.7	1923.1; 6492.5
$n_{K\pm}$	0.63; 2.00	1.16; 4.61	39.02; 153.83	168.35; 678.84
n_p	1.26; 1.54	1.67; 2.45	38.50; 58.99	156.46; 247.25
n_n	0.69; 0.93	1.68; 2.62	45.74; 66.57	180.02; 271.38
$n_{\bar{p}}$	0.058; 0.331	0.084; 0.691	1.740; 20.99	6.860; 91.840
n_Λ	0.157; 0.340	0.247; 0.714	5.544; 18.21	21.64; 74.89
$y_0 - y_{max}$	0.92; 1.18	2.32; 2.78		1.74; 2.08

It should be stressed that in our model, as in most string models, the produced strings do not interact, and decay independently. The introduction of some collective effects, like string fusion[9], can change the predictions essentially, specially for rapidity plateau heights and high multiplicity distribution tails.

Nuclear stopping power

In Table 3, predictions of the model for the relative position of the maxima of the rapidity net proton distributions are presented, to illustrate the so-called nuclear stopping power. The rapidity shift is determined by how the total nucleon momentum is divided among the partons sitting on the ends of the strings, i. e., by the quark and diquark structure functions and their number. In our case, the number of partons increases with the initial energy. But the rapidity shift is weakly dependent on energy for a given parton number, being the structure function a product of $1/\sqrt{x}$. In the case of a different sea quark structure function, as in the DPM[12] ($1/x$), the rapidity shift decreases with the energy, due to the introduced x–cut[36]. The rapidity shift is also dependent on the fragmentation function of a diquark to produce a proton. As it follows from our calculations, the rapidity shift per soft parton–parton collision is approximately $\sim 0.40 - 0.45$. We want to stress that these distributions are obtained in the model without secondary parton or hadron rescatterings.

Table 4. Mean multiplicities and maximum rapidity densities of all charmed particles and $D/\bar{D}$ mesons in pp, pAu, $AuAu$ and central $AuAu$ collisions at $E_{cm} = 200\ AGeV$. The number of simulated events is shown in Table 3.

Reaction	pp	pAu	AuAu	$AuAu_c$
n_{charm}	0.0231	0.0503	1.379	5.92
dn_{charm}/dy	0.0042	0.0097	0.33	1.12
$n_{D/\bar{D}}$	0.0072	0.0167	0.502	2.06
$dn_{D/\bar{D}}/dy$	0.0011	0.0037	0.11	0.41

Meson density evolution

It is very interesting to obtain information not only about measurable observables, but also to have a performance about the evolution of the colliding system. Here, it is done with help of a very simple idea[34]. In the Lund model[35], the time t_i and coordinate z_i of the hadron i, produced in string decay, are defined by:

$$t_i = (1/2\kappa)[M_s - 2\sum_{j=1}^{i-1} p_{z_j}] + E_i - p_{z_i} \ , \tag{8}$$

$$z_i = (1/2\kappa)[M_s - 2\sum_{j=1}^{i-1} E_j] + p_{z_i} - E_i \ , \tag{9}$$

in the string center of mass. Here index $i = 1, 2, ...$ orders the string break–up points, M_s is the string mass, $\kappa \sim 1\ GeV/fm$ is the string tension, and p_{z_i} and E_i are the hadron longitudinal momentum and energy respectively. Besides, quarks are assumed to be massless. To find hadron time and coordinate in the observer frame, Lorentz boosts and rotations are performed. If interactions between hadrons, after their production through string decay, are neglected, the time evolution of the particle and energy density can be calculated, since times, coordinates and momenta of the produced hadrons are known. Predictions are very limited by the absence of information about the longitudinal coordinates and times of the string formation points. They are assumed to be equal zero. It is not a realistic assumption, because the low momentum initial partons, whose interaction leads to the string formation, should be smeared along the longitudinal direction; but it gives the possibility to see clearly hadron production in finite time. In our model, mesons are produced only after $0.5 - 1\ fm/c$. Since the hadron formation time in this Lund definition[35] is proportional to the transverse hadron mass, baryons are produced later than mesons, and after $\sim 2\ fm/c$. It is a long time, compared to the temporal evolution extracted from models based on only hard parton scatterings. Particularly, at $RHIC$ energy in a hard parton collision model[37], already after 2 fm/c an equilibrated parton gas can be established.

To estimate the local meson density as a function of time, in the center of mass of the colliding nuclei, we choose a box of size: $L_x = 2\ fm$, $L_y = 15.0\ fm$ and $L_z = 24\ fm$. Space grids are also introduced: $\Delta x = 2\ fm$, $\Delta y = 2.5\ fm$, $\Delta z = 1\ fm$. At both energies the model predicts approximatelly equal maxima: $\sim 3 - 3.5$ mesons per fm^3, which can be reached after time $t_{cm} \sim 3\ fm/c$, but the energy and meson density decreases much slower at $RHIC$ than at $Sp\bar{p}S$ energy. After the maxima, the meson densities fall to ~ 1 meson per fm^3, at $\sim 3\ fm/c$ at $Sp\bar{p}S$ energy, and at $\sim 6\ fm/c$ at $RHIC$ energy respectively.

At ultra–relativistic energies (specially at $RHIC$), where the produced mesons are concentrated along the light cones, a more suitable evolution parameter is the proper time $\tau = \sqrt{t_{cm}^2 - z_{cm}^2}$, where t_{cm} and z_{cm} are the center of mass time and longitudinal coordinate of the produced hadron. Using Bjorken's formula[38] for the meson density:

$$\rho_m = 1/\tau A_{trans} dN/d\eta, \tag{10}$$

where A_{trans} is the Au transverse area, N is the number of produced mesons, and η is the space–time meson rapidity, we can calculate the τ evolution of the meson and also energy density, by specifying the transverse mass $m_t \approx 0.5\ GeV$. The meson density evolution calculated at both energies looks similar. It reaches a sharp maximum: ~ 3.7 mesons per fm^3 at $\tau \sim 2.0\ fm/c$, and ~ 4 mesons per fm^3 at $\tau \sim 2.5\ fm/c$, at $Sp\bar{p}S$ and $RHIC$ respectively. After the maxima, the meson densities decrease quickly, following a longitudinal ($1/\tau$) expansion.

ACKNOWLEDGEMENTS

The authors are grateful to the Alexander von Humboldt Foundation and the CICYT of Spain for financial support.

REFERENCES

1. Proceedings of "Quark matter 91", Nucl. Phys. **A525** (1992) 1.
2. B. Andersson, G. Gustafson and B. Nilsson–Almqvist, Nucl. Phys. **B281** (1987) 289; B. Andersson, G. Gustafson and Hong Pi, Z. Phys. **C57** (1993) 485.
3. H. Sorge, this Proceedings.
4. K. Werner, this Proceedings.
5. H.–J. Möhring, A. Capella, J. Ranft, J. Tran Thahn Van and C. Merino, Nucl. Phys. **A525** (1991) 493c.
6. N. S. Amelin, K. K. Gudima and V. D. Toneev, The Nuclear Equation of State, Part B, Ed. W. Greiner and H. Stöcker, NATO ASI Series **A216**, Plenum 1989, p. 473.
7. J. Bartke *et al.*, Z. Phys. **C48** (1990) 191; R. Stock *et al.*, Nucl. Phys. **A525** (1991) 221c.
8. E. Andersen *et al.*, Phys. Lett. **B316** (1993) 603.
9. N. S. Amelin, M. A. Braun and C. Pajares, Phys. Lett. **B306** (1993) 312; Santiago preprint US–FT/92–18 (submitted to Z. Phys. **C**); Proceedings of the XXII Int.

Symp. on Multiparticle Dynamics, Santiago de Compostela, July 1992 , Ed. C. Pajares, World Scientific 1993, p. 482.

10. L. V. Gribov, E. M. Levin and M. G. Ryskin, Phys. Rep. **100** (1983) 1.

11. V. A. Abramovsky, E. V. Gedalin, E. G. Gurvich and O. V. Kancheli, Sov. J. Nucl. Phys. **53** (1991) 271.

12. A. Capella, U. P. Sukhatme, C.-I. Tan and J. Tran Thanh Van, Phys. Lett. **B81** (1979) 68; Orsay preprint LPTHE 92/38 (to appear in Phys. Rep.).

13. A. B. Kaidalov, Nucl. Phys. **A525** (1991) 39c; A. B. Kaidalov and O. I. Piskunova, Z. Phys. **C30** (1985) 145.

14. R. D. Field and R. P. Feynman, Nucl. Phys. **B136** (1978) 1.

15. C. De Marzo *et al.*, Phys. Rev. **D26** (1982) 1019.

16. T. Akesson *et al.*, CERN preprint CERN–EP 111 (1989).

17. G. I. Lykasov and M. N. Sergeenko, Z. Phys. **C56** (1992) 697.

18. Yu. M. Shabelskii, Sov. J. Nucl. Phys. **55** (1992) 1399; Z. Phys. **C47** (1993) 409.

19. M. Aguilar–Benítez *et al.*, Phys. Lett. **B201** (1988) 176.

20. R. E. Ansorge *et al.*, Z. Phys. **C43** (1989) 357.

21. G. J. Alner *et al.*, Z. Phys. **C33** (1986) 1.

22. C. Albajar *et al.*, Nucl. Phys. **B335** (1990) 261; F. Abe *et al.*, Phys. Rev. Lett. **61** (1988) 1819.

23. R. E. Ansorge *et al.*, Z. Phys. **C41** (1988) 179.

24. B. Andersson, G. Gustafson and B. Söderberg, Nucl. Phys. **B264** (1986) 29.

25. T. Sjöstrand and M. van Zijl, Phys. Rev. **D36** (1987) 2019.

26. I. Kawrakov, H.-J. Möhring and J. Ranft, Leipzig preprint UL–HEP–92–11.

27. X.-N. Wang and M. Gyulassy, Phys. Rev. **D45** (1992) 844.

28. N. S. Amelin, E. F. Staubo and L. P. Csernai, Phys. Rev. **D46** (1992) 4873.

29. E. Eichten, I. Hinchliffe, K. Lane and C. Quigg, Rev. Mod. Phys. **56** (1984) 579.

30. T. Sjöstrand, Comput. Phys. Commun. **39** (1986) 347; T. Sjöstrand and M. Bengtsson, Comput. Phys. Commun. **43** (1987) 367; T. Sjöstrand, CERN preprint CERN–TH 6488 (1992).

31. B. Andersson, G. Gustafson, L. Lönnblad and U. Pettersson, Z. Phys. **C43** (1989) 625; U. Pettersson, Lund preprint LU TP 88–5; L. Lönnblad, Lund preprint LU TP 89–10.

32. T. Sjöstrand, Nucl. Phys. **B248** (1984) 469.

33. A. De Rújula, E. Fernández and J. J. Gómez–Cadenas, CERN preprint CERN–TH 6452 (1992).

34. K. Werner, Phys. Lett. **B219** (1989) 111.

35. B. Andersson, G. Gustafson, G. Ingelman and T. Sjöstrand, Phys. Rep. **97** (1983) 31.

36. J. A. Casado, Phys. Lett. **B309** (1993) 431.

37. K. Geiger and B. Müller, Nucl Phys. **B369** (1992) 600.

38. J. D. Bjorken, Phys. Rev. **D27** (1983) 140.

LEPTON PAIRS FROM THERMAL MESONS

Charles Gale

Physics Department, McGill University
3600 University St., Montréal QC
Canada H3A–2T8

INTRODUCTION

For a while now, electromagnetic signals have been known as ideal probes of strongly interacting matter at high temperatures and densities [1]. This owes to the fact that once they are produced, they will travel relatively unscathed from their point of origin to the detector. Since production rates are rapidly increasing functions of temperature and density, these electromagnetic signals provide valuable information on the hot and dense phases of the reaction. It is hoped that, because of these facts, those signals should constitute precious aids in the process of analyzing the behaviour of hot quark–gluon matter [2]. As with any possible experimental signature of the QGP, a great deal of care must go into the calculation of a corresponding "purely hadronic" signal, that is a contribution to the same experimental observables from sources other than the deconfined, chiral–symmetric phase. As far as the quark–gluon plasma is concerned, one may refer to these sources as the "background".

In this paper, we are concerned with the thermal rate of dielectron emission only but our treatment is completely general. The source is a hot environment of several meson species: for the first time, we use a rather complete set of mesons, rather than restricting ourselves to the usual pion gas approximation. The equilibrium assumptions inherent to the approaches similar to the one been used here have to be carried to their logical conclusion: in such scenarios, once the temperature has been set one can clearly calculate the population of species present. These mesons can then interact among themselves, or even decay, to produce lepton pairs in the final state. It is important to realize that we deliberately make no attempt here to connect with experiment because our calculation is rather meant to answer a well defined theoretical question: what is the electromagnetic emissivity (in the dilepton channel) of a hot hadron gas? To answer this question, we shall proceed along the lines of a similar calculation for photon rates [3].

Hot and Dense Nuclear Matter, Edited by
W. Greiner *et al.*, Plenum Press, New York, 1994

We estimate the rates of producing lepton pairs using relativistic kinetic theory. The mesonic interactions are modelled with an effective Lagrangian and the coupling of radiation to hadronic matter coupling is done in the vector meson dominance (VMD) approach. The values of the coupling constants involved are adjusted so that the experimentally measured radiative decay widths are reproduced.

THE MODEL

Our starting point is an ensemble of mesons in thermal equilibrium. We consider the lightest and thus most abundant strange and non–strange mesons together with their main interaction channels. This means we shall include: $\pi, \eta, \rho, \omega, \eta', \phi, K$ and K^*. The charge states are not labelled but all of them are present. This collection can be further divided in two categories: pseudoscalar (PS) and vector (V) particles. From this hot meson gas, how do we calculate what is the lepton pair radiation? It has been shown [4] that the thermal production rate for electron–positron pairs is related to the imaginary part of the retarded photon self energy by

$$E_+E_-\frac{dR}{d^3p_+\,d^3p_-} = \frac{2e^2}{(2\pi)^6}\frac{1}{M^4}(p_+^\mu p_-^\nu + p_+^\nu p_-^\mu - p_+\cdot p_- g^{\mu\nu})\,Im\Pi^{R}_{(\gamma)\mu\nu}(k) \qquad (1)$$
$$\times\frac{1}{e^{E/T}-1}.$$

Here p_+ and p_- are the positron and electron momenta, $k^\mu = (E,\vec{k})$ is the virtual photon momentum, T is the temperature, and we have set the electron mass to zero (nonzero lepton mass is easy to include). R is the number of times per unit four-volume an e^+e^- pair of invariant mass M is produced with the specified momentum configuration. Note that the above equation is perturbative in the electromagnetic interaction only; it is a completely non–perturbative expression in the strong interaction. Furthermore, we shall make use of the VMD model, which states that the hadronic electromagnetic current operator is given by the current–field identity[5].

The above expression tells us how the electromagnetic radiation couples to hadronic (in our case mesonic) matter: by first coupling to one of the vector mesons with some coupling constant. In the above, we have kept the ρ, ω and ϕ fields, but in some cases we shall tacitly include also higher vector mesons by using phenomenological form factors inspired by data. We further need a model for how the mesons interact among themselves. For this, we shall use a simple phenomenological approach, inspired by the chiral properties of low energy QCD. Such classes of phenomenological Lagrangians have been quite successful in the past in the description of low energy hadronic physics [6]. We are explicitly interested in the interaction between the different possible combinations of vector (V) and pseudoscalar (φ) fields. For reasons that will become clear shortly we restrict our discussion to the following interaction Lagrangians [6]:

$$\mathcal{L}^{\rm int}_{VV\varphi} = g_{VV\varphi}\,\epsilon_{\mu\nu\alpha\beta}\,\partial^\mu V^\nu \partial^\alpha V^\beta \varphi\,, \qquad (2)$$

and

$$\mathcal{L}^{\rm int}_{V\varphi\varphi} = g_{V\varphi\varphi}\,V_\mu\varphi\,\overleftrightarrow{\partial}\,\varphi\,. \qquad (3)$$

In the above, the coupling constants are fitted for each field combination, in a procedure we now describe. We have a model for how mesons interact among themselves and how they interact with the electromagnetic field. With this approach, let us study a simple radiative process like the decay of a vector meson into a pseudoscalar meson and a photon like $e.g.$ $\omega \rightarrow \pi^0\gamma$. In this model, the process goes via the $\omega\rho\pi$ vertex, owing to G parity conservation at the strong vertex, and the ρ^0 couples to the photon in virtue of the current–field identity. The coupling constants are adjusted so that the correct experimental radiative decay width [7] $\Gamma(\omega \rightarrow \pi^0\gamma)$ is obtained. Our Lagrangians are then "calibrated" through all the following processes: $\rho \rightarrow \pi\gamma$, $K^{*\pm} \rightarrow K^\pm\gamma$, $K^{*0}(\bar{K}^{*0}) \rightarrow K^0(\bar{K}^0)\gamma$, $\omega \rightarrow \pi^0\gamma$, $\rho^0 \rightarrow \eta\gamma$, $\eta' \rightarrow \rho^0\gamma$, $\eta' \rightarrow \omega\gamma$, $\phi \rightarrow \eta\gamma$, $\phi \rightarrow \eta'\gamma$, $\phi \rightarrow \pi^0\gamma$.

We now integrate our model for interacting mesons with a dilepton radiation calculation. If we keep a calculation of the photon self–energy at the one–loop level an evaluation of its imaginary part, as instructed in Eq. (1), will yield processes of the type V (PS) $\rightarrow$ PS (V) γ^*, PS + PS $\rightarrow \gamma^*$, V + PS $\rightarrow \gamma^*$ and V + V $\rightarrow \gamma^*$. Since such tree–level amplitudes can be readily computed and that our general field–theoretic treatment for dilepton emission has been shown to agree with relativistic kinetic calculations (up to temperature–dependent effects in the form factors, which have been shown to be small [4]) we use the latter approach. Finally note that the two–body channels listed above will kinematically dominate the contributions of the type V + PS $\rightarrow$ PS + γ^*, which we shall neglect. The inclusion of such processes would correspond to evaluation of the photon self–energy beyond the one–loop level. The first attempt to investigate the role of processes with more than two mesons involved has recently been made recently[8].

The basic relativistic kinetic expression for the dilepton production rate from processes of the type $a + b \rightarrow e^+e^-$ and $a \rightarrow b + e^+e^-$ are known[9] and we do not bother reproducing them here.

We also include the direct decay channels of the form V $\rightarrow e^+ e^-$. As we will show, their contributions are non–negligible. This is especially true in the case of $\rho \rightarrow e^+e^-$. One can show that for such decays

$$\frac{dR_{V \rightarrow e^+e^-}}{dM^2} = \frac{3}{2\pi^2} \frac{\Gamma_{V \rightarrow e^+e^-}}{\tilde{N}} \frac{m_V^3}{M^2} B(M^2) \int_M^\infty dE\, f(E)\, \sqrt{E^2 - M^2} \, , \quad (4)$$

where

$$B(M^2) = \beta \frac{\Gamma_{\text{tot}}}{(M^2 - m_V^2)^2 + (m_V\Gamma_{\text{tot}})^2} \, . \quad (5)$$

The constant β fixes the normalization of the Breit-Wigner probability density function. Its value is not important here as it enters also the factor

$$\tilde{N} = \int dM^2 \left(\frac{m_V}{M}\right)^3 B(M^2) \, , \quad (6)$$

which ensures the correct overall normalization based on the experimental value of the partial decay width into the dielectron channel, $\Gamma_{V \rightarrow e^+e^-}$. The integral runs over the allowed mass range.

In the above equations, m_V is the vector meson mass and Γ_{tot} is its total decay width. For the narrow resonances (ω, ϕ) the latter is taken constant but the ρ^0 width is given its proper mass dependence.

RESULTS AND DISCUSSION

The decay channels considered have already been listed: they are the same radiative decay reactions V (PS) → PS (V) + γ, as used to fix the couplings constants of our Lagrangians, with the obvious substitution: $\gamma \to \gamma^*$. The V + PS → e^+e^- amplitudes can all be obtained from the decay reaction amplitudes by crossing symmetry. We list the entrance channels anyway for completeness. They are: $\omega \pi^0$, $\rho \pi$, $\phi \pi^0$, $\omega \eta$, $\phi \eta$, $\rho^0 \eta$, $\omega \eta'$, $\phi \eta'$, $\rho \eta'$, $\bar{K}^* K$ and $K^* \bar{K}$. For each of the PS + PS and V + V reaction, we follow the following approach: their "bare" amplitude is calculated, squared, and finally multiplied by a form factor obtainable from experimental data on e^+e^- annihilation.

The topic of form factors deserves here a short discussion. Of course no information on time–like form factors is available through the analysis of meson radiative decays into real photons. With respect to this issue, we have followed a simple prescription. The time–like electromagnetic form factor of charged pion is experimentally very well known [12] and some experimental information exists also about those of both charged and neutral kaons [13]. In our calculations of $\pi^+\pi^-$, K^+K^-, and $K^0\bar{K}^0$ annihilation rates we have used a recent parametrization [14] of these quantities. The vector mesons annihilation channels have been given the same form factors as their corresponding (by strangeness and isospin) pseudoscalar counterparts. In the case of decays and V + PS reactions, whenever the G parity and isospin conservation laws allowed a coupling only to the ρ^0 and its recurrences, the charged pion electromagnetic form factor [14] was used. In the other cases, we have stuck with a form factor equivalent to a simple pole corresponding to the lightest permitted vector meson. Our way of normalizing coupling constants by means of the radiative decay radiative widths leads us to a belief that this conservative choice of form factors does not introduce too much uncertainty. We made only one exception from the simple rules sketched above. In the case of the reaction $\rho + \pi \to e^+e^-$ the rules would lead to a simple ω–pole. It would be a rather bad approximation because the threshold of this reaction lies below the position of the ϕ–resonance, which thus becomes extremely important. We take therefore a two–pole formula with the relative weight between the ω and ϕ contributions same as in the kaon isoscalar form factor $F_S = (F_{K+} + F_{K-})/2$ [14].

We have performed our thermal hadronic calculations at three temperatures: 100, 150 and 200 MeV. We feel that those reflect a range of energies that is somewhat reasonable, by current theoretical standards.

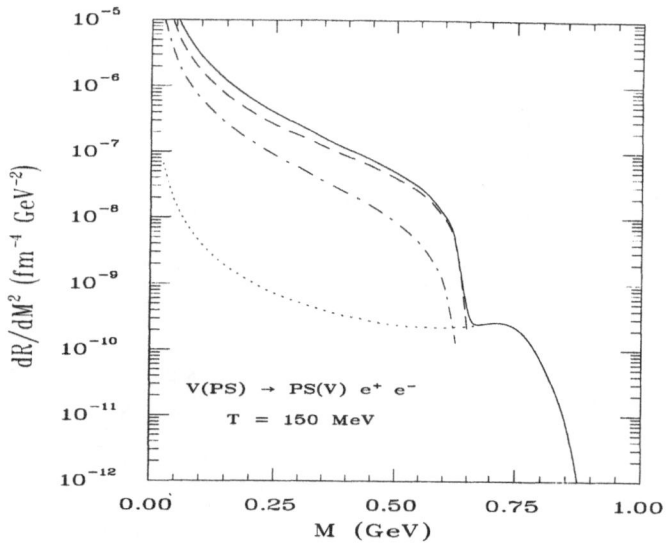

Figure 1: Differential rate for lepton pair production via vector or pseudoscalar meson decay. The dashed line represents the contribution from $\omega \rightarrow \pi^0 e^+ e^-$, the dashed–dotted line is the rate from $\rho \rightarrow \pi e^+ e^-$. The dotted line is the process $\phi \rightarrow \pi^0 e^+ e^-$. The structure in the latter channel is due to the $\rho(770)$. The solid line is the sum of all the decay processes, including those not listed in this caption but enumerated in the main text.

The results for V (PS) $\rightarrow$ PS (V) $e^+ e^-$ at a temperature of 150 MeV are shown on Fig. 1. Not all the decays are shown, but only the dominant ones. Coupling constants arguments aside, the largest contributions will come from the radiative channels where a heavy meson decays into a light one and a lepton pair. This is precisely what is observed on Fig. 1. The largest contribution up to invariant mass $\approx$ 0.65 GeV is from $\omega \rightarrow \pi^0 e^+ e^-$. Over this range, $\rho \rightarrow \pi e^+ e^-$ represents the next–to–leading contribution and the other decays are at least an order of magnitude lower. Above 0.65 GeV invariant mass, the only decay with phase space left is $\phi \rightarrow \pi^0 e^+ e^-$. Note that the widths for the radiative decays of the ω and ρ^0 are comparable and are two orders of magnitude larger than that for $\phi \rightarrow \pi^0 \gamma$ [7]. Dalitz decay (*e.g.* $\eta \rightarrow \gamma e^+ e^-$) is of higher order in α and can thus be neglected. However, this argument alone is not totally convincing as one could imagine that the η could be massively produced at such temperatures. We have therefore performed a calculation of the contribution from η Dalitz decay to thermal electron pair yield, using the VMD prescription for $d\Gamma/dM^2$ [15] with updated coupling constants. We have found it in fact to be orders of magnitude smaller than the channels discussed above.

For the pseudoscalar–pseudoscalar reactions, on Fig. 2 we display a plot of all the contributions, again at $T = 150$ MeV. The different contributions add up to a signal in which the only apparent structures are associated with the $\rho(770)$ and the ϕ, with a slight shoulder at the $\rho(2150)$. The peak in the pion form factor at the $\rho(1700)$ is washed out by the kaon contributions.

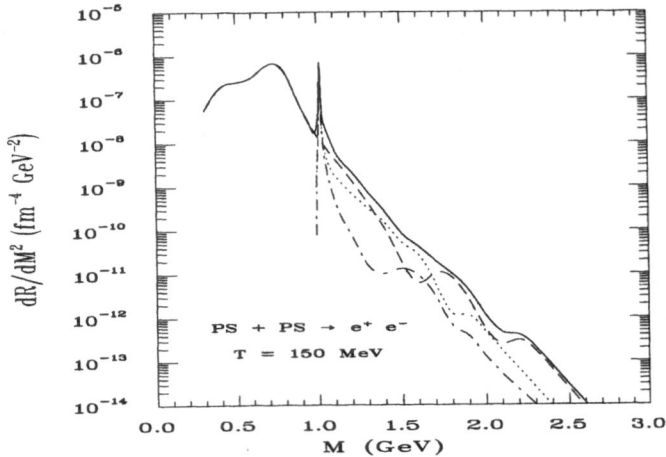

Figure 2: Rate from PS + PS type reactions. The dashed line is the rate for the pion annihilation process. The dotted curve represents the contribution from $K^+ + K^-$. The dashed–dotted line is the rate from $K^0 \bar{K}^0$ annihilation. The solid line is the sum of the PS + PS processes.

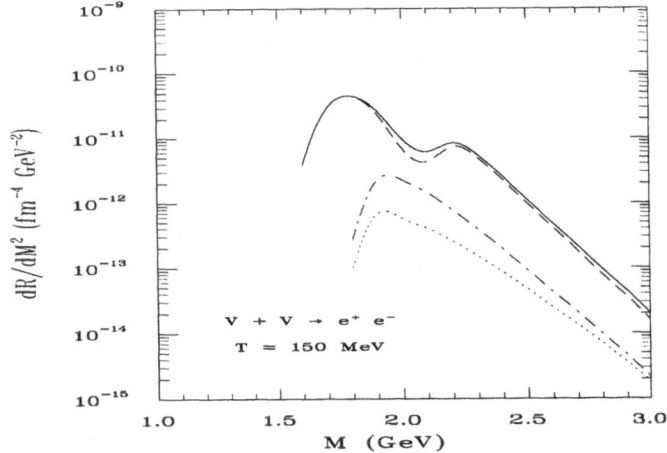

Figure 3: Rate from V + V type reactions. The dashed curve is the $\rho^+ + \rho^-$ contribution. The dashed–dotted and dotted curves represent charged and neutral K^* annihilation, respectively. The solid curve is the sum of the V + V contributions.

We show the V + V contributions on Fig. 3. Above threshold, the sum of these processes outshine the PS + PS ones by roughly an order of magnitude. The structure at $M = 2.15$ GeV owes to the corresponding excitation of the ρ.

The V + PS reactions are quite numerous, we show the brighter dilepton sources on Fig. 4, again for $T = 150$ MeV. The dominant channels are $\omega + \pi^0$, $\rho + \pi$ and

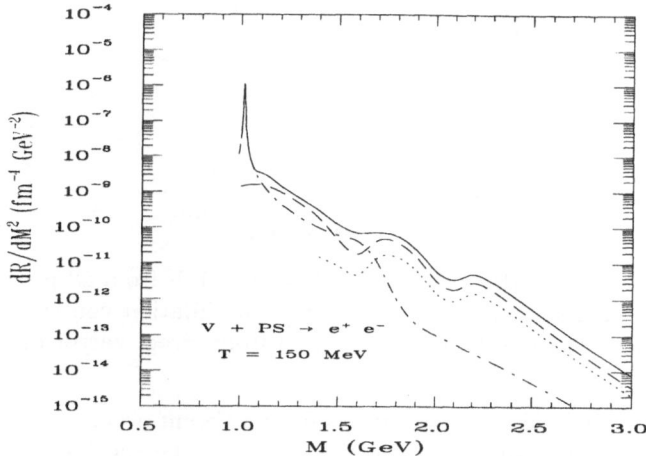

Figure 4: Rate from V + PS type reactions. The dashed curve is the rate from $\omega + \pi^0$. The dashed–dotted curve is the contribution from $\rho + \pi$ and the dotted curve is the rate from $\rho^0 + \eta$. Again, the solid line is a sum of all V + PS processes, as enumerated in the text.

$\rho^0 + \eta$. The kaon channels are not shown but are roughly the size of the $\pi + \rho$ contribution. The strongest signal is from $\omega + \pi^0$, over the entire invariant mass range considered here. Recall from our discussion of the decays that the radiative decay widths of the ρ and the ω are quite large.

Finally, the total rate corresponding to the sum of all processes discussed so far is shown on Fig. 5, along with a curve representing the $\pi^+\pi^-$ contribution only. We also show the net direct decay contribution, summing $\rho \rightarrow e^+e^-$, $\omega \rightarrow e^+e^-$ and $\phi \rightarrow e^+e^-$. The radiation from these channels turns out to be quite important. The signal from the decay reaction $\rho \rightarrow e^+e^-$ closely resembles the pion annihilation spectrum, which in retrospect is quite reasonable. In all cases (decays, PS + PS, V + V, V + PS) our findings at $T = 100$ and 200 MeV are qualitatively similar, with a global shift in the rate.

Up to now, thermal calculations of the variety discussed in this paper have rarely gone beyond a pure pion gas approximation, usually concentrating on the annihilation channel [16]. The contribution from thermal meson decays has been considered previously [17]. To our knowledge it is the first time that extensive mesonic reactions have been included, together with direct decays.

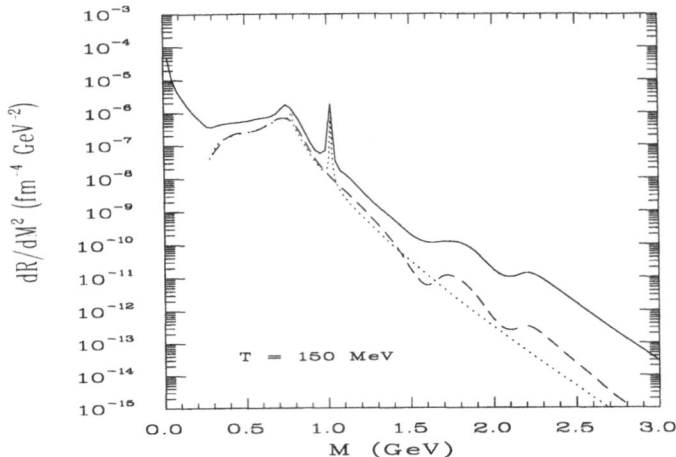

Figure 5: The solid line is the total rate at $T = 150$ MeV from all processes discussed in the text. The dashed line is the pion–pion annihilation contribution only. The short–dashed curve represents the contribution from direct vector meson decays.

Comparing the individual contributions from different processes to the total dilepton rate (Fig. 5) one sees that the dilepton invariant mass spectrum naturally divides in several parts. At low masses, the decay channels clearly dominate the entire spectrum. The crossover to the pion–pion annihilation and direct decay signal occurs just above 0.5 GeV (at the lower temperature, $T = 100$ MeV, this crossing point is shifted closer to the two–pion threshold). Already at $M \approx 1$ GeV, the total rate dominates over the pion gas approximation result by an approximate factor of 3. At $M = 1.5$ GeV, those rates differ by a little more than an order of magnitude. The difference increases with larger invariant masses. One also sees that the net rate at the vector meson positions is also larger than in the straight $\pi^+ - \pi^-$ scenario, owing principally to direct decays and also form factor effects. Probably the most striking conclusion of our work is that the "usual" pion results for lepton pair production calculation holds rather poorly over all regions of invariant masses considered in this work. This statement is true for all temperatures studied here.

Thus, the rate for $M > 1$ GeV is approximately one order of magnitude larger in our calculation than in "conventional" meson background calculations. This enhancement is also present in the momentum structure of the lepton signal[9]. These findings should have important implications in connection with the plasma signal identification. The conventional window for thermal lepton pairs of plasma origin is $m_\phi < M < m_{J/\psi}$[2], precisely the range discussed here. An observation of a signal from an exotic source can only be claimed if all other sources are under control. Here these would be identified with the Drell–Yan mechanism, open charm decay[18] and the thermal background we have considered in this work. However, before any more quantitative statements can be made, it is imperative to complement our calculations with a dynamical model of some sort, in order to make contact with genuine observables. Work in this direction is in progress.

It is of interest to compare the rates obtained with other similar calculations. Some recent interest has been devoted to the emission of lepton pairs from pionic bremsstrahlung processes [19]. It was concluded that the radiation from the external pion lines in pion–pion collisions would be a dominant contribution to the low mass lepton spectrum. Comparing with pion bremsstrahlung calculations at $T = 150$ MeV, we realize that the low mass signal is the same magnitude as the net meson decay contribution. Correcting the pion–pion bremsstrahlung rate for the Landau–Pomeranchuk effect [20] will cut this pion signal by some factor. This factor is only ≈ 2 for low invariant masses and $T = 150$ MeV [21]. This correction also goes down as invariant mass grows. This will then leave the pion bremsstrahlung to compete with the decay channels contribution, up to the two–pion annihilation threshold.

In this inquiry, we have pursued the same goals as a similar photon production calculation [3]. Our main aim has been to identify the most important dilepton production processes which operate in a hadron gas. We considered only the decays and reactions with the minimal possible number of hadrons: one in the decay final states, none in the final states of two-initial-hadron reactions. These processes are believed, on the basis of the phase-space and order-of-interaction arguments, to be dominant here. The reactions of this kind $(2 \rightarrow 0$ hadrons$)$ do not operate in real photon production due to restrictions from energy-momentum conservation. However, the dominant reactions for photon production $a + b \rightarrow c + \gamma$ can produce virtual photons as well. It is clear that they would populate preferably the low-mass region. Even there they would be probably negligible, as it was shown for the case of $\pi + \pi \rightarrow \pi +$ dilepton in [8]. But one cannot exclude surprises. The latter process amplifies, together with the three pion annihilation channel, the omega peak in dilepton spectrum. This may in turn serve as a signature of a hadron gas creation [8]. It has also been pointed out that the a_1 meson could have a significant influence on the real photon yield [22], through the process $\pi \rho \rightarrow a_1 \rightarrow \pi \gamma$. This conjecture has been carefully analyzed in a recent paper [23]. The reflection in the thermal dilepton sector is certainly worth studying as well and this has been addressed recently[24]. Three body initial state processes $a + b + c \rightarrow e^+ e^-$ may also contribute significantly in the high invariant mass region [8].

ACKNOWLEDGEMENTS

It is a pleasure to acknowledge that this work is the result of a fruitful collaboration with Peter Lichard. We both would like to acknowledge the warm hospitality of the Theoretical Physics Institute of the University of Minnesota, where this work was started. This work was supported in part by the Natural Sciences and Engineering Research Council of Canada, by the FCAR fund of the Québec Government and by a NATO Collaborative Research grant. The stay of P.L. at the University of Minnesota was supported by the U.S. Department of Energy under Contract No. DOE/DE-FG02-87ER-40328; travel expenses were borne by the grant MŠMŠ SR 01/35.

REFERENCES

1. E. Feinberg, Nuovo Cimento A **34**, 39 (1976); E. Shuryak, Phys. Lett. **79B**, 135 (1978); E. Shuryak, Phys. Rep. **67** , 71 (1980); G. Domokos and J. I. Goldman, Phys. Rev. D **23**, 203 (1981); G. Domokos, *ibid* **28**, 123 (1983); K. Kajantie and

H. I. Miettenen, Z. Phys. C **9**, 341 (1981); **14** 357, (1982); L. McLerran and T. Toimela, Phys. Rev. D **31**, 545 (1985).

2. For a recent review, see P.V. Ruuskanen in *Particle Production in Highly Excited Matter*, H.H. Gutbrod ed., Plenum, New York (1993), and references therein.

3. J. Kapusta, P. Lichard and D. Seibert, Phys. Rev. D **44**, 2774 (1991).

4. C. Gale and J.I. Kapusta, Nucl. Phys. **B357**, 65 (1991).

5. J. Sakurai, *Current and Mesons* (University of Chicago Press, 1969); N. Kroll, T. D. Lee and W. Zumino, Phys. Rev. **157**, 1376 (1967).

6. U.-G. Meißner, Phys. Rep. **161**, 213 (1988), and references therein.

7. Review of Particle Properties, Phys. Rev. D **45**, (1992).

8. P. Lichard, University of Minnesota preprint TPI–MINN–92/51–T, October 1992 (unpublished).

9. Charles Gale and Peter Lichard, Phys. Rev. D, in press.

10. P. Lichard and L. Van Hove, Phys. Lett. **245B**, 605 (1990).

11. P. Lichard and J.A. Thompson, Phys. Rev. D **44**, 668 (1991).

12. D. Bisello et al., Phys. Lett. **220B**, 321 (1989), and references therein.

13. N. Albrecht et al., Phys. Lett. **185B**, 223 (1987), and references therein.

14. M.E. Biagini, S. Dubnička, E. Etim and P. Kolář, Nuovo Cimento A **104**, 363 (1991). The paper does not contain all the necessary parameters to reconstruct the form factors. We are indebted to Dr. Dubnička for providing us with a more detailed information in the form of ready–to–use computer codes.

15. V.M. Budnev and V. A. Karnakov, Pisma Zh. Eks. Teor. Fiz. **29**, 439 (1979).

16. K. Kajantie, J. Kapusta, L. McLerran and A. Mekjian, Phys. Rev. D **34**, 2476 (1986).

17. K.K. Gudima, A.I. Titov and V.D. Toneev, Phys. Lett. **287B**, 302 (1992); P. Koch, Z. Phys. C **57**, 283 (1993).

18. A. Shor, Phys. Lett. **233B**, 231 (1989).

19. J. Cleymans, K. Redlich, and H. Satz, Z. Phys. C **52**, 517 (1991); K. Haglin, C. Gale, and V. Emel'yanov, Phys. Rev. D **46**, 4082 (1992); **47**, 973 (1993).

20. L. Landau and I. Pomeranchuk, Dokl. Akad. Nauk SSSR **92**, 535 (1953); **92**, 735 (1953).

21. J. Cleymans, V.V. Goloviznin, and K. Redlich, Phys. Rev. D **47**, 989 (1993).

22. L. Xiong, E.V. Shuryak and G.E. Brown, Phys. Rev. D **46**, 3798 (1992).

23. Chungsik Song, Phys. Rev. C **47**, 2861 (1993).

24. Chungsik Song, Che Ming Ko, and Charles Gale, TAMU/McGill preprint, November 1993.

DIMUON AND VECTOR MESON PRODUCTION IN P-W AND S-W COLLISIONS AT 200 GeV/c/A

I. Králik[1] for HELIOS-3 collaboration[2]

[1]Inst. of Exp. Physics Košice
[2]University of Bari and INFN, CERN,
Inst. of Exp. Physics Košice, Univ. of Montreal,
Moscow Lebedev Inst. of Physics, Moscow Phys. Eng. Inst.,
Univ. of Rome "La Sapienza" and INFN, DAPNIA CEN Saclay,
University of Torino and INFN

ABSTRACT

The Helios-3 experiment was designed for the measuring of dimuons with low $M_\perp$ (low mass and low $p_\perp$) in S-W collisions at the SPS energy. The data cover the mass range up to J/ψ. The ratio $\phi/(\rho+\omega)$ was found to be increasing with the centrality due to different behaviour of the $(\rho+\omega)/\pi$ and ϕ/π. The data statistics is good enough for some investigation of the dimuon continuum in the mass range $1 < M_{\mu\mu} < 3$ GeV/c^2. It is possible to explain the dimuon production in this mass range by resonance production, Drell-Yan process and charmed particle decays in p-W collisions, but in S-W there remains some unexplained excess over the contributions of the expected sources.

INTRODUCTION

According to lattice QCD calculations, a phase transition to a state in which quarks and gluons are deconfined should occur at high energy densities. The collisions of ultrarelativistic nuclei are the way how to obtain the critical conditions necessary to produce the plasma phase. Several signatures of possible phase transition[1] can be studied via the detection of the lepton pairs. The HELIOS-3 experiment was designed to detect the dimuons at low transverse mass (low mass and low $p_\perp$) up to the J/ψ mass. For the physical analysis, the data from interactions of p and ^{32}S at 200 GeV/c per nucleon were collected at the CERN SPS. The comparison of the p-W (which is believed to be understood) with S-W can show some evidence of the presence of collective phenomena. The resolution of the dimuon spectrometer allows the study of the production of ρ, ω and ϕ mesons as the function of the charged particle multiplicity (which is closely related to the collision centrality). The enhancement of the $\phi/(\rho + \omega)$ ratio was proposed[2] as a possible signature of the quark gluon plasma formation.

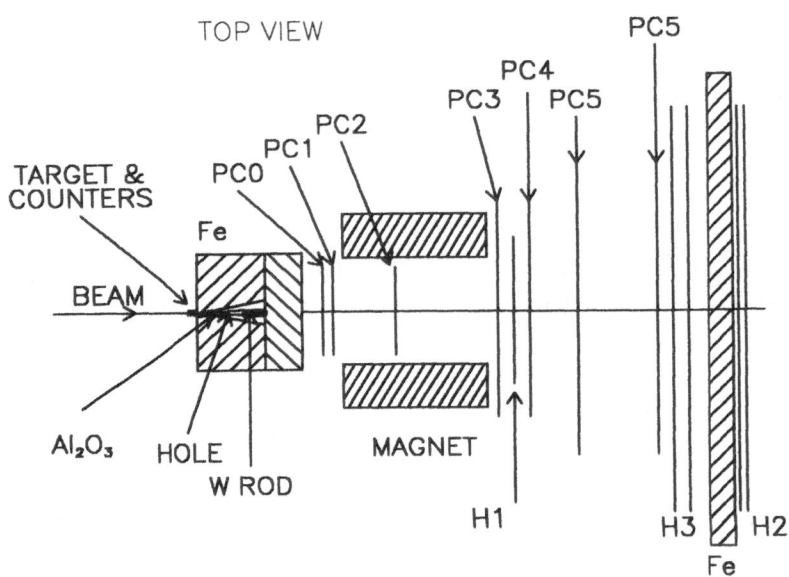

Figure 1: The top view of the Helios-3 dimuon spectrometer.

Another region where some possible QGP signatures could be found[3] is the continuum in the mass region above 1 GeV/c^2 which is also accessible to our experiment.

EXPERIMENTAL SET-UP

The apparatus (Figure 1) was designed for the measurement of muon pairs and charged particle multiplicity in approximately the same acceptance region in nucleus-nucleus collisions.

The beam particle (p or ^{32}S) was identified by a set of scintillator and quartz counters. An interaction in the W target was triggered by six scintillator counters arranged in hexagonal geometry.

After the 5 mm W rod target there were two silicon ring counters (covering $1.6 < \eta < 3.7$ and $2.6 < \eta < 5.5$, respectively) used for the charged particle multiplicity measurement. The forward part $3.7 < \eta < 5.5$ of the second ring detector provided a multiplicity trigger in the dimuon spectrometer acceptance region.

The hadron absorber, placed 25 cm after the target was optimized for the low $M_\perp$ dimuon measurement. It consisted of a 180 cm long Al_2O_3 cone followed by 100 cm of iron blocks. A 121 cm long hole in the centre of the alumina cone was followed by a W rod ensuring that the projectile fragments interact far from the target.

The muon spectrometer consisted of 7 MWPC's PC0 ... PC6 (32 X,Y,U,V planes total) and a 4.1 Tm dipole magnet. Two scintillator hodoscopes H3 and H2 and an 80 cm thick iron wall placed between them were used for muon identification. The dimuon trigger required 2 coincidences in the hodoscopes and at least 2 reconstructed tracks (in X-projections) in the chambers behind the magnet.

DATA REDUCTION AND BACKGROUND SUBTRACTION

During proton and sulphur runs 2.4×10^5 dimuons from p-W and 2.4×10^5 dimuons from S-W collisions were reconstructed. The muon momenta were reconstructed taking into account the energy loss in the absorber and the multiple scattering (the muon momentum direction is reconstructed using the Branson plane method[6]).

The number muons not originated in the target were reduced using the cut $Z_\mu/\sigma_Z < 3$ and $Z_\mu < 250$ cm, where Z_μ is the coordinate of the point of the closest approach of a single muon track to the beam axis, and σ_Z is the error of Z_μ due to the multiple scattering.

After applying all cuts 2.4×10^5 dimuons from S-W ($71\%\mu^+\mu^-$, $12\%\mu^+\mu^+$ and $17\%\mu^-\mu^-$) and 4.5×10^4 dimuons from p-W collisions remained.

Dimuon Kinematic Region and Charged Particle Multiplicity Classes

The experimental arrangement with the off-line cuts on data in general rejects events with small momenta and small $p_\perp$ of single muons. In the dimuon $(M_\perp, y)$ plane most of the non zero acceptance region can be expressed in the form

$$M_\perp \geq \sqrt{(2m_\mu)^2 + \left(\frac{2p_{min}}{\cosh y}\right)^2}, \tag{1}$$

$$M_\perp \geq 4(7 - 2y),$$

where the $p_{min} = 7.5$ GeV/c (given by the energy loss in the absorber), y is rapidity in lab. frame.

The charged particle multiplicity measured by the silicon ring detectors, reconstructed (using a method utilizing the VENUS[8] and the GEANT programs[9]) in the pseudorapidity range $3.5 < \eta < 5.2$ is used to split S-W data into five classes. For each class also parameters describing the collision centrality (like the impact parameter, number of projectile and target participants etc ...) were determined. The definition of classes as well as some centrality parameters are summarized in Table 1. The class 1 is not used separately, since the statistics is not sufficient to make some reasonable analysis.

Table 1: The multiplicity classes definition.

class	multiplicity $3.5 < \eta < 5.2$	Nr. of proj. part.	Impact param. [fm]
2	$20 < MULT \leq 95$	12.40 ± 5.81	7.72 ± 1.22
3	$95 < MULT \leq 130$	24.11 ± 4.78	5.23 ± 1.42
4	$130 < MULT \leq 160$	28.19 ± 3.60	3.96 ± 1.51
5	$MULT > 160$	30.39 ± 2.02	3.02 ± 1.35

The charged particle multiplicity measured by the silicon ring detectors, reconstructed (using a method utilizing the VENUS[8] and the GEANT programs[9]) in the pseudorapidity range $3.5 < \eta < 5.2$ is used to split S-W data into five classes. For each class also parameters describing the collision centrality (like the impact parameter, number of projectile and target participants etc ...) were determined. The definition of classes as well as some centrality parameters are summarized in Table 1. The class 1 is not used separately, since the statistics is not sufficient to make some reasonable analysis.

Before the dimuon spectra may be studied, two sources of background pairs must be subtracted:

- *combinatorial background* having its origin in simultaneous $\pi \to \mu\nu$ and K $\to \mu\nu$ decays,
- *dump background* having its origin in dimuons pair produced in the absorber and not having been removed by the off-line cuts.

Combinatorial Background

Since the "dimuons" from pion and kaon decays originate in front of the absorber, the combinatorial background cannot be reduced by means of some cuts. In spite of the fact that the free space in front of the absorber was as small as possible, the presence of $\mu^+\mu^+$ and $\mu^-\mu^-$ pairs (being $\sim 30\%$ of sulphur data) indicates considerable contribution from π and K decays which can be estimated using the relation[7]

$$N_{+-} = 2R\sqrt{N_{++}N_{--}}, \tag{2}$$

where N_{++}, N_{--} are the numbers of $\mu^+\mu^+$ and $\mu^-\mu^-$ pairs and the R factor was estimated using Monte Carlo method based on VENUS and GEANT ($R = 1.57 \pm 0.10$ for proton and $R = 1.09 \pm 0.02$ for sulphur data). Statistical fluctuations are minimized using the mixed event technique : muons from different events were combined into pairs and corrected for the acceptance.

Dump Background

It has already been mentioned, that secondary particles from the interaction in the target can produce dimuons in the absorber. A part of these unwanted pairs is removed by cuts. The contribution not removed by cuts is estimated using Monte Carlo programs tuned to fit the results of special runs with π beams at 25, 50 and 100 GeV/c focused on the absorber and with 200 GeV/c protons focused on the tungsten rod.

RESULTS

In the following, the results reflecting the current status of the HELIOS-3 analysis will be presented. The attention is given to the study of $\phi/(\rho+\omega)$ ratio, to the absolute normalization to charged particles (i.e. $(\rho + \omega)/\pi$, ϕ/π) and to the analysis of the intermediate mass region ($1 < M_{\mu\mu} < 3$ GeV/c).

Vector Meson Production

To obtain the $\phi/(\rho + \omega)$ ratio from the data, the dimuon mass spectra were fitted with the following function

$$f(M) = P_1\left[0.6B_\rho(m) + G_\omega(m) + P_2G_\phi(m)\right] + P_3\exp(P_4m) + P_5\exp(P_6m), \tag{3}$$

where G_i are gaussians describing the signals from ω and ϕ, B_ρ is a convolution of the experimental resolution function with the function describing the ρ meson shape taken in the form[10]

$$\frac{dN}{dm}(m) = d\sigma(m)\frac{\pi^{-1}m^2\Gamma_{\rho\to\mu\mu}(m)}{(m^2 - M_\rho^2)^2 + m^2\Gamma_{\rho\to\pi\pi}(m)}, \tag{4}$$

and $P_1, \ldots, P_6$ are fitted parameters.

Table 2: The ratios of $\phi/(\rho + \omega)$ in different $p_\perp$ bins.

	p-W	S-W
$(low\ p_\perp)/(high\ p_\perp)$	0.83 ± 0.10	0.95 ± 0.17
$(intermediate\ p_\perp)/(high\ p_\perp)$	0.98 ± 0.11	0.70 ± 0.11

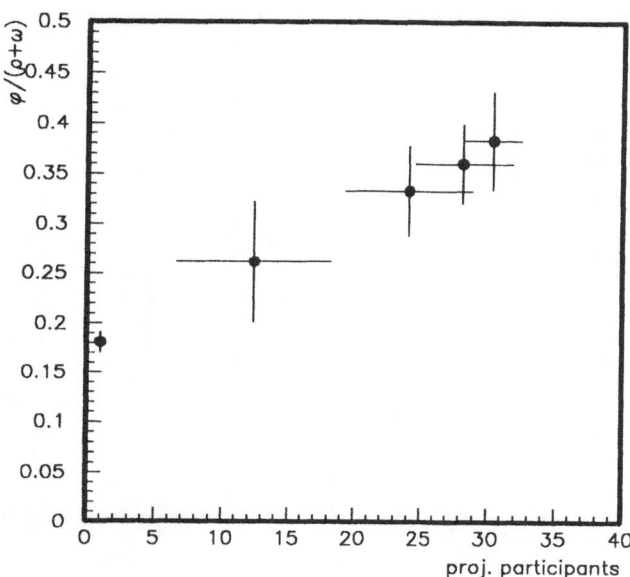

Figure 2: The strangeness enhancement as a function of the number of projectile participants.

The ρ/ω ratio was fixed at the value 0.6 measured in p-Be interactions.[11] The dependence of the $\phi/(\rho+\omega)$ ratio on the number of the projectile participants is shown in Fig 2.

It is clearly seen, that the $\phi/(\rho+\omega)$ ratio increases with centrality. To see, whether this increase is not just a kinematic effect, the data were divided into three $p_\perp$ bins:

low $p_\perp$: $p_\perp \leq 0.35$ GeV/c,

intermediate $p_\perp$: $0.35 < p_\perp \leq 0.6$ GeV/c,

high $p_\perp$: $p_\perp > 0.6$ GeV/c.

Table 2 shows, that within the errors, the $\phi/(\rho+\omega)$ ratio does not depend on $p_\perp$.

Vector Meson Production Normalized to Number of Charged Particles

Another information on the ϕ production, compared to $\rho+\omega$, provides the normalization of the resonance yield to the number of charged particles as a function of centrality.

In Fig 3 one can s ee that while $(\rho+\omega)/\pi$ decreases with centrality, ϕ/π seems to exhibit a different trend. This result is to be taken as very preliminary.

Intermediate Mass Region

In the mass region $1. < M_{\mu\mu} < 3$ GeV/c^2 there are three sources expected to contribute to the dimuon continuum:

i. the tails of resonances (ρ, ω, ϕ),

ii. the Drell-Yan process,

iii. simultaneous decays of charmed particles.

Since neither of these processes is measured under the same conditions as the HELIOS-3 data (beam,kinematic region, energy ...), it is necessary to use the existing experimental information and extrapolate it. In this analysis we extrapolate the known sources to get the dimuon mass shape, rather than the cross section (i.e. the final cocktail was

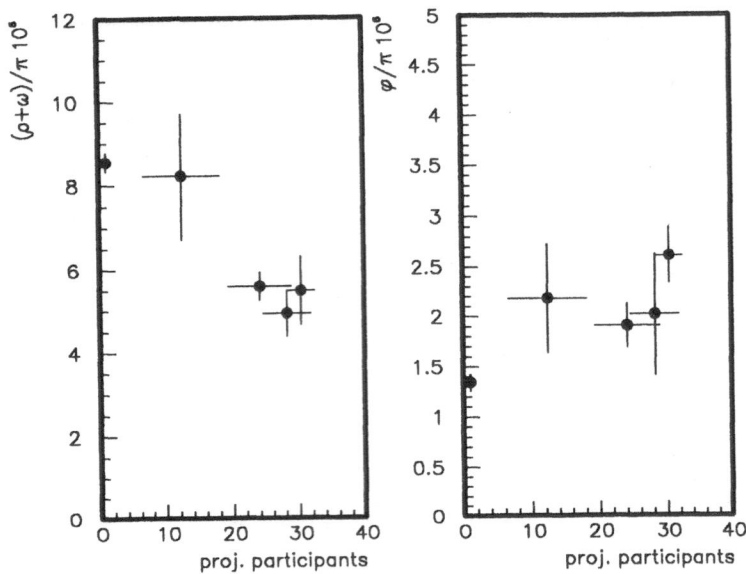

Figure 3: The resonance yield normalized to the number of charged particles.

normalized to the number of ρ mesons found by fitting the data). For the change of the cross sections going from p-p to p-W and S-W collisions (in kinematic region (2), in given impact parameter range etc...) the following assumptions were used:

- ρ *meson production* cross section dependence is estimated using the formula

$$\sigma_\rho^{SW}(kin.reg.) = \left(\frac{<n_\rho^{SW}>}{<n_\rho^{pp}>}\right)_{kin.reg} \frac{\sigma_{inel}^{SW}}{\sigma_{inel}^{pp}} \sigma_\rho^{pp}(kin.reg), \qquad (5)$$

where $<n_\rho^{SW}>$ and $<n_\rho^{pp}>$ are mean ρ multiplicities in the region (2) in p-W/S-W (for S-W in given multiplicity class selected using the impact parameter range from Table 1) and p-p collisions, respectively, predicted by the Monte Carlo event generator QGSM,[12] σ_{inel}^{SW} and σ_{inel}^{pp} are inelastic cross sections in p-W/S-W and p-p collisions, respectively, (predicted by the QGSM model) and $\sigma_\rho^{pp}(kin.reg)$ is experimental value of ρ production cross section in p-p collisions at 200 GeV/c in the region (2).

- *Drell-Yan process* cross section depends linearly on A and the cross section corresponding to some impact parameter range $b_{min} \le b \le b_{max}$ can be expressed using formula

$$\sigma^{SW}(b_{min}, b_{max}) = A_S A_W \sigma_0 \left(2\pi \int_{b_{min}}^{b_{max}} T_{SW}(b)bdb\right), \qquad (6)$$

where T_{SW} is the thickness function for the S-W collision, σ_0 the nucleon-nucleon cross section corrected for the isospin. Drell-Yan events generated using the Lund Monte Carlo event generator PYTHIA[13] (Duke-Owens structure functions,[14] $p_\perp$ adjusted to reproduce experimental value) were used as input for the HELIOS-3 simulation program based on GEANT. The K factor was chosen to be $K = 2.5 \pm 0.5$ not contradicting experiments.

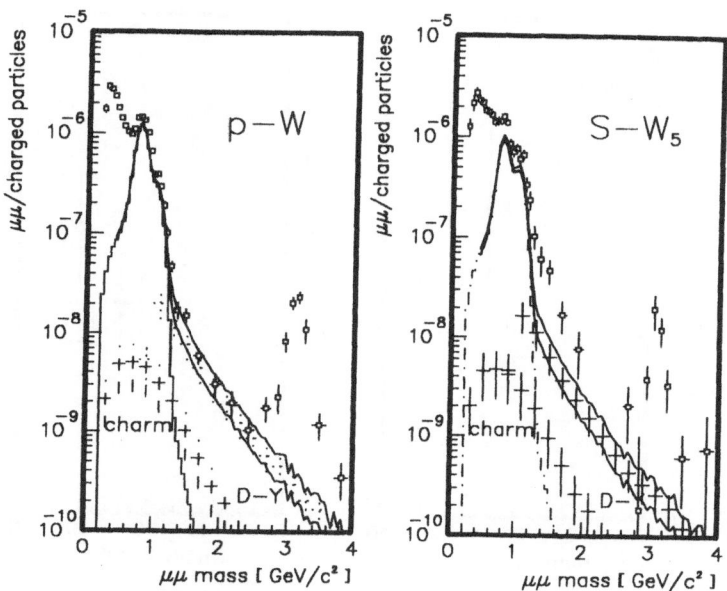

Figure 4: The sources expected to contribute to the intermediate mass region superimposed onto the measured dimuon mass spectra in p-W and S-W (multiplicity class 5) collisions. The histogram represents the resonance contributions. The solid lines show the expected signal given by the expected sources.

- *charm production* cross section was supposed to depend on A the same way as the drell-Yan process (i.e. as A^1) and (6) is used. The charm events generated using PYTHIA (tuned to reproduce the experimental $D\bar{D}$ distributions) are processed in the same way as the Drell-Yan ones. As σ_0 the NA-25 value for $D\bar{D}$ cross section in p-p at 200 GeV/c is taken.[15]

The mix of the contributions described above superimposed over the p-W dimuon mass spectrum is shown in Fig 4.

It can be seen that our mix can explain the dimuon signal in the proton data. To have some measure showing how the expected sources can explain observed signal, the ratio *data/sources* was chosen. Fig 5 shows, how the *data/sources* ratio for masses $1.35 < M_{\mu\mu} < 1.6$ GeV/c^2 and $1.35 < M_{\mu\mu} < 2.5$ GeV/c^2 depends on centrality. It is clearly seen, that the expected sources (extrapolated using the aforementioned assumptions) can explain p-W data, but there is an indication of an excess over the expected contributions.

CONCLUSIONS

The results of the analysis that just have been described, lead to the following conclusions:

- an increase of the $\phi/(\rho + \omega)$ ratio is observed when going from p-W to S-W collisions,
- in S-W collisions $\phi/(\rho + \omega)$ increases with centrality,
- no visible $p_\perp$ dependence of $\phi/(\rho + \omega)$ is observed,
- *preliminary:* $(\rho + \omega)/\pi$ decreases with centrality more than ϕ/π,

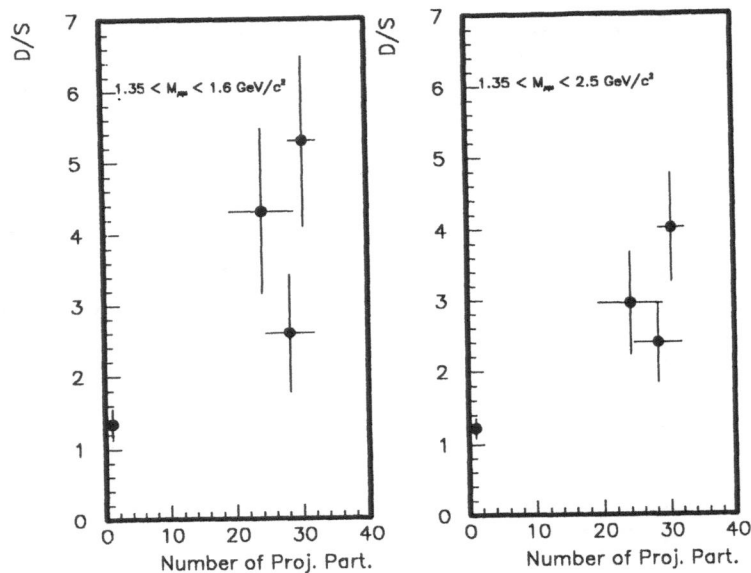

Figure 5: The ratio of experimentally measured dimuon production over the yield predicted by the extrapolation of the sources expected to contribute to the intermediate mass region shown as a function of the number of projectile participants. Two different mass regions are shown.

- dimuon continuum in the intermediate mass region measured in S-W cannot be explained by the contributions of known sources, extrapolated using available experimental knowledge sufficient to describe the observed signal in p-W data.

References

[1] P.V. Ruuskanen, Electromagnetic Probes of Quark-Gluon Plasma in Relativistic Heavy-Ion Collisions, preprint JYFL 28/91 (1991).

[2] A. Shor, Phys. Rev. Lett. 54:1122(1985).

[3] J. Masarik, N. Pišútová and J. Pišút, A Signature of the Mixed Phase or Quark Gluon Plasma In Sulphur-Tungsten Ion Collisions, preprint PRE 33853, (1993).

[4] M. Masera, Dimuon and Vector Meson Production in p-W and S-W Collisions at 200 GeV/c/A, Proceedings of Hadron Structure '92, Stará Lesná, Czechoslovakia, September 6-11, (1992).

[5] P. Giubellino et. al., Nucl. Instr. Meth. A275:89 (1989).

[6] J.G. Branson et. al., Phys. Rev. Lett. 38:1334 (1977)

[7] M.A. Mazzoni, HELIOS-3 internal note Nr. 50 (1991).

[8] K. Werner, Phys. Rev. Lett. 62:2460 (1989),
K. Werner, Strings, Pomerons, and the VENUS Model of Hadronic Interactions at Ultrarelativistic Energies, Preprint HD-TPV-93-1, (1993).

[9] R. Brun et. al., GEANT 3, CERN DD/EE/84-1, 1987.

[10] J.D. Jackson, Il Nuovo Cimento 34:1644 (1964)
Z.Y. Fang, G. Lopez Castro and J. Pestieau, Use and Misuse of the Breight-Wigner Formula, preprint UCL-IPT 87-07, (1987).

[11] T. Åkesson et. al., First Measurement of Branching Ratio $\omega \to \mu^+\mu^-$, to be published.

[12] N.S. Amelin, K.K. Gudima and V.D. Toneev, NATO ASI series B: Physics 2166:473 (1989), and preprint GSI 89-52.

[13] T. Sjöstrand and M. Bengtsson, Comp. Phys. Comm. 43:367 (1975),
T. Sjöstrand, PYTHIA 5.6 and JETSET 7.3 Physics and Manual. CERN-TH.6488/92, (1992).

[14] D.W. Duke and J.F. Owens, Phys. Rev. D26:1600 (1982).

[15] O. Erriquez et. al., Physica Scripta 33:202 (1986).

MESON PRODUCTION IN HEAVY-ION
COLLISIONS AT 1 GEV/NUCLEON

H. Oeschler for the KaoS Collaboration*

Institut für Kernphysik, Technische Hochschule Darmstadt
D - 64289 Darmstadt, Germany

INTRODUCTION

Relativistic heavy-ion collisions provide a unique tool to study the properties of hot and dense nuclear matter. The high stopping power of nuclei in nuclear matter leads to a rapid heating and to a compression of the medium. Two classes of observables are available. One of them tests the dynamics of the nuclear matter expressed by pictorial words like flow, squeeze out and side splash, whereas the other class of observables explores the properties of hot and dense nuclear matter via particle production.

This talk is focussed on the second observable and amoung the various species produced we concentrate on pions and kaons:

- **positive pions**. The energy needed for the production is the rest mass of 139 MeV plus the kinetic energy of the pion. This energy can be supplied by individual NN collisions at the beam energy of 1 GeV/nucleon. An amount of 447 MeV is available in the center-of-mass frame for the particle production in the very first collisions. This exceeds the above mentioned minimum and this production mechanism is called "above threshold". However, pions with total energies above this value also are observed. Their properties will be discussed at the end of this report.

- **positive kaons**. They are produced together with the transformation of a nucleon into a lambda. The corresponding energy needed in the center-of-mass frame is 671 MeV plus the kinetic energies. Hence, the beam energy of 1 GeV/nucleon is not sufficient to produce K^+ in NN collisions (neglecting the Fermi motion) and the production is called "subthreshold". This fact causes a great sensitivity of the K^+ yield to collective effects and it allows one to extract information on the properties of the nuclear equation of state (EOS)[1, 2, 3, 4, 5, 6].

Kaons carry strangeness, whereas pions do not. Concerning the production mechanism the property strangeness does not play a role (in contrast to ultrarelativistic energies). The production yield is simply related to the Q value of the reaction.

However, as soon as these mesons are produced, the interaction with matter is strongly affected by the property "strange" – "non-strange". Since **positive kaons** carry an antistrange quark, they hardly interact with nuclear matter due to strangeness and energy conservation. Therefore, the positive kaons constitute **direct probes of the hot collision zone**.

The situation for the **pions** is completely different. At the chosen incident energy the excitation of the Δ_{33} resonance is the dominating inelastic channel. By the decay of Δ resonances pions are continuously created. Due to their strong interaction with matter pions are continuously disappearing by the Δ formation via $\pi + N \rightarrow \Delta$. Most of the observed pions are therefore not primordial. The measured pion emission represents the "time integral" of the above mentioned processes. A systematics of the pion study allows us to extract informations about the **dynamics of Δ in matter**. In the following the shape of the π^+ spectra and the azimuthal distribution of π^+ will be discussed in this context.

During the discussion on K^+ production it will be shown that again the Δ resonance plays a key role. Therefore, information extracted from the π^+ spectra is essential for a quantitative understanding of the K^+ production.

THE EXPERIMENT

The experiments are performed with the **Kaon Spectrometer (KaoS)** installed at SIS/GSI.

This spectrometer [7] (Figure 1) has been designed to identify kaons over a wide range of momenta and angles in the presence of a high background of protons and pions. It consists of a quadrupole and a dipole magnet. KaoS combines a compact geometry to minimize the decay in flight, a large solid angle ($\Omega = 15 - 35$ msr) and a broad momentum range ($p_{max}/p_{min} \approx 2$ up to 1.7 GeV/c). The intrinsic momentum resolution without tracking is $\delta p/p \simeq 1\%$. A time-of-flight detector array of 50 plastic scintillator paddles is positioned along the focal plane of the spectrometer. Another time-of-flight detector (16 paddles) is located in between the dipole and quadrupole magnet. This arrangement allows for a very fast, mass-selective time-of-flight trigger which is indispensable for the efficient detection of rare particles. Two 120×35 cm^2 multi-wire proportional chambers – one located at the exit of the dipole and the other close to the focal plane – are used for offline tracking. The pions, kaons, protons and heavier particles are identified by the time-of-flight information together with a tracking analysis.

In the first experiments performed in 1991 the SIS accelerator has provided beam intensities of 2×10^5 Au and 6×10^7 Ne projectiles per spill. Targets of 1.93 g/cm^2 and 0.45 g/cm^2 for Au and NaF have been used, respectively.

In the analysis mesons from central and peripheral collisions are separated by means of the hit multiplicity of charged particles in the Large Angle Hodoscope (LAH). This hodoscope consists of a 96-fold segmented detector close to the target at angles of 12–48 degrees. In this angular range participating protons are the most abundant particles. The impact-parameter selection has been controlled[8] by the correlation of this multiplicity with the summed nuclear charge of projectile fragments observed in the Small Angle Hodoscope (SAH). This 380-fold segmented detector covers polar angles between 0.5 and 11 degrees. This hodoscope is located 7 m downstream of the target.

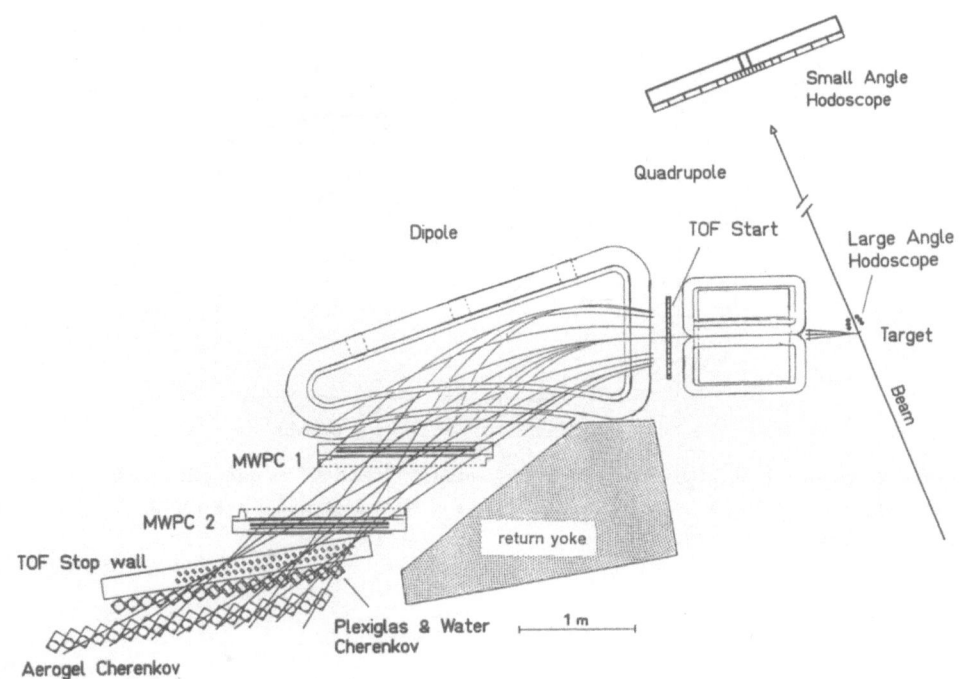

Figure 1. The Kaon Spectrometer and the various detector components. For the particle identification: Time-of-flight (TOF) start, time-of-flight stop, two multiwire proportional counters (MWPC), and Cherenkov detectors. For the centrality selection: Large Angle Hodoscope (LAH) and Small Angle Hodoscope (SAH).

KAON PRODUCTION

Figure 2 shows the double differential cross sections $d^2\sigma/dpd\Omega$ for protons, pions and kaons which have been measured within a polar angular range of $40 - 48$ degrees for the two collision systems Ne + NaF and Au +Au at 1 GeV/nucleon incident energy. The measured kaon spectra can be compared to microscopic model calculations like BUU[5] and QMD[6]. These calculations demonstrate that the compressibility parameter of the EOS, indeed, influences the kaon yield for the heavy system whereas the yield of the low-mass system remains nearly unchanged (see talk given by J. Aichelin). It is, however, too early to draw conclusions about the compressibility on the basis of the present data and the present calculations. Several problems concerning the input of the calculations exist as discussed in J. Aichelin's talk. Uncertainties in the

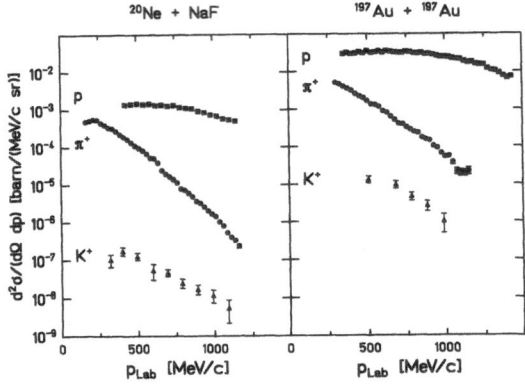

Figure 2. Double differential cross sections for protons, π^+ and K^+ as a function of the laboratory momentum measured at $40 - 48$ degrees in collisons of Ne + NaF and Au + Au at 1 GeV/nucleon incident energy.

elementary cross sections partly cancel out when the ratio of the results from the two mass systems is plotted as shown in Figure 3 (Reference[9]). This figure demonstrates nicely that QMD calculations predict an increasing sensitivity on the compressibility of the EOS at lower incident energies. Better statistics and experiments at lower beam energies are needed to progress further. It can also be seen from Figure 3 that the K^+ yield increases by a factor of 130 while the masses of the collision partners are only a factor of 10 higher. This will be discussed in the next section.

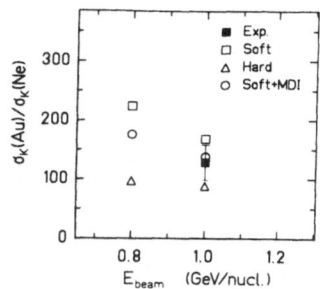

Figure 3. Excitation function of the ratio of the K^+ cross section for the two mass systems. The full symbol shows the experimental value[9]. The open symbols represent QMD calculations [6] for various values of the compressibility parameter (hard – 380 MeV, soft – 200 MeV and soft with momentum-dependent forces (MDI)).

Mass Dependence

As can be seen in Figure 2, the yield of protons and of positive pions varies by the same factor when comparing the two mass systems. The yield of the K^+ increases much stronger. The pion and proton yields scale roughly with $A^{5/3}$ with A the mass of the colliding nuclei, whereas the kaon yield scales with $A^{2.13 \pm 0.15}$ (Reference[9]). This mass dependence allows for an interpretation of the underlying mechanism. Particle production homogeneously distributed over the whole participant volume leads to a $A^{5/3}$ dependence in agreement with the trends observed for p and π^+ (for details see Reference[10]). The higher exponent for the K^+ yield indicates a collective process. Calculations[6] show that the channel $\Delta + N \rightarrow K^+ N \Lambda$ is the dominant process in producing K^+. When neglecting this channel the calculated yield drops by a factor of 6. The dominance of this two-step process for the K^+ production seems to be very convincing at 1 GeV/nucleon incident energy. In the first step a Δ resonance is formed with this process being above threshold. Subsequently the Δ collides with a nucleon producing the K^+. The Δ serves as an energy storage for the second step. The probability for this two-step mechanism is the product of the individual probabilities per step

$$P_1 \propto \rho_N \, \sigma(NN \rightarrow \Delta N)$$

and

$$P_2 \propto \rho_N \, \sigma(\Delta N \rightarrow K^+ N \Lambda)$$

giving

$$P_{K^+} \propto \rho_N^2 \, \sigma(NN \rightarrow \Delta N) \, \sigma(\Delta N \rightarrow K^+ N \Lambda).$$

This simplified consideration explains why a higher exponent for the K^+ production is observed than for the π^+ production, the number of collision and hence the density plays a role. The very same conclusion will be drawn from the impact-parameter dependence of the K$^+$ yield in the next paragraph.

Impact-Parameter Dependence

In order to examine the meson yield as a function of the centrality we study in Figure 4 the ratio of mesons and protons, both measured in the spectrometer under the same centrality condition. The centrality of the collision is determined by chosing four bins in the multiplicity range of the LAH. For these classes the quantity Z_{SAH}^{sum} is measured in the Small Angle Hodoscope in the region of beam rapidity and the number of participating nucleons A_{part} is obtained via the relation

$$A_{part} = A_{proj}/Z_{proj} \times 2 \times (Z_{proj} - Z_{SAH}^{sum}).$$

The upper part of Figure 4 shows the π^+/p ratio which does not vary with centrality. Since the number of protons is proportional to A_{part}, this dependence evidences that the pion rate increases linearly with the number of participating nucleons

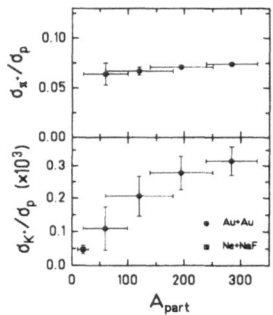

Figure 4. π/p and K$^+$/p ratios as a function of the participating nucleons for both collision systems.

(see also Reference[11]). This trend is in clear contrast to the observation for the Kaons. The ratio K$^+$/p increases drastically with centrality. A scaling with $A^{1.3\pm.3}$ is deduced[12]. Since the ordinate represents a multiplicity one has to multiply it with $A^{2/3}$ to obtain a cross section. Hence, a scaling with A^2 results which is just the same conclusion as obtained in the preceding chapter. The data point obtained for Ne + NaF follows the trend of the Au + Au system.

We conclude that multiple collisions, most likely the $NN \rightarrow \Delta N$ followed by $\Delta N \rightarrow K^+ N \Lambda$, are the dominating processes in producing K$^+$ at this incident energy. The yield of these multistep collisions is very sensitive to the number of collisions and hence to the density.

PION PRODUCTION

The Shape of the Pion Spectra

Figure 5 shows the inclusive, acceptance-corrected invariant production cross sections of positive pions observed in the symmetric mass systems Au+Au and Ne+NaF at 1 GeV/nucleon incident energy as a function of the total pion energy in the c.m. system. The experiment has been performed within a polar angular range of 40 − 48 degrees. This corresponds to normalized rapidities of $0.5 < y/y_{beam} < 0.75$. The measured shapes cannot be described by a single thermal pion source at midrapidity.

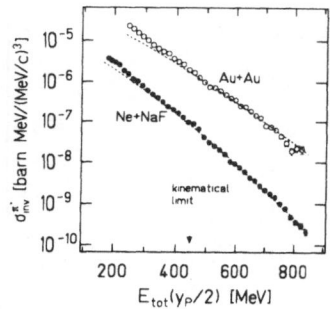

Figure 5. Invariant production cross section of positive pions for Au + Au (open circles) and Ne + NaF (full circles) reactions at 1 GeV/nucleon incident energy as a function of the total pion energy in the midrapidity system. The dashed lines are Maxwell-Boltzmann fits to the high-energy tails of the spectra.

For example, the dashed lines in Figure 5 indicate Maxwell-Boltzmann distributions $\sigma_{inv} \propto E \exp(-E/T)$ fitted to the high-energy tails of the spectra (above the free kinematical limit in the NN system of 447 MeV total pion energy, marked by an arrow). Similar observations have been made in References[13, 14, 15]. The deviations at low energies indicate the non-thermal behaviour of the spectra. The slope parameters of the high-energy part are $T = 76 \pm 3$ MeV for Au (top) and $T = 61 \pm 3$ MeV for Ne (bottom). The latter value is significantly lower than the one of the heavy mass system[10].

849

In order to study the variation of the spectra with impact parameter pion spectra from central and peripheral collisions are compared in Figure 6. They are distinguished by the multiplicity of charged particles Z_{LAH} in the Large Angle Hodoscope and adjusted to match in the low-energy part. In central collisions the relative yield of high-energy pions is enhanced. This dependence is reflected by a small but significant variation of the slope parameter of the asymptotic high-energy tails by 5 ± 2 MeV $(3 \pm 1$ MeV) for Au+Au (Ne+NaF) reactions. These results are in contrast to the proton spectra measured in the same runs; they exhibit a more pronounced variation with the reaction centrality by 30 ± 5 MeV $(13 \pm 5$ MeV). For details see Reference[10].

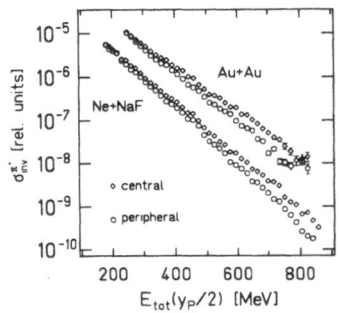

Figure 6. Spectra of positive pions from central and peripheral collisions from Au+Au and Ne+NaF reactions at an incident energy of 1 GeV/nucleon.

High-Energy Pions: A Probe for the Hot and Dense Collision Zone?

To obtain further insight into the mechanism of pion emission Figure 7 represents the ratio of the energy-integrated production cross sections (open data points) for positive pions to the proton cross sections for the Au+Au reaction as a function of the reaction centrality given by the average number of participating nucleons A_{part}. The proton cross sections are the integral over a Maxwell-Boltzmann distribution fitted to the high-energy protons with 800 MeV/c $\leq p_{Lab} \leq$ 1440 MeV/c emitted close to midrapidity. They define the size of the reaction volume, i.e. the number of participating nucleons. The pion cross sections are the results of the integral over the sum of two Maxwell-Boltzmann distributions which represents a proper fit to the pion spectra[10]. The open data points exhibit no significant dependence of the energy-integrated pion to proton ratio on the centrality. This result – as mentioned before – demonstrates that the number of pions, which is dominated by the low-energy part, exhibits a linear increase with the number of participating nucleons as already reported in Reference [11].

A totally different behaviour is observed if one selects high-energy pions as represented by the full data points in Figure 7. Now, the pion spectra are only integrated above the free kinematical limit of 447 MeV total energy in the center-of-mass frame. These high-energy pions are subthreshold particles at 1 GeV/nucleon incident energy. In contrast to the previous result the high-energy π to p ratio now increases for more violent reactions. Absorption in the spectator matter does not seem to be the reason for this behaviour since the low-energy pions do not exhibit such a trend although their absorption cross section is even higher. Such a dependence on the centrality has also been observed for the simultaneously measured positive kaons (see Figure 4 and Reference[12]). It has been interpreted as an experimental signature that subthreshold kaons are produced by preference in central collisions where multiple NN collisions as well as secondary ΔN and πN collisions happen more frequently. This scenario seems to be valid also for high-energy pions and is confirmed in Figure 2 where the production of high-energy pions in the low-mass system Ne+NaF is strongly reduced.

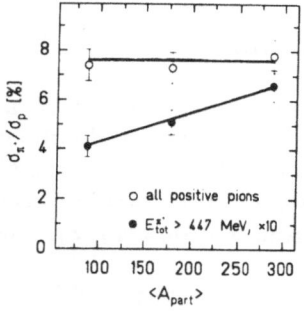

Figure 7. The ratio of the energy-integrated cross sections for positive pions to protons from Au+Au reactions as a function of the average number of participating nucleons A_{part}. The open symbols represent the integral over all pion energies while for the full symbols only pions with $E_{tot} > 447$ MeV are taken into account. The error bars are dominated by errors due to normalization, statistical errors can be neglected.

The interpretation of the results presented above is related to the understanding of the observed spectral shapes. In heavy-ion reactions in the energy regime of 1 GeV/nucleon pions are preferentially produced by the decay of Δ_{33} resonances. These resonances are excited in nucleon-nucleon collisions by $N\,N \rightarrow N\,\Delta$ or, in later stages of the reaction, when pions are more abundant, by the "absorption" of pions via $\pi N \rightarrow \Delta$. Hence, the Δ excitation and the decay kinematics play a key role in the interpretation of the pion spectra[13]. As the Q value of the Δ decay is comparable to the pion rest mass, the pion momentum depends strongly on the "effective" Δ mass whereas the momentum of the Δ resonance has a small effect. Therefore, we do not expect thermal spectra but those which are determined from the decay kinematics. This explains why the shape of the pion spectra vary little with impact parameter in contrast to the behaviour of the proton spectra. However, the measured pion spectra can hardly be explained by assuming decaying thermalized Δ resonances with a mass distribution of the known Breit-Wigner form centered at 1232 MeV/c^2: The low- and the high-energy part of the spectrum is underpredicted [10]. Due to the permanent creation and decay of these resonances via $NN \rightleftharpoons N\Delta$ and $\pi N \rightleftharpoons \Delta$ the mass distribution is affected by the Δ-formation cross section and the relative energies available in these elementary nucleon-nucleon and nucleon-pion collisions. At a later stage of the collision these energies are lower and the mass distribution is populated preferentially at lower Δ masses. Hence, the experimental pion spectrum can be interpreted as a superposition of pions from decaying Δ resonances at various stages of the collision[16]. It reflects the decrease of the average available energy in NN and Nπ collisions due to the continuous cooling of the reaction zone. The high-energy part of the resonance mass distribution is preferentially excited in the early hot stage of the collision whereas the low-energy part is filled by preference in a later stage of the collision. This dynamical reduction of the mean Δ mass with time has already been reported in INC calculations[17] as well as in Isospin-QMD[18] and BUU[19] calculations.

Azimuthal Anisotropic Distribution

The same physics, the strong interaction of pions with matter can be seen in a more exclusive observable. The azimuthal distribution of the pions with respect to the reaction plane. An enhanced emission of pions perpendicular to the reaction plane is observed in semi-central collisions (see talks given by D. Brill and S. Bass and References[20, 21, 22]). This azimuthal anisotropy is most pronounced for high-energy pions.

This experimental result can be interpreted as a shadowing effect of pions in the surrounding spectator matter. Pions which are emitted in the reaction plane undergo more reabsorptive collisions ($\pi N \rightleftharpoons \Delta$) in the cold spectator matter than pions emitted perpendicular to the reaction plane. High-energy pions are assumed to be emitted at an early stage of collision. For them the effect is most pronounced. Low-energy pions are emitted during the whole collision process but mainly towards its end. As a consequence the geometrical separation between hot and cold matter is washed out and the observed effect is reduced.

The high-energy pions emitted perpendicular to the reaction plane represent a new and promising probe of the hot and compressed collision zone.

SUMMARY AND OUTLOOK

Positive pions and kaons have been measured with the **Kaon Spectrometer** at SIS for the two symmetric mass systems Ne + NaF and Au + Au at an incident energy of 1 GeV/nucleon.

- The measured **K$^+$ yield** rises both with increasing mass and with increasing centrality stronger than the pion yield. This is an experimental signature that **multistep processes** – likely involving the Δ – are dominating in the production mechanism in agreement with theoretical descriptions.

- For the interpretation of the **shape of the pion spectra** the importance of the dynamics of the Δ_{33} formation and decay is pointed out. The high-energy pions are assumed to be emitted from decaying Δ_{33} resonances which are populated during the early high-density phase of the collision whereas the low-energy pions are predominantly emitted at later stages of the collision.

  The yield of high-energy pions (produced above the kinematical limit) rises faster with centrality than the total pion yield and exhibits a similarity to the kaon yield. This similarity suggests that also for the high-energy pions multiple collisions contribute to the yield.

- The **azimuthally anisotropic emission,** mostly pronounced for the high-energy pions, is interpreted as **shadowing** of pions **by the spectator matter.** This explanation agrees well with the assumption of the temporal scenario of the pion emission.

The perspectives can be summarized as:

- For the extraction of the **equation of state** the kaon yield at lower incident energies will be measured. More information is needed about the elementary processes used as input in the calculations and a devoted experimental research program is performed using proton and deuteron beams.

- The **high-energy pions** emitted perpendicular to the reaction plane represent **a new and promising probe of the hot and compressed collision zone.**

- The **future theoretical description** of these precisely measured shapes will test our knowledge of the Δ **dynamics in nuclear matter.** The calculations have to describe both the pion and the kaon spectra simultaneously since the kaon yield is strongly related to the number of Δ resonances present during the collision.

* The members of the **KaoS Collaboration**:

W. Ahner, R. Barth, M. Cieślak, M. Debowski, E. Grosse, W. Henning[1], P. Koczoń, M. Mang, D. Miśkowiec, R. Schicker, E. Schwab, P. Senger, (*Gesellschaft für Schwerionenforschung, D-64220 Darmstadt, Germany*)

P. Baltes, C. Müntz, H. Oeschler, A. Sartorius, C. Sturm, A. Wagner (*Technische Hochschule Darmstadt, D-64289 Darmstadt, Germany*)

P. Beckerle, C. Bormann, D. Brill, Y. Shin, J. Stein, K. Stiebing, R. Stock, H. Ströbele, (*Johann Wolfgang Goethe-Universität, D-60325 Frankfurt/Main, Germany*)

B. Kohlmeyer, H. Pöppl, F. Pühlhofer, B. Schlei, J. Speer, K. Völkel, (*Philipps-Universität, D-35037 Marburg, Germany*)

W. Waluś (*Jagiellonian University, PL-30-059 Kraków, Poland*)

REFERENCES

1. R. Stock, Phys. Rep. 135(1986)259.
2. J. Aichelin et al., Phys. Rev. Lett. 58(1987)1926.
3. W. Cassing et al., Phys. Rep. 188(1990)363.
4. J. Aichelin, Phys. Rep. 202(1991)233.
5. A. Lang, W. Cassing, U. Mosel, K. Weber, Nucl. Phys. A 541(1992)507.
6. C. Hartnack, H. Jaenicke, J. Aichelin, Preprint LPN - 93 - 11, submitted to Phys. Rev. C.
7. P. Senger et al., Nucl. Instr. Meth. A327(1993)393.
8. W. Ahner et al., Z.Phys A 341(1991)123.
9. W. Ahner, Ph.D. Thesis, University Heidelberg (1993); and to be published.
10. C. Müntz, Ph.D. Thesis, Technische Hochschule Darmstadt (1993); and to be published.
11. J.W. Harris et al., Phys. Rev. Lett. 58(1987)463.
12. D. Miśkowiec et al., to be published.
13. R. Brockmann et al., Phys. Rev. Lett. 53(1984)2012.
14. S. Backović et al., Phys. Rev. C46(1992)1501.
15. V. Metag et al., Nucl. Phys. A553(1993)283c.
16. B.A. Li, W. Bauer, Phys.Rev. C44(1991)450; B.A. Li, Nucl. Phys. A in press.
17. J. Cugnon et al., Nucl. Phys. A379(1982)553.
18. S.A. Bass, C. Hartnack, H. Stöcker and W. Greiner, submitted to Phys. Rev. Lett.
19. W. Ehehalt et al, Phys. Rev. C 47(1993)R2467.
20. D. Brill et al., Phys. Rev. Lett. 71(1993)336; D. Brill, Ph.D. Thesis, University Frankfurt (1993).
21. S.A. Bass, C. Hartnack, H. Stöcker, W. Greiner, Phys. Rev. Lett. 71(1993)1144.
22. L. Venema et al., Phys. Rev. Lett. 71(1993)835.

NUCLEONIC FLOW, KAON PRODUCTION AND THE NUCLEAR EQUATION OF STATE

Ch. Hartnack[a,b], J. Aichelin[b†], H. Stöcker[c], W. Greiner[c]

[a] Gesellschaft für Schwerionenforschung, Darmstadt, Germany
[b] Laboratoire de Physique Nucléaire, Univ. Nantes, France
[c] Institut f. Theoretische Physik, Univ. Frankfurt, Germany

INTRODUCTION

A major goal of studying heavy ion reactions is the unique opportunity to create hot and dense nuclear matter in the laboratory.[1] Unfortunately, this novel state of nuclear matter exists only for a very short time and expands afterwards. In order to gain information about nuclear matter under these extreme conditions, one must find probes which are most sensitive to the properties of this dense matter at the time of maximum density. The key mechanism for creating dense matter is the compression of nuclei due to inertial confinement[2]. Early hydrodynamical calculations [3] succeeded in describing the dynamics of the collisions of heavy nuclei. Mainly, there are three predictions of the hydrodynamical model which have now been confirmed by experiment [4, 5, 6] namely

- nuclear stopping of heavy systems at central collisions,

- transverse flow (in the reaction plane), the so-called bounce-off effect,

- emission of high energetic particles perpendicular to the reaction plane - the so-called squeeze out.

The production of secondary particles has recently gained much attention since they are expected to yield direct information about the hot high density region. Pions had been proposed as direct messengers from the high density region [2, 7, 8] since they are produced during the time of maximum compression. However, pions have a large cross section for a reabsorption by a nucleon forming a delta. This delta may decay, reemitting another pion: Thus, most of the observed pions have interacted several times with rather 'cold' nuclear matter and the signals from the high density region may have been washed out. [9]

---

† invited speaker

This understanding led to the idea to look for dileptons which interact only weakly with the surrounding matter. While dilepton pairs of low invariant masses are mainly produced by pn bremsstrahlung and Dalitz decays, the production of dilepton pairs of high invariant masses is dominated by the annihilation of $\pi^+\pi^-$ forming an intermediate ρ meson.[10] Therefore l^+l^- test predominantly the pionic densities. However, during the state of maximum compression a large fraction of the pions resides in the deltas, therefore dileptons do also reflect to a large extend the pion density at mid and low baryon density. [11]

Kaons are assumed to give information about the high density region in analogy to the pions.[12, 13, 14] They may be produced predominantly in the first high energy collisions. However, in contrast to the pions, the reabsorption cross sections of Kaons (K^+) are rather small. Therefore, one might expect that Kaons are direct messengers from the high density region.

THE NUCLEAR EQUATION OF STATE

For all observables listed up above there exists a more or less strong dependence on the nuclear equation of state (eos). The nuclear eos describes the amount of energy stored in compressed nuclear matter at a given density.[1, 2] Following the Skyrme ansatz the following density dependent potential is used.

$$U(\rho) = \alpha \left(\frac{\rho}{\rho_0}\right) + \beta \left(\frac{\rho}{\rho_0}\right)^\gamma, \qquad \rho_0 \approx 0.17 \; fm^{-3} \text{ (ground state density)} \qquad (1)$$

For fixing the parameters three conditions are needed, namely that the resulting mean energy per nucleon should have a minimum at $\rho = \rho_0$ with $E/A = -16$ MeV and that the corresponding compression modulus K should be within the range of several hundred MeV. For our calculations we commonly use the following parameter sets (see also Figure 1):

$$\begin{array}{lllll}
\alpha = -124 \, MeV & \beta = 70.5 \, MeV & \gamma = 2 & K = 380 \, MeV & H(ard \; eos) \\
\alpha = -356 \, MeV & \beta = 303 \, MeV & \gamma = 7/6 & K = 200 \, MeV & S(oft \; eos)
\end{array} \qquad (2)$$

Thus the stiffness of the eos, i.e. the repulsion at high energies, is directly connected to the compression modulus. It should be noted that this is artificially due to an ansatz with three parameters fixed. One could also introduce equations of state with a fixed compression modulus and different behaviour at higher densities, e.g.[15]

$$U = a_1 \left(\frac{\rho}{\rho_0}\right) + a_2 \left(\frac{\rho}{\rho_0}\right)^2 + a_3 \left(\frac{\rho}{\rho_0}\right)^3 + a_4 \left(\frac{\rho}{\rho_0}\right)^4 + a_5 \left(\frac{\rho}{\rho_0}\right)^5 \qquad (3)$$

As example we used two parametrizations, first one (Hs) with $K = 380$ MeV (like a hard eos) and a weak repulsion at higher densities. The other one has a strong repulsion with $K = 200$ MeV (Sh), see also Figure 1.

$$
\begin{array}{c|c|c|c|c|c|c}
 & a_1 & a_2 & a_3 & a_4 & a_5 & K \\
\hline
Hs & -120 MeV & 45.6 MeV & 36.8 MeV & -19 MeV & 2.2 MeV & 380 MeV \\
\hline
Sh & -152 MeV & 161 MeV & -80 MeV & 16.7 MeV & 0 MeV & 200 MeV
\end{array} \qquad (4)
$$

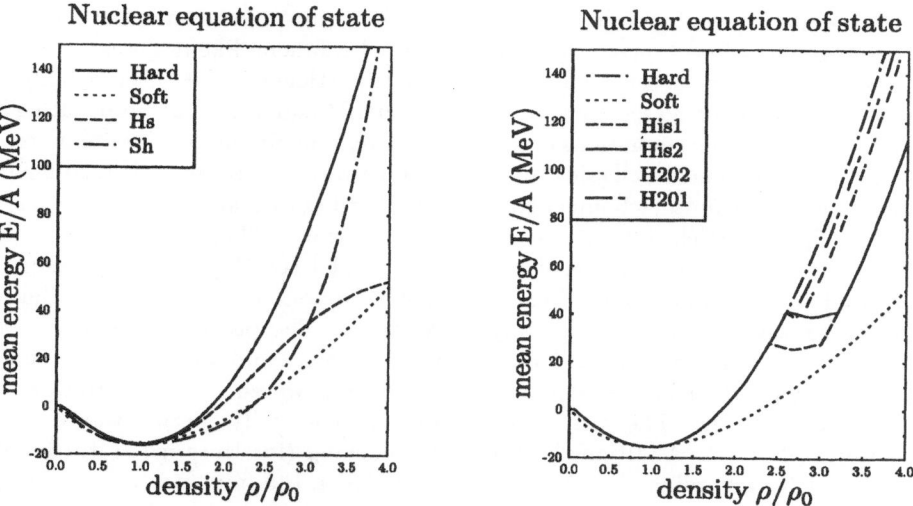

Figure 1. Equations of state with different parametrizations

The effect of the eos described above will be discussed later. Finally, it should be noted that it might also be possible to have additional local minima in the potential energy function. This might e.g. be due to nonlinear scalar meson terms in the Lagrangian or zero degree spin-isospin modes in nuclear matter.[16] Relativistic mean field models with included Delta resonances yielded second minima with its position strongly depending on the scalar and vector coupling constants of the Deltas.[17] To discuss its effect, as it will be done in a later section, we used a simplified parametization, namely a hard eos with a dip of 2 MeV depth and $0.6\rho_0$ width starting at $\rho = 2.4\rho_0$ (His1) resp. $\rho = 2.6\rho_0$ (His2) and two sets with dip widthes of $\rho = 0.2\rho_0$ (H202) and $\rho = 0.1\rho_0$ (H201) starting at $\rho = 2.6\rho_0$. The corresponding curves are shown in the right part of Figure 1.

THE SIMULATION MODELS

For our calculations we used the Vlasov-Uehling-Uhlenbeck (VUU) approach [18] and the Quantum Molecular Dynamics (QMD) model [19, 20, 21]. The VUU model solves the VUU equation which governs the time evolution of the single particle distribution function f in phase space. It is a combination of the Vlasov equation, which describes the motion of classical particles in a mean field and the Nordheim-Uehling Uhlenbeck collision integral. The latter one is a Boltzmann integral modified by additional factors $(1 - f)$ in order to take into account the Pauli blocking of the final states.

$$\frac{\partial f}{\partial t} + \vec{v} \cdot \nabla_r f - \nabla_r U \cdot \nabla_p f = -\int \frac{d^3 p_2 \, d^3 p_1' \, d^3 p_2'}{(2\pi)^6} \sigma v_{12}$$
$$\cdot \left[f f_2 (1 - f_1')(1 - f_2') - f_1' f_2' (1 - f)(1 - f_2) \right] \delta^3 (p + p_2 - p_1' - p_2') \quad (5)$$

The VUU equation contains single particle dynamics (mean field) and two body collisions but no many body correlations. Therefore, phenomena like fragment formation can only doubtfully be described directly by single particle theories. The N-body correlations can be investigated by solving the Liouville equation and propagating the particles under classical equations of motion with reasonable interactions between the nucleons, like it is done in classical molecular dynamics. [22] However, it is difficult to include some quantum mechanical features like hard core scattering with $\sigma(\sqrt{s}, \Omega)$, Pauli principle, particle production and Fermi motion. These additional features are found in the quantum molecular dynamics model (QMD). Here Gaussians are propagated by Hamilton's equations of motion. Smooth phase space distributions are obtained without loosing two body correlations. The particles interact via two (and three) body interactions of local Skyrme-, Yukawa- and Coulomb type and by momentum dependent interactions. For the collisions we employ the collision term of the VUU equation. The collisions are done stochastically, similar way as in the cascade models [7, 23]. Additionally, the Pauli blocking (for the final state) is taken into account by regarding the phase space densities in the final states. The calculations presented here are performed with IQMD, a version with included isospin for nucleons, deltas and pions.[9, 15]

NUCLEONIC FLOW

Much attention has been given to the transverse (in plane) flow in central heavy ion collisions. The projectile and target remnants are scattered off the dense participant region. This effect has been analysed using the sphericity method[24] and transverse momentum analysis[25]. It has frequently been quoted that a hard eos yields stronger flow than a soft one.[20, 21, 26] Momentum dependent interactions (mdi) also enlarge the

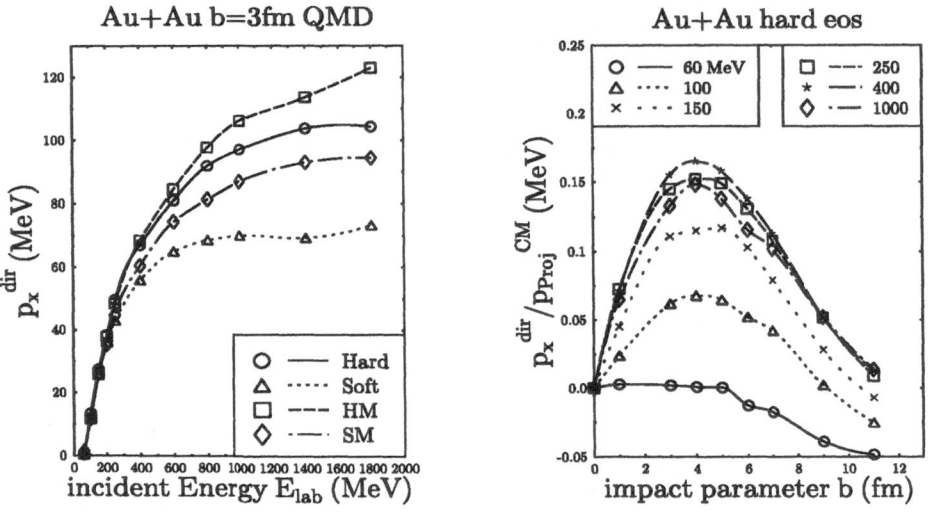

Figure 2. Left: Energy dependence of the momentum transfer p_x^{dir} for different equations of state. Right: Dependence of the scaled flow p_x^{dir} on the impact parameter for different energies.

flow.[13] The left part of Figure 2 illustrates this for the system Au+Au at semicentral collisions in the excitation function of the momentum transfer $p_x^{dir} = < p_x\ sign\ y_{CM} >$. Strongest flow is obtained with a hard equation of state with mdi (HM), while a soft equation of state without mdi shows the smallest values. The values of p_x^{dir} increases for energies higher than about 200 MeV almost linearly with the mean momentum of the incoming momentum of the projectile.[27] This can be seen in the right part of Figure 2 where we scaled p_x^{dir} with the projectile momentum p_P taken in the cm-system. We find that the calculations for 200 MeV and 1 GeV show nearly same curves. The maximum flow is obtained for semicentral collisions in the range b=3-6 fm. For very central and very peripheral collisions p_x^{dir} approaches 0.

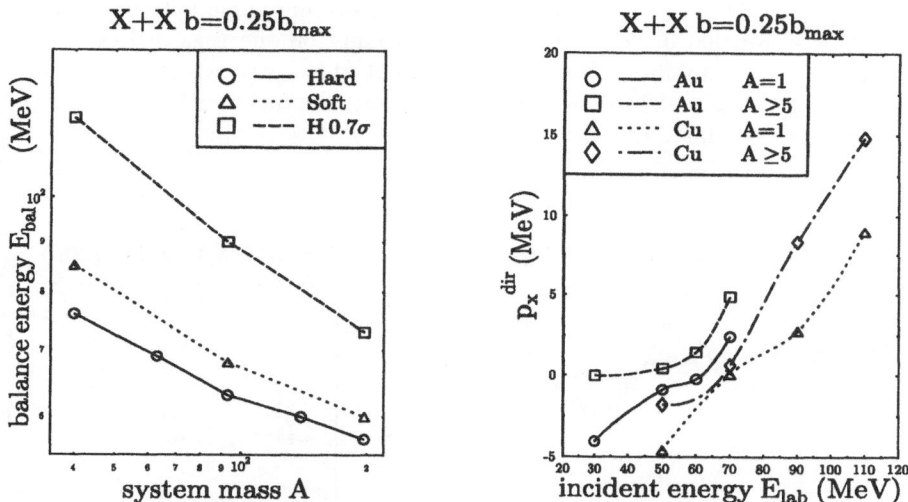

Figure 3. Left: Balance energies of different systems at same geometrical impact. Right: Energy dependence of p_x^{dir} for single particles and IMF's.

Onset of flow

For lower energies the behaviour of the (scaled) flow changes. Here the compression plays a minor role – one can see this in nearly vanishing differences between hard and soft eos (left part of Figure 2) – and the influence of the surface terms in the interaction increases. The flow is strongly going down with energy and may finally be inverted to a negative value. The point of zero flow (balance energy) depends strongly on the centrality of the collision. Also the treatment of long range interactions (especially the Coulomb forces) show strong effects on the onset of flow[31].

Figure 3 shows the dependence of the balance energy on the system size.[15] For this the impact of the incoming nuclei was scaled to the same geometrical overlap $b = 0.25 b_{max} = 0.5 R_{Nuc.}$. The balance energy shows a power law behaviour, which agrees well with the $A^{1/3}$ dependence reported by the Michigan group.[28] Hard and soft eos show small differences in the balances points whereas a change of the cross section has a much stronger effect.[29] The right part of Figure 3 shows the dependence of the flow on the fragment size. It is known that for higher energies the fragments exhibit a stronger flow than single particles [6, 20]. This remains valid up to energies of the balance point. We see that the flow of the fragments is always larger than that of the single particles. It may also happen (esspecially for large systems) that the fragments still exhibit a positive flow while the flow of the single particles has already changed its sign. [30] However it should be noted that this effect is very small.

Squeeze-out

The transverse momentum in plane has its maximum at projectile and target rapidities and is dominated by rather 'cold' matter.[31] Now it is interesting to learn something about the behaviour of hot participant matter stopped to $Y = Y_{CM}$, where the transverse flow vanishes $< p_x > \approx 0$ by construction. Let us consider the azimuthal (φ) distribution of the particles where φ is the inclination angle between $\vec{p}_T$ and the x-axis. The azimuthal distribution depends strongly on the regarded rapidity bin. While at projectile rapidity a strong enhancement at $\phi = 0$ degrees is found and at target rapidity the $\phi = 180$ degrees bin is dominant, the distribution at CM rapidity shows a completely different behaviour.[21]

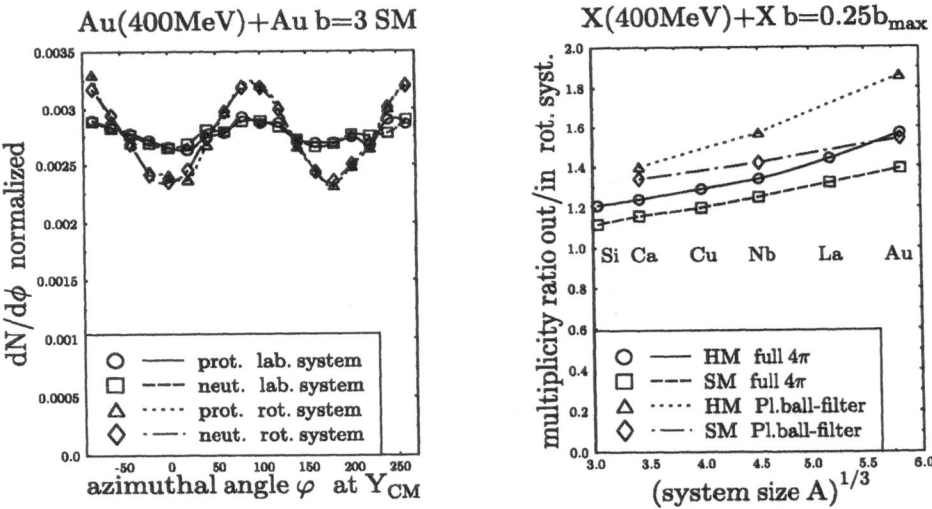

Figure 4. Left: azimuthal distribution at CM rapidity for protons and neutrons in the lab and in the rotated system.
Right: out-of-plane to in-plane ratio in the rotated system as the function of the system size.

Figure 4 presents the distribution of the azimuthal angle φ taken at CM rapidity for the system Au+Au at $b = 3$fm using a soft eos with included momentum dependence. A clear peak at 90 degree (resp. 270 degree), i.e. out of plane is visible, which becomes even more prominent, if we rotate the system from the z=beam-axis (lab system) to the eigensystem of the sphericity tensor (rotated system). The significance of this peak will be even more enlarged, if we fold the multiplicity distribution with the transverse energy of the particles, i.e. if we regard the total transverse energy per azimuthal bin.[30, 33] This finding is in agreement with data obtained by the Plasticball collaboration.[5] Recent experiments performed at the SIS also confirmed the squeeze-out effect as well for charged particles as for neutrons.[32] In our calculation protons and neutrons show very similar behaviour which is in agreement with experimental findings. Fragments show an enhanced squeeze-out effect at CM-rapidities.[30]

The right hand side of Figure 4 shows the out-of-plane to in-plane ratio in the rotated system as a function of the system size. The x-axis has been scaled by a power of one third. The curves for the hard eos and for the soft eos, both with mdi, show a nearly linear behaviour, as well for the full 4π event and for the events filtered with the Plasticball filter routine. A scaling with $A^{1/3}$ might refer to the influence of shadowing effects. However, the significance is very weak, so that a plotting versus A could also be fitted linearly. Furthermore the importance of collectivity should not be neglected, as it can be seen from a linear mass dependence, if we fold the given ratioswith the transverse energy, i.e. if we examine the out-of-plane to in-plane ratio of the total transverse energy.[33]

KAON PRODUCTION

Let us now turn to the production of kaons. In our calculation the kaons are treated perturbatively [12, 13]: the kaons are assumed to leave the system after they are produced without any further interaction with the surrounding matter. The kaons are assumed as isospin degenerate for the K^+, K^0 channel. The K^-, $\bar{K}^0$ channel has been neglected, since its threshold is so high that their direct pair production K^+K^- plays a minor role at 1 GeV/n. However $\pi - \Lambda$ charge exchange may alter this statement. In our calculation we only regarded baryon-baryon channels of the following kind:

$$NN \to NYK, \qquad N\Delta \to NYK, \qquad \Delta\Delta \to NYK$$

Kaons have been assumed to be produced predominantly in the very first pp collisions at the beginning of the reaction. In our calculation we find that the nucleon-nucleon channel plays a minor role in the Kaon production. For 1GeV incident energy this channel yields only about a quarter of the Kaons, about 60% of the Kaon multiplicity is obtained from $N-\Delta$ collisions. The values, which can be found when regarding the contributions of the different channels in Figure 5, are in well agreement with previous calculations.[34] Figure 5 shows the distribution of densities found at the processes producing kaons. Obviously a large number of kaons stem from the high density region.

Also the term 'very first collisions' has to be taken with caution. The Kaon production channel of those NN- collisions where both collision partners collided first time has a marginal contribution of about 3%. The contribution of collisions (NN or $N\Delta$) where only one of the partners collides for the first time (but the other particle does not) is about considerable 25%. In nearly 80% of all collisions at least one of the partners has undergone more than 3 collisions including the actual one. This corresponds to the picture that only few first-first collisions of projectile and target particles (both having $N_{Coll} = 0$) occur in the reaction. The particles of these very first collisions are stopped to midrapidity and form a participant region. All particles of projectile and target matter will predominently collide with these stopped particles and enlarge the region of stopped matter. Therefore the contribution of high energy projectile-target particle collisions is suppressed [35]. This also corresponds to the

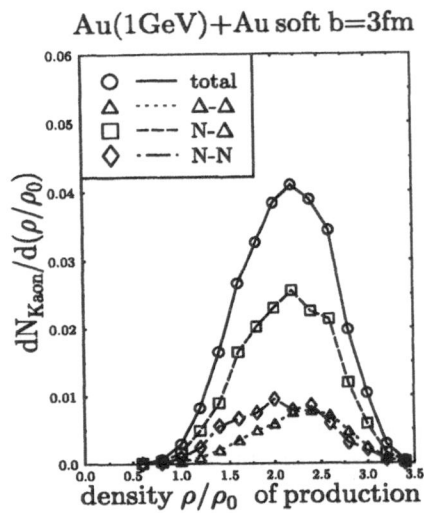

Au(1GeV)+Au soft b=3fm

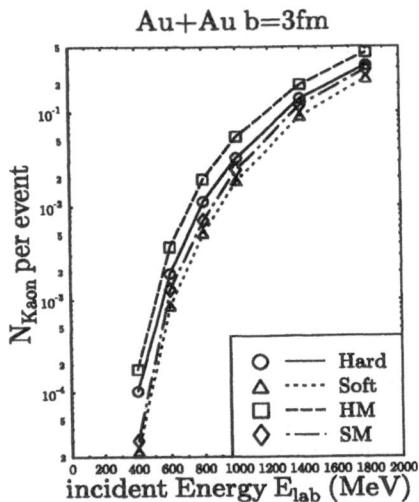

Au+Au b=3fm

Figure 5. Left: Density distribution of the kaon production channels for Au(1GeV)+Au, using a soft eos.
Right: Excitation function of the kaon production for hard and soft eos, with and without mdi.

high amount of kaons coming from rather high densities. Furthermore this underlines the importance of two-step processes for the kaon production, which causes a strong dependence of the kaon production on the system size.[36]

If a lot of kaons are testing high densities, the total production should be strongly influenced by the nuclear eos. The right hand side of Figure 5 confirms this assumption with the excitation function of the kaon multiplicity for the system Au+Au. A soft eos yields higher multiplicities than a hard one. Momentum dependent interactions (SM resp. HM) are decreasing the kaon number further, as it has been already reported.[13, 14] Recent data of the KaoS collaboration seem to be in qualitative agreement with our calculations.[37]

DENSITY ISOMERS

In the previous sections the dependence of flow and particle production on the nuclear equation of state has been analysed. Two common parametrizations, optionally combined with momentum dependent interactions have been used. Quite often these parametrizations have only been identified with their compression modulus K and therefore the behaviour of the equation of state at high compression has been connected to low density behaviour. In this section it shall be demonstrated that this identification can be misleading.

For our analysis we used the VUU model and compared the flow of Au+Au for four equations of state: H(ard), S(oft), Hs and Sh. H and Hs (resp. S and Sh) have the same compression modulus (380 resp. 200 MeV) but different repulsion at high densities (see Figure 1). Thus the regarded parametrizations will yield different effects at different densities. Especially the excitation function will show a different behaviour.

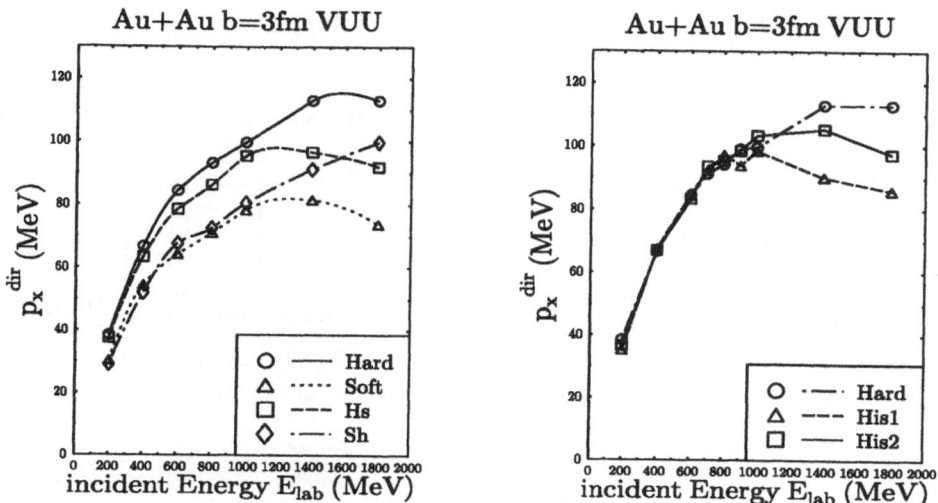

Figure 6. Left: Excitation function of Au+Au for hard and soft eos, Hs and Sh. Right: Excitation function for Au+Au, hard eos, with and without density isomer.

Figure 6 shows on the left hand side the excitation function of the flow p_x^{dir} for a hard and a soft eos in comparison with Hs and Sh. First it should be noted that the flow of hard and soft eos obtained with VUU yield similar values than the corresponding calculation performed with QMD (see Figure 2). For low energies the values obtained with hard and Hs (resp. soft and Sh) are nearly the same. For higher energies the differences increase and thus result different energy dependences of the parametrizations. It should be noted that the flow does not so much test the high density region but more the density gradient between low density and high density region.[31] Therefore observables which are stronger related to the high density region yield much more significant differences between hard and Hs (resp. soft and Sh). For example the kaon spectra of Au+Au at 1 GeV show the same shape for a hard eos and Sh, although they have different compression moduli.[15]

Let us now examine the influence of the density isomers on the observables. From the right part of Figure 6 one does not find significant differences between a hard eos without isomer (hard), hard eos with an isomer starting at $2.4\rho_0$ (His1) and with an isomer starting at $2.6\rho_0$, if one regards the flow values beyond an incident energy of 1GeV/u. For higher energies we find a decrease of the flow for the parametrizations with density isomers. At these energies a large fraction of the nucleons reaches maximum densities larger than $2.4\rho_0$ and $2.6\rho_0$ respectively.

If now regard the excitation function of the kaon production, as depicted in Figure 7 we find a very strong influence of the density isomers. At a certain energy the multiplicity of the kaons raise up by about one order of magnitude. The corresponding

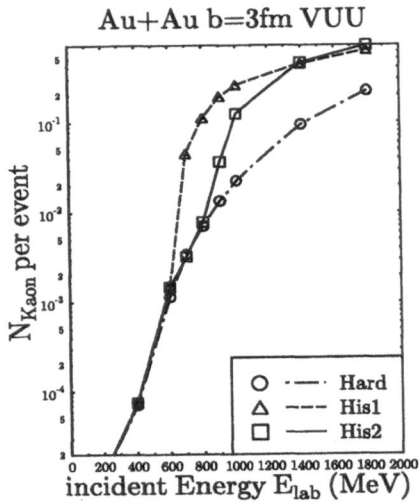

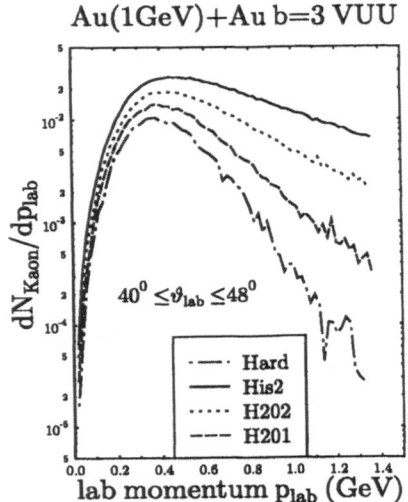

Figure 7. Left: excitation function of the kaon production for a hard eos, His1 and His2. Right: Dependence of the kaon spectra on the width of the gap in the eos.

incident energy depends on the starting point of the gap in the eos. Also the number of pions raises at this point, but only in the order of several 10%.[38]. The density isomers effects especially the number of high energetic kaons, as it can be taken from the right hand side of Figure 7. The raise of high energy kaons depends strongly on the width of the gap, the spectra approach continously from that of His2 (dip width $=0.6\rho_0$) to that of the hard eos (width=0), as it can be seen from the spectra of parametrizations with widthes of $0.2\rho_0$ (H202) and $0.1\rho_0$ (H201).

We conclude that the excitation function of the kaon production and the analysis of the spectra in combination with the excitation function of the flow might be a viable tool to analyse the behaviour of the nuclear eos at higher densities.

REFERENCES

1. For a review and further references see:
   H. Stöcker and W. Greiner, Phys. Rep. 137 (1986).
   G.F. Bertsch, S. Das Gupta, Phys. Rep. 160 (1988) 189.
   W. Cassing et al. Phys. Rep. 188 (1990) 361.
   J. Aichelin, Phys. Rep. 202 (1991) 233

2. W. Scheid, H. Müller, and W. Greiner, Phys. Rev. Lett. 32 (1974) 741

3. C.Y. Wong and T. Welton, Phys. Lett B49 (1974) 243;
   A.A. Amsden et al. Phys. Lett. 35 (1975) 905;

H.G. Baumgardt et al. Z. Phys. A273 (1975) 359.

G. Buchwald et al. Phys. Rev. Lett 52 (1984) 1594.

4. H.A. Gustafsson et al., Phys. Rev. Lett. 52 (1984) 1592.
D. Beauvis et al., Phys. Rev. C27 (1983) 2443;
H.G. Ritter et al., Nucl. Phys. A447 (1985) 3c;
K.G.R. Doss et al., Phys. Rev. Lett. 57 (1986) 302.

5. H.H. Gutbrod et al., Phys. Lett. 216B (1989) 267.

6. K.H. Kampert, J. Phys. G15 (1989) 691.
H.H. Gutbrod et al. Rep. Prog. Phys. 52 (1989) 1267.

7. J. Cugnon, Phys. Rev C22 (1980) 1885.

8. R. Stock et al., Phys. Rev. Lett. 49 (1982) 1236.
J.J. Molitoris and H. Stöcker Phys. Rev. C32 (1985) 346.
J. Harris et al., Phys. Rev. Lett. 58 (1987) 463.

9. C. Hartnack, H. Stöcker and W. Greiner, Proc. Gross Properties of Nuclear Matter
XVI, Hirschegg, Austria, 1988, 138.
M. Berenguer et al., Proceedings of the NATO Advanced Research Workshop on
Nuclear Matter and Heavy Ion Collisions, NATO ASI Series B 205, Plenum, NY,
1990, 343.

10. C. Gale, J. Kapusta, Phys. Rev. C35 (1987) 2107.
C.L. Korpa, L. Xiong, C.M. Ko, P.J. Siemens, Phys. Lett. B246 (1990) 333
Gy. Wolf et al., Nucl. Phys. A517 (1990) 615.

11. L. Winckelmann GSI Report GSI-91-01 p.103, and to be published.

12. J. Aichelin and C.M. Ko, Phys. Rev. Lett. 55 (1985) 2661.
B. Schürmann and W. Zwermann, Phys. Lett. B158 (1985) 366.
A. Lang et al., Nucl. Phys. A41 (1992) 507.

13. J. Aichelin, A. Rosenhauer, G. Peilert, H. Stöcker, and W. Greiner, Phys. Rev.
Lett. 58 (1987) 1926

14. A. Rosenhauer, PhD-Thesis, GSI-Report GSI–88–09.

15. Ch. Hartnack, GSI report 93-05

16. E. Feenberg and H. Primakoff, Phys. Rev. 70 (1946) 980.
A.R. Bodmer, Phys. Rev. D4 (1971) 1601.
A.B. Migdal, JETP 34 (1972) 1184.
T.D. Lee and C.G. Wick, Phys. Rev. D9 (1974) 2291.
W. Weise and G.E. Brown, Phys. Rep. C27 (1976) 1.

17. J. Boguta Phys. Lett. B109 (1982) 251.
B. Waldhauser et al., Zeitschr.. Physik A328 (1987) 19.

18. H. Kruse, B.V. Jacak and H. Stöcker, Phys. Rev. Lett. 54 (1985) 289.
The BUU model is described in: G.F. Bertsch, H. Kruse and S.Das Gupta, Phys.
Rev. C29 (1984) 673;   J. Aichelin, G.F. Bertsch, Phys. Rev. C31 (1985) 1730.
The Landau Vlasov model is described in: B. Remaud, C. Gregoire, F. Sebille,
L. Vinet, Y. Raffray, Nucl. Phys. A447 (1985) 555.

19. J. Aichelin and H. Stöcker, Phys. Lett. B176(1986) 14.

20. G. Peilert, H. Stöcker, W. Greiner, A. Rosenhauer, A. Bohnet, J. Aichelin, Phys. Rev. C39 (1989) 1402

21. C. Hartnack et al. Nucl. Phys. A495 (1989) 303c.

22. A.R. Bodmer und C.N. Panos, Phys. Rev. C15 (1977) 1342 and Phys. Rev. C22 (1980) 1023.
    L. Wilets, Y. Yariv und R. Chestnut, Nucl. Phys. A301 (1978) 359.
    S.M. Kiselev und Y.E. Pokrovskil, Sov. J. Nucl. Phys. 38(1) (1983) 46.
    J.J. Molitoris, J.B. Hoffer, H. Kruse und H. Stöcker, Phys. Rev. Lett. 53 (1984) 899.

23. Y. Yariv and Z.Fraenkel, Phys. Rev. C20 (1979) 2227 and Phys. Rev. C24 (1981) 488;

24. M. Gyulassy, K.A. Frankel and H. Stöcker, Phys. Lett. 110B (1982) 185

25. P. Danielewicz and G. Odyniec, Phys. Lett 157B (1985) 146.

26. J.J. Molitoris, H. Stöcker and B.L. Winer, Phys. Rev. C36 (1987) 220.

27. A. Bonasera, L.P. Csernai, B. Schürmann, Nucl. Phys. A476 (1988) 159.

28. G. Westfall et al., Phys. Rev. Lett. 71 (1993) 1986.

29. V. de la Mota et al., Phys. Rev. C46 (1992) 677.

30. Ch. Hartnack et al. Proceedings of the XXIX International Winter Meeting on Nuclear Physics, Bormio (Italy), ed. I.Iori, 1991, p.527.

31. C. Hartnack, H. Stöcker and W. Greiner, Proceedings on the NATO Advanced Research Course on The Nuclear Equation of State, NATO ASI B 216, Vol. A, Plenum, NY, 1990, 239.

32. Y. Leifels et al., Phys. Rev. Lett. 71 (1993) 963
    L.B. Venema et al. Phys. Rev. Lett. 71 (1993) 835.

33. C. Hartnack et al., Nucl. Phys. A538 (1992) 53c.

34. J. Aichelin and J. Jänicke, Proceedings of the workshop "Meson Production and Decay" Crakow, May 1991, ed. A. Magiera, World Scientific, Singapore.

35. M. Berenguer et al. J. Phys. G 18 (1992) 655.

36. Ch. Hartnack et al., Proceedings of the International Workshop on Gross Properties of Nuclei and Nuclear Exctations XXI, Hirschegg (Austria), 1993, p. 36.
    Ch. Hartnack et al., LPN preprint 93-11, submitted to Nucl. Phys. A

37. P. Senger et al., Nucl. Phys. A553 (1993) 757c.
    E. Grosse, Proceedings of the International Workshop on Gross Properties of Nuclei and Nuclear Exctations XXI, Hirschegg (Austria), 1993, p. 1.

38. Ch. Hartnack, J. Aichelin, H. Stöcker, W. Greiner, LPN preprint 93-12, submitted to Phys. Lett. B

ENTROPY AND ANCILLARY CONCEPTS IN QUANTUM PHYSICS

Michael Danos*

Physics Laboratory
National Institute of Standards and Technology
Technology Administration
Department of Commerce
Gaithersburg, MD U.S.A.

INTRODUCTION

In classical physics the entropy is a derived quantity; it must be computed from the characteristics of the system under consideration. The characteristics themselves are determined by the dynamics which is governed by a time-reversal invariant Hamiltonian. The same is true for the quantum physics entropy. In the limit of a "large" system it must go over into the classical entropy. I shall here give an outline of the physical concepts and quantities of quantum physics which are required for the definition of the entropy and give its quantum physics expression. For specificity I shall give the discussion in terms of the simple example of a dilute gas.[1]

The line of reasoning will be as follows. I begin by defining the phase space in terms of localized states directly appropriate to the description of reactions, in that the S-matrix (more precisely the U-matrix) can be directly labelled in terms of these states. The eigenstates of the U-matrix, commonly called the "eigenchannels", are the only time-reversal invariant states. They have probability measure zero and cannot be generated "in the laboratory". All other states turn out to be irreversible. Next I show that the quantum physics dynamics leads in a straight-forward manner to uniform probability density over the complete phase space; this way the classical axiom of equal *a-priory* probability for all phase space cells turns out to be a consequence of quantum dynamics. Herewith it is shown that without any auxiliary assumptions, or any averaging, commonly denoted as "coarse graining", quantum dynamics leads direcctly to classical statistical physics. At this point it is very easy to define an operator, the expectation value of which, when inserted in the Boltzmann expression, yields a quantity which has the characteristics of the entropy. Of course, being a quantum physics object, it gives not only the entropy itself but it also yields the fluctuations the entropy must show, even under equilibrium conditions. Furthermore, as behooves a quantum observable, it also responds to the characteristics of the measuring apparatus. As the last point I elucidate the process of measurement in quantum physics.

THE PHASE SPACE

Both in order to describe a collision, and to construct the phase space of a system one requires states which have a well-defined momentum and position;[2] furthermore they should be minimum-uncertaity states. We shall use states where the single-particle wave function is given by the Weyl prescription[3]

$$\Psi_{p,q}(x,t) = \mathcal{N} \int_{E_p-\epsilon}^{E_p+\epsilon} dE \; \chi_n e^{ik(x-q)-Et} \qquad (1)$$

which describes a particle of momentum around p; the wave function has a peak centered at $t = 0$ around $x = q$, which propagates with the classical velocity $v = p/m$. $\mathcal{N}$ is the normalization constant, and χ_n is the internal wave function of the particle where n specifies the state of the particle. This way p and q are the phase space co-ordinates. The system phase space is the product of the single-particle phase spaces.

THE SYSTEM U-MATRIX

The above states at the same time define a quasi-interaction-picture representation valid for a dilute gas: in between the collisions the particles are "free" owing to their large distance to the other particles; as two particles approach each other the interaction Hamiltonian, which of course is present all the time, acquires an interparticle energy, and leads to elastic or inelastic scattering between the particles. The evolution of the system thus is described by the modified Tomonaga – Schwinger evolution operator $U(t,t_0)$ which obeys the equation[4]

$$dU(t,t_0)/dt \; = \; -i \, (H \, - E)_W \, U(t,t_0) \qquad (2)$$

Here $(H \, - \, E)_W$ is the system Hamiltonian, E is the asymptotic total energy of the system, and the subscript W indicates that the parentheses should be inserted inside the integral, Eq. (1). This equation can be integrated to arrive at $U(t_0 + \tau, t_0)$, and as long as neither at t_0 nor at $t_0 + \tau$ an actual collision takes place, the matrix U can be labelled by the system phase space cell indices: the system phase space generates a representation for the U-matrix. We shall call this representation "the asymptotic representation". This so constructed U-matrix is unitary; that would not be the case if anywhere in the system at either $t_0 + \tau$, or at t_0 a collision would be in progress. In the time interval between two collisions the value of the U-matrix is constant.

THE U-MATRIX EIGENSTATES

Consider now a particular (pure) state of the system at time t (hence describable by a factorizable density matrix, i.e., by a wave function). In general it will have non-vanishing probability amplitudes in all phase space cells. Assume that before the considered time step, i.e., after the last collision, the amplitudes of the system wave function in the asymptotic representation are a_j. Then the amplitudes after the next time intervall τ are given by multiplication of the original amplitudes by the system U-matrix, $U(t + \tau, t)$. Denoting the system U-matrix elements by U_{ij}, then the new amplitudes, $\bar{a}_j$, will be

$$\bar{a}_1 = U_{11} a_1 + U_{12} a_2 + U_{13} a_3 + \cdots \tag{3}$$

$$\bar{a}_2 = U_{21} a_1 + U_{22} a_2 + U_{23} a_3 + \cdots \tag{4}$$

$$\cdots$$

$$\bar{a}_j = U_{j1} a_1 \cdots + U_{jj} a_j \cdots + U_{jk} a_k + \cdots \tag{5}$$

$$\cdots$$

$$\bar{a}_k = U_{k1} a_1 \cdots + U_{kj} a_j + \cdots + U_{kk} a_k + \cdots \tag{6}$$

$$\cdots$$

Of course, the system state has remained pure.

Being unitary, the U-matrix can be diagonalized. Denoting the eigenvectors of the U–matrix, called "eigenchannels", by $V_i^{(n)}$, we have

$$U_{ij}^{(n)} V_j^{(n)} = e^{-2i\delta^{(n)}} V_j^{(n)} \tag{7}$$

with real $\delta^{(n)}$. Each eigenchannel, and only each eigenchannel by itself, is invariant under time translation, in that after M time steps all asymptotic-representation amplitudes $V_j^{(n)}$ acquire the common phase factor $e^{-2iM\delta^{(n)}}$; hence these states are also time-reversal invariant. However, these time-reversal invariant states can not be generated in the "laboratory", they have probability measure zero: unless the system consists of non-interacting subsystems the eigenstates of the U-matrix have phase- and magnitude-related probability amplitudes in all channels. Any breach in these relations mixes in one or more eigenchannel states, which then destroys the time-reversal invariance, as we now discuss.

POINCARE TIME

For all states except for single eigenchannels the time evolution is irreversible. This is seen as follows. Expanding the above amplitudes a_j in terms of the eigenvectors we have

$$a_j = \sum C_n V_j^{(n)} \tag{8}$$

then after the time interval τ there will hold

$$\bar{a}_j = \sum C_n \, e^{-2i\delta^{(n)}} \, V_j^{(n)} \tag{9}$$

Thus unless all differences between pairs of eigenphases are rational multiples of π, the system never returns to its initial state: after each time step τ the phase differences between the components continue to accummulate. This is so already if the sum in Eq. (8) has only very few, even only two, terms. Thus the Poincaré time in quantum physics is infinite. This can be formulated in a provocative manner as the statement that in quantum physics all physically realizable systems are chaotic.

EQUIPARTITION

We now show that quantum dynamics leads to relaxation of the system to a state with uniform density in phase space. To that end we introduce as the measure of the pairwise non-uniformity of the phase space density the expression

$$\Delta_{jk} = |a_j|^2 - |a_k|^2 \tag{10}$$

and $\bar{\Delta}_{jk}$ for it after the time step. From (5) and (6) we compute the change in the non-uniformity

$$\bar{\Delta}_{jk} - \Delta_{jk} = \left(|\bar{a}_j|^2 - |\bar{a}_k|^2\right) - \left(|a_j|^2 - |a_k|^2\right) \tag{11}$$

or, explicitly

$$\bar{\Delta}_{jk} - \Delta_{jk} = -|U_{jk}|^2 \, \Delta_{jk} + [F_{jk}] \tag{12}$$

In the absence of the term $[F_{jk}]$, the non-uniformity Δ_{jk} would decrease exponentially with "time constant" $|U_{jk}|^2$. This is is true for every pair, j, k. The terms, denoted by $[F_{jk}]$ in Eq. (12), contain factors of the form $U_{js} \, U_{kr}^\dagger - U_{ks} \, U_{jr}^\dagger$ and thus are phase-sensitive. These terms induce fluctuations in Δ_{jk}, which are damped by Eq. (12). The time evolution of both the fluctuations and the relaxation can be followed by repeated matrix multiplication. Furthermore, the terms $[F_{jk}]$ are responsible for maintaining (unstably) the amplitude and phase relations between the channels in the eigenchannel solutions.

The above observations rest on the earlier result concerning the chaotic nature of quantum systems, in particular that the Poincaré time for chaotic quantum systems is infinite. Otherwise these observations would be unsupported; one would have to rely on some assumptions like random phases to control the influence of the fluctuation-inducing terms.

STATISTICAL THERMODYNAMICS

Above we have obtained the important result that in quantum theory all physically realizable states of a closed system relax towards equal probability density in phase space. To elucidate the physical significance of that result we recall the treatment in statistical thermodynamics.

In classical statistical thermodynamics one investigates the properties of a system at some fixed time in terms of the Gibbs microcanonical ensemble, which is the ensemble of an infinite number of replicas of the system under consideration; each of these replicas is in one of the phase space cells. As starting point one postulates the axiom of equal *a priori* probability for each phase space cell for thermal equilibrium. Further, all results are deduced by counting of phase space cells compatible with the characteristics of the system, to arrive at the probability of that state. In particular one computes the thermodynamic functions in terms of the so defined probability of the conditions of the system. Deviations from the equal probability distribution signify non-equilibrium conditions.

Our results show that quantum dynamics leads directly to relaxation towards uniform density in phase space, i.e., it follows from quantum physics that every system relaxes to the state which in classical physics had to be postulated by an axiom. Said differently, the equal *a priory* probability axiom of classical statistics is an inescapable consequence of quantum dynamics. Further, for non-equilibrium conditions the direction of the relaxation can be used to define the arrow of time.

THE QUANTUM ENTROPY

As a member of quantum physics the entropy must emerge as the result of a computation in terms of matrix elements of an observable, i.e., of the matrix elements of an operator, in terms of the state wave function, allowing for interferences, and allowing for the characteristics of the measurement device. Furthermore, for macroscopic systems the expression for the entropy must go over into the classical Boltzmann form,

$$S_B = - \sum_j p_j \, log p_j \qquad (13)$$

where p_j is the probability for the system being in state j. The corresponding expression in quantum physics is obtained by inserting the quantum value for the probability:

$$p_j = \sum_{kl} a_k^*(t_0) \, U_{kj}^\dagger(t,t_0) \, U_{jl}(t,t_0) \, a_l(t_0) \qquad (14)$$

Here a_k^*, a_l are the asymptotic-representation amplitudes of the system at time $t = t_0$; the expression (14) evidently incorporates all interference effects.[5] Defining the abbreviation

$$Q_{ki,jl}(t,t_0) = U_{kj}^\dagger(t,t_0) \, U_{jl}(t,t_0) \qquad (15)$$

and the asymptotic-representation density matrix

$$\rho_{kl}(t) = a_k^*(t_0) \, a_l(t_0) \qquad (16)$$

Eq. (14) can be written as

$$p_j = \sum_{kl} Q_{kj,jl}(t,t_0) \, \rho_{kl}(t_0) \qquad (17)$$

valid for impure states. The question of the purity of the state evidently is irrelevent for the entropy.

As an aside we note that by summing over the "similar", i.e., indistinguishable final states one obtains Fermi's Golden Rule

$$P = \sum_j p_j = \nu \, p_j \, \delta(E - E_j) \tag{18}$$

where ν is the final state level density.

To incorporate the characteristics of the measuring device one only must provide an appropriate projection operator in Eq. (13), say M_j,

$$S_B = \sum_j M_j \, p_j \, logp_j \tag{19}$$

with p_j given by (17). A particular use would be to obtain the entropy of a subsystem, for example, of the pions emitted in a relativistic heavy-ion collision.

We emphasize the important, actually inescapable, point that the entropy can be defined only for asymptotic states. During a collision the system contains off-the-mass-shell particles; states of that kind do not permit a probability measure.

MEASUREMENT

We have introduced the quantum entropy as an observable, and have given above a prescription how to take account of the characteristics ot the measuring apparatus, slighting the description of the measuring process itself; we now rectify that omission. Since the measurement process involves interactions, and interactions involve the emission or absorption of particles, say photons, the measurement process must be formulated in field theory.[6]

Consider an electron which has passed through a suitable configuration of collimators, electric, magnetic fields, shutters or what not. Consequently we know that every electron which emerges from the apparatus is in a well-defined state. The apparatus acts as a filter. To describe the electron it is best to use a normal mode expansion in terms of solutions of the Dirac equation involving the boundary conditions and field configurations of the apparatus:

$$\psi^{(A)}(x,t) = \sum f_r^{(A)}(x,t) \, b_r^{(A)} \tag{20}$$

where $f_r^{(A)}(x,t)$ are the normal modes in the form of Weyl functions, specified by the index r, while $b_r^{(A)}$ are the corresponding annihilation operators. The state vector describing a particular situation then is

$$S = \sum C_r \, b_r^{(A)\dagger} \tag{21}$$

with c-number amplitudes C_r. To ascertain whether an electron indeed has emerged from the apparatus at time, say t_1, one must perform a measurement, i.e., place a detector at the output of the apparatus. Its action is described by

$$\mathbf{D} = \sum \psi^{(D)\dagger}(x_D, t_D) \, D \, \psi^{(D)}(x_D, t_D) \tag{22}$$

with

$$\psi^{(D)}(x,t) = \sum f_s^{(D)}(x,t) b_s^{(D)} \tag{23}$$

where $f_s^{(D)}(x,t)$ and $b_s^{(D)}$ are the solutions and operators appropriate to the boundary conditions of the detector; D in (22) is the detector function, and x_D and t_D are the position of the detector and the time of the registered count, respectively. Owing to completeness there holds

$$f_s^{(D)}(x,t) \, b_s^{(D)} = \sum K_{s,r} \, f_r^{(A)}(x,t) \, b_r^{(A)} \tag{24}$$

and the result of the measurement is given by

$$R \rangle = \mathbf{D} \, S \, | \, 0 \, \rangle \tag{25}$$

or, in detail

$$R \rangle = \sum_s f_s^{(D)*}(x,t) \, b_s^{(D)\dagger} \, D \, \sum_r K_{s,r} \, f_r^{(A)}(x,t) \, b_r^{(A)} \, \sum_{r'} C_{r'} \, b_{r'}^{(A)\dagger} \, | \, 0 \, \rangle \tag{26}$$

Introducing the quantity

$$M_{s,r}(x_D, t_D) = f_s^{(D)*}(x_D, t_D) \, D \, K_{s,r} \, f_r^{(A)}(x_D, t_D) \, C_r \tag{27}$$

we find

$$R \rangle = \sum_{s,r} b_s^{(D)\dagger} \, | \, 0 \, \rangle \, M_{s,r}(x_D, t_D) \tag{28}$$

The wave function after the measurement is

$$\varphi(x,t) = \langle \, 0 \, | \psi^{(D)}(x,t) \, R \, \rangle \tag{29}$$

which yields

$$\varphi(x,t) = \sum_{s,r} M_{s,r}(x_D, t_D) \, f_s^{(D)}(x,t) \tag{30}$$

We recognize in the quantity $M_{s,r}(x_D, t_D)$ the projection operator M_j of Eq. (19) above.

We observe: (i) after interaction with the detector independently of the existence of an "observer" the state of the electron is changed from the initial state; if the state (20) represented an interference pattern as would be the case if the apparatus was that of a two-slit experiment, the state after the measurement, Eq. (26.2), would have lost this pattern; (ii) without any "collapse of the wave function" no further detector would detect a particle, as can be seen by evaluating the expression $\mathbf{D} \, \mathbf{D} \, S \, | \, 0 \, \rangle$ which vanishes identically; this is non-trivial but still straightforward in the case of the Einstein – Rosen – Podolski experiment where one is dealing with two-particle states. The analysis of that experiment along the present lines yields directly the quantum limit of Bell's inequalities.[7]

Acknowledgements

I would like to thank the Alexander von Humboldt Foundation for a grant. Further, I would like to acknowledge many useful discussions about the concepts of the present paper with many colleagues, including H. Bers, S. Chandrashekar, U. Fano, H. Fritsche, D. Mueller, Y. Nambu, D. Oxtoby, and R. Wald. I am particularly grateful for very helpful and stimulating discussions with P. Carruthers and especially with J. Rafelski, who in addition not only contributed essentially in clarifying my thinking but actually provided the initial stimulus which lead to this work.

REFERENCES

* Visiting Scholar, Enrico Fermi Institute, University of Chicago, Chicago, Illinois.

1. M. Danos, *Chaos, Dissipation. Arrow of time in Quantum Physics*; NIST Technical Note 1403 (1993)

2. E. Merzbacher, *Quantum Mechanics*, 2nd edition, Chapter 12; John Wiley and Sons, New York, London, Sydney, Toronto (1970).

3. A. Sommerfeld, *Partielle Differentialgleichungen der Physik*, Chapter 5, Appendix 1; Akademische Verlagsgesellschaft, Leipzig (1948).

4. M. Danos, *The System S-Matrix in Statistical Thermodynamics*; to be published

5. This definition differs from that of v. Neumann in J.von Neumann, *Mathematische Grundlagen der Quantenmechanik*; Springer (1932); translation *Mathematical Foundations of Quantum Mechanics*; Princeton University Press, Princeton, NJ (1955).

6. D. Lurie, *Particles and Fields*, Chapter 6; Interscience Publishers (1968); P. Roman, *Introduction to Quantum Field Theory*, Chapter 4; John Wiley and Sons, New York, London, Sydney, Toronto (1969).

7. J. S. Bell, Physics **1**, 195 (1964)

EQUILIBRATION AND MULTIFRAGMENTATION
IN HEAVY ION REACTIONS

G. Peilert,[1] M. G. Mustafa,[1] M. Blann,[1] and A. Botvina[2]

[1] Lawrence Livermore National Laboratory
Livermore, CA 94550

[2] Institute for Nuclear Research, Russian Academy of Sciences,
117312 Moscow, Russia

INTRODUCTION

Modeling of multifragmentation measurements from heavy ion reactions generally requires separate treatment of the initial fast part of the reaction, during which energetic nucleons are emitted, and of a quasi-equilibrated system where sufficient degrees of freedom have been excited, so that statistical approaches may be applied. Some of the more sophisticated fast cascade models, e.g., Quantum molecular dynamics (QMD), might also produce fragment yields, however, transport models have not yet been able to satisfactorily reproduce fragmentation properties of nuclear reactions.

In this work we consider the interactions of ^{36}Ar with ^{197}Au at incident energies of 35 to 110 MeV•A, which was investigated by de Souza et al.[1] We will first look at two dynamic models which may be used to estimate the excitation remaining for quasi-equilibrated systems following the fast nucleonic cascade, specifically the Boltzmann master equation (BME)[2-5] and Quantum molecular dynamics (QMD) approaches.[6,7] Using excitations from the BME model to estimate values for quasi-equilibrated nuclei, we will explore two approaches to statistical multifragmentation calculations, one of sequential binary decay,[8-10] the other a simultaneous multifragmentation model (SMM).[11-13] We will consider central collisions in our calculations, and experimental results gated on the highest total charged particle emissions - those deduced to be central collisions.

EQUILIBRATION (FAST CASCADE) MODELS

The QMD model is well known, and is still being improved for applications in relativistic heavy ion reactions. We refer to published references for details.[6,7]

Hot and Dense Nuclear Matter, Edited by
W. Greiner *et al.*, Plenum Press, New York, 1994

In this work we will show results with a form of 'Pauli potential'. The BME, unlike many models, follows reactions in energy space under the assumption that the geometric aspects — nucleon directions — roughly average out in the many collisions involving incident beam velocities and random Fermi velocities, and that we may make use of the well known isotropic energy distribution in nucleon-nucleon scattering.

The method of dynamics in the BME for heavy ion reactions is as follows: as the nuclei interact, we assume that the projectile (lighter partner) puts nucleons into the well of the heavier partner which mainly determines the center of mass of the reaction. We assume that every energy conserving partition is equally likely due to coupling of Fermi and beam momenta. In this case an exciton state density expression may be used to partition the energy.[5]

The BME then follows the rate of nucleon-nucleon scattering, (using energy and isospin dependent cross sections) in competition with an emission rate for the population in each energy bin.[2,3] Due to the simplicity of the model, the time dependent properties of many variables are easily followed for the two component gas, and even for the recoil momenta of the residues once a parameterization for emitted nucleon angular distributions is selected.[14] The time step in the BME may be varied; by default we use time increments of 2×10^{-23} sec, usually following 50 - 100 steps.

Model Intercomparisons

In Figures 1 and 2 we compare the time dependence of the QMD and BME models at incident energies of 35, 80 and 110 MeV•A for ^{36}Ar on ^{197}Au. Due to dimension limits in the BME, 100 MeV•A was used rather than 110. We show the E_{CM} + Q for each reaction as a horizontal bar; this would be the excitation of a compound nucleus if one were formed.

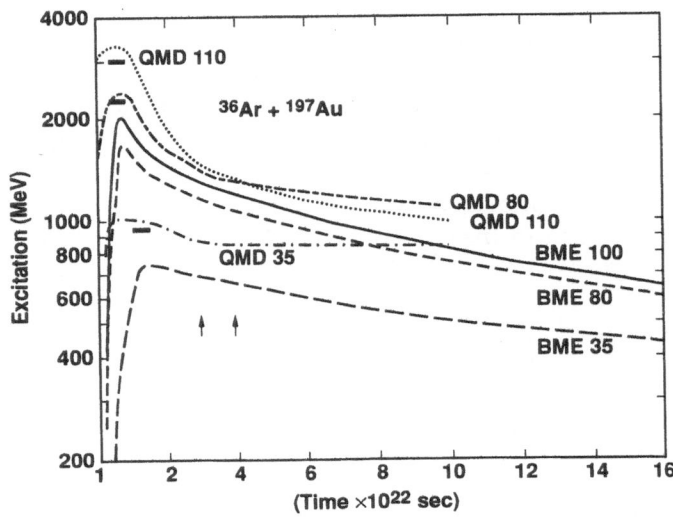

Figure 1. Time dependence of total excitation versus time for the interaction of 35, 80 and 110 MeV•A ^{36}Ar with ^{197}Au. The horizontal bars indicate the E_{CM} + Q for each reaction. The vertical arrows at 3 and 4 $\times 10^{-22}$ sec indicate the times at which excitations were taken for subsequent equilibrium calculations.

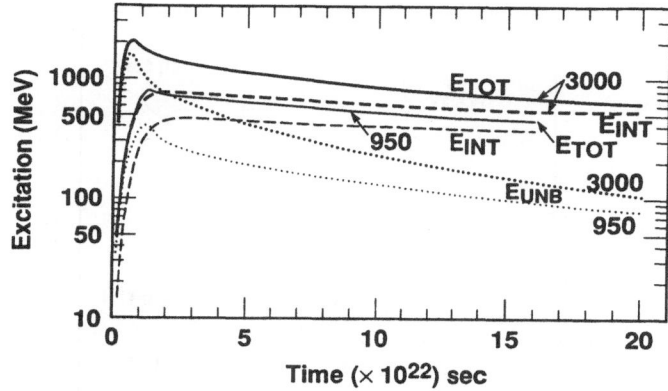

Figure 2. Time dependence of excitation for 35 and 110 MeV•A ^{36}Ar + ^{197}Au reactions, according to the BME model. The reactions are designated by corresponding $E_{CM} + Q$ values, 950 and 3000 respectively. The total excitation versus time, as shown in Figure 1, has been divided into particle (and hole) energy of quasi-bound nucleons (E_{int}) and energy carried by neutrons above the binding energy plus protons above the binding plus Coulomb energies (E_{UNB}).

In the BME, we see that the $E_{CM} + Q$ is never available to the two interacting nuclei; this is because nucleon emission during coalescence removes a substantial part of the reaction energy. The higher the incident energy per nucleon, the larger the fraction removed, making it increasingly difficult to gain high internal excitation. The QMD has qualitative similarities, but at these relatively low incident energies some of the longer time behavior is perhaps open to large dependencies on details of the calculation which are still under development.

Arrows in Figure 1 show times of 3 and 4 x 10^{-22} sec; typically we judge systems to be quasi-equilibrated within that time range. The greater the incident energy, the shorter the equilibration period.

Figure 2 shows some more detailed aspects of the 100 MeV•A and 35 MeV•A systems. We show the total excitation, as on the previous figure. We also show the energy of unbound nucleons (above the Coulomb barrier for protons) and of bound nucleons plus holes. The unbound nucleons will have a large probability of emission in a single transit of the nuclear diameter, while the bound excitation — nucleons and holes — have a much higher probability of undergoing multiple collisions and producing a quasi-equilibrated system. We see in Figure 2 that the internal (bound/quasi-bound) energy exceeds the unbound energy for the two cases shown at around 2 x 10^{-23} sec. We also note that the unbound energy monotonically decreases after this time, as does the total excitation, rendering the question of the 'time' and therefore the excitation of the 'equilibrated' system a subjective issue. We will show the sensitivity of multifragmentation yields to excitation in a few cases to be presented; first we show a few further comparisons of the BME and QMD approaches.

In Figures 3 and 4 we compare emitted nucleon multiplicity vs. time at 35 and 110 MeV•A. Qualitatively at 110 MeV•A the results are similar, but many more nucleons are emitted in the QMD than in the BME model. At 35 MeV•A the same is true up to about 90 fm/c (3 x 10^{-22} sec), after which emission in the QMD model suddenly drops to a very low instantaneous value.

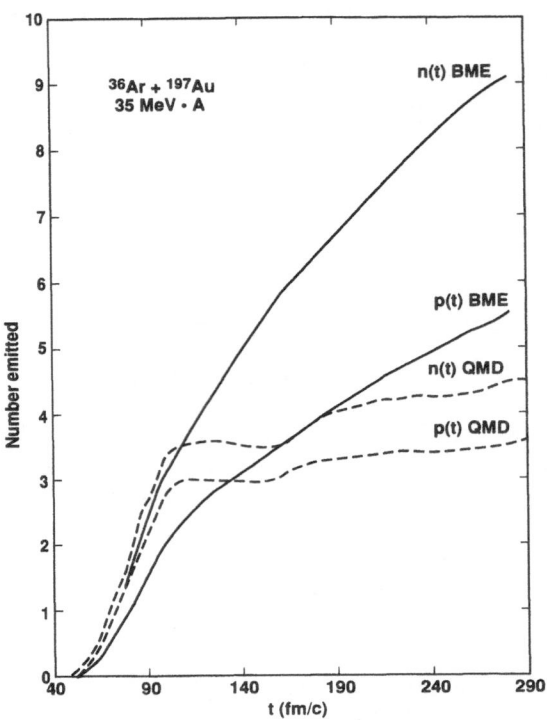

Figure 3. Emitted neutron multiplicity versus time in the BME and QMD models for 35 MeV•A
^{36}Ar + ^{197}Au. The values shown are the integrals up to time 't'.

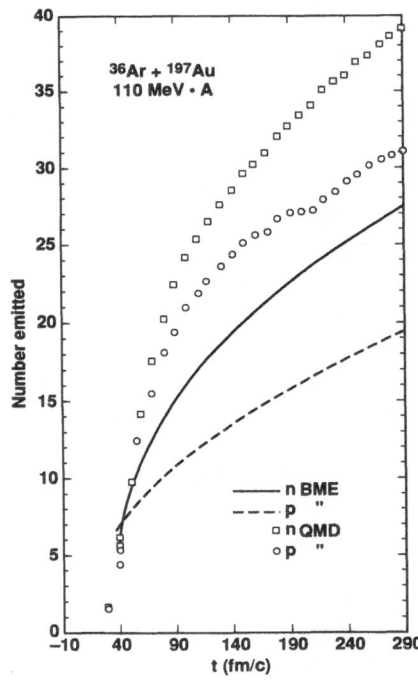

Figure 4. As in Figure 3 for 110 MeV•A incident energy.

The relationship of kinetic energies between the two models is displayed in Figures 5 and 6. For the BME the lower curves show the instantaneous kinetic energy per time step (Figure 6); all other points and curves are the results averaged up to time 't'. The two models are qualitatively similar, the greatest difference being in the Coulomb boost observed for proton kinetic energies in the BME which is not reflected in the QMD results.

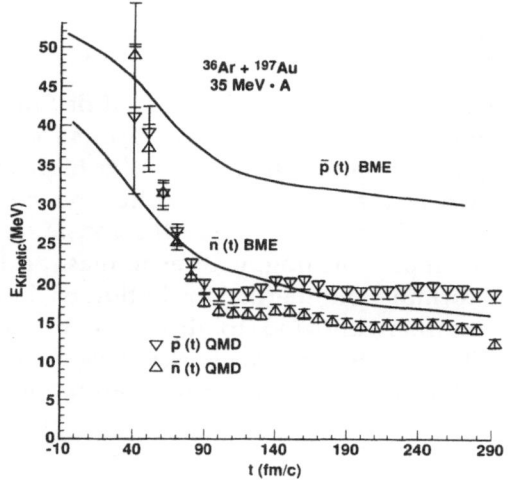

Figure 5. Kinetic energy of emitted neutrons and protons versus time in the BME and QMD approaches. The emission energies represent the kinetic energy of the neutron (proton) averaged up to the time 't' shown on the abscissa. Incident ^{36}Ar energy was 35 MeV•A.

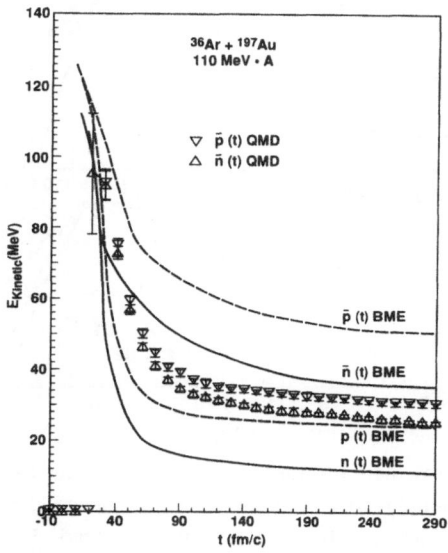

Figure 6. Kinetic energy of emitted neutrons and protons versus time as predicted in the BME and QMD models. These results are for 110 MeV•A ^{36}Ar on a ^{197}Au target. Results are for kinetic energies averaged up to time 't'; for the BME model we also show the kinetic energies of emission at time 't', (p (t), n (t)).

We have seen that there are uncertainties and ambiguities as to how we extract an 'excitation at equilibrium' following the fast cascade — which becomes monotonically lower with time in a continuous fashion. We have used the BME results at 4×10^{-23} sec for calculating the exclusive multiplicities for $^{36}Ar + ^{197}Au$ at 35, 50, 80 and 100 MeV•A; for 80 and 110 MeV•A we also calculated results for successive binary decay using the excitation at 3×10^{-22} sec, in order to demonstrate the sensitivity of final results to the assumption of excitation remaining in an 'equilibrated' system.

Multifragmentation Yields

In Figure 7 we show the experimental results of de Souza et al.[1] for incident energies between 35 and 110 MeV•A. We show results for gates set on either the highest total charged particle multiplicities, or the highest two multiplicities. These presumably represent the more central collisions, corresponding to those used for the BME and QMD models. The SMM considers the partition of the equilibrated system into all possible fragments as to mass and multiplicities, based on a liquid drop mass formula and thermal excitation of the equilibrated system. The hot fragments are then allowed to de-excite to ground via a simple evaporation model. The fragmenting nucleus is assumed to come from a low density nucleon gas; for this work a density of half saturation value was assumed.

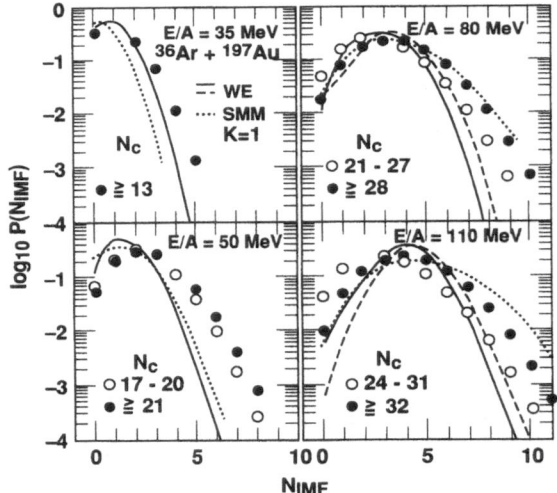

Figure 7. Experimental and calculated fragmentation multiplicities for 35, 50, 80 and 100 MeV•A $^{36}Ar + ^{197}Au$. The experimental results given by open and closed circles are those of de Souza et al. The total charged particle multiplicity gates used are indicated. Results using excitations from the BME at 4×10^{-22} sec are shown by solid lines for the sequential binary decay result (for primary fragments) and by dotted lines for the statistical (simultaneous) multifragmentation model. The dashed curves at 80 and 100 MeV•A are the sequential binary model results using excitations at 3×10^{-23} sec.

For the statistical calculations presented in Figure 7 for incident ^{36}Ar energies of 35, 50, 80 and 110 MeV•A with E_{CM} + Q values of 952, 1409, 2322 and 2950 MeV, we used the BME predicted total excitations of 651, 827, 1076 and 1200 MeV, respectively at 4 x 10^{-22} sec after contact. The number of emitted neutrons and protons, used to deduce the equilibrated nuclei, were respectively 5.9n, 3.7p; 9.2n, 6.0p; 14.9n, 10.2p; and 18.3n, 12.8p. At 80 and 110 MeV•A incident energy we also produce results for sequential binary decay assuming equilibration takes place at 3 x 10^{-22} sec; for these cases the excitations were 1171 and 1320 MeV respectively, with 12.5n, 8.7p and 15.5n, 10.9p being emitted.

The Weisskopf-Ewing (W-E) model has been extended in the ALICE code to include level densities for the emitted fragments,[9] with a sum over all partitions of internal excitation and channel energy between light and heavy fragments. The ALICE code was modified both to include emission of up to 20 clusters (up to ^{48}Ca), and to output the internal spectral excitations on an exclusive basis, i.e., clusters coming from a nucleus which had emitted no clusters before, one cluster before, etc.[9,10] However, at the higher excitations of Figure 7 (higher incident ^{36}Ar energies), some fraction of the clusters may themselves emit clusters in decaying to ground, while others will decay by nucleon and alpha emission to non-cluster residues. These decays will somewhat broaden calculated primary distributions, but they are not easily treated in an exclusive mode in the ALICE code. We therefore present primary distributions for the sequential binary decay (W-E) case in Figure 7.

The solid curves in Figure 7 are the successive binary results using excitations from the BME at 4 x 10^{-23} sec after first contact, as were used in the SMM calculations. The dashed curves result from using the excitations at 3 x 10^{-23} sec after contact. The two models (W-E and SMM) are rather close in their predictions, particularly at the two lower energies. At the higher energies the SMM results are in somewhat better agreement with the data; however, the W-E results may be expected to broaden upon de-excitation to ground. This would bring about a closer agreement between the two models.

CONCLUSIONS

The BME and QMD formulations give qualitatively similar behaviors for relaxation during heavy ion collisions. Results differ on a quantitative basis. Systems progress monotonically toward equilibrium following initiation of a reaction; the selection of a time, and therefore the excitation of an equilibrated system is subjective. This puts an uncertainty on the input into any equilibrium model. In comparing a sequential binary model with a simultaneous multifragmentation model, we get results which are reasonably consistent within the uncertainties of the choice of initial excitation, and effects of decay of primary fragments in the sequential binary model. Selection between models will therefore require tests other than exclusive multiplicities.

REFERENCES

1. R. T. de Souza et al., Phys. Lett. B268, 6 (1991).

2. G. D. Harp, J. M. Miller, and B. J. Berne, Phys. Rev. 165, 1166 (1968).

3. G. D. Harp and J. M. Miller, Phys. Rev. C3, 1847 (1971).

4. M. Blann, Nucl. Phys. A235, 211 (1974).

5. M. Blann, Phys. Rev. C31, 1245 (1985).

6. G. Peilert, H. Stöcker, W. Greiner, A. Rosenhauer, A. Bohnet, J. Aichelin: Mod. Phys. Lett. A3, 459 (1988); Phys. Rev. C39, 1402 (1989).

7. J. Aichelin, G. Peilert, A. Bohnet, A. Rosenhauer, H. Stöcker, W. Greiner: Phys. Rev. C37, 2451 (1988).

8. M. Blann, T. Komoto, and I. Tserruya, Phys. Rev. C40, 2498 (1989).

9. M. Blann and M. G. Mustafa, Phys. Rev. C44, R590 (1991).

10. A. Botvina, I. N. Mishustin, M. Blann, M. G. Mustafa, G. Peilert, H. Stöcker, and W. Greiner: Z. Physik A345, 297 (1993).

11. A. S. Botvina, A. S. Iljinov, I. N. Mishustin: Sov. J. Nucl. Phys. 42, 712 (1985).

12. A. S. Botvina, A. S. Iljinov, I. N. Mishustin, J. P. Bondorf, R. Donangelo, K. Sneppen: Nucl. Phys. A475, 663 (1987).

13. A. S. Botvina, A. S. Iljinov, I. N. Mishustin: Nucl. Phys. A507, 649 (1990).

14. M. Blann, Phys. Rev. C31, 295 (1985).

HOT AND DENSE NUCLEAR MATTER: SUMMARY TALK

R. L. Thews

Department of Physics
University of Arizona
Tucson, AZ 85721 USA

INTRODUCTION

I think my first mistake was to sit by the door on Monday of the opening session, taking notes, as Walter Greiner walked past after his introductory remarks. He stopped and asked if I would give the summary talk, and, almost without thinking, I agreed. The realization of what I had done came only seconds later, but it was too late. Since I have been only peripherally involved in the research fields which contribute to the subject of this meeting, my summary will necessarily be from a very narrow, but possibly unbiased point of view. I view this talk not as the traditional conference summary, where the speaker attempts to place into historical perspective the various accomplishments of the dinosaurs and promotes the recent accomplishments of the young turks. Rather I will more closely follow the pattern I remember in elementary school, where on the first day of classes in the fall everyone was expected to stand up in sequence and present an impromptu talk on: "What I Did on my Summer Holiday".

This meeting was subdivided by the organizers into eight general categories:

1. Collective Effects
2. Multifragmentation
3. Particle Production
4. Strangeness
5. Electromagnetic Probes
6. Astrophysics and Special Topics

Hot and Dense Nuclear Matter, Edited by
W. Greiner *et al.*, Plenum Press, New York, 1994

7. QCD

8. Theoretical Foundations

Altogether 91 talks were presented, and although I attended all of them, I am obviously not able to include every one in this summary. My choices of topics to follow is dictated partially by my own preparation (or lack thereof) which leads me to occasionally emphasize the areas with which I am somewhat familiar. I will also perhaps be too elementary in the presentation of some topics to which this is my first exposure, and consequently not place some of the contributions in the correct historical perspective. I thank the speakers who allowed me to show their transparencies in the oral version of this summary, and will refer to them as contained in the individual contributions in this written version.

A common theme running throughout all of the meeting is certainly the use of collisions between heavy nuclei as a production mechanism for hot and dense nuclear matter. I found it useful to make a table of the various experimental facilities which contribute to this enterprise. Listed below are both existing and future accelerators, with their (equivalent) laboratory energy per nucleon and the available (cm) energy per nucleon. It is noteworthy that in such a list one would normally expect the lowest energy facilities to be the next candidates for termination in the natural progression to higher energies. The exception occurs here in the recent (unfortunate) shutdown of the Bevalac, whose energy range was at the high end of the "low energy group". The large range of nine orders of magnitude in the equivalent lab energy is mirrored by the large span of background of the attendees at this meeting. It is indeed fortunate that the common interests of investigating hot and dense nuclear matter have brought together for perhaps the first time in recent history large segments of the particle and nuclear physics communities. We have a unique opportunitiy for fruitful collaborations and the cross-fertilization of ideas and concepts in this field, and should make a special effort to insure the full participation and support for every segment of our expanded community.

Facility	Lab E/A (MeV)	Available E/A (GeV)
GANIL	30	.015
MSU	.	.
GSI	.	.
LBL	1200	0.3
AGS	$11 - 14 \times 10^3$	3-3.5
SPS	$1 - 2 \times 10^5$	14-20
RHIC	2×10^7	200
LHC	2.5×10^{10}	7000

One of the most-frequently shown figures in the meeting is the canonical temperature-density phase diagram of nuclear matter. Of the many versions, I would like to give "first prize" to the one by Dr. K. H. Kampert[1]. Shown on these diagrams are possible time evolution paths that various systems might follow. Of primary interest in this meeting are the heavy ion collisions, but we can also consider the cases of neutron star formation and evolution of the early universe on this same diagram. In all of the analyses, one searches for signals of new phenomena due to the formation of hot and dense nuclear matter. One of the main tasks is to understand well enough the contribution of ordinary phenomena from merely superposing individual nucleon-

nucleon interactions, to ensure that one is seeing truly new effects. An (incomplete) list of the signals discusssed in this meeting includes: Collective effects (Flow, Correlations); Phase Transitions (Fluctuations, Critical Behaviour); In-Medium Particle Properties (Chiral Symmetry Restoration, Color Screening of Resonances); Thermal Properties (Transverse Mass Slopes, Direct Photons, Dileptons), Chemical Properties (Particle Ratios, Strangelets, MEMOS), Space-Time Properties (HBT Measurements of Source Sizes). This summary will attempt to place the results of selected contributions into some sort of perspective with respect to these categories.

QUARK MATTER IN ASTROPHYSICS AND COSMOLOGY

I would like to start with some remarks on three very nice presentations outside the direct area of heavy ion collisions. Prof. Vin Mau brought us up to date on the situation of the feasibility of producing the observed baryon asymmetry of the universe via processes at the electroweak phase transition. The minimal standard model turns out to proce an effect, but it is about seven orders of magnitude too small. Work is underway to extend to nonminimal models.

Prof. Kajino presented calculations on a very interesting scenario in which stable hot Strange Quark Matter (SQM) could survive the phase transition in the early universe. Investigation of the phase transition process via nucleation and supercooling leads to a possible spatial inhomogeneity in the baryon phase. Using as input the critical temperature and surface tension from QCD-inspired calculations, he found a range of mass number A between 10^{38} and 10^{40} for which survival to the present time is allowed. If this occurred, then it is possible that these lumps of SQM could be that required to make of the baryon density of the universe up to Ω_B of $0.2 \sim 0.4$. The resulting mass number abundances are greatly enhanced at high-A.[2]

Fridolin Weber presented two lectures on the scenarios of neutron star formation when a quark matter phase transition is included. An impressive program, it includes consideration of seventeen different nuclear equations of state. A new ingredient is the more general phase transition constraints which follow when there are two or more conserved quantities. This is then combined with the stability constraints of general relativity for a massive rotating system. The end result is a quite restrictive region of available parameter space in rotation period and mass. The lower limit on rotation period for a pulsar would be about 1.6 msec, and observations of periods almost at this lower limit have been reported. One can then ask, how would one interpret an observation of a pulsar with period below this limit? Weber argues that this could be an indication of a strange quark star, and proceeds to calculate its structure and mass-radius trajectories.[3] One might expect to see these strange quark pulsars at very low masses but overlapping the same range of radii as normal neutron stars.

COLLECTIVE EFFECTS AND MULTIFRAGMENTATION

This subject was represented by the largest number of experimental talks:

1. Ritter - Experimental overview.
2. Hildenbrand - FOPI Central Events - Radial Flow.
3. Pochodzalla - Multifragmentation.
4. Gelbke - Intermediate Mass Fragments.

Hans Georg Ritter presented a comprehensive overview of the experimental situation in the "low energy" ion collisions experiments where the final states involve a wide range of intermediate mass fragments. In the search for collective effects, a standard definition of flow is extracted from each event. The event plane is defined by the beam axis and the primary axis of the distribution of final particle momenta. The rapidity dependence of the projection of final particle transverse momenta into the reaction plane is then a signal for collective flow in the event. One finds a range at midrapidity for which the dependence is linear, and the slope of this linear rise is called the reduced flow. One can then examine the dependence of the flow signal on mass number, centrality of collision, energy, etc.

Gary Westfall gave a very nice talk on the energy dependence of the flow.[5] The data show that there is a disappearance of flow at beam energies in the range $50 - 100 MeV/A$, independent of the mass of the final fragments. This is nicely interpreted as a competition between the low energy attractive and the high energy repulsive nuclear force. Due to the surface vs. volume nature of these effects, one predicts that the resulting balance energy would vary with mass as $A^{-1/3}$, which is shown to be consistent with the data.

Klaus Hildenbrand showed additional evidence for collective flow from the particle energy spectra.[6] One finds that a fit to the data using only thermal velocities plus coulomb corrections grossly underestimates the high energy part. If one adds an isotropic radial flow the fit can be made acceptable. The magnitude of flow needed accounts for about 50% of the average particle energy. The need for this contribution is evident at large A, where the average energy per unit mass flattens to a constant, rather than continuing a $1/A$ thermal behavior.

J. P. Coffin showed data extracted on entropy production and compared with various hydrodynamic and thermal models.[7] New results on incident energy show a very flat dependence on S/A for intermediate impact parameter collisions, and lack a satisfactory explanation.

IMF PRODUCTION - MODELS AND THEORY

6. Ayik - Stochastic Collision Term.
7. Bonasera - Quantum Mechanics in Collision Term.
8. C. Greiner - Menory in Collision Term.
9. Li - Common Origin of Mean Field/Collision Term.
10. Belkacem - Reliability of Mean Field/Exact Classical Example.
11. Horiuchi - Antisymmetrization in QMD.
12. Bass - Pion Flow Correlations.
13. Wolter - Non-Equilibrium Effects in Kinetic Equations.
14. Aichelin - QMD Derivation and Applications.

Prof. Maruhn presented an excellent overview of the theoretical activities in this area. The main tools for connecting data with experiment are the hydrodynamic models and the microscopic cascade-type models. Their main properties and differences are listed below.[8]

	Hydrodynamics	Microscopic
Assumptions	Local equilibrium, continuum	Independent N-N Collisions mean field, diluteness
Input	Equation of state, dissipative terms	Mean field, Cross sections
Clusters	Chemical equilibrium	Correlations
Quantum effects	Partly in EOS	Pauli Blocking

One searches for effects which are sensitive to the input equations of state or in-medium effects in the two approaches. For example, one finds that the balance energy at which flow disappears has a dependence on the mass of the combined system which is quite sensitive to the stiffness of the equation of state. However, the slope of the curve also requires some modification of in-medium cross sections.

Many of the talks here dealt with modifications of the collision term in the microscopic models, to take into account effects which should appear in principle but have not yet been included due to their complexity. I have no time (or the ability) to summarize their many accomplishments, but recommend the reader refer to the contributions of those listed above in this section.

The second-most popular viewgraph in this meeting was certainly the classical picture of heavy nuclei colliding at nonzero impact parameters and low energy, showing how the phenomena of particle production might give rise to the generic in-plane bounce-off and off-plane squeeze-out. I recommend the version of Jens Konopka for your viewing pleasure.[9] This picture is especially relevant for the very nice analysis of pion flow correlations by Steffen Bass[10] applied to data described at this meeting by Dieter Brill[11].The analysis finds an effective anticorrelation between the patterns of emission of pions vs. nucleons in both rapidity and azimuthal dependence. For in-plane bounce-off, they interpret this as rescattering effects which are sensitive to in-medium cross sections. For the out of plane squeeze-out, the determining factor is the absorption by in-plane nucleons, thus favors high transverse momentum pions.

STRANGE QUARK/HADRON MATTER

1. C. Greiner - Overview: Strangelets and MEMOs.
2. Libby - AGS Searches for Anomalous Z/A.
3. Schaffner - Strange Hadronic Matter/Mean Field Model.
4. Gal - MEMO Production and Stability.

Carsten Greiner presented a thorough overview of this subject, including both theoretical motivations for the existence of Strangelets and MEMOs, and also experimental possibilities for verification. Some initial results from AGS were presented by Libby, along with expectations for the future. Model calculations by Schaffner and Gal indicate that strange hadron matter in the form of MEMOs could also exist in anomalously small (and even negative) regions of Z/A, and as such would have very similar properties to Strangelets.

COHERENT/NON-PERTURBATIVE QED

1. Soff - Coherent photon processes in Heavy Ion Collisions and New Particle Production at LHC.
   In high energy heavy ion colliders, the electromagnetic fields become intense enough that coherent photon -induced processes can be competitive. Soff has done a variety of calculations for the LHC energy region, and made estimates of the possibilities for detection of new particles (higgs, top, SUSY, etc.) It appears that production of b-quark pairs could be competitive for the CP violation studies.
2. Oberacker, Wells, Scheid - Nonperturbative electron-positron production calculations.
   Very impressive nonperturbative QED calculations by Volker Oberacker and Jack Wells, and independently by Prof. W. Scheid, have been carried through. They find that the nonperturbative effects can increase the cross section for pair production plus capture of one by the beam ion, in some cases by factors of 100. This has significant impact on the beam lifetimes for machines like RHIC, where the present estimates are about 10 hours.

FUTURE FACILITIES

Valuable and current updates on future facilities and detectors for the heavy ion program were presented as listed below.

1. Ludlam - Transitions from AGS to SPS to RHIC.
2. Nagamiya - Phenix Detector at RHIC.
3. Harris - Star Detector at RHIC.
4. Specht - Alice Detector at LHC.

THEORETICAL FOUNDATIONS

1. Carruthers - Correlations/Fluctuations in Data.
2. M. Greiner - Wavelets.

3. Eisenberg - Schwinger Mechanism, Back Reaction, Source Term in Transport.
4. Kluger - Quantum Simulation of Pre-Equilibrium QGP.
5. Wilets - Collapse of Flux Tubes.
6. Weigel - Nuclear Matter EOS, Relativistic BHF.

I have used some arbitrariness in placing contributions in the QCD or Theoretical Foundation categories, since there is, in this field, an inevitable degree of overlap. Peter Carruthers presented a view of multiparticle production from a somewhat historical point of view, and reminded us that we should not stray too far from the data in application of various theoretical concepts to physical situations. Martin Greiner gave an introduction to the theory of wavelets, a new (to physicists) method of analysis of phenomena at variable scales. This may be applicable to some of the self-similar cascades which we use in the various models to simulate parton branching and fragmentation. A new potential application resulted from interactions in this meeting, that of the quite nonuniform electric field spatial distributions in heavy ion collisions for which numerial QED calculations are in progress.

Judah Eisenberg summarized some results of his group on the QED calculation of back-reaction for particle production in strong fields. This type of calculation is intended to investigate the role of an effective source term in transport model descriptions of quark-gluon plasma formation. This connection opens the way for extension of the phenomenological models to include more realistic conditions. A related calculation by Larry Wilets on the collapse of flux tubes in QCD included an interesting homework problem for the students. The answer involves realizing that a displacement current is set up by the decaying color electric field. Yuval Kluger presented preliminary results on his very ambitious program for a full quantum dynamical simulation of the pre-equilibrium phase of a quark-gluon plasma. He used the $1/N_c$ expansion in $1 + 1$ dimensions, including the color mean field and radiation/collision terms to describe the dense system of interaction gluons in the initial stages. Very interesting results on time scales and equilibrium times are possible to extract.

ANTIPROTON PRODUCTION

1. Jahns - Overview.
2. König - Subthreshold Production.
3. Jacak - SPS NA44 results.
4. Röhrich - SPS NA35 results.

Andre Jahns presented an enlightening overview of this subject. [22] Consider the predeictions of the so-called first collision model, where p̄'s are produced just as in N-N collisions for the first collision only of each target or projectile nucleon. At low energies, the p-A data is described satisfactorily, but in A-A interations the yield is underestimated by 3 orders of magnitude. This has given rise to the modification of model calculations to include multi-step processes, typically via secondary collisions, color ropes, high mass flux tubes, etc. (see section on models). At low energy where subthreshold production is the only possibility, this is obviously of primary importance. Wolfgang König showed us results in this region are dominated by Δ interactions.[23]

Another effect to be considered is absorption of the p̄'s on the way out of the nuclear environment, which should act opposite to the enhanced production mechanisms. Previous data in the AGS energy region have been fit well by models such as RQMD, and show no definite signal of annihilation. However, there is probably a complicated interplay in the region between the threshold production behavior, enhanced production, and annihilation.

In this meeting, new results were presented from two SPS experiments. The energy should be high enough to avoid threshold complications, and the midrapidity baryon density is lower such that annihilation should be less important.

Barbara Jacak [21] showed preliminary data from NA44 on $\bar{p}$ production. This was compared with predictions of RQMD both with and without the enhanced interactions. The initial data points near mid-rapidity lie about 30% below the enhanced model predictions, but the systematic errors are still quite large. Overall, there is clearly some evidence for excess $\bar{p}$ production over first collision models. In this experiment, there were also four anti-deuteron candidate events.

Dr. D. Röhrich presented preliminary data on $\bar{p}$ production from NA35. Previous data on anti-Λ production had seen an excess over the first collision model, and successfully fitted by RQMD including the enhanced interactions. The model had then made predictions for the $\bar{p}$ yield. The preliminary data point at one value of rapidity agrees with these model predictions.[24]

PARTICLE SPECTRA AND RATIOS

1. Nagamiya - Overview, AGS E802 K/π.
2. Braun-Munzinger - AGS E814, π and K Spectra.
3. Röhrich - SPS NA35 K_s^o Enhancement.
4. Ukonov - Dubna Λ, π, K at $4 GeV/c$.
5. DiBari - WA85 Antibaryon Results.
6. Purschke - WA80 $< p_T >$.

Prof. S. Nagamiya presented two excellent lectures on the general nature of A-A collisions and proposed signals for QGP which were both informative and instructive.

Many new results were presented by Peter Braun-Munzinger from the E814 collaboration. The previously-reported low transverse mass enhancement in the pion spectra has been very well isolated. If interpreted in terms of decays from produced Δ's, it is required that about 1/3 of the pions come from decays. It is found that the Δ production directly measured in this experiment is consistent with this amount. New preliminary data on low transverse mass $K^{\pm}$ spectra were also presented. These data also show a sharp excess (a fit to the low momentum peaks would require an equivalent Boltzmann temperature of about $20 MeV$!). It is hard to see how resonance decays could account for this effect.[17]

New particle spectra data were also presented for the NA35 collaboration. [24] They found a very nice scaling relation for the pion spectra over the entire rapidity range. In going from p-A to A-A, a factor of about 30 was observed. However, for the K_s^o spectra, an additional (above the 30) factor of about 1.7 was required, indicative of a different effect for strangeness.

Domenico Di Bari presented the new results from WA85 on antihyperon production. [19] These are of intense current interest for comparison with thermal-type models of both hadron gas and QGP. A previous result of NA36 indicated that the ratio of anticascade to antilambda was 0.127 ± 0.022, about a factor of two above that in p-p interactions. The new WA80 measurement is 0.20 ± 0.03. Care must be taken when comparing the two since the kinematic regions are not identical, but one can certainly conclude that this ratio is enhanced in the A-A interactions. This experiment also has observed anti-omega baryons for the first time, and presented preliminary data at this meeting.

Martin Purschke presented a very interesting analysis of the average transverse momentum dependence on entropy (inferred from rapidity distributions scaled by $A^{2/3}$) for neutral pions from WA80. There is some hint that the increase is followed by a plateau and then another increase, as one would expect if the system were undergoing a first-order phase transition. Attempts to increase the statistics by using the single-photon spectra look promising but not definitive.[28]

SOURCE SIZES AND HBT

1. Ströbele - Rapidity Dependence.
2. Nagamiya - E859, Difference for K's and π's.
3. Pelte - Clustering Phase Space Method.
4. Barz - Expanding Pion Gas Model, HBT Size vs Physical.
5. Braun-Munzinger - $x - p_x$ Correlations and p_T cuts.

I have no time to review this subject, and must refer to the excellent contributions listed above.

MODELS

1. Werner - Overview.
2. Umar - Real Partons on Strings.
3. Amelin - Inclusion of Hard QCD Processes.
4. Neise - Off-Mass-Shell Effects.
5. Sorge - RQMD.
6. Katscher - Three-Fluid Model.
7. Czernai - Relativistic Fluid Models and Phase Transitions.

Klaus Werner presented a comprehensive overview of relativistic models which are used to simulate particle production, rescattering, and fragmentation in N-N and A-A collisions. [26] The main additional ingredient for A-A is the addition of some sort of rescattering or enhanced scattering process due to the dense nuclear medium. Heinz Sorge explained some details in the RQMD model, which was used with great regularity in the presentations at this meeting.[25] I was particularly impressed with the detailed calculations of Czernai and collaborators using a relativistic hydrodynamical model.[18] They have new results which allow predictions of the nucleation rates and supercooling for QGP to hadron gas phase transitions. One possible scenario includes a sudden timelike freezeout from a supercooled QGP, which may fit in nicely with the newly-measured particle ratios in A-A collisions at SPS energies.

ELECTROMAGNETIC PROBES

1. Kampert - Overview, WA80 γ/π^o.

2. Roche - Dileptons at LBL, GSI.
3. Specht - Direct Photons at Low P_T, Dileptons at NA34, NA45.
4. Gale - Hadron Gas Background for Direct Photons and Dileptons.
5. Winckelmann - Hadron Bremstrahlung of Low Mass Dielectrons.
6. Dumitru - Thermal Photons from QGP for Nonzero Chemical Potential.
7. Gerschel - NA38 Dimuons.
8. Kralik - Helios-3 ρ, ω, ϕ Production.
9. Cotanch - CEBAF Program.

This is another very large area of activity, as evidenced by the number of presentations. The overview by Kampert [1] emphasized that it is very difficult experimentally to isolate the direct photons from decay products in A-A collisions. Preliminary data from WA80 shows a very nice separation at low P_T however, and we can look forward to more in the future.

The aim of this, of course, is to determine a possible window through which direct photons or dileptons from a QGP might be observable. To that end, there is intense theoretical and experimental activity on hadronic decay backgrounds at low momenta and hard QCD processes at high momenta. Of course there is also substantial uncertainty in predicting the pure QGP signals.

A parallel approach involves looking for the in-medium effects on the vector meson spectra. New results from NA38 [20] on Ψ and Ψ' suppression patterns are compatible with original expectations in a QGP scenario, but there are also hadronic scenarios with initial and final state effects which remain viable. A new effect, seen also by Helios-3[27], is the increase from p-A to A-A of the $\phi/(\rho + \omega)$ yields. This increase is also fit nicely in a pure RQMD calculation.[25]

QCD

1. Fingberg - Overview of Lattice Results.
2. Weise - In-Medium Effects and Chiral Symmetry.
3. Schramm - Higher-Number $Q\bar{Q}$ Condensates.
4. Sailer - Model Y-M Field Theory with Deconfining Phase Transition.
5. McLerran - Calculation of Structure Function for Very Large Nucleus.
6. Quack - Nuclear Stopping Effects on x_F Distributions.
7. Gorenstein - QGP-Hadron Transition not in Chemical Equilibrium.
8. Gyulassy - Minijet Gluons Damping and Color Conductivity.
9. Rischke - QGP above T_c, Gluon Clusters.
10. Polonyi - QGP above T_c, Screening by Gluons.

J. Fingberg presented two excellent lectures reviewing the status of lattice calculations in QCD.[14] Some of the main conclusions include: (a) The order of the transition depends critically on the number of light quark species; (b) The effective thermal mass for quarks is still large above T_c; and (c) Meson correlators above T_c clearly reflect properties of the fermionic substructure.

He reminded us that even in the pure gauge theory, in the region above T_c the screened gluons by no means act like an ideal noninteracting gas. This is supported by calculations which show that the Stefan-Boltzmann limit is reached for neither pressure nor energy density, and the relation between them is very far from the canonical factor of three. This has been exploited in the work of Rischke and collaborators[16], who are able to reproduce this behavior by insisting that the low-momentum gluon modes are confined. The physical picture which emerges is that a QGP might contain significant fractions of confined gluon clusters.

Larry McLerran presented a preliminary version of an interesting idea on the possibility of a first-principles calculation of a structure function. [12] The method relies on the existence of a large energy scale in a very large nucleus, provided by the parton rapidity density and the transverse size. If this size is much larger than the energy scale of QCD, one can use a perturbative expansion.

Miklos Gyulassy showed an estimate of the damping of gluons in a medium with color conductivity. [13] The application would be in the RHIC energy range, where initially produced gluons would exist in a backgound produced by the co-existing soft processes. Because of the possibility to change color in each interaction of the gluons, the damping rate is much reduced. This in turn reduces the quark pair production rate, delaying the onset of chemical equilibrium and reducing the expected number of thermal dileptons.

Finally, J. Polonyi gave two very exciting lectures on his work on non-perturbative QCD above T_c. [15] He finds that the screening of quark charge must be accomplished by a collective effect of an infinite number of gluons which can simulate the quantum numbers of an antiquark. This collection can be either bosonic or ferminonic, leading to some interesting possibilities for the quasiparticle spectrum of a QGP.

CONCLUDING REMARKS

Finally, I would like to acknowledge the outstanding work done by our hosts and organizers, Walter Greiner, Horst Stöcker, and Andre Gallmann. In particular, the scheduling of the presentations such that free time for discussions was avilable in the afternoons was certainly to our benefit. Some of us, of course, occassionaly used that time to become acquainted with the local area, beaches, castle, shopping, etc. In the course of these excursions, I suspect there may have been some conversations with local residents or tourists on the nature of our meeting here. A cartoon appeared this week in the International Herald Tribune of Snoopy and his friend the bird, and contained an exchange which seems ideal for adapting to a typical situation. (I thank my wife Johnnie for sketching some pictures illustrating the conversation which were used in the oral summary presentation.) The comments by Snoopy the Physicist were as follows:

"What, you didn't know that QCD has a phase transition!?"

"Well, I suppose there are a lot of things you don't have to know if you're a (insert your favorite target group here)."

"Life is more difficult for physicists . . . We're required to know everything."

"Yes, most of us remain quite humble."

REFERENCES

1. K. Kampert, These proceedings.

2. T. Kajino, These proceedings.

3. F. Weber, These proceedings.

4. H.-G. Ritter, These proceedings.

5. G. Westfall, These proceedings.

6. K. Hildenbrand, These proceedings.

7. J. P. Coffin, These proceedings.

8. J. A. Maruhn, These proceedings.

9. J. Konopka, These proceedings.

10. S. Bass, These proceedings.

11. D. Brill, These proceedings.

12. L. McLerran, These proceedings.

13. M. Gyulassy, These proceedings.

14. J. Fingberg, These proceedings.

15. J. Polonyi, These proceedings.

16. D. Rischke, These proceedings.

17. P. Braun-Munzinger, These proceedings.

18. L. Czernai, These proceedings.

19. D. Di Bari, These proceedings.

20. C. Gerschel, These proceedings.

21. B. Jacak, These proceedings.

22. A. Jahns, These proceedings.

23. W. König, These proceedings.

24. D. Röhrich, These proceedings.

25. H. Sorge, These proceedings.

26. K. Werner, These proceedings.

27. I. Kralik, These proceedings.

28. M. Purschke, These proceedings.

INDEX

Spatial string tension, 73
Speckle pattern, 589
Spectrometer, 383
Speed of sound, 12
Sphaleron, 759
Spinodal region, 671
Squeeze-out, 102, 183, 194, 554, 549, 680, 860
Star experiment, 402
Statistical decay, 48
Statistical model, 48, 133, 139, 167
Statistical multifragmentation models, 39, 235, 875
Statistical thermodynamics, 870
Stefan-Boltzmann gas, 111, 278
Stopping power, 769
Stopping, 4, 182, 193, 296, 431, 769, 790, 818, 856
Strange dwarfs, 145
Strange hadronic matter, 53, 647
Strange matter, 145
Strange quark matter, 55, 153, 156, 159, 387, 647
Strange stars, 145
Strangelet distillation, 58
Strangelet formation, 53
Strangelet, 160, 259
Strangeness distillation, 58
Strangeness enhancement, 160, 833
Strangeness production, 383, 621
Strangeness separation, 53, 57, 58
Strangeness suppression, 626
Strangeness, 355, 383, 621, 787
Streamer chamber, 243
String excitation, 622, 810
String fragmentation, 518, 625, 751, 811
String model, 517, 809
String-parton model, 374, 752, 809
Strings, 374, 621
String tension, 73, 485, 747
Strong field Quantum Electrodynamics (QED), 777
Strong interaction, 73, 109
Subsaturation density, 39, 193
Surface instabilities, 5

TAPS, 463
Thermal model, 7
Thermal photons, 25, 227

Three-fluid model, 697
Time dependent cluster model, 216
Time projection chamber, 243, 505, 560
Time-dependent Hartree-Fock, 475
Top quark, 321
Transport equations, 333
Transport theory, 687
Transversal flow, 175, 316, 737
Transverse momentum, 4, 99, 207, 561, 709
Two-fragment correlation function, 39, 46

U-Matrix, 869
Ultrarelativistic heavy ion collisions, 73, 697, 709, 719
Ultrarelativistic heavy ion collisions, 709

Van der Waals gas, 39, 209
Vaporization, 41
Vector meson dominance, 27, 123, 827
Vector mesons, 122, 621, 719, 824, 833
Velocity correlations, 687
Venus, 378, 395, 524
Virtual photons, 24, 123
Viscosity, 13, 567
Viscous hydrodynamics, 11
Vlasov equation, 477
Vlasov-Uehling-Uhlenbeck Model (VUU), 4, 193, 221, 233, 449, 862
Volume emission, 48

WA80 experiment, 31, 378, 709
WA85 experiment, 356
WA93 experiment, 31
WA95 experiment, 189
W boson, 326
Walecka model, 738
Wavelet transformation, 265
Wavelet, 65, 265
Weizsäcker Williams method, 324, 441, 570, 604
White dwarf, 155
Wigner representation, 476
Wilson loop, 82

X-distribution, 769

Yang-Mills theory, 741

Z^0 boson, 322, 327